Jussi Behrndt · Seppo Hassi · Henk de Snoo

Boundary Value Problems, Weyl Functions, and Differential Operators

Jussi Behrndt
Institut für Angewandte Mathematik
Technische Universität Graz
Graz, Austria

Seppo Hassi
Mathematics and Statistics
University of Vaasa
Vaasa, Finland

Henk de Snoo
Bernoulli Institute for Mathematics
Computer Science and Artificial Intelligence
University of Groningen
Groningen, The Netherlands

https://doi.org/10.1007/978-3-030-36714-5

Mathematics Subject Classification (2010): 47A, 47B, 47E, 47F, 34B, 34L, 35P, 81C, 93B

Contents

Preface

This monograph is about boundary value problems, Weyl functions, and differential operators. It grew out of a number of courses and seminars on functional analysis, operator theory, and differential equations, which the authors have given over a long period of time at various institutions. The project goes back to 2005 with a course on extension theory of symmetric operators, boundary triplets, and Weyl functions given at TU Berlin, while an extended form of the course was presented in 2006/2007 at the University of Groningen. Many more such courses and seminars, often on special topics, would follow at TU Berlin, Jagiellonian University in Kraków, and, since 2011, at TU Graz.

The authors wish to thank all the students, PhD students, and postdocs who have attended these lectures; their critical questions and comments have led to numerous improvements. They have shown that lectures at the blackboard provide the ultimate test for the quality of the material. In particular, we mention Bernhard Gsell, Markus Holzmann, Christian Kühn, Vladimir Lotoreichik, Jonathan Rohleder, Peter Schlosser, Philipp Schmitz, Simon Stadler, Alef Sterk, and Rudi Wietsma. It is our experience that the individual chapters of this monograph can be used (with small additions from some of the other chapters) for independent courses on the respective topics.

The book has benefited from our collaboration with many different colleagues. We would like to single out our friends and faithful coauthors Yuri Arlinskiĭ, Vladimir Derkach, Peter Jonas, Matthias Langer, Annemarie Luger, Mark Malamud, Hagen Neidhardt, Franek Szafraniec, Carsten Trunk, Henrik Winkler, and Harald Woracek. Special thanks go to Fritz Gesztesy, Gerd Grubb, Heinz Langer, and James Rovnyak, who have responded to our queries concerning historical developments and references.

We gratefully acknowledge the support of the following institutions: Deutsche Forschungsgemeinschaft, Jagiellonian University, TU Berlin, and TU Graz. We would like to thank the Mathematisches Forschungsinstitut Oberwolfach and the Mittag-Leffler Institute in Djursholm for their hospitality during the final stages of the preparation of this book. Finally, we are indebted to the Austrian Science Fund (Grant PUB 683-Z) and the University of Vaasa for funding the open access publication of this monograph.

Jussi Behrndt, Seppo Hassi, and Henk de Snoo

Introduction

In this monograph the theory of boundary triplets and their Weyl functions is developed and applied to the analysis of boundary value problems for differential equations and general operators in Hilbert spaces. Concrete illustrations by means of weighted Sturm–Liouville differential operators, canonical systems of differential equations, and multidimensional Schrödinger operators are provided. The abstract notions of boundary triplets and Weyl functions have their roots in the theory of ordinary differential operators; they appear in a slightly different context also in the treatment of partial differential operators.

Before describing the contents of the monograph it may be helpful to explain the ideas in this text by means of the following simple Sturm–Liouville differential expression

$$L = -\frac{d^2}{dx^2} + V, \tag{1}$$

where it is assumed that the potential V is a real measurable function. The context in which this differential expression will be placed serves as an example as well as a motivation. The first step is to associate with L some differential operators in a suitable Hilbert space. Assume, e.g., that (1) is given on the positive half-line $\mathbb{R}^+ = (0, \infty)$ and assume for simplicity that the real function V is bounded. Define the linear space $\mathfrak{D}_{\max}$ by

$$\mathfrak{D}_{\max} = \big\{ f \in L^2(\mathbb{R}^+) : f, f' \text{ absolutely continuous}, \ Lf \in L^2(\mathbb{R}^+) \big\}$$

and define the *minimal operator* S associated with L by

$$Sf = -f'' + Vf, \qquad \operatorname{dom} S = \big\{ f \in \mathfrak{D}_{\max} : f(0) = f'(0) = 0 \big\}.$$

Then S is a closed densely defined symmetric operator $L^2(\mathbb{R}^+)$; in fact, it is the closure of (the graph of) the restriction of S to $C_0^\infty(\mathbb{R}^+)$. It can be shown that the adjoint operator S^* is given by

$$S^* f = -f'' + Vf, \qquad \operatorname{dom} S^* = \mathfrak{D}_{\max},$$

which is usually called the *maximal operator* associated with L. Roughly speaking, S is a two-dimensional restriction of S^* by means of the boundary conditions

J. Behrndt et al., *Boundary Value Problems, Weyl Functions, and Differential Operators*, Monographs in Mathematics 108,
https://doi.org/10.1007/978-3-030-36714-5_1

$f(0) = 0$ and $f'(0) = 0$. Note that the maximal domain $\mathfrak{D}_{\max}$ coincides with the second-order Sobolev space $H^2(\mathbb{R}^+)$.

The notion of *boundary triplet* will now be explained in the present situation. For this consider $f, g \in \operatorname{dom} S^*$ and observe that integration by parts leads to

$$\begin{aligned}(S^*f, g)_{L^2(\mathbb{R}^+)} - (f, S^*g)_{L^2(\mathbb{R}^+)} &= -f'(x)\overline{g(x)}\Big|_0^\infty + f(x)\overline{g'(x)}\Big|_0^\infty \\ &= f'(0)\overline{g(0)} - f(0)\overline{g'(0)},\end{aligned}$$

where it was used that the products $f'\bar{g}$ and $f\bar{g}'$ vanish at ∞. Inspired by the above identity, define boundary mappings

$$\Gamma_0, \Gamma_1 : \operatorname{dom} S^* \to \mathbb{C}, \quad f \mapsto \Gamma_0 f := f(0) \quad \text{and} \quad f \mapsto \Gamma_1 f := f'(0), \tag{2}$$

so that for all $f, g \in \operatorname{dom} S^*$ one has

$$(S^*f, g)_{L^2(\mathbb{R}^+)} - (f, S^*g)_{L^2(\mathbb{R}^+)} = (\Gamma_1 f, \Gamma_0 g)_{\mathbb{C}} - (\Gamma_0 f, \Gamma_1 g)_{\mathbb{C}}, \tag{3}$$

which is the so-called *abstract Green identity* in the definition of a boundary triplet; note that on the right-hand side of (3) the scalar product in the (boundary) Hilbert space $\mathbb{C}$ is used. This abstract Green identity is the key feature in the notion of a boundary triplet and it is primarily responsible for the succesful functioning of the whole theory. Note also that the combined boundary mapping

$$(\Gamma_0, \Gamma_1)^\top : \operatorname{dom} S^* \to \mathbb{C}^2$$

is surjective, which is understood as a maximality condition in the sense that the image space of the boundary maps is not unnecessarily large. Observe that one has $\operatorname{dom} S = \ker \Gamma_0 \cap \ker \Gamma_1$. The operator realizations A of the Sturm–Liouville differential expression L which are intermediate extensions, that is, $S \subset A \subset S^*$, can be described by boundary conditions expressed via the boundary maps. More precisely, for $\tau \in \mathbb{C} \cup \{\infty\}$ the operator A_τ is defined by

$$A_\tau f = S^* f, \qquad \operatorname{dom} A_\tau = \ker(\Gamma_1 - \tau\Gamma_0), \tag{4}$$

which in a more explicit form reads

$$A_\tau f = -f'' + Vf, \qquad \operatorname{dom} A_\tau = \big\{ f \in \mathfrak{D}_{\max} : f'(0) = \tau f(0) \big\};$$

the case $\tau = \infty$ is understood as the boundary condition $\ker \Gamma_0$, that is,

$$A_\infty f = -f'' + Vf, \qquad \operatorname{dom} A_\infty = \big\{ f \in \mathfrak{D}_{\max} : f(0) = 0 \big\}. \tag{5}$$

In the definition (4) the quantity τ plays the role of a boundary parameter that links the boundary values $\Gamma_0 f = f(0)$ and $\Gamma_1 f = f'(0)$ of the functions $f \in \operatorname{dom} S^*$, which determine the Dirichlet and Neumann boundary conditions, respectively. The properties of the boundary parameter are directly connected with the properties of the corresponding operator A_τ; in particular, the realization A_τ is self-adjoint in $L^2(\mathbb{R}^+)$ if and only if $\tau \in \mathbb{R} \cup \{\infty\}$.

The next main goal is to motivate and illustrate the definition of the *Weyl function* as an analytic object corresponding to a boundary triplet, which is indispensable in the spectral theory of the intermediate extensions. For this, let $\lambda \in \mathbb{C}$ and consider first the unique solutions φ_λ and ψ_λ of the boundary value problems

$$\begin{aligned} -\varphi_\lambda'' + V\varphi_\lambda &= \lambda\varphi_\lambda, \qquad & \varphi_\lambda(0) = 1, \quad \varphi_\lambda'(0) = 0, \\ -\psi_\lambda'' + V\psi_\lambda &= \lambda\psi_\lambda, \qquad & \psi_\lambda(0) = 0, \quad \psi_\lambda'(0) = 1, \end{aligned} \tag{6}$$

and note that in general $\varphi_\lambda, \psi_\lambda \notin L^2(\mathbb{R}^+)$. It was shown by H. Weyl more than a century ago that for $\lambda \in \mathbb{C} \setminus \mathbb{R}$ there exists $m(\lambda) \in \mathbb{C}$ such that

$$x \mapsto f_\lambda(x) = \varphi_\lambda(x) + m(\lambda)\psi_\lambda(x) \in L^2(\mathbb{R}^+), \tag{7}$$

and it turned out that the function $m : \mathbb{C} \setminus \mathbb{R} \to \mathbb{C}$ is holomorphic and has a positive imaginary part in the upper half-plane $\mathbb{C}^+$. This function and its interplay with spectral theory were later studied extensively by E.C. Titchmarsh; hence the frequently used terminology Titchmarsh–Weyl m-function. It plays a key role in the spectral analysis of Sturm–Liouville differential operators. E.g., the (real) poles of m coincide with the isolated eigenvalues of the self-adjoint Dirichlet operator A_∞ in (5) and the absolutely continuous spectrum of A_∞ is, roughly speaking, given by those $\lambda \in \mathbb{R}$ for which $\operatorname{Im} m(\lambda + i0) > 0$. In a similar way one can also characterize the continuous spectrum, the embedded eigenvalues, and exclude singular continuous spectrum of A_∞.

Observe that for each $\lambda \in \mathbb{C} \setminus \mathbb{R}$ the function $x \mapsto f_\lambda(x)$ in (7) belongs to $\operatorname{dom} S^* = \mathfrak{D}_{\max}$ and that, in fact, $-f_\lambda'' + Vf_\lambda = \lambda f_\lambda$ for $\lambda \in \mathbb{C} \setminus \mathbb{R}$; in other words, $f_\lambda \in \ker(S^* - \lambda)$. Let $\{\mathbb{C}, \Gamma_0, \Gamma_1\}$ be the boundary triplet for S^* with the boundary mappings defined in (2). From the choice of φ_λ and ψ_λ in (6) it is clear that

$$m(\lambda)\Gamma_0 f_\lambda = m(\lambda) f_\lambda(0) = m(\lambda) = \Gamma_1 f_\lambda, \quad f_\lambda \in \ker(S^* - \lambda). \tag{8}$$

In the general theory this identity is used as the definition of the *Weyl function* corresponding to a boundary triplet. In other words, the Weyl function corresponding to the boundary triplet $\{\mathbb{C}, \Gamma_0, \Gamma_1\}$ is defined as the function m that satisfies (8) for all $\lambda \in \mathbb{C} \setminus \mathbb{R}$ (and even for the possibly larger set of λ belonging to the resolvent set of the self-adjoint Dirichlet operator A_∞) and hence coincides with the Titchmarsh–Weyl m-function introduced via (7). Here the Weyl function maps Dirichlet boundary values of L^2-solutions of the equation $-f_\lambda'' + Vf_\lambda = \lambda f_\lambda$ onto the corresponding Neumann boundary values and therefore $m(\lambda)$ acts formally like a *Dirichlet-to-Neumann map*. Besides the Weyl function, one associates to the boundary triplet $\{\mathbb{C}, \Gamma_0, \Gamma_1\}$ the so-called *γ-field* as the mapping $\gamma(\lambda) : \mathbb{C} \to L^2(\mathbb{R}^+)$ that assigns to a prescribed boundary value $c \in \mathbb{C}$ the solution $h_\lambda \in \operatorname{dom} S^*$ of the boundary value problem

$$-h_\lambda'' + Vh_\lambda = \lambda h_\lambda, \qquad \Gamma_0 h_\lambda = h_\lambda(0) = c.$$

Since $\gamma(\lambda)c = h_\lambda = cf_\lambda$, it is clear that $m(\lambda) = \Gamma_1\gamma(\lambda)$. Moreover, one can show with the help of the abstract Green identity that the adjoint $\gamma(\lambda)^* : L^2(\mathbb{R}^+) \to \mathbb{C}$ is given by $\gamma(\lambda)^* = \Gamma_1(A_\infty - \bar{\lambda})^{-1}$. The Weyl function and γ-field associated to the boundary triplet $\{\mathbb{C}, \Gamma_0, \Gamma_1\}$ appear in the perturbation term in Kreĭn's formula

$$(A_\tau - \lambda)^{-1} = (A_\infty - \lambda)^{-1} + \gamma(\lambda)(\tau - m(\lambda))^{-1}\gamma(\bar{\lambda})^*,$$

where, for simplicity, it is assumed that A_τ is a self-adjoint realization of L as in (4) corresponding to some boundary parameter $\tau \in \mathbb{R}$ and $\lambda \in \rho(A_\tau) \cap \rho(A_\infty)$. Kreĭn's formula in this particular case provides a description of the resolvent difference of A_τ and the fixed self-adjoint extension A_∞. It is important to note that $\gamma(\lambda)$ and $\gamma(\bar{\lambda})^*$ in the perturbation term provide a link between the original Hilbert space $L^2(\mathbb{R}^+)$ and the boundary space $\mathbb{C}$, but do not affect the resolvents of A_∞ and A_τ. Therefore, if $\lambda \in \rho(A_0)$, then the singularities of the resolvent $\lambda \mapsto (A_\tau - \lambda)^{-1}$ are reflected in the singularities of the term $\lambda \mapsto (\tau - m(\lambda))^{-1}$ and vice versa. In fact, the function $\lambda \mapsto (\tau - m(\lambda))^{-1}$ is connected with the spectrum of A_τ in the same way as the function $\lambda \mapsto m(\lambda)$ is connected with the spectrum of A_∞.

There is another efficient technique to associate differential operators with the differential expression L, which is based on the *sesquilinear form* $\mathfrak{t}$ corresponding to L,

$$\mathfrak{t}[f,g] = (f',g')_{L^2(\mathbb{R}^+)} + (Vf,g)_{L^2(\mathbb{R}^+)}, \tag{9}$$

defined on, e.g.,

$$\mathfrak{D} = \{f \in L^2(\mathbb{R}^+) : f \text{ absolutely continuous, } f' \in L^2(\mathbb{R}^+)\}, \tag{10}$$

and the first representation theorem for sesquilinear forms. In fact, one verifies that $\mathfrak{t}$ in (9)–(10) is a densely defined closed semibounded form in $L^2(\mathbb{R}^+)$, and hence there exists a uniquely determined self-adjoint operator S_1 with $\operatorname{dom} S_1 \subset \mathfrak{D}$ such that

$$(S_1 f, g)_{L^2(\mathbb{R}^+)} = \mathfrak{t}[f,g], \qquad f \in \operatorname{dom} S_1,\ g \in \operatorname{dom} \mathfrak{D}. \tag{11}$$

Note that here the form domain $\mathfrak{D}$ coincides with the first-order Sobolev space $H^1(\mathbb{R}^+)$. It can be shown that the self-adjoint operator S_1 is actually an extension of the minimal operator S. Instead of the domain $\mathfrak{D}$ in (10) one may consider the sesquilinear form $\mathfrak{t}$ on the smaller domain $\mathfrak{D}_0 = \{f \in \mathfrak{D} : f(0) = 0\}$, which also leads to a densely defined closed semibounded form in $L^2(\mathbb{R}^+)$. Again, via the first representation theorem, there is a corresponding self-adjoint operator S_0 with $\operatorname{dom} S_0 \subset \mathfrak{D}_0$ determined by

$$(S_0 f, g)_{L^2(\mathbb{R}^+)} = \mathfrak{t}[f,g], \qquad f \in \operatorname{dom} S_0,\ g \in \operatorname{dom} \mathfrak{D}_0. \tag{12}$$

One verifies that the self-adjoint operator S_1 in (11) coincides with the self-adjoint realization of L determined by the boundary condition $\ker \Gamma_1$ and that the self-adjoint operator S_0 in (12) coincides with the self-adjoint realization of L determined by the boundary condition $\ker \Gamma_0$ in (4), that is, S_1 corresponds to the boundary parameter $\tau = 0$ and S_0 is the Dirichlet operator corresponding to the boundary parameter $\tau = \infty$. Furthermore, in the situation discussed here the self-adjoint operator S_0 in (12) is the *Friedrichs extension* of the minimal (or preminimal) operator associated to L.

The concept of boundary triplet is supplemented by the notion of *boundary pair*, which is inspired by the form approach indicated above. More precisely, in the present situation it turns out that $\{\mathcal{G}, \Lambda\}$, where $\mathcal{G} = \mathbb{C}$ and

$$\Lambda : \mathfrak{D} \to \mathbb{C}, \qquad f \mapsto \Lambda f := f(0), \tag{13}$$

is a boundary pair for the minimal operator S (corresponding to S_1). For this, one has to ensure that the mapping Λ defined on the form domain of S_1 is continuous with respect to the Hilbert space topology generated by the closed form $\mathfrak{t}$ on $\mathfrak{D}$, and that $\ker \Lambda$ coincides with the form domain corresponding to the Friedrichs extension of S. Note also that in the present situation the mapping Λ in (13) is an extension of the boundary mapping $\Gamma_0 : \operatorname{dom} S^* \to \mathbb{C}$ to the form domain $\mathfrak{D}$. With the help of the boundary pair $\{\mathbb{C}, \Lambda\}$ one can parametrize all densely defined closed semibounded forms corresponding to semibounded self-adjoint extensions of S via

$$\mathfrak{t}_\tau[f,g] = \mathfrak{t}[f,g] + (\tau \Lambda f, \Lambda g)_{\mathbb{C}}, \qquad f,g \in \mathfrak{D}, \tag{14}$$

where $\tau \in \mathbb{R} \cup \{\infty\}$, and the case $\tau = \infty$ corresponds to the boundary condition $\Lambda f = 0$ in $\mathfrak{D}_0$. The boundary pair and the boundary triplet are connected via the first Green identity

$$(S^* f, g)_{L^2(\mathbb{R}^+)} = \mathfrak{t}[f,g] + (\Gamma_1 f, \Lambda g)_{\mathbb{C}}, \quad f \in \operatorname{dom} S^*,\ g \in \mathfrak{D}.$$

The first Green identity makes it possible to identify the closed semibounded forms in (14) with the corresponding self-adjoint operator realizations A_τ of L described via boundary conditions in (4). For $f \in \operatorname{dom} A_\tau$ and $g \in \mathfrak{D}$, the first Green identity reduces to

$$(A_\tau f, g)_{L^2(\mathbb{R}^+)} = \mathfrak{t}[f,g] + (\tau \Gamma_0 f, \Lambda g)_{\mathbb{C}} = \mathfrak{t}[f,g] + (\tau \Lambda f, \Lambda g)_{\mathbb{C}},$$

and the expression $(\tau \Lambda f, \Lambda g)_{\mathbb{C}}$ on the right-hand side can also be interpreted as a sesquilinear form in the boundary space $\mathbb{C}$. In this sense the theory of boundary pairs for semibounded symmetric operators complements the theory of boundary triplets in a natural way: it provides a description of the closed semibounded forms corresponding to semibounded self-adjoint extensions of the minimal operator S.

Methods to treat Sturm–Liouville problems such as the one discussed above go back to H. Weyl [758, 759, 760], whose papers on this topic appeared in 1910/1911; see also [761]. The interpretation of a Sturm–Liouville expression as an operator in a Hilbert space can already be found in the 1932 book of M.H. Stone [724]. In this monograph Stone gave an abstract treatment of operators in a Hilbert space including the work of J. von Neumann [610, 611] from 1929 and 1932, who had also introduced the extension theory of densely defined symmetric operators and found the formulas which carry his name: self-adjoint extensions correspond to unitary mappings between the defect spaces. The von Neumann formulas are abstract, since they are formulated in terms of the defect spaces of the symmetric operator, and they needed to be related to concrete boundary value problems. With this in mind another approach involving abstract boundary conditions was developed by J.W. Calkin [187] in his 1937 Harvard doctoral dissertation, which was written under the direction of Stone, who suggested the topic. Calkin was also advised by von Neumann. Calkin's work on boundary value problems did not receive the attention it might have deserved. It seems that he never returned to it; his later mathematical work was related to World War II and the Manhattan project in Los Alamos.

Another way to deal with the self-adjoint extensions of a symmetric operator is via Kreĭn's resolvent formula. The early background of this formula can be found in the idea of perturbation of self-adjoint operators. Kreĭn's formula describes the resolvent of a self-adjoint extension in terms of the resolvent of a fixed self-adjoint extension and a perturbation term which involves a so-called Q-function and a parameter describing the self-adjoint extension. The Q-function uniquely determines the underlying symmetry and the fixed self-adjoint extension, up to unitary equivalence, and thus reflects their spectral properties. The original Kreĭn formula for equal finite defect numbers goes back to M.G. Kreĭn [491, 492] in the middle of the 1940s; only in 1965 it was finally established for the case of equal infinite defect numbers by S.N. Saakyan [679]. In fact, the self-adjoint extensions were allowed to be in a Hilbert space which contains the original Hilbert space as a closed subspace. This type of extension appeared after 1940 in papers by M.G. Kreĭn and M.A. Naĭmark [605, 606, 607]. Later A.V. Štraus in the 1950s and 1960s described such exit space extensions in the framework of the von Neumann formulas via holomorphic contractions between the defect spaces [731]. The Q-function in Kreĭn's formula can be seen as an abstract analog of the Titchmarsh–Weyl function in the above Sturm–Liouville example; it was extensively studied in the 1960s and 1970s by M.G. Kreĭn and H. Langer [497]–[504], also in the context of Pontryagin spaces.

From the early 1940s on E.C. Titchmarsh turned his attention to the singular Sturm–Liouville equation. He put aside Weyl's method of handling the Sturm–Liouville problem on the basis of integral equations and also bypassed the use of the general theory of linear operators in Hilbert spaces as in Stone's book [739]. Instead, Titchmarsh used contour integration and the Cauchy calculus of residues, influenced by the work of E. Hilb [417, 418, 419], a contemporary of Weyl. In this way he found a simple formula to determine the spectral measure; this last formula was also discovered by K. Kodaira around the same time [469, 470]. A complete survey of the work of Titchmarsh, both for ordinary and partial differential operators, is given in his two books on eigenfunction expansions [740, 741]. A different approach, followed by B.M. Levitan [541, 542], N. Levinson [539, 540], and K. Yosida [780, 781], is based on the fact that the resolvent operator of the self-adjoint realization of a singular differential operator can be approximated by compact resolvents corresponding to Sturm–Liouville problems for proper closed subintervals. Closely connected with this is an abstract approach to eigenfunction expansions generated by differential operators that was introduced by Kreĭn [495] in the form of directing functionals.

Influenced by questions from mathematical physics, von Neumann posed the following problem in the middle of the 1930s: can one extend a densely defined semibounded symmetric operator to a self-adjoint operator with the same lower bound? There were contributions by M.H. Stone [724] and K.O. Friedrichs [310] (whose work was simplified by H. Freudenthal [309]). The Friedrichs extension was the solution to von Neumann's problem. For Sturm–Liouville operators the Friedrichs extension was determined in various cases by K.O. Friedrichs [311] in 1935 and by F. Rellich [654] in 1950. Another semibounded extension, the so-called Kreĭn–von Neumann extension (going back to Stone) has particularly interesting properties. It was Kreĭn [493, 494] who established a complete theory of semibounded extensions. In the middle of the 1950s this circle of ideas was carried forward, and it inspired contributions by M.S. Birman [139], and also M.I. Vishik [747], who was particularly interested in the case of elliptic partial differential operators. Building on the work of J.L. Lions and E. Magenes [544] on Sobolev spaces and trace mappings G. Grubb [352, 353] gave a characterization of all closed extensions of a minimal elliptic operator by nonlocal boundary conditions in her 1966 Stanford doctoral dissertation, written under the direction of R.S. Phillips.

The context of symmetric operators which are densely defined was soon felt to be too restrictive. Already in 1949 M.A. Krasnoselskiĭ [490] described all self-adjoint operator extensions of a not necessarily densely defined symmetric operator. The appearance of the work on linear relations by R. Arens [42] in 1961 made all the difference. B.C. Orcutt [619] in a 1969 dissertation written under the direction of J. Rovnyak treated the spectral theory of canonical systems of differential equations in terms of linear relations. Subsequently, E.A. Coddington [202] in 1973 gave a description of all self-adjoint relation extensions of a symmetric relation. In fact, it turned out that many of the earlier results concerning extensions of symmetric operators could be put in the framework of relations. The new context made it also possible to consider nonstandard boundary conditions (involving integrals, for instance). Furthermore, in terms of relations the Kreĭn–von Neumann extension of a semibounded relation could be simply expressed in terms of the Friedrichs extension. There has been an abundance of papers devoted to linear relations in Hilbert spaces, and later also to linear relations in indefinite inner product spaces.

In the middle of the 1970s boundary triplets were introduced independently by V.M. Bruk [176] and A.N. Kochubei [466] as a convenient tool for the description of boundary values of abstract Hilbert space operators; they applied them to, e.g., Sturm–Liouville operators with an operator-valued potential. The main feature is that under a given boundary triplet there is a natural correspondence between self-adjoint extensions of a symmetric operator and self-adjoint relations in the parameter space. An overview of the theory with applications to differential operators is contained in the 1984 book by M.L. Gorbachuk and V.I. Gorbachuk [346]. Around the same time V.A. Derkach and M.M. Malamud [244, 246] continued the work on boundary triplets by associating the notion of Weyl function to a boundary triplet; their later work was written in the context of symmetric operators that are not necessarily densely defined. The Weyl function is a very useful tool in spectral analysis; it turns out to be a special choice of a Q-function (which is uniquely determined by the boundary triplet) and hence the analytic properties and the limit behavior of the Weyl function towards the real line reflect the spectral properties of the self-adjoint extensions. Broadly speaking, boundary triplets and Weyl functions placed the work of Titchmarsh, and others, in a more abstract setting while retaining the flavor of concrete boundary value problems. The link to form methods and the Birman–Kreĭn–Vishik approach to semibounded self-adjoint extensions is made with the help of so-called boundary pairs. The origin of the concept of boundary pair lies in the work of Kreĭn and Vishik; it was formalized and studied by V.E. Lyantse and O.G. Storozh [552] in the early 1980s. Its connection with boundary triplets was later established by Yu.M. Arlinskiĭ [44].

It is the main objective of this monograph to present the theory of boundary triplets and Weyl functions in an easily accessible and self-contained manner. The exposition is detailed and kept as simple as possible; the reader is only assumed to be familiar with the basic principles of functional analysis and some fundamentals of the spectral theory of self-adjoint operators in Hilbert spaces. The monograph is divided into the abstract part Chapters 1–5, the applied part Chapters 6–8, and Appendices A–D. The heart of the monograph is Chapter 2 and it is complemented by Chapter 5; for a rough idea on the general techniques the reader may first look through these chapters and examine one of the applications (which may also be read independently) afterwards: Sturm–Liouville operators, canonical systems, or Schrödinger operators – up to personal taste and preferences.

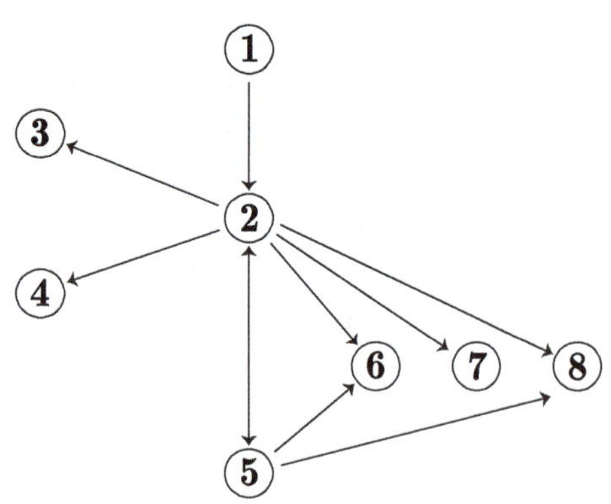

The monograph opens in Chapter 1 with a detailed introduction to the theory of linear operators and relations in Hilbert spaces. A large part of this material is preparatory and may be used for reference purposes in the rest of the text.

The heart of the matter in this book is contained in Chapter 2, where boundary value problems are presented as extension problems of symmetric operators or relations. Here the notions of boundary triplets and their Weyl functions are introduced, and the fundamental properties of these objects are provided. Particular attention is paid to the question of existence and uniqueness of boundary triplets. Closely connected with a boundary triplet is Kreĭn's resolvent formula for canonical extensions and self-adjoint extensions in larger Hilbert spaces.

Chapter 3 is a continuation and further refinement of the techniques in the previous chapter. Here the main objective is to give a detailed description of the complete spectrum of the self-adjoint extensions of a symmetric relation in terms of the Weyl function. The connection between the limit properties of the Weyl function and the spectrum of the self-adjoint extension is explained via the Borel transform of the spectral measure.

Most of the topics in Chapter 4 are supplementary to the main text as they are concerned with a certain type of inverse problem. More precisely, it will be shown that any (uniformly strict) operator-valued Nevanlinna function can be realized as the Weyl function corresponding to a boundary triplet for a symmetric relation in a reproducing kernel Hilbert model space. Of independent interest is the discussion around the orthogonal coupling of boundary triplets with a view to exit space extensions.

Another central theme in this monograph is presented in Chapter 5, where the important case of semibounded symmetric relations is treated in more detail; here the general methods from Chapter 2 are further developed. The chapter starts with an introduction to closed semibounded forms and the corresponding representation theorems, and continues with the Friedrichs extension, the so-called Kreĭn type extensions, and the Kreĭn–von Neumann extension. The ultimate result is a description of the semibounded self-adjoint extensions of a semibounded relation via the notions of a boundary triplet and a boundary pair; this establishes the connection with the Kreĭn–Birman–Vishik theory.

The general theory is applied to boundary value problems for differential operators in Chapters 6–8 in three different situations. In each case the presentation follows a similar scheme: After the necessary preparations to keep these chapters mostly self-contained, explicit boundary triplets and Weyl functions for the particular operators or relations under consideration, are provided. A further spectral analysis, depending on the nature of problem is presented. The class of Sturm–Liouville operators that is discussed in Chapter 6 covers also the example given earlier in this introduction. A good deal of preparation is needed to construct closed semibounded forms and corresponding boundary pairs in the singular situation. Chapter 7 deals with 2×2 canonical systems of differential equations and also illustrates the role of linear relations in the analysis of such systems. Finally, in Chapter 8 Schrödinger operators on bounded domains $\Omega \subset \mathbb{R}^n$ are treated, where one of the main challenges is to construct Dirichlet and Neumann traces on the maximal domain.

For the reader's convenience a number of appendices have been added: they contain material concerning Nevanlinna functions and some useful elementary observations on operators and subspaces in Hilbert spaces. At the end of the text a few notes and some (historical) comments, as well as a list of recent and earlier references, can be found. Here the reader is also referred to some recent literature for topics that go beyond this monograph.

Chapter 1

Linear Relations in Hilbert Spaces

A linear relation from one Hilbert space to another Hilbert space is a linear subspace of the product of these spaces. In this chapter some material about such linear relations is presented and it is shown how linear operators, whether densely defined or not, fit in this context. The basic terminology is provided in Section 1.1 and afterwards the spectrum, resolvent set, the adjoint, and operator decompositions of linear relations are discussed in Section 1.2 and Section 1.3. Linear relations with special properties, such as symmetric, self-adjoint, dissipative, and accumulative relations, are investigated in Sections 1.4, 1.5, and 1.6. More details on self-adjoint and semibounded relations can be found in Chapter 3 and Chapter 5. Intermediate extensions and the classical von Neumann formulas describing self-adjoint extensions of symmetric operators and relations can be found in Section 1.7. In Section 1.8 it is shown that there is a natural indefinite inner product by means of which the notion of adjoint relation corresponds to the notion of orthogonal companion. Strong graph convergence and strong resolvent convergence of sequences of linear relations are discussed in Section 1.9 and parametric representations of linear relations are studied in Section 1.10. Finally, in Section 1.11 some useful properties of a resolvent-type operator of a linear relation are given, and in Section 1.12 the class of so-called Nevanlinna families, a natural extension of the class of Nevanlinna functions (see Appendix A) is studied.

1.1 Elementary facts about linear relations

Let $\mathfrak{H}$ and $\mathfrak{K}$ be Hilbert spaces over $\mathbb{C}$. The Hilbert space inner product and the corresponding norm are usually denoted by $(\cdot,\cdot)$ and $\|\cdot\|$, respectively, and sometimes a subindex will be used in order to avoid confusion. The inner product is linear in the first entry and antilinear in the second entry. The orthogonal complement will be denoted by $\perp$, sometimes a subindex will be used to indicate the relevant space. The product $\mathfrak{H} \times \mathfrak{K}$ will often be regarded as a Hilbert space with the standard inner product $(\cdot,\cdot)_{\mathfrak{H}} + (\cdot,\cdot)_{\mathfrak{K}}$ and all topological notions in $\mathfrak{H} \times \mathfrak{K}$

J. Behrndt et al., *Boundary Value Problems, Weyl Functions, and Differential Operators*, Monographs in Mathematics 108,
https://doi.org/10.1007/978-3-030-36714-5_2

are understood with respect to the topology induced by the corresponding norm. The product space $\mathfrak{H} \times \mathfrak{K}$ will also be written as $\mathfrak{H} \oplus \mathfrak{K}$, and $\mathfrak{H}$ and $\mathfrak{K}$ are then regarded as closed linear subspaces in $\mathfrak{H} \oplus \mathfrak{K}$ which are orthogonal to each other.

A linear subspace of $\mathfrak{H} \times \mathfrak{K}$ is called a *linear relation* from $\mathfrak{H}$ to $\mathfrak{K}$. If H is a linear relation from $\mathfrak{H}$ to $\mathfrak{K}$ the elements $\widehat{h} \in H$ will in general be written as pairs $\{h, h'\}$ with components $h \in \mathfrak{H}$ and $h' \in \mathfrak{K}$. If $\mathfrak{K} = \mathfrak{H}$ one speaks simply of a linear relation in $\mathfrak{H}$. After this introductory section the adjective linear is usually omitted and one speaks of relations when linear relations are meant.

The *domain*, *range*, *kernel*, and *multivalued part* of a linear relation H from $\mathfrak{H}$ to $\mathfrak{K}$ are defined by

$$\begin{aligned} \operatorname{dom} H &= \big\{h \in \mathfrak{H} : \{h, h'\} \in H \text{ for some } h' \in \mathfrak{K}\big\}, \\ \operatorname{ran} H &= \big\{h' \in \mathfrak{K} : \{h, h'\} \in H \text{ for some } h \in \mathfrak{H}\big\}, \\ \ker H &= \big\{h \in \mathfrak{H} : \{h, 0\} \in H\big\}, \\ \operatorname{mul} H &= \big\{h' \in \mathfrak{K} : \{0, h'\} \in H\big\}, \end{aligned}$$

respectively. The closure of the linear space $\operatorname{dom} H$ will be denoted by $\overline{\operatorname{dom}}\, H$ and, likewise, the closure of the linear space $\operatorname{ran} H$ will be denoted by $\overline{\operatorname{ran}}\, H$. Note that each linear operator H from $\mathfrak{H}$ to $\mathfrak{K}$ is a linear relation if the operator is identified with its graph,

$$H = \big\{\{h, Hh\} : h \in \operatorname{dom} H\big\},$$

and that a linear relation H is (the graph of) an operator if and only if the multivalued part of H is trivial, $\operatorname{mul} H = \{0\}$. The *inverse* H^{-1} of a linear relation H from $\mathfrak{H}$ to $\mathfrak{K}$ is defined by

$$H^{-1} = \big\{\{h', h\} : \{h, h'\} \in H\big\},$$

so that H^{-1} is a linear relation from $\mathfrak{K}$ to $\mathfrak{H}$. In the next lemma some obvious facts concerning the inverse relation are collected.

Lemma 1.1.1. *Let H be a linear relation from $\mathfrak{H}$ to $\mathfrak{K}$. Then the following identities hold:*

$$\begin{aligned} \operatorname{dom} H^{-1} &= \operatorname{ran} H, & \operatorname{ran} H^{-1} &= \operatorname{dom} H, \\ \ker H^{-1} &= \operatorname{mul} H, & \operatorname{mul} H^{-1} &= \ker H. \end{aligned}$$

There is a linear structure on the collection of linear relations from $\mathfrak{H}$ to $\mathfrak{K}$. For linear relations H and K from $\mathfrak{H}$ to $\mathfrak{K}$ the *componentwise sum* is the linear relation from $\mathfrak{H}$ to $\mathfrak{K}$ defined by

$$H \mathbin{\widehat{+}} K = \big\{\{h + k, h' + k'\} : \{h, h'\} \in H, \{k, k'\} \in K\big\}, \tag{1.1.1}$$

while the *product* λH of H with a scalar $\lambda \in \mathbb{C}$ is the linear relation from $\mathfrak{H}$ to $\mathfrak{K}$ defined by

$$\lambda H = \big\{\{h, \lambda h'\} : \{h, h'\} \in H\big\}.$$

Note that the componentwise sum $H \mathbin{\widehat{+}} K$ is the linear span of the graphs of H and K, and

$$\operatorname{dom}(H \mathbin{\widehat{+}} K) = \operatorname{dom} H + \operatorname{dom} K, \quad \operatorname{ran}(H \mathbin{\widehat{+}} K) = \operatorname{ran} H + \operatorname{ran} K.$$

Likewise, if $\lambda \in \mathbb{C}$, one has

$$\operatorname{dom} \lambda H = \operatorname{dom} H \quad \text{and for} \quad \lambda \neq 0 \quad \operatorname{ran} \lambda H = \operatorname{ran} H.$$

Note that by definition $0\,H = O_{\operatorname{dom} H}$, where $O_{\operatorname{dom} H}$ stands for the zero operator on $\operatorname{dom} H$. It is useful to note that

$$(H \mathbin{\widehat{+}} K)^{-1} = H^{-1} \mathbin{\widehat{+}} K^{-1}, \qquad (\lambda H)^{-1} = \frac{1}{\lambda} H^{-1}, \quad \lambda \neq 0.$$

Let H and K be linear relations from $\mathfrak{H}$ to $\mathfrak{K}$. If $H \subset K$, then H is called a *restriction* of K and K is an *extension* of H.

Proposition 1.1.2. *Let H and K be linear relations from $\mathfrak{H}$ to $\mathfrak{K}$ and assume that $H \subset K$. Then*

$$\operatorname{dom} H = \operatorname{dom} K \quad \Leftrightarrow \quad K = H \mathbin{\widehat{+}} \big(\{0\} \times \operatorname{mul} K\big), \tag{1.1.2}$$

and, analogously,

$$\operatorname{ran} H = \operatorname{ran} K \quad \Leftrightarrow \quad K = H \mathbin{\widehat{+}} \big(\ker K \mathbin{\widehat{+}} \{0\}\big). \tag{1.1.3}$$

Proof. Note that $H \subset K$ is equivalent to $H^{-1} \subset K^{-1}$. Hence, in order to prove (1.1.3) one just applies (1.1.2) with H and K replaced by H^{-1} and K^{-1}, respectively. Thus it suffices to show (1.1.2). The implication ($\Leftarrow$) is trivial. To show ($\Rightarrow$), observe that $H \subset K$ yields $H \mathbin{\widehat{+}} (\{0\} \times \operatorname{mul} K) \subset K$ and hence it suffices to show that $K \subset H \mathbin{\widehat{+}} (\{0\} \times \operatorname{mul} K)$. Let $\{h, h'\} \in K$. Since $h \in \operatorname{dom} K = \operatorname{dom} H$, there exists an element $k' \in \mathfrak{K}$ such that $\{h, k'\} \in H$ and from $H \subset K$ it follows that also $\{h, k'\} \in K$. Hence, with $\varphi' = h' - k'$ one has

$$\{h, h'\} = \{h, k'\} + \{0, \varphi'\},$$

and thus $\{0, \varphi'\} \in K$, i.e., $\varphi' \in \operatorname{mul} K$. □

Corollary 1.1.3. *Let H and K be linear relations from $\mathfrak{H}$ to $\mathfrak{K}$ and assume that $H \subset K$. Then*

$$\operatorname{dom} H = \operatorname{dom} K \quad \textit{and} \quad \operatorname{mul} H = \operatorname{mul} K \quad \Leftrightarrow \quad H = K, \tag{1.1.4}$$

and, analogously,

$$\operatorname{ran} H = \operatorname{ran} K \quad \textit{and} \quad \ker H = \ker K \quad \Leftrightarrow \quad H = K. \tag{1.1.5}$$

Proof. It suffices to show (1.1.4), as (1.1.5) follows by taking inverses in (1.1.4). Clearly, the implication ($\Leftarrow$) is trivial. For the implication ($\Rightarrow$) apply (1.1.2). Then $\operatorname{dom} H = \operatorname{dom} K$ and $\operatorname{mul} H = \operatorname{mul} K$ give successively

$$K = H \,\widehat{+}\, (\{0\} \times \operatorname{mul} K) = H \,\widehat{+}\, (\{0\} \times \operatorname{mul} H) \subset H,$$

which together with $H \subset K$ implies $H = K$. □

Let H and K be linear relations from $\mathfrak{H}$ to $\mathfrak{K}$. The usual (operatorwise) *sum* $H + K$ is defined by

$$H + K = \{\{h, h' + h''\} : \{h, h'\} \in H,\ \{h, h''\} \in K\},$$

where $\operatorname{dom}(H+K) = \operatorname{dom} H \cap \operatorname{dom} K$. Note that $\operatorname{mul}(H+K) = \operatorname{mul} H + \operatorname{mul} K$. If H is a linear relation in $\mathfrak{H}$, then for $\lambda \in \mathbb{C}$ the sum $H + \lambda I$, where I denotes the identity operator in $\mathfrak{H}$, is usually simply written as $H + \lambda$ and has the form

$$H + \lambda = \{\{h, h' + \lambda h\} : \{h, h'\} \in H\},$$

with $\operatorname{dom}(H + \lambda) = \operatorname{dom} H$. Note that $\operatorname{mul}(H + \lambda) = \operatorname{mul} H$.

Let H be a linear relation from $\mathfrak{H}$ to $\mathfrak{K}$ and let K be a linear relation from $\mathfrak{K}$ to $\mathfrak{G}$, where $\mathfrak{G}$ is another Hilbert space. Then the *product* KH of K and H is the linear relation from $\mathfrak{H}$ to $\mathfrak{G}$ defined by

$$KH = \{\{h, h''\} : \{h, h'\} \in H,\ \{h', h''\} \in K\}.$$

Note that for $\lambda \in \mathbb{C}$ the notation λH agrees with $(\lambda I)H$, where I denotes the identity operator in $\mathfrak{K}$. It is straightforward to check that $(KH)^{-1} = H^{-1}K^{-1}$.

The following lemma shows an important feature of sums and products of linear relations. The notation $I_{\mathfrak{M}}$ stands for the identity operator on the linear subspace $\mathfrak{M}$, while $O_{\mathfrak{M}}$ stands for the zero operator on $\mathfrak{M}$.

Lemma 1.1.4. *Let H be a linear relation from $\mathfrak{H}$ to $\mathfrak{K}$. Then*

$$H + (-H) = O_{\operatorname{dom} H} \,\widehat{+}\, (\{0\} \times \operatorname{mul} H), \tag{1.1.6}$$

where the sum is direct. Moreover, the identities

$$HH^{-1} = I_{\operatorname{ran} H} \,\widehat{+}\, (\{0\} \times \operatorname{mul} H) \tag{1.1.7}$$

and

$$H^{-1}H = I_{\operatorname{dom} H} \,\widehat{+}\, (\{0\} \times \ker H) \tag{1.1.8}$$

hold, and both sums are direct.

Proof. First the identity (1.1.6) will be shown. For an element on the left-hand side of (1.1.6) one has

$$\{h, h' - h''\} = \{h, 0\} + \{0, h' - h''\},$$

where $\{h, h'\}, \{h, h''\} \in H$, so that $\{h, 0\} \in O_{\operatorname{dom} H}$ and $\{0, h' - h''\} \in \{0\} \times \operatorname{mul} H$.

Conversely, let $\{h,k\} \in O_{\mathrm{dom}\,H} \,\widehat{+}\, (\{0\}\times \mathrm{mul}\, H)$. Then $\{h,k\} = \{h,0\}+\{0,k\}$ with $h \in \mathrm{dom}\, H$ and $k \in \mathrm{mul}\, H$. Hence, $\{h,h'\} \in H$ for some $h' \in \mathfrak{K}$ so that also $\{h, h'-k\} \in H$. Consequently,

$$\{h,k\} = \{h, h' - (h'-k)\} \in H + (-H),$$

which completes the proof of (1.1.6).

The assertion (1.1.8) follows from (1.1.7) by replacing H with H^{-1}. Hence, only the identity in (1.1.7) has to be proved. By definition, the linear relation HH^{-1} is given by

$$HH^{-1} = \bigl\{\{h,h''\} : \{h,h'\} \in H^{-1},\ \{h',h''\} \in H\bigr\}.$$

Therefore, if $\{h,h''\} \in HH^{-1}$ with some $\{h,h'\} \in H^{-1}$ and $\{h',h''\} \in H$, then

$$\{h,h''\} = \{h,h\} + \{0, h''-h\}.$$

As $\{h',h\} \in H$, it follows that $h \in \mathrm{ran}\, H$ and

$$\{0, h''-h\} = \{h',h''\} - \{h',h\} \in H,$$

i.e., $h''-h \in \mathrm{mul}\, H$. Thus, $\{h,h''\} \in I_{\mathrm{ran}\,H} \,\widehat{+}\, (\{0\}\times \mathrm{mul}\, H)$.

Conversely, given an element $\{h,h\} + \{0,k\} \in I_{\mathrm{ran}\,H} \,\widehat{+}\, (\{0\}\times \mathrm{mul}\, H)$ with $h \in \mathrm{ran}\, H$ and $k \in \mathrm{mul}\, H$, there exists $h' \in \mathrm{dom}\, H$ such that $\{h',h\} \in H$ or, equivalently, $\{h,h'\} \in H^{-1}$. Since $\{0,k\} \in H$ it follows $\{h', h+k\} \in H$, so that $\{h, h+k\} \in HH^{-1}$. □

Thus far the Hilbert space structure of the spaces has not been used; only the linear space structure played a role. Now an interpretation of the componentwise sum $H \,\widehat{+}\, K$ in (1.1.1) will be given as an orthogonal componentwise sum. Let $\mathfrak{H}_1$, $\mathfrak{H}_2$, $\mathfrak{K}_1$, and $\mathfrak{K}_2$ be Hilbert spaces and let $\mathfrak{H} = \mathfrak{H}_1 \oplus \mathfrak{H}_2$ and $\mathfrak{K} = \mathfrak{K}_1 \oplus \mathfrak{K}_2$. Here and in the following $\mathfrak{H}_1$ and $\mathfrak{H}_2$ are viewed as closed linear subspaces of $\mathfrak{H}$, and $\mathfrak{K}_1$ and $\mathfrak{K}_2$ are viewed as closed linear subspaces of $\mathfrak{K}$. Assume that H is a linear relation from $\mathfrak{H}_1$ to $\mathfrak{K}_1$ and that K is a linear relation from $\mathfrak{H}_2$ to $\mathfrak{K}_2$. The *orthogonal sum* $H \,\widehat{\oplus}\, K$ is defined as

$$H \,\widehat{\oplus}\, K = \bigl\{\{h+k, h'+k'\} : \{h,h'\} \in H,\ \{k,k'\} \in K\bigr\}.$$

In other words, $H \,\widehat{\oplus}\, K$ is just the componentwise sum $H \,\widehat{+}\, K$ of H and K, when these linear relations H and K are interpreted as linear relations from $\mathfrak{H} = \mathfrak{H}_1 \oplus \mathfrak{H}_2$ to $\mathfrak{K} = \mathfrak{K}_1 \oplus \mathfrak{K}_2$. If $\mathfrak{H} = \mathfrak{K}$ and $\mathfrak{H}_1 = \mathfrak{K}_1$, $\mathfrak{H}_2 = \mathfrak{K}_2$, then this definition implies

$$(H \,\widehat{\oplus}\, K)^2 = H^2 \,\widehat{\oplus}\, K^2. \tag{1.1.9}$$

A linear relation H from $\mathfrak{H}$ to $\mathfrak{K}$ is called *bounded* if there is a constant $C \geq 0$ such that $\|h'\|_{\mathfrak{K}} \leq C\|h\|_{\mathfrak{H}}$ for all $\{h,h'\} \in H$. In this case it is clear that

$\operatorname{mul} H = \{0\}$, so that H is a bounded operator. Thus, there is no distinction between bounded linear relations or bounded linear operators. The set of everywhere defined bounded linear operators from $\mathfrak{H}$ to $\mathfrak{K}$ will be denoted by $\mathbf{B}(\mathfrak{H}, \mathfrak{K})$. If $\mathfrak{H} = \mathfrak{K}$ the notation $\mathbf{B}(\mathfrak{H})$ is used instead of $\mathbf{B}(\mathfrak{H}, \mathfrak{H})$.

A linear relation from $\mathfrak{H}$ to $\mathfrak{K}$ is called *closed* if it is closed as a linear subspace of $\mathfrak{H} \times \mathfrak{K}$. The *closure* $\overline{H}$ of the linear relation H as a linear subspace of $\mathfrak{H} \times \mathfrak{K}$ is itself a closed linear relation. It follows that $\operatorname{mul} H \subset \operatorname{mul} \overline{H}$; if $\operatorname{mul} H = \{0\}$ implies that $\operatorname{mul} \overline{H} = \{0\}$, then the operator H is called *closable* (as an operator). The following useful observations are easily verified.

Lemma 1.1.5. *Let H be a linear operator from $\mathfrak{H}$ to $\mathfrak{K}$. Then the following statements hold:*

(i) *Let H be closable. If* $\operatorname{dom} H$ *is closed, then H is closed.*

(ii) *Let H be bounded. Then H is closable.*

(iii) *Let H be bounded. Then* $\operatorname{dom} H$ *is closed if and only if H is closed.*

A linear relation H from $\mathfrak{H}$ to $\mathfrak{K}$ is called *contractive* if $\|h'\|_{\mathfrak{K}} \leq \|h\|_{\mathfrak{H}}$ for all $\{h, h'\} \in H$ and it is called *isometric* if $\|h'\|_{\mathfrak{K}} = \|h\|_{\mathfrak{H}}$ for all $\{h, h'\} \in H$. In each case $\operatorname{mul} H = \{0\}$ and H is an operator which is bounded and thus closable; cf. Lemma 1.1.5. Hence, there is no distinction between contractive relations or operators. Likewise, there is no distinction between isometric relations or operators. Clearly, the closure of a contractive or isometric operator is again contractive or isometric. Recall that a contraction H has the following useful property: if $\|Hk\|_{\mathfrak{K}} = \|k\|_{\mathfrak{H}}$ for some $k \in \operatorname{dom} H$, then

$$(Hh, Hk)_{\mathfrak{K}} = (h, k)_{\mathfrak{H}} \quad \text{for all} \quad h \in \operatorname{dom} H. \tag{1.1.10}$$

To see this, note that for all $\lambda \in \mathbb{C}$

$$\begin{aligned} 0 &\leq \|h + \lambda k\|_{\mathfrak{H}}^2 - \|H(h + \lambda k)\|_{\mathfrak{K}}^2 \\ &= \|h\|_{\mathfrak{H}}^2 - \|Hh\|_{\mathfrak{K}}^2 - 2\mathrm{Re}\left(\bar{\lambda}\left[(Hh, Hk)_{\mathfrak{K}} - (h, k)_{\mathfrak{H}}\right]\right), \end{aligned}$$

which implies that (1.1.10) holds.

For many combinations of linear relations the closedness is preserved. For instance, if H is a closed linear relation from $\mathfrak{H}$ to $\mathfrak{K}$, then H^{-1} is a closed linear relation from $\mathfrak{K}$ to $\mathfrak{H}$. Likewise, for $\lambda \neq 0$ the product λH is closed. If H and K are closed linear relations from $\mathfrak{H}$ to $\mathfrak{K}$, then the componentwise sum $H \mathbin{\widehat{+}} K$ is not necessarily closed (see Appendix C), while the orthogonal componentwise sum $H \mathbin{\widehat{\oplus}} K$ of H and K is closed. The sum $H + K$ of two closed linear relations H and K is not necessarily closed. However, in the special case that H is closed and $K \in \mathbf{B}(\mathfrak{H}, \mathfrak{K})$ the sum

$$H + K = \big\{\{h, h' + Kh\} : \{h, h'\} \in H\big\}$$

is also closed. In particular, the linear relation H in $\mathfrak{H}$ is closed if and only if $H + \lambda$ is closed for some, and hence for all $\lambda \in \mathbb{C}$. The product KH of closed linear relations K and H is not necessarily closed. However, in the special case that K is closed and $H \in \mathbf{B}(\mathfrak{H}, \mathfrak{K})$ the product

$$KH = \big\{\{h, h''\} : \{Hh, h''\} \in K\big\}$$

is also closed.

The above material will be used throughout the text. The rest of this section will be devoted to two specific items, namely, a discussion of questions around the so-called resolvent identity, and one involving Möbius transformations of linear relations.

For a linear relation H in $\mathfrak{H}$ and $\lambda \in \mathbb{C}$, the *resolvent relation* is defined by $(H-\lambda)^{-1}$. Clearly, H is closed if and only if $(H-\lambda)^{-1}$ is closed for some, and hence for all $\lambda \in \mathbb{C}$. The resolvent relation has a number of properties which will now be explored. First the λ-independence of $\ker(H-\lambda)^{-1}$ and $\mathrm{mul}\,(H-\lambda)$ is stated.

Lemma 1.1.6. *Let H be a linear relation in $\mathfrak{H}$ and let $\lambda \in \mathbb{C}$. Then*

$$\ker(H-\lambda)^{-1} = \mathrm{mul}\,(H-\lambda) = \mathrm{mul}\,H.$$

For practical purposes it is worthwhile mentioning the analogs of (1.1.7) and (1.1.8) for the resolvent relation of H. Using Lemma 1.1.6 one sees that

$$(H-\lambda)(H-\lambda)^{-1} = I_{\mathrm{ran}\,(H-\lambda)} \,\widehat{+}\, \big(\{0\} \times \mathrm{mul}\,H\big),$$

and, likewise,

$$(H-\lambda)^{-1}(H-\lambda) = I_{\mathrm{dom}\,H} \,\widehat{+}\, \big(\{0\} \times \ker(H-\lambda)\big).$$

In particular, when $\ker(H-\lambda) = \{0\}$ for some $\lambda \in \mathbb{C}$, one has

$$(H-\lambda)^{-1}(H-\lambda) = I_{\mathrm{dom}\,H}.$$

The *resolvent identity* in the next proposition involves a combination of the sum and the product of the resolvent relations $(H-\lambda)^{-1}$ and $(H-\mu)^{-1}$.

Proposition 1.1.7. *Let H be a linear relation in $\mathfrak{H}$ and let $\lambda, \mu \in \mathbb{C}$. Then*

$$(H-\lambda)^{-1} - (H-\mu)^{-1} = (H-\lambda)^{-1}(\lambda-\mu)(H-\mu)^{-1}. \tag{1.1.11}$$

If $\ker(H-\lambda) = \{0\}$ and $\ker(H-\mu) = \{0\}$, then $(H-\lambda)^{-1}$ and $(H-\mu)^{-1}$ are linear operators defined on $\mathrm{ran}\,(H-\lambda)$ and $\mathrm{ran}\,(H-\mu)$, respectively, with the same kernel $\mathrm{mul}\,H$. Moreover, if $\lambda \neq \mu$, then (1.1.11) may be written as

$$(H-\lambda)^{-1} - (H-\mu)^{-1} = (\lambda-\mu)(H-\lambda)^{-1}(H-\mu)^{-1}. \tag{1.1.12}$$

Proof. For the inclusion ($\subset$) in (1.1.11) let

$$\{h, h' - h''\} \in (H-\lambda)^{-1} - (H-\mu)^{-1},$$

with $\{h, h'\} \in (H-\lambda)^{-1}$ and $\{h, h''\} \in (H-\mu)^{-1}$. This gives

$$\{h', h + \lambda h'\} \in H \quad \text{and} \quad \{h'', h + \mu h''\} \in H,$$

which shows $\{h' - h'', \lambda h' - \mu h''\} \in H$, and thus $\{h' - h'', (\lambda - \mu)h''\} \in H - \lambda$ and

$$\big\{(\lambda - \mu)h'', h' - h''\big\} \in (H-\lambda)^{-1}.$$

Since $\{h, h''\} \in (H-\mu)^{-1}$, one sees that $\{h, (\lambda-\mu)h''\} \in (\lambda-\mu)(H-\mu)^{-1}$, as $\{h'', (\lambda-\mu)h''\} \in (\lambda-\mu)I$. Hence, the element $\{h, h'-h''\}$ belongs to the linear relation $(H-\lambda)^{-1}(\lambda-\mu)(H-\mu)^{-1}$, which shows the inclusion.

For the inclusion ($\supset$) in (1.1.11), let $\{h, h'\} \in (H-\lambda)^{-1}(\lambda-\mu)(H-\mu)^{-1}$. Then by definition there exists $k \in \mathfrak{H}$ such that

$$\{h, k\} \in (H-\mu)^{-1} \quad \text{and} \quad \{(\lambda-\mu)k, h'\} \in (H-\lambda)^{-1},$$

as $\{k, (\lambda-\mu)k\} \in (\lambda-\mu)I$. In addition, it is clear from $\{k, h\} \in H - \mu$ that $\{k, h + (\mu-\lambda)k\} \in H - \lambda$ and

$$\{h + (\mu-\lambda)k, k\} \in (H-\lambda)^{-1}.$$

Thus, it follows that $\{h, h'+k\} \in (H-\lambda)^{-1}$. Hence, $\{h, h'\} = \{h, h'+k-k\}$ belongs to $(H-\lambda)^{-1} - (H-\mu)^{-1}$, which shows the inclusion. This completes the proof of (1.1.11). If $\lambda \neq \mu$ this leads to (1.1.12).

The remaining statements follow directly from Lemma 1.1.6. $\square$

Note that in general the identity in (1.1.12) is not valid for $\lambda = \mu$. In this case the right-hand side of (1.1.12) clearly equals $O_{\mathrm{dom}\,(H-\lambda)^{-2}}$, while by (1.1.6) the left-hand side equals $O_{\mathrm{dom}\,(H-\lambda)^{-1}} \widehat{+} (\{0\} \times \mathrm{mul}\,(H-\lambda)^{-1})$. Hence, in (1.1.12) the right-hand side is contained in the left-hand side.

The following result shows that every linear relation H can be represented by means of a pair of operators expressed in terms of its resolvent operator $(H-\lambda)^{-1}$. This kind of representation of a linear relation will be considered in this text in various situations.

Lemma 1.1.8. *Let H be a linear relation in $\mathfrak{H}$ and assume that* $\ker(H-\lambda) = \{0\}$ *for some $\lambda \in \mathbb{C}$. Then*

$$H = \big\{\{(H-\lambda)^{-1}k, (I + \lambda(H-\lambda)^{-1})k\} : k \in \mathrm{ran}\,(H-\lambda)\big\}, \tag{1.1.13}$$

where the right-hand side is well defined since $\mathrm{dom}\,(H-\lambda)^{-1} = \mathrm{ran}\,(H-\lambda)$.

Proof. Denote the linear relation on the right-hand side of (1.1.13) by K. To see that $H \subset K$, let $\{h, h'\} \in H$. Then $\{h' - \lambda h, h\} \in (H-\lambda)^{-1}$ and from the assumption $\operatorname{mul}(H-\lambda)^{-1} = \ker(H-\lambda) = \{0\}$ it follows that

$$h = (H-\lambda)^{-1}(h' - \lambda h).$$

Therefore,

$$\begin{aligned}\{h, h'\} &= \{h, h' - \lambda h + \lambda h\} \\ &= \{(H-\lambda)^{-1}(h' - \lambda h), (I + \lambda(H-\lambda)^{-1})(h' - \lambda h)\},\end{aligned}$$

where $h' - \lambda h \in \operatorname{ran}(H-\lambda)$. Hence, $\{h, h'\} \in K$, so that $H \subset K$. Now the equality follows from Corollary 1.1.3, since

$$\operatorname{dom} K = \operatorname{ran}(H-\lambda)^{-1} = \operatorname{dom}(H-\lambda) = \operatorname{dom} H,$$

while

$$\operatorname{mul} K = \ker(H-\lambda)^{-1} = \operatorname{mul}(H-\lambda) = \operatorname{mul} H.$$

This completes the proof. □

Another algebraic identity involving the resolvent relations $(H-\lambda)^{-1}$ and $(H-\mu)^{-1}$ is contained in the next lemma; see also Corollary 1.2.8 in the next section. The formula in the lemma can also be checked via the Möbius transform to be defined below.

Lemma 1.1.9. *Let H be a linear relation in $\mathfrak{H}$ and let $\lambda, \mu \in \mathbb{C}$. Then*

$$\left((I + (\lambda - \mu)(H-\lambda)^{-1}\right)^{-1} = I + (\mu - \lambda)(H-\mu)^{-1}. \tag{1.1.14}$$

Proof. It is easy to see that

$$I + (\lambda - \mu)(H-\lambda)^{-1} = \{\{h' - \lambda h, h' - \mu h\} : \{h, h'\} \in H\},$$

and by symmetry

$$I + (\mu - \lambda)(H-\mu)^{-1} = \{\{h' - \mu h, h' - \lambda h\} : \{h, h'\} \in H\}.$$

This yields (1.1.14). □

Next, Möbius transformations of linear relations will be defined. For a Hilbert space $\mathfrak{H}$ and a 2×2 matrix

$$\mathcal{M} = \begin{pmatrix} \alpha & \beta \\ \gamma & \delta \end{pmatrix}, \qquad \alpha, \beta, \gamma, \delta \in \mathbb{C}, \tag{1.1.15}$$

the scalar *Möbius transform* $\mathcal{M}$ in $\mathfrak{H}^2 = \mathfrak{H} \times \mathfrak{H}$ is given by

$$\mathcal{M} : \mathfrak{H}^2 \to \mathfrak{H}^2, \qquad \{h, h'\} \mapsto \{\alpha h + \beta h', \gamma h + \delta h'\}.$$

The meaning of $\mathcal{M}$, either as a matrix or as a transformation, will be clear from the context. The scalar Möbius transform of a linear relation is defined as follows.

Definition 1.1.10. Let H be a linear relation in $\mathfrak{H}$ and let $\mathcal{M}$ be a 2×2 matrix as in (1.1.15). Then the *scalar Möbius transform* of H is the linear relation $\mathcal{M}[H]$ in $\mathfrak{H}$ defined by

$$\mathcal{M}[H] = \big\{\{\alpha h + \beta h', \gamma h + \delta h'\} : \{h, h'\} \in H\big\}. \tag{1.1.16}$$

Note that the domain and range of the scalar Möbius transform $\mathcal{M}[H]$ are given by

$$\begin{aligned} \operatorname{dom} \mathcal{M}[H] &= \big\{\alpha h + \beta h' : \{h, h'\} \in H\big\}, \\ \operatorname{ran} \mathcal{M}[H] &= \big\{\gamma h + \delta h' : \{h, h'\} \in H\big\}. \end{aligned}$$

If the 2×2 matrix $\mathcal{M}$ in Definition 1.1.10 is multiplied by a constant $\eta \in \mathbb{C} \setminus \{0\}$, then the corresponding Möbius transform $\mathcal{M}[H]$ and $(\eta\mathcal{M})[H]$ coincide.

Let $\mathcal{M}$ and $\mathcal{N}$ be 2×2 matrices. Then the identity

$$\mathcal{N}[\mathcal{M}[H]] = (\mathcal{N} \circ \mathcal{M})[H] \tag{1.1.17}$$

holds for any linear relation H in $\mathfrak{H}$. If $\det \mathcal{M} \neq 0$, then

$$\mathcal{M}^{-1} = \frac{1}{\alpha\delta - \beta\gamma} \begin{pmatrix} \delta & -\beta \\ -\gamma & \alpha \end{pmatrix}$$

and the Möbius transform corresponding to $\mathcal{M}^{-1}$ is given by

$$\mathcal{M}^{-1}[H] = \big\{\{\delta h - \beta h', -\gamma h + \alpha h'\} : \{h, h'\} \in H\big\}.$$

Thus, for any linear relation H one has

$$\mathcal{M}^{-1}[\mathcal{M}[H]] = H = \mathcal{M}[\mathcal{M}^{-1}[H]];$$

cf. (1.1.17). Note that in general $\mathcal{M}^{-1}[H]$ and $\mathcal{M}[H]^{-1}$ are different relations. In the case $\det \mathcal{M} \neq 0$ it clearly follows that

$$\mathcal{M}[H] \text{ is closed if and only if } H \text{ is closed.} \tag{1.1.18}$$

Observe that the linear relations λH, $H - \lambda$, H^{-1} correspond to the Möbius transforms determined by the following matrices

$$\begin{pmatrix} 1 & 0 \\ 0 & \lambda \end{pmatrix}, \quad \begin{pmatrix} 1 & 0 \\ -\lambda & 1 \end{pmatrix}, \quad \begin{pmatrix} 0 & 1 \\ 1 & 0 \end{pmatrix},$$

respectively. Thus, for instance, the linear relations $I + (\lambda - \mu)(H - \lambda)^{-1}$ and $I + (\mu - \lambda)(H - \mu)^{-1}$ correspond to Möbius transforms of H determined by the matrices

$$\begin{pmatrix} 1 & 0 \\ 1 & 1 \end{pmatrix} \begin{pmatrix} 1 & 0 \\ 0 & \lambda - \mu \end{pmatrix} \begin{pmatrix} 0 & 1 \\ 1 & 0 \end{pmatrix} \begin{pmatrix} 1 & 0 \\ -\lambda & 1 \end{pmatrix} = \begin{pmatrix} -\lambda & 1 \\ -\mu & 1 \end{pmatrix}$$

and

$$\begin{pmatrix} 1 & 0 \\ 1 & 1 \end{pmatrix} \begin{pmatrix} 1 & 0 \\ 0 & \mu - \lambda \end{pmatrix} \begin{pmatrix} 0 & 1 \\ 1 & 0 \end{pmatrix} \begin{pmatrix} 1 & 0 \\ -\mu & 1 \end{pmatrix} = \begin{pmatrix} -\mu & 1 \\ -\lambda & 1 \end{pmatrix},$$

respectively. This also confirms the identity (1.1.14).

For a 2×2 matrix $\mathcal{M}$ as in (1.1.15) with $\det \mathcal{M} \neq 0$ define the function

$$\lambda \mapsto \mathcal{M}[\lambda] = \frac{\gamma + \lambda\delta}{\alpha + \lambda\beta}, \qquad \alpha + \lambda\beta \neq 0. \tag{1.1.19}$$

Since the linear relation $\mathcal{M}[H] - \mathcal{M}[\lambda]$ corresponds to the matrix

$$\begin{pmatrix} 1 & 0 \\ -\mathcal{M}[\lambda] & 1 \end{pmatrix} \begin{pmatrix} \alpha & \beta \\ \gamma & \delta \end{pmatrix} = \begin{pmatrix} \alpha & \beta \\ \frac{-\lambda \det \mathcal{M}}{\alpha+\lambda\beta} & \frac{\det \mathcal{M}}{\alpha+\lambda\beta} \end{pmatrix},$$

one sees from (1.1.16) that for $\alpha + \lambda\beta \neq 0$,

$$\mathcal{M}[H] - \mathcal{M}[\lambda] = \Big\{ \big\{ \alpha h + \beta h', \tfrac{\det \mathcal{M}}{\alpha+\lambda\beta}(h' - \lambda h) \big\} : \{h, h'\} \in H \Big\}.$$

This identity yields, in particular, for $\alpha + \beta\lambda \neq 0$, that

$$\begin{aligned} \ker (H - \lambda) &= \ker \big(\mathcal{M}[H] - \mathcal{M}[\lambda]\big), \\ \operatorname{ran} (H - \lambda) &= \operatorname{ran} \big(\mathcal{M}[H] - \mathcal{M}[\lambda]\big). \end{aligned} \tag{1.1.20}$$

If, in addition, $\beta \neq 0$, then it follows from (1.1.16) that

$$\operatorname{mul} \mathcal{M}[H] = \ker (H + \alpha\beta^{-1}), \qquad \operatorname{mul} H = \ker \big(\mathcal{M}[H] - \delta\beta^{-1}\big),$$

and in the case $\beta = 0$ it is easy to see that $\operatorname{mul} \mathcal{M}[H] = \operatorname{mul} H$.

Proposition 1.1.11. *Let H be a linear relation in $\mathfrak{H}$ and let $\mathcal{M}$ be a 2×2 matrix as in (1.1.15) with $\det \mathcal{M} \neq 0$. Then for $\alpha + \lambda\beta \neq 0$*

$$\big(\mathcal{M}[H] - \mathcal{M}[\lambda]\big)^{-1} = \frac{(\alpha + \lambda\beta)\beta}{\det \mathcal{M}} + \frac{(\alpha + \lambda\beta)^2}{\det \mathcal{M}}(H - \lambda)^{-1}. \tag{1.1.21}$$

Proof. Use the abbreviation $\Delta = \det \mathcal{M}$. It suffices to see that the left-hand side corresponds to the matrix

$$\begin{pmatrix} 0 & 1 \\ 1 & 0 \end{pmatrix} \begin{pmatrix} 1 & 0 \\ -\mathcal{M}[\lambda] & 1 \end{pmatrix} \begin{pmatrix} \alpha & \beta \\ \gamma & \delta \end{pmatrix} = \begin{pmatrix} \frac{-\lambda\Delta}{\alpha+\lambda\beta} & \frac{\Delta}{\alpha+\lambda\beta} \\ \alpha & \beta \end{pmatrix},$$

while the right-hand side corresponds to the matrix

$$\begin{pmatrix} 1 & 0 \\ \frac{(\alpha+\lambda\beta)\beta}{\Delta} & 1 \end{pmatrix} \begin{pmatrix} 1 & 0 \\ 0 & \frac{(\alpha+\lambda\beta)^2}{\Delta} \end{pmatrix} \begin{pmatrix} 0 & 1 \\ 1 & 0 \end{pmatrix} \begin{pmatrix} 1 & 0 \\ -\lambda & 1 \end{pmatrix} = \begin{pmatrix} -\lambda & 1 \\ \frac{\alpha(\alpha+\lambda\beta)}{\Delta} & \frac{\beta(\alpha+\lambda\beta)}{\Delta} \end{pmatrix}.$$

Since these matrices coincide up to a nonzero multiplicative constant the assertion follows. □

It is clear that the following useful consequence of Proposition 1.1.11 is obtained by means of the special choice

$$\mathcal{M} = \begin{pmatrix} 0 & 1 \\ 1 & 0 \end{pmatrix},$$

so that $\det \mathcal{M} = -1$, $\mathcal{M}[H] = H^{-1}$, and $\mathcal{M}[\lambda] = 1/\lambda$, $\lambda \neq 0$.

Corollary 1.1.12. *Let H be a linear relation in $\mathfrak{H}$ and let $\lambda \in \mathbb{C} \setminus \{0\}$. Then*

$$(H^{-1} - \lambda^{-1})^{-1} = -\lambda - \lambda^2 (H - \lambda)^{-1}. \tag{1.1.22}$$

Next the Cayley transform and inverse Cayley transform of a linear relation will be introduced. These special Möbius transforms will be used later in Sections 1.5, 1.6, and 1.7.

Definition 1.1.13. Let H and V be linear relations in $\mathfrak{H}$ and let $\mu \in \mathbb{C} \setminus \mathbb{R}$. Then the *Cayley transform* $\mathcal{C}_\mu$ of H and the *inverse Cayley transform* $\mathcal{F}_\mu$ of V are defined by

$$\begin{aligned} \mathcal{C}_\mu[H] &= \big\{\{h' - \mu h, h' - \bar{\mu} h\} : \{h, h'\} \in H\big\}, \\ \mathcal{F}_\mu[V] &= \big\{\{k - k', \bar{\mu} k - \mu k'\} : \{k, k'\} \in V\big\}. \end{aligned} \tag{1.1.23}$$

Notice that the domain and range of the Cayley transform $\mathcal{C}_\mu$ and the inverse Cayley transform $\mathcal{F}_\mu$ are given by

$$\begin{aligned} \operatorname{dom} \mathcal{C}_\mu[H] &= \operatorname{ran}(H - \mu), & \operatorname{ran} \mathcal{C}_\mu[H] &= \operatorname{ran}(H - \bar{\mu}), \\ \operatorname{dom} \mathcal{F}_\mu[V] &= \operatorname{ran}(I - V), & \operatorname{ran} \mathcal{F}_\mu[V] &= \operatorname{ran}(\bar{\mu} - \mu V). \end{aligned} \tag{1.1.24}$$

It is clear that the Cayley transform $\mathcal{C}_\mu$ and the inverse Cayley transform $\mathcal{F}_\mu$ are Möbius transforms corresponding to the matrices

$$\mathcal{C}_\mu = \begin{pmatrix} -\mu & 1 \\ -\bar{\mu} & 1 \end{pmatrix} \quad \text{and} \quad \mathcal{F}_\mu = \begin{pmatrix} 1 & -1 \\ \bar{\mu} & -\mu \end{pmatrix} = (\bar{\mu} - \mu)\mathcal{C}_\mu^{-1}, \tag{1.1.25}$$

where $\det \mathcal{C}_\mu = \bar{\mu} - \mu$ was used. Note also that

$$\mathcal{C}_\mu[\lambda] = \frac{\lambda - \bar{\mu}}{\lambda - \mu}, \quad \lambda \neq \mu.$$

Thus, Proposition 1.1.11 leads to the following result.

Corollary 1.1.14. *Let H be a linear relation in $\mathfrak{H}$ and let $\mu \in \mathbb{C} \setminus \mathbb{R}$. Then*

$$\big(\mathcal{C}_\mu[H] - \mathcal{C}_\mu[\lambda]\big)^{-1} = \frac{\lambda - \mu}{\bar{\mu} - \mu} + \frac{(\lambda - \mu)^2}{\bar{\mu} - \mu}(H - \lambda)^{-1}, \quad \lambda \neq \mu. \tag{1.1.26}$$

1.2 Spectra, resolvent sets, and points of regular type

The resolvent set, spectrum, point, continuous, and residual spectrum, and the points of regular type of a linear relation or operator are defined. A priori it is not assumed that the linear relation is closed. Here and in the rest of the text linear relations will be referred to simply as relations and linear subspaces as subspaces.

Definition 1.2.1. Let H be a relation in $\mathfrak{H}$. Then $\lambda \in \mathbb{C}$ is said to be a *point of regular type* of H if $(H-\lambda)^{-1}$ is a (in general not everywhere defined) bounded operator. The set of points of regular type of H is denoted by $\gamma(H)$.

Some straightforward consequences of Definition 1.2.1 are presented in the next lemma.

Lemma 1.2.2. *Let H be a relation in $\mathfrak{H}$. Then $\lambda \in \gamma(H)$ if and only if there exists a positive constant c, depending on λ, such that*

$$\|h\| \leq c\|h' - \lambda h\|, \quad \{h, h'\} \in H. \tag{1.2.1}$$

Moreover, if $\gamma(H) \neq \emptyset$, then H is closed if and only if $\operatorname{ran}(H-\lambda)$ *is closed for some, and hence for all $\lambda \in \gamma(H)$.*

Proof. Assume that $\lambda \in \gamma(H)$, so that $(H-\lambda)^{-1}$ is a bounded operator. Let $\{h, h'\} \in H$; then $\{h' - \lambda h, h\} \in (H-\lambda)^{-1}$ and

$$\|h\| = \|(H-\lambda)^{-1}(h' - \lambda h)\| \leq c\|h' - \lambda h\|,$$

which gives (1.2.1). Conversely, assume that (1.2.1) holds. To see that $(H-\lambda)^{-1}$ is a bounded operator let $\{f, f'\} \in (H-\lambda)^{-1}$. Then $\{f, f'\} = \{h' - \lambda h, h\}$ for some $\{h, h'\} \in H$ and (1.2.1) shows $\|f'\| \leq c\|f\|$ for all $\{f, f'\} \in (H-\lambda)^{-1}$. This implies that $(H-\lambda)^{-1}$ is an operator that is bounded or, equivalently, $\lambda \in \gamma(H)$.

Assume that H is closed, so that also $(H-\lambda)^{-1}$ is closed. Then the relation $(H-\lambda)^{-1}$ is a closed and bounded operator for all $\lambda \in \gamma(H)$. This immediately implies that $\operatorname{ran}(H-\lambda) = \operatorname{dom}(H-\lambda)^{-1}$ is closed; cf. Lemma 1.1.5. Conversely, if $\operatorname{ran}(H-\lambda) = \operatorname{dom}(H-\lambda)^{-1}$ is closed for some $\lambda \in \gamma(H)$, then $(H-\lambda)^{-1}$ is a bounded operator defined on a closed subspace. It follows that $(H-\lambda)^{-1}$ is closed, cf. Lemma 1.1.5, and hence H is closed. □

Definition 1.2.3. Let H be a relation in $\mathfrak{H}$. A point $\lambda \in \mathbb{C}$ is said to belong to the *resolvent set* $\rho(H)$ of H if $(H-\lambda)^{-1}$ is a bounded operator and $\overline{\operatorname{ran}}(H-\lambda) = \mathfrak{H}$. The *spectrum* $\sigma(H)$ of H is the complement of $\rho(H)$ in $\mathbb{C}$. The spectrum $\sigma(H)$ decomposes into three disjoint components: the *point spectrum* $\sigma_p(H)$, *continuous spectrum* $\sigma_c(H)$, and *residual spectrum* $\sigma_r(H)$, defined by

$$\begin{aligned}
\sigma_p(H) &= \big\{\lambda \in \mathbb{C} : \ker(H-\lambda) \neq \{0\}\big\}, \\
\sigma_c(H) &= \big\{\lambda \in \mathbb{C} : \ker(H-\lambda) = \{0\},\ \overline{\operatorname{ran}}(H-\lambda) = \mathfrak{H},\ \lambda \notin \rho(H)\big\}, \\
\sigma_r(H) &= \big\{\lambda \in \mathbb{C} : \ker(H-\lambda) = \{0\},\ \overline{\operatorname{ran}}(H-\lambda) \neq \mathfrak{H}\big\}.
\end{aligned}$$

Let H be a relation in $\mathfrak{H}$. It follows from Definition 1.2.1 and Definition 1.2.3 that $\rho(H) \subset \gamma(H)$. Moreover, it follows from (1.2.1) that $\gamma(H) = \gamma(\overline{H})$ and the equivalence

$$\overline{\operatorname{ran}}\,(H - \lambda) = \mathfrak{H} \quad \Leftrightarrow \quad \overline{\operatorname{ran}}\,(\overline{H} - \lambda) = \mathfrak{H}$$

implies $\rho(H) = \rho(\overline{H})$.

The following state diagram is useful when discussing the spectral subsets and the resolvent set of H. The top row shows all possibilities for the range of $H-\lambda$. The first (second) rows show all possibilities for points λ such that $(H-\lambda)^{-1}$ is a bounded (unbounded) operator and the bottom row shows all possibilities for eigenvalues λ.

	$\operatorname{ran}(H-\lambda)$ $= \mathfrak{H}$	$\operatorname{ran}(H-\lambda)$ dense, $\neq \mathfrak{H}$	$\operatorname{ran}(H-\lambda)$ not dense
$(H-\lambda)^{-1}$ bounded operator	$\rho(H)$	$\rho(H)$	$\gamma(H) \cap \sigma_{\mathrm{r}}(H)$
$(H-\lambda)^{-1}$ unbounded operator	$\sigma_{\mathrm{c}}(H)$	$\sigma_{\mathrm{c}}(H)$	$\sigma_{\mathrm{r}}(H)$
$(H-\lambda)^{-1}$ not operator	$\sigma_{\mathrm{p}}(H)$	$\sigma_{\mathrm{p}}(H)$	$\sigma_{\mathrm{p}}(H)$

Now assume that H is a closed relation. Then it follows with the help of the closed graph theorem and Lemma 1.1.5 applied to the operator $(H-\lambda)^{-1}$ that two cases (marked by **X** below) in the above state diagram are not possible:

H closed relation	$\operatorname{ran}(H-\lambda)$ $= \mathfrak{H}$	$\operatorname{ran}(H-\lambda)$ dense, $\neq \mathfrak{H}$	$\operatorname{ran}(H-\lambda)$ not dense
$(H-\lambda)^{-1}$ bounded operator	$\rho(H)$	**X**	$\gamma(H) \cap \sigma_{\mathrm{r}}(H)$
$(H-\lambda)^{-1}$ unbounded operator	**X**	$\sigma_{\mathrm{c}}(H)$	$\sigma_{\mathrm{r}}(H)$
$(H-\lambda)^{-1}$ not operator	$\sigma_{\mathrm{p}}(H)$	$\sigma_{\mathrm{p}}(H)$	$\sigma_{\mathrm{p}}(H)$

In particular, for a closed relation the continuous spectrum is given by

$$\sigma_{\mathrm{c}}(H) = \bigl\{\lambda \in \mathbb{C} : \ker(H-\lambda) = \{0\},\ \overline{\operatorname{ran}}\,(H-\lambda) = \mathfrak{H},\ \operatorname{ran}(H-\lambda) \neq \mathfrak{H}\bigr\}.$$

Lemma 1.2.4. *A relation H in $\mathfrak{H}$ is closed if and only if* $\operatorname{ran}(H-\lambda) = \mathfrak{H}$ *for some, and hence for all $\lambda \in \rho(H)$. In this case the following statements hold:*

(i) $\rho(H) = \{\lambda \in \mathbb{C} : (H - \lambda)^{-1} \in \mathbf{B}(\mathfrak{H})\}$;

(ii) $H = \{\{(H - \lambda)^{-1} f, (I + \lambda(H - \lambda)^{-1})f\} : f \in \mathfrak{H}\}$ *for* $\lambda \in \rho(H)$.

Proof. If H is closed, then for all $\lambda \in \mathbb{C}$ also $(H-\lambda)^{-1}$ is closed. Hence, if $\lambda \in \rho(H)$, then $(H - \lambda)^{-1}$ is a bounded and closed operator, and therefore

$$\operatorname{dom}(H - \lambda)^{-1} = \operatorname{ran}(H - \lambda)$$

is closed and coincides with $\mathfrak{H}$; cf. Lemma 1.1.5. Conversely, if $\lambda \in \rho(H)$ and $\operatorname{ran}(H - \lambda) = \mathfrak{H}$, then $(H - \lambda)^{-1}$ is a bounded operator defined on $\mathfrak{H}$ and hence $(H - \lambda)^{-1}$ is closed by Lemma 1.1.5. This implies that also H is closed. Assertion (i) is now immediate and assertion (ii) follows from (1.1.13) in Lemma 1.1.8. □

In the next theorem the so-called defect of a relation H is studied. The proof uses the notions of opening and gap of closed subspaces from Appendix C.

Theorem 1.2.5. *Let H be a relation in $\mathfrak{H}$. Then the set $\gamma(H)$ of points of regular type of H is an open subset of $\mathbb{C}$ and the defect*

$$n_\lambda(H) := \dim \big(\operatorname{ran}(H - \lambda)\big)^\perp \tag{1.2.2}$$

of H is constant for all λ in a connected component of $\gamma(H)$.

Proof. Step 1. Let $\mu \in \gamma(H)$ and let $c_\mu > 0$ be any positive constant such that

$$\|h\| \le c_\mu \|h' - \mu h\|, \qquad \{h, h'\} \in H; \tag{1.2.3}$$

cf. Lemma 1.2.2. Hence, if $\lambda \in \mathbb{C}$ and $|\lambda - \mu| c_\mu < 1$, then

$$|\lambda - \mu| \|h\| \le |\lambda - \mu| c_\mu \|h' - \mu h\| < \|h' - \mu h\|.$$

In this case $h' - \lambda h = h' - \mu h - (\lambda - \mu)h$ yields

$$\|h' - \lambda h\| \ge \|h' - \mu h\| - |\lambda - \mu| \|h\| > 0,$$

and together with (1.2.3) this leads to

$$\begin{aligned} c_\mu \|h' - \lambda h\| &\ge c_\mu \|h' - \mu h\| - |\lambda - \mu| c_\mu \|h\| \\ &\ge \|h\| - |\lambda - \mu| c_\mu \|h\| \\ &= \big(1 - |\lambda - \mu| c_\mu\big) \|h\|. \end{aligned}$$

Since all elements of $(H - \lambda)^{-1}$ are of the form $\{h' - \lambda h, h\}$ with $\{h, h'\} \in H$, it follows from this inequality that $(H - \lambda)^{-1}$ is a bounded operator. In fact, one has for $|\lambda - \mu| c_\mu < 1$

$$\|(H - \lambda)^{-1} g\| \le \frac{c_\mu}{1 - |\lambda - \mu|\, c_\mu} \|g\|, \qquad g \in \operatorname{dom}(H - \lambda)^{-1}. \tag{1.2.4}$$

In particular, it follows that $\lambda \in \gamma(H)$ for $|\lambda - \mu| c_\mu < 1$. Therefore, $\gamma(H)$ is an open subset of $\mathbb{C}$.

Step 2. Let $\mu \in \gamma(H)$ and let P_μ be the orthogonal projection onto $\overline{\operatorname{ran}}\,(H-\mu)$. For each $f \in \mathfrak{H}$ one obtains

$$\|P_\mu f\| = \sup_{g\in\overline{\operatorname{ran}}\,(H-\mu)\setminus\{0\}} \frac{|(P_\mu f, g)|}{\|g\|} = \sup_{\{h,h'\}\in H\setminus\{0,0\}} \frac{|(f, h'-\mu h)|}{\|h'-\mu h\|},$$

since $\operatorname{ran}(H-\mu) = \{h'-\mu h : \{h,h'\} \in H\}$. Now choose $\lambda \in \mathbb{C}$ and write

$$h' - \mu h = h' - \lambda h + (\lambda - \mu)h.$$

If, in particular, $f \in \operatorname{ran}(H-\lambda)^\perp$, then $|(f, h'-\mu h)| = |\lambda-\mu||(f,h)|$ and it follows that

$$\begin{aligned}\|P_\mu f\| &= |\lambda-\mu| \sup_{\{h,h'\}\in H\setminus\{0,0\}} \frac{|(f,h)|}{\|h'-\mu h\|}\\ &\le |\lambda-\mu|\,\|f\| \sup_{\{h,h'\}\in H\setminus\{0,0\}} \frac{\|h\|}{\|h'-\mu h\|}.\end{aligned}$$

Let c_μ be as in Step 1, so that $\|h\| \le c_\mu\|h'-\mu h\|$ for $\{h,h'\} \in H$; cf. (1.2.3). Thus, for any $\lambda \in \mathbb{C}$ one has

$$\|P_\mu f\| \le |\lambda-\mu|c_\mu\|f\|, \quad f \in \operatorname{ran}(H-\lambda)^\perp. \tag{1.2.5}$$

Step 3. Let $\mu \in \gamma(H)$ and $|\lambda-\mu|c_\mu < 1$. By Step 1, $\lambda \in \gamma(H)$. Therefore, by symmetry, one obtains from Step 2 that

$$\|P_\lambda g\| \le |\lambda-\mu|c_\lambda\|g\|, \quad g \in \operatorname{ran}(H-\mu)^\perp, \tag{1.2.6}$$

where c_λ is any positive constant such that

$$\|(H-\lambda)^{-1}k\| \le c_\lambda\|k\| \quad \text{for} \quad k \in \operatorname{dom}(H-\lambda)^{-1}.$$

Due to the estimate (1.2.4) one may take

$$c_\lambda = \frac{c_\mu}{1-|\lambda-\mu|\,c_\mu},$$

and then one concludes from the estimate (1.2.6) that

$$\|P_\lambda g\| \le \frac{|\lambda-\mu|c_\mu}{1-|\lambda-\mu|c_\mu}\|g\|, \quad g \in \operatorname{ran}(H-\mu)^\perp, \tag{1.2.7}$$

for $|\lambda-\mu|c_\mu < 1$.

Step 4. Let $\mu \in \gamma(H)$ and assume that $|\lambda-\mu|c_\mu < C$ for some number $0 < C < \frac{1}{2}$. Then $\lambda \in \gamma(H)$ by Step 1, and

$$\begin{aligned}\|P_\mu f\| &\le C\|f\|, \quad f \in \operatorname{ran}(H-\lambda)^\perp,\\ \|P_\lambda g\| &\le \frac{C}{1-C}\|g\|, \quad g \in \operatorname{ran}(H-\mu)^\perp,\end{aligned}$$

by (1.2.5) and (1.2.7) in Step 2 and Step 3. Therefore,

$$\omega\big(\overline{\mathrm{ran}}\,(H-\mu), \mathrm{ran}\,(H-\lambda)^{\perp}\big) = \|P_\mu(I-P_\lambda)\| \le C < 1$$

and

$$\omega\big(\overline{\mathrm{ran}}\,(H-\lambda), \mathrm{ran}\,(H-\mu)^{\perp}\big) = \|P_\lambda(I-P_\mu)\| \le \frac{C}{1-C} < 1,$$

where ω stands for the opening between closed linear subspaces; cf. Definition C.5. For the gap in Definition C.9 one obtains

$$g\big(\overline{\mathrm{ran}}\,(H-\mu), \overline{\mathrm{ran}}\,(H-\lambda)\big) < 1$$

from Proposition C.10, and hence Theorem C.12 applied to the closed linear subspaces $\mathfrak{M} = \overline{\mathrm{ran}}\,(H-\mu)$ and $\mathfrak{N} = \mathrm{ran}\,(H-\lambda)^{\perp}$ implies

$$\dim\big(\mathrm{ran}\,(H-\lambda)\big)^{\perp} = \dim\big(\mathrm{ran}\,(H-\mu)\big)^{\perp} \tag{1.2.8}$$

for $\mu \in \gamma(H)$ and $|\lambda-\mu|c_\mu < C$ for some $0 < C < \frac{1}{2}$.

Step 5. Now let Γ be a connected open component of $\gamma(H)$. Then Γ is arcwise connected and each pair of points $\{\lambda_1, \lambda_2\}$ in Γ can be connected by a (piecewise) connected compact curve. Each point μ of the curve is the center of an open disc such that (1.2.8) holds for all λ in the disc. By compactness, finitely many such open discs form a cover of the curve and hence

$$\dim\big(\mathrm{ran}\,(H-\lambda_1)\big)^{\perp} = \dim\big(\mathrm{ran}\,(H-\lambda_2)\big)^{\perp},$$

that is, the defect of H is constant in each connected component of $\gamma(H)$. □

The next theorem is concerned with the properties of points in the resolvent set of a relation. This leads to the resolvent identity.

Theorem 1.2.6. *Let H be a relation in $\mathfrak{H}$. The resolvent set $\rho(H)$ is an open subset of $\mathbb{C}$. The resolvent identity*

$$(H-\lambda)^{-1} - (H-\mu)^{-1} = (H-\lambda)^{-1}(\lambda-\mu)(H-\mu)^{-1} \tag{1.2.9}$$

holds for $\lambda, \mu \in \rho(H)$; here $(H-\lambda)^{-1}$ and $(H-\mu)^{-1}$ are bounded operators defined on ran $(H-\lambda)$ *and* ran $(H-\mu)$*, respectively. If, in addition, H is closed, then $(H-\lambda)^{-1}, (H-\mu)^{-1} \in \mathbf{B}(\mathfrak{H})$ for $\lambda, \mu \in \rho(H)$, and (1.2.9) can be written as*

$$(H-\lambda)^{-1} - (H-\mu)^{-1} = (\lambda-\mu)(H-\lambda)^{-1}(H-\mu)^{-1} \tag{1.2.10}$$

for all $\lambda, \mu \in \rho(H)$.

Proof. Recall that the inclusion $\rho(H) \subset \gamma(H)$ holds. In fact, the resolvent set $\rho(H)$ of H is made up of the components of $\gamma(H)$ where the defect $n_\lambda(H)$ in (1.2.2) is zero. It follows in the same way as in Step 1 of the proof of Theorem 1.2.5 that for $\mu \in \rho(H)$ and $\lambda \in \mathbb{C}$ such that $|\lambda-\mu|\|(H-\mu)^{-1}\| < 1$ one has $\lambda \in \rho(H)$, and hence $\rho(H)$ is open. The identity (1.2.9) follows from Proposition 1.1.7. □

Corollary 1.2.7. *Let H be a closed relation in $\mathfrak{H}$ and assume that $\mu \in \rho(H)$ and $|\lambda - \mu| \|(H-\mu)^{-1}\| < 1$. Then $\lambda \in \rho(H)$ and*

$$(H-\lambda)^{-1} = \sum_{n=0}^{\infty} (\lambda - \mu)^n (H-\mu)^{-(n+1)}, \tag{1.2.11}$$

where the series converges in $\mathbf{B}(\mathfrak{H})$. In particular, the mapping

$$\lambda \mapsto (H-\lambda)^{-1}$$

is holomorphic on $\rho(H)$ and the limit

$$\lim_{\lambda \to \mu} \frac{(H-\lambda)^{-1} - (H-\mu)^{-1}}{\lambda - \mu} = (H-\mu)^{-2}$$

exists in $\mathbf{B}(\mathfrak{H})$.

Proof. With the notation $R(\lambda) = (H-\lambda)^{-1}$ it follows from the resolvent identity (1.2.10) and induction that

$$R(\lambda) = \sum_{n=0}^{k} (\lambda - \mu)^n R(\mu)^{n+1} + (\lambda - \mu)^{k+1} R(\lambda) R(\mu)^{k+1}. \tag{1.2.12}$$

The last term on the right-hand side of (1.2.12) obeys the estimate

$$\|(\lambda - \mu)^{k+1} R(\lambda) R(\mu)^{k+1}\| \leq \|R(\lambda)\| \big(|\lambda - \mu| \|R(\mu)\|\big)^{k+1},$$

and hence the condition $|\lambda - \mu| \|R(\mu)\| < 1$ implies that it tends to 0 in $\mathbf{B}(\mathfrak{H})$ as $k \to \infty$. This implies (1.2.11) and the holomorphy of $\lambda \mapsto (H-\lambda)^{-1}$. The last assertion follows from (1.2.10). □

Corollary 1.2.8. *Let H be a closed relation in $\mathfrak{H}$ and let $\lambda, \mu \in \rho(H)$. Then the operator $I + (\lambda - \mu)(H-\lambda)^{-1} \in \mathbf{B}(\mathfrak{H})$ is invertible and*

$$\big(I + (\lambda - \mu)(H-\lambda)^{-1}\big)^{-1} = I + (\mu - \lambda)(H-\mu)^{-1}.$$

Proof. The formal identity in terms of relations follows from Lemma 1.1.9. Since both $I + (\lambda - \mu)(H-\lambda)^{-1}$ and $I + (\mu - \lambda)(H-\mu)^{-1}$ belong to $\mathbf{B}(\mathfrak{H})$, the assertion is clear. □

The resolvent identity in (1.2.10) characterizes the closed relation H in a specific way.

Proposition 1.2.9. *Let $\mathcal{E} \subset \mathbb{C}$ be a nonempty set and assume that the mapping $\lambda \mapsto B(\lambda)$ from $\mathcal{E}$ to $\mathbf{B}(\mathfrak{H})$ satisfies the identity*

$$B(\lambda) - B(\mu) = (\lambda - \mu) B(\lambda) B(\mu), \quad \lambda, \mu \in \mathcal{E}. \tag{1.2.13}$$

Then there exists a closed relation H in $\mathfrak{H}$ such that $\mathcal{E} \subset \rho(H)$ and

$$B(\lambda) = (H-\lambda)^{-1}, \quad \lambda \in \mathcal{E}.$$

Proof. Define for $\lambda \in \mathcal{E}$ the relation $H(\lambda)$ by

$$H(\lambda) = B(\lambda)^{-1} + \lambda.$$

Since $B(\lambda) \in \mathbf{B}(\mathfrak{H})$, one sees that $B(\lambda)$ and thus also $B(\lambda)^{-1}$ are closed. Hence, also the relation $H(\lambda)$ is closed. Note that

$$\operatorname{ran}(H(\lambda) - \lambda)) = \operatorname{ran} B(\lambda)^{-1} = \operatorname{dom} B(\lambda) = \mathfrak{H},$$

while

$$\ker(H(\lambda) - \lambda)) = \ker B(\lambda)^{-1} = \operatorname{mul} B(\lambda) = \{0\},$$

so that $\lambda \in \rho(H(\lambda))$.

Now let $\lambda, \mu \in \mathcal{E}$ and let $\{h, h'\} \in H(\lambda)$. Then $h = B(\lambda)(h' - \lambda h)$ and due to the identity (1.2.13) (with μ and λ interchanged) one gets

$$\begin{aligned} h &= B(\lambda)(h' - \lambda h) \\ &= B(\mu)(h' - \lambda h) - (\mu - \lambda)B(\mu)B(\lambda)(h' - \lambda h) \\ &= B(\mu)(h' - \mu h). \end{aligned}$$

This implies $\{h, h'\} \in H(\mu)$. Therefore, $H(\lambda) \subset H(\mu)$ which, by symmetry, leads to $H(\lambda) = H(\mu)$. One concludes that $H(\lambda)$ does not depend on $\lambda \in \mathcal{E}$. Thus, one sees that

$$(H - \lambda)^{-1} = B(\lambda) \quad \text{and} \quad \lambda \in \rho(H),$$

which completes the proof. □

Let again H be a relation in $\mathfrak{H}$, let $\mathcal{M}$ be a 2×2 matrix as in (1.1.15) such that $\det \mathcal{M} \neq 0$, and let

$$\mathcal{M}[H] = \bigl\{\{\alpha h + \beta h', \gamma h + \delta h'\} : \{h, h'\} \in H\bigr\}$$

be the corresponding Möbius transform of H in Definition 1.1.10. The question is how the spectrum of H behaves under the Möbius transformation. Let the function $\mathcal{M}[\lambda]$ be defined by (1.1.19).

Proposition 1.2.10. *Let H be a relation in $\mathfrak{H}$ and let $\mathcal{M}$ be a 2×2 matrix as in (1.1.15) with $\det \mathcal{M} \neq 0$. Then the following statements hold for $\alpha + \lambda\beta \neq 0$:*

(i) *$\lambda \in \rho(H)$ if and only if $\mathcal{M}[\lambda] \in \rho(\mathcal{M}[H])$;*

(ii) *$\lambda \in \gamma(H)$ if and only if $\mathcal{M}[\lambda] \in \gamma(\mathcal{M}[H])$;*

(iii) *$\lambda \in \sigma_i(H)$ if and only if $\mathcal{M}[\lambda] \in \sigma_i(\mathcal{M}[H])$, where $i = \mathrm{p}, \mathrm{c}, \mathrm{r}$.*

If the relation H is closed and the equivalent assertions in (i) *hold, then the operators in the identity* (1.1.21) *belong to $\mathbf{B}(\mathfrak{H})$.*

Proof. (i) Assume that $\lambda \in \rho(H)$, that is, $\operatorname{ran}(H-\lambda)$ is dense in $\mathfrak{H}$ and $(H-\lambda)^{-1}$ is a bounded operator. Then the identities in (1.1.20) imply that $\operatorname{ran}(\mathcal{M}[H]-\mathcal{M}[\lambda])$ is dense in $\mathfrak{H}$ and that $(\mathcal{M}[H]-\mathcal{M}[\lambda])^{-1}$ is an operator. It follows with (1.1.21) that $(\mathcal{M}[H]-\mathcal{M}[\lambda])^{-1}$ is a bounded operator. This shows $\mathcal{M}[\lambda] \in \rho(\mathcal{M}[H])$. The converse statement follows by applying $\mathcal{M}^{-1}$.

(ii) and (iii) are now straightforward consequences from (1.1.20), (1.1.21), and the above considerations. □

Let H be a relation in $\mathfrak{H}$ and let $\lambda \in \mathbb{C}$. Then it follows from Proposition 1.2.10 that for $\lambda \neq 0$

$$\lambda \in \rho(H) \quad \Leftrightarrow \quad \lambda^{-1} \in \rho(H^{-1}), \tag{1.2.14}$$

in which case the resolvent operators in (1.1.22) belong to $\mathbf{B}(\mathfrak{H})$. Likewise, it follows from Proposition 1.2.10 that for $\lambda \neq \mu$

$$\lambda \in \rho(H) \quad \Leftrightarrow \quad \mathcal{C}_\mu[\lambda] \in \rho(\mathcal{C}_\mu[H]),$$

in which case the resolvent operators in (1.1.26) belong to $\mathbf{B}(\mathfrak{H})$.

1.3 Adjoint relations

Here the adjoint of a relation will be introduced, again as a relation, which will be automatically linear and closed. If the original relation is the graph of an operator, its adjoint will be the graph of an operator precisely when the original operator is densely defined.

Definition 1.3.1. Let H be a relation from $\mathfrak{H}$ to $\mathfrak{K}$. The *adjoint* H^* of H is defined as a relation from $\mathfrak{K}$ to $\mathfrak{H}$ by

$$H^* := \big\{\{f,f'\} \in \mathfrak{K}\times\mathfrak{H} : (f',h)_{\mathfrak{H}} = (f,h')_{\mathfrak{K}} \ \text{ for all } \ \{h,h'\} \in H\big\}.$$

Let J be the flip-flop operator from $\mathfrak{H}\times\mathfrak{K}$ to $\mathfrak{K}\times\mathfrak{H}$ defined by

$$J\{f,f'\} = \{f',-f\}, \qquad \{f,f'\} \in \mathfrak{H}\times\mathfrak{K}. \tag{1.3.1}$$

Then it is clear from Definition 1.3.1 that

$$H^* = (JH)^\perp = JH^\perp, \tag{1.3.2}$$

where the orthogonal complements refer to the componentwise inner product in $\mathfrak{K}\times\mathfrak{H}$ and $\mathfrak{H}\times\mathfrak{K}$, respectively. Note that

$$\mathfrak{K}\times\mathfrak{H} = \overline{JH} \oplus (JH)^\perp \quad \text{and} \quad \mathfrak{H}\times\mathfrak{K} = \overline{H} \oplus H^\perp.$$

Clearly, if H and K are relations with $H \subset K$, then $K^* \subset H^*$. It also follows from (1.3.2) that H^* is a closed linear relation from $\mathfrak{K}$ to $\mathfrak{H}$. Note that (1.3.2) gives $J^{-1}H^* = H^\perp$, i.e.,

$$(J^{-1}H^*)^\perp = H^{\perp\perp} = \overline{H}.$$

Since J^{-1} is the flip-flop operator from $\mathfrak{K}\times\mathfrak{H}$ to $\mathfrak{H}\times\mathfrak{K}$, the left-hand side coincides with H^{**} and hence

$$H^{**} = \overline{H},$$

so that the double adjoint of H gives the closure of H in $\mathfrak{H}\times\mathfrak{K}$. As a byproduct, one obtains $H^* = (\overline{H})^*$. It follows directly from the definition that

$$(H^*)^{-1} = (H^{-1})^*, \tag{1.3.3}$$

and sometimes the notation $H^{-*} := (H^*)^{-1} = (H^{-1})^*$ will be used. These facts and some further elementary properties of adjoint relations are collected in the next proposition.

Proposition 1.3.2. *Let H be a relation from $\mathfrak{H}$ to $\mathfrak{K}$. Then the following statements hold:*

(i) *H^* is a closed linear relation, $(\overline{H})^* = H^*$, and $\overline{H} = H^{**}$;*
(ii) $(\operatorname{dom} H)^\perp = \operatorname{mul} H^*$ *and* $(\operatorname{dom} H^*)^\perp = \operatorname{mul} \overline{H}$*;*
(iii) $\ker H^* = (\operatorname{ran} H)^\perp$ *and* $(\ker H^*)^\perp = \overline{\operatorname{ran}}\, H$.

It is a direct consequence of Proposition 1.3.2 that

$$\overline{\operatorname{dom}}\, H = \overline{\operatorname{dom}}\, H^{**} \quad \text{and} \quad \overline{\operatorname{ran}}\, H = \overline{\operatorname{ran}}\, H^{**}.$$

The domain and range of the adjoint relation can be characterized as follows.

Lemma 1.3.3. *Let H be a relation from $\mathfrak{H}$ to $\mathfrak{K}$. Then* $\operatorname{dom} H^* \subset \mathfrak{K}$ *and* $\operatorname{ran} H^* \subset \mathfrak{H}$ *are characterized by*

$$\operatorname{dom} H^* = \bigl\{ f \in (\operatorname{mul} \overline{H})^\perp : |(f,h')| \le M_f \|h\| \text{ for all } \{h,h'\} \in H \bigr\},$$

and

$$\operatorname{ran} H^* = \bigl\{ f' \in (\ker \overline{H})^\perp : |(f',h)| \le M_{f'} \|h'\| \text{ for all } \{h,h'\} \in H \bigr\},$$

where M_f and $M_{f'}$ are nonnegative constants depending on f and f', respectively.

Proof. The first identity will be proved; the second identity follows from the first one by using H^{-1} instead of H, and (1.3.3). So let $f \in \operatorname{dom} H^*$. Then there exists an element $f' \in \mathfrak{H}$ with $\{f,f'\} \in H^*$. For $\{0,h'\} \in \overline{H}$ there exists $\{h_n,h_n'\} \in H$ with $\{h_n,h_n'\} \to \{0,h'\}$. Hence, it follows from

$$(f',h_n) = (f,h_n')$$

that $(f,h') = 0$. Thus, $f \perp \operatorname{mul} \overline{H}$. Furthermore, for all $\{h,h'\} \in H$ it follows that $|(f,h')| = |(f',h)| \le M_f \|h\|$. Hence, $\operatorname{dom} H^*$ is contained in the right-hand side.

To prove the converse inclusion, let f belong to the right-hand side. Since $f \in (\operatorname{mul} \overline{H})^\perp$, the linear relation from $\mathfrak{H}$ to $\mathbb{C}$ given by

$$\Phi = \bigl\{ \{h,(h',f)\} : \{h,h'\} \in H \bigr\}, \quad \operatorname{dom} \Phi = \operatorname{dom} H,$$

is the graph of a linear functional, which is bounded because

$$|(h', f)| \leq M_f \|h\| \text{ for all } \{h, h'\} \in H.$$

Its closure $\overline{\Phi}$ is a bounded linear functional on $\overline{\mathrm{dom}}\, H$, and by the Riesz representation theorem there exists an element $f' \in \overline{\mathrm{dom}}\, H$ such that

$$\overline{\Phi} h = (h, f'), \quad h \in \overline{\mathrm{dom}}\, H.$$

In particular, this shows that $(h', f) = (h, f')$ for all $\{h, h'\} \in H$, which means that $\{f, f'\} \in H^*$. □

Proposition 1.3.2 and Lemma 1.3.3 immediately yield the following corollary.

Corollary 1.3.4. *Let H be a relation from $\mathfrak{H}$ to $\mathfrak{K}$. Then the following statements hold:*

(i) *H^* is an operator if and only if* $\mathrm{dom}\, H$ *is dense in $\mathfrak{H}$;*

(ii) *$\overline{H}$ is an operator if and only if* $\mathrm{dom}\, H^*$ *is dense in $\mathfrak{K}$;*

(iii) *if $H \in \mathbf{B}(\mathfrak{H}, \mathfrak{K})$, then $H^* \in \mathbf{B}(\mathfrak{K}, \mathfrak{H})$.*

Proof. Items (i) and (ii) are immediate from Proposition 1.3.2. To prove (iii), assume $H \in \mathbf{B}(\mathfrak{H}, \mathfrak{K})$. Since $\mathrm{dom}\, H = \mathfrak{H}$ it follows that H^* is a (closed) operator. Moreover, since $\mathrm{mul}\, \overline{H} = \{0\}$ and H is bounded, Lemma 1.3.3 shows that $\mathrm{dom}\, H^* = \mathfrak{K}$. Now the closed graph theorem implies $H^* \in \mathbf{B}(\mathfrak{K}, \mathfrak{H})$. □

Occasionally the following situation comes up. Let $\mathfrak{M} \subset \mathfrak{H}$ and $\mathfrak{N} \subset \mathfrak{K}$ be (not necessarily closed) linear subspaces and let $H = \mathfrak{M} \times \mathfrak{N}$. Then

$$H^* = \big(J(\mathfrak{M} \times \mathfrak{N})\big)^{\perp} = (\mathfrak{N} \times \mathfrak{M})^{\perp} = \mathfrak{N}^{\perp} \times \mathfrak{M}^{\perp}. \tag{1.3.4}$$

Note that by the same argument $H^{**} = \mathfrak{M}^{\perp\perp} \times \mathfrak{N}^{\perp\perp} = \overline{\mathfrak{M}} \times \overline{\mathfrak{N}}$, which is of course clear from $H^{**} = \overline{H}$.

Let H and K be closed linear relations from $\mathfrak{H}$ to $\mathfrak{K}$. Then the componentwise sum $H \mathbin{\widehat{+}} K$ is closed if and only if $H^{\perp} \mathbin{\widehat{+}} K^{\perp}$ is closed (see Theorem C.3). Since $H^* = JH^{\perp}$ and $K^* = JK^{\perp}$, this implies

$$H \mathbin{\widehat{+}} K \text{ closed} \quad \Leftrightarrow \quad H^* \mathbin{\widehat{+}} K^* \text{ closed}. \tag{1.3.5}$$

The next theorem is a variant of the closed range theorem in the general context of linear relations.

Theorem 1.3.5. *Let H be a closed relation from $\mathfrak{H}$ to $\mathfrak{K}$. Then the following statements hold:*

(i) $\mathrm{dom}\, H$ *is closed if and only if* $\mathrm{dom}\, H^*$ *is closed;*

(ii) $\mathrm{ran}\, H$ *is closed if and only if* $\mathrm{ran}\, H^*$ *is closed.*

Proof. Since H and $\{0\} \times \mathfrak{K}$ are closed linear subspaces in $\mathfrak{H} \times \mathfrak{K}$, it follows from the equivalence (1.3.5) that

$$H \mathbin{\widehat{+}} (\{0\} \times \mathfrak{K}) = \operatorname{dom} H \times \mathfrak{K}$$

is closed if and only if

$$H^* \mathbin{\widehat{+}} (\{0\} \times \mathfrak{H}) = \operatorname{dom} H^* \times \mathfrak{H}$$

is closed; cf. (1.3.4). This implies that $\operatorname{dom} H$ is closed if and only if $\operatorname{dom} H^*$ is closed, that is, (i) holds. Assertion (ii) follows immediately by applying (i) to the inverse H^{-1}. □

An operator H from $\mathfrak{H}$ to $\mathfrak{K}$ is *unitary* if H is isometric and $\operatorname{dom} H = \mathfrak{H}$ and $\operatorname{ran} H = \mathfrak{K}$. The next result gives criteria for a relation from $\mathfrak{H}$ to $\mathfrak{K}$ in terms of its adjoint to be the graph of an isometric or unitary operator.

Lemma 1.3.6. *Let H be a relation from $\mathfrak{H}$ to $\mathfrak{K}$. Then the following statements hold:*

(i) *$H^{-1} \subset H^*$ if and only if H is an isometric operator;*

(ii) *$H^{-1} = H^*$ if and only if H is a unitary operator.*

Proof. (i) Assume that $H^{-1} \subset H^*$. For $\{h, h'\} \in H$ one has $\{h', h\} \in H^{-1} \subset H^*$ which implies $\|h\| = \|h'\|$ for $\{h, h'\} \in H$. This shows that H is an isometric operator. Conversely, let H be an isometric operator and $\{h', h\} \in H^{-1}$. Then $\{h, h'\} \in H$ and one has $(h, k) = (h', k')$ for all $\{k, k'\} \in H$ by polarization. This implies $\{h', h\} \in H^*$ and hence $H^{-1} \subset H^*$.

(ii) Assume that $H^{-1} = H^*$. Then H is closed and by (i) H is an isometric operator. Therefore, $\operatorname{dom} H$ is closed by Lemma 1.1.5, and

$$(\operatorname{dom} H)^\perp = \operatorname{mul} H^* = \operatorname{mul} H^{-1} = \ker H = \{0\}$$

implies $\operatorname{dom} H = \mathfrak{H}$. Note that H^{-1} satisfies $(H^{-1})^{-1} = (H^*)^{-1} = (H^{-1})^*$ and hence by the above argument $\operatorname{dom} H^{-1} = \mathfrak{K}$. This implies $\operatorname{ran} H = \mathfrak{K}$ and it follows that H is a unitary operator. Conversely, assume that the operator H is unitary. Then $H \in \mathbf{B}(\mathfrak{H}, \mathfrak{K})$, $H^{-1} \in \mathbf{B}(\mathfrak{K}, \mathfrak{H})$, and $H^* \in \mathbf{B}(\mathfrak{K}, \mathfrak{H})$ by Corollary 1.3.4. Since H is isometric, one has $H^{-1} \subset H^*$ by (i) and equality follows as both H^{-1} and H^* belong to $\mathbf{B}(\mathfrak{K}, \mathfrak{H})$. □

Unitary operators are often used to identify different relations or Hilbert spaces.

Definition 1.3.7. Let H be a relation in $\mathfrak{H}$ and let K be a relation in $\mathfrak{K}$. Then H and K are said to be *unitarily equivalent* if there exists a unitary operator $U \in \mathbf{B}(\mathfrak{H}, \mathfrak{K})$ such that $K = UHU^*$ or, equivalently,

$$K = \big\{\{Uh, Uh'\} : \{h, h'\} \in H\big\}. \tag{1.3.6}$$

Assume that the relations H in $\mathfrak{H}$ and K in $\mathfrak{K}$ satisfy (1.3.6). Then one has $\{k,k'\} \in K^*$ if and only if $(U^*k',h) = (U^*k,h')$ for all $\{h,h'\} \in H$, that is, $\{U^*k,U^*k'\} \in H^*$. By setting $\{k,k'\} = \{Uh,Uh'\}$ it also follows from this that $\{Uh,Uh'\} \in K^*$ if and only if $\{h,h'\} \in H^*$. Hence, one has

$$H^* = \big\{\{U^*k,U^*k'\} : \{k,k'\} \in K^*\big\}$$

and

$$K^* = \big\{\{Uh,Uh'\} : \{h,h'\} \in H^*\big\}. \tag{1.3.7}$$

Lemma 1.3.8. *Let H be a closed relation in $\mathfrak{H}$, let K be a closed relation in $\mathfrak{K}$, assume that $\rho(H) \cap \rho(K) \neq \emptyset$, and that $U \in \mathbf{B}(\mathfrak{H},\mathfrak{K})$ is unitary. Then H and K are unitarily equivalent if and only if*

$$(K-\lambda)^{-1} = U(H-\lambda)^{-1}U^* \tag{1.3.8}$$

for some, and hence for all $\lambda \in \rho(H) \cap \rho(K)$.

Proof. Assume that $K = UHU^*$. Then for all $\lambda \in \rho(H) \cap \rho(K)$ one has

$$K - \lambda = U(H-\lambda)U^*.$$

Taking inverses yields (1.3.8). Conversely, assume that the identity (1.3.8) holds for some $\lambda \in \rho(H) \cap \rho(K)$. Then

$$H = \big\{\{(H-\lambda)^{-1}f, (I+\lambda(H-\lambda)^{-1})f\} : f \in \mathfrak{H}\big\}$$

and

$$K = \big\{\{(K-\lambda)^{-1}g, (I+\lambda(K-\lambda)^{-1})g\} : g \in \mathfrak{K}\big\}$$

by Lemma 1.2.4. Therefore,

$$\begin{aligned} K &= \big\{\{(K-\lambda)^{-1}Uf, (I+\lambda(K-\lambda)^{-1})Uf\} : f \in \mathfrak{H}\big\} \\ &= \big\{\{U(H-\lambda)^{-1}f, U(I+\lambda(H-\lambda)^{-1})f\} : f \in \mathfrak{H}\big\} \\ &= \big\{\{Uh,Uh'\} : \{h,h'\} \in H\big\} \\ &= UHU^*, \end{aligned}$$

which completes the argument. □

The next proposition concerns the adjoint of the sum and of the product of relations in Hilbert spaces.

Proposition 1.3.9. *Let H and K be relations from $\mathfrak{H}$ to $\mathfrak{K}$, and let L be a relation from $\mathfrak{K}$ to $\mathfrak{G}$. Then the following statements hold:*

(i) *$H^* + K^* \subset (H+K)^*$, and if $K \in \mathbf{B}(\mathfrak{H},\mathfrak{K})$, then $H^* + K^* = (H+K)^*$;*
(ii) *$H^*L^* \subset (LH)^*$, and if $L \in \mathbf{B}(\mathfrak{K},\mathfrak{G})$, then $H^*L^* = (LH)^*$.*

Proof. (i) To show the inclusion $H^* + K^* \subset (H+K)^*$ assume that

$$\{f, f' + g'\} \in H^* + K^*, \quad \text{where} \quad \{f, f'\} \in H^*, \{f, g'\} \in K^*.$$

Next consider $\{h, h' + k'\} \in H + K$, where $\{h, h'\} \in H$ and $\{h, k'\} \in K$. Then $(f', h) = (f, h')$ and $(g', h) = (f, k')$, and hence

$$(f' + g', h) - (f, h' + k') = (f', h) - (f, h') + (g', h) - (f, k') = 0,$$

that is, $\{f, f' + g'\} \in (H+K)^*$. Now it will be shown that $K \in \mathbf{B}(\mathfrak{H}, \mathfrak{K})$ implies the inclusion $(H+K)^* \subset H^* + K^*$. Let $\{f, f'\} \in (H+K)^*$. Then $(f', h) = (f, h' + k')$ for all $\{h, h'\} \in H$ and $\{h, k'\} \in K$. Since $K \in \mathbf{B}(\mathfrak{H}, \mathfrak{K})$ and $K^* \in \mathbf{B}(\mathfrak{K}, \mathfrak{H})$, it follows that $k' = Kh$ and

$$(f', h) = (f, h') + (f, Kh) = (f, h') + (K^* f, h),$$

and hence $(f' - K^* f, h) = (f, h')$ holds for all $\{h, h'\} \in H$. Therefore, one sees $\{f, f' - K^* f\} \in H^*$ and $\{f, f'\} \in H^* + K^*$.

(ii) First the inclusion $H^* L^* \subset (LH)^*$ will be shown. Let $\{f, f'\} \in H^* L^*$, so that $\{f, g'\} \in L^*$ and $\{g', f'\} \in H^*$ for some $g' \in \mathfrak{K}$. Consider $\{h, l'\} \in LH$, where $\{h, h'\} \in H$ and $\{h', l'\} \in L$ for some $h' \in \mathfrak{K}$. Then $(g', h') = (f, l')$ and $(f', h) = (g', h')$ and hence

$$(f', h) - (f, l') = (g', h') - (g', h') = 0$$

for any $\{h, l'\} \in LH$. This shows $\{f, f'\} \in (LH)^*$. Assume now that $L \in \mathbf{B}(\mathfrak{K}, \mathfrak{G})$ and hence $L^* \in \mathbf{B}(\mathfrak{G}, \mathfrak{K})$. In order to show the inclusion $(LH)^* \subset H^* L^*$, let $\{f, f'\} \in (LH)^*$. For $\{h, h'\} \in H$ one has $\{h, Lh'\} \in LH$ and hence

$$(f', h) = (f, Lh') = (L^* f, h').$$

This implies $\{L^* f, f'\} \in H^*$ and together with $\{f, L^* f\} \in L^*$ one concludes $\{f, f'\} \in H^* L^*$. ☐

Let H be a relation from $\mathfrak{H}$ to $\mathfrak{K}$ and $\lambda \in \mathbb{C}$. The following consequences of Proposition 1.3.9 will prove useful:

$$(\lambda H)^* = \bar{\lambda} H^*,$$

and for $\mathfrak{H} = \mathfrak{K}$,

$$(H - \lambda)^* = H^* - \bar{\lambda}.$$

Hence, according to Proposition 1.3.2 (iii) one has

$$\ker (H^* - \bar{\lambda}) = \big(\operatorname{ran}(H - \lambda)\big)^{\perp} \quad \text{and} \quad \overline{\operatorname{ran}}\,(H - \lambda) = \big(\ker (H^* - \bar{\lambda})\big)^{\perp}. \tag{1.3.9}$$

Furthermore, by (1.3.3),

$$\big((H - \lambda)^{-1}\big)^* = (H^* - \bar{\lambda})^{-1}.$$

In the next proposition the connection between the spectra of H and H^* is discussed.

Proposition 1.3.10. *Let H be a relation in $\mathfrak{H}$ and let $\lambda \in \mathbb{C}$. Then the following statements hold:*

(i) $\lambda \in \rho(H) \Leftrightarrow \overline{\lambda} \in \rho(H^*)$;

(ii) $\lambda \in \sigma(H) \Leftrightarrow \overline{\lambda} \in \sigma(H^*)$.

If, in addition, the relation H is closed, then

(iii) $\lambda \in \sigma_{\mathrm{p}}(H)$ *and* $\overline{\mathrm{ran}}\,(H-\lambda) \neq \mathfrak{H} \Leftrightarrow \overline{\lambda} \in \sigma_{\mathrm{p}}(H^*)$ *and* $\overline{\mathrm{ran}}\,(H^*-\overline{\lambda}) \neq \mathfrak{H}$;

(iv) $\lambda \in \sigma_{\mathrm{p}}(H)$ *and* $\overline{\mathrm{ran}}\,(H-\lambda) = \mathfrak{H} \Leftrightarrow \overline{\lambda} \in \sigma_{\mathrm{r}}(H^*)$;

(v) $\lambda \in \sigma_{\mathrm{c}}(H) \Leftrightarrow \overline{\lambda} \in \sigma_{\mathrm{c}}(H^*)$.

Proof. (i) & (ii) If $\lambda \in \rho(H)$, then $(H-\lambda)^{-1}$ is a bounded operator with dense domain $\mathrm{ran}\,(H-\lambda)$, and hence it admits a continuous extension

$$\overline{(H-\lambda)^{-1}} = (\overline{H}-\lambda)^{-1} \in \mathbf{B}(\mathfrak{H}). \tag{1.3.10}$$

Thus, also $(H^*-\overline{\lambda})^{-1} = ((\overline{H}-\lambda)^{-1})^* \in \mathbf{B}(\mathfrak{H})$ and $\overline{\lambda} \in \rho(H^*)$ follows. Conversely, for $\overline{\lambda} \in \rho(H^*)$ one has $(H^*-\overline{\lambda})^{-1} \in \mathbf{B}(\mathfrak{H})$ since H^* is closed. Hence, also (1.3.10) holds and from this it is clear that $(H-\lambda)^{-1}$ is a bounded operator with dense domain $\mathrm{ran}\,(H-\lambda)$. This gives (i), and (ii) follows immediately from (i).

(iii)–(v) are direct consequences of (1.3.9). □

In the next lemma it turns out that the scalar Möbius transform in Definition 1.1.10 behaves under adjoints as scalar multiplication does. In order to formulate the result, let the conjugate of a 2×2 matrix $\mathcal{M}$ be defined by

$$\overline{\mathcal{M}} = \begin{pmatrix} \overline{\alpha} & \overline{\beta} \\ \overline{\gamma} & \overline{\delta} \end{pmatrix} \quad \text{when} \quad \mathcal{M} = \begin{pmatrix} \alpha & \beta \\ \gamma & \delta \end{pmatrix}.$$

The scalar Möbius transform corresponding to $\overline{\mathcal{M}}$ will be denoted by $\overline{\mathcal{M}}$. The special case of the following lemma for the Cayley transform is particularly useful.

Lemma 1.3.11. *Let H be a relation in $\mathfrak{H}$ and let $\mathcal{M}$ be a 2×2 matrix as in (1.1.15), and assume that $\mathcal{M}$ is invertible. Then*

$$(\mathcal{M}[H])^* = \overline{\mathcal{M}}[H^*].$$

In particular, for any $\mu \in \mathbb{C} \setminus \mathbb{R}$,

$$(\mathcal{C}_\mu[H])^* = \mathcal{C}_{\overline{\mu}}[H^*].$$

Proof. First observe that $(\mathcal{M}[H])^* \subset \overline{\mathcal{M}}[H^*]$. To see this let, $\{f, f'\} \in (\mathcal{M}[H])^*$. Then, by definition, one has for all $\{h, h'\} \in H$

$$0 = (f', \alpha h + \beta h') - (f, \gamma h + \delta h') = (-\overline{\gamma} f + \overline{\alpha} f', h) - (\overline{\delta} f - \overline{\beta} f', h'),$$

which shows that

$$\begin{pmatrix} \bar{\delta} & -\bar{\beta} \\ -\bar{\gamma} & \bar{\alpha} \end{pmatrix} \begin{pmatrix} f \\ f' \end{pmatrix} \in H^*.$$

Multiplication by $\overline{\mathcal{M}}$ leads to

$$\begin{pmatrix} f \\ f' \end{pmatrix} \in \begin{pmatrix} \bar{\alpha} & \bar{\beta} \\ \bar{\gamma} & \bar{\delta} \end{pmatrix} [H^*] = \overline{\mathcal{M}}[H^*],$$

and so $(\mathcal{M}[H])^* \subset \overline{\mathcal{M}}[H^*]$.

To see the reverse inclusion $\overline{\mathcal{M}}[H^*] \subset (\mathcal{M}[H])^*$, let $\{f, f'\} \in \overline{\mathcal{M}}[H^*]$, so that for some $\{\varphi, \varphi'\} \in H^*$

$$\{f, f'\} = \{\bar{\alpha}\varphi + \bar{\beta}\varphi', \bar{\gamma}\varphi + \bar{\delta}\varphi'\}.$$

Then for all $\{h, h'\} \in H$ one has that

$$(f', \alpha h + \beta h') - (f, \gamma h + \delta h') = (\bar{\alpha}\bar{\delta} - \bar{\beta}\bar{\gamma})\big[(\varphi', h) - (\varphi, h')\big] = 0.$$

This implies that $\overline{\mathcal{M}}[H^*] \subset (\mathcal{M}[H])^*$.

The statement about the Cayley transform follows with the special choice $\alpha = -\mu$, $\gamma = -\bar{\mu}$, and $\beta = \delta = 1$; cf. (1.1.25). □

The adjoint of the componentwise sum of linear relations is determined in the following proposition. Here the notation $\operatorname{clos} H$ is used for the closure of a relation H. Recall that if H and K are closed, then $H \mathbin{\widehat{+}} K$ is closed if and only if $H^* \mathbin{\widehat{+}} K^*$ is closed; cf. (1.3.5).

Proposition 1.3.12. *Let H and K be relations from $\mathfrak{H}$ to $\mathfrak{K}$. Then one has*

$$\big(H \mathbin{\widehat{+}} K\big)^* = H^* \cap K^* \quad \textit{and} \quad \operatorname{clos}\big(H \mathbin{\widehat{+}} K\big) = \big(H^* \cap K^*\big)^*. \tag{1.3.11}$$

Proof. To verify the inclusion $(H \mathbin{\widehat{+}} K)^* \subset H^* \cap K^*$, let $\{f, f'\} \in (H \mathbin{\widehat{+}} K)^*$. Then for every $\{h, h'\} \in H$ and $\{k, k'\} \in K$ one has

$$(f', h + k) = (f, h' + k').$$

In particular, $(f', h) = (f, h')$ for all $\{h, h'\} \in H$ and $(f', k) = (f, k')$ for all $\{k, k'\} \in K$. It follows that $\{f, f'\} \in H^* \cap K^*$. Conversely, if $\{f, f'\} \in H^* \cap K^*$, then

$$(f', h) = (f, h') \qquad \text{and} \qquad (f', k) = (f, k')$$

hold for all $\{h, h'\} \in H$ and $\{k, k'\} \in K$. Adding these two identities one obtains $(f', h+k) = (f, h'+k')$ and hence $\{f, f'\} \in (H \mathbin{\widehat{+}} K)^*$. This shows the first identity in (1.3.11). The second identity in (1.3.11) follows from the first identity by taking adjoints. □

The adjoint of an orthogonal sum of relations behaves like the orthogonal complement of a sum of orthogonal subspaces.

Proposition 1.3.13. *Let H be a relation from $\mathfrak{H}_1$ to $\mathfrak{K}_1$, let K be a relation from $\mathfrak{H}_2$ to $\mathfrak{K}_2$, and let $H \mathbin{\widehat{\oplus}} K$ be their orthogonal sum. Then*

$$(H \mathbin{\widehat{\oplus}} K)^* = H^* \mathbin{\widehat{\oplus}} K^*,$$

where the adjoint in each case is taken in the corresponding Hilbert spaces.

Let H be a relation from $\mathfrak{H}$ to $\mathfrak{K}$. Recall that the closure of H is given by H^{**} and that H is a closable operator if and only if $\operatorname{mul} H^{**} = \{0\}$. The orthogonal decomposition

$$\mathfrak{K} = \overline{\operatorname{dom}} H^* \oplus \operatorname{mul} H^{**}$$

implies a related range decomposition of the relation H itself.

Theorem 1.3.14. *Let H be a relation from $\mathfrak{H}$ to $\mathfrak{K}$ and let Q be the orthogonal projection in $\mathfrak{K}$ onto $\overline{\operatorname{dom}} H^*$. Then H admits the sum decomposition*

$$H = QH + (I - Q)H, \tag{1.3.12}$$

where the relations QH and $(I - Q)H$ have the following properties:

(i) *QH is a closable operator;*

(ii) $\operatorname{clos}((I - Q)H) = \overline{\operatorname{dom}} H \times \operatorname{mul} H^{**}$.

Proof. As to the decomposition (1.3.12) it is clear that $H \subset QH + (I - Q)H$. For the converse, consider $\{h, Qh' + (I - Q)h''\}$ for some $\{h, h'\}, \{h, h''\} \in H$. Observe that $\{0, h' - h''\} \in H$, i.e., $h' - h'' \in \operatorname{mul} H \subset \operatorname{mul} H^{**} = \ker Q$. Hence, $Q(h' - h'') = 0$ and this leads to

$$\big\{h, Qh' + (I - Q)h''\big\} = \big\{h, Q(h' - h'') + h''\big\} = \{h, h''\} \in H.$$

Hence, also $QH + (I - Q)H \subset H$. Thus, (1.3.12) holds.

(i) By Corollary 1.3.4, it suffices to show that $\operatorname{dom}(QH)^*$ is dense in $\mathfrak{K}$. Observe that $(QH)^* = H^*Q$ by Proposition 1.3.9, and hence

$$\operatorname{dom}(QH)^* = \operatorname{dom} H^*Q = \operatorname{dom} H^* \oplus \ker Q. \tag{1.3.13}$$

To see the last identity in (1.3.13) first observe that $h \in \operatorname{dom} H^*Q$ if and only if $Qh \in \operatorname{dom} H^*$. Hence, if $h \in \operatorname{dom} H^*Q$, then $h = Qh + (I - Q)h$ shows that $h \in \operatorname{dom} H^* \oplus \ker Q$. Conversely, if $h \in \operatorname{dom} H^* \oplus \ker Q$, then $h = f + g$, where $f \in \operatorname{dom} H^*$ and $g \in \ker Q$. Hence, $Qh = f \in \operatorname{dom} H^*$ and thus $h \in \operatorname{dom} H^*Q$. This shows the last identity in (1.3.13). Now observe that $\ker Q = (\operatorname{dom} H^*)^\perp$ and the identity (1.3.13) takes the form

$$\operatorname{dom}(QH)^* = \operatorname{dom} H^* \oplus (\operatorname{dom} H^*)^\perp,$$

which implies that $\operatorname{dom}(QH)^*$ is dense in $\mathfrak{K}$.

(ii) First it will be shown that

$$H^*(I-Q) = \overline{\operatorname{dom}}\, H^* \times \operatorname{mul} H^*. \tag{1.3.14}$$

For the inclusion ($\subset$), let $\{h,h'\} \in H^*(I-Q)$. Then $\{(I-Q)h,h'\} \in H^*$ and since $(I-Q)h \in (\overline{\operatorname{dom}}\, H^*)^\perp$, it follows that $(I-Q)h = 0$. Thus, $h = Qh \in \overline{\operatorname{dom}}\, H^*$ and $h' \in \operatorname{mul} H^*$. For the inclusion ($\supset$) in (1.3.14), let $h \in \overline{\operatorname{dom}}\, H^*$ and $h' \in \operatorname{mul} H^*$. Then $(I-Q)h = 0$ and hence $\{(I-Q)h,h'\} = \{0,h'\} \in H^*$. This implies that $\{h,h'\} \in H^*(I-Q)$.

It follows from Proposition 1.3.9 that $((I-Q)H)^* = H^*(I-Q)$ and together with (1.3.14) one obtains

$$\begin{aligned} \operatorname{clos}\big((I-Q)H\big) &= \big((I-Q)H\big)^{**} \\ &= \big(H^*(I-Q)\big)^* \\ &= \big(\overline{\operatorname{dom}}\, H^* \times \operatorname{mul} H^*\big)^* \\ &= (\operatorname{mul} H^*)^\perp \times (\overline{\operatorname{dom}}\, H^*)^\perp \\ &= \overline{\operatorname{dom}}\, H \times \operatorname{mul} H^{**}; \end{aligned}$$

here (1.3.4) was used in the last but one step. This completes the proof of (ii). □

The sum decomposition in (1.3.12) is called the *Lebesgue decomposition* of the relation H into the *regular part* QH and the *singular part* $(I-Q)H$. The closure of the regular part QH is (the graph of) an operator, while the closure of the singular part $(I-Q)H$ is a product of closed subspaces. This decomposition is the abstract variant of the Lebesgue decomposition of a measure.

The Lebesgue decomposition (1.3.12) for a relation H from $\mathfrak{H}$ to $\mathfrak{K}$ gives rise to a componentwise direct sum decomposition when $\operatorname{mul} H = \operatorname{mul} H^{**}$.

Theorem 1.3.15. *Let H be a relation from $\mathfrak{H}$ to $\mathfrak{K}$ and let Q be the orthogonal projection in $\mathfrak{K}$ onto $\overline{\operatorname{dom}}\, H^*$. Assume that*

$$\operatorname{mul} H = \operatorname{mul} H^{**}, \tag{1.3.15}$$

so that $\mathfrak{K}$ can be decomposed as $\mathfrak{K} = \overline{\operatorname{dom}}\, H^* \oplus \operatorname{mul} H$. Then $QH \subset H$ and the relation H has the direct sum decomposition

$$H = QH \,\widehat{+}\, \big(\{0\} \times \operatorname{mul} H\big), \tag{1.3.16}$$

where QH is a closable operator from $\mathfrak{H}$ to $\mathfrak{K}$ and $\{0\} \times \operatorname{mul} H$ is a purely multivalued relation in $\operatorname{mul} H$. Moreover, if the relation H is closed, then (1.3.15) is automatically satisfied and the operator QH is closed.

Proof. Note that any element $\{h,h'\} \in H$ can be written as

$$\{h,h'\} = \{h,Qh'\} + \{0,(I-Q)h'\}. \tag{1.3.17}$$

Under the assumption (1.3.15) the orthogonal projection $I - Q$ maps onto $\operatorname{mul} H$ and hence the relation H is contained in the right-hand side of (1.3.16). The identity (1.3.17) also implies $QH \subset H$ and it follows that the right-hand side of (1.3.16) is contained in H. According to Theorem 1.3.14, QH is a closable operator and hence the sum in (1.3.16) is direct.

Now assume that the relation H is closed. In order to show that QH is closed, let $\{h_n, h_n'\} \in H$ be a sequence such that $\{h_n, Qh_n'\} \to \{\varphi, \psi\}$. Since $QH \subset H$, it follows that $\{\varphi, \psi\} \in H$. Moreover, $Qh_n' \to \psi$ implies $\psi = Q\psi$ and hence $\{\varphi, \psi\} = \{\varphi, Q\psi\} \in QH$. □

According to the above theorem, the closable operator QH acts as an *operator part* of the relation H in the direct sum decomposition (1.3.16). Note that

$$\operatorname{dom} QH = \operatorname{dom} H \quad \text{and} \quad \operatorname{ran} QH \subset \overline{\operatorname{dom}}\, H^*. \tag{1.3.18}$$

The following theorem continues this line of thought in the special but useful situation where $\mathfrak{K} = \mathfrak{H}$, i.e., when H is a relation in $\mathfrak{H}$. Recall from Theorem 1.3.15 that if the relation H is closed then actually the operator QH is closed.

Theorem 1.3.16. *Let H be a relation in $\mathfrak{H}$, let Q be the orthogonal projection onto $\overline{\operatorname{dom}}\, H^*$, and assume $\operatorname{mul} H = \operatorname{mul} H^{**}$. Suppose, in addition, that*

$$\operatorname{dom} H \subset \overline{\operatorname{dom}}\, H^* \quad \textit{or, equivalently,} \quad \operatorname{mul} H \subset \operatorname{mul} H^*. \tag{1.3.19}$$

Then the closable operator QH acts in the Hilbert space $\overline{\operatorname{dom}}\, H^*$ and H has the orthogonal sum decomposition

$$H = QH \,\widehat{\oplus}\, \big(\{0\} \times \operatorname{mul} H\big). \tag{1.3.20}$$

Moreover, QH is densely defined in $\overline{\operatorname{dom}}\, H^*$ if and only if $\operatorname{mul} H = \operatorname{mul} H^*$.

Proof. Since the condition (1.3.15) is assumed, Theorem 1.3.15 applies, and so the direct sum decomposition (1.3.16) holds, where QH is a closable operator in $\mathfrak{H}$ and $\{0\} \times \operatorname{mul} H$ is a purely multivalued relation in $\operatorname{mul} H$.

Now the equivalence in (1.3.19) will be shown. If $\operatorname{dom} H \subset \overline{\operatorname{dom}}\, H^*$, it follows by taking orthogonal complements that $\operatorname{mul} H = \operatorname{mul} H^{**} \subset \operatorname{mul} H^*$. Conversely, if $\operatorname{mul} H = \operatorname{mul} H^{**} \subset \operatorname{mul} H^*$, it follows by taking orthogonal complements that $\overline{\operatorname{dom}}\, H^{**} \subset \overline{\operatorname{dom}}\, H^*$ and, in particular, $\operatorname{dom} H \subset \overline{\operatorname{dom}}\, H^*$.

The conditions (1.3.19) and (1.3.18) imply that the closable operator QH acts in the Hilbert space $\overline{\operatorname{dom}}\, H^*$ and hence the componentwise decomposition of H in (1.3.16) is actually a componentwise orthogonal sum, i.e., (1.3.20) holds. Furthermore, since $\operatorname{dom} QH = \operatorname{dom} H$ by (1.3.18), it follows that the operator QH is densely defined in $\overline{\operatorname{dom}}\, H^*$ if and only if $\overline{\operatorname{dom}}\, H = \overline{\operatorname{dom}}\, H^*$, which is equivalent to $\operatorname{mul} H = \operatorname{mul} H^*$. □

The message of this theorem is that when $\operatorname{mul} H = \operatorname{mul} H^{**}$, the Hilbert space decomposes in $\mathfrak{H} = \overline{\operatorname{dom}}\, H^* \oplus \operatorname{mul} H$, and the regular part of the relation H serves as a not necessarily densely defined (orthogonal) operator part of H in the Hilbert space $\overline{\operatorname{dom}}\, H^*$. In the rest of this text the following notation will be used:

$$\mathfrak{H}_{\rm op} = \overline{\operatorname{dom}}\, H^*, \quad \mathfrak{H}_{\rm mul} = \operatorname{mul} H^{**} = \operatorname{mul} H,$$

and, similarly,

$$H_{\rm op} = QH, \quad H_{\rm mul} = \{0\} \times \operatorname{mul} H.$$

With these notations one has

$$\mathfrak{H} = \mathfrak{H}_{\rm op} \oplus \mathfrak{H}_{\rm mul}, \quad H = H_{\rm op} \,\widehat{\oplus}\, H_{\rm mul};$$

cf. Theorem 1.4.11, Theorem 1.5.1, and Theorem 1.6.12. The relation $H_{\rm mul}$ is purely multivalued and self-adjoint in the Hilbert space $\mathfrak{H}_{\rm mul}$ by (1.3.4), that is,

$$H_{\rm mul} = (H_{\rm mul})^*.$$

From Proposition 1.3.13 one then obtains

$$H^* = (H_{\rm op})^* \,\widehat{\oplus}\, H_{\rm mul}$$

and hence the adjoint $(H_{\rm op})^*$ of $H_{\rm op}$ in $\mathfrak{H}_{\rm op}$ satisfies

$$(H_{\rm op})^* = H^* \cap (\mathfrak{H}_{\rm op} \times \mathfrak{H}_{\rm op})$$

and its multivalued part in $\mathfrak{H}_{\rm op}$ is $\operatorname{mul} H^* \cap \mathfrak{H}_{\rm op}$. Note that

$$\operatorname{mul} H^* = \operatorname{mul}(H_{\rm op})^* \oplus \operatorname{mul} H_{\rm mul} = \bigl(\operatorname{mul} H^* \cap \mathfrak{H}_{\rm op}\bigr) \oplus \operatorname{mul} H.$$

This section ends by introducing the Moore–Penrose inverse of a relation; cf. Appendix D.

Definition 1.3.17. Let H be a relation from $\mathfrak{H}$ to $\mathfrak{K}$. Then the *Moore–Penrose inverse* $H^{(-1)}$ of H from $\mathfrak{K}$ to $\mathfrak{H}$ is defined as the relation

$$H^{(-1)} = P_{(\ker H)^\perp} H^{-1} = P_{(\operatorname{mul} H^{-1})^\perp} H^{-1}.$$

In fact, the Moore–Penrose inverse $H^{(-1)}$ of H is an operator. To see this, let $\{0, k\} \in H^{(-1)}$. Then $\{0, h\} \in H^{-1}$ and $\{h, k\} \in P_{(\ker H)^\perp}$ for some $h \in \mathfrak{H}$. Since $k = P_{(\ker H)^\perp} h$ and $h \in \ker H$, it follows that $k = 0$. Furthermore, if H is closed, then Theorem 1.3.15 applied to H^{-1} (in which case $Q = P_{(\ker H)^\perp}$) shows that

$$H^{-1} = P_{(\ker H)^\perp} H^{-1} \,\widehat{+}\, \bigl(\{0\} \times \ker H\bigr).$$

Hence, the Moore–Penrose inverse $H^{(-1)}$ coincides with the operator part of H^{-1}. Moreover, if H is closed then $\operatorname{ran} H$ is closed if and only if $H^{(-1)}$ takes $\operatorname{ran} H$ boundedly into $(\ker H)^\perp$, in which case $H^{(-1)} \in \mathbf{B}(\operatorname{ran} H, (\ker H)^\perp)$. Note that for $H \in \mathbf{B}(\mathfrak{H}, \mathfrak{K})$ the Moore–Penrose inverse coincides with the usual Moore–Penrose inverse, see Appendix D.

Example 1.3.18. Let T be a closed relation in $\mathfrak{H}$ and assume that $\lambda \in \rho(T)$. Then $H = (T-\lambda)^{-1} \in \mathbf{B}(\mathfrak{H})$ with $\ker H = \operatorname{mul} T$ and $\operatorname{ran} H = \operatorname{dom} T$, so that the Moore–Penrose inverse of H is the operator given by

$$H^{(-1)} = T_{\rm op} - \lambda,$$

which maps $\operatorname{dom} T$ into $(\operatorname{mul} T)^{\perp}$.

1.4 Symmetric relations

Symmetric relations are the building stones of this text. Here the basic properties of such relations will be developed. The special case of self-adjoint relations will be treated in more detail in the next section.

Definition 1.4.1. A relation S in $\mathfrak{H}$ is called *symmetric* if $S \subset S^*$, and *self-adjoint* if $S = S^*$. A symmetric relation S in $\mathfrak{H}$ is said to be *maximal symmetric* if every symmetric extension S' of S in $\mathfrak{H}$ satisfies $S' = S$.

It follows immediately from the definition of the adjoint relation that a relation S is symmetric if and only if

$$(f',g) = (f,g') \qquad \text{for all} \quad \{f,f'\},\ \{g,g'\} \in S. \tag{1.4.1}$$

The following lemma provides a slightly stronger statement and an easily verifiable condition for the symmetry of a relation.

Lemma 1.4.2. *A relation S in $\mathfrak{H}$ is symmetric if and only if*

$$\operatorname{Im}(f',f) = 0 \qquad \textit{for all} \quad \{f,f'\} \in S. \tag{1.4.2}$$

Proof. If $S \subset S^*$, then (1.4.1) implies (1.4.2). Conversely, assume that (1.4.2) holds. Let $\{f,f'\}, \{g,g'\} \in S$ and let $\lambda \in \mathbb{C}$. Then $\{f+\lambda g, f'+\lambda g'\} \in S$ and it follows from

$$(f'+\lambda g', f+\lambda g) = (f',f) + \bar\lambda(f',g) + \lambda(g',f) + |\lambda|^2(g',g),$$

and the assumption $\operatorname{Im}(f'+\lambda g', f+\lambda g) = 0$ that

$$\operatorname{Im}\big[\bar\lambda(f',g) + \lambda(g',f)\big] = 0.$$

By putting $\lambda = 1$ and $\lambda = i$, respectively, one obtains

$$\operatorname{Im}(f',g) = -\operatorname{Im}(g',f), \quad \operatorname{Re}(f',g) = \operatorname{Re}(g',f),$$

which leads to the equality $(f',g) = \overline{(g',f)} = (f,g')$. Hence, the relation S is symmetric by (1.4.1). □

If S is symmetric, then clearly also $\overline{S} \subset S^*$, since S^* is closed. Hence, the closure $\overline{S}$ is also symmetric. In particular, if S is maximal symmetric, then S is closed. Thus, every self-adjoint relation is maximal symmetric.

Lemma 1.4.3. *Let S be a symmetric relation in $\mathfrak{H}$. Then $\operatorname{mul} S \subset \operatorname{mul} S^*$. If S is maximal symmetric, then $\operatorname{mul} S = \operatorname{mul} S^*$.*

Proof. Let S be symmetric. Then it follows directly from Definition 1.4.1 that $\operatorname{mul} S \subset \operatorname{mul} S^*$.

Now assume S is maximal symmetric. It suffices to show $\operatorname{mul} S^* \subset \operatorname{mul} S$. If $k \in \operatorname{mul} S^* = (\operatorname{dom} S)^\perp$, then

$$S \,\widehat{+}\, \operatorname{span}\{0,k\} = \bigl\{\{h, h'+k\} : \{h,h'\} \in S\bigr\}$$

is a symmetric extension of S, as $\operatorname{Im}(h'+k,h) = \operatorname{Im}(h',h) = 0$ for all $\{h,h'\} \in S$. Since S is maximal symmetric, it follows that $\{0,k\} \in S$ and $k \in \operatorname{mul} S$. Thus, $\operatorname{mul} S^* \subset \operatorname{mul} S$. □

As an example consider the relation S defined by $S = \{0\} \times \mathfrak{N}$, where $\mathfrak{N} \subset \mathfrak{H}$ is a linear subspace. It follows from (1.3.4) that $S^* = \mathfrak{N}^\perp \times \mathfrak{H}$ and $S^{**} = \{0\} \times \overline{\mathfrak{N}}$. Hence, S is symmetric, while

$$\operatorname{mul} S = \mathfrak{N}, \qquad \operatorname{mul} \overline{S} = \overline{\mathfrak{N}}, \qquad \operatorname{mul} S^* = \mathfrak{H},$$

which shows that if S is closed the inclusion $\operatorname{mul} S \subset \operatorname{mul} S^*$ in Lemma 1.4.3 is in general strict. Moreover, in the present example S is self-adjoint if and only if $\mathfrak{N} = \mathfrak{H}$. If S is maximal symmetric then according to Lemma 1.4.3 one has $\mathfrak{N} = \mathfrak{H}$, so that S is self-adjoint.

In the rest of this text the interest will often be in extensions that are closed; in particular, in relations H that are self-adjoint extensions of a given symmetric relation S,

$$S \subset H = H^* \subset S^*.$$

Observe that H is a self-adjoint extension of S if and only if H is a self-adjoint extension of the closure $\overline{S}$. In that sense it will often be assumed without loss of generality that S is closed; recall that $\gamma(S) = \gamma(\overline{S})$.

Proposition 1.4.4. *Let S be a symmetric relation in $\mathfrak{H}$. Then $\mathbb{C} \setminus \mathbb{R}$ is contained in $\gamma(S)$ and, in particular, the defect $n_\lambda(S) = \dim(\operatorname{ran}(S-\lambda))^\perp$ is constant for all $\lambda \in \mathbb{C}^+$ and all $\lambda \in \mathbb{C}^-$. Furthermore, $\sigma_{\mathrm{p}}(S) \cup \sigma_{\mathrm{c}}(S) \subset \mathbb{R}$ and*

$$\|(S-\lambda)^{-1}h\| \leq \frac{1}{|\operatorname{Im}\lambda|}\,\|h\| \tag{1.4.3}$$

for all $h \in \operatorname{dom}(S-\lambda)^{-1} = \operatorname{ran}(S-\lambda)$ and $\lambda \in \mathbb{C} \setminus \mathbb{R}$.

Proof. Let $\lambda \in \mathbb{C} \setminus \mathbb{R}$ and $\{f, f'\} \in S$, so that $\{f' - \lambda f, f\} \in (S - \lambda)^{-1}$. As S is symmetric, one has $\operatorname{Im}(f', f) = 0$ by Lemma 1.4.2, and hence

$$0 \leq |\operatorname{Im} \lambda|(f, f) = |\operatorname{Im}(f' - \lambda f, f)| \leq \|f' - \lambda f\| \|f\|$$

and for $f \neq 0$ this implies

$$0 \leq |\operatorname{Im} \lambda| \|f\| \leq \|f' - \lambda f\|.$$

Therefore, $(S - \lambda)^{-1}$ is an operator and (1.4.3) holds for all $h \in \operatorname{dom}(S - \lambda)^{-1}$ and $\lambda \in \mathbb{C} \setminus \mathbb{R}$. This also shows $\mathbb{C} \setminus \mathbb{R} \subset \gamma(S)$ and it follows from Theorem 1.2.5 that the defect $n_\lambda(S)$ is constant on $\mathbb{C}^+$ and $\mathbb{C}^-$. It is clear that the point spectrum $\sigma_{\mathrm{p}}(S)$ is contained in $\mathbb{R}$, and since $(S - \lambda)^{-1}$ is bounded for $\lambda \in \mathbb{C} \setminus \mathbb{R}$, also the continuous spectrum $\sigma_{\mathrm{c}}(S)$ is contained in $\mathbb{R}$. □

The *defect numbers* $n_\pm(S)$ of a symmetric relation S are defined as

$$n_\pm(S) := \dim \big(\operatorname{ran}(S \mp i)\big)^\perp = \dim \big(\ker (S^* \pm i)\big), \tag{1.4.4}$$

where according to Proposition 1.4.4 the point $\pm i$ in (1.4.4) can be replaced by any $\lambda \in \mathbb{C}^\pm$.

In the case where the symmetric relation S is bounded from below in the sense of the next definition it follows that $\gamma(S) \cap \mathbb{R} \neq \emptyset$ and thus $\gamma(S)$ consists of one component only; cf. Proposition 1.4.6. In particular, the defect numbers $n_+(S)$ and $n_-(S)$ coincide in this case.

Definition 1.4.5. Let S be a relation in $\mathfrak{H}$. Then S is said to be *bounded from below* if there exists a number $\eta \in \mathbb{R}$ such that

$$(f', f) \geq \eta(f, f) \quad \text{for all} \quad \{f, f'\} \in S. \tag{1.4.5}$$

The *lower bound* $m(S)$ of S is the largest number $\eta \in \mathbb{R}$ for which (1.4.5) holds:

$$m(S) = \inf \left\{ \frac{(f', f)}{(f, f)} : \{f, f'\} \in S, \, f \neq 0 \right\}.$$

The inequality (1.4.5) will be written as $S \geq \eta I$, which will be further abbreviated to $S \geq \eta$. If $S \geq 0$, then S is called *nonnegative* (and S may have a positive lower bound).

For a relation S that is bounded from below also the terminology *semibounded relation* will be used. If S is bounded from below, then it follows directly from Lemma 1.4.2 that S is symmetric. Moreover, if (1.4.5) is satisfied for all $\eta \in \mathbb{R}$, then $\eta \|f\|^2 \leq \|f'\| \|f\|$ for $\{f, f'\} \in S$ and $\eta \in \mathbb{R}$ shows that $\operatorname{dom} S = \{0\}$ and hence S is a purely multivalued relation.

Proposition 1.4.6. *Let S be a symmetric relation in $\mathfrak{H}$ which is bounded from below with lower bound $m(S) \in \mathbb{R}$. Then $\mathbb{C}\setminus[m(S),\infty)$ is contained in $\gamma(S)$ and the defect $n_\lambda(S) = \dim(\operatorname{ran}(S-\lambda))^\perp$ is constant for all $\lambda \in \mathbb{C}\setminus[m(S),\infty)$. Furthermore, $\sigma_p(S)\cup\sigma_c(S) \subset [m(S),\infty)$ and*

$$\|(S-\nu)^{-1}h\| \leq \frac{1}{m(S)-\nu}\|h\| \tag{1.4.6}$$

for all $h \in \operatorname{dom}(S-\nu)^{-1}$ and $\nu < m(S)$.

Proof. For $\{f,f'\} \in S$ and $\nu < m(S)$ the assumption $(f'-m(S)f,f) \geq 0$ implies

$$\begin{aligned}(m(S)-\nu)(f,f) &\leq \bigl(f'-m(S)f+(m(S)-\nu)f,f\bigr) = (f'-\nu f,f)\\ &\leq \|f'-\nu f\|\|f\|.\end{aligned}$$

Hence, if $f \neq 0$ it follows that $(m(S)-\nu)\|f\| \leq \|f'-\nu f\|$ holds for all $\{f,f'\} \in S$ and $\nu < m(S)$. This shows that $(S-\nu)^{-1}$ is an operator and (1.4.6) holds. Recalling Proposition 1.4.4, one has $\mathbb{C}\setminus[m(S),\infty) \subset \gamma(S)$, and Theorem 1.2.5 implies that the defect $n_\lambda(S)$ is constant on $\mathbb{C}\setminus[m(S),\infty)$, and $\sigma_p(S)\cup\sigma_c(S) \subset [m(S),\infty)$ holds. □

Lemma 1.4.7. *Let S be a closed symmetric relation in $\mathfrak{H}$ and let $\lambda \in \gamma(S)$. Then* $\operatorname{ran}(S^*-\lambda) = \mathfrak{H}$.

Proof. Let $\lambda \in \gamma(S)$, then also $\bar\lambda \in \gamma(S)$ (for $\lambda \in \mathbb{C}\setminus\mathbb{R}$ this follows from Proposition 1.4.4). This implies that $\operatorname{ran}(S-\bar\lambda)$ is closed according to Lemma 1.2.2. Hence, $\operatorname{ran}(S^*-\lambda)$ is closed by Theorem 1.3.5. Moreover, $\bar\lambda \in \gamma(S)$ implies $\ker(S-\bar\lambda) = \{0\}$ and therefore

$$\bigl(\operatorname{ran}(S^*-\lambda)\bigr)^\perp = \ker(S-\bar\lambda) = \{0\},$$

that is, $\operatorname{ran}(S^*-\lambda)$ is dense in $\mathfrak{H}$. It follows that $\operatorname{ran}(S^*-\lambda) = \mathfrak{H}$. □

In the next proposition the Cayley transform and the inverse Cayley transform of symmetric relations are considered.

Proposition 1.4.8. *Let $\mu \in \mathbb{C}\setminus\mathbb{R}$ and let $\mathcal{C}_\mu$ and $\mathcal{F}_\mu$ be the Cayley transform and inverse Cayley transform in Definition 1.1.13. Let S and V be relations in $\mathfrak{H}$ such that $V = \mathcal{C}_\mu[S]$ or, equivalently, $S = \mathcal{F}_\mu[V]$. Then the following statements hold:*

(i) *S is a (closed) symmetric relation if and only if V is a (closed) isometric operator;*

(ii) *S is a maximal symmetric relation if and only if V is an isometric operator such that $\operatorname{dom} V = \mathfrak{H}$ or $\operatorname{ran} V = \mathfrak{H}$.*

Proof. (i) Let S and V be relations such that $V = \mathcal{C}_\mu[S]$ with $\mu \in \mathbb{C}\setminus\mathbb{R}$. Then (1.1.23), (1.1.25), and Lemma 1.3.11 show that

$$V^{-1} = \mathcal{C}_{\bar\mu}[S] \quad\text{and}\quad V^* = \mathcal{C}_{\bar\mu}[S^*].$$

These equalities and an application of the inverse Cayley transform give

$$V^{-1} \subset V^* \quad \Leftrightarrow \quad \mathcal{C}_{\bar{\mu}}[S] \subset \mathcal{C}_{\bar{\mu}}[S^*] \quad \Leftrightarrow \quad S \subset S^*.$$

Now Lemma 1.3.6 shows that S is a symmetric relation if and only if V is an isometric operator. Moreover, by (1.1.18), S is closed if and only if $V^{-1} = \mathcal{C}_{\bar{\mu}}[S]$ is closed, which completes the proof of (i).

(ii) Let S be maximal symmetric and let $V = \mathcal{C}_\mu[S]$. Assume that V' is an isometric extension of V in $\mathfrak{H}$. Then $S' = \mathcal{F}_\mu[V']$ is a symmetric extension of S and hence $S' = S$. This implies $V' = V$ and hence $\operatorname{dom} V = \mathfrak{H}$ or $\operatorname{ran} V = \mathfrak{H}$. The converse statement is proved by the same argument. □

It follows from Proposition 1.4.8 (ii) and (1.1.24) that a symmetric relation S in $\mathfrak{H}$ is maximal symmetric if and only if

$$\operatorname{ran}(S - \mu) = \mathfrak{H} \tag{1.4.7}$$

for some, and hence for all $\mu \in \mathbb{C}^+$ or for some, and hence for all $\mu \in \mathbb{C}^-$.

Definition 1.4.9. Let S be a symmetric relation in $\mathfrak{H}$ and let $\lambda \in \mathbb{C}$. The spaces

$$\mathfrak{N}_\lambda(S^*) := \ker(S^* - \lambda) \quad \text{and} \quad \widehat{\mathfrak{N}}_\lambda(S^*) = \big\{\{f_\lambda, \lambda f_\lambda\} : f_\lambda \in \mathfrak{N}_\lambda(S^*)\big\}$$

are called *defect subspaces* of S at the point $\lambda \in \mathbb{C}$.

Note that the defect numbers $n_\pm(S)$ in (1.4.4) satisfy

$$n_\pm(S) = \dim \mathfrak{N}_{\mp i}(S^*) = \dim \mathfrak{N}_\lambda(S^*), \qquad \lambda \in \mathbb{C}^\mp.$$

Since the adjoint relation S^* is closed by Proposition 1.3.2, the defect subspaces $\mathfrak{N}_\lambda(S^*) \subset \operatorname{dom} S^*$ and $\widehat{\mathfrak{N}}_\lambda(S^*) \subset S^*$ are closed subspaces of $\mathfrak{H}$ and $\mathfrak{H}^2$, respectively. The notation in Definition 1.4.9 will be used throughout the text. Besides the defect subspaces $\mathfrak{N}_\lambda(S^*)$ and $\widehat{\mathfrak{N}}_\lambda(S^*)$ also the spaces

$$\mathfrak{N}_\lambda(S) := \ker(S - \lambda) \quad \text{and} \quad \widehat{\mathfrak{N}}_\lambda(S) := \big\{\{f_\lambda, \lambda f_\lambda\} : f_\lambda \in \mathfrak{N}_\lambda(S)\big\}$$

for a symmetric relation S will be used. Moreover, let

$$\mathfrak{N}_\infty(S) := \operatorname{mul} S \quad \text{and} \quad \widehat{\mathfrak{N}}_\infty(S) := \big\{\{0, f'\} : f' \in \mathfrak{N}_\infty(S)\big\}.$$

Lemma 1.4.10. *Let S be a closed symmetric relation in $\mathfrak{H}$ and let $H \subset S^*$ be a closed extension of S such that $\rho(H) \neq \emptyset$. Then for $\lambda, \mu \in \rho(H)$*

$$I + (\lambda - \mu)(H - \lambda)^{-1} \tag{1.4.8}$$

is boundedly invertible with inverse $I + (\mu - \lambda)(H - \mu)^{-1}$. For $\mu \in \rho(H)$ the mapping (1.4.8) is holomorphic in λ. Moreover, for $\lambda, \mu \in \rho(H)$ the operator in (1.4.8) maps $\mathfrak{N}_\mu(S^*)$ bijectively onto $\mathfrak{N}_\lambda(S^*)$.

Proof. Let $\lambda, \mu \in \rho(H)$. Then the first assertion follows from Corollary 1.2.8. The holomorphy of $\lambda \mapsto I + (\lambda - \mu)(H - \lambda)^{-1}$ follows from the holomorphy of the resolvent; cf. Corollary 1.2.7. As to the defect spaces, it is first verified that for $f_\mu \in \mathfrak{N}_\mu(S^*)$ one has

$$f_\lambda := \big(I + (\lambda - \mu)(H - \lambda)^{-1}\big) f_\mu \in \mathfrak{N}_\lambda(S^*) = \operatorname{ran}(S - \bar{\lambda})^\perp. \tag{1.4.9}$$

To see this, let $\{g, g'\} \in S$ and consider

$$\begin{aligned}(f_\lambda, g' - \bar{\lambda} g) &= \big((I + (\lambda - \mu)(H - \lambda)^{-1}) f_\mu, g' - \bar{\lambda} g\big) \\ &= \big(f_\mu, \big(I + (\bar{\lambda} - \bar{\mu})(H^* - \bar{\lambda})^{-1}\big)(g' - \bar{\lambda} g)\big).\end{aligned}$$

Since $\bar{\lambda} \in \rho(H^*)$ according to Proposition 1.3.10 and $S \subset H^*$ it follows that $(H^* - \bar{\lambda})^{-1}(g' - \bar{\lambda} g) = g$, and so

$$(f_\lambda, g' - \bar{\lambda} g) = (f_\mu, g' - \bar{\lambda} g + (\bar{\lambda} - \bar{\mu}) g) = (f_\mu, g' - \bar{\mu} g) = 0.$$

Hence, (1.4.9) is clear. It follows that the operator in (1.4.8) maps $\mathfrak{N}_\mu(S^*)$ to $\mathfrak{N}_\lambda(S^*)$. The same reasoning with λ and μ interchanged shows that the map is in fact onto. □

Closed symmetric relations can be written as orthogonal sums of closed symmetric operators and self-adjoint purely multivalued linear relations. This is a straightforward consequence of Theorem 1.3.16.

Theorem 1.4.11. *Let S be a closed symmetric relation in $\mathfrak{H}$. Decompose $\mathfrak{H}$ as $\mathfrak{H} = \mathfrak{H}_{\rm op} \oplus \mathfrak{H}_{\rm mul}$, $\mathfrak{H}_{\rm op} := (\operatorname{mul} S)^\perp$ and $\mathfrak{H}_{\rm mul} := \operatorname{mul} S$, and denote the orthogonal projection from $\mathfrak{H}$ onto $\mathfrak{H}_{\rm op}$ by $P_{\rm op}$. Then S is the direct orthogonal sum $S_{\rm op} \widehat{\oplus} S_{\rm mul}$ of the closed symmetric operator*

$$S_{\rm op} = \big\{\{f, P_{\rm op} f'\} : \{f, f'\} \in S\big\}$$

in $\mathfrak{H}_{\rm op}$ and the self-adjoint purely multivalued relation

$$S_{\rm mul} = \big\{\{0, f'\} : f' \in \mathfrak{H}_{\rm mul}\big\} = \big\{\{0, (I - P_{\rm op}) f'\} : \{f, f'\} \in S\big\}$$

in $\mathfrak{H}_{\rm mul}$. Moreover, the operator $S_{\rm op}$ is densely defined in $\mathfrak{H}_{\rm op}$ if and only if $\operatorname{mul} S = \operatorname{mul} S^*$. If the relation S is maximal symmetric, then $S_{\rm op}$ is a densely defined maximal symmetric operator in $\mathfrak{H}_{\rm op}$.

Proof. By assumption the relation S is closed and symmetric, which implies that $\operatorname{mul} S = \operatorname{mul} S^{**}$ and $\operatorname{dom} S \subset \operatorname{dom} S^*$. Thus, Theorem 1.3.16 applies and yields the indicated decomposition and the criterion for the denseness of $S_{\rm op}$ follows.

If S is maximal symmetric, then $\operatorname{mul} S = \operatorname{mul} S^*$ by Lemma 1.4.3. Hence, in this case the operator $S_{\rm op}$ is densely defined. Assume that S_1 is a symmetric extension of $S_{\rm op}$ in $\mathfrak{H}_{\rm op}$. Then $S_1 \widehat{\oplus} S_{\rm mul}$ is a symmetric extension of S and hence coincides with S. This implies $S_{\rm op} = S_1$ and therefore $S_{\rm op}$ is a maximal symmetric operator in $\mathfrak{H}_{\rm op}$. □

1.5 Self-adjoint relations

For self-adjoint relations there is always an orthogonal decomposition into a self-adjoint operator and a self-adjoint purely multivalued part. This reduction allows one to apply the spectral theory for self-adjoint operators in the present context, see also Chapter 3. This section contains a brief introduction and a number of consequences of this approach; in particular, nonnegative and semibounded self-adjoint relations will be considered.

The following reduction result is a specialization of Theorem 1.4.11 for self-adjoint relations.

Theorem 1.5.1. *Let H be a self-adjoint relation in $\mathfrak{H}$. Decompose the space $\mathfrak{H}$ as $\mathfrak{H} = \mathfrak{H}_{\rm op} \oplus \mathfrak{H}_\infty$, where*

$$\mathfrak{H}_{\rm op} := \overline{\operatorname{dom}} \, H = (\operatorname{mul} H)^\perp \quad \text{and} \quad \mathfrak{H}_{\rm mul} := \operatorname{mul} H,$$

and denote the orthogonal projection from $\mathfrak{H}$ onto $\mathfrak{H}_{\rm op}$ by $P_{\rm op}$. Then H is the direct orthogonal sum $H_{\rm op} \mathbin{\widehat{\oplus}} H_{\rm mul}$ of the (densely defined) self-adjoint operator

$$H_{\rm op} = \bigl\{\{f, P_{\rm op} f'\} : \{f, f'\} \in H\bigr\}$$

in $\mathfrak{H}_{\rm op}$ and the self-adjoint purely multivalued relation

$$H_{\rm mul} = \bigl\{\{0, f'\} : f' \in \mathfrak{H}_{\rm mul}\bigr\} = \bigl\{\{0, (I - P_{\rm op})f'\} : \{f, f'\} \in H\bigr\}$$

in $\mathfrak{H}_{\rm mul}$.

Observe that for $\lambda \in \mathbb{C} \setminus \mathbb{R}$ the resolvent of H in Theorem 1.5.1 admits the matrix representation

$$(H - \lambda)^{-1} = \begin{pmatrix} P_{\rm op}(H - \lambda)^{-1}\iota_{\rm op} & 0 \\ 0 & 0 \end{pmatrix} = \begin{pmatrix} (H_{\rm op} - \lambda)^{-1} & 0 \\ 0 & 0 \end{pmatrix} \tag{1.5.1}$$

with respect to the space decomposition $\mathfrak{H} = \mathfrak{H}_{\rm op} \oplus \mathfrak{H}_{\rm mul}$. Here $\iota_{\rm op}$ denotes the canonical embedding of $\mathfrak{H}_{\rm op}$ in $\mathfrak{H}$.

In the following it is explained how the spectral theory and the functional calculus for self-adjoint operators extend via Theorem 1.5.1 to self-adjoint relations. First of all it is clear that the finite (real) spectrum of a self-adjoint relation $H = H_{\rm op} \mathbin{\widehat{\oplus}} H_{\rm mul}$ is the same as that of the self-adjoint operator part $H_{\rm op}$. Note that $\rho(H_{\rm mul}) = \mathbb{C}$ and that $\sigma(H_{\rm mul}^{-1})$ consists only of the eigenvalue 0. The *essential spectrum* $\sigma_{\rm ess}(H)$ and *discrete spectrum* $\sigma_{\rm d}(H)$ of a self-adjoint relation H are defined as the essential spectrum and discrete spectrum of its operator part $H_{\rm op}$, respectively. Recall that the discrete spectrum consists of all isolated eigenvalues of finite multiplicity and the essential spectrum is the remaining part of the spectrum; it consists of the continuous spectrum, eigenvalues embedded in the continuous spectrum, and isolated eigenvalues of infinite multiplicity. It is useful to observe that $\lambda \in \sigma_{\rm d}(H)$ if and only if $\dim \ker (H - \lambda) < \infty$ and $\operatorname{ran} (H - \lambda)$ is closed.

The spectral measure $E_{\rm op}(\cdot)$ of the self-adjoint operator $H_{\rm op}$ is defined for the Borel sets in $\mathbb{R}$ with orthogonal projections in $\mathfrak{H}_{\rm op}$ as values, and the corresponding spectral function is defined as $t \mapsto E_{\rm op}((-\infty, t))$. Then one has

$$\begin{aligned} H_{\rm op} f &= \int_{\mathbb{R}} t \, dE_{\rm op}(t) f, \\ \operatorname{dom} H_{\rm op} &= \left\{ f \in \mathfrak{H}_{\rm op} : \int_{\mathbb{R}} t^2 \, d(E_{\rm op}(t) f, f) < \infty \right\}. \end{aligned}$$

Furthermore, for a bounded measurable function $h : \mathbb{R} \to \mathbb{C}$ the bounded operator $h(H_{\rm op}) \in \mathbf{B}(\mathfrak{H}_{\rm op})$ is defined via the functional calculus for self-adjoint operators in $\mathfrak{H}_{\rm op}$:

$$h(H_{\rm op}) = \int_{\mathbb{R}} h(t) \, dE_{\rm op}(t). \tag{1.5.2}$$

In particular, the spectral calculus leads to the formula

$$(H_{\rm op} - \lambda)^{-1} = \int_{\mathbb{R}} \frac{1}{t - \lambda} dE_{\rm op}(t), \quad \lambda \in \rho(H_{\rm op}). \tag{1.5.3}$$

The spectral projection $E_{\rm op}((a, b))$ can be obtained with the help of the resolvent of $H_{\rm op}$ and Stone's formula

$$\lim_{\varepsilon \to +0} \lim_{\delta \to +0} \frac{1}{2\pi i} \int_{a+\varepsilon}^{b-\varepsilon} \left(\big(H_{\rm op} - (t + i\delta)\big)^{-1} - \big(H_{\rm op} - (t - i\delta)\big)^{-1} \right) dt, \tag{1.5.4}$$

where the limits exist in the strong sense.

The spectral measure of a self-adjoint relation will be defined on $\mathbb{R}$ as the orthogonal sum of the spectral measure of $H_{\rm op}$ and the zero operator in $\mathfrak{H}_{\rm mul}$.

Definition 1.5.2. Let $H = H_{\rm op} \,\widehat{\oplus}\, H_{\rm mul}$ be a self-adjoint relation in the Hilbert space $\mathfrak{H} = \mathfrak{H}_{\rm op} \oplus \mathfrak{H}_{\rm mul}$ and denote the spectral measure of the self-adjoint operator $H_{\rm op}$ in $\mathfrak{H}_{\rm op}$ by $E_{\rm op}(\cdot)$. Then the *spectral measure* $E(\cdot)$ of H is defined as

$$E(\cdot) = \begin{pmatrix} E_{\rm op}(\cdot) & 0 \\ 0 & 0 \end{pmatrix}$$

with respect to the decomposition $\mathfrak{H} = \mathfrak{H}_{\rm op} \oplus \mathfrak{H}_{\rm mul}$.

Now the functional calculus for the self-adjoint operator $H_{\rm op}$ yields the *functional calculus* for the self-adjoint relation $H = H_{\rm op} \,\widehat{\oplus}\, H_{\rm mul}$. More precisely, for a bounded measurable function $h : \mathbb{R} \to \mathbb{C}$ one defines

$$h(H) = \int_{\mathbb{R}} h(t) \, dE(t)$$

in accordance with (1.5.2). It follows directly from Definition 1.5.2 and (1.5.2) that

$$h(H) = h(H_{\rm op}) \,\widehat{\oplus}\, O_{\operatorname{mul} H} \tag{1.5.5}$$

is an everywhere defined bounded operator in $\mathfrak{H}$. In particular, for the resolvent of H in (1.5.1) one has

$$(H-\lambda)^{-1} = \int_{\mathbb{R}} \frac{1}{t-\lambda} dE(t), \quad \lambda \in \rho(H); \tag{1.5.6}$$

cf. (1.5.3). Note that the spectral projection $E((a,b))$ of H is also given by *Stone's formula*

$$\lim_{\varepsilon \to +0} \lim_{\delta \to +0} \frac{1}{2\pi i} \int_{a+\varepsilon}^{b-\varepsilon} \left(\left(H-(t+i\delta)\right)^{-1} - \left(H-(t-i\delta)\right)^{-1} \right) dt, \tag{1.5.7}$$

which again follows from the decomposition of the resolvent of H in (1.5.1); as in (1.5.4), the limits are understood in the strong sense. A proof of Stone's formula (in the weak sense) can also be found in Example A.1.4.

The next lemma on the strong convergence follows from the properties of the functional calculus of self-adjoint operators; cf. [649, Theorem VIII.5] and (1.5.5). It will be used in Chapter 3.

Lemma 1.5.3. *Let $h_n : \mathbb{R} \to \mathbb{C}$ be a sequence of bounded measurable functions which converge pointwise to h such that $\|h_n\|_\infty$ is bounded. Then*

$$\lim_{n\to\infty} h_n(H_{\text{op}})f = h(H_{\text{op}})f, \quad f \in \mathfrak{H}_{\text{op}},$$

and

$$\lim_{n\to\infty} h_n(H)g = h(H)g, \qquad g \in \mathfrak{H}.$$

The next proposition on the Cayley transform of self-adjoint relations complements Proposition 1.4.8.

Proposition 1.5.4. *Let $\mu \in \mathbb{C} \setminus \mathbb{R}$ and let $\mathcal{C}_\mu$ and $\mathcal{F}_\mu$ be the Cayley transform and inverse Cayley transform defined in (1.1.23). Let S and V be relations such that $V = \mathcal{C}_\mu[S]$ or, equivalently, $S = \mathcal{F}_\mu[V]$. Then S is a self-adjoint relation if and only if V is a unitary operator.*

Proof. As in the proof of Proposition 1.4.8 one obtains for $V = \mathcal{C}_\mu[S]$ and $\mu \in \mathbb{C}\setminus\mathbb{R}$ that

$$V^{-1} = V^* \quad \Leftrightarrow \quad \mathcal{C}_{\bar\mu}[S] = \mathcal{C}_{\bar\mu}[S^*] \quad \Leftrightarrow \quad S = S^*,$$

and now the assertion follows from Lemma 1.3.6. □

The following theorem is useful when one needs to prove that a given relation is self-adjoint. It is often easy to check that a relation is symmetric and hence it is convenient to have equivalent conditions for a symmetric relation to be self-adjoint available.

Theorem 1.5.5. *Let S be a closed symmetric relation in $\mathfrak{H}$. Then the following statements are equivalent:*

(i) $S = S^*$;

(ii) $\ker(S^* - \lambda) = \{0\} = \ker(S^* - \mu)$ *for some, and hence for all* $\lambda \in \mathbb{C}^+$ *and* $\mu \in \mathbb{C}^-$;

(iii) $\operatorname{ran}(S - \lambda) = \mathfrak{H} = \operatorname{ran}(S - \mu)$ *for some, and hence for all* $\lambda \in \mathbb{C}^+$ *and* $\mu \in \mathbb{C}^-$;

(iv) $\mathbb{C} \setminus \mathbb{R} \subset \rho(S)$.

If the closed symmetric relation S *is bounded from below by* $m(S) \in \mathbb{R}$ *or, more generally,* $\gamma(S) \cap \mathbb{R} \neq \emptyset$, *then* λ, μ *in* (ii) *and* (iii) *can also be chosen in* $(-\infty, m(S))$ *or* $\gamma(S) \cap \mathbb{R}$, *respectively, such that* $\lambda = \mu$. *In the case* $S \geq m(S)$ *item* (iv) *can be replaced by* $\mathbb{C} \setminus [m(S), \infty) \subset \rho(S)$.

Proof. (i) $\Rightarrow$ (ii) From Proposition 1.4.4 it follows that $\ker(S - \lambda) = \{0\}$ for $\lambda \in \mathbb{C} \setminus \mathbb{R}$, and as $S = S^*$ one concludes (ii).

(ii) $\Leftrightarrow$ (iii) follows from the identity $(\operatorname{ran}(S - \lambda))^\perp = \ker(S^* - \bar{\lambda})$ and the fact that $\operatorname{ran}(S - \lambda)$ is closed for $\lambda \in \mathbb{C} \setminus \mathbb{R}$ by Proposition 1.4.4 and Lemma 1.2.2. Note also that $\operatorname{ran}(S - \lambda) = \mathfrak{H}$ for some $\lambda \in \mathbb{C}^\pm$ implies $\operatorname{ran}(S - \lambda) = \mathfrak{H}$ for all $\lambda \in \mathbb{C}^\pm$ by Theorem 1.2.5 and $\mathbb{C}^\pm \subset \gamma(S)$.

(iii) $\Rightarrow$ (iv) Since $\mathbb{C} \setminus \mathbb{R} \subset \gamma(S)$ by Proposition 1.4.4 and $\operatorname{ran}(S - \lambda) = \mathfrak{H}$, $\lambda \in \mathbb{C} \setminus \mathbb{R}$, is closed, it follows that $(S - \lambda)^{-1} \in \mathbf{B}(\mathfrak{H})$, $\lambda \in \mathbb{C} \setminus \mathbb{R}$. Now Lemma 1.2.4 implies $\mathbb{C} \setminus \mathbb{R} \subset \rho(S)$.

(iv) $\Rightarrow$ (i) It suffices to show $S^* \subset S$. For this let $\{f, f'\} \in S^*$ and $\lambda \in \mathbb{C} \setminus \mathbb{R}$. As $\lambda \in \rho(S)$, there exists $\{g, g'\} \in S$ such that

$$f' - \lambda f = g' - \lambda g.$$

Hence, $\{f - g, \lambda(f - g)\} = \{f, f'\} - \{g, g'\} \in S^*$ and

$$f - g \in \ker(S^* - \lambda) = (\operatorname{ran}(S - \bar{\lambda}))^\perp.$$

Since with $\lambda \in \rho(S)$ also $\bar{\lambda} \in \rho(S)$, it follows that $f = g$ and hence $f' = g'$, that is, $\{f, f'\} \in S$.

If S is bounded from below with lower bound $m(S)$, then $\mathbb{C} \setminus [m(S), \infty) \subset \gamma(S)$ by Proposition 1.4.6. Hence, if $\lambda = \mu < m(S)$, then $\operatorname{ran}(S - \lambda) = \mathfrak{H}$ implies $\operatorname{ran}(S - \lambda) = \mathfrak{H}$ for all $\lambda \in \mathbb{C} \setminus [m(S), \infty)$ by Theorem 1.2.5. It is also clear that for $\lambda = \mu < m(S)$ one has $(\operatorname{ran}(S - \lambda))^\perp = \ker(S^* - \lambda)$, and that $\operatorname{ran}(S - \lambda)$ is closed. This shows the equivalence of (ii) and (iii), and the argument remains true in the more general situation $\gamma(S) \cap \mathbb{R} \neq \emptyset$. As above one concludes in the case $S \geq m(S)$ from $\mathbb{C} \setminus [m(S), \infty) \subset \gamma(S)$ that (iii) implies $\mathbb{C} \setminus [m(S), \infty) \subset \rho(S)$. □

Note also that if a closed symmetric relation S is self-adjoint and bounded from below with lower bound $m(S) \in \mathbb{R}$, then one has $\sigma(S) \subset [m(S), \infty)$ by

Theorem 1.5.5 (iv). In fact, one verifies with the help of the spectral measure of S or of its operator part $S_{\rm op}$ that

$$m(S) = \min \sigma(S).$$

In some cases it is useful to have the following variant of the equivalence of (i) and (iii) in Theorem 1.5.5 in which S is not assumed to be closed.

Proposition 1.5.6. *Let S be a symmetric relation in $\mathfrak{H}$. Then S is self-adjoint if and only if* $\operatorname{ran}(S-\lambda) = \mathfrak{H} = \operatorname{ran}(S-\mu)$ *for some, and hence for all $\lambda \in \mathbb{C}^+$ and $\mu \in \mathbb{C}^-$. If S is semibounded with lower bound $m(S)$ or, more generally, $\gamma(S) \cap \mathbb{R} \neq \emptyset$, then λ and μ can also be chosen in $(-\infty, m(S))$ or $\gamma(S) \cap \mathbb{R}$, respectively, such that $\lambda = \mu$.*

Proof. Note that by Lemma 1.2.2 the condition $\operatorname{ran}(S-\lambda) = \mathfrak{H}$ for some $\lambda \in \gamma(S)$ implies that S is closed. Now the assertions follow from Theorem 1.5.5. □

Let S be a symmetric relation in $\mathfrak{H}$. Then it is easy to see that

$$S \,\widehat{+}\, \widehat{\mathfrak{N}}_x(S^*), \quad x \in \mathbb{R}, \quad \text{and} \quad S \,\widehat{+}\, \widehat{\mathfrak{N}}_\infty(S^*)$$

are also symmetric relations in $\mathfrak{H}$. The next lemma provides a necessary and sufficient condition for the self-adjointness of these relations, which applies, in particular, when $\gamma(S) \cap \mathbb{R} \neq \emptyset$.

Lemma 1.5.7. *Let S be a symmetric relation in $\mathfrak{H}$ and let $x \in \mathbb{R}$.*

(i) *The symmetric relation $S \,\widehat{+}\, \widehat{\mathfrak{N}}_x(S^*)$ is self-adjoint if and only if*

$$\operatorname{ran}(S-x) = \overline{\operatorname{ran}}\,(S-x) \cap \operatorname{ran}(S^*-x).$$

In particular, if $\operatorname{ran}(S-x)$ *is closed, then $S \,\widehat{+}\, \widehat{\mathfrak{N}}_x(S^*)$ is self-adjoint.*

(ii) *The symmetric relation $S \,\widehat{+}\, \widehat{\mathfrak{N}}_\infty(S^*)$ is self-adjoint if and only if*

$$\operatorname{dom} S = \overline{\operatorname{dom} S} \cap \operatorname{dom} S^*. \tag{1.5.8}$$

In particular, if $\operatorname{dom} S$ *is closed, then $S \,\widehat{+}\, \widehat{\mathfrak{N}}_\infty(S^*)$ is self-adjoint.*

Proof. (i) This assertion is a consequence of item (ii). In fact, consider the symmetric relation $T = (S-x)^{-1}$, note that $T^* = (S^*-x)^{-1}$ and $\operatorname{mul} T^* = \mathfrak{N}_x(S^*)$, and observe that the following statements (a)–(c) are equivalent

(a) $S \,\widehat{+}\, \widehat{\mathfrak{N}}_x(S^*)$ is self-adjoint;

(b) $(S-x) \,\widehat{+}\, \big(\mathfrak{N}_x(S^*) \times \{0\}\big)$ is self-adjoint;

(c) $(S-x)^{-1} \,\widehat{+}\, \big(\{0\} \times \mathfrak{N}_x(S^*)\big) = T \,\widehat{+}\, \big(\{0\} \times \operatorname{mul} T^*\big)$ is self-adjoint.

Now (ii) shows that (a)–(c) are equivalent to

(d) $\operatorname{dom} T = \overline{\operatorname{dom}}\, T \cap \operatorname{dom} T^*$;

(e) $\operatorname{ran}(S - x) = \overline{\operatorname{ran}}\,(S - x) \cap \operatorname{ran}(S^* - x)$,

which implies (i).

(ii) Observe first that the relation $S \,\widehat{+}\, \widehat{\mathfrak{N}}_\infty(S^*) = S \,\widehat{+}\, (\{0\} \times \operatorname{mul} S^*)$ is symmetric and by Proposition 1.3.12 and (1.3.4) its adjoint is given by

$$\left(S \,\widehat{+}\, \widehat{\mathfrak{N}}_\infty(S^*)\right)^* = \left(S \,\widehat{+}\, (\{0\} \times \operatorname{mul} S^*)\right)^* = S^* \cap \left(\overline{\operatorname{dom}}\, S \times \mathfrak{H}\right). \tag{1.5.9}$$

Now assume that $S \,\widehat{+}\, \widehat{\mathfrak{N}}_\infty(S^*)$ is self-adjoint. Then

$$S \,\widehat{+}\, \left(\{0\} \times \operatorname{mul} S^*\right) = S^* \cap \left(\overline{\operatorname{dom}}\, S \times \mathfrak{H}\right),$$

which implies (1.5.8). Conversely, assume that (1.5.8) holds. Since $S \,\widehat{+}\, \widehat{\mathfrak{N}}_\infty(S^*)$ is symmetric, it suffices to show that

$$S^* \cap \left(\overline{\operatorname{dom}}\, S \times \mathfrak{H}\right) \subset S \,\widehat{+}\, \left(\{0\} \times \operatorname{mul} S^*\right); \tag{1.5.10}$$

cf. (1.5.9). Let $\{f, f'\}$ belong to the left-hand side of (1.5.10). Then it follows from (1.5.8) that $f \in \operatorname{dom} S$, so that $\{f, g\} \in S \subset S^*$ for some $g \in \mathfrak{H}$. Therefore, $\{0, f' - g\} = \{f, f'\} - \{f, g\} \in S^*$, so that $f' - g \in \operatorname{mul} S^*$ and

$$\{f, f'\} = \{f, g\} + \{0, f' - g\} \in S \,\widehat{+}\, \left(\{0\} \times \operatorname{mul} S^*\right).$$

This shows that (1.5.10) holds. Hence, $S \,\widehat{+}\, \widehat{\mathfrak{N}}_\infty(S^*)$ is self-adjoint. □

The rest of this section is devoted to self-adjoint relations that are semibounded; cf. Chapter 5. In particular, the square root of a nonnegative self-adjoint relation is constructed. The material presented here will play an essential role later in the text.

Lemma 1.5.8. *Let H be a closed relation from $\mathfrak{H}$ to $\mathfrak{K}$. Then the relations H^*H and HH^* are nonnegative and self-adjoint in $\mathfrak{H}$ and $\mathfrak{K}$, respectively.*

Proof. In order to see that $H^*H \geq 0$, let $\{h, h'\} \in H^*H$. Then $\{h, l\} \in H$ and $\{l, h'\} \in H^*$ for some $l \in \mathfrak{K}$, so that

$$(h', h) = (l, l) \geq 0.$$

Hence, H^*H is nonnegative and, in particular, symmetric.

In order to show that H^*H is self-adjoint in $\mathfrak{H}$ it suffices to verify the identity $\operatorname{ran}(H^*H + I) = \mathfrak{H}$; cf. Proposition 1.5.6. For this let $f \in \mathfrak{H}$ and note that $\mathfrak{H} \times \mathfrak{K} = H \oplus_{\mathfrak{H} \times \mathfrak{K}} H^\perp$, as H is closed. It follows that there is a unique decomposition

$$\{f, 0\} = \{h, h'\} + \{k, k'\}, \quad \{h, h'\} \in H, \quad \{k, k'\} \in H^\perp.$$

Hence,

$$f = h + k, \quad 0 = h' + k',$$

which leads to $\{-k, h'\} = \{-k, -k'\} \in H^\perp$ and $\{h', k\} = J\{-k, h'\} \in JH^\perp = H^*$, where J is the flip-flop operator in (1.3.1). Thus, $\{h, h'\} \in H$ and $\{h', k\} \in H^*$ imply $\{h, k\} \in H^*H$ and

$$\{h, f\} = \{h, k + h\} \in H^*H + I,$$

so that $f \in \operatorname{ran}(H^*H + I)$. Thus, $\operatorname{ran}(H^*H + I) = \mathfrak{H}$.

Applying what was established above to H^* (instead of H) and taking into account Proposition 1.3.2 (i) one concludes that $HH^* = H^{**}H^*$ is also nonnegative and self-adjoint. □

In the next theorem it will be shown that a nonnegative self-adjoint relation possesses a unique nonnegative *square root.*

Theorem 1.5.9. *Let H be a nonnegative self-adjoint relation in $\mathfrak{H}$. Then there exists a unique nonnegative self-adjoint relation K in $\mathfrak{H}$, denoted by $K = H^{\frac{1}{2}}$, such that $K^2 = H$. Moreover, $H^{\frac{1}{2}}$ has the representation*

$$H^{\frac{1}{2}} = (H_{\mathrm{op}})^{\frac{1}{2}} \mathbin{\widehat{\oplus}} H_{\mathrm{mul}}. \tag{1.5.11}$$

Proof. If H is a self-adjoint relation in $\mathfrak{H}$, then by Theorem 1.5.1 one has the orthogonal decomposition

$$H = H_{\mathrm{op}} \mathbin{\widehat{\oplus}} H_{\mathrm{mul}}. \tag{1.5.12}$$

Since H is assumed to be nonnegative, it follows that H_{op} is a nonnegative self-adjoint operator in $\mathfrak{H}_{\mathrm{op}}$ which possesses a unique nonnegative square root $(H_{\mathrm{op}})^{\frac{1}{2}}$ in $\mathfrak{H}_{\mathrm{op}}$. Now clearly K defined by the right-hand side of (1.5.11) is a nonnegative self-adjoint relation with $\operatorname{mul} K = \operatorname{mul} H$. Since $\operatorname{dom} H_{\mathrm{mul}} = \{0\}$, it is clear that $(H_{\mathrm{mul}})^2 = H_{\mathrm{mul}}$. It follows from (1.1.9) that

$$K^2 = ((H_{\mathrm{op}})^{\frac{1}{2}})^2 \mathbin{\widehat{\oplus}} (H_{\mathrm{mul}})^2 = H_{\mathrm{op}} \mathbin{\widehat{\oplus}} H_{\mathrm{mul}} = H.$$

In order to show uniqueness, let K be a nonnegative self-adjoint relation in $\mathfrak{H}$ such that $K^2 = H$. Then $\operatorname{mul} K = \operatorname{mul} H$. In fact, the inclusion $\operatorname{mul} K \subset \operatorname{mul} H$ is clear as $\{0, \varphi\} \in K$ and $\{0, 0\} \in K$ show $\{0, \varphi\} \in K^2 = H$. To show that $\operatorname{mul} H \subset \operatorname{mul} K$, let $\{0, \varphi\} \in H = K^2$. Then $\{0, \psi\} \in K$ and $\{\psi, \varphi\} \in K$ for some $\psi \in \mathfrak{H}$. As K is self-adjoint, $(\psi, \psi) = (0, \varphi) = 0$, that is, $\psi = 0$ and $\{0, \varphi\} \in K$, as needed. Therefore,

$$\operatorname{mul} K = \operatorname{mul} H \quad \text{and} \quad \overline{\operatorname{dom}}\, K = \overline{\operatorname{dom}}\, H.$$

Decompose the self-adjoint relation K as in Theorem 1.5.1

$$K = K_{\mathrm{op}} \mathbin{\widehat{\oplus}} K_{\mathrm{mul}}, \tag{1.5.13}$$

where $K_{\rm op}$ is a nonnegative self-adjoint operator in $\overline{\rm dom}\, K = \overline{\rm dom}\, H$. Furthermore, observe from (1.5.13) and (1.1.9) that

$$H = K^2 = (K_{\rm op})^2 \mathbin{\widehat{\oplus}} (K_{\rm mul})^2 = (K_{\rm op})^2 \mathbin{\widehat{\oplus}} K_{\rm mul}. \tag{1.5.14}$$

Moreover, comparing (1.5.14) with (1.5.12) shows that

$$H_{\rm op} = (K_{\rm op})^2, \quad H_{\rm mul} = K_{\rm mul},$$

and since the square root of a nonnegative self-adjoint operator is uniquely determined, it follows that $K_{\rm op} = (H_{\rm op})^{\frac{1}{2}}$. □

Let H be a nonnegative self-adjoint relation in $\mathfrak{H}$. Since Theorem 1.5.9 implies that $\operatorname{mul} H^{\frac{1}{2}} = \operatorname{mul} H$, it follows that

$$(H^{\frac{1}{2}})_{\rm op} = (H_{\rm op})^{\frac{1}{2}},$$

so that the notation $H_{\rm op}^{\frac{1}{2}}$ is unambiguous.

In the next lemma the square root of $H - x$ for a semibounded relation H is considered.

Lemma 1.5.10. *Let H be a semibounded self-adjoint relation in $\mathfrak{H}$ with lower bound $\eta = m(H)$ and let $x \le \eta$. Then the following statements hold:*

(i) $\operatorname{dom} H_{\rm op}$ *is a core for the operator* $(H_{\rm op} - x)^{\frac{1}{2}}$, *that is, the closure of the restriction*

$$(H_{\rm op} - x)^{\frac{1}{2}} \upharpoonright \operatorname{dom} H_{\rm op} \tag{1.5.15}$$

coincides with $(H_{\rm op} - x)^{\frac{1}{2}}$;

(ii) $\operatorname{dom}(H - x)^{\frac{1}{2}} = \operatorname{dom}(H - \eta)^{\frac{1}{2}}$;

(iii) *for all* $h \in \operatorname{dom}(H - x)^{\frac{1}{2}} = \operatorname{dom}(H - \eta)^{\frac{1}{2}}$,

$$\|(H_{\rm op} - x)^{\frac{1}{2}} h\|^2 + x\|h\|^2 = \|(H_{\rm op} - \eta)^{\frac{1}{2}} h\|^2 + \eta\|h\|^2. \tag{1.5.16}$$

Proof. (i) First observe that $H_{\rm op}$ is a self-adjoint operator in $\mathfrak{H}_{\rm op}$ with the same lower bound as H, and hence $H_{\rm op} - x$, $x \le \eta$, is a nonnegative operator in $\mathfrak{H}_{\rm op}$. It suffices to show that the graph of the operator in (1.5.15) is dense in the graph of the operator $(H_{\rm op} - x)^{\frac{1}{2}}$. Therefore, assume that for some $k \in \operatorname{dom}(H_{\rm op} - x)^{\frac{1}{2}}$ and all $h \in \operatorname{dom} H_{\rm op}$ one has

$$0 = (h,k) + \big((H_{\rm op} - x)^{\frac{1}{2}} h, (H_{\rm op} - x)^{\frac{1}{2}} k\big) = (h,k) + \big((H_{\rm op} - x)h, k\big).$$

Then k is orthogonal to $\operatorname{ran}((H_{\rm op} - x) + I)$ and as $x - 1 < \eta$, it follows that $\operatorname{ran}((H_{\rm op} - x) + I) = \mathfrak{H}_{\rm op}$. Hence, $k = 0$. This implies (i).

(ii) & (iii) Note first that for $h \in \operatorname{dom} H = \operatorname{dom} H_{\rm op}$ the identity

$$((H_{\rm op} - x)h, h) + x(h,h) = ((H_{\rm op} - \eta)h, h) + \eta(h,h)$$

can be rewritten in the form

$$\|(H_{\rm op} - x)^{\frac{1}{2}}h\|^2 + x\|h\|^2 = \|(H_{\rm op} - \eta)^{\frac{1}{2}}h\|^2 + \eta\|h\|^2, \quad h \in \operatorname{dom} H, \tag{1.5.17}$$

which coincides with (1.5.16) on $\operatorname{dom} H$.

In order to show the inclusion ($\subset$) in (ii), assume that

$$h \in \operatorname{dom}(H - x)^{\frac{1}{2}} = \operatorname{dom}(H_{\rm op} - x)^{\frac{1}{2}}.$$

According to (i), there exists a sequence $h_n \in \operatorname{dom} H_{\rm op}$ such that

$$h_n \to h \quad \text{and} \quad (H_{\rm op} - x)^{\frac{1}{2}}h_n \to (H_{\rm op} - x)^{\frac{1}{2}}h. \tag{1.5.18}$$

Therefore, it follows from (1.5.17) that $(H_{\rm op} - \eta)^{\frac{1}{2}}h_n$ is a Cauchy sequence in $\mathfrak{H}_{\rm op}$. Since $h_n \to h$ and the operator $(H_{\rm op} - \eta)^{\frac{1}{2}}$ is closed, one concludes that $h \in \operatorname{dom}(H_{\rm op} - \eta)^{\frac{1}{2}} = \operatorname{dom}(H - \eta)^{\frac{1}{2}}$. The other inclusion in (ii) is shown in the same way.

It remains to verify (1.5.16). Choose $h \in \operatorname{dom}(H - x)^{\frac{1}{2}} = \operatorname{dom}(H - \eta)^{\frac{1}{2}}$ and use (i) to get a sequence $h_n \in \operatorname{dom} H_{\rm op}$ as in (1.5.18). Then (1.5.17) shows that $(H_{\rm op} - \eta)^{\frac{1}{2}}h_n$ is a Cauchy sequence in $\mathfrak{H}_{\rm op}$, and as $(H_{\rm op} - \eta)^{\frac{1}{2}}$ is closed one has

$$(H_{\rm op} - \eta)^{\frac{1}{2}}h_n \to (H_{\rm op} - \eta)^{\frac{1}{2}}h, \quad n \to \infty. \tag{1.5.19}$$

From (1.5.17) applied with h_n, together with (1.5.18) and (1.5.19) one obtains (1.5.16). □

In the next proposition two semibounded self-adjoint relations are considered. It turns out that the inclusion of the square root domains implies a strong norm inequality for the operator parts.

Proposition 1.5.11. *Let H_1 and H_2 be semibounded self-adjoint relations in $\mathfrak{H}$ with lower bounds $m(H_1)$ and $m(H_2)$ and let $x < \min\{m(H_1), m(H_2)\}$. Then the inclusion*

$$\operatorname{dom}(H_2 - x)^{\frac{1}{2}} \subset \operatorname{dom}(H_1 - x)^{\frac{1}{2}}, \tag{1.5.20}$$

together with the inequality

$$\|(H_{1,\rm op} - x)^{\frac{1}{2}}\varphi\| \leq \rho\|(H_{2,\rm op} - x)^{\frac{1}{2}}\varphi\|, \quad \varphi \in \operatorname{dom}(H_2 - x)^{\frac{1}{2}}, \tag{1.5.21}$$

where $\rho > 0$, are equivalent to the inequality

$$(H_2 - x)^{-1} \leq \rho^2(H_1 - x)^{-1}. \tag{1.5.22}$$

Moreover, if the inclusion (1.5.20) holds, then there exists $\rho > 0$ for which the inequality (1.5.21) is satisfied.

Proof. Let $x < \min\{m(H_1), m(H_2)\}$, so that

$$A = (H_2 - x)^{-1} \quad \text{and} \quad B = (H_1 - x)^{-1}$$

are nonnegative operators in $\mathbf{B}(\mathfrak{H})$. Note that their square roots are given by $A^{\frac{1}{2}} = (H_2 - x)^{-\frac{1}{2}}$ and $B^{\frac{1}{2}} = (H_1 - x)^{-\frac{1}{2}}$. Hence, the Moore–Penrose inverses of $A^{\frac{1}{2}}$ and $B^{\frac{1}{2}}$ are given by

$$A^{(-\frac{1}{2})} = (H_{2,\mathrm{op}} - x)^{\frac{1}{2}}, \quad B^{(-\frac{1}{2})} = (H_{1,\mathrm{op}} - x)^{\frac{1}{2}},$$

cf. Definition 1.3.17 and Example 1.3.18, where it was used that

$$\ker A^{\frac{1}{2}} = \ker A = \operatorname{mul} H_2 \quad \text{and} \quad \ker B^{\frac{1}{2}} = \ker B = \operatorname{mul} H_1.$$

In terms of the operators A and B, the statements in (1.5.20) and (1.5.21) mean that

$$\operatorname{ran} A^{\frac{1}{2}} \subset \operatorname{ran} B^{\frac{1}{2}} \quad \text{and} \quad \|B^{(-\frac{1}{2})}\varphi\| \leq \rho \|A^{(-\frac{1}{2})}\varphi\|, \quad \varphi \in \operatorname{ran} A^{\frac{1}{2}}, \tag{1.5.23}$$

while the statement in (1.5.22) means that

$$A \leq \rho^2 B. \tag{1.5.24}$$

The equivalence of (1.5.23) and (1.5.24) follows from Proposition D.8. Moreover, Proposition D.8 also shows that (1.5.20) implies (1.5.21) for some $\rho > 0$, which completes the proof. □

The next lemma characterizes the domain of the square root of a nonnegative self-adjoint relation.

Lemma 1.5.12. *Let H be a nonnegative self-adjoint relation in $\mathfrak{H}$ and let $\varphi \in \mathfrak{H}$. Then the function*

$$x \mapsto \left((H^{-1} - x)^{-1}\varphi, \varphi\right), \quad x \in (-\infty, 0),$$

is nondecreasing, and

$$\lim_{x \uparrow 0} \left((H^{-1} - x)^{-1}\varphi, \varphi\right) = \begin{cases} \|H_{\mathrm{op}}^{\frac{1}{2}}\varphi\|^2, & \varphi \in \operatorname{dom} H^{\frac{1}{2}}, \\ \infty, & \text{otherwise}. \end{cases} \tag{1.5.25}$$

Proof. Since H is a nonnegative self-adjoint relation, so is H^{-1}, and each resolvent operator in the identity

$$(H^{-1} - x)^{-1} = -\frac{1}{x} - \frac{1}{x^2}\left(H - \frac{1}{x}\right)^{-1}, \quad x < 0, \tag{1.5.26}$$

belongs to $\mathbf{B}(\mathfrak{H})$ by Corollary 1.1.12 and (1.2.14). Since $\ker (H-1/x)^{-1} = \operatorname{mul} H$, it follows from (1.5.26) that for each $x<0$ and $\varphi \in \mathfrak{H}$

$$\begin{aligned}\big((H^{-1}-x)^{-1}\varphi,\varphi\big) = &-\frac{1}{x}\,\|(I-P_{\rm op})\varphi\|^2 - \frac{1}{x}\,\|P_{\rm op}\varphi\|^2 \\ &-\frac{1}{x^2}\left(\left(H-\frac{1}{x}\right)^{-1} P_{\rm op}\varphi, P_{\rm op}\varphi\right),\end{aligned} \tag{1.5.27}$$

where $P_{\rm op}$ is the orthogonal projection from $\mathfrak{H}$ onto $\overline{\operatorname{dom}}\, H$. Let $E(\cdot)$ be the spectral measure of H, so that $H_{\rm op} = \int_0^\infty t\, dE_{\rm op}(t)$. Then the formula (1.5.27) can be rewritten for each $x<0$ and $\varphi \in \mathfrak{H}$ as

$$\begin{aligned}\big((H^{-1}-x)^{-1}\varphi,\varphi\big) = &-\frac{1}{x}\,\|(I-P_{\rm op})\varphi\|^2 \\ &-\int_0^\infty \frac{t}{tx-1}\, d(E_{\rm op}(t)P_{\rm op}\varphi, P_{\rm op}\varphi).\end{aligned} \tag{1.5.28}$$

In particular, (1.5.28) shows that the function in (1.5.25) is nondecreasing for $x \in (-\infty, 0)$.

Furthermore, by the nonnegativity of the terms in (1.5.28), the limit as $x \uparrow 0$ of the left-hand side in (1.5.28) is finite if and only if the limit of each of the terms on the right-hand side of (1.5.28) is finite. The first limit is finite if and only if $(I-P_{\rm op})\varphi = 0$, i.e., $P_{\rm op}\varphi = \varphi$ and hence $\varphi \in \overline{\operatorname{dom}}\, H$. By the monotone convergence theorem, the limit of the second term is equal to $\int_0^\infty t\, d(E_{\rm op}(t)\varphi,\varphi)$, which is finite and equal to

$$\|H_{\rm op}^{\frac{1}{2}}\varphi\|^2$$

if and only if $\varphi \in \operatorname{dom} H^{\frac{1}{2}}$. □

1.6 Maximal dissipative and accumulative relations

In this section the basic properties of dissipative and accumulative relations are discussed. Of special interest are dissipative and accumulative relations which are maximal with this property. Such relations admit an orthogonal decomposition into a maximal dissipative or maximal accumulative operator and a self-adjoint purely multivalued part.

Definition 1.6.1. A relation H in $\mathfrak{H}$ is said to be *dissipative* (*accumulative*) if $\operatorname{Im}(f',f) \geq 0$ ($\operatorname{Im}(f',f) \leq 0$) for all $\{f,f'\} \in H$. The relation H is said to be *maximal dissipative* (*maximal accumulative*) if every dissipative (accumulative) extension H' of H in $\mathfrak{H}$ satisfies $H' = H$.

It is easy to see that if a relation H in $\mathfrak{H}$ dissipative or accumulative, then so is the closure $\overline{H}$. Hence, maximal dissipative or maximal accumulative relations are automatically closed.

Note that a linear relation H is dissipative (maximal dissipative) if and only if the relation $-H$ is accumulative (maximal accumulative). Thus, it suffices to state results for dissipative relations; the corresponding results for accumulative relations follow immediately.

Lemma 1.6.2. *Let H be a dissipative relation in $\mathfrak{H}$. Then* $\operatorname{mul} H \subset \operatorname{mul} H^*$. *If H is maximal dissipative, then* $\operatorname{mul} H = \operatorname{mul} H^*$.

Proof. Let $k \in \operatorname{mul} H$ and let $\{h, h'\} \in H$. Since $\{0, \lambda k\} \in H$ for every $\lambda \in \mathbb{C}$, one has $\{h, h' + \lambda k\} \in H$. Since H is dissipative one has

$$\operatorname{Im}(h', h) + \operatorname{Im}\big(\lambda(k, h)\big) = \operatorname{Im}(h' + \lambda k, h) \geq 0, \qquad \lambda \in \mathbb{C}.$$

In this inequality $\lambda \in \mathbb{C}$ is arbitrary and hence one concludes $(k, h) = 0$. Therefore, $\operatorname{mul} H \subset (\operatorname{dom} H)^\perp = \operatorname{mul} H^*$.

If H is dissipative and $k \in \operatorname{mul} H^* = (\operatorname{dom} H)^\perp$, then it follows that

$$H \,\widehat{+}\, \operatorname{span}\{0, k\}$$

is a dissipative extension of H. Hence, if H is maximal dissipative, then $\{0, k\} \in H$ and $k \in \operatorname{mul} H$. □

The next proposition is the analog of Proposition 1.4.4. Its proof is almost the same, and depends on the estimate

$$0 \leq -\operatorname{Im} \lambda (f, f) \leq \operatorname{Im}(f' - \lambda f, f) \leq \|f' - \lambda f\| \|f\|,$$

which is valid for $\lambda \in \mathbb{C}^-$ and $\{f, f'\} \in H$ such that $\operatorname{Im}(f', f) \geq 0$.

Proposition 1.6.3. *Let H be a dissipative relation in $\mathfrak{H}$. Then $\mathbb{C}^-$ is contained in $\gamma(H)$ and, in particular, the defect $n_\lambda(H) = \dim(\operatorname{ran}(H - \lambda)^\perp)$ is constant for all $\lambda \in \mathbb{C}^-$. Furthermore,*

$$\|(H - \lambda)^{-1} h\| \leq \frac{1}{-\operatorname{Im} \lambda} \|h\|$$

for all $h \in \operatorname{dom}(H - \lambda)^{-1} = \operatorname{ran}(H - \lambda)$ and $\lambda \in \mathbb{C}^-$.

For dissipative relations Theorem 1.5.5 reads as follows.

Theorem 1.6.4. *Let H be a closed dissipative relation in $\mathfrak{H}$. Then the following statements are equivalent:*

(i) *H is maximal dissipative;*

(ii) $\ker(H^* - \bar{\lambda}) = \{0\}$ *for some, and hence for all $\lambda \in \mathbb{C}^-$;*

(iii) $\operatorname{ran}(H - \lambda) = \mathfrak{H}$ *for some, and hence for all $\lambda \in \mathbb{C}^-$;*

(iv) $\mathbb{C}^- \subset \rho(H)$.

Proof. (ii) $\Leftrightarrow$ (iii) $\Rightarrow$ (iv) follow with the same arguments as in the proof of Theorem 1.5.5.

(i) $\Rightarrow$ (iii) As H is closed, also $\operatorname{ran}(H-\lambda)$ is closed for all $\lambda \in \mathbb{C}^-$ by Lemma 1.2.2 since $\mathbb{C}^- \subset \gamma(H)$ by Proposition 1.6.3. Now assume that $\operatorname{ran}(H-\lambda)$ is a proper subspace of $\mathfrak{H}$ for some $\lambda \in \mathbb{C}^-$ and define the relation H' in $\mathfrak{H}$ by

$$H' := \bigl\{\{f + f_{\bar\lambda}, f' + \bar\lambda f_{\bar\lambda}\} : \{f, f'\} \in H,\ f_{\bar\lambda} \in \mathfrak{N}_{\bar\lambda}(H^*)\bigr\},$$

where $\mathfrak{N}_{\bar\lambda}(H^*) = \ker(H^* - \bar\lambda) = (\operatorname{ran}(H-\lambda))^\perp$. Then clearly $H \subset H'$, and as

$$\begin{aligned}\operatorname{ran}(H'-\lambda) &= \bigl\{f' - \lambda f + (\bar\lambda - \lambda) f_{\bar\lambda} : \{f, f'\} \in H,\ f_{\bar\lambda} \in \mathfrak{N}_{\bar\lambda}(H^*)\bigr\}\\ &= \operatorname{ran}(H-\lambda) \oplus \mathfrak{N}_{\bar\lambda}(H^*) = \mathfrak{H}\end{aligned}$$

and $\mathfrak{N}_{\bar\lambda}(H^*) \neq \{0\}$, it follows that H' is a proper extension of H in $\mathfrak{H}$, $H \neq H'$. Since $f_{\bar\lambda} \in \mathfrak{N}_{\bar\lambda}(H^*)$ implies $\{f_{\bar\lambda}, \bar\lambda f_{\bar\lambda}\} \in H^*$, one sees that

$$(f', f_{\bar\lambda}) = (f, \bar\lambda f_{\bar\lambda}) = \lambda (f, f_{\bar\lambda})$$

for all $\{f, f'\} \in H$. Hence,

$$\begin{aligned}(f' + \bar\lambda f_{\bar\lambda}, f + f_{\bar\lambda}) &= (f', f) + (f', f_{\bar\lambda}) + \bar\lambda (f_{\bar\lambda}, f) + \bar\lambda (f_{\bar\lambda}, f_{\bar\lambda})\\ &= (f', f) + 2\mathrm{Re}\,\bigl(\lambda (f, f_{\bar\lambda})\bigr) + \bar\lambda (f_{\bar\lambda}, f_{\bar\lambda})\end{aligned}$$

and from the assumptions that H is dissipative and $\lambda \in \mathbb{C}^-$ one concludes that

$$\mathrm{Im}\,\bigl(f' + \bar\lambda f_{\bar\lambda}, f + f_{\bar\lambda}\bigr) = \mathrm{Im}\,(f', f) + \mathrm{Im}\,\bar\lambda (f_{\bar\lambda}, f_{\bar\lambda}) \geq 0,$$

i.e., H' is a proper dissipative extension of H in $\mathfrak{H}$. Thus, H is not maximal dissipative. This proves (iii) for all $\lambda \in \mathbb{C}^-$.

(iv) $\Rightarrow$ (i) Suppose that H' is a dissipative extension of H, and let $\{f, f'\} \in H'$ and $\lambda \in \mathbb{C}^-$. As $\mathbb{C}^- \subset \rho(H)$, there exists $\{g, g'\} \in H$ such that

$$f' - \lambda f = g' - \lambda g.$$

This implies $\{f - g, f' - g'\} \in H'$ and hence $f - g \in \ker(H' - \lambda)$. As H' is dissipative and $\lambda \in \mathbb{C}^-$, it follows from Proposition 1.6.3 that $f = g$, which also gives $f' = g'$. This shows $\{f, f'\} \in H$ and hence $H' = H$. Therefore, H is maximal dissipative. □

Corollary 1.6.5. *Let H be a relation in $\mathfrak{H}$. Then the following statements hold:*

(i) *If H is maximal dissipative (maximal accumulative), then $(H-\lambda)^{-1} \in \mathbf{B}(\mathfrak{H})$ is accumulative (dissipative) for each $\lambda \in \mathbb{C}^-$ ($\lambda \in \mathbb{C}^+$).*

(ii) *If H is closed, $\mathbb{C}^- \subset \rho(H)$ ($\mathbb{C}^+ \subset \rho(H)$), and $(H-\lambda)^{-1} \in \mathbf{B}(\mathfrak{H})$ is accumulative (dissipative) for all $\lambda \in \mathbb{C}^-$ ($\lambda \in \mathbb{C}^+$), then H is maximal dissipative (maximal accumulative).*

Proof. (i) Assume that H is maximal dissipative, which implies $\mathbb{C}^- \subset \rho(H)$. Let $\{f, f'\} \in H$. Then for $\lambda \in \mathbb{C}^-$ the identity

$$\begin{aligned}\operatorname{Im}\big((H-\lambda)^{-1}(f'-\lambda f), f'-\lambda f)\big) &= \operatorname{Im}(f, f'-\lambda f)\\ &= -\operatorname{Im}(f'-\lambda f, f)\\ &= -\operatorname{Im}(f', f) + \operatorname{Im}\lambda(f,f)\end{aligned} \tag{1.6.1}$$

shows that $(H-\lambda)^{-1} \in \mathbf{B}(\mathfrak{H})$ is accumulative.

(ii) Assume that H is closed, $\mathbb{C}^- \subset \rho(H)$, and $(H-\lambda)^{-1} \in \mathbf{B}(\mathfrak{H})$ is accumulative for all $\lambda \in \mathbb{C}^-$. Then $(H-\lambda)^{-1}(f'-\lambda f) = f$ for all $\{f, f'\} \in H$ and $\lambda \in \mathbb{C}^-$, and hence the identity (1.6.1) shows that

$$\operatorname{Im}(f', f) \geq \operatorname{Im}\lambda(f,f) \quad \text{for all} \quad \lambda \in \mathbb{C}^-.$$

This implies that H is dissipative and, since H is closed and $\mathbb{C}^- \subset \rho(H)$, it follows from Theorem 1.6.4 that H is maximal dissipative. □

In the next proposition the Cayley transform and the inverse Cayley transform of accumulative, dissipative, maximal accumulative, and maximal dissipative relations are considered. This proposition is the counterpart of Proposition 1.4.8 and Proposition 1.5.4.

Proposition 1.6.6. *Let $\mu \in \mathbb{C} \setminus \mathbb{R}$ and let $\mathcal{C}_\mu$ and $\mathcal{F}_\mu$ be the Cayley transform and inverse Cayley transform in Definition 1.1.13. Let H and V be relations in $\mathfrak{H}$ such that $V = \mathcal{C}_\mu[H]$ or, equivalently, $H = \mathcal{F}_\mu[V]$. Then the following statements hold:*

(i) *If $\mu \in \mathbb{C}^-$ ($\mu \in \mathbb{C}^+$), then H is a dissipative (accumulative) relation if and only if V is a, in general not everywhere defined, contractive operator.*
(ii) *If $\mu \in \mathbb{C}^-$ ($\mu \in \mathbb{C}^+$), then H is a maximal dissipative (maximal accumulative) relation if and only if V is a contractive operator defined on $\mathfrak{H}$.*

Proof. (i) Let H be dissipative or accumulative and for $\mu \in \mathbb{C} \setminus \mathbb{R}$ define

$$V = \mathcal{C}_\mu[H] = \big\{\{f'-\mu f, f'-\bar{\mu} f\} : \{f, f'\} \in H\big\}. \tag{1.6.2}$$

Then a straightforward computation shows that

$$\begin{aligned}\|f'-\bar{\mu}f\|^2 - \|f'-\mu f\|^2 &= 2\operatorname{Re}\big((\bar{\mu}-\mu)(f', f)\big)\\ &= 4(\operatorname{Im}\mu)\operatorname{Im}(f', f)\end{aligned}$$

for all $\{f, f'\} \in H$. Hence, $\|f'-\bar{\mu}f\| \leq \|f'-\mu f\|$ when $\mu \in \mathbb{C}^-$ and H is dissipative, or when $\mu \in \mathbb{C}^+$ and H is accumulative. This implies that V in (1.6.2) is a contractive operator.

Conversely, let V be a contractive operator and for $\mu \in \mathbb{C} \setminus \mathbb{R}$ define

$$H = \mathcal{F}_\mu[V] = \big\{\{k-k', \bar{\mu}k - \mu k'\} : \{k, k'\} \in V\big\}. \tag{1.6.3}$$

Then a computation shows

$$\big(\overline{\mu}k - \mu k', k - k'\big) = \overline{\mu}(k,k) - 2\mathrm{Re}\,\big(\mu(k',k)\big) + \mu(k',k'),$$

and consequently

$$\mathrm{Im}\,\big(\overline{\mu}k - \mu k', k - k'\big) = -\mathrm{Im}\,\mu\big(\|k\|^2 - \|k'\|^2\big). \tag{1.6.4}$$

Since $\|k'\| \leq \|k\|$ for $\{k,k'\} \in V$, it follows from (1.6.4) that H in (1.6.3) is dissipative for $\mu \in \mathbb{C}^-$ and accumulative for $\mu \in \mathbb{C}^+$.

(ii) If H is maximal dissipative or maximal accumulative, then $\mathrm{ran}\,(H - \mu) = \mathfrak{H}$ for $\mu \in \mathbb{C}^-$ or $\mu \in \mathbb{C}^+$, respectively; cf. Theorem 1.6.4. Hence, V in (1.6.2) satisfies $\mathrm{dom}\,V = \mathfrak{H}$.

Conversely, if $\mathrm{dom}\,V = \mathfrak{H}$, then (1.6.3) implies $\mathrm{ran}\,(H - \mu) = \mathfrak{H}$, and Theorem 1.6.4 then implies that H is maximal dissipative or maximal accumulative. □

The following result is sometimes useful.

Proposition 1.6.7. *Let H be a closed relation in $\mathfrak{H}$. Then H is maximal dissipative if and only if H^* is maximal accumulative.*

Proof. Let H be a maximal dissipative relation. Then Proposition 1.6.6 shows that for $\mu \in \mathbb{C}^-$ the Cayley transform $\mathcal{C}_\mu[H]$ is a contractive operator defined on $\mathfrak{H}$. But then also the adjoint is a contractive operator defined on all of $\mathfrak{H}$. Observe that by Lemma 1.3.11

$$(\mathcal{C}_\mu[H])^* = \mathcal{C}_{\overline{\mu}}[H^*],$$

which, since $\overline{\mu} \in \mathbb{C}^+$, implies by Proposition 1.6.6 that H^* is maximal accumulative. The converse is proved in the same way. □

The next proposition is of a slightly different nature than the previous results and complements Lemma 1.5.7. It shows that a closed symmetric relation always admits maximal dissipative and maximal accumulative extensions.

Proposition 1.6.8. *Let S be a closed symmetric relation in $\mathfrak{H}$, let $\lambda \in \mathbb{C} \setminus \mathbb{R}$, and let*

$$H = S \,\widehat{+}\, \widehat{\mathfrak{N}}_\lambda(S^*). \tag{1.6.5}$$

Then H is a closed extension of S and the sum is direct. Moreover,

(i) *if $\lambda \in \mathbb{C}^+$, then H is maximal dissipative;*

(ii) *if $\lambda \in \mathbb{C}^-$, then H is maximal accumulative.*

Proof. Since the eigenvalues of S are real, the sum in (1.6.5) is direct. In order to verify (i) and (ii) note that a typical element in H is of the form $\{f + f_\lambda, f' + \lambda f_\lambda\}$ with $\{f, f'\} \in S$ and $\{f_\lambda, \lambda f_\lambda\} \in S^*$. Therefore, as S is symmetric,

$$(f', f_\lambda) = (f, \lambda f_\lambda).$$

Hence, the identity

$$(f' + \lambda f_\lambda, f + f_\lambda) = (f', f) + 2\mathrm{Re}\,\big(\lambda(f_\lambda, f)\big) + \lambda(f_\lambda, f_\lambda)$$

together with $\mathrm{Im}\,(f', f) = 0$ shows that

$$\mathrm{Im}\,(f' + \lambda f_\lambda, f + f_\lambda) = (\mathrm{Im}\,\lambda)(f_\lambda, f_\lambda).$$

Therefore, H is dissipative for $\lambda \in \mathbb{C}^+$ and accumulative for $\lambda \in \mathbb{C}^-$. Finally, observe that (1.6.5) implies

$$\mathrm{ran}\,(H - \overline{\lambda}) = \mathrm{ran}\,(S - \overline{\lambda}) \oplus \ker\,(S^* - \lambda) = \mathfrak{H},$$

and Theorem 1.6.4 shows that H is maximal dissipative for $\lambda \in \mathbb{C}^+$ and maximal accumulative for $\lambda \in \mathbb{C}^-$. In particular, H is closed. □

The next proposition provides a direct sum decomposition of $\mathfrak{H}$ based on the construction in Proposition 1.6.8.

Proposition 1.6.9. *Let S be a closed symmetric relation in $\mathfrak{H}$ and let $\mu \in \mathbb{C} \setminus \mathbb{R}$. Then for λ in the same half-plane as μ there is the direct sum decomposition*

$$\mathfrak{H} = \mathrm{ran}\,(S - \lambda) + \ker\,(S^* - \overline{\mu}). \tag{1.6.6}$$

Proof. Let the relation $H(\overline{\mu})$ be defined by $H(\overline{\mu}) = S \,\widehat{+}\, \widehat{\mathfrak{N}}_{\overline{\mu}}(S^*)$. A straightforward calculation involving the Cayley transform

$$\mathcal{C}_\mu[H(\overline{\mu})] = \big\{\{h' - \mu h + \eta, h' - \overline{\mu} h\} : \{h, h'\} \in S,\ \eta \in \mathfrak{N}_{\overline{\mu}}(S^*)\big\}$$

yields the identity

$$\begin{aligned} I - &\frac{\lambda - \mu}{\lambda - \overline{\mu}}\,\mathcal{C}_\mu[H(\overline{\mu})] \\ &= \left\{\left\{h' - \mu h + \eta, \frac{\mu - \overline{\mu}}{\lambda - \overline{\mu}}(h' - \lambda h) + \eta\right\} : \{h, h'\} \in S,\ \eta \in \mathfrak{N}_{\overline{\mu}}(S^*)\right\}. \end{aligned} \tag{1.6.7}$$

Note that the domain and the range of this relation are given by

$$\mathrm{ran}\,(S - \mu) \oplus \ker\,(S^* - \overline{\mu}) = \mathfrak{H} \quad \text{and} \quad \mathrm{ran}\,(S - \lambda) + \ker\,(S^* - \overline{\mu}), \tag{1.6.8}$$

respectively.

Now observe that by Proposition 1.6.8 the relation $H(\overline{\mu})$ is maximal dissipative for $\mu \in \mathbb{C}^-$ or maximal accumulative for $\mu \in \mathbb{C}^+$, and thus the Cayley transform $\mathcal{C}_\mu[H(\overline{\mu})]$ is a contraction defined on $\mathfrak{H}$; cf. Proposition 1.6.6. Due to the assumption about λ, one has $|\lambda - \mu| < |\lambda - \overline{\mu}|$ and hence the left-hand side of (1.6.7) is a bijection from $\mathfrak{H}$ onto $\mathfrak{H}$. Therefore, the decomposition (1.6.6) in the lemma is now concluded from the second identity in (1.6.8). To see that the decomposition (1.6.6) is direct, assume that the second component on the right-hand side of (1.6.7) is zero. Then the first component must be zero, so that $h' = \mu h$ and $\eta = 0$. Since S is symmetric, it follows from $\{h, h'\} \in S$ and $h' = \mu h$ that $h = 0$ and $h' = 0$. Thus, indeed, the sum in (1.6.6) is direct. □

The next assertion is a special case of Lemma 1.4.10 for maximal dissipative and maximal accumulative extensions.

Lemma 1.6.10. *Let S be a closed symmetric relation in $\mathfrak{H}$ and let $H \subset S^*$ be a maximal dissipative (maximal accumulative) extension of S. Then for $\lambda, \mu \in \mathbb{C}^-$ ($\lambda, \mu \in \mathbb{C}^+$)*

$$I + (\lambda - \mu)(H - \lambda)^{-1} \tag{1.6.9}$$

is boundedly invertible with inverse $I + (\mu - \lambda)(H - \mu)^{-1}$. For fixed $\mu \in \mathbb{C}^-$ ($\mu \in \mathbb{C}^+$), the mapping (1.6.9) is holomorphic in $\lambda \in \mathbb{C}^-$ ($\lambda \in \mathbb{C}^+$). Moreover, the operator in (1.6.9) maps $\mathfrak{N}_\mu(S^*)$ bijectively onto $\mathfrak{N}_\lambda(S^*)$.

The following useful fact about the closed span (denoted by $\overline{\text{span}}$) of the defect spaces of a symmetric relation will be used in Chapter 3.

Lemma 1.6.11. *Let S be a closed symmetric relation in $\mathfrak{H}$. Let $\mathcal{U}^- \subset \mathbb{C}^-$ be a set which has an accumulation point in $\mathbb{C}^-$, and let $\mathcal{U}^+ \subset \mathbb{C}^+$ be a set which has an accumulation point in $\mathbb{C}^+$. Then*

$$\overline{\text{span}}\,\{\mathfrak{N}_\lambda(S^*) : \lambda \in \mathcal{U}^-\} = \overline{\text{span}}\,\{\mathfrak{N}_\lambda(S^*) : \lambda \in \mathbb{C}^-\} \tag{1.6.10}$$

and

$$\overline{\text{span}}\,\{\mathfrak{N}_\lambda(S^*) : \lambda \in \mathcal{U}^+\} = \overline{\text{span}}\,\{\mathfrak{N}_\lambda(S^*) : \lambda \in \mathbb{C}^+\}. \tag{1.6.11}$$

Proof. The equality (1.6.10) will be shown, the proof of (1.6.11) is analogous. Note also that the inclusion ($\subset$) in (1.6.10) is clear and hence it remains to verify the inclusion ($\supset$) in (1.6.10). It is sufficient to show that

$$\big(\text{span}\,\{\mathfrak{N}_\lambda(S^*) : \lambda \in \mathcal{U}^-\}\big)^\perp \subset \big(\text{span}\,\{\mathfrak{N}_\lambda(S^*) : \lambda \in \mathbb{C}^-\}\big)^\perp \tag{1.6.12}$$

holds. Fix a maximal dissipative extension H of S, see Proposition 1.6.8, and assume that $f \in \mathfrak{H}$ belongs to the left-hand side of (1.6.12). Then for all $\lambda \in \mathcal{U}^-$ and $f_\lambda \in \mathfrak{N}_\lambda(S^*)$ one has $(f_\lambda, f) = 0$. By Lemma 1.6.10, for $\mu \in \mathbb{C}^-$ the mapping

$$\lambda \mapsto I + (\lambda - \mu)(H - \lambda)^{-1} \tag{1.6.13}$$

is holomorphic in $\mathbb{C}^-$ and the operator in (1.6.13) maps $\mathfrak{N}_\mu(S^*)$ bijectively onto $\mathfrak{N}_\lambda(S^*)$. Fix $\mu \in \mathbb{C}^-$ and $f_\mu \in \mathfrak{N}_\mu(S^*)$, and consider the element

$$f_\lambda = (I + (\lambda - \mu)(H - \lambda)^{-1})f_\mu.$$

Then for all $\lambda \in \mathcal{U}^-$ one has

$$\big((I + (\lambda - \mu)(H - \lambda)^{-1})f_\mu, f\big) = (f_\lambda, f) = 0.$$

Since the function $\lambda \mapsto ((I + (\lambda - \mu)(H - \lambda)^{-1})f_\mu, f) = (f_\lambda, f)$ is holomorphic on $\mathbb{C}^-$ and vanishes on $\mathcal{U}^-$, one must have for all $\lambda \in \mathbb{C}^-$

$$\big((I + (\lambda - \mu)(H - \lambda)^{-1})f_\mu, f\big) = (f_\lambda, f) = 0.$$

Since $f_\mu \in \mathfrak{N}_\mu(S^*)$ was arbitrary and (1.6.13) maps $\mathfrak{N}_\mu(S^*)$ bijectively onto $\mathfrak{N}_\lambda(S^*)$ it follows that $(f_\lambda, f) = 0$ for all $f_\lambda \in \mathfrak{N}_\lambda(S^*)$ and all $\lambda \in \mathbb{C}^-$. Therefore, f belongs to the right-hand side of (1.6.12). □

Here is a variant of the decomposition in Theorem 1.3.16 for closed dissipative (accumulative) relations; cf. Theorem 1.4.11 and Theorem 1.5.1.

Theorem 1.6.12. *Let H be a closed dissipative (accumulative) relation in $\mathfrak{H}$ and decompose $\mathfrak{H}$ as $\mathfrak{H} = \mathfrak{H}_{\rm op} \oplus \mathfrak{H}_{\rm mul}$, $\mathfrak{H}_{\rm op} := (\operatorname{mul} H)^\perp$ and $\mathfrak{H}_{\rm mul} := \operatorname{mul} H$, and denote the orthogonal projection from $\mathfrak{H}$ onto $\mathfrak{H}_{\rm op}$ by $P_{\rm op}$. Then H is the direct orthogonal sum $H_{\rm op} \,\widehat{\oplus}\, H_{\rm mul}$ of the closed dissipative (accumulative) operator*

$$H_{\rm op} = \bigl\{\{f, P_{\rm op} f'\} : \{f, f'\} \in H\bigr\}$$

in $\mathfrak{H}_{\rm op}$ and the self-adjoint purely multivalued relation

$$H_{\rm mul} = \bigl\{\{0, f'\} : f' \in \mathfrak{H}_{\rm mul}\bigr\} = \bigl\{\{0, (I - P_{\rm op})f'\} : \{f, f'\} \in H\bigr\}$$

in $\mathfrak{H}_{\rm mul}$. Moreover, the operator $H_{\rm op}$ is densely defined in $\mathfrak{H}_{\rm op}$ if and only if $\operatorname{mul} H = \operatorname{mul} H^*$. *If the relation H is maximal dissipative (maximal accumulative), then $H_{\rm op}$ is a densely defined maximal dissipative (maximal accumulative) operator in $\mathfrak{H}_{\rm op}$.*

Proof. The proof follows the proof of Theorem 1.4.11 for symmetric relations. In order to apply Theorem 1.3.16 now one has to recall that $\operatorname{mul} H = \operatorname{mul} H^{**}$ and the inclusion $\operatorname{mul} H \subset \operatorname{mul} H^*$ holds by Lemma 1.6.2. The assertion about maximality follows from Lemma 1.6.2. □

1.7 Intermediate extensions and von Neumann's formulas

In this section intermediate extensions of a symmetric relation will be studied, with special attention paid to disjoint and transversal extensions. Furthermore, some important decompositions of intermediate extensions and the adjoint relation will be discussed. In particular, these investigations lead to the von Neumann formulas in the context of relations, which provide a description of accumulative, dissipative, symmetric, and self-adjoint extensions in terms of contractive, isometric, and unitary operators between the defect spaces of the symmetric relation.

The first result is a decomposition of a relation in a Hilbert space which has a closed restriction with nonempty resolvent set. As in Definition 1.4.9, the following notations are associated with the eigenspace of a relation T at $\lambda \in \mathbb{C}$:

$$\mathfrak{N}_\lambda(T) = \ker (T - \lambda) \quad \text{and} \quad \widehat{\mathfrak{N}}_\lambda(T) = \bigl\{\{f_\lambda, \lambda f_\lambda\} : f_\lambda \in \mathfrak{N}_\lambda(T)\bigr\}.$$

Theorem 1.7.1. *Let T be a relation in $\mathfrak{H}$ and let the relation H be a restriction of T. If* $\operatorname{ran}(H-\lambda)=\mathfrak{H}$ *for some $\lambda\in\mathbb{C}$, then*

$$T = H \,\widehat{+}\, \widehat{\mathfrak{N}}_\lambda(T) \quad \textit{and} \quad H\cap\widehat{\mathfrak{N}}_\lambda(T)=\widehat{\mathfrak{N}}_\lambda(H). \tag{1.7.1}$$

Assume, in addition, that H is closed and $\lambda\in\rho(H)$. Then the decomposition in (1.7.1) holds, the sum is direct, and

$$T \textit{ is closed if and only if } \mathfrak{N}_\lambda(T) \textit{ is closed.}$$

Proof. Since $\widehat{\mathfrak{N}}_\lambda(T)\subset T$, one has the inclusion $H\,\widehat{+}\,\widehat{\mathfrak{N}}_\lambda(T)\subset T$. To see the opposite inclusion, let $\{f,f'\}\in T$. Since $\operatorname{ran}(H-\lambda)=\mathfrak{H}$, there exists an element $\{h,h'\}\in H\subset T$ such that $f'-\lambda f=h'-\lambda h$. It follows that

$$\{f,f'\}-\{h,h'\}=\{f-h,f'-h'\}=\{f-h,\lambda(f-h)\}\in\widehat{\mathfrak{N}}_\lambda(T),$$

which shows that $\{f,f'\}\in H\,\widehat{+}\,\widehat{\mathfrak{N}}_\lambda(T)$ and thus $T\subset H\,\widehat{+}\,\widehat{\mathfrak{N}}_\lambda(T)$. The statement $H\cap\widehat{\mathfrak{N}}_\lambda(T)=\widehat{\mathfrak{N}}_\lambda(H)$ is immediate.

Now assume that the relation H is closed and $\lambda\in\rho(H)$. Then the conditions $\operatorname{ran}(H-\lambda)=\mathfrak{H}$ and $\ker(H-\lambda)=\{0\}$ are satisfied and hence the decomposition in (1.7.1) holds and the sum is direct. If T is closed, then clearly $\mathfrak{N}_\lambda(T)$ is closed. To prove the converse implication consider the linear mapping $B:\mathfrak{H}\times\mathfrak{H}\to\mathfrak{H}$ defined by $B\{f,f'\}=f'-\lambda f$. Clearly, B is a bounded operator with $\operatorname{ran}B=\mathfrak{H}$ and

$$\ker B=\bigl\{\{f,\lambda f\}:f\in\mathfrak{H}\bigr\}.$$

Consider the relation H as a closed subspace of $\mathfrak{H}\times\mathfrak{H}$. Then

$$BH=\operatorname{ran}(H-\lambda)=\mathfrak{H} \quad\text{and}\quad H\cap\ker B=\widehat{\mathfrak{N}}_\lambda(H)=\{0\}$$

since $\lambda\in\rho(H)$. Hence, BH is closed, which implies that the sum $H\,\widehat{+}\,\ker B$ is closed by Lemma C.4. Moreover, the sum $H\,\widehat{+}\,\ker B$ is direct. By assumption, $\widehat{\mathfrak{N}}_\lambda(T)$ is a closed subspace of $\ker B$, which implies that also $H\,\widehat{+}\,\widehat{\mathfrak{N}}_\lambda(T)=T$ is closed; cf. Corollary C.7. □

The next preliminary lemma contains some useful observations.

Lemma 1.7.2. *Let H and K be closed relations in $\mathfrak{H}$. Then, for all $\lambda\in\rho(H)\cap\rho(K)$,*

$$\operatorname{ran}\bigl((H\cap K)-\lambda\bigr)=\ker\bigl((K-\lambda)^{-1}-(H-\lambda)^{-1}\bigr)$$

and

$$\ker\bigl((H\,\widehat{+}\,K)-\lambda\bigr)=\operatorname{ran}\bigl((K-\lambda)^{-1}-(H-\lambda)^{-1}\bigr).$$

Proof. In order to prove the first equality let $\lambda\in\rho(H)\cap\rho(K)$ and assume that $g\in\operatorname{ran}((H\cap K)-\lambda)$. Then $g=h'-\lambda h$ for some $\{h,h'\}\in H\cap K$ and it follows that

$$(H-\lambda)^{-1}(h'-\lambda h)=h \quad\text{and}\quad (K-\lambda)^{-1}(h'-\lambda h)=h.$$

Hence, $((K-\lambda)^{-1}-(H-\lambda)^{-1})g=0$ and this shows the inclusion

$$\operatorname{ran}\big((H\cap K)-\lambda\big)\subset\ker\big((K-\lambda)^{-1}-(H-\lambda)^{-1}\big).$$

To prove the opposite inclusion, let $g\in\ker((K-\lambda)^{-1}-(H-\lambda)^{-1})$. Then with $k=(H-\lambda)^{-1}g=(K-\lambda)^{-1}g$ it follows that $\{k,g+\lambda k\}\in H\cap K$. Consequently, $g\in\operatorname{ran}((H\cap K)-\lambda)$ and therefore

$$\ker\big((K-\lambda)^{-1}-(H-\lambda)^{-1}\big)\subset\operatorname{ran}\big((H\cap K)-\lambda\big).$$

As to the second equality, let $\lambda\in\rho(H)\cap\rho(K)$ and $\{f,f'\}\in H\,\widehat{+}\,K$. Then according to Lemma 1.2.4 one has the representation

$$\begin{aligned}\{f,f'\}=\big\{(K-\lambda)^{-1}h,h+\lambda(K-\lambda)^{-1}h\big\}\\+\big\{(H-\lambda)^{-1})k,k+\lambda(H-\lambda)^{-1})k\big\}\end{aligned}$$

for some $h,k\in\mathfrak{H}$. Clearly, $\{f,\lambda f\}\in H\,\widehat{+}\,K$ if and only if $h+k=0$ or, equivalently, $f=(K-\lambda)^{-1}h-(H-\lambda)^{-1}h$ for some $h\in\mathfrak{H}$. This proves the second equality. □

Let H and K be closed relations in $\mathfrak{H}$. Then the intersection $H\cap K$ is closed, but the componentwise sum $H\,\widehat{+}\,K$ is in general not closed; cf. Proposition 1.3.12 and (1.3.5). An application of Theorem 1.7.1 and Lemma 1.7.2 gives the following characterization.

Theorem 1.7.3. *Let H and K be closed relations in $\mathfrak{H}$ such that $\rho(H)\cap\rho(K)\neq\emptyset$. Then the sum $H\,\widehat{+}\,K$ is closed if and only if*

$$\operatorname{ran}\big((K-\lambda)^{-1}-(H-\lambda)^{-1}\big)$$

is closed for some, and hence for all $\lambda\in\rho(H)\cap\rho(K)$.

Proof. Let $\lambda\in\rho(H)\cap\rho(K)$. Then by Lemma 1.7.2

$$\operatorname{ran}\big((K-\lambda)^{-1}-(H-\lambda)^{-1}\big)$$

is closed if and only if $\ker(H\,\widehat{+}\,K-\lambda)$ is closed. Now note that the relation H is closed, that $\lambda\in\rho(H)$, and that H is a restriction of $H\,\widehat{+}\,K$, so that by Theorem 1.7.1

$$H\,\widehat{+}\,K=H\,\widehat{+}\,\widehat{\mathfrak{N}}_\lambda(H\,\widehat{+}\,K).$$

Moreover, it follows from Theorem 1.7.1 that $\ker(H\,\widehat{+}\,K-\lambda)$ is closed if and only if $H\,\widehat{+}\,K$ is closed. □

Next follow some consequences of Theorem 1.7.1 and Theorem 1.7.3 in the context of closed symmetric relations. They are stated in terms of intermediate extensions.

Definition 1.7.4. Let S be a closed symmetric relation in $\mathfrak{H}$. A relation H is said to be an *intermediate extension* of S if $S \subset H \subset S^*$.

For instance, H defined in (1.6.5) is an intermediate extension of S. In general, an extension H of S need not be a restriction of S^*. However, if H is symmetric, then $H \subset H^*$, and it follows from $S \subset H$ and $H^* \subset S^*$ that $S \subset H \subset S^*$. Hence, symmetric and self-adjoint extensions of S are intermediate.

For intermediate extensions with nonempty resolvent set one obtains the following decomposition from Theorem 1.7.1.

Corollary 1.7.5. *Let S be a closed symmetric relation in $\mathfrak{H}$. If H is a closed intermediate extension of S such that $\rho(H) \neq \emptyset$, then*

$$S^* = H \mathbin{\widehat{+}} \widehat{\mathfrak{N}}_\lambda(S^*) \tag{1.7.2}$$

for all $\lambda \in \rho(H)$, and the sum in the decomposition (1.7.2) is direct. Furthermore, if $H \in \mathbf{B}(\mathfrak{H})$, then

$$S^* = H \mathbin{\widehat{+}} \widehat{\mathfrak{N}}_\infty(S^*), \tag{1.7.3}$$

and the sum in the decomposition (1.7.3) is direct.

Proof. The direct sum decomposition (1.7.2) follows from Theorem 1.7.1. In order to prove (1.7.3), note that the inclusion ($\supset$) is clear. For the inclusion ($\subset$) take $\{f, f'\} \in S^*$. Then $\{f, Hf\} \in H \subset S^*$ and hence $\{0, f' - Hf\} \in S^*$. Thus, $\{f, f'\} = \{f, Hf\} + \{0, f' - Hf\} \in H \mathbin{\widehat{+}} \widehat{\mathfrak{N}}_\infty(S^*)$. □

Next the notions of disjointness and transversality of two intermediate extensions are defined.

Definition 1.7.6. Let S be a closed symmetric relation in $\mathfrak{H}$. If H and K are closed intermediate extensions of S, then they are called *disjoint* if $H \cap K = S$, and they are called *transversal* if they are disjoint and $H \mathbin{\widehat{+}} K = S^*$.

Let H and K be closed intermediate extensions of S. By Proposition 1.3.12, $(H \cap K)^* = \operatorname{clos}(H^* \mathbin{\widehat{+}} K^*)$ and hence H and K are disjoint if and only if $S^* = \operatorname{clos}(H^* \mathbin{\widehat{+}} K^*)$. In the next lemma self-adjoint intermediate extensions are considered.

Lemma 1.7.7. *Let S be a closed symmetric relation in $\mathfrak{H}$ and let H and K be self-adjoint extensions of S. Then the following statements hold:*

(i) *H and K are disjoint if and only if $S^* = \operatorname{clos}(H \mathbin{\widehat{+}} K)$;*
(ii) *H and K are transversal if and only if $S^* = H \mathbin{\widehat{+}} K$.*

Consequently, if H and K are disjoint, then they are transversal if and only if $H \mathbin{\widehat{+}} K$ is closed.

Proof. (i) follows from the discussion before the lemma and the assumption that H and K are self-adjoint.

(ii) The implication ($\Rightarrow$) is clear and hence only the implication ($\Leftarrow$) has to be checked. But if $S^* = H \,\widehat{+}\, K$, then Proposition 1.3.12 implies

$$S = \big(H \,\widehat{+}\, K\big)^* = H^* \cap K^* = H \cap K,$$

and hence H and K are disjoint. Together with $S^* = H \,\widehat{+}\, K$ this shows that H and K are transversal. □

The next theorem provides useful criteria for disjointness and transversality of self-adjoint extensions.

Theorem 1.7.8. *Let S be a closed symmetric relation in $\mathfrak{H}$ and let H and K be self-adjoint extensions of S. Then the following statements hold:*

(i) *H and K are disjoint if and only if*

$$\operatorname{ran}(S-\lambda) = \ker\big((K-\lambda)^{-1} - (H-\lambda)^{-1}\big) \tag{1.7.4}$$

for some, and hence for all $\lambda \in \rho(H) \cap \rho(K)$;

(ii) *H and K are transversal if and only if*

$$\ker(S^*-\lambda) = \operatorname{ran}\big((K-\lambda)^{-1} - (H-\lambda)^{-1}\big) \tag{1.7.5}$$

for some, and hence for all $\lambda \in \rho(H) \cap \rho(K)$.

Proof. (i) If $S = H \cap K$, then (1.7.4) holds for all $\lambda \in \rho(H) \cap \rho(K)$ by Lemma 1.7.2. Conversely, if (1.7.4) holds for some $\lambda \in \rho(H) \cap \rho(K)$, then Lemma 1.7.2 shows that

$$\operatorname{ran}(S-\lambda) = \operatorname{ran}\big((H\cap K)-\lambda\big).$$

Since $\lambda \in \rho(H)$ and both $H \cap K$ and S are restrictions of H, one has

$$\ker\big((H\cap K)-\lambda\big) = \{0\} = \ker(S-\lambda).$$

Clearly, $S - \lambda \subset (H \cap K) - \lambda$ and now the equality $S - \lambda = (H \cap K) - \lambda$ follows from Corollary 1.1.3. This implies $S = H \cap K$.

(ii) If $S^* = H \,\widehat{+}\, K$, then (1.7.5) holds for all $\lambda \in \rho(H) \cap \rho(K)$ by Lemma 1.7.2. Conversely, assume that (1.7.5) holds for some $\lambda \in \rho(H) \cap \rho(K)$ and let

$$T = H \,\widehat{+}\, K.$$

Since $H \subset S^*$ and $H \subset T$, it follows from Theorem 1.7.1 that

$$S^* = H \,\widehat{+}\, \widehat{\mathfrak{N}}_\lambda(S^*) \quad \text{and} \quad T = H \,\widehat{+}\, \widehat{\mathfrak{N}}_\lambda(T).$$

By Lemma 1.7.2, the assumption (1.7.5) means that $\ker(S^*-\lambda) = \ker(T-\lambda)$. Therefore, $\widehat{\mathfrak{N}}_\lambda(S^*) = \widehat{\mathfrak{N}}_\lambda(T)$ and $S^* = T = H \,\widehat{+}\, K$. Now Lemma 1.7.7 (ii) implies that H and K are transversal. □

In the next result a closed intermediate extension H with $\rho(H) \neq \emptyset$ of S is decomposed into the direct sum of S and another closed subspace in H. Recall that in Proposition 1.6.8 it was shown that there always exist closed intermediate extensions with this property.

Proposition 1.7.9. *Let S be a closed symmetric relation in $\mathfrak{H}$ and let $\lambda \in \mathbb{C} \setminus \mathbb{R}$. Let H be a closed intermediate extension of S with $\lambda \in \rho(H)$. Then*

$$H = S \,\widehat{+}\, \{\{(H-\lambda)^{-1} f_{\bar\lambda}, (I + \lambda(H-\lambda)^{-1}) f_{\bar\lambda}\} : f_{\bar\lambda} \in \mathfrak{N}_{\bar\lambda}(S^*)\} \tag{1.7.6}$$

and the sum is direct.

Proof. In order to show (1.7.6) observe that by Lemma 1.2.4 and $S \subset H$ the right-hand side is contained in H. To see the opposite inclusion, let $\{h, h'\} \in H$. Since $\lambda \in \mathbb{C} \setminus \mathbb{R}$, one has

$$\mathfrak{H} = \operatorname{ran}(S - \lambda) \oplus \ker(S^* - \bar\lambda) = \operatorname{ran}(S - \lambda) \oplus \mathfrak{N}_{\bar\lambda}(S^*).$$

Due to this decomposition, there exist $\{f, f'\} \in S$ and $f_{\bar\lambda} \in \mathfrak{N}_{\bar\lambda}(S^*)$ such that

$$h' - \lambda h = f' - \lambda f + f_{\bar\lambda}.$$

Hence, it follows from $\{h - f, h' - f'\} \in H$ that $\{h - f, f_{\bar\lambda}\} \in H - \lambda$,

$$h - f = (H-\lambda)^{-1} f_{\bar\lambda}, \quad \text{and} \quad h' - f' = f_{\bar\lambda} + \lambda(H-\lambda)^{-1} f_{\bar\lambda},$$

and therefore

$$\{h, h'\} - \{f, f'\} = \{h - f, h' - f'\} = \{(H-\lambda)^{-1} f_{\bar\lambda}, (I + \lambda(H-\lambda)^{-1} f_{\bar\lambda}\}.$$

Thus, $\{h, h'\}$ belongs to the right-hand side of (1.7.6).

In order to show that the sum in (1.7.6) is direct, assume that

$$\{(H-\lambda)^{-1} f_{\bar\lambda}, (I + \lambda(H-\lambda)^{-1}) f_{\bar\lambda}\} \in S$$

for some $f_{\bar\lambda} \in \mathfrak{N}_{\bar\lambda}(S^*)$. Then since $\{f_{\bar\lambda}, \bar\lambda f_{\bar\lambda}\} \in S^*$ it follows that

$$\big((I + \lambda(H-\lambda)^{-1}) f_{\bar\lambda}, f_{\bar\lambda}\big) = \big((H-\lambda)^{-1} f_{\bar\lambda}, \bar\lambda f_{\bar\lambda}\big),$$

which leads to $f_{\bar\lambda} = 0$. $\square$

The next statement is a consequence of Corollary 1.7.5 and Proposition 1.7.9.

Corollary 1.7.10. *Let S be a closed symmetric relation in $\mathfrak{H}$ and let $\lambda \in \mathbb{C} \setminus \mathbb{R}$. Let H be a closed intermediate extension of S with $\lambda \in \rho(H)$. Then*

$$S^* = S \,\widehat{+}\, \{\, \{(H-\lambda)^{-1} f_{\bar\lambda}, (I + \lambda(H-\lambda)^{-1} f_{\bar\lambda}\} : \, f_{\bar\lambda} \in \mathfrak{N}_{\bar\lambda}(S^*) \,\} \,\widehat{+}\, \widehat{\mathfrak{N}}_{\lambda}(S^*)$$

and all sums are direct.

The following result is *von Neumann's first formula*, stated in the context of a closed symmetric relation S. This decomposition of S^* into the direct sum of S and two defect subspaces corresponding to two points in the upper and lower half-plane can be viewed as a consequence of Corollary 1.7.5 and Proposition 1.6.8.

Theorem 1.7.11. *Let S be a closed symmetric relation in $\mathfrak{H}$ and let $\lambda, \mu \in \mathbb{C} \setminus \mathbb{R}$ be in the same half-plane. Then*

$$S^* = S \mathbin{\widehat{+}} \widehat{\mathfrak{N}}_\lambda(S^*) \mathbin{\widehat{+}} \widehat{\mathfrak{N}}_{\bar{\mu}}(S^*), \quad \textit{direct sums.} \tag{1.7.7}$$

The sums are orthogonal in $\mathfrak{H}^2$ when $\lambda = \mu = \pm i$.

Proof. Assume that $\lambda, \mu \in \mathbb{C}^+$. By Proposition 1.6.8, the relation

$$H = S \mathbin{\widehat{+}} \widehat{\mathfrak{N}}_{\bar{\mu}}(S^*)$$

is a maximal accumulative intermediate extension of S, and the sum is direct. Since $\mathbb{C}^+ \subset \rho(H)$ by Theorem 1.6.4 and $\lambda \in \mathbb{C}^+$, it follows from Corollary 1.7.5 that $S^* = H \mathbin{\widehat{+}} \widehat{\mathfrak{N}}_\lambda(S^*)$ and the sum is direct. Hence, (1.7.7) follows. The case where $\lambda, \mu \in \mathbb{C}^-$ is completely similar.

It is a simple calculation to show that $\widehat{\mathfrak{N}}_i(S^*)$ and $\widehat{\mathfrak{N}}_{-i}(S^*)$ are orthogonal in $\mathfrak{H}^2$. The orthogonality of S and $\widehat{\mathfrak{N}}_{\pm i}(S^*)$ in $\mathfrak{H}^2$ follows from

$$(f, f_{\pm i}) + (f', \pm i f_{\pm i}) = (f, f_{\pm i}) \mp i(f', f_{\pm i}) = (f, f_{\pm i}) \mp i(f, \pm i f_{\pm i}) = 0,$$

where it was used that $\{f, f'\} \in S$ and $\widehat{\mathfrak{N}}_{\pm i}(S^*) \subset S^*$. □

The next result is *von Neumann's second formula*, stated in the context of a closed symmetric relation S. It describes all symmetric extensions of S in terms of isometric operators between the defect spaces $\mathfrak{N}_{\bar{\mu}}(S^*)$ and $\mathfrak{N}_\mu(S^*)$ appearing in Theorem 1.7.11. The following notation will be useful. Let $\lambda \in \mathbb{C} \setminus \mathbb{R}$ and let $\mathfrak{M}_\lambda$ be a closed linear subspace of $\mathfrak{N}_\lambda(S^*)$. Then $\widehat{\mathfrak{M}}_\lambda$ denotes the closed linear subspace of $\widehat{\mathfrak{N}}_\lambda(S^*)$ defined by

$$\widehat{\mathfrak{M}}_\lambda = \big\{\{f_\lambda, \lambda f_\lambda\} \in S^* : f_\lambda \in \mathfrak{M}_\lambda\big\}.$$

Now let $\mu \in \mathbb{C} \setminus \mathbb{R}$ and let W be a bounded linear mapping from a closed linear subspace $\mathfrak{M}_{\bar{\mu}}$ of $\mathfrak{N}_{\bar{\mu}}(S^*)$ to $\mathfrak{N}_\mu(S^*)$. Then W induces a linear mapping $\widehat{W}$ from $\widehat{\mathfrak{M}}_{\bar{\mu}}$ to $\widehat{\mathfrak{N}}_\mu(S^*)$ by

$$\widehat{W}\{f_{\bar{\mu}}, \bar{\mu} f_{\bar{\mu}}\} = \{W f_{\bar{\mu}}, \mu W f_{\bar{\mu}}\}.$$

Clearly, $\widehat{W}$ is bounded and $\|\widehat{W}\| = \|W\|$. In fact, every bounded linear mapping from $\widehat{\mathfrak{M}}_{\bar{\mu}}$ to $\widehat{\mathfrak{N}}_\mu(S^*)$ is of this form. To see this, it suffices to observe that if $\widehat{W}\{f_{\bar{\mu}}, \bar{\mu} f_{\bar{\mu}}\} = \{g_\mu, \mu g_\mu\}$, then the mapping $f_{\bar{\mu}} \in \mathfrak{M}_{\bar{\mu}} \mapsto g_\mu \in \mathfrak{N}_\mu(S^*)$ is linear. Moreover, this mapping is also bounded since

$$\sqrt{1 + |\mu|^2}\, \|g_\mu\| \leq \|\widehat{W}\| \sqrt{1 + |\bar{\mu}|^2}\, \|f_{\bar{\mu}}\|,$$

thanks to the boundedness of $\widehat{W}$ and the structure of the standard inner product.

Theorem 1.7.12. *Let S be a closed symmetric relation in $\mathfrak{H}$ and let $\mu \in \mathbb{C} \setminus \mathbb{R}$. Then H is a closed symmetric extension of S if and only if there exists an isometric operator W mapping a closed subspace $\mathfrak{M}_{\bar\mu} \subset \mathfrak{N}_{\bar\mu}(S^*)$ onto a closed subspace $\mathfrak{M}_\mu \subset \mathfrak{N}_\mu(S^*)$, such that*

$$H = S \,\widehat{+}\, (I - \widehat{W})\widehat{\mathfrak{M}}_{\bar\mu}. \tag{1.7.8}$$

The closed symmetric extension H is maximal if and only if $\mathfrak{M}_{\bar\mu} = \mathfrak{N}_{\bar\mu}(S^*)$ or $\mathfrak{M}_\mu = \mathfrak{N}_\mu(S^*)$ holds. Furthermore, the extension H is self-adjoint if and only if $\mathfrak{M}_{\bar\mu} = \mathfrak{N}_{\bar\mu}(S^*)$ and $\mathfrak{M}_\mu = \mathfrak{N}_\mu(S^*)$ hold.

Let $\mathfrak{M}_{\bar\mu} \subset \mathfrak{N}_{\bar\mu}(S^*)$ and $\mathfrak{M}_\mu \subset \mathfrak{N}_\mu(S^*)$ be closed subspaces and observe that isometric operators W from $\mathfrak{M}_{\bar\mu}$ onto $\mathfrak{M}_\mu$ exist if and only if the dimensions of the spaces $\mathfrak{M}_{\bar\mu}$ and $\mathfrak{M}_\mu$ coincide. This implies the following statement.

Corollary 1.7.13. *Let S be a closed symmetric relation in $\mathfrak{H}$. Then S admits self-adjoint extensions H in $\mathfrak{H}$ if and only if*

$$\dim \mathfrak{N}_\mu(S^*) = \dim \mathfrak{N}_{\bar\mu}(S^*)$$

for some, and hence for all $\mu \in \mathbb{C} \setminus \mathbb{R}$.

Proof of Theorem 1.7.12. ($\Rightarrow$) Let H be a closed symmetric extension of S, let $\mu \in \mathbb{C} \setminus \mathbb{R}$, and consider the Cayley transforms

$$V := \mathcal{C}_\mu[S] = \{\{f' - \mu f, f' - \bar\mu f\} : \{f, f'\} \in S\}$$

and

$$U := \mathcal{C}_\mu[H] = \{\{h' - \mu h, h' - \bar\mu h\} : \{h, h'\} \in H\}$$

of S and H, respectively. According to Proposition 1.4.8, V is a closed isometric operator from the closed subspace $\operatorname{ran}(S - \mu)$ onto the closed subspace $\operatorname{ran}(S - \bar\mu)$, and U is an isometric extension of V from the closed subspace $\operatorname{ran}(H - \mu)$ onto the closed subspace $\operatorname{ran}(H - \bar\mu)$. It follows that there exist closed subspaces

$$\mathfrak{M}_{\bar\mu} \subset \mathfrak{N}_{\bar\mu}(S^*) = \operatorname{ran}(S - \mu)^\perp \quad \text{and} \quad \mathfrak{M}_\mu \subset \mathfrak{N}_\mu(S^*) = \operatorname{ran}(S - \bar\mu)^\perp,$$

such that

$$\operatorname{ran}(H - \mu) = \operatorname{ran}(S - \mu) \oplus \mathfrak{M}_{\bar\mu} \quad \text{and} \quad \operatorname{ran}(H - \bar\mu) = \operatorname{ran}(S - \bar\mu) \oplus \mathfrak{M}_\mu.$$

Let W be the restriction of U to $\mathfrak{M}_{\bar\mu}$. Then W maps $\mathfrak{M}_{\bar\mu}$ isometrically onto $\mathfrak{M}_\mu$ and

$$U = \begin{pmatrix} V & 0 \\ 0 & W \end{pmatrix} : \begin{pmatrix} \operatorname{ran}(S - \mu) \\ \mathfrak{M}_{\bar\mu} \end{pmatrix} \to \begin{pmatrix} \operatorname{ran}(S - \bar\mu) \\ \mathfrak{M}_\mu \end{pmatrix}. \tag{1.7.9}$$

Taking the inverse Cayley transform leads to

$$\begin{aligned} H = \mathcal{F}_\mu[U] &= \mathcal{F}_\mu[V] \,\widehat{+}\, \mathcal{F}_\mu[W] \\ &= S \,\widehat{+}\, \{\{(I - W)f_{\bar\mu}, (\bar\mu - \mu W)f_{\bar\mu}\} : \; f_{\bar\mu} \in \mathfrak{M}_{\bar\mu}\} \\ &= S \,\widehat{+}\, \{\{f_{\bar\mu}, \bar\mu f_{\bar\mu}\} - \{W f_{\bar\mu}, \mu W f_{\bar\mu}\} : \; f_{\bar\mu} \in \mathfrak{M}_{\bar\mu}\}, \end{aligned}$$

which implies (1.7.8). In the case where the closed symmetric extension H is maximal, $\operatorname{ran}(H-\mu)=\mathfrak{H}$ or $\operatorname{ran}(H-\overline{\mu})=\mathfrak{H}$ by (1.4.7), and hence one has $\mathfrak{M}_{\overline{\mu}}=\mathfrak{N}_{\overline{\mu}}(S^*)$ or $\mathfrak{M}_\mu=\mathfrak{N}_\mu(S^*)$, respectively. If H is self-adjoint, then it is clear that $\operatorname{ran}(H-\mu)=\mathfrak{H}=\operatorname{ran}(H-\overline{\mu})$ by Theorem 1.5.5 and therefore $\mathfrak{M}_{\overline{\mu}}=\mathfrak{N}_{\overline{\mu}}(S^*)$ and $\mathfrak{M}_\mu=\mathfrak{N}_\mu(S^*)$.

($\Leftarrow$) Let $\mu\in\mathbb{C}\setminus\mathbb{R}$, assume that W is an isometric operator from a closed subspace $\mathfrak{M}_{\overline{\mu}}\subset\mathfrak{N}_{\overline{\mu}}(S^*)$ onto a closed subspace $\mathfrak{M}_\mu\subset\mathfrak{N}_\mu(S^*)$, and consider the relation $H=S\,\widehat{+}\,(I-\widehat{W})\widehat{\mathfrak{M}}_{\overline{\mu}}$. Let V be the Cayley transform of S, define the operator U as in (1.7.9), and note that

$$H=S\,\widehat{+}\,(I-\widehat{W})\widehat{\mathfrak{M}}_{\overline{\mu}}=\mathcal{F}_\mu[V]\,\widehat{+}\,\mathcal{F}_\mu[W]=\mathcal{F}_\mu[U]$$

holds. Since U is isometric and closed, it follows from Proposition 1.4.8 that H is a closed symmetric relation. If $\mathfrak{M}_{\overline{\mu}}=\mathfrak{N}_{\overline{\mu}}(S^*)$ or $\mathfrak{M}_\mu=\mathfrak{N}_\mu(S^*)$, then one has $\operatorname{dom}U=\mathfrak{H}$ or $\operatorname{ran}U=\mathfrak{H}$, respectively, and therefore one sees that $\operatorname{ran}(H-\mu)=\mathfrak{H}$ or $\operatorname{ran}(H-\overline{\mu})=\mathfrak{H}$, respectively, so that H is a maximal symmetric relation. Finally, if $\mathfrak{M}_{\overline{\mu}}=\mathfrak{N}_{\overline{\mu}}(S^*)$ and $\mathfrak{M}_\mu=\mathfrak{N}_\mu(S^*)$, then U is unitary and hence H is self-adjoint; cf. Proposition 1.4.8. □

The second von Neumann formula in Theorem 1.7.12 has a natural extension, which describes all accumulative (dissipative) extensions of S in terms of contractive operators between the defect spaces.

Theorem 1.7.14. *Let S be a closed symmetric relation in $\mathfrak{H}$. Then H is a closed accumulative (closed dissipative) extension of S if and only if for some $\mu\in\mathbb{C}^+$ ($\mu\in\mathbb{C}^-$), there exists a contraction W mapping a closed subspace $\mathfrak{M}_{\overline{\mu}}\subset\mathfrak{N}_{\overline{\mu}}(S^*)$ to $\mathfrak{N}_\mu(S^*)$, such that*

$$H=S\,\widehat{+}\,(I-\widehat{W})\widehat{\mathfrak{M}}_{\overline{\mu}}. \tag{1.7.10}$$

The closed accumulative (closed dissipative) extension H is maximal if and only if $\mathfrak{M}_{\overline{\mu}}=\mathfrak{N}_{\overline{\mu}}(S^*)$ holds for $\mu\in\mathbb{C}^+$ ($\mu\in\mathbb{C}^-$).

Proof. ($\Rightarrow$) Let H be a closed accumulative extension of S and let $\mu\in\mathbb{C}^+$. Define the Cayley transform $V=\mathcal{C}_\mu[S]$ of S, so that V is a closed isometry from the closed subspace $\operatorname{ran}(S-\mu)$ onto the closed subspace $\operatorname{ran}(S-\overline{\mu})$. By Proposition 1.6.6 and (1.1.18), the Cayley transform $U=\mathcal{C}_\mu[H]$ of H is a closed contractive extension of V from the closed subspace $\operatorname{ran}(H-\mu)$ onto the subspace $\operatorname{ran}(H-\overline{\mu})$. Then there exists a closed subspace $\mathfrak{M}_{\overline{\mu}}\subset\mathfrak{N}_{\overline{\mu}}(S^*)$ such that

$$\operatorname{ran}(H-\mu)=\operatorname{ran}(S-\mu)\oplus\mathfrak{M}_{\overline{\mu}}.$$

Let W be the restriction of U to $\mathfrak{M}_{\overline{\mu}}$. It will be shown that U is of the form

$$U=\begin{pmatrix} V & 0\\ 0 & W\end{pmatrix}:\begin{pmatrix}\operatorname{ran}(S-\mu)\\ \mathfrak{M}_{\overline{\mu}}\end{pmatrix}\to\begin{pmatrix}\operatorname{ran}(S-\overline{\mu})\\ \mathfrak{N}_\mu(S^*)\end{pmatrix}. \tag{1.7.11}$$

For this it suffices to verify that the contractive operator W is a mapping from $\mathfrak{M}_{\bar\mu}$ to $\mathfrak{N}_\mu(S^*) = \ker(S^* - \mu)$. To see this, note that the restriction V of U is isometric. Hence, by (1.1.10),

$$(V\varphi, U\psi) = (\varphi, \psi), \quad \varphi \in \operatorname{dom} V \subset \operatorname{dom} U, \quad \psi \in \operatorname{dom} U. \tag{1.7.12}$$

Observe that if $\psi \in \mathfrak{M}_{\bar\mu}$, then $(\varphi, \psi) = 0$ for all $\varphi \in \operatorname{dom} V$. Thus, (1.7.12) implies that $W\psi = U\psi \in (\operatorname{ran} V)^\perp = \mathfrak{N}_\mu(S^*)$. This yields (1.7.11).

Taking the inverse Cayley transform of U in (1.7.11) leads (in the same way as in the proof of Theorem 1.7.12) to

$$\begin{aligned} H &= \mathcal{F}_\mu[U] = \mathcal{F}_\mu[V] \,\widehat{+}\, \mathcal{F}_\mu[W] \\ &= S \,\widehat{+}\, \{\{f_{\bar\mu}, \bar\mu f_{\bar\mu}\} - \{W f_{\bar\mu}, \mu W f_{\bar\mu}\} : f_{\bar\mu} \in \mathfrak{M}_{\bar\mu}\}, \end{aligned}$$

which implies (1.7.10). If H is maximal accumulative, then $\operatorname{ran}(H - \mu) = \mathfrak{H}$ and hence $\mathfrak{M}_{\bar\mu} = \mathfrak{N}_{\bar\mu}(S^*)$.

($\Leftarrow$) Let $\mu \in \mathbb{C}^+$ and assume that W is a contractive operator from a closed subspace $\mathfrak{M}_{\bar\mu} \subset \mathfrak{N}_{\bar\mu}(S^*)$ to $\mathfrak{N}_\mu(S^*)$ and consider the relation $H = S \,\widehat{+}\, (I - \widehat{W})\widehat{\mathfrak{M}}_{\bar\mu}$. Let V be the Cayley transform of S, define the operator U as in (1.7.11), and note that

$$H = S \,\widehat{+}\, (I - \widehat{W})\widehat{\mathfrak{M}}_{\bar\mu} = \mathcal{F}_\mu[V] \,\widehat{+}\, \mathcal{F}_\mu[W] = \mathcal{F}_\mu[U].$$

Since U is a closed contractive operator, it follows from Proposition 1.6.6 and (1.1.18) that H is a closed accumulative extension of S. If $\mathfrak{M}_{\bar\mu} = \mathfrak{N}_{\bar\mu}(S^*)$, then $\operatorname{dom} U = \mathfrak{H}$ and hence $\operatorname{ran}(H - \mu) = \mathfrak{H}$, so that H is maximal accumulative. □

1.8 Adjoint relations and indefinite inner products

The adjoint of a relation in a Hilbert space $\mathfrak{H}$ has a natural interpretation in terms of a certain indefinite inner product $[\![\cdot,\cdot]\!]_{\mathfrak{H}^2}$ on the product space $\mathfrak{H} \times \mathfrak{H}$. It will be shown that surjective operators which are isometric with respect to such indefinite inner products and have a closed domain are automatically bounded. Furthermore, some geometric transformation properties of operator-valued Möbius transformations which are unitary with respect to indefinite inner products will be studied.

Let H be a relation in $\mathfrak{H}$. The adjoint relation H^* in Definition 1.3.1 satisfies $H^* = (JH)^\perp = JH^\perp$, where J is the flip-flop operator in (1.3.1) and the orthogonal complement refers to the componentwise inner product in the product space $\mathfrak{H} \times \mathfrak{H}$; cf. (1.3.2). Define the operator $\mathcal{J}$ on the product space $\mathfrak{H}^2$ as $\mathcal{J} = -iJ$, where J is the flip-flop operator:

$$\mathcal{J} := -i \begin{pmatrix} 0 & I_{\mathfrak{H}} \\ -I_{\mathfrak{H}} & 0 \end{pmatrix} = \begin{pmatrix} 0 & -iI_{\mathfrak{H}} \\ iI_{\mathfrak{H}} & 0 \end{pmatrix}. \tag{1.8.1}$$

Sometimes the notation $\mathfrak{J}_{\mathfrak{H}}$ is used to indicate the underlying Hilbert space. Clearly, the operator $\mathfrak{J}$ in (1.8.1) has the properties

$$\mathfrak{J} = \mathfrak{J}^* = \mathfrak{J}^{-1} \in \mathbf{B}(\mathfrak{H}^2),$$

so that $\mathfrak{J}$ is unitary and self-adjoint. The operator $\mathfrak{J}$ gives rise to an inner product $[\![\cdot,\cdot]\!]$ on $\mathfrak{H}^2$ as follows

$$[\![\widehat{h},\widehat{k}]\!] := (\mathfrak{J}\widehat{h},\widehat{k})_{\mathfrak{H}^2}, \qquad \widehat{h} = \begin{pmatrix} h \\ h' \end{pmatrix}, \ \widehat{k} = \begin{pmatrix} k \\ k' \end{pmatrix} \in \mathfrak{H}^2, \tag{1.8.2}$$

where for convenience $\widehat{h}$ and $\widehat{k}$ are written in vector notation. In the following sometimes an index is used to indicate in which space the indefinite inner product is defined, e.g., $[\![\cdot,\cdot]\!]_{\mathfrak{H}^2}$. Explicitly the new inner product is given by

$$[\![\widehat{h},\widehat{k}]\!] = -i\big((h',k) - (h,k')\big), \qquad \widehat{h} = \begin{pmatrix} h \\ h' \end{pmatrix}, \ \widehat{k} = \begin{pmatrix} k \\ k' \end{pmatrix} \in \mathfrak{H}^2, \tag{1.8.3}$$

and note that

$$[\![\widehat{h},\widehat{h}]\!] = 2\,\mathrm{Im}\,(h',h), \qquad \widehat{h} = \begin{pmatrix} h \\ h' \end{pmatrix} \in \mathfrak{H}^2. \tag{1.8.4}$$

This shows that the new inner product on $\mathfrak{H}^2$ is indefinite; and, in fact, $(\mathfrak{H}^2, [\![\cdot,\cdot]\!])$ is a so-called Kreĭn space. It follows from (1.8.2) that the inner product $[\![\cdot,\cdot]\!]$ is continuous: if $\widehat{h}_n \to \widehat{h}$ and $\widehat{k}_m \to \widehat{k}$ in $\mathfrak{H}^2$ in the usual sense, then clearly

$$[\![\widehat{h}_n,\widehat{k}_m]\!] \to [\![\widehat{h},\widehat{k}]\!], \quad \text{as} \quad m,n \to \infty.$$

For a linear subspace H of $\mathfrak{H}^2$, the $[\![\cdot,\cdot]\!]$-*orthogonal companion* is given by

$$H^{[\![\perp]\!]} = \big\{\widehat{h} \in \mathfrak{H}^2 : [\![\widehat{h},\widehat{k}]\!] = 0 \ \text{ for all } \widehat{k} \in H\big\}.$$

Hence, it follows from (1.8.3) that the adjoint H^* (with respect to the standard inner product) of the relation H in $\mathfrak{H}$ coincides with the orthogonal companion $H^{[\![\perp]\!]}$ (with respect to the indefinite inner product $[\![\cdot,\cdot]\!]$) of the subspace H in $\mathfrak{H}^2$:

$$H^* = H^{[\![\perp]\!]}. \tag{1.8.5}$$

The indefinite inner product $[\![\cdot,\cdot]\!]$ on $\mathfrak{H}^2$ provides an appropriate tool to describe certain fundamental notions and identities. A linear subspace H in the space $(\mathfrak{H}^2, [\![\cdot,\cdot]\!])$ is said to be

(i) *nonnegative* if $[\![\widehat{h},\widehat{h}]\!] \geq 0$ for all $\widehat{h} \in H$;

(ii) *nonpositive* if $[\![\widehat{h},\widehat{h}]\!] \leq 0$ for all $\widehat{h} \in H$;

(iiii) *neutral* if $[\![\widehat{h},\widehat{h}]\!] = 0$ for all $\widehat{h} \in H$ or, equivalently, $H \subset H^{[\![\perp]\!]}$;

(iv) *hypermaximal neutral* if $H = H^{[\![\perp]\!]}$;

the equivalence in (iii) follows from (1.8.4), (1.8.5), and Lemma 1.4.2. A linear subspace H in the space $(\mathfrak{H}^2, [\![\cdot,\cdot]\!])$ is *maximal nonnegative*, *maximal nonpositive*, or *maximal neutral* if the existence of a linear subspace with $H \subset H'$, where H' is nonnegative, nonpositive, or neutral, respectively, implies that $H' = H$.

By considering a relation H as a subspace of $\mathfrak{H}^2$ with the usual inner product or as a subspace of $\mathfrak{H}^2$ with the inner product $[\![\cdot,\cdot]\!]$, the following correspondence is an immediate consequence of (1.8.4) and (1.8.5):

(i) H is a (maximal) dissipative relation in $\mathfrak{H}$ if and only if H is a (maximal) nonnegative subspace of $(\mathfrak{H}^2, [\![\cdot,\cdot]\!])$;

(ii) H is a (maximal) accumulative relation in $\mathfrak{H}$ if and only if H is a (maximal) nonpositive subspace of $(\mathfrak{H}^2, [\![\cdot,\cdot]\!])$;

(iii) H is a (maximal) symmetric relation in $\mathfrak{H}$ if and only if H is a (maximal) neutral subspace of $(\mathfrak{H}^2, [\![\cdot,\cdot]\!])$;

(iv) H is a self-adjoint relation in $\mathfrak{H}$ if and only if H is a hypermaximal neutral subspace of $(\mathfrak{H}^2, [\![\cdot,\cdot]\!])$.

Let U be a linear *operator* from $\mathfrak{H}^2$ to $\mathfrak{K}^2$. Then U is said to be *isometric* from $(\mathfrak{H}^2, [\![\cdot,\cdot]\!])$ to $(\mathfrak{K}^2, [\![\cdot,\cdot]\!])$ if

$$[\![U\widehat{h}, U\widehat{k}]\!]_{\mathfrak{K}^2} = [\![\widehat{h}, \widehat{k}]\!]_{\mathfrak{H}^2} \quad \text{for all} \quad \widehat{h}, \widehat{k} \in \operatorname{dom} U. \tag{1.8.6}$$

In addition, U is said to be *unitary* from $(\mathfrak{H}^2, [\![\cdot,\cdot]\!])$ to $(\mathfrak{K}^2, [\![\cdot,\cdot]\!])$ if U is isometric from $(\mathfrak{H}^2, [\![\cdot,\cdot]\!])$ to $(\mathfrak{K}^2, [\![\cdot,\cdot]\!])$ and $\operatorname{dom} U = \mathfrak{H}^2$ and $\operatorname{ran} U = \mathfrak{K}^2$.

Lemma 1.8.1. *Let $\mathfrak{H}$ and $\mathfrak{K}$ be Hilbert spaces and let U be an isometric operator from $(\mathfrak{H}^2, [\![\cdot,\cdot]\!])$ to $(\mathfrak{K}^2, [\![\cdot,\cdot]\!])$. Assume that* $\operatorname{dom} U$ *is closed and that U is surjective. Then U is bounded.*

Proof. To see that the operator U is bounded, it suffices to show that U is closed and to apply the closed graph theorem. Let $(\widehat{h}_n)$ be a sequence in $\operatorname{dom} U$ such that

$$\widehat{h}_n \to \widehat{h}, \quad U\widehat{h}_n \to \widehat{\varphi}$$

for some $\widehat{h} \in \mathfrak{H}^2$ and $\widehat{\varphi} \in \mathfrak{K}^2$. Since $\operatorname{dom} U$ is closed, it follows that $\widehat{h} \in \operatorname{dom} U$. As U is surjective, one can choose for each $\widehat{\psi} \in \mathfrak{K}^2$ an element $\widehat{k} \in \operatorname{dom} U$ such that $U\widehat{k} = \mathfrak{J}_{\mathfrak{K}}\widehat{\psi}$; here $\mathfrak{J}_{\mathfrak{K}}$ is defined in the same way as in (1.8.1) and is a unitary and self-adjoint operator in $\mathfrak{K}^2$. Then it follows from the identity (1.8.6) and the continuity of $[\![\cdot,\cdot]\!]_{\mathfrak{H}^2}$ that

$$\begin{aligned}(\widehat{\varphi}, \widehat{\psi})_{\mathfrak{K}^2} &= \lim_{n\to\infty} (U\widehat{h}_n, \mathfrak{J}_{\mathfrak{K}}^{-1}U\widehat{k})_{\mathfrak{K}^2} = \lim_{n\to\infty} [\![U\widehat{h}_n, U\widehat{k}]\!]_{\mathfrak{K}^2} = \lim_{n\to\infty} [\![\widehat{h}_n, \widehat{k}]\!]_{\mathfrak{H}^2} \\ &= [\![\widehat{h}, \widehat{k}]\!]_{\mathfrak{H}^2} = [\![U\widehat{h}, U\widehat{k}]\!]_{\mathfrak{K}^2} = [\![U\widehat{h}, \mathfrak{J}_{\mathfrak{K}}\widehat{\psi}]\!]_{\mathfrak{K}^2} = (U\widehat{h}, \widehat{\psi})_{\mathfrak{K}^2}.\end{aligned}$$

Since $\widehat{\psi} \in \mathfrak{K}^2$ is arbitrary, this gives $\widehat{\varphi} = U\widehat{h}$. It follows that the operator U is closed, and since $\operatorname{dom} U$ is closed, one sees that U is bounded. □

Proposition 1.8.2. *Let $\mathfrak{H}$ and $\mathfrak{K}$ be Hilbert spaces and let U be an operator from $\mathfrak{H}^2$ to $\mathfrak{K}^2$. Then the following statements are equivalent:*

(i) U *is unitary from* $(\mathfrak{H}^2, [\![\cdot,\cdot]\!])$ *to* $(\mathfrak{K}^2, [\![\cdot,\cdot]\!])$*;*

(ii) $U \in \mathbf{B}(\mathfrak{H}^2, \mathfrak{K}^2)$ *satisfies the identities*

$$U^* \mathfrak{J}_{\mathfrak{K}} U = \mathfrak{J}_{\mathfrak{H}} \quad \textit{and} \quad U \mathfrak{J}_{\mathfrak{H}} U^* = \mathfrak{J}_{\mathfrak{K}}; \tag{1.8.7}$$

(iii) $U \in \mathbf{B}(\mathfrak{H}^2, \mathfrak{K}^2)$ *is surjective and* $U^* \mathfrak{J}_{\mathfrak{K}} U = \mathfrak{J}_{\mathfrak{H}}$ *holds.*

Proof. (i) $\Rightarrow$ (ii) It follows from Lemma 1.8.1 that the operator U is bounded and hence $U \in \mathbf{B}(\mathfrak{H}^2, \mathfrak{K}^2)$. Moreover, (1.8.6) implies

$$\big(U^* \mathfrak{J}_{\mathfrak{K}} U \widehat{\varphi}, \widehat{\psi}\big)_{\mathfrak{H}^2} = \big[\!\big[U\widehat{\varphi}, U\widehat{\psi}\big]\!\big]_{\mathfrak{K}^2} = \big[\!\big[\widehat{\varphi}, \widehat{\psi}\big]\!\big]_{\mathfrak{H}^2} = \big(\mathfrak{J}_{\mathfrak{H}} \widehat{\varphi}, \widehat{\psi}\big)_{\mathfrak{H}^2} \tag{1.8.8}$$

for all $\widehat{\varphi}, \widehat{\psi} \in \mathfrak{H}^2$, which yields the first identity in (1.8.7). To prove the second identity in (1.8.7), let $\widehat{\varphi}, \widehat{\psi} \in \mathfrak{K}^2$ and choose $\widehat{\eta} \in \mathfrak{H}^2$ such that $\widehat{\varphi} = \mathfrak{J}_{\mathfrak{K}} U \widehat{\eta}$, which is possible as U is surjective. It follows from the first identity in (1.8.7), and the identities $\mathfrak{J}_{\mathfrak{H}} = \mathfrak{J}_{\mathfrak{H}}^{-1}$ and $\mathfrak{J}_{\mathfrak{K}} = \mathfrak{J}_{\mathfrak{K}}^{-1}$, that

$$\big(U \mathfrak{J}_{\mathfrak{H}} U^* \widehat{\varphi}, \widehat{\psi}\big)_{\mathfrak{K}^2} = \big(U \mathfrak{J}_{\mathfrak{H}} U^* \mathfrak{J}_{\mathfrak{K}} U \widehat{\eta}, \widehat{\psi}\big)_{\mathfrak{K}^2} = \big(U \widehat{\eta}, \widehat{\psi}\big)_{\mathfrak{K}^2} = \big(\mathfrak{J}_{\mathfrak{K}} \widehat{\varphi}, \widehat{\psi}\big)_{\mathfrak{K}^2}.$$

This implies the second identity in (1.8.7).

(ii) $\Rightarrow$ (iii) The second identity in (1.8.7) yields that U is surjective, and hence (iii) holds.

(iii) $\Rightarrow$ (i) The identity $U^* \mathfrak{J}_{\mathfrak{K}} U = \mathfrak{J}_{\mathfrak{H}}$ and the reasoning in (1.8.8) show that (1.8.6) holds, and hence U is isometric from $(\mathfrak{H}^2, [\![\cdot,\cdot]\!])$ to $(\mathfrak{K}^2, [\![\cdot,\cdot]\!])$. As U is surjective, it follows that U is unitary from $(\mathfrak{H}^2, [\![\cdot,\cdot]\!])$ to $(\mathfrak{K}^2, [\![\cdot,\cdot]\!])$. □

An important feature of operators which are isometric or unitary in the present sense is the way they transform certain classes of subspaces. Let H be a linear subspace of $(\mathfrak{H}^2, [\![\cdot,\cdot]\!])$ which is nonnegative, nonpositive, or neutral. If U is a linear operator from $\mathfrak{H}^2$ to $\mathfrak{K}^2$ which is isometric from $(\mathfrak{H}^2, [\![\cdot,\cdot]\!])$ to $(\mathfrak{K}^2, [\![\cdot,\cdot]\!])$, then it follows directly from the definition that U maps $H \cap \operatorname{dom} U$ into a nonnegative, nonpositive, or neutral subspace of $(\mathfrak{K}^2, [\![\cdot,\cdot]\!])$, respectively.

Lemma 1.8.3. *Let* U *be a unitary operator from* $(\mathfrak{H}^2, [\![\cdot,\cdot]\!])$ *to* $(\mathfrak{K}^2, [\![\cdot,\cdot]\!])$*. Then* U *provides a one-to-one correspondence between (maximal) nonnegative, (maximal) nonpositive, (maximal) neutral, and hypermaximal neutral subspaces in* $(\mathfrak{H}^2, [\![\cdot,\cdot]\!])$ *and* $(\mathfrak{K}^2, [\![\cdot,\cdot]\!])$*, respectively.*

Proof. Only the statement about hypermaximal neutral subspaces needs attention. For it, one observes that for any subspace H of $\mathfrak{H}^2$ one has

$$(UH)^{[\![\perp]\!]} = U(H^{[\![\perp]\!]}).$$

Thus, $H = H^{[\![\perp]\!]}$ if and only if $UH = U(H^{[\![\perp]\!]}) = (UH)^{[\![\perp]\!]}$. □

Let $\mathfrak{H}$ and $\mathfrak{K}$ be Hilbert spaces and let $\mathcal{W} \in \mathbf{B}(\mathfrak{H} \times \mathfrak{H}, \mathfrak{K} \times \mathfrak{K})$ have the matrix decomposition

$$\mathcal{W} = \begin{pmatrix} W_{11} & W_{12} \\ W_{21} & W_{22} \end{pmatrix}. \tag{1.8.9}$$

In much the same way as the scalar Möbius transform in Definition 1.1.10, the operator $\mathcal{W}$ induces the transformation

$$\mathcal{W} : \mathfrak{H} \times \mathfrak{H} \to \mathfrak{K} \times \mathfrak{K}, \quad \{h, h'\} \mapsto \{W_{11}h + W_{12}h', W_{21}h + W_{22}h'\}.$$

The meaning of $\mathcal{W}$, either as a matrix of operators or as a transformation, will be clear from the context.

Definition 1.8.4. Let $\mathcal{W} \in \mathbf{B}(\mathfrak{H} \times \mathfrak{H}, \mathfrak{K} \times \mathfrak{K})$ have the matrix decomposition as in (1.8.9) and let H be a relation in $\mathfrak{H}$. Then the *Möbius transform* of H is the relation $\mathcal{W}[H]$ in $\mathfrak{K}$ defined by

$$\mathcal{W}[H] = \big\{\{W_{11}h + W_{12}h', W_{21}h + W_{22}h'\} : \{h, h'\} \in H\big\}.$$

Note that the domain and range of the Möbius transform are given by

$$\operatorname{dom} \mathcal{W}[H] = \big\{W_{11}h + W_{12}h' : \{h, h'\} \in H\big\},$$
$$\operatorname{ran} \mathcal{W}[H] = \big\{W_{21}h + W_{22}h' : \{h, h'\} \in H\big\}.$$

Moreover, if $\mathfrak{G}$ is a further Hilbert space and $\mathcal{V} \in \mathbf{B}(\mathfrak{K} \times \mathfrak{K}, \mathfrak{G} \times \mathfrak{G})$, then one has $\mathcal{V}[\mathcal{W}[H]] = (\mathcal{V} \circ \mathcal{W})[H]$. For the case where $\mathfrak{H} = \mathfrak{K}$ and $\mathcal{W}^{-1} \in \mathbf{B}(\mathfrak{H} \times \mathfrak{H})$ it follows that the inverse Möbius transform exists and is given by the inverse of $\mathcal{W}$. In this case it also follows that

$$\mathcal{W}[H] \text{ is closed if and only if } H \text{ is closed.} \tag{1.8.10}$$

If $\mathcal{W}$ in Definition 1.8.4 is unitary with respect to the indefinite inner products $[\![\cdot,\cdot]\!]$ in $\mathfrak{H}^2$ and $\mathfrak{K}^2$ (see Proposition 1.8.2), then the corresponding Möbius transform has useful additional geometric properties.

Theorem 1.8.5. *Let $\mathcal{W} \in \mathbf{B}(\mathfrak{H} \times \mathfrak{H}, \mathfrak{K} \times \mathfrak{K})$ have the matrix decomposition in (1.8.9) and assume that $\mathcal{W}$ satisfies the identities*

$$\mathcal{W}^* \mathcal{J}_{\mathfrak{K}} \mathcal{W} = \mathcal{J}_{\mathfrak{H}} \quad \text{and} \quad \mathcal{W} \mathcal{J}_{\mathfrak{H}} \mathcal{W}^* = \mathcal{J}_{\mathfrak{K}}. \tag{1.8.11}$$

Then $\mathcal{W}$ provides a one-to-one correspondence between (maximal) dissipative, (maximal) accumulative, (maximal) symmetric, and self-adjoint relations in $\mathfrak{H}$ and (maximal) dissipative, (maximal) accumulative, (maximal) symmetric, and self-adjoint relations in $\mathfrak{K}$, respectively.

Proof. By Proposition 1.8.2, the operator $\mathcal{W}$ is unitary from $(\mathfrak{H}^2, [\![\cdot,\cdot]\!])$ to $(\mathfrak{K}^2, [\![\cdot,\cdot]\!])$. Recall that the notions of (maximal) dissipative, (maximal) accumulative, (maximal) symmetric, and self-adjoint relations correspond to the notions of (maximal) nonnegative, (maximal) nonpositive, (maximal) neutral, and hypermaximal neutral relations, respectively. Therefore, the asserted results follow from Lemma 1.8.3. □

Note that in the case where $\mathcal{W} \in \mathbf{B}(\mathfrak{H} \times \mathfrak{H}, \mathfrak{K} \times \mathfrak{K})$ satisfies (1.8.11) the inverse $\mathcal{W}^{-1} \in \mathbf{B}(\mathfrak{K} \times \mathfrak{K}, \mathfrak{H} \times \mathfrak{H})$ is given by

$$\mathcal{W}^{-1} = \begin{pmatrix} W_{22}^* & -W_{12}^* \\ -W_{21}^* & W_{11}^* \end{pmatrix}.$$

1.9 Convergence of sequences of relations

This section is devoted to the convergence of sequences of relations in a Hilbert space. There are two notions to be discussed: strong graph convergence and strong resolvent convergence. It will be shown via the uniform boundedness principle that under certain circumstances these notions are equivalent. In particular, the equivalence holds for sequences of self-adjoint or maximal accretive (dissipative) relations.

First recall the following well-known result for bounded linear operators. Let $H_n \in \mathbf{B}(\mathfrak{H}, \mathfrak{K})$ be a sequence of bounded linear operators and assume that $\lim_{n\to\infty} H_n h$ exists in $\mathfrak{K}$ for all $h \in \mathfrak{H}$. An application of the uniform boundedness principle shows that there is a uniform bound: $\|H_n\| \leq C$ for some $C \geq 0$. Moreover,

$$H_\infty h = \lim_{n\to\infty} H_n h, \quad h \in \mathfrak{H}, \tag{1.9.1}$$

defines an operator $H_\infty \in \mathbf{B}(\mathfrak{H}, \mathfrak{K})$ and $\|H_\infty\| \leq C$. A sequence of operators $H_n \in \mathbf{B}(\mathfrak{H}, \mathfrak{K})$ is said to converge *strongly* to $H_\infty \in \mathbf{B}(\mathfrak{H}, \mathfrak{K})$ if $H_n h \to H_\infty h$ for all $h \in \mathfrak{H}$; in this case there is a uniform bound $\|H_n\| \leq C$ for some $C \geq 0$. These results will be used frequently in this section. In the special case $\mathfrak{K} = \mathfrak{H}$ the limit result (1.9.1) leads to the identity

$$(H_\infty h, h) = \lim_{n\to\infty} (H_n h, h), \quad h \in \mathfrak{H}. \tag{1.9.2}$$

Hence, if all $H_n \in \mathbf{B}(\mathfrak{H})$ are self-adjoint (dissipative, accumulative), then (1.9.2) shows $H_\infty \in \mathbf{B}(\mathfrak{H})$ is self-adjoint (dissipative, accumulative, respectively).

Also recall the following situation. Let H_n be a nondecreasing sequence of nonnegative operators in $\mathbf{B}(\mathfrak{H})$ bounded above by $H' \in \mathbf{B}(\mathfrak{H})$:

$$0 \leq (H_m h, h) \leq (H_n h, h) \leq (H' h, h), \quad h \in \mathfrak{H}, \quad n > m. \tag{1.9.3}$$

Then clearly $\|H_n\| \leq \|H'\|$ and it follows from the Cauchy–Schwarz inequality for the nonnegative inner product $((H_n - H_m)\cdot, \cdot)$ that

$$\begin{aligned} \|(H_n - H_m)h\|^2 &\leq \|H_n - H_m\|((H_n - H_m)h, h) \\ &\leq 2\|H'\|((H_n - H_m)h, h) \end{aligned} \tag{1.9.4}$$

for all $h \in \mathfrak{H}$. Consequently, there exists an operator $H_\infty \in \mathbf{B}(\mathfrak{H})$ such that $0 \leq H_n \leq H_\infty \leq H'$ and $H_n h \to H_\infty h$ for all $h \in \mathfrak{H}$ as $n \to \infty$. Note that a similar observation is valid for a nonincreasing sequence of nonnegative operators in $\mathbf{B}(\mathfrak{H})$.

Now one introduces two notions of convergence for relations from $\mathfrak{H}$ to $\mathfrak{K}$: strong graph convergence and strong resolvent convergence. First one defines the notion of strong graph limit.

Definition 1.9.1. Let H_n be a sequence of relations from $\mathfrak{H}$ to $\mathfrak{K}$. The *strong graph limit* is the linear relation H_∞ consisting of all $\{h,h'\} \in \mathfrak{H} \times \mathfrak{K}$ for which there exists a sequence $\{h_n,h'_n\} \in H_n$ such that $\{h_n,h'_n\} \to \{h,h'\}$ in $\mathfrak{H} \times \mathfrak{K}$. The sequence H_n is said to converge to H_∞ in the *strong graph sense* if H_∞ is the strong graph limit of H_n.

By definition, the strong graph limit H_∞ always exists, it is a uniquely determined relation from $\mathfrak{H}$ to $\mathfrak{K}$, and H_∞ is closed. In fact, let $\{h,h'\}$ be the limit of $\{k_n,k'_n\} \in H_\infty$. Then for each $n \in \mathbb{N}$ there exist elements $\{h_n,h'_n\} \in H_n$ with

$$\|\{k_n,k'_n\} - \{h_n,h'_n\}\| \leq \frac{1}{n}.$$

Clearly, $\{h_n,h'_n\} \to \{h,h'\}$ and it follows that $\{h,h'\} \in H_\infty$. Thus, H_∞ is closed. A similar argument shows that the strong graph limits of a sequence H_n and its closures $\overline{H}_n$ coincide. Note also that the strong graph limit may coincide with the zero set $\{0,0\}$ in $\mathfrak{H} \times \mathfrak{K}$. Furthermore, if H_∞ is the strong graph limit of H_n, then $(H_\infty)^{-1}$ is the strong graph limit of $(H_n)^{-1}$. Finally, note that if $H_\infty \in \mathbf{B}(\mathfrak{H},\mathfrak{K})$ is the strong limit of $H_n \in \mathbf{B}(\mathfrak{H},\mathfrak{K})$, then (the graph of) H_∞ is the strong graph limit of H_n

In general, the strong graph convergence $H_n \to H_\infty$ does not imply the strong graph convergence of the adjoints $(H_n)^*$ to $(H_\infty)^*$. But there is the following observation.

Lemma 1.9.2. *Let H_n and H_∞ be relations from $\mathfrak{H}$ to $\mathfrak{K}$. Assume that H_n converges to H_∞ in the strong graph sense. Let K be the strong graph limit in $\mathfrak{K} \times \mathfrak{H}$ of the sequence $(H_n)^*$. Then*

$$K \subset (H_\infty)^*.$$

Proof. Assume that $\{f,f'\} \in K$. Then there exist $\{f_n,f'_n\} \in (H_n)^*$ such that $\{f_n,f'_n\} \to \{f,f'\}$. Now let $\{h,h'\} \in H_\infty$, so that there exist $\{h_n,h'_n\} \in H_n$ such that $\{h_n,h'_n\} \to \{h,h'\}$. In particular, one sees that $(f'_n,h_n) = (f_n,h'_n)$, which in the limit gives

$$(f',h) = (f,h'), \quad \{h,h'\} \in H_\infty.$$

In other words, $\{f,f'\} \in (H_\infty)^*$ and thus $K \subset (H_\infty)^*$. □

In order to define strong resolvent convergence of a sequence of relations H_n in $\mathfrak{H}$ to a relation H_∞ in $\mathfrak{H}$ the following set is needed:

$$\rho_\infty = \rho(H_\infty) \cap \bigcap_{n=1}^{\infty} \rho(H_n),$$

and, whenever it is used, it is tacitly assumed that it is nonempty. Next the notion of strong resolvent limit is defined.

Definition 1.9.3. A sequence of closed relations H_n in $\mathfrak{H}$ is said to converge to a closed linear relation H_∞ in $\mathfrak{H}$ in the *strong resolvent sense* at the point $\lambda \in \rho_\infty$ if for all $h \in \mathfrak{H}$

$$(H_n - \lambda)^{-1}h \to (H_\infty - \lambda)^{-1}h. \tag{1.9.5}$$

In the case of strong resolvent convergence there is also an interplay between the convergence of H_n and that of $(H_n)^{-1}$. Let the closed relations H_n converge in the strong resolvent sense to the closed relation H_∞ at the point $\lambda \in \rho_\infty$. Then it follows from (1.2.14) that for $\lambda \neq 0$

$$\frac{1}{\lambda} \in \rho\big((H_\infty)^{-1}\big) \cap \bigcap_{n=1}^{\infty} \rho\big((H_n)^{-1}\big),$$

and Corollary 1.1.12 implies that $(H_n)^{-1}$ converges to $(H_\infty)^{-1}$ in the strong resolvent sense at $1/\lambda$. Of course, when $\lambda \in \rho_\infty$ and $\lambda = 0$ the operators $(H_n)^{-1} \in \mathbf{B}(\mathfrak{H})$ converge strongly to $(H_\infty)^{-1} \in \mathbf{B}(\mathfrak{H})$.

Strong graph convergence and strong resolvent convergence are closely related in the presence of a uniform bound, as described below.

Theorem 1.9.4. *Let H_n and H_∞ be closed linear relations in $\mathfrak{H}$. Then the following statements hold:*

(i) *Assume that H_n converges to H_∞ in the strong resolvent sense at the point $\lambda \in \rho_\infty$. Then H_n converges to H_∞ in the strong graph sense and there exists $C_\lambda > 0$ such that for all $n \in \mathbb{N}$*

$$\|(H_n - \lambda)^{-1}\| \leq C_\lambda. \tag{1.9.6}$$

(ii) *Assume that H_n converges to H_∞ in the strong graph sense. Let $\lambda \in \rho_\infty$ be any point for which there exists $C_\lambda > 0$ such that (1.9.6) holds for all $n \in \mathbb{N}$. Then H_n converges to H_∞ in the strong resolvent sense at the point λ.*

Proof. (i) Assume that (1.9.5) holds for some $\lambda \in \rho_\infty$. In particular, then one has $(H_n - \lambda)^{-1} \in \mathbf{B}(\mathfrak{H})$. Recall that (1.9.5) implies that the uniform estimate (1.9.6) holds, via the uniform boundedness principle.

Let Γ be the strong graph limit of the sequence H_n. Let $\{h, h'\} \in H_\infty$. Then the sequence

$$\big\{(H_n - \lambda)^{-1}(h' - \lambda h), (I + \lambda(H_n - \lambda)^{-1})(h' - \lambda h)\big\} \in H_n$$

converges to

$$\big\{(H_\infty - \lambda)^{-1}(h' - \lambda h), (I + \lambda(H_\infty - \lambda)^{-1})(h' - \lambda h)\big\} = \{h, h'\}.$$

Hence, $\{h, h'\} \in \Gamma$, which shows that $H_\infty \subset \Gamma$.

Conversely, let $\{h, h'\} \in \Gamma$ and let $\{h_n, h_n'\} \in H_n$ be a sequence such that $\{h_n, h_n'\} \to \{h, h'\}$. Then

$$\begin{aligned}(H_\infty - \lambda)^{-1}&(h_n' - \lambda h_n) - h_n \\ &= (H_\infty - \lambda)^{-1}(h_n' - \lambda h_n) - (H_n - \lambda)^{-1}(h_n' - \lambda h_n) \\ &= \big[(H_\infty - \lambda)^{-1} - (H_n - \lambda)^{-1}\big]\big((h_n' - \lambda h_n) - (h' - \lambda h)\big) \\ &\quad + \big[(H_\infty - \lambda)^{-1} - (H_n - \lambda)^{-1}\big](h' - \lambda h),\end{aligned}$$

and the terms on the right-hand side tend to 0 as $n \to \infty$ due to the pointwise bound $\|(H_n - \lambda)^{-1}\| \leq C_\lambda$ and the strong resolvent convergence. Hence, it follows that

$$(H_\infty - \lambda)^{-1}(h' - \lambda h) = h,$$

so that $\{h, h'\} \in H_\infty$. This shows that $\Gamma \subset H_\infty$.

(ii) Assume that H_∞ is the strong graph limit of the sequence H_n. Let $\lambda \in \rho_\infty$ and let $h \in \mathfrak{H}$. Then, since $\lambda \in \rho(H_\infty)$, there is an element $\{f, f'\} \in H_\infty$ with $f' - \lambda f = h$, so that

$$(H_\infty - \lambda)^{-1} h = (H_\infty - \lambda)^{-1}(f' - \lambda f) = f.$$

Since H_∞ is the strong graph limit of the sequence H_n, there exists a sequence $\{f_n, f_n'\} \in H_n$ with the property that $\{f_n, f_n'\} \to \{f, f'\}$. Then

$$\begin{aligned}(H_n - \lambda)^{-1}&h - (H_\infty - \lambda)^{-1} h \\ &= (H_n - \lambda)^{-1}\big((f' - \lambda f) - (f_n' - \lambda f_n)\big) \\ &\quad + (H_n - \lambda)^{-1}(f_n' - \lambda f_n) - (H_\infty - \lambda)^{-1}(f' - \lambda f) \\ &= (H_n - \lambda)^{-1}\big((f' - \lambda f) - (f_n' - \lambda f_n)\big) + f_n - f,\end{aligned}$$

and, since for $\lambda \in \rho_\infty$ there is the bound (1.9.6), the right-hand side tends to 0 as $n \to \infty$. Hence, H_n converges to H_∞ in the strong resolvent sense at λ. □

The following result is a useful consequence of Theorem 1.9.4.

Corollary 1.9.5. *Let H_n and H_∞ be closed relations in $\mathfrak{H}$ and let H_n satisfy the uniform bound*

$$\|(H_n - \lambda)^{-1}\| \leq C_\lambda \tag{1.9.7}$$

for some $\lambda \in \rho_\infty$. Assume that the relation H is a restriction of H_∞ which satisfies

(i) $\operatorname{ran}(H - \lambda)$ *is dense in* $\mathfrak{H}$;

(ii) *for each $\{h, h'\} \in H$ there exists $h_n' \in \mathfrak{H}$ such that $\{h, h_n'\} \in H_n$ and $h_n' \to h'$ in $\mathfrak{H}$.*

Then H is dense in H_∞ and H_n converges to H_∞ in the strong graph sense or, equivalently, in the strong resolvent sense at $\lambda \in \rho_\infty$.

Proof. Let $\{h, h'\} \in H \subset H_\infty$ and $\{h, h'_n\} \in H_n$ such that $h'_n \to h'$. Then for $\lambda \in \rho_\infty$ one has

$$\{h' - \lambda h, h\} \in (H_\infty - \lambda)^{-1} \quad \text{and} \quad \{h'_n - \lambda h, h\} \in (H_n - \lambda)^{-1},$$

so that

$$(H_\infty - \lambda)^{-1}(h' - \lambda h) = (H_n - \lambda)^{-1}(h'_n - \lambda h).$$

Consequently,

$$\begin{aligned}(H_n - \lambda)^{-1}&(h' - \lambda h) - (H_\infty - \lambda)^{-1}(h' - \lambda h) \\ &= (H_n - \lambda)^{-1}(h' - \lambda h) - (H_n - \lambda)^{-1}(h'_n - \lambda h) \\ &= (H_n - \lambda)^{-1}(h' - h'_n),\end{aligned}$$

and therefore, by the uniform bound,

$$\|(H_n - \lambda)^{-1}(h' - \lambda h) - (H_\infty - \lambda)^{-1}(h' - \lambda h)\| \leq C_\lambda \|h' - h'_n\|$$

for all $\{h, h'\} \in H$. Since $\operatorname{ran}(H - \lambda)$ is dense in $\mathfrak{H}$, it follows from (1.9.7) that H_n converges to H_∞ in the strong resolvent sense at $\lambda \in \rho_\infty$, and hence also in the strong graph sense.

It remains to show that H is dense in H_∞. Observe that $H \subset H_\infty$ implies $(H - \lambda)^{-1} \subset (H_\infty - \lambda)^{-1}$. Since $\lambda \in \rho_\infty$ and $\operatorname{ran}(H - \lambda)$ is dense in $\mathfrak{H}$, it follows that $(H - \lambda)^{-1}$ is a densely defined bounded operator in $\mathfrak{H}$. Thus, its closure coincides with $(H_\infty - \lambda)^{-1}$, which gives $\overline{H} = H_\infty$. □

Let H_n and H_∞ be closed relations in $\mathfrak{H}$. When all these relations are self-adjoint or maximal dissipative (accumulative), then there is automatically a uniform bound of the form (1.9.6).

Corollary 1.9.6. *Let H_n and H_∞ be relations in $\mathfrak{H}$. Then the following statements hold:*

(i) *Assume that H_n and H_∞ are self-adjoint. Then H_n converges to H_∞ in the strong resolvent sense for some, and hence for all $\lambda \in \mathbb{C} \setminus \mathbb{R}$ if and only if H_n converges to H_∞ in the strong graph sense.*

(ii) *Assume that H_n and H_∞ are semibounded and self-adjoint in $\mathfrak{H}$ and assume that γ is a common lower bound. Then H_n converges to H_∞ in the strong resolvent sense for some, and hence for all $\lambda \in \mathbb{C} \setminus [\gamma, \infty)$ if and only if H_n converges to H_∞ in the strong graph sense.*

(iii) *Assume that H_n and H_∞ are maximal dissipative (maximal accumulative). Then H_n converges to H_∞ in the strong resolvent sense for some, and hence for all $\lambda \in \mathbb{C}^-$ ($\lambda \in \mathbb{C}^+$) if and only if H_n converges to H_∞ in the strong graph sense.*

Proof. The proof follows from Theorem 1.9.4 when one recalls that in case (i) one has

$$\|(H_n - \lambda)^{-1}\| \leq \frac{1}{|\mathrm{Im}\,\lambda|}, \qquad \lambda \in \mathbb{C} \setminus \mathbb{R},$$

while in case (ii) one has in addition

$$\|(H_n - \lambda)^{-1}\| \leq \frac{1}{\gamma - \lambda}, \qquad \lambda < \gamma;$$

cf. Proposition 1.4.4 and Proposition 1.4.6. In case (iii) with maximal dissipative relations H_n one has

$$\|(H_n - \lambda)^{-1}\| \leq \frac{1}{-\mathrm{Im}\,\lambda}, \qquad \lambda \in \mathbb{C}^-;$$

cf. Proposition 1.6.3. The case of maximal accumulative relations is analogous. □

In the definition of convergence in strong resolvent sense the limit relation is included. There are situations when the limit is not known beforehand.

Theorem 1.9.7. *Let H_n be a sequence of closed relations. Let*

$$\mathcal{E} \subset \bigcap_{n=1}^{\infty} \rho(H_n) \tag{1.9.8}$$

be a nonempty set such that for all $\lambda \in \mathcal{E}$ and all $h \in \mathfrak{H}$ the sequence $(H_n - \lambda)^{-1}h$ converges. Then there is a closed relation H_∞ with $\mathcal{E} \subset \rho(H_\infty)$ such that H_n converges to H_∞ in the strong resolvent sense for each $\lambda \in \mathcal{E}$.

Proof. Let $\lambda \in \mathcal{E}$, so that the sequence $(H_n - \lambda)^{-1}h$ converges for all $h \in \mathfrak{H}$. Since $(H_n - \lambda)^{-1} \in \mathbf{B}(\mathfrak{H})$, it follows from (1.9.1) (with H_n in (1.9.1) replaced by $(H_n - \lambda)^{-1}$) that there exists an operator $B(\lambda) \in \mathbf{B}(\mathfrak{H})$ such that for all $h \in \mathfrak{H}$

$$(H_n - \lambda)^{-1}h \to B(\lambda)h.$$

Define the relation $H_\infty(\lambda)$ by

$$H_\infty(\lambda) = B(\lambda)^{-1} + \lambda.$$

Then $H_\infty(\lambda)$ is closed, $B(\lambda) = (H_\infty(\lambda) - \lambda)^{-1}$, and $\lambda \in \rho(H_\infty(\lambda))$. In other words, H_n in $\mathfrak{H}$ converges to $H_\infty(\lambda)$ in $\mathfrak{H}$ in the strong resolvent sense at the point $\lambda \in \mathcal{E}$. By Theorem 1.9.4, $H_\infty(\lambda)$ is the strong graph limit of H_n. Hence, $H_\infty(\lambda)$ is independent of the choice of $\lambda \in \mathcal{E}$. □

Theorem 1.9.7 gives rise to the following weakening of Corollary 1.9.6.

Corollary 1.9.8. *Let H_n be a sequence of relations in $\mathfrak{H}$. Then the following statements hold:*

(i) *Assume that all* H_n *are self-adjoint and that for some* $\lambda = \lambda^+ \in \mathbb{C}^+$ *and for some* $\lambda = \lambda^- \in \mathbb{C}^-$ *and all* $h \in \mathfrak{H}$ *the sequence* $(H_n - \lambda)^{-1}h$ *converges. Then there exists a self-adjoint relation* H_∞ *such that* H_n *converges to* H_∞ *in the strong resolvent sense on* $\mathbb{C} \setminus \mathbb{R}$.

(ii) *Assume that all* H_n *are semibounded and self-adjoint with lower bound* γ, *and that for some* $\lambda \in \mathbb{C} \setminus [\gamma, \infty)$ *and all* $h \in \mathfrak{H}$ *the sequence* $(H_n - \lambda)^{-1}h$ *converges. Then there exists a semibounded self-adjoint relation* H_∞ *bounded below by* γ *such that* H_n *converges to* H_∞ *in the strong resolvent sense on* $\mathbb{C} \setminus [\gamma, \infty)$.

(iii) *Assume that all* H_n *are maximal dissipative (maximal accumulative) and that for some* $\lambda \in \mathbb{C}^-$ $(\lambda \in \mathbb{C}^+)$ *and all* $h \in \mathfrak{H}$ *the sequence* $(H_n - \lambda)^{-1}h$ *converges. Then there exists a maximal dissipative (maximal accumulative) relation* H_∞ *such that* H_n *converges to* H_∞ *in the strong resolvent sense on* $\mathbb{C}^-$ $(\mathbb{C}^+)$.

Proof. Let H_n be any sequence of closed relations in $\mathfrak{H}$ such that for all $h \in \mathfrak{H}$ the sequence $(H_n - \lambda)^{-1}h$ converges for each $\lambda \in \mathcal{E}$, where $\mathcal{E}$ is a nonempty set satisfying (1.9.8). According to Theorem 1.9.7, there is a closed relation H_∞, which is the limit of H_n in the strong resolvent sense on $\mathcal{E}$, such that

$$\operatorname{ran}(H_\infty - \lambda) = \mathfrak{H}, \quad \lambda \in \mathcal{E}. \tag{1.9.9}$$

If there is a uniform bound as in (1.9.6), then Theorem 1.9.4 implies that the limit H_∞ is also the strong graph limit of H_n. Hence, every $\{h, h'\} \in H_\infty$ can be approximated by $\{h_n, h_n'\} \in H_n$, which implies that

$$(h', h) = \lim_{n\to\infty} (h_n', h_n), \tag{1.9.10}$$

and thus also

$$\operatorname{Im}(h', h) = \lim_{n\to\infty} \operatorname{Im}(h_n', h_n). \tag{1.9.11}$$

(i) Assume that all H_n are self-adjoint. Then the set $\mathcal{E} = \{\lambda^+, \lambda^-\}$ with some $\lambda^\pm \in \mathbb{C}^\pm$ satisfies (1.9.8) and since all H_n are symmetric, it follows from (1.9.11) that the closed relation H_∞ is symmetric. Hence, (1.9.9) with $\mathcal{E} = \{\lambda^+, \lambda^-\}$ shows that H_∞ is self-adjoint; cf. Theorem 1.5.5. Due to Corollary 1.9.6, one sees that H_n converges to H_∞ in the strong resolvent sense on $\mathbb{C} \setminus \mathbb{R}$.

(ii) Assume that all H_n are semibounded and self-adjoint with common lower bound γ. Then the set $\mathcal{E} = \{\lambda\}$ with some $\lambda \in \mathbb{C} \setminus [\gamma, \infty)$ satisfies (1.9.8) and, since $(h_n', h_n) \geq \gamma(h_n, h_n)$ for $\{h_n, h_{n'}\} \in H_n$, it follows from (1.9.10) that the closed relation H_∞ is bounded below with lower bound γ. Hence, (1.9.9) shows that H_∞ is self-adjoint; cf. Theorem 1.5.5. In view of Corollary 1.9.6, H_n converges to H_∞ in the strong resolvent sense on $\mathbb{C} \setminus [\gamma, \infty)$.

(iii) Assume that all H_n are maximal dissipative. Then the set $\mathcal{E} = \{\lambda\}$ with some $\lambda \in \mathbb{C}^-$ satisfies (1.9.8) and since all H_n are dissipative it follows from

(1.9.11) that the closed relation H_∞ is dissipative. Hence, (1.9.9) shows that H_∞ is maximal dissipative; cf. Theorem 1.6.4. Due to Corollary 1.9.6, one sees that H_n converges to H_∞ in the strong resolvent sense on $\mathbb{C}^-$. The case where H_n is maximal accumulative is treated analogously. □

The following result is an illustration of the methods involving convergence in the graph sense and in the resolvent sense. In Chapter 5 this result will be used extensively.

Proposition 1.9.9. *Let H_n be a sequence of semibounded self-adjoint relations with common lower bound γ in $\mathfrak{H}$. Assume that for $m > n$ and some $\lambda < \gamma$*

$$0 \leq (H_m - \lambda)^{-1} \leq (H_n - \lambda)^{-1}. \tag{1.9.12}$$

Then there exists a semibounded self-adjoint relation H_∞ with lower bound γ such that H_n converges to H_∞ in the strong resolvent sense on $\mathbb{C} \setminus [\gamma, \infty)$, and

$$0 \leq (H_\infty - \lambda)^{-1} \leq (H_n - \lambda)^{-1}. \tag{1.9.13}$$

Proof. Let $\lambda < \gamma$ and let $h \in \mathfrak{H}$. By (1.9.12),

$$0 \leq \big((H_m - \lambda)^{-1}h, h\big) \leq \big((H_n - \lambda)^{-1}h, h\big)$$

for $m > n$ and now it follows in the same way as in (1.9.3)–(1.9.4) that the sequence $(H_n - \lambda)^{-1}h$ converges for $h \in \mathfrak{H}$. Then by Corollary 1.9.8 there is a self-adjoint relation H_∞ bounded below by γ, such that H_n converges to H_∞ in the strong resolvent sense on $\mathbb{C} \setminus [\gamma, \infty)$. It follows from (1.9.12) that (1.9.13) holds. □

Before moving to a corollary of Proposition 1.9.9, recall the following simple antitonicity result. Let $A, B \in \mathbf{B}(\mathfrak{H})$ satisfy $0 \leq A \leq B$ and let A be boundedly invertible. Then B is boundedly invertible, $0 \leq B^{-1}$, and $B^{-1} \leq A^{-1}$. To see the last inequality, note that $(A\cdot, \cdot)$ is a nonnegative semi-inner product, thus one obtains for any $\varphi, \psi \in \mathfrak{H}$:

$$|(A\varphi, \psi)|^2 \leq (A\varphi, \varphi)(A\psi, \psi) \leq (A\varphi, \varphi)(B\psi, \psi).$$

Let $h \in \mathfrak{H}$ and choose $\varphi = A^{-1}h$ and $\psi = B^{-1}h$. Then this inequality leads to $0 \leq B^{-1} \leq A^{-1}$.

The following corollary deals with the situation from the beginning of this section. However, now the nondecreasing sequence of self-adjoint operators in $\mathbf{B}(\mathfrak{H})$ does not necessarily have an upper bound.

Corollary 1.9.10. *Let $H_n \in \mathbf{B}(\mathfrak{H})$ be a sequence of self-adjoint operators which is nondecreasing, i.e., for all $h \in \mathfrak{H}$*

$$(H_n h, h) \leq (H_m h, h), \quad n < m, \tag{1.9.14}$$

and let $\gamma \in \mathbb{R}$ be the lower bound of H_1. Then there exists a semibounded self-adjoint relation H_∞, bounded below by γ, such that H_n converges to H_∞ in the strong resolvent sense on $\mathbb{C} \setminus [\gamma, \infty)$, and

$$0 \le (H_\infty - \lambda)^{-1} \le (H_n - \lambda)^{-1}, \quad \lambda < \gamma. \tag{1.9.15}$$

Proof. Let $\gamma \in \mathbb{R}$ be the lower bound for H_1. Then γ is a common lower bound for all H_n, i.e., $\gamma(h,h) \le (H_n h, h)$ for all $h \in \mathfrak{H}$. Furthermore, (1.9.14) gives for $n < m$:

$$0 \le ((H_n - \lambda)h, h) \le ((H_m - \lambda)h, h), \quad \lambda < \gamma.$$

Since $H_n - \lambda$ with $\lambda < \gamma$ is boundedly invertible for all $n \in \mathbb{N}$, this implies by antitonicity that (1.9.12) holds. Thus, (1.9.15) follows from Proposition 1.9.9. □

1.10 Parametric representations for relations

The discussion in this section is centered on the question when a relation from $\mathfrak{H}$ to $\mathfrak{K}$ can be seen as the range of a bounded column operator or as the kernel of a bounded row operator. The results will be used in the description of boundary value problems in Chapter 2.

Let $\mathfrak{H}$, $\mathfrak{K}$, and $\mathfrak{E}$ be Hilbert spaces and let $\mathcal{A} \in \mathbf{B}(\mathfrak{E}, \mathfrak{H})$, $\mathcal{B} \in \mathbf{B}(\mathfrak{E}, \mathfrak{K})$. Then H defined by

$$H = \big\{\{\mathcal{A}e, \mathcal{B}e\} : e \in \mathfrak{E}\big\} \tag{1.10.1}$$

is a relation from $\mathfrak{H}$ to $\mathfrak{K}$. The representation of the relation H in (1.10.1) is called a *parametric representation* and is denoted by $H = \{\mathcal{A}, \mathcal{B}\}$. It is sometimes convenient to rewrite (1.10.1) as

$$H = \operatorname{ran} \begin{pmatrix} \mathcal{A} \\ \mathcal{B} \end{pmatrix}, \tag{1.10.2}$$

that is, H is the range of the corresponding bounded column operator from $\mathfrak{E}$ to $\mathfrak{H} \times \mathfrak{K}$. Not all relations from $\mathfrak{H}$ to $\mathfrak{K}$ can be represented in the form (1.10.1); below the ones that do will be characterized.

An interesting feature of parametric representations is how they show up in adjoints. Namely, if H is given by (1.10.1) or, equivalently, by (1.10.2), then the adjoint H^* of H satisfies

$$H^* = \big\{\{f, f'\} \in \mathfrak{K} \times \mathfrak{H} : \mathcal{B}^* f = \mathcal{A}^* f'\big\}, \tag{1.10.3}$$

or, equivalently,

$$H^* = \ker \begin{pmatrix} \mathcal{B}^* & -\mathcal{A}^* \end{pmatrix},$$

that is, H^* can be written as the kernel of the corresponding bounded row operator from $\mathfrak{K} \times \mathfrak{H}$ to $\mathfrak{E}$.

The following theorem uses the notion of operator range. An *operator range* $\mathfrak{R}$ in a Hilbert space $\mathfrak{X}$ is defined as the range of a bounded everywhere defined operator from some Hilbert space $\mathfrak{Y}$ to $\mathfrak{X}$.

Theorem 1.10.1. *A relation H from $\mathfrak{H}$ to $\mathfrak{K}$ is of the form (1.10.1) with $\mathcal{A} \in \mathbf{B}(\mathfrak{E}, \mathfrak{H})$ and $\mathcal{B} \in \mathbf{B}(\mathfrak{E}, \mathfrak{K})$ if and only if H is an operator range. In particular, every closed relation H from $\mathfrak{H}$ to $\mathfrak{K}$ is of the form (1.10.1).*

Proof. Let H be given by (1.10.1) and define the column operator R by

$$R = \begin{pmatrix} \mathcal{A} \\ \mathcal{B} \end{pmatrix} : \mathfrak{E} \to \begin{pmatrix} \mathfrak{H} \\ \mathfrak{K} \end{pmatrix},$$

where the column on the right stands for the Hilbert space $\mathfrak{H} \times \mathfrak{K}$. Then R belongs to $\mathbf{B}(\mathfrak{E}, \mathfrak{H} \times \mathfrak{K})$ and clearly $\operatorname{ran} R$ coincides with (the graph of) the relation H; i.e., H is an operator range in $\mathfrak{H} \times \mathfrak{K}$.

Conversely, assume that H is an operator range in $\mathfrak{H} \times \mathfrak{K}$, so that (the graph of) H coincides with $\operatorname{ran} R$ for some $R \in \mathbf{B}(\mathfrak{E}, \mathfrak{H} \times \mathfrak{K})$. Let $P_{\mathfrak{H}}$ and $P_{\mathfrak{K}}$ be the orthogonal projections from $\mathfrak{H} \times \mathfrak{K}$ onto $\mathfrak{H}$ and $\mathfrak{K}$, respectively. Then

$$\mathcal{A} = P_{\mathfrak{H}} R \quad \text{and} \quad \mathcal{B} = P_{\mathfrak{K}} R$$

define a pair of bounded operators $\mathcal{A} \in \mathbf{B}(\mathfrak{E}, \mathfrak{H})$ and $\mathcal{B} \in \mathbf{B}(\mathfrak{E}, \mathfrak{K})$ such that

$$H = \big\{ Rf : f \in \mathfrak{E} \big\} = \big\{ \{\mathcal{A}f, \mathcal{B}f\} : f \in \mathfrak{E} \big\}.$$

Hence, H has the form (1.10.1).

Finally, every closed relation H from $\mathfrak{H}$ to $\mathfrak{K}$ is of the form (1.10.1), since it coincides with the range of the orthogonal projection from $\mathfrak{H} \times \mathfrak{K}$ onto (the closed graph of) H. □

In the general operator representation (1.10.1) there clearly exists some redundancy: the closed linear subspace

$$\ker \begin{pmatrix} \mathcal{A} \\ \mathcal{B} \end{pmatrix} = \ker \mathcal{A} \cap \ker \mathcal{B} \subset \mathfrak{E}$$

does not contribute to H. Thus, one can restrict the operators $\mathcal{A}$ and $\mathcal{B}$ to the orthogonal complement of $\ker \mathcal{A} \cap \ker \mathcal{B}$ in $\mathfrak{E}$. The representing pair $H = \{\mathcal{A}, \mathcal{B}\}$ is called *tight* if $\ker \mathcal{A} \cap \ker \mathcal{B} = \{0\}$. All tight representations $H = \{\mathcal{A}, \mathcal{B}\}$ are easily characterized.

Lemma 1.10.2. *Let H_j, $j = 1, 2$, be relations from $\mathfrak{H}$ to $\mathfrak{K}$. Assume that the representations*

$$H_j = \big\{ \{\mathcal{A}_j e, \mathcal{B}_j e\} : e \in \mathfrak{E}_j \big\}, \quad j = 1, 2,$$

where $\mathcal{A}_j \in \mathbf{B}(\mathfrak{E}_j, \mathfrak{H})$, $\mathcal{B}_j \in \mathbf{B}(\mathfrak{E}_j, \mathfrak{K})$, and $\mathfrak{E}_j$ are Hilbert spaces, are tight. Then the equality $H_1 = H_2$ holds if and only if there exists a bounded bijective operator $X \in \mathbf{B}(\mathfrak{E}_1, \mathfrak{E}_2)$ such that

$$\mathcal{A}_1 = \mathcal{A}_2 X, \quad \mathcal{B}_1 = \mathcal{B}_2 X.$$

Proof. Since the representations of H_j, $j = 1, 2$, are tight, one has

$$\overline{\operatorname{ran}} \begin{pmatrix} \mathcal{A}_1 \\ \mathcal{B}_1 \end{pmatrix}^* = \left(\ker \begin{pmatrix} \mathcal{A}_1 \\ \mathcal{B}_1 \end{pmatrix} \right)^\perp = \mathfrak{E}_1 \quad \text{and} \quad \overline{\operatorname{ran}} \begin{pmatrix} \mathcal{A}_2 \\ \mathcal{B}_2 \end{pmatrix}^* = \left(\ker \begin{pmatrix} \mathcal{A}_2 \\ \mathcal{B}_2 \end{pmatrix} \right)^\perp = \mathfrak{E}_2.$$

Now assume that $H_1 = H_2$, so that

$$\operatorname{ran} \begin{pmatrix} \mathcal{A}_1 \\ \mathcal{B}_1 \end{pmatrix} = \operatorname{ran} \begin{pmatrix} \mathcal{A}_2 \\ \mathcal{B}_2 \end{pmatrix}.$$

Then Corollary D.4 and the discussion preceding it show that there exists a boundedly invertible operator $X \in \mathbf{B}(\mathfrak{E}_1, \mathfrak{E}_2)$ such that

$$\begin{pmatrix} \mathcal{A}_1 \\ \mathcal{B}_1 \end{pmatrix} = \begin{pmatrix} \mathcal{A}_2 \\ \mathcal{B}_2 \end{pmatrix} X.$$

The converse is clear. □

The question comes up when relations of the form (1.10.1) are closed. The following result gives a necessary and sufficient condition.

Proposition 1.10.3. *Let H be a relation from $\mathfrak{H}$ to $\mathfrak{K}$ of the form (1.10.1) with $\mathcal{A} \in \mathbf{B}(\mathfrak{E}, \mathfrak{H})$ and $\mathcal{B} \in \mathbf{B}(\mathfrak{E}, \mathfrak{K})$. Then H is closed if and only if*

$$\mathfrak{E}' = \operatorname{ran} (\mathcal{A}^*\mathcal{A} + \mathcal{B}^*\mathcal{B})$$

is closed in $\mathfrak{E}$. In this case there exists a tight representation $\{\mathcal{A}', \mathcal{B}'\}$ of H, where $\mathcal{A}' \in \mathbf{B}(\mathfrak{E}', \mathfrak{H})$, $\mathcal{B}' \in \mathbf{B}(\mathfrak{E}', \mathfrak{K})$, such that

$$(\mathcal{A}')^*\mathcal{A}' + (\mathcal{B}')^*\mathcal{B}' = I_{\mathfrak{E}'}.$$

Proof. The identity

$$\operatorname{ran} \begin{pmatrix} \mathcal{A} \\ \mathcal{B} \end{pmatrix}^* \begin{pmatrix} \mathcal{A} \\ \mathcal{B} \end{pmatrix} = \operatorname{ran} (\mathcal{A}^*\mathcal{A} + \mathcal{B}^*\mathcal{B})$$

together with Lemma D.1 and Lemma D.2 shows that $\operatorname{ran} (\mathcal{A}^*\mathcal{A} + \mathcal{B}^*\mathcal{B})$ is closed if and only if

$$H = \operatorname{ran} \begin{pmatrix} \mathcal{A} \\ \mathcal{B} \end{pmatrix}$$

is closed. Now assume that H is closed or, equivalently, that $\operatorname{ran} (\mathcal{A}^*\mathcal{A} + \mathcal{B}^*\mathcal{B})$ is closed. Since $\mathcal{A}^*\mathcal{A} + \mathcal{B}^*\mathcal{B}$ is self-adjoint the space $\mathfrak{E}$ has the orthogonal decomposition

$$\mathfrak{E} = \operatorname{ran} (\mathcal{A}^*\mathcal{A} + \mathcal{B}^*\mathcal{B}) \oplus \ker (\mathcal{A}^*\mathcal{A} + \mathcal{B}^*\mathcal{B}).$$

It follows from the identity

$$\ker (\mathcal{A}^*\mathcal{A} + \mathcal{B}^*\mathcal{B}) = \ker \begin{pmatrix} \mathcal{A} \\ \mathcal{B} \end{pmatrix}^* \begin{pmatrix} \mathcal{A} \\ \mathcal{B} \end{pmatrix} = \ker \begin{pmatrix} \mathcal{A} \\ \mathcal{B} \end{pmatrix},$$

that the restrictions $\mathcal{A}_0$ and $\mathcal{B}_0$ of $\mathcal{A}$ and $\mathcal{B}$ to $\mathfrak{E}' = \operatorname{ran}(\mathcal{A}^*\mathcal{A} + \mathcal{B}^*\mathcal{B})$ form a tight representation of H. Moreover, it can be seen that $\mathcal{A}_0^*\mathcal{A}_0 + \mathcal{B}_0^*\mathcal{B}_0$ coincides with the restriction of $\mathcal{A}^*\mathcal{A} + \mathcal{B}^*\mathcal{B}$ onto $\mathfrak{E}'$ and hence it follows that $\mathcal{A}_0^*\mathcal{A}_0 + \mathcal{B}_0^*\mathcal{B}_0$ is a bounded bijective nonnegative operator in $\mathfrak{E}'$. Now define

$$X = (\mathcal{A}_0^*\mathcal{A}_0 + \mathcal{B}_0^*\mathcal{B}_0)^{-\frac{1}{2}} \in \mathbf{B}(\mathfrak{E}'),$$

and set $\mathcal{A}' = \mathcal{A}_0 X \in \mathbf{B}(\mathfrak{E}', \mathfrak{H})$, $\mathcal{B}' = \mathcal{B}_0 X \in \mathbf{B}(\mathfrak{E}', \mathfrak{K})$. Then it follows that H can be represented in the form

$$H = \big\{\{\mathcal{A}'e, \mathcal{B}'e\} : e \in \mathfrak{E}'\big\},$$

where the pair $\{\mathcal{A}', \mathcal{B}'\}$ is normalized by

$$(\mathcal{A}')^*\mathcal{A}' + (\mathcal{B}')^*\mathcal{B}' = X^*\big(\mathcal{A}_0^*\mathcal{A}_0 + \mathcal{B}_0^*\mathcal{B}_0\big)X = I_{\mathfrak{E}'}.$$

Since the representing pair $H = \{\mathcal{A}_0, \mathcal{B}_0\}$ is tight, so is the representing pair $H = \{\mathcal{A}', \mathcal{B}'\}$. □

A direct by-product of Theorem 1.10.1 is the following representation of a relation in terms of the kernel of a bounded row operator.

Proposition 1.10.4. *A relation H from $\mathfrak{H}$ to $\mathfrak{K}$ is of the form*

$$H = \big\{\{f, f'\} \in \mathfrak{H} \times \mathfrak{K} : \mathcal{M}f = \mathcal{N}f'\big\}, \tag{1.10.4}$$

where $\mathcal{M} \in \mathbf{B}(\mathfrak{H}, \mathfrak{F})$, $\mathcal{N} \in \mathbf{B}(\mathfrak{K}, \mathfrak{F})$, and $\mathfrak{F}$ is a Hilbert space if and only if H is closed. In this case the Hilbert space $\mathfrak{F}$ can be chosen such that

$$\mathfrak{F} = \overline{\operatorname{span}}\,\big\{\overline{\operatorname{ran}}\,\mathcal{M}, \overline{\operatorname{ran}}\,\mathcal{N}\big\}, \tag{1.10.5}$$

where $\mathcal{M}$ and $\mathcal{N}$ are uniquely determined up to left-multiplication by a bounded bijective operator.

Proof. Note first that for any relation H from $\mathfrak{H}$ to $\mathfrak{K}$ the adjoint H^* is a closed relation from $\mathfrak{K}$ to $\mathfrak{H}$. Hence, by Theorem 1.10.1, there exist a Hilbert space $\mathfrak{F}$ and a pair $\mathcal{C} \in \mathbf{B}(\mathfrak{F}, \mathfrak{K})$ and $\mathcal{D} \in \mathbf{B}(\mathfrak{F}, \mathfrak{H})$, such that

$$H^* = \big\{\{\mathcal{C}e, \mathcal{D}e\} : e \in \mathfrak{F}\big\}.$$

Then it follows from (1.10.3) that

$$H^{**} = \big\{\{f, f'\} \in \mathfrak{H} \times \mathfrak{K} : \mathcal{D}^*f = \mathcal{C}^*f'\big\}.$$

Now assume that H is a closed relation from $\mathfrak{H}$ to $\mathfrak{K}$. Then $H = H^{**}$ and hence (1.10.4) is valid with $\mathcal{M} = \mathcal{D}^* \in \mathbf{B}(\mathfrak{H}, \mathfrak{F})$ and $\mathcal{N} = \mathcal{C}^* \in \mathbf{B}(\mathfrak{K}, \mathfrak{F})$. For the converse assume that H has the form (1.10.4), where $\mathcal{M} \in \mathbf{B}(\mathfrak{H}, \mathfrak{F})$ and $\mathcal{N} \in \mathbf{B}(\mathfrak{K}, \mathfrak{F})$. Then it follows directly that H is closed.

If H is given by (1.10.4), then it follows that H^* has the representation

$$H^* = \{\{\mathcal{N}^* e, \mathcal{M}^* e\} : e \in \mathfrak{F}\}$$

with $\mathcal{N}^* \in \mathbf{B}(\mathfrak{F}, \mathfrak{K})$ and $\mathcal{M}^* \in \mathbf{B}(\mathfrak{F}, \mathfrak{H})$. By Proposition 1.10.3, this representation can be assumed to be tight. Then one has

$$\{0\} = \ker \begin{pmatrix} \mathcal{N}^* \\ \mathcal{M}^* \end{pmatrix} = \left(\operatorname{ran} \begin{pmatrix} \mathcal{N}^* \\ \mathcal{M}^* \end{pmatrix}^* \right)^{\perp} = \left(\operatorname{ran} (\mathcal{N} \ \mathcal{M}) \right)^{\perp},$$

which gives (1.10.5).

Likewise, assume that H is given by

$$H = \{\{f, f'\} \in \mathfrak{H} \times \mathfrak{K} : \mathcal{M}_1 f = \mathcal{N}_1 f'\},$$

where $\mathfrak{F}_1$ is a Hilbert space, $\mathcal{M}_1 \in \mathbf{B}(\mathfrak{H}, \mathfrak{F}_1)$, and $\mathcal{N}_1 \in \mathbf{B}(\mathfrak{K}, \mathfrak{F}_1)$, and that the condition $\mathfrak{F}_1 = \overline{\operatorname{span}}\, \{\overline{\operatorname{ran}}\, \mathcal{M}_1, \overline{\operatorname{ran}}\, \mathcal{N}_1\}$ holds. Then H^* also has the following tight representation

$$H^* = \{\{(\mathcal{N}_1)^* e, (\mathcal{M}_1)^* e\} : e \in \mathfrak{F}_1\}.$$

By Lemma 1.10.2, there exists a bounded bijective operator $X \in \mathbf{B}(\mathfrak{F}_1, \mathfrak{F})$ such that

$$(\mathcal{M}_1)^* = (\mathcal{M})^* X, \quad (\mathcal{N}_1)^* = (\mathcal{N})^* X,$$

or

$$\mathcal{M}_1 = X^* \mathcal{M}, \quad \mathcal{N}_1 = X^* \mathcal{N},$$

with $X^* \in \mathbf{B}(\mathfrak{F}, \mathfrak{F}_1)$ is bijective. This completes the proof. □

Let H be a closed relation from $\mathfrak{H}$ to $\mathfrak{K}$. Then it has a representation as in (1.10.1) and a representation as in (1.10.4). The interest is now in explicitly connecting these representations. The first main result concerns the case when the resolvent set of the relation is nonempty.

Theorem 1.10.5. *The relation H in $\mathfrak{H}$ is closed with $\mu \in \rho(H)$ if and only if H has a representation*

$$H = \{\{\mathcal{A}e, \mathcal{B}e\} : e \in \mathfrak{H}\} \tag{1.10.6}$$

with $\mathcal{A}, \mathcal{B} \in \mathbf{B}(\mathfrak{H})$, such that $(\mathcal{B} - \mu\mathcal{A})^{-1} \in \mathbf{B}(\mathfrak{H})$. This representation is automatically tight. Moreover, in this case the pair $\{\mathcal{A}, \mathcal{B}\}$ may be chosen such that H^* has the tight representation

$$H^* = \{\{\mathcal{A}^* e, \mathcal{B}^* e\} : e \in \mathfrak{H}\}, \tag{1.10.7}$$

so that H can also be written as

$$H = \{\{f, f'\} \in \mathfrak{H} \times \mathfrak{H} : \mathcal{B}f = \mathcal{A}f'\}. \tag{1.10.8}$$

Proof. Let H be the relation in (1.10.6) and assume that $(\mathcal{B} - \mu\mathcal{A})^{-1} \in \mathbf{B}(\mathfrak{H})$. Then it is clear that

$$(H - \mu)^{-1} = \{\{(\mathcal{B} - \mu\mathcal{A})e, \mathcal{A}e\} : e \in \mathfrak{H}\} = \mathcal{A}(\mathcal{B} - \mu\mathcal{A})^{-1},$$

which implies that $\mu \in \rho(H)$ and that H is closed. The representation is tight, since $\mathcal{A}e = 0$ and $\mathcal{B}e = 0$ imply $(\mathcal{B} - \mu\mathcal{A})e = 0$, and hence $e = 0$.

Conversely, let H be closed and assume that $\mu \in \rho(H)$. Then, by Lemma 1.2.4, H has the representation

$$H = \{\{(H - \mu)^{-1}f, (I + \mu(H - \mu)^{-1})f\} : f \in \mathfrak{H}\}.$$

Hence, one gets (1.10.6) by taking $\mathcal{A} = (H - \mu)^{-1}$ and $\mathcal{B} = I + \mu(H - \mu)^{-1}$, in which case $\mathcal{B} - \mu\mathcal{A} = I$ is boundedly invertible. Since $\bar{\mu} \in \rho(H^*)$, one also has, by Lemma 1.2.4,

$$H^* = \{\{(H^* - \bar{\mu})^{-1}g, (I + \bar{\mu}(H - \bar{\mu})^{-1})g\} : g \in \mathfrak{H}\},$$

which then leads to (1.10.7). It also follows that this representation is tight. The assertion (1.10.8) follows from (1.10.7), (1.10.3), and $H = H^{**}$. □

Note that a possible choice for (1.10.6) and (1.10.7) (and hence also (1.10.8)) to hold is given by

$$\mathcal{A} = (H - \mu)^{-1} \quad \text{and} \quad \mathcal{B} = I + \mu(H - \mu)^{-1}, \qquad \mu \in \rho(H). \tag{1.10.9}$$

In the next statement, starting from an arbitrary representing pair $\{\mathcal{A}, \mathcal{B}\}$ for H in (1.10.6) a representing pair $\{X^{-*}\mathcal{A}^*, X^{-*}\mathcal{B}^*\}$ for H^* as in (1.10.7) is obtained. In fact, Corollary 1.10.6 is an immediate consequence of (1.10.9), Lemma 1.10.2 and Theorem 1.10.5.

Corollary 1.10.6. *Let H be a closed relation in $\mathfrak{H}$ with $\mu \in \rho(H)$ given in the form (1.10.6) with $\mathcal{A}, \mathcal{B} \in \mathbf{B}(\mathfrak{H})$, such that $(\mathcal{B} - \mu\mathcal{A})^{-1} \in \mathbf{B}(\mathfrak{H})$. Then*

$$\mathcal{A} = (H - \mu)^{-1}X, \quad \mathcal{B} = \big(I + \mu(H - \mu)^{-1}\big)X,$$

for some bijective $X \in \mathbf{B}(\mathfrak{H})$, and the pair $\{X^{-*}\mathcal{A}^*, X^{-*}\mathcal{B}^*\}$ represents the adjoint H^* as in (1.10.7). In particular, H is given by

$$H = \{\{f, f'\} \in \mathfrak{H} : \mathcal{B}X^{-1}f = \mathcal{A}X^{-1}f'\}.$$

For a given representation $H = \{\mathcal{A}, \mathcal{B}\}$ as in (1.10.6) and some bijective operator $X \in \mathbf{B}(\mathfrak{H})$, Lemma 1.10.2 shows that also

$$\mathcal{A}' = \mathcal{A}X, \quad \mathcal{B}' = \mathcal{B}X,$$

is a tight representation of H. With $\{\mathcal{A}, \mathcal{B}\}$ in (1.10.9) and $X = \bar{\mu} - \mu$, where $\mu \in \rho(H) \cap \mathbb{C} \setminus \mathbb{R}$, one gets the following representing pair $\{\mathcal{A}', \mathcal{B}'\}$ for H in terms of the Cayley transform in Definition 1.1.13:

$$\begin{aligned} \mathcal{A}' &= (\bar{\mu} - \mu)(H - \mu)^{-1} = I - \mathcal{C}_\mu[H], \\ \mathcal{B}' &= (\bar{\mu} - \mu)\bigl(I + \mu(H - \mu)^{-1}\bigr) = \bar{\mu} - \mu\mathcal{C}_\mu[H]. \end{aligned} \tag{1.10.10}$$

The next proposition is also closely related to Theorem 1.10.5.

Proposition 1.10.7. *Let H be a closed relation in $\mathfrak{H}$ of the form*

$$H = \bigl\{\{f, f'\} \in \mathfrak{H} \times \mathfrak{H} : \mathcal{M}f = \mathcal{N}f'\bigr\}, \tag{1.10.11}$$

where $\mathfrak{F}$ is a Hilbert space and $\mathcal{M}, \mathcal{N} \in \mathbf{B}(\mathfrak{H}, \mathfrak{F})$, and assume that (1.10.5) is satisfied. Then $\mu \in \rho(H)$ if and only if $\mathcal{M} - \mu\mathcal{N} \in \mathbf{B}(\mathfrak{H}, \mathfrak{F})$ is bijective. In this case H has the parametrization (1.10.6) with

$$\{\mathcal{A}, \mathcal{B}\} = \bigl\{(\mathcal{M} - \mu\mathcal{N})^{-1}\mathcal{N}, (\mathcal{M} - \mu\mathcal{N})^{-1}\mathcal{M}\bigr\}. \tag{1.10.12}$$

Proof. Assume that $\mu \in \rho(H)$, so that also $\bar{\mu} \in \rho(H^*)$ and hence one has the parametrization $H^* = \{(H^* - \bar{\mu})^{-1}, I + \bar{\mu}(H^* - \bar{\mu})^{-1}\}$. Define the relation K in $\mathfrak{H}$ by

$$K = \bigl\{\{\mathcal{N}^*e, \mathcal{M}^*e\} : e \in \mathfrak{F}\bigr\}.$$

Then it follows from (1.10.3) that $H = K^*$. Furthermore, one sees that the representation of $K = H^*$ is tight due to (1.10.5). Hence, there exists a bijective operator $X \in \mathbf{B}(\mathfrak{F}, \mathfrak{H})$ such that

$$\mathcal{N}^* = (H^* - \bar{\mu})^{-1}X \quad \text{and} \quad \mathcal{M}^* = \bigl(I + \bar{\mu}(H^* - \bar{\mu})^{-1}\bigr)X,$$

and therefore

$$\mathcal{M} = X^*\bigl(I + \mu(H - \mu)^{-1}\bigr) \quad \text{and} \quad \mathcal{N} = X^*(H - \mu)^{-1}. \tag{1.10.13}$$

It follows that

$$\mathcal{M} - \mu\mathcal{N} = X^*, \tag{1.10.14}$$

and hence $\mathcal{M} - \mu\mathcal{N} \in \mathbf{B}(\mathfrak{H}, \mathfrak{F})$ is bijective.

Conversely, assume that $\mathcal{M} - \mu\mathcal{N} \in \mathbf{B}(\mathfrak{H}, \mathfrak{F})$ is bijective. It follows from (1.10.11) that

$$H - \mu = \bigl\{\{f, f' - \mu f\} \in \mathfrak{H} \times \mathfrak{H} : \mathcal{M}f = \mathcal{N}f'\bigr\}.$$

Then it is clear that $\ker(H - \mu) = \{0\}$. To show that $\operatorname{ran}(H - \mu) = \mathfrak{H}$, let $h \in \mathfrak{H}$ and define

$$f = (\mathcal{M} - \mu\mathcal{N})^{-1}\mathcal{N}h \quad \text{and} \quad f' = \mu f + h.$$

From this definition one sees that

$$\mathcal{N}f' = \mu\mathcal{N}f + \mathcal{N}h = \mu\mathcal{N}f + (\mathcal{M} - \mu\mathcal{N})f = \mathcal{M}f,$$

which shows $\{f, f'\} \in H$. Furthermore one sees that $f' - \mu f = h$, which then implies that $\operatorname{ran}(H - \mu) = \mathfrak{H}$. Hence, $\mu \in \rho(H)$.

It remains to show the parametrization (1.10.12) if $\mu \in \rho(H)$ or, equivalently, $(\mathcal{M} - \mu\mathcal{N})^{-1} \in \mathbf{B}(\mathfrak{F}, \mathfrak{H})$. For this note that $H = \{(H - \mu)^{-1}, I + \mu(H - \mu)^{-1}\}$, and that (1.10.13) and (1.10.14) imply

$$(H - \mu)^{-1} = (\mathcal{M} - \mu\mathcal{N})^{-1}\mathcal{N} \quad \text{and} \quad I + \mu(H - \mu)^{-1} = (\mathcal{M} - \mu\mathcal{N})^{-1}\mathcal{M}.$$

This gives (1.10.12). □

In the next corollary the self-adjoint, maximal dissipative, and maximal accumulative relations are treated.

Corollary 1.10.8. *Let H be a relation in $\mathfrak{H}$. Then the following statements hold:*

(i) *H is self-adjoint if and only if there exist $\mathcal{A}, \mathcal{B} \in \mathbf{B}(\mathfrak{H})$, such that*

$$H = \big\{\{\mathcal{A}e, \mathcal{B}e\} : e \in \mathfrak{H}\big\} \tag{1.10.15}$$

holds with

$$\operatorname{Im}(\mathcal{A}^*\mathcal{B}) = 0 \quad \text{and} \quad (\mathcal{B} - \mu\mathcal{A})^{-1} \in \mathbf{B}(\mathfrak{H})$$

for some, and hence for all $\mu \in \mathbb{C}^+$ and for some, and hence for all $\mu \in \mathbb{C}^-$.

(ii) *H is maximal dissipative if and only if there exist $\mathcal{A}, \mathcal{B} \in \mathbf{B}(\mathfrak{H})$, such that* (1.10.15) *holds with*

$$\operatorname{Im}(\mathcal{A}^*\mathcal{B}) \geq 0 \quad \text{and} \quad (\mathcal{B} - \mu\mathcal{A})^{-1} \in \mathbf{B}(\mathfrak{H})$$

for some, and hence for all $\mu \in \mathbb{C}^-$.

(iii) *H is maximal accumulative if and only if there exist $\mathcal{A}, \mathcal{B} \in \mathbf{B}(\mathfrak{H})$, such that* (1.10.15) *holds with*

$$\operatorname{Im}(\mathcal{A}^*\mathcal{B}) \leq 0 \quad \text{and} \quad (\mathcal{B} - \mu\mathcal{A})^{-1} \in \mathbf{B}(\mathfrak{H})$$

for some, and hence for all $\mu \in \mathbb{C}^+$.

If $\mathcal{A}, \mathcal{B} \in \mathbf{B}(\mathfrak{H})$ are chosen such that also (1.10.7) *is satisfied, then H has the representation*

$$H = \big\{\{f, f'\} \in \mathfrak{H} \times \mathfrak{H} : \mathcal{B}f = \mathcal{A}f'\big\}.$$

Proof. It has been shown in Theorem 1.10.5 that a relation H is closed with $\mu \in \rho(H)$ if and only if it admits the representation (1.10.15) with $\mathcal{A}, \mathcal{B} \in \mathbf{B}(\mathfrak{H})$ such that $(\mathcal{B} - \mu\mathcal{A})^{-1} \in \mathbf{B}(\mathfrak{H})$. Note that when H is given in this way, then $\{f, f'\} \in H$ if and only if $\{f, f'\} = \{\mathcal{A}e, \mathcal{B}e\}$ for some $e \in \mathfrak{H}$. The identity

$$(f', f) = (\mathcal{B}e, \mathcal{A}e) = (\mathcal{A}^*\mathcal{B}e, e)$$

shows that H is dissipative, accumulative, or symmetric if and only if

$$\operatorname{Im}(\mathcal{A}^*\mathcal{B}) \geq 0, \quad \operatorname{Im}(\mathcal{A}^*\mathcal{B}) \leq 0, \quad \text{or} \quad \operatorname{Im}(\mathcal{A}^*\mathcal{B}) = 0, \quad \text{respectively.}$$

Furthermore, it is clear that H is maximal dissipative if $\mu \in \mathbb{C}^-$, maximal accumulative if $\mu \in \mathbb{C}^+$, and self-adjoint if $\mu \in \mathbb{C}^+$ and $\mu \in \mathbb{C}^-$. □

The next corollary provides a special representing pair $\{\mathcal{A}, \mathcal{B}\}$ for a self-adjoint relation H.

Corollary 1.10.9. *Let H be a relation in $\mathfrak{H}$. Then H is self-adjoint if and only if there exist $\mathcal{A}, \mathcal{B} \in \mathbf{B}(\mathfrak{H})$ such that*

$$\mathcal{A}^*\mathcal{B} = \mathcal{B}^*\mathcal{A}, \quad \mathcal{A}\mathcal{B}^* = \mathcal{B}\mathcal{A}^*, \quad \mathcal{A}^*\mathcal{A} + \mathcal{B}^*\mathcal{B} = I = \mathcal{A}\mathcal{A}^* + \mathcal{B}\mathcal{B}^*. \tag{1.10.16}$$

Proof. Assume that H is self-adjoint and define

$$\mathcal{A} = \frac{1}{2}(I - \mathcal{C}_{-i}[H]) \quad \text{and} \quad \mathcal{B} = \frac{1}{2}(i + i\mathcal{C}_{-i}[H]),$$

where $\mathcal{C}_{-i}[H]$ denotes the Cayley transform of H (with respect to the point $\mu = -i$) in Definition 1.1.13. A straightforward calculation using the identity $(\mathcal{C}_{-i}[H])^{-1} = C_i[H] = (\mathcal{C}_{-i}[H])^*$ (see Lemma 1.3.11) shows that the properties in (1.10.16) are satisfied.

Conversely, it suffices to remark that for $\mu = \pm i$

$$(\mathcal{B} + \mu\mathcal{A})^*(\mathcal{B} + \mu\mathcal{A}) = I = (\mathcal{B} + \mu\mathcal{A})(\mathcal{B} + \mu\mathcal{A})^*$$

follows from (1.10.16). This shows $(\mathcal{B} + \mu\mathcal{A})^{-1} \in \mathbf{B}(\mathfrak{H})$ for $\mu = \pm i$. Furthermore, the first condition in (1.10.16) shows $\operatorname{Im}(\mathcal{A}^*\mathcal{B}) = 0$ and now Corollary 1.10.8 (i) implies that H is self-adjoint. □

In Theorem 1.10.5 and afterwards special attention was paid to representations of H of the form (1.10.6) and (1.10.8) under the assumption that $\rho(H) \neq \emptyset$. In the next proposition this assumption is dropped.

Proposition 1.10.10. *Let the relation $H = \{\mathcal{A}, \mathcal{B}\}$ from $\mathfrak{H}$ to $\mathfrak{K}$ be given by (1.10.1) with $\mathcal{A} \in \mathbf{B}(\mathfrak{E}, \mathfrak{H})$, $\mathcal{B} \in \mathbf{B}(\mathfrak{E}, \mathfrak{K})$, and assume that*

$$\mathcal{A}^*\mathcal{A} + \mathcal{B}^*\mathcal{B} = I. \tag{1.10.17}$$

Then the adjoint H^* from $\mathfrak{K}$ to $\mathfrak{H}$ has the parametrization

$$H^* = \bigl\{\{(I - \mathcal{B}\mathcal{B}^*)\varphi + \mathcal{B}\mathcal{A}^*\psi, \, \mathcal{A}\mathcal{B}^*\varphi + (I - \mathcal{A}\mathcal{A}^*)\psi\} : \varphi \in \mathfrak{K}, \, \psi \in \mathfrak{H}\bigr\}.$$

Consequently, H is given by all $\{f, f'\} \in \mathfrak{H} \times \mathfrak{K}$ for which

$$(I - \mathcal{A}\mathcal{A}^*)f = \mathcal{A}\mathcal{B}^*f', \quad \mathcal{B}\mathcal{A}^*f = (I - \mathcal{B}\mathcal{B}^*)f'. \tag{1.10.18}$$

Proof. The assumption (1.10.17) and Proposition 1.10.3 imply that the relation $H = \{\mathcal{A}, \mathcal{B}\}$ is closed. Let $J\{h, k\} = \{k, -h\}$ be the flip-flop operator from $\mathfrak{H} \times \mathfrak{K}$ to $\mathfrak{K} \times \mathfrak{H}$ in (1.3.1). Then $JH = \{\mathcal{B}, -\mathcal{A}\}$ is a closed relation from $\mathfrak{K}$ to $\mathfrak{H}$. It follows from (1.10.17) that

$$\begin{pmatrix} \mathcal{B} \\ -\mathcal{A} \end{pmatrix}^* \begin{pmatrix} \mathcal{B} \\ -\mathcal{A} \end{pmatrix} = I, \quad \text{and hence ran} \begin{pmatrix} \mathcal{B} \\ -\mathcal{A} \end{pmatrix}^* = \mathfrak{E}.$$

This implies that the orthogonal projection P_{JH} in $\mathfrak{K} \times \mathfrak{H}$ onto JH has the form

$$P_{JH} = \begin{pmatrix} \mathcal{B} \\ -\mathcal{A} \end{pmatrix} \begin{pmatrix} \mathcal{B} \\ -\mathcal{A} \end{pmatrix}^* = \begin{pmatrix} \mathcal{B}\mathcal{B}^* & -\mathcal{B}\mathcal{A}^* \\ -\mathcal{A}\mathcal{B}^* & \mathcal{A}\mathcal{A}^* \end{pmatrix}.$$

Since $H^* = (JH)^\perp$ by (1.3.2), the orthogonal projection onto H^* is given by

$$P_{H^*} = I - P_{JH} = \begin{pmatrix} I - \mathcal{B}\mathcal{B}^* & \mathcal{B}\mathcal{A}^* \\ \mathcal{A}\mathcal{B}^* & I - \mathcal{A}\mathcal{A}^* \end{pmatrix},$$

and this leads to the form of H^* in the proposition. It then follows from (1.10.3) that $H^{**} = H$ consists of all $\{f, f'\} \in \mathfrak{H} \times \mathfrak{K}$ for which (1.10.18) holds. □

1.11 Resolvent operators with respect to a bounded operator

Many of the results in this chapter are phrased for $(H - \lambda)^{-1}$, where H is a relation in $\mathfrak{H}$ and $\lambda \in \mathbb{C}$. In the rest of this text there will be several occasions to use similar results phrased for $(H - R)^{-1}$ when $R \in \mathbf{B}(\mathfrak{H})$. A brief survey is offered.

For a relation H in $\mathfrak{H}$ recall that the difference $H - R$ is a well-defined relation in $\mathfrak{H}$ given by

$$H - R = \big\{\{h, h' - Rh\} : \{h, h'\} \in H\big\},$$

and that $H - R$ is closed whenever H is closed. It is clear that

$$\ker (H - R)^{-1} = \operatorname{mul} (H - R) = \operatorname{mul} H. \tag{1.11.1}$$

The next lemma is a variant of Lemma 1.1.8 in the present context. The proof is not repeated.

Lemma 1.11.1. *Let H be a relation in $\mathfrak{H}$. If $R \in \mathbf{B}(\mathfrak{H})$ and* $\ker (H - R) = \{0\}$, *then*

$$H = \big\{\{(H - R)^{-1} f, (I + R(H - R)^{-1}) f\} : f \in \operatorname{ran} (H - R)\big\}.$$

The next proposition is concerned with the *resolvent identity* as in Proposition 1.1.7, but in the present context.

Lemma 1.11.2. *Let H be a relation in $\mathfrak{H}$ and let $R, S \in \mathbf{B}(\mathfrak{H})$. Then*

$$(H-R)^{-1}-(H-S)^{-1}=(H-R)^{-1}(R-S)(H-S)^{-1}. \tag{1.11.2}$$

If $\ker(H-R)=\{0\}$ *and* $\ker(H-S)=\{0\}$, *then* $(H-R)^{-1}$ *and* $(H-S)^{-1}$ *are linear operators with the same kernel* $\operatorname{mul} H$.

Proof. For the inclusion ($\subset$), let $\{h, h'-h''\} \in (H-R)^{-1}-(H-S)^{-1}$ with $\{h,h'\} \in (H-R)^{-1}$ and $\{h,h''\} \in (H-S)^{-1}$. This gives

$$\{h', h+Rh'\} \in H \quad \text{and} \quad \{h'', h+Sh''\} \in H,$$

which shows that $\{h'-h'', Rh'-Sh''\} \in H$, and thus

$$\{(R-S)h'', h'-h''\} \in (H-R)^{-1}.$$

Since $\{h,h''\} \in (H-S)^{-1}$ and $\{h'', (R-S)h''\} \in R-S$, one concludes that $\{h,(R-S)h''\} \in (R-S)(H-S)^{-1}$. Hence, the element $\{h, h'-h''\}$ belongs to the relation $(H-R)^{-1}(R-S)(H-S)^{-1}$, which shows the inclusion.

For the inclusion ($\supset$), let $\{h,h'\} \in (H-R)^{-1}(R-S)(H-S)^{-1}$. Then by definition there exists $k \in \mathfrak{H}$ such that

$$\{h,k\} \in (H-S)^{-1} \quad \text{and} \quad \{(R-S)k, h'\} \in (H-R)^{-1},$$

as $\{k,(R-S)k\} \in R-S$. It is clear from $\{k,h\} \in H-S$ that

$$\{h+(S-R)k, k\} \in (H-R)^{-1}.$$

Thus, it follows that $\{h, h'+k\} \in (H-R)^{-1}$. Hence, $\{h,h'\} = \{h, h'+k-k\}$ belongs to $(H-R)^{-1}-(H-S)^{-1}$, which shows the inclusion.

The last statements follow directly from (1.11.1). □

Observe that if $(H-R)^{-1}$ and $(H-S)^{-1}$ belong to $\mathbf{B}(\mathfrak{H})$, then the resolvent identity (1.11.2) involves only operators from $\mathbf{B}(\mathfrak{H})$ defined on all of $\mathfrak{H}$. Hence, the following lemma can be verified by direct computation.

Lemma 1.11.3. *Let H be a closed relation in $\mathfrak{H}$, let $R, S \in \mathbf{B}(\mathfrak{H})$, and assume that*

$$(H-R)^{-1} \quad \textit{and} \quad (H-S)^{-1} \in \mathbf{B}(\mathfrak{H}).$$

Then the operator $I+(H-R)^{-1}(R-S) \in \mathbf{B}(\mathfrak{H})$ is boundedly invertible, with inverse given by

$$\big[I+(H-R)^{-1}(R-S)\big]^{-1} = I-(H-S)^{-1}(R-S). \tag{1.11.3}$$

Likewise, the operator $I-(R-S)(H-S)^{-1} \in \mathbf{B}(\mathfrak{H})$ is boundedly invertible, with inverse given by

$$\big[I-(R-S)(H-S)^{-1}\big]^{-1} = I+(R-S)(H-R)^{-1}. \tag{1.11.4}$$

Under the conditions of Lemma 1.11.3 it follows by rewriting the resolvent identity (1.11.2) that

$$(H - R)^{-1} = \big[I + (H - R)^{-1}(R - S)\big](H - S)^{-1} \tag{1.11.5}$$

and

$$(H - R)^{-1}\big[I - (R - S)(H - S)^{-1}\big] = (H - S)^{-1}; \tag{1.11.6}$$

here the factors are bounded and boundedly invertible by Lemma 1.11.3. Hence, the identities (1.11.3) and (1.11.4) lead to the following useful result, expressing the resolvent difference $(H_1 - R)^{-1} - (H_2 - R)^{-1}$ in terms of the resolvent difference $(H_1 - S)^{-1} - (H_2 - S)^{-1}$.

Lemma 1.11.4. *Let H_1 and H_2 be closed relations in $\mathfrak{H}$, and let R and S be operators in $\mathbf{B}(\mathfrak{H})$. For $i = 1, 2$ assume that $(H_i - R)^{-1}$ and $(H_i - S)^{-1}$ belong to $\mathbf{B}(\mathfrak{H})$. Then the bounded operators*

$$I - (H_1 - S)^{-1}(R - S), \quad I - (R - S)(H_2 - S)^{-1}$$

are boundedly invertible, and

$$\begin{aligned}
&(H_1 - R)^{-1} - (H_2 - R)^{-1} \\
&\quad = \big[I - (H_1 - S)^{-1}(R - S)\big]^{-1} \\
&\qquad\qquad \big[(H_1 - S)^{-1} - (H_2 - S)^{-1}\big]\big[I - (R - S)(H_2 - S)^{-1}\big]^{-1}.
\end{aligned}$$

Proof. It follows from the identities (1.11.5)–(1.11.6) and Lemma 1.11.3 that

$$(H_1 - R)^{-1} = \big[I - (H_1 - S)^{-1}(R - S)\big]^{-1}(H_1 - S)^{-1}$$

and

$$(H_2 - R)^{-1} = (H_2 - S)^{-1}\big[I - (R - S)(H_2 - S)^{-1}\big]^{-1}.$$

Subtracting these identities yields the desired result. □

The question arises for what relations H in $\mathfrak{H}$ and $R \in \mathbf{B}(\mathfrak{H})$ one can conclude that $(H - R)^{-1} \in \mathbf{B}(\mathfrak{H})$. The following lemma presents some sufficient conditions.

Lemma 1.11.5. *Let H be a closed relation in $\mathfrak{H}$ and let $R \in \mathbf{B}(\mathfrak{H})$ with $\operatorname{Im} R \geq \varepsilon$ for some $\varepsilon > 0$. Then the following statements hold:*

(i) *if H is maximal accumulative, then $(H - R)^{-1} \in \mathbf{B}(\mathfrak{H})$ is dissipative;*
(ii) *if H is maximal dissipative, then $(H - R^*)^{-1} \in \mathbf{B}(\mathfrak{H})$ is accumulative;*
(iii) *if H is self-adjoint, then $(H - R)^{-1} \in \mathbf{B}(\mathfrak{H})$ and $(H - R^*)^{-1} \in \mathbf{B}(\mathfrak{H})$ are accumulative and dissipative, respectively.*

Proof. (i) Since the relation H is maximal accumulative, it follows that H is closed, which implies that the relation $(H-R)^{-1}$ is closed. In order to show that $(H-R)^{-1}$ is a bounded operator, let $\{f,f'\} \in (H-R)^{-1}$. Then $\{f', f+Rf'\} \in H$ and, since H is accumulative and $\operatorname{Im} R \geq \varepsilon$, this shows that

$$\operatorname{Im}(f,f') + \varepsilon\|f'\|^2 \leq \operatorname{Im}(f,f') + ((\operatorname{Im} R)f',f') = \operatorname{Im}(f+Rf',f') \leq 0, \quad (1.11.7)$$

which leads to

$$\varepsilon\|f'\|^2 \leq -\operatorname{Im}(f,f') = \operatorname{Im}(f',f) \leq \|f'\|\,\|f\|.$$

This implies that the closed relation $(H-R)^{-1}$ is a bounded operator. Note also that (1.11.7) implies $\operatorname{Im}(f,f') \leq 0$ for $\{f,f'\} \in (H-R)^{-1}$. Hence, $\operatorname{Im}(f',f) \geq 0$ and $(H-R)^{-1}$ is dissipative.

To show that $(H-R)^{-1} \in \mathbf{B}(\mathfrak{H})$ it therefore suffices to verify that $\operatorname{ran}(H-R)$ is dense in $\mathfrak{H}$. Note that $(\operatorname{ran}(H-R))^{\perp} = \ker(H^*-R^*)$ by Proposition 1.3.2 and Proposition 1.3.9. Now observe that $\ker(H^*-R^*) = \{0\}$. To see this, assume that $f \in \ker(H^*-R^*)$ or, equivalently, $\{f, R^*f\} \in H^*$. Since H^* is maximal dissipative by Proposition 1.6.7 one obtains that

$$0 \leq \operatorname{Im}(R^*f,f) = \operatorname{Im}(f,Rf) = -((\operatorname{Im} R)f,f) \leq -\varepsilon\|f\|^2,$$

which gives $f=0$.

(ii) & (iii) The proofs are similar. □

Let H be a closed relation in $\mathfrak{H}$ and let $\mathcal{A} \in \mathbf{B}(\mathfrak{E},\mathfrak{H})$, $\mathcal{B} \in \mathbf{B}(\mathfrak{E},\mathfrak{H})$ be a tight representing pair for H, that is,

$$H = \bigl\{\{\mathcal{A}e, \mathcal{B}e\} : e \in \mathfrak{E}\bigr\} \quad (1.11.8)$$

and $\ker \mathcal{A} \cap \ker \mathcal{B} = \{0\}$; cf. Theorem 1.10.1 and Proposition 1.10.3. Note that if for some $\mu \in \mathbb{C}$ one has $(\mathcal{B}-\mu\mathcal{A})^{-1} \in \mathbf{B}(\mathfrak{E})$, then the tightness condition $\ker \mathcal{A} \cap \ker \mathcal{B} = \{0\}$ is automatically satisfied.

Lemma 1.11.6. *Let H be a closed relation in $\mathfrak{H}$ and assume that H has the tight representation (1.11.8), where $\mathcal{A},\mathcal{B} \in \mathbf{B}(\mathfrak{E},\mathfrak{H})$. Then for any $R \in \mathbf{B}(\mathfrak{H})$ one has that*

$$(H-R)^{-1} \in \mathbf{B}(\mathfrak{H}) \quad \Leftrightarrow \quad (\mathcal{B}-R\mathcal{A})^{-1} \in \mathbf{B}(\mathfrak{H},\mathfrak{E}), \quad (1.11.9)$$

in which case

$$(H-R)^{-1} = \mathcal{A}(\mathcal{B}-R\mathcal{A})^{-1}. \quad (1.11.10)$$

Proof. One sees by the definition of $H-R$ that

$$H-R = \bigl\{\{\mathcal{A}e, (\mathcal{B}-R\mathcal{A})e\} : e \in \mathfrak{E}\bigr\}, \quad (1.11.11)$$

and thus it follows directly that

$$\operatorname{ran}(H-R) = \operatorname{ran}(\mathcal{B}-R\mathcal{A}) \quad \text{and} \quad \ker(H-R) = \mathcal{A}\ker(\mathcal{B}-R\mathcal{A}).$$

Furthermore, it is clear that in general

$$\ker(\mathcal{B} - R\mathcal{A}) = \{0\} \quad \Rightarrow \quad \ker(H - R) = \{0\}.$$

Moreover, under the assumption $\ker \mathcal{A} \cap \ker \mathcal{B} = \{0\}$ one concludes that

$$\ker(H - R) = \{0\} \quad \Rightarrow \quad \ker(\mathcal{B} - R\mathcal{A}) = \{0\}.$$

To see this, let $h \in \ker(\mathcal{B} - R\mathcal{A})$, which means that $\mathcal{B}h = R\mathcal{A}h$. Due to

$$\ker(H - R) = \mathcal{A} \ker(\mathcal{B} - R\mathcal{A})$$

one has $\mathcal{A}h \in \ker(H - R) = \{0\}$ and thus also $\mathcal{B}h = 0$. The tightness condition now implies that $h = 0$.

In order to prove the equivalence (1.11.9), assume that H has the tight representation (1.11.8). Then it is clear that $\ker(H - R) = \{0\}$ and $\operatorname{ran}(H - R) = \mathfrak{H}$ if and only if $\ker(\mathcal{B} - R\mathcal{A}) = \{0\}$ and $\operatorname{ran}(\mathcal{B} - R\mathcal{A}) = \mathfrak{H}$. This implies (1.11.9). Moreover, (1.11.11) yields (1.11.10). □

1.12 Nevanlinna families and their representations

Let $\mathfrak{H}$ be a Hilbert space and let $N : \mathbb{C} \setminus \mathbb{R} \to \mathbf{B}(\mathfrak{H})$ be a holomorphic function. Then N is a Nevanlinna function (or $\mathbf{B}(\mathfrak{H})$-valued Nevanlinna function) if

$$(\operatorname{Im} \lambda)(\operatorname{Im} N(\lambda)) \geq 0, \qquad \lambda \in \mathbb{C} \setminus \mathbb{R}, \tag{1.12.1}$$

and N satisfies the symmetry condition $N(\lambda) = N(\bar{\lambda})^*$ for all $\lambda \in \mathbb{C} \setminus \mathbb{R}$; see Definition A.4.1. If, in addition, the imaginary part $\operatorname{Im} N(\lambda)$ is boundedly invertible for some, and hence for all $\lambda \in \mathbb{C} \setminus \mathbb{R}$, then the Nevanlinna function N is said to be uniformly strict; cf. Definition A.4.7. Observe that by (1.12.1) the operators $N(\lambda) \in \mathbf{B}(\mathfrak{H})$ are dissipative (accumulative) for $\lambda \in \mathbb{C}^+$ ($\lambda \in \mathbb{C}^-$). In this section the notion of a Nevanlinna function is extended to a so-called Nevanlinna family, that is, a family of relations $Z(\lambda)$, $\lambda \in \mathbb{C} \setminus \mathbb{R}$, in $\mathfrak{H}$ which are maximal dissipative or maximal accumulative for $\lambda \in \mathbb{C}^+$ or $\mathbb{C}^-$, respectively, and satisfy a symmetry condition and a holomorphy condition.

Definition 1.12.1. A family of relations $Z(\lambda)$, $\lambda \in \mathbb{C} \setminus \mathbb{R}$, in $\mathfrak{H}$ is called a *Nevanlinna family* if the following conditions are satisfied:

(i) $Z(\lambda)$ is maximal dissipative (maximal accumulative) for $\lambda \in \mathbb{C}^+$ ($\lambda \in \mathbb{C}^-$);

(ii) $Z(\lambda) = Z(\bar{\lambda})^*$, $\lambda \in \mathbb{C} \setminus \mathbb{R}$;

(iii) there exists $\mu \in \mathbb{C}^+$ such that $\lambda \mapsto (Z(\lambda) + \mu)^{-1}$ is holomorphic on $\mathbb{C}^+$ and $\lambda \mapsto (Z(\lambda) + \bar{\mu})^{-1}$ is holomorphic on $\mathbb{C}^-$.

Note that condition (i) in this definition and Theorem 1.6.4 ensure that

$$\mathbb{C}^- \subset \rho(Z(\lambda)), \ \lambda \in \mathbb{C}^+, \qquad \text{and} \qquad \mathbb{C}^+ \subset \rho(Z(\lambda)), \ \lambda \in \mathbb{C}^-. \tag{1.12.2}$$

In particular, one has $(Z(\lambda)+\mu)^{-1} \in \mathbf{B}(\mathfrak{H})$ in (iii). The condition $Z(\lambda) = Z(\bar{\lambda})^*$ in (ii) leads to the following conclusions. First of all, it follows from Proposition 1.6.7 that $Z(\lambda)$ is maximal accumulative for $\lambda \in \mathbb{C}^-$ if and only if $Z(\lambda)$ is maximal dissipative for $\lambda \in \mathbb{C}^+$. Secondly, $\lambda \mapsto (Z(\lambda) + \bar{\mu})^{-1}$ is holomorphic on $\mathbb{C}^-$ if and only if $\lambda \mapsto (Z(\lambda) + \mu)^{-1}$ is holomorphic on $\mathbb{C}^+$; this follows from the fact that for a $\mathbf{B}(\mathfrak{H})$-valued function H one has that $\lambda \mapsto H(\lambda)$ is holomorphic if and only if $\bar{\lambda} \mapsto H(\bar{\lambda})^*$ is holomorphic. Furthermore, each element $Z(\lambda)$ is a closed relation in $\mathfrak{H}$ and therefore it has a tight operator representation by Theorem 1.10.1 and Proposition 1.10.3. In particular, one has the following general representation result.

Proposition 1.12.2. *Let $Z(\lambda)$, $\lambda \in \mathbb{C} \setminus \mathbb{R}$, be a Nevanlinna family in $\mathfrak{H}$. Then $Z(\lambda)$ has the tight representation*

$$Z(\lambda) = \big\{\{A(\lambda)h, B(\lambda)h\} : h \in \mathfrak{H}\big\}, \quad \lambda \in \mathbb{C} \setminus \mathbb{R}. \tag{1.12.3}$$

Here $\{A, B\}$ is a pair of $\mathbf{B}(\mathfrak{H})$-valued functions on $\mathbb{C} \setminus \mathbb{R}$ which satisfies:

(a) *the mappings $\lambda \mapsto A(\lambda)$ and $\lambda \mapsto B(\lambda)$ are holomorphic on $\mathbb{C} \setminus \mathbb{R}$;*

(b) $(\operatorname{Im} \lambda) \operatorname{Im} (A(\lambda)^* B(\lambda)) \geq 0$, $\lambda \in \mathbb{C} \setminus \mathbb{R}$;

(c) $A(\bar{\lambda})^* B(\lambda) = B(\bar{\lambda})^* A(\lambda)$, $\lambda \in \mathbb{C} \setminus \mathbb{R}$;

(d) *there exists $\mu \in \mathbb{C}^+$ such that $(B(\lambda) + \mu A(\lambda))^{-1} \in \mathbf{B}(\mathfrak{H})$ for $\lambda \in \mathbb{C}^+$ and $(B(\lambda) + \bar{\mu} A(\lambda))^{-1} \in \mathbf{B}(\mathfrak{H})$ for $\lambda \in \mathbb{C}^-$.*

If the pair $\{C, D\}$ is another tight representation of the Nevanlinna family $Z(\lambda)$, $\lambda \in \mathbb{C} \setminus \mathbb{R}$, with the above properties, then there exists a bounded and boundedly invertible holomorphic operator family $X(\lambda)$, $\lambda \in \mathbb{C} \setminus \mathbb{R}$, such that

$$C(\lambda) = A(\lambda)X(\lambda), \quad D(\lambda) = B(\lambda)X(\lambda), \quad \lambda \in \mathbb{C} \setminus \mathbb{R}. \tag{1.12.4}$$

Proof. Let $Z(\lambda)$, $\lambda \in \mathbb{C} \setminus \mathbb{R}$, be a Nevanlinna family, choose $\mu \in \mathbb{C}^+$ as in Definition 1.12.1, and define $A(\lambda)$ and $B(\lambda)$ by

$$A(\lambda) = \begin{cases} (Z(\lambda) + \mu)^{-1}, & \lambda \in \mathbb{C}^+, \\ (Z(\lambda) + \bar{\mu})^{-1}, & \lambda \in \mathbb{C}^-, \end{cases} \tag{1.12.5}$$

and

$$B(\lambda) = \begin{cases} I - \mu(Z(\lambda) + \mu)^{-1}, & \lambda \in \mathbb{C}^+, \\ I - \bar{\mu}(Z(\lambda) + \bar{\mu})^{-1}, & \lambda \in \mathbb{C}^-. \end{cases} \tag{1.12.6}$$

Then it follows from (1.12.2) that $A(\lambda)$ and $B(\lambda)$ belong to $\mathbf{B}(\mathfrak{H})$, and Lemma 1.2.4 shows that $Z(\lambda)$ has the representation (1.12.3). Definition 1.12.1 (iii) implies that the mappings $\lambda \mapsto A(\lambda)$ and $\lambda \mapsto B(\lambda)$ are holomorphic, which shows (a). Furthermore, it follows from (1.12.5) and (1.12.6) that $B(\lambda)+\mu A(\lambda) = I$, $\lambda \in \mathbb{C}^+$, and $B(\lambda) + \bar{\mu}A(\lambda) = I$, $\lambda \in \mathbb{C}^-$, which shows (d). Since by (i) $Z(\lambda)$ is dissipative (accumulative) for $\lambda \in \mathbb{C}^+$ ($\lambda \in \mathbb{C}^-$) it follows from

$$\operatorname{Im}\big(A(\lambda)^*B(\lambda)h, h\big) = \operatorname{Im}\big(B(\lambda)h, A(\lambda)h\big), \quad \lambda \in \mathbb{C}\setminus\mathbb{R}, \quad h \in \mathfrak{H}, \tag{1.12.7}$$

that (b) holds. It is a direct consequence of (1.12.3) that

$$Z(\bar{\lambda})^* = \big\{\{k, k'\} \in \mathfrak{H} \times \mathfrak{H} : B(\bar{\lambda})^*k = A(\bar{\lambda})^*k'\big\};$$

cf. (1.10.2) and (1.10.3). Since by (ii) $Z(\lambda) = Z(\bar{\lambda})^*$ it follows from (1.12.3) that (c) holds.

The representation in (1.12.3) with $A(\lambda)$ and $B(\lambda)$ in (1.12.5)–(1.12.6) is tight for each $\lambda \in \mathbb{C}\setminus\mathbb{R}$, that is,

$$\ker A(\lambda) \cap \ker B(\lambda) = \{0\}, \qquad \lambda \in \mathbb{C}\setminus\mathbb{R}.$$

In order to see this, assume that $A(\lambda)g = 0$ and $B(\lambda)g = 0$ for some $\lambda \in \mathbb{C}^+$ with some $g \in \mathfrak{H}$. Then $(B(\lambda) + \mu A(\lambda))g = 0$ and hence (d) implies that $g = 0$. Likewise, the same conclusion holds when $\lambda \in \mathbb{C}^-$.

Now assume that $\{C, D\}$ is another tight representation of the same Nevanlinna family $Z(\lambda)$, $\lambda \in \mathbb{C}\setminus\mathbb{R}$, i.e., assume that $Z(\lambda)$ is also given by

$$Z(\lambda) = \big\{\{C(\lambda)h, D(\lambda)h\} : h \in \mathfrak{H}\big\}.$$

It follows from Lemma 1.10.2 that there exist bounded bijective operators $X(\lambda)$, $\lambda \in \mathbb{C}\setminus\mathbb{R}$, such that (1.12.4) holds. In particular, for $\lambda \in \mathbb{C}^+$ one has

$$D(\lambda) + \mu C(\lambda) = \big(B(\lambda) + \mu A(\lambda)\big)X(\lambda)$$

and hence the function

$$\lambda \mapsto X(\lambda) = \big(B(\lambda) + \mu A(\lambda)\big)^{-1}\big(D(\lambda) + \mu C(\lambda)\big), \quad \lambda \in \mathbb{C}^+,$$

is holomorphic on $\mathbb{C}^+$. Similarly, on $\mathbb{C}^-$ the function X has the form

$$\lambda \mapsto X(\lambda) = \big(B(\lambda) + \bar{\mu}A(\lambda)\big)^{-1}\big(D(\lambda) + \bar{\mu}C(\lambda)\big), \quad \lambda \in \mathbb{C}^-,$$

and is holomorphic. □

Definition 1.12.3. Let $\{A, B\}$ be a pair of $\mathbf{B}(\mathfrak{H})$-valued functions on $\mathbb{C}\setminus\mathbb{R}$. Then $\{A, B\}$ is called a *Nevanlinna pair* if it satisfies the properties (a), (b), (c), and (d) in Proposition 1.12.2.

Hence, by Proposition 1.12.2 each Nevanlinna family is represented by a Nevanlinna pair. The converse is also true: each Nevanlinna pair defines a Nevanlinna family.

Proposition 1.12.4. *Let $\{A, B\}$ be a Nevanlinna pair in $\mathfrak{H}$. Then $Z(\lambda)$, $\lambda \in \mathbb{C} \setminus \mathbb{R}$, defined by (1.12.3) is a Nevanlinna family in $\mathfrak{H}$.*

Proof. Let $\{A, B\}$ be a Nevanlinna pair and define the family $Z(\lambda)$, $\lambda \in \mathbb{C} \setminus \mathbb{R}$, by (1.12.3). Then (b) implies that $Z(\lambda)$ is dissipative (accumulative) for $\lambda \in \mathbb{C}^+$ ($\lambda \in \mathbb{C}^-$); cf. (1.12.7). It follows from (d) and the definition of $Z(\lambda)$ that one has

$$\begin{aligned}(Z(\lambda) + \mu)^{-1} &= A(\lambda)\big(B(\lambda) + \mu A(\lambda)\big)^{-1} \in \mathbf{B}(\mathfrak{H}), \quad \lambda \in \mathbb{C}^+,\\ (Z(\lambda) + \bar{\mu})^{-1} &= A(\lambda)\big(B(\lambda) + \bar{\mu} A(\lambda)\big)^{-1} \in \mathbf{B}(\mathfrak{H}), \quad \lambda \in \mathbb{C}^-,\end{aligned} \tag{1.12.8}$$

for some $\mu \in \mathbb{C}^+$. With (a) this shows that the holomorphy condition (iii) is satisfied. Moreover, from (1.12.8) and Theorem 1.6.4 it is now clear that $Z(\lambda)$ is maximal dissipative (maximal accumulative) for $\lambda \in \mathbb{C}^+$ ($\lambda \in \mathbb{C}^-$), which is (i). From (c) one concludes

$$A(\lambda)^*\big(B(\bar{\lambda}) + \bar{\mu} A(\bar{\lambda})\big) = \big(B(\lambda)^* + \bar{\mu} A(\lambda)^*\big)A(\bar{\lambda}), \quad \lambda \in \mathbb{C} \setminus \mathbb{R}.$$

For $\lambda \in \mathbb{C}^+$ and $\mu \in \mathbb{C}^+$ this reads

$$\big(B(\lambda)^* + \bar{\mu} A(\lambda)^*\big)^{-1} A(\lambda)^* = A(\bar{\lambda})\big(B(\bar{\lambda}) + \bar{\mu} A(\bar{\lambda})\big)^{-1},$$

so that

$$\begin{aligned}(Z(\lambda) + \mu)^{-*} &= \big(B(\lambda)^* + \bar{\mu} A(\lambda)^*\big)^{-1} A(\lambda)^*\\ &= A(\bar{\lambda})\big(B(\bar{\lambda}) + \bar{\mu} A(\bar{\lambda})\big)^{-1} = (Z(\bar{\lambda}) + \bar{\mu})^{-1}.\end{aligned}$$

However, the left-hand side is equal to $(Z(\lambda)^* + \bar{\mu})^{-1}$, and hence it follows that $Z(\lambda)^* = Z(\bar{\lambda})$ for $\lambda \in \mathbb{C}^+$. A similar reasoning is valid for $\lambda \in \mathbb{C}^-$. Hence, (ii) follows and therefore $Z(\lambda)$, $\lambda \in \mathbb{C} \setminus \mathbb{R}$, defined by (1.12.3) is a Nevanlinna family in $\mathfrak{H}$. □

In the next lemma it turns out that the conditions (iii) in the definition of a Nevanlinna family and the conditions (d) in the definition of a Nevanlinna pair hold for all $\mu \in \mathbb{C} \setminus \mathbb{R}$.

Lemma 1.12.5. *Let $Z(\lambda)$, $\lambda \in \mathbb{C} \setminus \mathbb{R}$, be a Nevanlinna family in $\mathfrak{H}$ and let $\{A, B\}$ be a Nevanlinna pair in $\mathfrak{H}$. Then the following statements hold:*

(i) *for all $\mu \in \mathbb{C}^+$ the mapping $\lambda \mapsto (Z(\lambda) + \mu)^{-1}$ is holomorphic on $\mathbb{C}^+$ and for all $\mu \in \mathbb{C}^-$ the mapping $\lambda \mapsto (Z(\lambda) + \mu)^{-1}$ is holomorphic on $\mathbb{C}^-$;*

(ii) *for all $\mu \in \mathbb{C}^+$ one has $(B(\lambda) + \mu A(\lambda))^{-1} \in \mathbf{B}(\mathfrak{H})$ on $\mathbb{C}^+$ and for all $\mu \in \mathbb{C}^-$ one has $(B(\lambda) + \mu A(\lambda))^{-1} \in \mathbf{B}(\mathfrak{H})$ on $\mathbb{C}^-$.*

Proof. (i) Assume that $Z(\lambda)$ satisfies (i) in Definition 1.12.1, so that, (1.12.2) holds. Fix ν, μ in the same half-plane as $\lambda \in \mathbb{C} \setminus \mathbb{R}$. Then one has $-\nu \in \rho(Z(\lambda))$ and $-\mu \in \rho(Z(\lambda))$ According to the resolvent formula one has

$$\begin{aligned}(Z(\lambda) + \mu)^{-1} - (Z(\lambda) + \nu)^{-1} &= (\nu - \mu)(Z(\lambda) + \nu)^{-1}(Z(\lambda) + \mu)^{-1}\\ &= (\nu - \mu)(Z(\lambda) + \mu)^{-1}(Z(\lambda) + \nu)^{-1}.\end{aligned} \tag{1.12.9}$$

Assume that for some $\mu \in \mathbb{C}^+$ the mapping $\lambda \mapsto (Z(\lambda) + \mu)^{-1}$ is holomorphic on $\mathbb{C}^+$. Let $\nu \in \mathbb{C}^+$. Then it follows from the resolvent formula (1.12.9) that

$$(Z(\lambda) + \mu)^{-1} = \big(I + (\nu - \mu)(Z(\lambda) + \mu)^{-1}\big)(Z(\lambda) + \nu)^{-1}.$$

The factor $I + (\nu - \mu)(Z(\lambda) + \mu)^{-1}$ is holomorphic in $\lambda \in \mathbb{C}^+$ and according to Lemma 1.6.10 it is boundedly invertible. Hence, its inverse is also holomorphic in $\lambda \in \mathbb{C}^+$ and one obtains for $\lambda \in \mathbb{C}^+$

$$(Z(\lambda) + \nu)^{-1} = \big(I + (\nu - \mu)(Z(\lambda) + \mu)^{-1}\big)^{-1}(Z(\lambda) + \mu)^{-1}.$$

Hence, $\lambda \mapsto (Z(\lambda) + \nu)^{-1}$ is holomorphic on $\mathbb{C}^+$. The corresponding statement for the half-plane $\mathbb{C}^-$ follows from the symmetry of the Nevanlinna family.

(ii) Assume that for some $\mu \in \mathbb{C}^+$ one has $(B(\lambda) + \mu A(\lambda))^{-1} \in \mathbf{B}(\mathfrak{H})$. Define $Z(\lambda)$, $\lambda \in \mathbb{C}^+$, by (1.12.3), and note that this representation is tight. Then $Z(\lambda)$, $\lambda \in \mathbb{C}^+$, is maximal dissipative and hence $(Z(\lambda) + \mu)^{-1} \in \mathbf{B}(\mathfrak{H})$ for all $\mu \in \mathbb{C}^+$. Now Lemma 1.11.6 yields $(B(\lambda) + \mu A(\lambda))^{-1} \in \mathbf{B}(\mathfrak{H})$ for all $\mu \in \mathbb{C}^+$. A similar reasoning holds for $\mu \in \mathbb{C}^-$. □

Now the conditions (iii) in Definition 1.12.1 and, likewise, the conditions (d) in Definition 1.12.3 will be further relaxed in a useful way.

Proposition 1.12.6. *Let $Z(\lambda)$, $\lambda \in \mathbb{C} \setminus \mathbb{R}$, be a Nevanlinna family and let $\{A, B\}$ be a Nevanlinna pair in $\mathfrak{H}$ such that the representation* (1.12.3) *holds. Let N be a uniformly strict Nevanlinna function with values in $\mathbf{B}(\mathfrak{H})$. Then the conditions in* (iii) *in Definition 1.12.1 may be replaced by*

$$\lambda \mapsto \big(Z(\lambda) + N(\lambda)\big)^{-1} \textit{is holomorphic on } \mathbb{C} \setminus \mathbb{R} \textit{ with values in } \mathbf{B}(\mathfrak{H}).$$

Moreover, the conditions in (d) *in Definition 1.12.3 may be replaced by*

$$\big(B(\lambda) + N(\lambda)A(\lambda)\big)^{-1} \in \mathbf{B}(\mathfrak{H}), \quad \lambda \in \mathbb{C} \setminus \mathbb{R}.$$

In particular, the choice $N(\lambda) = \lambda$ is allowed for these statements. Moreover,

$$-\big(Z(\lambda) + N(\lambda)\big)^{-1} = -A(\lambda)\big(B(\lambda) + N(\lambda)A(\lambda)\big)^{-1}, \quad \lambda \in \mathbb{C} \setminus \mathbb{R}, \tag{1.12.10}$$

defines a Nevanlinna function with values in $\mathbf{B}(\mathfrak{H})$.

Proof. The proof will be given in three steps. In Step 1 it is shown how condition (iii) in Definition 1.12.1 and condition (d) in Definition 1.12.3 give rise to the stated conditions in the proposition. In Step 2 and Step 3 the reverse direction is traversed for Nevanlinna families and Nevanlinna pairs, respectively.

Step 1. Let $Z(\lambda)$ be a Nevanlinna family in $\mathfrak{H}$ and let $\{A, B\}$ be a Nevanlinna pair in $\mathfrak{H}$ as in Definition 1.12.1 and Definition 1.12.3, respectively, such that (1.12.3) holds. Then for $\lambda \in \mathbb{C} \setminus \mathbb{R}$

$$\big(Z(\lambda) + N(\lambda)\big)^{-1} \in \mathbf{B}(\mathfrak{H}) \quad \text{and} \quad \big(B(\lambda) + N(\lambda)A(\lambda)\big)^{-1} \in \mathbf{B}(\mathfrak{H}), \tag{1.12.11}$$

and the identity (1.12.10) holds and defines a Nevanlinna function with values in $\mathbf{B}(\mathfrak{H})$. In fact, for $\lambda \in \mathbb{C}^+$ the first assertion in (1.12.11) follows from Lemma 1.11.5 (i) with $H = -Z(\lambda)$ and $R = N(\lambda)$. Similarly, for $\lambda \in \mathbb{C}^-$ Lemma 1.11.5 (ii) yields the first assertion in (1.12.11). The second assertion in (1.12.11) and the identity in (1.12.10) follow from Lemma 1.11.6. Since the functions A, B, and N are holomorphic, the identity in (1.12.10) defines a holomorphic function and Lemma 1.11.5 implies

$$(\operatorname{Im}\lambda)\operatorname{Im}\big(-(Z(\lambda)+N(\lambda))^{-1}h,h\big)\geq 0, \qquad \lambda \in \mathbb{C}\setminus\mathbb{R},\ h \in \mathfrak{H}.$$

Furthermore, for $\lambda \in \mathbb{C}\setminus\mathbb{R}$ one has $-(Z(\bar\lambda)+N(\bar\lambda))^{-*} = -(Z(\lambda)+N(\lambda))^{-1}$ by Definition 1.12.1 (ii) and the fact that N is a Nevanlinna function. Now it follows that the function in (1.12.10) is a Nevanlinna function with values in $\mathbf{B}(\mathfrak{H})$.

Step 2. Let $Z(\lambda)$, $\lambda \in \mathbb{C}\setminus\mathbb{R}$, satisfy (i) and (ii) of Definition 1.12.1, and assume that

$$\lambda \mapsto \big(Z(\lambda)+N(\lambda)\big)^{-1} \text{ is holomorphic with values in } \mathbf{B}(\mathfrak{H}).$$

Define $A(\lambda)$ and $B(\lambda)$ for $\lambda \in \mathbb{C}\setminus\mathbb{R}$ by

$$A(\lambda) = \big(Z(\lambda)+N(\lambda)\big)^{-1} \quad\text{and}\quad B(\lambda) = I - N(\lambda)\big(Z(\lambda)+N(\lambda)\big)^{-1};$$

then it follows from Lemma 1.2.4 that the family $Z(\lambda)$, $\lambda \in \mathbb{C}\setminus\mathbb{R}$, has the representation (1.12.3). Note that by assumption $A(\lambda)$ and $B(\lambda)$ belong to $\mathbf{B}(\mathfrak{H})$ and that each of the mappings $\lambda \mapsto A(\lambda)$ and $\lambda \mapsto B(\lambda)$ is holomorphic. Since $Z(\lambda)$ is maximal dissipative (maximal accumulative) for $\lambda \in \mathbb{C}^+$ ($\lambda \in \mathbb{C}^-$) it follows from Lemma 1.11.6 that for $\mu \in \mathbb{C}^+$ the operator $(B(\lambda)+\mu A(\lambda))^{-1}$ belongs to $\mathbf{B}(\mathfrak{H})$ when $\lambda \in \mathbb{C}^+$, the operator $(B(\lambda)+\bar\mu A(\lambda))^{-1}$ belongs to $\mathbf{B}(\mathfrak{H})$ when $\lambda \in \mathbb{C}^-$, and

$$\begin{aligned}(Z(\lambda)+\mu)^{-1} &= A(\lambda)\big(B(\lambda)+\mu A(\lambda)\big)^{-1}, \quad \lambda \in \mathbb{C}^+,\\ (Z(\lambda)+\bar\mu)^{-1} &= A(\lambda)\big(B(\lambda)+\bar\mu A(\lambda)\big)^{-1}, \quad \lambda \in \mathbb{C}^-.\end{aligned}$$

Since $\lambda \mapsto A(\lambda)$ and $\lambda \mapsto B(\lambda)$ are holomorphic, it follows that the mapping $\lambda \mapsto (Z(\lambda)+\mu)^{-1}$ is holomorphic on $\mathbb{C}^+$ with values in $\mathbf{B}(\mathfrak{H})$ and the mapping $\lambda \mapsto (Z(\lambda)+\bar\mu)^{-1}$ is holomorphic on $\mathbb{C}^-$ with values in $\mathbf{B}(\mathfrak{H})$. Hence, (iii) in Definition 1.12.1 is satisfied.

Step 3. Let $\{A,B\}$ satisfy (a), (b), and (c) of Definition 1.12.3, and assume that

$$\big(B(\lambda)+N(\lambda)A(\lambda)\big)^{-1} \in \mathbf{B}(\mathfrak{H}).$$

Define the family $Z(\lambda)$, $\lambda \in \mathbb{C}\setminus\mathbb{R}$, by (1.12.3). It will be shown first that $Z(\lambda)$, $\lambda \in \mathbb{C}\setminus\mathbb{R}$, is a Nevanlinna family. In fact, it follows from the definition that

$$\big(Z(\lambda)+N(\lambda)\big)^{-1} = A(\lambda)\big(B(\lambda)+N(\lambda)A(\lambda)\big)^{-1} \in \mathbf{B}(\mathfrak{H}),$$

and via (a) one sees that

$$\lambda \mapsto \big(Z(\lambda) + N(\lambda)\big)^{-1} \text{ is holomorphic with values in } \mathbf{B}(\mathfrak{H}). \tag{1.12.12}$$

Note that (b) shows that $Z(\lambda)$ is dissipative (accumulative) for $\lambda \in \mathbb{C}^+$ ($\lambda \in \mathbb{C}^-$). In fact, $Z(\lambda)$ is maximal dissipative (maximal accumulative) for $\lambda \in \mathbb{C}^+$ ($\lambda \in \mathbb{C}^-$). To see this, let $Z'(\lambda)$ be an extension of $Z(\lambda)$ which is dissipative for $\lambda \in \mathbb{C}^+$. Then clearly

$$\big(Z(\lambda) + N(\lambda)\big)^{-1} \subset \big(Z'(\lambda) + N(\lambda)\big)^{-1}, \tag{1.12.13}$$

the left-hand side is an operator in $\mathbf{B}(\mathfrak{H})$, and the right-hand side defines an operator. In fact, if $\{0, k\} \in (Z'(\lambda) + N(\lambda))^{-1}$, then $\{k, -N(\lambda)k\} \in Z'(\lambda)$ and as $Z'(\lambda)$ is dissipative, it follows that $\operatorname{Im}(-N(\lambda)k, k) \geq 0$. On the other hand, one has $\operatorname{Im}(N(\lambda)k, k) \geq 0$ as N is a Nevanlinna function. Hence, $k = 0$ and $(Z'(\lambda) + N(\lambda))^{-1}$ is an operator. It follows that the inclusion in (1.12.13) is an equality and therefore $Z'(\lambda) = Z(\lambda)$ for $\lambda \in \mathbb{C}^+$. Thus, $Z(\lambda)$ is maximal dissipative for $\lambda \in \mathbb{C}^+$. A similar argument shows that $Z(\lambda)$ is maximal accumulative for $\lambda \in \mathbb{C}^-$. Hence, (i) in Definition 1.12.1 has been shown. It clearly follows from (c) that $Z(\lambda) \subset Z(\bar{\lambda})^*$, which implies that

$$\big(Z(\lambda) + N(\lambda)\big)^{-1} \subset \big(Z(\bar{\lambda})^* + N(\lambda)\big)^{-1} = \big(Z(\bar{\lambda}) + N(\bar{\lambda})\big)^{-*},$$

where in the last step it was used that $N(\lambda) = N(\bar{\lambda})^*$. The above inclusion is in fact an equality, since the operators on the left and on the right belong to $\mathbf{B}(\mathfrak{H})$. Therefore, $Z(\lambda) = Z(\bar{\lambda})^*$, and hence (ii) in Definition 1.12.1 holds. Now it follows from (1.12.12) and Step 2 of this lemma that also (iii) in Definition 1.12.1 holds. Therefore, one concludes that $Z(\lambda)$, $\lambda \in \mathbb{C} \setminus \mathbb{R}$, defined by (1.12.3) is a Nevanlinna family.

Now it follows from Proposition 1.12.2 that $\{A, B\}$ is a Nevanlinna pair and thus, in particular, condition (d) holds. □

Chapter 2

Boundary Triplets and Weyl Functions

The basic properties of boundary triplets for closed symmetric operators or relations in Hilbert spaces are presented. These triplets give rise to a parametrization of the intermediate extensions of symmetric relations, in particular of the self-adjoint extensions. Closely related is the Kreĭn formula which describes the resolvent operators of such intermediate extensions. The introduction of boundary triplets and a discussion of corresponding boundary value problems can be found in Section 2.1 and Section 2.2. Associated with a boundary triplet are the γ-field and the Weyl function, and these analytic objects are treated in Section 2.3. The existence and construction of boundary triplets is discussed in Section 2.4; their transformations are the contents of Section 2.5. Section 2.6 on Kreĭn's resolvent formula for canonical extensions and a description of their spectra is central in this chapter. Furthermore, a discussion of self-adjoint exit space extensions, Štraus families, and the Kreĭn–Naĭmark formula can be found Section 2.7. Some related perturbation problems are treated in Section 2.8.

2.1 Boundary triplets

The following definition introduces a boundary triplet, one of the key objects in this text. It is based on the well-known Green or Lagrange formula together with an additional maximality condition.

Definition 2.1.1. Let S be a closed symmetric relation in a Hilbert space $\mathfrak{H}$. Then $\{\mathcal{G}, \Gamma_0, \Gamma_1\}$ is a *boundary triplet* for S^* if $\mathcal{G}$ is a Hilbert space and $\Gamma_0, \Gamma_1 : S^* \to \mathcal{G}$ are linear mappings such that the mapping $\Gamma : S^* \to \mathcal{G} \times \mathcal{G}$ defined by

$$\Gamma \widehat{f} = \{\Gamma_0 \widehat{f}, \Gamma_1 \widehat{f}\}, \quad \widehat{f} = \{f, f'\} \in S^*,$$

is surjective and the identity

$$(f', g)_{\mathfrak{H}} - (f, g')_{\mathfrak{H}} = (\Gamma_1 \widehat{f}, \Gamma_0 \widehat{g})_{\mathcal{G}} - (\Gamma_0 \widehat{f}, \Gamma_1 \widehat{g})_{\mathcal{G}} \tag{2.1.1}$$

holds for all $\widehat{f} = \{f, f'\}$, $\widehat{g} = \{g, g'\} \in S^*$.

J. Behrndt et al., *Boundary Value Problems, Weyl Functions, and Differential Operators*, Monographs in Mathematics 108,
https://doi.org/10.1007/978-3-030-36714-5_3

Note that a symmetric relation S is densely defined if and only if S^* is an operator. In this case the boundary mappings Γ_0 and Γ_1 can be defined on $\operatorname{dom} S^*$ instead of (the graph of) S^*. More precisely, if $\{\mathcal{G}, \Gamma_0, \Gamma_1\}$ is a boundary triplet for S^*, then one defines boundary mappings Γ_0 and Γ_1 on $\operatorname{dom} S^*$ by the following identifications

$$\Gamma_0 f = \Gamma_0 \widehat{f}, \quad \Gamma_1 f = \Gamma_1 \widehat{f}, \quad \widehat{f} = \{f, f'\} \in S^*.$$

In the following treatment whenever S is a densely defined operator, boundary mappings defined on S^* and on $\operatorname{dom} S^*$ will be identified in this sense. After this identification (2.1.1) turns into

$$(S^* f, g)_{\mathfrak{H}} - (f, S^* g)_{\mathfrak{H}} = (\Gamma_1 f, \Gamma_0 g)_{\mathcal{G}} - (\Gamma_0 f, \Gamma_1 g)_{\mathcal{G}}, \tag{2.1.2}$$

where $f, g \in \operatorname{dom} S^*$. This formalism will be used in Chapter 6 and Chapter 8 in the treatment of ordinary and partial differential operators.

The identity (2.1.1) or the identity (2.1.2) is sometimes called the *abstract Green identity* or the *abstract Lagrange identity*; in this text mostly the terminology abstract Green identity will be used. This identity has a geometric interpretation which is best expressed in terms of the indefinite inner products

$$\begin{aligned} [\![\cdot,\cdot]\!]_{\mathfrak{H}^2} &:= (J_{\mathfrak{H}}\cdot,\cdot)_{\mathfrak{H}^2}, \qquad J_{\mathfrak{H}} = \begin{pmatrix} 0 & -iI_{\mathfrak{H}} \\ iI_{\mathfrak{H}} & 0 \end{pmatrix}, \\ [\![\cdot,\cdot]\!]_{\mathcal{G}^2} &:= (J_{\mathcal{G}}\cdot,\cdot)_{\mathcal{G}^2}, \qquad J_{\mathcal{G}} = \begin{pmatrix} 0 & -iI_{\mathcal{G}} \\ iI_{\mathcal{G}} & 0 \end{pmatrix}, \end{aligned} \tag{2.1.3}$$

where $J_{\mathfrak{H}} = J_{\mathfrak{H}}^* = J_{\mathfrak{H}}^{-1} \in \mathbf{B}(\mathfrak{H}^2)$ and $J_{\mathcal{G}} = J_{\mathcal{G}}^* = J_{\mathcal{G}}^{-1} \in \mathbf{B}(\mathcal{G}^2)$; cf. Section 1.8. By means of these inner products, the identity (2.1.1) can be rewritten as

$$[\![\widehat{f}, \widehat{g}]\!]_{\mathfrak{H}^2} = [\![\Gamma\widehat{f}, \Gamma\widehat{g}]\!]_{\mathcal{G}^2} \tag{2.1.4}$$

for $\widehat{f} = \{f, f'\}$, $\widehat{g} = \{g, g'\} \in S^*$. Later the scalar products in (2.1.1), (2.1.2), and (2.1.4) will be used without indices $\mathfrak{H}$ and $\mathcal{G}$, respectively, when there is no danger of confusion. Recall that the adjoint A^* of a relation A in $\mathfrak{H}$ can be written as an orthogonal complement with respect to the inner product $[\![\cdot,\cdot]\!]$, that is $A^* = A^{[\![\perp]\!]}$; cf. Section 1.8.

Some elementary but important properties of the boundary mappings are collected in the following proposition.

Proposition 2.1.2. *Let S be a closed symmetric relation in $\mathfrak{H}$ and assume that $\{\mathcal{G}, \Gamma_0, \Gamma_1\}$ is a boundary triplet for S^*. Then the following statements hold:*

(i) *the mappings $\Gamma : S^* \to \mathcal{G} \times \mathcal{G}$ and $\Gamma_0, \Gamma_1 : S^* \to \mathcal{G}$ are surjective and continuous;*

(ii) $\ker \Gamma = S$.

Proof. (i) The continuity of $\Gamma : S^* \to \mathcal{G} \times \mathcal{G}$ is essentially a consequence of the fact that Γ is isometric in the sense of (2.1.4). More precisely, by definition the mapping Γ is surjective, and since $\operatorname{dom} \Gamma = S^*$ is closed it follows from Lemma 1.8.1 that Γ is continuous. Clearly, the mappings Γ_0 and Γ_1 are also surjective and continuous.

(ii) In order to show that $\ker \Gamma \subset S$, let $\widehat{f} \in \ker \Gamma$. Then it follows from (2.1.4) that $[\![\widehat{f}, \widehat{g}]\!] = [\![\Gamma \widehat{f}, \Gamma \widehat{g}]\!] = 0$ for all $\widehat{g} \in S^*$, which implies $\widehat{f} \in (S^*)^{[\![\perp]\!]} = S^{**} = S$, since S is closed. Hence, $\ker \Gamma \subset S$ has been shown. To show that $S \subset \ker \Gamma$, let $\widehat{f} \in S$. Since Γ is surjective, for arbitrary fixed $\widehat{\varphi} \in \mathcal{G} \times \mathcal{G}$ one can choose $\widehat{g} \in S^* = S^{[\![\perp]\!]}$ such that $\Gamma \widehat{g} = \mathcal{J}_{\mathcal{G}} \widehat{\varphi}$, with $\mathcal{J}_{\mathcal{G}}$ as in (2.1.3). Since $\widehat{f} \in S$ and $\widehat{g} \in S^*$ it follows from (2.1.4) that

$$(\Gamma \widehat{f}, \widehat{\varphi})_{\mathcal{G}^2} = \big(\Gamma \widehat{f}, \mathcal{J}_{\mathcal{G}}^{-1} \Gamma \widehat{g}\big)_{\mathcal{G}^2} = \big[\!\big[\Gamma \widehat{f}, \Gamma \widehat{g}\big]\!\big]_{\mathcal{G}^2} = \big[\!\big[\widehat{f}, \widehat{g}\big]\!\big]_{\mathfrak{H}^2} = 0$$

for all $\widehat{\varphi} \in \mathcal{G}^2$, which leads to $\Gamma \widehat{f} = 0$. Thus, $S \subset \ker \Gamma$ has been shown. □

By means of a boundary triplet $\{\mathcal{G}, \Gamma_0, \Gamma_1\}$ for S^* the intermediate extensions of S defined in Section 1.7 can be described via relations in the space $\mathcal{G}$. In particular, the one-to-one correspondence in the next theorem preserves adjoints, which is a consequence of the abstract Green identity (2.1.4).

Theorem 2.1.3. *Let S be a closed symmetric relation in $\mathfrak{H}$ and let $\{\mathcal{G}, \Gamma_0, \Gamma_1\}$ be a boundary triplet for S^*. Then the following statements hold:*

(i) *there is a bijective correspondence between the set of intermediate extensions A_Θ of S and the set of relations Θ in $\mathcal{G}$, via*

$$A_\Theta := \big\{\widehat{f} \in S^* : \Gamma \widehat{f} \in \Theta\big\}; \tag{2.1.5}$$

(ii) $\overline{A_\Theta} = A_{\overline{\Theta}}$ *and, in particular, the relation A_Θ is closed if and only if the relation Θ is closed;*

(iii) $A_\Theta = \ker \big(\Gamma_1 - \Theta \Gamma_0\big)$;

(iv) $(A_\Theta)^* = A_{\Theta^*}$ *for every relation Θ in $\mathcal{G}$;*

(v) $A_\Theta \subset A_{\Theta'}$ *if and only if $\Theta \subset \Theta'$, when Θ and Θ' are relations in $\mathcal{G}$;*

(vi) A_Θ *is an operator if and only if S is an operator and*

$$\Theta \cap \Gamma\big(\{0\} \times \operatorname{mul} S^*\big) = \{0, 0\}. \tag{2.1.6}$$

Proof. (i) & (ii) The relation $S^* \subset \mathfrak{H}^2$ is equipped with the Hilbert space inner product of $\mathfrak{H}^2$. Now let $\mathfrak{M} \subset \mathfrak{H}^2$ be the orthogonal complement of S in S^*, so that $S \oplus \mathfrak{M} = S^*$. Since $\ker \Gamma = S$, the restriction Γ' of Γ to $\mathfrak{M}$ is an isomorphism between $\mathfrak{M}$ and $\mathcal{G} \times \mathcal{G}$. Hence Γ' gives a one-to-one correspondence between the subspaces H' of $\mathfrak{M}$ and the subspaces Θ of $\mathcal{G} \times \mathcal{G}$ via

$$\Theta = \Gamma' H' \quad \text{or, equivalently,} \quad (\Gamma')^{-1} \Theta = H'. \tag{2.1.7}$$

Clearly, this gives rise to a one-to-one correspondence between all intermediate extensions H of S and all subspaces H' of $\mathfrak{M}$ via $H = S \oplus H'$, which is expressed in (2.1.5). Moreover, since Γ' is an isomorphism it also follows from (2.1.7) that $\overline{\Theta} = \overline{\Gamma' H'} = \Gamma' \overline{H'}$ and hence the closure $\overline{H}$ of H corresponds to the closure $\overline{\Theta}$ of Θ. This implies via (2.1.5) that $\overline{A_\Theta} = A_{\overline{\Theta}}$.

(iii) Let A_Θ be defined by (2.1.5). It will be verified that

$$A_\Theta = \ker\big(\Gamma_1 - \Theta\Gamma_0\big) \tag{2.1.8}$$

holds for any relation Θ in $\mathcal{G}$. Note that (2.1.8) is clear in the special case that Θ is an operator, since $\Gamma\widehat{f} = \{\Gamma_0\widehat{f}, \Gamma_1\widehat{f}\} \in \Theta$ means that $\Theta\Gamma_0\widehat{f} = \Gamma_1\widehat{f}$. Now assume that Θ is a relation.

First the inclusion ($\subset$) in (2.1.8) will be shown. For this consider $\widehat{f} \in A_\Theta$. Hence, $\widehat{f} \in S^*$ and $\{\Gamma_0\widehat{f}, \Gamma_1\widehat{f}\} \in \Theta$. Then $\{\widehat{f}, \Gamma_0\widehat{f}\} \in \Gamma_0$ gives $\{\widehat{f}, \Gamma_1\widehat{f}\} \in \Theta\Gamma_0$. Since $\{\widehat{f}, \Gamma_1\widehat{f}\} \in \Gamma_1$ one finds $\{\widehat{f}, 0\} \in \Gamma_1 - \Theta\Gamma_0$. In other words $\widehat{f} \in \ker(\Gamma_1 - \Theta\Gamma_0)$.

For the inclusion ($\supset$) in (2.1.8) consider $\widehat{f} \in \ker(\Gamma_1 - \Theta\Gamma_0)$. Then one has $\{\widehat{f}, 0\} \in \Gamma_1 - \Theta\Gamma_0$ and hence there exists an element $\widehat{\psi}$ such that $\{\widehat{f}, \widehat{\psi}\} \in \Gamma_1$ and $\{\widehat{f}, \widehat{\psi}\} \in \Theta\Gamma_0$. Thus $\{\widehat{f}, \widehat{\varphi}\} \in \Gamma_0$ and $\{\widehat{\varphi}, \widehat{\psi}\} \in \Theta$ for some $\widehat{\varphi}$. Since both Γ_0 and Γ_1 are operators, one has $\widehat{\psi} = \Gamma_1\widehat{f}$ and $\widehat{\varphi} = \Gamma_0\widehat{f}$, and therefore $\{\Gamma_0\widehat{f}, \Gamma_1\widehat{f}\} \in \Theta$, that is, $\widehat{f} \in A_\Theta$.

(iv) To show that $(A_\Theta)^* \subset A_{\Theta^*}$, let $\widehat{g} \in (A_\Theta)^*$. Let $\widehat{\varphi} \in \Theta$ and choose $\widehat{f} \in A_\Theta$ such that $\Gamma\widehat{f} = \widehat{\varphi}$. Then one has

$$\big[\widehat{\varphi}, \Gamma\widehat{g}\big]_{\mathcal{G}^2} = \big[\Gamma\widehat{f}, \Gamma\widehat{g}\big]_{\mathcal{G}^2} = \big[\widehat{f}, \widehat{g}\big]_{\mathfrak{H}^2} = 0,$$

which implies $\Gamma\widehat{g} \in \Theta^*$, that is, $\widehat{g} \in A_{\Theta^*}$. One concludes $(A_\Theta)^* \subset A_{\Theta^*}$. Since A_{Θ^*} is closed by (ii), the inclusion $A_{\Theta^*} \subset (A_\Theta)^*$ follows together with (ii) from

$$A_{\Theta^*} = (A_{\Theta^*})^{**} \subset (A_{\Theta^{**}})^* = (A_{\overline{\Theta}})^* = (\overline{A_\Theta})^* = (A_\Theta)^*.$$

(v) This assertion is obvious from the correspondence in (2.1.5).

(vi) Let A_Θ be an operator. Then clearly S is an operator. Assume that $\Gamma\widehat{f} \in \Theta$ for some element $\widehat{f} = \{0, f'\} \in S^*$. Then $\widehat{f} \in A_\Theta$ and hence $f' = 0$, so that (2.1.6) holds.

Conversely, assume now that S is an operator and that (2.1.6) holds. If $\widehat{f} = \{0, f'\} \in A_\Theta$, then $f' \in \operatorname{mul} S^*$ and $\Gamma\{0, f'\} \in \Theta$. Hence, $\Gamma\{0, f'\} = \{0, 0\}$ and as $S = \ker\Gamma$, it follows that $\{0, f'\} \in S$. This implies $f' = 0$. Therefore, A_Θ is an operator. □

Due to the abstract Green identity (2.1.1), (2.1.2), or (2.1.4) some properties of intermediate extensions are preserved in the corresponding relations in $\mathcal{G}$.

Corollary 2.1.4. *Let S be a closed symmetric relation in $\mathfrak{H}$, let $\{\mathcal{G}, \Gamma_0, \Gamma_1\}$ be a boundary triplet for S^*, and let A_Θ be the extension of S in $\mathfrak{H}$ corresponding to the relation Θ in $\mathcal{G}$ via (2.1.5). Then the following statements hold:*

(i) *A_Θ is dissipative (accumulative) if and only if Θ is dissipative (accumulative);*

(ii) *A_Θ is maximal dissipative (maximal accumulative) if and only if Θ is maximal dissipative (maximal accumulative);*

(iii) *A_Θ is symmetric if and only if Θ is symmetric;*

(iv) *A_Θ is maximal symmetric if and only if Θ is maximal symmetric;*

(v) *A_Θ is self-adjoint if and only if Θ is self-adjoint.*

Proof. (i) This assertion follows immediately from the identity (see (1.8.4))

$$\mathrm{Im}\,(f', f) = \frac{1}{2}\big[\widehat{f}, \widehat{f}\big]_{\mathfrak{H}^2} = \frac{1}{2}\big[\Gamma\widehat{f}, \Gamma\widehat{f}\big]_{\mathcal{G}^2} = \mathrm{Im}\,(\Gamma_1\widehat{f}, \Gamma_0\widehat{f}),$$

where $\widehat{f} = \{f, f'\} \in S^*$, and (2.1.5).

(ii) According to Theorem 2.1.3 (v), for any two extensions A_Θ and $A_{\Theta'}$ of S one has $A_\Theta \subset A_{\Theta'}$ if and only if $\Theta \subset \Theta'$ holds. Therefore, if A_Θ is maximal dissipative, then Θ is dissipative because of (i) and if Θ' is a dissipative extension of Θ in $\mathcal{G}$, then $A_{\Theta'}$ is a dissipative extension of A_Θ, so that $A_\Theta = A_{\Theta'}$. Hence, $\Theta = \Theta'$ and Θ is maximal dissipative. The converse direction is proved in exactly the same way. The statement for maximal accumulative extensions follows analogously.

(iii)–(v) These assertions follow from the previous items, and the fact that a relation is symmetric (self-adjoint) if and only if it is (maximal) dissipative and (maximal) accumulative. □

Let H and H' be two closed intermediate extensions of S in $\mathfrak{H}$. Recall that H and H' are disjoint if $H \cap H' = S$, and that H and H' are transversal if they are disjoint and $H \widehat{+} H' = S^*$; cf. Definition 1.7.6. If, in addition, the extensions H and H' are self-adjoint, then disjointness implies

$$S^* = \mathrm{clos}\,(H \widehat{+} H');$$

and in this case H and H' are transversal if and only if $H \widehat{+} H'$ is closed; cf. Lemma 1.7.7. In a similar way the closed relations Θ and Θ' in $\mathcal{G}$, as intermediate extensions of the trivial symmetric relation $\{0, 0\}$, are disjoint if $\Theta \cap \Theta' = \{0, 0\}$ and transversal if they are disjoint and $\Theta \widehat{+} \Theta' = \mathcal{G}^2$.

Lemma 2.1.5. *Let S be a closed symmetric relation in $\mathfrak{H}$ and let $\{\mathcal{G}, \Gamma_0, \Gamma_1\}$ be a boundary triplet for S^*. Let Θ and Θ' be relations in $\mathcal{G}$. Then*

$$A_\Theta \cap A_{\Theta'} = A_{\Theta \cap \Theta'} \tag{2.1.9}$$

and

$$A_\Theta \widehat{+} A_{\Theta'} = A_{\Theta \widehat{+} \Theta'}. \tag{2.1.10}$$

In particular, if A_Θ and $A_{\Theta'}$ are closed or, equivalently, Θ and Θ' are closed, then the following statements hold:

(i) *A_Θ and $A_{\Theta'}$ are disjoint if and only if Θ and Θ' are disjoint;*

(ii) *A_Θ and $A_{\Theta'}$ are transversal if and only if Θ and Θ' are transversal.*

Proof. The identity (2.1.9) follows from

$$\begin{aligned} A_\Theta \cap A_{\Theta'} &= \{\widehat{f} \in S^* : \Gamma\widehat{f} \in \Theta\} \cap \{\widehat{f} \in S^* : \Gamma\widehat{f} \in \Theta'\} \\ &= \{\widehat{f} \in S^* : \Gamma\widehat{f} \in \Theta \cap \Theta'\} \\ &= A_{\Theta\cap\Theta'}, \end{aligned}$$

while the identity (2.1.10) follows from

$$\Gamma(A_\Theta \,\widehat{+}\, A_{\Theta'}) = \Gamma(A_\Theta) \,\widehat{+}\, \Gamma(A_{\Theta'}) = \Theta \,\widehat{+}\, \Theta'.$$

In particular, (2.1.9) together with $S = \ker \Gamma$ shows that $A_\Theta \cap A_{\Theta'} = S$ if and only if $\Theta \cap \Theta' = \{0, 0\}$, while (2.1.9) and (2.1.10) show that $A_\Theta \cap A_{\Theta'} = S$ and $A_\Theta \,\widehat{+}\, A_{\Theta'} = S^*$ if and only if $\Theta \cap \Theta' = \{0, 0\}$ and $\Theta \,\widehat{+}\, \Theta' = \mathcal{G}^2$. This completes the proof. □

Corollary 2.1.6. *Let S be a closed symmetric relation in $\mathfrak{H}$, let $\{\mathcal{G}, \Gamma_0, \Gamma_1\}$ be a boundary triplet for S^* and assume that $\dim \mathcal{G} < \infty$. If A_Θ and $A_{\Theta'}$ are self-adjoint extensions of S which are disjoint, then they are transversal.*

Let S be a closed symmetric relation in $\mathfrak{H}$ and let $\{\mathcal{G}, \Gamma_0, \Gamma_1\}$ be a boundary triplet for S^*. There are two special extensions of S which will be frequently used in the following; they are defined by

$$A_0 := \ker \Gamma_0 \quad \text{and} \quad A_1 := \ker \Gamma_1. \tag{2.1.11}$$

It is clear that A_0 and A_1 are self-adjoint extensions of S, since they correspond to the self-adjoint parameters Θ in $\mathcal{G}$ in (2.1.5) given by

$$\Theta = \{0\} \times \mathcal{G} \quad \text{and} \quad \Theta = \mathcal{G} \times \{0\}, \tag{2.1.12}$$

respectively. Furthermore, the representations in (2.1.12) show that the self-adjoint extensions A_0 and A_1 are transversal; cf. Lemma 2.1.5. Note also in this context that a boundary triplet $\{\mathcal{G}, \Gamma_0, \Gamma_1\}$ for S^* can only exist if the defect numbers of the closed symmetric relation S coincide (since it admits the self-adjoint extensions A_0 and A_1 in (2.1.11)); a more detailed discussion on the existence and uniqueness of boundary triplets will be provided in Section 2.4 and Section 2.5.

As $S = \ker \Gamma$, it follows from von Neumann's decomposition Theorem 1.7.11 that Γ is an isomorphism from $\widehat{\mathfrak{N}}_\lambda(S^*) \,\widehat{+}\, \widehat{\mathfrak{N}}_{\bar\lambda}(S^*)$, $\lambda \in \mathbb{C} \setminus \mathbb{R}$, onto $\mathcal{G}^2$. Due to the definitions in (2.1.11) a similar observation can be made for the components Γ_0 and Γ_1.

Lemma 2.1.7. *Let S be a closed symmetric relation in $\mathfrak{H}$, let $\{\mathcal{G}, \Gamma_0, \Gamma_1\}$ be a boundary triplet for S^*, and let $A_0 = \ker \Gamma_0$ and $A_1 = \ker \Gamma_1$. Then the adjoint S^* admits the direct sum decompositions*

$$\begin{aligned} S^* &= A_0 \,\widehat{+}\, \widehat{\mathfrak{N}}_\lambda(S^*), \quad \lambda \in \rho(A_0), \\ S^* &= A_1 \,\widehat{+}\, \widehat{\mathfrak{N}}_\lambda(S^*), \quad \lambda \in \rho(A_1). \end{aligned} \tag{2.1.13}$$

In particular, the restrictions $\Gamma_0 \upharpoonright \widehat{\mathfrak{N}}_\lambda(S^*)$ and $\Gamma_1 \upharpoonright \widehat{\mathfrak{N}}_\lambda(S^*)$ are isomorphisms from $\widehat{\mathfrak{N}}_\lambda(S^*)$ onto $\mathcal{G}$ for $\lambda \in \rho(A_0)$ and $\lambda \in \rho(A_1)$, respectively.

Proof. As A_0 and A_1 are self-adjoint, the direct sum decompositions (2.1.13) hold by Corollary 1.7.5. Since Γ_0 and Γ_1 map S^* onto $\mathcal{G}$, and A_0 and A_1 are their respective kernels, it is clear that the restrictions $\Gamma_0 \upharpoonright \widehat{\mathfrak{N}}_\lambda(S^*)$ and $\Gamma_1 \upharpoonright \widehat{\mathfrak{N}}_\lambda(S^*)$ are isomorphisms from $\widehat{\mathfrak{N}}_\lambda(S^*)$ onto $\mathcal{G}$. □

In the rest of the text the self-adjoint extension $A_0 = \ker \Gamma_0$ will often serve as a point of reference due to the corresponding representation $\{0\} \times \mathcal{G}$ in the parameter space $\mathcal{G}$. In the next proposition it is shown that A_0 and a given closed extension A_Θ are disjoint (transversal) if and only if the parameter Θ is a (bounded everywhere defined) operator.

Proposition 2.1.8. *Let S be a closed symmetric relation in $\mathfrak{H}$, let $\{\mathcal{G}, \Gamma_0, \Gamma_1\}$ be a boundary triplet for S^*, let $A_0 = \ker \Gamma_0$, and let A_Θ be the closed intermediate extension of S in $\mathfrak{H}$ corresponding to the closed relation Θ in $\mathcal{G}$ via (2.1.5). Then the following statements hold:*

(i) *$A_\Theta \cap A_0 = S$ if and only if Θ is a closed operator in $\mathcal{G}$;*

(ii) *$A_\Theta \cap A_0 = S$ and $A_\Theta \widehat{+} A_0 = S^*$ if and only if $\Theta \in \mathbf{B}(\mathcal{G})$.*

Proof. Apply Lemma 2.1.5 to the self-adjoint extension $A_0 = \ker \Gamma_0$ that corresponds to $\{0\} \times \mathcal{G}$.

(i) A_Θ and A_0 are disjoint if and only if $\Theta \cap (\{0\} \times \mathcal{G}) = \{0, 0\}$, which is the same as saying that $\operatorname{mul} \Theta = \{0\}$.

(ii) A_Θ and A_0 are transversal if and only if

$$\Theta \cap (\{0\} \times \mathcal{G}) = \{0, 0\} \quad \text{and} \quad \Theta \,\widehat{+}\, (\{0\} \times \mathcal{G}) = \mathcal{G} \times \mathcal{G},$$

which is the same as saying that $\operatorname{mul} \Theta = \{0\}$ and $\operatorname{dom} \Theta = \mathcal{G}$. By the closed graph theorem, the last two conditions are equivalent to $\Theta \in \mathbf{B}(\mathcal{G})$. □

Let S be a closed symmetric relation in $\mathfrak{H}$ with equal defect numbers and let H be a self-adjoint extension of S. Later it will be shown that there exists a boundary triplet $\{\mathcal{G}, \Gamma_0, \Gamma_1\}$ for S^* such that $\ker \Gamma_0$ concides with H; cf. Theorem 2.4.1. Furthermore, it will be shown that for a pair of self-adjoint extensions of S which are transversal, there exists a boundary triplet $\{\mathcal{G}, \Gamma_0, \Gamma_1\}$ for S^* such that $\ker \Gamma_0$

and $\ker \Gamma_1$ coincide with this pair; cf. Theorem 2.5.9. The notion of boundary triplet is not unique; in fact, a parametrization of all possible boundary triplets will be provided in Section 2.5.

The following theorem is of a different nature. It can be used to prove that a given relation T is the adjoint of a symmetric relation S.

Theorem 2.1.9. *Let T be a relation in $\mathfrak{H}$, let $\mathcal{G}$ be a Hilbert space, and assume that*

$$\Gamma = \begin{pmatrix} \Gamma_0 \\ \Gamma_1 \end{pmatrix} : T \to \mathcal{G} \times \mathcal{G}$$

is a linear mapping such that the following conditions are satisfied:

(i) $\ker \Gamma_0$ *contains a self-adjoint relation* A_0;

(ii) $\operatorname{ran} \Gamma = \mathcal{G} \times \mathcal{G}$;

(iii) *for all* $\widehat{f}, \widehat{g} \in T$,

$$(f', g)_{\mathfrak{H}} - (f, g')_{\mathfrak{H}} = (\Gamma_1 \widehat{f}, \Gamma_0 \widehat{g})_{\mathcal{G}} - (\Gamma_0 \widehat{f}, \Gamma_1 \widehat{g})_{\mathcal{G}}.$$

Then $S := \ker \Gamma$ is a closed symmetric relation in $\mathfrak{H}$ such that $S^* = T$ and $\{\mathcal{G}, \Gamma_0, \Gamma_1\}$ is a boundary triplet for S^* with $A_0 = \ker \Gamma_0$.

Proof. First note that condition (iii) implies that $\ker \Gamma_0$ is symmetric. To see this, let $\widehat{f}, \widehat{g} \in \ker \Gamma_0$. Then, by condition (iii),

$$[\widehat{f}, \widehat{g}] = [\![\Gamma \widehat{f}, \Gamma \widehat{g}]\!] = 0,$$

and hence $\ker \Gamma_0$ is a symmetric relation in $\mathfrak{H}$. Then (i) gives $A_0 = A_0^* \subset \ker \Gamma_0$, which implies

$$\ker \Gamma_0 \subset (\ker \Gamma_0)^* \subset A_0^* = A_0 \subset \ker \Gamma_0.$$

Therefore, $\ker \Gamma_0 = A_0$ is self-adjoint in $\mathfrak{H}$. Moreover, $S := \ker \Gamma \subset \ker \Gamma_0$ is a symmetric relation in $\mathfrak{H}$.

It will be shown that

$$S = T^*, \tag{2.1.14}$$

so that, in particular, S is closed. To see ($\subset$) in (2.1.14), let $\widehat{f} \in S = \ker \Gamma$. For any $\widehat{g} \in T$ one has $[\![\widehat{f}, \widehat{g}]\!] = [\![\Gamma \widehat{f}, \Gamma \widehat{g}]\!] = 0$, so that $\widehat{f} \in T^*$. To see ($\supset$) in (2.1.14), let $\widehat{f} \in T^*$. Since A_0 is self-adjoint and $A_0 \subset T$, it follows that $T^* \subset A_0 = \ker \Gamma_0$, so that $\Gamma_0 \widehat{f} = 0$. For arbitrary $\widehat{g} \in T$ it therefore follows that

$$0 = [\![\widehat{f}, \widehat{g}]\!] = [\![\Gamma \widehat{f}, \Gamma \widehat{g}]\!] = -i(\Gamma_1 \widehat{f}, \Gamma_0 \widehat{g}).$$

From condition (ii) one concludes $\operatorname{ran} \Gamma_0 = \mathcal{G}$ and this leads to $\Gamma_1 \widehat{f} = 0$. Hence, $\widehat{f} \in \ker \Gamma_0 \cap \ker \Gamma_1 = S$. Therefore, (2.1.14) is proved.

It follows from $S = T^*$ that $S^* = T^{**} = \overline{T}$. Hence, it remains to show that T is closed. Let $(\widehat{f}_n)$ be a sequence in T converging to $\widehat{f}$. It suffices to show that $\widehat{f} \in T$. Let $\widehat{\psi} \in \mathcal{G}^2$ and let $\widehat{g} \in T$ be such that $\widehat{\psi} = \mathcal{J}_{\mathcal{G}}^{-1}\Gamma\widehat{g}$ (here condition (ii) is being used). Using the continuity of the indefinite inner product $[\![\cdot,\cdot]\!]$ (see Section 1.8) one obtains

$$(\Gamma\widehat{f}_n, \widehat{\psi}) = (\Gamma\widehat{f}_n, \mathcal{J}_{\mathcal{G}}^{-1}\Gamma\widehat{g}) = [\![\Gamma\widehat{f}_n, \Gamma\widehat{g}]\!] = [\![\widehat{f}_n, \widehat{g}]\!] \to [\![\widehat{f}, \widehat{g}]\!]. \tag{2.1.15}$$

This shows that $\Gamma\widehat{f}_n$ is a weak Cauchy sequence in $\mathcal{G}$, hence weakly bounded and thus bounded. It follows that there exists a subsequence, again denoted by $\Gamma\widehat{f}_n$, which converges weakly to some $\widehat{\varphi} \in \mathcal{G} \times \mathcal{G}$. Now let $\widehat{h} \in T$ be such that $\Gamma\widehat{h} = \widehat{\varphi}$ (again condition (ii) is being used). Choose $\widehat{g} \in T$ and let, as above, $\widehat{\psi} = \mathcal{J}_{\mathcal{G}}^{-1}\Gamma\widehat{g}$, so that (2.1.15) remains valid. Then (2.1.15) implies

$$[\![\widehat{f}, \widehat{g}]\!] = \lim_{n\to\infty} (\Gamma\widehat{f}_n, \widehat{\psi}) = (\widehat{\varphi}, \widehat{\psi}) = (\Gamma\widehat{h}, \mathcal{J}_{\mathcal{G}}^{-1}\Gamma\widehat{g}) = [\![\Gamma\widehat{h}, \Gamma\widehat{g}]\!] = [\![\widehat{h}, \widehat{g}]\!],$$

and therefore $[\![\widehat{f} - \widehat{h}, \widehat{g}]\!] = 0$. Since $\widehat{g} \in T$, one concludes that $\widehat{f} - \widehat{h} \in T^* = S \subset T$. Now $\widehat{h} \in T$ implies that $\widehat{f} \in T$. Therefore, T is closed and it follows that $S^* = T$.

By conditions (ii) and (iii) $\{\mathcal{G}, \Gamma_0, \Gamma_1\}$ is a boundary triplet for S^*. Above it was also shown that $A_0 = \ker \Gamma_0$. □

2.2 Boundary value problems

Let S be a closed symmetric relation in $\mathfrak{H}$ and let $\{\mathcal{G}, \Gamma_0, \Gamma_1\}$ be a boundary triplet for S^*. Due to Theorem 2.1.3 one may think of the intermediate extensions of S being parametrized by the relations in the space $\mathcal{G}$; for this reason the space $\mathcal{G}$ will often be called the *boundary space* or *parameter space* associated with the boundary triplet. Let Θ be a closed relation in $\mathcal{G}$ and let A_Θ be the corresponding closed extension of S in $\mathfrak{H}$ via (2.1.5):

$$A_\Theta = \{\widehat{f} \in S^* : \Gamma\widehat{f} \in \Theta\} = \ker(\Gamma_1 - \Theta\Gamma_0). \tag{2.2.1}$$

Recall from Section 1.10 that any closed relation Θ in $\mathcal{G}$ has a parametric representation of the form $\Theta = \{\mathcal{A}, \mathcal{B}\}$, i.e.,

$$\Theta = \{\{\mathcal{A}e, \mathcal{B}e\} : e \in \mathcal{E}\} \tag{2.2.2}$$

with some operators $\mathcal{A}, \mathcal{B} \in \mathbf{B}(\mathcal{E}, \mathcal{G})$ and a Hilbert space $\mathcal{E}$. Likewise, since Θ^* is closed, it has a representation of the form

$$\Theta^* = \{\{\mathcal{C}e', \mathcal{D}e'\} : e' \in \mathcal{E}'\} \tag{2.2.3}$$

with some operators $\mathcal{C}, \mathcal{D} \in \mathbf{B}(\mathcal{E}', \mathcal{G})$ and a Hilbert space $\mathcal{E}'$. Thus, (2.2.3) gives

$$\Theta = \{\{\varphi, \varphi'\} \in \mathcal{G} \times \mathcal{G} : \mathcal{D}^*\varphi = \mathcal{C}^*\varphi'\}. \tag{2.2.4}$$

Therefore, it follows that A_Θ in (2.2.1) can be written as

$$A_\Theta = \{\widehat{f} \in S^* : \mathcal{D}^*\Gamma_0\widehat{f} = \mathcal{C}^*\Gamma_1\widehat{f}\}. \tag{2.2.5}$$

In the following it will be shown how the pair $\{\mathcal{C}, \mathcal{D}\}$ in (2.2.3) and (2.2.5) can be expressed in terms of the original pair $\{\mathcal{A}, \mathcal{B}\}$ in (2.2.2). The main result is contained in the next proposition.

Recall that the condition that $\Theta = \{\mathcal{A}, \mathcal{B}\}$ is closed with some $\mathcal{A}, \mathcal{B} \in \mathbf{B}(\mathcal{E}, \mathcal{G})$ is equivalent to the condition that $\operatorname{ran}(\mathcal{A}^*\mathcal{A} + \mathcal{B}^*\mathcal{B})$ is closed in $\mathcal{E}$; cf. Proposition 1.10.3. In fact, in the case where Θ is closed one may assume that the representing pair $\{\mathcal{A}, \mathcal{B}\}$ satisfies the normalization condition $\mathcal{A}^*\mathcal{A} + \mathcal{B}^*\mathcal{B} = I$; cf. Proposition 1.10.3.

Proposition 2.2.1. *Let S be a closed symmetric relation in $\mathfrak{H}$, let $\{\mathcal{G}, \Gamma_0, \Gamma_1\}$ be a boundary triplet for S^*, and let A_Θ be the closed extension of S in $\mathfrak{H}$ corresponding to the closed relation Θ in $\mathcal{G}$ via (2.1.5). Assume that Θ has the representation $\Theta = \{\mathcal{A}, \mathcal{B}\}$ with $\mathcal{A}, \mathcal{B} \in \mathbf{B}(\mathcal{E}, \mathcal{G})$ such that*

$$\mathcal{A}^*\mathcal{A} + \mathcal{B}^*\mathcal{B} = I. \tag{2.2.6}$$

Then the intermediate extension A_Θ in (2.2.1) can be described as

$$A_\Theta = \left\{ \widehat{f} \in S^* : \begin{pmatrix} \mathcal{B}\mathcal{A}^* \\ I - \mathcal{A}\mathcal{A}^* \end{pmatrix} \Gamma_0\widehat{f} = \begin{pmatrix} I - \mathcal{B}\mathcal{B}^* \\ \mathcal{A}\mathcal{B}^* \end{pmatrix} \Gamma_1\widehat{f} \right\}. \tag{2.2.7}$$

Proof. By Proposition 1.10.10, condition (2.2.6) implies that the relation Θ is given by

$$\Theta = \left\{ \{\varphi, \varphi'\} \in \mathcal{G}^2 : \begin{pmatrix} \mathcal{B}\mathcal{A}^* \\ I - \mathcal{A}\mathcal{A}^* \end{pmatrix} \varphi = \begin{pmatrix} I - \mathcal{B}\mathcal{B}^* \\ \mathcal{A}\mathcal{B}^* \end{pmatrix} \varphi' \right\}.$$

Then (2.2.7) follows from (2.1.5). □

In the next proposition it will be assumed, in addition, that $\rho(\Theta) \neq \emptyset$. The following result is a reformulation of Theorem 1.10.5 and formula (2.1.5).

Proposition 2.2.2. *Let S be a closed symmetric relation in $\mathfrak{H}$, let $\{\mathcal{G}, \Gamma_0, \Gamma_1\}$ be a boundary triplet for S^*, and let A_Θ be the closed extension of S in $\mathfrak{H}$ corresponding to the closed relation Θ in $\mathcal{G}$ via (2.1.5). Then $\mu \in \rho(\Theta)$ if and only if Θ has the representation $\Theta = \{\mathcal{A}, \mathcal{B}\}$ with $\mathcal{A}, \mathcal{B} \in \mathbf{B}(\mathcal{G})$ such that $(\mathcal{B} - \mu\mathcal{A})^{-1} \in \mathbf{B}(\mathcal{G})$. Moreover, the pair $\{\mathcal{A}, \mathcal{B}\}$ may be chosen such that $\Theta^* = \{\mathcal{A}^*, \mathcal{B}^*\}$. In this case the intermediate extension A_Θ in (2.2.1) can be described as*

$$A_\Theta = \{\widehat{f} \in S^* : \mathcal{B}\Gamma_0\widehat{f} = \mathcal{A}\Gamma_1\widehat{f}\}. \tag{2.2.8}$$

For $\mu \in \rho(\Theta)$ it follows from (1.10.9) that in Proposition 2.2.2 one can choose

$$\mathcal{A} = (\Theta - \mu)^{-1} \quad \text{and} \quad \mathcal{B} = I + \mu(\Theta - \mu)^{-1}. \tag{2.2.9}$$

In the case that $\mu \in \mathbb{C} \setminus \mathbb{R}$ one may also choose

$$\mathcal{A} = I - \mathcal{C}_\mu[\Theta] \quad \text{and} \quad \mathcal{B} = \bar{\mu} - \mu \mathcal{C}_\mu[\Theta] \tag{2.2.10}$$

by (1.10.10), where $\mathcal{C}_\mu$ denotes the Cayley transform.

The next corollary is a translation of Corollary 2.1.4 and Corollary 1.10.8. In each of the cases in this corollary one may apply Proposition 2.2.2 by choosing the pair $\{\mathcal{A}, \mathcal{B}\}$ as in (2.2.9) or (2.2.10) with $\mu \in \mathbb{C} \setminus \mathbb{R}$, $\mu \in \mathbb{C}^+$, or $\mu \in \mathbb{C}^-$, respectively.

Corollary 2.2.3. *Let S be a closed symmetric relation in $\mathfrak{H}$, let $\{\mathcal{G}, \Gamma_0, \Gamma_1\}$ be a boundary triplet for S^*, and let A_Θ be the closed extension of S in $\mathfrak{H}$ corresponding to the closed relation Θ in $\mathcal{G}$ via (2.1.5). Assume that $\Theta = \{\mathcal{A}, \mathcal{B}\}$. Then the following statements hold:*

(i) *A_Θ is self-adjoint if and only if*

$$\operatorname{Im}(\mathcal{A}^*\mathcal{B}) = 0 \quad \textit{and} \quad (\mathcal{B} - \mu\mathcal{A})^{-1} \in \mathbf{B}(\mathcal{G})$$

for some, and hence for all $\mu \in \mathbb{C}^+$ and for some, and hence for all $\mu \in \mathbb{C}^-$;

(ii) *A_Θ is maximal dissipative if and only if*

$$\operatorname{Im}(\mathcal{A}^*\mathcal{B}) \geq 0 \quad \textit{and} \quad (\mathcal{B} - \mu\mathcal{A})^{-1} \in \mathbf{B}(\mathcal{G})$$

for some, and hence for all $\mu \in \mathbb{C}^-$;

(iii) *A_Θ is maximal accumulative if and only if*

$$\operatorname{Im}(\mathcal{A}^*\mathcal{B}) \leq 0 \quad \textit{and} \quad (\mathcal{B} - \mu\mathcal{A})^{-1} \in \mathbf{B}(\mathcal{G})$$

for some, and hence for all $\mu \in \mathbb{C}^+$.

In the case that the representation $\Theta = \{\mathcal{A}, \mathcal{B}\}$ is chosen so that $\Theta^* = \{\mathcal{A}^*, \mathcal{B}^*\}$, the extension A_Θ is given by (2.2.8).

Now the converse question will be addressed. Let A be a closed extension of S given in terms of boundary conditions. The problem is to determine a corresponding parameter Θ in $\mathcal{G}$ such that $A = A_\Theta$.

Proposition 2.2.4. *Let S be a closed symmetric relation in $\mathfrak{H}$, and let $\{\mathcal{G}, \Gamma_0, \Gamma_1\}$ be a boundary triplet for S^*. Assume that $\mathfrak{F}$ is a Hilbert space, $\mathcal{M}, \mathcal{N} \in \mathbf{B}(\mathcal{G}, \mathfrak{F})$, and that, without loss of generality, the space $\mathfrak{F}$ is minimal:*

$$\mathfrak{F} = \overline{\operatorname{span}}\,\{\overline{\operatorname{ran}}\,\mathcal{M}, \overline{\operatorname{ran}}\,\mathcal{N}\}.$$

Furthermore, assume that $\mathcal{M} - \mu\mathcal{N} \in \mathbf{B}(\mathcal{G}, \mathfrak{F})$ is bijective for some $\mu \in \mathbb{C}$ and let A be an intermediate extension of S of the form

$$A = \{\widehat{f} \in S^* : \mathcal{M}\Gamma_0\widehat{f} = \mathcal{N}\Gamma_1\widehat{f}\}. \tag{2.2.11}$$

Then A is closed and $A = A_\Theta$, where the parameter $\Theta = \{\mathcal{A}, \mathcal{B}\}$ is given by

$$\{\mathcal{A}, \mathcal{B}\} = \{(\mathcal{M} - \mu\mathcal{N})^{-1}\mathcal{N}, (\mathcal{M} - \mu\mathcal{N})^{-1}\mathcal{M}\}.$$

Proof. First observe that the intermediate extension A in (2.2.11) is closed since $\mathcal{M}, \mathcal{N} \in \mathbf{B}(\mathcal{G}, \mathfrak{F})$. Moreover, A corresponds to the closed relation Θ in $\mathcal{G}$ given by

$$\Theta = \big\{\{\varphi, \varphi'\} \in \mathcal{G} \times \mathcal{G} : \mathcal{M}\varphi = \mathcal{N}\varphi'\big\}.$$

Now the assertion follows from Proposition 1.10.7. □

Let again Θ be a closed relation in $\mathcal{G}$ and let A_Θ be the corresponding closed extension in $\mathfrak{H}$ via (2.1.5). Assume, in addition, that Θ admits an orthogonal decomposition

$$\Theta = \Theta_{\mathrm{op}} \,\widehat{\oplus}\, \Theta_{\mathrm{mul}}, \qquad \mathcal{G} = \mathcal{G}_{\mathrm{op}} \oplus \mathcal{G}_{\mathrm{mul}},$$

into a (not necessarily densely defined) operator part Θ_{op} acting in the Hilbert space $\mathcal{G}_{\mathrm{op}} = \overline{\mathrm{dom}}\,\Theta^* = (\mathrm{mul}\,\Theta)^\perp$ and a multivalued part $\Theta_{\mathrm{mul}} = \{0\} \times \mathrm{mul}\,\Theta$ in the Hilbert space $\mathcal{G}_{\mathrm{mul}} = \mathrm{mul}\,\Theta$; cf. Theorem 1.3.16 and the discussion following it. Recall from Theorem 1.4.11, Theorem 1.5.1, and Theorem 1.6.12 that any closed symmetric, self-adjoint, (maximal) dissipative, or (maximal) accumulative relation Θ in $\mathcal{G}$ gives rise to such a decomposition. If P_{op} denotes the orthogonal projection in $\mathcal{G}$ onto $\mathcal{G}_{\mathrm{op}}$, then the closed extension A_Θ in (2.1.5) has the form

$$A_\Theta = \big\{\widehat{f} \in S^* : \Theta_{\mathrm{op}} P_{\mathrm{op}} \Gamma_0 \widehat{f} = P_{\mathrm{op}} \Gamma_1 \widehat{f},\, (I_{\mathcal{G}} - P_{\mathrm{op}})\Gamma_0 \widehat{f} = 0\big\}. \tag{2.2.12}$$

Note that this abstract boundary condition also requires $P_{\mathrm{op}}\Gamma_0\widehat{f} \in \mathrm{dom}\,\Theta_{\mathrm{op}}$.

2.3 Associated γ-fields and Weyl functions

Let S be a closed symmetric relation in the Hilbert space $\mathfrak{H}$ and let $\{\mathcal{G}, \Gamma_0, \Gamma_1\}$ be a boundary triplet for S^*. Recall from Lemma 2.1.7 that Γ_0 maps $\widehat{\mathfrak{N}}_\lambda(S^*)$ bijectively onto $\mathcal{G}$ when $\lambda \in \rho(A_0)$. Hence, the inverse mapping

$$\widehat{\gamma}(\lambda) := \big(\Gamma_0 \upharpoonright \widehat{\mathfrak{N}}_\lambda(S^*)\big)^{-1}, \quad \lambda \in \rho(A_0),$$

maps $\mathcal{G}$ bijectively onto $\widehat{\mathfrak{N}}_\lambda(S^*)$. Let π_1 be the orthogonal projection from $\mathfrak{H} \times \mathfrak{H}$ onto $\mathfrak{H} \times \{0\}$. Then π_1 maps $\widehat{\mathfrak{N}}_\lambda(S^*)$ bijectively onto $\mathfrak{N}_\lambda(S^*)$.

Definition 2.3.1. Let S be a closed symmetric relation in $\mathfrak{H}$, let $\{\mathcal{G}, \Gamma_0, \Gamma_1\}$ be a boundary triplet for S^*, and let $A_0 = \ker \Gamma_0$. Then

$$\rho(A_0) \ni \lambda \mapsto \gamma(\lambda) = \big\{\{\Gamma_0 \widehat{f}_\lambda, f_\lambda\} : \widehat{f}_\lambda \in \widehat{\mathfrak{N}}_\lambda(S^*)\big\} \tag{2.3.1}$$

or, equivalently,

$$\rho(A_0) \ni \lambda \mapsto \gamma(\lambda) = \pi_1 \widehat{\gamma}(\lambda) = \pi_1 \big(\Gamma_0 \upharpoonright \widehat{\mathfrak{N}}_\lambda(S^*)\big)^{-1},$$

is called the *γ-field* associated with the boundary triplet $\{\mathcal{G}, \Gamma_0, \Gamma_1\}$.

The main properties of the γ-field will now be discussed.

Proposition 2.3.2. *Let S be a closed symmetric relation in $\mathfrak{H}$, let $\{\mathcal{G},\Gamma_0,\Gamma_1\}$ be a boundary triplet for S^*, and let $A_0 = \ker\Gamma_0$. Then the following statements hold for the corresponding γ-field γ:*

(i) *$\gamma(\lambda) \in \mathbf{B}(\mathcal{G},\mathfrak{H})$ for all $\lambda \in \rho(A_0)$ and, in fact, $\gamma(\lambda)$ maps $\mathcal{G}$ isomorphically onto $\mathfrak{N}_\lambda(S^*) \subset \mathfrak{H}$;*

(ii) *for all $\lambda, \mu \in \rho(A_0)$ the operators $\gamma(\lambda)$ and $\gamma(\mu)$ are connected via*

$$\gamma(\lambda) = \big(I + (\lambda-\mu)(A_0-\lambda)^{-1}\big)\gamma(\mu);$$

(iii) *the operator function $\gamma : \rho(A_0) \to \mathbf{B}(\mathcal{G},\mathfrak{H})$, $\lambda \mapsto \gamma(\lambda)$, is holomorphic, i.e., the limit*

$$\frac{d}{d\mu}\gamma(\mu) = \lim_{\lambda\to\mu} \frac{\gamma(\lambda)-\gamma(\mu)}{\lambda-\mu}$$

exists for all $\mu \in \rho(A_0)$ in $\mathbf{B}(\mathcal{G},\mathfrak{H})$;

(iv) *for all $\lambda \in \rho(A_0)$ the operator $\gamma(\lambda)^* \in \mathbf{B}(\mathfrak{H},\mathcal{G})$ is given by*

$$\gamma(\lambda)^*h = \Gamma_1\big\{(A_0-\bar\lambda)^{-1}h, \big(I+\bar\lambda(A_0-\bar\lambda)^{-1}\big)h\big\}, \quad h \in \mathfrak{H}, \tag{2.3.2}$$

and $\ker\gamma(\lambda)^* = (\mathfrak{N}_\lambda(S^*))^\perp = \operatorname{ran}(S-\bar\lambda)$ holds. Moreover, one has

$$\Gamma\big\{(A_0-\bar\lambda)^{-1}h, \big(I+\bar\lambda(A_0-\bar\lambda)^{-1}\big)h\big\} = \{0,\gamma(\lambda)^*h\}, \quad h \in \mathfrak{H}. \tag{2.3.3}$$

Proof. (i) Let $\lambda \in \rho(A_0)$. Since the restriction of Γ_0 to $\widehat{\mathfrak{N}}_\lambda(S^*)$ is an isomorphism from $\widehat{\mathfrak{N}}_\lambda(S^*)$ onto $\mathcal{G}$ (see Lemma 2.1.7), while π_1 is an isomorphism from $\widehat{\mathfrak{N}}_\lambda(S^*)$ onto $\mathfrak{N}_\lambda(S^*)$, it follows from Definition 2.3.1 that the mapping $\gamma(\lambda)$ is an isomorphism from $\mathcal{G}$ onto $\mathfrak{N}_\lambda(S^*)$. From this it is also clear that $\gamma(\lambda) \in \mathbf{B}(\mathcal{G},\mathfrak{H})$.

(ii) Let $\lambda,\mu \in \rho(A_0)$ and let $\varphi \in \mathcal{G}$. Then there exists $\widehat{f}_\mu = \{f_\mu, \mu f_\mu\} \in \widehat{\mathfrak{N}}_\mu(S^*)$ such that $\varphi = \Gamma_0\widehat{f}_\mu$ and hence $f_\mu = \gamma(\mu)\varphi$. Due to $S^* = A_0 \,\widehat{+}\, \widehat{\mathfrak{N}}_\lambda(S^*)$ there exist $\widehat{h} \in A_0$ and $\widehat{f}_\lambda = \{f_\lambda, \lambda f_\lambda\} \in \widehat{\mathfrak{N}}_\lambda(S^*)$ such that

$$\widehat{f}_\mu = \widehat{h} + \widehat{f}_\lambda.$$

Observe that $\widehat{f}_\lambda - \widehat{f}_\mu = -\widehat{h} \in A_0$, which gives $\Gamma_0\widehat{f}_\lambda = \Gamma_0\widehat{f}_\mu$, so that $\Gamma_0\widehat{f}_\lambda = \varphi$ and $f_\lambda = \gamma(\lambda)\varphi$. Moreover, this observation also shows that for some $g \in \mathfrak{H}$

$$\{f_\lambda, \lambda f_\lambda\} = \{f_\mu, \mu f_\mu\} + \big\{(A_0-\lambda)^{-1}g, \big(I+\lambda(A_0-\lambda)^{-1}\big)g\big\}.$$

Hence, $\{f_\lambda, 0\} = \{f_\mu, (\mu-\lambda)f_\mu\} + \{(A_0-\lambda)^{-1}g, g\}$, so that $g = (\lambda-\mu)f_\mu$. Therefore, $f_\lambda = f_\mu + (\lambda-\mu)(A_0-\lambda)^{-1}f_\mu$, which implies

$$\gamma(\lambda)\varphi = \big(I + (\lambda-\mu)(A_0-\lambda)^{-1}\big)\gamma(\mu)\varphi.$$

(iii) Fix some $\mu \in \rho(A_0)$. Then it follows from (ii) and the fact that the mapping $\lambda \mapsto (A_0 - \lambda)^{-1}$ is a holomorphic operator function with values in $\mathbf{B}(\mathfrak{H})$ that $\lambda \mapsto \gamma(\lambda)$ is a holomorphic operator function on $\rho(A_0)$ with values in $\mathbf{B}(\mathcal{G}, \mathfrak{H})$.

(iv) Fix $\lambda \in \rho(A_0)$ and let $h \in \mathfrak{H}$. Then there exists $\widehat{k} = \{k, k'\} \in A_0$ with $h = k' - \bar{\lambda} k$. Let $\varphi \in \mathcal{G}$; then $\gamma(\lambda)\varphi = f_\lambda$ for some $f_\lambda \in \mathfrak{N}_\lambda(S^*)$. Hence, with the abstract Green identity for $\widehat{k} = \{k, k'\}$ and $\widehat{f}_\lambda = \{f_\lambda, \lambda f_\lambda\}$ it follows from $\Gamma_0 \widehat{k} = 0$ that

$$\begin{aligned}
(\varphi, \gamma(\lambda)^* h) &= \big(\gamma(\lambda)\varphi, k' - \bar{\lambda} k\big) \\
&= (f_\lambda, k' - \bar{\lambda} k) \\
&= -\big((\lambda f_\lambda, k) - (f_\lambda, k')\big) \\
&= -\big((\Gamma_1 \widehat{f}_\lambda, \Gamma_0 \widehat{k}) - (\Gamma_0 \widehat{f}_\lambda, \Gamma_1 \widehat{k})\big) \\
&= (\Gamma_0 \widehat{f}_\lambda, \Gamma_1 \widehat{k}) \\
&= (\varphi, \Gamma_1 \widehat{k}),
\end{aligned}$$

which implies

$$\gamma(\lambda)^* h = \Gamma_1 \widehat{k} = \Gamma_1 \big\{(A_0 - \bar{\lambda})^{-1} h, \big(I + \bar{\lambda}(A_0 - \bar{\lambda})^{-1}\big) h\big\}.$$

The identity $\ker \gamma(\lambda)^* = (\mathfrak{N}_\lambda(S^*))^\perp$ follows from $\operatorname{ran} \gamma(\lambda) = \mathfrak{N}_\lambda(S^*)$. Furthermore, the identity $(\mathfrak{N}_\lambda(S^*))^\perp = \operatorname{ran}(S - \bar{\lambda})$ is clear and (2.3.3) follows from (2.3.2) and

$$\big\{(A_0 - \bar{\lambda})^{-1} h, \big(I + \bar{\lambda}(A_0 - \bar{\lambda})^{-1}\big) h\big\} \in A_0 = \ker \Gamma_0.$$

This completes the proof. □

In the case where the symmetric relation S is a densely defined symmetric operator and $\{\mathcal{G}, \Gamma_0, \Gamma_1\}$ is a boundary triplet for S^* with boundary mappings Γ_0 and Γ_1 defined on $\operatorname{dom} S^*$ (see the text below Definition 2.1.1 and (2.1.2)) the formula for the adjoint $\gamma(\lambda)^*$ of the corresponding γ-field in Proposition 2.3.2 (iv) has the simpler form

$$\gamma(\lambda)^* h = \Gamma_1 (A_0 - \bar{\lambda})^{-1} h, \qquad \lambda \in \rho(A_0), \ h \in \mathfrak{H}.$$

According to Proposition 2.3.2 (iv), the action of Γ_1 on a general element of A_0 is expressed in terms of the operator $\gamma(\lambda)^*$. The form of this action is particularly simple on eigenelements of A_0.

Corollary 2.3.3. *Let $\lambda \in \rho(A_0)$ and assume that $\{h, xh\} \in A_0$ with $x \in \mathbb{R}$. Then*

$$\Gamma_1 \{h, xh\} = (x - \bar{\lambda}) \gamma(\lambda)^* h.$$

If $\{0, h\} \in A_0$, then

$$\Gamma_1 \{0, h\} = \gamma(\lambda)^* h.$$

Proof. Let $\lambda \in \rho(A_0)$ and let $\{h, xh\} \in A_0$ with $x \in \mathbb{R}$. Then

$$h = (A_0 - \bar{\lambda})^{-1}(x - \bar{\lambda})h,$$

which together with Proposition 2.3.2 (iv) leads to

$$\begin{aligned}(x - \bar{\lambda})\gamma(\lambda)^* h &= (x - \bar{\lambda})\Gamma_1\{(A_0 - \bar{\lambda})^{-1}h, \big(I + \bar{\lambda}(A_0 - \bar{\lambda})^{-1}\big)h\} \\ &= \Gamma_1\{h, xh\}.\end{aligned}$$

If $\{0, h\} \in A_0$, then $h \in \ker(A_0 - \bar{\lambda})^{-1}$ and the expression for $\Gamma_1\{0, h\}$ follows directly from Proposition 2.3.2 (iv). □

The definition and properties of the γ-field now give rise to the notion of Weyl function. It is defined, as in the case of the γ-field, for a closed symmetric relation S in terms of the boundary triplet for S^* and the eigenspaces of S^*.

Definition 2.3.4. Let S be a closed symmetric relation in $\mathfrak{H}$, let $\{\mathcal{G}, \Gamma_0, \Gamma_1\}$ be a boundary triplet for S^*, and let $A_0 = \ker \Gamma_0$. Then

$$\rho(A_0) \ni \lambda \mapsto M(\lambda) = \big\{\{\Gamma_0 \widehat{f}_\lambda, \Gamma_1 \widehat{f}_\lambda\} : \widehat{f}_\lambda \in \widehat{\mathfrak{N}}_\lambda(S^*)\big\} \tag{2.3.4}$$

or, equivalently,

$$\rho(A_0) \ni \lambda \mapsto M(\lambda) = \Gamma_1 \widehat{\gamma}(\lambda) = \Gamma_1 \big(\Gamma_0 \restriction \widehat{\mathfrak{N}}_\lambda(S^*)\big)^{-1},$$

is called the *Weyl function* associated with the boundary triplet $\{\mathcal{G}, \Gamma_0, \Gamma_1\}$.

Here is a simple example of a Weyl function for a trivial symmetric relation S in $\mathfrak{H}$. Note that in this example one has $\mathcal{G} = \mathfrak{H}$, i.e., the corresponding boundary triplet maps onto $\mathfrak{H} \times \mathfrak{H}$; this situation is not typical in standard applications; cf. Chapters 6, 7, and 8.

Example 2.3.5. Let $S = \{0, 0\}$ be the trivial symmetric relation in $\mathfrak{H}$. It is clear that $S^* = \mathfrak{H} \times \mathfrak{H}$ and $\mathfrak{N}_\lambda(S^*) = \mathfrak{H}$ for $\lambda \in \mathbb{C}$. Now define

$$\Gamma_0 \widehat{f} = f' \quad \text{and} \quad \Gamma_1 \widehat{f} = -f, \quad \widehat{f} = \{f, f'\} \in S^*,$$

so that $\Gamma : S^* \to \mathfrak{H} \times \mathfrak{H}$ is surjective and (2.1.1) is satisfied. Hence, $\{\mathfrak{H}, \Gamma_0, \Gamma_1\}$ is a boundary triplet for S^*. Note that

$$A_0 = \ker \Gamma_0 = \mathfrak{H} \times \{0\}$$

is a self-adjoint extension of S with $\rho(A_0) = \mathbb{C}\setminus\{0\}$, $\sigma(A_0) = \{0\}$, and $\mathfrak{N}_0(A_0) = \mathfrak{H}$. It follows from Definition 2.3.1 and Definition 2.3.4 that the γ-field and the Weyl function are given by $\gamma(\lambda) = (1/\lambda)I$ and $M(\lambda) = -(1/\lambda)I$, respectively.

Next some elementary properties of the Weyl function are discussed. Recall that the *real part* and *imaginary part* of a bounded operator $T \in \mathbf{B}(\mathcal{G})$ are defined as $\operatorname{Re} T = \frac{1}{2}(T + T^*)$ and $\operatorname{Im} T = \frac{1}{2i}(T - T^*)$, respectively.

Proposition 2.3.6. *Let S be a closed symmetric relation in $\mathfrak{H}$, let $\{\mathcal{G}, \Gamma_0, \Gamma_1\}$ be a boundary triplet for S^*, and let $A_0 = \ker \Gamma_0$. Then the following statements hold* *for the corresponding γ-field γ and Weyl function M:*

(i) $M(\lambda) \in \mathbf{B}(\mathcal{G})$ *for all* $\lambda \in \rho(A_0)$;

(ii) *for all* $\lambda \in \rho(A_0)$ *one has* $M(\lambda)\Gamma_0 \widehat{f}_\lambda = \Gamma_1 \widehat{f}_\lambda$ *for every* $\widehat{f}_\lambda \in \widehat{\mathfrak{N}}_\lambda(S^*)$ *or, equivalently, one has*

$$\Gamma\widehat{\gamma}(\lambda)\varphi = \{\Gamma_0\widehat{\gamma}(\lambda)\varphi, \Gamma_1\widehat{\gamma}(\lambda)\varphi\} = \{\varphi, M(\lambda)\varphi\},$$

for every $\varphi \in \mathcal{G}$;

(iii) *for all* $\lambda, \mu \in \rho(A_0)$ *the identity*

$$M(\lambda) - M(\mu)^* = (\lambda - \bar{\mu})\gamma(\mu)^*\gamma(\lambda)$$

holds, and, in particular, the symmetry condition $M(\lambda)^* = M(\bar{\lambda})$ *holds for all* $\lambda \in \rho(A_0)$;

(iv) $\operatorname{Im} M(\lambda) \in \mathbf{B}(\mathcal{G})$ *is a nonnegative (nonpositive) self-adjoint operator for all* $\lambda \in \mathbb{C}^+$ $(\lambda \in \mathbb{C}^-)$ *and* $0 \in \rho(\operatorname{Im} M(\lambda))$ *for all* $\lambda \in \mathbb{C} \setminus \mathbb{R}$;

(v) *for any fixed* $\lambda_0 \in \rho(A_0)$ *and all* $\lambda \in \rho(A_0)$ *the resolvent of* A_0 *and the function* M *are connected via*

$$M(\lambda) = \operatorname{Re} M(\lambda_0) + \gamma(\lambda_0)^*\bigl[\lambda - \operatorname{Re}\lambda_0 + (\lambda - \lambda_0)(\lambda - \bar{\lambda}_0)(A_0 - \lambda)^{-1}\bigr]\gamma(\lambda_0);$$

(vi) *the identity*

$$\gamma(\bar{\mu})^*(A_0 - \lambda)^{-1}\gamma(\nu) = \frac{M(\lambda)}{(\lambda - \nu)(\lambda - \mu)} + \frac{M(\mu)}{(\mu - \lambda)(\mu - \nu)} + \frac{M(\nu)}{(\nu - \lambda)(\nu - \mu)}$$

holds for $\lambda, \mu, \nu \in \rho(A_0)$ *such that* $\lambda \neq \nu$, $\lambda \neq \mu$, *and* $\nu \neq \mu$.

Proof. (i) Let $\lambda \in \rho(A_0)$. By Lemma 2.1.7, the restriction of Γ_0 to $\widehat{\mathfrak{N}}_\lambda(S^*)$ is an isomorphism between $\widehat{\mathfrak{N}}_\lambda(S^*)$ and $\mathcal{G}$. Hence, the inverse $\widehat{\gamma}(\lambda)$ is an isomorphism between $\mathcal{G}$ and $\widehat{\mathfrak{N}}_\lambda(S^*)$, and since the operator $\Gamma_1 : S^* \to \mathcal{G}$ is continuous by Proposition 2.1.2 (i), it follows from Definition 2.3.4 that $M(\lambda) = \Gamma_1\widehat{\gamma}(\lambda) \in \mathbf{B}(\mathcal{G})$.

(ii) It is clear from (i) and the definition of $M(\lambda)$ that $M(\lambda)\Gamma_0\widehat{f}_\lambda = \Gamma_1\widehat{f}_\lambda$ for every $\widehat{f}_\lambda \in \widehat{\mathfrak{N}}_\lambda(S^*)$. Now $\widehat{\gamma}(\lambda)\varphi$ belongs to $\widehat{\mathfrak{N}}_\lambda(S^*)$ for $\varphi \in \mathcal{G}$, so that

$$\{\Gamma_0\widehat{\gamma}(\lambda)\varphi, \Gamma_1\widehat{\gamma}(\lambda)\varphi\} = \{\Gamma_0\widehat{\gamma}(\lambda)\varphi, M(\lambda)\Gamma_0\widehat{\gamma}(\lambda)\varphi\} = \{\varphi, M(\lambda)\varphi\}.$$

Conversely, assume that $\{\Gamma_0\widehat{\gamma}(\lambda)\varphi, \Gamma_1\widehat{\gamma}(\lambda)\varphi\} = \{\varphi, M(\lambda)\varphi\}$ for all $\varphi \in \mathcal{G}$ and let $\widehat{f}_\lambda \in \widehat{\mathfrak{N}}_\lambda(S^*)$. Then $\widehat{f}_\lambda = \widehat{\gamma}(\lambda)\varphi$ for some $\varphi \in \mathcal{G}$ and hence

$$\{\Gamma_0\widehat{f}_\lambda, \Gamma_1\widehat{f}_\lambda\} = \{\Gamma_0\widehat{\gamma}(\lambda)\varphi, \Gamma_1\widehat{\gamma}(\lambda)\varphi\} = \{\varphi, M(\lambda)\varphi\}$$

yields $M(\lambda)\Gamma_0\widehat{f}_\lambda = \Gamma_1\widehat{f}_\lambda$.

(iii) Let $\lambda, \mu \in \rho(A_0)$. For given $\varphi, \psi \in \mathcal{G}$ one can choose

$$\widehat{h}_\lambda = \{h_\lambda, \lambda h_\lambda\} \in \widehat{\mathfrak{N}}_\lambda(S^*) \quad \text{and} \quad \widehat{k}_\mu = \{k_\mu, \mu k_\mu\} \in \widehat{\mathfrak{N}}_\mu(S^*),$$

such that $\varphi = \Gamma_0 \widehat{h}_\lambda$ and $\psi = \Gamma_0 \widehat{k}_\mu$. Clearly, $\gamma(\lambda)\varphi = h_\lambda$, $\gamma(\mu)\psi = k_\mu$, and the abstract Green identity applied to $\widehat{h}_\lambda$ and $\widehat{k}_\mu$ shows that

$$\begin{aligned}
\big((M(\lambda) - M(\mu)^*)\varphi, \psi\big) &= (M(\lambda)\varphi, \psi) - (\varphi, M(\mu)\psi) \\
&= \big(M(\lambda)\Gamma_0 \widehat{h}_\lambda, \Gamma_0 \widehat{k}_\mu\big) - \big(\Gamma_0 \widehat{h}_\lambda, M(\mu)\Gamma_0 \widehat{k}_\mu\big) \\
&= (\Gamma_1 \widehat{h}_\lambda, \Gamma_0 \widehat{k}_\mu) - (\Gamma_0 \widehat{h}_\lambda, \Gamma_1 \widehat{k}_\mu) \\
&= (\lambda h_\lambda, k_\mu) - (h_\lambda, \mu k_\mu) \\
&= (\lambda - \overline{\mu})(h_\lambda, k_\mu) \\
&= \big((\lambda - \overline{\mu})\gamma(\lambda)\varphi, \gamma(\mu)\psi\big).
\end{aligned}$$

Thus, one has the identity $M(\lambda) - M(\mu)^* = (\lambda - \overline{\mu})\gamma(\mu)^*\gamma(\lambda)$. Setting $\mu = \overline{\lambda}$ it follows that $M(\lambda) = M(\overline{\lambda})^*$ and therefore $M(\lambda)^* = M(\overline{\lambda})$, $\lambda \in \rho(A_0)$.

(iv) The assertion (iii) gives for $\lambda \in \mathbb{C} \setminus \mathbb{R}$

$$\frac{(\operatorname{Im} M(\lambda)\varphi, \varphi)}{\operatorname{Im} \lambda} = (\gamma(\lambda)^*\gamma(\lambda)\varphi, \varphi) = \|\gamma(\lambda)\varphi\|^2, \quad \varphi \in \mathcal{G}.$$

Hence, for $\lambda \in \mathbb{C}^+$ or $\lambda \in \mathbb{C}^-$ the operator $\operatorname{Im} M(\lambda)$ is nonnegative or nonpositive, respectively. As $\gamma(\lambda)$ is an isomorphism from $\mathcal{G}$ onto $\mathfrak{N}_\lambda(S^*)$ it follows that for all $\lambda \in \mathbb{C} \setminus \mathbb{R}$ the operator $\operatorname{Im} M(\lambda)$ is boundedly invertible.

(v) Let $\lambda_0 \in \rho(A_0)$ be fixed. Then assertion (iii) implies

$$\operatorname{Im} M(\lambda_0) = (\operatorname{Im} \lambda_0)\gamma(\lambda_0)^*\gamma(\lambda_0),$$

while $\gamma(\lambda) = (I + (\lambda - \lambda_0)(A_0 - \lambda)^{-1})\gamma(\lambda_0)$, $\lambda \in \rho(A_0)$, by Proposition 2.3.2. Using (iii) this leads to

$$\begin{aligned}
M(\lambda) &= M(\lambda_0)^* + (\lambda - \overline{\lambda}_0)\gamma(\lambda_0)^*\gamma(\lambda) \\
&= \operatorname{Re} M(\lambda_0) - i \operatorname{Im} M(\lambda_0) + (\lambda - \overline{\lambda}_0)\gamma(\lambda_0)^*\big[I + (\lambda - \lambda_0)(A_0 - \lambda)^{-1}\big]\gamma(\lambda_0) \\
&= \operatorname{Re} M(\lambda_0) + \gamma(\lambda_0)^*\big[(\lambda - \operatorname{Re} \lambda_0) + (\lambda - \lambda_0)(\lambda - \overline{\lambda}_0)(A_0 - \lambda)^{-1}\big]\gamma(\lambda_0)
\end{aligned}$$

for all $\lambda \in \rho(A_0)$.

(vi) It follows from item (iii) and $\gamma(\lambda) = (I + (\lambda - \nu)(A_0 - \lambda)^{-1})\gamma(\nu)$ in Proposition 2.3.2 (ii) that

$$\begin{aligned}
\gamma(\overline{\mu})^*(A_0 - \lambda)^{-1}\gamma(\nu) &= \gamma(\overline{\mu})^* \,\frac{\gamma(\lambda) - \gamma(\nu)}{\lambda - \nu} \\
&= \frac{1}{\lambda - \nu}\Big(\gamma(\overline{\mu})^*\gamma(\lambda) - \gamma(\overline{\mu})^*\gamma(\nu)\Big) \\
&= \frac{1}{\lambda - \nu}\left(\frac{M(\lambda) - M(\overline{\mu})^*}{\lambda - \mu} - \frac{M(\nu) - M(\overline{\mu})^*}{\nu - \mu}\right),
\end{aligned}$$

and a simple calculation using $M(\overline{\mu})^* = M(\mu)$ then yields the assertion. □

In the next corollary it turns out that the Weyl function M is a uniformly strict Nevanlinna function; cf. Definition A.4.1 and Definition A.4.7.

Corollary 2.3.7. *The Weyl function M in Definition 2.3.4 is a uniformly strict $\mathbf{B}(\mathcal{G})$-valued Nevanlinna function. Its values $M(\lambda)$ are maximal dissipative (maximal accumulative) operators for $\lambda \in \mathbb{C}^+$ ($\lambda \in \mathbb{C}^-$), and $-\lambda \in \rho(M(\lambda))$ for all $\lambda \in \mathbb{C} \setminus \mathbb{R}$.*

Proof. According to Proposition 2.3.2 (iii), the function $\lambda \mapsto \gamma(\lambda)$ is holomorphic on $\rho(A_0)$. Hence, it follows from Proposition 2.3.6 (iii) with fixed $\mu \in \rho(A_0)$ that the function $\lambda \mapsto M(\lambda)$ is holomorphic on $\rho(A_0)$ and hence, in particular, on the possibly smaller subset $\mathbb{C} \setminus \mathbb{R}$. Clearly, according to Proposition 2.3.6 (iii) and (iv) one has $M(\lambda)^* = M(\bar{\lambda})$ and $(\operatorname{Im} \lambda)(\operatorname{Im} M(\lambda)) \geq 0$ for $\lambda \in \mathbb{C} \setminus \mathbb{R}$, and hence M is a $\mathbf{B}(\mathcal{G})$-valued Nevanlinna function. It follows from Proposition 2.3.6 (iv) that M is uniformly strict. □

Corollary 2.3.8. *Let M be the Weyl function in Definition 2.3.4. Then the following statements hold:*

(i) *if $x \in \rho(A_0) \cap \mathbb{R}$, then $M(x) \in \mathbf{B}(\mathcal{G})$ is self-adjoint;*

(ii) *if $x \in \rho(A_0) \cap \mathbb{R}$, then the derivative $M'(x) \in \mathbf{B}(\mathcal{G})$ is a nonnegative self-adjoint operator and $0 \in \rho(M'(x))$;*

(iii) *if $(a, b) \subset \mathbb{R}$ belongs to $\rho(A_0)$, then for all $\varphi \in \mathcal{G}$ the function*

$$x \mapsto (M(x)\varphi, \varphi)$$

is nondecreasing on (a, b);

(iv) *if $(a, b) \subset \mathbb{R}$ belongs to $\rho(A_0)$, then there exist self-adjoint relations $M(a)$ and $M(b)$ in $\mathcal{G}$ such that*

$$M(b) = \lim_{x \uparrow b} M(x) \quad \text{and} \quad M(a) = \lim_{x \downarrow a} M(x)$$

in the strong graph sense or, equivalently, in the strong resolvent sense on $\mathbb{C} \setminus \mathbb{R}$.

Proof. (i) It follows from $M(\lambda)^* = M(\bar{\lambda})$, $\lambda \in \rho(A_0)$, that for $x \in \rho(A_0) \cap \mathbb{R}$ one has that $M(x)^* = M(x)$, i.e., $M(x) \in \mathbf{B}(\mathcal{G})$ is self-adjoint.

(ii) Since M is holomorphic on $\rho(A_0)$, it is clear that the derivative $M'(x) \in \mathbf{B}(\mathcal{G})$ exists. Moreover, for all $\varphi \in \mathcal{G}$ and $y \neq x$ Proposition 2.3.6 (iii) shows that

$$\begin{aligned}(M'(x)\varphi, \varphi) &= \lim_{y \to x} \frac{(M(x)\varphi, \varphi) - (M(y)\varphi, \varphi)}{x - y} \\ &= \lim_{y \to x} (\gamma(x)\varphi, \gamma(y)\varphi) = \|\gamma(x)\varphi\|^2\end{aligned}$$

and hence $M'(x) \geq 0$ is self-adjoint. Since $\gamma(x)$ maps $\mathcal{G}$ isomorphically onto $\mathfrak{N}_x(S^*)$ it also follows that $0 \in \rho(M'(x))$.

(iii) If $(a, b) \subset \mathbb{R}$ belongs to $\rho(A_0)$, then for all $\varphi \in \mathcal{G}$ the mapping $x \mapsto (M(x)\varphi, \varphi)$ is differentiable and (ii) implies that $x \mapsto (M(x)\varphi, \varphi)$ is nondecreasing on (a, b).

(iv) For $a < y < x < b$ it follows from (iii) that $(M(y)\varphi, \varphi) \leq (M(x)\varphi, \varphi)$ for all $\varphi \in \mathcal{G}$. If γ_y is a lower bound for $M(y)$, then Corollary 1.9.10 implies that there exists a semibounded self-adjoint relation $M(b)$ such that $M(x)$ converges in the strong resolvent sense to $M(b)$ on $\mathbb{C} \setminus [\gamma_y, \infty)$ when x tends to b. According to Corollary 1.9.6 (i), this is equivalent to strong graph convergence of $M(x)$ to $M(b)$.

The same considerations as above show that $(-M(y)\varphi, \varphi) \leq (-M(x)\varphi, \varphi)$ for $a < x < y < b$ and $\varphi \in \mathcal{G}$, and hence $-M(x)$ converges in the strong resolvent sense to a semibounded self-adjoint relation $-M(a)$ on $\mathbb{C} \setminus [\gamma_y, \infty)$ when x tends to a; here γ_y is a lower bound for $-M(y)$. This implies that $M(x)$ tends to $M(a)$ in the strong graph sense and in the strong resolvent sense. □

It is known that every isolated spectral point of a self-adjoint operator or relation A_0 is an eigenvalue and a pole of first order of the resolvent $\lambda \mapsto (A_0 - \lambda)^{-1}$. As a consequence of Proposition 2.3.6 (v), the isolated singularities of the Weyl function M are poles of first order. This is formulated in the next corollary, which can also be regarded as a simple example of the connection between the properties of the Weyl function M and the spectrum of A_0. The full connection between these objects is studied in detail in Section 3.5 and Section 3.6.

Corollary 2.3.9. *If $x \in \mathbb{R}$ is an isolated singularity of M and B_x is a disc centered at x such that M is holomorphic in $\overline{B}_x \setminus \{x\}$, then M admits a norm convergent Laurent series expansion of the form*

$$M(\lambda) = \frac{M_{-1}}{\lambda - x} + \sum_{k=0}^{\infty} M_k (\lambda - x)^k, \quad M_{-1}, M_0, M_1, \ldots \in \mathbf{B}(\mathcal{G}).$$

In particular,

$$\lim_{\lambda \to x} (\lambda - x) M(\lambda) = M_{-1} = \frac{1}{2\pi i} \int_{\mathcal{C}} M(\lambda)\, d\lambda,$$

where $\mathcal{C}$ denotes the boundary of B_x.

In the next remark it is explained that a self-adjoint part of a symmetric relation has, roughly speaking, no influence on the corresponding boundary triplet, γ-field, and Weyl function.

Remark 2.3.10. Let S be a closed symmetric relation in $\mathfrak{H}$, let $\{\mathcal{G}, \Gamma_0, \Gamma_1\}$ be a boundary triplet for S^*, and let $A_0 = \ker \Gamma_0$. Assume that $\mathfrak{H}$ admits an orthogonal decomposition $\mathfrak{H} = \mathfrak{H}' \oplus \mathfrak{H}''$ and that S has the orthogonal decomposition

$$S = S' \mathbin{\widehat{\oplus}} A, \tag{2.3.5}$$

where S' is a closed symmetric relation in $\mathfrak{H}'$ and A is a self-adjoint relation in $\mathfrak{H}''$. Then it follows from (2.3.5) and Proposition 1.3.13 that

$$S^* = (S')^* \,\widehat{\oplus}\, A, \tag{2.3.6}$$

where $(S')^*$ stands for the adjoint of S' in the space $\mathfrak{H}'$. Observe that according to (2.3.6) every element $\{f, f'\} \in S^*$ has the decomposition

$$\{f, f'\} = \{h, h'\} + \{k, k'\}, \quad \{h, h'\} \in (S')^*, \ \{k, k'\} \in A. \tag{2.3.7}$$

Since $A \subset S$, (2.3.7) shows that

$$\Gamma_0\{f, f'\} = \Gamma_0\{h, h'\} \quad \text{and} \quad \Gamma_1\{f, f'\} = \Gamma_1\{h, h'\}.$$

Hence, if Γ_0' and Γ_1' denote the restrictions of Γ_0 and Γ_1 to $(S')^*$, then it is easily seen that $\{\mathcal{G}, \Gamma_0', \Gamma_1'\}$ is a boundary triplet for $(S')^*$ such that

$$A_0' = \ker \Gamma_0', \quad A_0 = \ker \Gamma_0 = A_0' \,\widehat{\oplus}\, A.$$

Moreover, note that (2.3.6) shows

$$\widehat{\mathfrak{N}}_\lambda(S^*) = \widehat{\mathfrak{N}}_\lambda((S')^*), \quad \lambda \in \rho(A_0) = \rho(A_0') \cap \rho(A),$$

which implies that the Weyl function M' and the γ-field γ' satisfy

$$M'(\lambda) = M(\lambda), \quad \gamma'(\lambda) = \gamma(\lambda), \quad \lambda \in \rho(A_0).$$

For completeness observe that if H is a closed intermediate extension of S and $H' = H \cap (S')^*$, then $S' \subset H' \subset (S')^*$ and $H = H' \,\widehat{\oplus}\, A$. Hence, one may discard the self-adjoint part A in $\mathfrak{H}''$ without disturbing the boundary triplet structure.

2.4 Existence and construction of boundary triplets

Here the existence and possible constructions of boundary triplets based on decompositions in Section 1.7 are addressed. Recall first from Corollary 1.7.13 that a closed symmetric relation S in a Hilbert space $\mathfrak{H}$ admits self-adjoint extensions in $\mathfrak{H}$ if and only if the defect numbers

$$\dim \widehat{\mathfrak{N}}_\lambda(S^*) \quad \text{and} \quad \dim \widehat{\mathfrak{N}}_\mu(S^*) \tag{2.4.1}$$

of S are equal for some, and hence for all $\lambda \in \mathbb{C}^+$ and some, and hence for all $\mu \in \mathbb{C}^-$. Since any boundary triplet for S^* induces two self-adjoint extensions A_0 and A_1 of S it is clear that a boundary triplet can only exist if the defect numbers are equal. It turns out that this condition is also sufficient.

The following main result makes explicit how to construct a boundary triplet in terms of a given self-adjoint extension of a closed symmetric relation S (which exists if and only if the defect numbers in (2.4.1) coincide). The following notation will be used. For $\mu \in \mathbb{C}$ the natural embedding of $\mathfrak{N}_\mu(S^*)$ into $\mathfrak{H}$ is denoted by $\iota_{\mathfrak{N}_\mu(S^*)}$ and its adjoint is the orthogonal projection $P_{\mathfrak{N}_\mu(S^*)}$ from $\mathfrak{H}$ onto $\mathfrak{N}_\mu(S^*)$.

Theorem 2.4.1. *Let S be a closed symmetric relation in $\mathfrak{H}$ and assume that H is a self-adjoint extension of S in $\mathfrak{H}$. Fix $\mu \in \rho(H)$ and decompose S^* as*

$$S^* = H \,\widehat{+}\, \widehat{\mathfrak{N}}_\mu(S^*), \quad \text{direct sum.} \tag{2.4.2}$$

Let $\widehat{f} = \{f, f'\} \in S^*$ have the corresponding decomposition

$$\widehat{f} = \widehat{f}_0 + \widehat{f}_\mu, \tag{2.4.3}$$

with $\widehat{f}_0 = \{f_0, f_0'\} \in H$ and $\widehat{f}_\mu = \{f_\mu, \mu f_\mu\} \in \widehat{\mathfrak{N}}_\mu(S^*)$. Then

$$\Gamma_0 \widehat{f} := f_\mu \quad \text{and} \quad \Gamma_1 \widehat{f} := P_{\mathfrak{N}_\mu(S^*)}(f_0' - \bar{\mu} f_0 + \mu f_\mu) \tag{2.4.4}$$

define a boundary triplet $\{\mathfrak{N}_\mu(S^*), \Gamma_0, \Gamma_1\}$ for S^* such that $H = \ker \Gamma_0$. Moreover, for $\lambda \in \rho(H)$ the corresponding γ-field γ is given by

$$\gamma(\lambda) = \bigl(I + (\lambda - \mu)(H - \lambda)^{-1}\bigr)\iota_{\mathfrak{N}_\mu(S^*)} \tag{2.4.5}$$

and the corresponding Weyl function M is given by

$$M(\lambda) = \lambda + (\lambda - \mu)(\lambda - \bar{\mu}) P_{\mathfrak{N}_\mu(S^*)}(H - \lambda)^{-1}\iota_{\mathfrak{N}_\mu(S^*)}. \tag{2.4.6}$$

Proof. Since H is a self-adjoint extension of S, the direct sum decomposition (2.4.2) with $\mu \in \rho(H)$ follows from Corollary 1.7.5. Hence, for every $\widehat{f} \in S^*$ there is a unique decomposition as in (2.4.3). Let $\widehat{g} \in S^*$ have a corresponding decomposition

$$\widehat{g} = \widehat{g}_0 + \widehat{g}_\mu, \tag{2.4.7}$$

where $\widehat{g}_0 = \{g_0, g_0'\} \in H$ and $\widehat{g}_\mu = \{g_\mu, \mu g_\mu\} \in \widehat{\mathfrak{N}}_\mu(S^*)$. Then it follows directly from the decompositions (2.4.3), (2.4.7), and $(f_0', g_0) = (f_0, g_0')$ that

$$\begin{aligned} (f', g) - (f, g') &= (f_0' + \mu f_\mu, g_0 + g_\mu) - (f_0 + f_\mu, g_0' + \mu g_\mu) \\ &= (f_0' + \mu f_\mu, g_\mu) + (\mu f_\mu, g_0) - (f_\mu, g_0' + \mu g_\mu) - (f_0, \mu g_\mu) \\ &= (f_0' - \bar{\mu} f_0 + \mu f_\mu, g_\mu) - (f_\mu, g_0' - \bar{\mu} g_0 + \mu g_\mu). \end{aligned} \tag{2.4.8}$$

Moreover, it follows from the definition (2.4.4) applied to $\widehat{f}$ and $\widehat{g}$ that

$$\begin{aligned} &(\Gamma_1 \widehat{f}, \Gamma_0 \widehat{g})_{\mathfrak{N}_\mu(S^*)} - (\Gamma_0 \widehat{f}, \Gamma_1 \widehat{g})_{\mathfrak{N}_\mu(S^*)} \\ &\quad = \bigl(P_{\mathfrak{N}_\mu(S^*)}(f_0' - \bar{\mu} f_0 + \mu f_\mu), g_\mu\bigr)_{\mathfrak{N}_\mu(S^*)} \\ &\qquad\qquad - \bigl(f_\mu, P_{\mathfrak{N}_\mu(S^*)}(g_0' - \bar{\mu} g_0 + \mu g_\mu)\bigr)_{\mathfrak{N}_\mu(S^*)} \\ &\quad = (f_0' - \bar{\mu} f_0 + \mu f_\mu, g_\mu) - (f_\mu, g_0' - \bar{\mu} g_0 + \mu g_\mu). \end{aligned} \tag{2.4.9}$$

A combination of (2.4.8) and (2.4.9) shows that the abstract Green identity (2.1.1) holds.

It will now be verified that $\Gamma = (\Gamma_0, \Gamma_1)^\top : S^* \to \mathfrak{N}_\mu(S^*) \times \mathfrak{N}_\mu(S^*)$ is surjective. To see this, let $\varphi, \varphi' \in \mathfrak{N}_\mu(S^*)$. Since $\bar\mu \in \rho(H)$, one can choose $\{f_0, f_0'\} \in H$ such that

$$f_0' - \bar\mu f_0 = \varphi' - \mu\varphi. \tag{2.4.10}$$

It is clear from (2.4.2) that

$$\widehat{f} := \{f_0, f_0'\} + \{\varphi, \mu\varphi\} \in S^*,$$

and therefore (2.4.4) shows that

$$\Gamma_0 \widehat{f} = \varphi, \quad \Gamma_1 \widehat{f} = P_{\mathfrak{N}_\mu(S^*)}(f_0' - \bar\mu f_0 + \mu\varphi) = \varphi',$$

where (2.4.10) was used in the last equality. Hence, $\operatorname{ran}\Gamma = \mathfrak{N}_\mu(S^*) \times \mathfrak{N}_\mu(S^*)$ and thus $\{\mathfrak{N}_\mu(S^*), \Gamma_0, \Gamma_1\}$ is a boundary triplet for S^*. It follows from the definition of Γ_0 and the decomposition (2.4.2) that $H = \ker\Gamma_0$.

Now (2.4.5) and (2.4.6) will be verified. Let $\widehat{f}_\mu = \{f_\mu, \mu f_\mu\} \in \widehat{\mathfrak{N}}_\mu(S^*)$. Then (2.4.4) gives

$$\Gamma_0 \widehat{f}_\mu = f_\mu \quad \text{and} \quad \Gamma_1 \widehat{f}_\mu = \mu f_\mu.$$

Therefore, Definition 2.3.1 leads to

$$\gamma(\mu) = \bigl\{\{\Gamma_0 \widehat{f}_\mu, f_\mu\} : \widehat{f}_\mu \in \widehat{\mathfrak{N}}_\mu(S^*)\bigr\} = \bigl\{\{f_\mu, f_\mu\} : \widehat{f}_\mu \in \widehat{\mathfrak{N}}_\mu(S^*)\bigr\}$$

or, equivalently, $\gamma(\mu) : \mathfrak{N}_\mu(S^*) \to \mathfrak{H}$ acts as $f_\mu \mapsto f_\mu$. Thus, $\gamma(\mu)$ is the canonical embedding of $\mathfrak{N}_\mu(S^*)$ into $\mathfrak{H}$,

$$\gamma(\mu) = \iota_{\mathfrak{N}_\mu(S^*)}, \tag{2.4.11}$$

and $\gamma(\mu)^* : \mathfrak{H} \to \mathfrak{N}_\mu(S^*)$ is the orthogonal projection onto $\mathfrak{N}_\mu(S^*)$, that is, $\gamma(\mu)^* = P_{\mathfrak{N}_\mu(S^*)}$. Proposition 2.3.2 (ii) and (2.4.11) imply that the γ-field of $\{\mathfrak{N}_\mu(S^*), \Gamma_0, \Gamma_1\}$ is of the required form. Furthermore, Definition 2.3.4 leads to

$$M(\mu) = \bigl\{\{\Gamma_0 \widehat{f}_\mu, \Gamma_1 \widehat{f}_\mu\} : \widehat{f}_\mu \in \widehat{\mathfrak{N}}_\mu(S^*)\bigr\} = \bigl\{\{f_\mu, \mu f_\mu\} : \widehat{f}_\mu \in \widehat{\mathfrak{N}}_\mu(S^*)\bigr\}$$

or, equivalently,

$$M(\mu) = \mu.$$

Hence, Proposition 2.3.6 (v) with $\lambda_0 = \mu$ gives the desired result. □

It is interesting to see what Theorem 2.4.1 means in the simple case when the underlying symmetric relation is trivial; cf. Example 2.3.5, which is opposite in the sense that there $\ker\Gamma_0 = \mathfrak{H} \times \{0\}$ and $M(\lambda) = -(1/\lambda)I$. Again this example is not typical, since in standard applications $\mathcal{G} \neq \mathfrak{H}$; cf. Chapters 6, 7, and 8.

Example 2.4.2. Let $S = \{0,0\}$ be the trivial symmetric relation in $\mathfrak{H}$. Note that

$$H = \{0\} \times \mathfrak{H}$$

is a self-adjoint extension of S with $0 \in \rho(H)$. It is clear that $S^* = \mathfrak{H} \times \mathfrak{H}$ and that $\widehat{\mathfrak{N}}_0(S^*) = \mathfrak{H} \times \{0\}$. Therefore, one has the direct sum decomposition

$$S^* = H \mathbin{\widehat{+}} \widehat{\mathfrak{N}}_0(S^*)$$

and any $\widehat{f} \in S^*$ has the corresponding decomposition

$$\widehat{f} = \{f, f'\} = \{0, f'\} + \{f, 0\}, \quad \{0, f'\} \in H, \quad \{f, 0\} \in \widehat{\mathfrak{N}}_0(S^*).$$

According to (2.4.4), one sees that

$$\Gamma_0 \widehat{f} = f \quad \text{and} \quad \Gamma_1 \widehat{f} = f', \quad \widehat{f} = \{f, f'\} \in S^*,$$

defines a boundary triplet $\{\mathfrak{H}, \Gamma_0, \Gamma_1\}$ for S^* with $\ker \Gamma_0 = H$. Note that $\rho(H) = \mathbb{C}$ and that for every $\lambda \in \mathbb{C}$ the resolvent $(H - \lambda)^{-1}$ is the zero operator. Hence, the γ-field is given by $\gamma(\lambda) = I$ and the Weyl function is given by $M(\lambda) = \lambda I$.

There is an addendum to Theorem 2.4.1 when the decomposition (2.4.2) is replaced by a decomposition involving

$$\widehat{\mathfrak{N}}_\infty(S^*) = \bigl\{\{0, f'\} : f' \in \operatorname{mul} S^*\bigr\}.$$

In fact, the following result may be seen as a limit result obtained from (2.4.2) with $\mu \to \infty$. The embedding of $\mathfrak{N}_\infty(S^*) = \operatorname{mul} S^*$ into $\mathfrak{H}$ is denoted by $\iota_{\mathfrak{N}_\infty(S^*)}$ and its adjoint is the orthogonal projection $P_{\mathfrak{N}_\infty(S^*)}$ from $\mathfrak{H}$ onto $\mathfrak{N}_\infty(S^*)$. The proof of Proposition 2.4.3 is straightforward. Observe that Example 2.3.5 is an illustration of the following proposition.

Proposition 2.4.3. *Let S be a closed symmetric operator in $\mathfrak{H}$ and assume that H is a self-adjoint extension of S which belongs to $\mathbf{B}(\mathfrak{H})$. Then S^* can be decomposed as*

$$S^* = H \mathbin{\widehat{+}} \widehat{\mathfrak{N}}_\infty(S^*), \quad \textit{direct sum}. \tag{2.4.12}$$

Let $\widehat{f} = \{f, f'\} \in S^*$ have the corresponding decomposition

$$\widehat{f} = \widehat{f}_0 + \widehat{f}_\infty,$$

with $\widehat{f}_0 = \{f_0, Hf_0\} \in H$ and $\widehat{f}_\infty = \{0, f_\infty\} \in \widehat{\mathfrak{N}}_\infty(S^*)$. Then

$$\Gamma_0 \widehat{f} := f_\infty \quad \textit{and} \quad \Gamma_1 \widehat{f} := -P_{\mathfrak{N}_\infty(S^*)} f_0 \tag{2.4.13}$$

define a boundary triplet $\{\mathfrak{N}_\infty(S^*), \Gamma_0, \Gamma_1\}$ for S^* such that $H = \ker \Gamma_0$. Moreover, for $\lambda \in \rho(H)$ the corresponding γ-field γ is given by

$$\gamma(\lambda) = -(H - \lambda)^{-1} \iota_{\mathfrak{N}_\infty(S^*)},$$

and the corresponding Weyl function M is given by

$$M(\lambda) = P_{\mathfrak{N}_\infty(S^*)} (H - \lambda)^{-1} \iota_{\mathfrak{N}_\infty(S^*)}.$$

Remark 2.4.4. Let S be a closed symmetric operator in $\mathfrak{H}$ and let $H \in \mathbf{B}(\mathfrak{H})$ be a self-adjoint extension of S as in Proposition 2.4.3. Then S is bounded and hence $\operatorname{dom} S$ is closed. Decompose $\mathfrak{H} = \operatorname{dom} S \oplus \mathfrak{N}_\infty(S^*)$ and note that

$$S = \begin{pmatrix} S_{11} \\ S_{21} \end{pmatrix} : \operatorname{dom} S \to \begin{pmatrix} \operatorname{dom} S \\ \mathfrak{N}_\infty(S^*) \end{pmatrix},$$

and in a similar way

$$H = \begin{pmatrix} H_{11} & H_{21}^* \\ H_{21} & H_{22} \end{pmatrix} : \begin{pmatrix} \operatorname{dom} S \\ \mathfrak{N}_\infty(S^*) \end{pmatrix} \to \begin{pmatrix} \operatorname{dom} S \\ \mathfrak{N}_\infty(S^*) \end{pmatrix}.$$

It follows that $H_{11} = S_{11}$, $H_{21} = S_{21}$, and $H_{21}^* = S_{21}^*$. Relative to the decomposition (2.4.12) of S^*, the boundary triplet in (2.4.13) can be written as

$$\Gamma_0 \widehat{f} = f_\infty \quad \text{and} \quad \Gamma_1 \widehat{f} = -f_2, \quad \text{where } \widehat{f} = \left\{ \begin{pmatrix} f_1 \\ f_2 \end{pmatrix}, \begin{pmatrix} f_1' \\ f_2' \end{pmatrix} \right\} \in S^*.$$

Let Θ be a closed relation in $\mathcal{G} = \mathfrak{N}_\infty(S^*)$. Then the corresponding extension A_Θ of S is given by

$$A_\Theta = \left\{ \left\{ \begin{pmatrix} f_1 \\ f_2 \end{pmatrix}, \begin{pmatrix} S_{11} f_1 + S_{21}^* f_2 \\ S_{21} f_1 + H_{22} f_2 + f_\infty \end{pmatrix} \right\} : \{f_\infty, -f_2\} \in \Theta \right\},$$

which can formally be written as

$$A_\Theta = \begin{pmatrix} S_{11} & S_{21}^* \\ S_{21} & H_{22} - \Theta^{-1} \end{pmatrix}.$$

Therefore, the extensions of S may be interpreted as solutions of the *completion problem* posed by the incomplete 2×2 operator matrix

$$\begin{pmatrix} S_{11} & S_{21}^* \\ S_{21} & * \end{pmatrix}.$$

Theorem 2.4.1 has some variations when the self-adjoint extension H in (2.4.2) is further decomposed; cf. Section 1.7. The most straightforward results are presented in the following corollaries. In the next result the direct sum decomposition from Corollary 1.7.10 (with $\mu = \overline{\lambda}$) is used.

Corollary 2.4.5. *Let S be a closed symmetric relation in $\mathfrak{H}$, assume that H is a self-adjoint extension of S in $\mathfrak{H}$, and fix $\mu \in \mathbb{C} \setminus \mathbb{R}$. Then*

$$S^* = S \,\widehat{+}\, \{\{(H-\overline{\mu})^{-1}k, (I + \overline{\mu}(H-\overline{\mu})^{-1})k\} : k \in \mathfrak{N}_\mu(S^*)\} \,\widehat{+}\, \widehat{\mathfrak{N}}_\mu(S^*),$$

where the sums are direct. Let $\widehat{f} = \{f, f'\} \in S^*$ have the corresponding decomposition

$$\{f, f'\} = \{h, h'\} + \{(H-\overline{\mu})^{-1}k, (I + \overline{\mu}(H-\overline{\mu})^{-1})k\} + \{f_\mu, \mu f_\mu\}, \tag{2.4.14}$$

where $\widehat{h} = \{h, h'\} \in S$, $k \in \mathfrak{N}_\mu(S^*)$, *and* $f_\mu \in \mathfrak{N}_\mu(S^*)$. *Then*

$$\Gamma_0 \widehat{f} := f_\mu \quad \textit{and} \quad \Gamma_1 \widehat{f} := k + \mu f_\mu \tag{2.4.15}$$

define a boundary triplet $\{\mathfrak{N}_\mu(S^*), \Gamma_0, \Gamma_1\}$ *for* S^* *such that* $H = \ker \Gamma_0$. *The corresponding* γ*-field and Weyl function are given by* (2.4.5) *and* (2.4.6).

Proof. It follows from Corollary 1.7.10 that every $\widehat{f} = \{f, f'\} \in S^*$ can be written as

$$\{f, f'\} = \{f_0, f_0'\} + \{f_\mu, \mu f_\mu\},$$

where $\{f_0, f_0'\} \in H$, $\{f_\mu, \mu f_\mu\} \in \widehat{\mathfrak{N}}_\mu(S^*)$, and

$$\{f_0, f_0'\} = \{h, h'\} + \{(H - \overline{\mu})^{-1} k, (I + \overline{\mu}(H - \overline{\mu})^{-1}) k\},$$

with $\{h, h'\} \in S$ and $k \in \mathfrak{N}_\mu(S^*)$. The boundary mappings in Theorem 2.4.1 then have the form $\Gamma_0 \widehat{f} = f_\mu$ and

$$\begin{aligned} \Gamma_1 \widehat{f} &= P_{\mathfrak{N}_\mu(S^*)}(f_0' - \overline{\mu} f_0 + \mu f_\mu) \\ &= P_{\mathfrak{N}_\mu(S^*)}\big(h' - \overline{\mu} h + k + \mu f_\mu\big) \\ &= k + \mu f_\mu, \end{aligned}$$

where $h' - \overline{\mu} h \in \operatorname{ran}(S - \overline{\mu}) = \mathfrak{N}_\mu(S^*)^\perp$ and $k + \mu f_\mu \in \mathfrak{N}_\mu(S^*)$ was used in the last step. This shows that the mappings in (2.4.15) form a boundary triplet with the same γ-field and Weyl function as in Theorem 2.4.1. □

In Theorem 2.4.1, Proposition 2.4.3, and Corollary 2.4.5 the boundary triplets were based on decompositions of S^* in Section 1.7. The following result gives a boundary triplet for a decomposition of S^* which is a mixture of the above decompositions.

Corollary 2.4.6. *Let* S *be a closed symmetric relation in* $\mathfrak{H}$, *assume that* H *is a self-adjoint extension of* S *in* $\mathfrak{H}$, *and fix* $\mu \in \mathbb{C} \setminus \mathbb{R}$. *Every* $\widehat{f} = \{f, f'\} \in S^*$ *has the unique decomposition*

$$\begin{aligned} \{f, f'\} = \{h, h'\} &+ \{(I + \overline{\mu}(H - \overline{\mu})^{-1})\psi, (\mu + \overline{\mu} + \overline{\mu}^2 (H - \overline{\mu})^{-1})\psi\} \\ &+ \{(H - \overline{\mu})^{-1}\varphi, (I + \overline{\mu}(H - \overline{\mu})^{-1})\varphi\}, \end{aligned}$$

with $\widehat{h} = \{h, h'\} \in S$ *and* $\psi, \varphi \in \mathfrak{N}_\mu(S^*)$. *Then*

$$\Gamma_0 \widehat{f} = \psi \quad \textit{and} \quad \Gamma_1 \widehat{f} = \varphi + 2(\operatorname{Re} \mu)\psi \tag{2.4.16}$$

define a boundary triplet $\{\mathfrak{N}_\mu(S^*), \Gamma_0, \Gamma_1\}$ *for* S^* *such that* $H = \ker \Gamma_0$. *The corresponding* γ*-field and Weyl function are given by* (2.4.5) *and* (2.4.6).

Proof. Let $\widehat{f} = \{f, f'\} \in S^*$. Then according to (2.4.14) there is the decomposition

$$\{f, f'\} = \{h, h'\} + \{(H - \overline{\mu})^{-1}k, (I + \overline{\mu}(H - \overline{\mu})^{-1})k\} + \{\psi, \mu\psi\},$$

where $\widehat{h} = \{h, h'\} \in S$, $k \in \mathfrak{N}_\mu(S^*)$, and $\psi \in \mathfrak{N}_\mu(S^*)$ are uniquely determined. Define the element φ by $k = \overline{\mu}\psi + \varphi$, so that $\varphi \in \mathfrak{N}_\mu(S^*)$ and the right-hand side of the above decomposition can be rewritten as

$$\{h, h'\} + \{(H - \overline{\mu})^{-1}(\overline{\mu}\psi + \varphi), (I + \overline{\mu}(H - \overline{\mu})^{-1})(\overline{\mu}\psi + \varphi)\} + \{\psi, \mu\psi\}.$$

This yields the decomposition for $\{f, f'\}$ in the statement. The boundary triplet in (2.4.15) now reads as (2.4.16); the corresponding γ-field and Weyl function are given by (2.4.5) and (2.4.6). □

By von Neumann's second formula (see Theorem 1.7.12) one can describe the self-adjoint extension H in Theorem 2.4.1 by means of an isometric operator from $\mathfrak{N}_{\overline{\mu}}(S^*)$ onto $\mathfrak{N}_\mu(S^*)$. This observation also gives rise to the construction of a boundary triplet, where the parameter space is given by $\mathfrak{N}_\mu(S^*)$.

Theorem 2.4.7. *Let S be a closed symmetric relation in $\mathfrak{H}$, let H be a self-adjoint extension of S, and fix some $\mu \in \mathbb{C} \setminus \mathbb{R}$. Let W be the isometric mapping from $\mathfrak{N}_{\overline{\mu}}(S^*)$ onto $\mathfrak{N}_\mu(S^*)$ such that*

$$H = S \,\widehat{+}\, (I - \widehat{W})\widehat{\mathfrak{N}}_{\overline{\mu}}(S^*) = S \,\widehat{+}\, \{f_{\overline{\mu}} - W f_{\overline{\mu}}, \overline{\mu} f_{\overline{\mu}} - \mu W f_{\overline{\mu}}\} \tag{2.4.17}$$

and decompose $\widehat{f} = \{f, f'\} \in S^*$ according to von Neumann's first formula:

$$\widehat{f} = \{h, h'\} + \{f_\mu, \mu f_\mu\} + \{f_{\overline{\mu}}, \overline{\mu} f_{\overline{\mu}}\}, \tag{2.4.18}$$

where $\widehat{h} = \{h, h'\} \in S$, $\widehat{f}_\mu = \{f_\mu, \mu f_\mu\} \in \widehat{\mathfrak{N}}_\mu(S^*)$, and $\widehat{f}_{\overline{\mu}} = \{f_{\overline{\mu}}, \overline{\mu} f_{\overline{\mu}}\} \in \widehat{\mathfrak{N}}_{\overline{\mu}}(S^*)$. Then

$$\Gamma_0 \widehat{f} = f_\mu + W f_{\overline{\mu}} \quad \text{and} \quad \Gamma_1 \widehat{f} = \mu f_\mu + \overline{\mu} W f_{\overline{\mu}} \tag{2.4.19}$$

define a boundary triplet $\{\mathfrak{N}_\mu(S^*), \Gamma_0, \Gamma_1\}$ for S^* such that $H = \ker \Gamma_0$. The corresponding γ-field and the Weyl function are given by (2.4.5) and (2.4.6).

Proof. Let $\widehat{f} = \{f, f'\} \in S^*$ be decomposed as in (2.4.18) and let $\widehat{g} = \{g, g'\} \in S^*$ be decomposed in the analogous form

$$\widehat{g} = \{k, k'\} + \{g_\mu, \mu g_\mu\} + \{g_{\overline{\mu}}, \overline{\mu} g_{\overline{\mu}}\},$$

where $\widehat{k} = \{k, k'\} \in S$, $\widehat{g}_\mu = \{g_\mu, \mu g_\mu\} \in \widehat{\mathfrak{N}}_\mu(S^*)$, and $\widehat{g}_{\overline{\mu}} = \{g_{\overline{\mu}}, \overline{\mu} g_{\overline{\mu}}\} \in \widehat{\mathfrak{N}}_{\overline{\mu}}(S^*)$. Since $\{h, h'\}, \{k, k'\} \in S$, one has $(h', k) = (h, k')$,

$$\begin{aligned} (h', g_\mu + g_{\overline{\mu}}) - (h, \mu g_\mu + \overline{\mu} g_{\overline{\mu}}) &= 0, \\ (\mu f_\mu + \overline{\mu} f_{\overline{\mu}}, k) - (f_\mu + f_{\overline{\mu}}, k') &= 0, \end{aligned}$$

and therefore

$$\begin{aligned}(f',g)-(f,g') &= \big(h'+\mu f_\mu+\bar\mu f_{\bar\mu}, k+g_\mu+g_{\bar\mu}\big)-\big(h+f_\mu+f_{\bar\mu}, k'+\mu g_\mu+\bar\mu g_{\bar\mu}\big)\\ &= \big(\mu f_\mu+\bar\mu f_{\bar\mu}, g_\mu+g_{\bar\mu}\big)-\big(f_\mu+f_{\bar\mu}, \mu g_\mu+\bar\mu g_{\bar\mu}\big)\\ &= (\mu-\bar\mu)(f_\mu,g_\mu)+(\bar\mu-\mu)(f_{\bar\mu},g_{\bar\mu})\\ &= (\mu-\bar\mu)(f_\mu,g_\mu)_{\mathfrak{N}_\mu(S^*)}+(\bar\mu-\mu)(Wf_{\bar\mu},Wg_{\bar\mu})_{\mathfrak{N}_\mu(S^*)}.\end{aligned}$$

On the other hand it follows from (2.4.19) that

$$\begin{aligned}&(\Gamma_1\widehat f,\Gamma_0\widehat g)_{\mathfrak{N}_\mu(S^*)}-(\Gamma_0\widehat f,\Gamma_1\widehat g)_{\mathfrak{N}_\mu(S^*)}\\ &\quad= \big(\mu f_\mu+\bar\mu Wf_{\bar\mu}, g_\mu+Wg_{\bar\mu}\big)_{\mathfrak{N}_\mu(S^*)}-\big(f_\mu+Wf_{\bar\mu}, \mu g_\mu+\bar\mu Wg_{\bar\mu}\big)_{\mathfrak{N}_\mu(S^*)}\\ &\quad= (\mu-\bar\mu)(f_\mu,g_\mu)_{\mathfrak{N}_\mu(S^*)}+(\bar\mu-\mu)(Wf_{\bar\mu},Wg_{\bar\mu})_{\mathfrak{N}_\mu(S^*)},\end{aligned}$$

i.e., the abstract Green identity (2.1.1) holds.

In order to see that $\Gamma=(\Gamma_0,\Gamma_1)^\top: S^*\to\mathfrak{N}_\mu(S^*)\times\mathfrak{N}_\mu(S^*)$ is surjective consider $\varphi,\psi\in\mathfrak{N}_\mu(S^*)$ and define $\widehat f=\{f_\mu,\mu f_\mu\}+\{f_{\bar\mu},\bar\mu f_{\bar\mu}\}\in S^*$ by

$$\widehat f=\frac{1}{\bar\mu-\mu}\{\bar\mu\varphi-\psi,\mu(\bar\mu\varphi-\psi)\}+\frac{1}{\mu-\bar\mu}\{W^*(\mu\varphi-\psi),\bar\mu W^*(\mu\varphi-\psi)\}.$$

Then

$$\Gamma_0\widehat f=f_\mu+Wf_{\bar\mu}=\varphi\quad\text{and}\quad\Gamma_1\widehat f=\mu f_\mu+\bar\mu Wf_{\bar\mu}=\psi,$$

and therefore $\Gamma=(\Gamma_0,\Gamma_1)^\top$ maps onto $\mathfrak{N}_\mu(S^*)\times\mathfrak{N}_\mu(S^*)$. This implies that $\{\mathfrak{N}_\mu(S^*),\Gamma_0,\Gamma_1\}$ is a boundary triplet for S^*. Note that $\widehat f$ in (2.4.18) is in $\ker\Gamma_0$ if and only if $f_\mu=-Wf_{\bar\mu}$ and from (2.4.17) it then follows that $H=\ker\Gamma_0$.

Finally, to describe the γ-field and the Weyl function, consider the decomposition

$$S^*=H\,\widehat{+}\,\widehat{\mathfrak{N}}_\mu(S^*)=\ker\Gamma_0\,\widehat{+}\,\widehat{\mathfrak{N}}_\mu(S^*)$$

and note that if $\widehat f$ in (2.4.18) belongs to $\widehat{\mathfrak{N}}_\mu(S^*)$, then $\widehat f=\widehat f_\mu=\{f_\mu,\mu f_\mu\}$. Hence, (2.4.19) gives

$$\Gamma_0\widehat f_\mu=f_\mu\quad\text{and}\quad\Gamma_1\widehat f_\mu=\mu f_\mu.$$

In the same way as in the proof of Theorem 2.4.1 one concludes that $\gamma(\mu)=\iota_{\mathfrak{N}_\mu(S^*)}$ and $M(\mu)=\mu$. Now Proposition 2.3.2 (ii) yields (2.4.5) and Proposition 2.3.6 (v) implies (2.4.6). □

Note that the strategy in the proof of Theorem 2.4.7 is different from the strategy in the two previous results. The connection will be sketched now. In Theorem 2.4.7 the isometric mapping W from $\mathfrak{N}_{\bar\mu}(S^*)$ onto $\mathfrak{N}_\mu(S^*)$ determines the boundary triplet $\{\mathfrak{N}_\mu(S^*),\Gamma_0,\Gamma_1\}$ for S^* in (2.4.19). The self-adjoint extension H of S determined by W in (2.4.17) then satisfies $H=\ker\Gamma_0$. Now apply

Theorem 2.4.1 with this particular self-adjoint extension. Hence, $\widehat{f} \in S^*$ in Theorem 2.4.1 is decomposed in the form

$$\widehat{f} = \{f_0, f_0'\} + \{\varphi_\mu, \mu\varphi_\mu\}, \tag{2.4.20}$$

where $\{f_0, f_0'\} \in H$ and $\{\varphi_\mu, \mu\varphi_\mu\} \in \widehat{\mathfrak{N}}_\mu(S^*)$. Making use of the decomposition (2.4.17) of H it follows that

$$\{f_0, f_0'\} = \{h, h'\} + \{-W\psi_{\bar\mu}, -\mu W\psi_{\bar\mu}\} + \{\psi_{\bar\mu}, \bar\mu\psi_{\bar\mu}\} \tag{2.4.21}$$

with $\{h, h'\} \in S$ and $\psi_{\bar\mu} \in \mathfrak{N}_{\bar\mu}(S^*)$. Therefore, $\widehat{f}$ in (2.4.20) is given by

$$\widehat{f} = \{h, h'\} + \{f_\mu, \mu f_\mu\} + \{f_{\bar\mu}, \bar\mu f_{\bar\mu}\},$$

where $\{f_\mu, \mu f_\mu\} = \{\varphi_\mu - W\psi_{\bar\mu}, \mu\varphi_\mu - \mu W\psi_{\bar\mu}\}$ and $\{f_{\bar\mu}, \bar\mu f_{\bar\mu}\} = \{\psi_{\bar\mu}, \bar\mu\psi_{\bar\mu}\}$. Now the identity

$$\varphi_\mu = f_\mu + W\psi_{\bar\mu} = f_\mu + W f_{\bar\mu}$$

shows that the boundary maps Γ_0 in Theorem 2.4.1 and Theorem 2.4.7 coincide. Moreover, as $P_{\mathfrak{N}_\mu(S^*)}(h' - \bar\mu h) = 0$, the identity

$$P_{\mathfrak{N}_\mu(S^*)}\big(f_0' - \bar\mu f_0 + \mu\varphi_\mu\big) = -\mu W\psi_{\bar\mu} + \bar\mu W\psi_{\bar\mu} + \mu\varphi_\mu = \mu f_\mu + \bar\mu W f_{\bar\mu}$$

follows from (2.4.21), and shows that the boundary maps Γ_1 in Theorem 2.4.1 and Theorem 2.4.7 are the same.

2.5 Transformations of boundary triplets

Let S be a closed symmetric relation in $\mathfrak{H}$ with equal defect numbers. Then S admits self-adjoint extensions in $\mathfrak{H}$ and each self-adjoint extension gives rise to a boundary triplet as in Theorem 2.4.1. Hence, boundary triplets for S^* are not uniquely determined, with the exception of the trivial case $S = S^*$. A complete description of all boundary triplets for S^* will be given with the help of block operator matrices that are unitary with respect to the indefinite inner products in Section 1.8; cf. (2.1.3). The transformation properties of the corresponding boundary parameters, γ-fields, and Weyl functions are discussed afterwards.

The main result on the description of all boundary triplets for S^* is the following theorem. It describes the *transformation* of boundary triplets.

Theorem 2.5.1. *Let S be a closed symmetric relation in $\mathfrak{H}$, assume that $\{\mathcal{G}, \Gamma_0, \Gamma_1\}$ is a boundary triplet for S^*, and let $\mathcal{G}'$ be a Hilbert space. Then the following statements hold:*

(i) *Let $\mathcal{W} \in \mathbf{B}(\mathcal{G} \times \mathcal{G}, \mathcal{G}' \times \mathcal{G}')$ satisfy*

$$\mathcal{W}^* \mathcal{J}_{\mathcal{G}'} \mathcal{W} = \mathcal{J}_{\mathcal{G}} \quad \textit{and} \quad \mathcal{W} \mathcal{J}_{\mathcal{G}} \mathcal{W}^* = \mathcal{J}_{\mathcal{G}'}, \tag{2.5.1}$$

and define

$$\begin{pmatrix}\Gamma_0'\\ \Gamma_1'\end{pmatrix} = \mathcal{W}\begin{pmatrix}\Gamma_0\\ \Gamma_1\end{pmatrix} = \begin{pmatrix}W_{11} & W_{12}\\ W_{21} & W_{22}\end{pmatrix}\begin{pmatrix}\Gamma_0\\ \Gamma_1\end{pmatrix}. \tag{2.5.2}$$

Then $\{\mathcal{G}', \Gamma_0', \Gamma_1'\}$ *is a boundary triplet for* S^*.

(ii) *Let* $\{\mathcal{G}', \Gamma_0', \Gamma_1'\}$ *be a boundary triplet for* S^*. *Then there exists a unique operator* $\mathcal{W} \in \mathbf{B}(\mathcal{G}\times\mathcal{G}, \mathcal{G}'\times\mathcal{G}')$ *satisfying* (2.5.1) *such that* (2.5.2) *holds.*

Proof. (i) Note that the operator $\mathcal{W} \in \mathbf{B}(\mathcal{G}\times\mathcal{G}, \mathcal{G}'\times\mathcal{G}')$ is unitary from $(\mathcal{G}^2, [\![\cdot,\cdot]\!]_{\mathcal{G}^2})$ to $(\mathcal{G}'^2, [\![\cdot,\cdot]\!]_{\mathcal{G}'^2})$; cf. Proposition 1.8.2. Hence, for $\widehat{f}, \widehat{g} \in S^*$ one has

$$[\![\Gamma'\widehat{f}, \Gamma'\widehat{g}]\!]_{\mathcal{G}'^2} = [\![\mathcal{W}\Gamma\widehat{f}, \mathcal{W}\Gamma\widehat{g}]\!]_{\mathcal{G}'^2} = [\![\Gamma\widehat{f}, \Gamma\widehat{g}]\!]_{\mathcal{G}^2} = [\![\widehat{f}, \widehat{g}]\!]_{\mathfrak{H}^2}.$$

Therefore, Γ_0' and Γ_1' satisfy the abstract Green identity (2.1.1); cf. (2.1.4). Since $\mathcal{W}$ is surjective by Proposition 1.8.2, $\Gamma' = \mathcal{W}\Gamma$ is also surjective thanks to the surjectivity of Γ. It follows that $\{\mathcal{G}', \Gamma_0', \Gamma_1'\}$ is a boundary triplet for S^*.

(ii) Assume that $\{\mathcal{G}', \Gamma_0', \Gamma_1'\}$ is a boundary triplet for S^* and define a linear relation $\mathcal{W} \subset \mathcal{G}^2 \times \mathcal{G}'^2$ by

$$\mathcal{W} := \bigl\{\{\Gamma\widehat{f}, \Gamma'\widehat{f}\} : \widehat{f} \in S^*\bigr\}. \tag{2.5.3}$$

It follows from Proposition 2.1.2 (ii) that $\mathcal{W}$ is an operator. Indeed, if $\Gamma\widehat{f} = 0$, then $\widehat{f} \in S$ and thus $\Gamma'\widehat{f} = 0$. For the operator $\mathcal{W}$ one has $\operatorname{dom}\mathcal{W} = \mathcal{G}\times\mathcal{G}$ and $\operatorname{ran}\mathcal{W} = \mathcal{G}'\times\mathcal{G}'$ since $\operatorname{ran}\Gamma = \mathcal{G}\times\mathcal{G}$ and $\operatorname{ran}\Gamma' = \mathcal{G}'\times\mathcal{G}'$.

Define the inner product $[\![\cdot,\cdot]\!]_{\mathcal{G}'^2}$ on $\mathcal{G}'^2$ as in (2.1.3). Let $\widehat{\varphi}, \widehat{\psi} \in \mathcal{G}\times\mathcal{G}$ and let $\widehat{f}, \widehat{g} \in S^*$ be such that $\Gamma\widehat{f} = \widehat{\varphi}$ and $\Gamma\widehat{g} = \widehat{\psi}$. Then one has

$$\begin{aligned}[\![\mathcal{W}\widehat{\varphi}, \mathcal{W}\widehat{\psi}]\!]_{\mathcal{G}'^2} &= [\![\mathcal{W}\Gamma\widehat{f}, \mathcal{W}\Gamma\widehat{g}]\!]_{\mathcal{G}'^2} = [\![\Gamma'\widehat{f}, \Gamma'\widehat{g}]\!]_{\mathcal{G}'^2}\\ &= [\![\widehat{f}, \widehat{g}]\!]_{\mathfrak{H}^2} = [\![\Gamma\widehat{f}, \Gamma\widehat{g}]\!]_{\mathcal{G}^2} = [\![\widehat{\varphi}, \widehat{\psi}]\!]_{\mathcal{G}^2},\end{aligned}$$

and hence $\mathcal{W}$ is an isometric operator from $(\mathcal{G}^2, [\![\cdot,\cdot]\!]_{\mathcal{G}^2})$ to $(\mathcal{G}'^2, [\![\cdot,\cdot]\!]_{\mathcal{G}'^2})$. This implies that the first identity in (2.5.1) is satisfied and from Lemma 1.8.1 it follows that $\mathcal{W} \in \mathbf{B}(\mathcal{G}\times\mathcal{G}, \mathcal{G}'\times\mathcal{G}')$. Furthermore, as $\mathcal{W}$ is surjective, Proposition 1.8.2 implies that also the second identity in (2.5.1) holds.

By the definition (2.5.3) of the operator $\mathcal{W}$, the boundary triplets $\{\mathcal{G}, \Gamma_0, \Gamma_1\}$ and $\{\mathcal{G}', \Gamma_0', \Gamma_1'\}$ are connected via (2.5.2). Moreover, the operator $\mathcal{W}$ is unique. Indeed, if $\Gamma' = \mathcal{W}\Gamma$ and $\Gamma' = \widetilde{\mathcal{W}}\Gamma$, then $(\mathcal{W} - \widetilde{\mathcal{W}})\Gamma\widehat{f} = 0$ for all $\widehat{f} \in S^*$ and as $\operatorname{ran}\Gamma = \mathcal{G}\times\mathcal{G}$ it follows that $\mathcal{W} = \widetilde{\mathcal{W}}$. □

The transformation of a boundary triplet $\{\mathcal{G}, \Gamma_0, \Gamma_1\}$ in Theorem 2.5.1 induces a transformation of the closed relations in the parameter space. Assume that $\mathcal{W} \in \mathbf{B}(\mathcal{G}\times\mathcal{G}, \mathcal{G}'\times\mathcal{G}')$ satisfies the identities in (2.5.1) and let $\{\mathcal{G}', \Gamma_0', \Gamma_1'\}$ be the corresponding transformed boundary triplet in (2.5.2). Let Θ be a relation in $\mathcal{G}$ and define Θ' in $\mathcal{G}'$ as a Möbius transform of Θ by

$$\Theta' = \mathcal{W}[\Theta] = \bigl\{\{W_{11}\varphi + W_{12}\varphi', W_{21}\varphi + W_{22}\varphi'\} : \{\varphi, \varphi'\} \in \Theta\bigr\}; \tag{2.5.4}$$

cf. Definition 1.8.4. As $\mathcal{W}$ is bijective, it follows that $\Theta' = \mathcal{W}[\Theta]$ is closed in $\mathcal{G}'$ if and only if Θ is closed in $\mathcal{G}$; cf. (1.8.10).

Proposition 2.5.2. *Let S be a closed symmetric relation in $\mathfrak{H}$ and assume that $\{\mathcal{G}, \Gamma_0, \Gamma_1\}$ and $\{\mathcal{G}', \Gamma_0', \Gamma_1'\}$ are boundary triplets for S^* connected via $\Gamma' = \mathcal{W}\Gamma$ in Theorem 2.5.1. Let Θ be a closed relation in $\mathcal{G}$ and let Θ' be defined by (2.5.4). Then the closed intermediate extensions*

$$A_\Theta = \ker(\Gamma_1 - \Theta\Gamma_0) \quad \text{and} \quad A'_{\Theta'} = \ker(\Gamma_1' - \Theta'\Gamma_0')$$

coincide, that is, for $\widehat{f} \in S^*$ one has

$$\Gamma'\widehat{f} \in \Theta' \quad \Leftrightarrow \quad \Gamma\widehat{f} \in \Theta.$$

Proof. Let $\widehat{f} \in S^*$. Then the transformation formulas (2.5.2), (2.5.4), and the fact that $\mathcal{W}$ is bijective imply

$$\Gamma'\widehat{f} \in \Theta' \quad \Leftrightarrow \quad \mathcal{W}\Gamma\widehat{f} \in \mathcal{W}[\Theta] \quad \Leftrightarrow \quad \Gamma\widehat{f} \in \Theta.$$

Hence, $\ker(\Gamma_1 - \Theta\Gamma_0)$ and $\ker(\Gamma_1' - \Theta'\Gamma_0')$ coincide; cf. Theorem 2.1.3 (iii). □

Likewise, the transformation of the boundary triplet leads to a transformation of the γ-field and the Weyl function.

Proposition 2.5.3. *Let S be a closed symmetric relation in $\mathfrak{H}$ and assume that $\{\mathcal{G}, \Gamma_0, \Gamma_1\}$ and $\{\mathcal{G}', \Gamma_0', \Gamma_1'\}$ are boundary triplets for S^* connected via $\Gamma' = \mathcal{W}\Gamma$ in Theorem 2.5.1. Let $A_0 = \ker\Gamma_0$, $A_0' = \ker\Gamma_0'$, and let γ, γ' and M, M' be the γ-fields and Weyl functions corresponding to $\{\mathcal{G}, \Gamma_0, \Gamma_1\}$ and $\{\mathcal{G}', \Gamma_0', \Gamma_1'\}$, respectively. Then for all $\lambda \in \rho(A_0) \cap \rho(A_0')$ the operator*

$$W_{11} + W_{12}M(\lambda)$$

is an isomorphism from $\mathcal{G}$ onto $\mathcal{G}'$, and the identities

$$\gamma'(\lambda) = \gamma(\lambda)\big(W_{11} + W_{12}M(\lambda)\big)^{-1} \tag{2.5.5}$$

and

$$M'(\lambda) = \big(W_{21} + W_{22}M(\lambda)\big)\big(W_{11} + W_{12}M(\lambda)\big)^{-1} \tag{2.5.6}$$

hold.

Proof. For $\lambda \in \rho(A_0)$ and $\widehat{f}_\lambda \in \widehat{\mathfrak{N}}_\lambda(S^*)$ one has

$$\Gamma_0'\widehat{f}_\lambda = W_{11}\Gamma_0\widehat{f}_\lambda + W_{12}\Gamma_1\widehat{f}_\lambda = \big(W_{11} + W_{12}M(\lambda)\big)\Gamma_0\widehat{f}_\lambda,$$

which leads to

$$\Gamma_0' \upharpoonright \widehat{\mathfrak{N}}_\lambda(S^*) = \big(W_{11} + W_{12}M(\lambda)\big)\big(\Gamma_0 \upharpoonright \widehat{\mathfrak{N}}_\lambda(S^*)\big). \tag{2.5.7}$$

If, in addition, $\lambda \in \rho(A_0) \cap \rho(A_0')$, then, by Lemma 2.1.7, Γ_0 and Γ_0' are isomorphisms from $\widehat{\mathfrak{N}}_\lambda(S^*)$ onto $\mathcal{G}$ and $\mathcal{G}'$, respectively. Hence, it follows from (2.5.7) that $W_{11} + W_{12}M(\lambda)$ is an isomorphism from $\mathcal{G}$ onto $\mathcal{G}'$, and therefore

$$\big(\Gamma_0' \upharpoonright \widehat{\mathfrak{N}}_\lambda(S^*)\big)^{-1} = \big(\Gamma_0 \upharpoonright \widehat{\mathfrak{N}}_\lambda(S^*)\big)^{-1}\big(W_{11} + W_{12}M(\lambda)\big)^{-1}.$$

If one applies the orthogonal projection π_1 from $\mathfrak{H} \times \mathfrak{H}$ onto $\mathfrak{H} \times \{0\}$ to both sides, then (2.5.5) follows. Similarly, for $\lambda \in \rho(A_0) \cap \rho(A_0')$ one finds that

$$\begin{aligned}\big(W_{21} + W_{22}M(\lambda)\big)\big(W_{11} + W_{12}M(\lambda)\big)^{-1}\Gamma_0'\widehat{f}_\lambda &= \big(W_{21} + W_{22}M(\lambda)\big)\Gamma_0\widehat{f}_\lambda \\ &= W_{21}\Gamma_0\widehat{f}_\lambda + W_{22}\Gamma_1\widehat{f}_\lambda \\ &= \Gamma_1'\widehat{f}_\lambda,\end{aligned}$$

which yields (2.5.6). □

Next some special transformations of boundary triplets and Weyl functions will be discussed. In the first corollary the boundary mappings are interchanged via a flip-flop, which leads to the Weyl function $-M^{-1}$.

Corollary 2.5.4. *Let S be a closed symmetric relation in $\mathfrak{H}$ and assume that $\{\mathcal{G}, \Gamma_0, \Gamma_1\}$ is a boundary triplet for S^* with γ-field γ and Weyl function M, and define*

$$\Gamma_0' = \Gamma_1 \quad \textit{and} \quad \Gamma_1' = -\Gamma_0.$$

Then $\{\mathcal{G}, \Gamma_0', \Gamma_1'\}$ is a boundary triplet for S^* and $\ker \Gamma_0' = \ker \Gamma_1$. Moreover, for $\lambda \in \rho(A_0) \cap \rho(A_1)$ the corresponding γ-field γ' and the Weyl function M' are given by

$$\gamma'(\lambda) = \gamma(\lambda)M(\lambda)^{-1} \quad \textit{and} \quad M'(\lambda) = -M(\lambda)^{-1},$$

respectively.

Proof. The operator

$$\mathcal{W} = \begin{pmatrix} 0 & I \\ -I & 0 \end{pmatrix} \in \mathbf{B}(\mathcal{G} \times \mathcal{G})$$

satisfies both identities in (2.5.1). Now the assertions follow from Theorem 2.5.1 and Proposition 2.5.3. □

The second corollary treats the situation in which a bijective operator D dilates the Weyl function M and a self-adjoint operator P produces a shift of the dilated Weyl function D^*MD.

Corollary 2.5.5. *Let S be a closed symmetric relation in $\mathfrak{H}$ and assume that $\{\mathcal{G}, \Gamma_0, \Gamma_1\}$ is a boundary triplet for S^* with γ-field γ and Weyl function M. Let $\mathcal{G}'$ be a Hilbert space, let $D \in \mathbf{B}(\mathcal{G}', \mathcal{G})$ be boundedly invertible, let $P \in \mathbf{B}(\mathcal{G}')$ be self-adjoint, and define*

$$\Gamma_0' = D^{-1}\Gamma_0 \quad \textit{and} \quad \Gamma_1' = D^*\Gamma_1 + PD^{-1}\Gamma_0.$$

Then $\{\mathcal{G}', \Gamma_0', \Gamma_1'\}$ *is a boundary triplet for* S^* *and* $\ker \Gamma_0' = \ker \Gamma_0$*. Moreover, for* $\lambda \in \rho(A_0)$ *the corresponding* γ*-field* γ' *and the Weyl function* M' *are given by*

$$\gamma'(\lambda) = \gamma(\lambda)D \quad \text{and} \quad M'(\lambda) = D^* M(\lambda) D + P, \tag{2.5.8}$$

respectively.

Proof. It is not difficult to check that the operator

$$\mathcal{W} = \begin{pmatrix} D^{-1} & 0 \\ PD^{-1} & D^* \end{pmatrix} \in \mathbf{B}(\mathcal{G} \times \mathcal{G}, \mathcal{G}' \times \mathcal{G}')$$

satisfies both identities in (2.5.1). Now the assertions follow from Theorem 2.5.1 and Proposition 2.5.3. □

The next corollary complements Corollary 2.5.5.

Corollary 2.5.6. *Let* S *be a closed symmetric relation in* $\mathfrak{H}$ *and assume that* $\{\mathcal{G}, \Gamma_0, \Gamma_1\}$ *and* $\{\mathcal{G}', \Gamma_0', \Gamma_1'\}$ *are boundary triplets for* S^* *such that*

$$\ker \Gamma_0 = \ker \Gamma_0'.$$

Then there exist a boundedly invertible operator $D \in \mathbf{B}(\mathcal{G}', \mathcal{G})$ *and a self-adjoint operator* $P \in \mathbf{B}(\mathcal{G}')$ *such that*

$$\Gamma_0' = D^{-1}\Gamma_0 \quad \text{and} \quad \Gamma_1' = D^*\Gamma_1 + PD^{-1}\Gamma_0. \tag{2.5.9}$$

In particular, the γ*-fields and Weyl functions corresponding to the boundary triplets* $\{\mathcal{G}, \Gamma_0, \Gamma_1\}$ *and* $\{\mathcal{G}, \Gamma_0', \Gamma_1'\}$ *satisfy* (2.5.8).

Proof. It follows from Theorem 2.5.1 that there exists $\mathcal{W} \in \mathbf{B}(\mathcal{G} \times \mathcal{G}, \mathcal{G}' \times \mathcal{G}')$ with the properties (2.5.1) such that

$$\begin{pmatrix} \Gamma_0' \\ \Gamma_1' \end{pmatrix} = \mathcal{W} \begin{pmatrix} \Gamma_0 \\ \Gamma_1 \end{pmatrix} = \begin{pmatrix} W_{11} & W_{12} \\ W_{21} & W_{22} \end{pmatrix} \begin{pmatrix} \Gamma_0 \\ \Gamma_1 \end{pmatrix}. \tag{2.5.10}$$

The assumption $\ker \Gamma_0 = \ker \Gamma_0'$ implies $W_{12} = 0$. In fact, if $\widehat{f} \in \ker \Gamma_0 = \ker \Gamma_0'$, then $W_{12}\Gamma_1 \widehat{f} = 0$ by (2.5.10) and hence Proposition 2.1.2 (i) implies $W_{12} = 0$. Therefore, the first identity $\mathcal{W}^* \mathcal{J}_{\mathcal{G}'} \mathcal{W} = \mathcal{J}_{\mathcal{G}}$ in (2.5.1) means that

$$W_{11}^* W_{22} = I_{\mathcal{G}}, \quad W_{22}^* W_{11} = I_{\mathcal{G}}, \quad W_{11}^* W_{21} = W_{21}^* W_{11}.$$

Likewise, the second equality $\mathcal{W} \mathcal{J}_{\mathcal{G}} \mathcal{W}^* = \mathcal{J}_{\mathcal{G}'}$ in (2.5.1) means that

$$W_{11} W_{22}^* = I_{\mathcal{G}'}, \quad W_{22} W_{11}^* = I_{\mathcal{G}'}, \quad W_{21} W_{22}^* = W_{22} W_{21}^*.$$

It follows that $D := W_{22}^* \in \mathbf{B}(\mathcal{G}', \mathcal{G})$ is boundedly invertible with $D^{-1} = W_{11}$, the operator $P := W_{21} D \in \mathbf{B}(\mathcal{G}')$ is self-adjoint and (2.5.9) is satisfied. □

A combination of Corollary 2.5.5 with $\mathcal{G} = \mathcal{G}'$, $D = I$, $P = -\Theta$, and Corollary 2.5.4 leads to the followings statement.

Corollary 2.5.7. *Let S be a closed symmetric relation in $\mathfrak{H}$ and assume that $\{\mathcal{G}, \Gamma_0, \Gamma_1\}$ is a boundary triplet for S^* with γ-field γ and Weyl function M. Let $\Theta \in \mathbf{B}(\mathcal{G})$ be self-adjoint, and define*

$$\Gamma_0' = \Gamma_1 - \Theta\Gamma_0 \quad \textit{and} \quad \Gamma_1' = -\Gamma_0.$$

Then $\{\mathcal{G}, \Gamma_0', \Gamma_1'\}$ is a boundary triplet for S^* and $\ker \Gamma_0' = \ker(\Gamma_1 - \Theta\Gamma_0) = A_\Theta$ holds. Moreover, for $\lambda \in \rho(A_0) \cap \rho(A_\Theta)$ the corresponding γ-field γ' and the Weyl function M' are given by

$$\gamma'(\lambda) = -\gamma(\lambda)(\Theta - M(\lambda))^{-1} \quad \textit{and} \quad M'(\lambda) = (\Theta - M(\lambda))^{-1},$$

respectively.

The following statement is also a direct consequence of Corollary 2.5.5.

Corollary 2.5.8. *Let S be a closed symmetric relation in $\mathfrak{H}$ and assume that $\{\mathcal{G}, \Gamma_0, \Gamma_1\}$ is a boundary triplet for S^* with γ-field γ and Weyl function M. Let $Q(\lambda) \in \mathbf{B}(\mathcal{G})$, $\lambda \in \mathbb{C} \setminus \mathbb{R}$, be a family of operators which satisfy*

$$\frac{Q(\lambda) - Q(\mu)^*}{\lambda - \bar{\mu}} = \gamma(\mu)^*\gamma(\lambda), \qquad \lambda, \mu \in \mathbb{C} \setminus \mathbb{R}. \tag{2.5.11}$$

Let $\lambda_0 \in \mathbb{C} \setminus \mathbb{R}$ and define the self-adjoint operator $P \in \mathbf{B}(\mathcal{G})$ by

$$P = \operatorname{Re} Q(\lambda_0) - \operatorname{Re} M(\lambda_0).$$

Then Q is the Weyl function corresponding to the boundary triplet $\{\mathcal{G}, \Gamma_0', \Gamma_1'\}$, where

$$\Gamma_0' = \Gamma_0 \quad \textit{and} \quad \Gamma_1' = \Gamma_1 + P\Gamma_0,$$

and $Q(\lambda) = M(\lambda) + P$ holds for all $\lambda \in \rho(A_0)$.

Proof. Due to Proposition 2.3.6 (iii) it follows from the identity (2.5.11) that

$$Q(\lambda) - Q(\lambda_0)^* = M(\lambda) - M(\lambda_0)^*, \qquad \lambda, \lambda_0 \in \mathbb{C} \setminus \mathbb{R},$$

and, in particular, $\operatorname{Im} Q(\lambda_0) = \operatorname{Im} M(\lambda_0)$. Hence, one obtains

$$Q(\lambda) - M(\lambda) = Q(\lambda_0)^* - M(\lambda_0)^* = \operatorname{Re} Q(\lambda_0) - \operatorname{Re} M(\lambda_0) = P.$$

With the choice $D = I$ and P as above, the result follows from Corollary 2.5.5. □

Now it will be shown that a pair of transversal self-adjoint extensions induces a boundary triplet $\{\mathcal{G}, \Gamma_0, \Gamma_1\}$ which determines these extensions via the boundary conditions $\ker \Gamma_0$ and $\ker \Gamma_1$. The following theorem is a consequence of Theorem 2.4.1 and Corollary 2.5.5.

Theorem 2.5.9. *Let S be a closed symmetric relation in $\mathfrak{H}$ and assume that H and H' are transversal self-adjoint extensions of S in $\mathfrak{H}$, that is,*

$$S^* = H \mathbin{\widehat{+}} H'$$

holds; cf. Lemma 1.7.7. Then there exists a boundary triplet $\{\mathcal{G}, \Gamma_0, \Gamma_1\}$ for S^* such that

$$H = \ker \Gamma_0 \quad \textit{and} \quad H' = \ker \Gamma_1. \tag{2.5.12}$$

Proof. As H is a self-adjoint extension of S, there is a boundary triplet $\{\mathcal{G}, \Upsilon_0, \Upsilon_1\}$ for S^* such that $H = \ker \Upsilon_0$; cf. Theorem 2.4.1. Since H' is a self-adjoint extension of S it follows from Corollary 2.1.4 (v) that there exists a self-adjoint relation Θ in $\mathcal{G}$ such that

$$H' = \ker (\Upsilon_1 - \Theta\Upsilon_0).$$

Furthermore, since H' and H are transversal, it follows from Proposition 2.1.8 (ii) that $\Theta \in \mathbf{B}(\mathcal{G})$. Now define

$$\begin{pmatrix} \Gamma_0 \\ \Gamma_1 \end{pmatrix} = \begin{pmatrix} I & 0 \\ -\Theta & I \end{pmatrix} \begin{pmatrix} \Upsilon_0 \\ \Upsilon_1 \end{pmatrix},$$

so that $\{\mathcal{G}, \Gamma_0, \Gamma_1\}$ is a boundary triplet for S^* by Corollary 2.5.5. By construction, the boundary triplet $\{\mathcal{G}, \Gamma_0, \Gamma_1\}$ has the properties (2.5.12). □

Corollary 2.5.10. *Let S be a closed symmetric relation in $\mathfrak{H}$ and assume that $\{\mathcal{G}, \Gamma_0, \Gamma_1\}$ is a boundary triplet for S^*. Let $A_\Theta = \ker (\Gamma_1 - \Theta\Gamma_0)$ be a self-adjoint extension of S corresponding to the self-adjoint relation Θ in $\mathcal{G}$ via (2.1.5). Then there exists a boundary triplet $\{\mathcal{G}, \Gamma_0', \Gamma_1'\}$ such that*

$$\ker \Gamma_0' = \ker \big(\Gamma_1 - \Theta\Gamma_0\big) \quad \textit{and} \quad \ker \Gamma_1' = \ker \big(\Gamma_1 + \Theta^{-1}\Gamma_0\big), \tag{2.5.13}$$

that is, $A_0' = A_\Theta$ and $A_1' = A_{-\Theta^{-1}}$.

Proof. Recall that $\Theta^* = (J\Theta)^\perp$, where J denotes the flip-flop operator in $\mathcal{G}^2$ from (1.3.1). Since Θ is self-adjoint it follows that $\Theta = (J\Theta)^\perp$ or, in other words,

$$\mathcal{G}^2 = \Theta \oplus J\Theta.$$

In particular, one sees that Θ and $J\Theta = -\Theta^{-1}$ are transversal in $\mathcal{G}$. Therefore, the self-adjoint extensions $\ker (\Gamma_1 - \Theta\Gamma_0)$ and $\ker (\Gamma_1 + \Theta^{-1}\Gamma_0)$ are transversal; cf. Lemma 2.1.5 (ii). Now the assertion follows from Theorem 2.5.9. □

Corollary 2.5.10 is concerned with the existence of the boundary triplet $\{\mathcal{G}, \Gamma_0', \Gamma_1'\}$ with the properties (2.5.13). In fact, it is possible to explicitly construct such a boundary triplet via the choice of an appropriate operator $\mathcal{W}$ such that $\Gamma' = \mathcal{W}\Gamma$; cf. Corollary 2.5.7 for the special case $\Theta \in \mathbf{B}(\mathcal{G})$.

Corollary 2.5.11. *Let S be a closed symmetric relation in $\mathfrak{H}$ and assume that $\{\mathcal{G}, \Gamma_0, \Gamma_1\}$ is a boundary triplet for S^*. Let Θ be a self-adjoint relation in $\mathcal{G}$ and choose $\mathcal{A}, \mathcal{B} \in \mathbf{B}(\mathcal{G})$ such that*

$$\Theta = \big\{\{\mathcal{A}\varphi, \mathcal{B}\varphi\} : \varphi \in \mathcal{G}\big\}$$

and the identities

$$\mathcal{A}^*\mathcal{B} = \mathcal{B}^*\mathcal{A}, \quad \mathcal{A}\mathcal{B}^* = \mathcal{B}\mathcal{A}^*, \quad \mathcal{A}^*\mathcal{A} + \mathcal{B}^*\mathcal{B} = I = \mathcal{A}\mathcal{A}^* + \mathcal{B}\mathcal{B}^*, \tag{2.5.14}$$

hold; cf. Corollary 1.10.9. Then $\{\mathcal{G}, \Gamma_0', \Gamma_1'\}$, where

$$\Gamma_0' = \mathcal{B}^*\Gamma_0 - \mathcal{A}^*\Gamma_1 \quad \textit{and} \quad \Gamma_1' = \mathcal{A}^*\Gamma_0 + \mathcal{B}^*\Gamma_1, \tag{2.5.15}$$

is a boundary triplet for S^* such that both identities in (2.5.13) hold, that is, $A_0' = A_\Theta$ and $A_1' = A_{-\Theta^{-1}}$.

Proof. It is not difficult to check that

$$\mathcal{W} = \begin{pmatrix} \mathcal{B}^* & -\mathcal{A}^* \\ \mathcal{A}^* & \mathcal{B}^* \end{pmatrix} \in \mathbf{B}(\mathcal{G} \times \mathcal{G}) \tag{2.5.16}$$

satisfies (2.5.1) and it is clear from (2.5.15) that $\{\mathcal{G}, \Gamma_0', \Gamma_1'\}$ and $\{\mathcal{G}, \Gamma_0, \Gamma_1\}$ are connected via $\mathcal{W}$ as in Theorem 2.5.1 (i). Hence, $\{\mathcal{G}, \Gamma_0', \Gamma_1'\}$ is a boundary triplet for S^*. It follows from (2.5.14) that

$$\begin{aligned} \mathcal{W}[\Theta] &= \big\{\{\mathcal{B}^*\mathcal{A}\varphi - \mathcal{A}^*\mathcal{B}\varphi, \mathcal{A}^*\mathcal{A}\varphi + \mathcal{B}^*\mathcal{B}\varphi\} : \{\mathcal{A}\varphi, \mathcal{B}\varphi\} \in \Theta\big\} \\ &= \{0\} \times \mathcal{G}, \end{aligned}$$

and since $-\Theta^{-1} = \{\{\mathcal{B}\varphi, -\mathcal{A}\varphi\} : \varphi \in \mathcal{G}\}$, it follows in the same way that

$$\begin{aligned} \mathcal{W}[-\Theta^{-1}] &= \big\{\{\mathcal{B}^*\mathcal{B}\varphi + \mathcal{A}^*\mathcal{A}\varphi, \mathcal{A}^*\mathcal{B}\varphi - \mathcal{B}^*\mathcal{A}\varphi\} : \{\mathcal{B}\varphi, -\mathcal{A}\varphi\} \in -\Theta^{-1}\big\} \\ &= \mathcal{G} \times \{0\}. \end{aligned}$$

Recall from Proposition 2.5.2 that

$$\Gamma'\widehat{f} \in \mathcal{W}[\Xi] \quad \Leftrightarrow \quad \Gamma\widehat{f} \in \Xi, \qquad \widehat{f} \in S^*,$$

holds for any closed relation Ξ in $\mathcal{G}$. With $\Xi = \Theta$ and $\Xi = -\Theta^{-1}$ one then has

$$\Gamma'\widehat{f} \in \{0\} \times \mathcal{G} \quad \Leftrightarrow \quad \Gamma\widehat{f} \in \Theta, \qquad \widehat{f} \in S^*,$$

and

$$\Gamma'\widehat{f} \in \mathcal{G} \times \{0\} \quad \Leftrightarrow \quad \Gamma\widehat{f} \in -\Theta^{-1}, \qquad \widehat{f} \in S^*,$$

respectively. Now (2.1.11)–(2.1.12) imply

$$A_0' = \ker \Gamma_0' = \ker \big(\Gamma_1 - \Theta\Gamma_0\big) = A_\Theta$$

and

$$A_1' = \ker \Gamma_1' = \ker \big(\Gamma_1 + \Theta^{-1}\Gamma_0\big) = A_{-\Theta^{-1}}. \qquad \square$$

Assume that the boundary triplets $\{\mathcal{G},\Gamma_0,\Gamma_1\}$ and $\{\mathcal{G},\Gamma_0',\Gamma_1'\}$ are as in Corollary 2.5.11 and let γ and M, and γ' and M', be the corresponding γ-fields and Weyl functions, respectively. Then it follows from Proposition 2.5.3 that for all $\lambda \in \rho(A_0)\cap\rho(A_0')$ one has

$$\gamma'(\lambda) = \gamma(\lambda)\bigl(\mathcal{B}^* - \mathcal{A}^*M(\lambda)\bigr)^{-1} \tag{2.5.17}$$

and

$$M'(\lambda) = \bigl(\mathcal{A}^* + \mathcal{B}^*M(\lambda)\bigr)\bigl(\mathcal{B}^* - \mathcal{A}^*M(\lambda)\bigr)^{-1}. \tag{2.5.18}$$

In the special case where the defect numbers of S are $(1,1)$ one may choose

$$\mathcal{A} = \frac{1}{\sqrt{s^2+1}} \quad \text{and} \quad \mathcal{B} = \frac{s}{\sqrt{s^2+1}}, \qquad s \in \mathbb{R}\cup\{\infty\},$$

where $\mathcal{A}=0$ and $\mathcal{B}=1$ if $s=\infty$; this interpretation will be used also in the following. With this choice of $\mathcal{A}$ and $\mathcal{B}$ the operator in (2.5.16) reduces to the 2×2-matrix

$$\mathcal{W} = \frac{1}{\sqrt{s^2+1}}\begin{pmatrix} s & -1 \\ 1 & s\end{pmatrix}, \qquad s\in\mathbb{R}\cup\{\infty\}.$$

In this case

$$\Gamma_0' = \frac{1}{\sqrt{s^2+1}}(s\Gamma_0 - \Gamma_1), \quad \Gamma_1' = \frac{1}{\sqrt{s^2+1}}(\Gamma_0 + s\Gamma_1), \quad s\in\mathbb{R}\cup\{\infty\}, \tag{2.5.19}$$

and for $\lambda\in\rho(A_0)\cap\rho(A_0')$ the corresponding γ-field and Weyl function are given by

$$\gamma'(\lambda) = \frac{\sqrt{s^2+1}}{s-M(\lambda)}\gamma(\lambda) \quad \text{and} \quad M'(\lambda) = \frac{1+sM(\lambda)}{s-M(\lambda)}, \qquad s\in\mathbb{R}\cup\{\infty\}. \tag{2.5.20}$$

Now let S be a closed symmetric relation in $\mathfrak{H}$ and assume that $\{\mathcal{G},\Gamma_0,\Gamma_1\}$ is a boundary triplet for S^*. Consider a closed symmetric extension S' of S with the property $S'\subset A_0 = \ker\Gamma_0$. Then the boundary triplet $\{\mathcal{G},\Gamma_0,\Gamma_1\}$ can be restricted to $(S')^*\subset S^*$ and A_0 coincides with the kernel of the restriction of Γ_0. The Weyl function corresponding to this *restricted boundary triplet* is a compression of the original Weyl function onto a subspace of $\mathcal{G}$. In the following proposition this is made precise from the point of view of an orthogonal decomposition of $\mathcal{G}$.

Proposition 2.5.12. *Let S be a closed symmetric relation in $\mathfrak{H}$ and assume that $\{\mathcal{G},\Gamma_0,\Gamma_1\}$ is a boundary triplet for S^* with $A_0=\ker\Gamma_0$, and corresponding γ-field γ and Weyl function M. Assume that $\mathcal{G}$ has the orthogonal decomposition*

$$\mathcal{G} = \mathcal{G}'\oplus\mathcal{G}'' \tag{2.5.21}$$

with corresponding orthogonal projections P' and P'' and canonical embedding ι'. Then the following statements hold:

(i) *The relation*

$$S' = \{\widehat{f} \in S^* : \Gamma_0\widehat{f} = 0,\ P'\Gamma_1\widehat{f} = 0\} \tag{2.5.22}$$

is closed and symmetric with $S \subset S' \subset A_0$.

(ii) *The adjoint* $(S')^*$ *of* S' *is given by*

$$(S')^* = \{\widehat{f} \in S^* : P''\Gamma_0\widehat{f} = 0\}.$$

(iii) *The triplet* $\{\mathcal{G}', \Gamma_0', \Gamma_1'\}$, *where*

$$\Gamma_0' = \Gamma_0 \restriction (S')^* \quad \textit{and} \quad \Gamma_1' = P'\Gamma_1 \restriction (S')^*,$$

is a boundary triplet for $(S')^*$ *such that* $A_0 = \ker \Gamma_0'$.

(iv) *The* γ*-field* γ' *and Weyl function* M' *corresponding to the boundary triplet* $\{\mathcal{G}', \Gamma_0', \Gamma_1'\}$ *are given by*

$$\gamma'(\lambda) = \gamma(\lambda)\iota' \quad \textit{and} \quad M'(\lambda) = P'M(\lambda)\iota', \quad \lambda \in \rho(A_0).$$

Moreover, for every closed symmetric extension S' *with* $S \subset S' \subset A_0$ *there exists an orthogonal decomposition* (2.5.21) *of* $\mathcal{G}$ *such that* (2.5.22) *holds.*

Proof. (i) & (ii) It is clear from the definition that $S \subset S' \subset A_0$ and that S' can be written as

$$S' = \{\widehat{f} \in S^* : \Gamma\widehat{f} \in \{0\} \times \mathcal{G}''\}.$$

Hence, $S' = A_\Theta$ when $\Theta = \{0\} \times \mathcal{G}''$. It follows that S' is closed, and by (1.3.4) one has $\Theta^* = \mathcal{G}' \times \mathcal{G}$, so that Theorem 2.1.3 (iv) shows

$$(S')^* = \{\widehat{f} \in S^* : \Gamma\widehat{f} \in \mathcal{G}' \times \mathcal{G}\} = \{\widehat{f} \in S^* : P''\Gamma_0\widehat{f} = 0\}.$$

(iii) With the choice $\widehat{f}, \widehat{g} \in (S')^*$ one has $\Gamma_0\widehat{f} = P'\Gamma_0\widehat{f}$ and $\Gamma_0\widehat{g} = P'\Gamma_0\widehat{g}$. Then (2.1.1) yields

$$\begin{aligned}(f', g)_{\mathfrak{H}} - (f, g')_{\mathfrak{H}} &= (\Gamma_1\widehat{f}, \Gamma_0\widehat{g})_{\mathcal{G}} - (\Gamma_0\widehat{f}, \Gamma_1\widehat{g})_{\mathcal{G}} \\ &= (\Gamma_1\widehat{f}, P'\Gamma_0\widehat{g})_{\mathcal{G}'} - (P'\Gamma_0\widehat{f}, \Gamma_1\widehat{g})_{\mathcal{G}'} \\ &= (\Gamma_1'\widehat{f}, \Gamma_0'\widehat{g})_{\mathcal{G}'} - (\Gamma_0'\widehat{f}, \Gamma_1'\widehat{g})_{\mathcal{G}'}.\end{aligned} \tag{2.5.23}$$

It follows from the surjectivity of Γ and the identity $\Gamma_0\widehat{f} = P'\Gamma_0\widehat{f}$ when $\widehat{f} \in (S')^*$, that

$$\Gamma' = \begin{pmatrix} \Gamma_0 \restriction (S')^* \\ P'\Gamma_1 \restriction (S')^* \end{pmatrix} = \begin{pmatrix} P'\Gamma_0 \restriction (S')^* \\ P'\Gamma_1 \restriction (S')^* \end{pmatrix} : (S')^* \to \begin{pmatrix} \mathcal{G}' \\ \mathcal{G}' \end{pmatrix}$$

maps $(S')^*$ onto $\mathcal{G}' \times \mathcal{G}'$. Together with (2.5.23) this shows that $\{\mathcal{G}', \Gamma_0', \Gamma_1'\}$ is a boundary triplet for $(S')^*$. It is clear that $A_0 = \ker \Gamma_0'$ holds.

(iv) Since Γ_0 maps $\widehat{\mathfrak{N}}_\lambda(S^*)$, $\lambda \in \rho(A_0)$, one-to-one onto $\mathcal{G}$, the restriction Γ_0' maps $\widehat{\mathfrak{N}}_\lambda((S')^*)$, $\lambda \in \rho(A_0)$, one-to-one onto $\mathcal{G}'$. Hence, (2.3.1) implies that

$$\rho(A_0) \ni \lambda \mapsto \gamma'(\lambda) = \bigl\{\{\Gamma_0'\widehat{f}_\lambda, f_\lambda\} : \widehat{f}_\lambda \in \widehat{\mathfrak{N}}_\lambda((S')^*)\bigr\}$$

and therefore $\gamma'(\lambda) = \gamma(\lambda)\iota'$. It follows from (2.3.4) that

$$\rho(A_0) \ni \lambda \mapsto M'(\lambda) = \bigl\{\{\Gamma_0'\widehat{f}_\lambda, \Gamma_1'\widehat{f}_\lambda\} : \widehat{f}_\lambda \in \widehat{\mathfrak{N}}_\lambda((S')^*)\bigr\},$$

which shows that $M'(\lambda) = P'M(\lambda)\iota'$.

Finally, if S' is a closed symmetric extension of S with the property $S' \subset A_0$, then $S' = A_\Theta$ for some closed symmetric relation Θ in $\mathcal{G}$ such that $\Theta \subset \{0\} \times \mathcal{G}$ by Theorem 2.1.3 (v). As Θ is closed, there exists a closed subspace $\mathcal{G}'' \subset \mathcal{G}$ such that $\Theta = \{0\} \times \mathcal{G}''$. With $\mathcal{G}' = (\mathcal{G}'')^\perp$ it is clear that the orthogonal decomposition (2.5.21) of $\mathcal{G}$ holds and S' is of the form (2.5.22). □

In the situation of Proposition 2.5.12 the intermediate extensions of S' can also be interpreted as intermediate extensions of S. In the next corollary the connection between these extensions relative to the appropriate boundary triplets is explained.

Corollary 2.5.13. *Assume that the parameter space $\mathcal{G}$ has the orthogonal decomposition (2.5.21)and let S' be as in Proposition 2.5.12. Let Θ' be a closed relation in $\mathcal{G}'$ and let Θ be the closed linear relation in $\mathcal{G}$ defined by*

$$\Theta = \Theta' \,\widehat{\oplus}\, (\{0\} \times \mathcal{G}''). \tag{2.5.24}$$

For the intermediate extensions induced by Θ and Θ' one has

$$\bigl\{\widehat{f} \in S^* : \Gamma\widehat{f} \in \Theta\bigr\} = \bigl\{\widehat{f} \in (S')^* : \Gamma'\widehat{f} \in \Theta'\bigr\} \tag{2.5.25}$$

or, equivalently, $\ker(\Gamma_1 - \Theta\Gamma_0) = \ker(\Gamma_1' - \Theta'\Gamma_0')$.

Proof. It is clear that the relation Θ defined in (2.5.24) is a closed relation in $\mathcal{G}$. The identity in (2.5.25) now follows from

$$\begin{aligned}\bigl\{\widehat{f} \in S^* : \Gamma\widehat{f} \in \Theta\bigr\} &= \bigl\{\widehat{f} \in S^* : \Gamma\widehat{f} \in \Theta' \,\widehat{\oplus}\, (\{0\} \times \mathcal{G}'')\bigr\} \\ &= \bigl\{\widehat{f} \in S^* : \{P'\Gamma_0\widehat{f}, P'\Gamma_1\widehat{f}\} \in \Theta',\ P''\Gamma_0\widehat{f} = 0\bigr\} \\ &= \bigl\{\widehat{f} \in (S')^* : \{\Gamma_0\widehat{f}, P'\Gamma_1\widehat{f}\} \in \Theta'\bigr\} \\ &= \bigl\{\widehat{f} \in (S')^* : \Gamma'\widehat{f} \in \Theta'\bigr\},\end{aligned}$$

where (2.5.24) has been used in conjunction with the boundary triplet in Proposition 2.5.12. □

Let S and S' be closed symmetric relations in $\mathfrak{H}$ and $\mathfrak{H}'$ which are unitarily equivalent, and let $\{\mathcal{G}, \Gamma_0, \Gamma_1\}$ and $\{\mathcal{G}', \Gamma_0', \Gamma_1'\}$ be boundary triplets for S^* and $(S')^*$, respectively. The notion of unitary equivalence for these boundary triplets will now be introduced which leads to unitary equivalence of the corresponding extensions, γ-fields and Weyl functions.

Let $\mathfrak{H}$ and $\mathfrak{H}'$ be Hilbert spaces and let $U \in \mathbf{B}(\mathfrak{H}, \mathfrak{H}')$ be a unitary operator from $\mathfrak{H}$ onto $\mathfrak{H}'$. Let S and S' be closed symmetric relations in $\mathfrak{H}$ and $\mathfrak{H}'$, respectively, such that they are unitarily equivalent by means of U, that is,

$$S' = \big\{\{Uf, Uf'\} : \{f, f'\} \in S\big\} \tag{2.5.26}$$

in the sense of Definition 1.3.7. It follows from (1.3.7) that this assumption is equivalent to S^* and $(S')^*$ being equivalent under U,

$$(S')^* = \big\{\{Uf, Uf'\} : \{f, f'\} \in S^*\big\}.$$

Then U maps $\mathfrak{N}_\lambda(S^*)$ unitarily onto $\mathfrak{N}_\lambda((S')^*)$, and hence

$$\widehat{\mathfrak{N}}_\lambda((S')^*) = \big\{\{Uf_\lambda, \lambda Uf_\lambda\} : \{f, \lambda f_\lambda\} \in S^*\big\}.$$

Furthermore, let $V \in \mathbf{B}(\mathcal{G}, \mathcal{G}')$ be a unitary mapping from $\mathcal{G}$ onto $\mathcal{G}'$. Then the closed relations Θ in $\mathcal{G}$ and Θ' in $\mathcal{G}'$ are unitarily equivalent if

$$\Theta' = \big\{\{Vf, Vf'\} : \{f, f'\} \in \Theta\big\}. \tag{2.5.27}$$

The notion of unitary equivalence of two boundary triplets involves not only the unitary equivalence between $\mathfrak{H}$ and $\mathfrak{H}'$, but also the unitary equivalence between $\mathcal{G}$ and $\mathcal{G}'$.

Definition 2.5.14. Let S and S' be closed symmetric relations in $\mathfrak{H}$ and $\mathfrak{H}'$, and let $\{\mathcal{G}, \Gamma_0, \Gamma_1\}$ and $\{\mathcal{G}', \Gamma_0', \Gamma_1'\}$ be boundary triplets for S^* and $(S')^*$, respectively. Then $\{\mathcal{G}, \Gamma_0, \Gamma_1\}$ and $\{\mathcal{G}', \Gamma_0', \Gamma_1'\}$ are said to be *unitarily equivalent* if there exist a unitary operator $U \in \mathbf{B}(\mathfrak{H}, \mathfrak{H}')$ and a unitary operator $V \in \mathbf{B}(\mathcal{G}, \mathcal{G}')$ such that

(i) S and S' are unitarily equivalent by means of U as in (2.5.26);

(ii) $\{\mathcal{G}, \Gamma_0, \Gamma_1\}$ and $\{\mathcal{G}', \Gamma_0', \Gamma_1'\}$ are connected via

$$\Gamma_0'\{Uf, Uf'\} = V\Gamma_0\{f, f'\} \quad \text{and} \quad \Gamma_1'\{Uf, Uf'\} = V\Gamma_1\{f, f'\} \tag{2.5.28}$$

for all $\{f, f'\} \in S^*$.

In the next proposition it will be shown that for unitarily equivalent boundary triplets the corresponding closed extensions, γ-fields, and Weyl functions are unitarily equivalent.

Proposition 2.5.15. *Let S and S' be closed symmetric relations in $\mathfrak{H}$ and $\mathfrak{H}'$, and let $\{\mathcal{G}, \Gamma_0, \Gamma_1\}$ and $\{\mathcal{G}', \Gamma_0', \Gamma_1'\}$ be boundary triplets for S^* and $(S')^*$, respectively, which are unitarily equivalent by means of the unitary operators $U \in \mathbf{B}(\mathfrak{H}, \mathfrak{H}')$ and $V \in \mathbf{B}(\mathcal{G}, \mathcal{G}')$. Then the following statements hold:*

(i) *For all closed relations* Θ *in* $\mathcal{G}$ *and* Θ' *in* $\mathcal{G}'$ *connected via* (2.5.27) *the closed extensions*

$$A_\Theta = \{\widehat{f} \in S^* : \Gamma\widehat{f} \in \Theta\} \quad \textit{and} \quad A'_{\Theta'} = \{\widehat{h} \in (S')^* : \Gamma'\widehat{h} \in \Theta'\}$$

are unitarily equivalent by means of $U \in \mathbf{B}(\mathfrak{H}, \mathfrak{H}')$, *that is,*

$$A'_{\Theta'} = \{\{Uf, Uf'\} : \{f, f'\} \in A_\Theta\}.$$

(ii) *The* γ*-fields* γ *and* γ' *corresponding to* $\{\mathcal{G}, \Gamma_0, \Gamma_1\}$ *and* $\{\mathcal{G}', \Gamma'_0, \Gamma'_1\}$ *are related by*

$$\gamma'(\lambda) = U\gamma(\lambda)V^{-1}, \quad \lambda \in \rho(A_0) = \rho(A'_0).$$

(iii) *The Weyl functions* M *and* M' *corresponding to* $\{\mathcal{G}, \Gamma_0, \Gamma_1\}$ *and* $\{\mathcal{G}', \Gamma'_0, \Gamma'_1\}$ *are related by*

$$M'(\lambda) = VM(\lambda)V^{-1}, \quad \lambda \in \rho(A_0) = \rho(A'_0).$$

Proof. (i) It follows from the definition that

$$A_\Theta = \{\{f, f'\} \in S^* : \Gamma\{f, f'\} \in \Theta\}$$

and

$$A'_{\Theta'} = \{\{g, g'\} \in (S')^* : \Gamma'\{g, g'\} \in \Theta'\}.$$

Since S and S' are unitarily equivalent, so are S^* and $(S')^*$, and hence one has $\{g, g'\} \in (S')^*$ if and only if $\{g, g'\} = \{Uf, Uf'\}$ for some $\{f, f'\} \in S^*$. Thus, by the unitary equivalence of the boundary triplets one obtains

$$\begin{aligned}
A'_{\Theta'} &= \{\{Uf, Uf'\} : \{f, f'\} \in S^*,\ \{\Gamma'_0\{Uf, Uf'\}, \Gamma'_1\{Uf, Uf'\}\} \in \Theta'\} \\
&= \{\{Uf, Uf'\} : \{f, f'\} \in S^*,\ \{V\Gamma_0\{f, f'\}, V\Gamma_1\{f, f'\}\} \in \Theta'\} \\
&= \{\{Uf, Uf'\} : \{f, f'\} \in S^*,\ \{\Gamma_0\{f, f'\}, \Gamma_1\{f, f'\}\} \in \Theta\} \\
&= \{\{Uf, Uf'\} : \{f, f'\} \in A_\Theta\}.
\end{aligned}$$

(ii) By item (i), the self-adjoint relations $A_0 = \ker \Gamma_0$ and $A'_0 = \ker \Gamma'_0$ are unitarily equivalent by means of U, which implies that $\rho(A_0) = \rho(A'_0)$, and hence the γ-fields γ and γ' are defined on the same subset of $\mathbb{C}$. For $\lambda \in \rho(A_0)$ one computes

$$\begin{aligned}
\gamma'(\lambda) &= \{\{\Gamma'_0\{g_\lambda, \lambda g_\lambda\}, g_\lambda\} : \{g_\lambda, \lambda g_\lambda\} \in \widehat{\mathfrak{N}}_\lambda((S')^*)\} \\
&= \{\{\Gamma'_0\{Uf_\lambda, \lambda Uf_\lambda\}, Uf_\lambda\} : \{f_\lambda, \lambda f_\lambda\} \in \widehat{\mathfrak{N}}_\lambda(S^*)\} \\
&= \{\{V\Gamma_0\{f_\lambda, \lambda f_\lambda\}, Uf_\lambda\} : \{f_\lambda, \lambda f_\lambda\} \in \widehat{\mathfrak{N}}_\lambda(S^*)\} \\
&= U\gamma(\lambda)V^{-1}.
\end{aligned}$$

(iii) Since $\rho(A_0) = \rho(A_0')$, the Weyl functions M and M' are defined on the same subset of $\mathbb{C}$. Fix $\lambda \in \rho(A_0)$, let $\psi \in \mathcal{G}'$ and choose $\{f_\lambda, \lambda f_\lambda\} \in \widehat{\mathfrak{N}}_\lambda(S^*)$ such that $\Gamma_0'\{Uf_\lambda, \lambda Uf_\lambda\} = \psi$. Since $\{Uf_\lambda, \lambda Uf_\lambda\} \in \widehat{\mathfrak{N}}_\lambda((S')^*)$, it follows from the definition of the Weyl function and (2.5.28) that

$$\begin{aligned} M'(\lambda)\psi &= M'(\lambda)\Gamma_0'\{Uf_\lambda, \lambda Uf_\lambda\} \\ &= \Gamma_1'\{Uf_\lambda, \lambda Uf_\lambda\} \\ &= V\Gamma_1\{f_\lambda, \lambda f_\lambda\} \\ &= VM(\lambda)\Gamma_0\{f_\lambda, \lambda f_\lambda\} \\ &= VM(\lambda)V^{-1}\Gamma_0'\{Uf_\lambda, \lambda Uf_\lambda\} \\ &= VM(\lambda)V^{-1}\psi. \end{aligned}$$

This yields $M'(\lambda) = VM(\lambda)V^{-1}$ for all $\lambda \in \rho(A_0)$. □

Let S and S' be closed symmetric relations in $\mathfrak{H}$ and $\mathfrak{H}'$ with boundary triplets $\{\mathcal{G}, \Gamma_0, \Gamma_1\}$ and $\{\mathcal{G}', \Gamma_0', \Gamma_1'\}$ that are unitarily equivalent by means of a unitary operator $U \in \mathbf{B}(\mathfrak{H}, \mathfrak{H}')$ and a unitary operator $V \in \mathbf{B}(\mathcal{G}, \mathcal{G}')$ as in Proposition 2.5.15. Then according to Proposition 2.5.15 one has that

$$A_0' = \big\{\{Uf, Uf'\} : \{f, f'\} \in A_0\big\};$$

cf. (2.5.28). In particular, this implies that the multivalued parts of A_0 and A_0' are connected by

$$\operatorname{mul} A_0' = U(\operatorname{mul} A_0),$$

and since U is unitary it also follows that

$$\overline{\operatorname{dom}}\, A_0' = U(\overline{\operatorname{dom}}\, A_0).$$

The following corollary is an immediate consequence of Proposition 2.5.15 (ii) and Proposition 2.3.2 (ii).

Corollary 2.5.16. *Let P and P' be the orthogonal projections in $\mathfrak{H}$ and $\mathfrak{H}'$ onto $(\operatorname{mul} A_0)^\perp$ and $(\operatorname{mul} A_0')^\perp$, respectively. Then*

$$\begin{pmatrix} P'\gamma'(\lambda) \\ (I - P')\gamma'(\lambda) \end{pmatrix} = U \begin{pmatrix} P\gamma(\lambda) \\ (I - P)\gamma(\lambda) \end{pmatrix} V^{-1}, \quad \lambda \in \rho(A_0) = \rho(A_0'),$$

where $(I - P)\gamma(\lambda) = (I - P)\gamma(\lambda_0)$ and $(I - P')\gamma'(\lambda) = (I - P')\gamma'(\lambda_0)$ are parts that do not depend on λ.

In Theorem 4.2.6 it will be shown that if S and S' are simple (see Section 3.4) and their Weyl functions are unitarily equivalent, then in fact the corresponding boundary triplets are unitarily equivalent.

2.6 Kreĭn's formula for intermediate extensions

Let S be a closed symmetric relation in a Hilbert space $\mathfrak{H}$ and assume that $\{\mathcal{G}, \Gamma_0, \Gamma_1\}$ is a boundary triplet for S^*. According to Theorem 2.1.3, the mapping $\Gamma = (\Gamma_0, \Gamma_1)^\top$ induces a bijective correspondence between the set of (closed) intermediate extensions A_Θ of S and the set of (closed) relations Θ in $\mathcal{G}$, via

$$\Theta \mapsto A_\Theta = \{\widehat{f} \in S^* : \Gamma \widehat{f} \in \Theta\} = \ker\big(\Gamma_1 - \Theta\Gamma_0\big);$$

and $A_0 = \ker \Gamma_0$ corresponds to $\Theta = \{0\} \times \mathcal{G}$. For $\lambda \in \rho(A_0)$ the relation $(A_\Theta - \lambda)^{-1}$ will be regarded as a perturbation of the resolvent of the self-adjoint extension A_0 of S. This fact is expressed by the formula provided in Theorem 2.6.1 and some variants under the additional assumption $\lambda \in \rho(A_\Theta)$ are discussed afterwards. In the special case $\lambda \in \rho(A_\Theta)$ one has $(A_\Theta - \lambda)^{-1} \in \mathbf{B}(\mathfrak{H})$ and the resolvent difference, and hence also the perturbation term, are bounded operators. Moreover, it is shown later how the different types of spectral points $\lambda \in \sigma(A_\Theta)$ which are contained in $\rho(A_0)$ are related to the Weyl function and the parameter Θ. A more in-depth treatment of the connection of the spectrum and the Weyl function can be found in Chapter 3.

In the next theorem the difference of $(A_\Theta - \lambda)^{-1}$ and $(A_0 - \lambda)^{-1}$, $\lambda \in \rho(A_0)$, is expressed in a perturbation term which involves the Weyl function M and the parameter Θ. This results in a general version of *Kreĭn's formula* for intermediate extensions.

Theorem 2.6.1. *Let S be a closed symmetric relation in $\mathfrak{H}$, let $\{\mathcal{G}, \Gamma_0, \Gamma_1\}$ be a boundary triplet for S^*, $A_0 = \ker \Gamma_0$, and let γ and M be the corresponding γ-field and Weyl function, respectively. Moreover, let Θ be a closed relation in $\mathcal{G}$ and let*

$$A_\Theta = \{\widehat{f} \in S^* : \Gamma \widehat{f} \in \Theta\} \tag{2.6.1}$$

be the corresponding extension via (2.1.5). Then for all $\lambda \in \rho(A_0)$ one has the equality

$$(A_\Theta - \lambda)^{-1} = (A_0 - \lambda)^{-1} + \gamma(\lambda)\big(\Theta - M(\lambda)\big)^{-1}\gamma(\bar{\lambda})^*, \tag{2.6.2}$$

where the inverses in the first and the last term are taken in the sense of relations. Moreover, if $\lambda \in \rho(A_0) \cap \rho(A_\Theta)$, then $(\Theta - M(\lambda))^{-1} \in \mathbf{B}(\mathcal{G})$ and (2.6.2) holds in the sense of bounded linear operators.

Proof. Assume that $\lambda \in \rho(A_0)$. In order to establish the identity (2.6.2) it must be shown that the relations on the left-hand side and right-hand side coincide.

First the inclusion ($\subset$) in (2.6.2) will be shown. For this purpose, consider $\{g, g'\} \in (A_\Theta - \lambda)^{-1}$ so that, equivalently, $\widehat{g}_\Theta = \{g', g + \lambda g'\} \in A_\Theta$. Moreover, denote

$$\widehat{g}_0 = \big\{(A_0 - \lambda)^{-1}g, (I + \lambda(A_0 - \lambda)^{-1})g\big\} \in A_0.$$

Then

$$\widehat{g}_\Theta - \widehat{g}_0 = \{g' - (A_0 - \lambda)^{-1} g, \lambda(g' - (A_0 - \lambda)^{-1} g)\},$$

and hence $\widehat{g}_\Theta - \widehat{g}_0 \in \widehat{\mathfrak{N}}_\lambda(S^*)$. Since $\widehat{\gamma}(\lambda)$ maps $\mathcal{G}$ onto $\widehat{\mathfrak{N}}_\lambda(S^*)$ there exists an element $\varphi \in \mathcal{G}$ such that

$$\widehat{g}_\Theta = \widehat{g}_0 + \widehat{\gamma}(\lambda)\varphi. \tag{2.6.3}$$

By Proposition 2.3.6 (ii) one has $\Gamma\widehat{\gamma}(\lambda)\varphi = \{\varphi, M(\lambda)\varphi\}$ and, moreover, Proposition 2.3.2 (iv) shows that $\Gamma\widehat{g}_0 = \{0, \gamma(\bar{\lambda})^* g\}$. Since $\widehat{g}_\Theta \in A_\Theta$, an application of Γ to (2.6.3) yields

$$\{0, \gamma(\bar{\lambda})^* g\} + \{\varphi, M(\lambda)\varphi\} = \Gamma\widehat{g}_0 + \Gamma\widehat{\gamma}(\lambda)\varphi = \Gamma\widehat{g}_\Theta \in \Theta,$$

see (2.6.1). Thus, $\{\varphi, \gamma(\bar{\lambda})^* g + M(\lambda)\varphi\} \in \Theta$ and $\{\varphi, \gamma(\bar{\lambda})^* g\} \in \Theta - M(\lambda)$ or, equivalently, $\{g, \varphi\} \in (\Theta - M(\lambda))^{-1}\gamma(\bar{\lambda})^*$, which implies that

$$\{g, \gamma(\lambda)\varphi\} \in \gamma(\lambda)(\Theta - M(\lambda))^{-1}\gamma(\bar{\lambda})^*. \tag{2.6.4}$$

Now consider the first component $g' = (A_0 - \lambda)^{-1} g + \gamma(\lambda)\varphi$ in the identity (2.6.3). Then one has

$$\{g, g'\} = \{g, (A_0 - \lambda)^{-1} g + \gamma(\lambda)\varphi\}, \tag{2.6.5}$$

and due to $\{g, (A_0 - \lambda)^{-1} g\} \in (A_0 - \lambda)^{-1}$ and (2.6.4) it follows from (2.6.5) that

$$\{g, g'\} \in (A_0 - \lambda)^{-1} + \gamma(\lambda)\big(\Theta - M(\lambda)\big)^{-1}\gamma(\bar{\lambda})^*,$$

and hence the inclusion ($\subset$) in (2.6.2) holds.

Next the inclusion ($\supset$) in (2.6.2) will be shown. For this purpose, let

$$\{g, g'\} \in (A_0 - \lambda)^{-1} + \gamma(\lambda)\big(\Theta - M(\lambda)\big)^{-1}\gamma(\bar{\lambda})^*.$$

By the definition of the sum of relations, this means that

$$g' = (A_0 - \lambda)^{-1} g + h, \quad \text{where} \quad \{g, h\} \in \gamma(\lambda)\big(\Theta - M(\lambda)\big)^{-1}\gamma(\bar{\lambda})^*.$$

Recall from Proposition 2.3.2 (i) that $\gamma(\lambda) \in \mathbf{B}(\mathcal{G}, \mathfrak{H})$, since $\lambda \in \rho(A_0)$. Hence, $h = \gamma(\lambda)\varphi$, where $\{\gamma(\bar{\lambda})^* g, \varphi\} \in (\Theta - M(\lambda))^{-1}$. Consequently, it is clear that $\{\varphi, \gamma(\bar{\lambda})^* g + M(\lambda)\varphi\} \in \Theta$. Next observe that

$$\{g', g + \lambda g'\} = \{(A_0 - \lambda)^{-1} g, (I + \lambda(A_0 - \lambda)^{-1}) g\} + \{\gamma(\lambda)\varphi, \lambda\gamma(\lambda)\varphi\},$$

which implies that

$$\Gamma\{g', g + \lambda g'\} = \{0, \gamma(\bar{\lambda})^* g\} + \{\varphi, M(\lambda)\varphi\} \in \Theta$$

or $\{g', g + \lambda g'\} \in A_\Theta$. Thus, $\{g, g'\} \in (A_\Theta - \lambda)^{-1}$ and therefore

$$(A_0 - \lambda)^{-1} + \gamma(\lambda)\big(\Theta - M(\lambda)\big)^{-1}\gamma(\bar{\lambda})^* \subset (A_\Theta - \lambda)^{-1}.$$

Hence, the inclusion ($\supset$) in (2.6.2) has been shown.

To prove the last assertion in the theorem, assume that $\lambda \in \rho(A_0)\cap\rho(A_\Theta)$. It is first shown that $\ker(\Theta - M(\lambda)) = \{0\}$. To see this, let $\varphi \in \ker(\Theta - M(\lambda))$. Then clearly $\{\varphi, M(\lambda)\varphi\} \in \Theta$. Define $\widehat{f}_\lambda = \widehat{\gamma}(\lambda)\varphi$, so that $\widehat{f}_\lambda = \{f_\lambda, \lambda f_\lambda\} \in \widehat{\mathfrak{N}}_\lambda(S^*)$ and

$$\Gamma\widehat{f}_\lambda = \{\Gamma_0\widehat{f}_\lambda, \Gamma_1\widehat{f}_\lambda\} = \{\Gamma_0\widehat{f}_\lambda, M(\lambda)\Gamma_0\widehat{f}_\lambda\} = \{\varphi, M(\lambda)\varphi\} \in \Theta.$$

Thus, $\widehat{f}_\lambda \in A_\Theta$ and $f_\lambda = \gamma(\lambda)\varphi \in \ker(A_\Theta - \lambda)$. Since $\lambda \in \rho(A_\Theta)$, one concludes that $\gamma(\lambda)\varphi = 0$ and $\varphi = 0$. Hence, $\ker(\Theta - M(\lambda)) = \{0\}$.

Next it is shown that $(\Theta - M(\lambda))^{-1} \in \mathbf{B}(\mathcal{G})$. Since $\lambda \in \rho(A_0)$, the identity (2.6.2) holds and as it is assumed that $\lambda \in \rho(A_\Theta)$, one has $\operatorname{dom}(A_\Theta - \lambda)^{-1} = \mathfrak{H}$. Therefore,

$$\operatorname{dom}\left[(\Theta - M(\lambda))^{-1}\gamma(\bar\lambda)^*\right] = \operatorname{dom}\left[\gamma(\lambda)(\Theta - M(\lambda))^{-1}\gamma(\bar\lambda)^*\right] = \mathfrak{H}, \tag{2.6.6}$$

where the first identity is clear since $\gamma(\lambda) \in \mathbf{B}(\mathcal{G}, \mathfrak{H})$. As $\operatorname{ran}\gamma(\bar\lambda)^* = \mathcal{G}$, one concludes from (2.6.6) that $\operatorname{dom}(\Theta - M(\lambda))^{-1} = \mathcal{G}$. By assumption, Θ is closed and then $M(\lambda) \in \mathbf{B}(\mathcal{G})$ implies that $\Theta - M(\lambda)$ is closed. Then $(\Theta - M(\lambda))^{-1}$ is a closed operator and by the closed graph theorem $(\Theta - M(\lambda))^{-1} \in \mathbf{B}(\mathcal{G})$. □

Assume that $\lambda \in \rho(A_0)$. In Theorem 2.6.1 it is shown that then $\lambda \in \rho(A_\Theta)$ leads to $(\Theta - M(\lambda))^{-1} \in \mathbf{B}(\mathcal{G})$. In fact, there is a one-to-one correspondence between the part of the spectrum of A_Θ contained in $\rho(A_0)$ and the spectrum of $\Theta - M(\lambda)$ contained in $\rho(A_0)$. The following theorem and its corollary are direct consequences of the Kreĭn formula (2.6.2). A complete description of the spectrum of self-adjoint extensions A_Θ in terms of the singularities of the function $\lambda \mapsto (\Theta - M(\lambda))^{-1}$ is given in Section 3.8.

Theorem 2.6.2. *Let S be a closed symmetric relation in $\mathfrak{H}$, let $\{\mathcal{G}, \Gamma_0, \Gamma_1\}$ be a boundary triplet for S^*, $A_0 = \ker\Gamma_0$, and let γ and M be the corresponding γ-field and Weyl function, respectively. Moreover, let Θ be a closed relation in $\mathcal{G}$ and let*

$$A_\Theta = \{\widehat{f} \in S^* : \Gamma\widehat{f} \in \Theta\}$$

be the corresponding extension via (2.1.5). Then the following statements hold for all $\lambda \in \rho(A_0)$:

(i) $\lambda \in \sigma_p(A_\Theta) \Leftrightarrow 0 \in \sigma_p(\Theta - M(\lambda))$, *and in this case*

$$\ker(A_\Theta - \lambda) = \gamma(\lambda)\ker(\Theta - M(\lambda)); \tag{2.6.7}$$

(ii) $\lambda \in \sigma_r(A_\Theta) \Leftrightarrow 0 \in \sigma_r(\Theta - M(\lambda))$;

(iii) $\lambda \in \sigma_c(A_\Theta) \Leftrightarrow 0 \in \sigma_c(\Theta - M(\lambda))$;

(iv) $\lambda \in \rho(A_\Theta) \Leftrightarrow 0 \in \rho(\Theta - M(\lambda))$.

Proof. Assume that $\lambda \in \rho(A_0)$ and consider the right-hand side of (2.6.2) as the sum of the operator $(A_0 - \lambda)^{-1} \in \mathbf{B}(\mathfrak{H})$ and the relation

$$\gamma(\lambda)\big(\Theta - M(\lambda)\big)^{-1}\gamma(\bar\lambda)^*$$

in $\mathfrak{H}$. Hence, the domain of the right-hand side of (2.6.2) is given by

$$\mathrm{dom}\,\big[\gamma(\lambda)\big(\Theta - M(\lambda)\big)^{-1}\gamma(\bar\lambda)^*\big] = \mathrm{dom}\,\big[\big(\Theta - M(\lambda)\big)^{-1}\gamma(\bar\lambda)^*\big],$$

where it was used that $\gamma(\lambda) \in \mathbf{B}(\mathcal{G}, \mathfrak{H})$. Thus, it follows from (2.6.2) that

$$\mathrm{dom}\,(A_\Theta - \lambda)^{-1} = \mathrm{dom}\,\big(\Theta - M(\lambda)\big)^{-1}\gamma(\bar\lambda)^*. \tag{2.6.8}$$

Due to the definition of the sum of relations and $\mathrm{mul}\,(A_0 - \lambda)^{-1} = \{0\}$ the multivalued part of the right-hand side of (2.6.2) is given by

$$\begin{aligned}\mathrm{mul}\,\big[\gamma(\lambda)\big(\Theta - M(\lambda)\big)^{-1}\gamma(\bar\lambda)^*\big] &= \mathrm{mul}\,\big[\gamma(\lambda)\big(\Theta - M(\lambda)\big)^{-1}\big]\\ &= \gamma(\lambda)\,\mathrm{mul}\,\big(\Theta - M(\lambda)\big)^{-1}.\end{aligned}$$

Thus, it follows from (2.6.2) that

$$\mathrm{mul}\,(A_\Theta - \lambda)^{-1} = \gamma(\lambda)\,\mathrm{mul}\,\big(\Theta - M(\lambda)\big)^{-1}. \tag{2.6.9}$$

The proof of the theorem is based on the identities (2.6.8) and (2.6.9).

For the interpretation of (2.6.8) recall that $\gamma(\lambda)$, $\lambda \in \rho(A_0)$, maps $\mathcal{G}$ isomorphically onto $\mathfrak{N}_\lambda(S^*)$; see Proposition 2.3.2 (i). This implies that the restriction

$$\gamma(\bar\lambda)^* : \mathfrak{N}_{\bar\lambda}(S^*) \to \mathcal{G} \quad \text{is an isomorphism.}$$

In particular, $\gamma(\bar\lambda)^*$ is a bijection between closed or dense subspaces in $\mathfrak{N}_{\bar\lambda}(S^*)$ and closed or dense subspaces in $\mathcal{G}$, respectively. Now assume that V is a closed relation in $\mathcal{G}$. Since $\ker \gamma(\bar\lambda)^* = (\mathfrak{N}_{\bar\lambda}(S^*))^\perp$, it follows that

$$\mathrm{dom}\,V\gamma(\bar\lambda)^* \text{ is closed in } \mathfrak{H} \quad \Leftrightarrow \quad \mathrm{dom}\,V \text{ is closed in } \mathcal{G}, \tag{2.6.10}$$

and

$$\mathrm{dom}\,V\gamma(\bar\lambda)^* \text{ is dense in } \mathfrak{H} \quad \Leftrightarrow \quad \mathrm{dom}\,V \text{ is dense in } \mathcal{G}. \tag{2.6.11}$$

(i) The identity (2.6.7) follows from (2.6.9). It is clear that (2.6.7) implies the equivalence $\lambda \in \sigma_p(A_\Theta) \Leftrightarrow 0 \in \sigma_p(\Theta - M(\lambda))$.

(iii) It follows from (2.6.8) that $\mathrm{ran}\,(A_\Theta - \lambda)$ is a dense nonclosed subspace of $\mathfrak{H}$ if and only if $\mathrm{dom}\,(\Theta - M(\lambda))^{-1}\gamma(\bar\lambda)^*$ is a dense nonclosed subspace of $\mathfrak{H}$. By (2.6.10) and (2.6.11) with $V = (\Theta - M(\lambda))^{-1}$, this is equivalent to $\mathrm{ran}\,(\Theta - M(\lambda))$ being a dense nonclosed subspace of $\mathcal{G}$. In addition, it follows from (i) that $A_\Theta - \lambda$ is injective if and only if $\Theta - M(\lambda)$ is injective. This proves the assertion.

(iv) The implication ($\Rightarrow$) holds by Theorem 2.6.1. The implication ($\Leftarrow$) is easy to see. Indeed, assume that $0 \in \rho(\Theta - M(\lambda))$. Then $(\Theta - M(\lambda))^{-1} \in \mathbf{B}(\mathcal{G})$ and since $\gamma(\lambda) \in \mathbf{B}(\mathcal{G},\mathfrak{H})$ and $\gamma(\bar\lambda)^* \in \mathbf{B}(\mathfrak{H},\mathcal{G})$ for $\lambda \in \rho(A_0)$, one concludes from (2.6.2) that $(A_\Theta - \lambda)^{-1} \in \mathbf{B}(\mathfrak{H})$, i.e., $\lambda \in \rho(A_\Theta)$.

(ii) This assertion is a consequence of (i), (iii), and (iv). □

The Kreĭn formula (2.6.2) was formulated above in terms of the closed relation Θ in the Hilbert space $\mathcal{G}$. Now the form of the Kreĭn formula will be given when a tight parametric representation of Θ is chosen; cf. Section 1.10.

Corollary 2.6.3. *Let S be a closed symmetric relation in $\mathfrak{H}$, let $\{\mathcal{G},\Gamma_0,\Gamma_1\}$ be a boundary triplet for S^*, $A_0 = \ker\Gamma_0$, and let γ and M be the corresponding γ-field and Weyl function, respectively. Let the closed relation Θ have the parametric representation*

$$\Theta = \big\{\{\mathcal{A}e, \mathcal{B}e\} : e \in \mathcal{E}\big\}, \tag{2.6.12}$$

where $\mathcal{E}$ is a Hilbert space and $\mathcal{A},\mathcal{B} \in \mathbf{B}(\mathcal{E},\mathcal{G})$, and assume that this representation of Θ is tight, i.e., $\ker\mathcal{A} \cap \ker\mathcal{B} = \{0\}$ holds. Then for all $\lambda \in \rho(A_0)$ one has

$$\lambda \in \rho(A_\Theta) \quad\Leftrightarrow\quad \big(\mathcal{B} - M(\lambda)\mathcal{A}\big)^{-1} \in \mathbf{B}(\mathcal{G},\mathcal{E}),$$

and in this case

$$(A_\Theta - \lambda)^{-1} = (A_0 - \lambda)^{-1} + \gamma(\lambda)\mathcal{A}\big(\mathcal{B} - M(\lambda)\mathcal{A}\big)^{-1}\gamma(\bar\lambda)^*. \tag{2.6.13}$$

Proof. According to Theorem 2.6.2, for $\lambda \in \rho(A_0)$ one has

$$\lambda \in \rho(A_\Theta) \quad\Leftrightarrow\quad 0 \in \rho(\Theta - M(\lambda)),$$

that is, $\lambda \in \rho(A_\Theta)$ if and only if $(\Theta - M(\lambda))^{-1} \in \mathbf{B}(\mathcal{G})$. Due to the tightness of the representation (2.6.12), Lemma 1.11.6 shows that

$$\big(\Theta - M(\lambda)\big)^{-1} \in \mathbf{B}(\mathcal{G}) \quad\Leftrightarrow\quad \big(\mathcal{B} - M(\lambda)\mathcal{A}\big)^{-1} \in \mathbf{B}(\mathcal{G},\mathcal{E}),$$

as for all $\lambda \in \rho(A_0)$ one has that $M(\lambda) \in \mathbf{B}(\mathcal{G})$. In this case it follows that $(\Theta - M(\lambda))^{-1} = \mathcal{A}(\mathcal{B} - M(\lambda)\mathcal{A})^{-1}$. Furthermore, the resolvent formula (2.6.13) follows from (2.6.2). □

Let again Θ be a closed relation in $\mathcal{G}$ and assume, in the same way as at the end of Section 2.2, that Θ admits an orthogonal decomposition

$$\Theta = \Theta_{\rm op} \,\widehat{\oplus}\, \Theta_{\rm mul}, \qquad \mathcal{G} = \mathcal{G}_{\rm op} \oplus \mathcal{G}_{\rm mul}, \tag{2.6.14}$$

into a (not necessarily densely defined) operator part $\Theta_{\rm op}$ acting in the Hilbert space $\mathcal{G}_{\rm op} = \overline{\operatorname{dom}\Theta^*} = (\operatorname{mul}\Theta)^\perp$ and a multivalued part $\Theta_{\rm mul} = \{0\} \times \operatorname{mul}\Theta$ in the Hilbert space $\mathcal{G}_{\rm mul} = \operatorname{mul}\Theta$; cf. Section 1.3. Recall that, in particular, closed symmetric, self-adjoint, (maximal) dissipative, and (maximal) accumulative relations Θ in $\mathcal{G}$ admit such a decomposition.

Corollary 2.6.4. *Assume that the closed relation Θ in Theorem 2.6.1 has the orthogonal decomposition (2.6.14), let P_{op} be the orthogonal projection onto $\mathcal{G}_{\mathrm{op}}$, and denote the canonical embedding of $\mathcal{G}_{\mathrm{op}}$ into $\mathcal{G}$ by ι_{op}. Let*

$$A_\Theta = \{\widehat{f} \in S^* : \Gamma\widehat{f} \in \Theta\}$$

be the intermediate extension of S via (2.2.12). Then for all $\lambda \in \rho(A_0) \cap \rho(A_\Theta)$ one has $(\Theta_{\mathrm{op}} - P_{\mathrm{op}} M(\lambda)\iota_{\mathrm{op}})^{-1} \in \mathbf{B}(\mathcal{G}_{\mathrm{op}})$ and

$$(A_\Theta - \lambda)^{-1} = (A_0 - \lambda)^{-1} + \gamma(\lambda)\iota_{\mathrm{op}}\big(\Theta_{\mathrm{op}} - P_{\mathrm{op}}M(\lambda)\iota_{\mathrm{op}}\big)^{-1} P_{\mathrm{op}}\gamma(\bar\lambda)^*.$$

Proof. In view of (2.6.14), one sees that for $\lambda \in \rho(A_0) \cap \rho(A_\Theta)$

$$\Theta - M(\lambda) = \left\{\left\{\begin{pmatrix}\varphi\\0\end{pmatrix}, \begin{pmatrix}\Theta_{\mathrm{op}}\varphi - P_{\mathrm{op}}M(\lambda)\iota_{\mathrm{op}}\varphi\\ \psi - (I-P_{\mathrm{op}})M(\lambda)\iota_{\mathrm{op}}\varphi\end{pmatrix}\right\} : \begin{matrix}\varphi \in \operatorname{dom}\Theta_{\mathrm{op}},\\ \psi \in \mathcal{G}_{\mathrm{mul}}\end{matrix}\right\},$$

and hence

$$\big(\Theta - M(\lambda)\big)^{-1} = \left\{\left\{\begin{pmatrix}\Theta_{\mathrm{op}}\varphi - P_{\mathrm{op}}M(\lambda)\iota_{\mathrm{op}}\varphi\\ \chi\end{pmatrix}, \begin{pmatrix}\varphi\\0\end{pmatrix}\right\} : \begin{matrix}\varphi \in \operatorname{dom}\Theta_{\mathrm{op}},\\ \chi \in \mathcal{G}_{\mathrm{mul}}\end{matrix}\right\}.$$

Since $(\Theta - M(\lambda))^{-1} \in \mathbf{B}(\mathcal{G})$, one has

$$\ker\big(\Theta_{\mathrm{op}} - P_{\mathrm{op}}M(\lambda)\iota_{\mathrm{op}}\big) = \{0\} \quad \text{and} \quad \operatorname{ran}\big(\Theta_{\mathrm{op}} - P_{\mathrm{op}}M(\lambda)\iota_{\mathrm{op}}\big) = \mathcal{G}_{\mathrm{op}}.$$

This shows that $\Theta_{\mathrm{op}} - P_{\mathrm{op}}M(\lambda)\iota_{\mathrm{op}}$ is a bijective closed operator in $\mathcal{G}_{\mathrm{op}}$. Hence, $(\Theta_{\mathrm{op}} - P_{\mathrm{op}}M(\lambda)\iota_{\mathrm{op}})^{-1} \in \mathbf{B}(\mathcal{G}_{\mathrm{op}})$ and so

$$\big(\Theta - M(\lambda)\big)^{-1} = \begin{pmatrix}(\Theta_{\mathrm{op}} - P_{\mathrm{op}}M(\lambda)\iota_{\mathrm{op}})^{-1} & 0\\ 0 & 0\end{pmatrix}$$

with respect to the decomposition (2.6.14). Now the identity

$$\big(\Theta - M(\lambda)\big)^{-1} = \iota_{\mathrm{op}}\big(\Theta_{\mathrm{op}} - P_{\mathrm{op}}M(\lambda)\iota_{\mathrm{op}}\big)^{-1}P_{\mathrm{op}}$$

is an immediate consequence. This together with Theorem 2.6.1 implies the statement. □

If the closed relation Θ in $\mathcal{G}$ admits a decomposition of the form

$$\Theta = \Theta' \,\widehat{\oplus}\, (\{0\} \times \mathcal{G}'') \tag{2.6.15}$$

as in Corollary 2.5.13, where Θ' is a closed relation in the Hilbert space $\mathcal{G}'$ and $\mathcal{G} = \mathcal{G}' \oplus \mathcal{G}''$, then Kreĭn's formula can also be interpreted in the context of the intermediate symmetric extension S' of S in Proposition 2.5.12 and the corresponding restriction of the boundary triplet $\{\mathcal{G}, \Gamma_0, \Gamma_1\}$. More precisely, if Θ is of the form

(2.6.15) and S' and the boundary triplet $\{\mathcal{G}', \Gamma_0', \Gamma_1'\}$ are as in Proposition 2.5.12 with corresponding γ-field γ' and Weyl function M', and

$$A_{\Theta'} = \{\widehat{f} \in (S')^* : \Gamma'\widehat{f} \in \Theta'\} = \ker(\Gamma_1' - \Theta'\Gamma_0'),$$

then $A_{\Theta'} = \ker(\Gamma_1 - \Theta\Gamma_0) = A_\Theta$ and $A_0' = \ker \Gamma_0' = \ker \Gamma_0 = A_0$ hold by Corollary 2.5.13 and Proposition 2.5.12, respectively. Moreover, by Theorem 2.6.1 one has $(\Theta' - M'(\lambda))^{-1} \in \mathbf{B}(\mathcal{G}')$ for all $\lambda \in \rho(A_{\Theta'}) \cap \rho(A_0')$ and

$$(A_{\Theta'} - \lambda)^{-1} = (A_0' - \lambda)^{-1} + \gamma'(\lambda)\big(\Theta' - M'(\lambda)\big)^{-1}\gamma'(\bar{\lambda})^*.$$

In the special case where $\Theta' = \Theta_{\mathrm{op}}$ and $\{0\} \times \mathcal{G}'' = \Theta_{\mathrm{mul}}$ as in (2.6.14) one has

$$M'(\lambda) = P_{\mathrm{op}} M(\lambda)\iota_{\mathrm{op}} \quad \text{and} \quad \gamma'(\lambda) = \gamma(\lambda)\iota_{\mathrm{op}},$$

so that Kreĭn's formula in Corollary 2.6.4 can be rewritten in the form

$$(A_{\Theta_{\mathrm{op}}} - \lambda)^{-1} = (A_0' - \lambda)^{-1} + \gamma'(\lambda)\big(\Theta_{\mathrm{op}} - M'(\lambda)\big)^{-1}\gamma'(\bar{\lambda})^*.$$

The behavior of Kreĭn's formula under transformations of boundary triplets will be discussed next. To this end suppose that S is a closed symmetric relation in $\mathfrak{H}$, let $\{\mathcal{G}, \Gamma_0, \Gamma_1\}$ be a boundary triplet for S^*, $A_0 = \ker \Gamma_0$, and let γ and M be the corresponding γ-field and Weyl function, respectively. Consider a closed extension

$$A_\Theta = \{\widehat{f} \in S^* : \Gamma\widehat{f} \in \Theta\} = \ker\big(\Gamma_1 - \Theta\Gamma_0\big)$$

corresponding to a closed relation Θ in $\mathcal{G}$. Then for all $\lambda \in \rho(A_\Theta) \cap \rho(A_0)$ one has

$$(A_\Theta - \lambda)^{-1} = (A_0 - \lambda)^{-1} + \gamma(\lambda)\big(\Theta - M(\lambda)\big)^{-1}\gamma(\bar{\lambda})^*$$

according to Theorem 2.6.1. Let $\mathcal{G}'$ be a further Hilbert space and assume that $\mathcal{W} \in \mathbf{B}(\mathcal{G} \times \mathcal{G}, \mathcal{G}' \times \mathcal{G}')$ satisfies the identities in (2.5.1). Let $\{\mathcal{G}', \Gamma_0', \Gamma_1'\}$ be the corresponding transformed boundary triplet in (2.5.2) with γ-field γ' and Weyl function M' specified in Proposition 2.5.3. Let $A_0' = \ker \Gamma_0'$ and define the closed relation Θ' in $\mathcal{G}'$ by $\Theta' = \mathcal{W}[\Theta]$; cf. (2.5.4). By Proposition 2.5.2, one has

$$A_\Theta = \ker\big(\Gamma_1 - \Theta\Gamma_0\big) = \ker\big(\Gamma_1' - \Theta'\Gamma_0'\big) = A_{\Theta'}',$$

and hence for all $\lambda \in \rho(A_\Theta) \cap \rho(A_0')$ Kreĭn's formula in Theorem 2.6.1 has the form

$$(A_\Theta - \lambda)^{-1} = (A_0' - \lambda)^{-1} + \gamma'(\lambda)\big(\Theta' - M'(\lambda)\big)^{-1}\gamma'(\bar{\lambda})^* = (A_{\Theta'}' - \lambda)^{-1}.$$

In this sense Kreĭn's formula is invariant under transformations of boundary triplets.

Next, Theorem 2.6.2 will be complemented for the case where the extensions are self-adjoint. Recall from Section 1.5 that for a self-adjoint relation H a spectral point $\lambda \in \mathbb{R}$ belongs to the discrete spectrum $\sigma_{\mathrm{d}}(H)$ if λ is an eigenvalue with finite multiplicity which is an isolated point in $\sigma(H)$. It will be used that $\lambda \in \sigma_{\mathrm{d}}(H)$ if and only if

$$\dim \ker (H - \lambda) < \infty \quad \text{and} \quad \operatorname{ran} (H - \lambda) = \overline{\operatorname{ran}} (H - \lambda). \tag{2.6.16}$$

The complement of the discrete spectrum of H in $\sigma(H)$ is the essential spectrum, denoted by $\sigma_{\rm ess}(H)$.

Theorem 2.6.5. *Let S be a closed symmetric relation, let $\{\mathcal{G}, \Gamma_0, \Gamma_1\}$ be a boundary triplet for S^*, $A_0 = \ker \Gamma_0$, and let M be the corresponding Weyl function. Let Θ be a self-adjoint relation in $\mathcal{G}$ and let*

$$A_\Theta = \{\widehat{f} \in S^* : \Gamma \widehat{f} \in \Theta\}$$

be the corresponding self-adjoint extension via (2.1.5). Then the following statements hold for all $\lambda \in \rho(A_0)$:

(i) $\lambda \in \sigma_{\rm d}(A_\Theta) \Leftrightarrow 0 \in \sigma_{\rm d}(\Theta - M(\lambda))$;

(ii) $\lambda \in \sigma_{\rm ess}(A_\Theta) \Leftrightarrow 0 \in \sigma_{\rm ess}(\Theta - M(\lambda))$.

Proof. Here one relies on the observations made in the proof of Theorem 2.6.2. Assume that $\lambda \in \rho(A_0)$.

(i) It follows from Theorem 2.6.2 (i) that

$$0 < \dim \ker (A_\Theta - \lambda) < \infty \quad \Leftrightarrow \quad 0 < \dim \ker (\Theta - M(\lambda)) < \infty,$$

and it follows from (2.6.8) and (2.6.10) with $V = (\Theta - M(\lambda))^{-1}$ that

$$\operatorname{ran} (A_\Theta - \lambda) \text{ closed} \quad \Leftrightarrow \quad \operatorname{ran} (\Theta - M(\lambda)) \text{ closed}.$$

Now assertion (i) is a consequence of the above equivalences and the characterization (2.6.16) of discrete eigenvalues of self-adjoint relations.

(ii) Note that $\lambda \in \sigma(A_\Theta)$ if and only if $0 \in \sigma(\Theta - M(\lambda))$ by Theorem 2.6.2. Hence, this assertion is a consequence of item (i), $\sigma_{\rm ess}(A_\Theta) = \sigma(A_\Theta) \setminus \sigma_{\rm d}(A_\Theta)$, and $\sigma_{\rm ess}(\Theta - M(\lambda)) = \sigma(\Theta - M(\lambda)) \setminus \sigma_{\rm d}(\Theta - M(\lambda))$. □

2.7 Kreĭn's formula for exit space extensions

The Kreĭn formula in Theorem 2.6.1 holds for intermediate extensions of a symmetric relation S in a Hilbert space $\mathfrak{H}$. In particular, these intermediate extensions contain maximal dissipative, maximal accumulative, and self-adjoint extensions. Now consider larger Hilbert spaces $\mathfrak{K}$ which contain $\mathfrak{H}$ as a closed subspace and self-adjoint relations $\widetilde{A}$ in $\mathfrak{K}$ which extend S as studied by Kreĭn and Naĭmark. It will be shown that such self-adjoint extensions induce families of relations in $\mathfrak{H}$ which also extend S. For these families of relations there is a version of the Kreĭn formula, which will also be called Kreĭn–Naĭmark formula in this text.

The following notions are useful. Let $\mathfrak{H}$ and $\mathfrak{H}'$ be Hilbert spaces and let $\widetilde{A}$ be a self-adjoint relation in the Hilbert space $\mathfrak{H} \oplus \mathfrak{H}'$. The *Štraus family* $T(\lambda)$, $\lambda \in \mathbb{C}$, in $\mathfrak{H}$ corresponding to the self-adjoint relation $\widetilde{A}$ in $\mathfrak{H} \oplus \mathfrak{H}'$ is defined by

$$T(\lambda) = \left\{ \{f, f'\} \in \mathfrak{H} \times \mathfrak{H} : \left\{ \begin{pmatrix} f \\ h \end{pmatrix}, \begin{pmatrix} f' \\ h' \end{pmatrix} \right\} \in \widetilde{A},\, h' = \lambda h \right\}. \tag{2.7.1}$$

Here a vector notation is used for the elements

$$\begin{pmatrix} f \\ h \end{pmatrix} \in \operatorname{dom} \widetilde{A} \subset \mathfrak{H} \oplus \mathfrak{H}' \quad \text{and} \quad \begin{pmatrix} f' \\ h' \end{pmatrix} \in \operatorname{ran} \widetilde{A} \subset \mathfrak{H} \oplus \mathfrak{H}',$$

where $f, f' \in \mathfrak{H}$ and $h, h' \in \mathfrak{H}'$. This notation will be frequently used in the rest of this section. Closely associated with the Štraus family $T(\lambda)$ is the *compressed resolvent* $R(\lambda) \in \mathbf{B}(\mathfrak{H})$ of the self-adjoint relation $\widetilde{A}$ in $\mathfrak{H} \oplus \mathfrak{H}'$, defined by

$$R(\lambda) = P_{\mathfrak{H}} (\widetilde{A} - \lambda)^{-1} \iota_{\mathfrak{H}}, \quad \lambda \in \rho(\widetilde{A}); \tag{2.7.2}$$

here $P_{\mathfrak{H}} : \mathfrak{H} \oplus \mathfrak{H}' \to \mathfrak{H}$ denotes the orthogonal projection from $\mathfrak{H} \oplus \mathfrak{H}'$ onto $\mathfrak{H}$ and $\iota_{\mathfrak{H}} : \mathfrak{H} \to \mathfrak{H} \oplus \mathfrak{H}'$ is the canonical embedding of $\mathfrak{H}$ into $\mathfrak{H} \oplus \mathfrak{H}'$.

Lemma 2.7.1. *Let $\widetilde{A}$ be a self-adjoint relation in $\mathfrak{H} \oplus \mathfrak{H}'$. Then the following statements holds for the Štraus family $T(\lambda)$ in (2.7.1):*

(i) *$T(\lambda)$ is maximal accumulative (maximal dissipative) for $\lambda \in \mathbb{C}^+$ ($\lambda \in \mathbb{C}^-$);*
(ii) *$T(\lambda) = T(\bar{\lambda})^*$, $\lambda \in \mathbb{C} \setminus \mathbb{R}$;*
(iii) *$\lambda \mapsto (T(\lambda) - \lambda)^{-1}$ is holomorphic on $\mathbb{C} \setminus \mathbb{R}$ with values in $\mathbf{B}(\mathfrak{H})$.*

Moreover, the following statements hold for the compressed resolvent $R(\lambda) \in \mathbf{B}(\mathfrak{H})$ in (2.7.2):

(iv) *$\lambda \mapsto R(\lambda)$ is holomorphic on $\mathbb{C} \setminus \mathbb{R}$ with values in $\mathbf{B}(\mathfrak{H})$;*
(v) *$R(\lambda) = R(\bar{\lambda})^*$, $\lambda \in \mathbb{C} \setminus \mathbb{R}$;*
(vi) *for all $\lambda \in \mathbb{C} \setminus \mathbb{R}$,*

$$\frac{\operatorname{Im} R(\lambda)}{\operatorname{Im} \lambda} - R(\lambda) R(\lambda)^* \geq 0. \tag{2.7.3}$$

Furthermore, the Štraus family $T(\lambda)$ in (2.7.1) and the compressed resolvent in (2.7.2) are related via

$$R(\lambda) = (T(\lambda) - \lambda)^{-1}, \quad \lambda \in \mathbb{C} \setminus \mathbb{R}. \tag{2.7.4}$$

Proof. It is clear from (2.7.1) that for $\lambda \in \mathbb{C}$ the Štraus family $T(\lambda)$ satisfies

$$(T(\lambda) - \lambda)^{-1} = \left\{ \{f' - \lambda f, f\} \in \mathfrak{H} \times \mathfrak{H} : \left\{ \begin{pmatrix} f \\ h \end{pmatrix}, \begin{pmatrix} f' \\ h' \end{pmatrix} \right\} \in \widetilde{A},\, h' = \lambda h \right\}.$$

On the other hand, it is clear that

$$(\widetilde{A}-\lambda)^{-1}=\left\{\left\{\begin{pmatrix} f'-\lambda f\\ h'-\lambda h\end{pmatrix},\begin{pmatrix} f\\ h\end{pmatrix}\right\}:\left\{\begin{pmatrix} f\\ h\end{pmatrix},\begin{pmatrix} f'\\ h'\end{pmatrix}\right\}\in\widetilde{A}\right\},$$

so that the compressed resolvent of $\widetilde{A}$ in (2.7.2) is given for $\lambda\in\rho(\widetilde{A})$ by

$$\begin{aligned} R(\lambda)&=P_{\mathfrak{H}}(\widetilde{A}-\lambda)^{-1}\iota_{\mathfrak{H}}\\ &=\left\{\{f'-\lambda f,f\}\in\mathfrak{H}\times\mathfrak{H}:\left\{\begin{pmatrix} f\\ h\end{pmatrix},\begin{pmatrix} f'\\ h'\end{pmatrix}\right\}\in\widetilde{A},\, h'=\lambda h\right\}.\end{aligned}$$

Comparison of the above identities shows that (2.7.4) holds.

(iv) & (v) follow immediately from (2.7.2).

(i) Let $\{f,f'\}\in T(\lambda)$. Then there exists a pair $\{h,h'\}$ such that

$$\left\{\begin{pmatrix} f\\ h\end{pmatrix},\begin{pmatrix} f'\\ h'\end{pmatrix}\right\}\in\widetilde{A}\quad\text{and}\quad h'=\lambda h.$$

Since $\widetilde{A}$ is self-adjoint, it follows that

$$0=\operatorname{Im}\big((f',f)+(h',h)\big)=\operatorname{Im}(f',f)+(\operatorname{Im}\lambda)(h,h)$$

and this implies that $T(\lambda)$ is accumulative (dissipative) for $\lambda\in\mathbb{C}^+$ ($\lambda\in\mathbb{C}^-$). The maximality follows from (2.7.4) since $\operatorname{ran}(T(\lambda)-\lambda)=\mathfrak{H}$; cf. Theorem 1.6.4.

(ii) It follows from (v) and (2.7.4) that

$$(T(\lambda)-\lambda)^{-1}=(T(\bar\lambda)-\bar\lambda)^{-*},$$

and this implies $T(\lambda)=T(\bar\lambda)^*$.

(iii) This is clear from (iv) and (2.7.4).

(vi) Let $\varphi\in\mathfrak{H}$ and $\varphi'=R(\lambda)\varphi$. Then (2.7.4) implies $\{\varphi',\varphi+\lambda\varphi'\}\in T(\lambda)$. For $\lambda\in\mathbb{C}^+$ the relation $T(\lambda)$ is maximal accumulative and hence (v) implies

$$\begin{aligned} 0\geq\operatorname{Im}(\varphi+\lambda\varphi',\varphi')&=\operatorname{Im}(\varphi,R(\lambda)\varphi)+(\operatorname{Im}\lambda)\|R(\lambda)\varphi\|^2\\ &=\operatorname{Im}(R(\bar\lambda)\varphi,\varphi)-(\operatorname{Im}\bar\lambda)\|R(\bar\lambda)^*\varphi\|^2.\end{aligned}$$

Thus, it follows for $\lambda\in\mathbb{C}^+$ and $\varphi\in\mathfrak{H}$ that

$$\frac{\operatorname{Im}(R(\bar\lambda)\varphi,\varphi)}{\operatorname{Im}\bar\lambda}-(R(\bar\lambda)R(\bar\lambda)^*\varphi,\varphi)\geq 0,$$

which implies (2.7.3) on $\mathbb{C}^-$. A similar reasoning leads to (2.7.3) on $\mathbb{C}^+$. □

In Chapter 4 it will be shown how the properties of the Štraus family and the compressed resolvent in Lemma 2.7.1 determine the space $\mathfrak{H}'$ and the self-adjoint relation $\widetilde{A}$ in $\mathfrak{H}\oplus\mathfrak{H}'$.

In the present context the Štraus family and the compressed resolvent appear when one considers self-adjoint extensions of a closed symmetric relation S in a Hilbert spaces $\mathfrak{H}$. Let the Hilbert space $\mathfrak{H} \oplus \mathfrak{H}'$ be an extension of $\mathfrak{H}$, where the Hilbert space $\mathfrak{H}'$ is an *exit space*. Assume that the self-adjoint relation $\widetilde{A}$ in $\mathfrak{H} \oplus \mathfrak{H}'$ is an extension of the symmetric relation S in $\mathfrak{H}$, i.e., $S \subset \widetilde{A}$. The Štraus family and the compressed resolvent of $\widetilde{A}$ consist of relations in the closed subspace $\mathfrak{H}$ that extend S in the following sense.

Proposition 2.7.2. *Let S be a closed symmetric relation in $\mathfrak{H}$ and let $\widetilde{A}$ be a self-adjoint extension of S in $\mathfrak{H} \oplus \mathfrak{H}'$. Then the Štraus family $T(\lambda)$ in (2.7.1) satisfies*

$$S \subset T(\lambda) \subset S^*, \quad \lambda \in \mathbb{C} \setminus \mathbb{R}, \tag{2.7.5}$$

and the compressed resolvent $R(\lambda)$ in (2.7.2) satisfies

$$R(\lambda) \upharpoonright_{\operatorname{ran}(S-\lambda)} = (S-\lambda)^{-1}, \quad \lambda \in \mathbb{C} \setminus \mathbb{R}. \tag{2.7.6}$$

Proof. In order to prove (2.7.5), let $\{f, f'\} \in S$. Then one has

$$\left\{ \begin{pmatrix} f \\ 0 \end{pmatrix}, \begin{pmatrix} f' \\ 0 \end{pmatrix} \right\} \in \widetilde{A},$$

and hence $\{f, f'\} \in T(\lambda)$ for all $\lambda \in \mathbb{C} \setminus \mathbb{R}$. This shows $S \subset T(\lambda)$ for all $\lambda \in \mathbb{C} \setminus \mathbb{R}$ and making use of Lemma 2.7.1 (ii) one concludes that $T(\lambda) = T(\bar{\lambda})^* \subset S^*$. The identity (2.7.6) follows from the inclusion $S \subset T(\lambda)$ and (2.7.4). □

Assume in the context of Proposition 2.7.2, that $\{\mathcal{G}, \Gamma_0, \Gamma_1\}$ is a boundary triplet for S^*. Then each relation $T(\lambda)$, $\lambda \in \mathbb{C} \setminus \mathbb{R}$, in (2.7.5), being an intermediate extension of S, can be described by the relation $\Gamma(T(\lambda))$ in the parameter space $\mathcal{G}$. It follows from Lemma 2.7.1 and Proposition 1.12.6 that the family $-T(\lambda)$, $\lambda \in \mathbb{C} \setminus \mathbb{R}$, is a Nevanlinna family in $\mathfrak{H}$ in the sense of Definition 1.12.1. Note that for the holomorphy condition in Definition 1.12.1 it is necessary to apply Proposition 1.12.6. A similar reasoning will also be used in the proof of the next theorem, which relates a Nevanlinna family in $\mathcal{G}$ to the Štraus family $T(\lambda)$.

Theorem 2.7.3. *Let S be a closed symmetric relation in $\mathfrak{H}$ and let $\{\mathcal{G}, \Gamma_0, \Gamma_1\}$ be a boundary triplet for S^*. Let $\widetilde{A}$ be a self-adjoint extension of S in $\mathfrak{H} \oplus \mathfrak{H}'$ and let $T(\lambda)$, $\lambda \in \mathbb{C} \setminus \mathbb{R}$, be the corresponding Štraus family in (2.7.1). Then*

$$T(\lambda) = \{\widehat{f} \in S^* : \Gamma \widehat{f} \in -\tau(\lambda)\} = \ker (\Gamma_1 + \tau(\lambda)\Gamma_0), \tag{2.7.7}$$

where $\tau(\lambda)$, $\lambda \in \mathbb{C} \setminus \mathbb{R}$, is a Nevanlinna family in $\mathcal{G}$.

Proof. It follows from Lemma 2.7.1 and Proposition 2.7.2 that $T(\lambda)$ is a closed extension of S for each $\lambda \in \mathbb{C} \setminus \mathbb{R}$. According to Theorem 2.1.3, the extension $T(\lambda)$ of S can be written in the form (2.7.7), where $\tau(\lambda) = -\Gamma(T(\lambda))$.

It will be shown that $\tau(\lambda)$, $\lambda \in \mathbb{C} \setminus \mathbb{R}$, is a Nevanlinna family in $\mathcal{G}$. Since $T(\lambda)$ is maximal accumulative (maximal dissipative) for $\lambda \in \mathbb{C}^+$ ($\lambda \in \mathbb{C}^-$) it follows from Corollary 2.1.4 (ii) that $\tau(\lambda)$ is maximal dissipative (maximal accumulative) for $\lambda \in \mathbb{C}^+$ ($\lambda \in \mathbb{C}^-$). The property $T(\lambda) = T(\bar{\lambda})^*$ and Theorem 2.1.3 imply $\tau(\lambda) = \tau(\bar{\lambda})^*$. Denote the γ-field and the Weyl function corresponding to the boundary triplet $\{\mathcal{G}, \Gamma_0, \Gamma_1\}$ by γ and M. Then for $\lambda \in \mathbb{C}^\pm$ and $\mu \in \mathbb{C}^\pm$ it follows from Lemma 1.11.5 that $(-\tau(\lambda) - M(\mu))^{-1} \in \mathbf{B}(\mathcal{G})$ and

$$(T(\lambda) - \mu)^{-1} = (A_0 - \mu)^{-1} - \gamma(\mu)\big(\tau(\lambda) + M(\mu)\big)^{-1}\gamma(\bar{\mu})^* \tag{2.7.8}$$

holds by Theorem 2.6.1. According to Lemma 2.7.1, the mapping $\lambda \mapsto (T(\lambda) - \lambda)^{-1}$ is holomorphic and hence by Proposition 1.12.6 it follows that also the mapping $\lambda \mapsto (T(\lambda) - \mu)^{-1}$ is holomorphic. Now (2.7.8) shows that $\lambda \mapsto (\tau(\lambda) + M(\mu))^{-1}$ is holomorphic and another application of Proposition 1.12.6 finally gives that $\lambda \mapsto (\tau(\lambda) + \mu)^{-1}$ is also holomorphic. Therefore, $\tau(\lambda)$, $\lambda \in \mathbb{C} \setminus \mathbb{R}$, is a Nevanlinna family. □

The Kreĭn–Naĭmark formula in the following theorem is now an immediate consequence.

Theorem 2.7.4. *Let S be a closed symmetric relation in $\mathfrak{H}$, let $\{\mathcal{G}, \Gamma_0, \Gamma_1\}$ be a boundary triplet for S^*, $A_0 = \ker \Gamma_0$, and let γ and M be the corresponding γ-field and Weyl function, respectively. Let $\widetilde{A}$ be a self-adjoint extension of S in $\mathfrak{H} \oplus \mathfrak{H}'$. Then with the Nevanlinna family τ in $\mathcal{G}$ from Theorem 2.7.3 the compressed resolvent $R(\lambda)$ in (2.7.2) of $\widetilde{A}$ is given by the Kreĭn–Naĭmark formula*

$$R(\lambda) = (A_0 - \lambda)^{-1} - \gamma(\lambda)\big(M(\lambda) + \tau(\lambda)\big)^{-1}\gamma(\bar{\lambda})^*, \quad \lambda \in \mathbb{C} \setminus \mathbb{R}. \tag{2.7.9}$$

Proof. As in the proof of Theorem 2.7.3, it follows from Theorem 2.6.1 that for $\lambda \in \mathbb{C} \setminus \mathbb{R}$ one has

$$(T(\lambda) - \lambda)^{-1} = (A_0 - \lambda)^{-1} - \gamma(\lambda)\big(\tau(\lambda) + M(\lambda)\big)^{-1}\gamma(\bar{\lambda})^*.$$

Hence, the formula (2.7.9) follows from (2.7.4). □

In Chapter 4 the converse of Theorem 2.7.4 will be proved: for every Nevanlinna family in $\mathcal{G}$ there exists a self-adjoint exit space extension $\widetilde{A}$ of S such that (2.7.9) holds for the compressed resolvent of $\widetilde{A}$.

Just as in the case of Corollary 2.6.3, there is now a formulation of the Kreĭn formula for exit space extensions in terms of a parametric representation of the Nevanlinna family τ. Assume that the Nevanlinna pair $\{\mathcal{A}, \mathcal{B}\}$ is a tight representation of the Nevanlinna family τ; cf. Section 1.12. Then the next corollary can be shown in the same way as Corollary 2.6.3 by applying Proposition 1.12.6.

Corollary 2.7.5. *Let S be a closed symmetric relation in $\mathfrak{H}$, let $\{\mathcal{G}, \Gamma_0, \Gamma_1\}$ be a boundary triplet for S^*, $A_0 = \ker \Gamma_0$, and let γ and M be the corresponding γ-field and Weyl function, respectively. Let the Nevanlinna family τ in Theorem 2.7.4* *have the tight representation* $\tau = \{\mathcal{A}, \mathcal{B}\}$ *with the Nevanlinna pair* $\{\mathcal{A}, \mathcal{B}\}$. *Then the compressed resolvent* $R(\lambda)$ *of* $\widetilde{A}$ *has the form*

$$R(\lambda) = (A_0 - \lambda)^{-1} - \gamma(\lambda)\mathcal{A}(\lambda)\big(\mathcal{B}(\lambda) + M(\lambda)\mathcal{A}(\lambda)\big)^{-1}\gamma(\bar{\lambda})^*, \quad \lambda \in \mathbb{C} \setminus \mathbb{R}.$$

The Štraus family in Theorem 2.7.3 can also be described in terms of a representing Nevanlinna pair $\{\mathcal{A}, \mathcal{B}\}$.

Corollary 2.7.6. *Let* S *be a closed symmetric relation in* $\mathfrak{H}$, *let* $\{\mathcal{G}, \Gamma_0, \Gamma_1\}$ *be a boundary triplet for* S^*, *and let* $T(\lambda)$, $\lambda \in \mathbb{C} \setminus \mathbb{R}$, *be the Štraus family in Theorem 2.7.3. Let the corresponding Nevanlinna family* τ *have the tight representation* $\tau = \{\mathcal{A}, \mathcal{B}\}$ *with the Nevanlinna pair* $\{\mathcal{A}, \mathcal{B}\}$. *Then*

$$T(\lambda) = \big\{\widehat{f} \in S^* : \mathcal{B}(\bar{\lambda})^*\Gamma_0\widehat{f} = -\mathcal{A}(\bar{\lambda})^*\Gamma_1\widehat{f}\big\}. \tag{2.7.10}$$

Proof. By assumption $\tau(\lambda)$, $\lambda \in \mathbb{C} \setminus \mathbb{R}$, is given as

$$\tau(\lambda) = \big\{\{\mathcal{A}(\lambda)\varphi, \mathcal{B}(\lambda)\varphi\} : \varphi \in \mathcal{G}\big\},$$

with a Nevanlinna pair $\{\mathcal{A}, \mathcal{B}\}$ and this representation is tight. The symmetry property $\tau(\lambda)^* = \tau(\bar{\lambda})$ implies that

$$\tau(\lambda)^* = \big\{\{\mathcal{A}(\bar{\lambda})\varphi, \mathcal{B}(\bar{\lambda})\varphi\} : \varphi \in \mathcal{G}\big\},$$

so that $\tau(\lambda)$ can also be written as

$$\tau(\lambda) = \big\{\{\varphi, \varphi'\} \in \mathcal{G}^2 : \mathcal{B}(\bar{\lambda})^*\varphi = \mathcal{A}(\bar{\lambda})^*\varphi'\big\},$$

and hence

$$-\tau(\lambda) = \big\{\{\varphi, \varphi'\} \in \mathcal{G}^2 : \mathcal{B}(\bar{\lambda})^*\varphi = -\mathcal{A}(\bar{\lambda})^*\varphi'\big\};$$

cf. (2.2.3) and (2.2.4). Thus, (2.7.10) follows from (2.7.7). □

In the following a particular self-adjoint exit space extension of S will be studied. Here the exit space is the parameter space $\mathcal{G}$.

Proposition 2.7.7. *Let* S *be a closed symmetric relation in* $\mathfrak{H}$ *and let* $\{\mathcal{G}, \Gamma_0, \Gamma_1\}$ *be a boundary triplet for* S^*. *Then*

$$\widetilde{A} = \left\{\left\{\begin{pmatrix} f \\ \Gamma_0\widehat{f} \end{pmatrix}, \begin{pmatrix} f' \\ -\Gamma_1\widehat{f} \end{pmatrix}\right\} : \widehat{f} = \{f, f'\} \in S^*\right\} \tag{2.7.11}$$

is a self-adjoint extension of S *in* $\mathfrak{H} \oplus \mathcal{G}$. *The corresponding Štraus family* $T(\lambda)$, $\lambda \in \mathbb{C} \setminus \mathbb{R}$, *in* $\mathfrak{H}$ *has the form*

$$T(\lambda) = \big\{\widehat{f} \in S^* : -\Gamma_1\widehat{f} = \lambda\Gamma_0\widehat{f}\big\} = \ker\big(\Gamma_1 + \lambda\Gamma_0\big) \tag{2.7.12}$$

and the compressed resolvent $R(\lambda)$ *onto* $\mathfrak{H}$ *is given by*

$$R(\lambda) = (A_0 - \lambda)^{-1} - \gamma(\lambda)\big(M(\lambda) + \lambda\big)^{-1}\gamma(\bar{\lambda})^*.$$

Proof. Observe that $S \subset \widetilde{A}$. Indeed, for $\widehat{f} = \{f, f'\} \in S$ one has $\Gamma_0 \widehat{f} = \Gamma_1 \widehat{f} = 0$ by Proposition 2.1.2 (ii) and hence

$$\left\{ \begin{pmatrix} f \\ 0 \end{pmatrix}, \begin{pmatrix} f' \\ 0 \end{pmatrix} \right\} \in \widetilde{A}.$$

It follows from the abstract Green identity (2.1.1) and the definition of $\widetilde{A}$ in (2.7.11) that the relation $\widetilde{A}$ is symmetric, that is, $\widetilde{A} \subset (\widetilde{A})^*$. Now let the element

$$\left\{ \begin{pmatrix} g \\ \alpha \end{pmatrix}, \begin{pmatrix} g' \\ \alpha' \end{pmatrix} \right\}, \qquad g, g' \in \mathfrak{H}, \ \alpha, \alpha' \in \mathcal{G}, \tag{2.7.13}$$

belong to $(\widetilde{A})^*$. Then for all $\widehat{f} = \{f, f'\} \in S^*$ one has

$$\left(\begin{pmatrix} f' \\ -\Gamma_1 \widehat{f} \end{pmatrix}, \begin{pmatrix} g \\ \alpha \end{pmatrix} \right) = \left(\begin{pmatrix} f \\ \Gamma_0 \widehat{f} \end{pmatrix}, \begin{pmatrix} g' \\ \alpha' \end{pmatrix} \right)$$

or, equivalently,

$$(f', g) - (f, g') = (\Gamma_1 \widehat{f}, \alpha) + (\Gamma_0 \widehat{f}, \alpha'). \tag{2.7.14}$$

In particular, since $\ker \Gamma = S$, it follows from (2.7.14) that if $\widehat{f} \in S$, then $\widehat{g} \in S^*$. Therefore, the abstract Green identity (2.1.1) together with (2.7.14) imply

$$(\Gamma_1 \widehat{f}, \Gamma_0 \widehat{g}) - (\Gamma_0 \widehat{f}, \Gamma_1 \widehat{g}) = (\Gamma_1 \widehat{f}, \alpha) + (\Gamma_0 \widehat{f}, \alpha')$$

for all $\widehat{f} \in S^*$. By definition, the mapping $\Gamma : S^* \to \mathcal{G} \times \mathcal{G}$ is surjective, and consequently

$$\alpha = \Gamma_0 \widehat{g} \quad \text{and} \quad \alpha' = -\Gamma_1 \widehat{g}.$$

Thus, the element in (2.7.13) belongs to $\widetilde{A}$. Hence, $\widetilde{A}$ is a self-adjoint extension of S in $\mathfrak{H} \oplus \mathcal{G}$.

The Štraus family $T(\lambda)$, $\lambda \in \mathbb{C}$, in $\mathfrak{H}$ corresponding to the self-adjoint relation $\widetilde{A}$ in (2.7.11) has the form

$$T(\lambda) = \left\{ \widehat{f} = \{f, f'\} \in S^* : \left\{ \begin{pmatrix} f \\ \Gamma_0 \widehat{f} \end{pmatrix}, \begin{pmatrix} f' \\ -\Gamma_1 \widehat{f} \end{pmatrix} \right\} \in \widetilde{A}, \, -\Gamma_1 \widehat{f} = \lambda \Gamma_0 \widehat{f} \right\},$$

and hence is given by (2.7.12). The statement concerning the compressed resolvent of $\widetilde{A}$ onto $\mathfrak{H}$ follows from the Kreĭn–Naĭmark formula in Theorem 2.7.4. □

Finally, the Štraus family and the compressed resolvent of the self-adjoint relation $\widetilde{A}$ in $\mathfrak{H} \oplus \mathfrak{H}'$ can be regarded from a slightly different point of view. Thus far, the Štraus family and the compressed resolvent were given as notions in the Hilbert space $\mathfrak{H}$; more structure was added by considering a closed symmetric relation S in $\mathfrak{H}$ and assuming that $\widetilde{A}$ is a self-adjoint extension of S in $\mathfrak{H} \oplus \mathfrak{H}'$. Now the role of the original space and the exit space will be interchanged and a self-adjoint relation $\widetilde{A}$ in $\mathfrak{H} \oplus \mathfrak{H}'$ will be viewed as a self-adjoint extension of the trivial symmetric relation S' in $\mathfrak{H}'$. The Štraus family $T'(\lambda)$, $\lambda \in \mathbb{C}$, in $\mathfrak{H}'$ corresponding to $\widetilde{A}$ in $\mathfrak{H} \oplus \mathfrak{H}'$ is defined by

$$T'(\lambda) = \left\{ \{h, h'\} \in \mathfrak{H}' \times \mathfrak{H}' : \left\{ \begin{pmatrix} f \\ h \end{pmatrix}, \begin{pmatrix} f' \\ h' \end{pmatrix} \right\} \in \widetilde{A}, \, f' = \lambda f \right\}$$

and the corresponding compressed resolvent $R'(\lambda) \in \mathbf{B}(\mathfrak{H}')$ is given by

$$P_{\mathfrak{H}'}(\widetilde{A} - \lambda)^{-1} \iota_{\mathfrak{H}'} = (T'(\lambda) - \lambda)^{-1}, \quad \lambda \in \rho(\widetilde{A}); \tag{2.7.15}$$

here $P_{\mathfrak{H}'} : \mathfrak{H} \oplus \mathfrak{H}' \to \mathfrak{H}'$ denotes the orthogonal projection from $\mathfrak{H} \oplus \mathfrak{H}'$ onto $\mathfrak{H}'$ and $\iota_{\mathfrak{H}'} : \mathfrak{H}' \to \mathfrak{H} \oplus \mathfrak{H}'$ is the canonical embedding of $\mathfrak{H}'$ into $\mathfrak{H} \oplus \mathfrak{H}'$. The adjoint of the trivial symmetric relation $S' = \{0, 0\}$ in $\mathfrak{H}'$ is $(S')^* = \mathfrak{H}' \times \mathfrak{H}'$ and

$$\Gamma_0' \widehat{h} = h \quad \text{and} \quad \Gamma_1' \widehat{h} = h', \quad \widehat{h} = \{h, h'\} \in (S')^*,$$

defines a boundary triplet $\{\mathfrak{H}', \Gamma_0', \Gamma_1'\}$ for $(S')^*$. Then $A_0' = \{0\} \times \mathfrak{H}'$, so that $(A_0' - \lambda)^{-1} = 0$, $\lambda \in \mathbb{C}$, and the γ-field and the Weyl function are given by

$$\gamma'(\lambda) = I \quad \text{and} \quad M'(\lambda) = \lambda I;$$

cf. Example 2.4.2. In this situation the Štraus family $T'(\lambda)$, $\lambda \in \mathbb{C} \setminus \mathbb{R}$, in $\mathfrak{H}'$ induces a Nevanlinna family $\tau(\lambda)$, $\lambda \in \mathbb{C} \setminus \mathbb{R}$, in the same space $\mathfrak{H}'$ as in (2.7.7) via

$$T'(\lambda) = \ker (\Gamma_1' + \tau(\lambda)\Gamma_0'),$$

so that $\tau(\lambda) = -T'(\lambda)$. Then the compressed resolvent (2.7.15) takes the form

$$P_{\mathfrak{H}'}(\widetilde{A} - \lambda)^{-1} \iota_{\mathfrak{H}'} = -(\tau(\lambda) + \lambda)^{-1}, \tag{2.7.16}$$

which can be viewed as the Kreĭn–Naĭmark formula in $\mathfrak{H}'$ for the extension $\widetilde{A}$ of S'.

For the self-adjoint relation $\widetilde{A}$ in Proposition 2.7.7 it turns out in this new context that the corresponding Štraus family in $\mathcal{G}$ is given by the function $-M$.

Proposition 2.7.8. *Let S be a closed symmetric relation in $\mathfrak{H}$, let $\{\mathcal{G}, \Gamma_0, \Gamma_1\}$ be a boundary triplet for S^*, and let M be the corresponding Weyl function. Consider the self-adjoint relation*

$$\widetilde{A} = \left\{ \left\{ \begin{pmatrix} f \\ \Gamma_0 \widehat{f} \end{pmatrix}, \begin{pmatrix} f' \\ -\Gamma_1 \widehat{f} \end{pmatrix} \right\} : \widehat{f} = \{f, f'\} \in S^* \right\}$$

in $\mathfrak{H} \oplus \mathcal{G}$. Then the corresponding Štraus family in $\mathcal{G}$ is given by

$$\left\{ \{\Gamma_0 \widehat{f}, -\Gamma_1 \widehat{f}\} \in \mathcal{G} \times \mathcal{G} : \left\{ \begin{pmatrix} f \\ \Gamma_0 \widehat{f} \end{pmatrix}, \begin{pmatrix} f' \\ -\Gamma_1 \widehat{f} \end{pmatrix} \right\} \in \widetilde{A}, \, \widehat{f} \in \widehat{\mathfrak{N}}_\lambda(S^*) \right\} \tag{2.7.17}$$

and coincides with $-M(\lambda)$, $\lambda \in \mathbb{C}\setminus\mathbb{R}$. *Furthermore, the compressed resolvent of* $\widetilde{A}$ *to* $\mathcal{G}$ *is given by*

$$P_{\mathcal{G}}(\widetilde{A}-\lambda)^{-1}\iota_{\mathcal{G}} = -(M(\lambda)+\lambda)^{-1}, \quad \lambda \in \mathbb{C}\setminus\mathbb{R}; \tag{2.7.18}$$

here $P_{\mathcal{G}} : \mathfrak{H}\oplus\mathcal{G} \to \mathcal{G}$ *denotes the orthogonal projection from* $\mathfrak{H}\oplus\mathcal{G}$ *onto* $\mathcal{G}$ *and* $\iota_{\mathcal{G}} : \mathcal{G} \to \mathfrak{H}\oplus\mathcal{G}$ *is the canonical embedding of* $\mathcal{G}$ *into* $\mathfrak{H}\oplus\mathcal{G}$.

Proof. It follows from the definition of the Štraus family in (2.7.1) that the Štraus family corresponding to $\widetilde{A}$ in the Hilbert space $\mathcal{G}$ has the form (2.7.17). Since $\{\Gamma_0\widehat{f}, -\Gamma_1\widehat{f}\}$ belongs to (2.7.17) if and only if $\widehat{f} \in \widehat{\mathfrak{N}}_\lambda(S^*)$, it is also clear that for all $\lambda \in \mathbb{C}\setminus\mathbb{R}$ the Štraus family coincides with the values $-M(\lambda)$ of the Weyl function corresponding to the boundary triplet $\{\mathcal{G},\Gamma_0,\Gamma_1\}$. The formula (2.7.18) follows from (2.7.16) in this special case. □

2.8 Perturbation problems

Let A be a self-adjoint relation in the Hilbert space $\mathfrak{H}$, let $V \in \mathbf{B}(\mathfrak{H})$ be a bounded self-adjoint operator in $\mathfrak{H}$, and consider the self-adjoint relation

$$B = A + V. \tag{2.8.1}$$

For $\lambda \in \rho(A)\cap\rho(B)$ one can rewrite (2.8.1) in the form

$$(B-\lambda)^{-1} - (A-\lambda)^{-1} = -(B-\lambda)^{-1}V(A-\lambda)^{-1};$$

this follows from Lemma 1.11.2 with $H = A$, $R = \lambda - V$ and $S = \lambda$. In particular, if V in (2.8.1) belongs to some left-sided or right-sided operator ideal, then the same is true for the difference of the resolvents of A and B. From this point of view perturbation problems in the resolvent sense are more general than additive perturbations of the form (2.8.1). Such perturbation problems embed naturally in the framework of the extension theory that has been discussed in this chapter.

In the next theorem the particularly simple case of finite-rank perturbations is treated.

Theorem 2.8.1. *Let* A *and* B *be self-adjoint relations in* $\mathfrak{H}$ *and assume that*

$$\dim\operatorname{ran}\big((B-\lambda_0)^{-1} - (A-\lambda_0)^{-1}\big) = n < \infty \tag{2.8.2}$$

for some, and hence for all $\lambda_0 \in \rho(A)\cap\rho(B)$. *Then* $S = A\cap B$ *is a closed symmetric relation in* $\mathfrak{H}$ *and there exists a boundary triplet* $\{\mathbb{C}^n,\Gamma_0,\Gamma_1\}$ *for* S^* *such that*

$$A = \ker\Gamma_0 \quad \text{and} \quad B = \ker\Gamma_1. \tag{2.8.3}$$

If γ and M are the γ-field and the Weyl function, respectively, corresponding to $\{\mathbb{C}^n, \Gamma_0, \Gamma_1\}$, then

$$(B-\lambda)^{-1} - (A-\lambda)^{-1} = -\gamma(\lambda)M(\lambda)^{-1}\gamma(\bar{\lambda})^* \tag{2.8.4}$$

for all $\lambda \in \rho(A) \cap \rho(B)$. Moreover, if $\lambda \in \rho(A)$, then $\lambda \in \sigma_p(B)$ if and only if $0 \in \sigma_p(M(\lambda))$, and the multiplicities are at most n and coincide.

Proof. Let $\lambda_0 \in \rho(A) \cap \rho(B)$ be such that (2.8.2) holds and consider the closed symmetric relation $S = A \cap B$ in $\mathfrak{H}$. By construction, A and B are disjoint self-adjoint extensions of S, and hence

$$\operatorname{ran}(S - \bar{\lambda}_0) = \ker\big((B-\bar{\lambda}_0)^{-1} - (A-\bar{\lambda}_0)^{-1}\big)$$

by Theorem 1.7.8. This leads to

$$\ker(S^* - \lambda_0) = \big(\operatorname{ran}(S-\bar{\lambda}_0)\big)^\perp = \operatorname{ran}\big((B-\lambda_0)^{-1} - (A-\lambda_0)^{-1}\big),$$

where (2.8.2) was used in the last equality. Now Theorem 1.7.8 implies that A and B are transversal self-adjoint extensions of S. Theorem 2.5.9 shows that there exists a boundary triplet $\{\mathbb{C}^n, \Gamma_0, \Gamma_1\}$ such that (2.8.3) holds, and the formula (2.8.4) follows from Theorem 2.6.1. One also concludes from (2.8.4) and the fact that $M(\lambda)$ is bijective for $\lambda \in \rho(A) \cap \rho(B)$ (see Corollary 2.5.4) that the difference of the resolvents in (2.8.2) is of rank n for all $\lambda \in \rho(A) \cap \rho(B)$. The last statement on the eigenvalues of B follows from Theorem 2.6.2 (i). □

The following result is a generalization of Theorem 2.8.1 that applies to non-self-adjoint intermediate extensions B.

Theorem 2.8.2. *Let A be a self-adjoint relation in $\mathfrak{H}$ and let B be a closed relation in $\mathfrak{H}$ such that $\rho(B) \neq \emptyset$. Then $S = A \cap B$ is a closed symmetric relation in $\mathfrak{H}$ and there exist a boundary triplet $\{\mathcal{G}, \Gamma_0, \Gamma_1\}$ for S^* and a closed operator Θ in $\mathcal{G}$ such that*

$$A = \ker \Gamma_0 \qquad \textit{and} \qquad B = \ker(\Gamma_1 - \Theta\Gamma_0).$$

If γ and M are the γ-field and the Weyl function, respectively, corresponding to $\{\mathcal{G}, \Gamma_0, \Gamma_1\}$, then

$$(B-\lambda)^{-1} - (A-\lambda)^{-1} = \gamma(\lambda)\big(\Theta - M(\lambda)\big)^{-1}\gamma(\bar{\lambda})^*$$

for all $\lambda \in \rho(A) \cap \rho(B)$. Moreover, for all $\lambda \in \rho(A)$ one has $\lambda \in \sigma_i(B)$ if and only if $0 \in \sigma_i(\Theta - M(\lambda))$, $i = \mathrm{p}, \mathrm{c}, \mathrm{r}$, and for $i = \mathrm{p}$ the geometric multiplicities coincide.

Proof. It is clear that $S = A \cap B$ is a closed symmetric relation and hence there exists a boundary triplet $\{\mathcal{G}, \Gamma_0, \Gamma_1\}$ for S^* such that $A = \ker \Gamma_0$; cf. Theorem 2.4.1. Since B is a closed extension of S, there exists a closed relation Θ in $\mathcal{G}$ such that $B = \ker(\Gamma_1 - \Theta\Gamma_0)$. By construction, the relations A and B are disjoint and hence it follows from Proposition 2.1.8 (i) that Θ is a closed operator in $\mathcal{G}$. The resolvent formula and the assertion on the spectrum of B are immediate consequences of Theorem 2.6.1 and Theorem 2.6.2. □

Let $\mathfrak{K}$ and $\mathfrak{L}$ be Hilbert spaces and let $T \in \mathbf{B}(\mathfrak{K}, \mathfrak{L})$ be a compact operator. Recall that the *singular values* $s_k(T)$, $k \in \mathbb{N}$, of T are defined as the eigenvalues of the nonnegative compact operator $(T^*T)^{1/2} \in \mathbf{B}(\mathfrak{K})$ (enumerated in nonincreasing order). The *Schatten–von Neumann ideal* $\mathfrak{S}_p(\mathfrak{K}, \mathfrak{L})$, $1 \leq p < \infty$, consists of all compact operators $T \in \mathbf{B}(\mathfrak{K}, \mathfrak{L})$ such that the singular values are p-summable, that is,

$$\sum_{k=1}^{\infty} (s_k(T))^p < \infty.$$

If $\mathfrak{K} = \mathfrak{L}$ the notation $\mathfrak{S}_p(\mathfrak{K})$ is used instead of $\mathfrak{S}_p(\mathfrak{K}, \mathfrak{K})$. Observe that the nonzero singular values of T coincide with the nonzero singular values of the restriction of T to $(\ker T)^\perp$ as the corresponding restriction of T^*T is a nonnegative compact operator in the Hilbert space $(\ker T)^\perp$. This fact will be used in the proof of the following theorem.

Theorem 2.8.3. *Let S be a closed symmetric relation in $\mathfrak{H}$, let $\{\mathcal{G}, \Gamma_0, \Gamma_1\}$ be a boundary triplet for S^*, let A_{Θ_1} and A_{Θ_2} be closed extensions of S corresponding to closed relations Θ_1 and Θ_2 in $\mathcal{G}$ via (2.1.5), and assume that $\rho(A_{\Theta_1}) \cap \rho(A_{\Theta_2}) \neq \emptyset$ and $\rho(\Theta_1) \cap \rho(\Theta_2) \neq \emptyset$. Then*

$$(A_{\Theta_1} - \lambda)^{-1} - (A_{\Theta_2} - \lambda)^{-1} \in \mathfrak{S}_p(\mathfrak{H}) \tag{2.8.5}$$

for some, and hence for all $\lambda \in \rho(A_{\Theta_1}) \cap \rho(A_{\Theta_2})$ if and only if

$$(\Theta_1 - \xi)^{-1} - (\Theta_2 - \xi)^{-1} \in \mathfrak{S}_p(\mathcal{G}) \tag{2.8.6}$$

for some, and hence for all $\xi \in \rho(\Theta_1) \cap \rho(\Theta_2)$.

Proof. Let $A_0 = \ker \Gamma_0$ and let γ and M be the γ-field and the Weyl function corresponding to the boundary triplet $\{\mathcal{G}, \Gamma_0, \Gamma_1\}$. Then one has

$$\begin{aligned} (A_{\Theta_1} - \lambda)^{-1} &= (A_0 - \lambda)^{-1} + \gamma(\lambda)\big(\Theta_1 - M(\lambda)\big)^{-1}\gamma(\bar{\lambda})^*, \\ (A_{\Theta_2} - \lambda)^{-1} &= (A_0 - \lambda)^{-1} + \gamma(\lambda)\big(\Theta_2 - M(\lambda)\big)^{-1}\gamma(\bar{\lambda})^*, \end{aligned}$$

for all $\lambda \in \rho(A_{\Theta_1}) \cap \rho(A_{\Theta_2}) \cap \rho(A_0)$, and hence

$$\begin{aligned} &(A_{\Theta_1} - \lambda)^{-1} - (A_{\Theta_2} - \lambda)^{-1} \\ &\qquad = \gamma(\lambda) \Big[\big(\Theta_1 - M(\lambda)\big)^{-1} - \big(\Theta_2 - M(\lambda)\big)^{-1} \Big] \gamma(\bar{\lambda})^*. \end{aligned} \tag{2.8.7}$$

It will be shown that (2.8.5) holds if and only if

$$\big(\Theta_1 - M(\lambda)\big)^{-1} - \big(\Theta_2 - M(\lambda)\big)^{-1} \in \mathfrak{S}_p(\mathcal{G}). \tag{2.8.8}$$

In fact, it is clear that if (2.8.8) holds, then so does (2.8.5). Conversely, if (2.8.5) holds, then

$$\gamma(\lambda) \Big[\big(\Theta_1 - M(\lambda)\big)^{-1} - \big(\Theta_2 - M(\lambda)\big)^{-1} \Big] \gamma(\bar{\lambda})^* \in \mathfrak{S}_p(\mathfrak{H}) \tag{2.8.9}$$

follows directly from (2.8.7). Since $\gamma(\lambda)$ is an isomorphism from $\mathcal{G}$ onto $\mathfrak{N}_\lambda(S^*)$ and $\ker \gamma(\bar\lambda)^* = \mathfrak{N}_{\bar\lambda}(S^*)^\perp$, it follows that the restriction of $\gamma(\bar\lambda)^*$ to $\mathfrak{N}_{\bar\lambda}(S^*)$ is an isomorphism onto $\mathcal{G}$. Hence, the operator in (2.8.9) may also be viewed as a bounded operator from $\mathfrak{N}_{\bar\lambda}(S^*)$ to $\mathfrak{N}_\lambda(S^*)$ and thus belongs to the Schatten–von Neumann ideal $\mathfrak{S}_p(\mathfrak{N}_{\bar\lambda}(S^*), \mathfrak{N}_\lambda(S^*))$. In this context $\gamma(\lambda) : \mathcal{G} \to \mathfrak{N}_\lambda(S^*)$ and $\gamma(\bar\lambda)^* : \mathfrak{N}_{\bar\lambda}(S^*) \to \mathcal{G}$ are boundedly invertible and hence it follows that (2.8.8) holds. Therefore, if $\lambda \in \rho(A_{\Theta_1}) \cap \rho(A_{\Theta_2}) \cap \rho(A_0)$, then (2.8.5) is equivalent to (2.8.8). Note that if (2.8.5) holds for some $\lambda \in \rho(A_{\Theta_1}) \cap \rho(A_{\Theta_2})$, then it holds for all $\lambda \in \rho(A_{\Theta_1}) \cap \rho(A_{\Theta_2})$ by Lemma 1.11.4.

It remains to show that for all $\lambda \in \rho(A_{\Theta_1}) \cap \rho(A_{\Theta_2}) \cap \rho(A_0)$ (2.8.8) is equivalent to (2.8.6). By Lemma 1.11.4,

$$\begin{aligned}
&\big(\Theta_1 - M(\lambda)\big)^{-1} - \big(\Theta_2 - M(\lambda)\big)^{-1} \\
&\quad = \big[I - (\Theta_1 - \xi)^{-1}(M(\lambda) - \xi)\big]^{-1} \\
&\qquad\quad \big[(\Theta_1 - \xi)^{-1} - (\Theta_2 - \xi)^{-1}\big]\big[I - (M(\lambda) - \xi)(\Theta_2 - \xi)^{-1}\big]^{-1}
\end{aligned}$$

and since the factors around $(\Theta_1 - \xi)^{-1} - (\Theta_2 - \xi)^{-1}$ on the right-hand side are boundedly invertible by Lemma 1.11.3, this establishes the equivalence of (2.8.8) and (2.8.6). □

If Θ_1 and Θ_2 in Theorem 2.8.3 are bounded operators in $\mathcal{G}$ the condition $\rho(\Theta_1) \cap \rho(\Theta_2) \neq \emptyset$ is automatically satisfied and the identity

$$(\Theta_1 - \xi)^{-1} - (\Theta_2 - \xi)^{-1} = (\Theta_1 - \xi)^{-1}\big(\Theta_2 - \Theta_1\big)(\Theta_2 - \xi)^{-1}$$

shows that $(\Theta_1 - \xi)^{-1} - (\Theta_2 - \xi)^{-1} \in \mathfrak{S}_p(\mathcal{G})$ for $\xi \in \rho(\Theta_1) \cap \rho(\Theta_2)$ if and only if $\Theta_1 - \Theta_2 \in \mathfrak{S}_p(\mathcal{G})$. This leads to the following corollary.

Corollary 2.8.4. *Let S be a closed symmetric relation in $\mathfrak{H}$, let $\{\mathcal{G}, \Gamma_0, \Gamma_1\}$ be a boundary triplet for S^*, let A_{Θ_1} and A_{Θ_2} be closed extensions of S which correspond to bounded operators $\Theta_1, \Theta_2 \in \mathbf{B}(\mathcal{G})$ via (2.1.5), and assume that $\rho(A_{\Theta_1}) \cap \rho(A_{\Theta_2}) \neq \emptyset$. Then*

$$(A_{\Theta_1} - \lambda)^{-1} - (A_{\Theta_2} - \lambda)^{-1} \in \mathfrak{S}_p(\mathfrak{H})$$

for some, and hence for all $\lambda \in \rho(A_{\Theta_1}) \cap \rho(A_{\Theta_2})$ if and only if

$$\Theta_1 - \Theta_2 \in \mathfrak{S}_p(\mathcal{G}).$$

The following proposition is an addendum to Theorem 2.8.2 in the special case where B is an $\mathfrak{S}_p$-perturbation of A in the resolvent sense.

Proposition 2.8.5. *Let A be a self-adjoint relation in $\mathfrak{H}$, let B be a closed relation in $\mathfrak{H}$ with $\rho(B) \neq \emptyset$, assume that*

$$(B - \lambda_0)^{-1} - (A - \lambda_0)^{-1} \in \mathfrak{S}_p(\mathfrak{H}) \tag{2.8.10}$$

for some $\lambda_0 \in \rho(A) \cap \rho(B)$ *and that* (2.8.10) *is not a finite-rank operator. Let* $S = A \cap B$ *and* $\{\mathcal{G}, \Gamma_0, \Gamma_1\}$ *be a boundary triplet for* S^* *as in Theorem* 2.8.2 *such that*

$$A = \ker \Gamma_0 \qquad \textit{and} \qquad B = \ker(\Gamma_1 - \Theta\Gamma_0)$$

for some closed operator Θ *in* $\mathcal{G}$. *If* $\rho(\Theta) \neq \emptyset$, *then* Θ *is an unbounded closed operator and* $(\Theta - \xi)^{-1} \in \mathfrak{S}_p(\mathcal{G})$ *for all* $\xi \in \rho(\Theta)$.

Proof. Assume that (2.8.10) holds and that $\rho(\Theta) \neq \emptyset$. As $\Theta_0 = \{0\} \times \mathcal{G}$ is the self-adjoint relation in $\mathcal{G}$ which corresponds to $A = \ker \Gamma_0$ and $(\Theta_0 - \xi)^{-1} = 0$ for $\xi \in \mathbb{C}$, one concludes from Theorem 2.8.3 and (2.8.10) that

$$(\Theta - \xi)^{-1} = (\Theta - \xi)^{-1} - (\Theta_0 - \xi)^{-1} \in \mathfrak{S}_p(\mathcal{G}), \quad \xi \in \rho(\Theta).$$

Together with the assumption that (2.8.10) is of infinite rank, this implies that Θ is an unbounded closed operator in $\mathcal{G}$. □

Chapter 3

Spectra, Simple Operators, and Weyl Functions

In this chapter the spectrum of a self-adjoint operator or relation will be completely characterized in terms of the analytic behavior and the limit properties of the Weyl function. In order to be able to treat the different parts of the spectrum, a short introduction to finite Borel measures on $\mathbb{R}$ and the corresponding Borel transforms will be given in Section 3.1 and Section 3.2. The notions and some properties of the absolutely continuous, singular continuous, pure point, and other spectral subsets of a self-adjoint relation are recalled in Section 3.3. Moreover, the concepts of simplicity (or complete non-self-adjointness) and local simplicity of symmetric operators and relations will be explained in detail in Section 3.4. For a boundary triplet $\{\mathcal{G}, \Gamma_0, \Gamma_1\}$ with corresponding Weyl function M, the spectrum of the self-adjoint extension $A_0 = \ker \Gamma_0$ is then characterized. An analytic description for the point spectrum of A_0 in terms of M is given in Section 3.5, the rest of the spectrum and its different parts, namely absolutely continuous, singular, and continuous spectrum are studied in Section 3.6 under the additional condition that the underlying symmetric relation S is simple or locally simple. The limit properties of the Weyl function are also connected with defect elements belonging to the domain or range of A_0. This is discussed in Section 3.7. Finally, it is shown with the help of tranformation properties of boundary triplets and Weyl functions in Section 3.8 how the earlier results in this chapter extend to a description of the spectrum of an arbitrary self-adjoint extension A_Θ.

3.1 Analytic descriptions of minimal supports of Borel measures

A Borel measure on $\mathbb{R}$ can be decomposed with respect to the Lebesgue measure into an absolutely continuous measure and a singular measure. The minimal supports of the measure and its parts can be described by means of the derivative of

J. Behrndt et al., *Boundary Value Problems, Weyl Functions, and Differential Operators*, Monographs in Mathematics 108,
https://doi.org/10.1007/978-3-030-36714-5_4

the measure. The present interest is in an analytic description of these minimal supports in terms of the Borel transform. For the convenience of the reader, a brief review on Borel measures on $\mathbb{R}$ and some properties of their Borel transforms are recalled.

In the following let μ be a regular Borel measure on $\mathbb{R}$ and denote the Lebesgue measure on $\mathbb{R}$ by m. Recall that any Borel measure on $\mathbb{R}$ which is finite on compact sets is automatically regular. Associated with the regular Borel measure μ is the nondecreasing, left-continuous function

$$\nu_\mu(x) = \begin{cases} \mu([0,x)), & x > 0, \\ 0, & x = 0, \\ -\mu([x,0)), & x < 0, \end{cases} \tag{3.1.1}$$

on $\mathbb{R}$. Observe that ν_μ is bounded if and only if μ is a finite measure, that the derivative ν'_μ of the nondecreasing function ν_μ exists m-almost everywhere, and that

$$\mu([x,y)) = \nu_\mu(y) - \nu_\mu(x), \qquad x < y. \tag{3.1.2}$$

It is important to note that via (3.1.2) the function ν_μ induces a Lebesgue-Stieltjes measure on $\mathbb{R}$, which is a complete measure that coincides with the completion of μ. In the following it is often more convenient to work with this completion, which will also be denoted by μ, and the corresponding μ-measurable subsets of $\mathbb{R}$.

The regular Borel measure μ has a *Lebesgue decomposition* with respect to the Lebesgue measure m:

$$\mu = \mu_{\rm ac} + \mu_{\rm s},$$

where the measure $\mu_{\rm ac}$ is absolutely continuous and the measure $\mu_{\rm s}$ is singular, each with respect to the Lebesgue measure. The singular measure $\mu_{\rm s}$ is further decomposed into the singular continuous part $\mu_{\rm sc}$ and the pure point part $\mu_{\rm p}$, so that

$$\mu = \mu_{\rm ac} + \mu_{\rm sc} + \mu_{\rm p}.$$

The corresponding nondecreasing, left-continuous functions $\nu_{\mu_{\rm ac}}$, $\nu_{\mu_{\rm sc}}$, and $\nu_{\mu_{\rm p}}$ defined via (3.1.1), are absolutely continuous, continuous with $\nu'_{\mu_{\rm sc}} = 0$ m-almost everywhere, and a step function, respectively, and

$$\nu_\mu = \nu_{\mu_{\rm ac}} + \nu_{\mu_{\rm sc}} + \nu_{\mu_{\rm p}}.$$

Furthermore,

$$\mu_{\rm ac}(\mathfrak{B}) = \int_{\mathfrak{B}} \nu'_\mu(x)\, dm(x) \tag{3.1.3}$$

for all Borel sets $\mathfrak{B}$, and hence the derivative ν'_μ coincides with the Radon–Nikodým derivative of $\mu_{\rm ac}$ m-almost everywhere.

For $x \in \mathbb{R}$ the derivative $\mu'(x)$ of the Borel measure μ with respect to the Lebesgue measure m is defined by

$$\mu'(x) = \lim_{m(I_x)\downarrow 0} \left\{ \frac{\mu(I_x)}{m(I_x)} : I_x \text{ an interval containing } x \right\}, \tag{3.1.4}$$

whenever the limit exists and takes values in $[0, \infty]$. It can be shown that the sets

$$\mathfrak{E}_0 = \{x \in \mathbb{R} : \mu'(x) \text{ exists finitely}\} \tag{3.1.5}$$

and

$$\mathfrak{E} = \{x \in \mathbb{R} : \mu'(x) \text{ exists finitely or infinitely}\} \tag{3.1.6}$$

are Borel sets, and for the set $\mathbb{R} \setminus \mathfrak{E}_0$ on which the derivative μ' does not exist finitely one has that

$$m(\mathbb{R} \setminus \mathfrak{E}_0) = 0, \tag{3.1.7}$$

while for the set $\mathbb{R}\setminus\mathfrak{E}$ on which the derivative μ' does not exist finitely or infinitely one has that

$$m(\mathbb{R} \setminus \mathfrak{E}) = 0 \quad \text{and} \quad \mu(\mathbb{R} \setminus \mathfrak{E}) = 0; \tag{3.1.8}$$

note that $\mathbb{R} \setminus \mathfrak{E} \subset \mathbb{R} \setminus \mathfrak{E}_0$. Recall also that the derivative ν_μ' of the function ν_μ in (3.1.2) and the derivative μ' in (3.1.4) of the measure μ coincide m-almost everywhere.

A μ-measurable set $\mathfrak{S} \subset \mathbb{R}$ is called a *support* of μ if $\mu(\mathbb{R} \setminus \mathfrak{S}) = 0$. In particular, this implies that $\mu(\mathfrak{A}) = \mu(\mathfrak{A} \cap \mathfrak{S})$ for all μ-measurable sets $\mathfrak{A} \subset \mathbb{R}$. A support $\mathfrak{S} \subset \mathbb{R}$ of μ is called *minimal* if for subsets $\mathfrak{S}_0 \subset \mathfrak{S}$ that are μ-measurable and m-measurable, $\mu(\mathfrak{S}_0) = 0$ implies $m(\mathfrak{S}_0) = 0$. A minimal support is not uniquely defined. The next auxiliary lemma provides some useful properties of minimal supports.

Lemma 3.1.1. *Let μ be a Borel measure on $\mathbb{R}$ and let $\mathfrak{S}, \mathfrak{S}' \subset \mathbb{R}$ be sets that are measurable with respect to μ and m.*

(i) *If $\mathfrak{S}$ and $\mathfrak{S}'$ are minimal supports for μ, then the symmetric difference $\mathfrak{S}\Delta\mathfrak{S}'$ satisfies $\mu(\mathfrak{S}\Delta\mathfrak{S}') = 0$ and $m(\mathfrak{S}\Delta\mathfrak{S}') = 0$.*

(ii) *If $\mathfrak{S}$ is a minimal support for μ while $\mu(\mathfrak{S} \setminus \mathfrak{S}') = 0$ and $m(\mathfrak{S}' \setminus \mathfrak{S}) = 0$, then $\mathfrak{S}'$ is a minimal support of μ. In particular, if $\mathfrak{S}$ is a minimal support for μ and $\mathfrak{S} \subset \mathfrak{S}'$ is such that $m(\mathfrak{S}' \setminus \mathfrak{S}) = 0$, then $\mathfrak{S}'$ is a minimal support of μ.*

Proof. (i) Since $\mathfrak{S}\Delta\mathfrak{S}' \subset ((\mathbb{R} \setminus \mathfrak{S}) \cup (\mathbb{R} \setminus \mathfrak{S}'))$ and both $\mathfrak{S}$ and $\mathfrak{S}'$ are supports for μ, one has

$$\mu(\mathfrak{S}\Delta\mathfrak{S}') \leq \mu(\mathbb{R} \setminus \mathfrak{S}) + \mu(\mathbb{R} \setminus \mathfrak{S}') = 0.$$

In particular, $\mu(\mathfrak{S} \setminus \mathfrak{S}') = 0$. Now $\mathfrak{S} \setminus \mathfrak{S}' \subset \mathfrak{S}$ is μ-measurable and m-measurable, and since $\mathfrak{S}$ is a minimal support, it follows that $m(\mathfrak{S}\setminus\mathfrak{S}') = 0$. A similar argument shows that $m(\mathfrak{S}' \setminus \mathfrak{S}) = 0$. Hence, $m(\mathfrak{S}\Delta\mathfrak{S}') = 0$.

(ii) From $\mathbb{R}\setminus\mathfrak{S}' = ((\mathbb{R}\setminus\mathfrak{S})\cup(\mathfrak{S}\setminus\mathfrak{S}'))\setminus(\mathfrak{S}'\setminus\mathfrak{S})$ one concludes that

$$\mu(\mathbb{R}\setminus\mathfrak{S}') \leq \mu(\mathbb{R}\setminus\mathfrak{S}) + \mu(\mathfrak{S}\setminus\mathfrak{S}').$$

Since $\mathfrak{S}$ is a support of μ and it is assumed that $\mu(\mathfrak{S}\setminus\mathfrak{S}') = 0$, it follows that $\mu(\mathbb{R}\setminus\mathfrak{S}') = 0$. Hence, $\mathfrak{S}'$ is a support of μ.

To prove that $\mathfrak{S}'$ is a minimal support for μ, let $\mathfrak{S}_0 \subset \mathfrak{S}'$ be μ-measurable and m-measurable, and assume that $m(\mathfrak{S}_0) > 0$. Since

$$\mathfrak{S}_0 = (\mathfrak{S}_0\cap\mathfrak{S}) \cup \big(\mathfrak{S}_0\cap(\mathfrak{S}'\setminus\mathfrak{S})\big) \tag{3.1.9}$$

and $m(\mathfrak{S}'\setminus\mathfrak{S}) = 0$ by assumption, it follows that $m(\mathfrak{S}_0\cap\mathfrak{S}) = m(\mathfrak{S}_0) > 0$. As $\mathfrak{S}$ is a minimal support for μ, this implies $\mu(\mathfrak{S}_0\cap\mathfrak{S}) > 0$. Therefore, (3.1.9) leads to

$$\mu(\mathfrak{S}_0) = \mu(\mathfrak{S}_0\cap\mathfrak{S}) + \mu\big(\mathfrak{S}_0\cap(\mathfrak{S}'\setminus\mathfrak{S})\big) \geq \mu(\mathfrak{S}_0\cap\mathfrak{S}) > 0.$$

Thus, $\mathfrak{S}'$ is a minimal support for μ. □

Minimal supports for the parts of the spectrum in the Lebesgue decomposition can be expressed in terms of the behavior of the derivative μ'; cf. [335, Lemma 4] (see also [676, 682]).

Theorem 3.1.2. *Let μ be a regular Borel measure on $\mathbb{R}$. Then the following sets*

(i) $\{x\in\mathfrak{E} : 0 < \mu'(x) \leq \infty\}$;
(ii) $\{x\in\mathfrak{E} : 0 < \mu'(x) < \infty\}$;
(iii) $\{x\in\mathfrak{E} : \mu'(x) = \infty\}$;
(iv) $\{x\in\mathfrak{E} : \mu'(x) = \infty,\ \mu(\{x\}) = 0\}$;
(v) $\{x\in\mathfrak{E} : \mu'(x) = \infty,\ \mu(\{x\}) > 0\}$,

are minimal supports for μ, $\mu_{\rm ac}$, $\mu_{\rm s}$, $\mu_{\rm sc}$, and $\mu_{\rm p}$, respectively.

For practical reasons the attention is now restricted to finite Borel measures on $\mathbb{R}$. The properties of such measures are reflected by the boundary behavior of their so-called Borel transform in a sense to be made precise; cf. Appendix A.

Definition 3.1.3. Let μ be a finite Borel measure on $\mathbb{R}$. Then the *Borel transform* F of μ is the function F defined by

$$F(\lambda) = \int_{\mathbb{R}} \frac{1}{t-\lambda}\, d\mu(t), \quad \lambda\in\mathbb{C}\setminus\mathbb{R}. \tag{3.1.10}$$

If for some $x\in\mathbb{R}$ the limit $\lim_{y\downarrow 0} F(x+iy)$ exists and takes values in $[0,\infty]$, it will be denoted by $F(x+i0)$. The set of points in $\mathbb{R}$ where the limit of the imaginary part of F exists and takes values in $[0,\infty]$ is denoted by

$$\mathfrak{F} = \big\{x\in\mathbb{R} : \operatorname{Im} F(x+i0) \text{ exists finitely or infinitely}\big\}. \tag{3.1.11}$$

It follows from the integral representation (3.1.10) that

$$y \operatorname{Re} F(x+iy) = \int_{\mathbb{R}} \frac{(t-x)y}{(t-x)^2+y^2}\, d\mu(t),$$

$$y \operatorname{Im} F(x+iy) = \int_{\mathbb{R}} \frac{y^2}{(t-x)^2+y^2}\, d\mu(t),$$

and hence, by dominated convergence,

$$\lim_{y\downarrow 0} y \operatorname{Re} F(x+iy) = 0 \quad \text{and} \quad \lim_{y\downarrow 0} y \operatorname{Im} F(x+iy) = \mu(\{x\}) \tag{3.1.12}$$

for all $x \in \mathbb{R}$; cf. Lemma A.2.6. In particular,

$$\lim_{y\downarrow 0} y\, F(x+iy) = \lim_{y\downarrow 0} iy \operatorname{Im} F(x+iy) \tag{3.1.13}$$

for all $x \in \mathbb{R}$. Note also that the Borel transform F is a Nevanlinna function (see Definition A.2.3) and $\mu(\mathbb{R}) = \sup_{y>0} y \operatorname{Im} F(iy)$. Conversely, every Nevanlinna function F with

$$\sup_{y>0} y \operatorname{Im} F(iy) < \infty \quad \text{and} \quad \lim_{y\to\infty} F(iy) = 0$$

is the Borel transform of a finite Borel measure μ as in (3.1.10); cf. Proposition A.5.3.

An important observation concerning the boundary values $\operatorname{Im} F(x+i0)$ is contained in the following theorem, which is formulated in terms of the *symmetric derivative*

$$(D\mu)(x) = \lim_{\epsilon\downarrow 0} \frac{\mu((x-\epsilon, x+\epsilon))}{2\epsilon} \tag{3.1.14}$$

of μ. Here the limit is assumed to take values in $[0,\infty]$. Note that if for some $x \in \mathbb{R}$ the derivative $\mu'(x)$ in (3.1.4) exists with values in $[0,\infty]$, then the same is true for the symmetric derivative $(D\mu)(x)$.

Theorem 3.1.4. *Let μ be a finite Borel measure on $\mathbb{R}$, let F be its Borel transform, and let $x \in \mathbb{R}$. If the symmetric derivative $(D\mu)(x)$ exists with values in $[0,\infty]$, then also $\operatorname{Im} F(x+i0)$ exists with values in $[0,\infty]$ and*

$$\operatorname{Im} F(x+i0) = \pi (D\mu)(x) \quad (\in [0,\infty]). \tag{3.1.15}$$

In particular, the following statements hold:

(i) *$\operatorname{Im} F(x+i0)$ and $(D\mu)(x)$ exist simultaneously finitely m-almost everywhere and (3.1.15) holds;*

(ii) *$\operatorname{Im} F(x+i0)$ and $(D\mu)(x)$ exist simultaneously finitely or infinitely μ-almost everywhere and m-almost everywhere and (3.1.15) holds.*

Proof. Assume first that the symmetric derivative $(D\mu)(x)$ exists in $[0,\infty)$ for some $x \in \mathbb{R}$ and choose $c_-, c_+ \in \mathbb{R}$ with $c_- < (D\mu)(x) < c_+$. From the definition

(3.1.14) it follows that there exists $\delta > 0$ such that

$$2c_-\epsilon \le \mu(I_\epsilon) \le 2c_+\epsilon, \qquad I_\epsilon := (x-\epsilon, x+\epsilon), \tag{3.1.16}$$

holds for all $\epsilon \in (0,\delta]$. In the following set $K_y(s) := \frac{y}{s^2+y^2}$ for $y > 0$ and $s \in \mathbb{R}$. Then one has

$$\begin{aligned} \operatorname{Im} F(x+iy) &= \int_{\mathbb{R}} \frac{y}{(x-t)^2+y^2}\, d\mu(t) \\ &= \int_{\mathbb{R}} K_y(x-t)\, d\mu(t) \\ &= \int_{I_\delta} K_y(x-t)\, d\mu(t) + \int_{\mathbb{R}\setminus I_\delta} K_y(x-t)\, d\mu(t) \end{aligned} \tag{3.1.17}$$

for $y > 0$. First one estimates the second term on the right-hand side in (3.1.17). Since $t \in \mathbb{R}\setminus I_\delta$, one has $|t-x| \ge \delta$, so that $0 \le K_y(t-x) \le K_y(\delta)$. Then it is clear that

$$0 \le \int_{\mathbb{R}\setminus I_\delta} K_y(x-t)\, d\mu(t) \le K_y(\delta)\mu(\mathbb{R}) \to 0 \tag{3.1.18}$$

for $y \downarrow 0$. In order to estimate the first integral on the right-hand side in (3.1.17) one uses the identity

$$\int_{I_\delta} K_y(t-x)\, d\mu(t) = \mu(I_\delta)K_y(\delta) - \int_0^\delta K_y'(\epsilon)\mu(I_\epsilon)\, d\epsilon. \tag{3.1.19}$$

To prove (3.1.19), observe that

$$\begin{aligned} \int_0^\delta K_y'(\epsilon)\mu(I_\epsilon)\, d\epsilon &= \int_0^\delta \int_{x-\epsilon}^{x+\epsilon} K_y'(\epsilon)\, d\mu(t)\, d\epsilon \\ &= \int_{x-\delta}^{x} \int_{x-t}^{\delta} K_y'(\epsilon)\, d\epsilon\, d\mu(t) + \int_x^{x+\delta} \int_{t-x}^{\delta} K_y'(\epsilon)\, d\epsilon\, d\mu(t) \\ &= \mu(I_\delta)K_y(\delta) - \int_{x-\delta}^{x+\delta} K_y(t-x)\, d\mu(t), \end{aligned}$$

where Fubini's theorem on the triangle in the (t,ϵ)-plane given by $\epsilon = t-x$, $\epsilon = x-t$, with $0 \le \epsilon \le \delta$, was used. Now integration by parts, the fact that (3.1.16), $-K_y'(\epsilon) \ge 0$ for $\epsilon, y > 0$, and (3.1.19) give the estimate

$$\begin{aligned} 2c_- \arctan(\delta/y) &= 2c_- \int_0^\delta K_y(\epsilon)\, d\epsilon \\ &= 2c_-\delta K_y(\delta) + 2c_- \int_0^\delta (-\epsilon K_y'(\epsilon))\, d\epsilon \\ &\le \mu(I_\delta)K_y(\delta) - \int_0^\delta K_y'(\epsilon)\mu(I_\epsilon)\, d\epsilon \\ &= \int_{I_\delta} K_y(t-x)\, d\mu(t). \end{aligned}$$

In the same way one verifies the estimate

$$\int_{I_\delta} K_y(t-x)\, d\mu(t) \le 2c_+ \arctan(\delta/y).$$

It follows that

$$\pi c_- \le \liminf_{y\downarrow 0} \int_{I_\delta} K_y(t-x)\, d\mu(t) \le \limsup_{y\downarrow 0} \int_{I_\delta} K_y(t-x)\, d\mu(t) \le \pi c_+.$$

Now (3.1.18) and (3.1.17) imply

$$\pi c_- \le \liminf_{y\downarrow 0} \operatorname{Im} F(x+iy) \le \limsup_{y\downarrow 0} \operatorname{Im} F(x+iy) \le \pi c_+.$$

Letting $c_- \uparrow (D\mu)(x)$ and $c_+ \downarrow (D\mu)(x)$, one obtains

$$\lim_{y\downarrow 0} \operatorname{Im} F(x+iy) = \pi (D\mu)(x).$$

Next the case where the symmetric derivative $(D\mu)(x)$ exists and equals ∞ for some $x \in \mathbb{R}$ is discussed. In this situation the above reasoning leads to

$$\pi c_- \le \liminf_{y\downarrow 0} \operatorname{Im} F(x+iy)$$

for all $c_- > 0$. This yields $\lim_{y\downarrow 0} \operatorname{Im} F(x+iy) = \infty$.

It remains to show assertions (i) and (ii). Recall that if $\mu'(x)$ exists at some point $x \in \mathbb{R}$, then so does the symmetric derivative $(D\mu)(x)$ and

$$\mu'(x) = (D\mu)(x),$$

with equality in $[0,\infty]$. For (ii) the above reasoning implies that the set $\mathfrak{E}$ in (3.1.6) is contained in the set $\mathfrak{F}$ in (3.1.11) and hence $\mu(\mathbb{R}\setminus\mathfrak{F}) = 0$ and $m(\mathbb{R}\setminus\mathfrak{F}) = 0$ by (3.1.8). Assertion (i) follows in the same way from (3.1.5) and (3.1.7). □

It follows from Theorem 3.1.4 and (3.1.12) that Theorem 3.1.2 has a counterpart expressing minimal supports in terms of the Borel transform of μ.

Theorem 3.1.5. *Let μ be a finite Borel measure and let F be its Borel transform. Then the sets*

(i) $\{x \in \mathfrak{F} : 0 < \operatorname{Im} F(x+i0) \le \infty\}$;
(ii) $\{x \in \mathfrak{F} : 0 < \operatorname{Im} F(x+i0) < \infty\}$;
(iii) $\{x \in \mathfrak{F} : \operatorname{Im} F(x+i0) = \infty\}$;
(iv) $\{x \in \mathfrak{F} : \operatorname{Im} F(x+i0) = \infty,\ \lim_{y\downarrow 0} y \operatorname{Im} F(x+iy) = 0\}$;
(v) $\{x \in \mathfrak{F} : \operatorname{Im} F(x+i0) = \infty,\ \lim_{y\downarrow 0} y \operatorname{Im} F(x+iy) > 0\}$,

are minimal supports for μ, $\mu_{\rm ac}$, $\mu_{\rm s}$, $\mu_{\rm sc}$, and $\mu_{\rm p}$, respectively.

Proof. Only statement (i) will be proved. The proofs of the other statements are similar. Let

$$\mathfrak{M} = \{x \in \mathfrak{E} : 0 < \mu'(x) \leq \infty\},$$

and note that $\mathfrak{M}$ is a Borel set. Recall that, by Theorem 3.1.2 (i), $\mathfrak{M}$ is a minimal support for μ. Now introduce the set

$$\mathfrak{M}' = \{x \in \mathfrak{F} : 0 < \operatorname{Im} F(x + i0) \leq \infty\},$$

which is also a Borel set, as $\operatorname{Im} F(x + iy)$, $y > 0$, and hence $\operatorname{Im} F(x + i0)$ are Borel measurable functions in x. Then Theorem 3.1.4 shows that $\mathfrak{M} \subset \mathfrak{M}'$ and furthermore one has

$$\mathfrak{M}' \setminus \mathfrak{M} \subset \mathbb{R} \setminus \mathfrak{E}.$$

Since $m(\mathbb{R} \setminus \mathfrak{E}) = 0$ according to (3.1.8), it follows that $m(\mathfrak{M}' \setminus \mathfrak{M}) = 0$ and as $\mathfrak{M} \subset \mathfrak{M}'$, and $\mathfrak{M}$ is a minimal support for μ, one concludes from Lemma 3.1.1 (ii) that $\mathfrak{M}'$ is a minimal support for μ. □

Most of the results in this section have been stated in the context of finite Borel measures on $\mathbb{R}$ and their Borel transforms. They will be applied to study the spectrum of self-adjoint relations and operators in Section 3.6. However, it is also useful for later references to have similar results in the more general context of scalar Nevanlinna functions and the corresponding spectral functions; cf. Chapter 6 and Chapter 7. Let N be a scalar Nevanlinna function of the form

$$N(\lambda) = \alpha + \beta\lambda + \int_{\mathbb{R}} \left(\frac{1}{t - \lambda} - \frac{t}{t^2 + 1} \right) d\tau(t), \quad \lambda \in \mathbb{C} \setminus \mathbb{R}, \tag{3.1.20}$$

where $\alpha \in \mathbb{R}$, $\beta \geq 0$, and τ is a Borel measure on $\mathbb{R}$ which satisfies

$$\int_{\mathbb{R}} \frac{1}{t^2 + 1} \, d\tau(t) < \infty; \tag{3.1.21}$$

cf. Theorem A.2.5. Then the last condition implies that μ defined by

$$d\mu(t) = \frac{d\tau(t)}{t^2 + 1} \tag{3.1.22}$$

is a finite Borel measure on $\mathbb{R}$. Let F be the Borel transform of μ:

$$F(\lambda) = \int_{\mathbb{R}} \frac{1}{t - \lambda} \, d\mu(t), \quad \lambda \in \mathbb{C} \setminus \mathbb{R}. \tag{3.1.23}$$

The connection between N and F is given in the following lemma.

Lemma 3.1.6. *The Nevanlinna function N in (3.1.20) and the Borel transform F in (3.1.23) are connected by*

$$N(\lambda) = a + b\lambda + (\lambda^2 + 1)F(\lambda), \quad \lambda \in \mathbb{C} \setminus \mathbb{R}, \tag{3.1.24}$$

where $a, b \in \mathbb{R}$. If $x \in \mathbb{R}$, then the limits $\operatorname{Im} N(x+i0)$ and $\operatorname{Im} F(x+i0)$ exist simultaneously with values in $[0,\infty]$, and in that case

$$\operatorname{Im} N(x+i0) = (x^2+1)\operatorname{Im} F(x+i0) \quad (\in [0,\infty]). \tag{3.1.25}$$

Moreover, for each $x \in \mathbb{R}$,

$$\lim_{y \downarrow 0} y \operatorname{Re} N(x+iy) = 0 \tag{3.1.26}$$

and

$$\lim_{y \downarrow 0} y \operatorname{Im} N(x+iy) = (x^2+1) \lim_{y \downarrow 0} y \operatorname{Im} F(x+iy). \tag{3.1.27}$$

Proof. It is an immediate consequence of the integral representation (3.1.20) that N can be rewritten as

$$N(\lambda) = \alpha + \lambda \left(\beta + \int_{\mathbb{R}} d\mu(t) \right) + (\lambda^2+1) \int_{\mathbb{R}} \frac{1}{t-\lambda}\, d\mu(t), \quad \lambda \in \mathbb{C} \setminus \mathbb{R};$$

cf. Theorem A.2.4. This leads to (3.1.24). Note that for $\lambda = x+iy$ one has

$$N(x+iy) = a + b(x+iy) + ((x+iy)^2+1)F(x+iy),$$

whence

$$\operatorname{Im} N(x+iy) = by + (x^2+1-y^2)\operatorname{Im} F(x+iy) + 2xy \operatorname{Re} F(x+iy).$$

Now observe that for each $x \in \mathbb{R}$ one has $\lim_{y \downarrow 0} y \operatorname{Re} F(x+iy) = 0$ by (3.1.12). Together with the previous identity this proves the assertion in (3.1.25). Furthermore, now one sees (3.1.27) directly; cf. (3.1.12). Finally, note that

$$\operatorname{Re} N(x+iy) = a + bx + (x^2+1-y^2)\operatorname{Re} F(x+iy) - 2xy\operatorname{Im} F(x+iy),$$

which together with (3.1.12) leads to the identity (3.1.26). □

The next corollary deals with the existence of the limit $\lim_{\epsilon \downarrow 0} N(x+i\epsilon)$ for any scalar Nevanlinna function N.

Corollary 3.1.7. *Let N be a scalar Nevanlinna function. Then the limit $N(x+i0)$ exists finitely m-almost everywhere.*

Proof. It is clear from (3.1.25) and Theorem 3.1.4 that $\lim_{\epsilon \downarrow 0} \operatorname{Im} N(x+i\epsilon)$ exists finitely m-almost everywhere. Hence, it suffices to show that

$$\lim_{\epsilon \downarrow 0} \operatorname{Re} N(x+i\epsilon) \tag{3.1.28}$$

exists finitely m-almost everywhere. Denote by $\sqrt{\cdot}$ the branch of the square root fixed by $\operatorname{Im} \sqrt{\lambda} > 0$ for $\lambda \in \mathbb{C} \setminus [0,\infty)$ and $\sqrt{\lambda} \geq 0$ for $\lambda \in [0,\infty)$. Then it is easy to see that $\operatorname{Im}\sqrt{N(\lambda)} \geq 0$ and $\operatorname{Im}(i\sqrt{N(\lambda)}) \geq 0$ for $\lambda \in \mathbb{C}^+$ and hence $\lambda \mapsto \sqrt{N(\lambda)}$ and $\lambda \mapsto i\sqrt{N(\lambda)}$ are scalar Nevanlinna functions when they are extended to $\mathbb{C}^-$ by symmetry. It follows from (3.1.25) and Theorem 3.1.4 that the limits

$$\lim_{\epsilon\downarrow 0} \operatorname{Im}\sqrt{N(x+i\epsilon)} \quad \text{and} \quad \lim_{\epsilon\downarrow 0} \operatorname{Re}\sqrt{N(x+i\epsilon)} = \lim_{\epsilon\downarrow 0} \operatorname{Im}\big(i\sqrt{N(x+i\epsilon)}\big)$$

exist finitely m-almost everywhere. Since

$$\operatorname{Re} N(x+i\epsilon) = \big(\operatorname{Re}\sqrt{N(x+i\epsilon)}\big)^2 - \big(\operatorname{Im}\sqrt{N(x+i\epsilon)}\big)^2$$

it follows that the limit in (3.1.28) exists finitely m-almost everywhere. □

Let τ be the Borel measure on $\mathbb{R}$ in (3.1.20) which satisfies the condition (3.1.21). It has the Lebesgue decomposition

$$\tau = \tau_{\rm ac} + \tau_{\rm s}, \qquad \tau_{\rm s} = \tau_{\rm sc} + \tau_{\rm p},$$

where $\tau_{\rm ac}$ is absolutely continuous, $\tau_{\rm s}$ is singular, $\tau_{\rm sc}$ is singular continuous, and $\tau_{\rm p}$ is pure point. In the next corollary, which is a consequence of Theorem 3.1.5, (3.1.22), and (3.1.25), minimal supports for these measures are expressed in terms of the boundary behavior of N.

Corollary 3.1.8. *Let N be a Nevanlinna function with the integral representation (3.1.20). Then the sets*

(i) $\{x \in \mathfrak{F} : 0 < \operatorname{Im} N(x+i0) \leq \infty\}$;

(ii) $\{x \in \mathfrak{F} : 0 < \operatorname{Im} N(x+i0) < \infty\}$;

(iii) $\{x \in \mathfrak{F} : \operatorname{Im} N(x+i0) = \infty\}$;

(iv) $\{x \in \mathfrak{F} : \operatorname{Im} N(x+i0) = \infty, \lim_{y\downarrow 0} y \operatorname{Im} N(x+iy) = 0\}$;

(v) $\{x \in \mathfrak{F} : \operatorname{Im} N(x+i0) = \infty, \lim_{y\downarrow 0} y \operatorname{Im} N(x+iy) > 0\}$,

are minimal supports for τ, $\tau_{\rm ac}$, $\tau_{\rm s}$, $\tau_{\rm sc}$, and $\tau_{\rm p}$, respectively.

3.2 Growth points of finite Borel measures

Let μ be a finite Borel measure on $\mathbb{R}$. In this section the set of its *growth points* $\sigma(\mu)$, defined by

$$\sigma(\mu) = \big\{x \in \mathbb{R} : \mu\big((x-\epsilon, x+\epsilon)\big) > 0 \text{ for all } \epsilon > 0\big\}, \tag{3.2.1}$$

is studied. The growth points $\sigma(\mu)$ and the growth points $\sigma(\mu_{\rm ac})$, $\sigma(\mu_{\rm s})$, and $\sigma(\mu_{\rm sc})$ of the absolutely continuous, singular, and singular continuous part of μ will be located by means of the minimal supports expressed in terms of the Borel transform of μ.

There is an intimate connection between the set of growth points $\sigma(\mu)$ and supports for μ.

Lemma 3.2.1. *Let μ be a finite Borel measure on $\mathbb{R}$. Then the following statements hold:*

(i) *If $\mathfrak{S} \subset \mathbb{R}$ is a support of μ, then $\sigma(\mu) \subset \overline{\mathfrak{S}}$.*

(ii) *The set $\sigma(\mu)$ is closed and it is a support of μ.*

Proof. (i) Let $\mathfrak{S}$ be a support of μ, so that $\mu(\mathbb{R} \setminus \mathfrak{S}) = 0$. Assume that $x \in \sigma(\mu)$, so that for any $\epsilon > 0$ one has $\mu((x-\epsilon, x+\epsilon)) > 0$. Since $\mathfrak{S}$ is a support of μ, it follows that

$$0 < \mu\big((x-\epsilon, x+\epsilon)\big) = \mu\big((x-\epsilon, x+\epsilon) \cap \mathfrak{S}\big),$$

which implies that for any $\epsilon > 0$ the set $(x-\epsilon, x+\epsilon) \cap \mathfrak{S}$ is nonempty. Hence, there exists a sequence $x_n \in (x - 1/n, x + 1/n) \cap \mathfrak{S}$ converging to x from inside $\mathfrak{S}$. This shows that $\sigma(\mu) \subset \overline{\mathfrak{S}}$.

(ii) In order to show that $\sigma(\mu)$ is closed, let $x_n \in \sigma(\mu)$ converge to $x \in \mathbb{R}$. Assume that $x \notin \sigma(\mu)$. Then there is $\epsilon > 0$ such that $\mu((x-\epsilon, x+\epsilon)) = 0$. For this ϵ there exist $n_0 \in \mathbb{N}$ and $\epsilon_0 > 0$ with $(x_{n_0} - \epsilon_0, x_{n_0} + \epsilon_0) \subset (x-\epsilon, x+\epsilon)$, and hence

$$\mu\big((x_{n_0} - \epsilon_0, x_{n_0} + \epsilon_0)\big) \leq \mu\big((x-\epsilon, x+\epsilon)\big) = 0,$$

a contradiction, since $x_{n_0} \in \sigma(\mu)$. Therefore, $x \in \sigma(\mu)$ and $\sigma(\mu)$ is closed.

Next it will be verified that $\sigma(\mu)$ is a support for μ. For each $x \in \mathbb{R} \setminus \sigma(\mu)$ there is $\epsilon_x > 0$ such that $\mu((x-\epsilon_x, x+\epsilon_x)) = 0$. Since the set $\sigma(\mu)$ is closed, it follows that the open intervals $(x-\epsilon_x, x+\epsilon_x)$, $x \in \mathbb{R} \setminus \sigma(\mu)$, form an open cover for $\mathbb{R} \setminus \sigma(\mu)$. Then there is a countable subcover of open intervals I_n with $\mu(I_n) = 0$ for $\mathbb{R} \setminus \sigma(\mu)$. It follows that

$$\mu(\mathbb{R} \setminus \sigma(\mu)) \leq \sum_n \mu(I_n) = 0$$

and hence $\mu(\mathbb{R} \setminus \sigma(\mu)) = 0$, that is, $\sigma(\mu)$ is a support for μ. □

For completeness it is noted that in general the set $\sigma(\mu)$ is not a minimal support of μ. Observe also that, by Lemma 3.2.1, the set of growth points $\sigma(\mu)$ has the following minimality property: each closed support $\mathfrak{S} \subset \mathbb{R}$ of μ satisfies $\sigma(\mu) \subset \mathfrak{S}$. Therefore, one has the next corollary.

Corollary 3.2.2. *Let μ be a finite Borel measure on $\mathbb{R}$. Then $\sigma(\mu)$ is the smallest closed support of μ.*

The set of growth points of μ will now be described by means of the Borel transform of μ.

Theorem 3.2.3. *Let μ be a finite Borel measure on $\mathbb{R}$ and let F be its Borel transform. Then*

$$\sigma(\mu) = \overline{\big\{x \in \mathbb{R} : 0 < \liminf_{y \downarrow 0} \operatorname{Im} F(x+iy)\big\}}.$$

Proof. With the notation

$$\mathfrak{N} = \bigl\{x \in \mathbb{R} : 0 < \liminf_{y \downarrow 0} \operatorname{Im} F(x+iy)\bigr\}$$

it will be proved that $\sigma(\mu) = \overline{\mathfrak{N}}$. Recall first that, by Theorem 3.1.5 (i), the set

$$\mathfrak{M}' = \bigl\{x \in \mathfrak{F} : 0 < \operatorname{Im} F(x+i0) \leq \infty\bigr\}$$

is a (minimal) support for μ. Since $\mathfrak{M}' \subset \mathfrak{N}$, it follows that $\mathfrak{N}$ is also a support for μ. Hence, Lemma 3.2.1 (i) yields $\sigma(\mu) \subset \overline{\mathfrak{N}}$. For the inclusion $\overline{\mathfrak{N}} \subset \sigma(\mu)$ it suffices to show $\mathfrak{N} \subset \sigma(\mu)$, since $\sigma(\mu)$ is closed; cf. Lemma 3.2.1 (ii). Assume that $x \notin \sigma(\mu)$. Then there exists $\epsilon > 0$ such that $\mu((x-\epsilon, x+\epsilon)) = 0$ and it follows from

$$\operatorname{Im} F(x+iy) = \int_{\mathbb{R}\setminus(x-\epsilon,x+\epsilon)} \frac{y}{(t-x)^2+y^2}\, d\mu(t)$$

that $\operatorname{Im} F(x+i0) = 0$. This implies $x \notin \mathfrak{N}$ and hence $\mathfrak{N} \subset \sigma(\mu)$. □

Analogous to Theorem 3.2.3 there are also results for the parts of the finite Borel measure μ on $\mathbb{R}$ in its Lebesgue decomposition. In order to describe these results one needs the following notions of closure.

Definition 3.2.4. Let $\mathfrak{B} \subset \mathbb{R}$ be a Borel set. The *absolutely continuous closure* (or *essential closure*) of $\mathfrak{B}$ is defined by

$$\operatorname{clos}_{\mathrm{ac}}(\mathfrak{B}) := \bigl\{x \in \mathbb{R} : m\bigl((x-\epsilon, x+\epsilon) \cap \mathfrak{B}\bigr) > 0 \text{ for all } \epsilon > 0\bigr\}.$$

The *continuous closure* of $\mathfrak{B}$ is defined by

$$\operatorname{clos}_{\mathrm{c}}(\mathfrak{B}) := \bigl\{x \in \mathbb{R} : (x-\epsilon, x+\epsilon) \cap \mathfrak{B} \text{ is not countable for all } \epsilon > 0\bigr\}.$$

In general, $\mathfrak{B}$ is not a subset of $\operatorname{clos}_{\mathrm{ac}}(\mathfrak{B})$ since, e.g., isolated points in $\mathfrak{B}$ are not contained in $\operatorname{clos}_{\mathrm{ac}}(\mathfrak{B})$. Moreover, if $\mathfrak{B} \subset \mathfrak{B}'$ and $m(\mathfrak{B}' \setminus \mathfrak{B}) = 0$, then $\operatorname{clos}_{\mathrm{ac}}(\mathfrak{B}) = \operatorname{clos}_{\mathrm{ac}}(\mathfrak{B}')$.

Lemma 3.2.5. *Let $\mathfrak{B} \subset \mathbb{R}$ be a Borel set. Then the sets $\operatorname{clos}_{\mathrm{ac}}(\mathfrak{B})$ and $\operatorname{clos}_{\mathrm{c}}(\mathfrak{B})$ are both closed and*

$$\operatorname{clos}_{\mathrm{ac}}(\mathfrak{B}) \subset \operatorname{clos}_{\mathrm{c}}(\mathfrak{B}) \subset \overline{\mathfrak{B}}. \tag{3.2.2}$$

Moreover, the following statements hold:

(i) $\operatorname{clos}_{\mathrm{ac}}(\mathfrak{B}) = \emptyset$ *if and only if* $m(\mathfrak{B}) = 0$*;*

(ii) $\operatorname{clos}_{\mathrm{c}}(\mathfrak{B}) = \emptyset$ *if and only if* $\mathfrak{B}$ *is countable.*

Proof. First it will be shown that for any Borel set $\mathfrak{B} \subset \mathbb{R}$ both sets $\operatorname{clos}_{\mathrm{ac}}(\mathfrak{B})$ and $\operatorname{clos}_{\mathrm{c}}(\mathfrak{B})$ are closed.

In order to show that $\operatorname{clos}_{\mathrm{ac}}(\mathfrak{B})$ is closed, let $x_n \in \operatorname{clos}_{\mathrm{ac}}(\mathfrak{B})$ converge to $x \in \mathbb{R}$. Assume that $x \notin \operatorname{clos}_{\mathrm{ac}}(\mathfrak{B})$. Then there is $\epsilon > 0$ such that

$$m\bigl((x-\epsilon, x+\epsilon) \cap \mathfrak{B}\bigr) = 0. \tag{3.2.3}$$

For this ϵ there is $n_0 \in \mathbb{N}$ and $\epsilon_0 > 0$ with $(x_{n_0} - \epsilon_0, x_{n_0} + \epsilon_0) \subset (x - \epsilon, x + \epsilon)$. One then concludes from (3.2.3) that $m((x_{n_0} - \epsilon_0, x_{n_0} + \epsilon_0) \cap \mathfrak{B}) = 0$, a contradiction as $x_{n_0} \in \mathrm{clos}_{\mathrm{ac}}(\mathfrak{B})$. Therefore, $x \in \mathrm{clos}_{\mathrm{ac}}(\mathfrak{B})$ and $\mathrm{clos}_{\mathrm{ac}}(\mathfrak{B})$ is closed.

To show that $\mathrm{clos}_{\mathrm{c}}(\mathfrak{B})$ is closed, let $x_n \in \mathrm{clos}_{\mathrm{c}}(\mathfrak{B})$ converge to $x \in \mathbb{R}$. Assume that $x \notin \mathrm{clos}_{\mathrm{c}}(\mathfrak{B})$. Then there is $\epsilon > 0$ such that the set $(x - \epsilon, x + \epsilon) \cap \mathfrak{B}$ is countable. For this ϵ there exist $n_0 \in \mathbb{N}$ and $\epsilon_0 > 0$ with

$$(x_{n_0} - \epsilon_0, x_{n_0} + \epsilon_0) \subset (x - \epsilon_0, x + \epsilon_0),$$

so that

$$\big((x_{n_0} - \epsilon_0, x_{n_0} + \epsilon_0) \cap \mathfrak{B}\big) \subset \big((x - \epsilon, x + \epsilon) \cap \mathfrak{B}\big)$$

is countable, a contradiction, as $x_{n_0} \in \mathrm{clos}_{\mathrm{c}}(\mathfrak{B})$. Therefore, $x \in \mathrm{clos}_{\mathrm{c}}(\mathfrak{B})$ and $\mathrm{clos}_{\mathrm{c}}(\mathfrak{B})$ is closed.

To see the first inclusion in (3.2.2) assume that $x \in \mathrm{clos}_{\mathrm{ac}}(\mathfrak{B})$. Then one has $m((x-\epsilon, x+\epsilon) \cap \mathfrak{B}) > 0$ for all $\epsilon > 0$ and hence for all $\epsilon > 0$ the set $(x-\epsilon, x+\epsilon) \cap \mathfrak{B}$ is not countable. This implies $\mathrm{clos}_{\mathrm{ac}}(\mathfrak{B}) \subset \mathrm{clos}_{\mathrm{c}}(\mathfrak{B})$. Likewise, to see the second inclusion assume that $x \in \mathrm{clos}_{\mathrm{c}}(\mathfrak{B})$ and that $x \notin \overline{\mathfrak{B}}$. Then there is $\epsilon > 0$ such that $(x - \epsilon, x + \epsilon) \cap \mathfrak{B} = \emptyset$, a contradiction. Hence, $\mathrm{clos}_{\mathrm{c}}(\mathfrak{B}) \subset \overline{\mathfrak{B}}$.

(i) ($\Rightarrow$) Assume that $\mathrm{clos}_{\mathrm{ac}}(\mathfrak{B}) = \emptyset$. This implies that for all $x \in \mathbb{R}$ there exists $\epsilon_x > 0$ such that $m((x-\epsilon_x, x+\epsilon_x) \cap \mathfrak{B}) = 0$. First assume that $\mathfrak{B}$ is compact. Then for all $x \in \mathfrak{B}$ the open sets $(x - \epsilon_x, x + \epsilon_x)$ form an open cover for $\mathfrak{B}$. Therefore, there exists a finite subcover $(x_i - \epsilon_i, x_i + \epsilon_i)$ of $\mathfrak{B}$ such that

$$\mathfrak{B} \subset \bigcup_{i=1}^{n} (x_i - \epsilon_i, x_i + \epsilon_i) \cap \mathfrak{B},$$

and hence

$$m(\mathfrak{B}) \leq \sum_{i=1}^{n} m\big((x_i - \epsilon_i, x_i + \epsilon_i) \cap \mathfrak{B}\big) = 0.$$

For arbitrary Borel sets $\mathfrak{B}$ the (inner) regularity of the Lebesgue measure implies $m(\mathfrak{B}) = 0$.

($\Leftarrow$) If $m(\mathfrak{B}) = 0$, then $m((x - \epsilon, x + \epsilon) \cap \mathfrak{B}) = 0$ for all $x \in \mathbb{R}$ and all $\epsilon > 0$. Therefore, $\mathrm{clos}_{\mathrm{ac}}(\mathfrak{B}) = \emptyset$.

(ii) ($\Rightarrow$) Assume that $\mathrm{clos}_{\mathrm{c}}(\mathfrak{B}) = \emptyset$. This implies that for all $x \in \mathbb{R}$ there exists $\epsilon_x > 0$ such that $(x - \epsilon_x, x + \epsilon_x) \cap \mathfrak{B}$ is countable; in particular, this holds for all rational x_i. The countable many open sets $(x_i - \epsilon_{x_i}, x_i + \epsilon_{x_i})$ form an open cover for $\mathfrak{B}$ and this implies that $\mathfrak{B}$ is countable.

($\Leftarrow$) If $\mathfrak{B}$ is countable, then $(x - \epsilon, x + \epsilon) \cap \mathfrak{B}$ is countable for all $x \in \mathbb{R}$ and all $\epsilon > 0$. Therefore, $\mathrm{clos}_{\mathrm{c}}(\mathfrak{B}) = \emptyset$. □

Here is the promised treatment of the absolutely continuous, singular, and singular continuous parts of the Borel measure μ.

Theorem 3.2.6. *Let μ be a finite Borel measure on $\mathbb{R}$ and let F be its Borel transform. Then the following statements hold:*

(i) $\sigma(\mu_{\rm ac}) = {\rm clos}_{\rm ac}\big(\{x \in \mathfrak{F} : 0 < {\rm Im}\, F(x+i0) < \infty\}\big)$;

(ii) $\sigma(\mu_{\rm s}) \subset \overline{\{x \in \mathfrak{F} : {\rm Im}\, F(x+i0) = \infty\}}$;

(iii) $\sigma(\mu_{\rm sc}) \subset {\rm clos}_{\rm c}\big(\{x \in \mathfrak{F} : {\rm Im}\, F(x+i0) = \infty,\ \lim_{y\downarrow 0} yF(x+iy) = 0\}\big)$.

Proof. (i) Let

$$\mathfrak{M}'_{\rm ac} := \big\{x \in \mathfrak{F} : 0 < {\rm Im}\, F(x+i0) < \infty\big\}$$

and note that $\mathfrak{M}'_{\rm ac}$ is a Borel set. It is claimed that

$$\sigma(\mu_{\rm ac}) = {\rm clos}_{\rm ac}(\mathfrak{M}'_{\rm ac}). \tag{3.2.4}$$

To verify the inclusion ($\subset$) in (3.2.4), assume that $x \notin {\rm clos}_{\rm ac}(\mathfrak{M}'_{\rm ac})$. Then there exists $\epsilon > 0$ such that

$$m\big((x-\epsilon, x+\epsilon) \cap \mathfrak{M}'_{\rm ac}\big) = 0.$$

As $\mu_{\rm ac}$ is absolutely continuous with respect to the Lebesgue measure m, also

$$\mu_{\rm ac}\big((x-\epsilon, x+\epsilon) \cap \mathfrak{M}'_{\rm ac}\big) = 0. \tag{3.2.5}$$

Furthermore, by Theorem 3.1.5 (ii), the set $\mathfrak{M}'_{\rm ac}$ is a minimal support for $\mu_{\rm ac}$ and, in particular, $\mu_{\rm ac}(\mathbb{R} \setminus \mathfrak{M}'_{\rm ac}) = 0$. Hence,

$$\mu_{\rm ac}\big((x-\epsilon, x+\epsilon) \setminus \mathfrak{M}'_{\rm ac}\big) = 0 \tag{3.2.6}$$

and from (3.2.5)–(3.2.6) one obtains $\mu_{\rm ac}((x-\epsilon, x+\epsilon)) = 0$. Hence, $x \notin \sigma(\mu_{\rm ac})$. Thus, the inclusion ($\subset$) in (3.2.4) has been shown.

For the converse inclusion ($\supset$), let $x \notin \sigma(\mu_{\rm ac})$. Then there exists $\epsilon > 0$ such that

$$0 = \mu_{\rm ac}\big((x-\epsilon, x+\epsilon)\big) = \int_{(x-\epsilon, x+\epsilon)} (D\mu)(t)\, dm(t),$$

where in the last equality the Radon–Nikodým theorem was used; cf. (3.1.3) and note that $\nu'_\mu = \mu' = D\mu$ m-almost everywhere. Due to Theorem 3.1.4 and the fact that ${\rm Im}\, F(t+i0) \geq 0$ for all $t \in \mathfrak{F}$, one concludes that

$$\begin{aligned} 0 &= \frac{1}{\pi} \int_{(x-\epsilon, x+\epsilon)} {\rm Im}\, F(t+i0)\, dm(t) \\ &= \frac{1}{\pi} \int_{(x-\epsilon, x+\epsilon)\cap \mathfrak{M}'_{\rm ac}} {\rm Im}\, F(t+i0)\, dm(t). \end{aligned}$$

This implies $m((x-\epsilon, x+\epsilon) \cap \mathfrak{M}'_{\rm ac}) = 0$ since ${\rm Im}\, F(t+i0)$ is positive on $\mathfrak{M}'_{\rm ac}$. Hence, $x \notin {\rm clos}_{\rm ac}(\mathfrak{M}'_{\rm ac})$. Thus, the inclusion ($\supset$) in (3.2.4) has been shown. Therefore, the equality (3.2.4) has been established, which gives the assertion (i).

(ii) According to Theorem 3.1.5 (iii) the set $\{x \in \mathfrak{F} : \operatorname{Im} F(x + i0) = \infty\}$ is a minimal support for the singular part μ_s of μ. Since $\sigma(\mu_s)$ is contained in the closure of this set by Lemma 3.2.1 (i), the assertion follows.

(iii) By Theorem 3.1.5 (iv) and (3.1.13), the Borel set

$$\mathfrak{M}'_{sc} := \bigl\{x \in \mathfrak{F} : \operatorname{Im} F(x + i0) = \infty,\ \lim_{y \downarrow 0} yF(x + iy) = 0\bigr\}$$

is a minimal support for μ_{sc} and hence, in particular, $\mu_{sc}(\mathbb{R} \setminus \mathfrak{M}'_{sc}) = 0$. Let $\operatorname{clos}_c(\mathfrak{M}'_{sc})$ be the continuous closure of $\mathfrak{M}'_{sc}$, which is a Borel set, as it is closed; cf. Lemma 3.2.5. It will be shown that $\operatorname{clos}_c(\mathfrak{M}'_{sc})$ is a support for μ_{sc}, that is,

$$\mu_{sc}\bigl(\mathbb{R} \setminus \operatorname{clos}_c(\mathfrak{M}'_{sc})\bigr) = 0, \tag{3.2.7}$$

since this implies that $\sigma(\mu_{sc}) \subset \operatorname{clos}_c(\mathfrak{M}'_{sc})$; cf. Lemma 3.2.1 (i) and Lemma 3.2.5.

In fact, for $x \in \mathbb{R} \setminus \operatorname{clos}_c(\mathfrak{M}'_{sc})$ by definition there exists $\epsilon > 0$ such that $(x - \epsilon, x + \epsilon) \cap \mathfrak{M}'_{sc}$ is countable; thus $\mu_{sc}((x - \epsilon, x + \epsilon) \cap \mathfrak{M}'_{sc}) = 0$, as μ_{sc} is continuous. Consequently,

$$\mu_{sc}((x - \epsilon, x + \epsilon)) \leq \mu_{sc}\bigl((x - \epsilon, x + \epsilon) \cap \mathfrak{M}'_{sc}\bigr) + \mu_{sc}(\mathbb{R} \setminus \mathfrak{M}'_{sc}) = 0.$$

This yields $\mu_{sc}(K) = 0$ for each compact set $K \subset \mathbb{R} \setminus \operatorname{clos}_c(\mathfrak{M}'_{sc})$ and hence, by the (inner) regularity of the finite measure μ_{sc}, (3.2.7) follows. □

3.3 Spectra of self-adjoint relations

The spectrum of a self-adjoint relation or operator in a Hilbert space will be studied in terms of its spectral measure. In particular, a division of the spectrum into absolutely continuous and singular spectra will be introduced based on the Lebesgue decomposition of a finite Borel measure; cf. Section 3.1.

Let A be a self-adjoint relation in the Hilbert space $\mathfrak{H}$. Then $\sigma(A) \subset \mathbb{R}$ by Theorem 1.5.5 and $\sigma_r(A) = \emptyset$, and hence $\sigma(A) = \sigma_p(A) \cup \sigma_c(A)$; cf. Proposition 1.4.4. The spectral measure $E(\cdot)$ of A satisfies

$$(A - \lambda)^{-1} = \int_{\mathbb{R}} \frac{1}{t - \lambda}\, dE(t), \quad \lambda \in \mathbb{C} \setminus \mathbb{R},$$

cf. (1.5.6). First the parts $\sigma_p(A)$ and $\sigma_c(A)$ of the spectrum $\sigma(A)$ will be characterized in terms of the spectral measure $E(\cdot)$. These results will play an important role in the further development; cf. Section 3.5 and Section 3.6. The facts in Proposition 3.3.1 are immediate consequences of the orthogonal decomposition

$$\mathfrak{H} = \mathfrak{H}_{op} \oplus \mathfrak{H}_{mul}, \quad A = A_{op} \,\widehat{\oplus}\, A_{mul}, \tag{3.3.1}$$

where $\mathfrak{H}_{op} = \overline{\operatorname{dom} A}$ and $\mathfrak{H}_{mul} = \operatorname{mul} A$, of the self-adjoint relation A (see Theorem 1.5.1) and the properties of the spectral measure of A_{op}.

Proposition 3.3.1. *Let A be a self-adjoint relation in $\mathfrak{H}$ with spectral measure $E(\cdot)$. Then the following statements hold:*

(i) $\lambda \in \rho(A) \cap \mathbb{R}$ *if and only if* $E((\lambda - \epsilon, \lambda + \epsilon)) = 0$ *for some* $\epsilon > 0$;

(ii) $\lambda \in \sigma_{\rm p}(A)$ *if and only if* $E(\{\lambda\}) \neq 0$, *in which case* $\mathfrak{N}_\lambda(A) = \operatorname{ran} E(\{\lambda\})$ *and*

$$\widehat{\mathfrak{N}}_\lambda(A) = \bigl\{\{E(\{\lambda\})h, \lambda E(\{\lambda\})h\} : h \in \mathfrak{H}\bigr\};$$

(iii) $\lambda \in \sigma_{\rm c}(A)$ *if and only if* $E(\{\lambda\}) = 0$ *and* $E((\lambda - \epsilon, \lambda + \epsilon)) \neq 0$ *for all* $\epsilon > 0$.

A further subdivision of the spectrum will be introduced analogous to the Lebesgue decomposition of a finite Borel measure on $\mathbb{R}$; cf. Section 3.1. This requires another description of the spectrum via the introduction of a collection of finite Borel measures induced by the spectral function. Let A be a self-adjoint relation in $\mathfrak{H}$ with spectral measure $E(\cdot)$. For each $h \in \mathfrak{H}$, define μ_h by

$$\mu_h = (E(\cdot)h, h) = \bigl(E_{\rm op}(\cdot)P_{\rm op}h, P_{\rm op}h\bigr), \tag{3.3.2}$$

so that μ_h is a regular Borel measure on $\mathbb{R}$. Note that $\mu_h = 0$ for $h \in \mathfrak{H}_{\rm mul}$. The set of growth points $\sigma(\mu_h)$ of μ_h is given by

$$\sigma(\mu_h) = \bigl\{x \in \mathbb{R} : \mu_h((x - \epsilon, x + \epsilon)) > 0 \text{ for all } \epsilon > 0\bigr\}.$$

It will be shown that the spectrum of A is made up of the growth points of μ_h for a dense set of elements $h \in \mathfrak{H}$. Furthermore, the statement in the following proposition is in a local sense, namely, it concerns the spectrum of A relative to an open interval $\Delta \subset \mathbb{R}$; cf. Definition 3.4.9.

Proposition 3.3.2. *Let A be a self-adjoint relation in $\mathfrak{H}$, let $\Delta \subset \mathbb{R}$ be an open interval, and assume that $\mathcal{D}_\Delta$ is a subset of the closed subspace $E(\Delta)\mathfrak{H}$ such that*

$$\overline{\operatorname{span}}\, \mathcal{D}_\Delta = E(\Delta)\mathfrak{H}.$$

Then the following identities hold:

$$\overline{\sigma(A) \cap \Delta} = \overline{\bigcup_{h \in E(\Delta)\mathfrak{H}} \sigma(\mu_h)} = \overline{\bigcup_{h \in \mathcal{D}_\Delta} \sigma(\mu_h)}. \tag{3.3.3}$$

Proof. First it will be shown that

$$\overline{\sigma(A) \cap \Delta} \supset \overline{\bigcup_{h \in E(\Delta)\mathfrak{H}} \sigma(\mu_h)} \supset \overline{\bigcup_{h \in \mathcal{D}_\Delta} \sigma(\mu_h)}. \tag{3.3.4}$$

For this purpose assume that $x \notin \overline{\sigma(A) \cap \Delta}$. Then there exists $\epsilon > 0$ such that $(x - \epsilon, x + \epsilon) \cap \Delta$ contains no spectrum of A. By Proposition 3.3.1 (i), this yields

$$E\bigl((x - \epsilon, x + \epsilon) \cap \Delta\bigr) = 0$$

and for $h \in E(\Delta)\mathfrak{H}$ one obtains

$$\begin{aligned}\mu_h\big((x-\epsilon,x+\epsilon)\big) &= \big(E\big((x-\epsilon,x+\epsilon)\big)h,h\big)\\ &= \big(E\big((x-\epsilon,x+\epsilon)\big)E(\Delta)h,h\big)\\ &= \big(E\big((x-\epsilon,x+\epsilon)\cap\Delta\big)h,h\big)\\ &= 0.\end{aligned}$$

Therefore, $(x-\epsilon,x+\epsilon)\cap\sigma(\mu_h)=\emptyset$ for all $h\in E(\Delta)\mathfrak{H}$, and thus

$$x \notin \overline{\bigcup_{h\in E(\Delta)\mathfrak{H}} \sigma(\mu_h)}.$$

Hence, the inclusions (3.3.4) follow. Next it will be shown that

$$\overline{\bigcup_{h\in\mathfrak{D}_\Delta} \sigma(\mu_h)} \supset \overline{\sigma(A)\cap\Delta},$$

which, together with (3.3.4), yields (3.3.3). For this purpose, assume that

$$x \notin \overline{\bigcup_{h\in\mathfrak{D}_\Delta} \sigma(\mu_h)}.$$

Then there exists $\epsilon>0$ such that $(x-\epsilon,x+\epsilon)\subset\mathbb{R}\setminus\sigma(\mu_h)$ for all $h\in\mathfrak{D}_\Delta$, that is,

$$\big\|E\big((x-\epsilon,x+\epsilon)\big)h\big\|^2 = \mu_h\big((x-\epsilon,x+\epsilon)\big) = 0 \tag{3.3.5}$$

for all $h\in\mathfrak{D}_\Delta$, and hence for all $h\in\operatorname{span}\mathfrak{D}_\Delta$. Since by assumption $\operatorname{span}\mathfrak{D}_\Delta$ is dense in $E(\Delta)\mathfrak{H}$, it follows that (3.3.5) holds for all $h\in E(\Delta)\mathfrak{H}$ and hence again by Proposition 3.3.1 (i),

$$E\big((x-\epsilon,x+\epsilon)\cap\Delta\big)h = E\big((x-\epsilon,x+\epsilon)\big)E(\Delta)h = 0$$

for all $h\in\mathfrak{H}$. This shows that $(x-\epsilon,x+\epsilon)\cap\Delta$ does not contain spectrum of A, in particular, $x\notin\overline{\sigma(A)\cap\Delta}$. □

The collection of Borel measures μ_h, $h\in\mathfrak{H}$, as defined in (3.3.2), is now used to introduce a number of subspaces of $\mathfrak{H}$.

Definition 3.3.3. Let A be a self-adjoint relation in $\mathfrak{H}$. The *pure point subspace*, the *absolutely continuous subspace*, and the *singular continuous subspace* corresponding to $A_{\rm op}$ are defined by

$$\begin{aligned}\mathfrak{H}_{\rm p}(A_{\rm op}) &= \big\{h\in\mathfrak{H} : \mu_h \text{ is pure point}\big\},\\ \mathfrak{H}_{\rm ac}(A_{\rm op}) &= \big\{h\in\mathfrak{H} : \mu_h \text{ is absolutely continuous}\big\},\\ \mathfrak{H}_{\rm sc}(A_{\rm op}) &= \big\{h\in\mathfrak{H} : \mu_h \text{ is singular continuous}\big\},\end{aligned}$$

respectively.

In conjunction with the orthogonal decomposition (3.3.1), these subspaces span the original Hilbert space and lead to invariant parts of the self-adjoint relation, see, e.g., [649, Theorem VII.4] or [691, Proposition 9.3].

Theorem 3.3.4. *Let A be a self-adjoint relation in $\mathfrak{H}$. Then $\mathfrak{H}_{\rm p}(A_{\rm op})$, $\mathfrak{H}_{\rm ac}(A_{\rm op})$, and $\mathfrak{H}_{\rm sc}(A_{\rm op})$ are mutually orthogonal closed subspaces of $\mathfrak{H}$ and*

$$\mathfrak{H} = \mathfrak{H}_{\rm p}(A_{\rm op}) \oplus \mathfrak{H}_{\rm ac}(A_{\rm op}) \oplus \mathfrak{H}_{\rm sc}(A_{\rm op}) \oplus \mathfrak{H}_{\rm mul}\,.$$

Each of the Hilbert spaces $\mathfrak{H}_{\rm p}(A_{\rm op})$, $\mathfrak{H}_{\rm ac}(A_{\rm op})$, and $\mathfrak{H}_{\rm sc}(A_{\rm op})$ is invariant for the operator $A_{\rm op}$, and the restrictions

$$\begin{aligned} A_{\rm op}^{\rm p} &= A_{\rm op} \upharpoonright \mathfrak{H}_{\rm p}(A_{\rm op}),\\ A_{\rm op}^{\rm ac} &= A_{\rm op} \upharpoonright \mathfrak{H}_{\rm ac}(A_{\rm op}),\\ A_{\rm op}^{\rm sc} &= A_{\rm op} \upharpoonright \mathfrak{H}_{\rm sc}(A_{\rm op}), \end{aligned}$$

are self-adjoint operators in $\mathfrak{H}_{\rm p}(A_{\rm op})$, $\mathfrak{H}_{\rm ac}(A_{\rm op})$, and $\mathfrak{H}_{\rm sc}(A_{\rm op})$, respectively.

By means of these subspaces one defines, in analogy with the case of finite Borel measures, the *singular subspace* and the *continuous subspace* corresponding to $A_{\rm op}$ by

$$\mathfrak{H}_{\rm s}(A_{\rm op}) = \mathfrak{H}_{\rm p}(A_{\rm op}) \oplus \mathfrak{H}_{\rm sc}(A_{\rm op}) \quad \text{and} \quad \mathfrak{H}_{\rm c}(A_{\rm op}) = \mathfrak{H}_{\rm ac}(A_{\rm op}) \oplus \mathfrak{H}_{\rm sc}(A_{\rm op}),$$

respectively. The restrictions of $A_{\rm op}$ to these subspaces are denoted by $A_{\rm op}^{\rm s}$ and $A_{\rm op}^{\rm c}$, respectively, and it follows that

$$A_{\rm op}^{\rm s} = A_{\rm op}^{\rm p} \,\widehat{\oplus}\, A_{\rm op}^{\rm sc} \quad \text{and} \quad A_{\rm op}^{\rm c} = A_{\rm op}^{\rm ac} \,\widehat{\oplus}\, A_{\rm op}^{\rm sc}.$$

Definition 3.3.5. Let A be a self-adjoint relation in $\mathfrak{H}$. The *absolutely continuous spectrum* $\sigma_{\rm ac}(A)$, the *singular continuous spectrum* $\sigma_{\rm sc}(A)$, and the *singular spectrum* $\sigma_{\rm s}(A)$ of A are defined by

$$\sigma_{\rm ac}(A) = \sigma\big(A_{\rm op}^{\rm ac}\big), \quad \sigma_{\rm sc}(A) = \sigma\big(A_{\rm op}^{\rm sc}\big), \quad \text{and} \quad \sigma_{\rm s}(A) = \sigma\big(A_{\rm op}^{\rm s}\big),$$

respectively.

Note that for the pure point part $A_{\rm op}^{\rm p}$ one only has $\overline{\sigma_{\rm p}(A)} = \sigma(A_{\rm op}^{\rm p})$. The spectral measures of the self-adjoint operators $A_{\rm op}^{\rm ac}$, $A_{\rm op}^{\rm sc}$, and $A_{\rm op}^{\rm s}$ in the Hilbert spaces $\mathfrak{H}_{\rm ac}(A_{\rm op})$, $\mathfrak{H}_{\rm sc}(A_{\rm op})$, and $\mathfrak{H}_{\rm s}(A_{\rm op})$, are given by the corresponding restrictions of the spectral measure $E(\cdot)$ of A. These spectral measures will be denoted by $E_{\rm ac}(\cdot)$, $E_{\rm sc}(\cdot)$, and $E_{\rm s}(\cdot)$, respectively.

The following corollary relates the absolutely continuous, singular continuous, and singular spectrum of A in an open interval Δ with the growth points of the absolutely continuous, singular continuous, and singular parts of the measures μ_h.

Corollary 3.3.6. *Let A be a self-adjoint relation in $\mathfrak{H}$, let $\Delta \subset \mathbb{R}$ be an open interval, and assume that $\mathcal{D}_\Delta$ is a subset of the closed subspace $E(\Delta)\mathfrak{H}$ such that*

$$\overline{\operatorname{span}}\, \mathcal{D}_\Delta = E(\Delta)\mathfrak{H}.$$

Denote by $\mu_{h,\mathrm{ac}}$, $\mu_{h,\mathrm{sc}}$, and $\mu_{h,\mathrm{s}}$ the absolutely continuous, singular continuous, and singular part in the Lebesgue decomposition of the Borel measure μ_h in (3.3.2). Then the following identity holds:

$$\overline{\sigma_i(A) \cap \Delta} = \overline{\bigcup_{h \in \mathfrak{D}_\Delta} \sigma(\mu_{h,i})}, \qquad i = \mathrm{ac,\ sc,\ s}.$$

Proof. Observe first that the absolutely continuous, singular continuous, and singular part of the Borel measure μ_h, $h \in \mathfrak{H}$, are given by

$$\mu_{h,\mathrm{ac}} = \mu_{P_{\mathrm{ac}}h}, \quad \mu_{h,\mathrm{sc}} = \mu_{P_{\mathrm{sc}}h}, \quad \text{and} \quad \mu_{h,\mathrm{s}} = \mu_{P_{\mathrm{s}}h}, \tag{3.3.6}$$

respectively, where P_i denote the orthogonal projections onto the corresponding Hilbert spaces $\mathfrak{H}_i(A_{\mathrm{op}})$, $i = \mathrm{ac,\ sc,\ s}$. This follows from the uniqueness of the Lebesgue decomposition and Theorem 3.3.4. If $\mu^i_{h_i} = (E_i(\cdot)h_i, h_i)$, $h_i \in \mathfrak{H}_i(A_{\mathrm{op}})$, is the Borel measure defined with the help of the spectral measures $E_i(\cdot)$ of A^i_{op}, $i = \mathrm{ac,\ sc,\ s}$, then Definition 3.3.5, Proposition 3.3.2 and (3.3.6) yield

$$\overline{\sigma_i(A) \cap \Delta} = \overline{\bigcup_{h_i \in P_i\mathfrak{D}_\Delta} \sigma(\mu^i_{h_i})} = \overline{\bigcup_{h \in \mathfrak{D}_\Delta} \sigma(\mu_{P_i h})} = \overline{\bigcup_{h \in \mathfrak{D}_\Delta} \sigma(\mu_{h,i})}$$

for $i = \mathrm{ac,\ sc,\ s}$. Here it was also used that the linear span of the set $P_i\mathfrak{D}_\Delta$ is dense in $E_i(\Delta)\mathfrak{H}_i(A_{\mathrm{op}}) = P_i E(\Delta)\mathfrak{H}$. □

Example 3.3.7. Let μ be a Borel measure on $\mathbb{R}$ and consider the maximal multiplication operator by the independent variable in $L^2_\mu(\mathbb{R})$, given by

$$(Af)(t) = tf(t), \quad \mathrm{dom}\, A = \bigl\{ f \in L^2_\mu(\mathbb{R}) : t \mapsto tf(t) \in L^2_\mu(\mathbb{R}) \bigr\}.$$

The operator A is self-adjoint in $L^2_\mu(\mathbb{R})$ and for every Borel set $\mathfrak{B} \subset \mathbb{R}$ the spectral measure of A is given by

$$E(\mathfrak{B})h = \chi_{\mathfrak{B}} h, \qquad h \in L^2_\mu(\mathbb{R}),$$

where $\chi_{\mathfrak{B}}$ denotes the characteristic function of $\mathfrak{B}$. For $h \in L^2_\mu(\mathbb{R})$ the Borel measure in (3.3.2) satisfies

$$\mu_h(\mathfrak{B}) = (E(\mathfrak{B})h, h)_{L^2_\mu(\mathbb{R})} = \int_{\mathfrak{B}} |h(t)|^2 \, d\mu(t)$$

for all Borel sets $\mathfrak{B} \subset \mathbb{R}$. It is not difficult to check that $\sigma(A) = \sigma(\mu)$. Furthermore, the Lebesgue decomposition $\mu = \mu_{\mathrm{ac}} + \mu_{\mathrm{s}}$, where $\mu_{\mathrm{s}} = \mu_{\mathrm{sc}} + \mu_{\mathrm{p}}$, gives rise to the orthogonal decompositions

$$L^2_\mu(\mathbb{R}) = L^2_{\mu_{\mathrm{ac}}}(\mathbb{R}) \oplus L^2_{\mu_{\mathrm{s}}}(\mathbb{R}) \quad \text{and} \quad L^2_{\mu_{\mathrm{s}}}(\mathbb{R}) = L^2_{\mu_{\mathrm{sc}}}(\mathbb{R}) \oplus L^2_{\mu_{\mathrm{p}}}(\mathbb{R}).$$

For the spectral subspaces of A in Definition 3.3.3 one has $\mathfrak{H}_i(A) = L^2_{\mu_i}(\mathbb{R})$, $i = \mathrm{ac,\ sc,\ s}$, and this implies

$$\sigma_{\mathrm{ac}}(A) = \sigma(\mu_{\mathrm{ac}}), \quad \sigma_{\mathrm{sc}}(A) = \sigma(\mu_{\mathrm{sc}}), \quad \text{and} \quad \sigma_{\mathrm{s}}(A) = \sigma(\mu_{\mathrm{s}}).$$

3.4 Simple symmetric operators

It will be shown that any closed symmetric relation in a Hilbert space can be decomposed into the orthogonal componentwise sum of a closed simple, i.e., completely non-self-adjoint, symmetric operator, and a self-adjoint relation. Criteria for the absence of the self-adjoint relation in this decomposition will be given, and a local version of simplicity will be studied.

First some attention is paid to the notions of invariance and reduction. These notions appeared already in the self-adjoint case in the previous section when subdividing the spectrum, and are also important in the description of self-adjoint extensions of symmetric relations. Let S be a closed symmetric relation in the Hilbert space $\mathfrak{H}$. Decompose $\mathfrak{H}$ as $\mathfrak{H} = \mathfrak{H}' \oplus \mathfrak{H}''$, let P' and P'' be the orthogonal projections onto $\mathfrak{H}'$ and $\mathfrak{H}''$, and define

$$\widehat{P}'\{f,g\} = \{P'f, P'g\} \quad \text{and} \quad \widehat{P}''\{f,g\} = \{P''f, P''g\}, \quad f,g \in \mathfrak{H}.$$

The closed symmetric relation S gives rise to the restrictions

$$S' = S \cap (\mathfrak{H}')^2 \subset \widehat{P}'S \quad \text{and} \quad S'' = S \cap (\mathfrak{H}'')^2 \subset \widehat{P}''S, \tag{3.4.1}$$

which are closed symmetric relations and

$$S' \mathbin{\widehat{\oplus}} S'' \subset S. \tag{3.4.2}$$

In order to describe when S' and S'' span S the following notions are useful. The subspaces $\mathfrak{H}'$ and $\mathfrak{H}''$ are called *invariant* under the symmetric relation S if $S' = \widehat{P}'S$ or $S'' = \widehat{P}''S$, respectively. Clearly, the spaces $\mathfrak{H}'$ or $\mathfrak{H}''$ are invariant under S if

$$\widehat{P}'S \subset S \quad \text{or} \quad \widehat{P}''S \subset S,$$

respectively. In the next lemma it turns out that $\mathfrak{H}'$ is invariant under S if and only if $\mathfrak{H}''$ is invariant under S; in which case S' and S'' can be orthogonally split off from S, i.e., $S = S' \mathbin{\widehat{\oplus}} S''$.

Lemma 3.4.1. *Let S be a closed symmetric relation in $\mathfrak{H} = \mathfrak{H}' \oplus \mathfrak{H}''$ and let S' and S'' be as in (3.4.1). Then the following statements hold:*

(i) *$S' = \widehat{P}'S$ or, equivalently, $S'' = \widehat{P}''S$ implies that $S = S' \mathbin{\widehat{\oplus}} S''$.*

(ii) *If S' is self-adjoint in $\mathfrak{H}'$ or S'' is self-adjoint in $\mathfrak{H}''$, then $S' = \widehat{P}'S$ and $S'' = \widehat{P}''S$.*

Assume, in addition, that S is self-adjoint. Then

(iii) *$S' = \widehat{P}'S$ or, equivalently, $S'' = \widehat{P}''S$ implies that S' and S'' are self-adjoint in $\mathfrak{H}'$ and $\mathfrak{H}''$, respectively.*

Proof. (i) Assume that $S' = \widehat{P}'S$. Since $S' \,\widehat{\oplus}\, S'' \subset S$ by (3.4.2), it suffices to show that $S \subset S' \,\widehat{\oplus}\, S''$. Let $\{f, f'\} \in S$ and decompose $\{f, f'\}$ with respect to $\mathfrak{H} = \mathfrak{H}' \oplus \mathfrak{H}''$ as

$$\{f, f'\} = \{h, h'\} + \{k, k'\}, \quad h, h' \in \mathfrak{H}', \quad k, k' \in \mathfrak{H}''.$$

Then $\{h, h'\} \in \widehat{P}'S = S' \subset S$ and therefore $\{k, k'\} \in S \cap (\mathfrak{H}'')^2 = S''$. Hence, $S = S' \,\widehat{\oplus}\, S''$, which implies that $S'' = \widehat{P}''S$.

(ii) Assume that S' is self-adjoint in $\mathfrak{H}'$. To show that $\widehat{P}'S \subset S'$, let $\{f, f'\} \in S$ and consider $\{P'f, P'f'\} \in \widehat{P}'S$. Since S is symmetric it follows for all $\{h, h'\} \in S' \subset S$ that

$$(P'f', h)_{\mathfrak{H}'} - (P'f, h')_{\mathfrak{H}'} = (f', h)_{\mathfrak{H}} - (f, h')_{\mathfrak{H}} = 0.$$

The assumption that S' is self-adjoint in $\mathfrak{H}'$ implies $\{P'f, P'f'\} \in S'$. Therefore, $\widehat{P}'S \subset S'$. This implies $S' = \widehat{P}'S$ and (i) yields $S'' = \widehat{P}''S$.

(iii) According to (i), either of the conditions $S' = \widehat{P}'S$ or $S'' = \widehat{P}''S$ implies that $S = S' \,\widehat{\oplus}\, S''$. Since S is self-adjoint, this shows that S' is self-adjoint in $\mathfrak{H}'$ and that S'' is self-adjoint in $\mathfrak{H}''$. □

Before introducing the notion of simplicity in Definition 3.4.3 below, the following lemma on symmetric and self-adjoint extensions of symmetric relations that contain a self-adjoint part is discussed.

Lemma 3.4.2. *Let S be a closed symmetric relation in $\mathfrak{H}$ whose defect numbers are not necessarily equal and assume that there are orthogonal decompositions*

$$\mathfrak{H} = \mathfrak{H}' \oplus \mathfrak{H}'', \quad S = S' \,\widehat{\oplus}\, S'', \tag{3.4.3}$$

such that S' is closed and symmetric in $\mathfrak{H}'$ and S'' is self-adjoint in $\mathfrak{H}''$. Then every closed symmetric (self-adjoint) extension A of S in $\mathfrak{H}$ admits the decomposition

$$A = A' \,\widehat{\oplus}\, S'',$$

where A' is a closed symmetric (self-adjoint) extension of S' in $\mathfrak{H}'$.

Proof. Observe that the inclusion $S \subset A$ and the decomposition (3.4.3) imply that

$$S'' = S \cap (\mathfrak{H}'')^2 \subset A \cap (\mathfrak{H}'')^2.$$

Therefore, the assumption that S'' is self-adjoint in $\mathfrak{H}''$ shows that the closed symmetric relation $A \cap (\mathfrak{H}'')^2$ is actually self-adjoint in $\mathfrak{H}''$ and that $S'' = A \cap (\mathfrak{H}'')^2$. Hence, by Lemma 3.4.1 (i)–(ii) the relation A decomposes as $A = A' \,\widehat{\oplus}\, S''$, where $A' = A \cap (\mathfrak{H}')^2$ is a symmetric extension of S' in $\mathfrak{H}'$. Therefore,

$$S' \,\widehat{\oplus}\, S'' \subset A' \,\widehat{\oplus}\, S''.$$

If A is self-adjoint in $\mathfrak{H}$, then Lemma 3.4.1 (iii) implies that $A' = A \cap (\mathfrak{H}')^2$ is a self-adjoint extension of S' in $\mathfrak{H}'$. This completes the proof. □

The notion of simplicity or complete non-self-adjointness is defined next.

Definition 3.4.3. Let S be a closed symmetric relation in $\mathfrak{H}$ whose defect numbers are not necessarily equal. Then S is *simple* if there is no orthogonal decomposition

$$S = S' \,\widehat{\oplus}\, S'', \quad \text{where} \quad \mathfrak{H} = \mathfrak{H}' \oplus \mathfrak{H}'', \tag{3.4.4}$$

such that $\mathfrak{H}'' \neq \{0\}$ and S'' is self-adjoint in $\mathfrak{H}''$.

Every closed symmetric relation S in $\mathfrak{H}$ has the orthogonal componentwise decomposition $S = S_{\rm op} \,\widehat{\oplus}\, S_{\rm mul}$, where $S_{\rm mul}$ is a purely multivalued self-adjoint relation in the closed subspace $\mathfrak{H}_{\rm mul} = \operatorname{mul} S$; cf. Theorem 1.4.11. Hence, a closed simple symmetric relation is necessarily an operator. A similar argument shows that a closed simple symmetric relation does not have any eigenvalues; cf. Lemma 3.4.7.

Any closed symmetric relation S in $\mathfrak{H}$ has a decomposition as in (3.4.4), where S' is simple in $\mathfrak{H}'$ and S'' is self-adjoint in $\mathfrak{H}''$. To see this, define the closed subspace $\mathfrak{R} \subset \mathfrak{H}$ by

$$\mathfrak{R} := \bigcap_{\lambda \in \mathbb{C}\setminus\mathbb{R}} \operatorname{ran}(S - \lambda), \tag{3.4.5}$$

and the closed subspace $\mathfrak{K} = \mathfrak{R}^\perp$, so that

$$\mathfrak{K} = \overline{\operatorname{span}}\,\{\mathfrak{N}_\lambda(S^*) : \lambda \in \mathbb{C}\setminus\mathbb{R}\}, \quad \mathfrak{N}_\lambda(S^*) = \ker(S^* - \lambda). \tag{3.4.6}$$

It follows from Lemma 1.6.11 that the set $\mathbb{C}\setminus\mathbb{R}$ in (3.4.6), and hence in the intersection in (3.4.5), can be replaced by any subset of $\mathbb{C}\setminus\mathbb{R}$ which has an accumulation point in $\mathbb{C}^+$ and an accumulation point in $\mathbb{C}^-$.

Theorem 3.4.4. *Let S be a closed symmetric relation in $\mathfrak{H}$ whose defect numbers are not necessarily equal. Let $\mathfrak{H}$ be decomposed as $\mathfrak{H} = \mathfrak{K} \oplus \mathfrak{R}$, where the closed subspaces $\mathfrak{K}$ and $\mathfrak{R}$ are defined as in (3.4.5) and (3.4.6), and denote*

$$S' = S \cap \mathfrak{K}^2 \quad \textit{and} \quad S'' = S \cap \mathfrak{R}^2. \tag{3.4.7}$$

Then the relation S admits the orthogonal decomposition

$$S = S' \,\widehat{\oplus}\, S'', \tag{3.4.8}$$

where S' is a closed simple symmetric operator in $\mathfrak{K}$ and S'' is a self-adjoint relation in $\mathfrak{R}$.

Proof. Step 1. First it will be shown that $\mathfrak{R}$ satisfies the following invariance property: for any $\lambda_0 \in \mathbb{C}\setminus\mathbb{R}$

$$(S - \lambda_0)^{-1}\mathfrak{R} \subset \mathfrak{R}. \tag{3.4.9}$$

To see this, let $h \in \mathfrak{R}$ and $h' = (S - \lambda_0)^{-1}h$. Hence, $\{h', h + \lambda_0 h'\} \in S$ and thus

$$(h + \lambda_0 h', f_{\bar\lambda}) = (h', \bar\lambda f_{\bar\lambda}), \quad \{f_{\bar\lambda}, \bar\lambda f_{\bar\lambda}\} \in S^*, \ \lambda \in \mathbb{C}\setminus\mathbb{R}.$$

Since $h \in \operatorname{ran}(S-\lambda) = (\ker(S^*-\bar\lambda))^\perp$, $\lambda \in \mathbb{C}\setminus\mathbb{R}$, and $f_{\bar\lambda} \in \ker(S^*-\bar\lambda)$, this implies

$$0 = (h, f_{\bar\lambda}) = (\lambda-\lambda_0)(h', f_{\bar\lambda}),$$

that is, $h' \perp f_{\bar\lambda}$ for all $\lambda \in \mathbb{C}\setminus\mathbb{R}$, $\lambda \neq \lambda_0$. Hence, $h' \in \operatorname{ran}(S-\lambda)$ for all $\lambda \in \mathbb{C}\setminus\mathbb{R}$, $\lambda \neq \lambda_0$, and it follows from Lemma 1.6.11 that

$$h' \in \bigcap_{\lambda\in\mathbb{C}\setminus\mathbb{R},\, \lambda\neq\lambda_0} \operatorname{ran}(S-\lambda) = \bigcap_{\lambda\in\mathbb{C}\setminus\mathbb{R}} \operatorname{ran}(S-\lambda) = \mathfrak{R},$$

which proves the inclusion in (3.4.9).

Step 2. Next it will be shown that the relation $S \cap \mathfrak{R}^2$ is self-adjoint. Fix some $\lambda_0 \in \mathbb{C}\setminus\mathbb{R}$ and define the relation S'' first by

$$S'' = \bigl\{\{(S-\lambda_0)^{-1}h, (I+\lambda_0(S-\lambda_0)^{-1})h\} : h \in \mathfrak{R}\bigr\}. \tag{3.4.10}$$

It follows from (3.4.5) that $\mathfrak{R} \subset \operatorname{ran}(S-\lambda_0)$, and hence Lemma 1.1.8 implies $S'' \subset S$, so that in particular S'' is symmetric in $\mathfrak{H}$. It follows from (3.4.9) that $S'' \subset \mathfrak{R}^2$. Therefore, $S'' \subset S \cap \mathfrak{R}^2$. Next $S \cap \mathfrak{R}^2 \subset S''$ will be verified. Let $\{f, f'\} \in S \cap \mathfrak{R}^2$, so that by Lemma 1.1.8

$$\{f, f'\} = \{(S-\lambda_0)^{-1}h, (I+\lambda_0(S-\lambda_0)^{-1})h\}$$

for some $h \in \operatorname{ran}(S-\lambda_0)$. Since $\{f, f'\} \in \mathfrak{R}^2$, it follows that

$$(S-\lambda_0)^{-1}h \in \mathfrak{R} \quad \text{and} \quad (I+\lambda_0(S-\lambda_0)^{-1})h \in \mathfrak{R}.$$

Therefore, $h \in \mathfrak{R}$ and hence $\{f, f'\} \in S''$, so that $S \cap \mathfrak{R}^2 \subset S''$. This leads to the equality $S'' = S \cap \mathfrak{R}^2$ in (3.4.7); in particular, S'' in (3.4.10) does not depend on the choice of $\lambda_0 \in \mathbb{C}\setminus\mathbb{R}$.

From $S'' \subset S$ it follows that S'' is symmetric and from (3.4.10) one obtains that $\operatorname{ran}(S''-\lambda_0) = \mathfrak{R}$. Since S'' is independent of the choice of λ_0, it follows that $\operatorname{ran}(S''-\lambda) = \mathfrak{R}$ holds for every $\lambda \in \mathbb{C}\setminus\mathbb{R}$. Hence, $S'' = S \cap \mathfrak{R}^2$ is a self-adjoint relation in $\mathfrak{R}$ by Theorem 1.5.5. Now Lemma 3.4.1 (i)–(ii) imply (3.4.8).

Step 3. In order to show that $S' = S \cap \mathfrak{K}^2$ is simple in the Hilbert space $\mathfrak{K}$, assume that there is an orthogonal decomposition $\mathfrak{K} = \mathfrak{K}_1 \oplus \mathfrak{K}_2$ and a corresponding orthogonal decomposition $S' = S_1 \widehat{\oplus} S_2$ such that S_2 is self-adjoint in $\mathfrak{K}_2$. Then $\operatorname{ran}(S_2-\lambda) = \mathfrak{K}_2$ for all $\lambda \in \mathbb{C}\setminus\mathbb{R}$ and thus

$$\mathfrak{K}_2 = \operatorname{ran}(S_2-\lambda) \subset \operatorname{ran}(S'-\lambda) \subset \operatorname{ran}(S-\lambda), \quad \lambda \in \mathbb{C}\setminus\mathbb{R}.$$

According to (3.4.5), this implies $\mathfrak{K}_2 \subset \mathfrak{R}$ while $\mathfrak{K}_2 \subset \mathfrak{K} = \mathfrak{R}^\perp$. Thus, $\mathfrak{K}_2 = \{0\}$, so that S' is simple. □

Corollary 3.4.5. *Let S be a closed symmetric relation in $\mathfrak{H}$. Then S is simple if and only if*

$$\mathfrak{H} = \overline{\operatorname{span}}\,\bigl\{\mathfrak{N}_\lambda(S^*) : \lambda \in \mathbb{C}\setminus\mathbb{R}\bigr\}. \tag{3.4.11}$$

The set $\mathbb{C} \setminus \mathbb{R}$ on the right-hand side can be replaced by any set $\mathfrak{U}$ which has an accumulation point in $\mathbb{C}^+$ and in $\mathbb{C}^-$.

Proof. It follows from Theorem 3.4.4 and the definition of $\mathfrak{K}$ in (3.4.6) that the equality (3.4.11) holds if and only if S is simple. The last assertion in the corollary follows from Lemma 1.6.11. □

Corollary 3.4.6. *Let S be a closed symmetric relation in $\mathfrak{H}$. Then the following statements are equivalent:*

(i) $\mathfrak{H} = \overline{\operatorname{span}}\,\{\mathfrak{N}_\lambda(S^*) : \lambda \in \mathbb{C} \setminus \mathbb{R}\} \oplus \operatorname{mul} S$;

(ii) *$S_{\rm op}$ is a closed simple symmetric operator in $\mathfrak{H}_{\rm op} = \mathfrak{H} \ominus \operatorname{mul} S$.*

The set $\mathbb{C} \setminus \mathbb{R}$ on the right-hand side in (i) *can be replaced by any set $\mathfrak{U}$ which has an accumulation point in $\mathbb{C}^+$ and in $\mathbb{C}^-$.*

Proof. (i) $\Rightarrow$ (ii) The assumption implies in the context of Theorem 3.4.4 that $\mathfrak{R} = \operatorname{mul} S$, so that

$$S'' = S \cap \mathfrak{R}^2 = \{0\} \times \operatorname{mul} S,$$

which is a self-adjoint relation in $\operatorname{mul} S$. Hence, by the decomposition $S = S' \mathbin{\widehat{\oplus}} S''$ in Theorem 3.4.4 it follows that $S' = S_{\rm op}$ in $\mathfrak{H}_{\rm op} = \mathfrak{H} \ominus \operatorname{mul} S$.

(ii) $\Rightarrow$ (i) Recall that $\mathfrak{H} = \mathfrak{H}_{\rm op} \oplus \mathfrak{H}_{\rm mul}$. By Corollary 3.4.5 one has

$$\mathfrak{H}_{\rm op} = \overline{\operatorname{span}}\,\{\mathfrak{N}_\lambda(S^*_{\rm op}) : \lambda \in \mathbb{C} \setminus \mathbb{R}\}.$$

From the decomposition $S = S_{\rm op} \mathbin{\widehat{\oplus}} (\{0\} \times \operatorname{mul} S)$ and Proposition 1.3.13 one concludes that $S^* = S^*_{\rm op} \mathbin{\widehat{\oplus}} (\{0\} \times \operatorname{mul} S)$. Hence, $\mathfrak{N}_\lambda(S^*) = \mathfrak{N}_\lambda(S^*_{\rm op})$, which yields (i). □

Lemma 3.4.7. *Let S be a closed simple symmetric relation in $\mathfrak{H}$. Then S is an operator and it has no eigenvalues.*

Proof. Indeed, it follows from Definition 3.4.3 that also $S - x$ and $(S - x)^{-1}$, $x \in \mathbb{R}$, are closed simple symmetric relations in $\mathfrak{H}$. In particular, $(S - x)^{-1}$ is an operator; cf. the discussion following Definition 3.4.3. This implies $\ker (S - x) = \{0\}$ for all $x \in \mathbb{R}$ and hence S has no eigenvalues. □

In certain situations the assertion in Lemma 3.4.7 has a converse.

Proposition 3.4.8. *Let S be a closed symmetric relation in $\mathfrak{H}$ and assume that there exists a self-adjoint extension A of S in $\mathfrak{H}$ such that $\sigma(A) = \sigma_{\rm p}(A)$. If $\sigma_{\rm p}(S) = \emptyset$, then the operator part $S_{\rm op}$ of S is a closed simple symmetric operator in the Hilbert space $\mathfrak{H}_{\rm op} = (\operatorname{mul} S)^\perp$.*

Proof. By Lemma 3.4.2 and Theorem 1.4.11, it suffices to consider the case where S is a closed symmetric operator and A is a self-adjoint extension of S. Now assume that S is not simple, so that by Theorem 3.4.4 there are nontrivial decompositions $\mathfrak{H} = \mathfrak{H}' \oplus \mathfrak{H}''$ and $S = S' \widehat{\oplus} S''$ with S' closed, simple and symmetric in $\mathfrak{H}'$, and S'' self-adjoint in $\mathfrak{H}''$. Then A decomposes accordingly as $A = A' \widehat{\oplus} S''$ with A' self-adjoint in $\mathfrak{H}'$ by Lemma 3.4.2. Now $\sigma(A) = \sigma_{\rm p}(A)$ implies that S'' and thus S has a nontrivial point spectrum, which gives a contradiction. □

The notion of simplicity of a closed symmetric relation S in $\mathfrak{H}$ will now be specified in a local sense. This will be done relative to a Borel set $\Delta \subset \mathbb{R}$ and by means of a self-adjoint extension A of S and its spectral measure $E(\cdot)$. Then $\mathfrak{H}$ admits the orthogonal decomposition

$$\mathfrak{H} = E(\Delta)\mathfrak{H} \oplus (I - E(\Delta))\mathfrak{H},$$

which leads to the orthogonal componentwise decomposition of A into self-adjoint components:

$$A = \bigl[A \cap \bigl(E(\Delta)\mathfrak{H}\bigr)^2\bigr] \widehat{\oplus} \bigl[A \cap \bigl((I - E(\Delta))\mathfrak{H}\bigr)^2\bigr].$$

Note that $A \cap (E(\Delta)\mathfrak{H})^2$ is a self-adjoint operator in $E(\Delta)\mathfrak{H}$ which coincides with $A_{\rm op} \upharpoonright E_{\rm op}(\Delta)\mathfrak{H}_{\rm op}$; cf. Section 1.5.

Definition 3.4.9. Let S be a closed symmetric relation in $\mathfrak{H}$ and let A be a self-adjoint extension of S with spectral measure $E(\cdot)$. Let $\Delta \subset \mathbb{R}$ be a Borel set. Then S is said to be *simple with respect to* $\Delta \subset \mathbb{R}$ and the self-adjoint extension A if

$$E(\Delta)\mathfrak{H} = \overline{\operatorname{span}}\,\bigl\{E(\Delta)k : \, k \in \mathfrak{N}_\lambda(S^*), \, \lambda \in \mathbb{C} \setminus \mathbb{R}\bigr\}. \tag{3.4.12}$$

In the next proposition this local notion and some of its consequences are discussed.

Proposition 3.4.10. *Let S be a closed symmetric relation in $\mathfrak{H}$ and let A be a self-adjoint extension of S with spectral measure $E(\cdot)$. Assume that S is simple with respect to the Borel set $\Delta \subset \mathbb{R}$ and the self-adjoint extension A. Then the following statements hold:*

(i) *For every Borel set $\Delta' \subset \Delta$ one has*

$$E(\Delta')\mathfrak{H} = \overline{\operatorname{span}}\,\bigl\{E(\Delta')k : \, k \in \mathfrak{N}_\lambda(S^*), \, \lambda \in \mathbb{C} \setminus \mathbb{R}\bigr\}. \tag{3.4.13}$$

(ii) *There is no point spectrum of S in Δ:*

$$\Delta \cap \sigma_{\rm p}(S) = \emptyset.$$

(iii) *If $\mathfrak{U}$ is a subset of $\rho(A)$ with an accumulation point in each connected component of $\rho(A)$, then*

$$E(\Delta)\mathfrak{H} = \overline{\operatorname{span}}\,\bigl\{E(\Delta)k : \, k \in \mathfrak{N}_\lambda(S^*), \, \lambda \in \mathfrak{U}\bigr\}. \tag{3.4.14}$$

Proof. (i) First note that the inclusion ($\supset$) in (3.4.13) holds. To see the converse inclusion, let $f \in E(\Delta')\mathfrak{H}$. As $\Delta' \subset \Delta$, one has

$$E(\Delta')\mathfrak{H} \subset E(\Delta)\mathfrak{H},$$

and hence $f \in E(\Delta)\mathfrak{H}$. By assumption, the identity (3.4.12) holds and so, in the linear span of

$$\big\{E(\Delta)k : k \in \mathfrak{N}_\lambda(S^*),\, \lambda \in \mathbb{C}\setminus\mathbb{R}\big\}$$

there exists a sequence (f_n), that converges to f. Then $(E(\Delta')f_n)$ is a sequence in the linear span of

$$\big\{E(\Delta')k : k \in \mathfrak{N}_\lambda(S^*),\, \lambda \in \mathbb{C}\setminus\mathbb{R}\big\}$$

which converges to $E(\Delta')f = f$. This shows the inclusion ($\subset$) in (3.4.13).

(ii) Assume that $\{f, xf\} \in S$ for some $x \in \Delta$. Since $S \subset A$, it follows that $f \in E(\Delta)\mathfrak{H}$. Observe that for $k \in \mathfrak{N}_\lambda(S^*)$ with $\lambda \in \mathbb{C}\setminus\mathbb{R}$ one has $\{k, \lambda k\} \in S^*$ and hence $(\lambda k, f) = (k, xf)$. As $x \in \mathbb{R}$ and $\lambda \in \mathbb{C}\setminus\mathbb{R}$, it follows that $(k, f) = 0$. Further, since $f \in E(\Delta)\mathfrak{H}$, one concludes that

$$0 = (k, f) = (k, E(\Delta)f) = (E(\Delta)k, f)$$

for all $k \in \mathfrak{N}_\lambda(S^*)$ and $\lambda \in \mathbb{C}\setminus\mathbb{R}$. Hence, (3.4.12) implies that $f \in E(\Delta)\mathfrak{H}$ is orthogonal to $E(\Delta)\mathfrak{H}$, which shows that $f = 0$. Thus, S does not possess any eigenvalues in Δ.

(iii) The inclusion ($\supset$) in (3.4.14) is clear. In order to prove the identity, fix $\mu \in \mathcal{U}$ and recall from Lemma 1.4.10 that the operator $I + (\lambda-\mu)(A-\lambda)^{-1}$ maps $\mathfrak{N}_\mu(S^*)$ bijectively onto $\mathfrak{N}_\lambda(S^*)$ for all $\lambda \in \mathbb{C}\setminus\mathbb{R}$. It suffices to verify that the vectors $E(\Delta)k$, $k \in \mathfrak{N}_\lambda(S^*)$, $\lambda \in \mathcal{U}$, span a dense set in $E(\Delta)\mathfrak{H}$. Suppose that $E(\Delta)f$ is orthogonal to this set, that is,

$$0 = \big(E(\Delta)(I + (\lambda-\mu)(A-\lambda)^{-1})g_\mu, E(\Delta)f\big) \tag{3.4.15}$$

for all $g_\mu \in \mathfrak{N}_\mu(S^*)$ and all $\lambda \in \mathcal{U}$. Since for each $g_\mu \in \mathfrak{N}_\mu(S^*)$ the function

$$\lambda \mapsto \big(E(\Delta)(I + (\lambda-\mu)(A-\lambda)^{-1})g_\mu, E(\Delta)f\big)$$

is analytic on $\rho(A)$, it follows from (3.4.15) and the assumption that $\mathcal{U}$ has an accumulation point in each connected component of $\rho(A)$ that this function is identically equal to zero. Hence, $(E(\Delta)k, E(\Delta)f) = 0$ for all $k \in \mathfrak{N}_\lambda(S^*)$ and $\lambda \in \mathbb{C}\setminus\mathbb{R}$. Now (3.4.12) yields $E(\Delta)f = 0$ and (iii) follows. □

The connection with the global notion of simplicity is given in the following corollary.

Corollary 3.4.11. *Let S be a closed symmetric relation in $\mathfrak{H}$ and let A be a self-adjoint extension of S with spectral measure $E(\cdot)$. Then S is simple if and only if S is simple with respect to any Borel set $\Delta \subset \mathbb{R}$ and the self-adjoint extension A.*

Proof. Assume that S is simple. Then (3.4.11) holds and hence (3.4.12) holds with $\Delta = \mathbb{R}$. Then Proposition 3.4.10 (i) implies that S is simple with respect to any Borel set $\Delta \subset \mathbb{R}$ and the self-adjoint extension A. Conversely, if (3.4.12) holds for any Borel set $\Delta \subset \mathbb{R}$, then (3.4.12) also holds for $\Delta = \mathbb{R}$, and hence reduces to (3.4.11), that is, S is simple. □

In the following lemma the eigenspace of A corresponding to an eigenvalue x is described in the case where x is not an eigenvalue of S. In particular, this observation leads to a characterization of local simplicity if the Borel set $\Delta \subset \mathbb{R}$ in Definition 3.4.9 is a singleton; cf. Corollary 3.4.13.

Lemma 3.4.12. *Let S be a closed symmetric relation in $\mathfrak{H}$, let A be a self-adjoint extension of S with spectral measure $E(\cdot)$, and let $x \in \mathbb{R}$. Then*

$$E(\{x\})\mathfrak{H} = E(\{x\})\mathfrak{N}_\lambda(S^*) \tag{3.4.16}$$

for some, and hence for all $\lambda \in \mathbb{C} \setminus \mathbb{R}$, if and only if $x \notin \sigma_{\mathrm{p}}(S)$.

Proof. Assume first that (3.4.16) holds for some fixed $\lambda \in \mathbb{C} \setminus \mathbb{R}$. Assume that $\{f, xf\} \in S$, which implies $\{f, xf\} \in A$ and hence $f \in E(\{x\})\mathfrak{H}$. Moreover, one has $(xf, k_\lambda) = (f, \lambda k_\lambda)$ for all $k_\lambda \in \mathfrak{N}_\lambda(S^*)$ as $\{k_\lambda, \lambda k_\lambda\} \in S^*$. It follows that

$$\big(xf, E(\{x\})k_\lambda\big) = (xf, k_\lambda) = (f, \lambda k_\lambda) = \big(f, \lambda E(\{x\})k_\lambda\big)$$

and hence $(f, E(\{x\})k_\lambda) = 0$ for all $k_\lambda \in \mathfrak{N}_\lambda(S^*)$. Now (3.4.16) and $f \in E(\{x\})\mathfrak{H}$ yield $f = 0$, which implies $x \notin \sigma_{\mathrm{p}}(S)$.

Conversely, assume that $x \notin \sigma_{\mathrm{p}}(S)$ and let $\lambda \in \mathbb{C} \setminus \mathbb{R}$. The inclusion ($\supset$) in (3.4.16) is clear and since both subspaces in (3.4.16) are closed, it suffices to verify that $E(\{x\})\mathfrak{N}_\lambda(S^*)$ is dense in $E(\{x\})\mathfrak{H}$. Suppose that there exists $f \in E(\{x\})\mathfrak{H}$ such that

$$\big(f, E(\{x\})k_\lambda\big) = 0, \qquad k_\lambda \in \mathfrak{N}_\lambda(S^*).$$

As $f \in E(\{x\})\mathfrak{H}$, this implies $(f, k_\lambda) = 0$ and hence $f \in \operatorname{ran}(S - \bar{\lambda})$. Choose $\{g, g'\} \in S$ such that $g' - \bar{\lambda} g = f$. Then

$$g = (S - \bar{\lambda})^{-1} f = (A - \bar{\lambda})^{-1} f = \frac{1}{x - \bar{\lambda}} f$$

and

$$g' = f + \bar{\lambda} g = f + \frac{\bar{\lambda}}{x - \bar{\lambda}} f = \frac{x}{x - \bar{\lambda}} f,$$

and it follows that $\{f, xf\} \in S$. Since $x \notin \sigma_{\mathrm{p}}(S)$ by assumption this yields $f = 0$. Hence, $E(\{x\})\mathfrak{N}_\lambda(S^*)$ is dense in $E(\{x\})\mathfrak{H}$ and therefore (3.4.16) holds. □

The above lemma together with Proposition 3.4.10 (ii) implies that S is simple with respect to a point $x \in \mathbb{R}$ if and only if x is not an eigenvalue of S.

Corollary 3.4.13. *Let S be a closed symmetric relation in $\mathfrak{H}$, let A be a self-adjoint extension of S with spectral measure $E(\cdot)$, and let $x \in \mathbb{R}$. Then*

$$E(\{x\})\mathfrak{H} = \overline{\operatorname{span}}\,\big\{E(\{x\})k : \, k \in \mathfrak{N}_\lambda(S^*),\, \lambda \in \mathbb{C} \setminus \mathbb{R}\big\}$$

holds if and only if $x \notin \sigma_{\mathrm{p}}(S)$.

3.5 Eigenvalues and eigenspaces

Let S be a closed symmetric relation in a Hilbert space $\mathfrak{H}$ and let $\{\mathcal{G}, \Gamma_0, \Gamma_1\}$ be a boundary triplet for S^* with $A_0 = \ker \Gamma_0$, and corresponding γ-field γ and Weyl function M. The purpose of the present section is to characterize eigenvalues and the associated eigenspaces of the self-adjoint relation A_0 by means of the corresponding Weyl function M.

Recall that the Weyl function M can be expressed in terms of the γ-field and the resolvent of the self-adjoint relation A_0; cf. Proposition 2.3.6 (v). In particular, for $\lambda = x + iy$, $y > 0$, and $\lambda_0 \in \rho(A_0)$ one has

$$\begin{aligned} M(x+iy) = \operatorname{Re} M(\lambda_0) + \gamma(\lambda_0)^* \big[&(x+iy - \operatorname{Re}\lambda_0) \\ &+ (x+iy-\lambda_0)(x+iy-\bar{\lambda}_0)\big(A_0 - (x+iy)\big)^{-1}\big]\gamma(\lambda_0). \end{aligned} \tag{3.5.1}$$

This formula will be used to study the behavior of the Weyl function M at a point $x \in \mathbb{R}$. In the next proposition it turns out that the strong limit of $iyM(x+iy)$, $y \downarrow 0$, is closely connected with the eigenspace of A_0 at x. Here the spectral measure E of A_0 is not used explicitly in the assertion; the orthogonal projection onto the eigenspace $\mathfrak{N}_x(A_0) = \ker(A_0 - x)$ is denoted by $P_{\mathfrak{N}_x(A_0)}$ instead of $E(\{x\})$.

Proposition 3.5.1. *Let S be a closed symmetric relation in $\mathfrak{H}$, let $\{\mathcal{G}, \Gamma_0, \Gamma_1\}$ be a boundary triplet for S^*, let $A_0 = \ker \Gamma_0$, let M and γ be the corresponding Weyl function and γ-field, and let $x \in \mathbb{R}$. Then for each $\lambda_0 \in \rho(A_0)$ and all $\varphi \in \mathcal{G}$ one has*

$$\lim_{y \downarrow 0} iyM(x+iy)\varphi = -|x-\lambda_0|^2 \gamma(\lambda_0)^* P_{\mathfrak{N}_x(A_0)}\gamma(\lambda_0)\varphi. \tag{3.5.2}$$

Proof. For $x \in \mathbb{R}$ and $\lambda_0 \in \rho(A_0)$, it follows from (3.5.1) that

$$\begin{aligned} iyM(x+iy) &= iy \operatorname{Re} M(\lambda_0) + iy\gamma(\lambda_0)^*(x+iy-\operatorname{Re}\lambda_0)\gamma(\lambda_0) \\ &\quad + iy\,\gamma(\lambda_0)^*(x+iy-\lambda_0)(x+iy-\bar{\lambda}_0)\big(A_0-(x+iy)\big)^{-1}\gamma(\lambda_0). \end{aligned}$$

As the first and second terms on the right-hand side tend to 0 as $y \downarrow 0$, one obtains

$$\lim_{y\downarrow 0} iyM(x+iy)\varphi = |x-\lambda_0|^2\gamma(\lambda_0)^*\big[\lim_{y\downarrow 0} iy\big(A_0-(x+iy)\big)^{-1}\big]\gamma(\lambda_0)\varphi \tag{3.5.3}$$

for all $\varphi \in \mathcal{G}$. Since $x \in \mathbb{R}$ is fixed and $y \downarrow 0$, one has that

$$\frac{iy}{t-(x+iy)} \to -\mathbf{1}_x(t), \quad t \in \mathbb{R},$$

where the approximating functions are uniformly bounded by 1. The spectral calculus for the self-adjoint relation A_0 in Lemma 1.5.3 yields

$$\lim_{y\downarrow 0} iy\big(A_0-(x+iy)\big)^{-1}\gamma(\lambda_0)\varphi = -P_{\mathfrak{N}_x(A_0)}\gamma(\lambda_0)\varphi, \quad \varphi \in \mathcal{G}.$$

Now the assertion follows from (3.5.3). ◻

Definition 3.5.2. Let S be a closed symmetric relation in $\mathfrak{H}$, let $\{\mathcal{G}, \Gamma_0, \Gamma_1\}$ be a boundary triplet for S^*, let $A_0 = \ker \Gamma_0$, and let M be the corresponding Weyl function. For $x \in \mathbb{R}$ the operator $\mathcal{R}_x : \mathcal{G} \to \mathcal{G}$ is defined as the strong limit

$$\mathcal{R}_x \varphi = \lim_{y \downarrow 0} iyM(x+iy)\varphi, \qquad \varphi \in \mathcal{G}.$$

Observe that $\mathcal{R}_x$ in Definition 3.5.2 is a well-defined operator in $\mathbf{B}(\mathcal{G})$; indeed, this is clear from the identity (3.5.2). It also follows from (3.5.2) that $\mathcal{R}_x = 0$ when $x \in \rho(A_0) \cap \mathbb{R}$.

Remark 3.5.3. If $x \in \mathbb{R}$ is an isolated singularity of the function M, then x is a pole of first order of M; cf. Corollary 2.3.9. Moreover, in a sufficiently small punctured disc $B_x \setminus \{x\}$ centered at x such that M is holomorphic in $\overline{B}_x \setminus \{x\}$, one has a norm convergent Laurent series expansion of the form

$$M(\lambda) = \frac{M_{-1}}{\lambda - x} + \sum_{k=0}^{\infty} M_k (\lambda - x)^k, \quad M_{-1}, M_0, M_1, \ldots \in \mathbf{B}(\mathcal{G}).$$

It follows that $\mathcal{R}_x$ coincides with the residue of M at x, i.e.,

$$\mathcal{R}_x = \frac{1}{2\pi i} \int_{\mathcal{C}} M(\lambda)\, d\lambda = M_{-1},$$

where $\mathcal{C}$ denotes the boundary of B_x.

In the following let $x \in \mathbb{R}$ and recall that the corresponding eigenspaces of S and A_0 are given by

$$\widehat{\mathfrak{N}}_x(S) = \{\{f, xf\} : f \in \mathfrak{N}_x(S)\}, \quad \mathfrak{N}_x(S) = \ker (S - x),$$

and

$$\widehat{\mathfrak{N}}_x(A_0) = \{\{f, xf\} : f \in \mathfrak{N}_x(A_0)\}, \quad \mathfrak{N}_x(A_0) = \ker (A_0 - x).$$

The main interest will be in the closed linear subspace $\widehat{\mathfrak{N}}_x(A_0) \ominus \widehat{\mathfrak{N}}_x(S)$, which is the orthogonal complement of $\widehat{\mathfrak{N}}_x(S)$ in $\widehat{\mathfrak{N}}_x(A_0)$. Similarly, the orthogonal complement of $\mathfrak{N}_x(S)$ in $\mathfrak{N}_x(A_0)$ is denoted by $\mathfrak{N}_x(A_0) \ominus \mathfrak{N}_x(S)$.

Lemma 3.5.4. *Let $\lambda_0 \in \rho(A_0)$, let $x \in \mathbb{R}$, and let P_x be the orthogonal projection from $\mathfrak{H}$ onto $\mathfrak{N}_x(A_0) \ominus \mathfrak{N}_x(S)$. Then the operator $\mathcal{R}_x$ has the representation*

$$\mathcal{R}_x \varphi = (\lambda_0 - x)\Gamma_1 \{P_x \gamma(\lambda_0)\varphi, xP_x \gamma(\lambda_0)\varphi\}, \quad \varphi \in \mathcal{G}. \tag{3.5.4}$$

Proof. First, recall from Corollary 2.3.3 that for $x \in \mathbb{R}$ and $\{h, xh\} \in A_0$ one has

$$\Gamma_1\{h, xh\} = (x - \overline{\lambda}_0)\gamma(\lambda_0)^* h, \quad \lambda \in \rho(A_0).$$

Now let $\varphi \in \mathcal{G}$ and consider

$$h = (\lambda_0 - x) P_{\mathfrak{N}_x(A_0)}\, \gamma(\lambda_0)\varphi \in \ker (A_0 - x).$$

According to Proposition 3.5.1 and Definition 3.5.2,

$$\begin{aligned}\mathcal{R}_x\varphi &= -|x-\lambda_0|^2\gamma(\lambda_0)^* P_{\mathfrak{N}_x(A_0)}\,\gamma(\lambda_0)\varphi \\ &= (x-\overline{\lambda}_0)\gamma(\lambda_0)^* h \\ &= \Gamma_1\{h, xh\} \\ &= (\lambda_0 - x)\Gamma_1\{P_{\mathfrak{N}_x(A_0)}\gamma(\lambda_0)\varphi, xP_{\mathfrak{N}_x(A_0)}\gamma(\lambda_0)\varphi\}.\end{aligned} \tag{3.5.5}$$

Now observe that for $\varphi \in \mathcal{G}$

$$P_{\mathfrak{N}_x(A_0)}\gamma(\lambda_0)\varphi = P_x\gamma(\lambda_0)\varphi + P_{\mathfrak{N}_x(S)}\gamma(\lambda_0)\varphi.$$

Since $\{P_{\mathfrak{N}_x(S)}\gamma(\lambda_0)\varphi, xP_{\mathfrak{N}_x(S)}\gamma(\lambda_0)\varphi\} \in S$ and $S = \ker\Gamma_0 \cap \ker\Gamma_1$ by Proposition 2.1.2 (ii), it follows that

$$\Gamma_1\{P_{\mathfrak{N}_x(S)}\gamma(\lambda_0)\varphi, xP_{\mathfrak{N}_x(S)}\gamma(\lambda_0)\varphi\} = 0$$

and hence (3.5.5) leads to (3.5.4). □

In the following theorem the eigenvalue $x \in \mathbb{R}$ and the corresponding eigenspace of A_0 are characterized by means of the Weyl function M and the operator $\mathcal{R}_x$. Later it will be shown how to distinguish between isolated and embedded eigenvalues of A_0; cf. Theorem 3.6.1.

Theorem 3.5.5. *Let S be a closed symmetric relation in $\mathfrak{H}$, let $\{\mathcal{G}, \Gamma_0, \Gamma_1\}$ be a boundary triplet for S^*, let $A_0 = \ker\Gamma_0$, let M and γ be the corresponding Weyl function and γ-field, and let $x \in \mathbb{R}$. Then the mapping*

$$\tau : \widehat{\mathfrak{N}}_x(A_0) \ominus \widehat{\mathfrak{N}}_x(S) \to \overline{\operatorname{ran}}\,\mathcal{R}_x, \quad \widehat{f} \mapsto \Gamma_1\widehat{f}, \tag{3.5.6}$$

is an isomorphism. In particular,

$$x \in \sigma_p(A_0) \text{ and } \widehat{\mathfrak{N}}_x(A_0) \ominus \widehat{\mathfrak{N}}_x(S) \neq \{0\} \quad \Leftrightarrow \quad \mathcal{R}_x \neq 0.$$

Proof. Let $x \in \mathbb{R}$ and define $\mathfrak{K}_x = \widehat{\mathfrak{N}}_x(A_0) \ominus \widehat{\mathfrak{N}}_x(S)$. The mapping $\Gamma_1 : S^* \to \mathcal{G}$ is continuous and, in particular, its restriction to $\mathfrak{K}_x \subset S^*$ is continuous. The proof consists of three steps. In Step 1 it will be shown that the restriction of Γ_1 to $\mathfrak{K}_x$ is injective and in Step 2 it will be shown that it has closed range. Then it follows from Step 3 that τ in (3.5.6) is an isomorphism.

Step 1. The restriction of the mapping Γ_1 to $\mathfrak{K}_x$ is injective. Indeed, let $\widehat{f} \in \mathfrak{K}_x$ with $\Gamma_1\widehat{f} = 0$. The assumption $\widehat{f} \in \mathfrak{K}_x$ implies that $\widehat{f} \in A_0$ and hence $\Gamma_0\widehat{f} = 0$. Therefore, $\widehat{f} \in \ker\Gamma_0 \cap \ker\Gamma_1 = S$. Since $\widehat{f} = \{f, xf\} \in \widehat{\mathfrak{N}}_x(A_0) \ominus \widehat{\mathfrak{N}}_x(S)$, this implies $\widehat{f} = 0$.

Step 2. The range of the restriction of Γ_1 to $\mathfrak{K}_x$ is closed. In fact, let (φ_n) be a sequence in $\operatorname{ran}(\Gamma_1 \upharpoonright \mathfrak{K}_x)$ such that $\varphi_n \to \varphi \in \mathcal{G}$. Then there exists a sequence

$(\widehat{f}_n)$ in $\mathfrak{K}_x$ such that $\Gamma_1 \widehat{f}_n = \varphi_n$ and as $\widehat{f}_n \in A_0$ one has $\Gamma_0 \widehat{f}_n = 0$. Therefore, $\Gamma \widehat{f}_n = \{0, \varphi_n\} \to \{0, \varphi\}$. Recall from Proposition 2.1.2 that the restriction of Γ to $S^* \ominus S$ is an isomorphism onto $\mathcal{G} \times \mathcal{G}$. It follows that $\widehat{f}_n$ converge to some element $\widehat{f}$, which belongs to the closed subspace $\mathfrak{K}_x$. This yields $\Gamma_1 \widehat{f} = \varphi$ and hence $\operatorname{ran}(\Gamma_1 \upharpoonright \mathfrak{K}_x)$ is closed.

Step 3. The linear space

$$\{\{P_x\gamma(\lambda_0)\varphi, xP_x\gamma(\lambda_0)\varphi\} : \varphi \in \mathcal{G}\}$$

is dense in the Hilbert space $\mathfrak{K}_x = \widehat{\mathfrak{N}}_x(A_0) \ominus \widehat{\mathfrak{N}}_x(S)$. To see this, let $\widehat{f} \in \mathfrak{K}_x$ be orthogonal to all $\{P_x\gamma(\lambda_0)\varphi, xP_x\gamma(\lambda_0)\varphi\}$, $\varphi \in \mathcal{G}$. Then, since $\widehat{f} = \{f, xf\}$, Corollary 2.3.3 shows that for all $\varphi \in \mathcal{G}$ one has

$$\begin{aligned}
0 &= \big(\widehat{f}, \{P_x\gamma(\lambda_0)\varphi, xP_x\gamma(\lambda_0)\varphi\}\big) \\
&= (f, P_x\gamma(\lambda_0)\varphi) + (xf, xP_x\gamma(\lambda_0)\varphi) \\
&= (1+x^2)(f, \gamma(\lambda_0)\varphi) \\
&= (1+x^2)(\gamma(\lambda_0)^* f, \varphi) \\
&= (1+x^2)(x-\bar{\lambda}_0)^{-1}(\Gamma_1 \widehat{f}, \varphi),
\end{aligned}$$

so that $\Gamma_1 \widehat{f} = 0$, and hence $\widehat{f} = 0$ by Step 1.

Step 4. The mapping in (3.5.6) is an isomorphism. To see this, observe that

$$\operatorname{ran} \mathcal{R}_x \subset \operatorname{ran}(\Gamma_1 \upharpoonright \mathfrak{K}_x) \subset \overline{\operatorname{ran}}\, \mathcal{R}_x. \tag{3.5.7}$$

The first inclusion in (3.5.7) follows directly from (3.5.4). From the same identity one also sees that

$$\Gamma_1 \{P_x\gamma(\lambda_0)\varphi, xP_x\gamma(\lambda_0)\varphi\} = \frac{1}{\lambda_0 - x} \mathcal{R}_x\varphi \in \operatorname{ran} \mathcal{R}_x \subset \overline{\operatorname{ran}}\, \mathcal{R}_x.$$

Hence, the second inclusion in (3.5.7) follows from Step 3 and the boundedness of Γ_1. It is clear from (3.5.7) and Step 2 that

$$\operatorname{ran}(\Gamma_1 \upharpoonright \mathfrak{K}_x) = \overline{\operatorname{ran}}\, \mathcal{R}_x,$$

and hence, due to Step 1, the mapping in (3.5.6) is an isomorphism. □

The statement of Theorem 3.5.5 can be simplified if x is not an eigenvalue of the symmetric relation S, that is, S satisfies a local simplicity condition at $x \in \mathbb{R}$; cf. Corollary 3.4.13.

Corollary 3.5.6. *Assume that x is not an eigenvalue of the closed symmetric relation S in Theorem 3.5.5. Then*

$$x \in \sigma_{\mathrm{p}}(A_0) \quad \Leftrightarrow \quad \mathcal{R}_x \neq 0.$$

Now the behavior of M at ∞ will be considered and the multivalued part of A_0 will be described. First recall that the self-adjoint relation A_0 is decomposed into the orthogonal sum

$$A_0 = A_{0,\mathrm{op}} \,\widehat{\oplus}\, A_{0,\mathrm{mul}}, \tag{3.5.8}$$

where $A_{0,\mathrm{op}}$ is a self-adjoint operator in the Hilbert space

$$\mathfrak{H}_{\mathrm{op}} = (\mathrm{mul}\, A_0)^\perp = \overline{\mathrm{dom}\, A_0} \tag{3.5.9}$$

and $A_{0,\mathrm{mul}}$ is the purely multivalued self-adjoint relation in $\mathfrak{H}_{\mathrm{mul}} = \mathrm{mul}\, A_0$. Then the resolvent of A_0 has the form

$$(A_0 - \lambda)^{-1} = \begin{pmatrix} (A_{0,\mathrm{op}} - \lambda)^{-1} & 0 \\ 0 & 0 \end{pmatrix}, \qquad \lambda \in \rho(A_0), \tag{3.5.10}$$

with respect to the decomposition $\mathfrak{H} = \mathfrak{H}_{\mathrm{op}} \oplus \mathfrak{H}_{\mathrm{mul}}$; cf. (1.5.1).

The representation (3.5.1) of M in terms of A_0 gives for $\lambda_0 \in \rho(A_0)$ and $x = 0$ that

$$\begin{aligned} M(iy) &= \mathrm{Re}\, M(\lambda_0) \\ &\quad + \gamma(\lambda_0)^*\bigl[iy - \mathrm{Re}\,\lambda_0 + (iy-\lambda_0)(iy-\bar\lambda_0)(A_0 - iy)^{-1}\bigr]\gamma(\lambda_0). \end{aligned} \tag{3.5.11}$$

In order to use this formula for large y decompose the term $\gamma(\lambda_0)^*\gamma(\lambda_0)$ as

$$\gamma(\lambda_0)^*\gamma(\lambda_0) = \gamma(\lambda_0)^*(I - P_{\mathrm{op}})\gamma(\lambda_0) + \gamma(\lambda_0)^*\iota_{\mathrm{op}}P_{\mathrm{op}}\gamma(\lambda_0),$$

where P_{op} denotes the orthogonal projection from $\mathfrak{H}$ onto $\mathfrak{H}_{\mathrm{op}}$, ι_{op} is the canonical embedding of $\mathfrak{H}_{\mathrm{op}}$ into $\mathfrak{H}$, and $I - P_{\mathrm{op}}$ is viewed as an orthogonal projection in $\mathfrak{H}$. From the representation of the resolvent of A_0 in terms of the resolvent of $A_{0,\mathrm{op}}$ in (3.5.10) it follows that (3.5.11) may be rewritten as

$$\begin{aligned} M(iy) &= \mathrm{Re}\, M(\lambda_0) + (iy - \mathrm{Re}\,\lambda_0)\,\gamma(\lambda_0)^*(I - P_{\mathrm{op}})\gamma(\lambda_0) \\ &\quad + \gamma(\lambda_0)^*\iota_{\mathrm{op}}\bigl[iy - \mathrm{Re}\,\lambda_0 + (iy-\lambda_0)(iy-\bar\lambda_0)(A_{0,\mathrm{op}} - iy)^{-1}\bigr]P_{\mathrm{op}}\gamma(\lambda_0) \end{aligned} \tag{3.5.12}$$

for all $y > 0$. This formula will be used to study the behavior of M at ∞. It turns out that the strong limit $\frac{1}{iy}M(iy)$, $y \to +\infty$, is closely connected with the multivalued part of A_0.

Proposition 3.5.7. *Let S be a closed symmetric relation in $\mathfrak{H}$, let $\{\mathcal{G}, \Gamma_0, \Gamma_1\}$ be a boundary triplet for S^*, let $A_0 = \ker \Gamma_0$, and let M and γ be the corresponding Weyl function and γ-field. Then for each $\lambda_0 \in \rho(A_0)$ and $\varphi \in \mathcal{G}$ one has*

$$\lim_{y\to+\infty} \frac{1}{iy} M(iy)\varphi = \gamma(\lambda_0)^*(I - P_{\mathrm{op}})\gamma(\lambda_0)\varphi. \tag{3.5.13}$$

Proof. It follows from (3.5.12) with $\lambda_0 \in \rho(A_0)$ that

$$\frac{1}{iy}M(iy) = \frac{1}{iy}\mathrm{Re}\,M(\lambda_0) + \frac{iy - \mathrm{Re}\,\lambda_0}{iy}\gamma(\lambda_0)^*(I - P_{\mathrm{op}})\gamma(\lambda_0)$$
$$+ \frac{1}{iy}\gamma(\lambda_0)^*\iota_{\mathrm{op}}\big[iy - \mathrm{Re}\,\lambda_0 + (iy - \lambda_0)(iy - \bar{\lambda}_0)(A_{0,\mathrm{op}} - iy)^{-1}\big]P_{\mathrm{op}}\gamma(\lambda_0).$$

It suffices to show that the first and the third term on the right-hand side converge to 0 strongly. This is obvious for the first term on the right-hand side. For the third term note that for $y \to +\infty$ one has

$$\frac{iy - \mathrm{Re}\,\lambda_0}{iy} + \frac{(iy - \lambda_0)(iy - \bar{\lambda}_0)}{iy}\frac{1}{t - iy} \to 0, \qquad t \in \mathbb{R},$$

and hence the spectral calculus for $A_{0,\mathrm{op}}$ shows that for $y \to +\infty$

$$\frac{1}{iy}\gamma(\lambda_0)^*\iota_{\mathrm{op}}\big[iy - \mathrm{Re}\,\lambda_0 + (iy - \lambda_0)(iy - \bar{\lambda}_0)(A_{0,\mathrm{op}} - iy)^{-1}\big]P_{\mathrm{op}}\gamma(\lambda_0)\varphi$$

tends to zero for all $\varphi \in \mathcal{G}$; cf. Lemma 1.5.3. This leads to (3.5.13). □

Definition 3.5.8. Let S be a closed symmetric relation in $\mathfrak{H}$, let $\{\mathcal{G}, \Gamma_0, \Gamma_1\}$ be a boundary triplet for S^*, let $A_0 = \ker \Gamma_0$, and let M be the corresponding Weyl function. The operator $\mathcal{R}_\infty : \mathcal{G} \to \mathcal{G}$ is defined as the strong limit

$$\mathcal{R}_\infty \varphi = \lim_{y \to +\infty} \frac{1}{iy}M(iy)\varphi, \qquad \varphi \in \mathcal{G}.$$

It follows from Proposition 3.5.7 that $\mathcal{R}_\infty \in \mathbf{B}(\mathcal{G})$. For the following properties of $\mathcal{R}_\infty$ recall the notations

$$\widehat{\mathfrak{N}}_\infty(S) = \big\{\{0, f\} : f \in \mathfrak{N}_\infty(S)\big\}, \quad \mathfrak{N}_\infty(S) = \mathrm{mul}\, S,$$

and

$$\widehat{\mathfrak{N}}_\infty(A_0) = \big\{\{0, f\} : f \in \mathfrak{N}_\infty(A_0)\big\}, \quad \mathfrak{N}_\infty(A_0) = \mathrm{mul}\, A_0.$$

The next lemma can be viewed as a variant of Lemma 3.5.4 for $x = \infty$. Here the main interest is in the closed subspace $\widehat{\mathfrak{N}}_\infty(A_0) \ominus \widehat{\mathfrak{N}}_\infty(S)$, that is, the orthogonal complement of $\widehat{\mathfrak{N}}_\infty(S)$ in $\widehat{\mathfrak{N}}_\infty(A_0)$.

Lemma 3.5.9. *Let $\lambda_0 \in \rho(A_0)$ and let P_∞ be the orthogonal projection from $\mathfrak{H}$ onto $\mathfrak{N}_\infty(A_0) \ominus \mathfrak{N}_\infty(S)$. Then the operator $\mathcal{R}_\infty$ has the representation*

$$\mathcal{R}_\infty \varphi = \Gamma_1\big\{0, P_\infty \gamma(\lambda_0)\varphi\big\}, \qquad \varphi \in \mathcal{G}. \tag{3.5.14}$$

Proof. First recall from Corollary 2.3.3 that for $\{0, h'\} \in A_0$ one has

$$\Gamma_1\{0, h'\} = \gamma(\lambda_0)^* h', \qquad \lambda_0 \in \rho(A_0).$$

Now let $\varphi \in \mathcal{G}$ and consider $h' = (I - P_{\rm op})\gamma(\lambda_0)\varphi \in \operatorname{mul} A_0$. According to Proposition 3.5.7,

$$\begin{aligned} \mathcal{R}_\infty \varphi = \gamma(\lambda_0)^*(I - P_{\rm op})\gamma(\lambda_0)\varphi = \gamma(\lambda_0)^* h' = \Gamma_1\{0, h'\} \\ = \Gamma_1\{0, (I - P_{\rm op})\gamma(\lambda_0)\varphi\}. \end{aligned} \tag{3.5.15}$$

Now observe that for $\varphi \in \mathcal{G}$

$$(I - P_{\rm op})\gamma(\lambda_0)\varphi = P_\infty \gamma(\lambda_0)\varphi + P_{\mathfrak{N}_\infty(S)}\gamma(\lambda_0)\varphi.$$

Since $\{0, P_{\mathfrak{N}_\infty(S)}\gamma(\lambda_0)\varphi\} \in S$ and $S = \ker \Gamma_0 \cap \ker \Gamma_1$ by Proposition 2.1.2 (ii), it follows that

$$\Gamma_1\{0, P_{\mathfrak{N}_\infty(S)}\gamma(\lambda_0)\varphi\} = 0$$

and hence (3.5.15) leads to (3.5.14). □

In the next theorem the multivalued part of A_0 is characterized by means of the Weyl function M and the operator $\mathcal{R}_\infty$.

Theorem 3.5.10. *Let S be a closed symmetric relation in $\mathfrak{H}$, let $\{\mathcal{G}, \Gamma_0, \Gamma_1\}$ be a boundary triplet for S^*, let $A_0 = \ker \Gamma_0$, and let M and γ be the corresponding Weyl function and γ-field. Then the mapping*

$$\tau : \widehat{\mathfrak{N}}_\infty(A_0) \ominus \widehat{\mathfrak{N}}_\infty(S) \to \overline{\operatorname{ran}}\, \mathcal{R}_\infty, \quad \widehat{f} \mapsto \Gamma_1 \widehat{f}, \tag{3.5.16}$$

is an isomorphism. In particular,

$$\operatorname{mul} A_0 \ominus \operatorname{mul} S \neq \{0\} \quad \Leftrightarrow \quad \mathcal{R}_\infty \neq 0.$$

Proof. The proof follows a strategy similar to the one used in the proof of Theorem 3.5.5. To simplify notation, set

$$\mathfrak{K}_\infty := \widehat{\mathfrak{N}}_\infty(A_0) \ominus \widehat{\mathfrak{N}}_\infty(S) = \big\{\{0, f'\} : f' \in \operatorname{mul} A_0 \ominus \operatorname{mul} S\big\}.$$

The mapping $\Gamma_1 : S^* \to \mathcal{G}$ is continuous and, in particular, its restriction to $\mathfrak{K}_\infty \subset S^*$ is continuous.

Step 1. The restriction of the mapping Γ_1 to $\mathfrak{K}_\infty$ is injective. Indeed, let $\widehat{f} \in \mathfrak{K}_\infty$ with $\Gamma_1 \widehat{f} = 0$. The assumption $\widehat{f} \in \mathfrak{K}_\infty$ implies that $\widehat{f} \in A_0$ and hence $\Gamma_0 \widehat{f} = 0$. Therefore, $\widehat{f} \in \ker \Gamma_0 \cap \ker \Gamma_1 = S$. Since $\widehat{f} = \{0, f'\} \in \widehat{\mathfrak{N}}_\infty(A_0) \ominus \widehat{\mathfrak{N}}_\infty(S)$, this implies $\widehat{f} = 0$.

Step 2. The range of the restriction of Γ_1 to $\mathfrak{K}_\infty$ is closed. In fact, let (φ_n) be a sequence in $\operatorname{ran}(\Gamma_1 \upharpoonright \mathfrak{K}_\infty)$ such that $\varphi_n \to \varphi \in \mathcal{G}$. Then there exist $(\widehat{f}_n)$ in $\mathfrak{K}_\infty$ such that $\Gamma_1 \widehat{f}_n = \varphi_n$ and as $\widehat{f}_n \in A_0$ one has $\Gamma_0 \widehat{f}_n = 0$. Thus, $\Gamma \widehat{f}_n = \{0, \varphi_n\} \to \{0, \varphi\}$, and since the restriction of Γ to $S^* \ominus S$ is an isomorphism onto $\mathcal{G} \times \mathcal{G}$, it follows that $\widehat{f}_n$ converge to some element $\widehat{f}$, which belongs to the closed subspace $\mathfrak{K}_\infty$. Therefore, $\Gamma_1 \widehat{f} = \varphi$ and hence $\operatorname{ran}(\Gamma_1 \upharpoonright \mathfrak{K}_\infty)$ is closed.

Step 3. The linear space

$$\{\{0, P_\infty \gamma(\lambda_0)\varphi\} : \varphi \in \mathcal{G}\}$$

is dense in the Hilbert space $\mathfrak{K}_\infty = \widehat{\mathfrak{N}}_\infty(A_0) \ominus \widehat{\mathfrak{N}}_\infty(S)$. To see this, let $\widehat{f} \in \mathfrak{K}_\infty$ be orthogonal to all elements $\{0, P_\infty \gamma(\lambda_0)\varphi\}$, $\varphi \in \mathcal{G}$. Then it follows from $\widehat{f} = \{0, f'\}$ and Corollary 2.3.3 that for all $\varphi \in \mathcal{G}$ one has

$$0 = \big(\widehat{f}, \{0, P_\infty \gamma(\lambda_0)\varphi\}\big) = (f', P_\infty \gamma(\lambda_0)\varphi) = (\gamma(\lambda_0)^* f', \varphi) = (\Gamma_1 \widehat{f}, \varphi),$$

so that $\Gamma_1 \widehat{f} = 0$, and hence $\widehat{f} = 0$ by Step 1.

Step 4. The mapping in (3.5.16) is an isomorphism. To see this, observe that

$$\operatorname{ran} \mathcal{R}_\infty \subset \operatorname{ran} (\Gamma_1 \restriction \mathfrak{K}_\infty) \subset \overline{\operatorname{ran}}\, \mathcal{R}_\infty. \tag{3.5.17}$$

The first inclusion in (3.5.17) follows from (3.5.14). From the same identity one also sees that

$$\Gamma_1 \{0, P_\infty \gamma(\lambda_0)\varphi\} = \mathcal{R}_\infty \varphi \in \operatorname{ran} \mathcal{R}_\infty \subset \overline{\operatorname{ran}}\, \mathcal{R}_\infty.$$

Hence, the second inclusion in (3.5.17) follows from Step 3 and the boundedness of Γ_1. It is clear from (3.5.17) and Step 2 that

$$\operatorname{ran} (\Gamma_1 \restriction \mathfrak{K}_\infty) = \overline{\operatorname{ran}}\, \mathcal{R}_\infty,$$

and hence, due to Step 1, the mapping in (3.5.16) is an isomorphism. □

Corollary 3.5.11. *Assume that the closed symmetric relation S in Theorem 3.5.10 is an operator. Then A_0 is an operator if and only if $\mathcal{R}_\infty = 0$.*

An equivalent statement is that A_0 is an operator if and only if for all $\varphi \in \mathcal{G}$

$$\lim_{y \to +\infty} \frac{1}{iy} M(iy)\varphi = 0. \tag{3.5.18}$$

3.6 Spectra and local minimality

As in Section 3.5, let S be a closed symmetric relation in $\mathfrak{H}$, let $\{\mathcal{G}, \Gamma_0, \Gamma_1\}$ be a boundary triplet for S^* with $A_0 = \ker \Gamma_0$, and corresponding γ-field γ and Weyl function M. The spectrum of the self-adjoint extension A_0 and its division into absolutely continuous and singular spectra (cf. Section 3.3) will now be discussed in detail in terms of the boundary behavior of M. For this purpose it is assumed that S either is simple or satisfies a local simplicity condition with respect to an open interval $\Delta \subset \mathbb{R}$ and the self-adjoint extension A_0; see Definition 3.4.9 for the notion of local simplicity.

The following theorem describes the point spectrum and the continuous spectrum of A_0 in terms of the boundary behavior of the Weyl function M; cf. Proposition 3.3.1.

Theorem 3.6.1. *Let S be a closed symmetric relation in $\mathfrak{H}$, let $\{\mathcal{G},\Gamma_0,\Gamma_1\}$ be a boundary triplet for S^* with $A_0 = \ker\Gamma_0$, let M and γ be the corresponding Weyl function and γ-field, and let $\mathcal{R}_x = \lim_{y\downarrow 0} iyM(x+iy)$, $x\in\mathbb{R}$, be the operator in Definition 3.5.2. Let $\Delta\subset\mathbb{R}$ be an open interval and assume that the condition*

$$E(\Delta)\mathfrak{H} = \overline{\operatorname{span}}\,\{E(\Delta)\gamma(\nu)\varphi : \nu\in\mathbb{C}\setminus\mathbb{R}, \varphi\in\mathcal{G}\} \tag{3.6.1}$$

is satisfied, where $E(\cdot)$ is the spectral measure of A_0. Then the following statements hold for each $x\in\Delta$:

(i) *$x\in\rho(A_0)$ if and only if M can be continued analytically to x;*

(ii) *$x\in\sigma_c(A_0)$ if and only if $\mathcal{R}_x = 0$ and M cannot be continued analytically to x;*

(iii) *x is an eigenvalue of A_0 if and only if $\mathcal{R}_x\neq 0$;*

(iv) *x is an isolated eigenvalue of A_0 if and only if x is a pole (of first order) of M; in this case $\mathcal{R}_x$ is the residue of M at x.*

Proof. (i) Recall first that by Proposition 2.3.6 (iii) or (v) the function $\lambda\mapsto M(\lambda)$ is holomorphic on $\rho(A_0)$, which proves the implication ($\Rightarrow$). In order to verify the other implication assume that M can be continued analytically to some $x\in\Delta$. Then there exists an open neighborhood $\mathcal{O}$ of x in $\mathbb{C}$ with $\mathcal{O}\cap\mathbb{R}\subset\Delta$ to which M can be continued analytically. Choose $a,b\in\mathbb{R}$ with $x\in(a,b)$, $[a,b]\subset\mathcal{O}$, and $a,b\notin\sigma_p(A_0)$. The spectral projection $E((a,b))$ of A_0 corresponding to the interval (a,b) is given by Stone's formula (1.5.7)

$$E((a,b)) = \lim_{\delta\downarrow 0}\frac{1}{2\pi i}\int_a^b \big((A_0-(t+i\delta))^{-1} - (A_0-(t-i\delta))^{-1}\big)\,dt,$$

where the integral on the right-hand side is understood in the strong sense. For $\nu\in\mathbb{C}\setminus\mathbb{R}$ and $\varphi\in\mathcal{G}$ this implies

$$\begin{aligned}\|E((a,b))\gamma(\nu)\varphi\|^2 &= \big(\gamma(\nu)^*E((a,b))\gamma(\nu)\varphi,\varphi\big)\\ &= \lim_{\delta\downarrow 0}\frac{1}{2\pi i}\int_a^b\Big(\big(\gamma(\nu)^*(A_0-(t+i\delta))^{-1}\gamma(\nu)\varphi,\varphi\big)\\ &\qquad\qquad - \big(\gamma(\nu)^*(A_0-(t-i\delta))^{-1}\gamma(\nu)\varphi,\varphi\big)\Big)dt\end{aligned} \tag{3.6.2}$$

and the identities

$$\begin{aligned}&\gamma(\nu)^*\big(A_0-(t\pm i\delta)\big)^{-1}\gamma(\nu)\\ &\quad = \frac{M(t\pm i\delta)}{|t\pm i\delta-\nu|^2} + \frac{M(\bar\nu)}{(\bar\nu-(t\pm i\delta))(\bar\nu-\nu)} + \frac{M(\nu)}{(\nu-(t\pm i\delta))(\nu-\bar\nu)}\end{aligned}$$

from Proposition 2.3.6 (vi) (with $\lambda = t\pm i\delta$ and $\bar\mu=\nu$) together with the holomorphy of M in $\mathcal{O}$ yield that the integral on the right-hand side of (3.6.2) is zero.

Hence, $E((a,b))\gamma(\nu)\varphi = 0$ for all $\nu \in \mathbb{C} \setminus \mathbb{R}$ and $\varphi \in \mathcal{G}$. On the other hand, since $(a,b) \subset \Delta$, the assumption (3.6.1) and Proposition 3.4.10 (i) yield

$$E((a,b))\mathfrak{H} = \overline{\operatorname{span}}\,\{E((a,b))\gamma(\nu)\varphi : \nu \in \mathbb{C} \setminus \mathbb{R}, \varphi \in \mathcal{G}\},$$

and hence one concludes from $E((a,b))\gamma(\nu)\varphi = 0$ for $\nu \in \mathbb{C} \setminus \mathbb{R}$ and $\varphi \in \mathcal{G}$ that $E((a,b)) = 0$. In particular, $x \in \rho(A_0)$ by Proposition 3.3.1 (i).

(ii)–(iii) According to Proposition 3.4.10 (ii), the condition (3.6.1) implies that S does not have eigenvalues in Δ. Hence, items (ii) and (iii) follow immediately from item (i) and Corollary 3.5.6.

(iv) Assume that $x \in \Delta$ is an isolated eigenvalue of A_0. Then by Proposition 2.3.6 (iii) or (v) there exists an open neighborhood $\mathcal{O}$ of x such that M is holomorphic on $\mathcal{O} \setminus \{x\}$. Since $x \notin \sigma_p(S)$ by Proposition 3.4.10 (ii), it follows from Corollary 3.5.6 that there exists $\varphi \in \mathcal{G}$ such that

$$\mathcal{R}_x\varphi = \lim_{y \downarrow 0} iyM(x+iy)\varphi \neq 0. \tag{3.6.3}$$

This implies that M has a pole at x, which is of first order; cf. Corollary 2.3.9. By Remark 3.5.3 the residue of M at x is given by $\mathcal{R}_x$. Conversely, if M has a pole (of first order) at x, then (3.6.3) holds for some $\varphi \in \mathcal{G}$. Thus, x is an eigenvalue of A_0 by Corollary 3.5.6 and from item (i) it follows that there exists an open neighborhood $\mathcal{O}$ of x in $\mathbb{C}$ such that $\mathcal{O} \setminus \{\lambda\} \subset \rho(A_0)$. Hence, x is an isolated point in the spectrum of A_0. □

Under the condition that S is simple the spectrum of A_0 can be described completely in terms of the Weyl function M.

Corollary 3.6.2. *Let S be a closed symmetric relation in $\mathfrak{H}$, let $\{\mathcal{G}, \Gamma_0, \Gamma_1\}$ be a boundary triplet for S^*, let $A_0 = \ker \Gamma_0$, and let M and γ be the corresponding Weyl function and γ-field. Assume that S is simple. Then the assertions* (i)–(iv) *in Theorem 3.6.1 hold for all $x \in \mathbb{R}$.*

To describe the absolutely continuous, singular, and singular continuous parts of the spectrum of A_0 in terms of the boundary behavior of the Weyl function M, some preliminary lemmas are needed.

Lemma 3.6.3. *Let $\lambda_0 \in \rho(A_0)$, $x \in \mathbb{R}$, and $\varphi \in \mathcal{G}$. Then the (possibly improper) limits*

$$\operatorname{Im}\,(M(x+i0)\varphi,\varphi) \quad \textit{and} \quad \operatorname{Im}\,\big((A_0-(x+i0))^{-1}\gamma(\lambda_0)\varphi,\gamma(\lambda_0)\varphi\big)$$

exist simultaneously, and they satisfy

$$\operatorname{Im}\,(M(x+i0)\varphi,\varphi) = |x-\lambda_0|^2\operatorname{Im}\,\big((A_0-(x+i0))^{-1}\gamma(\lambda_0)\varphi,\gamma(\lambda_0)\varphi\big).$$

Proof. It is no restriction to assume that $x \neq \lambda_0$, as otherwise $\lambda \mapsto M(\lambda)$ and $\lambda \mapsto (A_0 - \lambda)^{-1}$ are both holomorphic at $x = \lambda_0 \in \rho(A_0) \cap \mathbb{R}$, so that the above limits are zero and the identities hold.

For $x \neq \lambda_0$ it follows from (3.5.1) that

$$\begin{aligned}\operatorname{Im}\,(M(x+iy)\varphi,\varphi) &= y\|\gamma(\lambda_0)\varphi\|^2 \\ &\quad + \left(|x-\lambda_0|^2 - y^2\right)\operatorname{Im}\left(\left(A_0-(x+iy)\right)^{-1}\gamma(\lambda_0)\varphi,\gamma(\lambda_0)\varphi\right) \\ &\quad + 2(x-\operatorname{Re}\lambda_0)\,y\operatorname{Re}\left(\left(A_0-(x+iy)\right)^{-1}\gamma(\lambda_0)\varphi,\gamma(\lambda_0)\varphi\right).\end{aligned}$$

The first term on the right-hand side clearly goes to 0 as $y \downarrow 0$. For the third term on the right-hand side, observe that for $y \downarrow 0$ one has

$$y\operatorname{Re}\left(\frac{1}{t-(x+iy)}\right) = \frac{y(t-x)}{(t-x)^2+y^2} \to 0, \quad t \in \mathbb{R},$$

and since the approximating functions are uniformly bounded, the spectral calculus for A_0 (see Lemma 1.5.3) yields

$$\lim_{y\downarrow 0} y\operatorname{Re}\left((A_0-(x+iy))^{-1}\gamma(\lambda_0)\varphi,\gamma(\lambda_0)\varphi\right) = 0.$$

Hence, also the third term on the right-hand side goes to 0 as $y \downarrow 0$. Furthermore, $|x-\lambda_0|^2 - y^2 \to |x-\lambda_0|^2 > 0$ as $y \downarrow 0$. Therefore, $\operatorname{Im}\,(M(x+iy)\varphi,\varphi)$ converges as $y \downarrow 0$ if and only if

$$\operatorname{Im}\,\left((A_0-(x+iy))^{-1}\gamma(\lambda_0)\varphi,\gamma(\lambda_0)\varphi\right)$$

converges as $y \downarrow 0$. In addition, it is clear that the identity in the lemma for the limits is satisfied. □

Recall that the self-adjoint extension A_0 generates a collection of finite Borel measures on $\mathbb{R}$: for each $h \in \mathfrak{H}$ the finite Borel measure μ_h in (3.3.2) is defined by $\mu_h = (E(\cdot)h,h)$, where E is the spectral measure of A_0. Now the interest is in the Borel transform F_h of $\mu_h = (E(\cdot)h,h)$, that is

$$F_h(\lambda) = \int_{\mathbb{R}} \frac{1}{t-\lambda}\,d(E(t)h,h), \quad \lambda \in \mathbb{C}\setminus\mathbb{R};$$

cf. Definition 3.1.3. In particular, if $\lambda = x+iy$, where $x \in \mathbb{R}$ and $y > 0$, then one has

$$\operatorname{Im} F_h(x+iy) = \operatorname{Im}\left((A_0-(x+iy))^{-1}h,h\right) \tag{3.6.4}$$

and

$$yF_h(x+iy) = y\left((A_0-(x+iy))^{-1}h,h\right). \tag{3.6.5}$$

By means of Lemma 3.6.3 the boundary values of the Borel transform F_h for a class of elements $h \in \mathfrak{H}$ are expressed in terms of the boundary values of the Weyl function M.

Lemma 3.6.4. *Let $\Delta \subset \mathbb{R}$ be an open interval and let $\lambda_0 \in \mathbb{C} \setminus \mathbb{R}$. Then for elements of the form $h = E(\Delta)\gamma(\lambda_0)\varphi$, $\varphi \in \mathcal{G}$, the following statements hold:*

(i) *If $x \in \Delta$, then the (possibly improper) limits*

$$\operatorname{Im} F_h(x + i0) \quad \textit{and} \quad \operatorname{Im}(M(x + i0)\varphi, \varphi)$$

exist simultaneously, and

$$\operatorname{Im} F_h(x + i0) = |x - \lambda_0|^{-2} \operatorname{Im}(M(x + i0)\varphi, \varphi).$$

(ii) *If $x \notin \overline{\Delta}$, then $\operatorname{Im} F_h(x + i0) = |x - \lambda_0|^{-2} \operatorname{Im}(M(x + i0)\varphi, \varphi) = 0$.*

Proof. It follows from (3.6.4) that for all $h \in \mathfrak{H}$ the (possibly improper) limits $\operatorname{Im} F_h(x + i0)$ and $\operatorname{Im}((A_0 - (x + i0))^{-1}h, h)$ exist simultaneously and coincide. For the choice $h = \gamma(\lambda_0)\varphi$, $\varphi \in \mathcal{G}$, it follows from Lemma 3.6.3 that the (possibly improper) limits $\operatorname{Im} F_h(x + i0)$ and $\operatorname{Im}(M(x + i0)\varphi, \varphi)$ exist simultaneously, and

$$\begin{aligned} \operatorname{Im} F_h(x + i0) &= \operatorname{Im}\big((A_0 - (x + i0))^{-1}\gamma(\lambda_0)\varphi, \gamma(\lambda_0)\varphi\big) \\ &= |x - \lambda_0|^{-2} \operatorname{Im}(M(x + i0)\varphi, \varphi). \end{aligned}$$

If $h = E(\Delta)\gamma(\lambda_0)\varphi$, $\varphi \in \mathcal{G}$, then for $x \in \Delta$ the spectral calculus implies

$$\begin{aligned} \operatorname{Im} F_h(x + i0) &= \operatorname{Im}\big((A_0 - (x + i0))^{-1}E(\Delta)\gamma(\lambda_0)\varphi, E(\Delta)\gamma(\lambda_0)\varphi\big) \\ &= \operatorname{Im}\big((A_0 - (x + i0))^{-1}\gamma(\lambda_0)\varphi, \gamma(\lambda_0)\varphi\big) \\ &= |x - \lambda_0|^{-2} \operatorname{Im}(M(x + i0)\varphi, \varphi), \end{aligned}$$

while for $x \notin \overline{\Delta}$ it follows that

$$\begin{aligned} \operatorname{Im} F_h(x + i0) &= \operatorname{Im}\big((A_0 - (x + i0))^{-1}E(\Delta)\gamma(\lambda_0)\varphi, E(\Delta)\gamma(\lambda_0)\varphi\big) \\ &= 0. \end{aligned}$$

This shows the assertions in (i) and (ii). □

Now the absolutely continuous spectrum, the singular spectrum, and the singular continuous spectrum (cf. Section 3.3) of A_0 can be described in terms of the boundary behavior of the Weyl function M, still under the assumption of local simplicity. The results are essentially consequences of Theorem 3.2.3, Theorem 3.2.6, and Corollary 3.3.6.

Theorem 3.6.5. *Let S be a closed symmetric relation in $\mathfrak{H}$, let $\{\mathcal{G}, \Gamma_0, \Gamma_1\}$ be a boundary triplet for S^* with let $A_0 = \ker \Gamma_0$, and let M and γ be the corresponding Weyl function and γ-field. Let $\Delta \subset \mathbb{R}$ be an open interval and assume that the condition*

$$E(\Delta)\mathfrak{H} = \overline{\operatorname{span}}\,\big\{E(\Delta)\gamma(\nu)\varphi : \nu \in \mathbb{C} \setminus \mathbb{R}, \varphi \in \mathcal{G}\big\} \tag{3.6.6}$$

is satisfied, where $E(\cdot)$ is the spectral measure of A_0. Then the absolutely continuous spectrum of A_0 in Δ is given by

$$\overline{\sigma_{\mathrm{ac}}(A_0) \cap \Delta} = \overline{\bigcup_{\varphi \in \mathcal{G}} \mathrm{clos}_{\mathrm{ac}}\bigl(\{x \in \Delta : 0 < \mathrm{Im}\,(M(x+i0)\varphi, \varphi) < \infty\}\bigr)}. \tag{3.6.7}$$

If S is simple, then (3.6.7) holds for every open interval Δ, including $\Delta = \mathbb{R}$.

Proof. By assumption, the span of the set

$$\mathcal{D}_\Delta := \bigl\{E(\Delta)\gamma(\nu)\varphi : \nu \in \mathbb{C} \setminus \mathbb{R},\ \varphi \in \mathcal{G}\bigr\}$$

is dense in $E(\Delta)\mathfrak{H}$ and hence Corollary 3.3.6 implies the identity

$$\overline{\sigma_{\mathrm{ac}}(A_0) \cap \Delta} = \overline{\bigcup_{h \in \mathcal{D}_\Delta} \sigma(\mu_{h,\mathrm{ac}})}.$$

According to Theorem 3.2.6 (i) (where the set $\mathfrak{F}$ was replaced by $\mathbb{R}$),

$$\sigma(\mu_{h,\mathrm{ac}}) = \mathrm{clos}_{\mathrm{ac}}\bigl(\{x \in \mathbb{R} : 0 < \mathrm{Im}\, F_h(x+i0) < \infty\}\bigr),$$

which for $h = E(\Delta)\gamma(\nu)\varphi \in \mathcal{D}_\Delta$ is equivalent to

$$\sigma(\mu_{h,\mathrm{ac}}) = \mathrm{clos}_{\mathrm{ac}}\bigl(\{x \in \Delta : 0 < \mathrm{Im}\,(M(x+i0)\varphi, \varphi) < \infty\}$$

by Lemma 3.6.4. This yields (3.6.7). □

The next corollary gives a necessary and sufficient condition for the absence of absolutely continuous spectrum.

Corollary 3.6.6. *Let A_0 and M be as in Theorem 3.6.5 and let $\Delta \subset \mathbb{R}$ be an open interval such that the condition (3.6.6) is satisfied. Then*

$$\sigma_{\mathrm{ac}}(A_0) \cap \Delta = \emptyset$$

if and only if for all $\varphi \in \mathcal{G}$ and for almost all $x \in \Delta$

$$\mathrm{Im}\,(M(x+i0)\varphi, \varphi) = 0.$$

If S is simple, then the assertion holds for every open interval Δ, including $\Delta = \mathbb{R}$.

Proof. Since $\mathrm{clos}_{\mathrm{ac}}(\mathfrak{B}) = \emptyset$ if and only if $m(\mathfrak{B}) = 0$ for any Borel set $\mathfrak{B} \subset \mathbb{R}$ by Lemma 3.2.5 (i), it is clear that for $\varphi \in \mathcal{G}$

$$\mathrm{clos}_{\mathrm{ac}}\bigl(\{x \in \Delta : 0 < \mathrm{Im}\,(M(x+i0)\varphi, \varphi) < \infty\}\bigr) = \emptyset \tag{3.6.8}$$

if and only if

$$m\bigl(\{x \in \Delta : 0 < \mathrm{Im}\,(M(x+i0)\varphi, \varphi) < \infty\}\bigr) = 0. \tag{3.6.9}$$

Assume first that $\sigma_{\mathrm{ac}}(A_0) \cap \Delta = \emptyset$. Then (3.6.7) yields (3.6.8) for all $\varphi \in \mathcal{G}$, and hence (3.6.9) holds for all $\varphi \in \mathcal{G}$. Moreover, for $h = \gamma(\lambda_0)\varphi$, $\lambda_0 \in \mathbb{C} \setminus \mathbb{R}$, and $x \in \mathbb{R}$

one has

$$\operatorname{Im}(M(x+i0)\varphi,\varphi) = |x-\lambda_0|^2 \operatorname{Im} F_h(x+i0)$$

by Lemma 3.6.4 (with $\Delta = \mathbb{R}$), and according to Theorem 3.1.4 (i) this limit exists and is finite for m-almost all $x \in \mathbb{R}$. Hence, (3.6.9) implies $\operatorname{Im}(M(x+i0)\varphi,\varphi) = 0$ for all $\varphi \in \mathcal{G}$ and m-almost all $x \in \Delta$. For the converse implication assume that $\operatorname{Im}(M(x+i0)\varphi,\varphi) = 0$ for all $\varphi \in \mathcal{G}$ and for m-almost all $x \in \Delta$. Then (3.6.9) and hence also (3.6.8) hold for all $\varphi \in \mathcal{G}$. Thus, (3.6.7) yields $\sigma_{\mathrm{ac}}(A_0) \cap \Delta = \emptyset$. □

The next lemma is of similar nature as Lemma 3.6.4. Here the limits exist for all $x \in \mathbb{R}$ by (3.1.12)–(3.1.13) and Proposition 3.5.1.

Lemma 3.6.7. *Let $\Delta \subset \mathbb{R}$ be an open interval and let $\lambda_0 \in \mathbb{C} \setminus \mathbb{R}$. Then for elements of the form $h = E(\Delta)\gamma(\lambda_0)\varphi$, $\varphi \in \mathcal{G}$, one has*

$$\lim_{y \downarrow 0} y F_h(x+iy) = \begin{cases} |x-\lambda_0|^{-2} \lim_{y\downarrow 0} y(M(x+iy)\varphi,\varphi), & x \in \Delta, \\ 0, & x \notin \overline{\Delta}. \end{cases}$$

Proof. For $h = \gamma(\lambda_0)\varphi$, $\varphi \in \mathcal{G}$, it follows from (3.6.5) and (3.5.1) (cf. (3.5.3) in the proof of Proposition 3.5.1) that

$$\begin{aligned} \lim_{y \downarrow 0} y F_h(x+iy) &= \lim_{y \downarrow 0} y\big(\big(A_0 - (x+iy)\big)^{-1}\gamma(\lambda_0)\varphi, \gamma(\lambda_0)\varphi\big) \\ &= |x-\lambda_0|^{-2} \lim_{y \downarrow 0} y(M(x+iy)\varphi,\varphi) \end{aligned}$$

for all $x \in \mathbb{R}$. If $h = E(\Delta)\gamma(\lambda_0)\varphi$, $\varphi \in \mathcal{G}$, then for $x \in \Delta$ the spectral calculus shows that

$$\begin{aligned} \lim_{y \downarrow 0} y F_h(x+iy) &= \lim_{y \downarrow 0} y\big(\big(A_0 - (x+iy)\big)^{-1}E(\Delta)\gamma(\lambda_0)\varphi, E(\Delta)\gamma(\lambda_0)\varphi\big) \\ &= \lim_{y \downarrow 0} y\big(\big(A_0 - (x+iy)\big)^{-1}\gamma(\lambda_0)\varphi, \gamma(\lambda_0)\varphi\big) \\ &= |x-\lambda_0|^{-2} \lim_{y \downarrow 0} y(M(x+iy)\varphi,\varphi), \end{aligned}$$

while for $x \notin \overline{\Delta}$ one has

$$\begin{aligned} \lim_{y \downarrow 0} y F_h(x+iy) &= \lim_{y \downarrow 0} y\big(\big(A_0 - (x+iy)\big)^{-1}E(\Delta)\gamma(\lambda_0)\varphi, E(\Delta)\gamma(\lambda_0)\varphi\big) \\ &= 0. \end{aligned}$$

This completes the proof. □

Next some inclusions for the singular and singular continuous spectra of A_0 will be shown.

Theorem 3.6.8. *Let S be a closed symmetric relation in $\mathfrak{H}$, let $\{\mathcal{G}, \Gamma_0, \Gamma_1\}$ be a boundary triplet for S^* with $A_0 = \ker \Gamma_0$, and let M and γ be the corresponding Weyl function and γ-field. Let $\Delta \subset \mathbb{R}$ be an open interval and assume that* *the condition*

$$E(\Delta)\mathfrak{H} = \overline{\operatorname{span}}\,\{E(\Delta)\gamma(\nu)\varphi : \nu \in \mathbb{C}\setminus\mathbb{R}, \varphi \in \mathcal{G}\} \tag{3.6.10}$$

is satisfied, where $E(\cdot)$ is the spectral measure of A_0. Then the following statements hold:

(i) *The singular spectrum of A_0 in Δ satisfies*

$$\big(\sigma_s(A_0)\cap\Delta\big) \subset \overline{\bigcup_{\varphi\in\mathcal{G}} \{x\in\Delta : \operatorname{Im}(M(x+i0)\varphi,\varphi)=\infty\}}.$$

(ii) *The singular continuous spectrum of A_0 in Δ, i.e., $\sigma_{sc}(A_0)\cap\Delta$, is contained in the set*

$$\overline{\bigcup_{\varphi\in\mathcal{G}} \operatorname{clos}_c\Big(\big\{x\in\Delta : \operatorname{Im}(M(x+i0)\varphi,\varphi)=\infty, \lim_{y\downarrow 0} y(M(x+iy)\varphi,\varphi)=0\big\}\Big)}.$$

If S is simple, then (i) *and* (ii) *hold for every open interval Δ, including $\Delta=\mathbb{R}$.*

Proof. By assumption, the span of the set

$$\mathcal{D}_\Delta := \{E(\Delta)\gamma(\nu)\varphi : \nu\in\mathbb{C}\setminus\mathbb{R}, \varphi\in\mathcal{G}\}$$

is dense in $E(\Delta)\mathfrak{H}$.

(i) Recall that by Corollary 3.3.6 one has

$$\overline{\sigma_s(A_0)\cap\Delta} = \overline{\bigcup_{h\in\mathcal{D}_\Delta} \sigma(\mu_{h,s})} \tag{3.6.11}$$

and according to Theorem 3.2.6 (ii) (with $\mathfrak{F}$ replaced by $\mathbb{R}$)

$$\sigma(\mu_{h,s}) \subset \overline{\{x\in\mathbb{R} : \operatorname{Im} F_h(x+i0)=\infty\}}.$$

For $h = E(\Delta)\gamma(\nu)\varphi \in \mathcal{D}_\Delta$ this gives, via Lemma 3.6.4,

$$\sigma(\mu_{h,s}) \subset \overline{\{x\in\Delta : \operatorname{Im}(M(x+i0)\varphi,\varphi)=\infty\}}.$$

Hence, the set $\overline{\sigma_s(A_0)\cap\Delta}$ in (3.6.11) is contained in

$$\begin{aligned} &\overline{\bigcup_{h\in\mathcal{D}_\Delta} \overline{\{x\in\Delta : \operatorname{Im}(M(x+i0)\varphi,\varphi)=\infty\}}} \\ &\qquad = \overline{\bigcup_{h\in\mathcal{D}_\Delta} \{x\in\Delta : \operatorname{Im}(M(x+i0)\varphi,\varphi)=\infty\}}, \end{aligned}$$

which yields the assertion in (i).

(ii) Likewise, Corollary 3.3.6 implies

$$\overline{\sigma_{sc}(A_0)\cap\Delta} = \overline{\bigcup_{h\in\mathcal{D}_\Delta} \sigma(\mu_{h,sc})}. \tag{3.6.12}$$

By Theorem 3.2.6 (iii) (again with $\mathfrak{F}$ replaced by $\mathbb{R}$),

$$\sigma(\mu_{h,\mathrm{sc}}) \subset \mathrm{clos}_{\mathrm{c}}\big(\{x \in \mathbb{R} : \mathrm{Im}\, F_h(x+i0) = \infty,\ \lim_{y \downarrow 0} yF_h(x+iy) = 0\}\big),$$

and for $h = E(\Delta)\gamma(\nu)\varphi \in \mathfrak{D}_\Delta$ this gives, via Lemma 3.6.4 and Lemma 3.6.7, that $\sigma(\mu_{h,\mathrm{sc}})$ is contained in

$$\mathrm{clos}_{\mathrm{c}}\big(\{x \in \Delta : \mathrm{Im}\,(M(x+i0)\varphi,\varphi) = \infty,\ \lim_{y \downarrow 0} y(M(x+iy)\varphi,\varphi) = 0\}\big).$$

Hence, the assertion follows from (3.6.12). □

An immediate corollary of the previous theorem and Lemma 3.2.5 (ii) is a sufficient condition for the absence of the singular continuous spectrum in terms of the limit behavior of the function M.

Corollary 3.6.9. *Let A_0 and M be as in Theorem 3.6.8 and let $\Delta \subset \mathbb{R}$ be an open interval such that the condition (3.6.10) is satisfied. Assume that for each $\varphi \in \mathcal{G}$ there exist at most countably many $x \in \Delta$ such that*

$$\mathrm{Im}\,(M(x+iy)\varphi,\varphi) \to \infty \quad \textit{and} \quad y(M(x+iy)\varphi,\varphi) \to 0 \quad \textit{as} \quad y \downarrow 0.$$

Then

$$\sigma_{\mathrm{sc}}(A_0) \cap \Delta = \emptyset.$$

If S is simple, then the assertion holds for every open interval Δ, including $\Delta = \mathbb{R}$.

As a further corollary of the theorems of this section sufficient conditions are provided for the spectrum of A_0 to be purely absolutely continuous or purely singularly continuous, respectively, in some set.

Corollary 3.6.10. *Let A_0 and M be as in Theorem 3.6.5 or Theorem 3.6.8 and let $\Delta \subset \mathbb{R}$ be an open interval such that the condition (3.6.6) or (3.6.10) is satisfied. Assume that for all $\varphi \in \mathcal{G}$ and all $x \in \Delta$*

$$\lim_{y \downarrow 0} yM(x+iy)\varphi = 0. \tag{3.6.13}$$

Then the following statements hold:

(i) *If for each $\varphi \in \mathcal{G}$ there exist at most countably many $x \in \Delta$ such that $\mathrm{Im}\,(M(x+i0)\varphi,\varphi) = \infty$, then $\sigma(A_0) \cap \Delta = \sigma_{\mathrm{ac}}(A_0) \cap \Delta$.*

(ii) *If $\mathrm{Im}\,(M(x+i0)\varphi,\varphi) = 0$ holds for all $\varphi \in \mathcal{G}$ and almost all $x \in \Delta$, then $\sigma(A_0) \cap \Delta = \sigma_{\mathrm{sc}}(A_0) \cap \Delta$.*

If S is simple and Δ is an open interval such that (3.6.13) holds for all $\varphi \in \mathcal{G}$ and all $x \in \Delta$, then (i) and (ii) are satisfied.

Proof. Note first that the assumption (3.6.13) yields $\sigma_p(A_0) \cap \Delta = \emptyset$; this follows immediately from Corollary 3.5.6 and the fact that the condition (3.6.6) or (3.6.10) implies $\sigma_p(S) \cap \Delta = \emptyset$; cf. Proposition 3.4.10 (ii). The assumption in (i) and Corollary 3.6.9 imply $\sigma_{sc}(A_0) \cap \Delta = \emptyset$ and hence $\sigma(A_0) \cap \Delta = \sigma_{ac}(A_0) \cap \Delta$. Similarly, the assumption in (ii) and Corollary 3.6.6 imply $\sigma_{ac}(A_0) \cap \Delta = \emptyset$ and hence $\sigma(A_0) \cap \Delta = \sigma_{sc}(A_0) \cap \Delta$ follows. □

3.7 Limit properties of Weyl functions

Let S be a closed symmetric relation in a Hilbert space $\mathfrak{H}$, let $\{\mathcal{G}, \Gamma_0, \Gamma_1\}$ be a boundary triplet for S^*, let $A_0 = \ker \Gamma_0$, and let M be the corresponding Weyl function. The aim of this section is to relate limit properties of the imaginary part $\operatorname{Im} M$ of the Weyl function with defect elements in $\operatorname{dom} A_0$ and $\operatorname{dom} |A_0|^{\frac{1}{2}}$, and $\operatorname{ran}(A_0 - x)$ and $\operatorname{ran} |A_0 - x|^{\frac{1}{2}}$, $x \in \mathbb{R}$, respectively, where

$$A_0 = A_{0,\mathrm{op}} \,\widehat{\oplus}\, A_{0,\mathrm{mul}} \quad \text{and} \quad |A_0| = |A_{0,\mathrm{op}}| \,\widehat{\oplus}\, A_{0,\mathrm{mul}} \tag{3.7.1}$$

with respect to the usual decomposition $\mathfrak{H} = \mathfrak{H}_{\mathrm{op}} \oplus \mathfrak{H}_{\mathrm{mul}}$. This also leads to necessary and sufficient conditions for S to be a densely defined operator in terms of the Weyl function.

The first result connects limit properties of the Weyl function at ∞ with elements in $\operatorname{dom} A_0 \cap \ker(S^* - \lambda)$ and $\operatorname{dom} |A_0|^{\frac{1}{2}} \cap \ker(S^* - \lambda)$ for $\lambda \in \rho(A_0)$. Although the decomposition $S^* = A_0 \,\widehat{+}\, \widehat{\mathfrak{N}}_\lambda(S^*)$ is direct for all $\lambda \in \rho(A_0)$, it may happen that $\operatorname{dom} A_0 \cap \ker(S^* - \lambda) \neq \{0\}$ if S^* is multivalued. In fact, if $f \in \operatorname{dom} A_0 \cap \ker(S^* - \lambda)$, $f \neq 0$, then $\{f, f'\} \in A_0$ for some f' and hence $\{0, f' - \lambda f\} \in S^*$. Since $\lambda \in \rho(A_0)$, this yields $\operatorname{mul} S^* \neq \{0\}$.

The representation (3.5.12) of the Weyl function M in terms of the extension $A_0 = \ker \Gamma_0$ will now be used; cf. (3.5.8)–(3.5.9). For simplicity one takes $\lambda_0 = i$ in (3.5.12), which leads to the representation

$$\begin{aligned} M(iy) = \operatorname{Re} M(i) &+ iy\,\gamma(i)^*(I - P_{\mathrm{op}})\gamma(i) \\ &+ \gamma(i)^* \iota_{\mathrm{op}}\big[iy + (1 - y^2)(A_{0,\mathrm{op}} - iy)^{-1}\big] P_{\mathrm{op}}\gamma(i) \end{aligned} \tag{3.7.2}$$

for all $y > 0$. The spectral calculus for the self-adjoint operator $A_{0,\mathrm{op}}$ applied to (3.7.2) shows that for $\varphi \in \mathcal{G}$ and $y > 0$

$$\begin{aligned} \operatorname{Im}(M(iy)\varphi, \varphi) &= y\|(I - P_{\mathrm{op}})\gamma(i)\varphi\|^2 \\ &\quad + y \int_{\mathbb{R}} \frac{t^2 + 1}{t^2 + y^2}\, d\big(E_{\mathrm{op}}(t) P_{\mathrm{op}}\gamma(i)\varphi, P_{\mathrm{op}}\gamma(i)\varphi\big). \end{aligned} \tag{3.7.3}$$

Proposition 3.7.1. *Let S be a closed symmetric relation in $\mathfrak{H}$, let $\{\mathcal{G}, \Gamma_0, \Gamma_1\}$ be a boundary triplet for S^*, let $A_0 = \ker \Gamma_0$, and let M and γ be the corresponding Weyl function and γ-field. Then the following statements hold for $\varphi \in \mathcal{G}$:*

(i) $\gamma(\lambda)\varphi \in \operatorname{dom} A_0$ *for some, and hence for all* $\lambda \in \rho(A_0)$ *if and only if*

$$\lim_{y\to+\infty} y \operatorname{Im}(M(iy)\varphi,\varphi) < \infty;$$

(ii) $\gamma(\lambda)\varphi \in \operatorname{dom} |A_0|^{\frac{1}{2}}$ *for some, and hence for all* $\lambda \in \rho(A_0)$ *if and only if*

$$\int_1^\infty \frac{\operatorname{Im}(M(iy)\varphi,\varphi)}{y}\, dy < \infty. \tag{3.7.4}$$

Proof. (i) It suffices to prove the assertion for $\lambda = i$, since by Proposition 2.3.2 (ii)

$$\begin{aligned}\gamma(\lambda) &= \bigl(I + (\lambda - i)(A_0 - \lambda)^{-1}\bigr)\gamma(i),\\ \gamma(i) &= \bigl(I + (i-\lambda)(A_0 - i)^{-1}\bigr)\gamma(\lambda)\end{aligned} \tag{3.7.5}$$

for $\lambda \in \rho(A_0)$ and hence $\gamma(i)\varphi \in \operatorname{dom} A_0$ if and only if $\gamma(\lambda)\varphi \in \operatorname{dom} A_0$. Note first that (3.7.3) yields

$$\begin{aligned} y\operatorname{Im}(M(iy)\varphi,\varphi) &= y^2\|(I - P_{\mathrm{op}})\gamma(i)\varphi\|^2\\ &\quad + \int_{\mathbb{R}} \frac{y^2(t^2+1)}{t^2+y^2}\, d\bigl(E_{\mathrm{op}}(t)P_{\mathrm{op}}\gamma(i)\varphi, P_{\mathrm{op}}\gamma(i)\varphi\bigr).\end{aligned} \tag{3.7.6}$$

It is clear that the left-hand side of (3.7.6) has a finite limit for $y \to +\infty$ if and only if $(I - P_{\mathrm{op}})\gamma(i)\varphi = 0$ and

$$\int_{\mathbb{R}} t^2\, d\bigl(E_{\mathrm{op}}(t)P_{\mathrm{op}}\gamma(i)\varphi, P_{\mathrm{op}}\gamma(i)\varphi\bigr) < \infty,$$

which follows from the monotone convergence theorem. In other words, the left-hand side of (3.7.6) has a finite limit for $y \to +\infty$ if and only if $\gamma(i)\varphi \in \operatorname{dom} A_0$.

(ii) As in the proof of (i), it suffices to verify the assertion for $\lambda = i$. In fact, if $\gamma(i)\varphi \in \operatorname{dom} |A_0|^{\frac{1}{2}}$, then $\gamma(i)\varphi = (|A_0|^{\frac{1}{2}} - \mu)^{-1}g$ for some $\mu \in \mathbb{C}\setminus\mathbb{R}$ and $g \in \mathfrak{H}$. The first identity in (3.7.5) and the functional calculus for the self-adjoint operator $A_{0,\mathrm{op}}$ or self-adjoint relation A_0 (see Section 1.5) show

$$\begin{aligned}\gamma(\lambda)\varphi &= \bigl(I + (\lambda - i)(A_0 - \lambda)^{-1}\bigr)(|A_0|^{\frac{1}{2}} - \mu)^{-1}g\\ &= (|A_0|^{\frac{1}{2}} - \mu)^{-1}\bigl(I + (\lambda - i)(A_0 - \lambda)^{-1}\bigr)g \in \operatorname{dom} |A_0|^{\frac{1}{2}}.\end{aligned}$$

The same argument and the second identity in (3.7.5) show that $\gamma(\lambda)\varphi \in \operatorname{dom} |A_0|^{\frac{1}{2}}$ implies $\gamma(i)\varphi \in \operatorname{dom} |A_0|^{\frac{1}{2}}$.

It follows from (3.7.3) that

$$\begin{aligned}\frac{\operatorname{Im}(M(iy)\varphi,\varphi)}{y} &= \|(I - P_{\mathrm{op}})\gamma(i)\varphi\|^2\\ &\quad + \int_{\mathbb{R}} \frac{t^2+1}{t^2+y^2}\, d\bigl(E_{\mathrm{op}}(t)P_{\mathrm{op}}\gamma(i)\varphi, P_{\mathrm{op}}\gamma(i)\varphi\bigr).\end{aligned}$$

Hence, (3.7.4) holds if and only if $(I - P_{\mathrm{op}})\gamma(i)\varphi = 0$ and

$$\int_1^\infty \left(\int_{\mathbb{R}} \frac{t^2+1}{t^2+y^2}\, d\big(E_{\mathrm{op}}(t)P_{\mathrm{op}}\gamma(i)\varphi, P_{\mathrm{op}}\gamma(i)\varphi\big) \right) dy < \infty.$$

Change the order of integration in the last integral, note that

$$(t^2+1)\int_1^\infty \frac{1}{t^2+y^2}\, dy = (t^2+1)\frac{1}{|t|}\left(\frac{\pi}{2} - \arctan\frac{1}{|t|}\right), \qquad t \neq 0,$$

and observe that for large $|t|$ one has

$$(t^2+1)\frac{1}{|t|}\left(\frac{\pi}{2} - \arctan\frac{1}{|t|}\right) \sim |t|$$

and that on compact subsets of $\mathbb{R}$ the function

$$t \mapsto (t^2+1)\frac{1}{|t|}\left(\frac{\pi}{2} - \arctan\frac{1}{|t|}\right)$$

is bounded. Hence, (3.7.4) holds if and only if $(I - P_{\mathrm{op}})\gamma(i)\varphi = 0$ and

$$\int_{\mathbb{R}} |t|\, d\big(E_{\mathrm{op}}(t)P_{\mathrm{op}}\gamma(i)\varphi, P_{\mathrm{op}}\gamma(i)\varphi\big) < \infty.$$

In other words, (3.7.4) holds if and only if $\gamma(i)\varphi \in \operatorname{dom}|A_0|^{\frac{1}{2}}$. □

The following result is essentially a consequence of Proposition 3.7.1 (i).

Corollary 3.7.2. *Let S, A_0, and M be as in Proposition 3.7.1. Then* $\operatorname{dom} S$ *is dense in* $\overline{\operatorname{dom}}\, A_0$ *if and only if*

$$\lim_{y\to+\infty} y \operatorname{Im}(M(iy)\varphi,\varphi) = \infty \quad \textit{for all} \quad \varphi \in \mathcal{G},\ \varphi \neq 0.$$

Proof. Let $\lambda \in \rho(A_0)$ and note that $f \in (\operatorname{dom} S)^\perp$ if and only if for all $\{h, h'\} \in S$

$$0 = (f,h) = \big(f, (A_0 - \bar\lambda)^{-1}(h' - \bar\lambda h)\big) = \big((A_0-\lambda)^{-1}f, h' - \bar\lambda h\big).$$

Hence, $f \in (\operatorname{dom} S)^\perp$ if and only if $(A_0-\lambda)^{-1}f \in \ker(S^*-\lambda) = \operatorname{ran}\gamma(\lambda)$. Furthermore, $(A_0-\lambda)^{-1}f \neq 0$ if and only if $f \notin \operatorname{mul} A_0 = (\overline{\operatorname{dom}}\, A_0)^\perp$.

Now assume that $\operatorname{dom} S$ is not dense in $\overline{\operatorname{dom}}\, A_0$. Then there exists a nontrivial $f \in \overline{\operatorname{dom}}\, A_0$ such that $f \in (\operatorname{dom} S)^\perp$, and hence

$$(A_0-\lambda)^{-1}f \in \ker(S^*-\lambda).$$

Since $f \in \overline{\operatorname{dom}}\, A_0$ it follows that $(A_0-\lambda)^{-1}f = \gamma(\lambda)\varphi$ for a nontrivial $\varphi \in \mathcal{G}$. This means $\gamma(\lambda)\varphi \in \operatorname{dom} A_0$, and hence

$$\lim_{y\to+\infty} y \operatorname{Im}(M(iy)\varphi,\varphi) < \infty \tag{3.7.7}$$

by Proposition 3.7.1 (i). Conversely, if (3.7.7) holds for some nontrivial $\varphi \in \mathcal{G}$, then by Proposition 3.7.1 (i) it follows that $\gamma(\lambda)\varphi \in \overline{\operatorname{dom} A_0}$. Hence, there exists a nontrivial $f \in \overline{\operatorname{dom} A_0}$ such that $\gamma(\lambda)\varphi = (A_0 - \lambda)^{-1}f$. Therefore, one sees that $f \in (\operatorname{dom} S)^\perp$ and hence $\operatorname{dom} S$ is not dense in $\overline{\operatorname{dom} A_0}$. □

Corollary 3.7.3. *Let S, A_0, and M be as in Proposition 3.7.1. Then S is a densely defined operator if and only if the following conditions hold:*

(i) $\displaystyle\lim_{y\to+\infty} \frac{1}{iy}(M(iy)\varphi,\varphi) = 0$ *for all* $\varphi \in \mathcal{G}$;

(ii) $\lim_{y\to+\infty} y \operatorname{Im}(M(iy)\varphi,\varphi) = \infty$ *for all* $\varphi \in \mathcal{G}$, $\varphi \neq 0$.

In this case, S^* is an operator and all intermediate extensions of S are operators.

Proof. Note that Proposition 3.5.7 and the fact that $\gamma(\lambda_0)^*(I - P_{\mathrm{op}})\gamma(\lambda_0)$ in (3.5.13) is a nonnegative operator in $\mathcal{G}$ show that condition (i) is equivalent to the condition

$$\lim_{y\to+\infty} \frac{1}{iy} M(iy)\varphi = 0, \qquad \varphi \in \mathcal{G}.$$

By (3.5.18), this condition is necessary and sufficient for A_0 to be an operator, which is the case if and only if $\overline{\operatorname{dom} A_0} = \mathfrak{H}$. Moreover, according to Corollary 3.7.2, the condition (ii) is necessary and sufficient for the equality $\overline{\operatorname{dom} S} = \overline{\operatorname{dom} A_0}$ to hold. Therefore, $\overline{\operatorname{dom} S} = \mathfrak{H}$ if and only if conditions (i) and (ii) hold. □

In the next result, which is parallel to Proposition 3.7.1, the limit properties of the Weyl function at $x \in \mathbb{R}$ will be connected with elements in

$$\ker(S^* - \lambda) \cap \operatorname{ran}(A_0 - x) \quad \text{and} \quad \ker(S^* - \lambda) \cap \operatorname{ran}|A_0 - x|^{\frac{1}{2}}.$$

For this reason the representation (3.5.1) expressing the Weyl function M in terms of the self-adjoint relation $A_0 = \ker \Gamma_0$ will be used. For simplicity one takes $\lambda_0 \in \mathbb{C} \setminus \mathbb{R}$ such that $\operatorname{Re} \lambda_0 = x$ in (3.5.1), which leads to

$$\begin{aligned} M(x+iy) &= \operatorname{Re} M(\lambda_0) \\ &\quad + \gamma(\lambda_0)^*\big[iy + (|\operatorname{Im}\lambda_0|^2 - y^2)\big(A_0 - (x+iy)\big)^{-1}\big]\gamma(\lambda_0). \end{aligned} \tag{3.7.8}$$

It follows by means of the spectral calculus applied to (3.7.8) that for $x \in \mathbb{R}$ and $\varphi \in \mathcal{G}$ one has

$$\begin{aligned} \frac{\operatorname{Im}(M(x+iy)\varphi,\varphi)}{y} &= \|\gamma(\lambda_0)\varphi\|^2 \\ &\quad + \big(|\operatorname{Im}\lambda_0|^2 - y^2\big)\int_{\mathbb{R}} \frac{1}{(t-x)^2+y^2}\, d(E(t)\gamma(\lambda_0)\varphi,\gamma(\lambda_0)\varphi). \end{aligned} \tag{3.7.9}$$

Proposition 3.7.4. *Let S be a closed symmetric relation in $\mathfrak{H}$, let $\{\mathcal{G},\Gamma_0,\Gamma_1\}$ be a boundary triplet for S^*, let $A_0 = \ker\Gamma_0$ be decomposed as in (3.7.1), and let M and γ be the corresponding Weyl function and γ-field. Then the following statements hold for $x \in \mathbb{R}$ and $\varphi \in \mathcal{G}$:*

(i) $\gamma(\lambda)\varphi \in \operatorname{ran}(A_0 - x)$ *for some, and hence for all* $\lambda \in \rho(A_0)$ *if and only if*

$$\lim_{y \downarrow 0} \frac{\operatorname{Im}(M(x+iy)\varphi, \varphi)}{y} < \infty; \tag{3.7.10}$$

(ii) $P_{\rm op}\gamma(\lambda)\varphi \in \operatorname{ran}|A_{0,\rm op} - x|^{\frac{1}{2}}$ *for some, and hence for all* $\lambda \in \rho(A_0)$ *if and only if*

$$\int_0^1 \frac{\operatorname{Im}(M(x+iy)\varphi, \varphi)}{y}\, dy < \infty. \tag{3.7.11}$$

Proof. (i) It will first be shown that for $\lambda, \lambda_0 \in \rho(A_0)$ one has $\gamma(\lambda)\varphi \in \operatorname{ran}(A_0 - x)$ if and only if $\gamma(\lambda_0)\varphi \in \operatorname{ran}(A_0 - x)$. Assume that $\gamma(\lambda_0)\varphi \in \operatorname{ran}(A_0 - x)$. Then there is $\{f, f'\} \in A_0$ such that $\gamma(\lambda_0)\varphi = f' - xf$. As

$$\{f' - xf, (A_0 - \lambda)^{-1}(f' - xf)\} \in (A_0 - \lambda)^{-1},$$

it follows that

$$\{(A_0 - \lambda)^{-1}(f' - xf), f' - xf + (\lambda - x)(A_0 - \lambda)^{-1}(f' - xf)\} \in A_0 - x.$$

Hence, $f' - xf + (\lambda - x)(A_0 - \lambda)^{-1}(f' - xf) \in \operatorname{ran}(A_0 - x)$ and

$$(A_0 - \lambda)^{-1}(f' - xf) \in \operatorname{ran}(A_0 - x).$$

From the identity $\gamma(\lambda) = (I + (\lambda - \lambda_0)(A_0 - \lambda)^{-1})\gamma(\lambda_0)$, established in Proposition 2.3.2 (ii), one finds that

$$\gamma(\lambda)\varphi = f' - xf + (\lambda - \lambda_0)(A_0 - \lambda)^{-1}(f' - xf) \in \operatorname{ran}(A_0 - x).$$

Thus, $\gamma(\lambda_0)\varphi \in \operatorname{ran}(A_0 - x)$ implies that $\gamma(\lambda)\varphi \in \operatorname{ran}(A_0 - x)$. Since λ_0 and λ in the above argument can be interchanged, it is clear that $\gamma(\lambda)\varphi \in \operatorname{ran}(A_0 - x)$ if and only if $\gamma(\lambda_0)\varphi \in \operatorname{ran}(A_0 - x)$.

To verify the remaining assertion in (i) with $\lambda = \lambda_0$, note first that the limit as $y \downarrow 0$ in (3.7.10) is finite if and only if the limit of the integral in the second term in (3.7.9) is finite. An application of the monotone convergence theorem shows that the limit as $y \downarrow 0$ of the integral in the second term in (3.7.9) is finite if and only if

$$\int_{\mathbb{R}} \frac{1}{(t-x)^2}\, d(E(t)\gamma(\lambda_0)\varphi, \gamma(\lambda_0)\varphi) < \infty,$$

that is, if and only if

$$\int_{\mathbb{R}} \frac{1}{(t-x)^2}\, d(E_{\rm op}(t)P_{\rm op}\gamma(\lambda_0)\varphi, P_{\rm op}\gamma(\lambda_0)\varphi) < \infty,$$

where the definition of the spectral measure $E(\cdot)$ of A_0 via the spectral measure $E_{\rm op}(\cdot)$ of $A_{0,\rm op}$ was used. Therefore, the limit as $y \downarrow 0$ in (3.7.10) is finite if and only if $P_{\rm op}\,\gamma(\lambda_0)\varphi \in \operatorname{dom}(A_{0,\rm op} - x)^{-1} = \operatorname{ran}(A_{0,\rm op} - x)$, that is, if and only if $\gamma(\lambda_0)\varphi \in \operatorname{ran}(A_0 - x)$.

(ii) As in (i), it will first be shown that $P_{\rm op}\gamma(\lambda)\varphi \in \operatorname{ran}|A_{0,\rm op} - x|^{\frac{1}{2}}$ if and only if $P_{\rm op}\gamma(\lambda_0)\varphi \in \operatorname{ran}|A_{0,\rm op} - x|^{\frac{1}{2}}$ for $\lambda, \lambda_0 \in \rho(A_0)$. Assume that

$$P_{\rm op}\gamma(\lambda_0)\varphi = |A_{0,\rm op} - x|^{\frac{1}{2}} f$$

for some $f \in \operatorname{dom}|A_{0,\rm op} - x|^{\frac{1}{2}}$. It follows from the functional calculus for unbounded self-adjoint operators that

$$\begin{aligned}\overline{(A_{0,\rm op} - \lambda)^{-1}|A_{0,\rm op} - x|^{\frac{1}{2}}} &= \overline{|A_{0,\rm op} - x|^{\frac{1}{2}}(A_{0,\rm op} - \lambda)^{-1}}\\ &= |A_{0,\rm op} - x|^{\frac{1}{2}}(A_{0,\rm op} - \lambda)^{-1}\end{aligned}$$

and hence, since $\gamma(\lambda) = (I + (\lambda - \lambda_0)(A_0 - \lambda)^{-1})\gamma(\lambda_0)$, one has that

$$\begin{aligned}P_{\rm op}\gamma(\lambda)\varphi &= P_{\rm op}\gamma(\lambda_0)\varphi + (\lambda - \lambda_0)(A_{0,\rm op} - \lambda)^{-1}P_{\rm op}\gamma(\lambda_0)\varphi\\ &= |A_{0,\rm op} - x|^{\frac{1}{2}} f + (\lambda - \lambda_0)(A_{0,\rm op} - \lambda)^{-1}|A_{0,\rm op} - x|^{\frac{1}{2}} f\\ &= |A_{0,\rm op} - x|^{\frac{1}{2}} f + (\lambda - \lambda_0)|A_{0,\rm op} - x|^{\frac{1}{2}}(A_{0,\rm op} - \lambda)^{-1} f,\end{aligned}$$

that is, $P_{\rm op}\gamma(\lambda)\varphi \in \operatorname{ran}|A_{0,\rm op} - x|^{\frac{1}{2}}$. Thus, $P_{\rm op}\gamma(\lambda_0)\varphi \in \operatorname{ran}|A_{0,\rm op} - x|^{\frac{1}{2}}$ implies $P_{\rm op}\gamma(\lambda)\varphi \in \operatorname{ran}|A_{0,\rm op} - x|^{\frac{1}{2}}$. Since λ_0 and λ in the above argument can be interchanged, it is clear that $P_{\rm op}\gamma(\lambda_0)\varphi \in \operatorname{ran}|A_{0,\rm op} - x|^{\frac{1}{2}}$ holds if and only if $P_{\rm op}\gamma(\lambda)\varphi \in \operatorname{ran}|A_{0,\rm op} - x|^{\frac{1}{2}}$ holds.

To verify the remaining assertion in (ii), it is convenient to fix $\lambda = \lambda_0 \in \mathbb{C} \setminus \mathbb{R}$ such that $|\operatorname{Im}\lambda_0| > 1$. One then concludes from (3.7.9) that the integral in (3.7.11) converges if and only if the integral

$$\int_0^1 \left(\int_{\mathbb{R}} \frac{1}{(t - x)^2 + y^2}\, d(E(t)\gamma(\lambda_0)\varphi, \gamma(\lambda_0)\varphi) \right) dy$$

converges. Changing the order of integration in the last integral and observing that

$$\int_0^1 \frac{1}{(t - x)^2 + y^2}\, dy = \frac{1}{|t - x|} \arctan \frac{1}{|t - x|}, \qquad t \neq x,$$

one sees that the integral in (3.7.11) converges if and only if

$$\int_{\mathbb{R}} \frac{1}{|t - x|} \arctan \frac{1}{|t - x|}\, d(E(t)\gamma(\lambda_0)\varphi, \gamma(\lambda_0)\varphi) < \infty. \tag{3.7.12}$$

Since the integrand in (3.7.12) is bounded on $\mathbb{R} \setminus (x - 1, x + 1)$, it follows that the integral in (3.7.11) converges if and only

$$\int_{x-1}^{x+1} \frac{1}{|t - x|}\, d(E(t)\gamma(\lambda_0)\varphi, \gamma(\lambda_0)\varphi) < \infty,$$

which is equivalent to

$$\int_{\mathbb{R}} \frac{1}{|t-x|}\, d(E(t)\gamma(\lambda_0)\varphi, \gamma(\lambda_0)\varphi) < \infty$$

and to

$$\int_{\mathbb{R}} \frac{1}{|t-x|}\, d(E_{\rm op}(t)P_{\rm op}\gamma(\lambda_0)\varphi, P_{\rm op}\gamma(\lambda_0)\varphi) < \infty.$$

Therefore, (3.7.11) holds if and only if

$$P_{\rm op}\gamma(\lambda_0)\varphi \in \operatorname{dom}|A_{0,{\rm op}} - x|^{-\frac{1}{2}} = \operatorname{ran}|A_{0,{\rm op}} - x|^{\frac{1}{2}},$$

that is, if and only if $\gamma(\lambda_0)\varphi \in \operatorname{ran}|A_0 - x|^{\frac{1}{2}}$. □

3.8 Spectra and local minimality for self-adjoint extensions

In this section the results on eigenvalues, eigenspaces, continuous, absolutely continuous and singular continuous spectra from Section 3.5 and Section 3.6 will be explicitly formulated for arbitrary self-adjoint extensions of a symmetric relation.

Let S be a closed symmetric relation in $\mathfrak{H}$ and let $\{\mathcal{G}, \Gamma_0, \Gamma_1\}$ be a boundary triplet for S^* with γ-field γ and Weyl function M. Consider a self-adjoint extension

$$A_\Theta = \{\widehat{f} \in S^* : \Gamma\widehat{f} \in \Theta\} = \ker(\Gamma_1 - \Theta\Gamma_0) \tag{3.8.1}$$

of S in $\mathfrak{H}$, where $\Theta = \Theta^*$ is a self-adjoint relation in $\mathcal{G}$. Recall from Corollary 1.10.9 that there exist operators $\mathcal{A}, \mathcal{B} \in \mathbf{B}(\mathcal{G})$ with the properties

$$\mathcal{A}^*\mathcal{B} = \mathcal{B}^*\mathcal{A}, \quad \mathcal{A}\mathcal{B}^* = \mathcal{B}\mathcal{A}^*, \quad \mathcal{A}^*\mathcal{A} + \mathcal{B}^*\mathcal{B} = I = \mathcal{A}\mathcal{A}^* + \mathcal{B}\mathcal{B}^*,$$

such that

$$\Theta = \{\{\mathcal{A}\varphi, \mathcal{B}\varphi\} : \varphi \in \mathcal{G}\} = \{\{\psi, \psi'\} \in \mathcal{G}^2 : \mathcal{A}^*\psi' = \mathcal{B}^*\psi\}.$$

According to Section 2.2, the self-adjoint extensions A_Θ in (3.8.1) can also be written in the form

$$A_\Theta = \{\widehat{f} \in S^* : \mathcal{A}^*\Gamma_1\widehat{f} = \mathcal{B}^*\Gamma_0\widehat{f}\}.$$

In order to describe the spectrum of A_Θ consider the boundary triplet $\{\mathcal{G}, \Gamma_0', \Gamma_1'\}$, where

$$\begin{pmatrix} \Gamma_0' \\ \Gamma_1' \end{pmatrix} = \begin{pmatrix} \mathcal{B}^* & -\mathcal{A}^* \\ \mathcal{A}^* & \mathcal{B}^* \end{pmatrix} \begin{pmatrix} \Gamma_0 \\ \Gamma_1 \end{pmatrix}; \tag{3.8.2}$$

cf. Corollary 2.5.11. Then one has

$$A_\Theta = \ker \Gamma_0', \tag{3.8.3}$$

and the corresponding Weyl function and γ-field will be denoted by M_Θ and γ_Θ. For $\lambda \in \rho(A_\Theta) \cap \rho(A_0)$ they are given by

$$M_\Theta(\lambda) = \big(\mathcal{A}^* + \mathcal{B}^* M(\lambda)\big)\big(\mathcal{B}^* - \mathcal{A}^* M(\lambda)\big)^{-1} \tag{3.8.4}$$

and

$$\gamma_\Theta(\lambda) = \gamma(\lambda)\big(\mathcal{B}^* - \mathcal{A}^* M(\lambda)\big)^{-1},$$

respectively; cf. (2.5.17) and (2.5.18). From (3.8.3) it is clear that the spectrum of A_Θ can be described by means of the Weyl function M_Θ. Therefore, the earlier results expressing the spectrum of A_0 in terms of the Weyl function M (and the γ-field γ) can now be simply translated to the present context. The main results will be listed below; it is left to the reader to formulate analogs of the results in Section 3.7 in the present setting.

First the analogs of Theorem 3.5.5 and Theorem 3.5.10 will be described. For this purpose define the operators $\mathcal{R}_x^\Theta$, $x \in \mathbb{R}$, and $\mathcal{R}_\infty^\Theta$ similar to Definition 3.5.2 and Definition 3.5.8:

$$\mathcal{R}_x^\Theta \varphi = \lim_{y \downarrow 0} iy M_\Theta(x+iy)\varphi, \qquad \varphi \in \mathcal{G},$$

and

$$\mathcal{R}_\infty^\Theta \varphi = \lim_{y \to +\infty} \frac{1}{iy} M_\Theta(iy)\varphi, \qquad \varphi \in \mathcal{G}.$$

As in Section 3.5, one has that $\mathcal{R}_x^\Theta, \mathcal{R}_\infty^\Theta \in \mathbf{B}(\mathcal{G})$. In terms of the boundary triplet $\{\mathcal{G}, \Gamma_0', \Gamma_1'\}$ in (3.8.2) and the corresponding Weyl function M_Θ in (3.8.4), Theorem 3.5.5 and Corollary 3.5.6 read as follows.

Corollary 3.8.1. *Let S, A_Θ, and M_Θ be as above and let $x \in \mathbb{R}$. Then the mapping*

$$\tau : \widehat{\mathfrak{N}}_x(A_\Theta) \ominus \widehat{\mathfrak{N}}_x(S) \to \overline{\operatorname{ran}}\, \mathcal{R}_x^\Theta, \quad \widehat{f} \mapsto \mathcal{A}^* \Gamma_0 \widehat{f} + \mathcal{B}^* \Gamma_1 \widehat{f},$$

is an isomorphism. In particular,

$$x \in \sigma_{\rm p}(A_\Theta) \text{ and } \widehat{\mathfrak{N}}_x(A_\Theta) \ominus \widehat{\mathfrak{N}}_x(S) \neq \{0\} \quad \Leftrightarrow \quad \mathcal{R}_x^\Theta \neq 0,$$

and if $x \notin \sigma_{\rm p}(S)$, then $x \in \sigma_{\rm p}(A_\Theta)$ if and only if $\mathcal{R}_x^\Theta \neq 0$.

Similarly, Theorem 3.5.10 and Corollary 3.5.11 take the following form.

Corollary 3.8.2. *Let S, A_Θ, and M_Θ be as above. Then the mapping*

$$\tau : \widehat{\mathfrak{N}}_\infty(A_\Theta) \ominus \widehat{\mathfrak{N}}_\infty(S) \to \overline{\operatorname{ran}}\, \mathcal{R}_\infty^\Theta, \quad \widehat{f} \mapsto \mathcal{A}^* \Gamma_0 \widehat{f} + \mathcal{B}^* \Gamma_1 \widehat{f},$$

is an isomorphism. In particular,

$$\operatorname{mul} A_\Theta \ominus \operatorname{mul} S \neq \{0\} \quad \Leftrightarrow \quad \mathcal{R}_\infty^\Theta \neq 0,$$

and if $\operatorname{mul} S = \{0\}$, then A_Θ is an operator if and only if $\mathcal{R}_\infty^\Theta = 0$.

For the next results the local simplicity condition appearing in many of the results in Section 3.6 has to be reformulated with respect to A_Θ. According to Definition 3.4.9, the closed symmetric relation S is simple with respect to $\Delta \subset \mathbb{R}$ and the self-adjoint extension A_Θ if

$$E_\Theta(\Delta)\mathfrak{H} = \overline{\operatorname{span}}\,\{E_\Theta(\Delta)\gamma_\Theta(\nu)\varphi : \nu \in \mathbb{C}\setminus\mathbb{R}, \varphi \in \mathcal{G}\}, \tag{3.8.5}$$

where $E_\Theta(\cdot)$ is the spectral measure of A_Θ.

Then Theorem 3.6.1 yields the following statement.

Corollary 3.8.3. *Let S, A_Θ, and M_Θ be as above, let $\Delta \subset \mathbb{R}$ be an open interval, and assume that the local simplicity condition* (3.8.5) *is satisfied. Then the following statements hold for each $x \in \Delta$:*

(i) *$x \in \rho(A_\Theta)$ if and only if M_Θ can be continued analytically to x;*

(ii) *$x \in \sigma_c(A_\Theta)$ if and only if $\mathcal{R}_x^\Theta = 0$ and M_Θ cannot be continued analytically to x;*

(iii) *x is an eigenvalue of A_Θ if and only if $\mathcal{R}_x^\Theta \neq 0$;*

(iv) *x is an isolated eigenvalue of A_Θ if and only if x is a pole* (*of first order*) *of M_Θ; in this case $\mathcal{R}_x^\Theta$ is the residue of M_Θ at x.*

If S is simple, then the statements (i)–(iv) *hold for all $x \in \mathbb{R}$.*

Finally, the corresponding results for the absolutely continuous, singular, and singular continuous spectra will be formulated; it is left to the reader to state the analogs of Corollaries 3.6.6, 3.6.9, and 3.6.10.

In the present situation Theorem 3.6.5 reads as follows.

Corollary 3.8.4. *Let S, A_Θ, and M_Θ be as above, let $\Delta \subset \mathbb{R}$ be an open interval, and assume that the local simplicity condition* (3.8.5) *is satisfied. Then the absolutely continuous spectrum of A_Θ in Δ is given by*

$$\overline{\sigma_{\rm ac}(A_\Theta) \cap \Delta} = \overline{\bigcup_{\varphi\in\mathcal{G}} \operatorname{clos}_{\rm ac}\big(\{x \in \Delta : 0 < \operatorname{Im}(M_\Theta(x+i0)\varphi,\varphi) < \infty\}\big)}. \tag{3.8.6}$$

If S is simple, then (3.8.6) *holds for every open interval Δ, including $\Delta = \mathbb{R}$.*

For the singular and singular continuous spectra one obtains the following version of Theorem 3.6.8.

Corollary 3.8.5. *Let S, A_Θ, and M_Θ be as above, let $\Delta \subset \mathbb{R}$ be an open interval, and assume that the local simplicity condition* (3.8.5) *is satisfied. Then the following statements hold:*

(i) *The singular spectrum of A_Θ in Δ satisfies*

$$\big(\sigma_{\rm s}(A_\Theta) \cap \Delta\big) \subset \overline{\bigcup_{\varphi\in\mathcal{G}} \{x \in \Delta : \operatorname{Im}(M_\Theta(x+i0)\varphi,\varphi) = \infty\}}.$$

(ii) *The singular continuous spectrum of* A_Θ *in* Δ, $\sigma_{\rm sc}(A_\Theta)\cap\Delta$, *is contained in the set*

$$\overline{\bigcup_{\varphi\in\mathcal{G}} \operatorname{clos}_c\Bigl(\bigl\{x\in\Delta : \operatorname{Im}\,(M_\Theta(x+i0)\varphi,\varphi)=\infty,\ \lim_{y\downarrow 0} y(M_\Theta(x+iy)\varphi,\varphi)=0\bigr\}\Bigr)}.$$

If S is simple, then (i) *and* (ii) *hold for every open interval* Δ, *including* $\Delta=\mathbb{R}$.

Finally, the special case where the self-adjoint relation Θ in (3.8.1) is a bounded self-adjoint operator will be briefly discussed. In this situation there is a more natural choice of the transformed boundary triplet $\{\mathcal{G},\Gamma_0',\Gamma_1'\}$ above. In fact, if S is a closed symmetric relation, $\{\mathcal{G},\Gamma_0,\Gamma_1\}$ is a boundary triplet for S^* with γ-field γ and Weyl function M, and $\Theta\in\mathbf{B}(\mathcal{G})$ is self-adjoint, then, by Corollary 2.5.7, the mappings

$$\Gamma_0' = \Gamma_1 - \Theta\Gamma_0 \quad\text{and}\quad \Gamma_1' = -\Gamma_0$$

lead to a boundary triplet $\{\mathcal{G},\Gamma_0',\Gamma_1'\}$ for S^* such that

$$\ker\Gamma_0' = \ker\bigl(\Gamma_1-\Theta\Gamma_0\bigr) = A_\Theta.$$

For $\lambda\in\rho(A_0)\cap\rho(A_\Theta)$ the corresponding γ-field γ_Θ and the Weyl function M_Θ are given by

$$\gamma_\Theta(\lambda) = -\gamma(\lambda)\bigl(\Theta-M(\lambda)\bigr)^{-1} \quad\text{and}\quad M_\Theta(\lambda)=\bigl(\Theta-M(\lambda)\bigr)^{-1}, \tag{3.8.7}$$

respectively. Then the above results in Corollaries 3.8.1–3.8.5 remain valid with the function M_Θ in (3.8.7) and the mapping $\widehat f\mapsto \mathcal{A}^*\Gamma_0\widehat f+\mathcal{B}^*\Gamma_1\widehat f$ in Corollary 3.8.1 and Corollary 3.8.2 replaced by $\widehat f\mapsto -\Gamma_0\widehat f$.

Chapter 4

Operator Models for Nevanlinna Functions

The classes of Weyl functions and more generally of Nevanlinna functions will be studied from the point of view of reproducing kernel Hilbert spaces. It is clear from Chapter 2 that every Weyl function is a uniformly strict Nevanlinna function and it is one of the main objectives here to show that also the converse is true: every uniformly strict Nevanlinna function is a Weyl function. The model space is built as a reproducing kernel Hilbert space of holomorphic functions. A brief introduction to reproducing kernel Hilbert spaces is given in Section 4.1. Using the Nevanlinna kernel in Section 4.2 multiplication operators by the independent variable are studied and a boundary triplet whose Weyl function is the original Nevanlinna function is constructed. For scalar Nevanlinna functions an alternative model in an L^2-space is given in Section 4.3. The uniqueness of these constructions will also be discussed in detail. An extension of the operator model in Section 4.2 to Nevanlinna functions which are not necessarily uniformly strict, and to Nevanlinna families is provided in Section 4.4. This also includes a discussion of generalized resolvents, and as a byproduct one obtains the Sz.-Nagy dilation theorem. The connection with extension theory is given via the compressed resolvents of self-adjoint relations in the Kreĭn–Naĭmark formula in Section 4.5. It will be shown that for every Nevanlinna family there is a self-adjoint exit space extension whose compressed resolvent is parametrized by the Nevanlinna family. Closely connected is the discussion about the orthogonal coupling of two boundary triplets in Section 4.6, which also complements the considerations in Section 2.7.

4.1 Reproducing kernel Hilbert spaces

The following discussion of reproducing kernel Hilbert spaces is focused on what is needed in this text. Within these bounds there is a complete treatment for the reader's convenience. In the first definition one restricts attention to open

J. Behrndt et al., *Boundary Value Problems, Weyl Functions, and Differential Operators*, Monographs in Mathematics 108,
https://doi.org/10.1007/978-3-030-36714-5_5

sets, as the emphasis will be on reproducing kernel Hilbert spaces of holomorphic functions.

Definition 4.1.1. Let $\Omega \subset \mathbb{C}$ be an open set and let $\mathcal{G}$ be a Hilbert space. A mapping

$$\mathsf{K}(\cdot,\cdot) : \Omega \times \Omega \to \mathbf{B}(\mathcal{G}) \tag{4.1.1}$$

is called a $\mathbf{B}(\mathcal{G})$-valued *kernel* on Ω. The kernel $\mathsf{K}(\cdot,\cdot)$, the kernel K for short, is said to be

(i) *nonnegative*, if for any finite set of points $\lambda_1, \dots, \lambda_n \in \Omega$ and any choice of vectors $\varphi_1, \dots, \varphi_n \in \mathcal{G}$ the $n \times n$ matrix

$$\big((\mathsf{K}(\lambda_i,\lambda_j)\varphi_j,\varphi_i)_{\mathcal{G}}\big)_{i,j=1}^n$$

is nonnegative;

(ii) *symmetric*, if $\mathsf{K}(\lambda,\mu)^* = \mathsf{K}(\mu,\lambda)$ for all $\lambda,\mu \in \Omega$;

(iii) *holomorphic* on Ω, if the mapping $\lambda \mapsto \mathsf{K}(\lambda,\mu)$ is holomorphic on Ω for each $\mu \in \Omega$;

(iv) *uniformly bounded on compact subsets of* Ω, if for any compact set $K \subset \Omega$ one has $\sup_{\lambda\in K} \|\mathsf{K}(\lambda,\lambda)\| < \infty$.

Note that the first two items in this definition are not independent. Nonnegativity is the stronger condition.

Lemma 4.1.2. *Let $\mathsf{K}(\cdot,\cdot)$ be a $\mathbf{B}(\mathcal{G})$-valued kernel on Ω as in (4.1.1). If $\mathsf{K}(\cdot,\cdot)$ is nonnegative, then $\mathsf{K}(\cdot,\cdot)$ is symmetric.*

Proof. Let $\varphi,\psi \in \mathcal{G}$ and $\lambda,\mu \in \Omega$. According to (i), the 2×2 matrix

$$\begin{pmatrix} (\mathsf{K}(\lambda,\lambda)\varphi,\varphi)_{\mathcal{G}} & (\mathsf{K}(\lambda,\mu)\psi,\varphi)_{\mathcal{G}} \\ (\mathsf{K}(\mu,\lambda)\varphi,\psi)_{\mathcal{G}} & (\mathsf{K}(\mu,\mu)\psi,\psi)_{\mathcal{G}} \end{pmatrix}$$

is nonnegative, and, hence hermitian. In particular, this implies that

$$(\mathsf{K}(\lambda,\mu)\psi,\varphi)_{\mathcal{G}} = \overline{(\mathsf{K}(\mu,\lambda)\varphi,\psi)_{\mathcal{G}}} = (\psi,\mathsf{K}(\mu,\lambda)\varphi)_{\mathcal{G}}$$

for all $\varphi,\psi \in \mathcal{G}$. This gives $\mathsf{K}(\lambda,\mu)^* = \mathsf{K}(\mu,\lambda)$, so that the kernel $\mathsf{K}(\cdot,\cdot)$ is symmetric. □

The kernels described in Definition 4.1.1 form the basis of the theory of reproducing kernel Hilbert spaces. They arise naturally in the following context. Let $\Omega \subset \mathbb{C}$ be an open set and let $(\mathfrak{H}, \langle\cdot,\cdot\rangle)$ be a Hilbert space of functions defined on Ω with values in a Hilbert space $\mathcal{G}$. The Hilbert space $\mathfrak{H}$ is called a *reproducing kernel Hilbert space* if for all $\mu \in \Omega$ the operation of point evaluation

$$f \in \mathfrak{H} \mapsto f(\mu) \in \mathcal{G}$$

is bounded. In other words, for each $\mu \in \Omega$ the linear operator $E(\mu) : \mathfrak{H} \to \mathcal{G}$, defined by $E(\mu)f = f(\mu)$, belongs to $\mathbf{B}(\mathfrak{H}, \mathcal{G})$.

In the next theorem a kernel is related to a Hilbert space of functions in which point evaluation is bounded.

Theorem 4.1.3. *Let $\mathcal{G}$ be a Hilbert space and assume that $(\mathfrak{H}, \langle \cdot, \cdot \rangle)$ is a Hilbert space of $\mathcal{G}$-valued functions on an open set $\Omega \subset \mathbb{C}$ such that point evaluation is bounded for all $\mu \in \Omega$. Define the corresponding kernel $\mathsf{K}(\cdot, \cdot)$ by*

$$\mathsf{K}(\lambda, \mu) = E(\lambda)E(\mu)^* \in \mathbf{B}(\mathcal{G}), \quad \lambda, \mu \in \Omega.$$

Then the following statements hold:

(i) *For $f \in \mathfrak{H}$ one has the reproducing kernel property*

$$\langle f, \mathsf{K}(\cdot, \mu)\varphi \rangle = (f(\mu), \varphi)_{\mathcal{G}}, \quad \varphi \in \mathcal{G}, \quad \mu \in \Omega. \tag{4.1.2}$$

(ii) *The identity*

$$\langle \mathsf{K}(\cdot, \nu)\eta, \mathsf{K}(\cdot, \mu)\varphi \rangle = (\mathsf{K}(\mu, \nu)\eta, \varphi)_{\mathcal{G}}$$

is valid for all $\nu, \mu \in \Omega$ and $\eta, \varphi \in \mathcal{G}$.

(iii) *$\mathsf{K}(\cdot, \cdot)$ is nonnegative and symmetric.*

(iv) *$\mathfrak{H} = \overline{\operatorname{span}}\,\{\mathsf{K}(\cdot, \mu)\varphi : \mu \in \Omega, \ \varphi \in \mathcal{G}\}$.*

(v) *If the $\mathcal{G}$-valued functions in $\mathfrak{H}$ are holomorphic on Ω, then $\mathsf{K}(\cdot, \cdot)$ is holomorphic and uniformly bounded on compact subsets of Ω.*

Proof. (i) & (ii) Note that for $f \in \mathfrak{H}$, $\varphi \in \mathcal{G}$, and $\mu \in \Omega$ one has

$$(f(\mu), \varphi)_{\mathcal{G}} = (E(\mu)f, \varphi)_{\mathcal{G}} = \langle f, E(\mu)^*\varphi \rangle$$

and observe that $E(\mu)^*\varphi$ is a function in $\mathfrak{H}$ whose value at $\lambda \in \Omega$ is given by

$$(E(\mu)^*\varphi)(\lambda) = E(\lambda)E(\mu)^*\varphi = \mathsf{K}(\lambda, \mu)\varphi.$$

This implies that $\mathsf{K}(\cdot, \cdot)$ has the reproducing kernel property (4.1.2). The identity in (ii) follows with the special choice $f(\cdot) = \mathsf{K}(\cdot, \nu)\eta$.

(iii) To see that $\mathsf{K}(\cdot, \cdot)$ is a nonnegative kernel it suffices to observe that the matrix

$$\begin{aligned}\big((\mathsf{K}(\lambda_i, \lambda_j)\varphi_j, \varphi_i)_{\mathcal{G}}\big)_{i,j=1}^n &= \big((E(\lambda_i)E(\lambda_j)^*\varphi_j, \varphi_i)_{\mathcal{G}}\big)_{i,j=1}^n \\ &= \big(\langle E(\lambda_j)^*\varphi_j, E(\lambda_i)^*\varphi_i \rangle\big)_{i,j=1}^n\end{aligned}$$

is nonnegative. Lemma 4.1.2 implies that $\mathsf{K}(\cdot, \cdot)$ is symmetric.

(iv) In order to see that the subspace $\operatorname{span}\{\mathsf{K}(\cdot, \mu)\varphi : \mu \in \Omega, \ \varphi \in \mathcal{G}\}$ is dense in $\mathfrak{H}$, assume that there is an element $f \in \mathfrak{H}$ such that $\langle f, \mathsf{K}(\cdot, \mu)\varphi \rangle = 0$ for all $\varphi \in \mathcal{G}$

and $\mu \in \Omega$. But then $(f(\mu), \varphi)_{\mathcal{G}} = 0$ for all $\varphi \in \mathcal{G}$ and $\mu \in \Omega$. Hence, $f(\mu) = 0$ for all $\mu \in \mathcal{G}$, i.e., f is the null function, which completes the argument.

(v) To see that $\mathsf{K}(\cdot, \cdot)$ is uniformly bounded on compact subsets of Ω, note first that

$$\|\mathsf{K}(\lambda, \lambda)\| = \|E(\lambda)E(\lambda)^*\| = \|E(\lambda)\|^2. \tag{4.1.3}$$

Now observe that for all $f \in \mathfrak{H}$ and $\varphi \in \mathcal{G}$,

$$(E(\lambda)f, \varphi)_{\mathcal{G}} = (f(\lambda), \varphi)_{\mathcal{G}}.$$

Since by assumption the function $\lambda \mapsto f(\lambda)$ from Ω to $\mathcal{G}$ is holomorphic, it follows that the mapping $\lambda \mapsto E(\lambda)$ from Ω to $\mathbf{B}(\mathfrak{H}, \mathcal{G})$ is holomorphic, which implies that $\lambda \mapsto \|E(\lambda)\|$ is continuous. Hence, for any compact set $K \subset \Omega$ there is some $M' \geq 0$ such that

$$\sup_{\lambda \in K} \|E(\lambda)\| \leq M'.$$

Therefore, (4.1.3) shows that the kernel $\mathsf{K}(\cdot, \cdot)$ is uniformly bounded on compact subsets of Ω. □

It has been shown in Theorem 4.1.3 that a Hilbert space of $\mathcal{G}$-valued functions in which point evaluation is bounded gives rise to a nonnegative kernel that possesses the reproducing kernel property in (4.1.2). Now it will be shown, conversely, that any nonnegative kernel $\mathsf{K}(\cdot, \cdot)$ gives rise to such a reproducing kernel Hilbert space. Assume that $\mathsf{K}(\cdot, \cdot)$ is some nonnegative kernel on Ω with values in $\mathbf{B}(\mathcal{G})$. Then $\mathsf{K}(\cdot, \cdot)$ is automatically symmetric by Lemma 4.1.2. Consider the linear space of functions from Ω into $\mathcal{G}$ generated by $\mathsf{K}(\cdot, \cdot)$ via

$$\mathring{\mathfrak{H}}(\mathsf{K}) := \operatorname{span}\bigl\{\lambda \mapsto \mathsf{K}(\lambda, \mu)\varphi : \mu \in \Omega, \varphi \in \mathcal{G}\bigr\}. \tag{4.1.4}$$

Define the form $\langle \cdot, \cdot \rangle$ on generating elements by

$$\langle \mathsf{K}(\cdot, \nu)\eta, \mathsf{K}(\cdot, \mu)\varphi \rangle := (\mathsf{K}(\mu, \nu)\eta, \varphi)_{\mathcal{G}}, \quad \nu, \mu \in \Omega, \quad \eta, \varphi \in \mathcal{G}, \tag{4.1.5}$$

and extend it to a form on $\mathring{\mathfrak{H}}(\mathsf{K})$ by

$$\Bigl\langle \sum\nolimits_{j=1}^{n} \alpha_j \mathsf{K}(\cdot, \nu_j)\varphi_j, \sum\nolimits_{i=1}^{m} \beta_i \mathsf{K}(\cdot, \mu_i)\psi_i \Bigr\rangle = \sum_{i,j=1}^{n,m} \bigl(\mathsf{K}(\mu_i, \nu_j)\alpha_j\varphi_j, \beta_i\psi_i\bigr)_{\mathcal{G}}, \tag{4.1.6}$$

where $\alpha_j, \beta_i \in \mathbb{C}$, $\nu_j, \mu_i \in \Omega$, and $\varphi_j, \psi_i \in \mathcal{G}$ for $j = 1, \dots, n$, $i = 1, \dots, m$.

In particular, one has by (4.1.5) for all $f \in \mathring{\mathfrak{H}}(\mathsf{K})$

$$\langle f, \mathsf{K}(\cdot, \mu)\varphi \rangle = (f(\mu), \varphi)_{\mathcal{G}}, \qquad \mu \in \Omega, \quad \varphi \in \mathcal{G}. \tag{4.1.7}$$

Thus, the definition of the form $\langle \cdot, \cdot \rangle$ in (4.1.6) implies the reproducing kernel property in (4.1.7); the kernel $\mathsf{K}(\cdot, \cdot)$ is called a reproducing kernel, relative to the linear space $\mathring{\mathfrak{H}}(\mathsf{K})$ in (4.1.4). It will now be shown that the form defined by (4.1.5) or (4.1.6) is actually a scalar product.

Lemma 4.1.4. *Let $\Omega \subset \mathbb{C}$ be an open set, let $\mathcal{G}$ be a Hilbert space, and let the kernel $\mathsf{K}(\cdot,\cdot)$ in (4.1.1) be nonnegative. Define the space $\mathring{\mathfrak{H}}(\mathsf{K})$ by (4.1.4) and define the form $\langle\cdot,\cdot\rangle$ on $\mathring{\mathfrak{H}}(\mathsf{K})$ as in (4.1.5) and (4.1.6). Then $\mathring{\mathfrak{H}}(\mathsf{K})$ is a pre-Hilbert space with the scalar product $\langle\cdot,\cdot\rangle$.*

Proof. A straightforward calculation shows that $\langle\cdot,\cdot\rangle$ is a well-defined sesquilinear form on $\mathring{\mathfrak{H}}(\mathsf{K})$. By Lemma 4.1.2, the kernel $\mathsf{K}(\cdot,\cdot)$ is symmetric and this yields that $\langle\cdot,\cdot\rangle$ is symmetric. In order to show that $\langle\cdot,\cdot\rangle$ is nonnegative on $\mathring{\mathfrak{H}}(\mathsf{K})$, observe that

$$\begin{aligned}\Big\langle\sum\nolimits_{j=1}^{n}\alpha_j\mathsf{K}(\cdot,\nu_j)\varphi_j,\sum\nolimits_{i=1}^{n}\alpha_i\mathsf{K}(\cdot,\nu_i)\varphi_i\Big\rangle &= \sum_{i,j=1}^{n}\big(\mathsf{K}(\nu_i,\nu_j)\alpha_j\varphi_j,\alpha_i\varphi_i\big)_{\mathcal{G}}\\ &= \sum_{i,j=1}^{n}\big((\mathsf{K}(\nu_i,\nu_j)\varphi_j,\varphi_i)_{\mathcal{G}}\alpha_j,\alpha_i\big),\end{aligned} \tag{4.1.8}$$

for all $\nu_1,\dots,\nu_n\in\Omega$, $\varphi_1,\dots,\varphi_n\in\mathcal{G}$, and $\alpha_1,\dots,\alpha_n\in\mathbb{C}$. Clearly, the last term is equal to

$$\left(\begin{pmatrix}(\mathsf{K}(\nu_1,\nu_1)\varphi_1,\varphi_1)_{\mathcal{G}} & \cdots & (\mathsf{K}(\nu_1,\nu_n)\varphi_n,\varphi_1)_{\mathcal{G}}\\ \vdots & & \vdots\\ (\mathsf{K}(\nu_n,\nu_1)\varphi_1,\varphi_n)_{\mathcal{G}} & \cdots & (\mathsf{K}(\nu_n,\nu_n)\varphi_n,\varphi_n)_{\mathcal{G}}\end{pmatrix}\begin{pmatrix}\alpha_1\\ \vdots\\ \alpha_n\end{pmatrix},\begin{pmatrix}\alpha_1\\ \vdots\\ \alpha_n\end{pmatrix}\right).$$

The assumption that the kernel $\mathsf{K}(\cdot,\cdot)$ is nonnegative means that the $n\times n$ matrix

$$\big[(\mathsf{K}(\nu_i,\nu_j)\varphi_j,\varphi_i)_{\mathcal{G}}\big]_{i,j=1}^{n}$$

is nonnegative. Thus for a typical element

$$f=\sum_{j=1}^{n}\alpha_j\mathsf{K}(\cdot,\nu_j)\varphi_j\in\mathring{\mathfrak{H}}(\mathsf{K})$$

one sees that $\langle f,f\rangle\geq 0$ and hence $\langle\cdot,\cdot\rangle$ is a nonnegative symmetric sesquilinear form on $\mathring{\mathfrak{H}}(\mathsf{K})$. In particular, $\langle\cdot,\cdot\rangle$ satisfies the Cauchy–Schwarz inequality. This implies that $\langle\cdot,\cdot\rangle$ is positive definite. In fact, if $\langle f,f\rangle=0$ for some $f\in\mathring{\mathfrak{H}}(\mathsf{K})$, then

$$|\langle f,g\rangle|^2\leq\langle f,f\rangle\langle g,g\rangle=0\quad\text{for all}\quad g\in\mathring{\mathfrak{H}}(\mathsf{K}).$$

Hence, with $g=\mathsf{K}(\cdot,\mu)\psi$, $\mu\in\Omega$, $\psi\in\mathcal{G}$, the reproducing kernel property (4.1.7) shows that

$$0=\langle f,\mathsf{K}(\cdot,\mu)\psi\rangle=(f(\mu),\psi)_{\mathcal{G}}.$$

Thus, $f(\mu)=0$ for all $\mu\in\Omega$ and so $f=0\in\mathring{\mathfrak{H}}(\mathsf{K})$.

Summing up, it has been shown that $\langle\cdot,\cdot\rangle$ is a positive definite symmetric sesquilinear form on $\mathring{\mathfrak{H}}(\mathsf{K})$, that is, $\langle\cdot,\cdot\rangle$ is a scalar product and $(\mathring{\mathfrak{H}}(\mathsf{K}),\langle\cdot,\cdot\rangle)$ is a pre-Hilbert space. □

In the following theorem it is shown that a nonnegative kernel $\mathsf{K}(\cdot,\cdot)$ on Ω produces a Hilbert space $\mathfrak{H}(\mathsf{K})$, as a completion of $\mathring{\mathfrak{H}}(\mathsf{K})$, of functions on Ω for which point evaluation is a continuous map. Moreover, if the kernel is holomorphic and uniformly bounded on compact subsets of Ω, then the functions in the resulting Hilbert space are holomorphic.

Theorem 4.1.5. *Let $\mathcal{G}$ be a Hilbert space, let $\mathsf{K}(\cdot,\cdot)$ be a nonnegative kernel on the open set $\Omega \subset \mathbb{C}$, and let the form $\langle\cdot,\cdot\rangle$ on $\mathring{\mathfrak{H}}(\mathsf{K})$ be defined as in (4.1.5) and (4.1.6). Then the following statements hold:*

(i) *The completion $\mathfrak{H}(\mathsf{K})$ of the pre-Hilbert space $(\mathring{\mathfrak{H}}(\mathsf{K}),\langle\cdot,\cdot\rangle)$ can be identified with a Hilbert space of $\mathcal{G}$-valued functions defined on Ω.*

(ii) *For $f \in \mathfrak{H}(\mathsf{K})$ one has the reproducing kernel property*

$$\langle f, \mathsf{K}(\cdot,\mu)\varphi\rangle = (f(\mu),\varphi)_{\mathcal{G}}, \qquad \mu \in \Omega, \quad \varphi \in \mathcal{G}.$$

(iii) *For $\lambda \in \Omega$ the point evaluation $E(\lambda) : \mathfrak{H}(\mathsf{K}) \to \mathcal{G}$, $f \mapsto E(\lambda)f = f(\lambda)$ is a continuous linear mapping and*

$$\mathsf{K}(\lambda,\mu) = E(\lambda)E(\mu)^*, \quad \lambda,\mu \in \Omega.$$

(iv) *If the kernel $\mathsf{K}(\cdot,\cdot)$ is holomorphic and uniformly bounded on every compact subset of Ω, then the functions in $\mathfrak{H}(\mathsf{K})$ are holomorphic on Ω.*

Proof. (i) Let $(\mathfrak{H}(\mathsf{K}),\langle\cdot,\cdot\rangle)$ be the Hilbert space that is obtained when one completes the pre-Hilbert space $(\mathring{\mathfrak{H}}(\mathsf{K}),\langle\cdot,\cdot\rangle)$. It will be shown that the elements in $\mathfrak{H}(\mathsf{K})$ can be identified with $\mathcal{G}$-valued functions on Ω. For this let $f \in \mathfrak{H}(\mathsf{K})$ and fix some $\lambda \in \Omega$. Consider the functional

$$\Psi_{f,\lambda} : \mathcal{G} \to \mathbb{C}, \qquad \varphi \mapsto \langle \mathsf{K}(\cdot,\lambda)\varphi, f\rangle.$$

Then an application of the Cauchy–Schwarz inequality shows that

$$\begin{aligned}
|\Psi_{f,\lambda}(\varphi)|^2 &= |\langle \mathsf{K}(\cdot,\lambda)\varphi, f\rangle|^2 \\
&\leq \|\mathsf{K}(\cdot,\lambda)\varphi\|^2\|f\|^2 \\
&= \langle \mathsf{K}(\cdot,\lambda)\varphi, \mathsf{K}(\cdot,\lambda)\varphi\rangle\|f\|^2 \\
&= (\mathsf{K}(\lambda,\lambda)\varphi,\varphi)_{\mathcal{G}}\|f\|^2 \\
&\leq \|\mathsf{K}(\lambda,\lambda)\|\|f\|^2\|\varphi\|_{\mathcal{G}}^2,
\end{aligned}$$

and hence $\Psi_{f,\lambda}$ is continuous. By the Riesz representation theorem, there is a unique vector $\psi_{f,\lambda} \in \mathcal{G}$ such that

$$(\varphi,\psi_{f,\lambda})_{\mathcal{G}} = \Psi_{f,\lambda}(\varphi) = \langle \mathsf{K}(\cdot,\lambda)\varphi, f\rangle, \qquad \varphi \in \mathcal{G}.$$

Let $\mathfrak{F}(\Omega, \mathcal{G})$ be the space of all $\mathcal{G}$-valued functions defined on Ω, and consider the mapping

$$\iota : \mathfrak{H}(\mathsf{K}) \to \mathfrak{F}(\Omega, \mathcal{G}), \quad f \mapsto \iota(f), \quad \text{where} \quad \iota(f)(\lambda) := \psi_{f,\lambda}. \tag{4.1.9}$$

It follows from the definition of ι and $\psi_{f,\lambda}$ that

$$\big(\iota(f)(\lambda), \varphi\big)_{\mathcal{G}} = (\psi_{f,\lambda}, \varphi)_{\mathcal{G}} = \langle f, \mathsf{K}(\cdot, \lambda)\varphi\rangle, \tag{4.1.10}$$

and this equality also shows that ι is a linear mapping.

The mapping ι in (4.1.9) is injective. To see this, assume that $\iota(f) = 0$ for some $f \in \mathfrak{H}(\mathsf{K})$. This means $\iota(f)(\lambda) = 0$ for all $\lambda \in \Omega$, and (4.1.10) implies $\langle f, \mathsf{K}(\cdot, \lambda)\varphi\rangle = 0$ for all $\lambda \in \Omega$ and $\varphi \in \mathcal{G}$. Since the linear span of the functions $\mathsf{K}(\cdot, \lambda)\varphi$ forms the dense subspace $\mathring{\mathfrak{H}}(\mathsf{K})$ of $\mathfrak{H}(\mathsf{K})$, it follows that $f = 0$, that is, ι is injective.

Observe that for $f \in \mathring{\mathfrak{H}}(\mathsf{K})$ it follows from (4.1.10) and the reproducing kernel property (4.1.7) that

$$\big(\iota(f)(\lambda), \varphi\big)_{\mathcal{G}} = \langle f, \mathsf{K}(\cdot, \lambda)\varphi\rangle = (f(\lambda), \varphi)_{\mathcal{G}}, \qquad \varphi \in \mathcal{G},$$

for all $\lambda \in \Omega$, and hence $\iota(f) = f$ for $f \in \mathring{\mathfrak{H}}(\mathsf{K})$. In other words, ι restricted to the dense subspace $\mathring{\mathfrak{H}}(\mathsf{K})$ is the identity, so that $\iota(\mathring{\mathfrak{H}}(\mathsf{K})) = \mathring{\mathfrak{H}}(\mathsf{K})$.

Finally, item (i) follows when the subspace $\operatorname{ran} \iota$ of $\mathfrak{F}(\Omega, \mathcal{G})$ is equipped with the scalar product induced by $\mathfrak{H}(\mathsf{K})$, that is, for $\tilde{f}, \tilde{g} \in \operatorname{ran} \iota$ define

$$\langle \tilde{f}, \tilde{g}\rangle_{\sim} := \langle \iota^{-1}\tilde{f}, \iota^{-1}\tilde{g}\rangle.$$

Then ι is a unitary mapping from the Hilbert space $(\mathfrak{H}(\mathsf{K}), \langle \cdot, \cdot\rangle)$ onto the Hilbert space $(\operatorname{ran} \iota, \langle \cdot, \cdot\rangle_{\sim})$.

(ii) After identifying $\iota(f)$ and $f \in \mathfrak{H}(\mathsf{K})$ as in (i), the reproducing kernel property is immediate from (4.1.10).

(iii) With the identification from (ii) observe that for all $\lambda \in \Omega$ and $\varphi \in \mathcal{G}$ the mapping

$$f \mapsto (f(\lambda), \varphi)_{\mathcal{G}} \tag{4.1.11}$$

is continuous on $\mathfrak{H}(\mathsf{K})$. In fact, this follows from (ii) and the computation

$$\begin{aligned} |(f(\lambda), \varphi)_{\mathcal{G}}|^2 &= |\langle f, \mathsf{K}(\cdot, \lambda)\varphi\rangle|^2 \\ &\leq \langle f, f\rangle \langle \mathsf{K}(\cdot, \lambda)\varphi, \mathsf{K}(\cdot, \lambda)\varphi\rangle \\ &= (\mathsf{K}(\lambda, \lambda)\varphi, \varphi)_{\mathcal{G}} \|f\|^2. \end{aligned}$$

For a fixed $\lambda \in \Omega$ the mapping

$$E(\lambda) : \mathfrak{H}(\mathsf{K}) \to \mathcal{G}, \quad f \mapsto E(\lambda)f = f(\lambda),$$

is closed. To see this, suppose that $f_n \to f$ in $\mathfrak{H}(\mathsf{K})$ and $E(\lambda)f_n \to \psi$ in $\mathcal{G}$. As $\operatorname{dom} E(\lambda) = \mathfrak{H}(\mathsf{K})$, it follows that $f \in \operatorname{dom} E(\lambda)$ and the continuity of (4.1.11) then yields

$$\begin{aligned}(\psi, \varphi)_{\mathcal{G}} &= \lim_{n\to\infty} (E(\lambda)f_n, \varphi)_{\mathcal{G}} = \lim_{n\to\infty} (f_n(\lambda), \varphi)_{\mathcal{G}} \\ &= (f(\lambda), \varphi)_{\mathcal{G}} = (E(\lambda)f, \varphi)_{\mathcal{G}}\end{aligned}$$

for all $\varphi \in \mathcal{G}$. This shows $E(\lambda)f = \psi$ and hence $E(\lambda)$ is a closed operator. Since $\operatorname{dom} E(\lambda) = \mathfrak{H}(\mathsf{K})$, the closed graph theorem implies that $E(\lambda)$ is continuous.

It remains to check the identity $\mathsf{K}(\lambda, \mu) = E(\lambda)E(\mu)^*$ for $\lambda, \mu \in \Omega$. For this let $\varphi, \psi \in \mathcal{G}$, $\lambda, \mu \in \Omega$, and note that $E(\mu)^*\varphi \in \mathfrak{H}(\mathsf{K})$ is a function in the variable λ. Hence, $E(\lambda)E(\mu)^*\varphi = (E(\mu)^*\varphi)(\lambda)$, the reproducing kernel property, and the symmetry of the kernel $\mathsf{K}(\cdot, \cdot)$ imply

$$\begin{aligned}(E(\lambda)E(\mu)^*\varphi, \psi)_{\mathcal{G}} &= ((E(\mu)^*\varphi)(\lambda), \psi)_{\mathcal{G}} \\ &= \langle E(\mu)^*\varphi, \mathsf{K}(\cdot, \lambda)\psi\rangle \\ &= (\varphi, E(\mu)\mathsf{K}(\cdot, \lambda)\psi)_{\mathcal{G}} \\ &= (\varphi, \mathsf{K}(\mu, \lambda)\psi)_{\mathcal{G}} \\ &= (\mathsf{K}(\lambda, \mu)\varphi, \psi)_{\mathcal{G}},\end{aligned}$$

which shows that $\mathsf{K}(\lambda, \mu) = E(\lambda)E(\mu)^*$.

(iv) Let $f \in \mathfrak{H}(\mathsf{K})$ and choose a sequence $f_n \in \mathring{\mathfrak{H}}(\mathsf{K})$ such that $f_n \to f$ in $\mathfrak{H}(\mathsf{K})$. By assumption, the functions $\mathsf{K}(\cdot, \mu)\varphi$ are holomorphic on Ω, and hence so are the functions f_n. Now let $K \subset \Omega$ be a compact set and let $\sup_{\lambda\in K} \|\mathsf{K}(\lambda, \lambda)\| = M_K$. Then for $\lambda \in K$ one gets

$$\begin{aligned}|(f(\lambda) - f_n(\lambda), \varphi)_{\mathcal{G}}| &= |\langle f - f_n, \mathsf{K}(\cdot, \lambda)\varphi\rangle| \\ &\le \|f - f_n\| \|\mathsf{K}(\cdot, \lambda)\varphi\| \\ &\le \|f - f_n\| (\mathsf{K}(\lambda, \lambda)\varphi, \varphi)_{\mathcal{G}}^{1/2} \\ &\le M_K^{1/2} \|f - f_n\| \|\varphi\|_{\mathcal{G}},\end{aligned}$$

and hence $(f_n(\cdot), \varphi)_{\mathcal{G}} \to (f(\cdot), \varphi)_{\mathcal{G}}$ uniformly on K for all $\varphi \in \mathcal{G}$. As K is an arbitrary compact subset of Ω, it follows that the function $\lambda \mapsto (f(\lambda), \varphi)_{\mathcal{G}}$ is holomorphic on Ω for all $\varphi \in \mathcal{G}$. This implies that f is holomorphic. □

If the kernel $\mathsf{K}(\cdot, \cdot)$ is holomorphic and uniformly bounded on every compact subset of Ω, then the elements in the reproducing kernel Hilbert space $\mathfrak{H}(\mathsf{K})$ can be described as holomorphic functions from Ω to $\mathcal{G}$ which satisfy an additional boundedness condition involving the kernel $\mathsf{K}(\cdot, \cdot)$.

Theorem 4.1.6. *Let $\mathcal{G}$ be a Hilbert space, assume that $\mathsf{K}(\cdot, \cdot)$ is a nonnegative holomorphic kernel on the open set $\Omega \subset \mathbb{C}$ which is uniformly bounded on every compact subset of Ω, and let $(\mathfrak{H}(\mathsf{K}), \langle\cdot, \cdot\rangle)$ be the associated reproducing kernel*

Hilbert space. Then $f \in \mathfrak{H}(\mathsf{K})$ with $\|f\| \le \gamma$ if and only if $f : \Omega \to \mathcal{G}$ is holomorphic and the $n \times n$ matrix

$$\gamma^2 \big[(\mathsf{K}(\nu_i,\nu_j)\varphi_j,\varphi_i)_{\mathcal{G}}\big]_{i,j=1}^n - \big[(f(\nu_i),\varphi_i)_{\mathcal{G}}\overline{(f(\nu_j),\varphi_j)_{\mathcal{G}}}\,\big]_{i,j=1}^n \tag{4.1.12}$$

is nonnegative for all $n \in \mathbb{N}$, $\nu_1,\dots,\nu_n \in \Omega$, and $\varphi_1,\dots,\varphi_n \in \mathcal{G}$.

Proof. In order to prove the necessary and sufficient conditions it is helpful to note that the formulation of the condition (4.1.12) is based on the following identities. For the reproducing kernel $\mathsf{K}(\cdot,\cdot)$ one has, as in (4.1.8),

$$\sum_{i,j=1}^n \big((\mathsf{K}(\nu_i,\nu_j)\varphi_j,\varphi_i)_{\mathcal{G}}\alpha_j,\alpha_i\big) = \Big\|\sum\nolimits_{j=1}^n \alpha_j \mathsf{K}(\cdot,\nu_j)\varphi_j\Big\|^2 \tag{4.1.13}$$

for all $\nu_1,\dots,\nu_n \in \Omega$, $\varphi_1,\dots,\varphi_n \in \mathcal{G}$, and $\alpha_1,\dots,\alpha_n \in \mathbb{C}$. Furthermore, for a function $f : \Omega \to \mathcal{G}$ which is holomorphic one has

$$\begin{aligned}
&\sum_{i,j=1}^n \big((f(\nu_i),\varphi_i)_{\mathcal{G}}\overline{(f(\nu_j),\varphi_j)_{\mathcal{G}}}\,\alpha_j,\alpha_i\big) \\
&\quad = \sum_{i,j=1}^n (f(\nu_i),\alpha_i\varphi_i)_{\mathcal{G}}\overline{(f(\nu_j),\alpha_j\varphi_j)_{\mathcal{G}}} \\
&\quad = \Big(\sum_{i=1}^n (f(\nu_i),\alpha_i\varphi_i)_{\mathcal{G}}\Big)\Big(\sum_{j=1}^n \overline{(f(\nu_j),\alpha_j\varphi_j)_{\mathcal{G}}}\Big) \\
&\quad = \Big|\sum_{j=1}^n (f(\nu_j),\alpha_j\varphi_j)_{\mathcal{G}}\Big|^2 \\
&\quad = \Big|\sum_{j=1}^n (\alpha_j\varphi_j,f(\nu_j))_{\mathcal{G}}\Big|^2
\end{aligned} \tag{4.1.14}$$

for all $\nu_1,\dots,\nu_n \in \Omega$, $\varphi_1,\dots,\varphi_n \in \mathcal{G}$, and $\alpha_1,\dots,\alpha_n \in \mathbb{C}$.

Assume that the function $f : \Omega \to \mathcal{G}$ is holomorphic and there exists $\gamma > 0$ such that the $n\times n$ matrix (4.1.12) is nonnegative for all $n \in \mathbb{N}$, $\nu_1,\dots,\nu_n \in \Omega$, and $\varphi_1,\dots,\varphi_n \in \mathcal{G}$. Together with (4.1.13) and (4.1.14), this implies that the relation from $\mathfrak{H}(\mathsf{K})$ to $\mathbb{C}$, spanned by the elements

$$\left\{\sum_{j=1}^n \alpha_j \mathsf{K}(\cdot,\nu_j)\varphi_j, \sum_{j=1}^n (\alpha_j\varphi_j,f(\nu_j))_{\mathcal{G}}\right\},$$

where $\nu_1,\dots,\nu_n \in \Omega$, $\varphi_1,\dots,\varphi_n \in \mathcal{G}$, and $\alpha_1,\dots,\alpha_n \in \mathbb{C}$, is a bounded functional with bound γ. Furthermore, it is densely defined on $\mathfrak{H}(\mathsf{K})$, so that it admits a uniquely defined bounded linear extension defined on all of $\mathfrak{H}(\mathsf{K})$. This functional is represented by a unique element $F \in \mathfrak{H}(\mathsf{K})$ with $\|F\| \leq \gamma$ via the Riesz representation theorem. In particular this means that

$$(\varphi, f(\nu))_{\mathcal{G}} = \langle \mathsf{K}(\cdot,\nu)\varphi, F\rangle, \quad \nu \in \Omega, \ \varphi \in \mathcal{G},$$

whereas by the reproducing kernel property one has

$$\langle \mathsf{K}(\cdot,\nu)\varphi, F\rangle = \overline{\langle F, \mathsf{K}(\cdot,\nu)\varphi\rangle} = \overline{(F(\nu),\varphi)_{\mathcal{G}}} = (\varphi, F(\nu))_{\mathcal{G}}, \quad \nu \in \Omega, \ \varphi \in \mathcal{G}.$$

Combining the last two identities one concludes that $f = F$, which gives that $f \in \mathfrak{H}(\mathsf{K})$ and $\|f\| \leq \gamma$.

For the converse statement, assume that $f \in \mathfrak{H}(\mathsf{K})$ and $\|f\| \leq \gamma$. Then the function $f : \Omega \to \mathcal{G}$ is holomorphic and for all $\nu_1, \dots, \nu_n \in \Omega$, $\varphi_1, \dots, \varphi_n \in \mathcal{G}$, and $\alpha_1, \dots, \alpha_n \in \mathbb{C}$ one has by means of (4.1.14) and the fact that $f \in \mathfrak{H}(\mathsf{K})$

$$\begin{aligned}\sum_{i,j=1}^{n} &\left((f(\nu_i),\varphi_i)_{\mathcal{G}}\overline{(f(\nu_j),\varphi_j)_{\mathcal{G}}}\,\alpha_j, \alpha_i\right) \\ &= \left|\sum_{j=1}^{n} (f(\nu_j), \alpha_j\varphi_j)_{\mathcal{G}}\right|^2 \\ &= \left|\sum_{j=1}^{n} \langle f, \alpha_j \mathsf{K}(\cdot,\nu_j)\varphi_j\rangle\right|^2 \\ &= \left|\left\langle f, \sum_{j=1}^{n} \alpha_j \mathsf{K}(\cdot,\nu_j)\varphi_j\right\rangle\right|^2 \\ &\leq \|f\|^2 \left\|\sum_{j=1}^{n} \alpha_j \mathsf{K}(\cdot,\nu_j)\varphi_j\right\|^2.\end{aligned}$$

Together with (4.1.13) and $\|f\| \leq \gamma$ this gives (4.1.12). □

Due to the holomorphy it is sometimes convenient to consider a set of functions $\lambda \mapsto \mathsf{K}(\lambda,\mu)\varphi$, $\varphi \in \mathcal{G}$, on a determining set of points $\mu \in \Omega$.

Corollary 4.1.7. *Let $\mathsf{K}(\cdot,\cdot)$ be a nonnegative holomorphic kernel on an open set $\Omega \subset \mathbb{C}$ which is uniformly bounded on every compact subset of Ω, and let $\mathfrak{H}(\mathsf{K})$ be the associated reproducing kernel Hilbert space. Let $\mathcal{D} \subset \Omega$ be a set of points which has an accumulation point in each connected component of Ω. Then*

$$\mathfrak{H}(\mathsf{K}) = \overline{\operatorname{span}}\left\{\lambda \mapsto \mathsf{K}(\lambda,\mu)\varphi : \mu \in \mathcal{D}, \ \varphi \in \mathcal{G}\right\}.$$

Proof. The inclusion ($\supset$) is obvious from (4.1.4). To show the inclusion ($\subset$), it suffices to verify that the linear space

$$\operatorname{span}\left\{\lambda \mapsto \mathsf{K}(\lambda,\mu)\varphi : \mu \in \mathcal{D}, \ \varphi \in \mathcal{G}\right\}$$

is dense in $\mathfrak{H}(\mathsf{K})$. Therefore, let $f \in \mathfrak{H}(\mathsf{K})$ be orthogonal to this set. Then

$$0 = \langle f, \mathsf{K}(\cdot, \mu)\varphi \rangle = (f(\mu), \varphi)_{\mathcal{G}}$$

for all $\mu \in \mathcal{D}$ and $\varphi \in \mathcal{G}$, and hence $f(\mu) = 0$ for all $\mu \in \mathcal{D}$. Since $f \in \mathfrak{H}(\mathsf{K})$ is holomorphic on Ω, the assumption on $\mathcal{D}$ now implies that $f(\lambda) = 0$ for all $\lambda \in \Omega$. Hence, $f = 0$ and the proof is complete. □

Let $\mathsf{K}(\cdot,\cdot)$ be a nonnegative holomorphic kernel on an open set Ω. If $\Omega' \subset \mathbb{C}$ is an open set such that $\Omega \subset \Omega'$ and if $\mathsf{K}'(\cdot,\cdot)$ is a nonnegative holomorphic kernel on Ω' extending $\mathsf{K}(\cdot,\cdot)$, then the functions in the reproducing kernel Hilbert space $\mathfrak{H}(\mathsf{K})$ may be seen as restrictions to Ω of the functions in the reproducing kernel Hilbert space $\mathfrak{H}(\mathsf{K}')$.

Proposition 4.1.8. *Let $\mathsf{K}(\cdot,\cdot)$ be a nonnegative holomorphic kernel on an open set $\Omega \subset \mathbb{C}$ which is uniformly bounded on every compact subset of Ω. Assume that $\Omega' \subset \mathbb{C}$ is an open set such that $\Omega \subset \Omega'$ and that $\mathsf{K}'(\cdot,\cdot)$ is a nonnegative holomorphic kernel on Ω' which is uniformly bounded on every compact subset of Ω' and which is equal to $\mathsf{K}(\cdot,\cdot)$ on Ω. Then*

$$\mathfrak{H}(\mathsf{K}) = \big\{ f|_{\Omega} : f \in \mathfrak{H}(\mathsf{K}') \big\}.$$

Proof. Consider the linear space of functions from Ω' into $\mathcal{G}$ generated by $\mathsf{K}'(\cdot,\cdot)$ via

$$\mathring{\mathfrak{H}}(\mathsf{K}') := \operatorname{span} \big\{ \lambda \mapsto \mathsf{K}'(\lambda, \mu)\varphi : \, \mu \in \Omega', \, \varphi \in \mathcal{G} \big\}.$$

It is clear that the analogous linear space

$$\mathring{\mathfrak{H}}(\mathsf{K}) := \operatorname{span} \big\{ \lambda \mapsto \mathsf{K}(\lambda, \mu)\varphi : \, \mu \in \Omega, \, \varphi \in \mathcal{G} \big\}$$

is contained in $\mathring{\mathfrak{H}}(\mathsf{K}')$ in the sense that each function $\lambda \mapsto \mathsf{K}(\lambda,\mu)\varphi$ with $\mu \in \Omega$ is the restriction to Ω of the function $\lambda \mapsto \mathsf{K}'(\lambda,\mu)\varphi$. Hence, a continuity argument shows the inclusion

$$\mathfrak{H}(\mathsf{K}) \subset \big\{ f|_{\Omega} : f \in \mathfrak{H}(\mathsf{K}') \big\}.$$

For the opposite inclusion consider $f|_{\Omega} : \Omega \to \mathbb{C}$ for some $f \in \mathfrak{H}(\mathsf{K}')$, and set $\gamma = \|f\|$. Then, by Theorem 4.1.6, the matrix

$$\gamma^2 \big[(\mathsf{K}'(\nu_i, \nu_j)\varphi_j, \varphi_i)_{\mathcal{G}} \big]_{i,j=1}^{n} - \big[(f(\nu_i), \varphi_i)_{\mathcal{G}} \, \overline{(f(\nu_j), \varphi_j)_{\mathcal{G}}} \big]_{i,j=1}^{n}$$

is nonnegative for all $n \in \mathbb{N}$, $\nu_1, \dots, \nu_n \in \Omega'$, and $\varphi_1, \dots, \varphi_n \in \mathcal{G}$. In particular,

$$\gamma^2 \big[(\mathsf{K}(\nu_i, \nu_j)\varphi_j, \varphi_i)_{\mathcal{G}} \big]_{i,j=1}^{n} - \big[(f(\nu_i), \varphi_i)_{\mathcal{G}} \, \overline{(f(\nu_j), \varphi_j)_{\mathcal{G}}} \big]_{i,j=1}^{n}$$

is nonnegative for all $n \in \mathbb{N}$, $\nu_1, \dots, \nu_n \in \Omega$, and $\varphi_1, \dots, \varphi_n \in \mathcal{G}$. Another application of Theorem 4.1.6 implies $f|_{\Omega} \in \mathfrak{H}(\mathsf{K})$. □

Under suitable circumstances multiplication of a given reproducing kernel by an operator function gives rise to a new reproducing kernel. In the following proposition this fact and the relation between the corresponding reproducing kernel Hilbert spaces are explained.

Proposition 4.1.9. *Let $\mathcal{G}$ be a Hilbert space, assume that $\mathsf{K}(\cdot,\cdot)$ is a nonnegative kernel on Ω, and let $(\mathfrak{H}(\mathsf{K}),\langle\cdot,\cdot\rangle)$ be the associated reproducing kernel Hilbert space. Let $\Phi:\Omega\to\mathbf{B}(\mathcal{G})$ be such that $0\in\rho(\Phi(\lambda))$ for all $\lambda\in\Omega$. Then*

$$\mathsf{K}_\Phi(\lambda,\mu)=\Phi(\lambda)\mathsf{K}(\lambda,\mu)\Phi(\mu)^* \tag{4.1.15}$$

is a nonnegative kernel on Ω and the corresponding reproducing Hilbert space $(\mathfrak{H}(\mathsf{K}_\Phi),\langle\cdot,\cdot\rangle_\Phi)$ is unitarily equivalent to $\mathfrak{H}(\mathsf{K})$ via the mapping

$$\mathcal{M}_\Phi:\mathfrak{H}(\mathsf{K})\to\mathfrak{H}(\mathsf{K}_\Phi),\qquad f\mapsto\Phi f.$$

Moreover, if $\mathsf{K}(\cdot,\cdot)$ is holomorphic and uniformly bounded on every compact subset of Ω, and Φ is holomorphic, then also $\mathsf{K}_\Phi(\cdot,\cdot)$ is holomorphic and uniformly bounded on every compact subset of Ω.

Proof. The definition (4.1.15) leads to the identity

$$\big((\mathsf{K}_\Phi(\lambda_i,\lambda_j)\varphi_j,\varphi_i)_\mathcal{G}\big)_{i,j=1}^n=\big((\mathsf{K}(\lambda_i,\lambda_j)\Phi(\lambda_j)^*\varphi_j,\Phi(\lambda_i)^*\varphi_i)_\mathcal{G}\big)_{i,j=1}^n.$$

Hence, the nonnegativity of $\mathsf{K}(\cdot,\cdot)$ implies that $\mathsf{K}_\Phi(\cdot,\cdot)$ is a nonnegative kernel. Moreover, (4.1.15) shows that for all $\mu\in\Omega$ and $\varphi\in\mathcal{G}$

$$\Phi(\cdot)\mathsf{K}(\cdot,\mu)\varphi=\Phi(\cdot)\mathsf{K}(\cdot,\mu)\Phi(\mu)^*\Phi(\mu)^{-*}\varphi=\mathsf{K}_\Phi(\cdot,\mu)\Phi(\mu)^{-*}\varphi.$$

Hence, $\mathcal{M}_\Phi$ maps $\mathring{\mathfrak{H}}(\mathsf{K})$ onto $\mathring{\mathfrak{H}}(\mathsf{K}_\Phi)$. The identity

$$\begin{aligned}
\big\langle\Phi(\cdot)\mathsf{K}(\cdot,\mu)\varphi,\Phi(\cdot)\mathsf{K}(\cdot,\nu)\psi\big\rangle_\Phi&=\big\langle\mathsf{K}_\Phi(\cdot,\mu)\Phi(\mu)^{-*}\varphi,\mathsf{K}_\Phi(\cdot,\nu)\Phi(\nu)^{-*}\psi\big\rangle_\Phi\\
&=\big(\mathsf{K}_\Phi(\nu,\mu)\Phi(\mu)^{-*}\varphi,\Phi(\nu)^{-*}\psi\big)_\mathcal{G}\\
&=\big(\Phi(\nu)\mathsf{K}(\nu,\mu)\varphi,\Phi(\nu)^{-*}\psi\big)_\mathcal{G}\\
&=\big(\mathsf{K}(\nu,\mu)\varphi,\psi\big)_\mathcal{G}\\
&=\big\langle\mathsf{K}(\cdot,\mu)\varphi,\mathsf{K}(\cdot,\nu)\psi\big\rangle,
\end{aligned}$$

which is valid for all $\mu,\nu\in\Omega$ and all $\varphi,\psi\in\mathcal{G}$, shows that the mapping $\mathcal{M}_\Phi$ from $\mathring{\mathfrak{H}}(\mathsf{K})$ onto $\mathring{\mathfrak{H}}(\mathsf{K}_\Phi)$ is an isometry. Its unique bounded linear extension gives a unitary mapping from $\mathfrak{H}(\mathsf{K})$ onto $\mathfrak{H}(\mathsf{K}_\Phi)$. In order to see that this extension $\mathcal{M}_\Phi$ acts as multiplication by Φ on all functions in $\mathfrak{H}(\mathsf{K})$, let $f\in\mathfrak{H}(\mathsf{K})$ and choose a sequence $f_n\in\mathring{\mathfrak{H}}(\mathsf{K})$ such that $f_n\to f$ in $\mathfrak{H}(\mathsf{K})$. By isometry, the sequence $\mathcal{M}_\Phi f_n$ converges to $\mathcal{M}_\Phi f$ in $\mathfrak{H}(\mathsf{K}_\Phi)$. Observe that the approximating sequence (f_n) satisfies for all $\varphi\in\mathcal{G}$

$$\begin{aligned}
\big\langle(\mathcal{M}_\Phi f_n)(\cdot),\mathsf{K}_\Phi(\cdot,\mu)\varphi\big\rangle_\Phi&=\big\langle\Phi(\cdot)f_n(\cdot),\mathsf{K}_\Phi(\cdot,\mu)\varphi\big\rangle_\Phi\\
&=(\Phi(\mu)f_n(\mu),\varphi)_\mathcal{G}\\
&=\big(f_n(\mu),\Phi(\mu)^*\varphi\big)_\mathcal{G}\\
&=\big\langle f_n(\cdot),\mathsf{K}(\cdot,\mu)\Phi(\mu)^*\varphi\big\rangle.
\end{aligned}$$

Hence, taking limits one sees that

$$\big\langle (\mathcal{M}_\Phi f)(\cdot), \mathsf{K}_\Phi(\cdot,\mu)\varphi \big\rangle_\Phi = \big\langle f(\cdot), \mathsf{K}(\cdot,\mu)\Phi(\mu)^*\varphi \big\rangle,$$

and therefore

$$\begin{aligned} \big((\mathcal{M}_\Phi f)(\mu), \varphi\big)_\mathcal{G} &= \big\langle (\mathcal{M}_\Phi f)(\cdot), \mathsf{K}_\Phi(\cdot,\mu)\varphi \big\rangle_\Phi \\ &= \big\langle f(\cdot), \mathsf{K}(\cdot,\mu)\Phi(\mu)^*\varphi \big\rangle \\ &= \big(f(\mu), \Phi(\mu)^*\varphi \big)_\mathcal{G} \\ &= \big(\Phi(\mu) f(\mu), \varphi \big)_\mathcal{G} \end{aligned}$$

for all $\varphi \in \mathcal{G}$. This shows $(\mathcal{M}_\Phi f)(\mu) = \Phi(\mu)f(\mu)$ for all $\mu \in \Omega$.

The last assertion on the holomorphy and uniform boundedness of $\mathsf{K}_\Phi(\cdot,\cdot)$ on compact subsets of Ω is clear. □

4.2 Realization of uniformly strict Nevanlinna functions

The aim of this section is to show that every operator-valued uniformly strict Nevanlinna function can be realized as the Weyl function corresponding to a boundary triplet. The reproducing kernel Hilbert space associated with a given uniformly strict Nevanlinna function will serve as a model space. The uniqueness of the model is discussed as well.

Let $\mathcal{G}$ be a Hilbert space and let M be a $\mathbf{B}(\mathcal{G})$-valued Nevanlinna function. The associated *Nevanlinna kernel*

$$\mathsf{N}_M(\cdot,\cdot) : \Omega \times \Omega \to \mathbf{B}(\mathcal{G})$$

with $\Omega = \mathbb{C} \setminus \mathbb{R}$ is defined by

$$\mathsf{N}_M(\lambda,\mu) := \frac{M(\lambda) - M(\mu)^*}{\lambda - \bar{\mu}}, \quad \lambda,\mu \in \mathbb{C}\setminus\mathbb{R}, \quad \lambda \neq \bar{\mu}, \tag{4.2.1}$$

and $\mathsf{N}_M(\lambda,\bar{\lambda}) = M'(\lambda)$, $\lambda \in \mathbb{C}\setminus\mathbb{R}$. Then clearly the kernel N_M is symmetric. The kernel N_M is holomorphic, since

$$\lambda \mapsto \mathsf{N}_M(\lambda,\mu)$$

is holomorphic for each $\mu \in \mathbb{C}\setminus\mathbb{R}$. Moreover, from the definition of N_M one sees immediately that

$$\mathsf{N}_M(\lambda,\lambda) = \frac{\operatorname{Im} M(\lambda)}{\operatorname{Im}\lambda}, \quad \lambda \in \mathbb{C}\setminus\mathbb{R},$$

and hence $\mathsf{N}_M(\lambda,\lambda) \geq 0$, $\lambda \in \mathbb{C}\setminus\mathbb{R}$. In the next theorem it turns out that the kernel N_M is, in fact, nonnegative on $\mathbb{C}\setminus\mathbb{R}$. Note also that the kernel N_M is uniformly bounded on compact subsets of $\mathbb{C}\setminus\mathbb{R}$ since

$$\|\mathsf{N}_M(\lambda,\lambda)\| \leq \frac{\|M(\lambda)\|}{|\operatorname{Im}\lambda|}, \quad \lambda \in \mathbb{C}\setminus\mathbb{R}.$$

Theorem 4.2.1. *Let M be a $\mathbf{B}(\mathcal{G})$-valued Nevanlinna function. Then the kernel N_M in (4.2.1) is nonnegative.*

Proof. The function M has the integral representation

$$M(\lambda) = \alpha + \lambda\beta + \int_{\mathbb{R}} \left(\frac{1}{t-\lambda} - \frac{t}{1+t^2} \right) d\Sigma(t), \quad \lambda \in \mathbb{C} \setminus \mathbb{R}, \tag{4.2.2}$$

with self-adjoint operators $\alpha, \beta \in \mathbf{B}(\mathcal{G})$, $\beta \geq 0$, and a nondecreasing self-adjoint operator function $\Sigma : \mathbb{R} \to \mathbf{B}(\mathcal{G})$ such that

$$\int_{\mathbb{R}} \frac{d\Sigma(t)}{1+t^2} \in \mathbf{B}(\mathcal{G}),$$

where the integral in (4.2.2) converges in the strong topology; cf. Theorem A.4.2. For any $n \in \mathbb{N}$, points $\lambda_1, \dots, \lambda_n \in \mathbb{C} \setminus \mathbb{R}$, and elements $\varphi_1, \dots, \varphi_n \in \mathcal{G}$ it follows from (4.2.2) that

$$\begin{aligned} &\big((\mathsf{N}_M(\lambda_i, \lambda_j)\varphi_j, \varphi_i)_{\mathcal{G}} \big)_{i,j=1}^n \\ &\quad = \big((\beta\varphi_j, \varphi_i)_{\mathcal{G}} \big)_{i,j=1}^n + \left(\left(\left(\int_{\mathbb{R}} \frac{1}{t-\lambda_i} \frac{1}{t-\bar{\lambda}_j} \, d\Sigma(t) \right) \varphi_j, \varphi_i \right)_{\mathcal{G}} \right)_{i,j=1}^n . \end{aligned} \tag{4.2.3}$$

The first matrix on the right-hand side in (4.2.3) is nonnegative as for any vector $(x_1, \dots, x_n)^\top \in \mathbb{C}^n$ and $\varphi = \sum x_i \varphi_i$ the nonnegativity of the operator β implies

$$\left(\begin{pmatrix} (\beta\varphi_1, \varphi_1)_{\mathcal{G}} & \cdots & (\beta\varphi_n, \varphi_1)_{\mathcal{G}} \\ \vdots & & \vdots \\ (\beta\varphi_1, \varphi_n)_{\mathcal{G}} & \cdots & (\beta\varphi_n, \varphi_n)_{\mathcal{G}} \end{pmatrix} \begin{pmatrix} x_1 \\ \vdots \\ x_n \end{pmatrix}, \begin{pmatrix} x_1 \\ \vdots \\ x_n \end{pmatrix} \right) = \big(\beta\varphi, \varphi\big)_{\mathcal{G}} \geq 0.$$

To see that the second matrix on the right-hand side in (4.2.3) is also nonnegative, first use Proposition A.3.7 and Proposition A.3.4 to obtain

$$\begin{aligned} &\left(\left(\left(\left(\int_{\mathbb{R}} \frac{1}{t-\lambda_i} \frac{1}{t-\bar{\lambda}_j} \, d\Sigma(t) \right) \varphi_j, \varphi_i \right)_{\mathcal{G}} \right)_{i,j=1}^n \begin{pmatrix} x_1 \\ \vdots \\ x_n \end{pmatrix}, \begin{pmatrix} x_1 \\ \vdots \\ x_n \end{pmatrix} \right) \\ &\quad = \lim_{a \to -\infty} \lim_{b \to \infty} \sum_{i,j=1}^n \int_a^b \frac{1}{t-\lambda_i} \frac{1}{t-\bar{\lambda}_j} \, d\big(\Sigma(t) x_j\varphi_j, x_i\varphi_i\big)_{\mathcal{G}}. \end{aligned} \tag{4.2.4}$$

It is clear that for any finite partition $a = t_0 < t_1 < \cdots < t_k = b$ of a finite interval $[a, b]$ one has

$$\begin{aligned} &\sum_{l=1}^k \sum_{i,j=1}^n \frac{1}{t_l - \lambda_i} \frac{1}{t_l - \bar{\lambda}_j} \big((\Sigma(t_l) - \Sigma(t_{l-1})) x_j\varphi_j, x_i\varphi_i \big)_{\mathcal{G}} \\ &\quad = \sum_{l=1}^k \left((\Sigma(t_l) - \Sigma(t_{l-1})) \sum_{j=1}^n \frac{x_j\varphi_j}{t_l - \bar{\lambda}_j}, \sum_{i=1}^n \frac{x_i\varphi_i}{t_l - \bar{\lambda}_i} \right)_{\mathcal{G}} \geq 0 \end{aligned}$$

and when $\max|t_l - t_{l-1}|$ tends to zero these finite Riemann–Stieltjes sums converge to

$$\sum_{i,j=1}^{n} \int_a^b \frac{1}{t-\lambda_i} \frac{1}{t-\overline{\lambda}_j}\, d\big(\Sigma(t)x_j\varphi_j, x_i\varphi_i\big)_{\mathcal{G}} \geq 0.$$

Hence, also (4.2.4) is nonnegative; thus, both matrices on the right-hand side in (4.2.3) are nonnegative, and so is their sum, i.e., the kernel N_M is nonnegative. □

According to Theorem 4.1.5, the nonnegative kernel N_M gives rise to a Hilbert space of holomorphic $\mathcal{G}$-valued functions, which will be denoted by $\mathfrak{H}(\mathsf{N}_M)$, with inner product $\langle \cdot, \cdot \rangle$; cf. Section 4.1. Recall that the reproducing kernel property

$$\langle f, \mathsf{N}_M(\cdot, \mu)\varphi \rangle = (f(\mu), \varphi)_{\mathcal{G}}, \qquad \varphi \in \mathcal{G},\ \mu \in \mathbb{C} \setminus \mathbb{R}, \tag{4.2.5}$$

holds for all functions $f \in \mathfrak{H}(\mathsf{N}_M)$. The main results in this section concern a Nevanlinna function M and the construction of a self-adjoint relation which represents M in a sense to be explained. The construction will involve the associated reproducing kernel space $\mathfrak{H}(\mathsf{N}_M)$.

Let M be a (not necessarily uniformly strict) $\mathbf{B}(\mathcal{G})$-valued Nevanlinna function. The first main objective in this section is the contruction of a *minimal model* in which the function

$$\lambda \mapsto -(M(\lambda) + \lambda)^{-1}$$

is realized as the compressed resolvent of a self-adjoint relation. The uniqueness of the construction will be discussed after the theorem. Note that the definition of the self-adjoint relation involves multiplication by the independent variable; however, the resulting functions do not necessarily belong to $\mathfrak{H}(\mathsf{N}_M)$.

Theorem 4.2.2. *Let M be a $\mathbf{B}(\mathcal{G})$-valued Nevanlinna function and let $\mathfrak{H}(\mathsf{N}_M)$ be the associated reproducing kernel Hilbert space. Denote by $P_{\mathcal{G}}$ the orthogonal projection from $\mathfrak{H}(\mathsf{N}_M) \oplus \mathcal{G}$ onto $\mathcal{G}$ and let $\iota_{\mathcal{G}}$ be the canonical embedding of $\mathcal{G}$ into $\mathfrak{H}(\mathsf{N}_M) \oplus \mathcal{G}$. Then*

$$\widetilde{A} = \left\{ \left\{ \begin{pmatrix} f \\ \varphi \end{pmatrix}, \begin{pmatrix} f' \\ -\varphi' \end{pmatrix} \right\} : \begin{array}{l} f, f' \in \mathfrak{H}(\mathsf{N}_M),\ \varphi, \varphi' \in \mathcal{G}, \\ f'(\xi) - \xi f(\xi) = M(\xi)\varphi - \varphi' \end{array} \right\} \tag{4.2.6}$$

is a self-adjoint relation in the Hilbert space $\mathfrak{H}(\mathsf{N}_M) \oplus \mathcal{G}$ and the compressed resolvent of $\widetilde{A}$ onto $\mathcal{G}$ is given by

$$P_{\mathcal{G}}(\widetilde{A} - \lambda)^{-1}\iota_{\mathcal{G}} = -(M(\lambda) + \lambda)^{-1}, \qquad \lambda \in \mathbb{C} \setminus \mathbb{R}. \tag{4.2.7}$$

Furthermore, the self-adjoint relation $\widetilde{A}$ satisfies the following minimality condition:

$$\mathfrak{H}(\mathsf{N}_M) \oplus \mathcal{G} = \overline{\operatorname{span}}\, \big\{ \mathcal{G}, \operatorname{ran}(\widetilde{A} - \lambda)^{-1}\iota_{\mathcal{G}} : \lambda \in \mathbb{C} \setminus \mathbb{R} \big\}. \tag{4.2.8}$$

Proof. Step 1. The relation $\widetilde{A}$ in (4.2.6) contains an essentially self-adjoint relation. Indeed, define the relation B in $\mathfrak{H}(\mathsf{N}_M) \oplus \mathcal{G}$ by

$$B = \operatorname{span} \left\{ \left\{ \begin{pmatrix} \mathsf{N}_M(\cdot, \bar{\mu})\varphi \\ -\varphi \end{pmatrix}, \begin{pmatrix} \mu \mathsf{N}_M(\cdot, \bar{\mu})\varphi \\ M(\mu)\varphi \end{pmatrix} \right\} : \varphi \in \mathcal{G}, \, \mu \in \mathbb{C} \setminus \mathbb{R} \right\}.$$

It follows from the definition of N_M in (4.2.1) that

$$\mu \mathsf{N}_M(\xi, \bar{\mu})\varphi - \xi \mathsf{N}_M(\xi, \bar{\mu})\varphi = M(\mu)\varphi - M(\xi)\varphi, \quad \varphi \in \mathcal{G}.$$

Therefore, one sees from (4.2.6) that $B \subset \widetilde{A}$. It remains to show that B is essentially self-adjoint.

The symmetry of B is easily verified: it follows from the definition in (4.2.1) and the reproducing kernel property (4.2.5) that

$$\begin{aligned}
&\left(\begin{pmatrix} \mu \mathsf{N}_M(\cdot, \bar{\mu})\varphi \\ M(\mu)\varphi \end{pmatrix}, \begin{pmatrix} \mathsf{N}_M(\cdot, \bar{\nu})\psi \\ -\psi \end{pmatrix} \right) - \left(\begin{pmatrix} \mathsf{N}_M(\cdot, \bar{\mu})\varphi \\ -\varphi \end{pmatrix}, \begin{pmatrix} \nu \mathsf{N}_M(\cdot, \bar{\nu})\psi \\ M(\nu)\psi \end{pmatrix} \right) \\
&\quad = \langle \mu \mathsf{N}_M(\cdot, \bar{\mu})\varphi, \mathsf{N}_M(\cdot, \bar{\nu})\psi \rangle - (M(\mu)\varphi, \psi)_{\mathcal{G}} \\
&\qquad\qquad\qquad - \langle \mathsf{N}_M(\cdot, \bar{\mu})\varphi, \nu \mathsf{N}_M(\cdot, \bar{\nu})\psi \rangle + (\varphi, M(\nu)\psi)_{\mathcal{G}} \\
&\quad = (\mu - \bar{\nu})(\mathsf{N}_M(\bar{\nu}, \bar{\mu})\varphi, \psi)_{\mathcal{G}} - (M(\mu)\varphi, \psi)_{\mathcal{G}} + (\varphi, M(\nu)\psi)_{\mathcal{G}} = 0
\end{aligned}$$

for all $\varphi, \psi \in \mathcal{G}$ and all $\mu, \nu \in \mathbb{C} \setminus \mathbb{R}$. This identity implies that B is symmetric in $\mathfrak{H}(\mathsf{N}_M) \oplus \mathcal{G}$.

To see that B is essentially self-adjoint, it now suffices to establish that $\operatorname{ran}(B - \lambda_0)$ is dense in $\mathfrak{H}(\mathsf{N}_M) \oplus \mathcal{G}$ for arbitrary $\lambda_0 \in \mathbb{C} \setminus \mathbb{R}$. Observe that it follows from the definition that

$$\operatorname{ran}(B - \lambda_0) = \operatorname{span} \left\{ \begin{pmatrix} (\mu - \lambda_0)\mathsf{N}_M(\cdot, \bar{\mu})\varphi \\ (M(\mu) + \lambda_0)\varphi \end{pmatrix} : \varphi \in \mathcal{G}, \, \mu \in \mathbb{C} \setminus \mathbb{R} \right\}.$$

The choice $\mu = \lambda_0$ together with the fact $-\lambda_0 \in \rho(M(\lambda_0))$ (see Definition A.4.1) imply that $\operatorname{ran}(M(\lambda_0) + \lambda_0) = \mathcal{G}$ and hence

$$\{0\} \oplus \mathcal{G} \subset \operatorname{ran}(B - \lambda_0). \tag{4.2.9}$$

Therefore, also the elements of the form

$$\begin{pmatrix} \mathsf{N}_M(\cdot, \bar{\mu})\varphi \\ 0 \end{pmatrix}, \quad \varphi \in \mathcal{G}, \quad \mu \in \mathbb{C} \setminus \mathbb{R}, \quad \mu \neq \lambda_0, \tag{4.2.10}$$

belong to $\operatorname{ran}(B - \lambda_0)$. Moreover, since the set

$$\operatorname{span} \{ \mathsf{N}_M(\cdot, \mu)\varphi : \varphi \in \mathcal{G}, \, \mu \in \mathbb{C} \setminus \mathbb{R}, \, \mu \neq \lambda_0 \}$$

is dense in $\mathfrak{H}(\mathsf{N}_M)$, see Corollary 4.1.7, it follows from (4.2.9) and (4.2.10) that $\operatorname{ran}(B - \lambda_0)$ is dense in $\mathfrak{H}(\mathsf{N}_M) \oplus \mathcal{G}$ for all $\lambda_0 \in \mathbb{C} \setminus \mathbb{R}$.

Now let $\overline{B}$ be the closure of the symmetric relation B. It is clear that $\overline{B}$ is symmetric and that $\operatorname{ran}(\overline{B}-\lambda_0)$ is closed (see Proposition 1.4.4 and Lemma 1.2.2). Hence, it follows from the above considerations that $\operatorname{ran}(\overline{B}-\lambda_0) = \mathfrak{H}(\mathsf{N}_M) \oplus \mathcal{G}$, and Theorem 1.5.5 yields that $\overline{B}$ is self-adjoint in $\mathfrak{H}(\mathsf{N}_M) \oplus \mathcal{G}$.

Step 2. The relation $\widetilde{A}$ is self-adjoint. To prove this, it suffices to establish that the closure of B coincides with the relation $\widetilde{A}$.

First one shows that the relation $\widetilde{A}$ is closed. To see this, let

$$\left\{ \begin{pmatrix} f_n \\ \varphi_n \end{pmatrix}, \begin{pmatrix} f_n' \\ -\varphi_n' \end{pmatrix} \right\} \to \left\{ \begin{pmatrix} f \\ \varphi \end{pmatrix}, \begin{pmatrix} f' \\ -\varphi' \end{pmatrix} \right\} \quad \text{in} \quad \begin{pmatrix} \mathfrak{H}(\mathsf{N}_M) \\ \mathcal{G} \end{pmatrix} \times \begin{pmatrix} \mathfrak{H}(\mathsf{N}_M) \\ \mathcal{G} \end{pmatrix},$$

where the sequence on the left-hand side belongs to $\widetilde{A}$, so that

$$f_n'(\xi) - \xi f_n(\xi) = M(\xi)\varphi_n - \varphi_n', \qquad \xi \in \mathbb{C} \setminus \mathbb{R}.$$

Then taking limits in the last identity and using the continuity of point evaluation in $\mathfrak{H}(\mathsf{N}_M)$, see Theorem 4.1.5, leads to the identity

$$f'(\xi) - \xi f(\xi) = M(\xi)\varphi - \varphi', \qquad \xi \in \mathbb{C} \setminus \mathbb{R}.$$

Therefore, the relation $\widetilde{A}$ is closed.

Since $B \subset \widetilde{A}$ and $\widetilde{A}$ is closed, one has $\overline{B} \subset \widetilde{A}$. Hence, to see that $\widetilde{A} = \overline{B}$ it suffices to prove the inclusion $\widetilde{A} \subset \overline{B}$. As $\overline{B}$ is self-adjoint, it suffices to show $\widetilde{A} \subset B^*$. For this let

$$\left\{ \begin{pmatrix} f \\ \varphi \end{pmatrix}, \begin{pmatrix} f' \\ -\varphi' \end{pmatrix} \right\} \in \widetilde{A}.$$

Then $f, f' \in \mathfrak{H}(\mathsf{N}_M)$, $\varphi, \varphi' \in \mathcal{G}$, and

$$f'(\xi) - \xi f(\xi) = M(\xi)\varphi - \varphi', \qquad \xi \in \mathbb{C} \setminus \mathbb{R}. \tag{4.2.11}$$

For an element in B of the form

$$\left\{ \begin{pmatrix} \mathsf{N}_M(\cdot, \bar\mu)\psi \\ -\psi \end{pmatrix}, \begin{pmatrix} \mu \mathsf{N}_M(\cdot, \bar\mu)\psi \\ M(\mu)\psi \end{pmatrix} \right\}$$

with $\psi \in \mathcal{G}$ and some $\mu \in \mathbb{C} \setminus \mathbb{R}$ it follows that

$$\begin{aligned} \left(\begin{pmatrix} f' \\ -\varphi' \end{pmatrix}, \begin{pmatrix} \mathsf{N}_M(\cdot, \bar\mu)\psi \\ -\psi \end{pmatrix} \right) &- \left(\begin{pmatrix} f \\ \varphi \end{pmatrix}, \begin{pmatrix} \mu \mathsf{N}_M(\cdot, \bar\mu)\psi \\ M(\mu)\psi \end{pmatrix} \right) \\ &= \big(f'(\bar\mu) - \bar\mu f(\bar\mu) + \varphi' - M(\bar\mu)\varphi, \psi \big)_{\mathcal{G}} \\ &= \big(M(\bar\mu)\varphi - \varphi' + \varphi' - M(\bar\mu)\varphi, \psi \big)_{\mathcal{G}} = 0, \end{aligned}$$

where (4.2.11) with $\xi = \bar\mu$ was used. This implies $\widetilde{A} \subset B^*$ and thus $\widetilde{A} = \overline{B}$. Hence, $\widetilde{A}$ is self-adjoint in $\mathfrak{H}(\mathsf{N}_M) \oplus \mathcal{G}$.

Step 3. It remains to establish the identities (4.2.7) and (4.2.8). Both are direct consequences of (4.2.6). In fact, let $\lambda \in \mathbb{C} \setminus \mathbb{R}$ and note that

$$(\widetilde{A} - \lambda)^{-1} = \left\{ \left\{ \begin{pmatrix} f' - \lambda f \\ -\varphi' - \lambda\varphi \end{pmatrix}, \begin{pmatrix} f \\ \varphi \end{pmatrix} \right\} : \begin{matrix} f, f' \in \mathfrak{H}(\mathsf{N}_M), \, \varphi, \varphi' \in \mathcal{G}, \\ f'(\xi) - \xi f(\xi) = M(\xi)\varphi - \varphi' \end{matrix} \right\}$$

and hence

$$(\widetilde{A} - \lambda)^{-1}\iota_{\mathcal{G}} = \left\{ \left\{ -\varphi' - \lambda\varphi, \begin{pmatrix} f \\ \varphi \end{pmatrix} \right\} : \begin{matrix} f' = \lambda f, \, f \in \mathfrak{H}(\mathsf{N}_M), \, \varphi, \varphi' \in \mathcal{G}, \\ f'(\xi) - \xi f(\xi) = M(\xi)\varphi - \varphi' \end{matrix} \right\}.$$

The condition $f' = \lambda f$ yields $(\lambda - \xi)f(\xi) = M(\xi)\varphi - \varphi'$, $\xi \in \mathbb{C} \setminus \mathbb{R}$, and setting $\xi = \lambda$ one obtains $\varphi' = M(\lambda)\varphi$. Conversely, if $(\lambda - \xi)f(\xi) = (M(\xi) - M(\lambda))\varphi$ and $\varphi' = M(\lambda)\varphi$ and $f' = \lambda f$, then $f'(\xi) - \xi f(\xi) = M(\xi)\varphi - \varphi'$ for $\xi \in \mathbb{C} \setminus \mathbb{R}$. Therefore,

$$(\widetilde{A} - \lambda)^{-1}\iota_{\mathcal{G}} = \left\{ \left\{ -(M(\lambda) + \lambda)\varphi, \begin{pmatrix} f \\ \varphi \end{pmatrix} \right\} : \begin{matrix} f \in \mathfrak{H}(\mathsf{N}_M), \, \varphi \in \mathcal{G}, \\ (\lambda - \xi)f(\xi) = (M(\xi) - M(\lambda))\varphi \end{matrix} \right\}$$

and this yields $P_{\mathcal{G}}(\widetilde{A} - \lambda)^{-1}\iota_{\mathcal{G}} = -(M(\lambda) + \lambda)^{-1}$; recall that $-\lambda \in \rho(M(\lambda))$ for all $\lambda \in \mathbb{C} \setminus \mathbb{R}$. Hence, (4.2.7) is shown. Moreover, from (4.2.1) it follows that

$$\begin{aligned} \operatorname{ran} P_{\mathfrak{H}(\mathsf{N}_M)}(\widetilde{A} - \lambda)^{-1}\iota_{\mathcal{G}} &= \{ f \in \mathfrak{H}(\mathsf{N}_M) : (\lambda - \xi)f(\xi) = (M(\xi) - M(\lambda))\varphi, \varphi \in \mathcal{G} \} \\ &= \{ -\mathsf{N}_M(\xi, \bar{\lambda})\varphi : \varphi \in \mathcal{G}, \, \xi \in \mathbb{C} \setminus \mathbb{R}, \, \xi \neq \lambda \} \end{aligned}$$

and hence

$$\overline{\operatorname{span}} \{ \operatorname{ran} P_{\mathfrak{H}(\mathsf{N}_M)}(\widetilde{A} - \lambda)^{-1}\iota_{\mathcal{G}} : \lambda \in \mathbb{C} \setminus \mathbb{R} \} = \mathfrak{H}(\mathsf{N}_M)$$

by Theorem 4.1.5 and Corollary 4.1.7. This implies (4.2.8). □

The model and the self-adjoint relation in Theorem 4.2.2 are unique up to unitary equivalence. This is a consequence of the following general equivalence result.

Theorem 4.2.3. *Let $\mathcal{G}$, $\mathfrak{H}$, and $\mathfrak{H}'$ be Hilbert spaces and let $\widetilde{A}$ and $\widetilde{A}'$ be self-adjoint relations in the product spaces $\mathfrak{H} \oplus \mathcal{G}$ and $\mathfrak{H}' \oplus \mathcal{G}$, respectively. Denote by $P_{\mathcal{G}}$ and $P'_{\mathcal{G}}$ the orthogonal projections from $\mathfrak{H} \oplus \mathcal{G}$ and $\mathfrak{H}' \oplus \mathcal{G}$ onto $\mathcal{G}$, respectively, and let $\iota_{\mathcal{G}}$ and $\iota'_{\mathcal{G}}$ be the corresponding canonical embeddings. Assume that $\widetilde{A}$ satisfies the minimality condition*

$$\mathfrak{H} \oplus \mathcal{G} = \overline{\operatorname{span}} \{ \mathcal{G}, \operatorname{ran} (\widetilde{A} - \lambda)^{-1}\iota_{\mathcal{G}} : \lambda \in \mathbb{C} \setminus \mathbb{R} \} \tag{4.2.12}$$

and that $\widetilde{A}'$ satisfies the minimality condition

$$\mathfrak{H}' \oplus \mathcal{G} = \overline{\operatorname{span}} \{ \mathcal{G}, \operatorname{ran} (\widetilde{A}' - \lambda)^{-1}\iota'_{\mathcal{G}} : \lambda \in \mathbb{C} \setminus \mathbb{R} \}. \tag{4.2.13}$$

Furthermore, assume that

$$P_{\mathcal{G}}(\widetilde{A}-\lambda)^{-1}\iota_{\mathcal{G}} = P'_{\mathcal{G}}(\widetilde{A}'-\lambda)^{-1}\iota'_{\mathcal{G}}, \qquad \lambda \in \mathbb{C}\setminus\mathbb{R}. \tag{4.2.14}$$

Then $\widetilde{A}$ and $\widetilde{A}'$ are unitarily equivalent, that is, there exists a unitary operator $U \in \mathbf{B}(\mathfrak{H}\oplus\mathcal{G}, \mathfrak{H}'\oplus\mathcal{G})$ such that $\widetilde{A}' = U\widetilde{A}U^*$.

Proof. Note that the elements of the form

$$\sum_{j=1}^{n}\bigl(\alpha_j\varphi_j + \beta_j(\widetilde{A}-\lambda_j)^{-1}\psi_j\bigr), \tag{4.2.15}$$

where $\varphi_j, \psi_j \in \mathcal{G}$, $\alpha_j, \beta_j \in \mathbb{C}$, $\lambda_j \in \mathbb{C}\setminus\mathbb{R}$ for $j = 1, \dots, n$, and $n \in \mathbb{N}$ are arbitrarily chosen, form a dense subspace of the Hilbert space $\mathfrak{H}\oplus\mathcal{G}$ by the assumption (4.2.12). Likewise, the elements of the form

$$\sum_{j=1}^{n'}\bigl(\alpha'_j\varphi'_j + \beta'_j(\widetilde{A}'-\lambda'_j)^{-1}\psi'_j\bigr), \tag{4.2.16}$$

where $\varphi'_j, \psi'_j \in \mathcal{G}$, $\alpha'_j, \beta'_j \in \mathbb{C}$, $\lambda'_j \in \mathbb{C}\setminus\mathbb{R}$ for $j = 1, \dots, n'$, and $n' \in \mathbb{N}$ are arbitrarily chosen, form a dense subspace of the Hilbert space $\mathfrak{H}'\oplus\mathcal{G}$ by the assumption (4.2.13).

Define the linear relation U from $\mathfrak{H}\oplus\mathcal{G}$ to $\mathfrak{H}'\oplus\mathcal{G}$ as the linear span of all pairs of the form

$$\left\{\sum_{j=1}^{n}\bigl(\alpha_j\varphi_j + \beta_j(\widetilde{A}-\lambda_j)^{-1}\psi_j\bigr), \sum_{j=1}^{n}\bigl(\alpha_j\varphi_j + \beta_j(\widetilde{A}'-\lambda_j)^{-1}\psi_j\bigr)\right\},$$

where $\varphi_j, \psi_j \in \mathcal{G}$, $\alpha_j, \beta_j \in \mathbb{C}$, $\lambda_j \in \mathbb{C}\setminus\mathbb{R}$ for $j = 1, \dots, n$, and $n \in \mathbb{N}$ are arbitrarily chosen. Then according to (4.2.15) and (4.2.16) the relation U has a dense domain and a dense range. To show that the relation U is isometric, i.e., $\|h'\| = \|h\|$ for all $\{h, h'\} \in H$, one has to verify that

$$\begin{aligned}
&\left(\sum_{j=1}^{n}\bigl(\alpha_j\varphi_j + \beta_j(\widetilde{A}'-\lambda_j)^{-1}\psi_j\bigr), \sum_{i=1}^{n}\bigl(\alpha_i\varphi_i + \beta_i(\widetilde{A}'-\lambda_i)^{-1}\psi_i\bigr)\right)\\
&\quad = \left(\sum_{j=1}^{n}\bigl(\alpha_j\varphi_j + \beta_j(\widetilde{A}-\lambda_j)^{-1}\psi_j\bigr), \sum_{i=1}^{n}\bigl(\alpha_i\varphi_i + \beta_i(\widetilde{A}-\lambda_i)^{-1}\psi_i\bigr)\right).
\end{aligned}$$

To see this, it suffices to observe that (4.2.14) implies

$$\begin{aligned}
\bigl((\widetilde{A}'-\lambda_j)^{-1}\psi_j, \varphi_i\bigr)_{\mathcal{G}} &= \bigl(P'_{\mathcal{G}}(\widetilde{A}'-\lambda_j)^{-1}\psi_j, \varphi_i\bigr)_{\mathcal{G}}\\
&= \bigl(P_{\mathcal{G}}(\widetilde{A}-\lambda_j)^{-1}\psi_j, \varphi_i\bigr)_{\mathcal{G}}\\
&= \bigl((\widetilde{A}-\lambda_j)^{-1}\psi_j, \varphi_i\bigr)_{\mathcal{G}},
\end{aligned}$$

and, likewise, by symmetry,

$$\big(\varphi_j, (\widetilde{A}' - \lambda_i)^{-1}\psi_i\big)_{\mathcal{G}} = \big(\varphi_j, (\widetilde{A} - \lambda_i)^{-1}\psi_i\big)_{\mathcal{G}}.$$

Moreover, using the resolvent identity, one sees that for $\lambda_j \neq \overline{\lambda}_i$ (4.2.14) implies

$$\begin{aligned}
&\big((\widetilde{A}' - \lambda_j)^{-1}\psi_j, (\widetilde{A}' - \lambda_i)^{-1}\psi_i\big)_{\mathcal{G}} \\
&\quad = \big((\widetilde{A}' - \lambda_j)^{-1}(\widetilde{A}' - \overline{\lambda}_i)^{-1}\psi_j, \psi_i\big)_{\mathcal{G}} \\
&\quad = (\lambda_j - \overline{\lambda}_i)^{-1}\big[\big((\widetilde{A}' - \lambda_j)^{-1}\psi_j, \psi_i\big)_{\mathcal{G}} - \big((\widetilde{A}' - \overline{\lambda}_i)^{-1}\psi_j, \psi_i\big)_{\mathcal{G}}\big] \\
&\quad = (\lambda_j - \overline{\lambda}_i)^{-1}\big[\big((\widetilde{A} - \lambda_j)^{-1}\psi_j, \psi_i\big)_{\mathcal{G}} - \big((\widetilde{A} - \overline{\lambda}_i)^{-1}\psi_j, \psi_i\big)_{\mathcal{G}}\big] \\
&\quad = \big((\widetilde{A} - \lambda_j)^{-1}(\widetilde{A} - \overline{\lambda}_i)^{-1}\psi_j, \psi_i\big)_{\mathcal{G}} \\
&\quad = \big((\widetilde{A} - \lambda_j)^{-1}\psi_j, (\widetilde{A} - \lambda_i)^{-1}\psi_i\big)_{\mathcal{G}},
\end{aligned}$$

and a limit argument together with the continuity of the resolvent shows that the same is true in the case $\lambda_j = \overline{\lambda}_i$. Thus, the relation U is isometric; hence it is a well-defined isometric operator and the closure of U, denoted again by U, is a unitary operator from $\mathfrak{H} \oplus \mathcal{G}$ onto $\mathfrak{H}' \oplus \mathcal{G}$.

Next it will be shown that $\widetilde{A}$ and $\widetilde{A}'$ are unitarily equivalent under U. To see this, one needs to show

$$U(\widetilde{A} - \lambda)^{-1} = (\widetilde{A}' - \lambda)^{-1}U \tag{4.2.17}$$

for some $\lambda \in \mathbb{C} \setminus \mathbb{R}$; cf. Lemma 1.3.8. Since all operators involved are bounded, it suffices to check this identity on a dense set of $\mathfrak{H} \oplus \mathcal{G}$; thus, in fact, it suffices to check this only for the elements in (4.2.15). Observe that for elements in (4.2.15) with $\lambda \neq \lambda_j$ and $\gamma_j := \frac{\beta_j}{\lambda_j - \lambda}$ one has, again by the resolvent identity,

$$\begin{aligned}
&U(\widetilde{A} - \lambda)^{-1}\big(\alpha_j\varphi_j + \beta_j(\widetilde{A} - \lambda_j)^{-1}\psi_j\big) \\
&\quad = U\big((\widetilde{A} - \lambda)^{-1}(\alpha_j\varphi_j - \gamma_j\psi_j) + \gamma_j(\widetilde{A} - \lambda_j)^{-1}\psi_j\big) \\
&\quad = (\widetilde{A}' - \lambda)^{-1}(\alpha_j\varphi_j - \gamma_j\psi_j) + \gamma_j(\widetilde{A}' - \lambda_j)^{-1}\psi_j \\
&\quad = (\widetilde{A}' - \lambda)^{-1}\big(\alpha_j\varphi_j + \beta_j(\widetilde{A}' - \lambda_j)^{-1}\psi_j\big) \\
&\quad = (\widetilde{A}' - \lambda)^{-1}U\big(\alpha_j\varphi_j + \beta_j(\widetilde{A} - \lambda_j)^{-1}\psi_j\big),
\end{aligned}$$

and by the continuity of the resolvent the same relation holds also for $\lambda = \lambda_j$. Thus, (4.2.17) has been established. □

The second main result in this section concerns a minimal model for uniformly strict Nevanlinna functions. By means of Theorem 4.2.2 it will be shown that every uniformly strict Nevanlinna function M is the Weyl function corresponding to a boundary triplet of a simple symmetric operator in the Hilbert space $\mathfrak{H}(\mathsf{N}_M)$. After this result it will be shown that every boundary triplet producing the same Weyl function is unitarily equivalent to the boundary triplet in this construction. Note that the description of $(S_M)^*$ involves functions which do not necessarily belong to $\mathfrak{H}(\mathsf{N}_M)$; however the definition of S_M concerns functions which remain in $\mathfrak{H}(\mathsf{N}_M)$ after multiplication by the independent variable.

Theorem 4.2.4. *Let M be a uniformly strict $\mathbf{B}(\mathcal{G})$-valued Nevanlinna function and let $\mathfrak{H}(\mathsf{N}_M)$ be the associated reproducing kernel Hilbert space. Then*

$$S_M = \{\{f, f'\} \in \mathfrak{H}(\mathsf{N}_M)^2 : f'(\xi) = \xi f(\xi)\} \tag{4.2.18}$$

is a closed simple symmetric operator in $\mathfrak{H}(\mathsf{N}_M)$ and its adjoint is given by

$$(S_M)^* = \{\{f, f'\} \in \mathfrak{H}(\mathsf{N}_M)^2 : f'(\xi) - \xi f(\xi) = M(\xi)\varphi - \varphi',\ \varphi, \varphi' \in \mathcal{G}\}. \tag{4.2.19}$$

Moreover, the mappings

$$\Gamma_0 \widehat{f} = \varphi \quad \text{and} \quad \Gamma_1 \widehat{f} = \varphi', \quad \widehat{f} \in (S_M)^*, \tag{4.2.20}$$

are well defined and $\{\mathcal{G}, \Gamma_0, \Gamma_1\}$ is a boundary triplet for $(S_M)^*$. The corresponding γ-field is given by

$$\gamma(\lambda)\varphi = -\mathsf{N}_M(\cdot, \bar{\lambda})\varphi \tag{4.2.21}$$

and the corresponding Weyl function is given by M.

Proof. By Theorem 4.2.2, the relation

$$\widetilde{A} = \left\{ \left\{ \begin{pmatrix} f \\ \varphi \end{pmatrix}, \begin{pmatrix} f' \\ -\varphi' \end{pmatrix} \right\} : \begin{matrix} f, f' \in \mathfrak{H}(\mathsf{N}_M),\ \varphi, \varphi' \in \mathcal{G}, \\ f'(\xi) - \xi f(\xi) = M(\xi)\varphi - \varphi' \end{matrix} \right\}$$

is self-adjoint in $\mathfrak{H}(\mathsf{N}_M) \oplus \mathcal{G}$. The relations in (4.2.18) and (4.2.19) are defined in the component space $\mathfrak{H}(\mathsf{N}_M)$; the condition that M is uniformly strict makes it possible to connect them with $\widetilde{A}$.

Step 1. The relation S_M in (4.2.18) is a closed symmetric operator with adjoint $(S_M)^*$ given by (4.2.19). First observe that the relation S_M in (4.2.18) satisfies

$$S_M = \widetilde{A} \cap \left(\begin{pmatrix} \mathfrak{H}(\mathsf{N}_M) \\ \{0\} \end{pmatrix} \times \begin{pmatrix} \mathfrak{H}(\mathsf{N}_M) \\ \{0\} \end{pmatrix} \right), \tag{4.2.22}$$

when the space $\mathfrak{H}(\mathsf{N}_M) \times \{0\}$ is identified with $\mathfrak{H}(\mathsf{N}_M)$. Since $\widetilde{A}$ is self-adjoint, (4.2.22) implies that S_M is closed and symmetric, and it is clear from (4.2.18) that S_M is an operator. In order to find the adjoint of S_M, let the relation T_M be defined by

$$T_M = \{\{f, f'\} \in \mathfrak{H}(\mathsf{N}_M)^2 : f'(\xi) - \xi f(\xi) = M(\xi)\varphi - \varphi',\ \varphi, \varphi' \in \mathcal{G}\}.$$

Observe that

$$\{g, g'\} \in (T_M)^* \quad \Leftrightarrow \quad \left\{ \begin{pmatrix} g \\ 0 \end{pmatrix}, \begin{pmatrix} g' \\ 0 \end{pmatrix} \right\} \in \widetilde{A}^* = \widetilde{A} \quad \Leftrightarrow \quad \{g, g'\} \in S_M,$$

which leads to

$$(T_M)^{**} = (S_M)^*.$$

Hence, to conclude (4.2.19), it suffices to show that T_M is closed. To see this, assume that $\{f_n, f_n'\} \in T_M$ converges in $\mathfrak{H}(\mathsf{N}_M)^2$ to $\{f, f'\}$. Then there exists a sequence $\{\varphi_n, \varphi_n'\} \in \mathcal{G}^2$ such that

$$f_n'(\xi) - \xi f_n(\xi) = M(\xi)\varphi_n - \varphi_n', \quad \xi \in \mathbb{C} \setminus \mathbb{R}.$$

By Theorem 4.1.5 (iii), the point evaluation is continuous, so that

$$f_n(\xi) \to f(\xi) \quad \text{and} \quad f_n'(\xi) \to f'(\xi).$$

Hence,

$$M(\xi)\varphi_n - \varphi_n' \to f'(\xi) - \xi f(\xi)$$

for all $\xi \in \mathbb{C} \setminus \mathbb{R}$. Taking $\xi = \lambda_0$ and $\xi = \overline{\lambda}_0$ with $\lambda_0 \in \mathbb{C} \setminus \mathbb{R}$ one sees that

$$\big(M(\lambda_0) - M(\overline{\lambda}_0)\big)\varphi_n \to f'(\lambda_0) - \lambda_0 f(\lambda_0) - \big(f'(\overline{\lambda}_0) - \overline{\lambda}_0 f(\overline{\lambda}_0)\big)$$

and using that $\operatorname{Im} M(\lambda_0)$ is boundedly invertible (since M is assumed to be uniformly strict), it follows that

$$\varphi_n \to \varphi \quad \text{and hence also} \quad \varphi_n' \to \varphi'$$

for some $\varphi, \varphi' \in \mathcal{G}$. Therefore,

$$f'(\xi) - \xi f(\xi) = M(\xi)\varphi - \varphi', \quad \xi \in \mathbb{C} \setminus \mathbb{R}.$$

In other words, $\{f, f'\} \in T_M$, and hence T_M is closed.

Step 2. The mappings in (4.2.20) form a boundary triplet for $(S_M)^*$. First note that they are single-valued. Indeed, assume that $\{f, f'\} \in (S_M)^*$ is the trivial element, then $M(\xi)\varphi - \varphi' = 0$ for the corresponding elements $\varphi, \varphi' \in \mathcal{G}$ and all $\xi \in \mathbb{C} \setminus \mathbb{R}$. Taking $\xi = \lambda_0$ and $\xi = \overline{\lambda}_0$ with $\lambda_0 \in \mathbb{C} \setminus \mathbb{R}$ and using the fact that $\ker(\operatorname{Im} M(\lambda_0)) = \{0\}$, one concludes that $\varphi = 0$ and $\varphi' = 0$.

To verify the abstract Green identity, let $\widehat{f} = \{f, f'\}, \widehat{g} = \{g, g'\} \in (S_M)^*$. Then

$$f'(\xi) - \xi f(\xi) = M(\xi)\varphi - \varphi' \quad \text{and} \quad g'(\xi) - \xi g(\xi) = M(\xi)\psi - \psi'$$

for some $\varphi, \varphi', \psi, \psi'$ in $\mathcal{G}$ and some $\xi \in \mathbb{C} \setminus \mathbb{R}$; moreover it is clear that

$$\left\{ \begin{pmatrix} f \\ \varphi \end{pmatrix}, \begin{pmatrix} f' \\ -\varphi' \end{pmatrix} \right\}, \ \left\{ \begin{pmatrix} g \\ \psi \end{pmatrix}, \begin{pmatrix} g' \\ -\psi' \end{pmatrix} \right\} \in \widetilde{A}.$$

Since $\widetilde{A}$ is self-adjoint and thus symmetric, one sees that

$$\langle f', g \rangle - \langle f, g' \rangle = (\varphi', \psi)_{\mathcal{G}} - (\varphi, \psi')_{\mathcal{G}} = (\Gamma_1 \widehat{f}, \Gamma_0 \widehat{g})_{\mathcal{G}} - (\Gamma_0 \widehat{f}, \Gamma_1 \widehat{g})_{\mathcal{G}}.$$

Thus, the abstract Green identity is satisfied. It remains to show that Γ maps onto $\mathcal{G}^2$, which then implies that $\{\mathcal{G}, \Gamma_0, \Gamma_1\}$ is a boundary triplet for $(S_M)^*$.

First, it will be shown that $\operatorname{ran}\Gamma$ is dense in $\mathcal{G}^2$. Suppose that $\{\alpha', \alpha\} \in \mathcal{G}^2$ is orthogonal to $\operatorname{ran}\Gamma$, that is,

$$(\alpha', \varphi) + (\alpha, \varphi') = 0 \quad \text{for all} \quad \{\varphi, \varphi'\} \in \operatorname{ran}\Gamma.$$

It follows that

$$\left\{\begin{pmatrix} 0 \\ \alpha \end{pmatrix}, \begin{pmatrix} 0 \\ \alpha' \end{pmatrix}\right\} \in \widetilde{A}^* = \widetilde{A}$$

and hence $M(\xi)\alpha + \alpha' = 0$ for all $\xi \in \mathbb{C} \setminus \mathbb{R}$. Now as above one concludes that $\alpha = 0$ and $\alpha' = 0$. Therefore, $\operatorname{ran}\Gamma$ is dense in $\mathcal{G}^2$.

Next, it will be shown that $\operatorname{ran}\Gamma$ is closed. For this consider again the self-adjoint relation $\widetilde{A}$ as a subspace of $(\mathfrak{H}(\mathsf{N}_M) \oplus \mathcal{G})^2$ and define the orthogonal projections P and $I - P$ by

$$P : \left(\begin{pmatrix} \mathfrak{H}(\mathsf{N}_M) \\ \mathcal{G} \end{pmatrix} \times \begin{pmatrix} \mathfrak{H}(\mathsf{N}_M) \\ \mathcal{G} \end{pmatrix}\right) \to \left(\begin{pmatrix} \mathfrak{H}(\mathsf{N}_M) \\ \{0\} \end{pmatrix} \times \begin{pmatrix} \mathfrak{H}(\mathsf{N}_M) \\ \{0\} \end{pmatrix}\right)$$

and

$$I - P : \left(\begin{pmatrix} \mathfrak{H}(\mathsf{N}_M) \\ \mathcal{G} \end{pmatrix} \times \begin{pmatrix} \mathfrak{H}(\mathsf{N}_M) \\ \mathcal{G} \end{pmatrix}\right) \to \left(\begin{pmatrix} \{0\} \\ \mathcal{G} \end{pmatrix} \times \begin{pmatrix} \{0\} \\ \mathcal{G} \end{pmatrix}\right),$$

respectively. Then $P\widetilde{A} = T_M = (S_M)^*$ is closed and hence, by Lemma C.4,

$$\widetilde{A} \mathbin{\widehat{+}} \ker P = \widetilde{A} \mathbin{\widehat{+}} \left(\begin{pmatrix} \{0\} \\ \mathcal{G} \end{pmatrix} \times \begin{pmatrix} \{0\} \\ \mathcal{G} \end{pmatrix}\right)$$

is closed. Since $\widetilde{A}$ is self-adjoint, it follows from this and (1.3.5) that

$$\widetilde{A} \mathbin{\widehat{+}} \left(\begin{pmatrix} \mathfrak{H}(\mathsf{N}_M) \\ \{0\} \end{pmatrix} \times \begin{pmatrix} \mathfrak{H}(\mathsf{N}_M) \\ \{0\} \end{pmatrix}\right) = \widetilde{A} \mathbin{\widehat{+}} \ker (I - P)$$

is closed. By Lemma C.4, the relation $(I - P)\widetilde{A}$ is closed. Observe that $(I - P)\widetilde{A}$ is given by

$$\{\{\varphi, -\varphi'\} \in \mathcal{G}^2 : f'(\xi) - \xi f(\xi) = M(\xi)\varphi - \varphi',\ f, f' \in \mathfrak{H}(\mathsf{N}_M)\} \tag{4.2.23}$$

and hence (4.2.23) is closed. In other words, $\operatorname{ran}(\Gamma_0, -\Gamma_1)^\top$ is closed or, equivalently, $\operatorname{ran}(\Gamma_0, \Gamma_1)^\top$ is closed.

Step 3. The γ-field and Weyl function corresponding to the boundary triplet (4.2.20) are given by γ in (4.2.21) and M, respectively, and the symmetric operator S_M in (4.2.18) is simple.

To establish the assertion about M being the Weyl function, fix $\lambda \in \mathbb{C} \setminus \mathbb{R}$ and let $\widehat{f}_\lambda \in \widehat{\mathfrak{N}}_\lambda((S_M)^*)$. Then $f'_\lambda(\xi) = \lambda f_\lambda(\xi)$ for all $\xi \in \mathbb{C} \setminus \mathbb{R}$, and by (4.2.19)

$$(\lambda - \xi) f_\lambda(\xi) = M(\xi)\varphi_\lambda - \varphi'_\lambda, \quad \xi \in \mathbb{C} \setminus \mathbb{R}, \tag{4.2.24}$$

where $\varphi_\lambda = \Gamma_0 \widehat{f}_\lambda$ and $\varphi'_\lambda = \Gamma_1 \widehat{f}_\lambda$. The choice $\xi = \lambda$ in (4.2.24) shows $M(\lambda)\varphi_\lambda = \varphi'_\lambda$ and hence

$$M(\lambda)\Gamma_0 \widehat{f}_\lambda = \Gamma_1 \widehat{f}_\lambda.$$

As this is true for all $\widehat{f}_\lambda \in \widehat{\mathfrak{N}}_\lambda((S_M)^*)$ and all $\lambda \in \mathbb{C}\setminus\mathbb{R}$, one concludes that M is the Weyl function corresponding to the boundary triplet (4.2.20).

To compute the γ-field and to show that S_M is simple, assume again that $\widehat{f}_\lambda \in \widehat{\mathfrak{N}}_\lambda((S_M)^*)$. Then (4.2.24) with $\xi \neq \lambda$ implies that

$$f_\lambda(\xi) = \frac{M(\xi)\varphi_\lambda - \varphi'_\lambda}{\lambda - \xi} = -\frac{M(\xi) - M(\lambda)}{\xi - \lambda}\varphi_\lambda = -\mathsf{N}_M(\xi, \bar\lambda)\varphi_\lambda.$$

Hence, the γ-field corresponding to the boundary triplet (4.2.20) is given by (4.2.21), and since the elements $f_\lambda(\cdot) = \mathsf{N}_M(\cdot, \bar\lambda)\varphi_\lambda \in \ker((S_M)^* - \lambda)$ form a dense set in the Hilbert space $\mathfrak{H}(\mathsf{N}_M)$ (see Theorem 4.1.5 and Corollary 4.1.7), it follows from Corollary 3.4.5 that S_M is simple. □

Corollary 4.2.5. *Let $\{\mathcal{G}, \Gamma_0, \Gamma_1\}$ be the boundary triplet from Theorem 4.2.4 for $(S_M)^*$. Then*

$$A_0 = \ker \Gamma_0 = \bigl\{\{f, f'\} \in \mathfrak{H}(\mathsf{N}_M)^2 : f'(\xi) - \xi f(\xi) = \varphi',\ \varphi' \in \mathcal{G}\bigr\}$$

and

$$A_1 = \ker \Gamma_1 = \bigl\{\{f, f'\} \in \mathfrak{H}(\mathsf{N}_M)^2 : f'(\xi) - \xi f(\xi) = M(\xi)\varphi,\ \varphi \in \mathcal{G}\bigr\}$$

are self-adjoint relations in $\mathfrak{H}(\mathsf{N}_M)$.

It follows from Theorem 4.2.4 that $\operatorname{mul}(S_M)^*$ is the linear space spanned by all linear combinations $M(\cdot)\varphi - \varphi'$, $\varphi, \varphi' \in \mathcal{G}$, which belong to the Hilbert space $\mathfrak{H}(\mathsf{N}_M)$. Likewise, it follows from Corollary 4.2.5 that $\operatorname{mul} A_0$ consists of all constant functions in $\mathfrak{H}(\mathsf{N}_M)$, while $\operatorname{mul} A_1$ consists of all linear combinations $M(\cdot)\varphi$, $\varphi \in \mathcal{G}$, which belong to the Hilbert space $\mathfrak{H}(\mathsf{N}_M)$.

The construction of the boundary triplet in Theorem 4.2.4 is unique up to unitary equivalence. More precisely, if S is a simple symmetric operator in a Hilbert space $\mathfrak{H}$ and there is a boundary triplet for S^* with the same Weyl function M as in Theorem 4.2.4, then the boundary triplets are unitarily equivalent in the sense of Definition 2.5.14 (where $\mathcal{G} = \mathcal{G}'$). This is a consequence of the following general equivalence result, which is a further specification of Theorem 4.2.3.

Theorem 4.2.6. *Let S and S' be closed simple symmetric operators in Hilbert spaces $\mathfrak{H}$ and $\mathfrak{H}'$, respectively. Let $\{\mathcal{G}, \Gamma_0, \Gamma_1\}$ and $\{\mathcal{G}, \Gamma'_0, \Gamma'_1\}$ be boundary triplets for S^* and $(S')^*$ with γ-fields γ and γ', respectively. Assume that the corresponding Weyl functions M and M' coincide. Then the boundary triplets $\{\mathcal{G}, \Gamma_0, \Gamma_1\}$ and $\{\mathcal{G}, \Gamma'_0, \Gamma'_1\}$ are unitarily equivalent by means of a unitary operator $U : \mathfrak{H} \to \mathfrak{H}'$ which is determined by the property*

$$U\gamma(\lambda) = \gamma'(\lambda), \qquad \lambda \in \mathbb{C}\setminus\mathbb{R}. \tag{4.2.25}$$

Proof. The basic idea of the proof follows the proof of Theorem 4.2.3. By assumption, the Weyl functions M and M' of the boundary triplets $\{\mathcal{G},\Gamma_0,\Gamma_1\}$ and $\{\mathcal{G},\Gamma_0',\Gamma_1'\}$ coincide. It follows from Proposition 2.3.6 (iii) that the corresponding γ-fields γ and γ' satisfy the identity

$$\begin{aligned}\gamma(\mu)^*\gamma(\lambda) &= \frac{M(\lambda)-M(\mu)^*}{\lambda-\bar\mu}\\ &= \frac{M'(\lambda)-M'(\mu)^*}{\lambda-\bar\mu} = \gamma'(\mu)^*\gamma'(\lambda)\end{aligned}\tag{4.2.26}$$

for all $\lambda,\mu\in\mathbb{C}\setminus\mathbb{R}$, $\lambda\neq\bar\mu$, and $\gamma(\bar\lambda)^*\gamma(\lambda)=\gamma'(\bar\lambda)^*\gamma'(\lambda)$ follows by continuity. Define the linear relation U from $\mathfrak{H}$ to $\mathfrak{H}'$ as the linear set of all pairs of the form

$$\left\{\sum_{j=1}^n \alpha_j\gamma(\lambda_j)\varphi_j, \sum_{j=1}^n \alpha_j\gamma'(\lambda_j)\varphi_j\right\},$$

where $\varphi_j\in\mathcal{G}$, $\alpha_j\in\mathbb{C}$, $\lambda_j\in\mathbb{C}\setminus\mathbb{R}$ for $j=1,\dots,n$, and $n\in\mathbb{N}$ are arbitrarily chosen. It is clear from the definition of U that its domain is given by

$$\begin{aligned}\operatorname{dom}U &= \operatorname{span}\bigl\{\operatorname{ran}\gamma(\lambda):\lambda\in\mathbb{C}\setminus\mathbb{R}\bigr\}\\ &= \operatorname{span}\bigl\{\ker((S)^*-\lambda):\lambda\in\mathbb{C}\setminus\mathbb{R}\bigr\},\end{aligned}$$

and its range is given by

$$\begin{aligned}\operatorname{ran}U &= \operatorname{span}\bigl\{\operatorname{ran}\gamma'(\lambda):\lambda\in\mathbb{C}\setminus\mathbb{R}\bigr\}\\ &= \operatorname{span}\bigl\{\ker((S')^*-\lambda):\lambda\in\mathbb{C}\setminus\mathbb{R}\bigr\},\end{aligned}$$

which are dense in $\mathfrak{H}$ and $\mathfrak{H}'$, respectively, since S and S' are both simple by assumption; cf. Definition 3.4.3 and Corollary 3.4.5. From (4.2.26) it follows that the relation U is isometric; hence, it is a well-defined isometric operator. Therefore, U extends by continuity to a unitary operator from $\mathfrak{H}$ to $\mathfrak{H}'$, denoted again by U. From

$$U\gamma(\lambda)\varphi=\gamma'(\lambda)\varphi,\qquad \lambda\in\mathbb{C}\setminus\mathbb{R},\quad \varphi\in\mathcal{G},\tag{4.2.27}$$

it follows that for each $\lambda\in\mathbb{C}\setminus\mathbb{R}$ the restriction $U:\ker(S^*-\lambda)\to\ker((S')^*-\lambda)$ is unitary. Thus, (4.2.27) implies by Proposition 2.3.6 (ii) that

$$\begin{aligned}\Gamma\{\gamma(\lambda)\varphi,\lambda\gamma(\lambda)\varphi\} &= \{\varphi,M(\lambda)\varphi\}\\ &= \{\varphi,M'(\lambda)\varphi\}\\ &= \Gamma'\{\gamma'(\lambda)\varphi,\lambda\gamma'(\lambda)\varphi\}\\ &= \Gamma'\{U\gamma(\lambda)\varphi,\lambda U\gamma(\lambda)\varphi\},\end{aligned}$$

and, in particular,

$$\begin{aligned}\Gamma_0\{\gamma(\lambda)\varphi,\lambda\gamma(\lambda)\varphi\} &= \Gamma_0'\{U\gamma(\lambda)\varphi,\lambda U\gamma(\lambda)\varphi\},\\ \Gamma_1\{\gamma(\lambda)\varphi,\lambda\gamma(\lambda)\varphi\} &= \Gamma_1'\{U\gamma(\lambda)\varphi,\lambda U\gamma(\lambda)\varphi\},\end{aligned}\tag{4.2.28}$$

for all $\lambda\in\mathbb{C}\setminus\mathbb{R}$ and $\varphi\in\mathcal{G}$.

Now let $A_0 = \ker \Gamma_0$ and $A_0' = \ker \Gamma_0'$. Then the property (4.2.27) and Proposition 2.3.2 (ii) imply

$$\begin{aligned} U\big(I + (\lambda - \mu)(A_0 - \lambda)^{-1}\big)\gamma(\mu) &= U\gamma(\lambda) \\ &= \gamma'(\lambda) \\ &= \big(I + (\lambda - \mu)(A_0' - \lambda)^{-1}\big)\gamma'(\mu) \\ &= \big(I + (\lambda - \mu)(A_0' - \lambda)^{-1}\big)U\gamma(\mu) \end{aligned}$$

for all $\lambda, \mu \in \mathbb{C} \setminus \mathbb{R}$, and hence

$$U(A_0 - \lambda)^{-1}\gamma(\mu) = (A_0' - \lambda)^{-1}U\gamma(\mu), \qquad \lambda, \mu \in \mathbb{C} \setminus \mathbb{R}.$$

Since S is simple, one sees that $\operatorname{span}\{\operatorname{ran}\gamma(\mu) : \mu \in \mathbb{C} \setminus \mathbb{R}\}$ is a dense subspace of $\mathfrak{H}$ and hence

$$U(A_0 - \lambda)^{-1} = (A_0' - \lambda)^{-1}U, \qquad \lambda \in \mathbb{C} \setminus \mathbb{R},$$

and

$$U(A_0 - \lambda)^{-1}U^* = (A_0' - \lambda)^{-1}, \qquad \lambda \in \mathbb{C} \setminus \mathbb{R}, \tag{4.2.29}$$

follow. Therefore, by Lemma 1.3.8, the self-adjoint relations A_0' and A_0 are unitarily equivalent, that is,

$$A_0' = \big\{\{Uf, Uf'\} : \{f, f'\} \in A_0\big\}. \tag{4.2.30}$$

This immediately yields

$$\Gamma_0'\{Uf, Uf'\} = 0 \quad \text{and} \quad \Gamma_0\{f, f'\} = 0, \qquad \{f, f'\} \in A_0. \tag{4.2.31}$$

Furthermore, each $\{f, f'\} \in A_0$ can be written as

$$\{f, f'\} = \big\{(A_0 - \lambda)^{-1}U^*g, \big(I + \lambda(A_0 - \lambda)^{-1}\big)U^*g\big\}$$

for some $g \in \mathfrak{H}'$, so that by means of (4.2.29), Proposition 2.3.2 (iv), and (4.2.27) one obtains that

$$\begin{aligned} \Gamma_1'\{Uf, Uf'\} &= \Gamma_1'\big\{U(A_0 - \lambda)^{-1}U^*g, U\big(I + \lambda(A_0 - \lambda)^{-1}\big)U^*g\big\} \\ &= \Gamma_1'\big\{(A_0' - \lambda)^{-1}g, \big(I + \lambda(A_0' - \lambda)^{-1}\big)g\big\} \\ &= \gamma'(\bar{\lambda})^*g \\ &= \gamma(\bar{\lambda})^*U^*g \\ &= \Gamma_1\big\{(A_0 - \lambda)^{-1}U^*g, \big(I + \lambda(A_0 - \lambda)^{-1}\big)U^*g\big\}. \end{aligned}$$

Therefore,

$$\Gamma_1'\{Uf, Uf'\} = \Gamma_1\{f, f'\}, \qquad \{f, f'\} \in A_0. \tag{4.2.32}$$

To see that the boundary triplets are unitarily equivalent, first recall that A_0 and A_0' are unitarily equivalent, see (4.2.30), and that

$$\widehat{\mathfrak{N}}_\lambda((S')^*) = \big\{\{Uf_\lambda, \lambda Uf_\lambda\} : \{f_\lambda, \lambda f_\lambda\} \in S^*\big\};$$

cf. (4.2.27). The direct sum decompositions

$$(S')^* = A_0' \,\widehat{+}\, \mathfrak{N}_\lambda((S')^*) \quad \text{and} \quad S^* = A_0 \,\widehat{+}\, \mathfrak{N}_\lambda(S^*) \tag{4.2.33}$$

for $\lambda \in \rho(A_0) = \rho(A_0')$ from Theorem 1.7.1 now show that

$$(S')^* = \bigl\{\{Uf, Uf'\} : \{f, f'\} \in S^*\bigr\}, \tag{4.2.34}$$

so that S^* and $(S')^*$ are unitarily equivalent. It follows from (4.2.33) and the equalities (4.2.28), (4.2.31), and (4.2.32) that

$$\Gamma_0'\{Uf, Uf'\} = \Gamma_0\{f, f'\} \quad \text{and} \quad \Gamma_1'\{Uf, Uf'\} = \Gamma_1\{f, f'\}, \quad \{f, f'\} \in S^*.$$

Together with (4.2.34) this shows that the boundary triplets are unitarily equivalent. □

Let M be a uniformly strict $\mathbf{B}(\mathcal{G})$-valued Nevanlinna function and consider the corresponding model in Theorem 4.2.4. Denote the boundary triplet (4.2.20) for $(S_M)^*$ in this model by $\{\mathcal{G}, (\Gamma_M)_0, (\Gamma_M)_1\}$:

$$(\Gamma_M)_0\widehat{f} = \varphi \quad \text{and} \quad (\Gamma_M)_1\widehat{f} = \varphi', \qquad \widehat{f} = \{f, f'\} \in (S_M)^*. \tag{4.2.35}$$

According to Theorem 2.5.1 and Proposition 2.5.3 every operator matrix

$$\mathcal{W} = \begin{pmatrix} W_{11} & W_{12} \\ W_{21} & W_{22} \end{pmatrix} \in \mathbf{B}(\mathcal{G} \times \mathcal{G}, \mathcal{G} \times \mathcal{G}) \tag{4.2.36}$$

with the properties (2.5.1) gives rise to a boundary triplet $\{\mathcal{G}, (\Gamma_M)_0', (\Gamma_M)_1'\}$ for $(S_M)^*$ via $(\Gamma_M)' = \mathcal{W}\Gamma_M$, that is,

$$\begin{pmatrix} (\Gamma_M)_0' \\ (\Gamma_M)_1' \end{pmatrix} = \begin{pmatrix} W_{11} & W_{12} \\ W_{21} & W_{22} \end{pmatrix} \begin{pmatrix} (\Gamma_M)_0 \\ (\Gamma_M)_1 \end{pmatrix}, \tag{4.2.37}$$

and the corresponding γ-field and Weyl function are then given by

$$\gamma'(\lambda) = \gamma(\lambda)\bigl(W_{11} + W_{12}M(\lambda)\bigr)^{-1} \tag{4.2.38}$$

and

$$M'(\lambda) = \bigl(W_{21} + W_{22}M(\lambda)\bigr)\bigl(W_{11} + W_{12}M(\lambda)\bigr)^{-1}. \tag{4.2.39}$$

The function $M' = \mathcal{W}[M]$ in (4.2.39), being a Weyl function, is a uniformly strict $\mathbf{B}(\mathcal{G})$-valued Nevanlinna function. Let $\mathfrak{H}(\mathsf{N}_{M'})$ be the associated reproducing kernel Hilbert space. Then according to Theorem 4.2.4

$$S_{M'} = \bigl\{\{F, F'\} \in \mathfrak{H}(\mathsf{N}_{M'})^2 : \, F'(\xi) = \xi F(\xi)\bigr\}$$

is a closed simple symmetric operator in $\mathfrak{H}(\mathsf{N}_{M'})$ and its adjoint is given by

$$(S_{M'})^* = \bigl\{\{F, F'\} \in \mathfrak{H}(\mathsf{N}_{M'})^2 : F'(\xi) - \xi F(\xi) = M'(\xi)\psi - \psi', \, \psi, \psi' \in \mathcal{G}\bigr\}. \tag{4.2.40}$$

The corresponding boundary triplet $\{\mathcal{G}, (\Gamma_{M'})_0, (\Gamma_{M'})_1\}$ is given by

$$(\Gamma_{M'})_0\widehat{F} = \psi, \quad (\Gamma_{M'})_1\widehat{F} = \psi', \qquad \widehat{F} = \{F, F'\} \in (S_{M'})^*, \tag{4.2.41}$$

and according to Theorem 4.2.4 the corresponding Weyl function is M'. In the next proposition it will explained how this model for M' is connected with the space $\mathfrak{H}(\mathsf{N}_M)$ and the transformed boundary triplet $\{\mathcal{G}, (\Gamma_M)'_0, (\Gamma_M)'_1\}$ for $(S_M)^*$ in (4.2.37). The unitary map Φ in (4.2.43) below provides the unitary equivalence between the boundary triplets in the sense of Theorem 4.2.6.

Proposition 4.2.7. *Let M be the Weyl function in Theorem 4.2.4 with boundary triplet $\{\mathcal{G}, (\Gamma_M)_0, (\Gamma_M)_1\}$ and let $\mathcal{W}$ be of the form (4.2.36) with the properties in (2.5.1) so that $\{\mathcal{G}, (\Gamma_M)'_0, (\Gamma_M)'_1\}$ in (4.2.37) is a boundary triplet for $(S_M)^*$ with corresponding Weyl function M'. Then the kernels N_M and $\mathsf{N}_{M'}$ are connected via*

$$\mathsf{N}_{M'}(\lambda, \mu) = \Phi(\lambda)\mathsf{N}_M(\lambda, \mu)\Phi(\mu)^*, \tag{4.2.42}$$

where $\Phi : \mathbb{C} \setminus \mathbb{R} \to \mathbf{B}(\mathcal{G})$ is a holomorphic function given by

$$\Phi(\lambda) = \bigl(W_{11}^* + M(\lambda)W_{12}^*\bigr)^{-1}. \tag{4.2.43}$$

Furthermore, $\widehat{f} = \{f, f'\} \in (S_M)^*$ if and only if $\widehat{F} = \{\Phi f, \Phi f'\} \in (S_{M'})^*$, and the boundary triplets in (4.2.37) and (4.2.41) are connected via

$$(\Gamma'_M)_0\widehat{f} = (\Gamma_{M'})_0\widehat{F} \quad \text{and} \quad (\Gamma'_M)_1\widehat{f} = (\Gamma_{M'})_1\widehat{F} \tag{4.2.44}$$

for $\widehat{f} \in (S_M)^*$ and $\widehat{F} = \{\Phi f, \Phi f'\} \in (S_{M'})^*$.

Proof. To establish (4.2.42), note that

$$\mathsf{N}_{M'}(\lambda, \mu) = \frac{M'(\lambda) - M'(\mu)^*}{\lambda - \bar{\mu}} = \frac{M'(\bar{\mu}) - M'(\bar{\lambda})^*}{\bar{\mu} - \lambda} = \gamma'(\bar{\lambda})^*\gamma'(\bar{\mu}),$$

and hence, in view of (4.2.38) and (4.2.43),

$$\begin{aligned}\mathsf{N}_{M'}(\lambda, \mu) &= \bigl(W_{11}^* + M(\lambda)W_{12}^*\bigr)^{-1}\gamma(\bar{\lambda})^*\gamma(\bar{\mu})\bigl(W_{11} + W_{12}M(\bar{\mu})\bigr)^{-1} \\ &= \Phi(\lambda)\gamma(\bar{\lambda})^*\gamma(\bar{\mu})\Phi(\mu)^* \\ &= \Phi(\lambda)\mathsf{N}_M(\lambda, \mu)\Phi(\mu)^*\end{aligned}$$

for all $\lambda, \mu \in \mathbb{C} \setminus \mathbb{R}$.

Therefore, according to Proposition 4.1.9, each $\{F, F'\} \in \mathfrak{H}(\mathsf{N}_{M'})^2$ is of the form

$$\{F, F'\} = \{\Phi f, \Phi f'\}, \quad \{f, f'\} \in \mathfrak{H}(\mathsf{N}_M)^2, \tag{4.2.45}$$

and conversely. Let $\{F, F'\} \in \mathfrak{H}(\mathsf{N}_{M'})^2$ and $\{f, f'\} \in \mathfrak{H}(\mathsf{N}_M)^2$ be connected by (4.2.45), then

$$F'(\xi) - \xi F(\xi) = M'(\xi)\psi - \psi' \quad \Leftrightarrow \quad f'(\xi) - \xi f(\xi) = M(\xi)\varphi - \varphi', \tag{4.2.46}$$

where $\varphi, \varphi' \in \mathcal{G}$ and $\psi, \psi' \in \mathcal{G}$ are related by

$$\begin{pmatrix} \varphi \\ \varphi' \end{pmatrix} = \mathcal{W}^{-1} \begin{pmatrix} \psi \\ \psi' \end{pmatrix} = \begin{pmatrix} W_{22}^* & -W_{12}^* \\ -W_{21}^* & W_{11}^* \end{pmatrix} \begin{pmatrix} \psi \\ \psi' \end{pmatrix}. \tag{4.2.47}$$

In fact, if $F'(\xi) - \xi F(\xi) = M'(\xi)\psi - \psi'$, then it follows from (4.2.45), (4.2.39), and (4.2.43) that

$$\begin{aligned} f'(\xi) - \xi f(\xi) &= \Phi(\xi)^{-1}\big(F'(\xi) - \xi F(\xi)\big) \\ &= \Phi(\xi)^{-1}(M'(\xi)\psi - \psi') \\ &= \Phi(\xi)^{-1}(M'(\bar{\xi})^*\psi - \psi') \\ &= \Phi(\xi)^{-1}\big((W_{11}^* + M(\xi)W_{12}^*)^{-1}(W_{21}^* + M(\xi)W_{22}^*)\psi - \psi'\big) \\ &= (W_{21}^* + M(\xi)W_{22}^*)\psi - (W_{11}^* + M(\xi)W_{12}^*)\psi' \\ &= M(\xi)(W_{22}^*\psi - W_{12}^*\psi') - (-W_{21}^*\psi + W_{11}^*\psi') \\ &= M(\xi)\varphi - \varphi', \end{aligned}$$

where (4.2.47) was used in the last equality. Conversely, if $\{f, f'\} \in \mathfrak{H}(\mathsf{N}_M)^2$ and $f'(\xi) - \xi f(\xi) = M(\xi)\varphi - \varphi'$, then a similar computation shows that $\{F, F'\}$ in (4.2.45) satisfy $F'(\xi) - \xi F(\xi) = M'(\xi)\psi - \psi'$ with ψ, ψ' from (4.2.47).

Comparing (4.2.19) and (4.2.40), it follows from the equivalence (4.2.46) that

$$(S_{M'})^* = \big\{\{\Phi f, \Phi f'\} : \{f, f'\} \in (S_M)^*\big\}.$$

Moreover, from (4.2.35) and the model in Theorem 4.2.4 one then concludes

$$(\Gamma_M)_0 \widehat{f} = \varphi = W_{22}^*\psi - W_{12}^*\psi' \quad \text{and} \quad (\Gamma_M)_1 \widehat{f} = \varphi' = -W_{21}^*\psi + W_{11}^*\psi'$$

for $\widehat{f} \in (S_M)^*$, that is,

$$\begin{pmatrix} (\Gamma_M)_0 \widehat{f} \\ (\Gamma_M)_1 \widehat{f} \end{pmatrix} = \begin{pmatrix} W_{22}^* & -W_{12}^* \\ -W_{21}^* & W_{11}^* \end{pmatrix} \begin{pmatrix} \psi \\ \psi' \end{pmatrix}, \qquad \widehat{f} \in (S_M)^*,$$

or, equivalently,

$$\begin{pmatrix} W_{11} & W_{12} \\ W_{21} & W_{22} \end{pmatrix} \begin{pmatrix} (\Gamma_M)_0 \widehat{f} \\ (\Gamma_M)_1 \widehat{f} \end{pmatrix} = \begin{pmatrix} \psi \\ \psi' \end{pmatrix}, \qquad \widehat{f} \in (S_M)^*;$$

cf. (2.5.1). The result (4.2.44) now follows from (4.2.37) and (4.2.41). □

Finally, note that in Theorem 4.2.2 a model was constructed for a $\mathbf{B}(\mathcal{G})$-valued Nevanlinna function M. Under the extra condition that the Nevanlinna function M is uniformly strict, Theorem 4.2.4 provides by means of this model a boundary triplet for $(S_M)^*$ for which M is the Weyl function. Observe that with this boundary triplet the self-adjoint relation $\widetilde{A}$ in Theorem 4.2.2 in the Hilbert space $\mathfrak{H}(\mathsf{N}_M)\oplus\mathcal{G}$ can be written by means of (4.2.19) and (4.2.20) in the alternative way

$$\widetilde{A} = \left\{ \left\{ \begin{pmatrix} f \\ \Gamma_0 \widehat{f} \end{pmatrix}, \begin{pmatrix} f' \\ -\Gamma_1 \widehat{f} \end{pmatrix} \right\} : \widehat{f} = \{f, f'\} \in (S_M)^* \right\}.$$

This representation is in fact the counterpart of the observations concerning Weyl functions in Proposition 2.7.8.

4.3 Realization of scalar Nevanlinna functions via L^2-space models

In the case of a scalar Nevanlinna function M one may also construct a minimal model via the corresponding integral representation

$$M(\lambda) = \alpha + \beta\lambda + \int_{\mathbb{R}} \left(\frac{1}{t-\lambda} - \frac{t}{1+t^2} \right) d\sigma(t), \quad \lambda \in \mathbb{C}\setminus\mathbb{R}. \tag{4.3.1}$$

Here the constants α, β, and the nondecreasing function σ satisfy

$$\alpha \in \mathbb{R}, \quad \beta \geq 0, \quad \int_{\mathbb{R}} \frac{1}{1+t^2}\, d\sigma(t) < \infty. \tag{4.3.2}$$

It is a consequence of this representation that

$$\operatorname{Im} M(\lambda) = \beta + \int_{\mathbb{R}} \frac{1}{|t-\lambda|^2}\, d\sigma(t), \quad \lambda \in \mathbb{C}\setminus\mathbb{R}.$$

Therefore, a scalar Nevanlinna function M is equal to the real constant α if and only if M is not uniformly strict, i.e., if and only if $\operatorname{Im} M(\lambda) = 0$ for some, and hence for all $\lambda \in \mathbb{C}\setminus\mathbb{R}$. Under the assumption that the Nevanlinna function is not constant, a model involving the integral representation is constructed in this section. Moreover, a concrete natural isomorphism between the new model space and the reproducing kernel Hilbert space $\mathfrak{H}(\mathsf{N}_M)$ in Theorem 4.2.4 will be given.

The new model is build in the Hilbert space $L^2_{d\sigma}(\mathbb{R})$ consisting of all (equivalence classes of) complex $d\sigma$-measurable functions f such that $\int_{\mathbb{R}} |f|^2\, d\sigma < \infty$, equipped with the scalar product

$$(f,g)_{L^2_{d\sigma}(\mathbb{R})} := \int_{\mathbb{R}} f(t)\overline{g(t)}\, d\sigma(t), \qquad f,g \in L^2_{d\sigma}(\mathbb{R}).$$

The following observations will be used in the construction of the model. Under the integrability condition on σ in (4.3.2) one has that

$$\frac{t}{1+t^2}, \frac{1}{1+t^2} \in L^2_{d\sigma}(\mathbb{R}), \tag{4.3.3}$$

and, in addition,

$$f(t),\, tf(t) \in L^2_{d\sigma}(\mathbb{R}) \quad \Rightarrow \quad f(t) \in L^1_{d\sigma}(\mathbb{R}). \tag{4.3.4}$$

It is first assumed for convenience that the linear term in the integral representation (4.3.1) is absent, that is, $\beta = 0$. The general case $\beta \neq 0$ will be discussed afterwards in Theorem 4.3.4. The usual notation for general elements $\{f, f'\}$ will also be used here; the reader should be aware that f' is not the derivative here.

Theorem 4.3.1. *Let M be a scalar Nevanlinna function of the form (4.3.1) with $\beta = 0$ and assume that M is uniformly strict, that is, M is not identically equal to a constant. Then*

$$S = \left\{ \{f(t), tf(t)\} : f(t), tf(t) \in L^2_{d\sigma}(\mathbb{R}),\ \int_{\mathbb{R}} f(t)\, d\sigma(t) = 0 \right\} \tag{4.3.5}$$

is a closed simple symmetric operator in $L^2_{d\sigma}(\mathbb{R})$ and its adjoint is given by

$$S^* = \bigl\{ \{f(t), f'(t)\} : f(t), f'(t) \in L^2_{d\sigma}(\mathbb{R}),\ tf(t) - f'(t) = c \in \mathbb{C} \bigr\}. \tag{4.3.6}$$

Moreover, the mappings

$$\Gamma_0 \widehat{f} = c \quad \text{and} \quad \Gamma_1 \widehat{f} = \alpha c + \int_{\mathbb{R}} \frac{tf'(t) + f(t)}{1+t^2}\, d\sigma(t), \quad \widehat{f} = \{f, f'\} \in S^*, \tag{4.3.7}$$

are well defined and $\{\mathbb{C}, \Gamma_0, \Gamma_1\}$ is a boundary triplet for S^*. The corresponding γ-field is given by the mapping

$$c \mapsto f_\lambda(t) = \frac{c}{t-\lambda} \in \ker (S^* - \lambda), \tag{4.3.8}$$

and the corresponding Weyl function is M.

Proof. The proof consists of two steps. In Step 1 it will be shown that S in (4.3.5) is a closed symmetric operator and that $\{\mathbb{C}, \Gamma_0, \Gamma_1\}$ is a boundary triplet for its adjoint S^* in (4.3.6). Moreover, it will be shown that the γ-field is given by (4.3.8) and that M is the corresponding Weyl function. In Step 2 the simplicity of S is concluded from Corollary A.1.5.

Step 1. The right-hand side of (4.3.6) is a relation which satisfies the conditions in Theorem 2.1.9. To see this, denote the relation on the right-hand side of (4.3.6) by T and think of Γ_0 and Γ_1 as being defined on T. Observe that the mapping Γ_1 is well defined thanks to (4.3.3). Likewise, the operator S is well defined due to (4.3.4).

First, it is clear that $A_0 = \ker \Gamma_0 \subset T$ is the maximal multiplication operator by the independent variable in $L^2_{d\sigma}(\mathbb{R})$:

$$(A_0 f)(t) = tf(t), \quad \operatorname{dom} A_0 = \bigl\{ f(t) \in L^2_{d\sigma}(\mathbb{R}) : tf(t) \in L^2_{d\sigma}(\mathbb{R}) \bigr\}.$$

Since A_0 is a self-adjoint operator in $L^2_{d\sigma}(\mathbb{R})$, one sees that condition (i) in Theorem 2.1.9 is satisfied.

Next, one has that Γ is surjective. For this, note that, by (4.3.6), the defect subspace $\widehat{\mathfrak{N}}_\lambda(T)$, $\lambda \in \mathbb{C}\setminus\mathbb{R}$, of T consists of elements of the form

$$\widehat{f}_\lambda = \left\{\left\{\frac{c_\lambda}{t-\lambda}, \frac{\lambda c_\lambda}{t-\lambda}\right\} : c_\lambda \in \mathbb{C}\right\}, \tag{4.3.9}$$

which after a simple computation gives

$$\Gamma_0 \widehat{f}_\lambda = c_\lambda \quad \text{and} \quad \Gamma_1 \widehat{f}_\lambda = M(\lambda) c_\lambda,$$

or, in other words,

$$\Gamma_1 \widehat{f}_\lambda = M(\lambda)\Gamma_0 \widehat{f}_\lambda. \tag{4.3.10}$$

Using (4.3.10) one now observes that

$$\begin{aligned}\begin{pmatrix}\Gamma_0(\widehat{f}_\lambda c_\lambda + \widehat{f}_{\bar\lambda} c_{\bar\lambda})\\ \Gamma_1(\widehat{f}_\lambda c_\lambda + \widehat{f}_{\bar\lambda} c_{\bar\lambda})\end{pmatrix} &= \begin{pmatrix} c_\lambda + c_{\bar\lambda} \\ M(\lambda)c_\lambda + M(\bar\lambda)c_{\bar\lambda}\end{pmatrix}\\ &= \begin{pmatrix} 1 & 1 \\ M(\lambda) & M(\bar\lambda)\end{pmatrix}\begin{pmatrix} c_\lambda \\ c_{\bar\lambda}\end{pmatrix}.\end{aligned}$$

This shows that Γ is surjective; just note that $\lambda \in \mathbb{C}\setminus\mathbb{R}$ implies that $\operatorname{Im} M(\lambda) \neq 0$. Hence, condition (ii) in Theorem 2.1.9 is satisfied.

Finally, the abstract Green identity for T and Γ in Theorem 2.1.9 (iii) will be exhibited. For this purpose, let $\widehat{f} = \{f, f'\}, \widehat{g} = \{g, g'\} \in T$, and assume that $tf(t) - f'(t) = c$ and $tg(t) - g'(t) = d$ for some $c, d \in \mathbb{C}$. Then a calculation shows that

$$\begin{aligned}&\Gamma_1 \widehat{f}\, \overline{\Gamma_0 \widehat{g}} - \Gamma_0 \widehat{f}\, \overline{\Gamma_1 \widehat{g}}\\ &\quad = \left(\alpha c + \int_{\mathbb{R}} \frac{tf'(t) + f(t)}{1+t^2}\, d\sigma(t)\right)\bar d - c\,\overline{\left(\alpha d + \int_{\mathbb{R}} \frac{tg'(t) + g(t)}{1+t^2}\, d\sigma(t)\right)}\\ &\quad = \int_{\mathbb{R}} \frac{tf'(t) + f(t)}{1+t^2}\, \bar d \, d\sigma(t) - \int_{\mathbb{R}} c\, \frac{\overline{tg'(t) + g(t)}}{1+t^2}\, d\sigma(t).\end{aligned}$$

The last line gives, after substitution of c and d,

$$\begin{aligned}&\int_{\mathbb{R}} \frac{tf'(t) + f(t)}{1+t^2}\left(t\overline{g(t)} - \overline{g'(t)}\right) d\sigma(t) - \int_{\mathbb{R}} (tf(t) - f'(t))\, \frac{t\overline{g'(t)} + \overline{g(t)}}{1+t^2}\, d\sigma(t)\\ &\quad = \int_{\mathbb{R}} f'(t)\overline{g(t)}\, d\sigma(t) - \int_{\mathbb{R}} f(t)\overline{g'(t)}\, d\sigma(t)\\ &\quad = (f', g)_{L^2_{d\sigma}(\mathbb{R})} - (f, g')_{L^2_{d\sigma}(\mathbb{R})}.\end{aligned}$$

Hence, also condition (iii) in Theorem 2.1.9 is satisfied.

Therefore, all conditions of Theorem 2.1.9 have been verified. As a consequence the relation $\ker \Gamma_0 \cap \ker \Gamma_1$ is closed and symmetric. It coincides with S in (4.3.5), since for $\widehat{f} \in T$ one has

$$\Gamma_0 \widehat{f} = \Gamma_1 \widehat{f} = 0 \quad \text{if and only if} \quad f'(t) = tf(t) \text{ and } \int_{\mathbb{R}} f(t)\, d\sigma(t) = 0.$$

Thus, it follows from Theorem 2.1.9 that the adjoint of the closed symmetric operator S in (4.3.5) is given by T and hence has the form (4.3.6), and, moreover, $\{\mathbb{C}, \Gamma_0, \Gamma_1\}$ is a boundary triplet for S^*. As a byproduct of (4.3.9) and (4.3.10) one sees that the corresponding γ-field is given by (4.3.8) and that the corresponding Weyl function coincides with M.

Step 2. It remains to show that the operator S in (4.3.5) is simple. To see this, assume that there is an element $g \in L^2_{d\sigma}(\mathbb{R})$ which is orthogonal to all elements $f_\lambda \in \ker (S^* - \lambda)$, $\lambda \in \mathbb{C} \setminus \mathbb{R}$, that is,

$$\int_{\mathbb{R}} \frac{1}{t-\lambda}\, g(t)\, d\sigma(t) = 0, \qquad \lambda \in \mathbb{C} \setminus \mathbb{R}.$$

Then $g = 0$ in $L^2_{d\sigma}(\mathbb{R})$ by Corollary A.1.5. Thus, the linear span of the defect spaces $\ker (S^* - \lambda)$, $\lambda \in \mathbb{C} \setminus \mathbb{R}$, is dense in $L^2_{d\sigma}(\mathbb{R})$ and now Corollary 3.4.5 implies that the symmetric operator S is simple. This completes the proof of Theorem 4.3.1. □

Note that in the model in Theorem 4.3.1 the self-adjoint extension A_0 is equal to the operator of multiplication by the independent variable. The closed minimal operator S is not densely defined if and only if the constant functions belong to $L^2_{d\sigma}(\mathbb{R})$ or, equivalently, σ is a finite measure.

According to Theorem 4.2.6 the $L^2_{d\sigma}(\mathbb{R})$-space model for the function M and the model in Theorem 4.2.4 are unitarily equivalent thanks to the simplicity of the underlying symmetric operators. A concrete unitary map will be provided in the following proposition.

Proposition 4.3.2. *Let M be the Nevanlinna function in (4.3.1) with $\beta = 0$, and let $\mathfrak{H}(\mathsf{N}_M)$ be the associated reproducing kernel Hilbert space. Then the operator $V : L^2_{d\sigma}(\mathbb{R}) \to \mathfrak{H}(\mathsf{N}_M)$ defined by the rule*

$$f \mapsto -\int_{\mathbb{R}} \frac{1}{t-\xi} f(t)\, d\sigma(t), \quad \xi \in \mathbb{C} \setminus \mathbb{R}, \tag{4.3.11}$$

is unitary. Moreover, under this mapping the boundary triplets in Theorem 4.3.1 and Theorem 4.2.4 are unitarily equivalent.

Proof. It suffices to show that the operator in (4.3.11) satisfies (4.2.25); cf. Theorem 4.2.6. In fact, recall that the γ-field corresponding to the boundary triplet in Theorem 4.2.4 at a point $\lambda \in \mathbb{C} \setminus \mathbb{R}$ is given by the mapping

$$c \mapsto -c\, \mathsf{N}_M(\cdot, \bar{\lambda}) \in \ker ((S_M)^* - \lambda). \tag{4.3.12}$$

The γ-field corresponding to the boundary triplet in Theorem 4.3.1 at a point $\lambda \in \mathbb{C} \setminus \mathbb{R}$ is given by the mapping

$$c \mapsto f_\lambda(t) = \frac{c}{t-\lambda} \in \ker(S^* - \lambda). \tag{4.3.13}$$

It follows from the integral representation (4.3.1) with $\beta = 0$ that

$$\mathsf{N}_M(\xi, \bar{\lambda}) = \frac{M(\xi) - M(\bar{\lambda})}{\xi - \bar{\lambda}} = \int_{\mathbb{R}} \frac{1}{t-\xi} \frac{1}{t-\bar{\lambda}} \, d\sigma(t).$$

In view of (4.3.11) and (4.3.13), it follows that

$$(V f_\lambda)(\xi) = -\int_{\mathbb{R}} \frac{1}{t-\xi} \frac{c}{t-\lambda} \, d\sigma(t) = -c \, \mathsf{N}_M(\xi, \bar{\lambda}),$$

and taking into account (4.3.12) one concludes that (4.2.25) is satisfied. Hence, Theorem 4.2.6 ensures that the operator V in (4.3.11) is well defined and unitary, and the boundary triplets in Theorem 4.3.1 and Theorem 4.2.4 are unitarily equivalent. □

The special case of a rational Nevanlinna function serves as an illustration of Theorem 4.3.1. In this situation the measure $d\sigma$ in (4.3.1) has only finitely many point masses and the space $L^2_{d\sigma}(\mathbb{R})$ can be identified with $\mathbb{C}^n$.

Example 4.3.3. Let $\alpha_1 \in \mathbb{R}$, $n \in \mathbb{N}$, $\gamma_1, \dots, \gamma_n > 0$, and

$$-\infty < \delta_1 < \delta_2 < \cdots < \delta_n < \infty,$$

and consider the rational complex Nevanlinna function

$$N(\lambda) = \alpha_1 + \sum_{i=1}^{n} \frac{\gamma_i}{\delta_i - \lambda}, \qquad \lambda \neq \delta_i, \; i = 1, \dots, n. \tag{4.3.14}$$

Define a nondecreasing step function $\sigma : \mathbb{R} \to [0, \infty)$ by

$$\sigma(t) = \begin{cases} 0, & t \in (-\infty, \delta_1], \\ \gamma_1, & t \in (\delta_1, \delta_2], \\ \gamma_1 + \gamma_2, & t \in (\delta_2, \delta_3], \\ \dots & \\ \gamma_1 + \gamma_2 + \cdots + \gamma_n, & t \in (\delta_n, \infty), \end{cases}$$

and consider the corresponding L^2-space $L^2_{d\sigma}(\mathbb{R})$ with the scalar product

$$(f, g)_{L^2_{d\sigma}(\mathbb{R})} = \int_{\mathbb{R}} f(t) \overline{g(t)} \, d\sigma(t) = \sum_{i=1}^{n} \gamma_i \, f(\delta_i) \overline{g(\delta_i)}.$$

The rational Nevanlinna function N in (4.3.14) admits the integral representation

$$N(\lambda) = \alpha + \int_{\mathbb{R}} \left(\frac{1}{t-\lambda} - \frac{t}{1+t^2} \right) d\sigma(t)$$

as in (4.3.1), where

$$\alpha := \alpha_1 + \int_{\mathbb{R}} \frac{t}{1+t^2}\, d\sigma(t) = \alpha_1 + \sum_{i=1}^{n} \gamma_i \frac{\delta_i}{1+\delta_i^2}.$$

The Hilbert space $L^2_{d\sigma}(\mathbb{R})$ can be identified with $(\mathbb{C}^n, (\cdot,\cdot)_\gamma)$, where

$$(x,y)_\gamma := \sum_{i=1}^{n} \gamma_i\, x_i \bar{y}_i, \quad x = (x_1,\dots,x_n)^\top,\ y = (y_1,\dots,y_n)^\top \in \mathbb{C}^n,$$

via the unitary mapping

$$L^2_{d\sigma}(\mathbb{R}) \ni f \mapsto \begin{pmatrix} f(\delta_1) \\ \vdots \\ f(\delta_n) \end{pmatrix},$$

and the maximal operator of multiplication by the independent variable in $L^2_{d\sigma}(\mathbb{R})$ is unitarily equivalent to the diagonal matrix

$$\begin{pmatrix} \delta_1 & \cdots & 0 \\ \vdots & \ddots & \vdots \\ 0 & \cdots & \delta_n \end{pmatrix}. \tag{4.3.15}$$

As

$$\int_{\mathbb{R}} f(t)\, d\sigma(t) = \sum_{i=1}^{n} \gamma_i f(\delta_i) = \left(\begin{pmatrix} f(\delta_1) \\ \vdots \\ f(\delta_n) \end{pmatrix}, \begin{pmatrix} 1 \\ \vdots \\ 1 \end{pmatrix} \right)_\gamma,$$

the simple symmetric operator S in Theorem 4.3.1 is unitarily equivalent to the restriction of the diagonal matrix in (4.3.15) to the orthogonal complement of the subspace $\operatorname{span}(1,\dots,1)^\top$. Furthermore, S^* corresponds to the relation

$$\{\{f,\widehat{f}\} \in \mathbb{C}^n \times \mathbb{C}^n : \delta_i f(\delta_i) - f'(\delta_i) = c \in \mathbb{C},\ i = 1,\dots,n\}$$

and the boundary triplet $\{\mathbb{C}, \Gamma_0, \Gamma_1\}$ in Theorem 4.3.1 is of the form

$$\Gamma_0 \widehat{f} = c, \quad \Gamma_1 \widehat{f} = \alpha c + \sum_{i=1}^{n} \frac{\delta_i f'(\delta_i) + f(\delta_i)}{1+\delta_i^2}, \quad \widehat{f} = \{f, f'\} \in S^*.$$

According to Theorem 4.3.1, the corresponding Weyl function is the rational Nevanlinna function N in (4.3.14).

Now the general case of a scalar Nevanlinna function of the form (4.3.1) with $\beta > 0$ will be addressed.

Theorem 4.3.4. *Let M be a scalar Nevanlinna function of the form (4.3.1) with $\beta > 0$. Then*

$$S = \left\{ \left\{ \begin{pmatrix} f(t) \\ 0 \end{pmatrix}, \begin{pmatrix} tf(t) \\ h' \end{pmatrix} \right\} : f(t), tf(t) \in L^2_{d\sigma}(\mathbb{R}), \int_{\mathbb{R}} f(t)\, d\sigma(t) = -\beta^{1/2} h' \right\}$$

is a closed simple symmetric operator in $L^2_{d\sigma}(\mathbb{R}) \oplus \mathbb{C}$ and its adjoint is given by

$$S^* = \left\{ \left\{ \begin{pmatrix} f(t) \\ h \end{pmatrix}, \begin{pmatrix} f'(t) \\ h' \end{pmatrix} \right\} : \begin{matrix} f(t), f'(t) \in L^2_{d\sigma}(\mathbb{R}),\ h, h' \in \mathbb{C}, \\ tf(t) - f'(t) = \beta^{-1/2} h \end{matrix} \right\}.$$

Moreover, for $\widehat{f} \in S^*$ the mappings

$$\Gamma_0 \widehat{f} = \beta^{-1/2} h \quad \text{and} \quad \Gamma_1 \widehat{f} = \alpha \beta^{-1/2} h + \int_{\mathbb{R}} \frac{tf'(t) + f(t)}{1 + t^2}\, d\sigma(t) + \beta^{1/2} h'$$

are well defined and $\{\mathbb{C}, \Gamma_0, \Gamma_1\}$ is a boundary triplet for S^*. The corresponding γ-field is given by the mapping

$$c \mapsto f_\lambda(t) = \begin{pmatrix} \dfrac{c}{t - \lambda} \\ \beta^{1/2} c \end{pmatrix} \in \ker (S^* - \lambda) \tag{4.3.16}$$

and the corresponding Weyl function is given by M.

Proof. The proof is similar to the one of Theorem 4.3.1, thus only a brief sketch will be given. Denote the right-hand side of the formula for S^* by T and think of Γ_0 and Γ_1 as being defined on T. It is clear that $A_0 = \ker \Gamma_0 \subset T$ is the orthogonal componentwise sum of the maximal multiplication operator by the independent variable in $L^2_{d\sigma}(\mathbb{R})$ and the purely multivalued part $\{0\} \times \mathbb{C}$. Hence, A_0 is a self-adjoint relation. To show that Γ is surjective, note that the defect subspace $\widehat{\mathfrak{N}}_\lambda(T)$, $\lambda \in \mathbb{C} \setminus \mathbb{R}$, of T consists of elements of the form

$$\widehat{f}_\lambda = \left\{ \begin{pmatrix} \dfrac{c_\lambda}{t - \lambda} \\ \beta^{1/2} c_\lambda \end{pmatrix}, \begin{pmatrix} \dfrac{\lambda c_\lambda}{t - \lambda} \\ \lambda \beta^{1/2} c_\lambda \end{pmatrix} \right\} \in \widehat{\mathfrak{N}}_\lambda(S^*),$$

which gives

$$\Gamma_0 \widehat{f}_\lambda = c_\lambda \quad \text{and} \quad \Gamma_1 \widehat{f}_\lambda = M(\lambda) c_\lambda.$$

Again, as in the proof of Theorem 4.3.1 it follows that Γ is surjective. It can be checked by straightforward calculation as in the proof of Theorem 4.3.1 that the abstract Green identity is satisfied. Thus, by Theorem 2.1.9, one concludes that T is the adjoint of the closed symmetric relation S and that Γ_0 and Γ_1 define a boundary triplet for S^*. Hence, the statements about the γ-field and the Weyl function follow.

To show the simplicity of S, assume that there is an element orthogonal to $\mathfrak{N}_\lambda(S^*)$ for all $\lambda \in \mathbb{C}\setminus\mathbb{R}$, i.e., there exists an element $g \in L^2_{d\sigma}(\mathbb{R})$ and a constant $\gamma \in \mathbb{C}$ such that

$$\int_{\mathbb{R}} \frac{1}{t-\lambda}\, g(t)\, d\sigma(t) = \gamma, \qquad \lambda \in \mathbb{C}\setminus\mathbb{R}.$$

For $\lambda = iy$ and $y \to \infty$ it follows that $\gamma = 0$ and hence Corollary A.1.5 implies $g = 0$. Therefore, the closed linear span of all $\mathfrak{N}_\lambda(S^*)$, $\lambda \in \mathbb{C}\setminus\mathbb{R}$, is equal to $L^2_{d\sigma}(\mathbb{R}) \oplus \mathbb{C}$, and, as a consequence, the closed symmetric operator S is simple. □

In the situation of Theorem 4.3.4 one sees that S is a nondensely defined operator and that $\operatorname{mul} S^*$ is spanned by the vector

$$\begin{pmatrix} 0 \\ 1 \end{pmatrix} \in L^2_{d\sigma}(\mathbb{R}) \oplus \mathbb{C}. \tag{4.3.17}$$

Now A_0 is the only self-adjoint extension of S which is multivalued: it is the orthogonal sum of multiplication by the independent variable in $L^2_{d\sigma}$ and the space spanned by (4.3.17).

Proposition 4.3.5. *Let M be the Nevanlinna function in (4.3.1) with $\beta > 0$ and let $\mathfrak{H}(\mathsf{N}_M)$ be the associated reproducing kernel Hilbert space. Then the operator $V : L^2_{d\sigma}(\mathbb{R}) \oplus \mathbb{C} \to \mathfrak{H}(\mathsf{N}_M)$ given by the rule*

$$\begin{pmatrix} f \\ h \end{pmatrix} \mapsto -\beta^{1/2} h - \int_{\mathbb{R}} \frac{1}{t-\xi} f(t)\, d\sigma(t)$$

is unitary. Moreover, under this mapping the boundary triplets in Theorem 4.3.4 and Theorem 4.2.4 are unitarily equivalent.

Proof. The proof is similar to the one of Proposition 4.3.2 and will be sketched briefly. Recall that the γ-field corresponding to the boundary triplet in Theorem 4.2.4 at a point $\lambda \in \mathbb{C}\setminus\mathbb{R}$ is given by the mapping

$$c \mapsto -c\, \mathsf{N}_M(\cdot, \bar\lambda) \in \ker\big((S_M)^* - \lambda\big), \tag{4.3.18}$$

while the γ-field corresponding to the boundary triplet in Theorem 4.3.4 at a point $\lambda \in \mathbb{C}\setminus\mathbb{R}$ is given by the mapping

$$c \mapsto f_\lambda(t) = \begin{pmatrix} \dfrac{c}{t-\lambda} \\ \beta^{1/2} c \end{pmatrix} \in \ker\,(S^* - \lambda). \tag{4.3.19}$$

It follows from the integral representation (4.3.1) that

$$\mathsf{N}_M(\xi, \bar\lambda) = \frac{M(\xi) - M(\lambda)}{\xi - \lambda} = \beta + \int_{\mathbb{R}} \frac{1}{t-\xi}\frac{1}{t-\lambda}\, d\sigma(t).$$

Hence, (4.3.18)–(4.3.19) and the fact that

$$(Vf_\lambda)(\xi) = -c\beta - \int_{\mathbb{R}} \frac{1}{t-\xi}\frac{c}{t-\lambda}\, d\sigma(t) = -c\,\mathsf{N}_M(\xi,\bar\lambda),$$

imply that the property (4.2.25) holds. This implies that the boundary triplets in Theorem 4.3.4 and Theorem 4.2.4 are unitarily equivalent. □

In the following it is briefly explained how the self-adjoint multiplication operator in $L^2_{d\sigma}(\mathbb{R})$ and the model discussed in this section (in the case $\beta = 0$) are connected with the spectral theory and the limit properties of the Weyl function in Chapter 3. For this assume that $\sigma : \mathbb{R} \to \mathbb{R}$ is a nondecreasing function such that

$$\int_{\mathbb{R}} \frac{1}{1+t^2}\, d\sigma(t) < \infty$$

and consider the self-adjoint multiplication operator

$$(A_0 f)(t) = tf(t), \quad \operatorname{dom} A_0 = \bigl\{ f \in L^2_{d\sigma}(\mathbb{R}) : t \mapsto tf(t) \in L^2_{d\sigma}(\mathbb{R}) \bigr\}$$

in $L^2_{d\sigma}(\mathbb{R})$. Then it is known from Example 3.3.7 that the spectrum $\sigma(A_0)$ coincides with the set of growth points of the function σ, see (3.2.1), and the same is true for the absolutely continuous part $\sigma_{\rm ac}$, singular continuous part $\sigma_{\rm sc}$, and singular part $\sigma_{\rm s}$ of σ. On the other hand, the one-dimensional restriction

$$S = \left\{ \{f(t), tf(t)\} : f(t), tf(t) \in L^2_{d\sigma}(\mathbb{R}),\ \int_{\mathbb{R}} f(t)\, d\sigma(t) = 0 \right\}$$

of A_0 in Theorem 4.3.1 is a closed simple symmetric operator in $L^2_{d\sigma}(\mathbb{R})$ and $\{\mathbb{C}, \Gamma_0, \Gamma_1\}$ in (4.3.7) is a boundary triplet for S^* in (4.3.6) with $A_0 = \ker \Gamma_0$ and corresponding Weyl function

$$M(\lambda) = \alpha + \int_{\mathbb{R}} \left(\frac{1}{t-\lambda} - \frac{t}{1+t^2} \right) d\sigma(t), \quad \lambda \in \mathbb{C} \setminus \mathbb{R}, \tag{4.3.20}$$

where α is an arbitrary real number in the definition of the boundary map Γ_1 in (4.3.7). Hence, the results on the description of the spectrum of A_0 via the limit properties of the Weyl function from Section 3.5 and Section 3.6 apply in the present situation. For example, Theorem 3.6.5 shows that

$$\sigma_{\rm ac}(A_0) = \operatorname{clos}_{\rm ac}\bigl(\{ x \in \mathbb{R} : 0 < \operatorname{Im} M(x + i0) < \infty \} \bigr),$$

which is also clear from Theorem 3.2.6 (i), taking into account (3.1.25) and Corollary 3.1.8 (ii). Similar observations can be made for the other spectral subsets. In other words, in the special situation where A_0 is the self-adjoint multiplication operator in $L^2_{d\sigma}(\mathbb{R})$ the general description of the spectrum of A_0 and its subsets in Chapter 3 in terms of the limit properties of the associated Weyl function in (4.3.20) agrees with Example 3.3.7.

4.4 Realization of Nevanlinna pairs and generalized resolvents

In this section the model from Section 4.2 for Nevanlinna functions will be extended to the general setting of Nevanlinna pairs and of generalized resolvents. As a byproduct the extended model leads to the Sz.-Nagy dilation theorem.

Let $\mathcal{G}$ be a Hilbert space and let $\{A, B\}$ be a Nevanlinna pair of $\mathbf{B}(\mathcal{G})$-valued functions; cf. Section 1.12. The associated *Nevanlinna kernel* $\mathsf{N}_{A,B}$

$$\mathsf{N}_{A,B}(\cdot,\cdot) : \Omega \times \Omega \to \mathbf{B}(\mathcal{G})$$

is defined on $\Omega = \mathbb{C} \setminus \mathbb{R}$ by

$$\mathsf{N}_{A,B}(\lambda, \mu) = \frac{B(\bar{\lambda})^* A(\bar{\mu}) - A(\bar{\lambda})^* B(\bar{\mu})}{\lambda - \bar{\mu}}, \quad \lambda, \mu \in \mathbb{C} \setminus \mathbb{R}, \quad \lambda \neq \bar{\mu}, \tag{4.4.1}$$

and $\mathsf{N}_{A,B}(\lambda, \bar{\lambda}) = B'(\bar{\lambda})^* A(\lambda) - A'(\bar{\lambda})^* B(\lambda)$, $\lambda \in \mathbb{C} \setminus \mathbb{R}$. Then clearly the kernel $\mathsf{N}_{A,B}$ is symmetric. Recall that $\lambda \mapsto A(\lambda)$ and $\lambda \mapsto B(\lambda)$ are holomorphic mappings on $\mathbb{C} \setminus \mathbb{R}$. Hence,

$$\lambda \mapsto \mathsf{N}_{A,B}(\lambda, \mu)$$

is holomorphic for each $\mu \in \mathbb{C} \setminus \mathbb{R}$, that is, the kernel $\mathsf{N}_{A,B}$ is holomorphic. Moreover, it follows from (4.4.1) and Definition 1.12.3 that

$$\mathsf{N}_{A,B}(\lambda, \lambda) = \frac{\operatorname{Im}\left(A(\bar{\lambda})^* B(\bar{\lambda})\right)}{\operatorname{Im} \bar{\lambda}} \geq 0, \quad \lambda \in \mathbb{C} \setminus \mathbb{R}.$$

In the next theorem it is shown that the kernel $\mathsf{N}_{A,B}$ is, in fact, nonnegative on $\mathbb{C} \setminus \mathbb{R}$. Note also that the kernel $\mathsf{N}_{A,B}$ is uniformly bounded on compact subsets of $\mathbb{C} \setminus \mathbb{R}$ since

$$\|\mathsf{N}_{A,B}(\lambda, \lambda)\| \leq \frac{\|A(\bar{\lambda})^*\| \|B(\bar{\lambda})^*\|}{|\operatorname{Im} \lambda|}, \quad \lambda \in \mathbb{C} \setminus \mathbb{R}.$$

Theorem 4.4.1. *Let $\{A, B\}$ be a Nevanlinna pair in $\mathcal{G}$. Then the kernel $\mathsf{N}_{A,B}$ is nonnegative.*

Proof. To see this, let N be a uniformly strict $\mathbf{B}(\mathcal{G})$-valued Nevanlinna function and let $\varepsilon > 0$. Then εN is again a uniformly strict Nevanlinna function. Define the function S_ε by

$$S_\varepsilon(\lambda) = -A(\lambda)\big(\varepsilon N(\lambda) A(\lambda) + B(\lambda)\big)^{-1}, \quad \lambda \in \mathbb{C} \setminus \mathbb{R}.$$

By Proposition 1.12.6, S_ε is a Nevanlinna function. A calculation shows that the Nevanlinna kernel associated with the function S_ε is of the form

$$\begin{aligned} \mathsf{N}_{S_\varepsilon}(\lambda, \mu) &= \big(\varepsilon N(\bar{\lambda}) A(\bar{\lambda}) + B(\bar{\lambda})\big)^{-*} \cdot \\ &\quad \cdot \big[\mathsf{N}_{A,B}(\lambda, \mu) + \varepsilon A(\bar{\lambda})^* \mathsf{N}_N(\lambda, \mu) A(\bar{\mu})\big] \big(\varepsilon N(\bar{\mu}) A(\bar{\mu}) + B(\bar{\mu})\big)^{-1}. \end{aligned} \tag{4.4.2}$$

Observe that for any $\varepsilon > 0$ the kernel $\mathsf{N}_{S_\varepsilon}$ is nonnegative since S_ε is a Nevanlinna function. The identity (4.4.2) shows that the kernel

$$\mathsf{N}_{A,B}(\lambda,\mu) + \varepsilon A(\bar\lambda)^* \mathsf{N}_N(\lambda,\mu) A(\bar\mu) \tag{4.4.3}$$

is nonnegative for any $\varepsilon > 0$.

To show that the kernel $\mathsf{N}_{A,B}$ is nonnegative, assume the contrary, i.e., assume that $\mathsf{N}_{A,B}$ is not nonnegative. Then it follows from the definition of nonnegativity that there exist $n \in \mathbb{N}$, $\lambda_1, \dots, \lambda_n \in \mathbb{C} \setminus \mathbb{R}$, elements $\varphi_1, \dots, \varphi_n \in \mathcal{G}$, and a vector $c \in \mathbb{C}^n$, such that

$$\big(\big((\mathsf{N}_{A,B}(\lambda_i,\lambda_j)\varphi_j,\varphi_i)_{\mathcal{G}}\big)_{i,j=1}^n c, c\big) = x < 0.$$

Since $-x > 0$ and the kernel N_N is nonnegative, one can choose $\varepsilon > 0$ so small that

$$0 \le \varepsilon\big(\big((A(\bar\lambda_i)^* \mathsf{N}_N(\lambda_i,\lambda_j) A(\bar\lambda_j)\varphi_j,\varphi_i)_{\mathcal{G}}\big)_{i,j=1}^n c, c\big) < -x.$$

Combining these results one arrives at the inequality

$$\big(\big(((\mathsf{N}_{A,B}(\lambda_i,\lambda_j) + \varepsilon A(\bar\lambda_i)^* \mathsf{N}_N(\lambda_i,\lambda_j) A(\bar\lambda_j))\varphi_j,\varphi_i)_{\mathcal{G}}\big)_{i,j=1}^n c, c\big) < 0,$$

which contradicts the nonnegativity of the kernel in (4.4.3). Thus, the kernel $\mathsf{N}_{A,B}$ is nonnegative. □

Let $\{A, B\}$ be a Nevanlinna pair in $\mathcal{G}$. According to Theorem 4.1.5, with the nonnegative kernel $\mathsf{N}_{A,B}$ there is associated a Hilbert space of holomorphic $\mathcal{G}$-valued functions, which will be denoted by $\mathfrak{H}(\mathsf{N}_{A,B})$, with inner product $\langle\cdot,\cdot\rangle$; cf. Section 4.1. Recall that the reproducing kernel property

$$\langle f, \mathsf{N}_{A,B}(\cdot,\mu)\varphi\rangle = (f(\mu),\varphi)_{\mathcal{G}}, \qquad \varphi \in \mathcal{G},\ \mu \in \mathbb{C}\setminus\mathbb{R},$$

holds for all functions $f \in \mathfrak{H}(\mathsf{N}_{A,B})$. The following realization result extends Theorem 4.2.2 to the case of Nevanlinna pairs. It follows from Theorem 4.2.3 that this construction is unique up to unitary equivalence. Note in this context that a Nevanlinna function M always gives rise to a Nevanlinna pair $\{I, M\}$.

Theorem 4.4.2. *Let $\{A, B\}$ be a Nevanlinna pair in $\mathcal{G}$ and let $\tau = \{A, B\}$ be the corresponding Nevanlinna family. Let $\mathfrak{H}(\mathsf{N}_{A,B})$ be the associated reproducing kernel Hilbert space generated by $\{A, B\}$. Denote by $P_{\mathcal{G}}$ the orthogonal projection from $\mathfrak{H}(\mathsf{N}_M) \oplus \mathcal{G}$ onto $\mathcal{G}$ and let $\iota_{\mathcal{G}}$ be the canonical embedding of $\mathcal{G}$ into $\mathfrak{H}(\mathsf{N}_M) \oplus \mathcal{G}$. Then*

$$\widetilde{A}_{A,B} = \left\{\left\{\begin{pmatrix} f \\ \varphi \end{pmatrix}, \begin{pmatrix} f' \\ -\varphi' \end{pmatrix}\right\} : \begin{array}{l} f, f' \in \mathfrak{H}(\mathsf{N}_{A,B}),\ \varphi,\varphi' \in \mathcal{G}, \\ f'(\xi) - \xi f(\xi) = B(\bar\xi)^*\varphi - A(\bar\xi)^*\varphi' \end{array}\right\}$$

is a self-adjoint relation in the Hilbert space $\mathfrak{H}(\mathsf{N}_{A,B}) \oplus \mathcal{G}$ and the compressed resolvent of $\widetilde{A}_{A,B}$ onto $\mathcal{G}$ is given by

$$P_{\mathcal{G}}(\widetilde{A}_{A,B} - \lambda)^{-1}\iota_{\mathcal{G}} = -(\tau(\lambda) + \lambda)^{-1}, \qquad \lambda \in \mathbb{C}\setminus\mathbb{R}. \tag{4.4.4}$$

Furthermore, the self-adjoint relation $\widetilde{A}_{A,B}$ satisfies the following minimality condition:

$$\mathfrak{H}(\mathsf{N}_{A,B}) \oplus \mathcal{G} = \overline{\operatorname{span}}\,\{\mathcal{G}, \operatorname{ran}(\widetilde{A}_{A,B} - \lambda)^{-1}\iota_{\mathcal{G}} : \lambda \in \mathbb{C}\setminus\mathbb{R}\}. \tag{4.4.5}$$

Proof. The proof is almost the same as the proof of Theorem 4.2.2; therefore, only the main elements are recalled and the details are left to the reader.

Step 1. Use the Nevanlinna pair $\{A, B\}$ to define the auxiliary relation B in $\mathfrak{H}(\mathsf{N}_{A,B}) \oplus \mathcal{G}$ by

$$B = \operatorname{span}\left\{\left\{\begin{pmatrix}\mathsf{N}_{A,B}(\cdot,\bar\mu)\varphi \\ -A(\mu)\varphi\end{pmatrix}, \begin{pmatrix}\mu\mathsf{N}_{A,B}(\cdot,\bar\mu)\varphi \\ B(\mu)\varphi\end{pmatrix}\right\} : \varphi \in \mathcal{G},\, \mu \in \mathbb{C}\setminus\mathbb{R}\right\}.$$

It is a direct computation to show that $B \subset \widetilde{A}_{A,B}$. Likewise, by a similar computation one verifies that B is symmetric in $\mathfrak{H}(\mathsf{N}_{A,B}) \oplus \mathcal{G}$. Observe that for $\lambda_0 \in \mathbb{C}\setminus\mathbb{R}$ one has

$$\operatorname{ran}(B - \lambda_0) = \operatorname{span}\left\{\begin{pmatrix}(\mu-\lambda_0)\mathsf{N}_{A,B}(\cdot,\bar\mu)\varphi \\ (B(\mu) + \lambda_0 A(\mu))\varphi\end{pmatrix} : \varphi \in \mathcal{G},\, \mu \in \mathbb{C}\setminus\mathbb{R}\right\}.$$

Therefore, choosing $\mu = \lambda_0$ and taking into account that

$$\operatorname{ran}(B(\lambda_0) + \lambda_0 A(\lambda_0)) = \mathcal{G}$$

by Definition 1.12.3 and Lemma 1.12.5 it follows that $\{0\} \oplus \mathcal{G} \subset \operatorname{ran}(B - \lambda_0)$; hence also the elements of the form

$$\begin{pmatrix}\mathsf{N}_{A,B}(\cdot,\bar\mu)\varphi \\ 0\end{pmatrix}, \quad \varphi \in \mathcal{G}, \quad \mu \in \mathbb{C}\setminus\mathbb{R}, \quad \mu \neq \lambda_0,$$

belong to $\operatorname{ran}(B - \lambda_0)$. It follows from Corollary 4.1.7 that $\operatorname{ran}(B - \lambda_0)$ is dense in $\mathfrak{H}(\mathsf{N}_{A,B}) \oplus \mathcal{G}$, and thus B is essentially self-adjoint.

Step 2. One verifies in the same way as in the proof of Theorem 4.2.2 that $\widetilde{A}_{A,B}$ is closed and that $\widetilde{A}_{A,B} \subset B^*$. Since $\overline{B} \subset \widetilde{A}_{A,B}$ and $\overline{B}$ is self-adjoint, it follows that $\widetilde{A}_{A,B}$ is self-adjoint in $\mathfrak{H}(\mathsf{N}_{A,B}) \oplus \mathcal{G}$.

Step 3. The statement (4.4.4) follows in a similar way as in Theorem 4.2.2. In fact, observe that $(\widetilde{A}_{A,B} - \lambda)^{-1}$ consists of all the elements

$$\left\{\begin{pmatrix}f' - \lambda f \\ -\varphi' - \lambda\varphi\end{pmatrix}, \begin{pmatrix}f \\ \varphi\end{pmatrix}\right\}, \qquad f, f' \in \mathfrak{H}(\mathsf{N}_{A,B}), \quad \varphi, \varphi' \in \mathcal{G},$$

for which

$$f'(\xi) - \xi f(\xi) = B(\bar\xi)^*\varphi - A(\bar\xi)^*\varphi', \qquad \xi \in \mathbb{C}\setminus\mathbb{R}. \tag{4.4.6}$$

Hence, the compression $P_{\mathcal{G}}(\widetilde{A}_{A,B} - \lambda)^{-1}\iota_{\mathcal{G}}$ is formed by the pairs

$$\{-\varphi' - \lambda\varphi, \varphi\}, \qquad \varphi, \varphi' \in \mathcal{G}, \tag{4.4.7}$$

which satisfy (4.4.6) for some $f, f' \in \mathfrak{H}(\mathsf{N}_{A,B})$ and, in addition, $f'(\xi) = \lambda f(\xi)$ for $\xi \in \mathbb{C} \setminus \mathbb{R}$. This implies that (4.4.6) becomes

$$(\lambda - \xi)f(\xi) = B(\bar{\xi})^*\varphi - A(\bar{\xi})^*\varphi', \qquad \xi \in \mathbb{C} \setminus \mathbb{R},$$

and the choice $\xi = \lambda$ gives

$$B(\bar{\lambda})^*\varphi = A(\bar{\lambda})^*\varphi'. \tag{4.4.8}$$

On the other hand, as $\tau(\bar{\lambda}) = \{\{A(\bar{\lambda})\psi, B(\bar{\lambda})\psi\} : \psi \in \mathcal{G}\}$, one has by the symmetry property of the Nevanlinna family τ and (1.10.3) that

$$\tau(\lambda) = \tau(\bar{\lambda})^* = \bigl\{\{\psi, \psi'\} : B(\bar{\lambda})^*\psi = A(\bar{\lambda})^*\psi'\bigr\},$$

and since the pair $\{\varphi, \varphi'\}$ in (4.4.7) satisfies (4.4.8), it follows that $\{\varphi, \varphi'\} \in \tau(\lambda)$. Hence, $\{-\varphi' - \lambda\varphi, \varphi\} \in -(\tau(\lambda) + \lambda)^{-1}$, which yields the inclusion

$$P_{\mathcal{G}}(\widetilde{A}_{A,B} - \lambda)^{-1}\iota_{\mathcal{G}} \subset -(\tau(\lambda) + \lambda)^{-1}.$$

Since the compressed resolvent of $\widetilde{A}_{A,B}$ and $-(\tau(\lambda) + \lambda)^{-1}$ are both everywhere defined and bounded operators (4.4.4) follows.

Finally, the minimality condition (4.4.5) is shown in the same way as in the proof of Theorem 4.2.2. □

Theorem 4.4.2 provides a representation of the resolvent of the Nevanlinna family τ in terms of the model for the Nevanlinna pair $\{A, B\}$. Now let $\{C, D\}$ be a Nevanlinna pair which is equivalent to $\{A, B\}$:

$$C(\lambda) = A(\lambda)X(\lambda) \quad \text{and} \quad D(\lambda) = B(\lambda)X(\lambda), \tag{4.4.9}$$

where $X(\lambda)$, $\lambda \in \mathbb{C} \setminus \mathbb{R}$, is a bounded and boundedly invertible holomorphic operator function in $\mathbf{B}(\mathfrak{H})$; cf. Section 1.12. Then the kernels $\mathsf{N}_{A,B}$ and $\mathsf{N}_{C,D}$ of the Nevanlinna pairs in (4.4.9) are connected by

$$\mathsf{N}_{C,D}(\lambda, \mu) = X(\bar{\lambda})^*\mathsf{N}_{A,B}(\lambda, \mu)X(\bar{\mu}). \tag{4.4.10}$$

The following special case is of interest.

Lemma 4.4.3. *Let $\{A, B\}$ be a Nevanlinna pair in $\mathfrak{H}$ and consider the bounded and boundedly invertible holomorphic operator function $X(\lambda) = (B(\lambda) + \lambda A(\lambda))^{-1}$, $\lambda \in \mathbb{C} \setminus \mathbb{R}$. Then the Nevanlinna pair $\{C, D\}$ in (4.4.9) satisfies*

$$C(\lambda)^* = C(\bar{\lambda}), \quad D(\lambda)^* = D(\bar{\lambda}), \quad \text{and} \quad D(\lambda) + \lambda C(\lambda) = I,$$

and for $\lambda, \mu \in \mathbb{C} \setminus \mathbb{R}$ the corresponding Nevanlinna kernel can be written as

$$\mathsf{N}_{C,D}(\lambda, \mu) = \frac{D(\lambda)C(\mu)^* - C(\lambda)D(\mu)^*}{\lambda - \bar{\mu}} = \frac{C(\mu)^* - C(\lambda)}{\lambda - \bar{\mu}} - C(\lambda)C(\mu)^*.$$

Let again $\{A, B\}$ be a Nevanlinna pair, let τ be the corresponding Nevanlinna family, and consider a Nevanlinna pair $\{C, D\}$ which is equivalent to $\{A, B\}$ via (4.4.9), so that it generates the same Nevanlinna family τ. Then according to Theorem 4.4.2,

$$\widetilde{A}_{C,D} = \left\{ \left\{ \begin{pmatrix} F \\ \psi \end{pmatrix}, \begin{pmatrix} F' \\ -\psi' \end{pmatrix} \right\} : \begin{matrix} F, F' \in \mathfrak{H}(\mathsf{N}_{C,D}),\, \psi, \psi' \in \mathcal{G}, \\ F'(\xi) - \xi F(\xi) = D(\bar{\xi})^* \psi - C(\bar{\xi})^* \psi' \end{matrix} \right\} \tag{4.4.11}$$

is a self-adjoint relation in the Hilbert space $\mathfrak{H}(\mathsf{N}_{C,D}) \oplus \mathcal{G}$ and the compressed resolvent of $\widetilde{A}_{C,D}$ onto $\mathcal{G}$ is given by

$$P_{\mathcal{G}}(\widetilde{A}_{C,D} - \lambda)^{-1} \iota_{\mathcal{G}} = -(\tau(\lambda) + \lambda)^{-1}, \qquad \lambda \in \mathbb{C} \setminus \mathbb{R}.$$

Furthermore, the self-adjoint relation $\widetilde{A}_{C,D}$ satisfies the following minimality condition:

$$\mathfrak{H}(\mathsf{N}_{C,D}) \oplus \mathcal{G} = \overline{\operatorname{span}}\, \{\mathcal{G}, \operatorname{ran}(\widetilde{A}_{C,D} - \lambda)^{-1} \iota_{\mathcal{G}} : \lambda \in \mathbb{C} \setminus \mathbb{R}\}.$$

The explicit connection between the various models involving these kernels in Theorem 4.4.2 now depends on Proposition 4.1.9. The corresponding self-adjoint relations are then unitarily equivalent in the sense of Definition 1.3.7 and Lemma 1.3.8.

Lemma 4.4.4. *Let the Nevanlinna pairs $\{A, B\}$ and $\{C, D\}$ be equivalent in the sense of (4.4.9). Then the mapping U defined by*

$$U : \begin{pmatrix} \mathfrak{H}(\mathsf{N}_{A,B}) \\ \mathcal{G} \end{pmatrix} \to \begin{pmatrix} \mathfrak{H}(\mathsf{N}_{C,D}) \\ \mathcal{G} \end{pmatrix}, \qquad \begin{pmatrix} f(\xi) \\ \varphi \end{pmatrix} \mapsto \begin{pmatrix} X(\bar{\xi})^* f(\xi) \\ \varphi \end{pmatrix}, \tag{4.4.12}$$

is unitary. Moreover, the self-adjoint relation $\widetilde{A}_{A,B}$ in $\mathfrak{H}(\mathsf{N}_{A,B}) \oplus \mathcal{G}$ in Theorem 4.4.2 and the self-adjoint relation $\widetilde{A}_{C,D}$ in $\mathfrak{H}(\mathsf{N}_{C,D}) \oplus \mathcal{G}$ in (4.4.11) are unitarily equivalent under the mapping U, that is, $\widetilde{A}_{C,D} = U \widetilde{A}_{A,B} U^*$.

Proof. In the identity (4.4.10) set $\Phi(\lambda) = X(\bar{\lambda})^*$ with $X(\lambda)$ as in (4.4.9). Since $X(\lambda)$ is boundedly invertible one may apply Proposition 4.1.9 and hence U in (4.4.12) is unitary. Now consider an element

$$\left\{ \begin{pmatrix} f \\ \varphi \end{pmatrix}, \begin{pmatrix} f' \\ -\varphi' \end{pmatrix} \right\} \in \widetilde{A}_{A,B},$$

so that

$$f'(\xi) - \xi f(\xi) = B(\bar{\xi})^* \varphi - A(\bar{\xi})^* \varphi', \quad \xi \in \mathbb{C} \setminus \mathbb{R}.$$

Then with $F(\xi) = X(\bar{\xi})^* f(\xi)$ and $F'(\xi) = X(\bar{\xi})^* f'(\xi)$ it follows that

$$F'(\xi) - \xi F(\xi) = X(\bar{\xi})^* B(\bar{\xi})^* \varphi - X(\bar{\xi})^* A(\bar{\xi})^* \varphi' = D(\bar{\xi})^* \varphi - C(\bar{\xi})^* \varphi'.$$

This implies

$$\left\{ U \begin{pmatrix} f \\ \varphi \end{pmatrix}, U \begin{pmatrix} f' \\ -\varphi' \end{pmatrix} \right\} = \left\{ \begin{pmatrix} F \\ \varphi \end{pmatrix}, \begin{pmatrix} F' \\ -\varphi' \end{pmatrix} \right\} \in \widetilde{A}_{C,D}.$$

One verifies in the same way that every element

$$\left\{\begin{pmatrix} F \\ \varphi \end{pmatrix}, \begin{pmatrix} F' \\ -\varphi' \end{pmatrix}\right\} \in \widetilde{A}_{C,D}$$

can be written in the form

$$\left\{U\begin{pmatrix} f \\ \varphi \end{pmatrix}, U\begin{pmatrix} f' \\ -\varphi' \end{pmatrix}\right\} \quad \text{for some} \quad \left\{\begin{pmatrix} f \\ \varphi \end{pmatrix}, \begin{pmatrix} f' \\ -\varphi' \end{pmatrix}\right\} \in \widetilde{A}_{A,B}.$$

This shows that the self-adjoint relations $\widetilde{A}_{A,B}$ and $\widetilde{A}_{C,D}$ are unitarily equivalent under the mapping U; cf. Definition 1.3.7. □

The discussions in this section so far centered mainly on Nevanlinna pairs and will now be put in a slightly different context.

Definition 4.4.5. Let $\mathfrak{H}$ be a Hilbert space and let R be a $\mathbf{B}(\mathfrak{H})$-valued function defined on $\mathbb{C}\setminus\mathbb{R}$. Then R is a called a *generalized resolvent* if it has the following properties:

(i) $\lambda \mapsto R(\lambda)$ is holomorphic on $\mathbb{C}\setminus\mathbb{R}$;

(ii) $R(\lambda)^* = R(\bar\lambda)$, $\lambda \in \mathbb{C}\setminus\mathbb{R}$;

(iii) $\dfrac{\operatorname{Im} R(\lambda)}{\operatorname{Im}\lambda} - R(\lambda)R(\lambda)^* \geq 0$, $\lambda \in \mathbb{C}\setminus\mathbb{R}$.

With the function R one associates the kernel R_R

$$\mathsf{R}_R(\cdot,\cdot) : \Omega\times\Omega \to \mathbf{B}(\mathfrak{H}),$$

defined on $\Omega = \mathbb{C}\setminus\mathbb{R}$ by

$$\mathsf{R}_R(\lambda,\mu) = \frac{R(\lambda)-R(\mu)^*}{\lambda-\bar\mu} - R(\lambda)R(\mu)^*, \quad \lambda,\mu\in\mathbb{C}\setminus\mathbb{R}, \quad \lambda\neq\bar\mu, \tag{4.4.13}$$

and $\mathsf{R}(\lambda,\bar\lambda) = R'(\lambda) - R(\lambda)^2$, $\lambda\in\mathbb{C}\setminus\mathbb{R}$. Then clearly the kernel R_R is symmetric. Since $\lambda\mapsto R(\lambda)$ is holomorphic, the mapping $\lambda\mapsto\mathsf{R}_R(\lambda,\mu)$ is holomorphic for each $\mu\in\mathbb{C}\setminus\mathbb{R}$, that is, the kernel R_R is holomorphic. Note also that the kernel R_R is uniformly bounded on compact subsets of $\mathbb{C}\setminus\mathbb{R}$ since

$$\|\mathsf{R}_R(\lambda,\lambda)\| \leq \frac{\|R(\lambda)\|}{|\operatorname{Im}\lambda|} + \|R(\lambda)\|^2, \quad \lambda\in\mathbb{C}\setminus\mathbb{R}.$$

For R_R to be a reproducing kernel in the sense of Theorem 4.1.5 one needs nonnegativity.

Lemma 4.4.6. *Let $R : \mathbb{C}\setminus\mathbb{R} \to \mathbf{B}(\mathfrak{H})$ be a generalized resolvent. Then the kernel $\mathsf{R}_R(\cdot,\cdot)$ is nonnegative.*

Proof. Introduce the pair of $\mathbf{B}(\mathfrak{H})$-valued functions C and D by

$$C(\lambda) = -R(\lambda) \quad \text{and} \quad D(\lambda) = I + \lambda R(\lambda), \qquad \lambda \in \mathbb{C} \setminus \mathbb{R}.$$

Since R is a generalized resolvent, a straightforward computation shows that $\{C, D\}$ is a Nevanlinna pair and that the kernels satisfy

$$\mathsf{N}_{C,D}(\lambda, \mu) = \mathsf{R}_R(\lambda, \mu), \quad \lambda, \mu \in \mathbb{C} \setminus \mathbb{R}; \tag{4.4.14}$$

cf. (4.4.1) and (4.4.13). Now it follows from Theorem 4.4.1 (with $\mathcal{G} = \mathfrak{H}$) that the kernel $\mathsf{R}_R(\cdot, \cdot)$ is nonnegative. □

Let $R : \mathbb{C} \setminus \mathbb{R} \to \mathbf{B}(\mathfrak{H})$ be a generalized resolvent. By Theorem 4.1.5, the corresponding nonnegative kernel R_R induces a Hilbert space of holomorphic $\mathfrak{H}$-valued functions, which will be denoted by $\mathfrak{H}(\mathsf{R}_R)$, with inner product $\langle \cdot, \cdot \rangle$; cf. Section 4.1. Recall that the reproducing kernel property

$$\langle f, \mathsf{R}_R(\cdot, \mu)\varphi \rangle = (f(\mu), \varphi)_{\mathfrak{H}}, \qquad \varphi \in \mathfrak{H}, \ \mu \in \mathbb{C} \setminus \mathbb{R},$$

holds for all functions $f \in \mathfrak{H}(\mathsf{R}_R)$. The following result gives a representation of the function R.

Corollary 4.4.7. *Let $R : \mathbb{C} \setminus \mathbb{R} \to \mathbf{B}(\mathfrak{H})$ be a generalized resolvent and let $\mathfrak{H}(\mathsf{R}_R)$ be the associated reproducing kernel Hilbert space. Denote by $P_{\mathfrak{H}}$ the orthogonal projection from $\mathfrak{H}(\mathsf{R}_R) \oplus \mathfrak{H}$ onto $\mathfrak{H}$ and let $\iota_{\mathfrak{H}}$ be the canonical embedding of $\mathfrak{H}$ into $\mathfrak{H}(\mathsf{R}_R) \oplus \mathfrak{H}$. Then*

$$\widetilde{A}_R = \left\{ \left\{ \begin{pmatrix} f \\ h \end{pmatrix}, \begin{pmatrix} f' \\ -h' \end{pmatrix} \right\} : \begin{array}{c} f, f' \in \mathfrak{H}(\mathsf{R}_R),\ h, h' \in \mathfrak{H}, \\ f'(\xi) - \xi f(\xi) = (I + \xi R(\xi))h + R(\xi)h' \end{array} \right\}$$

is a self-adjoint relation in the Hilbert space $\mathfrak{H}(\mathsf{R}_R) \oplus \mathfrak{H}$ and the compressed resolvent of $\widetilde{A}_R$ onto $\mathfrak{H}$ is given by

$$P_{\mathfrak{H}}(\widetilde{A}_R - \lambda)^{-1}\iota_{\mathfrak{H}} = R(\lambda), \qquad \lambda \in \mathbb{C} \setminus \mathbb{R}.$$

Furthermore, the self-adjoint relation $\widetilde{A}_R$ satisfies the following minimality condition:

$$\mathfrak{H}(\mathsf{R}_R) \oplus \mathfrak{H} = \overline{\operatorname{span}}\, \{\mathfrak{H}, \operatorname{ran}(\widetilde{A}_R - \lambda)^{-1}\iota_{\mathfrak{H}} : \lambda \in \mathbb{C} \setminus \mathbb{R}\}. \tag{4.4.15}$$

Proof. Let $R : \mathbb{C} \setminus \mathbb{R} \to \mathbf{B}(\mathfrak{H})$ be a generalized resolvent and consider the Nevanlinna pair $\{C, D\}$ defined by

$$\{C(\lambda), D(\lambda)\} = \{-R(\lambda), I + \lambda R(\lambda)\};$$

cf. the proof of Lemma 4.4.6. Then the kernels $\mathsf{N}_{C,D}$ and R_R coincide by (4.4.14) and hence one has $\mathfrak{H}(\mathsf{N}_{C,D}) = \mathfrak{H}(\mathsf{R}_R)$. Now Theorem 4.4.2 (with $\mathcal{G} = \mathfrak{H}$) can be applied to the Nevanlinna family $\tau(\lambda) = \{C(\lambda), D(\lambda)\}$. It follows that $\widetilde{A}_R := \widetilde{A}_{C,D}$

is a self-adjoint relation in the Hilbert space $\mathfrak{H}(\mathsf{R}_R) \oplus \mathfrak{H}$ and that its compressed resolvent is given by

$$P_{\mathfrak{H}}(\widetilde{A}_R - \lambda)^{-1}\iota_{\mathfrak{H}} = -(\tau(\lambda) + \lambda)^{-1} = R(\lambda), \qquad \lambda \in \mathbb{C} \setminus \mathbb{R},$$

where the fact that $\tau(\lambda)+\lambda = \{-R(\lambda), I\}$ was used in the last equality. Moreover, the minimality condition (4.4.15) holds. □

By Corollary 4.4.7, every generalized resolvent can be interpreted as a compressed resolvent of a self-adjoint relation. Such compressed resolvents have been discussed briefly in the context of the Kreĭn formula in Section 2.7 and will be further studied in Section 4.5. The next theorem complements Corollary 4.4.7 by providing equivalent conditions. In particular, generalized resolvents or, equivalently, compressed resolvents, are characterized as Stieltjes transforms of nondecreasing families of nonnegative contractions. As a simple consequence one obtains the Sz.-Nagy dilation theorem in Corollary 4.4.9.

Theorem 4.4.8. *Let $\mathfrak{H}$ be a Hilbert space and let $R : \mathbb{C} \setminus \mathbb{R} \to \mathbf{B}(\mathfrak{H})$ be an operator function. Then the following statements are equivalent:*

(i) *The function R is a generalized resolvent.*

(ii) *There exist a Hilbert space $\mathfrak{K}$ and a self-adjoint relation $\widetilde{A}$ in the space $\mathfrak{H} \oplus \mathfrak{K}$ such that*

$$R(\lambda) = P_{\mathfrak{H}}(\widetilde{A} - \lambda)^{-1}\iota_{\mathfrak{H}}, \quad \lambda \in \mathbb{C} \setminus \mathbb{R}.$$

Furthermore, the self-adjoint relation $\widetilde{A}$ satisfies the following minimality condition:

$$\mathfrak{H} \oplus \mathfrak{K} = \overline{\operatorname{span}}\,\big\{\mathfrak{H}, \operatorname{ran}(\widetilde{A} - \lambda)^{-1}\iota_{\mathfrak{H}} : \lambda \in \mathbb{C} \setminus \mathbb{R}\big\}.$$

(iii) *There exists a nondecreasing function $\Sigma : \mathbb{R} \to \mathbf{B}(\mathfrak{H})$, whose values are nonnegative contractions, such that $\int_{\mathbb{R}} d\Sigma(t) \in \mathbf{B}(\mathfrak{H})$, $\| \int_{\mathbb{R}} d\Sigma(t)\| \leq 1$, and*

$$R(\lambda) = \int_{\mathbb{R}} \frac{1}{t-\lambda}\, d\Sigma(t), \quad \lambda \in \mathbb{C} \setminus \mathbb{R}.$$

Proof. (i) ⇒ (ii) This follows directly from Corollary 4.4.7.

(ii) ⇒ (iii) Since $\widetilde{A}$ is self-adjoint, one can write

$$(\widetilde{A} - \lambda)^{-1} = \int_{\mathbb{R}} \frac{1}{t-\lambda}\, dE(t), \quad \lambda \in \mathbb{C} \setminus \mathbb{R},$$

with the spectral measure $E(\cdot)$ of $\widetilde{A}$; cf. (1.5.6). The function $t \mapsto E((-\infty, t))$ is a nondecreasing family of orthogonal projections from $\mathbb{R}$ to $\mathbf{B}(\mathfrak{H} \oplus \mathfrak{K})$ and one has that

$$R(\lambda) = P_{\mathfrak{H}}(\widetilde{A} - \lambda)^{-1}\iota_{\mathfrak{H}} = \int_{\mathbb{R}} \frac{1}{t-\lambda}\, dP_{\mathfrak{H}}E(t)\iota_{\mathfrak{H}}.$$

Now define $\Sigma(t) = P_{\mathfrak{H}} E((-\infty, t))\iota_{\mathfrak{H}}$, which is a nondecreasing family of nonnegative contractions from $\mathbb{R}$ to $\mathbf{B}(\mathfrak{H})$ that satisfies $\int_{\mathbb{R}} d\Sigma(t) \in \mathbf{B}(\mathfrak{H})$ and the estimate $\| \int_{\mathbb{R}} d\Sigma(t)\| \leq 1$.

(iii) $\Rightarrow$ (i) It is clear that the function $R : \mathbb{C} \setminus \mathbb{R} \to \mathbf{B}(\mathfrak{H})$ is holomorphic and satisfies $R(\overline{\lambda}) = R(\lambda)^*$ for $\lambda \in \mathbb{C} \setminus \mathbb{R}$. Moreover, it follows from Proposition A.5.4 that

$$\frac{\operatorname{Im} R(\lambda)}{\operatorname{Im} \lambda} - R(\lambda)R(\lambda)^* = \frac{\operatorname{Im} R(\overline{\lambda})}{\operatorname{Im} \overline{\lambda}} - R(\overline{\lambda})^* R(\overline{\lambda}) \geq 0, \quad \lambda \in \mathbb{C} \setminus \mathbb{R},$$

which implies that R is a generalized resolvent. □

The next corollary is a variant of the dilation theorem, which goes back to M.A. Naĭmark and B. Sz.-Nagy; here it is obtained from Theorem 4.4.8 and the Stieltjes inversion formula.

Corollary 4.4.9. *Let $\Sigma : \mathbb{R} \to \mathbf{B}(\mathfrak{H})$ be a left-continuous nondecreasing function, whose values are nonnegative contractions, such that*

$$\int_{\mathbb{R}} d\Sigma(t) \in \mathbf{B}(\mathfrak{H}), \quad \left\| \int_{\mathbb{R}} d\Sigma(t) \right\| \leq 1, \quad \textit{and} \quad \Sigma(-\infty) = 0.$$

Then there exist a Hilbert space $\mathfrak{K}$ and a left-continuous nondecreasing function $E : \mathbb{R} \to \mathbf{B}(\mathfrak{H} \oplus \mathfrak{K})$, whose values are orthogonal projections, such that

$$\Sigma(t) = P_{\mathfrak{H}} E(t)\iota_{\mathfrak{H}}, \qquad t \in \mathbb{R}.$$

Proof. Associate with Σ the function

$$R(\lambda) = \int_{\mathbb{R}} \frac{1}{t-\lambda}\, d\Sigma(t), \quad \lambda \in \mathbb{C} \setminus \mathbb{R}. \tag{4.4.16}$$

By Theorem 4.4.8, there exists a Hilbert space $\mathfrak{K}$ and a self-adjoint relation $\widetilde{A}$ in $\mathfrak{H} \oplus \mathfrak{K}$ such that the compression of the resolvent of $\widetilde{A}$ onto $\mathfrak{H}$ is given by

$$P_{\mathfrak{H}}(\widetilde{A} - \lambda)^{-1}\iota_{\mathfrak{H}} = R(\lambda), \qquad \lambda \in \mathbb{C} \setminus \mathbb{R}. \tag{4.4.17}$$

Let $E(\cdot)$ be the spectral measure of $\widetilde{A}$ and let $t \mapsto E((-\infty, t))$ be the corresponding spectral function, which is left-continuous and satisfies $\lim_{t\to-\infty} E((-\infty,t)) = 0$. As in the proof of Theorem 4.4.8 one has

$$P_{\mathfrak{H}}(\widetilde{A} - \lambda)^{-1}\iota_{\mathfrak{H}} = P_{\mathfrak{H}}\left(\int_{\mathbb{R}} \frac{1}{t-\lambda}\, dE(t) \right)\iota_{\mathfrak{H}} = \int_{\mathbb{R}} \frac{1}{t-\lambda}\, dP_{\mathfrak{H}} E(t)\iota_{\mathfrak{H}}.$$

Taking into account (4.4.16) and (4.4.17), it follows that

$$\int_{\mathbb{R}} \frac{1}{t-\lambda} dP_{\mathfrak{H}} E(t)\iota_{\mathfrak{H}} = \int_{\mathbb{R}} \frac{1}{t-\lambda} d\Sigma(t), \qquad \lambda \in \mathbb{C} \setminus \mathbb{R},$$

and hence for all $h \in \mathfrak{H}$

$$\int_{\mathbb{R}} \frac{1}{t-\lambda} d(E(t)\iota_{\mathfrak{H}} h, \iota_{\mathfrak{H}} h) = \int_{\mathbb{R}} \frac{1}{t-\lambda} d(\Sigma(t)h, h), \qquad \lambda \in \mathbb{C} \setminus \mathbb{R}.$$

Since the functions $t \mapsto (E(t)\iota_{\mathfrak{H}} h, \iota_{\mathfrak{H}} h)$ and $t \mapsto (\Sigma(t)h, h)$ are left-continuous, and

$$\lim_{t \to -\infty} \big(E((-\infty, t))\iota_{\mathfrak{H}} h, \iota_{\mathfrak{H}} h\big) = 0 = (\Sigma(-\infty)h, h),$$

the Stieltjes inversion formula in Corollary A.1.2 yields $(E(t)\iota_{\mathfrak{H}} h, \iota_{\mathfrak{H}} h) = (\Sigma(t)h, h)$ for all $t \in \mathbb{R}$ and $h \in \mathfrak{H}$. This leads to the assertion. □

4.5 Kreĭn's formula for exit space extensions

Let S be a closed symmetric relation in the Hilbert space $\mathfrak{H}$, let $\{\mathcal{G}, \Gamma_0, \Gamma_1\}$ be a boundary triplet for S^*, $A_0 = \ker \Gamma_0$, and let γ and M be the corresponding γ-field and Weyl function, respectively. Suppose that $\widetilde{A}$ is a self-adjoint extension of S in $\mathfrak{H} \oplus \mathfrak{H}'$, where $\mathfrak{H}'$ is the exit space. It was shown in Theorem 2.7.4 that there exists a Nevanlinna family $\tau(\lambda)$, $\lambda \in \mathbb{C} \setminus \mathbb{R}$, in $\mathcal{G}$ such that

$$P_{\mathfrak{H}}(\widetilde{A} - \lambda)^{-1}\iota_{\mathfrak{H}} = (A_0 - \lambda)^{-1} - \gamma(\lambda)\big(M(\lambda) + \tau(\lambda)\big)^{-1}\gamma(\bar{\lambda})^*, \quad \lambda \in \mathbb{C} \setminus \mathbb{R},$$

holds. This is Kreĭn's formula for the compressed resolvents of self-adjoint exit space extensions (as studied by M. A. Naĭmark); it is also referred to as Kreĭn–Naĭmark formula in this text; cf. Section 2.7.

The goal of this section is to show the converse statement. More precisely, it will be proved that for every Nevanlinna family $\tau(\lambda)$, $\lambda \in \mathbb{C} \setminus \mathbb{R}$, in the Hilbert space $\mathcal{G}$ there exists a self-adjoint exit space extension $\widetilde{A}$ of S such that the compressed resolvent of $\widetilde{A}$ onto $\mathfrak{H}$ is given by the Kreĭn–Naĭmark formula. The following result is a first step.

Lemma 4.5.1. *Let S be a closed symmetric relation in $\mathfrak{H}$, let $\{\mathcal{G}, \Gamma_0, \Gamma_1\}$ be a boundary triplet for S^* with $A_0 = \ker \Gamma_0$, and let γ and M be the corresponding γ-field and Weyl function, respectively. Let $\tau = \{A, B\}$ be a Nevanlinna family in $\mathcal{G}$ and define $R(\lambda)$, $\lambda \in \mathbb{C} \setminus \mathbb{R}$, by*

$$R(\lambda) = (A_0 - \lambda)^{-1} - \gamma(\lambda)\big(M(\lambda) + \tau(\lambda)\big)^{-1}\gamma(\bar{\lambda})^*. \tag{4.5.1}$$

Then the kernel

$$\mathsf{R}_R(\lambda, \mu) = \frac{R(\lambda) - R(\mu)^*}{\lambda - \bar{\mu}} - R(\lambda)R(\mu)^*, \quad \lambda, \mu \in \mathbb{C} \setminus \mathbb{R}, \quad \lambda \neq \bar{\mu}, \tag{4.5.2}$$

satisfies

$$\mathsf{R}_R(\lambda, \mu) = W(\lambda)\mathsf{N}_{A,B}(\lambda, \mu)W(\mu)^*, \tag{4.5.3}$$

where

$$W(\lambda) = \gamma(\lambda)\big(M(\bar\lambda)A(\bar\lambda) + B(\bar\lambda)\big)^{-*}. \tag{4.5.4}$$

In particular, the kernel R_R *is nonnegative, symmetric, holomorphic, and uniformly bounded on compact subsets of* $\mathbb{C}\setminus\mathbb{R}$.

Proof. Step 1. For $\lambda\in\mathbb{C}\setminus\mathbb{R}$ introduce the following notations

$$R_0(\lambda) = (A_0-\lambda)^{-1} \quad\text{and}\quad Q(\lambda) = \big(M(\lambda)+\tau(\lambda)\big)^{-1},$$

so that R in (4.5.1) is given by

$$R(\lambda) = R_0(\lambda) - \gamma(\lambda)Q(\lambda)\gamma(\bar\lambda)^*.$$

Rewrite the kernel $\mathsf{R}_R(\cdot,\cdot)$ in (4.5.2) in terms of this notation:

$$\begin{aligned}
\mathsf{R}_R(\lambda,\mu) &= \frac{1}{\lambda-\bar\mu}\big(R_0(\lambda)-R_0(\mu)^* - \gamma(\lambda)Q(\lambda)\gamma(\bar\lambda)^* + \gamma(\bar\mu)Q(\mu)^*\gamma(\mu)^*\big)\\
&\qquad - \big(R_0(\lambda)-\gamma(\lambda)Q(\lambda)\gamma(\bar\lambda)^*\big)\big(R_0(\mu)^* - \gamma(\bar\mu)Q(\mu)^*\gamma(\mu)^*\big)\\
&= \frac{1}{\lambda-\bar\mu}\big(-\gamma(\lambda)Q(\lambda)\gamma(\bar\lambda)^* + \gamma(\bar\mu)Q(\mu)^*\gamma(\mu)^*\big)\\
&\qquad + R_0(\lambda)\gamma(\bar\mu)Q(\mu)^*\gamma(\mu)^* + \gamma(\lambda)Q(\lambda)\gamma(\bar\lambda)^*R_0(\mu)^*\\
&\qquad - \gamma(\lambda)Q(\lambda)\gamma(\bar\lambda)^*\gamma(\bar\mu)Q(\mu)^*\gamma(\mu)^*.
\end{aligned}$$

Recall that, by Proposition 2.3.2 (ii) and Proposition 2.3.6 (iii),

$$R_0(\lambda)\gamma(\bar\mu) = \frac{\gamma(\lambda)-\gamma(\bar\mu)}{\lambda-\bar\mu}, \quad \gamma(\bar\lambda)^*R_0(\mu)^* = \frac{\gamma(\bar\lambda)^*-\gamma(\mu)^*}{\lambda-\bar\mu},$$

and

$$\gamma(\bar\lambda)^*\gamma(\bar\mu) = \big(\gamma(\bar\mu)^*\gamma(\bar\lambda)\big)^* = \left(\frac{M(\bar\lambda)-M(\bar\mu)^*}{\bar\lambda-\mu}\right)^* = \frac{M(\lambda)-M(\mu)^*}{\lambda-\bar\mu}.$$

Therefore, the kernel $\mathsf{R}_R(\cdot,\cdot)$ has the form

$$\begin{aligned}
\mathsf{R}_R(\lambda,\mu) = \frac{1}{\lambda-\bar\mu}\gamma(\lambda)\big[&Q(\mu)^* - Q(\lambda)\\
&+ Q(\lambda)M(\mu)^*Q(\mu)^* - Q(\lambda)M(\lambda)Q(\mu)^*\big]\gamma(\mu)^*.
\end{aligned} \tag{4.5.5}$$

Step 2. Express the identity (4.5.5) in terms of the Nevanlinna pair $\{A,B\}$, representing the Nevanlinna family τ. For this, consider the equivalent Nevanlinna pair $\{C,D\}$ as in Lemma 4.4.3, that is,

$$C(\lambda) = A(\lambda)X(\lambda) \quad\text{and}\quad D(\lambda) = B(\lambda)X(\lambda),$$

where $X(\lambda) = (B(\lambda) + \lambda A(\lambda))^{-1}$, so that

$$\mathsf{N}_{C,D}(\lambda,\mu) = \frac{D(\lambda)C(\mu)^* - C(\lambda)D(\mu)^*}{\lambda - \bar{\mu}}. \tag{4.5.6}$$

Observe that $Q(\lambda)$ can be written in terms of $\tau = \{C, D\}$ as

$$Q(\lambda) = C(\lambda)\big(M(\lambda)C(\lambda) + D(\lambda)\big)^{-1};$$

cf. (1.12.10). It follows that

$$Q(\lambda) = Q(\bar{\lambda})^* = \big(C(\lambda)M(\lambda) + D(\lambda)\big)^{-1}C(\lambda) \tag{4.5.7}$$

and

$$Q(\mu)^* = Q(\bar{\mu}) = C(\mu)^*\big(C(\mu)M(\mu) + D(\mu)\big)^{-*}. \tag{4.5.8}$$

Inserting the expressions (4.5.7) and (4.5.8) in (4.5.5) one arrives after a straightforward computation at the identity

$$\mathsf{R}_R(\lambda,\mu) = Z(\lambda)\mathsf{N}_{C,D}(\lambda,\mu)Z(\mu)^*,$$

where the factor $Z(\lambda)$ is given by

$$Z(\lambda) = \gamma(\lambda)\big(C(\lambda)M(\lambda) + D(\lambda)\big)^{-1}. \tag{4.5.9}$$

Recall that $\mathsf{N}_{A,B}$ and $\mathsf{N}_{C,D}$ are related via (4.4.10). Therefore, with (4.5.6) and (4.5.9) one obtains the identity (4.5.3), where

$$W(\lambda) = \gamma(\lambda)\big(C(\lambda)M(\lambda) + D(\lambda)\big)^{-1}\big(B(\bar{\lambda}) + \bar{\lambda}A(\bar{\lambda})\big)^{-*}. \tag{4.5.10}$$

The proof is finished by writing the normalized pair $\{C, D\}$ in (4.5.10) in terms of the pair $\{A, B\}$. Observe that since $X(\lambda) = (B(\lambda) + \lambda A(\lambda))^{-1}$, the symmetry property $B(\bar{\lambda})^*A(\lambda) = A(\bar{\lambda})^*B(\lambda)$ of the Nevanlinna pair yields

$$\begin{aligned}
&\big(B(\bar{\lambda})^* + \lambda A(\bar{\lambda})^*\big)\big(C(\lambda)M(\lambda) + D(\lambda)\big)\\
&\quad = \big(B(\bar{\lambda})^* + \lambda A(\bar{\lambda})^*\big)\big(A(\lambda)X(\lambda)M(\lambda) + B(\lambda)X(\lambda)\big)\\
&\quad = A(\bar{\lambda})^*\big(B(\lambda) + \lambda A(\lambda)\big)X(\lambda)M(\lambda) + B(\bar{\lambda})^*\big(B(\lambda) + \lambda A(\lambda)\big)X(\lambda)\\
&\quad = A(\bar{\lambda})^*M(\lambda) + B(\bar{\lambda})^*\\
&\quad = \big(M(\bar{\lambda})A(\bar{\lambda}) + B(\bar{\lambda})\big)^*,
\end{aligned}$$

which gives (4.5.4).

It follows from (4.5.3) and (4.5.4) that the kernel R_R in (4.5.2) is nonnegative, symmetric, holomorphic, and uniformly bounded on compact subsets of $\mathbb{C} \setminus \mathbb{R}$. □

Lemma 4.5.1 shows that $\mathsf{R}_R(\cdot,\cdot)$ is a reproducing kernel. Therefore, one may apply Theorem 4.4.8.

Theorem 4.5.2. *Let S be a closed symmetric relation, let $\{\mathcal{G}, \Gamma_0, \Gamma_1\}$ be a boundary triplet for S^* with $A_0 = \ker \Gamma_0$, and let γ and M be the corresponding γ-field and Weyl function, respectively. Let τ be a Nevanlinna family in $\mathcal{G}$. Then there exist an exit Hilbert space $\mathfrak{H}'$ and a self-adjoint relation $\widetilde{A}$ in $\mathfrak{H} \oplus \mathfrak{H}'$ such that $\widetilde{A}$ is an extension of S and the compressed resolvent of $\widetilde{A}$ is given by the Kreĭn–Naĭmark formula:*

$$P_{\mathfrak{H}}(\widetilde{A}-\lambda)^{-1}\iota_{\mathfrak{H}} = (A_0-\lambda)^{-1} - \gamma(\lambda)\big(M(\lambda)+\tau(\lambda)\big)^{-1}\gamma(\bar{\lambda})^*, \quad \lambda \in \mathbb{C} \setminus \mathbb{R}. \tag{4.5.11}$$

Furthermore, the self-adjoint relation $\widetilde{A}$ satisfies the following minimality condition:

$$\mathfrak{H} \oplus \mathfrak{H}' = \overline{\operatorname{span}}\,\big\{\mathfrak{H}, \operatorname{ran}(\widetilde{A}-\mu)^{-1}\iota_{\mathfrak{H}} : \mu \in \mathbb{C} \setminus \mathbb{R}\big\}. \tag{4.5.12}$$

Proof. Define the function R as in Lemma 4.5.1. Then R is a $\mathbf{B}(\mathfrak{H})$-valued holomorphic function on $\mathbb{C} \setminus \mathbb{R}$ which satisfies $R(\bar{\lambda}) = R(\lambda)^*$ and so, by Lemma 4.5.1, R is a generalized resolvent. Hence, by Theorem 4.4.8, the function R is a compressed resolvent, that is, there exist a Hilbert space $\mathfrak{H}'$ and a self-adjoint relation $\widetilde{A}$ in $\mathfrak{H} \oplus \mathfrak{H}'$ such that

$$R(\lambda) = P_{\mathfrak{H}}(\widetilde{A}-\lambda)^{-1}\iota_{\mathfrak{H}}, \quad \lambda \in \mathbb{C} \setminus \mathbb{R};$$

this implies (4.5.11) Moreover, it follows from Theorem 4.4.8 that $\widetilde{A}$ satisfies the minimality condition (4.5.12).

It remains to prove that $S \subset \widetilde{A}$. Observe first that by (4.5.11) the Štraus family corresponding to $\widetilde{A}$ satisfies

$$\begin{aligned} T(\lambda) &= \big\{\{R(\lambda)h, (I+\lambda R(\lambda))h\} : h \in \mathfrak{H}\big\} \\ &\subset \big\{R_0(\lambda)h, (I+\lambda R_0(\lambda))h : h \in \mathfrak{H}\big\} \,\widehat{+}\, \big\{\{\gamma(\lambda)\varphi, \lambda\gamma(\lambda)\varphi\} : \varphi \in \mathcal{G}\big\}. \end{aligned}$$

Since each relation on the right-hand side is contained in S^*, so is the relation $T(\lambda)$. As $T(\lambda)^* = T(\bar{\lambda})$, it follows that $S \subset T(\lambda)$. Now let $\{f, f'\} \in S$ so that for each $\lambda \in \mathbb{C} \setminus \mathbb{R}$ there exists $h \in \mathfrak{H}'$ such that

$$\left\{ \begin{pmatrix} f \\ h \end{pmatrix}, \begin{pmatrix} f' \\ \lambda h \end{pmatrix} \right\} \in \widetilde{A}.$$

The relation $\widetilde{A}$ is self-adjoint and, in particular, symmetric. Therefore, one sees that $(f', f) + \lambda(h, h) \in \mathbb{R}$, while by definition $(f', f) \in \mathbb{R}$. Since $\lambda \in \mathbb{C} \setminus \mathbb{R}$, it follows that $h = 0$, and thus

$$\left\{ \begin{pmatrix} f \\ 0 \end{pmatrix}, \begin{pmatrix} f' \\ 0 \end{pmatrix} \right\} \in \widetilde{A}.$$

This shows that $S \subset \widetilde{A}$. □

4.6 Orthogonal coupling of boundary triplets

In this section a different look is taken at the Kreĭn–Naĭmark formula. By means of an abstract coupling method for direct orthogonal sums of symmetric relations and corresponding boundary triplets, a particular self-adjoint extension $\widetilde{A}$ of the direct sum is identified, and it is shown that the compressed resolvent of $\widetilde{A}$ is of the same form as in the Kreĭn–Naĭmark formula. When combined with Theorem 4.2.4, this coupling procedure provides a constructive approach to the exit space extension in Theorem 4.5.2 in the special case where the Nevanlinna family τ is a uniformly strict Nevanlinna function.

First a slightly more general, abstract point of view is adopted. In the following let S and T be closed symmetric relations in the Hilbert spaces $\mathfrak{H}$ and $\mathfrak{H}'$, respectively, and assume that the defect numbers of S and T coincide:

$$n_+(S) = n_-(S) = n_+(T) = n_-(T) \leq \infty.$$

Let $\{\mathcal{G}, \Gamma_0, \Gamma_1\}$ be a boundary triplet for S^* with $A_0 = \ker \Gamma_0$ and let $\{\mathcal{G}, \Gamma_0', \Gamma_1'\}$ be a boundary triplet for T^* with $B_0 = \ker \Gamma_0'$. Then it is easy to see that the direct orthogonal sum $S \,\widehat{\oplus}\, T$ is a closed symmetric relation in $\mathfrak{H} \oplus \mathfrak{H}'$ and $\{\mathcal{G} \oplus \mathcal{G}, \widetilde{\Gamma}_0, \widetilde{\Gamma}_1\}$, where

$$\widetilde{\Gamma}_0 \begin{pmatrix} \widehat{f} \\ \widehat{g} \end{pmatrix} = \begin{pmatrix} \Gamma_0 \widehat{f} \\ \Gamma_0' \widehat{g} \end{pmatrix} \quad \text{and} \quad \widetilde{\Gamma}_1 \begin{pmatrix} \widehat{f} \\ \widehat{g} \end{pmatrix} = \begin{pmatrix} \Gamma_1 \widehat{f} \\ \Gamma_1' \widehat{g} \end{pmatrix}, \quad \widehat{f} \in S^*, \, \widehat{g} \in T^*, \tag{4.6.1}$$

is a boundary triplet for $(S \,\widehat{\oplus}\, T)^* = S^* \,\widehat{\oplus}\, T^*$, and that

$$\widetilde{A}_0 := A_0 \,\widehat{\oplus}\, B_0 = \ker \widetilde{\Gamma}_0 \tag{4.6.2}$$

is a self-adjoint extension of $S \,\widehat{\oplus}\, T$ in $\mathfrak{H} \oplus \mathfrak{H}'$. Furthermore, if γ and γ' denote the γ-fields corresponding to the boundary triplets $\{\mathcal{G}, \Gamma_0, \Gamma_1\}$ and $\{\mathcal{G}, \Gamma_0', \Gamma_1'\}$, and M and τ are the Weyl functions corresponding to $\{\mathcal{G}, \Gamma_0, \Gamma_1\}$ and $\{\mathcal{G}, \Gamma_0', \Gamma_1'\}$, respectively, then it is clear that for $\lambda \in \rho(\widetilde{A}_0) = \rho(A_0) \cap \rho(B_0)$ the γ-field $\widetilde{\gamma}$ and the Weyl function $\widetilde{M}$ corresponding to the boundary triplet $\{\mathcal{G} \oplus \mathcal{G}, \widetilde{\Gamma}_0, \widetilde{\Gamma}_1\}$ have the forms

$$\widetilde{\gamma}(\lambda) = \begin{pmatrix} \gamma(\lambda) & 0 \\ 0 & \gamma'(\lambda) \end{pmatrix} \quad \text{and} \quad \widetilde{M}(\lambda) = \begin{pmatrix} M(\lambda) & 0 \\ 0 & \tau(\lambda) \end{pmatrix}. \tag{4.6.3}$$

Let $\widetilde{A}$ be a self-adjoint extension of $S \,\widehat{\oplus}\, T$ in $\mathfrak{H} \oplus \mathfrak{H}'$. Then Kreĭn's formula in Theorem 2.6.1 has the form

$$(\widetilde{A} - \lambda)^{-1} = (\widetilde{A}_0 - \lambda)^{-1} + \widetilde{\gamma}(\lambda)\big(\widetilde{\Theta} - \widetilde{M}(\lambda)\big)^{-1}\widetilde{\gamma}(\bar{\lambda})^*$$

for all $\lambda \in \rho(\widetilde{A}) \cap \rho(\widetilde{A}_0)$, where $\widetilde{\gamma}$ and $\widetilde{M}$ denote the γ-field and the Weyl function corresponding to the boundary triplet $\{\mathcal{G} \oplus \mathcal{G}, \widetilde{\Gamma}_0, \widetilde{\Gamma}_1\}$ in (4.6.3). If $\widetilde{\Theta} = \{\mathcal{A}, \mathcal{B}\}$ with $\mathcal{A}, \mathcal{B} \in \mathbf{B}(\mathcal{G} \oplus \mathcal{G})$, then

$$(\widetilde{A} - \lambda)^{-1} = (\widetilde{A}_0 - \lambda)^{-1} - \widetilde{\gamma}(\lambda)\mathcal{A}\big(\widetilde{M}(\lambda)\mathcal{A} - \mathcal{B}\big)^{-1}\widetilde{\gamma}(\bar{\lambda})^*,$$

see Corollary 2.6.3. Writing $\mathcal{A}$ and $\mathcal{B}$ as block operators

$$\mathcal{A} = \begin{pmatrix} \mathcal{A}_{11} & \mathcal{A}_{12} \\ \mathcal{A}_{21} & \mathcal{A}_{22} \end{pmatrix} \quad \text{and} \quad \mathcal{B} = \begin{pmatrix} \mathcal{B}_{11} & \mathcal{B}_{12} \\ \mathcal{B}_{21} & \mathcal{B}_{22} \end{pmatrix},$$

where $\mathcal{A}_{ij}, \mathcal{B}_{ij} \in \mathbf{B}(\mathcal{G})$, it follows that

$$\widetilde{M}(\lambda)\mathcal{A} - \mathcal{B} = \begin{pmatrix} M(\lambda)\mathcal{A}_{11} - \mathcal{B}_{11} & M(\lambda)\mathcal{A}_{12} - \mathcal{B}_{12} \\ \tau(\lambda)\mathcal{A}_{21} - \mathcal{B}_{21} & \tau(\lambda)\mathcal{A}_{22} - \mathcal{B}_{22} \end{pmatrix},$$

so that

$$\mathcal{A}\big(\widetilde{M}(\lambda)\mathcal{A} - \mathcal{B}\big)^{-1} = \begin{pmatrix} \mathcal{A}_{11} & \mathcal{A}_{12} \\ \mathcal{A}_{21} & \mathcal{A}_{22} \end{pmatrix} \begin{pmatrix} M(\lambda)\mathcal{A}_{11} - \mathcal{B}_{11} & M(\lambda)\mathcal{A}_{12} - \mathcal{B}_{12} \\ \tau(\lambda)\mathcal{A}_{21} - \mathcal{B}_{21} & \tau(\lambda)\mathcal{A}_{22} - \mathcal{B}_{22} \end{pmatrix}^{-1}.$$

The following proposition exhibits a particular self-adjoint extension of $S \,\widehat{\oplus}\, T$ in $\mathfrak{H} \oplus \mathfrak{H}'$.

Proposition 4.6.1. *Let S and T be closed symmetric relations in the Hilbert spaces $\mathfrak{H}$ and $\mathfrak{H}'$ with boundary triplets $\{\mathcal{G}, \Gamma_0, \Gamma_1\}$ and $\{\mathcal{G}, \Gamma_0', \Gamma_1'\}$ as above, respectively. Then*

$$\widetilde{A} = \left\{ \begin{pmatrix} \widehat{f} \\ \widehat{g} \end{pmatrix} : \widehat{f} \in S^*,\, \widehat{g} \in T^*,\, \Gamma_0 \widehat{f} = \Gamma_0' \widehat{g},\, \Gamma_1 \widehat{f} = -\Gamma_1' \widehat{g} \right\} \tag{4.6.4}$$

is a self-adjoint relation in $\mathfrak{H} \oplus \mathfrak{H}'$ and for all $\lambda \in \mathbb{C} \setminus \mathbb{R}$ the resolvent of $\widetilde{A}$ has the form

$$(\widetilde{A} - \lambda)^{-1} = (\widetilde{A}_0 - \lambda)^{-1} - \widetilde{\gamma}(\lambda) \begin{pmatrix} (M(\lambda) + \tau(\lambda))^{-1} & (M(\lambda) + \tau(\lambda))^{-1} \\ (M(\lambda) + \tau(\lambda))^{-1} & (M(\lambda) + \tau(\lambda))^{-1} \end{pmatrix} \widetilde{\gamma}(\bar{\lambda})^*,$$

where $\widetilde{A}_0$ and $\widetilde{\gamma}$ are as in (4.6.2), and M and τ denote the Weyl functions corresponding to $\{\mathcal{G}, \Gamma_0, \Gamma_1\}$ and $\{\mathcal{G}, \Gamma_0', \Gamma_1'\}$, respectively.

Proof. Consider the boundary triplet $\{\mathcal{G} \oplus \mathcal{G}, \widetilde{\Gamma}_0, \widetilde{\Gamma}_1\}$ in (4.6.1) and observe that the relation

$$\widetilde{\Theta} := \left\{ \left\{ \begin{pmatrix} \varphi \\ \varphi \end{pmatrix}, \begin{pmatrix} \psi \\ -\psi \end{pmatrix} \right\} : \varphi, \psi \in \mathcal{G} \right\} \tag{4.6.5}$$

is self-adjoint in $\mathcal{G} \oplus \mathcal{G}$. Hence, by Corollary 2.1.4 and (4.6.1),

$$\left\{ \begin{pmatrix} \widehat{f} \\ \widehat{g} \end{pmatrix} \in S^* \,\widehat{\oplus}\, T^* : \widetilde{\Gamma} \begin{pmatrix} \widehat{f} \\ \widehat{g} \end{pmatrix} = \left\{ \begin{pmatrix} \Gamma_0 \widehat{f} \\ \Gamma_0' \widehat{g} \end{pmatrix}, \begin{pmatrix} \Gamma_1 \widehat{f} \\ \Gamma_1' \widehat{g} \end{pmatrix} \right\} \in \widetilde{\Theta} \right\} \subset S^* \,\widehat{\oplus}\, T^* \tag{4.6.6}$$

is a self-adjoint relation $\mathfrak{H} \oplus \mathfrak{H}'$. Now it follows from the particular form of $\widetilde{\Theta}$ in (4.6.5) that the self-adjoint relation in (4.6.6) coincides with $\widetilde{A}$ in (4.6.4).

Next the resolvent of $\widetilde{A}$ will be computed. Recall first that Kreĭn's formula in Theorem 2.6.1 implies

$$(\widetilde{A} - \lambda)^{-1} = (\widetilde{A}_0 - \lambda)^{-1} - \widetilde{\gamma}(\lambda)\big(\widetilde{\Theta} - \widetilde{M}(\lambda)\big)^{-1} \widetilde{\gamma}(\bar{\lambda})^* \tag{4.6.7}$$

for all $\lambda \in \rho(\widetilde{A}) \cap \rho(\widetilde{A}_0)$, where $\widetilde{\gamma}$ and $\widetilde{M}$ denote the γ-field and the Weyl function corresponding to $\{\mathcal{G} \oplus \mathcal{G}, \widetilde{\Gamma}_0, \widetilde{\Gamma}_1\}$ in (4.6.3). From (4.6.5) and (4.6.3) one obtains

$$\big(\widetilde{\Theta} - \widetilde{M}(\lambda)\big)^{-1} = \left\{\left\{\begin{pmatrix}\psi - M(\lambda)\varphi \\ -\psi - \tau(\lambda)\varphi\end{pmatrix}, \begin{pmatrix}\varphi \\ \varphi\end{pmatrix}\right\} : \varphi, \psi \in \mathcal{G}\right\}.$$

Setting $\phi := \psi - M(\lambda)\varphi$ and $\chi := -\psi - \tau(\lambda)\varphi$ one has $\phi + \chi = -(M(\lambda) + \tau(\lambda))\varphi$. For $\lambda \in \mathbb{C} \setminus \mathbb{R}$ it follows from Lemma 1.11.5 (see also Proposition 1.12.6) that $(M(\lambda) + \tau(\lambda))^{-1} \in \mathbf{B}(\mathcal{G})$, and hence

$$\varphi = -(M(\lambda) + \tau(\lambda))^{-1}\phi - (M(\lambda) + \tau(\lambda))^{-1}\chi, \qquad \lambda \in \mathbb{C} \setminus \mathbb{R}.$$

This yields

$$\big(\widetilde{\Theta} - \widetilde{M}(\lambda)\big)^{-1} = -\begin{pmatrix}(M(\lambda) + \tau(\lambda))^{-1} & (M(\lambda) + \tau(\lambda))^{-1} \\ (M(\lambda) + \tau(\lambda))^{-1} & (M(\lambda) + \tau(\lambda))^{-1}\end{pmatrix},$$

and the statement about the resolvent of $\widetilde{A}$ follows from (4.6.7). □

The compressions of the resolvent of the self-adjoint relation $\widetilde{A}$ in (4.6.4) to $\mathfrak{H}$ and $\mathfrak{H}'$ are of interest. Note that the resolvent of $\widetilde{A}_0$ in (4.6.2) is given by the direct orthogonal sum of the resolvents of A_0 and B_0, and hence for $\lambda \in \rho(\widetilde{A}_0)$ the compressions to $\mathfrak{H}$ and $\mathfrak{H}'$ are

$$P_{\mathfrak{H}}(\widetilde{A}_0 - \lambda)^{-1}\iota_{\mathfrak{H}} = (A_0 - \lambda)^{-1} \quad \text{and} \quad P_{\mathfrak{H}'}(\widetilde{A}_0 - \lambda)^{-1}\iota_{\mathfrak{H}'} = (B_0 - \lambda)^{-1},$$

respectively. The next statement follows directly from Proposition 4.6.1 and (4.6.3).

Corollary 4.6.2. *Let S and T be closed symmetric relations in the Hilbert spaces $\mathfrak{H}$ and $\mathfrak{H}'$ with boundary triplets $\{\mathcal{G}, \Gamma_0, \Gamma_1\}$ and $\{\mathcal{G}, \Gamma_0', \Gamma_1'\}$, and corresponding γ-fields and Weyl functions γ, γ' and M, τ, respectively. Then for all $\lambda \in \mathbb{C} \setminus \mathbb{R}$ the following statements hold:*

(i) *The compression of the resolvent of the self-adjoint relation $\widetilde{A}$ in (4.6.4) to $\mathfrak{H}$ is given by*

$$P_{\mathfrak{H}}(\widetilde{A} - \lambda)^{-1}\iota_{\mathfrak{H}} = (A_0 - \lambda)^{-1} - \gamma(\lambda)\big(M(\lambda) + \tau(\lambda)\big)^{-1}\gamma(\bar{\lambda})^*.$$

(ii) *The compression of the resolvent of the self-adjoint relation $\widetilde{A}$ in (4.6.4) to $\mathfrak{H}'$ is given by*

$$P_{\mathfrak{H}'}(\widetilde{A} - \lambda)^{-1}\iota_{\mathfrak{H}'} = (B_0 - \lambda)^{-1} - \gamma'(\lambda)\big(M(\lambda) + \tau(\lambda)\big)^{-1}\gamma'(\bar{\lambda})^*.$$

Corollary 4.6.2 and Proposition 4.6.1 can also be viewed as an alternative approach to the Kreĭn–Naĭmark formula in the special case where the Nevanlinna family τ in Theorem 4.5.2 is a uniformly strict Nevanlinna function. In fact, according to Theorem 4.2.4 every uniformly strict $\mathbf{B}(\mathcal{G})$-valued Nevanlinna function can be realized as a Weyl function, that is, there exist a (reproducing kernel) Hilbert space $\mathfrak{H}'$ ($= \mathfrak{H}(\mathsf{N}_\tau)$), a closed simple symmetric operator T ($= S_\tau$) in $\mathfrak{H}'$, and a boundary triplet $\{\mathcal{G}, \Gamma_0', \Gamma_1'\}$ for the adjoint T^* such that τ is the corresponding Weyl function. In this situation the relation $\widetilde{A}$ in (4.6.4) is self-adjoint in $\mathfrak{H} \oplus \mathfrak{H}' = \mathfrak{H} \oplus \mathfrak{H}(\mathsf{N}_\tau)$ and its compressed resolvent in Corollary 4.6.2 coincides with the one in the Kreĭn–Naĭmark formula in Theorem 4.5.2. Summing up, the following special case of Theorem 4.5.2 is a consequence of the coupling method in Proposition 4.6.1 and Corollary 4.6.2.

Corollary 4.6.3. *Let S be a closed symmetric relation, let $\{\mathcal{G}, \Gamma_0, \Gamma_1\}$ be a boundary triplet for S^* with $A_0 = \ker \Gamma_0$, and let γ and M be the corresponding γ-field and Weyl function, respectively. Let τ be a uniformly strict $\mathbf{B}(\mathcal{G})$-valued Nevanlinna function. Then there exist an exit Hilbert space $\mathfrak{H}'$ and a self-adjoint relation $\widetilde{A}$ in $\mathfrak{H} \oplus \mathfrak{H}'$ such that $\widetilde{A}$ is an extension of S and the compressed resolvent of $\widetilde{A}$ is given by the Kreĭn–Naĭmark formula:*

$$P_{\mathfrak{H}}(\widetilde{A} - \lambda)^{-1}\iota_{\mathfrak{H}} = (A_0 - \lambda)^{-1} - \gamma(\lambda)\big(M(\lambda) + \tau(\lambda)\big)^{-1}\gamma(\bar{\lambda})^*, \quad \lambda \in \mathbb{C} \setminus \mathbb{R}.$$

Furthermore, the self-adjoint relation $\widetilde{A}$ satisfies the minimality condition

$$\mathfrak{H} \oplus \mathfrak{H}' = \overline{\operatorname{span}}\,\big\{\mathfrak{H}, \operatorname{ran}(\widetilde{A} - \mu)^{-1}\iota_{\mathfrak{H}} : \mu \in \mathbb{C} \setminus \mathbb{R}\big\}. \tag{4.6.8}$$

Proof. All statements except the minimality condition (4.6.8) follow from Proposition 4.6.1, Corollary 4.6.2, and Theorem 4.2.4 as explained above. For (4.6.8) recall first that the closed symmetric operator S_τ ($= T$) in Theorem 4.2.4 is simple, and hence

$$\mathfrak{H}' = \overline{\operatorname{span}}\,\big\{\ker(T^* - \mu) : \mu \in \mathbb{C} \setminus \mathbb{R}\big\} = \overline{\operatorname{span}}\,\big\{\operatorname{ran}\gamma'(\mu) : \mu \in \mathbb{C} \setminus \mathbb{R}\big\}. \tag{4.6.9}$$

It follows from Proposition 4.6.1 that

$$P_{\mathfrak{H}'}(\widetilde{A} - \mu)^{-1}\iota_{\mathfrak{H}} = -\gamma'(\mu)\big(M(\mu) + \tau(\mu)\big)^{-1}\gamma(\bar{\mu})^*,$$

and since $\operatorname{ran}\gamma(\bar{\mu})^* = \mathcal{G}$ and $\operatorname{dom}(M(\mu) + \tau(\mu)) = \mathcal{G}$, one sees that

$$\operatorname{ran}\big(P_{\mathfrak{H}'}(\widetilde{A} - \mu)^{-1}\iota_{\mathfrak{H}}\big) = \operatorname{ran}\gamma'(\mu), \qquad \mu \in \mathbb{C} \setminus \mathbb{R}.$$

With (4.6.9) one then concludes that

$$\mathfrak{H}' = \overline{\operatorname{span}}\,\big\{\operatorname{ran}\big(P_{\mathfrak{H}'}(\widetilde{A} - \mu)^{-1}\iota_{\mathfrak{H}}\big) : \mu \in \mathbb{C} \setminus \mathbb{R}\big\},$$

which in turn yields (4.6.8). □

In the next proposition a particular boundary triplet $\{\mathcal{G} \oplus \mathcal{G}, \widehat{\Gamma}_0, \widehat{\Gamma}_1\}$ is specified such that the self-adjoint relation $\widetilde{A}$ in (4.6.4) coincides with the kernel of the boundary mapping $\widehat{\Gamma}_0$. The corresponding Weyl function $\widehat{M}$ is useful for the spectral analysis of $\widetilde{A}$; cf. Chapter 6.

Proposition 4.6.4. *Let S and T be closed symmetric relations in the Hilbert spaces $\mathfrak{H}$ and $\mathfrak{H}'$ with boundary triplets $\{\mathcal{G}, \Gamma_0, \Gamma_1\}$ and $\{\mathcal{G}, \Gamma_0', \Gamma_1'\}$ and corresponding Weyl functions M and τ, respectively, as in the beginning of this section. Then $\{\mathcal{G} \oplus \mathcal{G}, \widehat{\Gamma}_0, \widehat{\Gamma}_1\}$, where*

$$\widehat{\Gamma}_0 \begin{pmatrix} \widehat{f} \\ \widehat{g} \end{pmatrix} = \begin{pmatrix} -\Gamma_1 \widehat{f} - \Gamma_1' \widehat{g} \\ \Gamma_0 \widehat{f} - \Gamma_0' \widehat{g} \end{pmatrix} \quad \text{and} \quad \widehat{\Gamma}_1 \begin{pmatrix} \widehat{f} \\ \widehat{g} \end{pmatrix} = \begin{pmatrix} \Gamma_0 \widehat{f} \\ -\Gamma_1' \widehat{g} \end{pmatrix}, \quad \widehat{f} \in S^*, \widehat{g} \in T^*,$$

is a boundary triplet for $S^* \mathbin{\widehat{\oplus}} T^*$ such that the self-adjoint relation $\widetilde{A}$ in (4.6.4) corresponds to the boundary mapping $\widehat{\Gamma}_0$, that is,

$$\widetilde{A} = \ker \widehat{\Gamma}_0.$$

The Weyl function of $\{\mathcal{G} \oplus \mathcal{G}, \widehat{\Gamma}_0, \widehat{\Gamma}_1\}$ is given by

$$\begin{aligned} \widehat{M}(\lambda) &= -\begin{pmatrix} M(\lambda) & -I \\ -I & -\tau(\lambda)^{-1} \end{pmatrix}^{-1} \\ &= \begin{pmatrix} -(M(\lambda)+\tau(\lambda))^{-1} & (M(\lambda)+\tau(\lambda))^{-1}\tau(\lambda) \\ \tau(\lambda)(M(\lambda)+\tau(\lambda))^{-1} & \tau(\lambda)(M(\lambda)+\tau(\lambda))^{-1}M(\lambda) \end{pmatrix} \end{aligned} \tag{4.6.10}$$

for $\lambda \in \mathbb{C} \setminus \mathbb{R}$.

Proof. Instead of a direct proof the assertions will be obtained as consequences of the results in Section 2.5. For this consider the boundary triplet $\{\mathcal{G} \oplus \mathcal{G}, \widetilde{\Gamma}_0, \widetilde{\Gamma}_1\}$ in (4.6.1) with the Weyl function $\widetilde{M}$ given in (4.6.3), let

$$\mathcal{A} = \frac{1}{\sqrt{2}} \begin{pmatrix} I & 0 \\ I & 0 \end{pmatrix} \quad \text{and} \quad \mathcal{B} = \frac{1}{\sqrt{2}} \begin{pmatrix} 0 & I \\ 0 & -I \end{pmatrix},$$

and observe that $\widetilde{\Theta} = \{\mathcal{A}, \mathcal{B}\}$, with $\widetilde{\Theta}$ in (4.6.5). It is easy to see that $\mathcal{A}$ and $\mathcal{B}$ satisfy the conditions in Corollary 2.5.11. Therefore, $\{\mathcal{G}^2, \breve{\Gamma}_0, \breve{\Gamma}_1\}$, where

$$\breve{\Gamma}_0 = \mathcal{B}^* \widetilde{\Gamma}_0 - \mathcal{A}^* \widetilde{\Gamma}_1 \quad \text{and} \quad \breve{\Gamma}_1 = \mathcal{A}^* \widetilde{\Gamma}_0 + \mathcal{B}^* \widetilde{\Gamma}_1,$$

is a boundary triplet with corresponding Weyl function

$$\breve{M}(\lambda) = \big(\mathcal{A}^* + \mathcal{B}^* \widetilde{M}(\lambda)\big)\big(\mathcal{B}^* - \mathcal{A}^* \widetilde{M}(\lambda)\big)^{-1}, \quad \lambda \in \mathbb{C} \setminus \mathbb{R}.$$

It follows that

$$\breve{\Gamma}_0 \begin{pmatrix} \widehat{f} \\ \widehat{g} \end{pmatrix} = \frac{1}{\sqrt{2}} \begin{pmatrix} -\Gamma_1 \widehat{f} - \Gamma_1' \widehat{g} \\ \Gamma_0 \widehat{f} - \Gamma_0' \widehat{g} \end{pmatrix}, \quad \widehat{f} \in S^*, \widehat{g} \in T^*,$$

and

$$\breve{\Gamma}_1 \begin{pmatrix} \widehat{f} \\ \widehat{g} \end{pmatrix} = \frac{1}{\sqrt{2}} \begin{pmatrix} \Gamma_0 \widehat{f} + \Gamma_0' \widehat{g} \\ \Gamma_1 \widehat{f} - \Gamma_1' \widehat{g} \end{pmatrix}, \quad \widehat{f} \in S^*, \widehat{g} \in T^*.$$

Furthermore, it is easily seen from the above that

$$\begin{aligned}\breve{M}(\lambda) &= \begin{pmatrix} 1 & 1 \\ M(\lambda) & -\tau(\lambda) \end{pmatrix} \begin{pmatrix} -M(\lambda) & -\tau(\lambda) \\ 1 & -1 \end{pmatrix}^{-1} \\ &= \begin{pmatrix} 1 & 1 \\ M(\lambda) & -\tau(\lambda) \end{pmatrix} \begin{pmatrix} -(M(\lambda)+\tau(\lambda))^{-1} & (M(\lambda)+\tau(\lambda))^{-1}\tau(\lambda) \\ -(M(\lambda)+\tau(\lambda))^{-1} & -(M(\lambda)+\tau(\lambda))^{-1}M(\lambda) \end{pmatrix},\end{aligned}$$

where the last step used the identity

$$M(\lambda)(M(\lambda)+\tau(\lambda))^{-1}\tau(\lambda) = \tau(\lambda)(M(\lambda)+\tau(\lambda))^{-1}M(\lambda).$$

Thus, it is clear that

$$\breve{M}(\lambda) = \begin{pmatrix} -2(M(\lambda)+\tau(\lambda))^{-1} & (M(\lambda)+\tau(\lambda))^{-1}(\tau(\lambda)-M(\lambda)) \\ (\tau(\lambda)-M(\lambda))(M(\lambda)+\tau(\lambda))^{-1} & 2M(\lambda)(M(\lambda+\tau(\lambda))^{-1}\tau(\lambda) \end{pmatrix}$$

holds for all $\lambda \in \mathbb{C} \setminus \mathbb{R}$. Now let

$$D = \frac{1}{\sqrt{2}} \begin{pmatrix} I & 0 \\ 0 & I \end{pmatrix} = D^* \quad \text{and} \quad P = \frac{1}{2} \begin{pmatrix} 0 & I \\ I & 0 \end{pmatrix},$$

and apply Corollary 2.5.5 to conclude that

$$\widehat{\Gamma}_0 = D^{-1}\breve{\Gamma}_0 \quad \text{and} \quad \widehat{\Gamma}_1 = D^*\breve{\Gamma}_1 + PD^{-1}\breve{\Gamma}_0$$

give a boundary triplet for $S^* \,\widehat{\oplus}\, T^*$. According to Corollary 2.5.5, the corresponding Weyl function is given by

$$\widehat{M}(\lambda) = D^*\breve{M}(\lambda)D + P, \qquad \lambda \in \mathbb{C} \setminus \mathbb{R},$$

and one verifies that the first identity in (4.6.10) holds. It is straightforward to check the second identity in (4.6.10). □

Chapter 5

Boundary Triplets and Boundary Pairs for Semibounded Relations

Semibounded relations in a Hilbert space automatically have equal defect numbers, so that there are always self-adjoint extensions. In this chapter the semibounded self-adjoint extensions of a semibounded relation will be investigated. Special attention will be paid to the Friedrichs extension, which is introduced with the help of closed semibounded forms. Section 5.1 provides a self-contained introduction to closed semibounded forms and their representations via semibounded self-adjoint relations. Closely related is the discussion of the ordering for closed semibounded forms and for semibounded self-adjoint relations in Section 5.2; this section also contains a general monotonicity principle about monotone sequences of semibounded relations. The Friedrichs extension of a semibounded relation is defined and its central properties are studied in Section 5.3. Particular attention is paid to semibounded self-adjoint extensions which are transversal to the Friedrichs extension. Section 5.4 is devoted to special semibounded extensions, namely the Kreĭn type extensions. In the nonnegative case these extensions include the well-known Kreĭn–von Neumann extension. The Friedrichs extension and the Kreĭn type extensions act as extremal elements to describe the semibounded self-adjoint extensions with a given lower bound. In Section 5.5 there is a return to boundary triplets and Weyl functions for symmetric relations which are semibounded. Of special interest is the case where the self-adjoint extensions determined by the boundary triplet are semibounded and one of them coincides with the Friedrichs extension. In particular, this leads to a useful abstract version of the first Green formula. The notion of a boundary pair for semibounded relations is developed in Section 5.6. In conjunction with the above first Green formula, this notion serves as a link between boundary triplet methods and form methods when semibounded self-adjoint extensions are described; in a wider sense it establishes the connection with the Birman–Kreĭn–Vishik method.

J. Behrndt et al., *Boundary Value Problems, Weyl Functions, and Differential Operators*, Monographs in Mathematics 108,
https://doi.org/10.1007/978-3-030-36714-5_6

5.1 Closed semibounded forms and their representations

A *sesquilinear form* $\mathfrak{t}[\cdot,\cdot]$ in a Hilbert space $\mathfrak{H}$ with inner product $(\cdot,\cdot)$ is a mapping from $\mathfrak{D} \times \mathfrak{D}$ to $\mathbb{C}$, where $\mathfrak{D}$ is a linear subspace of $\mathfrak{H}$, such that $\mathfrak{t}[\cdot,\cdot]$ is linear in the first entry and anti-linear in the second entry. The domain $\operatorname{dom}\mathfrak{t}$ is defined by $\operatorname{dom}\mathfrak{t} = \mathfrak{D}$. The form is said to be *symmetric* if $\mathfrak{t}[\varphi,\psi] = \overline{\mathfrak{t}[\psi,\varphi]}$ for all $\varphi,\psi \in \operatorname{dom}\mathfrak{t}$. The corresponding *quadratic form* $\mathfrak{t}[\cdot]$ is defined by $\mathfrak{t}[\varphi] = \mathfrak{t}[\varphi,\varphi]$, $\varphi \in \operatorname{dom}\mathfrak{t}$. The polarization formula

$$\mathfrak{t}[\varphi,\psi] = \frac{1}{4}\big(\mathfrak{t}[\varphi+\psi] - \mathfrak{t}[\varphi-\psi]\big) + \frac{i}{4}\big(\mathfrak{t}[\varphi+i\psi] - \mathfrak{t}[\varphi-i\psi]\big) \tag{5.1.1}$$

for $\varphi,\psi \in \operatorname{dom}\mathfrak{t}$ is easily checked. In the following the term sesquilinear will be dropped; whenever a form $\mathfrak{t}[\cdot,\cdot]$ is mentioned it is assumed to be sesquilinear and it will be denoted by $\mathfrak{t}$. For instance, the inner product $(\cdot,\cdot)$ is a form defined on all of $\mathfrak{H}$.

Definition 5.1.1. Let $\mathfrak{t}_1$ and $\mathfrak{t}_2$ be forms in $\mathfrak{H}$. Then the *inclusion* $\mathfrak{t}_2 \subset \mathfrak{t}_1$ means that

$$\operatorname{dom}\mathfrak{t}_2 \subset \operatorname{dom}\mathfrak{t}_1, \quad \mathfrak{t}_2[\varphi] = \mathfrak{t}_1[\varphi], \quad \varphi \in \operatorname{dom}\mathfrak{t}_2. \tag{5.1.2}$$

If $\mathfrak{t}_2 \subset \mathfrak{t}_1$, then $\mathfrak{t}_2$ is said to be a *restriction* of $\mathfrak{t}_1$ and $\mathfrak{t}_1$ is said to be an *extension* of $\mathfrak{t}_2$. The *sum* $\mathfrak{t}_1 + \mathfrak{t}_2$ is defined by

$$(\mathfrak{t}_1 + \mathfrak{t}_2)[\varphi,\psi] = \mathfrak{t}_1[\varphi,\psi] + \mathfrak{t}_2[\varphi,\psi], \quad \varphi,\psi \in \operatorname{dom}(\mathfrak{t}_1 + \mathfrak{t}_2),$$

where $\operatorname{dom}(\mathfrak{t}_1 + \mathfrak{t}_2) = \operatorname{dom}\mathfrak{t}_1 \cap \operatorname{dom}\mathfrak{t}_2$.

If $\alpha \in \mathbb{C}$ the sum $\mathfrak{t}[\cdot,\cdot] + \alpha(\cdot,\cdot)$ is given by

$$\mathfrak{t}[\varphi,\psi] + \alpha(\varphi,\psi), \quad \varphi,\psi \in \operatorname{dom}\mathfrak{t}.$$

This sum will be denoted by $\mathfrak{t}+\alpha$. It is symmetric when $\mathfrak{t}$ is symmetric and $\alpha \in \mathbb{R}$.

Definition 5.1.2. A symmetric form $\mathfrak{t}$ in $\mathfrak{H}$ is *bounded from below* if there exists a constant $c \in \mathbb{R}$ such that

$$\mathfrak{t}[\varphi] \geq c\|\varphi\|^2, \quad \varphi \in \operatorname{dom}\mathfrak{t}.$$

This inequality will be denoted by $\mathfrak{t} \geq c$. The *lower bound* $m(\mathfrak{t})$ is the largest of such numbers $c \in \mathbb{R}$:

$$m(\mathfrak{t}) = \inf\left\{\frac{\mathfrak{t}[\varphi]}{\|\varphi\|^2} : \varphi \in \operatorname{dom}\mathfrak{t},\ \varphi \neq 0\right\}.$$

If $m(\mathfrak{t}) \geq 0$, then $\mathfrak{t}$ is called *nonnegative*.

In the following the terminology *semibounded form* is used for a symmetric form which is bounded from below. Note that $\mathfrak{t}$ is a semibounded form if and only if for some, and hence for all $\alpha \in \mathbb{R}$ the form $\mathfrak{t} + \alpha$ is semibounded. For a semibounded form $\mathfrak{t}$ the lower bound $m(\mathfrak{t})$ will often be denoted by γ. Note that the form $\mathfrak{t} - \gamma$, $\gamma = m(\mathfrak{t})$, is nonnegative with lower bound 0. Therefore, one has the Cauchy–Schwarz inequality

$$|(\mathfrak{t} - \gamma)[\varphi, \psi]| \leq (\mathfrak{t} - \gamma)[\varphi]^{\frac{1}{2}} (\mathfrak{t} - \gamma)[\psi]^{\frac{1}{2}}, \quad \varphi, \psi \in \operatorname{dom} \mathfrak{t}, \tag{5.1.3}$$

and, hence the triangle inequality

$$(\mathfrak{t} - \gamma)[\varphi + \psi]^{\frac{1}{2}} \leq (\mathfrak{t} - \gamma)[\varphi]^{\frac{1}{2}} + (\mathfrak{t} - \gamma)[\psi]^{\frac{1}{2}}, \quad \varphi, \psi \in \operatorname{dom} \mathfrak{t}. \tag{5.1.4}$$

It follows from (5.1.4) that

$$\left|(\mathfrak{t} - \gamma)[\varphi]^{\frac{1}{2}} - (\mathfrak{t} - \gamma)[\psi]^{\frac{1}{2}}\right| \leq (\mathfrak{t} - \gamma)[\varphi - \psi]^{\frac{1}{2}}, \quad \varphi, \psi \in \operatorname{dom} \mathfrak{t}. \tag{5.1.5}$$

The following continuity property is a simple consequence of (5.1.5). For a sequence (φ_n) in $\operatorname{dom} \mathfrak{t}$ and $\varphi \in \operatorname{dom} \mathfrak{t}$ one has

$$(\mathfrak{t} - \gamma)[\varphi - \varphi_n] \to 0 \quad \Rightarrow \quad (\mathfrak{t} - \gamma)[\varphi_n] \to (\mathfrak{t} - \gamma)[\varphi]. \tag{5.1.6}$$

Let $\mathfrak{t}$ be a semibounded form in $\mathfrak{H}$ with lower bound γ and let $a < \gamma$. Equip the space $\operatorname{dom} \mathfrak{t} \subset \mathfrak{H}$ with the form

$$(\varphi, \psi)_{\mathfrak{t}-a} = \mathfrak{t}[\varphi, \psi] - a(\varphi, \psi), \quad \varphi, \psi \in \operatorname{dom} \mathfrak{t}. \tag{5.1.7}$$

By rewriting this definition as

$$(\varphi, \psi)_{\mathfrak{t}-a} = (\mathfrak{t} - \gamma)[\varphi, \psi] + (\gamma - a)(\varphi, \psi), \quad \varphi, \psi \in \operatorname{dom} \mathfrak{t}, \tag{5.1.8}$$

one sees that $(\cdot, \cdot)_{\mathfrak{t}-a}$ is the sum of the semidefinite inner product $\mathfrak{t} - \gamma$ and the inner product $(\gamma - a)(\cdot, \cdot)$. Hence, $(\cdot, \cdot)_{\mathfrak{t}-a}$ is an inner product on $\operatorname{dom} \mathfrak{t}$ and the corresponding norm $\| \cdot \|_{\mathfrak{t}-a}$ satisfies the inequality

$$\|\varphi\|^2_{\mathfrak{t}-a} \geq (\gamma - a)\|\varphi\|^2, \quad \varphi \in \operatorname{dom} \mathfrak{t}. \tag{5.1.9}$$

When $\operatorname{dom} \mathfrak{t}$ is equipped with the inner product $(\cdot, \cdot)_{\mathfrak{t}-a}$, the resulting inner product space will be denoted by $\mathfrak{H}_{\mathfrak{t}-a}$. Note that if $\gamma > 0$, then obviously $a = 0$ is a natural choice in the above and the following arguments.

Lemma 5.1.3. *Let $\mathfrak{t}$ be a semibounded form in $\mathfrak{H}$ with lower bound γ and let $a < \gamma$. Let (φ_n) be a sequence in $\operatorname{dom} \mathfrak{t}$. Then (φ_n) is a Cauchy sequence in $\mathfrak{H}_{\mathfrak{t}-a}$ if and only if*

$$\mathfrak{t}[\varphi_n - \varphi_m] \to 0 \quad \textit{and} \quad \|\varphi_n - \varphi_m\| \to 0. \tag{5.1.10}$$

Proof. According to (5.1.8), (φ_n) is a Cauchy sequence in $\mathfrak{H}_{\mathfrak{t}-a}$ if and only if

$$(\mathfrak{t}-\gamma)[\varphi_n - \varphi_m] \to 0 \quad \text{and} \quad \|\varphi_n - \varphi_m\|^2 \to 0. \tag{5.1.11}$$

Now assume that (φ_n) is a Cauchy sequence in $\mathfrak{H}_{\mathfrak{t}-a}$. Then it follows from (5.1.11) that (φ_n) is a Cauchy sequence in $\mathfrak{H}$ and that

$$\mathfrak{t}[\varphi_n - \varphi_m] = (\mathfrak{t}-\gamma)[\varphi_n - \varphi_m] + \gamma\|\varphi_n - \varphi_m\|^2 \to 0,$$

which shows (5.1.10). Conversely, if the sequence (φ_n) satisfies (5.1.10), then it follows from (5.1.7) that (φ_n) is a Cauchy sequence in $\mathfrak{H}_{\mathfrak{t}-a}$. □

Let (φ_n) be a Cauchy sequence in $\mathfrak{H}_{\mathfrak{t}-a}$. Since $\mathfrak{H}$ is a Hilbert space, it follows from Lemma 5.1.3 that there is an element $\varphi \in \mathfrak{H}$ such that $\varphi_n \to \varphi$ in $\mathfrak{H}$.

Definition 5.1.4. Let $\mathfrak{t}$ be a semibounded form in $\mathfrak{H}$. A sequence (φ_n) in $\operatorname{dom} \mathfrak{t}$ is said to be $\mathfrak{t}$-*convergent* to an element $\varphi \in \mathfrak{H}$, not necessarily belonging to $\operatorname{dom} \mathfrak{t}$, if

$$\varphi_n \to \varphi \text{ in } \mathfrak{H} \quad \text{and} \quad \mathfrak{t}[\varphi_n - \varphi_m] \to 0, \quad n, m \to \infty.$$

This type of convergence will be denoted by $\varphi_n \to_{\mathfrak{t}} \varphi$.

The following result is a direct consequence of Lemma 5.1.3 and the completeness of $\mathfrak{H}$.

Corollary 5.1.5. *Let $\mathfrak{t}$ be a semibounded form in $\mathfrak{H}$ with lower bound γ and let $a < \gamma$. Then any Cauchy sequence in $\mathfrak{H}_{\mathfrak{t}-a}$ is $\mathfrak{t}$-convergent. Conversely, any $\mathfrak{t}$-convergent sequence in* $\operatorname{dom} \mathfrak{t}$ *is a Cauchy sequence in $\mathfrak{H}_{\mathfrak{t}-a}$.*

If the sequence (φ_n) in $\operatorname{dom} \mathfrak{t}$ is $\mathfrak{t}$-convergent, then by Definition 5.1.4

$$(\mathfrak{t}-\gamma)[\varphi_n - \varphi_m] \to 0 \quad \text{and} \quad \|\varphi_n - \varphi_m\| \to 0.$$

Thus, one has the following result.

Corollary 5.1.6. *Let $\mathfrak{t}$ be a semibounded form in $\mathfrak{H}$ with lower bound γ and let the sequence (φ_n) in* $\operatorname{dom} \mathfrak{t}$ *be $\mathfrak{t}$-convergent. Then the sequences $((\mathfrak{t}-\gamma)[\varphi_n])$, $(\mathfrak{t}[\varphi_n])$, and $(\|\varphi_n\|)$ converge and, consequently, they are bounded.*

Proof. Since γ is the lower bound of $\mathfrak{t}$, one has $(\mathfrak{t}-\gamma)[\varphi_n - \varphi_m] \to 0$. Hence, (5.1.5) shows that $((\mathfrak{t}-\gamma)[\varphi_n])$ is a Cauchy sequence. Then the same is true for the sequence $(\mathfrak{t}[\varphi_n])$ and it is also clear that $(\|\varphi_n\|)$ is a Cauchy sequence. In particular, the sequences $((\mathfrak{t}-\gamma)[\varphi_n])$, $(\mathfrak{t}[\varphi_n])$, and $(\|\varphi_n\|)$ are bounded. □

The $\mathfrak{t}$-convergence is preserved when one takes a sum of sequences. To see this, let (φ_n) and (ψ_n) be sequences in $\operatorname{dom} \mathfrak{t}$ such that

$$\varphi_n \to_{\mathfrak{t}} \varphi \quad \text{and} \quad \psi_n \to_{\mathfrak{t}} \psi$$

for some $\varphi, \psi \in \mathfrak{H}$. Then clearly $\varphi_n + \psi_n \to \varphi + \psi$ in $\mathfrak{H}$ and

$$\begin{aligned}(\mathfrak{t}-\gamma)[\varphi_n + \psi_n - (\varphi_m + \psi_m)]^{\frac{1}{2}} \\ \leq (\mathfrak{t}-\gamma)[\varphi_n - \varphi_m]^{\frac{1}{2}} + (\mathfrak{t}-\gamma)[\psi_n - \psi_m]^{\frac{1}{2}},\end{aligned}$$

by the triangle inequality in (5.1.4). Therefore,

$$\varphi_n \to_{\mathfrak{t}} \varphi \quad \text{and} \quad \psi_n \to_{\mathfrak{t}} \psi \quad \Rightarrow \quad \varphi_n + \psi_n \to_{\mathfrak{t}} \varphi + \psi. \tag{5.1.12}$$

As a consequence, one sees that

$$\varphi_n \to_{\mathfrak{t}} \varphi \quad \text{and} \quad \psi_n \to_{\mathfrak{t}} \psi \quad \Rightarrow \quad \lim_{n\to\infty} \mathfrak{t}[\varphi_n, \psi_n] \quad \text{exists.} \tag{5.1.13}$$

This last implication follows easily from Corollary 5.1.6 and (5.1.12) by the polarization formula in (5.1.1).

Assume that $\varphi_n \in \operatorname{dom} \mathfrak{t}$ and that $\varphi_n \to_{\mathfrak{t}} \varphi$ for some $\varphi \in \mathfrak{H}$. Now the question is when $\varphi \in \operatorname{dom} \mathfrak{t}$ and, if this is the case, when $\mathfrak{t}[\varphi_n - \varphi] \to 0$? This question gives rise to the notions of closed form and closable form in Definition 5.1.7 and Definition 5.1.11.

Definition 5.1.7. A semibounded form $\mathfrak{t}$ in $\mathfrak{H}$ is said to be *closed* if

$$\varphi_n \to_{\mathfrak{t}} \varphi \quad \Rightarrow \quad \varphi \in \operatorname{dom} \mathfrak{t} \quad \text{and} \quad \mathfrak{t}[\varphi_n - \varphi] \to 0.$$

The statement in (5.1.13) can now be made more precise when the form is closed.

Lemma 5.1.8. *Let $\mathfrak{t}$ be a closed semibounded form in $\mathfrak{H}$. Then*

$$\varphi_n \to_{\mathfrak{t}} \varphi \quad \Rightarrow \quad \varphi \in \operatorname{dom} \mathfrak{t} \quad \textit{and} \quad \mathfrak{t}[\varphi_n] \to \mathfrak{t}[\varphi], \tag{5.1.14}$$

and, consequently,

$$\varphi_n \to_{\mathfrak{t}} \varphi, \quad \psi_n \to_{\mathfrak{t}} \psi \quad \Rightarrow \quad \varphi, \psi \in \operatorname{dom} \mathfrak{t} \quad \textit{and} \quad \mathfrak{t}[\varphi_n, \psi_n] \to \mathfrak{t}[\varphi, \psi]. \tag{5.1.15}$$

Proof. Assume that $\mathfrak{t}$ is a closed semibounded form and $\varphi_n \to_{\mathfrak{t}} \varphi$. Then $\varphi \in \operatorname{dom} \mathfrak{t}$ and $\mathfrak{t}[\varphi_n - \varphi] \to 0$, and since $\varphi_n \to \varphi$ it follows that $(\mathfrak{t}-\gamma)[\varphi_n - \varphi] \to 0$. Hence, $(\mathfrak{t}-\gamma)[\varphi_n] \to (\mathfrak{t}-\gamma)[\varphi]$ by (5.1.6), and therefore $\mathfrak{t}[\varphi_n] \to \mathfrak{t}[\varphi]$. This shows (5.1.14). Now polarization and (5.1.12) yield the assertion (5.1.15). □

Lemma 5.1.9. *Let $\mathfrak{t}$ be a semibounded form in $\mathfrak{H}$ with lower bound γ and let $a < \gamma$. Then the following statements are equivalent:*

(i) *$\mathfrak{H}_{\mathfrak{t}-a}$ is a Hilbert space;*

(ii) *$\mathfrak{t}$ is closed.*

In particular, $\mathfrak{t}$ is closed if and only if $\mathfrak{t} - x$ is closed for some, and hence for all $x \in \mathbb{R}$.

Proof. (i) $\Rightarrow$ (ii) Assume that $\mathfrak{H}_{\mathfrak{t}-a}$ is complete. To show that $\mathfrak{t}$ is closed, assume that $\varphi_n \to_{\mathfrak{t}} \varphi$, so

$$\varphi_n \to \varphi \quad \text{and} \quad \mathfrak{t}[\varphi_n - \varphi_m] \to 0.$$

In particular, this implies by Lemma 5.1.3 that $\|\varphi_n - \varphi_m\|_{\mathfrak{t}-a} \to 0$. Since $\mathfrak{H}_{\mathfrak{t}-a}$ is complete there is an element $\varphi_0 \in \mathfrak{H}_{\mathfrak{t}-a} = \operatorname{dom} \mathfrak{t}$ such that $\|\varphi_n - \varphi_0\|_{\mathfrak{t}-a} \to 0$. Hence, by (5.1.9),

$$\|\varphi_n - \varphi_0\| \to 0.$$

Thus $\varphi = \varphi_0 \in \operatorname{dom} \mathfrak{t}$. Therefore, $\|\varphi_n - \varphi\|_{\mathfrak{t}-a} \to 0$ and by (5.1.7) one sees that $\mathfrak{t}[\varphi_n - \varphi] \to 0$. This proves that $\mathfrak{t}$ is closed.

(ii) $\Leftarrow$ (i) Assume that $\mathfrak{t}$ is closed. To show that $\mathfrak{H}_{\mathfrak{t}-a}$ is complete, let (φ_n) be a Cauchy sequence in $\mathfrak{H}_{\mathfrak{t}-a}$. This implies that $\varphi_n \to_{\mathfrak{t}} \varphi$ for some $\varphi \in \mathfrak{H}$; cf. Corollary 5.1.5. The closedness of $\mathfrak{t}$ gives that $\varphi \in \operatorname{dom} \mathfrak{t} = \mathfrak{H}_{\mathfrak{t}-a}$ and $\mathfrak{t}[\varphi_n - \varphi] \to 0$. By (5.1.7) this leads to $\|\varphi_n - \varphi\|_{\mathfrak{t}-a} \to 0$, so that $\mathfrak{H}_{\mathfrak{t}-a}$ is complete.

Since $\mathfrak{t} - x$ is a semibounded form in $\mathfrak{H}$ with lower bound $\gamma - x$, the last statement follows from $\mathfrak{H}_{\mathfrak{t}-a} = \mathfrak{H}_{\mathfrak{t}-x-(a-x)}$ and the equivalence of (i) and (ii). □

Let $\mathfrak{t}$ be a semibounded form in $\mathfrak{H}$ with lower bound γ and let $\mathfrak{H}_{\mathfrak{t}-a}$ be the corresponding inner product space with $a < \gamma$. In general $\mathfrak{t}$ is not closed and hence $\mathfrak{H}_{\mathfrak{t}-a}$ is not complete; cf. Lemma 5.1.9. If $\mathfrak{t}_1$ is a semibounded form with lower bound γ_1, which extends the semibounded form $\mathfrak{t}$ with lower bound γ, then

$$\gamma_1 \leq \gamma.$$

Note that for $a < \gamma_1$ one has that $\mathfrak{t}_1$ is closed if and only if $\mathfrak{H}_{\mathfrak{t}_1-a}$ is a Hilbert space. The question is when such a closed extension $\mathfrak{t}_1$ exists and, if so, to determine the smallest such extension of $\mathfrak{t}$. In order to construct an extension of $\mathfrak{t}$, note that Lemma 5.1.8 suggests the following definition.

Definition 5.1.10. Let $\mathfrak{t}$ be a semibounded form in $\mathfrak{H}$. The linear subspace $\operatorname{dom} \widetilde{\mathfrak{t}}$ is the set of all $\varphi \in \mathfrak{H}$ for which there exists a sequence (φ_n) in $\operatorname{dom} \mathfrak{t}$ such that $\varphi_n \to_{\mathfrak{t}} \varphi$.

It is clear that $\operatorname{dom} \widetilde{\mathfrak{t}}$ is an extension of $\operatorname{dom} \mathfrak{t}$. To establish the linearity of $\operatorname{dom} \widetilde{\mathfrak{t}}$, recall the property (5.1.12). According to (5.1.13), it would now be natural to define the form $\widetilde{\mathfrak{t}}$ on $\operatorname{dom} \widetilde{\mathfrak{t}}$ as an extension of $\mathfrak{t}$ by

$$\widetilde{\mathfrak{t}}[\varphi, \psi] = \lim_{n\to\infty} \mathfrak{t}[\varphi_n, \psi_n] \quad \text{for any} \quad \varphi_n \to_{\mathfrak{t}} \varphi, \quad \psi_n \to_{\mathfrak{t}} \psi, \tag{5.1.16}$$

as the limit on the right-hand side exists. However, in general the limit on the right-hand side of (5.1.16) depends on the choice of the sequences (φ_n) and (ψ_n), so that $\widetilde{\mathfrak{t}}$ may not be well defined as a form.

Definition 5.1.11. A semibounded form $\mathfrak{t}$ in $\mathfrak{H}$ is said to be *closable* if for any sequence (φ_n) in $\operatorname{dom} \mathfrak{t}$

$$\varphi_n \to_{\mathfrak{t}} 0 \quad \Rightarrow \quad \mathfrak{t}[\varphi_n] \to 0.$$

It will be shown that the extension procedure in (5.1.16) defines a form extension of $\mathfrak{t}$ if $\mathfrak{t}$ is closable. In fact, in this case the resulting form $\widetilde{\mathfrak{t}}$ is unique, being the smallest closed extension, and will be called the *closure* of $\mathfrak{t}$.

Theorem 5.1.12. *Let $\mathfrak{t}$ be a semibounded form in $\mathfrak{H}$ with lower bound γ and let $a < \gamma$. Then $\mathfrak{t}$ has a closed extension if and only if $\mathfrak{t}$ is closable. In fact, if $\mathfrak{t}$ is closable, then*

(i) *the* closure $\widetilde{\mathfrak{t}}$ *in* (5.1.16) *is a well-defined form which extends* $\mathfrak{t}$*;*

(ii) $\widetilde{\mathfrak{t}}$ *has the same lower bound as* $\mathfrak{t}$*;*

(iii) $\widetilde{\mathfrak{t}}$ *is the smallest closed extension of* $\mathfrak{t}$*,*

and the inner product space $\mathfrak{H}_{\mathfrak{t}-a}$ is dense in the Hilbert space $\mathfrak{H}_{\widetilde{\mathfrak{t}}-a}$. Moreover, $\mathfrak{t}$ is closable if and only $\mathfrak{t}-x$ is closable for some, and hence for all $x \in \mathbb{R}$, in which case

$$\widetilde{\mathfrak{t}-x} = \widetilde{\mathfrak{t}} - x. \tag{5.1.17}$$

Proof. ($\Rightarrow$) Let $\mathfrak{t}_1$ be a closed extension of $\mathfrak{t}$. In order to show that $\mathfrak{t}$ is closable, assume that $\varphi_n \to_{\mathfrak{t}} 0$. The form $\mathfrak{t}_1$ is an extension of $\mathfrak{t}$ and this implies $\varphi_n \to_{\mathfrak{t}_1} 0$. Since $\mathfrak{t}_1$ is closed, it follows that

$$\mathfrak{t}[\varphi_n] = \mathfrak{t}_1[\varphi_n] \to 0.$$

Hence, $\mathfrak{t}$ is closable.

($\Leftarrow$) Assume that $\mathfrak{t}$ is closable. It will be shown that $\widetilde{\mathfrak{t}}$ in (5.1.16) is a well-defined form on $\operatorname{dom}\widetilde{\mathfrak{t}}$. It is clear from (5.1.13) that the limit on the right-hand side of (5.1.16) exists. To verify that this limit depends only on the elements φ, ψ and not on the particular sequences (φ_n), (ψ_n), let (φ_n'), (ψ_n') be other sequences such that $\varphi_n' \to_{\mathfrak{t}} \varphi$ and $\psi_n' \to_{\mathfrak{t}} \psi$. Then

$$\varphi_n' - \varphi_n \to_{\mathfrak{t}} 0 \quad \text{and} \quad \psi_n' - \psi_n \to_{\mathfrak{t}} 0;$$

cf. (5.1.12). In particular, this gives

$$\varphi_n' - \varphi_n \to 0 \quad \text{and} \quad \psi_n' - \psi_n \to 0,$$

while the closability of $\mathfrak{t}$ implies that

$$\mathfrak{t}[\varphi_n' - \varphi_n] \to 0 \quad \text{and} \quad \mathfrak{t}[\psi_n' - \psi_n] \to 0.$$

To see that the sequences $\mathfrak{t}[\varphi_n', \psi_n']$ and $\mathfrak{t}[\varphi_n, \psi_n]$ have the same limit, consider the inequalities

$$\begin{aligned}
&|(\mathfrak{t}-\gamma)[\varphi_n', \psi_n'] - (\mathfrak{t}-\gamma)[\varphi_n, \psi_n]| \\
&\quad = |(\mathfrak{t}-\gamma)[\varphi_n' - \varphi_n, \psi_n'] + (\mathfrak{t}-\gamma)[\varphi_n, \psi_n' - \psi_n]| \\
&\quad \leq |(\mathfrak{t}-\gamma)[\varphi_n' - \varphi_n, \psi_n']| + |(\mathfrak{t}-\gamma)[\varphi_n, \psi_n' - \psi_n]| \\
&\quad \leq (\mathfrak{t}-\gamma)[\varphi_n' - \varphi_n]^{\frac{1}{2}}(\mathfrak{t}-\gamma)[\psi_n']^{\frac{1}{2}} + (\mathfrak{t}-\gamma)[\varphi_n]^{\frac{1}{2}}(\mathfrak{t}-\gamma)[\psi_n' - \psi_n]^{\frac{1}{2}}.
\end{aligned}$$

Clearly, due to the closability assumption, the terms

$$(\mathfrak{t}-\gamma)[\varphi_n'-\varphi_n] \quad \text{and} \quad (\mathfrak{t}-\gamma)[\psi_n'-\psi_n]$$

converge to 0 as $n \to \infty$, while the terms

$$(\mathfrak{t}-\gamma)[\psi_n'] \quad \text{and} \quad (\mathfrak{t}-\gamma)[\varphi_n]$$

are bounded since $\psi_n' \to_{\mathfrak{t}} \psi$ and $\varphi_n \to_{\mathfrak{t}} \varphi$, respectively; cf. Corollary 5.1.6. It follows that $\mathfrak{t}[\varphi_n', \psi_n'] - \mathfrak{t}[\varphi_n, \psi_n] \to 0$ and hence $\widetilde{\mathfrak{t}}$ in (5.1.16) is a well-defined form. Moreover, it is clear that $\widetilde{\mathfrak{t}}$ extends $\mathfrak{t}$: $\mathfrak{t} \subset \widetilde{\mathfrak{t}}$.

The form $\widetilde{\mathfrak{t}}$ is semibounded. To see this, let $\varphi \in \operatorname{dom} \widetilde{\mathfrak{t}}$. Then there exists a sequence (φ_n) in $\operatorname{dom} \mathfrak{t}$ such that $\varphi_n \to_{\mathfrak{t}} \varphi$. In particular, $\varphi_n \to \varphi$ and hence $\|\varphi_n\| \to \|\varphi\|$. According to (5.1.16),

$$\widetilde{\mathfrak{t}}[\varphi] = \lim_{n\to\infty} \mathfrak{t}[\varphi_n],$$

where $\mathfrak{t}[\varphi_n] \geq \gamma \|\varphi_n\|^2$. Therefore,

$$\widetilde{\mathfrak{t}}[\varphi] \geq \gamma \|\varphi\|^2, \quad \varphi \in \operatorname{dom} \widetilde{\mathfrak{t}},$$

so that $\widetilde{\mathfrak{t}}$ is semibounded. Moreover, this argument shows that the lower bound of the extension is at least γ. Hence, $\widetilde{\mathfrak{t}}$ and $\mathfrak{t}$ have the same lower bound.

The argument to show that $\widetilde{\mathfrak{t}}$ is closed, is based on the observation that for the extension $\widetilde{\mathfrak{t}}$:

$$\varphi_n \to_{\mathfrak{t}} \varphi \quad \Rightarrow \quad \widetilde{\mathfrak{t}}[\varphi - \varphi_n] \to 0. \tag{5.1.18}$$

To see this, let $\varphi_n \to_{\mathfrak{t}} \varphi$, that is $\varphi_n \to \varphi$ and $\lim_{m,n\to\infty} \mathfrak{t}[\varphi_n - \varphi_m] = 0$. Now fix $n \in \mathbb{N}$, then $\varphi_m \to_{\mathfrak{t}} \varphi$ implies that

$$\varphi_m - \varphi_n \to_{\mathfrak{t}} \varphi - \varphi_n \quad \text{as} \quad m \to \infty,$$

so that, by definition,

$$\widetilde{\mathfrak{t}}[\varphi - \varphi_n] = \lim_{m\to\infty} \mathfrak{t}[\varphi_m - \varphi_n].$$

Now taking $n \to \infty$ gives (5.1.18).

The following three steps will establish that $\widetilde{\mathfrak{t}}$ is closed or, equivalently, that $\mathfrak{H}_{\widetilde{\mathfrak{t}}-a}$, $a < \gamma$, is complete.

Step 1. $\mathfrak{H}_{\mathfrak{t}-a}$ is dense in $\mathfrak{H}_{\widetilde{\mathfrak{t}}-a}$. Indeed, let $\varphi \in \mathfrak{H}_{\widetilde{\mathfrak{t}}-a} = \operatorname{dom} \widetilde{\mathfrak{t}}$. Then there is a sequence (φ_n) in $\mathfrak{H}_{\mathfrak{t}-a} = \operatorname{dom} \mathfrak{t}$ such that $\varphi_n \to_{\mathfrak{t}} \varphi$. It follows from this assumption and (5.1.18) that

$$\varphi_n \to \varphi \quad \text{and} \quad \widetilde{\mathfrak{t}}[\varphi - \varphi_n] \to 0,$$

in other words,

$$\|\varphi - \varphi_n\|^2_{\widetilde{\mathfrak{t}}-a} = \widetilde{\mathfrak{t}}[\varphi - \varphi_n] - a\|\varphi - \varphi_n\|^2 \to 0.$$

This shows that $\mathfrak{H}_{\mathfrak{t}-a}$ is dense in $\mathfrak{H}_{\widetilde{\mathfrak{t}}-a}$.

Step 2. Every Cauchy sequence in $\mathfrak{H}_{\mathfrak{t}-a}$ is convergent in $\mathfrak{H}_{\widetilde{\mathfrak{t}}-a}$. To see this, let (φ_n) be a Cauchy sequence in $\mathfrak{H}_{\mathfrak{t}-a}$. Then clearly there exists an element $\varphi \in \mathfrak{H}$ such that $\varphi_n \to_{\mathfrak{t}} \varphi$; cf. Corollary 5.1.5. Again by (5.1.18) it follows that

$$\|\varphi - \varphi_n\|_{\widetilde{\mathfrak{t}}-a} \to 0,$$

which now shows that the Cauchy sequence (φ_n) in $\mathfrak{H}_{\mathfrak{t}-a}$ is convergent in $\mathfrak{H}_{\widetilde{\mathfrak{t}}-a}$ to, in fact, $\varphi \in \operatorname{dom} \widetilde{\mathfrak{t}} = \mathfrak{H}_{\widetilde{\mathfrak{t}}-a}$.

Step 3. $\mathfrak{H}_{\widetilde{\mathfrak{t}}-a}$ is a Hilbert space. To see this, let (χ_n) be a Cauchy sequence in $\mathfrak{H}_{\widetilde{\mathfrak{t}}-a}$. By Step 1, there is an element $\varphi_n \in \mathfrak{H}_{\mathfrak{t}-a}$ such that

$$\|\chi_n - \varphi_n\|_{\widetilde{\mathfrak{t}}-a} \leq \frac{1}{n}.$$

Hence, the approximating sequence (φ_n) is a Cauchy sequence in $\mathfrak{H}_{\mathfrak{t}-a}$. By Step 2, (φ_n) converges in $\mathfrak{H}_{\widetilde{\mathfrak{t}}-a}$, which implies that the original sequence (χ_n) converges in $\mathfrak{H}_{\widetilde{\mathfrak{t}}-a}$.

Next it will be shown that $\widetilde{\mathfrak{t}}$ is the smallest closed extension of $\mathfrak{t}$. Assume that $\mathfrak{t}_1$ is a closed extension of $\mathfrak{t}$: $\mathfrak{t} \subset \mathfrak{t}_1$. Let $\varphi \in \operatorname{dom} \widetilde{\mathfrak{t}}$; then there exists a sequence (φ_n) in $\operatorname{dom} \mathfrak{t}$ with $\varphi_n \to_{\mathfrak{t}} \varphi$. Then also $\varphi_n \to_{\mathfrak{t}_1} \varphi$ and hence $\varphi \in \operatorname{dom} \mathfrak{t}_1$. Therefore, $\operatorname{dom} \widetilde{\mathfrak{t}} \subset \operatorname{dom} \mathfrak{t}_1$. For every $\varphi, \psi \in \operatorname{dom} \widetilde{\mathfrak{t}}$ it follows via corresponding sequences (φ_n), (ψ_n) in $\operatorname{dom} \mathfrak{t}$ with $\varphi_n \to_{\mathfrak{t}} \varphi$ and $\psi_n \to_{\mathfrak{t}} \psi$ that

$$\widetilde{\mathfrak{t}}[\varphi, \psi] = \lim_{n \to \infty} \mathfrak{t}[\varphi_n, \psi_n] = \lim_{n \to \infty} \mathfrak{t}_1[\varphi_n, \psi_n] = \mathfrak{t}_1[\varphi, \psi],$$

where the first equality follows from (5.1.16), the second equality is valid as $\mathfrak{t}_1$ extends $\mathfrak{t}$, and the third equality follows from (5.1.15). Therefore, $\widetilde{\mathfrak{t}} \subset \mathfrak{t}_1$, and $\widetilde{\mathfrak{t}}$ is the smallest closed extension of $\mathfrak{t}$.

As to the last statement, observe that Definition 5.1.11 implies that $\mathfrak{t}$ is closable if and only $\mathfrak{t} - x$ is closable for some, and hence for all $x \in \mathbb{R}$. Finally, (5.1.17) follows from (5.1.16). □

Thus, a closed semibounded form $\mathfrak{t}_1$ which extends $\mathfrak{t}$ contains the closure $\widetilde{\mathfrak{t}}$. The next corollary is a simple but useful description of the gap between $\mathfrak{t}_1$ and $\widetilde{\mathfrak{t}}$.

Corollary 5.1.13. *Let the semibounded form* $\mathfrak{t}$ *with lower bound* γ *be closable and let the closed form* $\mathfrak{t}_1$ *with lower bound* γ_1 *be an extension of* $\mathfrak{t}$, *so that* $\gamma_1 \leq \gamma$. *Assume that* $a < \gamma_1$, *then*

$$\mathfrak{H}_{\mathfrak{t}_1-a} = \left\{\varphi \in \mathfrak{H}_{\mathfrak{t}_1-a} : (\varphi, \psi)_{\mathfrak{t}_1-a} = 0, \psi \in \mathfrak{H}_{\widetilde{\mathfrak{t}}-a}\right\} \oplus_{\mathfrak{t}_1-a} \mathfrak{H}_{\widetilde{\mathfrak{t}}-a}.$$

Let $\mathfrak{t}$ be a closed semibounded form in $\mathfrak{H}$. Let $\mathfrak{D} \subset \operatorname{dom} \mathfrak{t}$ be a linear subspace and consider the restriction $\mathfrak{t}_{\mathfrak{D}}$ of $\mathfrak{t}$ to $\mathfrak{D}$,

$$\mathfrak{t}_{\mathfrak{D}}[\varphi, \psi] = \mathfrak{t}[\varphi, \psi], \quad \varphi, \psi \in \mathfrak{D}.$$

Since $\mathfrak{t}_{\mathfrak{D}}$ is a restriction of a closed form, it is closable, see Theorem 5.1.12. Let $\widetilde{\mathfrak{t}}_{\mathfrak{D}}$ be the closure of $\mathfrak{t}_{\mathfrak{D}}$. Then by definition $\operatorname{dom} \widetilde{\mathfrak{t}}_{\mathfrak{D}}$ is the set of all $\varphi \in \mathfrak{H}$ for which there exists a sequence (φ_n) in $\mathfrak{D}$ with $\varphi_n \to_{\mathfrak{t}_{\mathfrak{D}}} \varphi$, which means $\varphi_n \to_{\mathfrak{t}} \varphi$. Since $\mathfrak{t}$ is closed, one sees in particular that $\operatorname{dom} \widetilde{\mathfrak{t}}_{\mathfrak{D}} \subset \operatorname{dom} \mathfrak{t}$. Moreover, one has

$$\widetilde{\mathfrak{t}}_{\mathfrak{D}}[\varphi, \psi] = \lim_{n\to\infty} \mathfrak{t}_{\mathfrak{D}}[\varphi_n, \psi_n] = \lim_{n\to\infty} \mathfrak{t}[\varphi_n, \psi_n] = \mathfrak{t}[\varphi, \psi]$$

for $\varphi, \psi \in \operatorname{dom} \widetilde{\mathfrak{t}}_{\mathfrak{D}}$, where the first equality is by definition, and the third equality follows from Lemma 5.1.8. Hence, the closure $\widetilde{\mathfrak{t}}_{\mathfrak{D}}$ of $\mathfrak{t}_{\mathfrak{D}}$ is the restriction of $\mathfrak{t}$ to $\operatorname{dom} \widetilde{\mathfrak{t}}_{\mathfrak{D}}$. Since $\mathfrak{t}$ is closed, it follows that $\varphi \in \operatorname{dom} \widetilde{\mathfrak{t}}_{\mathfrak{D}}$ if and only if there is a sequence (φ_n) in $\mathfrak{D}$ such that

$$\varphi_n \to \varphi \quad \text{and} \quad \mathfrak{t}[\varphi_n - \varphi] \to 0.$$

Definition 5.1.14. Let $\mathfrak{t}$ be a closed semibounded form in $\mathfrak{H}$. A linear subspace $\mathfrak{D}$ of $\operatorname{dom} \mathfrak{t}$ is said to be a *core* of $\mathfrak{t}$ if the closure $\widetilde{\mathfrak{t}}_{\mathfrak{D}}$ of the restriction $\mathfrak{t}_{\mathfrak{D}}$ of $\mathfrak{t}$ to $\mathfrak{D}$ coincides with $\mathfrak{t}$.

Therefore, $\mathfrak{D} \subset \operatorname{dom} \mathfrak{t}$ is a core of $\mathfrak{t}$ if and only if for every $\varphi \in \operatorname{dom} \mathfrak{t}$ there is a sequence (φ_n) in $\mathfrak{D}$ such that

$$\varphi_n \to \varphi \quad \text{and} \quad \mathfrak{t}[\varphi_n - \varphi] \to 0. \tag{5.1.19}$$

This leads to the following corollary.

Corollary 5.1.15. *Let $\mathfrak{t}$ be a closed semibounded form in $\mathfrak{H}$ with lower bound γ, let $a < \gamma$, and let $\mathfrak{D} \subset \operatorname{dom} \mathfrak{t}$ be a linear subspace. Then $\mathfrak{D}$ is a core of $\mathfrak{t}$ if and only if $\mathfrak{D}$ is dense in the Hilbert space $\mathfrak{H}_{\mathfrak{t}-a}$.*

Note that in the situation of Theorem 5.1.12 the original domain $\operatorname{dom} \mathfrak{t}$ is a core of the closure $\widetilde{\mathfrak{t}}$ of $\mathfrak{t}$ (recall that the form $\widetilde{\mathfrak{t}}$ is closed). The following fact is useful: If $\mathfrak{t}$ and $\mathfrak{s}$ are closed semibounded forms in $\mathfrak{H}$ which coincide on $\mathfrak{D} \subset \operatorname{dom} \mathfrak{t} \cap \operatorname{dom} \mathfrak{s}$ and $\mathfrak{D}$ is a core of both $\mathfrak{t}$ and $\mathfrak{s}$, then $\mathfrak{t} = \mathfrak{s}$.

Recall the definition of the sum of two forms in Definition 5.1.1 and observe that a sum of semibounded forms is also semibounded. The following result is concerned with additive perturbations of forms: it provides a sufficient condition so that the sum of a closed semibounded form and a symmetric form remains closed and semibounded. Sometimes this result is referred to as KLMN theorem, named after Kato, Lions, Lax, Milgram, and Nelson. For a typical application to Sturm–Liouville operators, see, e.g., Lemma 6.8.3.

Theorem 5.1.16. *Assume that $\mathfrak{t}$ is a closed semibounded form in $\mathfrak{H}$ and let $\mathfrak{s}$ be a symmetric form in $\mathfrak{H}$ such that $\operatorname{dom} \mathfrak{t} \subset \operatorname{dom} \mathfrak{s}$ and*

$$|\mathfrak{s}[\varphi]| \leq a\|\varphi\|^2 + b\mathfrak{t}[\varphi], \qquad \varphi \in \operatorname{dom} \mathfrak{t}, \tag{5.1.20}$$

holds for some $a \geq 0$ *and* $b \in [0,1)$. *Then the symmetric form*

$$\mathfrak{t} + \mathfrak{s}, \qquad \operatorname{dom}(\mathfrak{t}+\mathfrak{s}) = \operatorname{dom}\mathfrak{t},$$

is closed and semibounded in $\mathfrak{H}$. *Furthermore, if* $\mathfrak{D}$ *is a core of* $\mathfrak{t}$, *then* $\mathfrak{D}$ *is also a core of* $\mathfrak{t}+\mathfrak{s}$.

Proof. Let γ be the lower bound of $\mathfrak{t}$. Fix some $a' < \gamma$ and assume $a' < 0$. For all $\varphi \in \operatorname{dom}\mathfrak{t}$, $\varphi \neq 0$, one obtains from (5.1.20) that

$$\mathfrak{s}[\varphi] \geq -a\|\varphi\|^2 - b\mathfrak{t}[\varphi], \tag{5.1.21}$$

and hence

$$(\mathfrak{t}+\mathfrak{s})[\varphi] \geq (1-b)\mathfrak{t}[\varphi] - a\|\varphi\|^2 > \big((1-b)a' - a\big)\|\varphi\|^2 = c'\|\varphi\|^2,$$

where $c' = (1-b)a' - a < 0$. This shows that $\mathfrak{t}+\mathfrak{s}$ is semibounded from below. Furthermore, the estimate (5.1.21) also shows that

$$\begin{aligned}(1-b)\|\varphi\|^2_{\mathfrak{t}-a'} &= (1-b)\mathfrak{t}[\varphi] - (1-b)a'\|\varphi\|^2 \\ &= \mathfrak{t}[\varphi] - b\mathfrak{t}[\varphi] - a\|\varphi\|^2 - \big((1-b)a' - a\big)\|\varphi\|^2 \\ &\leq (\mathfrak{t}+\mathfrak{s})[\varphi] - \big((1-b)a' - a\big)\|\varphi\|^2 \\ &= \|\varphi\|^2_{\mathfrak{t}+\mathfrak{s}-c'}.\end{aligned}$$

Using (5.1.20) one obtains

$$\begin{aligned}\|\varphi\|^2_{\mathfrak{t}+\mathfrak{s}-c'} &= \mathfrak{t}[\varphi] + \mathfrak{s}[\varphi] - c'\|\varphi\|^2 \\ &\leq (1+b)\mathfrak{t}[\varphi] - (c'-a)\|\varphi\|^2 \\ &\leq b'\|\varphi\|^2_{\mathfrak{t}-a'},\end{aligned}$$

where $b' = \max\{(1+b), (c'-a)/a'\}$. Therefore, the above estimates imply that the norms $\|\cdot\|^2_{\mathfrak{t}-a'}$ and $\|\cdot\|^2_{\mathfrak{t}+\mathfrak{s}-c'}$ are equivalent on $\operatorname{dom}\mathfrak{t} = \operatorname{dom}(\mathfrak{s}+\mathfrak{t})$. Since $\mathfrak{t}$ is closed, $\mathfrak{H}_{\mathfrak{t}-a'}$ is a Hilbert space and hence $\mathfrak{H}_{\mathfrak{t}+\mathfrak{s}-c'}$ is a Hilbert space, that is, the form $\mathfrak{t}+\mathfrak{s}$ is closed; cf. Lemma 5.1.9. The assertion about the core $\mathfrak{D}$ is clear from Corollary 5.1.15. □

Semibounded relations in a Hilbert space generate closable semibounded forms as will be shown in the following lemma. Note that if a relation is semibounded, then so is its closure, with the same lower bound; this follows directly from Definition 1.4.5. Furthermore, the closure will generate the same form. The particular situation of semibounded self-adjoint relations will be considered in detail in Theorem 5.1.18 and Proposition 5.1.19.

Lemma 5.1.17. *Let* S *be a semibounded relation in* $\mathfrak{H}$ *with lower bound* $m(S)$. *Then the form* $\mathfrak{t}_S$ *given by*

$$\mathfrak{t}_S[\varphi,\psi] = (\varphi',\psi), \quad \{\varphi,\varphi'\},\{\psi,\psi'\} \in S, \tag{5.1.22}$$

with $\operatorname{dom} \mathfrak{t}_S = \operatorname{dom} S$, *is well defined, semibounded with the lower bound* $m(S)$, *and closable. The closure* $\widetilde{\mathfrak{t}}_S$ *of* $\mathfrak{t}_S$ *is a semibounded closed form whose lower bound is equal to* $m(S)$, *and*

$$\operatorname{dom} \widetilde{\mathfrak{t}}_S \subset \overline{\operatorname{dom} S}. \tag{5.1.23}$$

Moreover, $\operatorname{dom} S = \operatorname{dom} \mathfrak{t}_S$ *is a core of* $\widetilde{\mathfrak{t}}_S$. *Furthermore, with the closure* $\overline{S}$ *of* S *one has*

$$\widetilde{\mathfrak{t}}_S = \widetilde{\mathfrak{t}}_{\overline{S}}. \tag{5.1.24}$$

Proof. As a semibounded relation S is automatically symmetric, it follows that $\operatorname{mul} S \subset \operatorname{mul} S^* = (\operatorname{dom} S)^\perp$, and hence

$$(\varphi', \psi) = (\varphi'', \psi), \quad \{\varphi, \varphi'\}, \{\varphi, \varphi''\}, \{\psi, \psi'\} \in S.$$

Thus, the form in (5.1.22) is well defined with $\operatorname{dom} \mathfrak{t}_S = \operatorname{dom} S$. By definition $\mathfrak{t}_S$ is semibounded and its lower bound is clearly equal to $\gamma = m(S)$.

In order to show that $\mathfrak{t}_S$ is closable, let $\varphi_n \to_{\mathfrak{t}_S} 0$. Then, equivalently,

$$\varphi_n \to 0 \quad \text{and} \quad (\mathfrak{t}_S - \gamma)[\varphi_n - \varphi_m] \to 0.$$

It suffices to verify that $(\mathfrak{t}_S - \gamma)[\varphi_n] \to 0$. Note that there exists $\varphi_n' \in \mathfrak{H}$ such that $\{\varphi_n, \varphi_n'\} \in S$. Then

$$(\mathfrak{t}_S - \gamma)[\varphi_n] = (\mathfrak{t}_S - \gamma)[\varphi_n, \varphi_n] = (\mathfrak{t}_S - \gamma)[\varphi_n, \varphi_n - \varphi_m] + (\mathfrak{t}_S - \gamma)[\varphi_n, \varphi_m],$$

and it follows with the help of the Cauchy–Schwarz inequality (5.1.3) for the nonnegative form $(\mathfrak{t}_S - \gamma)$ and (5.1.22) that

$$\begin{aligned} |(\mathfrak{t}_S - \gamma)[\varphi_n]| &\le |(\mathfrak{t}_S - \gamma)[\varphi_n, \varphi_n - \varphi_m]| + |(\mathfrak{t}_S - \gamma)[\varphi_n, \varphi_m]| \\ &\le (\mathfrak{t}_S - \gamma)[\varphi_n]^{\frac{1}{2}} (\mathfrak{t}_S - \gamma)[\varphi_n - \varphi_m]^{\frac{1}{2}} + |(\varphi_n' - \gamma\varphi_n, \varphi_m)|. \end{aligned}$$

By Corollary 5.1.6, the sequence $((\mathfrak{t}_S - \gamma)[\varphi_n])$ is bounded by M^2 for some $M > 0$. Moreover, for every $\varepsilon > 0$ there exists $N \in \mathbb{N}$ such that $(\mathfrak{t}_S - \gamma)[\varphi_n - \varphi_m] \le \varepsilon^2$ for $n, m \ge N$. Therefore,

$$|(\mathfrak{t}_S - \gamma)[\varphi_n]| \le M\varepsilon + |(\varphi_n' - \gamma\varphi_n, \varphi_m)|, \qquad n, m \ge N.$$

Fix $n \ge N$ and let $m \to \infty$. From $|(\varphi_n' - \gamma\varphi_n, \varphi_m)| \le \|\varphi_n' - \gamma\varphi_n\| \|\varphi_m\|$ and $\|\varphi_m\| \to 0$ it follows that $|(\mathfrak{t}_S - \gamma)[\varphi_n]| \le M\varepsilon$ for $n \ge N$. This shows that $(\mathfrak{t}_S - \gamma)[\varphi_n] \to 0$ as $n \to \infty$, and hence $\mathfrak{t}_S$ is closable.

By Theorem 5.1.12, it is clear that the closure $\widetilde{\mathfrak{t}}_S$ of $\mathfrak{t}_S$ is a semibounded closed form whose lower bound is equal to $m(S)$. It also follows from the definition of $\widetilde{\mathfrak{t}}$ that the inclusion (5.1.23) holds. Furthermore, $\operatorname{dom} S = \operatorname{dom} \mathfrak{t}_S$ is a core of $\widetilde{\mathfrak{t}}_S$.

It remains to show (5.1.24). The inclusion $\widetilde{\mathfrak{t}}_S \subset \widetilde{\mathfrak{t}}_{\overline{S}}$ is clear. For the opposite inclusion, let $\varphi \in \operatorname{dom} \mathfrak{t}_{\overline{S}} = \operatorname{dom} \overline{S}$ and $\varphi' \in \mathfrak{H}$ such that $\{\varphi, \varphi'\} \in \overline{S}$, in which case

$$\mathfrak{t}_{\overline{S}}[\varphi, \varphi] = (\varphi', \varphi).$$

Then there exists a sequence $(\{\varphi_n, \varphi_n'\})$ in S with $\varphi_n \to \varphi$ and $\varphi_n' \to \varphi'$, and hence

$$\mathfrak{t}_S[\varphi_n - \varphi_m] = (\varphi_n' - \varphi_m', \varphi_n - \varphi_m) \to 0.$$

Therefore, $\varphi_n \to_{\mathfrak{t}_S} \varphi$, so that $\varphi \in \operatorname{dom} \widetilde{\mathfrak{t}}_S$. Moreover,

$$\mathfrak{t}_{\overline{S}}[\varphi, \varphi] = (\varphi', \varphi) = \lim_{n\to\infty} (\varphi_n', \varphi_n) = \lim_{n\to\infty} \mathfrak{t}_S[\varphi_n, \varphi_n] = \widetilde{\mathfrak{t}}_S[\varphi, \varphi],$$

where in the last equality the definition of the closure in Theorem 5.1.12 was used. This implies $\mathfrak{t}_{\overline{S}} \subset \widetilde{\mathfrak{t}}_S$ and hence $\widetilde{\mathfrak{t}}_{\overline{S}} \subset \widetilde{\mathfrak{t}}_S$. Therefore, $\widetilde{\mathfrak{t}}_S = \widetilde{\mathfrak{t}}_{\overline{S}}$. □

In the next theorem it is shown that every closed semibounded form can be represented by a semibounded self-adjoint relation.

Theorem 5.1.18 (First representation theorem). *Assume that $\mathfrak{t}$ is a closed semibounded form in $\mathfrak{H}$. Then there exists a semibounded self-adjoint relation H in $\mathfrak{H}$ such that the following statements hold:*

(i) $\operatorname{dom} H \subset \operatorname{dom} \mathfrak{t}$ *and*
$$\mathfrak{t}[\varphi, \psi] = (\varphi', \psi) \tag{5.1.25}$$
for every $\{\varphi, \varphi'\} \in H$ *and* $\psi \in \operatorname{dom} \mathfrak{t}$;

(ii) $\operatorname{dom} H$ *is a core of* $\mathfrak{t}$;

(iii) *if* $\varphi \in \operatorname{dom} \mathfrak{t}$, $\varphi' \in \mathfrak{H}$, *and*
$$\mathfrak{t}[\varphi, \psi] = (\varphi', \psi) \tag{5.1.26}$$
for every ψ *in a core of* $\mathfrak{t}$, *then* $\{\varphi, \varphi'\} \in H$;

(iv) $\operatorname{mul} H = (\operatorname{dom} \mathfrak{t})^\perp$ *and*
$$\mathfrak{t}[\varphi, \psi] = (H_{\mathrm{op}}\, \varphi, \psi) \tag{5.1.27}$$
for every $\varphi \in \operatorname{dom} H$ *and* $\psi \in \operatorname{dom} \mathfrak{t}$.

The semibounded self-adjoint relation H is uniquely determined by (i). *The closed form $\mathfrak{t}$ and the corresponding semibounded self-adjoint relation H have the same lower bound: $m(\mathfrak{t}) = m(H)$. Moreover, for each $x \in \mathbb{R}$ the closed semibounded form $\mathfrak{t} - x$ corresponds to the semibounded self-adjoint relation $H - x$.*

Proof. (i) Let $m(\mathfrak{t}) = \gamma$ and choose $a < \gamma$. Then the assumption that $\mathfrak{t}$ is closed is equivalent to the inner product space $\mathfrak{H}_{\mathfrak{t}-a}$ being complete, where $\mathfrak{H}_{\mathfrak{t}-a} = \operatorname{dom} \mathfrak{t}$ is equipped with the inner product of $(\cdot,\cdot)_{\mathfrak{t}-a}$ as in (5.1.7)–(5.1.8); cf. Lemma 5.1.9. For any fixed $\omega \in \mathfrak{H}$ consider the linear functional

$$\psi \mapsto (\psi, \omega)$$

defined for all $\psi \in \mathfrak{H}_{\mathfrak{t}-a} = \operatorname{dom} \mathfrak{t} \subset \mathfrak{H}$. It follows from (5.1.9) that

$$|(\psi, \omega)| \le \|\psi\| \|\omega\| \le \left(\frac{1}{\sqrt{\gamma - a}} \|\omega\| \right) \|\psi\|_{\mathfrak{t}-a}, \quad \psi \in \mathfrak{H}_{\mathfrak{t}-a}.$$

Hence, the mapping $\psi \mapsto (\psi, \omega)$ from $\mathfrak{H}_{\mathfrak{t}-a}$ to $\mathbb{C}$ is bounded with bound at most $\|\omega\|/\sqrt{\gamma - a}$. Therefore, by the Riesz representation theorem, there exists an element $\widehat{\omega}$ in $\mathfrak{H}_{\mathfrak{t}-a}$ such that for all $\psi \in \mathfrak{H}_{\mathfrak{t}-a}$:

$$(\psi, \omega) = (\psi, \widehat{\omega})_{\mathfrak{t}-a}, \quad \|\widehat{\omega}\|_{\mathfrak{t}-a} \leq \frac{1}{\sqrt{\gamma - a}}\|\omega\|.$$

Taking conjugates for convenience, it follows from the definition (5.1.7) of $(\cdot, \cdot)_{\mathfrak{t}-a}$ that

$$(\omega, \psi) = (\widehat{\omega}, \psi)_{\mathfrak{t}-a} = \mathfrak{t}[\widehat{\omega}, \psi] - a(\widehat{\omega}, \psi), \tag{5.1.28}$$

or, in other words,

$$\mathfrak{t}[\widehat{\omega}, \psi] = (\omega + a\widehat{\omega}, \psi), \quad \psi \in \mathfrak{H}_{\mathfrak{t}-a}. \tag{5.1.29}$$

Note that the linear mapping A from $\mathfrak{H}$ to $\mathfrak{H}_{\mathfrak{t}-a}$ defined by $A\omega = \widehat{\omega}$ satisfies

$$\sqrt{\gamma - a}\,\|A\omega\| \leq \|A\omega\|_{\mathfrak{t}-a} \leq \frac{1}{\sqrt{\gamma - a}}\|\omega\|;$$

where in the first inequality (5.1.9) was used. In other words, if A is interpreted as a mapping from $\mathfrak{H}$ to $\mathfrak{H}$, then

$$\|A\omega\| \leq \frac{1}{\gamma - a}\|\omega\|.$$

By means of A define the linear relation H in $\mathfrak{H}$ by

$$H = \bigl\{\{A\omega, \omega + aA\omega\} : \omega \in \mathfrak{H}\bigr\},$$

so that

$$A = (H - a)^{-1}.$$

One sees that $\operatorname{dom} H = \operatorname{ran} A \subset \operatorname{dom} \mathfrak{t}$ and $\operatorname{mul} H = \ker A$. Moreover, every element $\{\varphi, \varphi'\} \in H$ can be written as $\{\varphi, \varphi'\} = \{\widehat{\omega}, \omega + a\widehat{\omega}\}$ for some ω, so that by the identity (5.1.29) one obtains

$$\mathfrak{t}[\varphi, \psi] = (\varphi', \psi), \quad \{\varphi, \varphi'\} \in H, \qquad \psi \in \mathfrak{H}_{\mathfrak{t}-a} = \operatorname{dom} \mathfrak{t}. \tag{5.1.30}$$

It follows from (5.1.30) with $\psi = \varphi$ that H is a semibounded relation with lower bound

$$m(H) \geq m(\mathfrak{t}) = \gamma. \tag{5.1.31}$$

It is clear that H is symmetric. According to the definition of H one sees that $\operatorname{ran}(H - a) = \mathfrak{H}$, which, since $a < \gamma$, implies that H is self-adjoint; cf. Proposition 1.5.6. Thus, (i) has been proved.

(ii) The statement that $\operatorname{dom} H$ is a core of $\mathfrak{t}$ is equivalent to the statement that $\operatorname{dom} H$ is dense in the Hilbert space $\mathfrak{H}_{\mathfrak{t}-a}$. To verify denseness, assume that the element $\psi \in \mathfrak{H}_{\mathfrak{t}-a}$ is orthogonal to $\operatorname{dom} H = \operatorname{ran} A$, i.e.,

$$0 = (A\omega, \psi)_{\mathfrak{t}-a} = (\widehat{\omega}, \psi)_{\mathfrak{t}-a} = (\omega, \psi),$$

for all $\omega \in \mathfrak{H}$; cf. (5.1.28). This leads to $\psi = 0$. Hence, $\operatorname{dom} H = \operatorname{ran} A$ is dense in the Hilbert space $\mathfrak{H}_{\mathfrak{t}-a}$, and the assertion follows from Corollary 5.1.15.

(iii) Let $\varphi \in \operatorname{dom} \mathfrak{t}$ and $\varphi' \in \mathfrak{H}$ satisfy (5.1.26) for every ψ in a core $\mathfrak{D}$ of the form $\mathfrak{t}$. Then (5.1.26) holds for all $\psi \in \operatorname{dom} \mathfrak{t}$. To see this, let $\psi \in \operatorname{dom} \mathfrak{t}$. Then there exists a sequence (ψ_n) in $\mathfrak{D}$ such that $\psi_n \to_{\mathfrak{t}} \psi$, which implies that $\mathfrak{t}[\psi_n - \psi] \to 0$. Since $\psi_n \in \mathfrak{D}$, the assumption yields

$$\mathfrak{t}[\varphi, \psi] = \lim_{n\to\infty} \mathfrak{t}[\varphi, \psi_n] = \lim_{n\to\infty} (\varphi', \psi_n) = (\varphi', \psi), \quad \psi \in \operatorname{dom} \mathfrak{t},$$

so that (5.1.26) holds for all $\psi \in \operatorname{dom} \mathfrak{t}$. Due to the symmetry of $\mathfrak{t}$ this result may also be written as

$$\mathfrak{t}[\psi, \varphi] = (\psi, \varphi'), \qquad \psi \in \operatorname{dom} \mathfrak{t}. \tag{5.1.32}$$

Now let $\{\psi, \psi'\} \in H$. Then $\psi \in \operatorname{dom} H \subset \operatorname{dom} \mathfrak{t}$ and, by (i),

$$\mathfrak{t}[\psi, \varphi] = (\psi', \varphi), \tag{5.1.33}$$

because $\varphi \in \operatorname{dom} \mathfrak{t}$. Comparing (5.1.32) and (5.1.33) gives

$$(\psi, \varphi') = (\psi', \varphi) \quad \text{for all} \quad \{\psi, \psi'\} \in H,$$

which leads to $\{\varphi, \varphi'\} \in H^* = H$. This proves (iii).

(iv) It follows from (i) that if $\{0, \varphi'\} \in H$, then $(\varphi', \psi) = 0$ for all $\psi \in \operatorname{dom} \mathfrak{t}$, and hence $\operatorname{mul} H \subset (\operatorname{dom} \mathfrak{t})^\perp$. Conversely, as $\operatorname{dom} H \subset \operatorname{dom} \mathfrak{t}$ by (i) and H is self-adjoint, $(\operatorname{dom} \mathfrak{t})^\perp \subset (\operatorname{dom} H)^\perp = \operatorname{mul} H$. This shows that $\operatorname{mul} H = (\operatorname{dom} \mathfrak{t})^\perp$.

To see (5.1.27), let $\{\varphi, \varphi'\} \in H$. Then $\varphi' = H_{\rm op}\, \varphi + \chi$, where $\chi \in \operatorname{mul} H$. Hence, from (5.1.25) one obtains

$$\mathfrak{t}[\varphi, \psi] = (\varphi', \psi) = (H_{\rm op}\, \varphi + \chi, \psi) = (H_{\rm op}\, \varphi, \psi),$$

which gives (5.1.27). This completes the proof of (iv).

To show uniqueness, assume that H' is a semibounded self-adjoint relation in $\mathfrak{H}$ such that $\operatorname{dom} H' \subset \operatorname{dom} \mathfrak{t}$ and

$$\mathfrak{t}[\varphi, \psi] = (\varphi', \psi)$$

for every $\{\varphi, \varphi'\} \in H'$ and $\psi \in \operatorname{dom} \mathfrak{t}$. Then, in particular, one concludes that $\varphi \in \operatorname{dom} H' \subset \operatorname{dom} \mathfrak{t}$ and $\varphi' \in \mathfrak{H}$, so that by (iii) it follows that $\{\varphi, \varphi'\} \in H$. Hence, $H' \subset H$ and one obtains equality as H' and H are both self-adjoint.

Recall that it has been shown in the proof of (i) that $m(H) \geq m(\mathfrak{t})$; cf. (5.1.31). The equality follows from the fact that $\operatorname{dom} H$ is a core of $\mathfrak{t}$; see (ii). In fact, if $\varphi \in \operatorname{dom} \mathfrak{t}$, then there exists a sequence (φ_n) in $\operatorname{dom} H$ such that $\varphi_n \to_{\mathfrak{t}} \varphi$. Therefore, if $\varphi \neq 0$, then

$$\frac{\mathfrak{t}[\varphi]}{\|\varphi\|^2} = \lim_{n\to\infty} \frac{\mathfrak{t}[\varphi_n]}{\|\varphi_n\|^2} = \lim_{n\to\infty} \frac{(H_{\rm op}\, \varphi_n, \varphi_n)}{\|\varphi_n\|^2} \geq m(H).$$

Since this inequality holds for every nontrivial $\varphi \in \operatorname{dom} \mathfrak{t}$ one concludes that

$$m(\mathfrak{t}) = \inf \left\{ \frac{\mathfrak{t}[\varphi]}{\|\varphi\|^2} : \varphi \in \operatorname{dom} \mathfrak{t},\ \varphi \neq 0 \right\} \geq m(H),$$

and so $m(\mathfrak{t}) = m(H)$.

Finally, note that for $x \in \mathbb{R}$ the form $\mathfrak{t} - x$ is semibounded and closed, and the relation $H - x$ is semibounded and self-adjoint. For $\{\varphi, \varphi'\} \in H$,

$$(\mathfrak{t} - x)[\varphi, \psi] = (\varphi', \psi) - x(\varphi, \psi) = (\varphi' - x\varphi, \psi) \tag{5.1.34}$$

for all $\psi \in \operatorname{dom} \mathfrak{t} = \operatorname{dom} \mathfrak{t} - x$. Observe from (iii) that $\{\varphi, \varphi' - x\varphi\} \in H - x$ belongs to the semibounded self-adjoint relation corresponding to $\mathfrak{t} - x$. As $H - x$ is self-adjoint and contained in the semibounded self-adjoint relation corresponding to $\mathfrak{t} - x$ both coincide, i.e., $H - x$ corresponds to the closed semibounded form $\mathfrak{t} - x$. □

The representation result in Theorem 5.1.18 gives assertions concerning the semibounded self-adjoint relation associated with a given semibounded form. In fact, every semibounded self-adjoint relation appears in such a context, as is shown in the following proposition; cf. Lemma 5.1.17.

Proposition 5.1.19. *Let A be a semibounded self-adjoint relation in $\mathfrak{H}$. Then the semibounded, closable form defined by*

$$\mathfrak{t}_A[\varphi, \psi] = (\varphi', \psi), \quad \{\varphi, \varphi'\}, \{\psi, \psi'\} \in A,$$

has a closure whose corresponding semibounded self-adjoint relation is given by A.

Proof. Since A is semibounded and self-adjoint, Lemma 5.1.17 shows that the form $\mathfrak{t}_A$ is well defined, semibounded, and closable. Moreover, $\operatorname{dom} A$ is a core of its closure $\widetilde{\mathfrak{t}}$. Let H be the semibounded self-adjoint relation corresponding to $\widetilde{\mathfrak{t}}$. Since $\widetilde{\mathfrak{t}}$ is an extension of $\mathfrak{t}$, one has

$$\widetilde{\mathfrak{t}}[\varphi, \psi] = \mathfrak{t}[\varphi, \psi] = (\varphi', \psi), \quad \{\varphi, \varphi'\}, \{\psi, \psi'\} \in A.$$

Therefore, Theorem 5.1.18 (iii) implies that $\{\varphi, \varphi'\} \in H$, since $\operatorname{dom} A$ is a core of $\widetilde{\mathfrak{t}}$. Consequently, $A \subset H$ and since A and H are both self-adjoint, one concludes $A = H$. □

The following observation, based on Theorem 5.1.18 and Proposition 5.1.19, is included for completeness.

Corollary 5.1.20. *There is a one-to-one correspondence between all closed semibounded forms and all closed semibounded self-adjoint relations via the identity (5.1.25) or, equivalently, via the identity (5.1.27) in the first representation theorem.*

The correspondence between closed semibounded forms and semibounded self-adjoint relations in Theorem 5.1.18 can be illuminated further in the context of nonnegative forms and nonnegative self-adjoint relations. As a preparation, observe that a typical way to define forms is via linear operators.

Lemma 5.1.21. *Let T be a linear operator from a Hilbert space $\mathfrak{H}$ to a Hilbert space $\mathfrak{K}$ and define a nonnegative form $\mathfrak{t}$ in $\mathfrak{H}$ by*

$$\mathfrak{t}[\varphi,\psi] = (T\varphi, T\psi), \quad \varphi,\psi \in \operatorname{dom}\mathfrak{t} = \operatorname{dom} T.$$

Then

$$\mathfrak{t} \text{ is a closable form} \quad \Leftrightarrow \quad T \text{ is a closable operator},$$

and in this case the closure of $\mathfrak{t}$ is given by

$$\widetilde{\mathfrak{t}}[\varphi,\psi] = (\overline{T}\varphi, \overline{T}\psi), \quad \varphi,\psi \in \operatorname{dom}\widetilde{\mathfrak{t}} = \operatorname{dom}\overline{T}. \tag{5.1.35}$$

Proof. ($\Rightarrow$) Assume that $\mathfrak{t}$ is closable. Let (φ_n) be a sequence in $\operatorname{dom} T$ such that $\varphi_n \to 0$ in $\mathfrak{H}$ and $T\varphi_n \to \psi$ in $\mathfrak{K}$. Then

$$\mathfrak{t}[\varphi_n - \varphi_m] = \|T(\varphi_n - \varphi_m)\|^2 \to 0,$$

which implies that $\varphi_n \to_{\mathfrak{t}} 0$. Since $\mathfrak{t}$ is closable, one obtains

$$\|T\varphi_n\|^2 = \mathfrak{t}[\varphi_n] \to 0,$$

so that $T\varphi_n \to 0$. It follows that T is closable.

($\Leftarrow$) Assume that T is closable. Let (φ_n) in $\operatorname{dom}\mathfrak{t}$ with $\varphi_n \to_{\mathfrak{t}} 0$. Then $\varphi_n \to 0$ in $\mathfrak{H}$ and $(T\varphi_n)$ is a Cauchy sequence in $\mathfrak{K}$. Hence, $T\varphi_n \to \psi$ for some $\psi \in \mathfrak{K}$ and since T is closable one sees that $\psi = 0$. Therefore, $\mathfrak{t}[\varphi_n] = \|T\varphi_n\|^2 \to 0$. It follows that $\mathfrak{t}$ is closable.

Finally, assume that $\mathfrak{t}$ or, equivalently, T is closable. Then one has

$$\operatorname{dom}\widetilde{\mathfrak{t}} = \operatorname{dom}\overline{T}. \tag{5.1.36}$$

Indeed, for the inclusion ($\subset$) in (5.1.36) consider $\varphi \in \operatorname{dom}\widetilde{\mathfrak{t}}$. Then there exists a sequence (φ_n) in $\operatorname{dom}\mathfrak{t}$ with $\varphi_n \to_{\mathfrak{t}} \varphi$; cf. Theorem 5.1.12. Hence, $\varphi_n \to \varphi$ in $\mathfrak{H}$ and $(T\varphi_n)$ is a Cauchy sequence in $\mathfrak{K}$. Thus, there exists $\varphi' \in \mathfrak{K}$ such that $T\varphi_n \to \varphi'$. Since T is closable it follows that $\varphi \in \operatorname{dom}\overline{T}$ and $\varphi' = \overline{T}\varphi$. Moreover, by Theorem 5.1.12 and (5.1.16) it follows that

$$\widetilde{\mathfrak{t}}[\varphi,\varphi] = \lim_{n\to\infty} \mathfrak{t}[\varphi_n,\varphi_n] = \lim_{n\to\infty} (T\varphi_n, T\varphi_n) = (\overline{T}\varphi, \overline{T}\varphi),$$

and polarization leads to the identity in (5.1.35). For the inclusion ($\supset$) in (5.1.36) let $\varphi \in \operatorname{dom} \overline{T}$. Then $\overline{T}\varphi = \varphi'$ for some $\varphi' \in \mathfrak{K}$, and there exists a sequence (φ_n) in $\operatorname{dom} T$ for which $\varphi_n \to \varphi$ while $T\varphi_n \to \varphi'$. In particular, it follows that $\varphi_n \to_{\mathfrak{t}} \varphi$. Therefore, $\varphi \in \operatorname{dom} \widetilde{\mathfrak{t}}$; this proves (5.1.36). □

The following result specializes the first representation theorem to closed nonnegative forms as in Lemma 5.1.21. For a class of closed nonnegative forms it identifies the associated self-adjoint relations. Recall that for a closed operator R a linear subspace $\mathfrak{D} \subset \operatorname{dom} R$ is a *core* if the closure of the restriction $R \upharpoonright_{\mathfrak{D}}$ of R to $\mathfrak{D}$ coincides with R; cf. Lemma 1.5.10.

Proposition 5.1.22. *Let T be a closed relation from a Hilbert space $\mathfrak{H}$ to a Hilbert space $\mathfrak{K}$ and let $T_{\rm op} = PT$ be the closed orthogonal operator part of T, where P is the orthogonal projection in $\mathfrak{K}$ onto* $(\operatorname{mul} T)^\perp$; *cf. Theorem 1.3.15. Then the rule*

$$\mathfrak{t}[\varphi,\psi] = (T_{\rm op}\varphi, T_{\rm op}\psi), \quad \varphi,\psi \in \operatorname{dom} \mathfrak{t} = \operatorname{dom} T_{\rm op} = \operatorname{dom} T, \tag{5.1.37}$$

defines a closed nonnegative form $\mathfrak{t}$ in $\mathfrak{H}$. The nonnegative self-adjoint relation corresponding to the form $\mathfrak{t}$ is given by T^*T. Moreover, a subset of $\operatorname{dom} \mathfrak{t} = \operatorname{dom} T$ *is a core of the form $\mathfrak{t}$ if and only if it is a core of the operator $T_{\rm op}$.*

Proof. Since the operator $T_{\rm op}$ is closed, the nonnegative form $\mathfrak{t}$ in (5.1.37) is closed, with $\operatorname{dom} \mathfrak{t} = \operatorname{dom} T_{\rm op} = \operatorname{dom} T$; cf. Lemma 5.1.21. Recall that T^*T is a nonnegative self-adjoint relation in $\mathfrak{H}$; cf. Lemma 1.5.8. Assume that $\varphi \in \operatorname{dom} T^*T$ and $\psi \in \operatorname{dom} T$. Let $\varphi' \in \mathfrak{H}$ be any element such that $\{\varphi, \varphi'\} \in T^*T$. This implies that $\{\varphi, \eta\} \in T$ and $\{\eta, \varphi'\} \in T^*$ for some $\eta \in \mathfrak{K}$. Clearly, $\eta = T_{\rm op}\varphi + \omega$ for some $\omega \in \operatorname{mul} T$. Since $\{T_{\rm op}\varphi + \omega, \varphi'\} \in T^*$ and $\{\psi, T_{\rm op}\psi\} \in T$, one sees that

$$0 = (\varphi', \psi) - (T_{\rm op}\varphi + \omega, T_{\rm op}\psi) = (\varphi', \psi) - (T_{\rm op}\varphi, T_{\rm op}\psi), \quad \psi \in \operatorname{dom} T,$$

i.e.,

$$\mathfrak{t}[\varphi, \psi] = (\varphi', \psi), \quad \{\varphi, \varphi'\} \in T^*T, \quad \psi \in \operatorname{dom} T.$$

Let H be the nonnegative self-adjoint relation associated with $\mathfrak{t}$ via Theorem 5.1.18. According to (iii) of Theorem 5.1.18, the nonnegative self-adjoint relation T^*T satisfies $T^*T \subset H$, which gives $T^*T = H$.

Now let $\mathfrak{D} \subset \operatorname{dom} \mathfrak{t} = \operatorname{dom} T$ be a linear subset. Then $\mathfrak{D}$ is a core of $\mathfrak{t}$ if and only if for every $\varphi \in \operatorname{dom} \mathfrak{t} = \operatorname{dom} T$ there is a sequence (φ_n) in $\mathfrak{D}$ such that

$$\varphi_n \to \varphi \quad \text{and} \quad \mathfrak{t}[\varphi_n - \varphi] \to 0;$$

cf. (5.1.19). In view of the definition of $\mathfrak{t}$, this condition reads as

$$\varphi_n \to \varphi \quad \text{and} \quad T_{\rm op}\varphi_n \to T_{\rm op}\varphi,$$

in other words $\mathfrak{D}$ is a core of $T_{\rm op}$. □

The so-called second representation theorem may be seen as a corollary of Theorem 5.1.18 and Proposition 5.1.22.

Theorem 5.1.23 (Second representation theorem). *Assume that the closed semibounded form* $\mathfrak{t}$ *and the semibounded self-adjoint relation* H *are connected as in Theorem 5.1.18, so that* $m(H) = m(\mathfrak{t}) = \gamma$, *and let* $x \leq \gamma$. *Then*

$$\operatorname{dom} \mathfrak{t} = \operatorname{dom}(H - x)^{\frac{1}{2}}$$

and the form $\mathfrak{t}$ *is represented by*

$$\mathfrak{t}[\varphi, \psi] = \big((H_{\mathrm{op}} - x)^{\frac{1}{2}}\varphi, (H_{\mathrm{op}} - x)^{\frac{1}{2}}\psi\big) + x(\varphi, \psi), \qquad \varphi, \psi \in \operatorname{dom} \mathfrak{t}.$$

Moreover, a subset of $\operatorname{dom} \mathfrak{t} = \operatorname{dom}(H - x)^{\frac{1}{2}}$ *is a core of the form* $\mathfrak{t}$ *if and only if it is a core of the operator* $(H_{\mathrm{op}} - x)^{\frac{1}{2}}$.

Proof. For $x \leq \gamma$ define the form $\mathfrak{s}_x$ by

$$\mathfrak{s}_x[\varphi, \psi] = \big((H_{\mathrm{op}} - x)^{\frac{1}{2}}\varphi, (H_{\mathrm{op}} - x)^{\frac{1}{2}}\psi\big), \qquad \varphi, \psi \in \operatorname{dom} \mathfrak{s}_x,$$

on the domain $\operatorname{dom} \mathfrak{s}_x = \operatorname{dom}(H_{\mathrm{op}} - x)^{1/2} = \operatorname{dom}(H - x)^{1/2}$. By Proposition 5.1.22, the form $\mathfrak{s}_x$ is closed and nonnegative. The corresponding nonnegative self-adjoint relation is given by

$$\big((H - x)^{\frac{1}{2}}\big)^*(H - x)^{\frac{1}{2}} = H - x,$$

and hence $\mathfrak{s}_x[\varphi, \psi] = (\varphi', \psi)$ holds for all $\{\varphi, \varphi'\} \in H - x$ and $\psi \in \operatorname{dom} \mathfrak{s}_x$. It follows as in the proof of Theorem 5.1.18 (see (5.1.34)) that the closed semibounded form

$$(\mathfrak{s}_x + x)[\varphi, \psi] = \big((H_{\mathrm{op}} - x)^{\frac{1}{2}}\varphi, (H_{\mathrm{op}} - x)^{\frac{1}{2}}\psi\big) + x(\varphi, \psi), \qquad \varphi, \psi \in \operatorname{dom} \mathfrak{s}_x,$$

is represented by the semibounded self-adjoint relation H. Furthermore,

$$(\mathfrak{s}_x + x)[\varphi, \psi] = ((H_{\mathrm{op}} - x)\varphi, \psi) + x(\varphi, \psi) = (H_{\mathrm{op}}\varphi, \psi)$$

for all $\varphi, \psi \in \operatorname{dom} H$, and hence the restrictions of the form $\mathfrak{s}_x + x$ and of the form $\mathfrak{t}$ to $\operatorname{dom} H_{\mathrm{op}}$ coincide; cf. Theorem 5.1.18 (iv). According to Proposition 5.1.22 and Lemma 1.5.10, $\operatorname{dom} H_{\mathrm{op}}$ is a core of $\mathfrak{s}_x$ and hence also of $\mathfrak{s}_x + x$. On the other hand, by Theorem 5.1.18 (ii), $\operatorname{dom} H_{\mathrm{op}} = \operatorname{dom} H$ is also a core of $\mathfrak{t}$. Hence, the forms $\mathfrak{s}_x + x$ and $\mathfrak{t}$ coincide on the common core $\operatorname{dom} H_{\mathrm{op}}$. This implies that the forms $\mathfrak{s}_x + x$ and $\mathfrak{t}$ coincide. Therefore,

$$\operatorname{dom} \mathfrak{t} = \operatorname{dom}(\mathfrak{s}_x + x) = \operatorname{dom}(H - x)^{\frac{1}{2}}, \qquad x \leq \gamma,$$

and

$$\mathfrak{t}[\varphi, \psi] = \big((H_{\mathrm{op}} - x)^{\frac{1}{2}}\varphi, (H_{\mathrm{op}} - x)^{\frac{1}{2}}\psi\big) + x(\varphi, \psi), \qquad \varphi, \psi \in \operatorname{dom} \mathfrak{t}.$$

Finally, Proposition 5.1.22 shows that a subset of $\operatorname{dom} \mathfrak{t}$ is a core of $\mathfrak{t}$ if and only if it is a core of the operator $(H_{\mathrm{op}} - x)^{\frac{1}{2}}$. This completes the proof. □

5.2 Ordering and monotonicity

In this section an ordering will be introduced for semibounded closed forms $\mathfrak{t}_1$ and $\mathfrak{t}_2$, and for semibounded self-adjoint relations H_1 and H_2 in a Hilbert space $\mathfrak{H}$. It will be shown that these orderings are compatible if $\mathfrak{t}_1$ and H_1, and $\mathfrak{t}_2$ and H_2 are related via the first representation theorem (Theorem 5.1.18), respectively. An alternative formulation of the ordering of semibounded self-adjoint relations will be given in terms of their resolvent operators. The last part of the section is devoted to a general monotonicity principle in the context of semibounded self-adjoint relations or, equivalently, of closed semibounded forms.

First an ordering will be defined for semibounded forms that are not necessarily closed.

Definition 5.2.1. Let $\mathfrak{t}_1$ and $\mathfrak{t}_2$ be semibounded forms in $\mathfrak{H}$ that are not necessarily closed. Then one writes $\mathfrak{t}_1 \leq \mathfrak{t}_2$, if

$$\operatorname{dom}\mathfrak{t}_2 \subset \operatorname{dom}\mathfrak{t}_1, \quad \mathfrak{t}_1[\varphi] \leq \mathfrak{t}_2[\varphi], \quad \varphi \in \operatorname{dom}\mathfrak{t}_2. \tag{5.2.1}$$

Note that if $\mathfrak{t}_1 \leq \mathfrak{t}_2$, then $\mathfrak{t}_2$-convergence implies $\mathfrak{t}_1$-convergence. Indeed, let $\varphi_n \to_{\mathfrak{t}_2} \varphi$. By Definition 5.1.4, this means that

$$\varphi_n \in \operatorname{dom}\mathfrak{t}_2, \quad \varphi_n \to \varphi, \quad \text{and} \quad \mathfrak{t}_2[\varphi_n - \varphi_m] \to 0.$$

Since $\mathfrak{t}_1 \leq \mathfrak{t}_2$, this implies that

$$\varphi_n \in \operatorname{dom}\mathfrak{t}_1 \quad \text{and} \quad \mathfrak{t}_1[\varphi_n - \varphi_m] \to 0,$$

which shows that $\varphi_n \to_{\mathfrak{t}_1} \varphi$. Definition 5.2.1 generates a number of simple but useful observations.

Lemma 5.2.2. *Let $\mathfrak{t}_1$, $\mathfrak{t}_2$, and $\mathfrak{t}_3$ be semibounded forms in $\mathfrak{H}$ that are not necessarily closed. Then the following statements hold:*

(i) $\mathfrak{t}_2 \subset \mathfrak{t}_1 \Rightarrow \mathfrak{t}_1 \leq \mathfrak{t}_2$;

(ii) $\mathfrak{t}_1 \leq \mathfrak{t}_2 \Rightarrow m(\mathfrak{t}_1) \leq m(\mathfrak{t}_2)$;

(iii) $\mathfrak{t}_1 \leq \mathfrak{t}_2$ *and* $\mathfrak{t}_2 \leq \mathfrak{t}_3 \Rightarrow \mathfrak{t}_1 \leq \mathfrak{t}_3$;

(iv) $\mathfrak{t}_1 \leq \mathfrak{t}_2$ *and* $\mathfrak{t}_2 \leq \mathfrak{t}_1 \Rightarrow \mathfrak{t}_1 = \mathfrak{t}_2$;

(v) $\mathfrak{t}_1 \leq \mathfrak{t}_2 \Rightarrow \widetilde{\mathfrak{t}}_1 \leq \widetilde{\mathfrak{t}}_2$, *when* $\mathfrak{t}_1$ *and* $\mathfrak{t}_2$ *are closable.*

Proof. (i) This follows from the definition of $\mathfrak{t}_2 \subset \mathfrak{t}_1$; cf. (5.1.2).

(ii) It follows from (5.2.1) that

$$\begin{aligned}\inf\left\{\frac{\mathfrak{t}_1[\varphi]}{\|\varphi\|^2} : \varphi \in \operatorname{dom}\mathfrak{t}_1,\, \varphi \neq 0\right\} &\leq \inf\left\{\frac{\mathfrak{t}_1[\varphi]}{\|\varphi\|^2} : \varphi \in \operatorname{dom}\mathfrak{t}_2,\, \varphi \neq 0\right\} \\ &\leq \inf\left\{\frac{\mathfrak{t}_2[\varphi]}{\|\varphi\|^2} : \varphi \in \operatorname{dom}\mathfrak{t}_2,\, \varphi \neq 0\right\}.\end{aligned}$$

Hence, Definition 5.1.2 implies that $m(\mathfrak{t}_1) \leq m(\mathfrak{t}_2)$.

(iii) This is an immediate consequence of Definition 5.2.1.

(iv) If $\mathfrak{t}_1 \leq \mathfrak{t}_2$ and $\mathfrak{t}_2 \leq \mathfrak{t}_1$, then it follows from (5.2.1) that $\operatorname{dom} \mathfrak{t}_1 = \operatorname{dom} \mathfrak{t}_2$ and that $\mathfrak{t}_1[\varphi] = \mathfrak{t}_2[\varphi]$ for all $\varphi \in \operatorname{dom} \mathfrak{t}_1 = \operatorname{dom} \mathfrak{t}_2$. The conclusion now follows by polarization; cf. (5.1.1).

(v) Assume that $\mathfrak{t}_1$ and $\mathfrak{t}_2$ are closable forms. Let $\varphi \in \operatorname{dom} \widetilde{\mathfrak{t}}_2$; then, by Definition 5.1.10, there exists a sequence (φ_n) in $\operatorname{dom} \mathfrak{t}_2$ such that $\varphi_n \to_{\mathfrak{t}_2} \varphi$. Recall that $\mathfrak{t}_2$-convergence implies $\mathfrak{t}_1$-convergence and thus $\varphi \in \operatorname{dom} \widetilde{\mathfrak{t}}_1$. This shows $\operatorname{dom} \widetilde{\mathfrak{t}}_2 \subset \operatorname{dom} \widetilde{\mathfrak{t}}_1$. Therefore, Theorem 5.1.12 implies that for $\varphi \in \operatorname{dom} \widetilde{\mathfrak{t}}_2$ one has

$$\widetilde{\mathfrak{t}}_1[\varphi] = \lim_{n \to \infty} \mathfrak{t}_1[\varphi_n] \leq \lim_{n \to \infty} \mathfrak{t}_2[\varphi_n] = \widetilde{\mathfrak{t}}_2[\varphi],$$

which shows (v). □

Next an ordering will be defined for semibounded self-adjoint relations. It will be shown in Proposition 5.2.6 below that this ordering is in agreement with the notation $\gamma \leq H$ for a semibounded self-adjoint relation H with lower bound γ; cf. Definition 1.4.5. Note that the following definition relies on Lemma 1.5.10.

Definition 5.2.3. Let H_1 and H_2 be semibounded self-adjoint relations in $\mathfrak{H}$, with lower bounds $m(H_1)$ and $m(H_2)$, respectively. Then the relations H_1 and H_2 are said to be *ordered*, and one writes $H_1 \leq H_2$, if

$$\begin{aligned} &\operatorname{dom} (H_2 - x)^{\frac{1}{2}} \subset \operatorname{dom} (H_1 - x)^{\frac{1}{2}}, \\ &\|(H_{1,\mathrm{op}} - x)^{\frac{1}{2}}\varphi\| \leq \|(H_{2,\mathrm{op}} - x)^{\frac{1}{2}}\varphi\|, \quad \varphi \in \operatorname{dom} (H_2 - x)^{\frac{1}{2}}, \end{aligned} \tag{5.2.2}$$

is satisfied for some, and hence for all $x \leq \min \{m(H_1), m(H_2)\}$.

In the next theorem it is shown that the ordering for semibounded forms in Definition 5.2.1 and the ordering for semibounded self-adjoint relations in Definition 5.2.3 are compatible. Here the second representation theorem (Theorem 5.1.23) plays an essential role.

Theorem 5.2.4. *Let $\mathfrak{t}_1$ and $\mathfrak{t}_2$ be closed semibounded forms in $\mathfrak{H}$ and let H_1 and H_2 be the corresponding semibounded self-adjoint relations. Then*

$$\mathfrak{t}_1 \leq \mathfrak{t}_2 \quad \Leftrightarrow \quad H_1 \leq H_2.$$

Proof. Assume first that $\mathfrak{t}_1 \leq \mathfrak{t}_2$. Then, by Definition 5.2.1,

$$\operatorname{dom} \mathfrak{t}_2 \subset \operatorname{dom} \mathfrak{t}_1, \quad \mathfrak{t}_1[\varphi] \leq \mathfrak{t}_2[\varphi], \quad \varphi \in \operatorname{dom} \mathfrak{t}_2,$$

and for all $x \leq \min \{m(\mathfrak{t}_1), m(\mathfrak{t}_2)\}$ it follows from Theorem 5.1.23 that (5.2.2) holds. Hence, $H_1 \leq H_2$ by Definition 5.2.3.

Conversely, assume that $H_1 \leq H_2$. Then, by Definition 5.2.3, (5.2.2) holds for all $x \leq \min \{m(H_1), m(H_2)\}$ and hence Theorem 5.1.23 implies $\mathfrak{t}_1 \leq \mathfrak{t}_2$. □

Lemma 5.2.5. *Let H_1, H_2, and H_3 be semibounded self-adjoint relations in $\mathfrak{H}$. Then the following statements hold:*

(i) $H_1 \leq H_2 \Rightarrow \operatorname{mul} H_1 \subset \operatorname{mul} H_2$;

(ii) $H_1 \leq H_2 \Rightarrow m(H_1) \leq m(H_2)$;

(iii) $H_1 \leq H_2$ *and* $H_2 \leq H_3 \Rightarrow H_1 \leq H_3$;

(iv) $H_1 \leq H_2$ *and* $H_2 \leq H_1 \Rightarrow H_1 = H_2$;

(v) $H_1 \leq H_2 \Leftrightarrow H_1 - x \leq H_2 - x$ *for every* $x \in \mathbb{R}$.

Proof. Let $\mathfrak{t}_i$ be the closed semibounded form corresponding to H_i, $i = 1, 2, 3$. For the proof of (i) it is sufficient to observe that

$$\overline{\operatorname{dom} H_2} = \overline{\operatorname{dom} \mathfrak{t}_2} \subset \overline{\operatorname{dom} \mathfrak{t}_1} = \overline{\operatorname{dom} H_1},$$

where Theorem 5.2.4 and Theorem 5.1.18 (iv) were used. Taking orthogonal complements then gives $\operatorname{mul} H_1 \subset \operatorname{mul} H_2$. For (ii) recall that

$$m(H_1) = m(\mathfrak{t}_1) \leq m(\mathfrak{t}_2) = m(H_2),$$

as follows from Lemma 5.2.2 and Theorem 5.1.18. Statements (iii) and (iv) are translations of similar statements in Lemma 5.2.2. The statement (v) is clear from Theorem 5.2.4. □

Assume that in Definition 5.2.3 the self-adjoint relation H_1 has a closed domain $\operatorname{dom} H_1$. Then the operator part $H_{1,\mathrm{op}}$ of H_1 is a bounded operator which implies that $\operatorname{dom} H_1 = \operatorname{dom} (H_1 - x)^{\frac{1}{2}}$. Thus, in this case $H_1 \leq H_2$ if and only if

$$\begin{aligned} &\operatorname{dom} (H_2 - x)^{\frac{1}{2}} \subset \operatorname{dom} H_1, \\ &((H_{1,\mathrm{op}} - x)\varphi, \varphi) \leq \|(H_{2,\mathrm{op}} - x)^{\frac{1}{2}}\varphi\|^2, \quad \varphi \in \operatorname{dom} (H_2 - x)^{\frac{1}{2}}. \end{aligned} \tag{5.2.3}$$

The following proposition gives an alternative version of this statement.

Proposition 5.2.6. *Let H_1 and H_2 be semibounded self-adjoint relations in $\mathfrak{H}$ and assume that* $\operatorname{dom} H_1$ *is closed. Then the following statements are equivalent:*

(i) $H_1 \leq H_2$;

(ii) $\operatorname{dom} H_2 \subset \operatorname{dom} H_1$, $(H_{1,\mathrm{op}}\, \varphi, \varphi) \leq (H_{2,\mathrm{op}}\, \varphi, \varphi)$, $\varphi \in \operatorname{dom} H_2$.

Moreover, if $H_1 \in \mathbf{B}(\mathfrak{H})$, then these statements are equivalent to

(iii) $(H_1\varphi, \varphi) \leq (H_{2,\mathrm{op}}\, \varphi, \varphi)$, $\varphi \in \operatorname{dom} H_2$;

(iv) $(H_1\varphi, \varphi) \leq (\varphi', \varphi)$, $\{\varphi, \varphi'\} \in H_2$;

and in the particular case that $H_1 = \gamma_1 I_{\mathfrak{H}}$ to

(v) $\gamma_1\|\varphi\|^2 \leq (H_{2,\mathrm{op}}\, \varphi, \varphi)$, $\varphi \in \operatorname{dom} H_2$;

(vi) $\gamma_1\|\varphi\|^2 \leq (\varphi', \varphi)$, $\{\varphi, \varphi'\} \in H_2$.

Proof. (i) ⇒ (ii) Let (i) be satisfied. Then $\operatorname{dom} H_2 \subset \operatorname{dom}(H_2 - x)^{\frac{1}{2}} \subset \operatorname{dom} H_1$ by (5.2.3), and for all $\varphi \in \operatorname{dom} H_2$ the inequality in (5.2.3) takes the form

$$((H_{1,\mathrm{op}} - x)\varphi, \varphi) \leq ((H_{2,\mathrm{op}} - x)\varphi, \varphi),$$

which implies (ii).

(ii) ⇒ (i) Let (ii) be satisfied and let $\varphi \in \operatorname{dom}(H_2 - x)^{\frac{1}{2}}$. Then there exists a sequence (φ_n) in $\operatorname{dom} H_2$ such that

$$\varphi_n \to \varphi \quad \text{and} \quad (H_{2,\mathrm{op}} - x)^{\frac{1}{2}}\varphi_n \to (H_{2,\mathrm{op}} - x)^{\frac{1}{2}}\varphi, \quad n \to \infty,$$

since $\operatorname{dom} H_2$ is a core of $(H_2 - x)^{\frac{1}{2}}$; see Lemma 1.5.10. Due to the assumption one has $\varphi_n \in \operatorname{dom} H_1$ and

$$((H_{1,\mathrm{op}} - x)\varphi_n, \varphi_n) \leq ((H_{2,\mathrm{op}} - x)\varphi_n, \varphi_n) = \|(H_{2,\mathrm{op}} - x)^{\frac{1}{2}}\varphi_n\|^2.$$

Since $\operatorname{dom} H_1$ is closed it follows by taking the limit that

$$((H_{1,\mathrm{op}} - x)\varphi, \varphi) \leq \|(H_{2,\mathrm{op}} - x)^{\frac{1}{2}}\varphi\|^2, \quad \varphi \in \operatorname{dom}(H_2 - x)^{\frac{1}{2}}.$$

Hence, (5.2.3) is satisfied or, equivalently, $H_1 \leq H_2$.

If $H_1 \in \mathbf{B}(\mathfrak{H})$, then $\operatorname{dom} H_1 = \mathfrak{H}$ and hence the rest of the statements is clear. □

In particular, the inequality in (v)–(vi) of Proposition 5.2.6 shows that the ordering $\gamma I_{\mathfrak{H}} \leq H$ is equivalent to H being semibounded with lower bound γ as defined in Definition 1.4.5. Furthermore, if both H_1 and H_2 are self-adjoint operators in $\mathbf{B}(\mathfrak{H})$, then they are semibounded and Proposition 5.2.6 (iii) shows that $H_1 \leq H_2$ in the sense of Definition 5.2.3 agrees with the usual definition $(H_1\varphi, \varphi) \leq (H_2\varphi, \varphi)$ for all $\varphi \in \mathfrak{H}$.

The ordering for semibounded relations H_1 and H_2 can also be expressed in terms of their resolvent operators. The next proposition is an immediate consequence of Proposition 1.5.11 (for the special case $\rho = 1$).

Proposition 5.2.7. *Let H_1 and H_2 be semibounded self-adjoint relations in $\mathfrak{H}$. Then the following statements are equivalent:*

(i) $H_1 \leq H_2$;

(ii) *for some, and hence for all* $x < \min\{m(H_1), m(H_2)\}$

$$(H_2 - x)^{-1} \leq (H_1 - x)^{-1}.$$

The next corollary slightly extends Proposition 5.2.7 and gives a further interpretation of the inequality $H_1 \leq H_2$ when $x \leq \min\{m(H_1), m(H_2)\}$. The equivalence in (5.2.4) below is an example of the antitonicity property.

Corollary 5.2.8. *Let H_1 and H_2 be semibounded self-adjoint relations in $\mathfrak{H}$. Then*

$$H_1 \leq H_2$$

if and only if for $\gamma \leq \min\{m(H_1), m(H_2)\}$ one has

$$(H_2-\gamma)^{-1} \leq (H_1-\gamma)^{-1}.$$

In particular, if H_1 and H_2 are nonnegative self-adjoint relations, then

$$H_1 \leq H_2 \quad \Leftrightarrow \quad H_2^{-1} \leq H_1^{-1}. \tag{5.2.4}$$

Proof. Let H be a semibounded self-adjoint relation with $\gamma \leq m(H)$. Then $H-\gamma$ is nonnegative and hence also $(H-\gamma)^{-1}$ is a nonnegative self-adjoint relation. Now write for $x<\gamma$

$$H-x = H-\gamma-(x-\gamma),$$

and apply Corollary 1.1.12 (with H replaced by $(H-\gamma)^{-1}$ and λ replaced by $(x-\gamma)^{-1}$), obtaining

$$(H-x)^{-1} = -\frac{1}{x-\gamma} - \frac{1}{(x-\gamma)^2}\left((H-\gamma)^{-1} - \frac{1}{x-\gamma}\right)^{-1}.$$

Hence, for the pair of semibounded self-adjoint relations H_1 and H_2 and with $\gamma \leq \min\{m(H_1), m(H_2)\}$ one obtains for each $x<\gamma$:

$$\begin{aligned}
&(H_1-x)^{-1} - (H_2-x)^{-1} \\
&= \frac{1}{(x-\gamma)^2}\left[\left((H_2-\gamma)^{-1} - \frac{1}{x-\gamma}\right)^{-1} - \left((H_1-\gamma)^{-1} - \frac{1}{x-\gamma}\right)^{-1}\right].
\end{aligned}$$

Since $x-\gamma<0$, a repeated application of Proposition 5.2.7 shows the equivalence. In fact, $H_1 \leq H_2$ if and only if $(H_2-x)^{-1} \leq (H_1-x)^{-1}$ by Proposition 5.2.7, which by the above formula is equivalent to

$$\left((H_1-\gamma)^{-1} - \frac{1}{x-\gamma}\right)^{-1} \leq \left((H_2-\gamma)^{-1} - \frac{1}{x-\gamma}\right)^{-1}. \tag{5.2.5}$$

Another application of Proposition 5.2.7 shows that the inequality (5.2.5) is equivalent to the inequality $(H_2-\gamma)^{-1} \leq (H_1-\gamma)^{-1}$. □

As a corollary to Proposition 5.2.7 it will be shown that in the case $H_1 \leq H_2$ the difference $(H_1-x)^{-1} - (H_2-x)^{-1}$, $x < \min\{m(H_1), m(H_2)\}$, can be used to describe the gap between the corresponding form domains

$$\operatorname{dom}(H_2-x)^{\frac{1}{2}} \subset \operatorname{dom}(H_1-x)^{\frac{1}{2}}.$$

Corollary 5.2.9. *Let H_1 and H_2 be semibounded self-adjoint relations in $\mathfrak{H}$ and assume that*

$$H_1 \leq H_2.$$

Then for all $x < \min\{m(H_1), m(H_2)\}$ the operator $(H_1-x)^{-1}-(H_2-x)^{-1} \in \mathbf{B}(\mathfrak{H})$ is nonnegative and

$$\operatorname{dom}(H_1 - x)^{\frac{1}{2}} = \operatorname{ran}\left((H_1 - x)^{-1} - (H_2 - x)^{-1}\right)^{\frac{1}{2}} + \operatorname{dom}(H_2 - x)^{\frac{1}{2}}.$$

Proof. Since $H_1 \leq H_2$, the operator $R(x) \in \mathbf{B}(\mathfrak{H})$, defined by

$$R(x) = (H_1 - x)^{-1} - (H_2 - x)^{-1},$$

is nonnegative for $x < \min\{m(H_1), m(H_2)\}$; cf. Proposition 5.2.7. Hence, one can write

$$\begin{aligned} (H_1 - x)^{-1} &= R(x) + (H_2 - x)^{-1} \\ &= \begin{pmatrix} R(x)^{\frac{1}{2}} & (H_2 - x)^{-\frac{1}{2}} \end{pmatrix} \begin{pmatrix} R(x)^{\frac{1}{2}} \\ (H_2 - x)^{-\frac{1}{2}} \end{pmatrix}. \end{aligned} \tag{5.2.6}$$

Now recall that if $T = (A \;\; B)$ is a row operator with $A, B \in \mathbf{B}(\mathfrak{H})$, then it follows from $\operatorname{ran}(TT^*)^{\frac{1}{2}} = \operatorname{ran}|T^*| = \operatorname{ran} T$, cf. Corollary D.6, that

$$\operatorname{ran}(AA^* + BB^*)^{\frac{1}{2}} = \operatorname{ran}(A \;\; B) = \operatorname{ran} A + \operatorname{ran} B. \tag{5.2.7}$$

Hence, taking square roots in the identity (5.2.6) and applying (5.2.7) shows that

$$\operatorname{ran}(H_1 - x)^{-\frac{1}{2}} = \operatorname{ran} R(x)^{\frac{1}{2}} + \operatorname{ran}(H_2 - x)^{-\frac{1}{2}},$$

which yields the desired decomposition

$$\operatorname{dom}(H_1 - x)^{\frac{1}{2}} = \operatorname{ran} R(x)^{\frac{1}{2}} + \operatorname{dom}(H_2 - x)^{\frac{1}{2}}$$

for $x < \min\{m(H_1), m(H_2)\}$. □

Now the ordering for semibounded self-adjoint relations and for semibounded closed forms will be used to reinterpret and extend the monotonicity result in Proposition 1.9.9

For the proof of the following theorem it is useful to have available an auxiliary result concerning the interchange of limits. Let (f_n) be a nondecreasing sequence of real nondecreasing functions defined on an open interval (a, b). Thus, for all $x \in (a, b)$ one has

$$f_m(x) \leq f_n(x), \quad m \leq n, \tag{5.2.8}$$

and for all $n \in \mathbb{N}$

$$f_n(x) \leq f_n(y), \quad a < x \leq y < b. \tag{5.2.9}$$

In view of (5.2.8) the pointwise limit

$$f_\infty(x) = \lim_{n\to\infty} f_n(x), \quad x \in (a, b), \tag{5.2.10}$$

gives a function $f_\infty : (a,b) \to \mathbb{R} \cup \{\infty\}$ that is nondecreasing, thanks to (5.2.9). This is clear when all $f_\infty(x)$ are finite, in which case $\lim_{x\to b} f_\infty(x)$ is proper or improper. However, if $f_\infty(x_0) = \infty$ for some $x_0 \in (a,b)$, then (5.2.9) shows that $f_\infty(x) = \infty$ for all $x_0 \le x < b$. In this case the function f_∞ is also called nondecreasing (in the sense of $\mathbb{R} \cup \{\infty\}$) and one defines $\lim_{x\to b} f_\infty(x) = \infty$. In view of (5.2.9) the limit

$$f_n(b) = \lim_{x\to b} f_n(x), \quad n \in \mathbb{N}, \tag{5.2.11}$$

gives a sequence with values in $\mathbb{R} \cup \{\infty\}$ that is nondecreasing, thanks to (5.2.8). This is again clear when all limits $f_n(b)$ are finite in which case $\lim_{n\to\infty} f_n(b)$ is proper or improper. However, if there exists some $m \in \mathbb{N}$ for which $f_m(b) = \infty$, then for all $n \ge m$ one has $f_n(b) = \infty$. In this case one defines $\lim_{n\to\infty} f_n(b) = \infty$.

Lemma 5.2.10. *Let (f_n) be a nondecreasing sequence of nondecreasing functions defined on some open interval (a,b). Let f_∞ be the nondecreasing limit function in (5.2.10) and let $(f_n(b))$ be the nondecreasing sequence of limits in (5.2.11). Then*

$$\lim_{x\to b} f_\infty(x) = \lim_{n\to\infty} f_n(b). \tag{5.2.12}$$

In particular, both limits in (5.2.12) are finite or infinite simultaneously.

Proof. Consider the case that all values of f_∞ are real. Since $f_n(x) \le f_\infty(x)$ for all $x \in (a,b)$, it follows that for any $n \in \mathbb{N}$

$$f_n(b) = \lim_{x\to b} f_n(x) \le \lim_{x\to b} f_\infty(x).$$

This implies

$$\lim_{n\to\infty} f_n(b) \le \lim_{x\to b} f_\infty(x), \tag{5.2.13}$$

where the limits may be infinite. Assume that there is strict inequality in (5.2.13). First consider the case $\lim_{x\to b} f_\infty(x) < \infty$. Then clearly there exists some $\delta > 0$ for which

$$\delta + \lim_{n\to\infty} f_n(b) < \lim_{x\to b} f_\infty(x). \tag{5.2.14}$$

Next consider the case $\lim_{x\to b} f_\infty(x) = \infty$. Then $\lim_{n\to\infty} f_n(b) < \infty$ (otherwise there would be equality in (5.2.13)) and (5.2.14) holds for any $\delta > 0$. In each case, there exists some $x \in (a,b)$ such that

$$\delta + \lim_{n\to\infty} f_n(b) < f_\infty(x).$$

From this one concludes

$$\delta + f_\infty(x) = \delta + \lim_{n\to\infty} f_n(x) \le \delta + \lim_{n\to\infty} f_n(b) < f_\infty(x);$$

a contradiction. Hence, there is equality in (5.2.13). It remains to consider the situation where $f_\infty(x_0) = \infty$ for some $a < x_0 < b$. In this case $f_\infty(x) = \infty$ for all $x_0 < x < b$ and $\lim_{x\to b} f_\infty(x) = \infty$. Assume that

$$L = \lim_{n\to\infty} f_n(b) < \infty.$$

For any $x_0 \le x < b$ one has

$$f_n(x) \le f_n(b) \le L,$$

which implies that $\lim_{n\to\infty} f_n(x) \le L$; a contradiction. Again, there is equality in (5.2.13). □

Theorem 5.2.11 (Monotonicity principle). *Let (H_n) be a nondecreasing sequence of semibounded self-adjoint relations in $\mathfrak{H}$ and let $\gamma \leq m(H_1)$. Then there exists a semibounded self-adjoint relation H_∞ with $\gamma \leq m(H_\infty)$ and $H_n \leq H_\infty$ such that $H_n \to H_\infty$ in the strong resolvent sense, i.e.,*

$$(H_n - \lambda)^{-1}\varphi \to (H_\infty - \lambda)^{-1}\varphi, \quad \varphi \in \mathfrak{H}, \quad \lambda \in \mathbb{C} \setminus [\gamma, \infty). \tag{5.2.15}$$

Furthermore, H_∞ satisfies

$$\begin{aligned}\operatorname{dom}(H_\infty - \gamma)^{\frac{1}{2}} \\ = \left\{ \varphi \in \bigcap_{n=1}^{\infty} \operatorname{dom}(H_n - \gamma)^{\frac{1}{2}} : \lim_{n\to\infty} \|(H_{n,\mathrm{op}} - \gamma)^{\frac{1}{2}}\varphi\| < \infty \right\}\end{aligned} \tag{5.2.16}$$

and for all $\varphi \in \operatorname{dom}(H_\infty - \gamma)^{\frac{1}{2}}$ it holds that

$$\|(H_{\infty,\mathrm{op}} - \gamma)^{\frac{1}{2}}\varphi\| = \lim_{n\to\infty} \|(H_{n,\mathrm{op}} - \gamma)^{\frac{1}{2}}\varphi\|. \tag{5.2.17}$$

Proof. The assumption $H_n \leq H_m$ for $n \leq m$ and Proposition 5.2.7 lead to

$$0 \leq (H_m - x)^{-1} \leq (H_n - x)^{-1}, \quad x < \gamma,$$

where $\gamma \leq m(H_1)$. Hence, by Proposition 1.9.14, there exists a semibounded self-adjoint relation H_∞ with $\gamma \leq m(H_\infty)$ such that

$$0 \leq (H_\infty - x)^{-1} \leq (H_n - x)^{-1}, \quad x < \gamma, \tag{5.2.18}$$

and H_n converges to H_∞ in the strong resolvent sense on $\mathbb{C} \setminus [\gamma, \infty)$, that is, (5.2.15) holds.

It remains to prove (5.2.16) and (5.2.17). It follows from Corollary 1.1.12 with H replaced by $H_n - \gamma$ and $H_\infty - \gamma$, respectively, that for $x < 0$ one has

$$\begin{aligned}&\left(\left((H_n - \gamma)^{-1} - x\right)^{-1}\varphi, \varphi\right) - \left(\left((H_\infty - \gamma)^{-1} - x\right)^{-1}\varphi, \varphi\right) \\ &\quad = \frac{1}{x^2}\left[\left(\left((H_\infty - \gamma) - \frac{1}{x}\right)^{-1}\varphi, \varphi\right) - \left(\left((H_n - \gamma) - \frac{1}{x}\right)^{-1}\varphi, \varphi\right)\right].\end{aligned}$$

Since $\gamma + 1/x < \gamma$, the right-hand side tends to zero monotonically from below for $n \to \infty$, as follows from (5.2.15) and (5.2.18); but then also the left-hand side tends to zero monotonically from below.

To complete the proof, consider the functions, defined for $\varphi \in \mathfrak{H}$ and $x < 0$ by

$$f_n(x) = \left(\left((H_n - \gamma)^{-1} - x\right)^{-1}\varphi, \varphi\right)$$

and

$$f_\infty(x) = \left(\left((H_\infty - \gamma)^{-1} - x\right)^{-1}\varphi, \varphi\right).$$

The above argument shows that the sequence f_n is nondecreasing with f_∞ as pointwise limit. It follows from Lemma 1.5.12 (with H replaced by $H_n - \gamma$ and $H_\infty - \gamma$, respectively), that both functions f_n and f_∞ are nondecreasing on the interval $(-\infty, 0)$ and that

$$\begin{aligned} f_n(0) &= \lim_{x \uparrow 0} \left(((H_n - \gamma)^{-1} - x)^{-1} \varphi, \varphi \right) \\ &= \begin{cases} \|(H_{n,\mathrm{op}} - \gamma)^{\frac{1}{2}} \varphi\|^2, & \varphi \in \mathrm{dom}\,(H_n - \gamma)^{\frac{1}{2}}, \\ \infty, & \text{otherwise}, \end{cases} \end{aligned} \tag{5.2.19}$$

while

$$\begin{aligned} f_\infty(0) &= \lim_{x \uparrow 0} \left(((H_\infty - \gamma)^{-1} - x)^{-1} \varphi, \varphi \right) \\ &= \begin{cases} \|(H_{\infty,\mathrm{op}} - \gamma)^{\frac{1}{2}} \varphi\|^2, & \varphi \in \mathrm{dom}\,(H_\infty - \gamma)^{\frac{1}{2}}, \\ \infty, & \text{otherwise}. \end{cases} \end{aligned} \tag{5.2.20}$$

Hence, by Lemma 5.2.10,

$$\lim_{n \to \infty} f_n(0) = f_\infty(0), \tag{5.2.21}$$

where the limits in (5.2.21) are finite or infinite simultaneously.

Assume that $\varphi \in \mathrm{dom}\,(H_\infty - \gamma)^{\frac{1}{2}}$. Then, by (5.2.20), $f_\infty(0) < \infty$, which, in view of (5.2.21), implies that all $f_n(0) < \infty$. Hence, $\varphi \in \bigcap_{n=1}^\infty \mathrm{dom}\,(H_n - \gamma)^{\frac{1}{2}}$ by (5.2.19), and (5.2.21) reads

$$\lim_{n \to \infty} \|(H_{n,\mathrm{op}} - \gamma)^{\frac{1}{2}} \varphi\|^2 = \|(H_{\infty,\mathrm{op}} - \gamma)^{\frac{1}{2}} \varphi\|^2. \tag{5.2.22}$$

Thus, φ belongs to the right-hand side of (5.2.16). This shows the inclusion $(\subset)$ in (5.2.16), and (5.2.22) gives (5.2.17).

Conversely, assume that φ belongs to the right-hand side of (5.2.16), that is, $\varphi \in \bigcap_{n=1}^\infty \mathrm{dom}\,(H_n - \gamma)^{\frac{1}{2}}$ and

$$\lim_{n \to \infty} \|(H_{n,\mathrm{op}} - \gamma)^{\frac{1}{2}} \varphi\| < \infty.$$

By (5.2.19) one sees that $f_n(0) < \infty$ and that $\lim_{n\to\infty} f_n(0) < \infty$. It follows from (5.2.21) that $f_\infty(0) < \infty$. Now apply (5.2.20) to conclude that $\varphi \in \mathrm{dom}\,(H_\infty - \gamma)^{\frac{1}{2}}$. This shows the inclusion $(\supset)$ in (5.2.16). □

Corollary 5.2.12. *Let (H_n) be a nondecreasing sequence of semibounded self-adjoint relations and let H_∞ be the strong resolvent limit as in Theorem 5.2.11. Then the following statements hold:*

(i) *If K is a semibounded self-adjoint relation such that $H_n \leq K$ for all $n \in \mathbb{N}$, then also $H_\infty \leq K$.*

(ii) *If S is a symmetric relation such that $S \subset H_n$ for all $n \in \mathbb{N}$, then also $S \subset H_\infty$.*

Proof. (i) Assume that $H_n \leq K$. Then for all $x < \gamma \leq m(H_1)$

$$0 \leq ((K-x)^{-1}\varphi, \varphi) \leq ((H_n - x)^{-1}\varphi, \varphi), \quad \varphi \in \mathfrak{H}.$$

By (5.2.15), $(H_n - x)^{-1}\varphi \to (H_\infty - x)^{-1}\varphi$ for $\varphi \in \mathfrak{H}$ and one concludes that

$$0 \leq ((K-x)^{-1}\varphi, \varphi) \leq ((H_\infty - x)^{-1}\varphi, \varphi), \quad \varphi \in \mathfrak{H}.$$

Hence, by Proposition 5.2.7 it follows that $H_\infty \leq K$.

(ii) Assume that $\{\varphi, \varphi'\} \in S$. Then $\{\varphi, \varphi'\} \in H_n$ by assumption and hence for all $n \in \mathbb{N}$ one has

$$(H_n - \lambda)^{-1}(\varphi' - \lambda\varphi) = \varphi, \quad \lambda \in \mathbb{C} \setminus [\gamma, \infty).$$

By (5.2.15), $(H_n - x)^{-1}\psi \to (H_\infty - x)^{-1}\psi$ for $\psi \in \mathfrak{H}$ and one concludes that

$$(H_\infty - \lambda)^{-1}(\varphi' - \lambda\varphi) = \lim_{n\to\infty} (H_n - \lambda)^{-1}(\varphi' - \lambda\varphi) = \varphi,$$

which gives $\{\varphi, \varphi'\} \in H_\infty$. Hence, $S \subset H_\infty$. □

Now consider the special case of a nondecreasing sequence of self-adjoint operators (H_n) in $\mathbf{B}(\mathfrak{H})$. Then it is clear that H_∞ is a semibounded self-adjoint relation which is an operator in $\mathbf{B}(\mathfrak{H})$ if and only if the sequence (H_n) is uniformly bounded; cf. Corollary 1.9.10 and the beginning of Section 1.9. The following corollary shows that the domain of the square root of $H_\infty - \gamma$, $\gamma \leq m(H_1)$, is given by those $\varphi \in \mathfrak{H}$ for which $(H_n\varphi, \varphi)$ has a finite limit as $n \to \infty$.

Corollary 5.2.13. *Let (H_n) be a nondecreasing sequence of self-adjoint operators in $\mathbf{B}(\mathfrak{H})$ with $\gamma \leq m(H_1)$ and define*

$$\mathfrak{E} = \left\{ \varphi \in \mathfrak{H} : \lim_{n\to\infty} (H_n\varphi, \varphi) < \infty \right\}.$$

Let H_∞ be the semibounded self-adjoint limit of the sequence H_n. Then

$$\mathfrak{E} = \operatorname{dom}(H_\infty - \gamma)^{\frac{1}{2}}.$$

In particular, one has

(i) $\mathfrak{E} = \mathfrak{H} \Leftrightarrow H_\infty \in \mathbf{B}(\mathfrak{H})$;
(ii) $\mathfrak{E}$ *is closed* $\Leftrightarrow H_{\infty,\mathrm{op}}$ *is a bounded operator;*
(iii) $\operatorname{clos} \mathfrak{E} = \mathfrak{H} \Leftrightarrow H_\infty$ *is an operator;*
(iv) $\mathfrak{E} = \{0\} \Leftrightarrow H_\infty = \{0\} \times \mathfrak{H}$.

A useful variant of Corollary 5.2.13 is concerned with a nondecreasing function $M : (a, b) \to \mathbf{B}(\mathfrak{H})$, whose values are self-adjoint operators; cf. Corollary 2.3.8. Then there exists a self-adjoint limit at the right endpoint b, which can be retrieved via sequences converging to b. For the existence of the self-adjoint limit at the left endpoint a consider the function $x \mapsto -M(x)$, which is nondecreasing when $x \in (a, b)$ tends to a.

Corollary 5.2.14. *Let $M : (a,b) \to \mathbf{B}(\mathfrak{H})$ be a nondecreasing function, whose values are self-adjoint operators. Then there exist self-adjoint relations $M(a)$ and $M(b)$ in $\mathfrak{H}$ such that $M(x) \to M(b)$ in the strong resolvent sense when $x \to b$ and $M(x) \to M(a)$ in the strong resolvent sense when $x \to a$. Furthermore,*

$$M(x) \leq M(b) \quad \text{and} \quad -M(x) \leq -M(a), \qquad x \in (a,b).$$

Define

$$\mathfrak{E}_b = \left\{ \varphi \in \mathfrak{H} : \lim_{x \uparrow b} (M(x)\varphi, \varphi) < \infty \right\}$$

and

$$\mathfrak{E}_a = \left\{ \varphi \in \mathfrak{H} : \lim_{x \downarrow a} (M(x)\varphi, \varphi) > -\infty \right\}.$$

Then for $c = a$ or $c = b$ one has

(i) $\mathfrak{E}_c = \mathfrak{H} \Leftrightarrow M(c) \in \mathbf{B}(\mathfrak{H})$;

(ii) $\mathfrak{E}_c$ *is closed* $\Leftrightarrow M(c)_{\mathrm{op}}$ *is a bounded operator;*

(iii) $\operatorname{clos} \mathfrak{E}_c = \mathfrak{H} \Leftrightarrow M(c)$ *is an operator;*

(iv) $\mathfrak{E}_c = \{0\} \Leftrightarrow M(c) = \{0\} \times \mathfrak{H}$.

Let $(\mathfrak{t}_n)$ be a nondecreasing sequence of closed semibounded forms in $\mathfrak{H}$ which satisfy $\gamma \leq m(\mathfrak{t}_1)$. By Theorem 5.1.18, there exist unique semibounded self-adjoint relations H_n, bounded from below by γ, which correspond to $\mathfrak{t}_n$. According to Theorem 5.2.4, the sequence (H_n) is nondecreasing. By the monotonicity principle in Theorem 5.2.11 the strong resolvent limit of the sequence (H_n) exists as a semibounded self-adjoint relation H_∞ with lower bound γ such that $H_n \leq H_\infty$. Let $\mathfrak{t}_\infty$ be the form corresponding to H_∞ by Proposition 5.1.19. Then $\mathfrak{t}_\infty$ is bounded below by γ and $\mathfrak{t}_n \leq \mathfrak{t}_\infty$ by Theorem 5.2.4. Therefore, the following theorem concerning a nondecreasing sequence of forms may be seen as a direct consequence of Theorem 5.2.11.

Theorem 5.2.15 (Monotonicity principle). *Let $(\mathfrak{t}_n)$ be a nondecreasing sequence of closed semibounded forms in $\mathfrak{H}$ and let $\gamma \leq m(\mathfrak{t}_1)$. Then there exists a closed semibounded form $\mathfrak{t}_\infty$ with $\gamma \leq m(\mathfrak{t}_\infty)$ such that $\mathfrak{t}_n \leq \mathfrak{t}_\infty$ and*

$$\operatorname{dom} \mathfrak{t}_\infty = \left\{ \varphi \in \bigcap_{n=1}^{\infty} \operatorname{dom} \mathfrak{t}_n : \lim_{n \to \infty} \mathfrak{t}_n[\varphi] < \infty \right\} \tag{5.2.23}$$

and

$$\mathfrak{t}_\infty[\varphi] = \lim_{n \to \infty} \mathfrak{t}_n[\varphi], \quad \varphi \in \operatorname{dom} \mathfrak{t}_\infty. \tag{5.2.24}$$

Moreover, the relations H_n corresponding to the forms $\mathfrak{t}_n$ converge in the strong resolvent sense to the relation H_∞ corresponding to the form $\mathfrak{t}_\infty$.

Proof. It is clear from Theorem 5.1.23 and the formulas (5.2.16) and (5.2.17) in Theorem 5.2.11 that the limit form $\mathfrak{t}_\infty$ satisfies (5.2.23) and (5.2.24). □

5.3 Friedrichs extensions of semibounded relations

A semibounded, not necessarily closed, relation S in a Hilbert space $\mathfrak{H}$ has equal defect numbers, and hence admits self-adjoint extensions in $\mathfrak{H}$. It will be shown that such a relation S has a distinguished semibounded self-adjoint extension $S_{\rm F}$, the so-called Friedrichs extension of S, with $m(S_{\rm F}) = m(S)$. The construction of this extension involves a closed semibounded form associated with S. The characteristic properties of this extension will be investigated in detail.

Let S be a semibounded relation in $\mathfrak{H}$. Recall from Lemma 5.1.17 that the form $\mathfrak{t}_S$ given by

$$\mathfrak{t}_S[f,g] = (f',g), \quad \{f,f'\},\{g,g'\} \in S, \tag{5.3.1}$$

is well defined and that it is semibounded with the lower bound $m(S)$. Moreover, it has been shown that $\mathfrak{t}_S$ is closable and that the closure $\widetilde{\mathfrak{t}}_S$ of $\mathfrak{t}_S$ is a semibounded closed form whose lower bound is equal to $m(S)$. Also, $\operatorname{dom} \mathfrak{t}_S = \operatorname{dom} S$ is a core of $\widetilde{\mathfrak{t}}_S$.

Lemma 5.3.1. *Let S be a semibounded relation in $\mathfrak{H}$ with lower bound $m(S)$. Let $\widetilde{\mathfrak{t}}_S$ be the closure of the form $\mathfrak{t}_S$ in (5.3.1). Then the unique relation $S_{\rm F}$ corresponding to $\widetilde{\mathfrak{t}}_S$ via Theorem 5.1.18 is a semibounded self-adjoint extension of S with the lower bound $m(S_{\rm F}) = m(S)$. In fact, $S_{\rm F}$ is a self-adjoint extension of the semibounded relation $S \,\widehat{+}\, \widehat{\mathfrak{N}}_\infty(S^*)$, so that*

$$S \subset S \,\widehat{+}\, \widehat{\mathfrak{N}}_\infty(S^*) \subset S_{\rm F}, \quad \text{where} \quad \widehat{\mathfrak{N}}_\infty(S^*) = \{0\} \times \operatorname{mul} S^*. \tag{5.3.2}$$

Moreover,

$$S_{\rm F} = (\overline{S})_{\rm F}. \tag{5.3.3}$$

Proof. By Theorem 5.1.18, the closed form $\widetilde{\mathfrak{t}}_S$ induces a unique semibounded self-adjoint relation $S_{\rm F}$ in $\mathfrak{H}$ such that

$$\widetilde{\mathfrak{t}}_S[f,g] = (f',g), \quad \{f,f'\} \in S_{\rm F}, \quad g \in \operatorname{dom} \widetilde{\mathfrak{t}}_S.$$

To show that $S_{\rm F}$ is an extension of S, let $\{f,f'\} \in S$. As $f \in \operatorname{dom} \mathfrak{t}_S$, it follows that for all $g \in \operatorname{dom} \mathfrak{t}_S$

$$\widetilde{\mathfrak{t}}_S[f,g] = \mathfrak{t}_S[f,g] = (f',g).$$

Since $\operatorname{dom} \mathfrak{t}_S = \operatorname{dom} S$ is a core of $\widetilde{\mathfrak{t}}_S$, one obtains $\{f,f'\} \in S_{\rm F}$ from Theorem 5.1.18 (iii). Hence, $S_{\rm F}$ is a self-adjoint extension of S with lower bound $m(S_{\rm F}) = m(S)$.

In order to verify (5.3.2) it now suffices to see that $\{0\} \times \operatorname{mul} S^* \subset S_{\rm F}$. Let $\varphi \in \operatorname{mul} S^*$ and let $g \in \operatorname{dom} S$, then clearly $\widetilde{\mathfrak{t}}_S[0,g] = 0$ and $(\varphi, g) = 0$. Therefore,

$$\widetilde{\mathfrak{t}}_S[0,g] = (\varphi, g) \quad \text{for all} \quad g \in \operatorname{dom} S.$$

Since $\operatorname{dom} S$ is a core of $\widetilde{\mathfrak{t}}_S$, it follows that $\{0,\varphi\} \in S_{\rm F}$.

To see that (5.3.3) holds, it suffices to recall from Lemma 5.1.17 that $\widetilde{\mathfrak{t}}_S = \widetilde{\mathfrak{t}}_{\overline{S}}$ holds for the closures of $\mathfrak{t}_S$ and $\mathfrak{t}_{\overline{S}}$. □

Definition 5.3.2. Let S be a semibounded relation in $\mathfrak{H}$. The semibounded self-adjoint relation S_{F} associated with the closure of the form $\mathfrak{t}_S$ in (5.3.1) is called the *Friedrichs extension* of S. The closure $\widetilde{\mathfrak{t}}_S$ of the form $\mathfrak{t}_S$ will be denoted by $\mathfrak{t}_{S_{\mathrm{F}}}$, so that $\mathfrak{t}_{S_{\mathrm{F}}} = \widetilde{\mathfrak{t}}_S$.

Let S be a semibounded relation in $\mathfrak{H}$ and let $a < m(S)$. Then $S - a$ is a nonnegative relation and it is a consequence of (5.3.1) that $\mathfrak{t}_{S-a} = \mathfrak{t}_S - a$. The translation invariance of the closures, cf. (5.1.17), leads to

$$\mathfrak{t}_{(S-a)_{\mathrm{F}}} = \widetilde{\mathfrak{t}_{S-a}} = \widetilde{\mathfrak{t}_S - a} = \widetilde{\mathfrak{t}}_S - a = \mathfrak{t}_{S_{\mathrm{F}}} - a.$$

The nonnegative self-adjoint relation $(S-a)_{\mathrm{F}}$ corresponding to the form on the left-hand side is equal to the nonnegative self-adjoint relation corresponding to the form on the right-hand side. Thus, one obtains

$$(S-a)_{\mathrm{F}} = S_{\mathrm{F}} - a, \quad a < m(S). \tag{5.3.4}$$

In other words, the Friedrichs extension is translation invariant.

By Lemma 5.3.1, the Friedrichs extension S_{F} is a semibounded self-adjoint extension of S. As a restriction of S^* the Friedrichs extensions can be characterized as follows.

Theorem 5.3.3. *Let S be a semibounded relation in $\mathfrak{H}$. The Friedrichs extension S_{F} of S admits the representation*

$$S_{\mathrm{F}} = \big\{\{f, f'\} \in S^* : f \in \operatorname{dom} \mathfrak{t}_{S_{\mathrm{F}}}\big\} \tag{5.3.5}$$

with $\operatorname{mul} S_{\mathrm{F}} = \operatorname{mul} S^*$. *Furthermore, if H is a self-adjoint extension of S, that is not necessarily semibounded, then*

$$\operatorname{dom} H \subset \operatorname{dom} \mathfrak{t}_{S_{\mathrm{F}}} \quad \Rightarrow \quad H = S_{\mathrm{F}}.$$

Proof. In order to show that S_{F} is contained in the right-hand side of (5.3.5), let $\{f, f'\} \in S_{\mathrm{F}}$. Clearly, $S \subset S_{\mathrm{F}}$ and since S_{F} is self-adjoint this implies $S_{\mathrm{F}} \subset S^*$, so that $\{f, f'\} \in S^*$. Note also that $f \in \operatorname{dom} S_{\mathrm{F}} \subset \operatorname{dom} \mathfrak{t}_{S_{\mathrm{F}}}$. Hence, S_{F} is contained in the right-hand side of (5.3.5).

To show the opposite inclusion, let $\{f, f'\} \in S^*$ be such that $f \in \operatorname{dom} \mathfrak{t}_{S_{\mathrm{F}}}$. Then there exists a sequence (f_n) in $\operatorname{dom} \mathfrak{t}_S = \operatorname{dom} S$ with

$$f_n \to_{\mathfrak{t}_{S_{\mathrm{F}}}} f.$$

Let $\{f_n, f_n'\}$ be corresponding elements in S and let $\{g, g'\} \in S$ be arbitrary. Then $\mathfrak{t}_S \subset \mathfrak{t}_{S_{\mathrm{F}}}$ and $S \subset S^*$ imply that

$$\mathfrak{t}_{S_{\mathrm{F}}}[f_n, g] = \mathfrak{t}_S[f_n, g] = (f_n', g) = (f_n, g').$$

Since $\mathfrak{t}_{S_F}$ is a closed form, it follows that

$$\mathfrak{t}_{S_F}[f,g] = (f,g').$$

As $\{f,f'\} \in S^*$ and $\{g,g'\} \in S$, one obtains $(f,g') = (f',g)$, so that

$$\mathfrak{t}_{S_F}[f,g] = (f',g).$$

This identity holds for an arbitrary element $g \in \operatorname{dom} S$. Since $\operatorname{dom} S = \operatorname{dom} \mathfrak{t}_S$ is a core of the form $\mathfrak{t}_{S_F}$ it follows from Theorem 5.1.18 (iii) that $\{f,f'\} \in S_F$.

Now let H be any self-adjoint extension of S with $\operatorname{dom} H \subset \operatorname{dom} \mathfrak{t}_{S_F}$. Hence, if $\{f,f'\} \in H$, then $\{f,f'\} \in S^*$ and $f \in \operatorname{dom} H \subset \operatorname{dom} \mathfrak{t}_{S_F}$. By (5.3.5), one has $\{f,f'\} \in S_F$. This shows $H \subset S_F$, and since both H and S_F are self-adjoint, it follows that $H = S_F$. □

According to Theorem 5.3.3, the inclusion $\operatorname{dom} H \subset \operatorname{dom} \mathfrak{t}_{S_F}$ for any self-adjoint extension H of S implies $H = S_F$. Note that in general a self-adjoint extension of a semibounded relation is not necessarily semibounded. The situation is different when S has finite defect numbers; cf. Proposition 5.5.8.

The construction of S_F in (5.3.5) results in a description of S_F by means of approximating elements from the graph of S.

Corollary 5.3.4. *Let S be a semibounded relation in $\mathfrak{H}$. Then S_F is the set of all elements $\{f,f'\} \in S^*$ for which there exists a sequence $(\{f_n,f_n'\})$ in S such that*

$$f_n \to f \quad \textit{and} \quad (f_n',f_n) \to (f',f).$$

Proof. By Theorem 5.3.3, S_F is the set of all elements $\{f,f'\} \in S^*$ for which $f \in \operatorname{dom} \mathfrak{t}_{S_F}$. Hence, there exists a sequence $(\{f_n,f_n'\})$ in S such that

$$f_n \to_{\mathfrak{t}_{S_F}} f.$$

In particular, $f_n \to f$ in $\mathfrak{H}$ and, moreover,

$$(f',f) = \mathfrak{t}_{S_F}[f,f] = \lim_{n\to\infty} \mathfrak{t}_{S_F}[f_n,f_n] = \lim_{n\to\infty} \mathfrak{t}_S[f_n,f_n] = \lim_{n\to\infty} (f_n',f_n).$$

Hence, S_F is contained in the relation

$$\bigl\{\{f,f'\} \in S^* : f_n \to f \text{ and } (f_n',f_n) \to (f',f) \text{ for some } \{f_n,f_n'\} \in S\bigr\}.$$

Observe that this relation is symmetric since $(f_n',f_n) \in \mathbb{R}$ implies $(f',f) \in \mathbb{R}$. Thus, the self-adjoint relation S_F is contained in the symmetric relation above, and therefore they coincide. □

Since $S \subset S_F \subset S^*$ and $\operatorname{dom} S_F \subset \overline{\operatorname{dom}} S$ by Corollary 5.3.4 one has that

$$\operatorname{dom} S \subset \operatorname{dom} S_F \subset \bigl(\overline{\operatorname{dom}} S \cap \operatorname{dom} S^*\bigr).$$

The next corollary shows when these inclusions are identities.

Corollary 5.3.5. *Let S be a semibounded relation in $\mathfrak{H}$. Then*

$$S \,\widehat{+}\, \widehat{\mathfrak{N}}_\infty(S^*) = S_{\mathrm{F}}, \quad \widehat{\mathfrak{N}}_\infty(S^*) = \{0\} \times \operatorname{mul} S^*, \tag{5.3.6}$$

if and only if

$$\operatorname{dom} S = \overline{\operatorname{dom} S} \cap \operatorname{dom} S^*.$$

In particular, if $\operatorname{dom} S$ *is closed, then* S_{F} *has the form* (5.3.6).

Proof. Recall from (5.3.2) that $S \,\widehat{+}\, \widehat{\mathfrak{N}}_\infty(S^*) \subset S_{\mathrm{F}}$. Note that there is equality if and only if $S \,\widehat{+}\, \widehat{\mathfrak{N}}_\infty(S^*)$ is self-adjoint. Hence, the assertion follows from Lemma 1.5.7. □

The construction of the Friedrichs extension S_{F} of a semibounded relation S via the form $\mathfrak{t}_S$ in (5.3.1) leads to an important characteristic property. First of all, recall that

$$\mathfrak{t}_{S_{\mathrm{F}}}[f,g] = (f',g), \quad \{f,f'\} \in S_{\mathrm{F}}, \quad g \in \operatorname{dom} \mathfrak{t}_{S_{\mathrm{F}}},$$

with lower bound $m(S_{\mathrm{F}}) = m(S)$. Now assume that H is another semibounded self-adjoint extension of S. Then clearly $m(H) \leq m(S)$, and according to Proposition 5.1.19, the relation H generates a closed semibounded form $\mathfrak{t}_H$ on $\mathfrak{H}$ with

$$\mathfrak{t}_H[f,g] = (f',g), \quad \{f,f'\} \in H, \quad g \in \operatorname{dom} \mathfrak{t}_H.$$

By specializing $\{f,f'\} \in S$ and $g \in \operatorname{dom} S$, it follows that

$$\mathfrak{t}_H[f,g] = (f',g) = \mathfrak{t}_S[f,g]$$

and hence $\mathfrak{t}_S \subset \mathfrak{t}_H$. By construction, $\mathfrak{t}_S \subset \widetilde{\mathfrak{t}} = \mathfrak{t}_{S_{\mathrm{F}}}$ and hence $\mathfrak{t}_{S_{\mathrm{F}}}$ is the smallest closed form extension of $\mathfrak{t}_S$; cf. Theorem 5.1.12. Therefore,

$$\mathfrak{t}_S \subset \mathfrak{t}_{S_{\mathrm{F}}} \subset \mathfrak{t}_H. \tag{5.3.7}$$

This leads to the extremality property of S_{F} stated in the next result.

Proposition 5.3.6. *Let S be a semibounded relation in $\mathfrak{H}$ and let H be a semibounded self-adjoint extension of S. Then $m(\mathfrak{t}_H) = m(H) \leq m(S_{\mathrm{F}}) = m(S)$ and*

$$\mathfrak{t}_H \leq \mathfrak{t}_{S_{\mathrm{F}}} \quad \text{or} \quad H \leq S_{\mathrm{F}};$$

or, equivalently, for some, and hence for all $a < m(H)$,

$$(S_{\mathrm{F}} - a)^{-1} \leq (H - a)^{-1}.$$

Proof. It is a consequence of (5.3.7) and Lemma 5.2.2 (i) that $\mathfrak{t}_H \leq \mathfrak{t}_{S_{\mathrm{F}}}$. The rest of the statements follow from Theorem 5.2.4 and Proposition 5.2.7. □

According to Proposition 5.3.6, the Friedrichs extension S_{F} is the largest semibounded self-adjoint extension of S in the sense of the ordering for forms or relations, and so it has the smallest form domain. Recall that for any semibounded self-adjoint extension H of S one has for $a < m(H) \leq m(S_{\mathrm{F}})$ that

$$\operatorname{dom}(H-a)^{\frac{1}{2}} = \operatorname{ran} R(a)^{\frac{1}{2}} + \operatorname{dom}(S_{\mathrm{F}}-a)^{\frac{1}{2}}, \tag{5.3.8}$$

where the nonnegative operator $R(a)$ is defined by

$$R(a) = (H-a)^{-1} - (S_{\mathrm{F}}-a)^{-1} \in \mathbf{B}(\mathfrak{H}); \tag{5.3.9}$$

cf. Corollary 5.2.9. The identity (5.3.8) will now be put in a geometric context. Recall that for $a < m(H)$ the closed nonnegative form $\mathfrak{t}_H - a$ on $\mathfrak{H}$ defines the following inner product on $\operatorname{dom}\mathfrak{t}_H$

$$(f,g)_{\mathfrak{t}_H-a} = \mathfrak{t}_H[f,g] - a(f,g), \quad f,g \in \operatorname{dom}\mathfrak{t}_H = \operatorname{dom}(H-a)^{\frac{1}{2}}, \tag{5.3.10}$$

which makes the space $\mathfrak{H}_{\mathfrak{t}_H-a} = \operatorname{dom}\mathfrak{t}_H = \operatorname{dom}(H-a)^{\frac{1}{2}}$ complete; cf. Lemma 5.1.8. Similarly, the closed nonnegative form $\mathfrak{t}_{S_{\mathrm{F}}} - a$ on $\mathfrak{H}$ defines the following inner product on $\operatorname{dom}\mathfrak{t}_{S_{\mathrm{F}}}$:

$$(f,g)_{\mathfrak{t}_{S_{\mathrm{F}}}-a} = \mathfrak{t}_{S_{\mathrm{F}}}[f,g] - a(f,g), \quad f,g \in \operatorname{dom}\mathfrak{t}_{S_{\mathrm{F}}} = \operatorname{dom}(S_{\mathrm{F}}-a)^{\frac{1}{2}}, \tag{5.3.11}$$

which makes the space $\mathfrak{H}_{\mathfrak{t}_{S_{\mathrm{F}}}-a} = \operatorname{dom}\mathfrak{t}_{S_{\mathrm{F}}} = \operatorname{dom}(S_{\mathrm{F}}-a)^{\frac{1}{2}}$ complete. Thus, in terms of inner product spaces one obtains

$$\mathfrak{H}_{\mathfrak{t}_{S_{\mathrm{F}}}-a} \subset \mathfrak{H}_{\mathfrak{t}_H-a},$$

and by (5.3.7) the restriction of the inner product in (5.3.10) to $\mathfrak{H}_{\mathfrak{t}_{S_{\mathrm{F}}}-a}$ coincides with the inner product in (5.3.11). Therefore,

$$\mathfrak{H}_{\mathfrak{t}_H-a} = \left(\mathfrak{H}_{\mathfrak{t}_H-a} \ominus_{\mathfrak{t}_H-a} \mathfrak{H}_{\mathfrak{t}_{S_{\mathrm{F}}}-a}\right) \oplus_{\mathfrak{t}_H-a} \mathfrak{H}_{\mathfrak{t}_{S_{\mathrm{F}}}-a}, \tag{5.3.12}$$

see Corollary 5.1.13. In terms of the spaces $\mathfrak{H}_{\mathfrak{t}_H-a}$ and $\mathfrak{H}_{\mathfrak{t}_{S_{\mathrm{F}}}-a}$ the sum decomposition in (5.3.8) may be rewritten as

$$\mathfrak{H}_{\mathfrak{t}_H-a} = \operatorname{ran} R(a)^{\frac{1}{2}} + \mathfrak{H}_{\mathfrak{t}_{S_{\mathrm{F}}}-a}. \tag{5.3.13}$$

The connection between the decompositions in (5.3.12) and (5.3.13) is discussed in the next proposition.

Proposition 5.3.7. *Let S be a semibounded relation in $\mathfrak{H}$, let S_{F} be its Friedrichs extension, and let H be a semibounded self-adjoint extension of S with lower bound $m(H)$. Furthermore, let $a < m(H)$ and let $R(a)$ be the nonnegative operator in (5.3.9). Then*

$$\mathfrak{H}_{\mathfrak{t}_H-a} \ominus_{\mathfrak{t}_H-a} \mathfrak{H}_{\mathfrak{t}_{S_{\mathrm{F}}}-a} = \ker(S^*-a) \cap \mathfrak{H}_{\mathfrak{t}_H-a} = \operatorname{ran} R(a)^{\frac{1}{2}}, \tag{5.3.14}$$

and, consequently, the Hilbert space $\mathfrak{H}_{\mathfrak{t}_H-a}$ *has the orthogonal decomposition*

$$\begin{aligned}\mathfrak{H}_{\mathfrak{t}_H-a} &= \bigl(\ker(S^*-a)\cap\mathfrak{H}_{\mathfrak{t}_H-a}\bigr)\oplus_{\mathfrak{t}_H-a}\mathfrak{H}_{\mathfrak{t}_{S_\mathrm{F}}-a} \\ &= \operatorname{ran} R(a)^{\frac{1}{2}}\oplus_{\mathfrak{t}_H-a}\mathfrak{H}_{\mathfrak{t}_{S_\mathrm{F}}-a}.\end{aligned}\tag{5.3.15}$$

In particular, the sum decomposition in (5.3.8) *is direct for every* $a<m(H)$.

Proof. First the identity

$$\mathfrak{H}_{\mathfrak{t}_H-a}\ominus_{\mathfrak{t}_H-a}\mathfrak{H}_{\mathfrak{t}_{S_\mathrm{F}}-a} = \ker(S^*-a)\cap\mathfrak{H}_{\mathfrak{t}_H-a}\tag{5.3.16}$$

in (5.3.14) will be shown. Recall from Lemma 5.1.17 and Theorem 5.1.12 that $\mathfrak{H}_{\mathfrak{t}_S-a}$ is a dense subspace of $\mathfrak{H}_{\mathfrak{t}_{S_\mathrm{F}}-a}$, and hence it suffices to verify that

$$\mathfrak{H}_{\mathfrak{t}_H-a}\ominus_{\mathfrak{t}_H-a}\mathfrak{H}_{\mathfrak{t}_S-a} = \ker(S^*-a)\cap\mathfrak{H}_{\mathfrak{t}_H-a}.\tag{5.3.17}$$

Assume first that g belongs to the left-hand side of (5.3.17). Then $(f,g)_{\mathfrak{t}_H-a}=0$ for all $f\in\operatorname{dom}S$. Hence,

$$0=(f,g)_{\mathfrak{t}_H-a}=\mathfrak{t}_H[f,g]-a(f,g)=(f',g)-a(f,g)=(f'-af,g)$$

for all $\{f,f'\}\in S\subset H$, where in the third equality the first representation theorem was used. This implies $g\in(\operatorname{ran}(S-a))^\perp\cap\mathfrak{H}_{\mathfrak{t}_H-a}=\ker(S^*-a)\cap\mathfrak{H}_{\mathfrak{t}_H-a}$. Conversely, assume that $g\in\ker(S^*-a)\cap\mathfrak{H}_{\mathfrak{t}_H-a}$. Then the same reasoning as above shows that g belongs to the left-hand side of (5.3.17).

In order to prove the second equality in (5.3.14), one first shows that

$$\operatorname{ran} R(a)^{\frac{1}{2}}\subset\ker(S^*-a).\tag{5.3.18}$$

To see this, let $\{f,f'\}\in S$. Then $\{f,f'\}\in H\cap S_\mathrm{F}$ and hence $(H-a)^{-1}(f'-af)=f$ and $(S_\mathrm{F}-a)^{-1}(f'-af)=f$, which implies that for $h\in\mathfrak{H}$

$$(f'-af,R(a)h)=\bigl((H-a)^{-1}(f'-af)-(S_\mathrm{F}-a)^{-1}(f'-af),h\bigr)=0.$$

Therefore,

$$\operatorname{ran}R(a)\subset(\operatorname{ran}(S-a))^\perp=\ker(S^*-a)$$

and hence also $\overline{\operatorname{ran}}\,R(a)\subset\ker(S^*-a)$. Moreover, since $R(a)$ is a nonnegative self-adjoint operator it follows from Corollary D.7 that

$$\operatorname{ran}R(a)\subset\operatorname{ran}R(a)^{\frac{1}{2}}\subset\overline{\operatorname{ran}}\,R(a)^{\frac{1}{2}}=\overline{\operatorname{ran}}\,R(a)\subset\ker(S^*-a),\tag{5.3.19}$$

which shows (5.3.18). By (5.3.8), one has $\operatorname{ran}R(a)^{\frac{1}{2}}\subset\operatorname{dom}(H-a)^{\frac{1}{2}}=\mathfrak{H}_{\mathfrak{t}_H-a}$. Together with (5.3.19) one concludes that $\operatorname{ran}R(a)^{\frac{1}{2}}\subset\ker(S^*-a)\cap\mathfrak{H}_{\mathfrak{t}_H-a}$. From (5.3.16) it is clear that

$$\operatorname{ran}R(a)^{\frac{1}{2}}\subset\mathfrak{H}_{\mathfrak{t}_H-a}\ominus_{\mathfrak{t}_H-a}\mathfrak{H}_{\mathfrak{t}_S-a}.$$

Comparing (5.3.8) with (5.3.12), one then concludes that

$$\operatorname{ran} R(a)^{\frac{1}{2}} = \mathfrak{H}_{\mathfrak{t}_H - a} \ominus_{\mathfrak{t}_H - a} \mathfrak{H}_{\mathfrak{t}_S - a}.$$

Together with (5.3.16) the identities in (5.3.14) follow. Furthermore, the above reasoning shows that the sum decomposition in (5.3.8) is direct. □

The semibounded self-adjoint extensions H of S for which the subspace $\ker(S^* - a) \cap \mathfrak{H}_{\mathfrak{t}_H - a}$ is not a proper subset of $\ker(S^* - a)$ are of special interest. In fact, they coincide with the semibounded self-adjoint extensions H for which H and S_F are transversal.

Theorem 5.3.8. *Let S be a semibounded relation in $\mathfrak{H}$ and let H be a semibounded self-adjoint extension of S. Then the following statements are equivalent:*

(i) *H and S_F are transversal, i.e., $S^* = H \,\widehat{+}\, S_\mathrm{F}$;*

(ii) *$\ker(S^* - a) \subset \operatorname{dom}(H - a)^{\frac{1}{2}}$ for some, and hence for all $a < m(H)$;*

(iii) *$\operatorname{dom}(H - a)^{\frac{1}{2}} = \ker(S^* - a) + \operatorname{dom}(S_\mathrm{F} - a)^{\frac{1}{2}}$ for some, and hence for all $a < m(H)$;*

(iv) *$\operatorname{dom} S^* \subset \operatorname{dom}(H - a)^{\frac{1}{2}}$ for some, and hence for all $a \le m(H)$.*

Proof. (i) ⇔ (ii) In general, the self-adjoint extensions H and S_F are transversal if and only if for some, and hence for all $a < m(H)$

$$\operatorname{ran} R(a) = \ker(S^* - a),$$

where $R(a) = (H - a)^{-1} - (S_\mathrm{F} - a)^{-1}$, which follows from Theorem 1.7.8. Since $R(a)$ is a nonnegative self-adjoint operator and since $\ker(S^* - a)$ is closed, the last statement is equivalent to

$$\operatorname{ran} R(a)^{\frac{1}{2}} = \ker(S^* - a);$$

cf. Corollary D.7. It follows from Proposition 5.3.7 that this condition is the same as

$$\ker(S^* - a) \cap \operatorname{dom}(H - a)^{\frac{1}{2}} = \ker(S^* - a).$$

(ii) ⇒ (iii) This follows immediately from the direct sum decomposition

$$\operatorname{dom}(H - a)^{\frac{1}{2}} = \big(\ker(S^* - a) \cap \operatorname{dom}(H - a)^{\frac{1}{2}}\big) + \operatorname{dom}(S_\mathrm{F} - a)^{\frac{1}{2}};$$

cf. (5.3.8) and Proposition 5.3.7.

(iii) ⇒ (ii) This implication is trivial.

(i) ⇒ (iv) The identity $S^* = S_\mathrm{F} \,\widehat{+}\, H$ shows that

$$\operatorname{dom} S^* = \operatorname{dom} S_\mathrm{F} + \operatorname{dom} H,$$

and note that $\operatorname{dom} H$ and $\operatorname{dom} S_\mathrm{F}$ are subsets of $\operatorname{dom}(H - a)^{\frac{1}{2}}$.

(iv) ⇒ (ii) This is clear. □

The following result is a consequence of Theorem 5.3.8; it describes the behavior of any semibounded self-adjoint extension H' of S in the presence of a semibounded self-adjoint extension H such that H and S_{F} are transversal.

Corollary 5.3.9. *Let S be a semibounded relation in $\mathfrak{H}$ and let H be a semibounded self-adjoint extension of S such that H and S_{F} are transversal. Then every semibounded self-adjoint extension H' of S satisfies*

$$\mathrm{dom}\,(H'-a)^{\frac{1}{2}} \subset \mathrm{dom}\,(H-a)^{\frac{1}{2}}, \quad a < \min\{m(H), m(H')\}, \tag{5.3.20}$$

in which case there exists $C > 0$ such that

$$\|(H_{\mathrm{op}}-a)^{\frac{1}{2}}\varphi\| \leq C\|((H')_{\mathrm{op}}-a)^{\frac{1}{2}}\varphi\| \tag{5.3.21}$$

for all $\varphi \in \mathrm{dom}\,(H'-a)^{\frac{1}{2}}$. Moreover, there is equality in (5.3.20) if and only if H' and S_{F} are transversal, in which case there exist $c > 0$ and $C > 0$ such that

$$c\|((H')_{\mathrm{op}}-a)^{\frac{1}{2}}\varphi\| \leq \|(H_{\mathrm{op}}-a)^{\frac{1}{2}}\varphi\| \leq C\|((H')_{\mathrm{op}}-a)^{\frac{1}{2}}\varphi\| \tag{5.3.22}$$

for all $\varphi \in \mathrm{dom}\,(H'-a)^{\frac{1}{2}} = \mathrm{dom}\,(H-a)^{\frac{1}{2}}$.

Proof. By Theorem 5.3.8, the semibounded self-adjoint extension H of S satisfies all of the equivalent conditions (i)–(iv) in Theorem 5.3.8. Let H' be another semibounded self-adjoint extension of S. Applying Proposition 5.3.7 to H' one sees that for $a < m(H')$

$$\mathrm{dom}\,(H'-a)^{\frac{1}{2}} = \big(\ker(S^*-a) \cap \mathfrak{H}_{\mathfrak{t}_{H'}-a}\big) + \mathrm{dom}\,(S_{\mathrm{F}}-a)^{\frac{1}{2}}.$$

Choosing $a < \min\{m(H), m(H')\}$ it follows from Theorem 5.3.8 (iii) for H that the inclusion (5.3.20) holds. The inequality (5.3.21) is a direct consequence of Proposition 1.5.11.

Assume that there is equality in (5.3.20). Then Theorem 5.3.8 (iii) implies that H' and S_{F} are transversal. Conversely, if H' and S_{F} are transversal, then it follows from Theorem 5.3.8 (iii) that there is equality in (5.3.20). If there is equality in (5.3.20), then (5.3.21) also holds for some $c > 0$ when H and H' are interchanged. Thus, the inequalities in (5.3.22) hold. □

The extreme case of equality of H and S_{F} is described in the following immediate corollary of Proposition 5.3.7 and (5.3.8).

Corollary 5.3.10. *Let S be a semibounded relation in $\mathfrak{H}$ and let H be a semibounded self-adjoint extension of S. Then $H = S_{\mathrm{F}}$ if and only if*

$$\ker(S^*-a) \cap \mathrm{dom}\,(H-a)^{\frac{1}{2}} = \{0\}$$

for some, and hence for all $a < m(H)$.

In the next corollary the form corresponding to a semibounded self-adjoint extension H which is transversal to S_{F} is specified.

Corollary 5.3.11. *Let S be a semibounded relation in $\mathfrak{H}$, let H be a semibounded self-adjoint extension of S such that H and S_{F} are transversal, and let $\mathfrak{t}_{S_{\mathrm{F}}}$ and $\mathfrak{t}_H$ be the corresponding closed semibounded forms in $\mathfrak{H}$. Then*

$$\operatorname{dom} \mathfrak{t}_H = \ker (S^* - a) \oplus_{\mathfrak{t}_H - a} \operatorname{dom} \mathfrak{t}_{S_{\mathrm{F}}}, \qquad a < m(H), \tag{5.3.23}$$

and the restriction of $\mathfrak{t}_H$ to $\mathfrak{N}_a(S^*) = \ker (S^* - a)$ is a closed form in $\mathfrak{N}_a(S^*)$ which is bounded from below by $m(H)$ and represented by a bounded self-adjoint operator $L_a \in \mathbf{B}(\mathfrak{N}_a(S^*))$. Furthermore, one has

$$\mathfrak{t}_H[f,g] - a(f,g) = \big((L_a - a)f_a, g_a\big) + \mathfrak{t}_{S_{\mathrm{F}}}[f_{\mathrm{F}}, g_{\mathrm{F}}] - a(f_{\mathrm{F}}, g_{\mathrm{F}}) \tag{5.3.24}$$

for all $f = f_a + f_{\mathrm{F}}, g = g_a + g_{\mathrm{F}} \in \operatorname{dom} \mathfrak{t}_H$, where $f_a, g_a \in \ker (S^* - a)$ and $f_{\mathrm{F}}, g_{\mathrm{F}} \in \operatorname{dom} \mathfrak{t}_{S_{\mathrm{F}}}$.

Proof. The orthogonal decomposition (5.3.23) of $\operatorname{dom} \mathfrak{t}_H$ follows from (5.3.15) in Proposition 5.3.7 and Theorem 5.3.8 (iii). Since the restriction of the form $\mathfrak{t}_H$ to $\mathfrak{N}_a(S^*)$ is a closed form which is bounded from below by $m(H)$, it follows from Theorem 5.1.18 that there exists a semibounded self-adjoint relation L_a in $\mathfrak{N}_a(S^*)$ which represents this form. Moreover, it follows from Theorem 5.1.23 that $\mathfrak{N}_a(S^*) = \operatorname{dom} (L_a - x)^{\frac{1}{2}}$, $x \leq m(H)$, and hence $(L_a - x)^{\frac{1}{2}}$ and L_a are bounded self-adjoint operators defined on $\mathfrak{N}_a(S^*)$.

For $f, g \in \operatorname{dom} \mathfrak{t}_H$ decomposed, with respect to (5.3.23), in

$$f = f_a + f_{\mathrm{F}} \quad \text{and} \quad g = g_a + g_{\mathrm{F}},$$

where $f_a, g_a \in \ker (S^* - a)$ and $f_{\mathrm{F}}, g_{\mathrm{F}} \in \operatorname{dom} \mathfrak{t}_{S_{\mathrm{F}}}$, one has $(f_a, g_{\mathrm{F}})_{\mathfrak{t}_H - a} = 0$ and $(f_{\mathrm{F}}, g_a)_{\mathfrak{t}_H - a} = 0$, and hence

$$\begin{aligned} \mathfrak{t}_H[f,g] - a(f,g) &= (f,g)_{\mathfrak{t}_H - a} = (f_a, g_a)_{\mathfrak{t}_H - a} + (f_{\mathrm{F}}, g_{\mathrm{F}})_{\mathfrak{t}_H - a} \\ &= \mathfrak{t}_H[f_a, g_a] - a(f_a, g_a) + \mathfrak{t}_{S_{\mathrm{F}}}[f_{\mathrm{F}}, g_{\mathrm{F}}] - a(f_{\mathrm{F}}, g_{\mathrm{F}}). \end{aligned}$$

Now (5.3.24) follows from $\mathfrak{t}_H[f_a, g_a] = (L_a f_a, g_a)$. □

5.4 Semibounded self-adjoint extensions and their lower bounds

Let S be a, not necessarily closed, semibounded relation in a Hilbert space $\mathfrak{H}$ with lower bound $m(S)$. The Friedrichs extension S_{F} is a self-adjoint extension of S whose lower bound $m(S_{\mathrm{F}})$ is equal to the lower bound $m(S)$ of S. If H is a semibounded self-adjoint extension of S, then necessarily $m(H) \leq m(S)$. In this section the so-called Kreĭn type extensions $S_{\mathrm{K},x}$ of S will be introduced. They can be viewed as generalizations of the Kreĭn–von Neumann extension of a nonnegative symmetric operator or relation. The Kreĭn type extension $S_{\mathrm{K},x}$ can be used to describe all semibounded self-adjoint extensions H of S whose lower bound satisfies $m(H) \in [x, m(S)]$ when $x \leq m(S)$.

Let S be a semibounded relation in $\mathfrak{H}$ with lower bound $\gamma = m(S)$. It is clear that $x \leq \gamma$ implies that $S - x \geq 0$. Hence, for $x \leq \gamma$ the relation $(S-x)^{-1}$ is nonnegative and one can define the Friedrichs extension $((S-x)^{-1})_{\mathrm{F}}$ of $(S-x)^{-1}$, which is nonnegative; cf. Definition 5.3.2.

Lemma 5.4.1. *Let S be a semibounded relation in $\mathfrak{H}$ with lower bound γ. For $x \leq \gamma$ the relation $S_{\mathrm{K},x}$ defined by*

$$S_{\mathrm{K},x} := \big(((S-x)^{-1})_{\mathrm{F}}\big)^{-1} + x \tag{5.4.1}$$

is a semibounded self-adjoint extension of S with lower bound $m(S_{\mathrm{K},x}) = x$. Moreover, $S_{\mathrm{K},x} = (\overline{S})_{\mathrm{K},x}$ for $x \leq \gamma$ and

$$S \subset S \,\widehat{+}\, \widehat{\mathfrak{N}}_x(S^*) \subset \overline{S} \,\widehat{+}\, \widehat{\mathfrak{N}}_x(S^*) \subset S_{\mathrm{K},x}, \quad x \leq \gamma, \tag{5.4.2}$$

while, in particular,

$$\overline{S} \,\widehat{+}\, \widehat{\mathfrak{N}}_x(S^*) = S_{\mathrm{K},x}, \quad x < \gamma. \tag{5.4.3}$$

Proof. Since for $x \leq \gamma$ the Friedrichs extension $((S-x)^{-1})_{\mathrm{F}}$ of the nonnegative relation $(S-x)^{-1}$ is nonnegative, it follows that

$$\big(((S-x)^{-1})_{\mathrm{F}}\big)^{-1}$$

is a nonnegative self-adjoint extension of $S - x$. Hence, $S_{\mathrm{K},x}$ defined by (5.4.1) is a self-adjoint extension of S and, clearly,

$$m(S_{\mathrm{K},x}) \geq x, \quad x \leq \gamma. \tag{5.4.4}$$

Since the closure of $(S-x)^{-1}$ is given by $(\overline{S}-x)^{-1}$, it follows from Lemma 5.3.1, with S replaced by $(S-x)^{-1}$, that

$$((\overline{S}-x)^{-1})_{\mathrm{F}} = ((S-x)^{-1})_{\mathrm{F}},$$

which leads to $S_{\mathrm{K},x} = (\overline{S})_{\mathrm{K},x}$ for $x \leq \gamma$.

Let $x \leq \gamma$ and note that the first and second inclusion in (5.4.2) are clear. It is also clear that $S \subset \overline{S} \subset S_{\mathrm{K},x}$; thus, to show the third inclusion in (5.4.2) it suffices to check that $\widehat{\mathfrak{N}}_x(S^*) \subset S_{\mathrm{K},x}$. Set $T = (S-x)^{-1}$, so that T is nonnegative and $\operatorname{mul} T^* = \mathfrak{N}_x(S^*)$. By Lemma 5.3.1, $\{0\} \times \operatorname{mul} T^* \subset T_{\mathrm{F}}$ or, equivalently,

$$\{0\} \times \mathfrak{N}_x(S^*) \subset ((S-x)^{-1})_{\mathrm{F}} \quad \text{or} \quad \mathfrak{N}_x(S^*) \times \{0\} \subset ((S-x)^{-1})_{\mathrm{F}})^{-1}.$$

Thus, $\widehat{\mathfrak{N}}_x(S^*) \subset S_{\mathrm{K},x}$ which completes the argument.

For $x < \gamma$ it follows from Proposition 1.4.6 and Lemma 1.2.2 that $\operatorname{ran}(\overline{S}-x)$ is closed. Hence, the relation $\overline{S} \,\widehat{+}\, \widehat{\mathfrak{N}}_x(S^*)$ is self-adjoint, cf. Lemma 1.5.7, and thus the equality (5.4.3) prevails.

It remains to show that $m(S_{K,x}) = x$. When $x < \gamma$ one concludes this from (5.4.3). When $x = \gamma$ observe that $S \subset S_{K,\gamma}$ implies $m(S_{K,\gamma}) \leq m(S) = \gamma$. On the other hand, from (5.4.4) it follows that $m(S_{K,\gamma}) \geq \gamma$. □

Definition 5.4.2. Let S be a semibounded relation in $\mathfrak{H}$ with lower bound γ and let $x \leq \gamma$. The semibounded self-adjoint extensions $S_{K,x}$ in (5.4.1) are called *Kreĭn type extensions* of S. In the case $\gamma \geq 0$ the nonnegative self-adjoint extension $S_{K,0}$ is called the *Kreĭn–von Neumann extension* of S.

The definition of the Kreĭn type extensions $S_{K,x}$ in (5.4.1) incorporates the lower bound $m(S) = \gamma$ of S. Note that $m(S - x) = \gamma - x$ for any $x \in \mathbb{R}$, so that $(S - x)_{K,\gamma - x}$ is well defined, and

$$(S - x)_{K,\gamma-x} = \big((((S - x) - (\gamma - x))^{-1})_F\big)^{-1} + \gamma - x, \quad x \in \mathbb{R},$$

which leads to

$$(S - x)_{K,\gamma-x} = S_{K,\gamma} - x, \quad x \in \mathbb{R}. \tag{5.4.5}$$

The identity (5.4.5) is the analog for the Kreĭn type extension $S_{K,\gamma}$ of the shift invariance property (5.3.4) of S_F. There are some more useful identities involving Kreĭn type extensions of S. First note the simple equality

$$\big(S \mathbin{\widehat{+}} \widehat{\mathfrak{N}}_x(S^*)\big)^{-1} = S^{-1} \mathbin{\widehat{+}} \widehat{\mathfrak{N}}_{1/x}(S^{-*}), \quad x \in \mathbb{R} \setminus \{0\}. \tag{5.4.6}$$

If $S \geq 0$ or, equivalently, $S^{-1} \geq 0$, then it follows from (5.4.6) and Lemma 5.4.1 that

$$(S_{K,x})^{-1} = (S^{-1})_{K,1/x}, \quad x < 0, \tag{5.4.7}$$

since $0 \leq \min\{m(S), m(S^{-1})\}$. In particular, one sees from (5.4.7) that

$$\big((S - \gamma)_{K,x}\big)^{-1} = \big((S - \gamma)^{-1}\big)_{K,1/x}, \quad x < 0. \tag{5.4.8}$$

Returning to the general case where S has lower bound γ, note the simple equality

$$\big(S \mathbin{\widehat{+}} \widehat{\mathfrak{N}}_{a+x}(S^*)\big) - a = (S - a) \mathbin{\widehat{+}} \widehat{\mathfrak{N}}_x((S - a)^*), \quad a, x \in \mathbb{R}.$$

For $a + x < \gamma$ this implies

$$S_{K,a+x} - a = (S - a)_{K,x}, \tag{5.4.9}$$

and taking $a = \gamma$ in (5.4.9) gives an analog of (5.4.5):

$$S_{K,\gamma+x} - \gamma = (S - \gamma)_{K,x}, \quad x < 0. \tag{5.4.10}$$

By Lemma 5.4.1, the Kreĭn type extensions $S_{K,x}$, $x \leq \gamma$, are semibounded self-adjoint extensions of S. As restrictions of S^* the Kreĭn type extensions can be characterized in a similar way as the Friedrichs extension S_F; cf. Theorem 5.3.3.

Theorem 5.4.3. *Let S be a semibounded relation in $\mathfrak{H}$ with lower bound γ. Then for each $x \leq \gamma$ the Kreĭn type extension $S_{\mathrm{K},x}$ of S has the representation*

$$S_{\mathrm{K},x} = \{\{f, f'\} \in S^* : f' - xf \in \operatorname{dom} \mathfrak{t}_{((S-x)^{-1})_{\mathrm{F}}}\} \tag{5.4.11}$$

with $\ker(S_{\mathrm{K},x} - x) = \ker(S^* - x)$. *Furthermore, if H is a self-adjoint extension of S, which is not necessarily semibounded, then*

$$\operatorname{ran}(H - x) \subset \operatorname{dom} \mathfrak{t}_{((S-x)^{-1})_{\mathrm{K}}} \quad \Rightarrow \quad H = S_{\mathrm{K},x}.$$

Proof. Let $x \leq \gamma$. Then by definition one has $(S_{\mathrm{K},x} - x)^{-1} = ((S-x)^{-1})_{\mathrm{F}}$ and $\{f, f'\} \in S_{\mathrm{K},x}$ if and only if $\{f' - xf, f\} \in (S_{\mathrm{K},x} - x)^{-1}$. Similarly, $\{f, f'\} \in S^*$ if and only if $\{f' - xf, f\} \in (S^* - x)^{-1}$. Hence, the description (5.4.11) follows from the representation of $((S-x)^{-1})_{\mathrm{F}}$ in Theorem 5.3.3 (with S now replaced by $(S-x)^{-1}$). Likewise,

$$\ker(S_{\mathrm{K},x} - x) = \operatorname{mul}((S-x)^{-1})_{\mathrm{F}} = \operatorname{mul}(S^* - x)^{-1} = \ker(S^* - x).$$

The last item also follows from Theorem 5.3.3. □

There is also an approximation of $S_{\mathrm{K},x}$ by elements in S as in Corollary 5.3.4; in particular, this gives a useful description of $\operatorname{mul} S_{\mathrm{K},x}$.

Corollary 5.4.4. *Let S be a semibounded relation in $\mathfrak{H}$ with lower bound γ. Then $S_{\mathrm{K},x}$, $x \leq \gamma$, is the set of all elements $\{f, f'\} \in S^*$ for which there exists a sequence $(\{f_n, f_n'\})$ in S such that*

$$f_n' - xf_n \to f' - xf \quad \text{and} \quad (f_n, f_n' - xf_n) \to (f, f' - xf).$$

In particular, $\operatorname{mul} S_{\mathrm{K},x}$ *is the set of all elements $f' \in \operatorname{mul} S^*$ for which there exists a sequence $(\{f_n, f_n'\})$ in S such that*

$$f_n' - xf_n \to f' \quad \text{and} \quad (f_n, f_n' - xf_n) \to 0.$$

As in the case of the Friedrichs extension, the Kreĭn type extension $S_{\mathrm{K},\gamma}$ can sometimes be explicitly given in terms of S and an eigenspace of S^*; cf. Lemma 1.5.7. The following result is the analog of Corollary 5.3.5. The special case where $\operatorname{ran}(S - \gamma)$ is closed is particularly useful.

Corollary 5.4.5. *Let S be a semibounded relation with lower bound γ. Then*

$$S_{\mathrm{K},\gamma} = S \,\widehat{+}\, \widehat{\mathfrak{N}}_\gamma(S^*) \tag{5.4.12}$$

if and only if

$$\operatorname{ran}(S - \gamma) = \overline{\operatorname{ran}}(S - \gamma) \cap \operatorname{ran}(S^* - \gamma).$$

In particular, if $\operatorname{ran}(S - \gamma)$ *is closed, then $S_{\mathrm{K},\gamma}$ has the form* (5.4.12).

The semibounded self-adjoint extensions $S_{K,x}$ with $x \le \gamma$ become extremal extensions of S when a lower bound $x \le \gamma$ for semibounded self-adjoint extensions of S is prescribed.

Theorem 5.4.6. *Let S be a semibounded relation in $\mathfrak{H}$ with lower bound γ. Let $x \le \gamma$ be fixed and let H be a semibounded self-adjoint relation in $\mathfrak{H}$. Then the following equivalence holds:*

$$S \subset H \quad \text{and} \quad x \le m(H) \quad \Leftrightarrow \quad S_{K,x} \le H \le S_F. \tag{5.4.13}$$

In particular, the class of semibounded self-adjoint extensions of S preserving the lower bound of S is characterized by

$$S \subset H \quad \text{and} \quad \gamma = m(H) \quad \Leftrightarrow \quad S_{K,\gamma} \le H \le S_F. \tag{5.4.14}$$

In fact, $S_{K,x} \le H \le S_F$, $x \le \gamma$, implies that $S \subset (S_F \cap S_{K,x}) \subset H$.

Proof. ($\Rightarrow$) Assume that H is a semibounded self-adjoint extension of S with lower bound $m(H) \ge x$. Then clearly $S - x \subset H - x$ and here both sides are nonnegative relations. But then also $(S-x)^{-1} \subset (H-x)^{-1}$, where both sides are nonnegative relations. Applying Proposition 5.3.6 to the relation $(S-x)^{-1}$ one obtains

$$(H-x)^{-1} \le ((S-x)^{-1})_F.$$

Since these relations are nonnegative, the antitonicity property of the inverse in Corollary 5.2.8 gives the inequality

$$\big(((S-x)^{-1})_F\big)^{-1} \le H - x$$

or, equivalently, $S_{K,x} \le H$. The inequality $H \le S_F$ holds by Proposition 5.3.6.

($\Leftarrow$) Let H be a semibounded self-adjoint relation such that $S_{K,x} \le H \le S_F$. Then $m(S_{K,x}) \le m(H) \le m(S_F)$ by Lemma 5.2.5 (ii) and, since $m(S_{K,x}) = x$ and $m(S_F) = \gamma$, one concludes that H is semibounded with $x \le m(H) \le \gamma$.

It remains to show that H is an extension of S. With $a < x$ the assumption on H is equivalent to

$$\big((S_F - a)^{-1}h, h\big) \le \big((H-a)^{-1}h, h\big) \le \big((S_{K,x} - a)^{-1}h, h\big), \quad h \in \mathfrak{H}; \tag{5.4.15}$$

cf. Proposition 5.2.7. For $R(a) = (H-a)^{-1} - (S_F - a)^{-1} \in \mathbf{B}(\mathfrak{H})$ it follows from (5.4.15) that

$$0 \le (R(a)h, h) \le \big((S_{K,x} - a)^{-1}h, h\big) - \big((S_F - a)^{-1}h, h\big), \quad h \in \mathfrak{H}. \tag{5.4.16}$$

Now, let $\{f, f'\} \in S$ and define $h = f' - af$. Then $h \in \operatorname{ran}(S-a)$ and $\{h, f\} \in (S-a)^{-1}$, and hence

$$\{h, f\} \in (S_F - a)^{-1} \cap (S_{K,x} - a)^{-1},$$

so that $(S_F - a)^{-1}h = (S_{K,x} - a)^{-1}h$. It follows from (5.4.16) that $(R(a)h, h) = 0$, and since $R(a) \ge 0$, one concludes that $R(a)h = 0$ for all $h \in \operatorname{ran}(S-a)$.

In other words,

$$(H-a)^{-1}(f'-af)=(S_{\mathrm{F}}-a)^{-1}(f'-af)=f, \quad \{f,f'\}\in S,$$

and it follows that $\{f,f'\}\in H$. This proves the claim $S\subset H$ and completes the proof of (5.4.13). The inclusion $S_{\mathrm{F}}\cap S_{\mathrm{K},x}\subset H$ follows similarly. □

If S is nonnegative, then Theorem 5.4.6 shows that the Kreĭn–von Neumann extension

$$S_{\mathrm{K},0}=\bigl((S^{-1})_{\mathrm{F}}\bigr)^{-1} \tag{5.4.17}$$

in Definition 5.4.2 is the smallest nonnegative self-adjoint extension.

Corollary 5.4.7. *Let S be a nonnegative relation in $\mathfrak{H}$ and let H be a semibounded self-adjoint relation in $\mathfrak{H}$. Then the following equivalence holds:*

$$S\subset H \quad \text{and} \quad m(H)\geq 0 \quad \Leftrightarrow \quad S_{\mathrm{K},0}\leq H\leq S_{\mathrm{F}}.$$

In fact, $S_{\mathrm{K},0}\leq H\leq S_{\mathrm{F}}$, implies that $S\subset(S_{\mathrm{F}}\cap S_{\mathrm{K},0})\subset H$.

For completeness it is observed that the inequalities $S_{\mathrm{K},x}\leq H\leq S_{\mathrm{F}}$ in (5.4.13) can also be expressed by inequalities for the corresponding forms:

$$\mathfrak{t}_{S_{\mathrm{K},x}}\leq\mathfrak{t}_H\leq\mathfrak{t}_{S_{\mathrm{F}}},$$

thanks to Theorem 5.2.4. Recall from (5.3.7) that the inequality $\mathfrak{t}_H\leq\mathfrak{t}_{S_{\mathrm{F}}}$ is in fact equivalent to the inclusion $\mathfrak{t}_{S_{\mathrm{F}}}\subset\mathfrak{t}_H$.

It is clear from Lemma 5.3.1 or Theorem 5.3.3 that the Friedrichs extension S_{F} is an operator if and only if S is densely defined, in which case all self-adjoint extensions of S are operators. If S is not densely defined, then S may not be closable as an operator, in which case all self-adjoint extensions of S are multivalued. The following result shows when semibounded self-adjoint operator extensions exist.

Corollary 5.4.8. *Let S be a semibounded operator in $\mathfrak{H}$ with lower bound γ. Then the following statements are equivalent:*

(i) *S has a semibounded self-adjoint operator extension;*
(ii) *for some $x\leq\gamma$ the self-adjoint extension $S_{\mathrm{K},x}$ is an operator;*
(iii) *there exists $x\leq\gamma$ such that for any sequence $(\{f_n,f_n'\})$ in S with the properties $f_n'-xf_n\to f'$ and $(f_n,f_n'-xf_n)\to 0$ one has $f'=0$.*

If any of these statements hold, then S is a closable operator. Furthermore, the following statements are equivalent:

(i′) *S has a bounded self-adjoint operator extension;*
(ii′) *for some $x\leq\gamma$ the self-adjoint extension $S_{\mathrm{K},x}$ belongs to $\mathbf{B}(\mathfrak{H})$;*
(iii′) *there exist $x\leq\gamma$ and $M\geq 0$ such that $\|f'-xf\|^2\leq M(f,f'-xf)$ for all $\{f,f'\}\in S$.*

If any of these statements hold, then S is a bounded operator.

Proof. (i) $\Rightarrow$ (ii) Let H be a semibounded self-adjoint operator extension of S. Then H is densely defined and $\operatorname{mul} H = \{0\}$. If $x = m(H)$, then Theorem 5.4.6 shows that $S_{\mathrm{K},x} \leq H$. Hence, $\operatorname{mul} S_{\mathrm{K},x} \subset \operatorname{mul} H = \{0\}$ by Lemma 5.2.5 (i), and therefore $S_{\mathrm{K},x}$ is an operator extension of S.

(ii) $\Rightarrow$ (i) This is clear.

(ii) $\Leftrightarrow$ (iii) This follows from Corollary 5.4.4.

The inclusion $S \subset H$ for a self-adjoint operator H shows that S is a closable operator; this holds when one of the equivalent conditions (i)–(iii) is satisfied.

(i′) $\Rightarrow$ (iii′) Let H be a bounded self-adjoint operator which extends S and let $m(H) = x$. Then $H \in \mathbf{B}(\mathfrak{H})$ and an application of the Cauchy–Schwarz inequality for the inner product $((H - x)\cdot, \cdot)$ yields

$$\|(H - x)f\|^2 \leq \|H - x\| \, (f, (H - x)f), \quad f \in \mathfrak{H}.$$

In particular, for $f \in \operatorname{dom} S$ one obtains (iii′) with $f' = Sf$ and $M = \|H - x\|$.

(iii′) $\Rightarrow$ (ii′) Assume that (iii′) holds for some x. To show (ii′) let $\{f, f'\} \in S_{\mathrm{K},x}$. Then according to Corollary 5.4.4 there exists a sequence $(\{f_n, f_n'\})$ in S such that $f_n' - xf_n \to f' - xf$ and $(f_n, f_n' - xf_n) \to (f, f' - xf)$. By assumption $\|f_n' - xf_n\|^2 \leq M(f_n, f_n' - xf_n)$ and taking limits gives

$$\|f' - xf\|^2 \leq M(f, f' - xf) \leq M\|f\|\|f' - xf\|.$$

Thus, if $\{f, f'\} \in S_{\mathrm{K},x}$, then $\|f' - xf\| \leq M\|f\|$, which gives $\operatorname{mul} S_{\mathrm{K},x} = \{0\}$. In addition, one now sees that $S_{\mathrm{K},x} - x$ is a bounded operator, and since $S_{\mathrm{K},x}$ is self-adjoint, it follows that $S_{\mathrm{K},x} \in \mathbf{B}(\mathfrak{H})$.

(ii′) $\Rightarrow$ (i′) This is clear.

The last statement follows from any of the statements (i′), (ii′), and (iii′). □

Let S be a semibounded relation in $\mathfrak{H}$ with lower bound $m(S) = \gamma$. By means of Theorem 5.4.6 it will be shown that the mapping $x \mapsto S_{\mathrm{K},x}$, $x < \gamma$, is nondecreasing.

Corollary 5.4.9. *Let S be a semibounded relation in $\mathfrak{H}$ with lower bound γ. Let $x \leq y < \gamma$, then*

$$S_{\mathrm{K},x} \leq S_{\mathrm{K},y} \leq S_{\mathrm{K},\gamma} \leq S_{\mathrm{F}}.$$

Proof. By construction, $S_{\mathrm{K},x}$ and $S_{\mathrm{K},y}$ are semibounded self-adjoint extensions of S with lower bounds $m(S_{\mathrm{K},x}) = x$ and $m(S_{\mathrm{K},y}) = y$. Hence, $m(S_{\mathrm{K},x}) \leq m(S_{\mathrm{K},y})$ and an application of (5.4.13) in Theorem 5.4.6 gives $S_{\mathrm{K},x} \leq S_{\mathrm{K},y}$. Similarly, $m(S_{\mathrm{K},y}) \leq m(S_{\mathrm{K},\gamma}) = \gamma$ leads to $S_{\mathrm{K},y} \leq S_{\mathrm{K},\gamma}$. The inequality $S_{\mathrm{K},\gamma} \leq S_{\mathrm{F}}$ also follows from Theorem 5.4.6. □

The Friedrichs extension S_F and the Kreĭn type extensions $S_{K,x}$ can be approximated in the strong resolvent sense by the semibounded self-adjoint relations $S_{K,t}$ with $t \in (-\infty, \gamma)$; cf. Theorem 5.2.11.

Theorem 5.4.10. *Let S be a semibounded relation in $\mathfrak{H}$ with lower bound γ. Then the Friedrichs extension S_F is given by the strong resolvent limit*

$$(S_F - \lambda)^{-1}h = \lim_{t \downarrow -\infty} (S_{K,t} - \lambda)^{-1}h, \quad h \in \mathfrak{H}, \tag{5.4.18}$$

where $\lambda \in \mathbb{C} \setminus [\gamma, \infty)$, and for each $x \leq \gamma$ the Kreĭn type extension $S_{K,x}$ is given by the strong resolvent limit

$$(S_{K,x} - \lambda)^{-1}h = \lim_{t \uparrow x} (S_{K,t} - \lambda)^{-1}h, \quad h \in \mathfrak{H}, \tag{5.4.19}$$

where $\lambda \in \mathbb{C} \setminus [x, \infty)$.

Proof. First the result in (5.4.19) will be shown. Let $x \leq \gamma$, let $\varepsilon > 0$ be arbitrary, and note that by Corollary 5.4.9

$$S_{K,x-\varepsilon} \leq S_{K,t} \leq S_{K,x}, \quad x - \varepsilon \leq t < x.$$

In particular, for $t \in [x-\varepsilon, x)$ the relations $S_{K,t}$ are bounded from below by $x-\varepsilon$. By the monotonicity of $S_{K,t}$ and Theorem 5.2.11, the strong resolvent limit of $S_{K,t}$ as $t \uparrow x$ exists for $\lambda \in \mathbb{C} \setminus [x - \varepsilon, \infty)$ and it is a semibounded self-adjoint relation S' with $x - \varepsilon \leq t \leq S_{K,t} \leq S'$. It will now be shown that

$$S' = S_{K,x}. \tag{5.4.20}$$

In fact, since $S_{K,t} \leq S_{K,x}$, there is a common upper bound and hence one has $S' \leq S_{K,x}$; see Corollary 5.2.12 (i). As $S \subset S_{K,t}$ for all $t < x$, this implies that $S \subset S'$ by Corollary 5.2.12 (ii). Thus, S' is a semibounded self-adjoint extension of S. Since $m(S_{K,t}) \leq m(S')$ for all $t < x$, it follows that $x \leq m(S')$ and, hence $S_{K,x} \leq S'$ by Theorem 5.4.6. Combining $S' \leq S_{K,x}$ and $S_{K,x} \leq S'$, it follows from Lemma 5.2.5 (iv) that (5.4.20) holds. This establishes (5.4.19) for $\lambda \in \mathbb{C} \setminus [x-\varepsilon, \infty)$. Since $\varepsilon > 0$ is arbitrary, one obtains (5.4.19).

Next, (5.4.18) will be shown. Apply the previous result (5.4.19) to the Kreĭn–von Neumann extension $((S-\gamma)^{-1})_{K,0}$ of the nonnegative relation $(S-\gamma)^{-1}$:

$$\big(((S-\gamma)^{-1})_{K,0} - \lambda\big)^{-1}h = \lim_{t \uparrow 0} \big(((S-\gamma)^{-1})_{K,t} - \lambda\big)^{-1}h, \quad h \in \mathfrak{H}, \tag{5.4.21}$$

where $\lambda \in \mathbb{C} \setminus [0, \infty)$. Then it follows from (5.4.1) (with $x = 0$ and S replaced by $(S-\gamma)^{-1}$) and the translation invariance property (5.3.4) of the Friedrichs extension that

$$((S-\gamma)^{-1})_{K,0} = ((S-\gamma)_F)^{-1} = (S_F - \gamma)^{-1}. \tag{5.4.22}$$

Likewise, using (5.4.8) and (5.4.10) one obtains for $t < 0$ that

$$((S-\gamma)^{-1})_{\mathrm{K},t} = ((S-\gamma)_{\mathrm{K},1/t})^{-1} = (S_{\mathrm{K},\gamma+1/t}-\gamma)^{-1}. \tag{5.4.23}$$

Substitute (5.4.22) and (5.4.23) into (5.4.21) and replace λ by $1/\lambda$:

$$\left((S_{\mathrm{F}}-\gamma)^{-1}-\frac{1}{\lambda}\right)^{-1} h = \lim_{t\uparrow 0}\left((S_{\mathrm{K},\gamma+1/t}-\gamma)^{-1}-\frac{1}{\lambda}\right)^{-1} h, \quad h \in \mathfrak{H}. \tag{5.4.24}$$

Now recall that for any relation H one has the identity

$$(H^{-1}-1/\lambda)^{-1} = -\lambda - \lambda^2(H-\lambda)^{-1}, \quad \lambda \neq 0,$$

see Corollary 1.1.12. Therefore, (5.4.24) yields

$$\left(S_{\mathrm{F}}-\gamma-\lambda\right)^{-1} h = \lim_{t\uparrow 0}\left(S_{\mathrm{K},\gamma+1/t}-\gamma-\lambda\right)^{-1},$$

where $\lambda \in \mathbb{C}\setminus[0,\infty)$, which is equivalent to (5.4.18). □

The next lemma and Proposition 5.4.12 below show that the convergence in (5.4.18) in Theorem 5.4.10 is uniform if the limit is a compact operator. First the case where the Kreĭn–von Neumann extension $S_{\mathrm{K},0}$ is compact is treated. There is in general no analog of Lemma 5.4.11 for the other Kreĭn type extensions $S_{\mathrm{K},x}$, $x \neq 0$, since the eigenspace $\ker(S_{\mathrm{K},x}-x) = \ker(S^*-x)$ for the eigenvalue $x \in \sigma_{\mathrm{p}}(S_{\mathrm{K},x})$ is infinite-dimensional whenever the defect numbers of S are infinite. Hence, $S_{\mathrm{K},x}$ cannot be compact for $x \neq 0$.

Lemma 5.4.11. *Let S be a bounded nonnegative operator in $\mathfrak{H}$ and assume that the Kreĭn–von Neumann extension $S_{\mathrm{K},0}$ is a compact operator. Then $S_{\mathrm{K},t} \in \mathbf{B}(\mathfrak{H})$ for $t < 0$ and*

$$\lim_{t\uparrow 0}\|S_{\mathrm{K},t}-S_{\mathrm{K},0}\| = 0.$$

Proof. Since $S_{\mathrm{K},0}$ is compact one has, in particular, $S_{\mathrm{K},0} \in \mathbf{B}(\mathfrak{H})$ and hence $S_{\mathrm{K},t} \in \mathbf{B}(\mathfrak{H})$ for $t < 0$; cf. Corollary 5.4.9 and Definition 5.2.3. By Theorem 5.4.10, the resolvents of $S_{\mathrm{K},t}$ converge in the strong sense to the resolvent of $S_{\mathrm{K},0}$ and since all operators belong to $\mathbf{B}(\mathfrak{H})$ it follows that $S_{\mathrm{K},t}$ converges strongly to $S_{\mathrm{K},0}$. In fact, strong resolvent convergence is equivalent to strong graph convergence by Corollary 1.9.6, and for operators in $\mathbf{B}(\mathfrak{H})$ this implies strong convergence. Now it will be shown that this convergence is uniform. Since $S_{\mathrm{K},0}$ is compact by assumption, for $\varepsilon > 0$ one can choose an orthogonal projection P_ε such that $\|S_{\mathrm{K},0}P_\varepsilon\| < \varepsilon$ and $I-P_\varepsilon$ is a finite-rank operator. Then it follows that the finite-rank operators

$$(S_{\mathrm{K},t}-S_{\mathrm{K},0})(I-P_\varepsilon) \quad \text{and} \quad (I-P_\varepsilon)(S_{\mathrm{K},t}-S_{\mathrm{K},0})P_\varepsilon \tag{5.4.25}$$

tend to zero uniformly as $t \uparrow 0$. For $t < 0$ one has $0 \leq S_{\mathrm{K},t}-t \leq S_{\mathrm{K},0}-t$ by Corollary 5.4.9, and hence

$$0 \leq P_\varepsilon(S_{\mathrm{K},t}-t)P_\varepsilon \leq P_\varepsilon(S_{\mathrm{K},0}-t)P_\varepsilon.$$

This implies

$$\begin{aligned}\|P_\varepsilon(S_{\mathrm{K},t}-S_{\mathrm{K},0})P_\varepsilon\| &\le \|P_\varepsilon(S_{\mathrm{K},t}-t)P_\varepsilon\| + \|P_\varepsilon(t-S_{\mathrm{K},0})P_\varepsilon\| \\ &\le 2\|P_\varepsilon(S_{\mathrm{K},0}-t)P_\varepsilon\| \\ &\le 2\varepsilon + 2|t|,\end{aligned}$$

and now the assertion follows together with (5.4.25) and the estimate

$$\begin{aligned}\|(S_{\mathrm{K},t}-S_{\mathrm{K},0})\| \le \|(S_{\mathrm{K},t}-S_{\mathrm{K},0})(I-P_\varepsilon)\| \\ + \|P_\varepsilon(S_{\mathrm{K},t}-S_{\mathrm{K},0})P_\varepsilon\| + \|(I-P_\varepsilon)(S_{\mathrm{K},t}-S_{\mathrm{K},0})P_\varepsilon\|.\end{aligned}$$

This completes the proof. □

The counterpart of Lemma 5.4.11 for the case where the Friedrichs extension S_{F} has a compact resolvent is provided next.

Proposition 5.4.12. *Let S be a semibounded relation in $\mathfrak{H}$ with lower bound γ and assume that the resolvent $(S_{\mathrm{F}}-\lambda)^{-1}$ of the Friedrichs extension S_{F} is compact for some, and hence for all $\lambda \in \mathbb{C}\setminus[\gamma,\infty)$. Then*

$$\lim_{t\downarrow-\infty} \|(S_{\mathrm{K},t}-\lambda)^{-1} - (S_{\mathrm{F}}-\lambda)^{-1}\| = 0.$$

Proof. It follows from the resolvent identity (see Theorem 1.2.6) that the resolvent of S_{F} is compact for all $\lambda \in \mathbb{C}\setminus[\gamma,\infty)$ if it is compact for some $\lambda \in \mathbb{C}\setminus[\gamma,\infty)$. Now let $x_0<\gamma$ and note that $(S-x_0)^{-1}$ is a bounded nonnegative operator. By (5.4.17) and (5.3.4) one has

$$\big((S-x_0)^{-1}\big)_{\mathrm{K},0} = \big((S-x_0)_{\mathrm{F}}\big)^{-1} = (S_{\mathrm{F}}-x_0)^{-1},$$

which is a compact operator by assumption. From Lemma 5.4.11 it follows that $((S-x_0)^{-1})_{\mathrm{K},x}$ converge uniformly to $((S-x_0)^{-1})_{\mathrm{K},0}$ when $x\uparrow 0$. This implies the assertion for $\lambda = x_0$, since

$$\begin{aligned}\lim_{x\uparrow 0}\big((S-x_0)^{-1}\big)_{\mathrm{K},x} &= \lim_{t\downarrow-\infty}\big((S-x_0)^{-1}\big)_{\mathrm{K},1/t} \\ &= \lim_{t\downarrow-\infty}\big((S-x_0)_{\mathrm{K},t}\big)^{-1} \\ &= \lim_{t\downarrow-\infty}\big(S_{\mathrm{K},t+x_0}-x_0\big)^{-1} \\ &= \lim_{t\downarrow-\infty}\big(S_{\mathrm{K},t}-x_0\big)^{-1},\end{aligned}$$

where (5.4.7) was used in the second equality and (5.4.9) was used in the third equality. The general case $\lambda \in \mathbb{C}\setminus[\gamma,\infty)$ follows with Lemma 1.11.4. □

Let S be a closed semibounded relation in $\mathfrak{H}$ with lower bound γ and let $x < \gamma$. Then $x \in \rho(S_{\rm F})$ is a point of regular type of S and one has that

$$S^* = S_{\rm F} \,\widehat{+}\, \widehat{\mathfrak{N}}_x(S^*) \quad \text{and} \quad S_{{\rm K},x} = S \,\widehat{+}\, \widehat{\mathfrak{N}}_x(S^*);$$

cf. Theorem 1.7.1 and (5.4.3). Therefore, it is clear that

$$S_{{\rm K},x} \,\widehat{+}\, S_{\rm F} = S^*, \quad x < \gamma; \tag{5.4.26}$$

in other words, the extensions $S_{{\rm K},x}$ and $S_{\rm F}$ are transversal for $x < \gamma$. For $x = \gamma$ the situation is different. Now it is possible that the extensions $S_{{\rm K},\gamma}$ and $S_{\rm F}$ are transversal, but it is also possible that they are not transversal because, for instance, the extensions $S_{{\rm K},\gamma}$ and $S_{\rm F}$ may even coincide. First the case of transversality is discussed.

Corollary 5.4.13. *Let S be a semibounded relation in $\mathfrak{H}$ with lower bound γ. Then the following statements hold:*

(i) *$S_{{\rm K},\gamma}$ and $S_{\rm F}$ are transversal if and only if* $\operatorname{dom} S^* \subset \operatorname{dom}(S_{{\rm K},\gamma} - \gamma)^{\frac{1}{2}}$;

(ii) *$S_{{\rm K},\gamma}$ and $S_{\rm F}$ are transversal and S is bounded if and only if $S_{{\rm K},\gamma} \in \mathbf{B}(\mathfrak{H})$.*

Proof. (i) This statement follows from Theorem 5.3.8.

(ii) Assume that $S_{\rm F}$ and $S_{{\rm K},\gamma}$ are transversal and that S is a bounded operator. Then part (i) shows that

$$\operatorname{dom} S^* \subset \operatorname{dom}(S_{{\rm K},\gamma} - \gamma)^{\frac{1}{2}}. \tag{5.4.27}$$

Since S^{**} is a bounded closed operator, $\operatorname{dom} S^{**}$ is closed, and hence so is $\operatorname{dom} S^*$; see Theorem 1.3.5. Moreover, $(\operatorname{dom} S^*)^\perp = \operatorname{mul} S^{**} = \{0\}$ implies that $\operatorname{dom} S^*$ is dense in $\mathfrak{H}$, so that $\operatorname{dom} S^* = \mathfrak{H}$. Then it follows from (5.4.27) that

$$\operatorname{dom}(S_{{\rm K},\gamma} - \gamma)^{\frac{1}{2}} = \mathfrak{H}.$$

Therefore, $\operatorname{dom} S_{{\rm K},\gamma} = \mathfrak{H}$ and hence $S_{{\rm K},\gamma} \in \mathbf{B}(\mathfrak{H})$.

Conversely, assume that $S_{{\rm K},\gamma} \in \mathbf{B}(\mathfrak{H})$. Then also S is bounded. Moreover, $\operatorname{dom} S^* \subset \mathfrak{H} = \operatorname{dom}(S_{{\rm K},\gamma} - \gamma)^{\frac{1}{2}}$, which together with (i) shows that $S_{\rm F}$ and $S_{{\rm K},\gamma}$ are transversal. □

The extreme case of equality of $S_{{\rm K},\gamma}$ and $S_{\rm F}$ is described in the following corollary.

Corollary 5.4.14. *Let S be a semibounded relation in $\mathfrak{H}$ with lower bound γ. Then the following statements hold:*

(i) *$S_{\rm F} = S_{{\rm K},\gamma}$ if and only if* $\ker(S^* - a) \cap \operatorname{dom}(S_{{\rm K},\gamma} - a)^{\frac{1}{2}} = \{0\}$ *for some, and hence for all $a < \gamma$;*

(ii) *$S_{\rm F} = S_{{\rm K},\gamma}$ and $S_{{\rm K},\gamma} \in \mathbf{B}(\mathfrak{H})$ if and only if $\overline{S} \in \mathbf{B}(\mathfrak{H})$.*

Proof. (i) This statement follows from Corollary 5.3.10.

(ii) Assume that $S_{\rm F} = S_{{\rm K},\gamma}$ and $S_{{\rm K},\gamma} \in \mathbf{B}(\mathfrak{H})$. The assumption $S_{{\rm K},\gamma} \in \mathbf{B}(\mathfrak{H})$ and Corollary 5.4.13 (ii) imply that $S_{\rm F}$ and $S_{{\rm K},\gamma}$ are transversal and S is bounded. Furthermore, $S_{\rm F} = S_{{\rm K},\gamma}$ implies that $S^* = S_{\rm F} \,\widehat{+}\, S_{{\rm K},\gamma} = S_{{\rm K},\gamma}$, so that $\overline{S} = S_{{\rm K},\gamma}$ is a self-adjoint operator in $\mathbf{B}(\mathfrak{H})$.

Conversely, if $\overline{S} \in \mathbf{B}(\mathfrak{H})$, then $\overline{S}$ is the only self-adjoint extension of S and hence $S_{\rm F} = S_{{\rm K},\gamma} = S^{**} \in \mathbf{B}(\mathfrak{H})$. □

In the next corollary, which is a special version of Corollary 5.3.11, the form $\mathfrak{t}_{S_{{\rm K},x}}$ for $x < \gamma$ corresponding to the Kreĭn type extensions $S_{{\rm K},x}$ is expressed in terms of the Friedrichs form $\mathfrak{t}_{S_{\rm F}}$.

Corollary 5.4.15. *Let S be a semibounded relation in $\mathfrak{H}$ with lower bound γ, let $x < \gamma$, and let $\mathfrak{t}_{S_{{\rm K},x}}$ be the form corresponding to $S_{{\rm K},x}$. Then*

$$\operatorname{dom} \mathfrak{t}_{S_{{\rm K},x}} = \ker (S^* - a) \oplus_{\mathfrak{t}_{S_{{\rm K},x}} - a} \operatorname{dom} \mathfrak{t}_{S_{\rm F}}, \qquad a < x, \tag{5.4.28}$$

and the restriction of $\mathfrak{t}_{S_{{\rm K},x}}$ to $\mathfrak{N}_a(S^*) = \ker (S^* - a)$ is represented by the bounded self-adjoint operator

$$L_a = P_{\mathfrak{N}_a(S^*)} \big(x + (x-a)^2 (S_{\rm F} - x)^{-1} \big) \iota_{\mathfrak{N}_a(S^*)} \in \mathbf{B}(\mathfrak{N}_a(S^*)), \tag{5.4.29}$$

where $\iota_{\mathfrak{N}_a(S^*)}$ is the canonical embedding of $\mathfrak{N}_a(S^*)$ into $\mathfrak{H}$ and $P_{\mathfrak{N}_a(S^*)}$ is the orthogonal projection onto $\mathfrak{N}_a(S^*)$. Furthermore,

$$\begin{aligned} \mathfrak{t}_{S_{{\rm K},x}}[f,g] - a(f,g) = (x-a) \big(\big(I + (x-a)(S_{\rm F} - x)^{-1} \big) f_a, g_a \big) \\ + \mathfrak{t}_{S_{\rm F}}[f_{\rm F}, g_{\rm F}] - a(f_{\rm F}, g_{\rm F}) \end{aligned} \tag{5.4.30}$$

holds for all $f = f_a + f_{\rm F}, g = g_a + g_{\rm F} \in \operatorname{dom} \mathfrak{t}_{S_{{\rm K},x}}$, where $f_a, g_a \in \ker (S^* - a)$ and $f_{\rm F}, g_{\rm F} \in \operatorname{dom} \mathfrak{t}_{S_{\rm F}}$.

Proof. The decomposition (5.4.28) is clear from Corollary 5.3.11, since $S_{{\rm K},x}$ and $S_{\rm F}$ are transversal for $x < \gamma$. Next it will be shown that the representing operator for the restriction of $\mathfrak{t}_{S_{{\rm K},x}}$ to $\mathfrak{N}_a(S^*)$ is given by (5.4.29); then (5.3.24) in Corollary 5.3.11 also leads to (5.4.30).

In order to verify (5.4.29) consider $f_a, g_a \in \mathfrak{N}_a(S^*)$ and let

$$f_x = \big(I + (x-a)(S_{\rm F} - x)^{-1} \big) f_a.$$

Then $f_x \in \mathfrak{N}_x(S^*)$ and $f_a = (I + (a-x)(S_{\rm F} - a)^{-1}) f_x$ by Lemma 1.4.10. Moreover, since $S_{{\rm K},x}$ is representing the form $\mathfrak{t}_{S_{{\rm K},x}}$ and $f_x \in \operatorname{dom} S_{{\rm K},x}$ one has $\mathfrak{t}_{S_{{\rm K},x}}[f_x, g_a] = (x f_x, g_a)$. Using $(S_{\rm F} - a)^{-1} f_x \in \operatorname{dom} \mathfrak{t}_{S_{\rm F}}$ and $g_a \in \mathfrak{N}_a(S^*)$ the orthogonal decomposition (5.4.28) yields $((S_{\rm F} - a)^{-1} f_x, g_a)_{\mathfrak{t}_{S_{{\rm K},x}} - a} = 0$.

Now one computes

$$\begin{aligned}
\mathfrak{t}_{S_{\mathrm{K},x}}[f_a, g_a] &= \mathfrak{t}_{S_{\mathrm{K},x}}[f_x, g_a] + (a-x)\mathfrak{t}_{S_{\mathrm{K},x}}\big[(S_{\mathrm{F}}-a)^{-1} f_x, g_a\big] \\
&= (x f_x, g_a) + (a-x)a\big((S_{\mathrm{F}}-a)^{-1} f_x, g_a\big) \\
&= \big(x f_x + a(f_a - f_x), g_a\big) \\
&= \big((x-a)\big(I + (x-a)(S_{\mathrm{F}}-x)^{-1}\big) f_a + a f_a, g_a\big) \\
&= \big(\big(x + (x-a)^2 (S_{\mathrm{F}}-x)^{-1}\big) f_a, g_a\big),
\end{aligned}$$

which implies (5.4.29). □

Finally, the decomposition (5.4.28) in the previous corollary is used to show a similar direct sum decomposition for $a = x$.

Corollary 5.4.16. *Let S be a semibounded relation in $\mathfrak{H}$ with lower bound γ, let $x < \gamma$, and let $\mathfrak{t}_{S_{\mathrm{K},x}}$ be the form corresponding to $S_{\mathrm{K},x}$. Then*

$$\mathrm{dom}\, \mathfrak{t}_{S_{\mathrm{K},x}} = \ker (S^* - x) + \mathrm{dom}\, \mathfrak{t}_{S_{\mathrm{F}}} \tag{5.4.31}$$

is a direct sum decomposition.

Proof. Let $a < x < \gamma$. Recall that the decomposition (5.4.28) holds since $S_{\mathrm{K},x}$ and S_{F} are transversal.

It is clear that the right-hand side of (5.4.31) is contained in the left-hand side. Observe for this that

$$\mathrm{dom}\, \mathfrak{t}_{S_{\mathrm{F}}} \subset \mathrm{dom}\, \mathfrak{t}_{S_{\mathrm{K},x}} \quad \text{and} \quad \ker (S^* - x) \subset \mathrm{dom}\, S_{\mathrm{K},x} \subset \mathrm{dom}\, \mathfrak{t}_{S_{\mathrm{K},x}}.$$

To show that the left-hand side of (5.4.31) is contained in the right-hand side, let $f \in \mathrm{dom}\, \mathfrak{t}_{S_{\mathrm{K},x}}$. According to (5.4.28) one has $f = f_a + f_{\mathrm{F}}$, with $f_a \in \ker (S^* - a)$ and $f_{\mathrm{F}} \in \mathrm{dom}\, \mathfrak{t}_{S_{\mathrm{F}}}$. Define

$$f_x = (I + (x-a)(S_{\mathrm{F}} - x)^{-1}) f_a.$$

Then

$$f_x \in \ker (S^* - x) \quad \text{and} \quad f = f_x + (f_a - f_x + f_{\mathrm{F}}),$$

where the last term is in $\mathrm{dom}\, t_{S_{\mathrm{F}}}$ since $f_a - f_x \in \mathrm{dom}\, S_{\mathrm{F}}$. Hence, the left-hand side of (5.4.31) belongs to the right-hand side. Thus, the sum decomposition (5.4.31) has been shown.

Finally, it will be shown that the sum decomposition (5.4.31) is direct. For this assume that $f_x \in \ker (S^* - x)$ is nontrivial and belongs to $\mathrm{dom}\, \mathfrak{t}_{S_{\mathrm{F}}}$. Then $\{f_x, x f_x\} \in S^*$ implies that $\{f_x, x f_x\} \in S_{\mathrm{F}}$; cf. Theorem 5.3.3. Since $x < \gamma$, this is a contradiction. □

5.5 Boundary triplets for semibounded relations

In this section semibounded self-adjoint extensions of semibounded symmetric relations are studied in the context of boundary triplets and their Weyl functions. The initial observations are general results about a closed symmetric relation S with a boundary triplet $\{\mathcal{G},\Gamma_0,\Gamma_1\}$, where $A_0 = \ker\Gamma_0$ is semibounded. In particular, the Friedrichs and the Kreĭn type extensions will be identified. In the remaining part of the section it will be assumed that S is a semibounded relation with a boundary triplet $\{\mathcal{G},\Gamma_0,\Gamma_1\}$, where $A_0 = S_{\rm F}$, and various specific properties are derived. The case where the self-adjoint extension $A_1 = \ker\Gamma_1$ is also semibounded is of specific interest. As a preparation for the main results in the following section it will be explained how the corresponding semibounded form is the first stepping stone to the notion of boundary pair.

Let S be a closed symmetric relation in a Hilbert space $\mathfrak{H}$ and let $\{\mathcal{G},\Gamma_0,\Gamma_1\}$ be a boundary triplet for S^*. Assume that $A_0 = \ker\Gamma_0$ is semibounded with lower bound $\gamma_0 = m(A_0)$. Then clearly S is semibounded and $\gamma_0 \le m(S) = \gamma$. Therefore, one may speak of the Friedrichs extension $S_{\rm F}$ so that $\gamma_0 \le m(S_{\rm F}) = \gamma$ and the Kreĭn type extensions $S_{{\rm K},x}$ of S with $x \le \gamma$. The corresponding Weyl function M is holomorphic on $\rho(A_0)$ and, in particular, on $\mathbb{C}\setminus[\gamma_0,\infty)$. Moreover, one has

$$M(x) = \Gamma(\widehat{\mathfrak{N}}_x(S^*)) = \Gamma(S \mathbin{\widehat{+}} \widehat{\mathfrak{N}}_x(S^*)) = \Gamma(S_{{\rm K},x}), \quad x < \gamma_0; \tag{5.5.1}$$

cf. Definition 2.3.4 and (5.4.3). By Corollary 2.3.8, the mapping $x \mapsto M(x)$ from $(-\infty,\gamma_0)$ to $\mathbf{B}(\mathcal{G})$ is nondecreasing. In particular, by Corollary 5.2.14 the limit $M(-\infty)$ exists in the strong resolvent sense,

$$\big(M(-\infty)-\lambda\big)^{-1} = \lim_{x\downarrow-\infty}\,(M(x)-\lambda)^{-1}, \tag{5.5.2}$$

and the limit $M(\gamma_0)$ exists in the strong resolvent sense,

$$\big(M(\gamma_0)-\lambda\big)^{-1} = \lim_{x\uparrow\gamma_0}\,(M(x)-\lambda)^{-1}, \tag{5.5.3}$$

where $\lambda\in\mathbb{C}\setminus[\gamma_0,\infty)$. Then $M(-\infty)$ and $M(\gamma_0)$ are self-adjoint relations in $\mathcal{G}$; cf. Theorem 5.2.11 and Corollary 5.2.14. In the following theorem the Friedrichs extension $S_{\rm F}$ and the Kreĭn type extension $S_{{\rm K},x}$ with $x \le \gamma_0$ will be characterized by means of the limits in (5.5.2) and (5.5.3).

Theorem 5.5.1. *Let S be a closed semibounded relation in $\mathfrak{H}$ with lower bound γ. Let $\{\mathcal{G},\Gamma_0,\Gamma_1\}$ be a boundary triplet for S^* and let M be the corresponding Weyl function. Assume that the self-adjoint extension $A_0 = \ker\Gamma_0$ is semibounded with $\gamma_0 = m(A_0) \le m(S_{\rm F}) = \gamma$. Then the Friedrichs extension $S_{\rm F}$ of S is given by*

$$S_{\rm F} = \big\{\widehat{f}\in S^* : \{\Gamma_0\widehat{f},\Gamma_1\widehat{f}\}\in M(-\infty)\big\} \tag{5.5.4}$$

and the Kreĭn type extension $S_{{\rm K},x}$ of S with $x \le \gamma_0$ is given by

$$S_{{\rm K},x} = \big\{\widehat{f}\in S^* : \{\Gamma_0\widehat{f},\Gamma_1\widehat{f}\}\in M(x)\big\}. \tag{5.5.5}$$

Proof. According to Theorem 5.4.10, the Friedrichs extension S_{F} of S is given by the strong resolvent limit

$$(S_{\mathrm{F}} - \lambda)^{-1}h = \lim_{t \downarrow -\infty} (S_{\mathrm{K},t} - \lambda)^{-1}h, \quad h \in \mathfrak{H}, \tag{5.5.6}$$

and for each $x \leq \gamma_0$ the Kreĭn type extension $S_{\mathrm{K},x}$ is given by the strong resolvent limit

$$(S_{\mathrm{K},x} - \lambda)^{-1}h = \lim_{t \uparrow x} (S_{\mathrm{K},t} - \lambda)^{-1}h, \quad h \in \mathfrak{H}, \tag{5.5.7}$$

where $\lambda \in \mathbb{C} \setminus [x, \infty)$. The idea behind the proof of the theorem is to connect the limit formulas in (5.5.2) and (5.5.3) with the limit formulas in (5.5.6) and (5.5.7). This in fact will be done by means of the Kreĭn formula. For this purpose observe that for $t < \gamma_0$ and $\lambda \in \mathbb{C} \setminus \mathbb{R}$ the resolvent formula in Theorem 2.6.1 for $S_{\mathrm{K},t}$ reads

$$(S_{\mathrm{K},t} - \lambda)^{-1} = (A_0 - \lambda)^{-1} + \gamma(\lambda)\big(M(t) - M(\lambda)\big)^{-1}\gamma(\bar{\lambda})^*, \tag{5.5.8}$$

due to (5.5.1). Here $(M(t) - M(\lambda))^{-1} \in \mathbf{B}(\mathcal{G})$ by Theorem 2.6.1 and Theorem 2.6.2.

First consider the Kreĭn type extension $S_{\mathrm{K},x}$ of S. If $x < \gamma_0$, then the formula (5.5.5) is a direct consequence of (5.5.1). To treat the case $x = \gamma_0$ let Θ_{K} be the self-adjoint relation in $\mathcal{G}$ which corresponds to S_{K,γ_0}, that is,

$$S_{\mathrm{K},\gamma_0} = \big\{ \widehat{f} \in S^* : \{\Gamma_0 \widehat{f}, \Gamma_1 \widehat{f}\} \in \Theta_{\mathrm{K}} \big\}. \tag{5.5.9}$$

Then again by the resolvent formula in Theorem 2.6.1 one has

$$(S_{\mathrm{K},\gamma_0} - \lambda)^{-1} = (A_0 - \lambda)^{-1} + \gamma(\lambda)\big(\Theta_{\mathrm{K}} - M(\lambda)\big)^{-1}\gamma(\bar{\lambda})^* \tag{5.5.10}$$

for $\lambda \in \mathbb{C} \setminus \mathbb{R}$. Here $(\Theta_{\mathrm{K}} - M(\lambda))^{-1} \in \mathbf{B}(\mathcal{G})$ by Theorem 2.6.1 and Theorem 2.6.2. Subtracting (5.5.10) from (5.5.8) leads to

$$\begin{aligned} &(S_{\mathrm{K},t} - \lambda)^{-1} - (S_{\mathrm{K},\gamma_0} - \lambda)^{-1} \\ &\qquad = \gamma(\lambda)\big[\big(M(t) - M(\lambda)\big)^{-1} - \big(\Theta_{\mathrm{K}} - M(\lambda)\big)^{-1}\big]\gamma(\bar{\lambda})^* \end{aligned} \tag{5.5.11}$$

with $t < \gamma_0$. Now take the strong limit for $t \uparrow \gamma_0$ and apply (5.5.7). Then for each $h \in \mathfrak{H}$

$$(S_{\mathrm{K},t} - \lambda)^{-1}h \to (S_{\mathrm{K},\gamma_0} - \lambda)^{-1}h \quad \text{as} \quad t \uparrow \gamma_0,$$

which, via (5.5.11), leads to

$$\gamma(\lambda)\big[\big(M(t) - M(\lambda)\big)^{-1} - \big(\Theta_{\mathrm{K}} - M(\lambda)\big)^{-1}\big]\gamma(\bar{\lambda})^* h \to 0 \quad \text{as} \quad t \uparrow \gamma_0.$$

Since $\gamma(\lambda)$ maps $\mathcal{G}$ isomorphically onto $\ker(S^* - \lambda)$ and $\gamma(\bar{\lambda})^* : \mathfrak{H} \to \mathcal{G}$ is surjective, see Proposition 2.3.2, it follows that for each $\varphi \in \mathcal{G}$

$$\big(M(t) - M(\lambda)\big)^{-1}\varphi \to \big(\Theta_{\mathrm{K}} - M(\lambda)\big)^{-1}\varphi \quad \text{as} \quad t \uparrow \gamma_0. \tag{5.5.12}$$

Next the parameter Θ_K will be identified with $M(\gamma_0)$. For this purpose, observe that for $\varphi \in \mathcal{G}$

$$\{(M(t) - M(\lambda))^{-1}\varphi, \varphi + M(\lambda)(M(t) - M(\lambda))^{-1}\varphi\} \in M(t). \tag{5.5.13}$$

As $t \uparrow \gamma_0$, the components on the left-hand side of (5.5.13) converge due to (5.5.12), while the bounded operators $M(t)$ converge in the strong resolvent sense, and hence in the graph sense (see Corollary 1.9.6) to the self-adjoint relation $M(\gamma_0)$. Hence, (5.5.13) implies

$$\{(\Theta_K - M(\lambda))^{-1}\varphi, \varphi + M(\lambda)(\Theta_K - M(\lambda))^{-1}\varphi\} \in M(\gamma_0) \tag{5.5.14}$$

for all $\varphi \in \mathcal{G}$, and thus $(\Theta_K - M(\lambda))^{-1} \subset (M(\gamma_0) - M(\lambda))^{-1}$. Since $M(\lambda) \in \mathbf{B}(\mathcal{G})$, it follows that $\Theta_K \subset M(\gamma_0)$, or, since both relations are self-adjoint, $\Theta_K = M(\gamma_0)$. Now (5.5.5) for $x = \gamma_0$ follows from (5.5.9).

Next consider the Friedrichs extension S_F of S. Let Θ_F be the self-adjoint relation in $\mathcal{G}$ which corresponds to S_F, that is,

$$S_F = \{\widehat{f} \in S^* : \{\Gamma_0\widehat{f}, \Gamma_1\widehat{f}\} \in \Theta_F\}.$$

Then again by the resolvent formula in Theorem 2.6.1 one has

$$(S_F - \lambda)^{-1} = (A_0 - \lambda)^{-1} + \gamma(\lambda)\big(\Theta_F - M(\lambda)\big)^{-1}\gamma(\bar{\lambda})^* \tag{5.5.15}$$

for $\lambda \in \mathbb{C} \setminus \mathbb{R}$. As above, $(\Theta_F - M(\lambda))^{-1} \in \mathbf{B}(\mathcal{G})$. Subtracting (5.5.15) from (5.5.8) and using the same reasoning as above involving Theorem 5.4.10 yields

$$\big(M(t) - M(\lambda)\big)^{-1}\varphi \to \big(\Theta_F - M(\lambda)\big)^{-1}\varphi \quad \text{as} \quad t \downarrow -\infty.$$

From the fact that $M(-\infty)$ is the strong resolvent limit, and hence the strong graph limit of $M(t)$ when $t \downarrow -\infty$, one concludes $\Theta_F = M(-\infty)$ in the same way as in (5.5.13)–(5.5.14). This shows (5.5.4). □

The statements in the following corollary are consequences of Theorem 5.5.1 and Proposition 2.1.8.

Corollary 5.5.2. *Let S be a closed symmetric relation, let $\{\mathcal{G}, \Gamma_0, \Gamma_1\}$ be a boundary triplet for S^*, and let M be the corresponding Weyl function. Assume that the self-adjoint extension $A_0 = \ker \Gamma_0$ is semibounded with lower bound γ_0. Then the following statements hold:*

(i) *$A_0 = S_F$ if and only if $M(-\infty) = \{0\} \times \mathcal{G}$;*
(ii) *$A_0 \cap S_F = S$ if and only if $M(-\infty)$ is a closed operator;*
(iii) *$A_0 \widehat{+} S_F = S^*$ if and only if $M(-\infty) \in \mathbf{B}(\mathcal{G})$,*

and, similarly

(iv) *$A_0 = S_{K,\gamma_0}$ if and only if $M(\gamma_0) = \{0\} \times \mathcal{G}$;*
(v) *$A_0 \cap S_{K,\gamma_0} = S$ if and only if $M(\gamma_0)$ is a closed operator;*
(vi) *$A_0 \widehat{+} S_{K,\gamma_0} = S^*$ if and only if $M(\gamma_0) \in \mathbf{B}(\mathcal{G})$.*

Moreover, for $A_1 = \ker \Gamma_1$ *one has*

(vii) $A_1 = S_{\rm F}$ *if and only if* $M(-\infty) = \mathcal{G} \times \{0\}$*;*

(viii) $A_1 = S_{{\rm K},\gamma_0}$ *if and only if* $M(\gamma_0) = \mathcal{G} \times \{0\}$.

Such facts can also be stated in terms of the limit behavior of $M(-\infty)$ and $M(\gamma_0)$ via Corollary 5.2.14 applied to the Weyl function M. For this purpose recall the notations

$$\mathfrak{E}_{\gamma_0} = \left\{ \varphi \in \mathcal{G} : \lim_{x \uparrow \gamma_0} (M(x)\varphi, \varphi) < \infty \right\},$$

$$\mathfrak{E}_{-\infty} = \left\{ \varphi \in \mathcal{G} : \lim_{x \downarrow -\infty} (M(x)\varphi, \varphi) > -\infty \right\}.$$

Corollary 5.5.3. *Let* S *be a closed symmetric relation, let* $\{\mathcal{G}, \Gamma_0, \Gamma_1\}$ *be a boundary triplet for* S^**, and let* M *be the corresponding Weyl function. Assume that the self-adjoint extension* $A_0 = \ker \Gamma_0$ *is semibounded with lower bound* γ_0*. Then the following statements hold:*

(i) $A_0 = S_{\rm F}$ *if and only if* $\mathfrak{E}_{-\infty} = \{0\}$*;*

(ii) $A_0 \cap S_{\rm F} = S$ *if and only if* $\operatorname{clos} \mathfrak{E}_{-\infty} = \mathcal{G}$*;*

(iii) $A_0 \widehat{+} S_{\rm F} = S^*$ *if and only if* $\mathfrak{E}_{-\infty} = \mathcal{G}$*,*

and, similarly

(iv) $A_0 = S_{{\rm K},\gamma_0}$ *if and only if* $\mathfrak{E}_{\gamma_0} = \{0\}$*;*

(v) $A_0 \cap S_{{\rm K},\gamma_0} = S$ *if and only if* $\operatorname{clos} \mathfrak{E}_{\gamma_0} = \mathcal{G}$*;*

(vi) $A_0 \widehat{+} S_{{\rm K},\gamma_0} = S^*$ *if and only if* $\mathfrak{E}_{\gamma_0} = \mathcal{G}$.

In the context of Theorem 5.5.1, the Weyl function of the boundary triplet is holomorphic on the interval $(-\infty, \gamma_0)$. In this situation the inverse result in Theorem 4.2.4 can be formulated as follows.

Proposition 5.5.4. *Let* $\mathcal{G}$ *be a Hilbert space and let* M *be a uniformly strict* $\mathbf{B}(\mathcal{G})$*-valued Nevanlinna function, which is holomorphic on* $\mathbb{C} \setminus [\gamma_0, \infty)$ *and not holomorphic at* γ_0*. Then there exist a Hilbert space* $\mathfrak{H}$*, a closed simple symmetric operator* S *in* $\mathfrak{H}$*, and a boundary triplet* $\{\mathcal{G}, \Gamma_0, \Gamma_1\}$ *for* S^* *such that* A_0 *is a semibounded self-adjoint relation with lower bound* $m(A_0) = \gamma_0$ *and* M *is the corresponding Weyl function.*

Proof. Let M be a uniformly strict Nevanlinna function with values in $\mathbf{B}(\mathcal{G})$. Let $\mathfrak{H}(\mathsf{N}_M)$ be the reproducing kernel Hilbert space associated with the Nevanlinna kernel

$$\frac{M(\lambda) - M(\mu)^*}{\lambda - \bar{\mu}}, \quad \lambda, \mu \in \mathbb{C} \setminus \mathbb{R}.$$

By Theorem 4.2.4, there exist a closed simple symmetric operator S in the reproducing kernel Hilbert space $\mathfrak{H}(\mathsf{N}_M)$ and a boundary triplet $\{\mathcal{G}, \Gamma_0, \Gamma_1\}$ for S^*

such that M is the corresponding Weyl function. The assumption that M is holomorphic on $(-\infty, \gamma_0)$ and the fact that S is simple imply, by Theorem 3.6.1, that $(-\infty, \gamma_0) \subset \rho(A_0)$. Moreover, since M is not holomorphic at γ_0, one has $\gamma_0 \in \sigma(A_0)$. Therefore, the self-adjoint relation A_0 is semibounded with lower bound $m(A_0) = \gamma_0$. □

The context of Theorem 5.5.1 will now be narrowed. Let S be a closed semibounded relation in $\mathfrak{H}$. Then the existence of a semibounded self-adjoint extension of S is guaranteed by the Friedrichs extension S_F of S. The interest in the rest of this section is in boundary triplets $\{\mathcal{G}, \Gamma_0, \Gamma_1\}$ for which $A_0 = S_\mathrm{F}$. The following result is a consequence of Theorem 2.4.1 since for any self-adjoint extension H of S there is a boundary triplet $\{\mathcal{G}, \Gamma_0, \Gamma_1\}$ for S^* such that $H = \ker \Gamma_0$.

Corollary 5.5.5. *Let S be a closed semibounded relation in $\mathfrak{H}$ with lower bound γ. Then there exists a boundary triplet $\{\mathcal{G}, \Gamma_0, \Gamma_1\}$ for S^* such that*

$$S_\mathrm{F} = \ker \Gamma_0.$$

The corresponding Weyl function M is holomorphic on $(-\infty, \gamma)$ and the mapping $x \mapsto M(x)$ from $(-\infty, \gamma)$ to $\mathbf{B}(\mathcal{G})$ is nondecreasing, while $M(-\infty) = \{0\} \times \mathcal{G}$.

The following result will be useful in treating the connection between semibounded self-adjoint extensions and the Weyl function.

Proposition 5.5.6. *Let S be a closed semibounded relation in $\mathfrak{H}$, let $\{\mathcal{G}, \Gamma_0, \Gamma_1\}$ be a boundary triplet for S^* such that $S_\mathrm{F} = \ker \Gamma_0$, and let M be the corresponding Weyl function. Let A_Θ be a self-adjoint extension of S corresponding to the self-adjoint relation Θ in $\mathcal{G}$ and assume that $x < m(S)$. Then $M(x) \in \mathbf{B}(\mathcal{G})$ and the following equivalence holds:*

$$x \leq A_\Theta \quad \Leftrightarrow \quad M(x) \leq \Theta.$$

In particular, if A_Θ is semibounded in $\mathfrak{H}$, then Θ is semibounded in $\mathcal{G}$.

Proof. The assumption $x < m(S) = m(S_\mathrm{F})$ implies that $x \in \rho(S_\mathrm{F})$ and hence $M(x) \in \mathbf{B}(\mathcal{G})$ is clear. The formula in Theorem 2.6.1, applied to A_Θ and S_F, gives

$$(A_\Theta - x)^{-1} - (S_\mathrm{F} - x)^{-1} = \gamma(x)(\Theta - M(x))^{-1}\gamma(x)^*, \tag{5.5.16}$$

since $(S_\mathrm{F} - x)^{-1} \in \mathbf{B}(\mathfrak{H})$. Recall that $\gamma(x)^*$ maps $\ker (S^* - x)$ onto $\mathcal{G}$; see Proposition 2.3.2. Note that if, in addition, $x \in \rho(A_\Theta)$, then $(\Theta - M(x))^{-1} \in \mathbf{B}(\mathcal{G})$.

($\Rightarrow$) Since A_Θ is a semibounded self-adjoint extension of S, it follows from Proposition 5.3.6 that $A_\Theta \leq S_\mathrm{F}$. Observe that for $x \in \rho(A_\Theta)$

$$0 \leq (S_\mathrm{F} - x)^{-1} \leq (A_\Theta - x)^{-1},$$

where both operators belong to $\mathbf{B}(\mathfrak{H})$ since $x \in \rho(S_{\mathrm{F}}) \cap \rho(A_\Theta)$. Hence, it then follows from (5.5.16) that

$$(\Theta - M(x))^{-1} \geq 0 \quad \text{or, equivalently,} \quad \Theta - M(x) \geq 0.$$

Since $M(x) \in \mathbf{B}(\mathcal{G})$ it follows that $M(x) \leq \Theta$; cf. Proposition 5.2.6.

Now let $x \leq A_\Theta$. First assume that $x < m(A_\Theta)$. In this case, $x \in \rho(A_\Theta)$ and thus $M(x) \leq \Theta$. Next, assume that $x = m(A_\Theta)$ and consider an increasing sequence x_n whose limit is x. Then clearly $M(x_n) \leq \Theta$ and thus $M(x) \leq \Theta$; cf. Corollary 5.2.12 (i).

($\Leftarrow$) Assume that $M(x) \leq \Theta$. Then $\Theta - M(x) \geq 0$ by Proposition 5.2.6 and hence $(\Theta - M(x))^{-1} \geq 0$. It is straightforward to see that the right-hand side of (5.5.16) is a nonnegative relation. Thus, the relation

$$(A_\Theta - x)^{-1} - (S_{\mathrm{F}} - x)^{-1}$$

on the left-hand side of (5.5.16) is also nonnegative and, in fact, this relation is also self-adjoint since $(A_\Theta - x)^{-1}$ is self-adjoint and $(S_{\mathrm{F}} - x)^{-1} \in \mathbf{B}(\mathfrak{H})$ is self-adjoint. Therefore, one concludes with the help of Proposition 5.2.6 and $x < m(S_{\mathrm{F}})$ that

$$0 \leq (S_{\mathrm{F}} - x)^{-1} \leq (A_\Theta - x)^{-1}.$$

In particular, this shows that $0 \leq A_\Theta - x$ or $x \leq A_\Theta$. □

From Proposition 5.5.6 one sees that if the self-adjoint extension A_Θ is semibounded in $\mathfrak{H}$, then the corresponding self-adjoint relation Θ is semibounded in $\mathcal{G}$. The converse is not true in general; cf. Remark 5.6.16. However, in Proposition 5.5.8 below it will be shown that the converse holds if S has finite defect numbers or S_{F} has a compact resolvent. The following result is a preliminary observation.

Lemma 5.5.7. *Let S be a closed semibounded relation in $\mathfrak{H}$ with lower bound γ and let $\{\mathcal{G}, \Gamma_0, \Gamma_1\}$ be a boundary triplet for S^* such that $S_{\mathrm{F}} = \ker \Gamma_0$. Let M be the corresponding Weyl function and assume that for any $C > 0$ there exists $x_1 < \gamma$ such that*

$$M(x) \leq -C, \qquad x \leq x_1. \tag{5.5.17}$$

Then for every semibounded self-adjoint relation Θ in $\mathcal{G}$ the corresponding self-adjoint extension A_Θ is semibounded from below.

Proof. Let Θ be a self-adjoint relation in $\mathcal{G}$ with lower bound ν and choose $C > 0$ in (5.5.17) such that $-C < \nu \leq \Theta$. For all $x \leq x_1$ one then has

$$0 < \nu + C \leq \Theta + C \leq \Theta - M(x),$$

which implies that $\Theta - M(x)$ is boundedly invertible for all $x \leq x_1$. From Theorem 2.6.2 one concludes $(-\infty, x_1) \in \rho(A_\Theta)$ and hence A_Θ is semibounded from below. This conclusion also follows from Proposition 5.5.6. □

Now it will be shown that the condition (5.5.17) holds when S has finite defect numbers or S_{F} has a compact resolvent.

Proposition 5.5.8. *Let S be a closed semibounded relation in $\mathfrak{H}$ with lower bound γ and let $\{\mathcal{G}, \Gamma_0, \Gamma_1\}$ be a boundary triplet for S^* such that $S_{\mathrm{F}} = \ker \Gamma_0$. Assume that one of the following conditions hold:*

(i) *$\mathcal{G}$ is finite-dimensional;*

(ii) *$(S_{\mathrm{F}} - \lambda)^{-1}$ is compact for some, and hence for all, $\lambda \in \rho(S_{\mathrm{F}})$.*

Then for any $C > 0$ there exists $x_1 < \gamma$ such that (5.5.17) holds. In particular, if (i) *holds, then all self-adjoint extensions of S in $\mathfrak{H}$ are semibounded from below, or if* (ii) *holds and Θ is a semibounded self-adjoint relation in $\mathcal{G}$, then the self-adjoint extension A_Θ of S is semibounded from below.*

Proof. (i) As $S_{\mathrm{F}} = \ker \Gamma_0$, one has $(M(x)\varphi, \varphi) \to -\infty$ for $x \to -\infty$ and all $\varphi \in \mathcal{G}$ by Corollary 5.5.3 (i). Since $\mathcal{G}$ is finite-dimensional, a compactness argument shows that there exists $x_1 < \gamma$ such that (5.5.17) holds. Every self-adjoint relation Θ in the finite-dimensional space $\mathcal{G}$ is semibounded and hence it follows from Lemma 5.5.7 that all self-adjoint extensions A_Θ are semibounded.

(ii) Recall from Proposition 5.4.12 that the resolvents of the Kreĭn type extensions $S_{\mathrm{K},t}$ converge uniformly to the resolvent of S_{F}, that is, for all $\lambda \in \mathbb{C} \setminus [\gamma, \infty)$ one has

$$\lim_{t \downarrow -\infty} \|(S_{\mathrm{K},t} - \lambda)^{-1} - (S_{\mathrm{F}} - \lambda)^{-1}\| = 0. \tag{5.5.18}$$

In the following fix some $\lambda = x_0 < m(S_{\mathrm{F}})$ and note that, by (5.5.18), there exists $t' < x_0$ such that $x_0 \in \rho(S_{\mathrm{K},t})$ for all $t \leq t'$. Using (5.5.1) it follows that the resolvent of $S_{\mathrm{K},t}$ has the form

$$(S_{\mathrm{K},t} - x_0)^{-1} - (S_{\mathrm{F}} - x_0)^{-1} = \gamma(x_0)\big(M(t) - M(x_0)\big)^{-1}\gamma(x_0)^*,$$

where $(M(t) - M(x_0))^{-1} \in \mathbf{B}(\mathcal{G})$ for all $t \leq t'$ by Theorem 2.6.1 and Theorem 2.6.2. Since $\gamma(x_0)$ maps $\mathcal{G}$ isomorphically to $\mathfrak{N}_{x_0}(S^*)$ and $\gamma(x_0)^*$ maps $\mathfrak{N}_{x_0}(S^*)$ isomorphically to $\mathcal{G}$, it follows together with (5.5.18) that

$$\lim_{t \downarrow -\infty} \big\|\big(M(t) - M(x_0)\big)^{-1}\big\| = 0. \tag{5.5.19}$$

This implies that for any $C > 0$ there exists $x_1 < \gamma$ such that (5.5.17) holds. In fact, otherwise there exists some $C_0 > 0$ and a sequence $s_n \to -\infty$ such that $s_n < t' < x_0$ and $M(s_n) > -C_0$. Then the estimate

$$-C_0 - \|M(x_0)\| \leq -C_0 - M(x_0) \leq M(s_n) - M(x_0) \leq 0$$

and $(M(s_n) - M(x_0))^{-1} \in \mathbf{B}(\mathcal{G})$ contradict (5.5.19). Therefore, the condition (5.5.17) is satisfied and if Θ is a semibounded self-adjoint relation in $\mathcal{G}$, then by Lemma 5.5.7 the corresponding self-adjoint extension A_Θ is semibounded. □

In the case of Corollary 5.5.5 the relationship between the Friedrichs extension S_F and the Kreĭn type extension $S_{K,\gamma}$ is described in the following corollary, which is a translation of (iv)–(vi) in Corollary 5.5.2; cf. Corollary 5.4.13 and Corollary 5.4.14.

Corollary 5.5.9. *Let S be a closed semibounded relation in $\mathfrak{H}$. Let $\{\mathcal{G}, \Gamma_0, \Gamma_1\}$ be a boundary triplet for S^* such that $S_F = \ker \Gamma_0$ and let M be the corresponding Weyl function. Let $\gamma = m(S_F)$. Then the following statements hold:*

(i) *S_F and $S_{K,\gamma}$ coincide if and only if $M(\gamma) = \{0\} \times \mathcal{G}$;*

(ii) *S_F and $S_{K,\gamma}$ are disjoint if and only if $M(\gamma)$ is a closed operator;*

(iii) *S_F and $S_{K,\gamma}$ are transversal if and only if $M(\gamma) \in \mathbf{B}(\mathcal{G})$.*

In general it may not be possible to simultaneously prescribe $\ker \Gamma_0$ as the Friedrichs extension S_F and $\ker \Gamma_1$ as the Kreĭn type extension $S_{K,\gamma}$, since $\ker \Gamma_0$ and $\ker \Gamma_1$ are necessarily transversal; cf. Section 2.1. However, note that the Friedrichs extension S_F and the Kreĭn type extension $S_{K,x}$ for $x < \gamma$ are automatically transversal; cf. (5.4.26).

Proposition 5.5.10. *Let S be a closed semibounded relation in $\mathfrak{H}$ with lower bound γ. Then the following statements hold:*

(i) *For $x < \gamma$ there exists a boundary triplet $\{\mathcal{G}, \Gamma_0, \Gamma_1\}$ for S^* such that*

$$S_F = \ker \Gamma_0 \quad \text{and} \quad S_{K,x} = \ker \Gamma_1. \tag{5.5.20}$$

(ii) *If S_F and $S_{K,\gamma}$ are transversal, then there exists a boundary triplet $\{\mathcal{G}, \Gamma_0, \Gamma_1\}$ for S^* such that*

$$S_F = \ker \Gamma_0 \quad \text{and} \quad S_{K,\gamma} = \ker \Gamma_1.$$

In both cases the corresponding Weyl function satisfies $M(-\infty) = \{0\} \times \mathcal{G}$ and $M(x) = \mathcal{G} \times \{0\}$, $x \le \gamma$. In particular, $M(t) \le 0$ for all $t \le x$, i.e., M belongs to the class $\mathbf{S}_{\mathcal{G}}^{-1}(-\infty, x)$ of inverse Stieltjes functions.

Proof. (i) The extensions S_F and $S_{K,x}$ for $x < \gamma$ are automatically transversal according to (5.4.26). Hence, it follows from Theorem 2.5.9 that there exists a boundary triplet $\{\mathcal{G}, \Gamma_0, \Gamma_1\}$ for S^* such that (5.5.20) holds.

(ii) Since it is assumed that S_F and $S_{K,\gamma}$ are transversal, Theorem 2.5.9 yields the statement.

The Weyl function M satisfies $M(-\infty) = \{0\} \times \mathcal{G}$ and $M(x) = \mathcal{G} \times \{0\}$ as a consequence of Corollary 5.5.2. Finally, that $M(t) \le 0$ for all $t \le x$ is a consequence of the monotonicity of the Weyl function M in Corollary 2.3.8. The assertion $M \in \mathbf{S}_{\mathcal{G}}^{-1}(-\infty, x)$ is immediate from the definition of the inverse Stieltjes class in Definition A.6.1. □

The following corollary is a consequence of Proposition 5.5.4 and Corollary 5.5.2.

Corollary 5.5.11. *Let $\mathcal{G}$ be a Hilbert space and let M be a uniformly strict $\mathbf{B}(\mathcal{G})$-valued Nevanlinna function, which is holomorphic on $\mathbb{C}\setminus[\gamma,\infty)$ and not holomorphic at γ. Assume, in addition, that*

$$M(-\infty)=\{0\}\times\mathcal{G} \quad \textit{and} \quad M(\gamma)=\mathcal{G}\times\{0\}. \tag{5.5.21}$$

Then there exist a Hilbert space $\mathfrak{H}$, a closed simple semibounded operator S in $\mathfrak{H}$ with lower bound γ, and a boundary triplet $\{\mathcal{G},\Gamma_0,\Gamma_1\}$ for S^* with $S_{\rm F}=\ker\Gamma_0$ and $S_{{\rm K},\gamma}=\ker\Gamma_1$, such that M is the corresponding Weyl function.

Proof. It follows from Proposition 5.5.4 that there exist a Hilbert space $\mathfrak{H}$, a closed simple symmetric operator S in $\mathfrak{H}$, and a boundary triplet $\{\mathcal{G},\Gamma_0,\Gamma_1\}$ for S^* such that $A_0=\ker\Gamma_0$ is a semibounded self-adjoint relation with lower bound $m(A_0)=\gamma$ and M is the corresponding Weyl function. The assumptions in (5.5.21) and Corollary 5.5.2 (i) and (viii) imply $S_{\rm F}=\ker\Gamma_0=A_0$ and $S_{{\rm K},\gamma}=\ker\Gamma_1$. Since $m(S_{\rm F})=m(A_0)=\gamma$, it is also clear that the symmetric operator S is semibounded with lower bound γ. □

In the next corollary a boundary triplet with the properties as in Proposition 5.5.10 (i) is exhibited.

Corollary 5.5.12. *Let S be a closed semibounded relation in $\mathfrak{H}$ with lower bound γ. Then*

$$S^*=S_{\rm F}\,\widehat{+}\,\widehat{\mathfrak{N}}_x(S^*),\qquad x<\gamma, \tag{5.5.22}$$

is a direct sum decomposition. Let $\widehat{f}=\{f,f'\}\in S^*$ have the unique decomposition

$$\widehat{f}=\widehat{f}_{\rm F}+\widehat{f}_x,$$

with $\widehat{f}_{\rm F}=\{f_{\rm F},f'_{\rm F}\}\in S_{\rm F}$ and $\widehat{f}_x=\{f_x,xf_x\}\in\widehat{\mathfrak{N}}_x(S^*)$. Then

$$\Gamma_0\widehat{f}=f_x \quad \textit{and} \quad \Gamma_1\widehat{f}=P_{\mathfrak{N}_x(S^*)}(f'_{\rm F}-xf_{\rm F})$$

defines a boundary triplet $\{\mathfrak{N}_x(S^*),\Gamma_0,\Gamma_1\}$ for S^* such that (5.5.20) holds. For $\lambda\in\rho(S_{\rm F})$ the corresponding γ-field γ is given by

$$\gamma(\lambda)=\bigl(I+(\lambda-x)(S_{\rm F}-\lambda)^{-1}\bigr)\iota_{\mathfrak{N}_x(S^*)}, \tag{5.5.23}$$

and the corresponding Weyl function M is given by

$$M(\lambda)=\lambda-x+(\lambda-x)^2P_{\mathfrak{N}_x(S^*)}(S_{\rm F}-\lambda)^{-1}\iota_{\mathfrak{N}_x(S^*)}. \tag{5.5.24}$$

Proof. It is clear from Theorem 1.7.1 that (5.5.22) is a direct sum decomposition. Now choose $\mu=x$ in Theorem 2.4.1 and modify the boundary triplet $\{\mathfrak{N}_x(S^*),\Gamma_0,\Gamma_1\}$ in Theorem 2.4.1 to $\{\mathfrak{N}_x(S^*),\Gamma_0,\Gamma_1-x\Gamma_0\}$. Then $S_{\rm F}=\ker\Gamma_0$

and the corresponding γ-field and Weyl function have the form (5.5.23) and (5.5.24); cf. Theorem 2.4.1 and Corollary 2.5.5. It is easy to see that

$$S \,\widehat{+}\, \widehat{\mathfrak{N}}_x(S^*) \subset \ker \Gamma_1,$$

and since $S_{\mathrm{K},x} = S \,\widehat{+}\, \widehat{\mathfrak{N}}_x(S^*)$ and $\ker \Gamma_1$ are both self-adjoint, one concludes that $S_{\mathrm{K},x} = \ker \Gamma_1$, so that (5.5.20) holds. □

The following example is an illustration of Proposition 5.5.10 (i) and Corollary 5.5.12 for the case where the semibounded relation S is uniformly positive. In this situation it is convenient to have a boundary triplet for which the Kreĭn–von Neumann extension $S_{\mathrm{K},0}$ corresponds to the boundary mapping Γ_1; cf. Chapter 8.

Example 5.5.13. Let S be a closed nonnegative symmetric relation in $\mathfrak{H}$ with lower bound $\gamma > 0$. In this case the Kreĭn–von Neumann extension $S_{\mathrm{K},0}$ is given by $S_{\mathrm{K},0} = S \,\widehat{+}\, \widehat{\mathfrak{N}}_0(S^*)$; cf. (5.4.3). Moreover, the Friedrichs extension S_{F} and the Kreĭn–von Neumann extension $S_{\mathrm{K},0}$ are transversal by (5.4.26). For $x = 0$ Corollary 5.5.12 shows that $\{\mathfrak{N}_0(S^*), \Gamma_0, \Gamma_1\}$, where

$$\Gamma_0 \widehat{f} = f_0 \quad \text{and} \quad \Gamma_1 \widehat{f} = P_{\mathfrak{N}_0(S^*)} f'_{\mathrm{F}}, \qquad \widehat{f} = \{f_{\mathrm{F}}, f'_{\mathrm{F}}\} + \{f_0, 0\},$$

is a boundary triplet for $S^* = S_{\mathrm{F}} \,\widehat{+}\, \widehat{\mathfrak{N}}_0(S^*)$ such that

$$S_{\mathrm{F}} = \ker \Gamma_0 \quad \text{and} \quad S_{\mathrm{K},0} = \ker \Gamma_1;$$

moreover, for $\lambda \in \rho(S_{\mathrm{F}})$ the corresponding γ-field γ is given by

$$\gamma(\lambda) = \big(I + \lambda(S_{\mathrm{F}} - \lambda)^{-1}\big)\iota_{\mathfrak{N}_0(S^*)},$$

and the corresponding Weyl function M is given by

$$M(\lambda) = \lambda + \lambda^2 P_{\mathfrak{N}_0(S^*)}(S_{\mathrm{F}} - \lambda)^{-1}\iota_{\mathfrak{N}_0(S^*)}.$$

Note that, in particular, $\gamma(0) = \iota_{\mathfrak{N}_0(S^*)}$ is the canonical embedding of $\mathfrak{N}_0(S^*)$ into $\mathfrak{H}$, $\gamma(0)^* = P_{\mathfrak{N}_0(S^*)}$ is the orthogonal projection onto $\mathfrak{N}_0(S^*)$, and $M(0) = 0$.

The last objective in this section is to derive an *abstract first Green identity.* For this consider a boundary triplet $\{\mathcal{G}, \Gamma_0, \Gamma_1\}$ for which $S_{\mathrm{F}} = \ker \Gamma_0$ and assume that also the self-adjoint extension corresponding to Γ_1 is semibounded. In the following the notation $S_1 = \ker \Gamma_1$ (instead of A_1) is used; this will turn out to be more convenient for the next section. As a first step rewrite the abstract Green identity (2.1.1) in the form

$$(f', g) - (\Gamma_1 \widehat{f}, \Gamma_0 \widehat{g}) = (f, g') - (\Gamma_0 \widehat{f}, \Gamma_1 \widehat{g}), \quad \widehat{f}, \widehat{g} \in S^*. \tag{5.5.25}$$

In the following theorem it will be shown that the expression on the left-hand side, and hence on the right-hand side, of (5.5.25) can be seen as a restriction of the form $\mathfrak{t}_{S_1}$.

Theorem 5.5.14. *Let S be a closed semibounded relation in $\mathfrak{H}$ and let $\{\mathcal{G}, \Gamma_0, \Gamma_1\}$ be a boundary triplet for S^* such that*

$$S_{\rm F} = \ker \Gamma_0 \quad and \quad S_1 = \ker \Gamma_1, \tag{5.5.26}$$

where $S_{\rm F}$ is the Friedrichs extension and S_1 is a semibounded self-adjoint extension of S. Moreover, let $\mathfrak{t}_{S_1}$ be the closed semibounded form corresponding to S_1. Then $\operatorname{dom} S^* \subset \operatorname{dom} \mathfrak{t}_{S_1}$ *and the following equality holds:*

$$(f', g) = \mathfrak{t}_{S_1}[f, g] + (\Gamma_1 \widehat{f}, \Gamma_0 \widehat{g}), \quad \widehat{f}, \widehat{g} \in S^*. \tag{5.5.27}$$

Proof. By the assumption (5.5.26), the extensions $S_{\rm F}$ and S_1 are transversal and hence it follows from Theorem 5.3.8 that $\operatorname{dom} S^* \subset \operatorname{dom} \mathfrak{t}_{S_1}$. Moreover, every $\widehat{f}, \widehat{g} \in S^*$ can be decomposed as

$$\widehat{f} = \widehat{f}_{\rm F} + \widehat{f}_1, \quad \widehat{g} = \widehat{g}_{\rm F} + \widehat{g}_1, \quad \widehat{f}_{\rm F}, \widehat{g}_{\rm F} \in S_{\rm F}, \quad \widehat{f}_1, \widehat{g}_1 \in S_1.$$

Using the conditions in (5.5.26) one sees that

$$\Gamma_0 \widehat{f}_{\rm F} = \Gamma_0 \widehat{g}_{\rm F} = 0 \quad \text{and} \quad \Gamma_1 \widehat{f}_1 = \Gamma_1 \widehat{g}_1 = 0, \tag{5.5.28}$$

and therefore the identity (5.5.25) can be rewritten as

$$\begin{aligned} (f', g) - (\Gamma_1 \widehat{f}, \Gamma_0 \widehat{g}) &= (f, g') - (\Gamma_0 \widehat{f}, \Gamma_1 \widehat{g}) \\ &= (f_{\rm F} + f_1, g_{\rm F}' + g_1') - (\Gamma_0 \widehat{f}_1, \Gamma_1 \widehat{g}_{\rm F}). \end{aligned} \tag{5.5.29}$$

In order to rewrite the last term on the right-hand side of (5.5.29) observe that $\widehat{f}_1, \widehat{g}_{\rm F} \in S^*$. Therefore, another application of the abstract Green identity for the boundary triplet $\{\mathcal{G}, \Gamma_0, \Gamma_1\}$ shows that

$$\begin{aligned} (f_1', g_{\rm F}) - (f_1, g_{\rm F}') &= (\Gamma_1 \widehat{f}_1, \Gamma_0 \widehat{g}_{\rm F}) - (\Gamma_0 \widehat{f}_1, \Gamma_1 \widehat{g}_{\rm F}) \\ &= -(\Gamma_0 \widehat{f}_1, \Gamma_1 \widehat{g}_{\rm F}), \end{aligned} \tag{5.5.30}$$

where (5.5.28) was used in the last equality. A combination of (5.5.29) and (5.5.30) gives

$$(f', g) - (\Gamma_1 \widehat{f}, \Gamma_0 \widehat{g}) = (f_{\rm F}, g_{\rm F}') + (f_{\rm F}, g_1') + (f_1, g_1') + (f_1', g_{\rm F}). \tag{5.5.31}$$

Since $\operatorname{dom} S^* \subset \operatorname{dom} \mathfrak{t}_{S_1}$, the last three terms on the right-hand side of (5.5.31) can be rewritten by means of Theorem 5.1.18:

$$(f_{\rm F}, g_1') = \mathfrak{t}_{S_1}[f_{\rm F}, g_1], \quad (f_1, g_1') = \mathfrak{t}_{S_1}[f_1, g_1], \quad (f_1', g_{\rm F}) = \mathfrak{t}_{S_1}[f_1, g_{\rm F}],$$

whereas the first term on the right-hand side of (5.5.31) can be rewritten by means of Theorem 5.1.18 and the inclusion $\mathfrak{t}_{S_{\rm F}} \subset \mathfrak{t}_{S_1}$ as follows:

$$(f_{\rm F}, g_{\rm F}') = \mathfrak{t}_{S_{\rm F}}[f_{\rm F}, g_{\rm F}] = \mathfrak{t}_{S_1}[f_{\rm F}, g_{\rm F}].$$

Combined with (5.5.31), the above rewriting leads to

$$\begin{aligned}(f',g)-(\Gamma_1\widehat{f},\Gamma_0\widehat{g}) &= \mathfrak{t}_{S_1}[f_{\mathrm{F}},g_{\mathrm{F}}]+\mathfrak{t}_{S_1}[f_{\mathrm{F}},g_1]+\mathfrak{t}_{S_1}[f_1,g_1]+\mathfrak{t}_{S_1}[f_1,g_{\mathrm{F}}]\\ &= \mathfrak{t}_{S_1}[f_{\mathrm{F}}+f_1,g_{\mathrm{F}}+g_1]\\ &= \mathfrak{t}_{S_1}[f,g],\end{aligned}$$

and hence (5.5.27) has been shown. □

Let Θ be a self-adjoint relation in $\mathcal{G}$ and consider the corresponding self-adjoint extension

$$H_\Theta=\big\{\widehat{f}\in S^*:\{\Gamma_0\widehat{f},\Gamma_1\widehat{f}\}\in\Theta\big\};\tag{5.5.32}$$

here the notation H_Θ (instead of A_Θ) is used, which is more convenient for the next section. For $\widehat{f}\in S^*$ the condition $\{\Gamma_0\widehat{f},\Gamma_1\widehat{f}\}\in\Theta$ is equivalent to

$$\Gamma_0\widehat{f}\in\operatorname{dom}\Theta_{\mathrm{op}}\quad\text{and}\quad P_{\mathrm{op}}\Gamma_1\widehat{f}=\Theta_{\mathrm{op}}\Gamma_0\widehat{f},\tag{5.5.33}$$

where P_{op} denotes the orthogonal projection from $\mathcal{G}$ onto $\mathcal{G}_{\mathrm{op}}=\overline{\operatorname{dom}\Theta}$ and Θ_{op} is the self-adjoint operator part of the self-adjoint relation Θ; cf. the end of Section 2.2. The following statement is an immediate consequence of Theorem 5.5.14.

Corollary 5.5.15. *Let S be a closed semibounded relation in $\mathfrak{H}$ and let $\{\mathcal{G},\Gamma_0,\Gamma_1\}$ be a boundary triplet for S^* such that $S_{\mathrm{F}}=\ker\Gamma_0$ and $S_1=\ker\Gamma_1$, where S_{F} is the Friedrichs extension and S_1 is a semibounded self-adjoint extension of S. Let $\mathfrak{t}_{S_1}$ be the closed semibounded form corresponding to S_1. Assume that H_Θ is a self-adjoint extension of S corresponding to the self-adjoint relation Θ in $\mathcal{G}$. Then*

$$(f',g)=\mathfrak{t}_{S_1}[f,g]+(\Theta_{\mathrm{op}}\Gamma_0\widehat{f},\Gamma_0\widehat{g}),\quad \widehat{f},\widehat{g}\in H_\Theta.\tag{5.5.34}$$

Under the assumption that H_Θ is semibounded, the left-hand side of (5.5.34) can be written as $\mathfrak{t}_{H_\Theta}[f,g]$. One may view the identity (5.5.34) as a perturbation of the form $\mathfrak{t}_{S_1}$ by means of the term $(\Theta_{\mathrm{op}}\Gamma_0\widehat{f},\Gamma_0\widehat{g})$. The proper interpretation of (5.5.34) in terms of quadratic forms requires an extension of the mapping Γ_0; this procedure will be taken up in detail in Section 5.6 with the introduction of the notion of a boundary pair.

5.6 Boundary pairs and boundary triplets

In this section the notion of a boundary pair for a semibounded symmetric relation in a Hilbert space $\mathfrak{H}$ is developed. It will turn out that there is an intimate connection between boundary pairs and boundary triplets. In fact, a boundary pair helps to express the closed semibounded form associated with a semibounded self-adjoint extension in terms of the parameter provided by the boundary triplet. The concept of a boundary pair is motivated by applications occurring in the study of semibounded differential operators.

Definition 5.6.1. Let S be a closed semibounded relation in $\mathfrak{H}$ and let S_1 be a semibounded self-adjoint extension of S such that S_1 and the Friedrichs extension S_F are transversal. Let $\mathfrak{t}_{S_1}$ be the closed form associated with S_1 and let

$$\mathfrak{H}_{\mathfrak{t}_{S_1}-a} = \bigl(\mathrm{dom}\,\mathfrak{t}_{S_1}, (\cdot,\cdot)_{\mathfrak{t}_{S_1}-a}\bigr), \quad a < m(S_1),$$

be the corresponding Hilbert space. A pair $\{\mathcal{G}, \Lambda\}$ is called a *boundary pair* for S corresponding to S_1 if $\mathcal{G}$ is a Hilbert space and $\Lambda \in \mathbf{B}(\mathfrak{H}_{\mathfrak{t}_{S_1}-a}, \mathcal{G})$ satisfies

$$\ker \Lambda = \mathrm{dom}\,\mathfrak{t}_{S_\mathrm{F}} \quad \text{and} \quad \mathrm{ran}\,\Lambda = \mathcal{G}.$$

Let S be a closed semibounded relation in $\mathfrak{H}$, let S_F be the Friedrichs extension of S, and assume that $\{\mathcal{G}, \Lambda\}$ is a boundary pair for S corresponding to a semibounded self-adjoint extension S_1 of S. Then, by Definition 5.6.1, the semibounded self-adjoint extensions S_1 and S_F are transversal, which, by Theorem 5.3.8, is equivalent to the inclusion

$$\mathrm{dom}\,S^* \subset \mathrm{dom}\,\mathfrak{t}_{S_1} = \mathrm{dom}\,\Lambda.$$

One also has the orthogonal decomposition

$$\mathfrak{H}_{\mathfrak{t}_{S_1}-a} = \ker(S^* - a) \oplus_{\mathfrak{t}_{S_1}-a} \mathfrak{H}_{\mathfrak{t}_{S_\mathrm{F}}-a}, \quad a < m(S_1) \le m(S_\mathrm{F});$$

cf. Proposition 5.3.7 and Theorem 5.3.8. Since $\ker \Lambda = \mathrm{dom}\,\mathfrak{t}_{S_\mathrm{F}}$ and $\mathrm{ran}\,\Lambda = \mathcal{G}$, one sees that the restriction of Λ to the space $\ker(S^* - a)$ equipped with the norm $\|\cdot\|_{\mathfrak{t}_{S_1}-a}$, is a bounded mapping from $\ker(S^* - a)$ to $\mathcal{G}$ such that

$$\mathrm{ran}\,\bigl(\Lambda\!\restriction\!\ker(S^* - a)\bigr) = \mathcal{G}.$$

Hence, the restriction $\Lambda\!\restriction\!\ker(S^* - a)$ has a bounded everywhere defined inverse.

Boundary pairs have a useful invariance property. To see this consider a pair of semibounded self-adjoint extensions S_1 and S_2 of S which are each transversal with S_F, i.e.,

$$S^* = S_1 \,\widehat{+}\, S_\mathrm{F} = S_2 \,\widehat{+}\, S_\mathrm{F}. \tag{5.6.1}$$

Lemma 5.6.2. *Let S be a closed semibounded relation in $\mathfrak{H}$, let S_1 and S_2 be semibounded self-adjoint extensions which satisfy the transversality conditions (5.6.1), and assume that $a < \min\{m(S_1), m(S_2)\}$. Then $\mathrm{dom}\,\mathfrak{t}_{S_1} = \mathrm{dom}\,\mathfrak{t}_{S_2}$ and the form topologies of $\mathfrak{t}_{S_1}$ and $\mathfrak{t}_{S_2}$ coincide. Consequently, $\{\mathcal{G}, \Lambda\}$ is a boundary pair for S corresponding to S_1 if and only if $\{\mathcal{G}, \Lambda\}$ is a boundary pair for S corresponding to S_2.*

Proof. It suffices to consider the boundedness property, as the other properties of a boundary pair in Definition 5.6.1 do not depend on the choice of the transversal extensions S_1 and S_2. In fact, according to Corollary 5.3.9, the transversality conditions in (5.6.1) imply that

$$\mathrm{dom}\,(S_1 - a)^{\frac{1}{2}} = \mathrm{dom}\,(S_2 - a)^{\frac{1}{2}}, \quad a < \min\{m(S_1), m(S_2)\}.$$

Moreover, again by Corollary 5.3.9, this implies that

$$c_1\|((S_1)_{\rm op}-a)^{\frac{1}{2}}\varphi\|^2 \le \|((S_2)_{\rm op}-a)^{\frac{1}{2}}\varphi\|^2 \le c_2\|((S_1)_{\rm op}-a)^{\frac{1}{2}}\varphi\|^2$$

for all $\varphi \in \operatorname{dom}(S_2-a)^{\frac{1}{2}} = \operatorname{dom}(S_1-a)^{\frac{1}{2}}$ and for some constants $c_1, c_2 > 0$. In other words, $c_1(\mathfrak{t}_{S_1}-a) \le \mathfrak{t}_{S_2}-a \le c_2(\mathfrak{t}_{S_1}-a)$ for some constants $c_1, c_2 > 0$. Therefore, the form topologies of $\mathfrak{t}_{S_1}$ and $\mathfrak{t}_{S_2}$ coincide and thus $\Lambda \in \mathbf{B}(\mathfrak{H}_{\mathfrak{t}_{S_1}-a}, \mathcal{G})$ if and only if $\Lambda \in \mathbf{B}(\mathfrak{H}_{\mathfrak{t}_{S_2}-a}, \mathcal{G})$. This implies that $\{\mathcal{G}, \Lambda\}$ is a boundary pair for S corresponding to S_1 if and only if $\{\mathcal{G}, \Lambda\}$ is a boundary pair for S corresponding to S_2. □

To explore the connection between boundary pairs and boundary triplets, the notion of extension in the next definition will be important.

Definition 5.6.3. Let S be a closed semibounded relation in $\mathfrak{H}$, let $\{\mathcal{G}, \Gamma_0, \Gamma_1\}$ be a boundary triplet for S^* with $A_0 = \ker \Gamma_0$ and assume that $\operatorname{mul} A_0 = \operatorname{mul} S^*$. Let S_1 be a semibounded self-adjoint extension of S such that S_1 and $S_{\rm F}$ are transversal, so that, $\operatorname{dom} S^* \subset \operatorname{dom} \mathfrak{t}_{S_1}$. Then an operator $\Lambda \in \mathbf{B}(\mathfrak{H}_{\mathfrak{t}_{S_1}-a}, \mathcal{G})$ is said to be an *extension* of Γ_0 if

$$\Gamma_0 \widehat{f} = \Lambda f \quad \text{for all } \widehat{f} = \{f, f'\} \in S^*. \tag{5.6.2}$$

It follows already from the assumption $\operatorname{mul} A_0 = \operatorname{mul} S^*$ that the mapping $f \mapsto \Gamma_0 \widehat{f}$ from $\operatorname{dom} S^*$ to $\mathcal{G}$ in (5.6.2) is an operator. In fact, if $\widehat{f} = \{0, f'\} \in S^*$, then $\widehat{f} \in A_0 = \ker \Gamma_0$, so that $\Gamma_0 \widehat{f} = 0$. Note also that in the case $A_0 = S_{\rm F}$ the condition $\operatorname{mul} A_0 = \operatorname{mul} S^*$ is satisfied by Theorem 5.3.3.

Next the notion of a compatible boundary pair will be introduced.

Definition 5.6.4. Let S be a closed semibounded relation in $\mathfrak{H}$ and let $\{\mathcal{G}, \Gamma_0, \Gamma_1\}$ be a boundary triplet for S^* with

$$A_0 = \ker \Gamma_0 \quad \text{and} \quad A_1 = \ker \Gamma_1.$$

Let S_1 be a semibounded self-adjoint extension of S such that S_1 and $S_{\rm F}$ are transversal and let $\{\mathcal{G}, \Lambda\}$ be a boundary pair for S corresponding to S_1. Then $\{\mathcal{G}, \Gamma_0, \Gamma_1\}$ and $\{\mathcal{G}, \Lambda\}$ are said to be *compatible* if Λ is an extension of Γ_0 and the self-adjoint relations A_1 and S_1 coincide.

The next lemma provides a sufficient condition for an extension Λ of Γ_0 such that $\{\mathcal{G}, \Lambda\}$ is a boundary pair, or compatible boundary pair, for S corresponding to S_1. In the special case where the defect numbers of S are finite this condition is automatically satisfied, which makes the lemma useful in applications to Sturm–Liouville operators in Chapter 6.

Lemma 5.6.5. *Let S be a closed semibounded relation in $\mathfrak{H}$ and let $\{\mathcal{G}, \Gamma_0, \Gamma_1\}$ be a boundary triplet for S^* with $A_0 = \ker \Gamma_0$ and $A_1 = \ker \Gamma_1$. Let S_1 be a semibounded self-adjoint extension of S such that S_1 and $S_{\rm F}$ are transversal or, equivalently, $\operatorname{dom} S^* \subset \operatorname{dom} \mathfrak{t}_{S_1}$, and let $a < m(S_1)$. Then the following statements hold:*

(i) *If* $\Lambda \in \mathbf{B}(\mathfrak{H}_{\mathfrak{t}_{S_1}-a}, \mathcal{G})$ *is an extension of* Γ_0, *then* $\operatorname{ran}\Lambda = \mathcal{G}$ *and*

$$\operatorname{dom}\mathfrak{t}_{S_{\mathrm{F}}} \subset \ker\Lambda.$$

(ii) *If* $\Lambda \in \mathbf{B}(\mathfrak{H}_{\mathfrak{t}_{S_1}-a}, \mathcal{G})$ *is an extension of* Γ_0 *and*

$$\operatorname{dom}\mathfrak{t}_{S_{\mathrm{F}}} = \ker\Lambda, \tag{5.6.3}$$

then $A_0 = \ker\Gamma_0 = S_{\mathrm{F}}$ *and* $\{\mathcal{G}, \Lambda\}$ *is a boundary pair for* S *corresponding to* S_1. *If, in addition,* $A_1 = S_1$, *then* $\{\mathcal{G}, \Gamma_0, \Gamma_1\}$ *and* $\{\mathcal{G}, \Lambda\}$ *are compatible.*

In particular, if $\Lambda \in \mathbf{B}(\mathfrak{H}_{\mathfrak{t}_{S_1}-a}, \mathcal{G})$ *is an extension of* Γ_0 *and the defect numbers of* S *are finite, then* $A_0 = \ker\Gamma_0 = S_{\mathrm{F}}$ *and* $\{\mathcal{G}, \Lambda\}$ *is a boundary pair for* S *corresponding to* S_1. *If, in this case, also* $A_1 = S_1$, *then* $\{\mathcal{G}, \Gamma_0, \Gamma_1\}$ *and* $\{\mathcal{G}, \Lambda\}$ *are compatible.*

Proof. (i) Let $\Lambda \in \mathbf{B}(\mathfrak{H}_{\mathfrak{t}_{S_1}-a}, \mathcal{G})$ be an extension of Γ_0. It follows from the extension property (5.6.2) that $\operatorname{ran}\Lambda = \mathcal{G}$ and that

$$\operatorname{dom} S \subset \operatorname{dom} A_0 \subset \ker\Lambda. \tag{5.6.4}$$

Since $\Lambda \in \mathbf{B}(\mathfrak{H}_{\mathfrak{t}_{S_1}-a}, \mathcal{G})$, it is clear that $\ker\Lambda$ is closed in $\mathfrak{H}_{\mathfrak{t}_{S_1}-a}$ and, by (5.6.4), the closure of $\operatorname{dom} S$ with respect to the inner product $(\cdot,\cdot)_{\mathfrak{t}_{S_1}-a}$ is contained in $\ker\Lambda$. On the other hand, $\operatorname{dom} S \subset \operatorname{dom} S_{\mathrm{F}} \subset \operatorname{dom}\mathfrak{t}_{S_{\mathrm{F}}}$ and the inner product on $\mathfrak{H}_{\mathfrak{t}_{S_1}-a}$ restricted to $\mathfrak{H}_{\mathfrak{t}_{S_{\mathrm{F}}}-a}$ coincides with the inner product $(\cdot,\cdot)_{\mathfrak{t}_{S_{\mathrm{F}}}-a}$ in $\mathfrak{H}_{\mathfrak{t}_{S_{\mathrm{F}}}-a}$ (see the discussion below (5.3.10) and (5.3.11)). As $\operatorname{dom} S$ is a core of $\mathfrak{t}_{S_{\mathrm{F}}}$, it follows from Corollary 5.1.15 that the closure of $\operatorname{dom} S$ with respect to the inner product $(\cdot,\cdot)_{\mathfrak{t}_{S_{\mathrm{F}}}-a}$ coincides with $\operatorname{dom}\mathfrak{t}_{S_{\mathrm{F}}}$. Thus, one concludes $\operatorname{dom}\mathfrak{t}_{S_{\mathrm{F}}} \subset \ker\Lambda$.

(ii) Assume that $\Lambda \in \mathbf{B}(\mathfrak{H}_{\mathfrak{t}_{S_1}-a}, \mathcal{G})$ is an extension of Γ_0 and that (5.6.3) holds. Then

$$A_0 = \bigl\{\widehat{f} \in S^* : \widehat{f} \in \ker\Gamma_0\bigr\} \subset \bigl\{\widehat{f} \in S^* : f \in \ker\Lambda\bigr\}. \tag{5.6.5}$$

Since $\ker\Lambda = \operatorname{dom}\mathfrak{t}_{S_{\mathrm{F}}}$, it follows from Theorem 5.3.3 that the right-hand side of (5.6.5) concides with S_{F}. Thus, $A_0 \subset S_{\mathrm{F}}$ and since both relations are self-adjoint, it follows that $A_0 = S_{\mathrm{F}}$. Moreover, one has $\operatorname{ran}\Lambda = \mathcal{G}$ by (i) and hence $\{\mathcal{G}, \Lambda\}$ is a boundary pair for S^* corresponding to S_1. It follows directly from Definition 5.6.4 that the additional assumption $A_1 = S_1$ yields compatibility of $\{\mathcal{G}, \Gamma_0, \Gamma_1\}$ and $\{\mathcal{G}, \Lambda\}$.

For the last statement assume that the defect numbers of S are finite and let $\Lambda \in \mathbf{B}(\mathfrak{H}_{\mathfrak{t}_{S_1}-a}, \mathcal{G})$ be an extension of Γ_0. Then $\operatorname{ran}\Lambda = \mathcal{G}$ and $\operatorname{dom}\mathfrak{t}_{S_{\mathrm{F}}} \subset \ker\Lambda$ by (i). Since S_1 and S_{F} are transversal, one has the orthogonal decomposition

$$\mathfrak{H}_{\mathfrak{t}_{S_1}-a} = \ker(S^* - a) \oplus_{\mathfrak{t}_{S_1}-a} \mathfrak{H}_{\mathfrak{t}_{S_{\mathrm{F}}}-a}$$

for $a < m(S_1)$ by Proposition 5.3.7 and Theorem 5.3.8. Then

$$\dim\operatorname{ran}\Lambda = \dim\mathcal{G} = \dim\ker(S^* - a) < \infty$$

together with $\operatorname{dom}\mathfrak{t}_{S_{\mathrm{F}}} \subset \ker\Lambda$ implies $\operatorname{dom}\mathfrak{t}_{S_{\mathrm{F}}} = \ker\Lambda$. Now the assertions follow from (ii). □

If the boundary triplet $\{\mathcal{G}, \Gamma_0, \Gamma_1\}$ and the boundary pair $\{\mathcal{G}, \Lambda\}$ are compatible, then automatically $\ker \Gamma_0 = S_{\mathrm{F}}$. In the next theorem it will be shown that for a boundary triplet for S^* such that $\ker \Gamma_0 = S_{\mathrm{F}}$ and $\ker \Gamma_1 = S_1$ is semibounded, there exists a compatible boundary pair $\{\mathcal{G}, \Lambda\}$ for S corresponding to S_1.

Theorem 5.6.6. *Let S be a closed semibounded relation in $\mathfrak{H}$ and let $\{\mathcal{G}, \Gamma_0, \Gamma_1\}$ be a boundary triplet for S^*. Assume that*

$$\ker \Gamma_0 = S_{\mathrm{F}} \quad \textit{and} \quad \ker \Gamma_1 = S_1,$$

where S_{F} is the Friedrichs extension and S_1 is a semibounded self-adjoint extension of S. Then, with $a < m(S_1)$ fixed, the mapping

$$\Lambda_0 = \left\{\{f, \Gamma_0 \widehat{f}\} : \widehat{f} \in S^*\right\} \subset \mathfrak{H}_{\mathfrak{t}_{S_1}-a} \times \mathcal{G}, \tag{5.6.6}$$

is (the graph of) a densely defined bounded operator. Its unique bounded extension Λ to all of $\mathfrak{H}_{\mathfrak{t}_{S_1}-a}$ induces a boundary pair $\{\mathcal{G}, \Lambda\}$ for S corresponding to S_1 which is compatible with the boundary triplet $\{\mathcal{G}, \Gamma_0, \Gamma_1\}$.

Proof. The assumption that $\{\mathcal{G}, \Gamma_0, \Gamma_1\}$ is a boundary triplet for S^* implies that S_{F} and S_1 are transversal extensions of S. By Theorem 5.3.8, this is equivalent to $\operatorname{dom} S^* \subset \operatorname{dom} \mathfrak{t}_{S_1}$, and hence $\operatorname{dom} \Lambda_0 = \operatorname{dom} S^*$ is contained in $\mathfrak{H}_{\mathfrak{t}_{S_1}-a}$, i.e., the relation Λ_0 in (5.6.6) is well defined from $\mathfrak{H}_{\mathfrak{t}_{S_1}-a}$ to $\mathcal{G}$.

In order to see that Λ_0 is an operator, assume that $\{f, \Gamma_0 \widehat{f}\} \in \Lambda_0$ with $\widehat{f} \in S^*$ satisfying $f = 0$. Hence, $\widehat{f} = \{0, f'\} \in S^*$, which by Theorem 5.3.3 shows that $\widehat{f} \in S_{\mathrm{F}}$. Now the identity $S_{\mathrm{F}} = \ker \Gamma_0$ implies that $\Gamma_0 \widehat{f} = 0$. Hence, $\operatorname{mul} \Lambda_0 = \{0\}$, that is, Λ_0 in (5.6.6) is an operator. Furthermore, it is clear that $S_{\mathrm{F}} = \ker \Gamma_0$ yields $\ker \Lambda_0 = \operatorname{dom} S_{\mathrm{F}}$.

Next it will be shown that the operator

$$\Lambda_0 : \mathfrak{H}_{\mathfrak{t}_{S_1}-a} \supset \operatorname{dom} S^* \to \mathcal{G}, \qquad f \mapsto \Lambda_0 f = \Gamma_0 \widehat{f}, \tag{5.6.7}$$

is bounded. By assumption, $a < m(S_1) \leq m(S_{\mathrm{F}})$, and hence one has the decomposition $S^* = S_{\mathrm{F}} \,\widehat{+}\, \widehat{\mathfrak{N}}_a(S^*)$; see Theorem 1.7.1. Let $\widehat{f} = \{f, f'\} \in S^*$ and decompose it accordingly,

$$\widehat{f} = \widehat{f}_{\mathrm{F}} + \widehat{f}_a, \quad \widehat{f}_{\mathrm{F}} \in S_{\mathrm{F}}, \ \widehat{f}_a = \{f_a, a f_a\} \in \widehat{\mathfrak{N}}_a(S^*).$$

From $S_{\mathrm{F}} = \ker \Gamma_0$ and the fact that $\Gamma_0 : S^* \to \mathcal{G}$ is bounded (with respect to the graph norm; cf. Proposition 2.1.2) it follows that

$$\|\Lambda_0 f\|^2 = \|\Gamma_0 \widehat{f}\|^2 = \|\Gamma_0 \widehat{f}_a\|^2 \leq M(\|f_a\|^2 + a^2 \|f_a\|^2) \leq M' \|f_a\|^2$$

holds for some constants $M, M' > 0$. Then is clear from (5.1.9) that there exists $M'' > 0$ such that

$$\|\Lambda_0 f\|^2 \leq M'' \|f_a\|^2_{\mathfrak{t}_{S_1}-a} \leq M''(\|f_{\mathrm{F}}\|^2_{\mathfrak{t}_{S_1}-a} + \|f_a\|^2_{\mathfrak{t}_{S_1}-a}) = M'' \|f\|^2_{\mathfrak{t}_{S_1}-a},$$

where $f_{\mathrm{F}} \in \operatorname{dom} S_{\mathrm{F}} \subset \operatorname{dom} \mathfrak{t}_{S_{\mathrm{F}}}$; here in the last equality one uses the orthogonal decomposition

$$\operatorname{dom} \mathfrak{t}_{S_1} = \ker (S^* - a) \oplus_{\mathfrak{t}_{S_1}-a} \operatorname{dom} \mathfrak{t}_{S_{\mathrm{F}}}$$

with respect to the inner product $(\cdot,\cdot)_{\mathfrak{t}_{S_1}-a}$ in the Hilbert space $\mathfrak{H}_{\mathfrak{t}_{S_1}-a}$, see Proposition 5.3.7 and Theorem 5.3.8. This shows that the operator Λ_0 in (5.6.7) is bounded.

By Theorem 5.1.18 (ii), $\operatorname{dom} S_1$ is a core of $\mathfrak{t}_{S_1}$ and hence $\operatorname{dom} S_1$ is dense in the Hilbert space $\mathfrak{H}_{\mathfrak{t}_{S_1}-a}$ by Corollary 5.1.15. As $\operatorname{dom} S_1 \subset \operatorname{dom} S^* \subset \operatorname{dom} \mathfrak{t}_{S_1}$, it follows that $\operatorname{dom} S^*$ is also a dense subspace of $\mathfrak{H}_{\mathfrak{t}_{S_1}-a}$. Therefore, the bounded operator Λ_0 in (5.6.7) is densely defined, and hence Λ_0 admits a unique extension Λ by continuity to all of $\mathfrak{H}_{\mathfrak{t}_{S_1}-a}$.

Since the restriction of the inner product in $\mathfrak{H}_{\mathfrak{t}_{S_1}-a}$ to $\mathfrak{H}_{\mathfrak{t}_{S_{\mathrm{F}}}-a}$ coincides with the inner product in $\mathfrak{H}_{\mathfrak{t}_{S_{\mathrm{F}}}-a}$ (see the discussion below (5.3.10) and (5.3.11)), the closure of $\operatorname{dom} S_{\mathrm{F}} = \ker \Lambda_0$ in $\mathfrak{H}_{\mathfrak{t}_{S_1}-a}$ is $\operatorname{dom} \mathfrak{t}_{S_{\mathrm{F}}}$. This implies $\ker \Lambda = \operatorname{dom} \mathfrak{t}_{S_{\mathrm{F}}}$. Finally, by definition $\operatorname{ran} \Lambda = \operatorname{ran} \Lambda_0 = \mathcal{G}$, and hence $\{\mathcal{G}, \Lambda\}$ is a boundary pair for S corresponding to S_1. It is also clear that the boundary pair $\{\mathcal{G}, \Lambda\}$ and the boundary triplet $\{\mathcal{G}, \Gamma_0, \Gamma_1\}$ are compatible. □

In the next corollary it is shown that in the context of Theorem 5.6.6 the continuity of $\Lambda : \mathfrak{H}_{\mathfrak{t}_{S_1}-a} \to \mathcal{G}$ makes it possible to extend the identity

$$(f', g) = \mathfrak{t}_{S_1}[f, g] + (\Gamma_1 \widehat{f}, \Gamma_0 \widehat{g}), \quad \widehat{f}, \widehat{g} \in S^*, \tag{5.6.8}$$

in Theorem 5.5.14 to $\widehat{f} \in S^*$ and $g \in \operatorname{dom} \mathfrak{t}_{S_1}$.

Corollary 5.6.7. *Let S be a closed semibounded relation in $\mathfrak{H}$, and let $\{\mathcal{G}, \Gamma_0, \Gamma_1\}$ and $\{\mathcal{G}, \Lambda\}$ be the boundary triplet and boundary pair in Theorem 5.6.6, respectively. Then the following equality holds:*

$$(f', g) = \mathfrak{t}_{S_1}[f, g] + (\Gamma_1 \widehat{f}, \Lambda g), \quad \widehat{f} \in S^*, \ g \in \operatorname{dom} \mathfrak{t}_{S_1}.$$

Proof. As $\operatorname{dom} S^*$ is a dense subspace of $\mathfrak{H}_{\mathfrak{t}_{S_1}-a}$, there exist $g_n \in \operatorname{dom} S^*$ and $\widehat{g}_n = \{g_n, g_n'\} \in S^*$ such that $g_n \to g$ in $\mathfrak{H}_{\mathfrak{t}_{S_1}-a}$, and hence $\Gamma_0 \widehat{g}_n = \Lambda g_n \to \Lambda g$ in $\mathcal{G}$. Furthermore, $g_n \to g$ in $\mathfrak{H}_{\mathfrak{t}_{S_1}-a}$ also implies $g_n \to g \in \mathfrak{H}$ and $\mathfrak{t}_{S_1}[f, g_n] \to \mathfrak{t}_{S_1}[f, g]$ for $\widehat{f} = \{f, f'\} \in S^*$. By (5.6.8), the identity

$$(f', g_n) = \mathfrak{t}_{S_1}[f, g_n] + (\Gamma_1 \widehat{f}, \Gamma_0 \widehat{g}_n) = \mathfrak{t}_{S_1}[f, g_n] + (\Gamma_1 \widehat{f}, \Lambda g_n)$$

holds for $\widehat{f} = \{f, f'\}$ and $\widehat{g}_n = \{g_n, g_n'\} \in S^*$. Now the assertion follows by taking limits. □

For the sake of completeness the existence of boundary pairs is stated in the following corollary as an addendum to Definition 5.6.1.

Corollary 5.6.8. *Let S be a closed semibounded relation in $\mathfrak{H}$. Then there exist a semibounded self-adjoint extension S_1 of S such that S_1 and $S_{\rm F}$ are transversal and a mapping $\Lambda \in \mathbf{B}(\mathfrak{H}_{\mathfrak{t}_{S_1}-a}, \mathcal{G})$, $a < m(S_1)$, such that $\{\mathcal{G}, \Lambda\}$ is a boundary pair for S.*

Proof. By Proposition 5.5.10, there exists a boundary triplet $\{\mathcal{G}, \Gamma_0, \Gamma_1\}$ for S^* such that $S_{\rm F} = \ker \Gamma_0$ and $S_{{\rm K},x} = \ker \Gamma_1$ for $x < m(S) = m(S_{\rm F})$. Now the statement follows with $S_1 = S_{{\rm K},x}$ and $a < m(S_1) = x$ from Theorem 5.6.6. □

Example 5.6.9. Let S be a closed semibounded relation in $\mathfrak{H}$ with lower bound γ, fix $x < \gamma$, and consider the boundary triplet $\{\mathfrak{N}_x(S^*), \Gamma_0, \Gamma_1\}$ for S^* in Corollary 5.5.12 with

$$\Gamma_0 \widehat{f} = f_x, \qquad \widehat{f} = \{f_{\rm F}, f_{\rm F}'\} + \{f_x, x f_x\} \in S_{\rm F} \,\widehat{+}\, \widehat{\mathfrak{N}}_x(S^*).$$

Then $S_{\rm F} = \ker \Gamma_0$ and $S_{{\rm K},x} = \ker \Gamma_1$ and for $a < x$ one has the direct sum decomposition

$$\mathfrak{H}_{\mathfrak{t}_{S_{{\rm K},x}}-a} = \operatorname{dom} \mathfrak{t}_{S_{{\rm K},x}} = \operatorname{dom} \mathfrak{t}_{S_{\rm F}} + \mathfrak{N}_x(S^*), \quad a < x < \gamma; \tag{5.6.9}$$

cf. Corollary 5.4.16. Then the mapping

$$\Lambda f = f_x, \qquad f = f_{\rm F} + f_x \in \operatorname{dom} \mathfrak{t}_{S_{\rm F}} + \mathfrak{N}_x(S^*),$$

belongs to $\mathbf{B}(\mathfrak{H}_{\mathfrak{t}_{S_{{\rm K},x}}-a}, \mathfrak{N}_x(S^*))$. In fact, let $f \in \mathfrak{H}_{\mathfrak{t}_{S_{{\rm K},x}}-a}$ have the decomposition $f = f_{\rm F} + f_x$ as in (5.6.9), and define $f_a = (I + (a - x)(S_{\rm F} - a)^{-1}) f_x$. Then $f = g_{\rm F} + f_a$, where $g_{\rm F} = f_x - f_a + f_{\rm F} \in \operatorname{dom} \mathfrak{t}_{S_{\rm F}}$ and $f_a \in \mathfrak{N}_a(S^*)$. Now observe that $f_x = (I + (x - a)(S_{\rm F} - x)^{-1}) f_a$, so that Proposition 1.4.6 leads to the estimate

$$\|f_x\| \le \frac{\gamma - a}{\gamma - x} \|f_a\|.$$

Recall from (5.1.9) (with $\mathfrak{t} = \mathfrak{t}_{S_{{\rm K},x}}$, $\gamma = x$, $\varphi = f_a$) and (5.4.28) that

$$(x - a)\|f_a\|^2 \le \|f_a\|^2_{\mathfrak{t}_{S_{{\rm K},x}}-a} \le \|f_a\|^2_{\mathfrak{t}_{S_{{\rm K},x}}-a} + \|g_{\rm F}\|^2_{\mathfrak{t}_{S_{{\rm K},x}}-a} = \|f\|^2_{\mathfrak{t}_{S_{{\rm K},x}}-a},$$

which proves that $\Lambda \in \mathbf{B}(\mathfrak{H}_{\mathfrak{t}_{S_{{\rm K},x}}-a}, \mathfrak{N}_x(S^*))$. Thus, Λ extends Γ_0 in the sense of Definition 5.6.3. It is clear that $\operatorname{dom} \mathfrak{t}_{S_{\rm F}} = \ker \Lambda$, and hence Lemma 5.6.5 (ii) implies that $\{\mathfrak{N}_x(S^*), \Lambda\}$ is a boundary pair for S corresponding to $S_{{\rm K},x}$ which is compatible with the boundary triplet $\{\mathfrak{N}_x(S^*), \Gamma_0, \Gamma_1\}$ in Corollary 5.5.12.

Theorem 5.6.6 admits a converse. If S is a semibounded relation and $\{\mathcal{G}, \Lambda\}$ is a boundary pair for S in the sense of Definition 5.6.1, then there exists a compatible boundary triplet $\{\mathcal{G}, \Gamma_0, \Gamma_1\}$ for S^*. The construction of the mapping $\Gamma_0 : S^* \to \mathcal{G}$ is inspired by Lemma 5.6.5 and the construction of $\Gamma_1 : S^* \to \mathcal{G}$ is inspired by the first Green formula in Theorem 5.5.14.

Theorem 5.6.10. *Let S be a closed semibounded relation in $\mathfrak{H}$ and let S_1 be a semibounded self-adjoint extension of S such that S_1 and $S_{\rm F}$ are transversal. Let $\{\mathcal{G}, \Lambda\}$ be a boundary pair for S corresponding to S_1. Then*

$$\Gamma_0 = \{\{\widehat{f}, \Lambda f\} : \widehat{f} \in S^*\} \tag{5.6.10}$$

is (the graph of) a linear operator from S^* to $\mathcal{G}$ and there exists a unique linear operator $\Gamma_1 : S^* \to \mathcal{G}$ such that $\{\mathcal{G}, \Gamma_0, \Gamma_1\}$ defines a boundary triplet for S^* which is compatible with the boundary pair $\{\mathcal{G}, \Lambda\}$ for S corresponding to S_1.

Proof. The relations S_1 and $S_{\rm F}$ are semibounded self-adjoint extensions of S and hence $m(S_1) \le m(S_{\rm F}) = m(S)$. There are the following decompositions of the relation S^*:

$$S^* = S_{\rm F} \,\widehat{+}\, \widehat{\mathfrak{N}}_a(S^*), \quad a < m(S_{\rm F}), \tag{5.6.11}$$

and, likewise,

$$S^* = S_1 \,\widehat{+}\, \widehat{\mathfrak{N}}_a(S^*), \quad a < m(S_1); \tag{5.6.12}$$

cf. Theorem 1.7.1. Recall that $\mathfrak{t}_{S_{\rm F}} \subset \mathfrak{t}_{S_1}$ and, since S_1 and $S_{\rm F}$ are transversal, there is the orthogonal decomposition

$$\operatorname{dom} \mathfrak{t}_{S_1} = \ker (S^* - a) \oplus_{\mathfrak{t}_{S_1} - a} \operatorname{dom} \mathfrak{t}_{S_{\rm F}}, \quad a < m(S_1), \tag{5.6.13}$$

of the Hilbert space $\mathfrak{H}_{\mathfrak{t}_{S_1} - a}$. Moreover, in this case one also has $\operatorname{dom} S^* \subset \operatorname{dom} \mathfrak{t}_{S_1}$; cf. Proposition 5.3.7 and Theorem 5.3.8. The proof will be given in a number of steps. The mapping Γ_0 is considered in Step 1. Step 2 and Step 3 are preparations for the construction of Γ_1 in Step 4. In the remaining steps the various properties of Γ_1 are established.

Step 1. This step concerns the properties of Γ_0 in (5.6.10). Since $\operatorname{dom} S^* \subset \operatorname{dom} \mathfrak{t}_{S_1}$ one sees from Definition 5.6.1 that the relation Γ_0 is well defined. It is clear that Γ_0 is the graph of an operator,

$$\Gamma_0 : S^* \to \mathcal{G}, \qquad \widehat{f} \mapsto \Gamma_0 \widehat{f} = \Lambda f, \tag{5.6.14}$$

and that $\{0\} \times \operatorname{mul} S^* \subset \ker \Gamma_0$. Furthermore,

$$S_{\rm F} = \bigl\{\widehat{f} \in S^* : f \in \operatorname{dom} \mathfrak{t}_{S_{\rm F}}\bigr\} = \bigl\{\widehat{f} \in S^* : f \in \ker \Lambda\bigr\} = \ker \Gamma_0, \tag{5.6.15}$$

where the first equality holds by Theorem 5.3.3, the second equality is due to $\operatorname{dom} \mathfrak{t}_{S_{\rm F}} = \ker \Lambda$, and the third equality follows from (5.6.14).

Since $\operatorname{ran} \Lambda = \mathcal{G}$ and $\ker \Lambda = \operatorname{dom} \mathfrak{t}_{S_{\rm F}}$, it follows from (5.6.13) that Λ maps $\ker (S^* - a)$ bijectively onto $\mathcal{G}$. Therefore,

$$\Gamma_0 \text{ is a bijection between } \widehat{\mathfrak{N}}_a(S^*) \text{ and } \mathcal{G}, \tag{5.6.16}$$

and, in particular,

$$\operatorname{ran} \Gamma_0 = \mathcal{G}. \tag{5.6.17}$$

Step 2. Now it will be shown that the identity

$$\mathfrak{t}_{S_1}[f, g_{\mathrm{F}}] = (f', g_{\mathrm{F}}), \quad \widehat{f} \in S^*, \; g_{\mathrm{F}} \in \operatorname{dom} \mathfrak{t}_{S_{\mathrm{F}}}, \tag{5.6.18}$$

holds. For this, assume that $\widehat{f}$ is decomposed as

$$\widehat{f} = \widehat{f}_{\mathrm{F}} + \widehat{h}_a, \qquad \widehat{f}_{\mathrm{F}} \in S_{\mathrm{F}}, \; \widehat{h}_a \in \widehat{\mathfrak{N}}_a(S^*); \tag{5.6.19}$$

cf. (5.6.11). Recall that $\operatorname{dom} S^* \subset \operatorname{dom} \mathfrak{t}_{S_1}$ and observe that with (5.6.19) one gets

$$\mathfrak{t}_{S_1}[f, g_{\mathrm{F}}] = \mathfrak{t}_{S_1}[f_{\mathrm{F}} + h_a, g_{\mathrm{F}}] = \mathfrak{t}_{S_{\mathrm{F}}}[f_{\mathrm{F}}, g_{\mathrm{F}}] + \mathfrak{t}_{S_1}[h_a, g_{\mathrm{F}}]. \tag{5.6.20}$$

The orthogonal decomposition in (5.6.13) gives

$$0 = (h_a, g_{\mathrm{F}})_{\mathfrak{t}_{S_1} - a} = \mathfrak{t}_{S_1}[h_a, g_{\mathrm{F}}] - a(h_a, g_{\mathrm{F}}).$$

Hence, (5.6.20) leads to the identity

$$\mathfrak{t}_{S_1}[f, g_{\mathrm{F}}] = \mathfrak{t}_{S_{\mathrm{F}}}[f_{\mathrm{F}}, g_{\mathrm{F}}] + a(h_a, g_{\mathrm{F}}) = (f'_{\mathrm{F}}, g_{\mathrm{F}}) + a(h_a, g_{\mathrm{F}}),$$

which shows (5.6.18).

Step 3. Next it will be shown that

$$\mathfrak{t}_{S_1}[f, g] - (f', g) = (f_a, g_a)_{\mathfrak{t}_{S_1} - a}, \quad \widehat{f}, \widehat{g} \in S^*, \tag{5.6.21}$$

where $\widehat{f}$ and $\widehat{g}$ are decomposed as

$$\widehat{f} = \widehat{f}_1 + \widehat{f}_a, \qquad \widehat{f}_1 \in S_1, \; \widehat{f}_a \in \widehat{\mathfrak{N}}_a(S^*), \tag{5.6.22}$$

and

$$\widehat{g} = \widehat{g}_{\mathrm{F}} + \widehat{g}_a, \qquad \widehat{g}_{\mathrm{F}} \in S_{\mathrm{F}}, \; \widehat{g}_a \in \widehat{\mathfrak{N}}_a(S^*); \tag{5.6.23}$$

cf. (5.6.12) and (5.6.11). For this note first that with (5.6.22) the identity (5.6.18) in Step 2 gives

$$\mathfrak{t}_{S_1}[f_1 + f_a, g_{\mathrm{F}}] = (f'_1 + a f_a, g_{\mathrm{F}}). \tag{5.6.24}$$

Furthermore, note that with g_a from (5.6.23) one has

$$\mathfrak{t}_{S_1}[f_1, g_a] = (f'_1, g_a) \tag{5.6.25}$$

due to $\widehat{f}_1 \in S_1$ and $g_a \in \mathfrak{N}_a(S^*) \subset \operatorname{dom} \mathfrak{t}_{S_1}$; cf. Theorem 5.1.18. A combination of (5.6.24) and (5.6.25) leads to

$$\begin{aligned} \mathfrak{t}_{S_1}[f, g] - (f', g) &= \mathfrak{t}_{S_1}[f_1 + f_a, g_{\mathrm{F}} + g_a] - (f'_1 + a f_a, g_{\mathrm{F}} + g_a) \\ &= \mathfrak{t}_{S_1}[f_1 + f_a, g_a] - (f'_1 + a f_a, g_a) \\ &= \mathfrak{t}_{S_1}[f_a, g_a] - a(f_a, g_a), \end{aligned}$$

which gives (5.6.21).

Step 4. In this step the operator $\Gamma_1 : S^* \to \mathcal{G}$ will be constructed. For this purpose fix $\widehat{f} \in S^*$ and consider the linear relation

$$\Phi_{\widehat{f}} = \big\{\{\Gamma_0 \widehat{g}, (g, f') - \mathfrak{t}_{S_1}[g, f]\} : \widehat{g} \in S^* \big\}. \tag{5.6.26}$$

It follows from $\operatorname{ran} \Gamma_0 = \mathcal{G}$ in (5.6.17) that $\operatorname{dom} \Phi_{\widehat{f}} = \mathcal{G}$. Next it will be shown that $\Phi_{\widehat{f}}$ is the graph of a bounded linear functional. If $\widehat{f}$ and $\widehat{g}$ are decomposed as in (5.6.22) and (5.6.23), then it follows from (5.6.21) in Step 3 that

$$\big|(g, f') - \mathfrak{t}_{S_1}[g, f]\big| = \big|\mathfrak{t}_{S_1}[f, g] - (f', g)\big| = |(f_a, g_a)_{\mathfrak{t}_{S_1}-a}| \leq \|f_a\|_{\mathfrak{t}_{S_1}-a} \|g_a\|_{\mathfrak{t}_{S_1}-a}.$$

Recall that the restriction of Λ to $\ker (S^* - a)$ has a bounded inverse (with respect to the norm $\|\cdot\|_{\mathfrak{t}_{S_1}-a}$ on $\ker (S^* - a)$). Therefore, by (5.6.10) and (5.6.15),

$$\|g_a\|_{\mathfrak{t}_{S_1}-a} \leq C\|\Lambda g_a\| = C\|\Gamma_0 \widehat{g}_a\| = C\|\Gamma_0 \widehat{g}\| \tag{5.6.27}$$

for some constant $C > 0$ and, as a consequence,

$$\big|(g, f') - \mathfrak{t}_{S_1}[g, f]\big| \leq C\|f_a\|_{\mathfrak{t}_{S_1}-a} \|\Gamma_0 \widehat{g}\|, \qquad \widehat{g} \in S^*.$$

This implies that the relation $\Phi_{\widehat{f}}$ in (5.6.26) is the graph of an everywhere defined bounded functional. Hence, by the Riesz representation theorem, there exists a unique $\varphi_{\widehat{f}} \in \mathcal{G}$ such that

$$\Phi_{\widehat{f}}(\Gamma_0 \widehat{g}) = \big(\Gamma_0 \widehat{g}, \varphi_{\widehat{f}}\big), \quad \widehat{g} \in S^*.$$

Define the mapping Γ_1 by

$$\Gamma_1 : S^* \to \mathcal{G}, \qquad \widehat{f} \mapsto \Gamma_1 \widehat{f} := \varphi_{\widehat{f}}. \tag{5.6.28}$$

By construction, Γ_1 is linear and it follows from (5.6.21) and (5.6.26) that

$$(\Gamma_1 \widehat{f}, \Gamma_0 \widehat{g}) = (f', g) - \mathfrak{t}_{S_1}[f, g] = -(f_a, g_a)_{\mathfrak{t}_{S_1}-a} \tag{5.6.29}$$

for all $\widehat{f} \in S^*$ and $\widehat{g} \in S^*$ decomposed in the forms (5.6.22) and (5.6.23), respectively.

Step 5. It will be shown that the operator Γ_1, constructed in Step 4, satisfies

$$S_1 = \ker \Gamma_1. \tag{5.6.30}$$

To show that $S_1 \subset \ker \Gamma_1$, assume that $\widehat{f} \in S_1$. Then (5.6.29) in Step 4 implies that $(\Gamma_1 \widehat{f}, \Gamma_0 \widehat{g}) = 0$ for all $\widehat{g} \in S^*$, and since Γ_0 is surjective (see (5.6.17)), one concludes that $\Gamma_1 \widehat{f} = 0$. Thus, $S_1 \subset \ker \Gamma_1$. To show the reverse inclusion, assume that $\widehat{f} = \{f, f'\} \in \ker \Gamma_1$. Then it follows from (5.6.29) that

$$\mathfrak{t}_{S_1}[f, g] = (f', g) \quad \text{for all} \quad g \in \operatorname{dom} S_1 \subset \operatorname{dom} S^*.$$

Since $\operatorname{dom} S_1$ is a core of $\mathfrak{t}_{S_1}$, it is a consequence of the first representation theorem (Theorem 5.1.18) that $\widehat{f} = \{f, f'\} \in S_1$. Thus, $\ker \Gamma_1 \subset S_1$, and so (5.6.30) has been proved.

Step 6. Next it will be shown that the operator Γ_1, constructed in Step 4, satisfies

$$\operatorname{ran}\Gamma_1 = \mathcal{G}. \tag{5.6.31}$$

For this purpose note first that $\overline{\operatorname{ran}}\,\Gamma_1 = \mathcal{G}$. In fact, if $\overline{\operatorname{ran}}\,\Gamma_1 \neq \mathcal{G}$, then in view of (5.6.16) there exists $\widehat{g}_a \in \widehat{\mathfrak{N}}_a(S^*)$ such that $\Gamma_0\widehat{g}_a \neq 0$ and

$$(\Gamma_1\widehat{f}, \Gamma_0\widehat{g}_a) = 0 \quad \text{for all} \quad \widehat{f} \in S^*. \tag{5.6.32}$$

Now apply (5.6.29) with $\widehat{f} = \widehat{f}_1 + \widehat{f}_a \in S^*$, $\widehat{f}_1 \in S_1$, $\widehat{f}_a \in \widehat{\mathfrak{N}}_a(S^*)$, and $\widehat{g}_a \in \widehat{\mathfrak{N}}_a(S^*)$. Then (5.6.32) implies

$$(f_a, g_a)_{\mathfrak{t}_{S_1}-a} = 0$$

for all $f_a \in \mathfrak{N}_a(S^*)$. Therefore, $g_a = 0$ and hence $\widehat{g}_a = 0$ and $\Gamma_0\widehat{g}_a = 0$, which is a contradiction. Thus, $\overline{\operatorname{ran}}\,\Gamma_1 = \mathcal{G}$.

To conclude (5.6.31), it suffices to show that $\operatorname{ran}\Gamma_1$ is closed. For this consider the restriction Γ_1' to $\widehat{\mathfrak{N}}_a(S^*)$. It follows from $S^* = S_1 \,\widehat{+}\, \widehat{\mathfrak{N}}_a(S^*)$ and (5.6.30) that Γ_1' is injective and that

$$\operatorname{ran}\Gamma_1' = \operatorname{ran}\Gamma_1. \tag{5.6.33}$$

With the inner product $(\cdot,\cdot)_{\mathfrak{t}_{S_1}-a}$ the space $\widehat{\mathfrak{N}}_a(S^*)$ is a closed subspace of the Hilbert space $\mathfrak{H}_{\mathfrak{t}_{S_1}-a} \times \mathfrak{H}_{\mathfrak{t}_{S_1}-a}$ (see (5.6.13)). Since

$$\big|(\Gamma_1'\widehat{f}_a, \Gamma_0\widehat{g})\big| = |(f_a, g_a)_{\mathfrak{t}_{S_1}-a}| \leq C\|f_a\|_{\mathfrak{t}_{S_1}-a}\|\Gamma_0\widehat{g}\|$$

by (5.6.29) and (5.6.27), it follows from

$$\|\Gamma_1'\widehat{f}_a\| = \sup_{\|\Gamma_0\widehat{g}\|=1} \big|(\Gamma_1\widehat{f}_a, \Gamma_0\widehat{g})\big| \leq C\|f_a\|_{\mathfrak{t}_{S_1}-a}$$

that the operator Γ_1' is bounded in the topology of $\mathfrak{H}_{\mathfrak{t}_{S_1}-a} \times \mathfrak{H}_{\mathfrak{t}_{S_1}-a}$. Hence, Γ_1' is closed and the same is true for the inverse operator

$$(\Gamma_1')^{-1} : \mathcal{G} \supset \operatorname{ran}\Gamma_1' \to \widehat{\mathfrak{N}}_a(S^*).$$

Assume that $(\Gamma_1')^{-1}$ is unbounded. Then there exists a sequence $(\widehat{g}_n)$ in $\widehat{\mathfrak{N}}_a(S^*)$ such that $\|g_n\|_{\mathfrak{t}_{S_1}-a} = 1$ and $\Gamma_1'\widehat{g}_n \to 0$ in $\mathcal{G}$. From (5.6.29) and the definition of Γ_0 one obtains

$$1 = (g_n, g_n)_{\mathfrak{t}_{S_1}-a} = -(\Gamma_1'\widehat{g}_n, \Gamma_0\widehat{g}_n) = -(\Gamma_1'\widehat{g}_n, \Lambda g_n) \leq \|\Gamma_1'\widehat{g}_n\|\|\Lambda g_n\|,$$

and as $\Lambda : \mathfrak{H}_{\mathfrak{t}_{S_1}-a} \to \mathcal{G}$ is bounded this yields

$$1 \leq C'\|\Gamma_1'\widehat{g}_n\|\|g_n\|_{\mathfrak{t}_{S_1}-a} = C'\|\Gamma_1'\widehat{g}_n\| \to 0;$$

a contradiction. Hence, the operator $(\Gamma_1')^{-1}$ is bounded. As $(\Gamma_1')^{-1}$ is closed, it follows that $\operatorname{ran}\Gamma_1' = \operatorname{dom}(\Gamma_1')^{-1}$ is closed, which together with (5.6.33) and $\overline{\operatorname{ran}}\,\Gamma_1 = \mathcal{G}$ shows (5.6.31).

Step 7. First it will be verified that the mappings Γ_0 and Γ_1 form a boundary triplet for S^*. Observe that (5.6.29) implies the Green identity

$$(f',g)-(f,g')=(\Gamma_1\widehat{f},\Gamma_0\widehat{g})-(\Gamma_0\widehat{f},\Gamma_1\widehat{g}),\quad \widehat{f},\widehat{g}\in S^*.$$

It remains to show that

$$\operatorname{ran}\begin{pmatrix}\Gamma_0\\ \Gamma_1\end{pmatrix}=\mathcal{G}\times\mathcal{G}. \tag{5.6.34}$$

For this, let $\varphi,\varphi'\in\mathcal{G}$. From $\operatorname{ran}\Gamma_0=\mathcal{G}$ in (5.6.17) and $\operatorname{ran}\Gamma_1=\mathcal{G}$ in (5.6.31) it is clear that there exist $\widehat{h},\widehat{k}\in S^*$ such that $\Gamma_0\widehat{h}=\varphi$ and $\Gamma_1\widehat{k}=\varphi'$. It follows from the transversality $S^*=S_{\rm F}\,\widehat{+}\,S_1$ that

$$\widehat{h}=\widehat{h}_{\rm F}+\widehat{h}_1\quad\text{and}\quad\widehat{k}=\widehat{k}_{\rm F}+\widehat{k}_1,\quad \widehat{h}_{\rm F},\widehat{k}_{\rm F}\in S_{\rm F},\ \widehat{h}_1,\widehat{k}_1\in S_1.$$

Define $\widehat{f}:=\widehat{h}_1+\widehat{k}_{\rm F}\in S^*$. Making use of the facts that $\ker\Gamma_0=S_{\rm F}$ in (5.6.15) and $\ker\Gamma_1=S_1$ in (5.6.30), one obtains

$$\begin{aligned}\Gamma_0\widehat{f}&=\Gamma_0\widehat{h}_1=\Gamma_0\widehat{h}=\varphi,\\ \Gamma_1\widehat{f}&=\Gamma_1\widehat{k}_{\rm F}=\Gamma_1\widehat{k}=\varphi',\end{aligned}$$

which shows (5.6.34). Therefore, $\{\mathcal{G},\Gamma_0,\Gamma_1\}$ is a boundary triplet for S^*.

Since $\ker\Gamma_0=S_{\rm F}$ and $\ker\Gamma_1=S_1$, and since Λ is an extension of Γ_0, see (5.6.10), one concludes that the boundary triplet $\{\mathcal{G},\Gamma_0,\Gamma_1\}$ and the boundary pair $\{\mathcal{G},\Lambda\}$ are compatible; see Definition 5.6.4.

It remains to check that Γ_1 constructed in (5.6.28)–(5.6.29) is uniquely determined. Note that the mapping Γ_0 and the kernel S_1 of Γ_1 are uniquely determined as the boundary triplet is required to be compatible with the boundary pair $\{\mathcal{G},\Lambda\}$. Under these circumstances the action of Γ_1 is uniquely determined by formula (5.5.27) in Theorem 5.5.14. □

The following result gives a connection via a boundary pair $\{\mathcal{G},\Lambda\}$ between closed semibounded forms $\mathfrak{t}_H$ corresponding to semibounded self-adjoint extensions H of S such that $S_1\le H\le S_{\rm F}$ and closed nonnegative forms ω in $\mathcal{G}$. A similar result also involving boundary triplets follows later.

Theorem 5.6.11. *Let S be a closed semibounded relation in $\mathfrak{H}$ and let S_1 be a semibounded self-adjoint extension of S such that S_1 and $S_{\rm F}$ are transversal. Let $\{\mathcal{G},\Lambda\}$ be a boundary pair for S corresponding to S_1. Then the following statements hold:*

(i) *If H is a semibounded self-adjoint extension of S such that $S_1\le H$, then there exists a closed nonnegative form ω in $\mathcal{G}$ defined on $\operatorname{dom}\omega=\Lambda(\operatorname{dom}\mathfrak{t}_H)$ such that*

$$\mathfrak{t}_H[f,g]=\mathfrak{t}_{S_1}[f,g]+\omega[\Lambda f,\Lambda g],\qquad f,g\in\operatorname{dom}\mathfrak{t}_H. \tag{5.6.35}$$

Moreover, the space $\Lambda(\operatorname{dom}H)$ is a core of the form ω.

(ii) *If ω is a closed nonnegative form in $\mathcal{G}$, then*

$$\begin{aligned} \mathfrak{t}[f,g] &= \mathfrak{t}_{S_1}[f,g] + \omega[\Lambda f, \Lambda g], \\ \operatorname{dom} \mathfrak{t} &= \{ f \in \operatorname{dom} \mathfrak{t}_{S_1} : \Lambda f \in \operatorname{dom} \omega \}, \end{aligned} \tag{5.6.36}$$

is a closed semibounded form in $\mathfrak{H}$ and the corresponding self-adjoint relation H is a semibounded self-adjoint extension of S which satisfies $S_1 \leq H$.

The formulas (5.6.35) and (5.6.36) establish a one-to-one correspondence between all closed nonnegative forms ω in $\mathcal{G}$ and all semibounded self-adjoint extensions H of S satisfying the inequalities $S_1 \leq H \leq S_{\mathrm{F}}$.

Proof. (i) Let H be a semibounded self-adjoint extension of S and let $\mathfrak{t}_H$ be the corresponding closed semibounded form. By assumption, $S_1 \leq H$ or, equivalently, $\mathfrak{t}_{S_1} \leq \mathfrak{t}_H$; cf. Theorem 5.2.4. Hence, $\operatorname{dom} \mathfrak{t}_H \subset \operatorname{dom} \mathfrak{t}_{S_1}$ and $\mathfrak{t}_{S_1}[f] \leq \mathfrak{t}_H[f]$ for all $f \in \operatorname{dom} \mathfrak{t}_H$. Recall that $\mathfrak{t}_{S_{\mathrm{F}}}$, as the closure of $\mathfrak{t}_S$, satisfies $\mathfrak{t}_{S_{\mathrm{F}}} \subset \mathfrak{t}_{S_1}$ and $\mathfrak{t}_{S_{\mathrm{F}}} \subset \mathfrak{t}_H$. Since $\ker \Lambda = \operatorname{dom} \mathfrak{t}_{S_{\mathrm{F}}}$, one concludes that the form

$$\omega[\Lambda f, \Lambda g] := \mathfrak{t}_H[f,g] - \mathfrak{t}_{S_1}[f,g], \quad \operatorname{dom} \omega = \Lambda(\operatorname{dom} \mathfrak{t}_H), \quad f,g \in \operatorname{dom} \mathfrak{t}_H, \tag{5.6.37}$$

is well defined and nonnegative in the Hilbert space $\mathcal{G}$. To see that it is well defined, just note that for $f, g \in \operatorname{dom} \mathfrak{t}_H$ the Cauchy–Schwarz inequality shows

$$\left| \mathfrak{t}_H[f,g] - \mathfrak{t}_{S_1}[f,g] \right| \leq \left| \mathfrak{t}_H[f,f] - \mathfrak{t}_{S_1}[f,f] \right|^{\frac{1}{2}} \left| \mathfrak{t}_H[g,g] - \mathfrak{t}_{S_1}[g,g] \right|^{\frac{1}{2}},$$

and hence $\mathfrak{t}_H[f,g] - \mathfrak{t}_{S_1}[f,g]$ in (5.6.37) vanishes when either f or g belongs to $\ker \Lambda = \operatorname{dom} \mathfrak{t}_{S_{\mathrm{F}}}$.

Next it will be shown that the form ω is closed in $\mathcal{G}$. To this end consider a sequence (φ_n) in $\operatorname{dom} \omega$ and assume that $\varphi_n \to_\omega \varphi$ for some $\varphi \in \mathcal{G}$, that is, (φ_n) is a sequence in $\operatorname{dom} \omega = \Lambda(\operatorname{dom} \mathfrak{t}_H)$, such that

$$\varphi_n \to \varphi \in \mathcal{G} \qquad \text{and} \qquad \omega[\varphi_n - \varphi_m] \to 0. \tag{5.6.38}$$

Since $\ker \Lambda = \operatorname{dom} \mathfrak{t}_{S_{\mathrm{F}}}$ and

$$\operatorname{dom} \mathfrak{t}_{S_{\mathrm{F}}} \subset \operatorname{dom} \mathfrak{t}_H \subset \operatorname{dom} \mathfrak{t}_{S_1} = \left(\operatorname{dom} \mathfrak{t}_{S_1} \ominus_{\mathfrak{t}_{S_1} - a} \operatorname{dom} \mathfrak{t}_{S_{\mathrm{F}}} \right) \oplus_{\mathfrak{t}_{S_1} - a} \operatorname{dom} \mathfrak{t}_{S_{\mathrm{F}}}$$

for $a < m(S_1)$, there exists a sequence (f_n) in $\operatorname{dom} \mathfrak{t}_H \ominus_{\mathfrak{t}_{S_1}-a} \operatorname{dom} \mathfrak{t}_{S_{\mathrm{F}}}$ such that $\Lambda f_n = \varphi_n$. Moreover, since $\operatorname{ran} \Lambda = \mathcal{G}$, there exists $f \in \operatorname{dom} \mathfrak{t}_{S_1} \ominus_{\mathfrak{t}_{S_1}-a} \operatorname{dom} \mathfrak{t}_{S_{\mathrm{F}}}$ such that $\Lambda f = \varphi$; see Proposition 5.3.7. Since the restriction of Λ to the space $\operatorname{dom} \mathfrak{t}_{S_1} \ominus_{\mathfrak{t}_{S_1}-a} \operatorname{dom} \mathfrak{t}_{S_{\mathrm{F}}}$ has a bounded inverse (see the discussion following Definition 5.6.1), it follows that $f_n \to f$ in $\mathfrak{H}_{\mathfrak{t}_{S_1}-a}$. In particular, $f_n \to f$ in $\mathfrak{H}$ and $\mathfrak{t}_{S_1}[f_n - f_m] \to 0$. Then (5.6.37) and (5.6.38) imply

$$\begin{aligned} \mathfrak{t}_H[f_n - f_m] &= \mathfrak{t}_{S_1}[f_n - f_m] + \omega[\Lambda f_n - \Lambda f_m] \\ &= \mathfrak{t}_{S_1}[f_n - f_m] + \omega[\varphi_n - \varphi_m] \to 0, \end{aligned}$$

and as $\mathfrak{t}_H$ is closed one concludes $f \in \operatorname{dom} \mathfrak{t}_H$ and $\mathfrak{t}_H[f_n - f] \to 0$. This implies $\varphi = \Lambda f \in \operatorname{dom} \omega$. Furthermore, as $\mathfrak{t}_{S_1}$ is closed, also $\mathfrak{t}_{S_1}[f_n - f] \to 0$, and hence

$$\omega[\varphi_n - \varphi] = \omega[\Lambda f_n - \Lambda f] = \mathfrak{t}_H[f_n - f] - \mathfrak{t}_{S_1}[f_n - f] \to 0,$$

so that ω is a closed form in $\mathcal{G}$. It is clear that the definition of ω in (5.6.37) implies the representation of $\mathfrak{t}_H$ in (i).

It remains to show that $\Lambda(\operatorname{dom} H)$ is a core of ω. For this let $\varphi \in \operatorname{dom} \omega$ and choose $f \in \operatorname{dom} \mathfrak{t}_H$ such that $\varphi = \Lambda f$. As $\operatorname{dom} H$ is a core of $\mathfrak{t}_H$, there exists a sequence (f_n) in $\operatorname{dom} H$ such that $f_n \to f$ in $\mathfrak{H}$ and $\mathfrak{t}_H[f_n - f] \to 0$. Then $0 \leq (\mathfrak{t}_{S_1} - a)[f_n - f] \leq (\mathfrak{t}_H - a)[f_n - f] \to 0$ and, in particular, one has $f_n \to f$ in $\mathfrak{H}_{\mathfrak{t}_{S_1}-a}$. Setting $\varphi_n := \Lambda f_n$ one has $\varphi_n \subset \Lambda(\operatorname{dom} H)$ and using the fact that Λ is bounded one concludes that

$$\varphi_n = \Lambda f_n \to \Lambda f = \varphi$$

and

$$\omega[\varphi_n - \varphi] = \omega[\Lambda f_n - \Lambda f] = \mathfrak{t}_H[f_n - f] - \mathfrak{t}_{S_1}[f_n - f] \to 0.$$

This shows that $\Lambda(\operatorname{dom} H)$ is a core of ω.

(ii) Assume that ω is a closed nonnegative form in $\mathcal{G}$. Then it is clear that the form

$$\mathfrak{t}[f, g] = \mathfrak{t}_{S_1}[f, g] + \omega[\Lambda f, \Lambda g] \tag{5.6.39}$$

defined on $\operatorname{dom} \mathfrak{t} = \Lambda^{-1}(\operatorname{dom} \omega) \subset \operatorname{dom} \mathfrak{t}_{S_1}$ is semibounded and $\mathfrak{t}[f] \geq \mathfrak{t}_{S_1}[f]$ holds for all $f \in \operatorname{dom} \mathfrak{t}$. To verify that $\mathfrak{t}$ is closed consider a sequence (f_n) in $\operatorname{dom} \mathfrak{t}$ such that $f_n \to_{\mathfrak{t}} f$ for some $f \in \mathfrak{H}$, that is, $f_n \to f$ in $\mathfrak{H}$ and $\mathfrak{t}[f_n - f_m] \to 0$. Since the forms $\mathfrak{t}_{S_1} - a$, $a < m(S_1)$, and ω are nonnegative, it follows from (5.6.39) and $(\mathfrak{t} - a)[f_n - f_m] \to 0$ that $0 \leq (\mathfrak{t}_{S_1} - a)[f_n - f_m] \to 0$ and $\omega[\Lambda f_n - \Lambda f_m] \to 0$. As $\mathfrak{t}_{S_1}$ is a closed form in $\mathfrak{H}$, one concludes that $f \in \operatorname{dom} \mathfrak{t}_{S_1}$ and $\mathfrak{t}_{S_1}[f_n - f] \to 0$. This shows that f_n converges to f in $\mathfrak{H}_{\mathfrak{t}_{S_1}-a}$ and as Λ is bounded one has $\Lambda f_n \to \Lambda f$ in $\mathcal{G}$. Moreover, since $\omega[\Lambda f_n - \Lambda f_m] \to 0$ and ω is closed in $\mathcal{G}$, one concludes that $\Lambda f \in \operatorname{dom} \omega$ and $\omega[\Lambda f_n - \Lambda f] \to 0$. Hence, $f \in \operatorname{dom} \mathfrak{t} = \Lambda^{-1}(\operatorname{dom} \omega)$ and $\mathfrak{t}[f_n - f] \to 0$, and $\mathfrak{t}$ is a closed form in $\mathfrak{H}$.

Let H be the semibounded self-adjoint relation associated with $\mathfrak{t}$ via the first representation theorem; see Theorem 5.1.18. Since $\operatorname{dom} \mathfrak{t}_{S_{\mathrm{F}}} = \ker \Lambda$, it follows from (5.6.36) that $\mathfrak{t}_{S_{\mathrm{F}}} \subset \mathfrak{t}$. Hence, $\mathfrak{t}_{S_1} \leq \mathfrak{t} \leq \mathfrak{t}_{S_{\mathrm{F}}}$ or, equivalently, $S_1 \leq H \leq S_{\mathrm{F}}$; see Theorem 5.2.4. One concludes from Theorem 5.4.6 (or its proof) that H is a self-adjoint extension of S. This completes the proof of (ii).

The indicated one-to-one correspondence is clear from (i) and (ii) by the uniqueness of the representing semibounded self-adjoint relation associated with a closed semibounded form. □

A combination of Theorem 5.6.11 with Theorem 5.4.6 leads to the following observations. Recall that the Kreĭn type extensions $S_{\mathrm{K},x}$ and S_{F} are transversal when $x < \gamma = m(S) = m(S_{\rm F})$ (see (5.4.26)) and that in the nonnegative case $\gamma \geq 0$ the Kreĭn–von Neumann extension is given by $S_{{\rm K},0}$; cf. Definition 5.4.2.

Corollary 5.6.12. *Let S be a closed semibounded relation in $\mathfrak{H}$ with lower bound γ.*

(i) *For $x < \gamma$ and $S_1 = S_{{\rm K},x}$ the formulas (5.6.35) and (5.6.36) establish a one-to-one correspondence between all closed nonnegative forms ω in $\mathcal{G}$ and all self-adjoint extensions H of S satisfying $m(H) \geq x$.*

(ii) *If $S_{{\rm K},\gamma}$ and $S_{\rm F}$ are transversal, then with $S_1 = S_{{\rm K},\gamma}$ the formulas (5.6.35) and (5.6.36) establish a one-to-one correspondence between all closed nonnegative forms ω in $\mathcal{G}$ and all self-adjoint extensions H of S satisfying $m(H) = \gamma$.*

(iii) *If $\gamma > 0$, then with the Kreĭn–von Neumann extension $S_1 = S_{{\rm K},0}$ the formulas (5.6.35) and (5.6.36) establish a one-to-one correspondence between all closed nonnegative forms ω in $\mathcal{G}$ and all nonnegative self-adjoint extensions H of S. If $\gamma = 0$ the same is true if the Kreĭn–von Neumann extension $S_1 = S_{{\rm K},0}$ and $S_{\rm F}$ are transversal.*

Theorem 5.6.11 is a first step towards a full description of all semibounded self-adjoint extensions and their associated forms. The following result is a continuation of Theorem 5.5.14 for semibounded self-adjoint extensions (see Corollary 5.5.15) and an extension of the first part of Theorem 5.6.11, in which also the boundary conditions of the extensions and the corresponding forms are connected.

Theorem 5.6.13. *Let S be a closed semibounded relation in $\mathfrak{H}$ and let S_1 be a semibounded self-adjoint extension of S such that S_1 and $S_{\rm F}$ are transversal. Let $\{\mathcal{G}, \Gamma_0, \Gamma_1\}$ be a boundary triplet for S^* and let $\{\mathcal{G}, \Lambda\}$ be a compatible boundary pair for S corresponding to S_1. Assume that H_Θ is a semibounded self-adjoint extension of S corresponding to the self-adjoint relation Θ in $\mathcal{G}$ as in (5.5.32)–(5.5.33). Then Θ is semibounded in $\mathcal{G}$ and the corresponding closed semibounded form ω_Θ in $\mathcal{G}$ and the closed semibounded form $\mathfrak{t}_{H_\Theta}$ corresponding to H_Θ are related by*

$$\begin{aligned} \mathfrak{t}_{H_\Theta}[f,g] &= \mathfrak{t}_{S_1}[f,g] + \omega_\Theta[\Lambda f, \Lambda g], \\ \operatorname{dom} \mathfrak{t}_{H_\Theta} &= \big\{ f \in \operatorname{dom} \mathfrak{t}_{S_1} : \Lambda f \in \operatorname{dom} \omega_\Theta \big\}. \end{aligned} \tag{5.6.40}$$

Proof. The proof of the theorem will rely on the results in Theorem 5.5.14 and Corollary 5.5.15, where H_Θ is now taken to be semibounded. In the first two steps of the proof the equality between the forms in (5.6.40) will be verified. In the last step the domain characterization in (5.6.40) will be shown.

Step 1. First recall from Corollary 5.5.15 the formula (5.5.34):

$$(f', g) = \mathfrak{t}_{S_1}[f,g] + (\Theta_{\rm op} \Gamma_0 \widehat{f}, \Gamma_0 \widehat{g}), \quad \widehat{f}, \widehat{g} \in H_\Theta. \tag{5.6.41}$$

Since H_Θ is assumed to be semibounded, it follows from Theorem 5.1.18 that $(f', g) = \mathfrak{t}_{H_\Theta}[f,g]$. As the boundary pair $\{\mathcal{G}, \Lambda\}$ is compatible with the boundary triplet $\{\mathcal{G},\Gamma_0,\Gamma_1\}$, the mapping Λ is an extension of Γ_0. Hence, (5.6.41) may now be rewritten as

$$\mathfrak{t}_{H_\Theta}[f,g]=\mathfrak{t}_{S_1}[f,g]+(\Theta_{\rm op}\Lambda f,\Lambda g),\quad f,g\in\operatorname{dom}H_\Theta. \tag{5.6.42}$$

Step 2. In this step it is shown that the formula (5.6.42) can be extended to the form domain of $\mathfrak{t}_{H_\Theta}$ as in (5.6.40). First observe that by Lemma 5.6.5 one has $A_0=S_{\rm F}$. Moreover, since H_Θ is a semibounded extension of S, it follows from Proposition 5.5.6 that Θ is semibounded from below. Hence, (5.6.42) can be written as

$$\mathfrak{t}_{H_\Theta}[f,g]=\mathfrak{t}_{S_1}[f,g]+\omega_\Theta[\Lambda f,\Lambda g],\quad f,g\in\operatorname{dom}H_\Theta, \tag{5.6.43}$$

where ω_Θ is the closed semibounded form corresponding to Θ in $\mathcal{G}$. It follows from Corollary 5.3.9 that

$$\operatorname{dom}(H_\Theta-a)^{\frac{1}{2}}\subset\operatorname{dom}(S_1-a)^{\frac{1}{2}} \tag{5.6.44}$$

and hence there is a constant $C>0$ such that

$$\|((S_1)_{\rm op}-a)^{\frac{1}{2}}\varphi\|\le C\|((H_\Theta)_{\rm op}-a)^{\frac{1}{2}}\varphi\| \tag{5.6.45}$$

for all $\varphi\in\operatorname{dom}(H_\Theta-a)^{\frac{1}{2}}$.

Now let $f\in\operatorname{dom}\mathfrak{t}_{H_\Theta}$. As $\operatorname{dom}H_\Theta$ is a core of $\mathfrak{t}_{H_\Theta}$, there exists a sequence (f_n) in $\operatorname{dom}H_\Theta$ such that $f_n\to f$ in $\mathfrak{H}$ and $\mathfrak{t}_{H_\Theta}[f_n-f]\to 0$. By (5.6.44)–(5.6.45) it follows that $f\in\operatorname{dom}\mathfrak{t}_{S_1}$ and $\mathfrak{t}_{S_1}[f_n-f]\to 0$, so that $f_n\to f$ in $\mathfrak{H}_{\mathfrak{t}_{S_1}-a}$. Since Λ is bounded, this shows that $\Lambda f_n\to\Lambda f$ in $\mathcal{G}$. Furthermore, from (5.6.43) one sees that

$$\omega_\Theta[\Lambda f_n-\Lambda f_m]=\mathfrak{t}_{H_\Theta}[f_n-f_m]-\mathfrak{t}_{S_1}[f_n-f_m]\to 0.$$

Since ω_Θ is closed, one obtains

$$\Lambda f\in\operatorname{dom}\omega_\Theta\quad\text{and}\quad\omega_\Theta[\Lambda f_n-\Lambda f]\to 0.$$

Therefore, the following inclusion has been shown

$$\operatorname{dom}\mathfrak{t}_{H_\Theta}\subset\bigl\{f\in\operatorname{dom}\mathfrak{t}_{S_1}:\Lambda f\in\operatorname{dom}\omega_\Theta\bigr\}. \tag{5.6.46}$$

Let $f,g\in\operatorname{dom}\mathfrak{t}_{H_\Theta}$ and choose $(f_n),(g_n)$ in $\operatorname{dom}H_\Theta$ as above. Then one has $\mathfrak{t}_{H_\Theta}[f_n,g_n]\to\mathfrak{t}_{H_\Theta}[f,g]$, $\mathfrak{t}_{S_1}[f_n,g_n]\to\mathfrak{t}_{S_1}[f,g]$, and $\omega_\Theta[\Lambda f_n,\Lambda g_n]\to\omega_\Theta[\Lambda f,\Lambda g]$ as $n\to\infty$ by Lemma 5.1.8, and hence (5.6.43) extends to

$$\mathfrak{t}_{H_\Theta}[f,g]=\mathfrak{t}_{S_1}[f,g]+\omega_\Theta[\Lambda f,\Lambda g],\quad f,g\in\operatorname{dom}\mathfrak{t}_{H_\Theta}. \tag{5.6.47}$$

Step 3. To complete the proof of the theorem the equality between the domains in (5.6.40) must be verified. Due to (5.6.46) it suffices to show that

$$\bigl\{f\in\operatorname{dom}\mathfrak{t}_{S_1}:\Lambda f\in\operatorname{dom}\omega_\Theta\bigr\}\subset\operatorname{dom}\mathfrak{t}_{H_\Theta}.$$

Let $f \in \operatorname{dom} \mathfrak{t}_{S_1}$ and assume that $\varphi = \Lambda f \in \operatorname{dom} \omega_\Theta$. Using the orthogonal decomposition

$$\operatorname{dom} \mathfrak{t}_{S_1} = \big(\operatorname{dom} \mathfrak{t}_{S_1} \ominus_{\mathfrak{t}_{S_1}-a} \operatorname{dom} \mathfrak{t}_{S_F}\big) \oplus_{\mathfrak{t}_{S_1}-a} \operatorname{dom} \mathfrak{t}_{S_F}, \quad a < m(S_1), \tag{5.6.48}$$

write f in the form $f = h + k$, where $h \in \operatorname{dom} \mathfrak{t}_{S_1} \ominus_{\mathfrak{t}_{S_1}-a} \operatorname{dom} \mathfrak{t}_{S_F}$ and $k \in \operatorname{dom} \mathfrak{t}_{S_F}$. Then $k \in \operatorname{dom} \mathfrak{t}_{H_\Theta}$, and since $\ker \Lambda = \operatorname{dom} \mathfrak{t}_{S_F}$, one has $\varphi = \Lambda h$. It remains to show that $h \in \operatorname{dom} \mathfrak{t}_{H_\Theta}$.

Recall that $\operatorname{dom} \Theta$ is a core of ω_Θ. Hence, there exists a sequence (φ_n) in $\operatorname{dom} \Theta$ such that $\varphi_n \to_{\omega_\Theta} \varphi$, that is,

$$\varphi_n \to \varphi \in \mathcal{G} \qquad \text{and} \qquad \omega_\Theta[\varphi_n - \varphi_m] \to 0.$$

Note that $\varphi_n \in \operatorname{dom} \Theta$ means $\{\varphi_n, \varphi_n'\} \in \Theta$ for some $\varphi_n' \in \mathcal{G}$ and there exists $\{f_n, f_n'\} \in H_\Theta$ such that $\Gamma\{f_n, f_n'\} = \{\varphi_n, \varphi_n'\}$. Hence, $\Lambda f_n = \Gamma_0\{f_n, f_n'\} = \varphi_n$, where $f_n \in \operatorname{dom} H_\Theta \subset \operatorname{dom} \mathfrak{t}_{H_\Theta} \subset \operatorname{dom} \mathfrak{t}_{S_1}$. Using (5.6.48), one can write f_n in the form

$$f_n = h_n + k_n, \quad h_n \in \operatorname{dom} \mathfrak{t}_{S_1} \ominus_{\mathfrak{t}_{S_1}-a} \operatorname{dom} \mathfrak{t}_{S_F}, \ k_n \in \operatorname{dom} \mathfrak{t}_{S_F}.$$

From $\ker \Lambda = \operatorname{dom} \mathfrak{t}_{S_F}$ it is clear that $\varphi_n = \Lambda f_n = \Lambda h_n$. Since the restriction of Λ to $\operatorname{dom} \mathfrak{t}_{S_1} \ominus_{\mathfrak{t}_{S_1}-a} \operatorname{dom} \mathfrak{t}_{S_F}$ has a bounded inverse it follows from $\varphi_n \to \varphi$ in $\mathcal{G}$ that $h_n \to h$ in $\mathfrak{H}_{\mathfrak{t}_{S_1}-a}$. In particular, $h_n \to h$ in $\mathfrak{H}$ and $\mathfrak{t}_{S_1}[h_n - h_m] \to 0$. Then it follows from (5.6.47) that

$$\begin{aligned} \mathfrak{t}_{H_\Theta}[h_n - h_m] &= \mathfrak{t}_{S_1}[h_n - h_m] + \omega_\Theta[\Lambda h_n - \Lambda h_m] \\ &= \mathfrak{t}_{S_1}[h_n - h_m] + \omega_\Theta[\varphi_n - \varphi_m] \to 0, \end{aligned}$$

and as $\mathfrak{t}_{H_\Theta}$ is closed, one concludes that $h \in \operatorname{dom} \mathfrak{t}_{H_\Theta}$. □

One may apply the second representation theorem (Theorem 5.1.23) to the closed form ω_Θ in Theorem 5.6.13. Assume that $\mu \le m(\Theta)$, then it follows that

$$\begin{aligned} \omega_\Theta[\Lambda f, \Lambda g] &= \big((\Theta_{\mathrm{op}} - \mu)^{\frac12} \Lambda f, (\Theta_{\mathrm{op}} - \mu)^{\frac12} \Lambda g\big) + \mu\,(\Lambda f, \Lambda g), \\ \operatorname{dom} \omega_\Theta &= \operatorname{dom} (\Theta_{\mathrm{op}} - \mu)^{\frac12}. \end{aligned}$$

Hence, one obtains the following result; cf. Corollary 5.5.15.

Corollary 5.6.14. *Let the assumptions be as in Theorem 5.6.13 and let $\mu \le m(\Theta)$. Then the closed semibounded form $\mathfrak{t}_{H_\Theta}$ corresponding to H_Θ is given by*

$$\begin{aligned} \mathfrak{t}_{H_\Theta}[f,g] &= \mathfrak{t}_{S_1}[f,g] + \big((\Theta_{\mathrm{op}} - \mu)^{\frac12} \Lambda f, (\Theta_{\mathrm{op}} - \mu)^{\frac12} \Lambda g\big) + \mu\,(\Lambda f, \Lambda g), \\ \operatorname{dom} \mathfrak{t}_{H_\Theta} &= \big\{ f \in \operatorname{dom} \mathfrak{t}_{S_1} : \Lambda f \in \operatorname{dom} (\Theta_{\mathrm{op}} - \mu)^{\frac12} \big\}. \end{aligned}$$

Furthermore, if $\Theta_{\mathrm{op}} \in \mathbf{B}(\mathcal{G}_{\mathrm{op}})$, then

$$\begin{aligned} \mathfrak{t}_{H_\Theta}[f,g] &= \mathfrak{t}_{S_1}[f,g] + \big(\Theta_{\mathrm{op}} \Lambda f, \Lambda g\big), \\ \operatorname{dom} \mathfrak{t}_{H_\Theta} &= \big\{ f \in \operatorname{dom} \mathfrak{t}_{S_1} : \Lambda f \in \operatorname{dom} \Theta_{\mathrm{op}} \big\}, \end{aligned} \tag{5.6.49}$$

and in the special case $\Theta \in \mathbf{B}(\mathcal{G})$

$$\mathfrak{t}_{H_\Theta}[f,g] = \mathfrak{t}_{S_1}[f,g] + \big(\Theta \Lambda f, \Lambda g\big), \qquad \operatorname{dom} \mathfrak{t}_{H_\Theta} = \operatorname{dom} \mathfrak{t}_{S_1}.$$

Example 5.6.15. Let S be a closed semibounded relation in $\mathfrak{H}$ with lower bound γ, fix $x < \gamma$, and consider the boundary triplet $\{\mathfrak{N}_x(S^*), \Gamma_0, \Gamma_1\}$ for S^* in Corollary 5.5.12 and the corresponding compatible boundary pair $\{\mathfrak{N}_x(S^*), \Lambda\}$ in Example 5.6.9. Assume that H_Θ is a semibounded self-adjoint extension of S corresponding to the self-adjoint relation Θ in $\mathfrak{N}_x(S^*)$ as in (5.5.32)–(5.5.33). Then Θ is semibounded in $\mathfrak{N}_x(S^*)$ and the corresponding closed semibounded form ω_Θ in $\mathfrak{N}_x(S^*)$ and the closed semibounded form $\mathfrak{t}_{H_\Theta}$ corresponding to H_Θ are related by

$$\begin{aligned} \mathfrak{t}_{H_\Theta}[f,g] &= \mathfrak{t}_{S_{\mathrm{K},x}}[f,g] + \omega_\Theta[f_x, g_x], \\ \operatorname{dom} \mathfrak{t}_{H_\Theta} &= \bigl\{ f = f_{\mathrm{F}} + f_x \in \operatorname{dom} \mathfrak{t}_{S_{\mathrm{F}}} + \mathfrak{N}_x(S^*) : f_x \in \operatorname{dom} \omega_\Theta \bigr\}. \end{aligned}$$

Let H_Θ be a semibounded self-adjoint extension of S corresponding to the self-adjoint relation Θ in $\mathcal{G}$ as in (5.5.32)–(5.5.33). The first boundary condition in (5.5.33) is the *essential boundary condition* given by $\Gamma_0 \widehat{f} \in \operatorname{dom} \Theta_{\mathrm{op}}$. Since H_Θ is now assumed to be semibounded, it follows from $f \in \operatorname{dom} S^* \subset \operatorname{dom} \Lambda$ that this condition can be written as

$$\Lambda f = \Gamma_0 \widehat{f} \in \operatorname{dom} \Theta_{\mathrm{op}} \subset \operatorname{dom} (\Theta_{\mathrm{op}} - \mu)^{\frac{1}{2}}, \quad \mu \le m(\Theta),$$

which implies that $f \in \operatorname{dom} \mathfrak{t}_{H_\Theta}$. The second boundary condition in (5.5.33) is the *natural boundary condition* given by $P_{\mathrm{op}} \Gamma_1 \widehat{f} = \Theta_{\mathrm{op}} \Gamma_0 \widehat{f}$. It is subsumed in the additive term in the structure of the form $\mathfrak{t}_{H_\Theta}$:

$$\bigl((\Theta_{\mathrm{op}} - \mu)^{\frac{1}{2}} \Lambda f, (\Theta_{\mathrm{op}} - \mu)^{\frac{1}{2}} \Lambda g\bigr) + \mu \, (\Lambda f, \Lambda g)\,;$$

cf. Corollary 5.5.15, which in case of a bounded operator part Θ_{op} simplifies to

$$\bigl(\Theta_{\mathrm{op}} \Lambda f, \Lambda g\bigr);$$

cf. (5.6.49). In particular, the elements in $\operatorname{dom} H_\Theta$ satisfy an essential boundary condition if and only if $\operatorname{dom} \Theta \neq \mathcal{G}$, that is, $\Theta \notin \mathbf{B}(\mathcal{G})$. Note that the extreme case $\operatorname{dom} \Theta = \{0\}$ corresponds to $\Lambda f = 0$ and $\Gamma_0 \widehat{f} = 0$, i.e., $f \in \operatorname{dom} \mathfrak{t}_{S_{\mathrm{F}}}$ and $\widehat{f} \in S_{\mathrm{F}}$.

Remark 5.6.16. In Theorem 5.6.11 a one-to-one correspondence between the closed nonnegative forms ω in $\mathcal{G}$ and the semibounded self-adjoint extensions H of S in $\mathfrak{H}$ satisfying $S_1 \le H \le S_{\mathrm{F}}$ is established. For closed semibounded forms ω in $\mathcal{G}$ the situation is different: Although Theorem 5.6.13 shows that for each semibounded self-adjoint extension $H = H_\Theta$ of S there exists a closed semibounded form $\omega = \omega_\Theta$ in $\mathcal{G}$ such that

$$\mathfrak{t}_H[f,g] = \mathfrak{t}_{S_1}[f,g] + \omega[\Lambda f, \Lambda g], \tag{5.6.50}$$

one can also see that for an arbitrary closed semibounded form ω in $\mathcal{G}$ the right-hand side in (5.6.50) is not necessarily bounded from below. However, if, e.g., ω is a symmetric form with $\operatorname{dom} \omega = \mathcal{G}$ such that for some $a \ge 0$ and $b \in [0,1)$

$$|\omega[\Lambda f]| \le a \|f\|^2 + b\,\mathfrak{t}_{S_1}[f], \qquad f \in \operatorname{dom} \mathfrak{t}_{S_1},$$

then Theorem 5.1.16 shows that $\mathfrak{t}_H$ in (5.6.50) is a closed semibounded form with $\operatorname{dom}\mathfrak{t}_H = \operatorname{dom}\mathfrak{t}_{S_1}$ in $\mathfrak{H}$. In particular, in this situation the corresponding self-adjoint extension H of S is semibounded.

Recall from Proposition 5.5.8 that in the case of finite defect numbers or in the case that S_F has a compact resolvent the implication

$$\Theta \text{ semibounded in } \mathcal{G} \quad \Rightarrow \quad H_\Theta \text{ semibounded in } \mathfrak{H}$$

holds. The following corollary supplements Theorem 5.6.13 and can be seen as an extension and completion of the second part of Theorem 5.6.11. When the defect numbers are not finite or the resolvent of S_F is not compact there is in general no analog of the second part of Theorem 5.6.11.

Corollary 5.6.17. *Let S be a closed semibounded relation in $\mathfrak{H}$ and let S_1 be a semibounded self-adjoint extension of S such that S_1 and S_F are transversal. Let $\{\mathcal{G},\Gamma_0,\Gamma_1\}$ be a boundary triplet for S^*, let $\{\mathcal{G},\Lambda\}$ be a compatible boundary pair for S corresponding to S_1 and assume, in addition, that one of the following conditions hold:*

(i) *the defect numbers of S are finite;*
(ii) *$(S_\mathrm{F}-\lambda)^{-1}$ is a compact operator for some $\lambda\in\rho(S_\mathrm{F})$.*

Let Θ be a semibounded self-adjoint relation in $\mathcal{G}$ and let H_Θ be the corresponding self-adjoint extension of S as in (5.5.32)–(5.5.33). Then H_Θ is semibounded and the closed semibounded forms $\mathfrak{t}_{H_\Theta}$ and ω_Θ are related by (5.6.40).

The following corollary complements Corollary 5.6.12 (iii). If the symmetric relation S is positive a natural choice for S_1 is the Kreĭn–von Neumann extension $S_{\mathrm{K},0}$. A possible explicit choice for the boundary triplet can be found in Example 5.5.13.

Corollary 5.6.18. *Let S be a closed semibounded relation in $\mathfrak{H}$ with lower bound $\gamma>0$, let $\{\mathcal{G},\Gamma_0,\Gamma_1\}$ be a boundary triplet for S^*, and let $\{\mathcal{G},\Lambda\}$ be a compatible boundary pair for S corresponding to the Kreĭn–von Neumann extension $S_{\mathrm{K},0}$. Then the formula*

$$\begin{aligned}\mathfrak{t}_{H_\Theta}[f,g] &= \mathfrak{t}_{S_{\mathrm{K},0}}[f,g] + \bigl(\Theta_{\mathrm{op}}^{\frac12}\Lambda f, \Theta_{\mathrm{op}}^{\frac12}\Lambda g\bigr),\\ \operatorname{dom}\mathfrak{t}_{H_\Theta} &= \bigl\{f\in\operatorname{dom}\mathfrak{t}_{S_{\mathrm{K},0}} : \Lambda f\in\operatorname{dom}\Theta_{\mathrm{op}}^{\frac12}\bigr\},\end{aligned} \tag{5.6.51}$$

establishes a one-to-one correspondence between all closed nonnegative forms $\mathfrak{t}_{H_\Theta}$ corresponding to nonnegative self-adjoint extensions H_Θ of S in $\mathfrak{H}$ and all closed nonnegative forms ω_Θ corresponding to nonnegative self-adjoint relations Θ in $\mathcal{G}$.

Proof. By assumption, one has $S_{\mathrm{K},0}=\ker\Gamma_1$ and hence the Weyl function M corresponding to $\{\mathcal{G},\Gamma_0,\Gamma_1\}$ satisfies $M(0)=0$ by Corollary 5.5.2 (viii). Assume that H_Θ is a nonnegative self-adjoint extension of S with corresponding closed nonnegative form $\mathfrak{t}_{H_\Theta}$. Since $\gamma > 0$, Proposition 5.5.6 with $x = 0$ shows that the self-adjoint relation Θ in $\mathcal{G}$ is nonnegative. Formula (5.6.51) follows from Theorem 5.6.13 and Corollary 5.6.14 with $\mu = 0$. Conversely, if Θ is a nonnegative self-adjoint relation in $\mathcal{G}$, then Theorem 5.6.11 (ii) shows that H_Θ is a nonnegative self-adjoint extension of S and (5.6.51) holds. □

In the next corollary the ordering of semibounded self-adjoint extensions is translated in the ordering of the corresponding parameters.

Corollary 5.6.19. *Let S be a closed semibounded relation in $\mathfrak{H}$ and let $\{\mathcal{G}, \Gamma_0, \Gamma_1\}$ be a boundary triplet for S^*. Assume that*

$$\ker \Gamma_0 = S_{\mathrm{F}} \quad \text{and} \quad \ker \Gamma_1 = S_1,$$

where S_{F} is the Friedrichs extension and S_1 is a semibounded self-adjoint extension of S. Let H_{Θ_1} and H_{Θ_2} be semibounded self-adjoint extensions of S corresponding to the semibounded self-adjoint relations Θ_1 and Θ_2. Then

$$H_{\Theta_1} \leq H_{\Theta_2} \quad \Leftrightarrow \quad \Theta_1 \leq \Theta_2. \tag{5.6.52}$$

In particular, $S_1 \leq H_{\Theta_2} \Leftrightarrow 0 \leq \Theta_2$.

Proof. Let $\{\mathcal{G}, \Lambda\}$ be a compatible boundary pair for S corresponding to S_1 as in Theorem 5.6.6. Then according to Theorem 5.6.13 one has the following identities

$$\begin{aligned} \mathfrak{t}_{H_{\Theta_1}}[f,g] &= \mathfrak{t}_{S_1}[f,g] + \omega_{\Theta_1}[\Lambda f, \Lambda g], \\ \operatorname{dom} \mathfrak{t}_{H_{\Theta_1}} &= \bigl\{ f \in \operatorname{dom} \mathfrak{t}_{S_1} : \Lambda f \in \operatorname{dom} \omega_{\Theta_1} \bigr\}, \end{aligned} \tag{5.6.53}$$

and

$$\begin{aligned} \mathfrak{t}_{H_{\Theta_2}}[f,g] &= \mathfrak{t}_{S_1}[f,g] + \omega_{\Theta_2}[\Lambda f, \Lambda g], \\ \operatorname{dom} \mathfrak{t}_{H_{\Theta_2}} &= \bigl\{ f \in \operatorname{dom} \mathfrak{t}_{S_1} : \Lambda f \in \operatorname{dom} \omega_{\Theta_2} \bigr\}. \end{aligned} \tag{5.6.54}$$

Recall from Theorem 5.2.4 that $H_{\Theta_1} \leq H_{\Theta_2}$ if and only if $\mathfrak{t}_{H_{\Theta_1}} \leq \mathfrak{t}_{H_{\Theta_2}}$. This last statement means by definition that

$$\operatorname{dom} \mathfrak{t}_{H_{\Theta_2}} \subset \operatorname{dom} \mathfrak{t}_{H_{\Theta_1}} \quad \text{and} \quad \mathfrak{t}_{H_{\Theta_1}}[f] \leq \mathfrak{t}_{H_{\Theta_2}}[f], \qquad f \in \operatorname{dom} \mathfrak{t}_{H_{\Theta_2}}, \tag{5.6.55}$$

which, via (5.6.53) and (5.6.54), is equivalent to

$$\operatorname{dom} \mathfrak{t}_{H_{\Theta_2}} \subset \operatorname{dom} \mathfrak{t}_{H_{\Theta_1}} \quad \text{and} \quad \omega_{\Theta_1}[\Lambda f] \leq \omega_{\Theta_2}[\Lambda f], \qquad f \in \operatorname{dom} \mathfrak{t}_{H_{\Theta_2}}. \tag{5.6.56}$$

Assume now that $H_{\Theta_1} \leq H_{\Theta_2}$, i.e., that (5.6.56) (and (5.6.55)) holds. First it will be shown that $\operatorname{dom} \mathfrak{t}_{H_{\Theta_2}} \subset \operatorname{dom} \mathfrak{t}_{H_{\Theta_1}}$ implies that

$$\operatorname{dom} \Theta_2 \subset \operatorname{dom} \omega_{\Theta_1}. \tag{5.6.57}$$

To see this, let $\varphi \in \operatorname{dom}\Theta_2$. Then $\{\varphi,\varphi'\} \in \Theta_2$ for some $\varphi' \in \mathcal{G}$. Now choose $\{f,f'\} \in H_{\Theta_2} \subset S^*$ with the property $\Gamma\{f,f'\} = \{\varphi,\varphi'\}$. Then it follows that $\Lambda f = \Gamma_0\{f,f'\} = \varphi$. Furthermore, since $f \in \operatorname{dom} S^* \subset \operatorname{dom} \mathfrak{t}_{S_1}$ and

$$\varphi = \Lambda f \in \operatorname{dom}\Theta_2 \subset \operatorname{dom}\omega_{\Theta_2},$$

it follows from $\operatorname{dom}\mathfrak{t}_{H_{\Theta_2}} \subset \operatorname{dom}\mathfrak{t}_{H_{\Theta_1}}$ that $\varphi = \Lambda f \in \operatorname{dom}\omega_{\Theta_1}$. Hence, (5.6.57) has been shown. Next observe that due to the previous reasoning the inequality in (5.6.56) gives

$$\omega_{\Theta_1}[\varphi] \leq \omega_{\Theta_2}[\varphi], \quad \varphi \in \operatorname{dom}\Theta_2. \tag{5.6.58}$$

Denote the restriction of the form ω_{Θ_2} to $\operatorname{dom}\Theta_2$ by $\mathring{\omega}_{\Theta_2}$. Then the inclusion (5.6.57) and the inequality (5.6.58) can be written as

$$\omega_{\Theta_1} \leq \mathring{\omega}_{\Theta_2}, \tag{5.6.59}$$

and, since $\operatorname{dom}\Theta_2$ is a core of ω_{Θ_2}, it follows from (5.6.59) and Lemma 5.2.2 (v) that

$$\omega_{\Theta_1} \leq \omega_{\Theta_2} \quad \text{or, equivalently,} \quad \Theta_1 \leq \Theta_2.$$

Hence, $H_{\Theta_1} \leq H_{\Theta_2}$ implies that $\Theta_1 \leq \Theta_2$.

For the converse statement assume $\Theta_1 \leq \Theta_2$ or, equivalently, $\omega_{\Theta_1} \leq \omega_{\Theta_2}$, i.e.,

$$\operatorname{dom}\omega_{\Theta_2} \subset \operatorname{dom}\omega_{\Theta_1} \quad \text{and} \quad \omega_{\Theta_1}[\varphi] \leq \omega_{\Theta_2}[\varphi], \qquad \varphi \in \operatorname{dom}\omega_{\Theta_2}. \tag{5.6.60}$$

It will be shown that (5.6.56) holds. Let $f \in \operatorname{dom}\mathfrak{t}_{H_{\Theta_2}}$, so that $f \in \operatorname{dom}\mathfrak{t}_{S_1}$ and $\Lambda f \in \operatorname{dom}\omega_{\Theta_2}$. Then it follows from (5.6.60) that also $\Lambda f \in \operatorname{dom}\omega_{\Theta_1}$. Hence, one sees that $\operatorname{dom}\mathfrak{t}_{H_{\Theta_2}} \subset \operatorname{dom}\mathfrak{t}_{H_{\Theta_1}}$. Furthermore, if $f \in \operatorname{dom}\mathfrak{t}_{H_{\Theta_2}}$, then it follows directly from (5.6.60) that $\omega_{\Theta_1}[\Lambda f] \leq \omega_{\Theta_2}[\Lambda f]$. Thus, (5.6.56) holds and one concludes that $H_1 \leq H_2$.

Finally, note that for the choice $\Theta_1 = 0$ one has $H_{\Theta_1} = \ker\Gamma_1 = S_1$ and hence the equivalence (5.6.52) takes the form $S_1 \leq H_{\Theta_2} \Leftrightarrow 0 \leq \Theta_2$. □

If S is a semibounded relation in $\mathfrak{H}$ with lower bound γ and one chooses H_{Θ_1} to be the Kreĭn type extension $S_{\mathrm{K},x}$ for some $x \leq \gamma$ in the previous corollary, then the next statement follows from (5.5.1).

Corollary 5.6.20. *Let the assumptions be as in Corollary 5.6.19 and let H_Θ be a semibounded self-adjoint extension of S corresponding to the self-adjoint relation Θ in $\mathcal{G}$ as in (5.5.32)–(5.5.33). Then for any $x \leq m(S)$*

$$S_{\mathrm{K},x} \leq H_\Theta \quad \Leftrightarrow \quad M(x) \leq \Theta.$$

Chapter 6

Sturm–Liouville Operators

Second-order Sturm–Liouville differential expressions generate self-adjoint differential operators in weighted L^2-spaces on an interval (a, b). A brief exposition of elementary properties of Sturm–Liouville expressions can be found in Section 6.1; this includes the limit-circle and limit-point terminology. The corresponding maximal and minimal operators associated with the Sturm–Liouville expression are introduced in Section 6.2; here one can find also a discussion of quasi-derivatives which are useful in limit-circle cases. In this chapter the boundary triplets and Weyl functions which can be associated with Sturm–Liouville operators will be studied. The case where the endpoints of (a, b) are regular or limit-circle is treated in Section 6.3, while the case where a is regular or limit-circle and b is limit-point is treated in Section 6.4. In each of these sections the spectrum is related to the limit properties of the Weyl function, as in Chapter 3. The case where both endpoints are limit-point can be found in Section 6.5. Here the useful technique of interface conditions is explained by means of the coupling concept in Section 4.6. Closely related is the case of exit space extensions resulting in boundary conditions depending on the eigenvalue parameter; such extensions are treated in Section 6.6. The characterization of the spectrum via subordinate solutions can be found in Section 6.7, where again the results in Chapter 3 play a central role. The rest of this chapter is devoted to boundary triplets and Weyl functions for Sturm–Liouville operators which are semibounded. Particular attention is paid to the corresponding semibounded forms and boundary pairs; cf. Section 5.6. The special case of regular endpoints is given in Section 6.8. In the singular case it is possible to construct semibounded closed forms by means of solutions which are nonoscillatory near the endpoints. This construction, which is suggested by a particular form of the Green formula, can be found in Section 6.9. Section 6.10 contains an overview of the necessary properties of the so-called nonprincipal and principal solutions which make these forms useful. The case where both endpoints a and b are limit-circle is treated in Section 6.11, while the case where a is limit-circle and b is limit-point is treated in Section 6.12. In each section the connection between the boundary triplet and the form is studied in detail as in Section 5.6.

J. Behrndt et al., *Boundary Value Problems, Weyl Functions, and Differential Operators*, Monographs in Mathematics 108,
https://doi.org/10.1007/978-3-030-36714-5_7

Finally, in Section 6.13 the particular case $L = -D^2 + q$, where q is a real integrable potential on a half-line is treated. Here again the spectral theory from Chapter 3 will be employed.

6.1 Sturm–Liouville differential expressions

This section offers a brief review of the properties of the Sturm–Liouville differential expression L defined by

$$L = \frac{1}{r}\left[-DpD + q\right], \quad D = d/dx, \tag{6.1.1}$$

where p, q, and r are assumed to be real functions on an open interval (a, b) with $-\infty \leq a < b \leq \infty$. Throughout the text the following minimal conditions will be imposed:

$$\begin{cases} p(x) \neq 0, \ r(x) > 0, & \text{for almost all } x \in (a, b), \\ 1/p, q, r \in L^1_{\mathrm{loc}}(a, b). \end{cases} \tag{6.1.2}$$

Here $L^1_{\mathrm{loc}}(a, b)$ stands for the linear space of all (equivalence classes of) complex functions which are integrable on each compact subset $K \subset (a, b)$.

For the reader's convenience some more notations which are used in the following are collected. The space of complex integrable functions on (a, b) will be denoted by $L^1(a, b)$. One denotes by $L^2_{r,\mathrm{loc}}(a, b)$ the set of all functions f for which $|f|^2 r \in L^1_{\mathrm{loc}}(a, b)$, while $L^2_r(a, b)$ stands for the set of all functions f for which $|f|^2 r \in L^1(a, b)$. The space $L^2_r(a, b)$ is a Hilbert space when equipped with the usual inner product

$$(f, g)_{L^2_r(a,b)} := \int_a^b f(x)\overline{g(x)} r(x)\, dx.$$

For $f \in L^2_{r,\mathrm{loc}}(a, b)$ the generic notations

$$f \in L^2_r(a, a') \quad \text{or} \quad f \in L^2_r(b', b)$$

indicate that $|f|^2 r$ is integrable on an interval (a, a') for some, and hence for all $a < a' < b$ or on an interval (b', b) for some, and hence for all $a < b' < b$, respectively. A complex function f is *absolutely continuous* on (a, b) if there exists $g \in L^1_{\mathrm{loc}}(a, b)$ such that

$$f(x) - f(y) = \int_y^x g(t)\, dt \tag{6.1.3}$$

for all $x, y \in (a, b)$. One denotes by $AC(a, b)$ the linear space of absolutely continuous functions on (a, b). Note that if $f \in AC(a, b)$, then f is differentiable almost everywhere on (a, b) and $f' = g$ almost everywhere, where g is as in (6.1.3). When $a \in \mathbb{R}$, then $AC[a, b)$ stands for the subclass of $f \in AC(a, b)$ for which

$g \in L^1_{\rm loc}(a,b)$ in (6.1.3) additionally belongs to $L^1(a,a')$ for some, and hence for all $a < a' < b$, in which case

$$f(x) - f(a) = \int_a^x g(t)\,dt$$

for all $x \in (a,b)$ and thus $f(a) = \lim_{x\to a} f(x)$. When $b \in \mathbb{R}$ there is a similar notation $AC(a,b]$ and for $f \in AC(a,b]$ one has $f(b) = \lim_{x\to b} f(x)$. The notation $AC[a,b]$ is analogous. If for some $a < c < b$ there are functions $f_{\rm l} \in AC(a,c]$ and $f_{\rm r} \in AC[c,b)$ with the property $f_{\rm l}(c) = f_{\rm r}(c)$, then the function $f : (a,b) \to \mathbb{C}$ defined by

$$f(x) = \begin{cases} f_{\rm l}(x), & a < x \leq c, \\ f_{\rm r}(x), & c < x < b, \end{cases}$$

belongs to the space $AC(a,b)$, as follows easily from the above observations.

In order to apply the differential expression L in (6.1.1) to a complex function f on (a,b) in a meaningful way one must first assume that $f \in AC(a,b)$, so that as a consequence f is differentiable almost everywhere. However, then the function pf' is only defined almost everywhere. The natural domain of the differential expression L in (6.1.1) is the linear space of all $f \in AC(a,b)$ for which the equivalence class $[pf']$ (in the sense of Lebesgue measure) contains an absolutely continuous function, which again will be denoted by pf'. For such functions f one defines Lf by

$$(Lf)(x) = \frac{1}{r(x)}\big[-(pf')'(x) + q(x)f(x)\big], \quad x \in (a,b),$$

so that $(Lf)(x)$ is well defined almost everywhere. This convention will be used tacitly: for $f \in AC(a,b)$ the assertion $pf' \in AC(a,b)$ means that the equivalence class $[pf']$ contains an absolutely continuous function, which is denoted by pf'. Sufficient conditions for the equivalence class $[pf']$ to have a representative in $AC(a,b)$ can be found in Theorem 6.1.2.

The following simple observation will be used frequently: for all functions $f \in AC(a,b)$ with $pf' \in AC(a,b)$ one has

$$L\overline{f} = \overline{Lf} \tag{6.1.4}$$

since the coefficient functions of L are assumed to be real.

Often the coefficient functions are integrable in a neighborhood of an endpoint. Hence, the following definition is presented.

Definition 6.1.1. Let the coefficient functions p, q, and r of the differential expression L in (6.1.1) satisfy the conditions (6.1.2). Then L is said to be *regular*

(i) *at the endpoint* a if one has $a \in \mathbb{R}$ and $1/p, q, r \in L^1(a, a')$ for one, and hence for all $a' \in (a, b)$.

(ii) *at the endpoint* b if one has $b \in \mathbb{R}$ and $1/p, q, r \in L^1(b', b)$ for one, and hence for all $b' \in (a, b)$.

The endpoint a or b is said to be *regular* if L is regular there and *singular* if a or b is not a regular endpoint, respectively.

Under the conditions in (6.1.2) there is an existence and uniqueness result for initial value problems involving the inhomogeneous equation $(L - \lambda)f = g$ when the initial values are posed at an interior point of (a, b). The initial value problem may also be posed at a finite endpoint when the differential expression L is regular there. The uniqueness and existence result can be proved by writing the Sturm–Liouville equation as a first-order system of differential equations:

$$\begin{pmatrix} f \\ pf' \end{pmatrix}' = \begin{pmatrix} 0 & 1/p \\ q - \lambda r & 0 \end{pmatrix} \begin{pmatrix} f \\ pf' \end{pmatrix} - \begin{pmatrix} 0 \\ rg \end{pmatrix}.$$

The new initial value problem is equivalent to a Volterra integral equation which can be solved in the usual way by successive approximations when all data are locally integrable; see, e.g., [754, Theorem 2.1]. Note also that the assumption $g \in L^2_{r,\mathrm{loc}}(a, b)$ in the next theorem implies that $rg \in L^1_{\mathrm{loc}}(a, b)$; this follows by means of the Cauchy–Schwarz inequality from the condition $r \in L^1_{\mathrm{loc}}(a, b)$.

Theorem 6.1.2. *Let $g \in L^2_{r,\mathrm{loc}}(a, b)$ and $c_1, c_2 \in \mathbb{C}$. Then for all $\lambda \in \mathbb{C}$ and $x_0 \in (a, b)$ the initial value problem*

$$(L - \lambda)f = g, \qquad f(x_0) = c_1, \quad (pf')(x_0) = c_2, \tag{6.1.5}$$

has a unique solution $f \in AC(a, b)$ for which $pf' \in AC(a, b)$. Moreover, if a or b is regular, then $x_0 = a$ or $x_0 = b$ is allowed and f, pf' belong to $AC[a, b)$ or $AC(a, b]$, respectively. In addition, the functions

$$\lambda \mapsto f(x_0, \lambda) \quad \text{and} \quad \lambda \mapsto (pf')(x_0, \lambda)$$

are entire for each $x_0 \in (a, b)$ and for $x_0 = a$ or $x_0 = b$ if the endpoint a or b is regular, respectively.

Let f, g be complex functions in $AC(a, b)$. Then the *Wronskian determinant* $W(f, g)$ is defined by

$$W(f, g) = p(fg' - f'g) = f(pg') - (pf')g. \tag{6.1.6}$$

The value at $x \in (a, b)$ of $W(f, g)$ will be denoted by $W_x(f, g)$. For complex functions f, g, h, k in $AC(a, b)$ one has the so-called *Plücker identity*

$$W(f, g)W(h, k) = W(f, h)W(g, k) - W(f, k)W(g, h). \tag{6.1.7}$$

This identity can be easily verified by writing out the various terms according to (6.1.6). The Wronskian determinant and the Plücker identity will be applied in conjunction with the differential expression L. Assume, in addition, that $pf', pg' \in AC(a,b)$ (in the sense that their equivalence classes contain an element in $AC(a,b)$). Then differentiation of the Wronskian gives an identity involving the differential expression L:

$$(W(f,g))' = r\,[(Lf)\,g - f\,(Lg)]. \tag{6.1.8}$$

When g in (6.1.8) is a solution of $(L-\mu)y = 0$ for some $\mu \in \mathbb{C}$ one obtains the useful formula

$$(W(f,g))' = r\,((L-\mu)f)\,g. \tag{6.1.9}$$

Furthermore, if for some $\lambda \in \mathbb{C}$ the function f is a solution of $(L-\lambda)y = 0$, then (6.1.9) gives

$$(W(f,g))' = r\,(\lambda-\mu) f\,g. \tag{6.1.10}$$

In particular, one sees that the Wronskian $x \mapsto W_x(f,g)$ is constant for solutions f,g of $(L-\lambda)y = 0$. Moreover, if the functions f,g are solutions of $(L-\lambda)y = 0$, then it is straightforward to verify that these functions are linearly independent if and only if $W(f,g) \neq 0$.

Integration by parts in (6.1.8) over a compact subinterval $[\alpha,\beta] \subset (a,b)$ leads to the Green or Lagrange identity:

$$\int_\alpha^\beta \left[(Lf)(x)\,\overline{g(x)} - f(x)\,\overline{(Lg)(x)}\,\right] r(x)\,dx = W_x(f,\bar{g})|_\alpha^\beta, \tag{6.1.11}$$

assuming that $f, pf', g, pg' \in AC(a,b)$ and $Lf, Lg \in L^2_r(\alpha,\beta)$. In particular, when $f = g$ are solutions of $(L-\lambda)y = 0$ for some $\lambda \in \mathbb{C}$ this gives

$$(\lambda - \bar{\lambda}) \int_\alpha^\beta |f(x)|^2\, r(x)\,dx = W_x(f,\bar{f})|_\alpha^\beta, \tag{6.1.12}$$

see also (6.1.10).

The interest in the present chapter is in solutions of $(L-\lambda)y = 0$ which are square-integrable with respect to the weight r near the endpoint a or the endpoint b. Thus, if f is a solution of $(L-\lambda)y = 0$ on (a,b) for some $\lambda \in \mathbb{C}$, then for $\lambda \in \mathbb{C}\setminus\mathbb{R}$ the identity (6.1.12) implies that

$$f \in L^2_r(a,a') \quad \Leftrightarrow \quad \lim_{x\to a} W_x(f,\bar{f}) \quad \text{exists},$$

$$f \in L^2_r(b',b) \quad \Leftrightarrow \quad \lim_{x\to b} W_x(f,\bar{f}) \quad \text{exists}.$$

Clearly, if these statements hold for some $a < a' < b$ or $a < b' < b$, then they hold for all $a < a' < b$ or $a < b' < b$, respectively.

It follows that under the circumstances of Theorem 6.1.2 for any $\lambda \in \mathbb{C}$ the homogeneous equation has a *fundamental system* of solutions, i.e., for any $\lambda \in \mathbb{C}$ there are two solutions $u_1(\cdot,\lambda)$ and $u_2(\cdot,\lambda)$ of $(L-\lambda)y = 0$ which are linearly independent. This can be seen by imposing the conditions

$$\begin{pmatrix} u_1(x_0,\lambda) & u_2(x_0,\lambda) \\ (pu_1')(x_0,\lambda) & (pu_2')(x_0,\lambda) \end{pmatrix} = \begin{pmatrix} 1 & 0 \\ 0 & 1 \end{pmatrix}$$

for a fixed $x_0 \in (a,b)$; when L is regular at the endpoint a or b, then $x_0 = a$ or $x_0 = b$, respectively, is also allowed. Note that then the Wronskian determinant satisfies $W_x(u_1(\cdot,\lambda), u_2(\cdot,\lambda)) = 1$ for all $x \in (a,b)$. Hence, for $g \in L^2_{r,\mathrm{loc}}(a,b)$ it is clear that the function

$$h(x) = u_1(x,\lambda)\int_{x_0}^{x} u_2(t,\lambda)g(t)r(t)\,dt - u_2(x,\lambda)\int_{x_0}^{x} u_1(t,\lambda)g(t)r(t)\,dt$$

belongs to $AC(a,b)$, while $(ph')(x)$ is equal almost everywhere to

$$(pu_1')(x,\lambda)\int_{x_0}^{x} u_2(t,\lambda)g(t)r(t)\,dt - (pu_2')(x,\lambda)\int_{x_0}^{x} u_1(t,\lambda)g(t)r(t)\,dt,$$

and the last expression belongs to $AC(a,b)$. Hence, h provides a solution of the inhomogeneous equation $(L-\lambda)h = g$ with $h(x_0) = 0$ and $(ph')(x_0) = 0$. Moreover, by adding $c_1u_1(\cdot,\lambda)+c_2u_2(\cdot,\lambda)$, $c_1, c_2 \in \mathbb{C}$, to the solution h one obtains a function

$$f = h + c_1u_1(\cdot,\lambda) + c_2u_2(\cdot,\lambda) \tag{6.1.13}$$

which is a solution of the initial value problem (6.1.5). The formula (6.1.13) is sometimes referred to as the *variation of constants formula*.

The following result is about smoothly cutting off the solution of an inhomogeneous Sturm–Liouville equation near an endpoint of the interval (a,b), so that it becomes trivial in a neighborhood of that endpoint.

Proposition 6.1.3. *Let $g \in L^2_{r,\mathrm{loc}}(a,b)$, $\lambda \in \mathbb{C}$, and let f be a solution of the inhomogeneous equation*

$$(L-\lambda)f = g,$$

with $f, pf' \in AC(a,b)$. Let $[\alpha,\beta] \subset (a,b)$ be a compact subinterval. Then there exist functions f_a and g_a with

$$f_a, pf_a' \in AC(a,b) \quad \text{and} \quad g_a \in L^2_{r,\mathrm{loc}}(a,b),$$

such that

$$(L-\lambda)f_a = g_a,$$

and, in addition,

$$f_a(t) = \begin{cases} f(t), & t \in (a,\alpha], \\ 0, & t \in [\beta,b), \end{cases} \qquad g_a(t) = \begin{cases} g(t), & t \in (a,\alpha], \\ 0, & t \in [\beta,b). \end{cases} \tag{6.1.14}$$

Likewise, there exist functions f_b *and* g_b *with*

$$f_b, pf_b' \in AC(a,b) \quad \textit{and} \quad g_b \in L^2_{r,\mathrm{loc}}(a,b),$$

such that

$$(L-\lambda)f_b = g_b,$$

and, in addition,

$$f_b(t) = \begin{cases} 0, & t \in (a,\alpha], \\ f(t), & t \in [\beta,b), \end{cases} \qquad g_b(t) = \begin{cases} 0, & t \in (a,\alpha], \\ g(t), & t \in [\beta,b). \end{cases} \tag{6.1.15}$$

Proof. The cut-off process at the endpoint a as exhibited in (6.1.15) will be shown; the other case of cutting off at the endpoint b as in (6.1.14) is treated in a similar way. Define the functions f_b and g_b as indicated on the interval (a,α) and on the interval (β,b). On the interval $[\alpha,\beta]$ choose any function $h \in L^2_r(\alpha,\beta)$ and define the function f_b on (α,β) by

$$f_b(x) = u_1(x,\lambda)\int_\alpha^x u_2(t,\lambda)h(t)r(t)\,dt - u_2(x,\lambda)\int_\alpha^x u_1(t,\lambda)h(t)r(t)\,dt,$$

where $u_1(\cdot,\lambda)$ and $u_2(\cdot,\lambda)$ form a fundamental system of $(L-\lambda)y=0$, fixed by standard initial conditions at α:

$$\begin{pmatrix} u_1(\alpha,\lambda) & u_2(\alpha,\lambda) \\ (pu_1')(\alpha,\lambda) & (pu_2')(\alpha,\lambda) \end{pmatrix} = \begin{pmatrix} 1 & 0 \\ 0 & 1 \end{pmatrix};$$

cf. Theorem 6.1.2. Then it is clear that on the interval $[\alpha,\beta]$ the function f_b satisfies $(L-\lambda)f_b = h$ and $f_b(\alpha)=0$, $(pf_b')(\alpha)=0$. Furthermore, one sees that

$$\begin{pmatrix} f_b(\beta) \\ (pf_b')(\beta) \end{pmatrix} = \begin{pmatrix} u_1(\beta,\lambda) & u_2(\beta,\lambda) \\ (pu_1')(\beta,\lambda) & (pu_2')(\beta,\lambda) \end{pmatrix} \begin{pmatrix} \int_\alpha^\beta u_2(t,\lambda)h(t)r(t)\,dt \\ -\int_\alpha^\beta u_1(t,\lambda)h(t)r(t)\,dt \end{pmatrix}.$$

Now one may choose the function $h \in L^2_r(\alpha,\beta)$ in such a way that

$$\begin{pmatrix} f_b(\beta) \\ (pf_b')(\beta) \end{pmatrix} = \begin{pmatrix} f(\beta) \\ (pf')(\beta) \end{pmatrix}.$$

To see this, note that the mapping from $L^2_r(\alpha,\beta)$ to $\mathbb{C}^2$ defined by

$$h \mapsto \begin{pmatrix} \int_\alpha^\beta u_2(t,\lambda)h(t)r(t)\,dt \\ -\int_\alpha^\beta u_1(t,\lambda)h(t)r(t)\,dt \end{pmatrix}$$

is surjective: any element in $\mathbb{C}^2$ which is orthogonal to its range is trivial.

Observe that with such a choice of h the above components of f_b belong to $AC(a,\alpha]$, $AC[\alpha,\beta]$, and $AC[\beta,b)$, respectively, and that there are no jumps. There is a similar statement for pf_b' and hence one concludes that $f_b, pf_b' \in AC(a,b)$. Since $h \in L^2_r(\alpha,\beta)$, one sees that the choice

$$g_b(t) = h(t), \quad t \in [\alpha,\beta],$$

implies $g_b \in L^2_{r,\mathrm{loc}}(a,b)$ and $(L-\lambda)f_b = g_b$. □

The following lemma is useful in the proof of Theorem 6.1.5.

Lemma 6.1.4. *Assume that $r \in L^1_{\mathrm{loc}}(b',b)$ is nonnegative almost everywhere and let $\varphi \in L^2_r(b',b)$ be a nonnegative function. If $u \in L^2_{r,\mathrm{loc}}[b',b)$ and there exist nonnegative constants A and B such that*

$$|u(x)|^2 \leq \varphi(x)^2 \left(A + B \int_{b'}^x |u(s)|^2 r(s)\,ds \right), \quad b' \leq x < b, \tag{6.1.16}$$

then $u \in L^2_r(b',b)$.

Proof. For $B = 0$ the statement is clear. In the following it will be assumed that $B > 0$. Since $\varphi \in L^2_r(b',b)$, one can choose $b' < c < b$ such that

$$2B \left(\int_c^b \varphi(x)^2 r(x)\,dx \right) < 1.$$

Let $y \in \mathbb{R}$ be an arbitrary number with $b' < c < y < b$. Then it is clear from the assumption (6.1.16) that

$$|u(x)|^2 \leq A\varphi(x)^2 + B\varphi(x)^2 \int_{b'}^y |u(s)|^2 r(s)\,ds, \quad c \leq x \leq y.$$

Multiply this inequality by $2r(x)$ and integrate over the interval $[c,y]$; then

$$\begin{aligned} 2\int_c^y |u(x)|^2 r(x)\,dx &\leq 2A \int_c^y \varphi(x)^2 r(x)\,dx \\ &\quad + 2B \left(\int_c^y \varphi(x)^2 r(x)\,dx \right) \left(\int_{b'}^y |u(s)|^2 r(s)\,ds \right) \\ &< \frac{A}{B} + \int_{b'}^y |u(s)|^2 r(s)\,ds. \end{aligned}$$

It follows that for any $c < y < b$

$$\int_c^y |u(x)|^2 r(x)\,dx \leq \frac{A}{B} + \int_{b'}^c |u(s)|^2 r(s)\,ds.$$

The monotone convergence theorem implies that $u \in L^2_r(c,b)$, and hence one concludes $u \in L^2_r(b',b)$. □

The next two theorems present fundamental results proved in a purely analytical way.

Theorem 6.1.5. *Assume that all solutions of* $(L - \lambda_0)y = 0$, $\lambda_0 \in \mathbb{R}$, *belong to* $L^2_r(a, a')$ *or to* $L^2_r(b', b)$ *for some* $\lambda_0 \in \mathbb{C}$, *respectively. Then for any* $\lambda \in \mathbb{C}$ *all solutions of* $(L - \lambda)y = 0$ *belong to* $L^2_r(a, a')$ *or to* $L^2_r(b', b)$, *respectively.*

Proof. It suffices to give the proof for the endpoint b. In the following, fix $\lambda_0 \in \mathbb{C}$ and let $u_1 := u_1(\cdot, \lambda_0)$ and $u_2 := u_2(\cdot, \lambda_0)$ be two linearly independent solutions of $(L - \lambda_0)y = 0$ such that

$$u_1 \in L^2_r(b', b) \quad \text{and} \quad u_2 \in L^2_r(b', b) \tag{6.1.17}$$

for some $a < b' < b$. Let $\lambda \in \mathbb{C}$ and let $u := u(\cdot, \lambda)$ be an arbitrary solution of $(L - \lambda)y = 0$. It will be shown that $u \in L^2_r(b', b)$, which proves the theorem.

Assume without loss of generality that

$$\begin{pmatrix} u_1(b', \lambda_0) & u_2(b', \lambda_0) \\ (pu_1')(b', \lambda_0) & (pu_2')(b', \lambda_0) \end{pmatrix} = \begin{pmatrix} 1 & 0 \\ 0 & 1 \end{pmatrix}.$$

Observe that $(L - \lambda_0)u = (\lambda - \lambda_0)u$, so that by the variation of constants formula there exist $\alpha_1, \alpha_2 \in \mathbb{C}$ such that

$$\begin{aligned} u(x) &= \alpha_1 u_1(x) + \alpha_2 u_2(x) \\ &\quad + (\lambda - \lambda_0) \left[u_1(x) \int_{b'}^x u_2(s)u(s)r(s)\, ds - u_2(x) \int_{b'}^x u_1(s)u(s)r(s)\, ds \right] \end{aligned}$$

for all $x \in (a, b)$. Define

$$\alpha = \max\{|\alpha_1|, |\alpha_2|\}, \quad \varphi(x) = \max\{|u_1(x)|, |u_2(x)|\}, \quad x \in (a, b),$$

and note that $\varphi \in L^2_r(b', b)$ by (6.1.17). It follows from the above representation of u that

$$|u(x)| \le 2 \left[\alpha\varphi(x) + |\lambda - \lambda_0|\varphi(x) \int_{b'}^x \varphi(s)|u(s)|r(s)\, ds \right],$$

which leads to

$$|u(x)|^2 \le 8\alpha^2\varphi(x)^2 + 8|\lambda - \lambda_0|^2\varphi(x)^2 \left(\int_{b'}^x \varphi(s)|u(s)|r(s)\, ds \right)^2.$$

An application of the Cauchy–Schwarz inequality gives

$$|u(x)|^2 \le 8\alpha^2\varphi(x)^2 + B\varphi(x)^2 \int_{b'}^x |u(s)|^2 r(s)\, ds,$$

where

$$B = 8|\lambda - \lambda_0|^2 \int_{b'}^b \varphi(s)^2 r(s)\, ds.$$

Since $\varphi \in L^2_r(b', b)$ one can apply Lemma 6.1.4 with $A = 8\alpha^2$ and B as above. This leads to $u \in L^2_r(b', b)$, as claimed. □

Limit-circle case and limit-point case. The following discussion is devoted to the construction of solutions of $(L-\lambda)y=0$ that belong to $L^2_r(a,a')$ or $L^2_r(b',b)$. Here the case for $L^2_r(b',b)$ will be considered; the treatment for the case $L^2_r(a,a')$ is entirely similar. For $\lambda \in \mathbb{C}\setminus\mathbb{R}$ let $u := u(\cdot,\lambda)$ and $v := v(\cdot,\lambda)$ be solutions of $(L-\lambda)y=0$, and denote

$$\{u,v\}_x = \frac{W_x(u,\bar{v})}{\lambda-\bar{\lambda}}.$$

It is clear that $u \mapsto \{u,v\}_x$ is linear and $v \mapsto \{u,v\}_x$ is anti-linear; in addition, $\{u,v\}_x = \overline{\{v,u\}_x}$. Fix $a<c<b$, then for any solution u of $(L-\lambda)y=0$ one has

$$\int_c^x |u(t)|^2\, r(t)\, dt = \{u,u\}_x - \{u,u\}_c; \tag{6.1.18}$$

cf. (6.1.12). Hence, the function $x \mapsto \{u,u\}_x$ is nondecreasing on (c,b). In the following let $u_1 := u_1(\cdot,\lambda)$ and $u_2 := u_2(\cdot,\lambda)$ be two linearly independent solutions of $(L-\lambda)y=0$ fixed by

$$\begin{pmatrix} u_1(c,\lambda) & u_2(c,\lambda) \\ (pu_1')(c,\lambda) & (pu_2')(c,\lambda)\end{pmatrix} = \begin{pmatrix} 1 & 0 \\ 0 & 1\end{pmatrix}, \tag{6.1.19}$$

so that $W(u_1,u_2)=1$. Then it is clear that $\{u_1,u_1\}_c=0$, $\{u_2,u_2\}_c=0$, and that for $x>c$:

$$\begin{aligned} \{u_1,u_1\}_x &= \int_c^x |u_1(t)|^2\, r(t)\, dt > 0, \\ \{u_2,u_2\}_x &= \int_c^x |u_2(t)|^2\, r(t)\, dt > 0; \end{aligned} \tag{6.1.20}$$

cf. (6.1.18). For each $\zeta \in \mathbb{C}$ define a solution $u = u(\cdot,\lambda)$ of $(L-\lambda)y=0$ in terms of the fundamental system u_1, u_2 by $u = \zeta u_1 + u_2$. Then for $c<x<b$ one has

$$\{u,u\}_x = \zeta\bar{\zeta}\{u_1,u_1\}_x + \zeta\{u_1,u_2\}_x + \bar{\zeta}\{u_2,u_1\}_x + \{u_2,u_2\}_x$$

and it easily follows from this identity that

$$\frac{\{u,u\}_x}{\{u_1,u_1\}_x} = \left(|\zeta-\zeta_x|^2 - \frac{\{u_1,u_2\}_x\{u_2,u_1\}_x - \{u_1,u_1\}_x\{u_2,u_2\}_x}{\{u_1,u_1\}_x\{u_1,u_1\}_x}\right),$$

where $\zeta_x \in \mathbb{C}$ is defined by

$$\zeta_x = -\frac{\{u_2,u_1\}_x}{\{u_1,u_1\}_x}. \tag{6.1.21}$$

It is a direct consequence of the definition, $(\lambda-\bar{\lambda})^2 = -4|\mathrm{Im}\,\lambda|^2$, and the Plücker identity (6.1.7) with $f = u_1$, $g = u_2$, $h = \bar{u}_1$, and $k = \bar{u}_2$, that

$$\begin{aligned}\frac{\{u_1,u_2\}_x\{u_2,u_1\}_x - \{u_1,u_1\}_x\{u_2,u_2\}_x}{\{u_1,u_1\}_x\{u_1,u_1\}_x}&\\ &= \frac{W(u_1,\bar{u}_1)W(u_2,\bar{u}_2) - W(u_1,\bar{u}_2)W(u_2,\bar{u}_1)}{4|\mathrm{Im}\,\lambda|^2(\{u_1,u_1\}_x)^2}\\ &= \frac{W(u_1,u_2)W(\bar{u}_1,\bar{u}_2)}{4|\mathrm{Im}\,\lambda|^2(\{u_1,u_1\}_x)^2}\\ &= r_x^2,\end{aligned}$$

where $r_x > 0$ is defined for $c < x < b$ by

$$r_x = \frac{1}{2|\mathrm{Im}\,\lambda|\{u_1,u_1\}_x}. \tag{6.1.22}$$

Consequently, for all $c < x < b$ the solution $u = \zeta u_1 + u_2$ of $(L-\lambda)y = 0$ satisfies the identity

$$\{u,u\}_x = \{u_1,u_1\}_x\left(|\zeta - \zeta_x|^2 - r_x^2\right), \tag{6.1.23}$$

where $\zeta_x \in \mathbb{C}$ and $r_x > 0$ are given by (6.1.21) and (6.1.22), respectively. Since $\{u_1,u_1\}_x > 0$ for all $x > c$ by (6.1.20), it follows from (6.1.23) that the equation $\{u,u\}_x = 0$ gives the circle with center ζ_x and radius r_x; and for $\zeta = \zeta_x$ one has $\{u,u\}_x = -\{u_1,u_1\}_x r_x^2 < 0$. Hence, by (6.1.23), one sees for $u = \zeta u_1 + u_2$ and $c < x < b$ that

$$\{u,u\}_x \le 0 \quad \Leftrightarrow \quad |\zeta - \zeta_x| \le r_x. \tag{6.1.24}$$

For each $c < x < b$ the closed disk with center ζ_x and radius r_x will be denoted by $\mathbb{D}(\zeta_x, r_x)$. Now let $c < x_1 < x_2 < b$ and assume that $\zeta \in \mathbb{D}(\zeta_{x_2}, r_{x_2})$. Then it follows from (6.1.24) that $\{u,u\}_{x_2} \le 0$ where $u = \zeta u_1 + u_2$. Recall that (6.1.18) with $u = \zeta u_1 + u_2$ implies that

$$\{u,u\}_{x_1} \le \{u,u\}_{x_2} \le 0.$$

By (6.1.24) this means that $\zeta \in \mathbb{D}(\zeta_{x_1}, r_{x_1})$. In other words,

$$c < x_1 < x_2 < b \quad \Rightarrow \quad \mathbb{D}(\zeta_{x_2}, r_{x_2}) \subset \mathbb{D}(\zeta_{x_1}, r_{x_1}). \tag{6.1.25}$$

Therefore, the disks $\mathbb{D}(\zeta_x, r_x)$ tend to a limit disk as $x \to b$ or shrink to exactly one point. These are the *limit-circle case* and the *limit-point case*, respectively. In the limit-circle case $r_b = \lim_{x\to b} r_x > 0$ is the radius of the limit circle and its center is given by $\zeta_b = \lim_{x\to b}\zeta_x$. In the limit-point case $r_b = \lim_{x\to b} r_x = 0$ and $\zeta_b = \lim_{x\to b}\zeta_x$ show that the limit circle degenerates into one point: the limit point.

The main fact which goes together with Theorem 6.1.5 is that at each endpoint and for all $\lambda \in \mathbb{C}\setminus\mathbb{R}$ there is at least one nontrivial solution in $L^2_r(a,a')$ or in $L^2_r(b',b)$. This leads to the following classification of the limit-circle and limit-point cases.

Theorem 6.1.6. *Let L be the differential expression in (6.1.1) on (a,b) such that the conditions in (6.1.2) are satisfied, and let $\lambda \in \mathbb{C}\setminus\mathbb{R}$. Then the following statements hold for the endpoint b:*

(i) *There exists a nontrivial solution u of $(L-\lambda)y = 0$ with $u \in L^2_r(b',b)$.*

(ii) *The limit-point case prevails at the endpoint b if and only if there exists a solution v of $(L-\lambda)y=0$ such that $v \notin L^2_r(b',b)$. In this case*

$$\lim_{x\to b} W_x(u,\bar{u}) = 0$$

holds for each solution u of $(L-\lambda)y=0$ such that $u \in L^2_r(b',b)$.

(iii) *The limit-circle case prevails at the endpoint b if and only if every solution of $(L-\lambda)y=0$ belongs to $L^2_r(b',b)$. In this case there exists a fundamental system u_1 and u_2 of $(L-\lambda)y=0$ and a disk with center ζ_b and radius $r_b>0$ so that with $u=\zeta u_1+u_2$*

$$\lim_{x\to b} W_x(u,\bar{u}) = 0$$

holds for all ζ with $|\zeta-\zeta_b| = r_b$.

There are similar statements for the endpoint a.

Proof. It suffices to give the proof for the endpoint b as the proof for the other endpoint is completely similar. Let $u_1 := u_1(\cdot,\lambda)$ and $u_2 := u_2(\cdot,\lambda)$ be two linearly independent solutions of $(L-\lambda)y=0$ fixed by (6.1.19).

Step 1. Assume that b is in the limit-circle case. Then $r_b = \lim_{x\to b} r_x > 0$ is the radius of the limit circle and $\zeta_b = \lim_{x\to b}\zeta_x$ is the center of the limit circle. It follows from (6.1.22) that

$$\lim_{x\to b}\{u_1,u_1\}_x = \frac{1}{2|\operatorname{Im}\lambda|\, r_b}$$

and by (6.1.20) and the monotone convergence theorem one therefore obtains

$$\int_c^b |u_1(t)|^2\, r(t)\, dt = \lim_{x\to b}\{u_1,u_1\}_x < \infty.$$

Moreover, for any ζ with $|\zeta-\zeta_b|\le r_b$ it follows from (6.1.18) with $u=\zeta u_1+u_2$ and (6.1.24) that

$$\int_c^x |\zeta u_1(t)+u_2(t)|^2\, r(t)\, dt = \{u,u\}_x - \{u,u\}_c \le -\{u,u\}_c.$$

Hence, the monotone convergence theorem implies that

$$\int_c^b |\zeta u_1(t)+u_2(t)|^2\, r(t)\, dt < \infty.$$

Therefore, u_1 and $\zeta u_1 + u_2$ form a fundamental system of $(L - \lambda)y = 0$ and both functions belong to $L^2_r(b', b)$.

Still assuming that b is in the limit-circle case, take the limit $x \to b$ in (6.1.23) to obtain

$$\{u, u\}_b = \{u_1, u_1\}_b \left(|\zeta - \zeta_b|^2 - r_b^2\right).$$

Since $\{u_1, u_1\}_b > 0$ it follows for $|\zeta - \zeta_b| = r_b$ that $\{u, u\}_b = 0$, so that

$$\lim_{x \to b} W_x(u, \overline{u}) = 0.$$

Step 2. Assume that b is in the limit-point case. Then $r_b = \lim_{x \to b} r_x = 0$ and $\zeta_b = \lim_{x \to b} \zeta_x$ show that the limit circle degenerates into the limit point. It follows from (6.1.22) that

$$\lim_{x \to b} \{u_1, u_1\}_x = \lim_{x \to b} \frac{1}{2|\mathrm{Im}\,\lambda|\, r_x} = \infty,$$

and therefore

$$\int_c^b |u_1(t)|^2\, r(t)\, dt = \lim_{x \to b} \{u_1, u_1\}_x = \infty.$$

Thus, there exists a solution v of $(L - \lambda)y = 0$ such that $v \notin L^2_r(b', b)$.

Still assuming that b is in the limit-point case, observe that by (6.1.25) ζ_b is inside all circles with center ζ_x and radius r_x, and therefore it follows from (6.1.24) that $\{u_{\zeta_b}, u_{\zeta_b}\}_x \leq 0$, where $u_{\zeta_b} = \zeta_b u_1 + u_2$. Hence, by (6.1.18),

$$\int_c^x |u_{\zeta_b}(t)|^2\, r(t)\, dt = \{u_{\zeta_b}, u_{\zeta_b}\}_x - \{u_{\zeta_b}, u_{\zeta_b}\}_c \leq -\{u_{\zeta_b}, u_{\zeta_b}\}_c,$$

and the monotone convergence theorem implies that

$$\int_c^b |u_{\zeta_b}(t)|^2\, r(t)\, dt < \infty,$$

that is, $u_{\zeta_b} \in L^2_r(b', b)$. Thus, u_1 and u_{ζ_b} form a fundamental system of $(L - \lambda)y = 0$ and every solution in $L^2_r(b', b)$ must be a multiple of u_{ζ_b}. It follows from (6.1.22), (6.1.20), and (6.1.23) that for $u_{\zeta_b} = \zeta_b u_1 + u_2$ and $c < x < b$

$$\begin{aligned} -\frac{1}{4|\mathrm{Im}\,\lambda|^2 \{u_1, u_1\}_x} &= -\{u_1, u_1\}_x\, r_x^2 \\ &\leq \{u_1, u_1\}_x \left(|\zeta_b - \zeta_x|^2 - r_x^2\right) = \{u_{\zeta_b}, u_{\zeta_b}\}_x \leq 0. \end{aligned}$$

Since $\{u_1, u_1\}_x \to \infty$ as $x \to b$, it follows from these inequalities that

$$\lim_{x \to b} \{u_{\zeta_b}, u_{\zeta_b}\}_x = 0.$$

In other words, $\lim_{x \to b} W_x(u_{\zeta_b}, \overline{u}_{\zeta_b}) = 0$ and since every solution $u \in L^2_r(b', b)$ of $(L - \lambda)y = 0$ is a multiple of u_{ζ_b} it follows that

$$\lim_{x \to b} W_x(u, \overline{u}) = 0.$$

Step 3. It follows from Step 1 and Step 2 that both in the limit-circle case and in the limit-point case there is a nontrivial solution of $(L-\lambda)y=0$ which belongs to $L^2_r(b',b)$. Therefore, (i) has been shown.

According to Step 1, in the limit-circle case every solution of $(L-\lambda)y=0$ belongs to $L^2_r(b',b)$. Conversely, if every solution of $(L-\lambda)y=0$ belongs to $L^2_r(b',b)$, then the limit-circle case must prevail. To see this, assume that the limit-point case prevails. Then by Step 2 there is a nontrivial solution of $(L-\lambda)y=0$ which does not belong to $L^2_r(b',b)$, which gives a contradiction. Therefore, (iii) has been shown.

According to Step 2, the limit-point case implies that there exists a nontrivial solution of $(L-\lambda)y=0$ which does not belong to $L^2_r(b',b)$. Conversely, if there exists a nontrivial solution of $(L-\lambda)y=0$ which does not belong to $L^2_r(b',b)$, then the limit-point case must prevail. To see this, assume that the limit-circle case prevails. Then by Step 1 all solutions of $(L-\lambda)y=0$ belong to $L^2_r(b',b)$, which gives a contradiction. Therefore, (ii) has been shown. □

By means of Theorem 6.1.5 the alternative in Theorem 6.1.6 is shown to be independent of $\lambda\in\mathbb{C}\setminus\mathbb{R}$. The existence of these two possibilities at an endpoint is known as *Weyl's alternative.*

Corollary 6.1.7. *Let L be the differential expression in (6.1.1) on (a,b) such that the conditions in (6.1.2) are satisfied. Then the following statements hold:*

(i) *The endpoint b is in the limit-circle case for some $\lambda\in\mathbb{C}\setminus\mathbb{R}$ if and only if it is in the limit-circle case for all $\lambda\in\mathbb{C}\setminus\mathbb{R}$. In this situation, for all $\lambda\in\mathbb{C}$ every solution of $(L-\lambda)y=0$ belongs to $L^2_r(b',b)$.*

(ii) *The endpoint b is in the limit-point case for some $\lambda\in\mathbb{C}\setminus\mathbb{R}$ if and only if it is in the limit-point case for all $\lambda\in\mathbb{C}\setminus\mathbb{R}$. In this situation, for all $\lambda\in\mathbb{C}\setminus\mathbb{R}$ there is, up to scalar multiples, exactly one nontrivial solution $(L-\lambda)y=0$ in $L^2_r(b',b)$ and for all $\lambda\in\mathbb{R}$ there is, up to scalar multiples, at most one nontrivial solution of $(L-\lambda)y=0$ in $L^2_r(b',b)$.*

There are similar statements for the endpoint a.

In addition to the assertions in Corollary 6.1.7 note that if for some $\lambda\in\mathbb{R}$ every solution of $(L-\lambda)y=0$ belongs to $L^2_r(b',b)$, then the limit-circle case prevails for all $\lambda\in\mathbb{C}\setminus\mathbb{R}$. Likewise, if for some $\lambda\in\mathbb{R}$ there is at most one nontrivial solution of $(L-\lambda)y=0$ which belongs to $L^2_r(b',b)$, then the limit-point prevails for all $\lambda\in\mathbb{C}\setminus\mathbb{R}$.

So far the Sturm–Liouville differential expression L in (6.1.1) was considered on the interval (a,b) under the conditions (6.1.2). At the end of this section some attention is paid to the following extra condition:

$$p(x)>0 \ \text{ for almost all } x\in(a,b). \tag{6.1.26}$$

This sign condition will play an important role later in this chapter. Here are a couple of useful remarks.

Let v be a real solution of $(L-\lambda)y=0$, $\lambda\in\mathbb{R}$, and let $\alpha<\beta$ be a pair of consecutive zeros of v, i.e., $v(\alpha)=v(\beta)=0$ and $v(x)\neq 0$ for $x\in(\alpha,\beta)$. Assume the sign condition (6.1.26). Then

$$v(t)>0 \quad \text{for} \quad \alpha<t<\beta \quad \Rightarrow \quad (pv')(\alpha)>0 \quad \text{and} \quad (pv')(\beta)<0,$$

and there is a similar implication when the signs are changed. It suffices to show the first inequality. By the uniqueness and existence theorem it is not possible to have $(pv')(\alpha)=0$. Now assume that $(pv')(\alpha)<0$; then since pv' is absolutely continuous, there exists $\delta>0$ such that $pv'<0$ on the interval $(\alpha-\delta,\alpha+\delta)$. As $p(x)>0$ almost everywhere on (a,b), one sees that $v'(x)<0$ almost everywhere on $(\alpha-\delta,\alpha+\delta)$. Since v is absolutely continuous one has $v(x)=\int_\alpha^x v'(t)\,dt$, which implies that $v(x)\leq 0$ for $\alpha<x<\alpha+\delta$; a contradiction. Hence, it has been shown that $(pv')(\alpha)>0$. The above implication will play a role in the following lemma, which is concerned with the Sturm comparison theory.

Lemma 6.1.8. *Assume the additional sign condition (6.1.26) and let u and v be real solutions of $(L-\lambda)y=0$ with $\lambda\in\mathbb{R}$. Then the following statements are equivalent:*

(i) *u and v are linearly independent;*

(ii) *u vanishes exactly once between consecutive zeros of v.*

Let w be a real solution of $(L-\mu)y=0$ with $\mu>\lambda$. Then between consecutive zeros of v there is at least one zero of w.

Proof. (i) $\Rightarrow$ (ii) Let u and v be real solutions which are linearly independent so that $x\mapsto W_x(u,v)$ is a nonzero constant. Let $\alpha<\beta$ be consecutive zeros of v and assume that $v(t)>0$ for $\alpha<t<\beta$. Then clearly

$$0<W_\alpha(u,v)W_\beta(u,v)=u(\alpha)(pv')(\alpha)u(\beta)(pv')(\beta). \tag{6.1.27}$$

The inequality (6.1.27) and $(pv')(\alpha)>0$ and $(pv')(\beta)<0$ together show that $u(\alpha)u(\beta)<0$. Hence, u has a zero between α and β. It is the only zero of u on this interval. For if not, a repetition of the argument applied to u would produce a zero of v between α and β, which is a contradiction.

(ii) $\Rightarrow$ (i) This implication is clear.

In order to see the last statement, assume again that $\alpha<\beta$ are consecutive zeros of v and that $v(t)>0$ for $\alpha<t<\beta$. Let w be a real solution of $(L-\mu)y=0$ with $\mu>\lambda$. Assume that w has no zeros between α and β, and that, in fact, $w(t)>0$ for $\alpha<t<\beta$. Then

$$W_\alpha(w,v)=w(\alpha)(pv')(\alpha)\geq 0, \quad W_\beta(w,v)=w(\beta)(pv')(\beta)\leq 0,$$

since $(pv')(\alpha) > 0$ and $(pv')(\beta) < 0$. Recall from (6.1.10) that on (α, β)

$$(W_x(w,v))' = r(x)\,(\mu - \lambda)w(x)\,v(x), \quad \alpha < x < \beta,$$

and the right-hand side is positive almost everywhere on (α, β) which gives a contradiction. Thus, w has at least one zero between α and β. □

For later use the following simple observation is included.

Lemma 6.1.9. *Assume that the sign condition (6.1.26) holds. Let f and g be real solutions of $(L - \lambda)y = 0$ with $\lambda \in \mathbb{R}$, and assume that f does not vanish on $(\alpha, \beta) \subset (a, b)$. Then the function g/f is monotone on (α, β) and for $x < y$ in (α, β) one has*

$$\frac{g(y)}{f(y)} - \frac{g(x)}{f(x)} = c \int_x^y \frac{1}{p(s)f(s)^2}\, ds,$$

where $c = W_x(f, g)$.

Proof. A straightforward calculation shows that

$$\left(\frac{g}{f}\right)'(x) = \frac{W_x(f,g)}{p(x)f(x)^2}, \tag{6.1.28}$$

where $W_x(f, g)$ is constant and the right-hand side has a constant sign on the interval (α, β), due to the condition (6.1.26). □

6.2 Maximal and minimal Sturm–Liouville differential operators

In Section 6.1 it has been shown that the differential expression L in (6.1.1) may be applied to a complex function f on (a, b) when $f, pf' \in AC(a, b)$. The conditions in (6.1.2) were used for the existence and uniqueness theorem for the corresponding initial value problems. In this section it will be shown that the differential expression L generates differential operators in the Hilbert space $L^2_r(a, b)$, where the weight function r is positive almost everywhere on the open interval (a, b).

The *maximal operator* $T_{\max}$ in $L^2_r(a, b)$ associated with the differential expression L is defined by

$$\begin{aligned} T_{\max} f &= Lf = \frac{1}{r}\big[-(pf')' + qf\big], \\ \operatorname{dom} T_{\max} &= \big\{ f \in L^2_r(a,b) : f, pf' \in AC(a,b),\ Lf \in L^2_r(a,b) \big\}. \end{aligned} \tag{6.2.1}$$

Recall from (6.1.6) that for $f, g \in AC(a, b)$ the Wronskian is defined as

$$W_x(f, \bar{g}) = f(x)\overline{(pg')(x)} - (pf')(x)\overline{g(x)}, \tag{6.2.2}$$

and it follows from (6.1.11) and the definition of $T_{\max}$ that for all $f, g \in \operatorname{dom} T_{\max}$ the limits

$$\lim_{x \to a} W_x(f, \bar{g}) \quad \text{and} \quad \lim_{x \to b} W_x(f, \bar{g})$$

exist separately. Therefore, it is a consequence of (6.1.11) that the following Green identity

$$(T_{\max} f, g)_{L^2_r(a,b)} - (f, T_{\max} g)_{L^2_r(a,b)} = \lim_{x \to b} W_x(f, \bar{g}) - \lim_{x \to a} W_x(f, \bar{g}) \qquad (6.2.3)$$

holds. The *preminimal operator* T_0 in $L^2_r(a,b)$ is defined by

$$T_0 f = Lf = \frac{1}{r}\big[-(pf')' + qf\big],$$
$$\operatorname{dom} T_0 = \big\{ f \in \operatorname{dom} T_{\max} : \operatorname{supp} f \text{ is compact in } (a,b) \big\}.$$

It follows from the Green formula (6.2.3) that the operator T_0 is symmetric. The following theorem concerning the *minimal operator* $T_{\min} = \overline{T_0}$ of the operator T_0 is based on the existence and uniqueness result in Theorem 6.1.2.

Theorem 6.2.1. *The closure $T_{\min} = \overline{T_0}$ of T_0 is a densely defined closed symmetric operator in $L^2_r(a,b)$ and it satisfies*

$$T_{\min} \subset (T_{\min})^* = T_{\max},$$

and, consequently, $T_{\min} = (T_{\max})^*$.

Proof. Step 1. This step is preparatory. Let $[\alpha, \beta] \subset (a,b)$ be a compact interval in which case the restriction of L to (α, β) is regular at α and β. Define the linear space $\mathfrak{D}_{[\alpha,\beta]}$ as

$$\mathfrak{D}_{[\alpha,\beta]} = \big\{ \varphi \in AC[\alpha,\beta] : p\varphi' \in AC[\alpha,\beta],\ L\varphi \in L^2_r(\alpha,\beta),$$
$$\varphi(\alpha) = \varphi(\beta) = 0,\ (p\varphi')(\alpha) = (p\varphi')(\beta) = 0 \big\},$$

and the operator $S_{[\alpha,\beta]}$ from $L^2_r(\alpha,\beta)$ into itself by

$$S_{[\alpha,\beta]}\varphi = L\varphi, \quad \varphi \in \mathfrak{D}_{[\alpha,\beta]}.$$

Denote by $\mathfrak{N}$ the two-dimensional space of all solutions of $Ly = 0$ on $[\alpha, \beta]$; thus if $h \in \mathfrak{N}$, then $h, ph' \in AC[\alpha,\beta]$ and $Lh = 0$. In particular, one has $\mathfrak{N} \subset L^2_r(\alpha,\beta)$. Now one shows that the Hilbert space $L^2_r(\alpha,\beta)$ admits the orthogonal decomposition

$$L^2_r(\alpha,\beta) = \operatorname{ran} S_{[\alpha,\beta]} \oplus \mathfrak{N}, \qquad (6.2.4)$$

so that $\operatorname{ran} S_{[\alpha,\beta]}$ is automatically closed. For this, let $g = S_{[\alpha,\beta]}\varphi$ with $\varphi \in \mathfrak{D}_{[\alpha,\beta]}$. Then for any solution u of $Ly = 0$ it follows from the Green identity that

$$\int_\alpha^\beta g(x)\overline{u(x)} r(x)\, dx = W_x(\varphi, \bar{u})\,\big|_\alpha^\beta = 0.$$

Hence, one concludes that $g \perp \mathfrak{N}$ in $L^2_r(\alpha,\beta)$, which shows $\operatorname{ran} S_{[\alpha,\beta]} \subset \mathfrak{N}^\perp$. Conversely, assume that $g \in \mathfrak{N}^\perp$. Let φ be the solution of the equation $L\varphi = g$ that is uniquely determined by the initial conditions $\varphi(\beta) = 0$ and $(p\varphi')(\beta) = 0$; cf. Theorem 6.1.2. For any solution u of $Ly = 0$ it follows from the Green identity that

$$0 = \int_\alpha^\beta g(x)\overline{u(x)}r(x)\,dx = -W_\alpha(\varphi, \bar{u}).$$

By choosing the right initial conditions at α for the solution u it follows that $\varphi(\alpha) = 0$ and $(p\varphi')(\alpha) = 0$. Hence, $g = L\varphi$ with $\varphi \in \mathfrak{D}_{[\alpha,\beta]}$, i.e., $g = S_{[\alpha,\beta]}\varphi$. Thus, $\mathfrak{N}^\perp \subset \operatorname{ran} S_{[\alpha,\beta]}$. Hence, (6.2.4) has been shown.

Step 2. The previous step will be used in conjunction with the following observation. Let $[\alpha,\beta] \subset (a,b)$ be a compact subinterval. Then one has the equivalence:

$$f \in \operatorname{dom} T_0 \quad \text{and} \quad \operatorname{supp} f \subset [\alpha,\beta] \quad \Leftrightarrow \quad f \in \mathfrak{D}_{[\alpha,\beta]}. \tag{6.2.5}$$

The implication ($\Rightarrow$) about the restriction of f to $[\alpha,\beta]$ is clear by definition. Conversely, if $f \in \mathfrak{D}_{[\alpha,\beta]}$, then f can be trivially extended to all of (a,b) and then the extension, also denoted by f, belongs to $\operatorname{dom} T_{\max}$ since f and pf' are continuous across α and β. Hence, the implication ($\Leftarrow$) in (6.2.5) follows.

Step 3. The symmetric operator T_0 is densely defined. To see this, assume that $g \in L^2_r(a,b)$ satisfies $(g,\varphi)_{L^2_r(a,b)} = 0$ for all $\varphi \in \operatorname{dom} T_0$ and let u be any solution of $Lu = g$. For any compact interval $[\alpha,\beta]$ one has $(g,\varphi)_{L^2_r(a,b)} = 0$ for all $\varphi \in \operatorname{dom} T_0$ with support in $[\alpha,\beta]$. Hence, for $[\alpha,\beta]$ fixed one has by means of Step 2 that

$$0 = (g,\varphi)_{L^2_r(a,b)} = \int_\alpha^\beta (Lu)(x)\overline{\varphi(x)}r(x)\,dx = \int_\alpha^\beta u(x)\overline{(L\varphi)(x)}r(x)\,dx$$

for all $\varphi \in \mathfrak{D}_{[\alpha,\beta]}$. According to Step 1 it follows that $u \in \mathfrak{N}$, i.e., $g = Lu = 0$ on $[\alpha,\beta]$. Since $[\alpha,\beta]$ is arbitrary, it follows that $g = 0$ in $L^2_r(a,b)$. Thus, T_0 is densely defined. In particular it follows from $T_0 \subset (T_0)^*$ that the operator T_0 is closable. Therefore, $T_{\min} = \overline{T_0}$ is an operator.

Step 4. Observe first that $T_{\max} \subset (T_0)^*$. In fact, if $f \in \operatorname{dom} T_{\max}$, then for all $\varphi \in \operatorname{dom} T_0$ one has

$$(T_{\max} f,\varphi)_{L^2_r(a,b)} - (f, T_0\varphi)_{L^2_r(a,b)} = (T_{\max} f,\varphi)_{L^2_r(a,b)} - (f, T_{\max}\varphi)_{L^2_r(a,b)} = 0$$

due to (6.2.3) and φ being zero in a neighborhood of a and of b; this proves the claim. It will now be shown that $(T_0)^* \subset T_{\max}$. For this let $\{f,g\} \in (T_0)^*$. Then for all $\varphi \in \operatorname{dom} T_0$ one has

$$(g,\varphi)_{L^2_r(a,b)} = (f, T_0\varphi)_{L^2_r(a,b)}.$$

Now let $[\alpha,\beta] \subset (a,b)$ be any compact interval. Then for all $\varphi \in \operatorname{dom} T_0$ with $\operatorname{supp}\varphi \subset [\alpha,\beta]$ or, by Step 2, for all $\varphi \in \mathfrak{D}_{[\alpha,\beta]}$ one has

$$\int_\alpha^\beta g(x)\overline{\varphi(x)}r(x)\,dx = \int_\alpha^\beta f(x)\overline{(L\varphi)(x)}r(x)\,dx.$$

On $[\alpha, \beta]$ choose u with $u, pu' \in AC[\alpha, \beta]$ such that $Lu = g$ almost everywhere. Then

$$\int_\alpha^\beta (Lu)(x)\overline{\varphi(x)}r(x)\,dx = \int_\alpha^\beta f(x)\overline{(L\varphi)(x)}r(x)\,dx$$

and integration by parts of the left-hand side yields

$$\int_\alpha^\beta u(x)\overline{(L\varphi)(x)}r(x)\,dx = \int_\alpha^\beta f(x)\overline{(L\varphi)(x)}r(x)\,dx,$$

so that $f - u \perp \operatorname{ran} S_{[\alpha,\beta]}$ in $L^2_r(\alpha, \beta)$. By Step 1 one sees that $f - u \in \mathfrak{N}$. It follows that f has the decomposition

$$f = h + u, \quad h = f - u,$$

where $h, ph', u, pu' \in AC[\alpha, \beta]$ and $Lh = 0$. Therefore, $f, pf' \in AC[\alpha, \beta]$ and $Lf = Lu = g$ almost everywhere on $[\alpha, \beta]$. This is true for each compact subinterval $[\alpha, \beta]$ of (a, b) and hence $f, pf' \in AC(a, b)$ and one has $g = Lf$ almost everywhere on (a, b). Therefore, $\{f, g\} \in T_{\max}$ and $(T_0)^* \subset T_{\max}$. □

As a consequence of Theorem 6.2.1 and Theorem 1.7.11 one sees that the graph of $T_{\max}$ has the componentwise sum decomposition

$$T_{\max} = T_{\min} \,\widehat{+}\, \widehat{\mathfrak{N}}_\lambda(T_{\max}) \,\widehat{+}\, \widehat{\mathfrak{N}}_{\bar\lambda}(T_{\max}), \quad \lambda \in \mathbb{C} \setminus \mathbb{R}, \quad \text{direct sums.}$$

Observe that $\mathfrak{N}_\lambda(T_{\max})$ consists of functions that solve the differential equation $(L - \lambda)y = 0$ and belong to $L^2_r(a, b)$. In particular, each of the defect numbers of $T_{\min}$ is at most 2 and since the coefficient functions are real, it follows that the defect numbers are equal; cf. (6.1.4). If both endpoints are in the limit-circle case, then every solution of $(L - \lambda)y = 0$, $\lambda \in \mathbb{C} \setminus \mathbb{R}$, belongs to $L^2_r(a, b)$ by Theorem 6.1.5. If one endpoint is in the limit-circle case and the other endpoint is in the limit-point case, there is, up to scalar multiples, only one solution that belongs to $L^2_r(a, b)$ by Theorem 6.1.6. This leads to the next corollary on the defect numbers of $T_{\min}$.

Corollary 6.2.2. *Let $T_{\min}$ be the minimal operator associated with the differential expression L in $L^2_r(a, b)$. Then the following statements hold:*

(i) *If both endpoints of (a, b) are in the limit-circle case, then the defect numbers of $T_{\min}$ are $(2, 2)$.*

(ii) *If one endpoint of (a, b) is in the limit-circle case and the other is in the limit-point case, then the defect numbers of $T_{\min}$ are $(1, 1)$.*

In addition to the cases in Corollary 6.2.2 there is the situation where both endpoints of (a, b) are in the limit-point case. Then the defect numbers of $T_{\min}$ are $(0, 0)$, which will become clear in Section 6.5; cf. Corollary 6.5.2.

The cases (i) and (ii) in Corollary 6.2.2 will be treated in Section 6.3 and Section 6.4, respectively, in terms of boundary triplets. These triplets will be defined by means of the Green formula

$$(T_{\max} f, g)_{L^2_r(a,b)} - (f, T_{\max} g)_{L^2_r(a,b)} = \lim_{x \to b} W_x(f, \bar{g}) - \lim_{x \to a} W_x(f, \bar{g}), \tag{6.2.6}$$

where for $f, g \in \operatorname{dom} T_{\max}$ each of the limits

$$\lim_{x \to a} W_x(f, \bar{g}) \quad \text{and} \quad \lim_{x \to b} W_x(f, \bar{g})$$

exists separately. These limits will be essential ingredients in defining "boundary values" of functions in $\operatorname{dom} T_{\max}$. In particular, observe that $T_{\min} = (T_{\max})^*$ implies that $\operatorname{dom} T_{\min}$ consists of all $f \in \operatorname{dom} T_{\max}$ for which

$$\lim_{x \to b} W_x(f, \bar{g}) = \lim_{x \to a} W_x(f, \bar{g})$$

for all $g \in \operatorname{dom} T_{\max}$. Hence, it follows from Proposition 6.1.3 that $\operatorname{dom} T_{\min}$ consists of all $f \in \operatorname{dom} T_{\max}$ for which the two separate limits must satisfy

$$\lim_{x \to b} W_x(f, \bar{g}) = 0 \quad \text{and} \quad \lim_{x \to a} W_x(f, \bar{g}) = 0 \tag{6.2.7}$$

for all $g \in \operatorname{dom} T_{\max}$. An essential ingredient is the behavior of the Wronskian $W(f, \bar{g})$ for $f, g \in \operatorname{dom} T_{\max}$ near an endpoint which is regular or in the limit-circle case. First the regular case is considered.

Lemma 6.2.3. *Assume that L is regular at a or b. Then a or b is in the limit-circle case, respectively. In particular, if $f \in \operatorname{dom} T_{\max}$, then the limits*

$$\begin{aligned} f(a) &= \lim_{x \to a} f(x), \quad (pf')(a) = \lim_{x \to a} (pf')(x), \\ f(b) &= \lim_{x \to b} f(x), \quad (pf')(b) = \lim_{x \to b} (pf')(x), \end{aligned} \tag{6.2.8}$$

exist, respectively. Moreover, for $f, g \in \operatorname{dom} T_{\max}$

$$\begin{aligned} \lim_{x \to a} W_x(f, \bar{g}) &= f(a)\overline{(pg')(a)} - (pf')(a)\overline{g(a)}, \\ \lim_{x \to b} W_x(f, \bar{g}) &= f(b)\overline{(pg')(b)} - (pf')(b)\overline{g(b)}, \end{aligned}$$

respectively.

Proof. Assume that L is regular at b. Then it follows from Theorem 6.1.2 that all solutions of $(L - \lambda)y = 0$, $\lambda \in \mathbb{C}$, belong to $AC(a, b]$ and hence are bounded near b. Since r is integrable at b, all solutions belongs to $L^2_r(b', b)$ and so the limit-circle case prevails at b by Theorem 6.1.6. Let $f \in \operatorname{dom} T_{\max}$ and let $g = T_{\max} f$. Then $f, g \in L^2_r(a, b)$ and $f, pf' \in AC(a, b]$ by Theorem 6.1.2. This implies the statements. □

If the endpoint is in the limit-circle case but not regular, then the existence of the individual limits in (6.2.8) is not guaranteed. In that case the notion of quasi-derivative proves useful.

Definition 6.2.4. Let u and v be linearly independent real solutions of the equation $(L-\lambda_0)y=0$ for some $\lambda_0 \in \mathbb{R}$ and assume that the solutions are normalized by $W(u,v)=1$. Let f be a complex function on (a,b) for which $f, pf' \in AC(a,b)$. Then the *quasi-derivatives* of f, induced by the normalized solutions u and v, are defined as complex functions on (a,b) given by

$$f^{[0]} := W(f,v) \quad \text{and} \quad f^{[1]} := -W(f,u). \tag{6.2.9}$$

Let f and g be functions on (a,b) for which $f, pf', g, pg' \in AC(a,b)$. Then it follows from the Plücker identity in (6.1.7) that

$$W_x(f,g) = f^{[0]}(x)g^{[1]}(x) - f^{[1]}(x)g^{[0]}(x). \tag{6.2.10}$$

Note that the right-hand side has an appearance which is similar to the right-hand side of (6.1.6) and (6.2.2), but both $f^{[0]}$ and $f^{[1]}$ are made up of f and pf'. In the limit-circle case also the individual factors have limits.

Lemma 6.2.5. *Let u and v be linearly independent real solutions of the equation $(L-\lambda_0)y=0$ for some $\lambda_0 \in \mathbb{R}$ which are normalized by $W(u,v)=1$, and let $f,g \in \operatorname{dom} T_{\max}$. If $u,v \subset L^2_r(a,a')$, then the limits*

$$f^{[0]}(a) = \lim_{x\to a} f^{[0]}(x) \quad \textit{and} \quad f^{[1]}(a) = \lim_{x\to a} f^{[1]}(x)$$

exist, and consequently

$$\lim_{x\to a} W_x(f,\bar{g})(x) = f^{[0]}(a)\overline{g^{[1]}(a)} - f^{[1]}(a)\overline{g^{[0]}(a)}.$$

Likewise, if $u,v \in L^2_r(b',b)$, then the limits

$$f^{[0]}(b) = \lim_{x\to b} f^{[0]}(x) \quad \textit{and} \quad f^{[1]}(b) = \lim_{x\to b} f^{[1]}(x)$$

exist, and consequently

$$\lim_{x\to b} W_x(f,\bar{g})(x) = f^{[0]}(b)\overline{g^{[1]}(b)} - f^{[1]}(b)\overline{g^{[0]}(b)}.$$

Proof. Let ϕ be any real solution of $(L-\lambda_0)y=0$. Since $f \in \operatorname{dom} T_{\max}$ by assumption, it is clear that $(L-\lambda_0)f \in L^2_r(a,b)$. It is easily seen from (6.1.9) that the following identity

$$W_x(f,\phi) - W_s(f,\phi) = \int_s^x ((L-\lambda_0)f)(t)\,\phi(t)\,r(t)\,dt$$

holds for all $a < s < x < b$. If, in addition, $\phi \in L^2_r(a, a')$, then

$$W_a(f, \phi) = \lim_{x \to a} W_x(f, \phi)$$

exists as can be seen from the dominated convergence theorem. The assertions in the lemma then follow by taking $\phi = v$ and $\phi = -u$, respectively. □

At the end of this section it is assumed that the coefficient functions satisfy (6.1.2) and that the endpoint a is in the limit-circle case. The following result is about solving the Sturm–Liouville equation $(L - \lambda)f = g$ with initial conditions in terms of quasi-derivatives. The proof of this result has the same background as the proof of Theorem 6.1.2. Just write the equation as a first-order system, but now involving quasi-derivatives:

$$\begin{pmatrix} f^{[0]} \\ f^{[1]} \end{pmatrix}' = (\lambda - \lambda_0) \begin{pmatrix} uvr & v^2 r \\ -u^2 r & -uvr \end{pmatrix} \begin{pmatrix} f^{[0]} \\ f^{[1]} \end{pmatrix} + \begin{pmatrix} vrg \\ -urg \end{pmatrix}.$$

The new initial value problem is equivalent to a Volterra integral equation which can be solved in the usual way by successive approximations when all data are locally integrable. Note that the condition $g \in L^2_r(a, b)$ in the proposition implies that $urg, vrg \in L^1(a, a')$. There is a similar statement when the endpoint b is in the limit-circle case.

Proposition 6.2.6. *Let a be in the limit-circle case. Let u, v be real solutions of $(L - \lambda_0)y = 0$, $\lambda_0 \in \mathbb{R}$, such that $W(u, v) = 1$, and $u, v \in L^2_r(a, a')$. Let $c_1, c_2 \in \mathbb{C}$ and let $g \in L^2_r(a, b)$. Then for all $\lambda \in \mathbb{C}$ the initial value problem*

$$(L - \lambda)f = g, \qquad f^{[0]}(a) = c_1, \quad f^{[1]}(a) = c_2,$$

has a unique solution $f \in AC(a, b)$ for which $pf' \in AC(a, b)$. In addition, the functions

$$\lambda \mapsto f^{[0]}(a, \lambda) \quad \text{and} \quad \lambda \mapsto f^{[1]}(a, \lambda)$$

are entire.

It follows that under the circumstances of Proposition 6.2.6 for any $\lambda \in \mathbb{C}$ the homogeneous equation $(L - \lambda)y = 0$ has a *fundamental system*, i.e., for any $\lambda \in \mathbb{C}$ there are two solutions $u_1(\cdot, \lambda)$ and $u_2(\cdot, \lambda)$ of $(L - \lambda)y = 0$ which are linearly independent when it is required that

$$\begin{pmatrix} u_1^{[0]}(a, \lambda) & u_2^{[0]}(a, \lambda) \\ u_1^{[1]}(a, \lambda) & u_2^{[1]}(a, \lambda) \end{pmatrix} = \begin{pmatrix} 1 & 0 \\ 0 & 1 \end{pmatrix}; \tag{6.2.11}$$

cf. (6.2.10). Moreover, each of the entries of the left-hand side gives an entire function in λ.

The quasi-derivatives of an element $f \in \operatorname{dom} T_{\max}$ may be interpreted as coefficients of f in terms of a local expansion involving the square-integrable solutions u and v as follows.

Lemma 6.2.7. *Let u and v be linearly independent real solutions of the equation $(L-\lambda_0)y=0$ for some $\lambda_0\in\mathbb{R}$ which are normalized by $W(u,v)=1$. Assume that $f\in\operatorname{dom}T_{\max}$ and let the quasi-derivatives $f^{[0]}$ and $f^{[1]}$ be defined as in (6.2.9). If $u,v\in L^2_r(a,a')$, then*

$$f(x)=u(x)\big(f^{[0]}(a)+o(1)\big)+v(x)\big(f^{[1]}(a)+o(1)\big),\quad x\to a.$$

Likewise, if $u,v\in L^2_r(b',b)$, then

$$f(x)=u(x)\big(f^{[0]}(b)+o(1)\big)+v(x)\big(f^{[1]}(b)+o(1)\big),\quad x\to b.$$

Proof. Consider the case of the endpoint a. Let the function g be defined by $g=(T_{\max}-\lambda_0)f$ and note that $g\in L^2_r(a,b)$. There exist unique $\alpha,\beta\in\mathbb{C}$ such that

$$f(x)=u(x)\left(\alpha+\int_a^x v(t)g(t)r(t)\,dt\right)+v(x)\left(\beta-\int_a^x u(t)g(t)r(t)\,dt\right);\quad (6.2.12)$$

cf. the variation of constants formula (6.1.13). One sees, via the Cauchy–Schwarz inequality, that

$$\int_a^x v(t)g(t)r(t)\,dt=o(1)\quad\text{and}\quad\int_a^x u(t)g(t)r(t)\,dt=o(1),\quad x\to a.$$

In order to identify the parameters α and β use that

$$f^{[0]}=f(pv')-(pf')v\quad\text{and}\quad f^{[1]}=(pf')u-f(pu').$$

Now substitute (6.2.12) and its differentiated form

$$\begin{aligned}(pf')(x)=(pu')(x)&\left(\alpha+\int_a^x v(t)g(t)r(t)\,dt\right)\\&+(pv')(x)\left(\beta-\int_a^x u(t)g(t)r(t)\,dt\right),\end{aligned}$$

so that with the normalization $W(u,v)=1$ it follows that

$$f^{[0]}(x)=\alpha+\int_a^x v(t)g(t)r(t)\,dt\quad\text{and}\quad f^{[1]}(x)=\beta-\int_a^x u(t)g(t)r(t)\,dt.$$

Hence, one obtains that $\alpha=f^{[0]}(a)$ and $\beta=f^{[1]}(a)$.

Next consider the case of the endpoint b. Likewise, if $u,v\in L^2_r(b',b)$, then there exist unique $\gamma,\delta\in\mathbb{C}$ such that

$$f(x)=u(x)\left(\gamma-\int_x^b v(t)g(t)r(t)\,dt\right)+v(x)\left(\delta+\int_x^b u(t)g(t)r(t)\,dt\right),$$

where

$$\int_x^b v(t)g(t)r(t)\,dt=o(1)\quad\text{and}\quad\int_x^b u(t)g(t)r(t)\,dt=o(1),\quad x\to b.$$

In a similar way one can show that $\gamma=f^{[0]}(b)$ and $\delta=f^{[1]}(b)$. □

6.3 Regular and limit-circle endpoints

Let $T_{\max} = (T_{\min})^*$ be the maximal operator associated with the Sturm–Liouville differential expression L in (6.1.1) on the interval (a,b). This situation will be considered first under the assumption that both endpoints a and b are regular and then in the end of this section the endpoints in the limit-circle case are treated.

Assume that the endpoints a and b are regular, i.e., $[a,b]$ is a compact interval and

$$\begin{cases} p(x) \neq 0, \ r(x) > 0, \ \text{for almost all } x \in (a,b), \\ 1/p, q, r \in L^1(a,b); \end{cases}$$

cf. Definition 6.1.1. It follows from Theorem 6.1.2 that there exists a fundamental system $(u_1(\cdot,\lambda); u_2(\cdot,\lambda))$ for the equation $(L-\lambda)y = 0$ with the initial conditions

$$\begin{pmatrix} u_1(a,\lambda) & u_2(a,\lambda) \\ (pu_1')(a,\lambda) & (pu_2')(a,\lambda) \end{pmatrix} = \begin{pmatrix} 1 & 0 \\ 0 & 1 \end{pmatrix}. \tag{6.3.1}$$

Then for $i = 1,2$ each of the mappings $\lambda \mapsto u_i(x,\lambda)$ and $\lambda \mapsto (pu_i')(x,\lambda)$ is entire for fixed $x \in [a,b]$. From (6.2.1) and Theorem 6.1.2 one also sees that every function $f \in \operatorname{dom} T_{\max}$ satisfies $f, pf' \in AC[a,b]$ and the quantities $f(a)$, $(pf')(a)$, $f(b)$, and $(pf')(b)$ are well defined.

Proposition 6.3.1. *Assume that the endpoints a and b are regular. Then $\{\mathbb{C}^2, \Gamma_0, \Gamma_1\}$, where*

$$\Gamma_0 f = \begin{pmatrix} f(a) \\ f(b) \end{pmatrix} \quad \text{and} \quad \Gamma_1 f = \begin{pmatrix} (pf')(a) \\ -(pf')(b) \end{pmatrix}, \quad f \in \operatorname{dom} T_{\max}, \tag{6.3.2}$$

is a boundary triplet for $(T_{\min})^* = T_{\max}$. The self-adjoint extension A_0 corresponding to Γ_0 is the restriction of $T_{\max}$ defined on

$$\operatorname{dom} A_0 = \{ f \in \operatorname{dom} T_{\max} : f(a) = f(b) = 0 \}$$

and the minimal operator $T_{\min}$ is the restriction of $T_{\max}$ defined on

$$\operatorname{dom} T_{\min} = \{ f \in \operatorname{dom} T_{\max} : f(a) = f(b) = (pf')(a) = (pf')(b) = 0 \}.$$

Moreover, for all $\lambda \in \rho(A_0)$ one has $u_2(b,\lambda) \neq 0$. The corresponding γ-field and Weyl function are given by

$$\gamma(\lambda) = \begin{pmatrix} u_1(\cdot,\lambda) & u_2(\cdot,\lambda) \end{pmatrix} \frac{1}{u_2(b,\lambda)} \begin{pmatrix} u_2(b,\lambda) & 0 \\ -u_1(b,\lambda) & 1 \end{pmatrix}, \quad \lambda \in \rho(A_0),$$

and

$$M(\lambda) = \frac{1}{u_2(b,\lambda)} \begin{pmatrix} -u_1(b,\lambda) & 1 \\ 1 & -(pu_2')(b,\lambda) \end{pmatrix}, \quad \lambda \in \rho(A_0).$$

Proof. Assume that the endpoints a and b are regular. First it will be shown that (6.3.2) defines a boundary triplet. For $f, g \in \operatorname{dom} T_{\max}$ one has by (6.2.6) and Lemma 6.2.3

$$\begin{aligned}(T_{\max} f, g) - (f, T_{\max} g) &= \lim_{x \to b} W_x(f, \bar{g}) - \lim_{x \to a} W_x(f, \bar{g}) \\ &= f(b)\overline{(pg')(b)} - (pf')(b)\overline{g(b)} - f(a)\overline{(pg')(a)} + (pf')(a)\overline{g(a)},\end{aligned}$$

which implies that the abstract Green identity is satisfied with the choice of Γ_0 and Γ_1 in (6.3.2). Furthermore, the mapping $(\Gamma_0, \Gamma_1)^\top : \operatorname{dom} T_{\max} \to \mathbb{C}^4$ is surjective. To see this, choose $\alpha \in \mathbb{C}^4$ and consider the initial value problem (6.1.5) with $\lambda = 0$, $g = 0$, and $x_0 = a$ with initial data $\alpha_1, \alpha_2 \in \mathbb{C}$. Then there is a unique solution f with $f, pf' \in AC[a, b]$. Now cut off this function so that it becomes a solution f_a of a corresponding inhomogeneous equation which is trivial in a neighborhood of b; see Proposition 6.1.3. It is clear that $f_a \in \operatorname{dom} T_{\max}$ and

$$\Gamma_0 f_a = \begin{pmatrix} \alpha_1 \\ 0 \end{pmatrix}, \quad \Gamma_1 f_a = \begin{pmatrix} \alpha_2 \\ 0 \end{pmatrix}.$$

A similar procedure at the other endpoint gives a function $f_b \in \operatorname{dom} T_{\max}$ and

$$\Gamma_0 f_b = \begin{pmatrix} 0 \\ \alpha_3 \end{pmatrix}, \quad \Gamma_1 f_b = \begin{pmatrix} 0 \\ \alpha_4 \end{pmatrix}.$$

Taking $h = f_a + f_b$ completes the argument. Thus, (6.3.2) defines a boundary triplet for $(T_{\min})^*$.

The description of the domain of the self-adjoint extension A_0 is trivial. Since $u_2(a, \lambda) = 0$ for all $\lambda \in \mathbb{C}$ it follows that $u_2(b, \lambda) \neq 0$ for $\lambda \in \rho(A_0)$, as otherwise $u_2(\cdot, \lambda)$ would be an eigenfunction for A_0, which contradicts $\lambda \in \rho(A_0)$. Furthermore, by Proposition 2.1.2 (ii) one has $\operatorname{dom} T_{\min} = \ker \Gamma_0 \cap \ker \Gamma_1$, which implies the description of the domain of the minimal operator.

In order to compute the γ-field and Weyl function corresponding to the boundary triplet $\{\mathbb{C}^2, \Gamma_0, \Gamma_1\}$ note that in terms of the fundamental system determined by (6.3.1) every element in $\mathfrak{N}_\lambda(T_{\max})$ has the form

$$f(\cdot, \lambda) = \begin{pmatrix} u_1(\cdot, \lambda) & u_2(\cdot, \lambda) \end{pmatrix} \begin{pmatrix} c_1 \\ c_2 \end{pmatrix}, \quad c_1, c_2 \in \mathbb{C}.$$

It follows from Definition 2.3.1 that $\gamma(\lambda)$ is given by

$$\left\{ \begin{pmatrix} 1 & 0 \\ u_1(b, \lambda) & u_2(b, \lambda) \end{pmatrix} \begin{pmatrix} c_1 \\ c_2 \end{pmatrix}, \begin{pmatrix} u_1(\cdot, \lambda) & u_2(\cdot, \lambda) \end{pmatrix} \begin{pmatrix} c_1 \\ c_2 \end{pmatrix} \right\}$$

for all pairs $c_1, c_2 \in \mathbb{C}$ and, likewise, it follows from Definition 2.3.4 that $M(\lambda)$ is given by

$$\left\{ \begin{pmatrix} 1 & 0 \\ u_1(b, \lambda) & u_2(b, \lambda) \end{pmatrix} \begin{pmatrix} c_1 \\ c_2 \end{pmatrix}, \begin{pmatrix} 0 & 1 \\ -(pu_1')(b, \lambda) & -(pu_2')(b, \lambda) \end{pmatrix} \begin{pmatrix} c_1 \\ c_2 \end{pmatrix} \right\}$$

for all pairs $c_1, c_2 \in \mathbb{C}$. Since $u_2(b,\lambda) \neq 0$ for $\lambda \in \rho(A_0)$ the stated results follow. In particular, the last result on the form of the Weyl function M follows as the Wronskian of u_1 and u_2 is constant and equal to one; cf. (6.3.1). □

Note that the self-adjoint operator A_0 in Proposition 6.3.1 corresponds to *Dirichlet boundary conditions* and the self-adjoint operator A_1 defined on $\ker \Gamma_1$ corresponds to *Neumann boundary conditions.* In the next corollary the boundary condition at the endpoint b is fixed as the Dirichlet condition $f(b) = 0$. The corresponding boundary triplet appears as a restriction of the boundary triplet in Proposition 6.3.1. Corollary 6.3.2 can be seen as an immediate consequence of Proposition 6.3.1 and Proposition 2.5.12 applied to the decomposition

$$\mathbb{C}^2 = \mathcal{G} = \mathcal{G}' \oplus \mathcal{G}'', \quad \text{with } \mathcal{G}' = \operatorname{span}\begin{pmatrix}1\\0\end{pmatrix} \text{ and } \mathcal{G}'' = \operatorname{span}\begin{pmatrix}0\\1\end{pmatrix}.$$

A short direct argument will be given.

Corollary 6.3.2. *Assume that the endpoints a and b are regular. Let the operator $T'_{\min}$ be the extension of $T_{\min}$ defined on*

$$\operatorname{dom} T'_{\min} = \bigl\{ f \in \operatorname{dom} T_{\max} : f(a) = (pf')(a) = f(b) = 0 \bigr\}.$$

Then $T'_{\min}$ is a densely defined closed symmetric operator with defect numbers $(1,1)$ and $T_{\min} \subset T'_{\min} \subset A_0$. The adjoint $(T'_{\min})^*$ is defined on

$$\operatorname{dom} (T'_{\min})^* = \bigl\{ f \in \operatorname{dom} T_{\max} : f(b) = 0 \bigr\}. \tag{6.3.3}$$

Then $\{\mathbb{C}, \Gamma'_0, \Gamma'_1\}$, where

$$\Gamma'_0 f = f(a) \quad \text{and} \quad \Gamma'_1 f = (pf')(a), \quad f \in \operatorname{dom} (T'_{\min})^*, \tag{6.3.4}$$

is a boundary triplet for $(T'_{\min})^*$. Moreover, for $\lambda \in \rho(A_0)$ the corresponding γ-field and the Weyl function are given by

$$\gamma'(\cdot,\lambda) = u_1(\cdot,\lambda) - \frac{u_1(b,\lambda)}{u_2(b,\lambda)} u_2(\cdot,\lambda) \quad \text{and} \quad M'(\lambda) = -\frac{u_1(b,\lambda)}{u_2(b,\lambda)}.$$

Proof. One verifies that the adjoint $(T'_{\min})^*$ is a restriction of $T_{\max}$ given by the boundary condition in (6.3.3) and that $\{\mathbb{C}, \Gamma'_0, \Gamma'_1\}$ (6.3.4) is a boundary triplet for $(T'_{\min})^*$. To see the last statement let

$$f(\cdot,\lambda) = \alpha(\lambda) u_1(\cdot,\lambda) + \beta(\lambda) u_2(\cdot,\lambda) \in \ker\bigl((T'_{\min})^* - \lambda\bigr),$$

where $\alpha(\lambda) u_1(b,\lambda) + \beta(\lambda) u_2(b,\lambda) = 0$. For $\lambda \in \rho(A_0)$ one obtains

$$\gamma'(\lambda) = u_1(\cdot,\lambda) + \frac{\beta(\lambda)}{\alpha(\lambda)} u_2(\cdot,\lambda) \quad \text{and} \quad M'(\lambda) = \frac{\beta(\lambda)}{\alpha(\lambda)}.$$

This completes the proof. □

Example 6.3.3. In the special case $r = p = 1$ and a constant $q \in \mathbb{R}$, the Sturm–Liouville expression is $Lf = -f'' + qf$. For $\lambda > q$ the fundamental system determined by (6.3.1) is given by

$$u_1(x,\lambda) = \cos\bigl[\sqrt{\lambda - q}\,(x-a)\bigr], \quad u_2(x,\lambda) = \frac{\sin\bigl[\sqrt{\lambda - q}\,(x-a)\bigr]}{\sqrt{\lambda - q}},$$

and if the square root $\sqrt{\cdot}$ is fixed such that $\operatorname{Im}\sqrt{\lambda} > 0$ for $\lambda \in \mathbb{C} \setminus [0,\infty)$ and $\sqrt{\lambda} \geq 0$ for $\lambda \in [0,\infty)$, the formula extends to $\lambda \in \mathbb{C} \setminus \{0\}$. Hence the Weyl function in Proposition 6.3.1 is

$$M(\lambda) = \frac{\sqrt{\lambda - q}}{\sin\bigl[\sqrt{\lambda - q}\,(b-a)\bigr]} \begin{pmatrix} -\cos\bigl[\sqrt{\lambda - q}\,(b-a)\bigr] & 1 \\ 1 & -\cos\bigl[\sqrt{\lambda - q}\,(b-a)\bigr] \end{pmatrix},$$

and the Weyl function in Corollary 6.3.2 is

$$M'(\lambda) = -\sqrt{\lambda - q}\cot\bigl[\sqrt{\lambda - q}\,(b-a)\bigr].$$

The poles of M and M' are given by

$$\left\{ \frac{(k\pi)^2}{(b-a)^2} + q \,:\, k \in \mathbb{N} \right\}$$

and coincide with the eigenvalues of the self-adjoint extension A_0.

Proposition 6.3.4. *For $\lambda \in \rho(A_0)$ the resolvent of A_0 is an integral operator of the form*

$$\bigl((A_0 - \lambda)^{-1} g\bigr)(t) = \int_a^b G_0(t,s,\lambda) g(s) r(s)\, ds, \quad g \in L^2_r(a,b),$$

where the Green function $G_0(t,s,\lambda)$ is given by

$$G_0(t,s,\lambda) = \begin{cases} (u_1(t,\lambda) + M'(\lambda)u_2(t,\lambda))u_2(s,\lambda), & a < s < t, \\ u_2(t,\lambda)(u_1(s,\lambda) + M'(\lambda)u_2(s,\lambda)), & t < s < b. \end{cases}$$

Here M' is the Weyl function in Corollary 6.3.2. The integral operator belongs to the Hilbert–Schmidt class. In particular, $\sigma(A_0) = \sigma_{\mathrm{p}}(A_0)$ and the multiplicity of each eigenvalue is 1.

Proof. A straightforward calculation shows that the function

$$\begin{aligned} f(t) &= \int_a^b G_0(t,s,\lambda) g(s) r(s)\, ds \\ &= (u_1(t,\lambda) + M'(\lambda)u_2(t,\lambda)) \int_a^t u_2(s,\lambda) g(s) r(s)\, ds \\ &\quad + u_2(t,\lambda) \int_t^b (u_1(s,\lambda) + M'(\lambda)u_2(s,\lambda)) g(s) r(s)\, ds \end{aligned}$$

satisfies the differential equation $(L - \lambda)f = g$. Since $u_2(a,\lambda) = 0$, it is clear that $f(a) = 0$. Moreover, since $M'(\lambda)u_2(b,\lambda) = -u_1(b,\lambda)$, one also has $f(b) = 0$. As

f is continuous on $[a,b]$ and $r \in L^1(a,b)$, it follows that $f \in L^2_r(a,b)$ and hence $f \in \operatorname{dom} A_0$. Therefore, $(A_0 - \lambda)f = g$ and for $\lambda \in \rho(A_0)$ the resolvent of A_0 is of the form as stated.

Furthermore, one has

$$\int_a^b \int_a^b |G_0(t,s,\lambda)|^2 r(s) r(t)\, ds\, dt < \infty$$

since $G_0(\cdot,\cdot,\lambda)$ is continuous for $a \le s \le t$ and $t \le s \le b$ and $r \in L^1(a,b)$. Thus, $(A_0 - \lambda)^{-1}$ is a Hilbert–Schmidt operator and, in particular, $\sigma(A_0) = \sigma_{\rm p}(A_0)$ holds. Since the eigenfunctions of A_0 are multiples of the solution $u_2(\cdot,\lambda)$ it also follows that the eigenvalues of A_0 have multiplicity one. □

It is easy to see that the closed symmetric operators $T_{\min}$ and $T'_{\min}$ in Proposition 6.3.1 and Corollary 6.3.2 do not have eigenvalues. Therefore, since the spectrum of A_0 is purely discrete according to Proposition 6.3.4, the next corollary is immediate from Proposition 3.4.8.

Corollary 6.3.5. *$T_{\min}$ and $T'_{\min}$ are simple symmetric operators in $L^2_r(a,b)$.*

It follows from Corollary 6.3.5 and the results in Section 3.5 that the spectrum of A_0 can be characterized with the help of the Weyl functions M and M' in Proposition 6.3.1 and Corollary 6.3.2. More precisely, in the present situation the functions

$$M(\lambda) = \frac{1}{u_2(b,\lambda)} \begin{pmatrix} -u_1(b,\lambda) & 1 \\ 1 & -(pu_2')(b,\lambda) \end{pmatrix} \quad \text{and} \quad M'(\lambda) = -\frac{u_1(b,\lambda)}{u_2(b,\lambda)}$$

are defined and holomorphic on $\rho(A_0)$, and one has $\lambda \in \sigma(A_0) = \sigma_{\rm p}(A_0)$ if and only if λ is a pole of M and M'. In particular, it follows that $\lambda \in \sigma_{\rm p}(A_0)$ if and only if $u_2(b,\lambda) = 0$; this extends the observation that $u_2(b,\lambda) \neq 0$ for all $\lambda \in \rho(A_0)$ in Proposition 6.3.1. The two linear maps

$$\tau : \ker(A_0 - \lambda) \to \operatorname{ran} \mathcal{R}_\lambda, \qquad f(\cdot,\lambda) \mapsto \Gamma_1 f(\cdot,\lambda) = \begin{pmatrix} (pf')(a,\lambda) \\ -(pf')(b,\lambda) \end{pmatrix},$$

and

$$\tau' : \ker(A_0 - \lambda) \to \operatorname{ran} \mathcal{R}'_\lambda, \qquad f(\cdot,\lambda) \mapsto \Gamma'_1 f(\cdot,\lambda) = (pf')(a,\lambda),$$

where

$$\mathcal{R}_\lambda \varphi = \lim_{\varepsilon \downarrow 0} i\varepsilon M(\lambda + i\varepsilon)\varphi, \quad \varphi \in \mathbb{C}^2, \quad \text{and} \quad \mathcal{R}'_\lambda = \lim_{\varepsilon \downarrow 0} i\varepsilon M'(\lambda + i\varepsilon)$$

coincide with the residues of M and M' at λ, are bijective.

In the following some classes of extensions of $T_{\min}$ and their spectral properties are briefly discussed. Let $\{\mathbb{C}^2, \Gamma_0, \Gamma_1\}$ be the boundary triplet in Proposition 6.3.1 with corresponding γ-field γ and Weyl function M. Recall first that the self-adjoint (maximal dissipative, maximal accumulative) extensions $A_\Theta \subset T_{\max}$ of $T_{\min}$ are in a one-to-one correspondence to the self-adjoint (maximal dissipative, maximal accumulative, respectively) relations Θ in $\mathbb{C}^2$ via

$$\begin{aligned} \operatorname{dom} A_\Theta &= \{f \in \operatorname{dom} T_{\max} : \{\Gamma_0 f, \Gamma_1 f\} \in \Theta\} \\ &= \left\{ f \in \operatorname{dom} T_{\max} : \left\{ \begin{pmatrix} f(a) \\ f(b) \end{pmatrix}, \begin{pmatrix} (pf')(a) \\ -(pf')(b) \end{pmatrix} \right\} \in \Theta \right\}. \end{aligned} \tag{6.3.5}$$

In the following assume that Θ is a self-adjoint relation in $\mathbb{C}^2$, so that the operator A_Θ is a self-adjoint realization of the Sturm–Liouville differential expression L in $L^2_r(a,b)$. By Corollary 1.10.9, the relation Θ in $\mathbb{C}^2$ can be represented by means of 2×2 matrices $\mathcal{A}$ and $\mathcal{B}$ satisfying the conditions $\mathcal{A}^*\mathcal{B} = \mathcal{B}^*\mathcal{A}$, $\mathcal{A}\mathcal{B}^* = \mathcal{B}\mathcal{A}^*$ and $\mathcal{A}^*\mathcal{A} + \mathcal{B}^*\mathcal{B} = I = \mathcal{A}\mathcal{A}^* + \mathcal{B}\mathcal{B}^*$, namely,

$$\Theta = \{\{\mathcal{A}\varphi, \mathcal{B}\varphi\} : \varphi \in \mathbb{C}^2\} = \{\{\psi, \psi'\} : \mathcal{A}^*\psi' = \mathcal{B}^*\psi\}.$$

In that case one has

$$\operatorname{dom} A_\Theta = \left\{ f \in \operatorname{dom} T_{\max} : \mathcal{A}^* \begin{pmatrix} (pf')(a) \\ -(pf')(b) \end{pmatrix} = \mathcal{B}^* \begin{pmatrix} f(a) \\ f(b) \end{pmatrix} \right\}.$$

Recall from Theorem 2.6.1 and Corollary 2.6.3 for $\lambda \in \rho(A_\Theta) \cap \rho(A_0)$ the Kreĭn formula for the corresponding resolvents

$$\begin{aligned} (A_\Theta - \lambda)^{-1} &= (A_0 - \lambda)^{-1} + \gamma(\lambda)\big(\Theta - M(\lambda)\big)^{-1}\gamma(\bar\lambda)^* \\ &= (A_0 - \lambda)^{-1} + \gamma(\lambda)\mathcal{A}\big(\mathcal{B} - M(\lambda)\mathcal{A}\big)^{-1}\gamma(\bar\lambda)^*. \end{aligned}$$

Since the spectrum of A_0 is discrete and the difference of the resolvents of A_0 and A_Θ is an operator of rank ≤ 2, it is clear that the spectrum of the self-adjoint operator A_Θ is discrete. Note that $\lambda \in \rho(A_0)$ is an eigenvalue of A_Θ if and only if $\ker(\Theta - M(\lambda))$ or, equivalently, $\ker(\mathcal{B} - M(\lambda)\mathcal{A})$ is nontrivial, and that

$$\ker(A_\Theta - \lambda) = \gamma(\lambda) \ker\big(\Theta - M(\lambda)\big) = \gamma(\lambda)\mathcal{A} \ker\big(\mathcal{B} - M(\lambda)\mathcal{A}\big).$$

For a complete description of the (discrete) spectrum of A_Θ recall that the symmetric operator $T_{\min}$ is simple and make use of a transform of the boundary triplet $\{\mathbb{C}^2, \Gamma_0, \Gamma_1\}$ as in Section 3.8. This reasoning implies that λ is an eigenvalue of A_Θ if and only if λ is a pole of the function

$$\lambda \mapsto M_\Theta(\lambda) = \big(\mathcal{A}^* + \mathcal{B}^* M(\lambda)\big)\big(\mathcal{B}^* - \mathcal{A}^* M(\lambda)\big)^{-1}.$$

It is important to note in this context that the multiplicity of the eigenvalues of A_Θ is at most 2 and that the dimension of the eigenspace $\ker(A_\Theta - \lambda)$ coincides with the dimension of the range of the residue of M_Θ at λ. In the special case where the self-adjoint relation Θ in $\mathbb{C}^2$ is a 2×2 matrix the boundary condition in (6.3.5) reads

$$\operatorname{dom} A_\Theta = \left\{ f \in \operatorname{dom} T_{\max} : \Theta \begin{pmatrix} f(a) \\ f(b) \end{pmatrix} = \begin{pmatrix} (pf')(a) \\ -(pf')(b) \end{pmatrix} \right\},$$

and according to Section 3.8 the spectral properties of the self-adjoint operator A_Θ can also be described with the help of the function

$$\lambda \mapsto \bigl(\Theta - M(\lambda)\bigr)^{-1}; \tag{6.3.6}$$

that is, the poles of the matrix function (6.3.6) coincide with the (discrete) spectrum of A_Θ and the dimension of the eigenspace $\ker(A_\Theta - \lambda)$ coincides with the dimension of the range of the residue of the function in (6.3.6) at λ.

In what follows some special types of boundary conditions will be discussed.

Example 6.3.6. Let $\{\mathbb{C}^2, \Gamma_0, \Gamma_1\}$ be the boundary triplet in Proposition 6.3.1 with corresponding γ-field γ and Weyl function M. Consider a 2×2 diagonal matrix

$$\Theta = \begin{pmatrix} \alpha & 0 \\ 0 & \beta \end{pmatrix}, \qquad \alpha, \beta \in \mathbb{R}.$$

The domain of the corresponding self-adjoint Sturm–Liouville operator A_Θ is given by

$$\operatorname{dom} A_\Theta = \bigl\{ f \in \operatorname{dom} T_{\max} : \alpha f(a) = (pf')(a),\, \beta f(b) = -(pf')(b) \bigr\}.$$

Such boundary conditions are often called *separated boundary conditions.* The eigenvalues of A_Θ have multiplicity one and they coincide with the poles of the function

$$\lambda \mapsto v(\lambda) \begin{pmatrix} \beta u_2(b,\lambda) + (pu_2')(b,\lambda) & 1 \\ 1 & u_1(b,\lambda) + \alpha u_2(b,\lambda) \end{pmatrix},$$

where

$$v(\lambda) = \frac{u_2(b,\lambda)}{(u_1(b,\lambda) + \alpha u_2(b,\lambda))(\beta u_2(b,\lambda) + (pu_2')(b,\lambda)) - 1}.$$

In the special case $\alpha = \beta = 0$ the operator A_Θ is defined on $\ker \Gamma_1$, which corresponds to Neumann boundary conditions. In this situation the poles of the function

$$\lambda \mapsto -M(\lambda)^{-1} = \frac{1}{(pu_1')(b,\lambda)} \begin{pmatrix} (pu_2')(b,\lambda) & 1 \\ 1 & u_1(b,\lambda) \end{pmatrix}$$

coincide with the Neumann eigenvalues.

Next the boundary triplet $\{\mathbb{C}^2, \Gamma_0, \Gamma_1\}$ in Proposition 6.3.1 will be transformed so that the so-called *periodic boundary conditions* can be treated in a convenient way. For this consider the matrix

$$\mathcal{W} = \frac{1}{\sqrt{2}} \begin{pmatrix} 1 & -1 & 0 & 0 \\ 0 & 0 & 1 & 1 \\ 0 & 0 & 1 & -1 \\ -1 & -1 & 0 & 0 \end{pmatrix},$$

and note that the condition (2.5.1) in Theorem 2.5.1 is satisfied. It then follows that $\{\mathbb{C}^2, \Upsilon_0, \Upsilon_1\}$, where

$$\Upsilon_0 f = \frac{1}{\sqrt{2}} \begin{pmatrix} f(a) - f(b) \\ (pf')(a) - (pf')(b) \end{pmatrix}, \qquad f \in \operatorname{dom} T_{\max}, \tag{6.3.7}$$

and

$$\Upsilon_1 f = \frac{1}{\sqrt{2}} \begin{pmatrix} (pf')(a) + (pf')(b) \\ -f(a) - f(b) \end{pmatrix}, \qquad f \in \operatorname{dom} T_{\max}, \tag{6.3.8}$$

is a boundary triplet for $T_{\max}$. In order to compute the corresponding γ-field γ_Υ and Weyl function M_Υ use Proposition 2.5.5 and note first that

$$\begin{pmatrix} 0 & 0 \\ -1 & -1 \end{pmatrix} + \begin{pmatrix} 1 & -1 \\ 0 & 0 \end{pmatrix} M(\lambda) = \frac{1}{u_2(b,\lambda)} \begin{pmatrix} -1 - u_1(b,\lambda) & 1 + (pu_2')(b,\lambda) \\ -u_2(b,\lambda) & -u_2(b,\lambda) \end{pmatrix}$$

and

$$\left(\begin{pmatrix} 1 & -1 \\ 0 & 0 \end{pmatrix} + \begin{pmatrix} 0 & 0 \\ 1 & 1 \end{pmatrix} M(\lambda) \right)^{-1} = w(\lambda) \begin{pmatrix} 1 - (pu_2')(b,\lambda) & u_2(b,\lambda) \\ u_1(b,\lambda) - 1 & u_2(b,\lambda) \end{pmatrix},$$

where

$$w(\lambda) = \frac{1}{2 - u_1(b,\lambda) - (pu_2')(b,\lambda)}$$

and M is the Weyl function corresponding to the boundary triplet $\{\mathbb{C}^2, \Gamma_0, \Gamma_1\}$ in Proposition 6.3.1. Now it follows that the γ-field γ_Υ and Weyl function M_Υ of $\{\mathbb{C}^2, \Upsilon_0, \Upsilon_1\}$ are given by

$$\gamma_\Upsilon(\lambda) = \begin{pmatrix} u_1(\cdot,\lambda) & u_2(\cdot,\lambda) \end{pmatrix} w(\lambda) \begin{pmatrix} 1 - (pu_2')(b,\lambda) & u_2(b,\lambda) \\ (pu_1')(b,\lambda) & 1 - u_1(b,\lambda) \end{pmatrix} \tag{6.3.9}$$

and

$$M_\Upsilon(\lambda) = w(\lambda) \begin{pmatrix} 2(pu_1')(b,\lambda) & -u_1(b,\lambda) + (pu_2')(b,\lambda) \\ -u_1(b,\lambda) + (pu_2')(b,\lambda) & -2u_2(b,\lambda) \end{pmatrix}. \tag{6.3.10}$$

As above, it follows from Corollary 6.3.5 and the considerations in Section 3.5 that the eigenvalues of the self-adjoint operator A_{Υ_0} which corresponds to the boundary condition $\ker \Upsilon_0$, that is,

$$\operatorname{dom} A_{\Upsilon_0} = \bigl\{ f \in \operatorname{dom} T_{\max} : f(a) = f(b),\ (pf')(a) = (pf')(b) \bigr\},$$

coincide with the isolated poles of M_Υ, that is, λ is an eigenvalue of A_{Υ_0} if and only if $u_1(b,\lambda) + (pu_2')(b,\lambda) = 2$. It is important to note that in this situation eigenvalues of multiplicity two may arise and hence a reduction of the spectral problem to a scalar Weyl function as in Corollary 6.3.2 is, in general, not possible. This is the case in the following example, where a symmetric extension of $T_{\min}$ appears which is not simple.

Example 6.3.7. Let $\{\mathbb{C}^2, \Upsilon_0, \Upsilon_1\}$ be the boundary triplet in (6.3.7)–(6.3.8) with corresponding γ-field γ_Υ and Weyl function M_Υ in (6.3.9) and (6.3.10), respectively. Consider the Sturm–Liouville expression $Lf = -f''$ in $L^2(0, 2\pi)$, that is, $r = p = 1$ and $q = 0$, and $(a, b) = (0, 2\pi)$. Then the positive eigenvalues k^2, $k \in \mathbb{N}$, of A_{Υ_0} are of multiplicity two and the eigenvalue 0 has multiplicity one. Consider the symmetric operator $T''_{\rm min}$ defined on

$$\operatorname{dom} T''_{\rm min} = \{f \in \operatorname{dom} T_{\rm max} : f(0) = f(2\pi),\ f'(0) = f'(2\pi) = 0\}$$

which arises in the same way as in Corollary 6.3.2, but now with the boundary triplet $\{\mathbb{C}^2, \Gamma_0, \Gamma_1\}$ from Proposition 6.3.1 replaced by the boundary triplet $\{\mathbb{C}^2, \Upsilon_0, \Upsilon_1\}$. As in Corollary 6.3.2, one obtains that $T''_{\rm min}$ is a closed symmetric operator with defect numbers $(1, 1)$ and adjoint $(T''_{\rm min})^*$ defined on

$$\operatorname{dom} (T''_{\rm min})^* = \{f \in \operatorname{dom} T_{\rm max} : f'(0) = f'(2\pi)\},$$

and one has $T_{\rm min} \subset T''_{\rm min} \subset A_{\Upsilon_0}$. Moreover, $\{\mathbb{C}, \Upsilon'_0, \Upsilon'_1\}$, where

$$\Upsilon'_0 f = \frac{1}{\sqrt{2}}\bigl(f(0) - f(2\pi)\bigr) \quad\text{and}\quad \Upsilon'_1 f = \frac{1}{\sqrt{2}}\bigl(f'(0) + f'(2\pi)\bigr),\ f \in \operatorname{dom} (T''_{\rm min})^*,$$

is a boundary triplet for $(T''_{\rm min})^*$ with corresponding Weyl function M'_Υ given by

$$M'_\Upsilon(\lambda) = \frac{2u'_1(2\pi, \lambda)}{2 - u_1(2\pi, \lambda) - u'_2(2\pi, \lambda)}.$$

However, in contrast to the situation in Corollary 6.3.2 and Corollary 6.3.5, here $T''_{\rm min}$ is not simple. In fact, since the eigenvalues k^2, $k \in \mathbb{N}$, of A_{Υ_0} have multiplicity two and $T''_{\rm min}$ is a one-dimensional restriction of A_{Υ_0} each k^2, $k \in \mathbb{N}$, is an eigenvalue of $T''_{\rm min}$ of multiplicity one. Therefore, the corresponding eigenfunctions span an infinite-dimensional subspace of $L^2(0, 2\pi)$ which reduces $T''_{\rm min}$ and in which $T''_{\rm min}$ is self-adjoint. Note however, that a further one-dimensional restriction of $T''_{\rm min}$ leads to the minimal operator $T_{\rm min}$ which is simple by Corollary 6.3.5.

This section is concluded with the limit-circle case. It is assumed that the coefficient functions satisfy (6.1.2) and that both endpoints a and b are in the limit-circle case; cf. Theorem 6.1.6. Now consider real solutions u_a, v_a and u_b, v_b of $(L - \lambda_0)y = 0$, $\lambda_0 \in \mathbb{R}$, which satisfy $W(u_a, v_a) = 1$ and $W(u_b, v_b) = 1$, while $u_a, v_a \in L^2_r(a, a')$ and $u_b, v_b \in L^2_r(b', b)$. In the following it is tacitly assumed that quasi-derivatives at a are defined in terms of u_a, v_a and that the quasi-derivatives at b are defined in terms of u_b, v_b.

Fix a fundamental system $(u_1(\cdot, \lambda); u_2(\cdot, \lambda))$ for the equation $(L - \lambda)y = 0$ by the initial conditions

$$\begin{pmatrix} u_1^{[0]}(a, \lambda) & u_2^{[0]}(a, \lambda) \\ u_1^{[1]}(a, \lambda) & u_2^{[1]}(a, \lambda) \end{pmatrix} = \begin{pmatrix} 1 & 0 \\ 0 & 1 \end{pmatrix}.$$

Recall that for each function $f \in \operatorname{dom} T_{\max}$ the quasi-derivaties $f^{[0]}(a)$, $f^{[1]}(a)$, $f^{[0]}(b)$, $f^{[1]}(b)$ are well defined; cf. Definition 6.2.4. The proof of the following proposition follows the lines of the proof of Proposition 6.3.1 in conjunction with Lemma 6.2.5 and Proposition 6.2.6.

Proposition 6.3.8. *Assume that the endpoints a and b are in the limit-circle case. Then $\{\mathbb{C}^2, \Gamma_0, \Gamma_1\}$, where*

$$\Gamma_0 f = \begin{pmatrix} f^{[0]}(a) \\ f^{[0]}(b) \end{pmatrix} \quad \text{and} \quad \Gamma_1 f = \begin{pmatrix} f^{[1]}(a) \\ -f^{[1]}(b) \end{pmatrix}, \quad f \in \operatorname{dom} T_{\max},$$

is a boundary triplet for $T_{\max}$. The self-adjoint extension A_0 corresponding to Γ_0 is the restriction of $T_{\max}$ defined on

$$\operatorname{dom} A_0 = \big\{ f \in \operatorname{dom} T_{\max} : f^{[0]}(a) = f^{[0]}(b) = 0 \big\}$$

and the minimal operator $T_{\min}$ is the restriction of $T_{\max}$ defined on

$$\operatorname{dom} T_{\min} = \big\{ f \in \operatorname{dom} T_{\max} : f^{[0]}(a) = f^{[0]}(b) = f^{[1]}(a) = f^{[1]}(b) = 0 \big\}.$$

Moreover, for all $\lambda \in \rho(A_0)$ one has $u_2^{[0]}(b, \lambda) \neq 0$. The corresponding γ-field and Weyl function are given by

$$\gamma(\lambda) = \big(u_1(\cdot, \lambda) \quad u_2(\cdot, \lambda) \big) \frac{1}{u_2^{[0]}(b, \lambda)} \begin{pmatrix} u_2^{[0]}(b, \lambda) & 0 \\ -u_1^{[0]}(b, \lambda) & 1 \end{pmatrix}, \quad \lambda \in \rho(A_0),$$

and

$$M(\lambda) = \frac{1}{u_2^{[0]}(b, \lambda)} \begin{pmatrix} -u_1^{[0]}(b, \lambda) & 1 \\ 1 & -u_2^{[1]}(b, \lambda) \end{pmatrix}, \quad \lambda \in \rho(A_0).$$

6.4 The case of one limit-point endpoint

Let $T_{\max} = (T_{\min})^*$ be the maximal operator associated with the Sturm–Liouville differential expression L in (6.1.1) on the interval (a, b). This situation will be considered under the assumption that the endpoint a is regular and the endpoint b is in the limit-point case. In the end of the section also the situation that a is in the limit-circle case is briefly discussed.

Recall that the assumption that the endpoint a is regular means

$$\begin{cases} p(x) \neq 0, \ r(x) > 0, \ \text{ for a.e. } x \in (a, b), \\ 1/p, q, r \in L^1(a, a'), \quad 1/p, q, r \in L^1_{\mathrm{loc}}(a', b); \end{cases}$$

cf. Definition 6.1.1. Let the fundamental system $(u_1(\cdot, \lambda); u_2(\cdot, \lambda))$ for the equation $(L - \lambda)y = 0$ be fixed by the initial conditions

$$\begin{pmatrix} u_1(a, \lambda) & u_2(a, \lambda) \\ (pu_1')(a, \lambda) & (pu_2')(a, \lambda) \end{pmatrix} = \begin{pmatrix} 1 & 0 \\ 0 & 1 \end{pmatrix}. \tag{6.4.1}$$

For each function $f \in \operatorname{dom} T_{\max}$ one has $f, pf' \in AC[a,b)$ and the quantities $f(a)$ and $(pf')(a)$ are well defined; cf. Theorem 6.1.2.

Proposition 6.4.1. *Assume that the endpoint a is regular and that the endpoint b is in the limit-point case. Then $\{\mathbb{C}, \Gamma_0, \Gamma_1\}$, where*

$$\Gamma_0 f = f(a) \quad \text{and} \quad \Gamma_1 f = (pf')(a), \quad f \in \operatorname{dom} T_{\max}, \tag{6.4.2}$$

is a boundary triplet for the operator $(T_{\min})^* = T_{\max}$. The self-adjoint extension A_0 corresponding to Γ_0 is the restriction of $T_{\max}$ defined on

$$\operatorname{dom} A_0 = \bigl\{ f \in \operatorname{dom} T_{\max} : f(a) = 0 \bigr\}$$

and the minimal operator $T_{\min}$ is the restriction of $T_{\max}$ defined on

$$\operatorname{dom} T_{\min} = \bigl\{ f \in \operatorname{dom} T_{\max} : f(a) = (pf')(a) = 0 \bigr\}.$$

Moreover, if $\lambda \in \mathbb{C} \setminus \mathbb{R}$ and $\chi(\cdot, \lambda)$ is a nontrivial element in $\mathfrak{N}_\lambda(T_{\max})$, then one has $\chi(a, \lambda) \neq 0$. For all $\lambda \in \mathbb{C} \setminus \mathbb{R}$ the corresponding γ-field and Weyl function are given by

$$\gamma(\cdot, \lambda) = u_1(\cdot, \lambda) + M(\lambda) u_2(\cdot, \lambda) \quad \text{and} \quad M(\lambda) = \frac{(p\chi')(a, \lambda)}{\chi(a, \lambda)}. \tag{6.4.3}$$

Proof. First it will be verified that the mapping $(\Gamma_0, \Gamma_1)^\top : \operatorname{dom} T_{\max} \to \mathbb{C}^2$ is surjective. Let $\alpha \in \mathbb{C}^2$. Then there exists $f \in \operatorname{dom} T_{\max}$ such that $f(a) = \alpha_1$, $(pf')(a) = \alpha_2$, which vanishes in a neighborhood of b. To see this, let h be a solution of $Ly = 0$ with $h, ph' \in AC[a,b)$ and $h(a) = \alpha_1$, $(ph')(a) = \alpha_2$. Now by cutting off the function h near b one obtains a function f which satisfies $Lf = g$ for some $g \in L^2_r(a,b)$ and which vanishes in a neighborhood of b; see Proposition 6.1.3. Hence, $f \in \operatorname{dom} T_{\max}$ and at a one has

$$\Gamma_0 f = f(a) = h(a) = \alpha_1 \quad \text{and} \quad \Gamma_1 f = (pf')(a) = (ph')(a) = \alpha_2.$$

This proves the claim.

Next one verifies the abstract Green identity. For this one shows first that $\lim_{x \to b} W_x(f, \bar{g}) = 0$ for all $f, g \in \operatorname{dom} T_{\max}$. In fact, since the endpoint b is in the limit-point case it follows from Corollary 6.2.2 that $T_{\min}$ has defect numbers $(1,1)$. Now choose $h_1, h_2 \in \operatorname{dom} T_{\max}$ such that

$$h_1(a) = 1, \quad (ph_1')(a) = 0, \quad h_2(a) = 0, \quad (ph_2')(a) = 1,$$

and such that h_1 and h_2 vanish in a neighborhood of b; cf. Proposition 6.1.3. Then $h_1, h_2 \notin \operatorname{dom} T_{\min}$, since otherwise

$$\lim_{x \to a} W_x(h_i, \bar{g}) = h_i(a)\overline{(pg')(a)} - (ph_i')(a)\overline{g(a)} = 0, \quad i = 1, 2,$$

for all $g \in \operatorname{dom} T_{\max}$ by (6.2.7) and Lemma 6.2.3, which is not possible by the considerations in the beginning of the proof. Thus, every function $f \in \operatorname{dom} T_{\max}$ can be written in the form

$$f = f_0 + c_1 h_1 + c_2 h_2, \qquad f_0 \in \operatorname{dom} T_{\min},$$

for some $c_1, c_2 \in \mathbb{C}$. Observe that therefore

$$W_x(f, \bar{g}) = W_x(f_0, \bar{g}) + W_x(c_1 h_1 + c_2 h_2, \bar{g})$$

for all $g \in \operatorname{dom} T_{\max}$, and since the last term vanishes in a neighborhood of b one obtains

$$\lim_{x \to b} W(f, \bar{g}) = \lim_{x \to b} W(f_0, \bar{g}) = 0$$

for all $g \in \operatorname{dom} T_{\max}$. Hence, by (6.2.6) and Lemma 6.2.3, for $f, g \in \operatorname{dom} T_{\max}$ one has

$$\begin{aligned}(T_{\max} f, g)_{L^2_r(a,b)} - (f, T_{\max} g)_{L^2_r(a,b)} &= - \lim_{x \to a} W(f, \bar{g}) \\ &= (pf')(a)\overline{g(a)} - f(a)\overline{(pg')(a)},\end{aligned}$$

which implies that the abstract Green identity is satisfied with the choice of Γ_0 and Γ_1 in (6.4.2). Thus, (6.4.2) defines a boundary triplet for $(T_{\min})^* = T_{\max}$.

The description of the domain of the self-adjoint extension A_0 is trivial. Furthermore, by Proposition 2.1.2 (ii) one has $\operatorname{dom} T_{\min} = \ker \Gamma_0 \cap \ker \Gamma_1$, which yields the stated description of the domain of the minimal operator.

Due to the assumption that the endpoint b is in the limit-point case, each eigenspace $\mathfrak{N}_\lambda(T_{\max})$, $\lambda \in \mathbb{C} \setminus \mathbb{R}$, has dimension one. Hence, if $\chi(\cdot, \lambda)$ is a nontrivial element which spans $\mathfrak{N}_\lambda(T_{\max})$, $\lambda \in \mathbb{C} \setminus \mathbb{R}$, then, by Definition 2.3.4,

$$M(\lambda) = \big\{\{\chi(a, \lambda)c, (p\chi')(a, \lambda)c\} : c \in \mathbb{C}\big\}, \qquad \lambda \in \mathbb{C} \setminus \mathbb{R}.$$

Observe that $\chi(a, \lambda) \neq 0$ for $\lambda \in \mathbb{C} \setminus \mathbb{R}$, since otherwise $\chi(\cdot, \lambda) \in \ker(A_0 - \lambda)$ and the fact that A_0 is self-adjoint would imply $\chi(\cdot, \lambda) = 0$. Consequently,

$$M(\lambda) = \frac{(p\chi')(a, \lambda)}{\chi(a, \lambda)}, \qquad \lambda \in \mathbb{C} \setminus \mathbb{R}.$$

Likewise, it follows from Definition 2.3.1 that

$$\gamma(\lambda) = \big\{\{\chi(a, \lambda)c, \chi(\cdot, \lambda)c\} : c \in \mathbb{C}\big\}, \qquad \lambda \in \mathbb{C} \setminus \mathbb{R},$$

and so

$$\gamma(\lambda) = \frac{\chi(\cdot, \lambda)}{\chi(a, \lambda)}, \qquad \lambda \in \mathbb{C} \setminus \mathbb{R}.$$

Writing $\chi(\cdot, \lambda)$ in terms of the fundamental system,

$$\chi(\cdot, \lambda) = \chi(a, \lambda) u_1(\cdot, \lambda) + (p\chi')(a, \lambda) u_2(\cdot, \lambda), \qquad \lambda \in \mathbb{C} \setminus \mathbb{R},$$

the form of the γ-field follows. □

Note that the γ-field and the Weyl function corresponding to the boundary triplet $\{\mathbb{C}, \Gamma_0, \Gamma_1\}$ in Proposition 6.4.1 are defined and analytic on the resolvent set of the self-adjoint operator A_0. The expressions in (6.4.3) extend from $\mathbb{C} \setminus \mathbb{R}$ to all of $\rho(A_0)$ when $\rho(A_0) \cap \mathbb{R} \neq \emptyset$. In fact, it follows from the direct sum decomposition $\operatorname{dom} T_{\max} = \operatorname{dom} A_0 \dotplus \mathfrak{N}_\lambda(T_{\max})$ that also for each $\lambda \in \rho(A_0) \cap \mathbb{R}$ there exists a nontrivial element $\chi(\cdot, \lambda)$ in $\mathfrak{N}_\lambda(T_{\max})$ such that $\chi(a, \lambda) \neq 0$. It is clear from (6.4.2) that also for these points $\lambda \in \rho(A_0) \cap \mathbb{R}$ the γ-field and Weyl function are given by (6.4.3).

Example 6.4.2. In the special case $r = p = 1$ on $\iota = (a, \infty)$ and a constant $q \in \mathbb{R}$, the Sturm–Liouville expression is $Lf = -f'' + qf$. Fix the square root $\sqrt{\cdot}$ such that $\operatorname{Im} \sqrt{\lambda} > 0$ for all $\lambda \in \mathbb{C} \setminus [0, \infty)$ and $\sqrt{\lambda} \geq 0$ for $\lambda \in [0, \infty)$. For all $\lambda \in \mathbb{C} \setminus [q, \infty)$, the function

$$x \mapsto \chi(x, \lambda) = e^{i\sqrt{\lambda - q}\,(x-a)} \in L^2(a, \infty)$$

spans the one-dimensional eigenspace $\mathfrak{N}_\lambda(T_{\max})$. Hence, the Weyl function M in Proposition 6.4.1 is given by

$$M(\lambda) = \frac{\Gamma_1 \chi(\cdot, \lambda)}{\Gamma_0 \chi(\cdot, \lambda)} = i\sqrt{\lambda - q}\,.$$

In terms of the fundamental system

$$u_1(x, \lambda) = \cos\bigl[\sqrt{\lambda - q}\,(x - a)\bigr], \quad u_2(x, \lambda) = \frac{\sin\bigl[\sqrt{\lambda - q}\,(x - a)\bigr]}{\sqrt{\lambda - q}},$$

one has $\chi(x, \lambda) = u_1(x, \lambda) + M(\lambda) u_2(x, \lambda) = e^{i\sqrt{\lambda - q}\,(x-a)}$ for $\lambda \in \mathbb{C} \setminus [q, \infty)$. Note also that M is holomorphic on $\mathbb{C} \setminus [q, \infty)$ and that $\sigma(A_0) = [q, \infty)$. Moreover, for $\lambda \in (q, \infty)$ one has

$$\lim_{\varepsilon \downarrow 0} \operatorname{Im} M(\lambda + i\varepsilon) = \sqrt{\lambda - q} > 0,$$

and for $\lambda \in [q, \infty)$

$$\lim_{\varepsilon \downarrow 0} i\varepsilon M(\lambda + i\varepsilon) = 0.$$

Below in Proposition 6.4.4 it is shown that $T_{\min}$ is simple and hence the results in Section 3.5 and Section 3.6 apply. In particular, it follows from Theorem 3.6.5 that $\sigma_{\mathrm{ac}}(A_0) = [q, \infty)$ and Corollary 3.5.6 shows $\sigma_{\mathrm{p}}(A_0) \cap [q, \infty) = \emptyset$ (see also Theorem 3.6.1).

Proposition 6.4.3. *For $\lambda \in \rho(A_0)$ the resolvent of the self-adjoint extension A_0 is an integral operator of the form*

$$\bigl((A_0 - \lambda)^{-1} g\bigr)(t) = \int_a^b G_0(t, s, \lambda) g(s) r(s) ds, \quad g \in L^2_r(a, b), \tag{6.4.4}$$

where the Green function $G_0(t, s, \lambda)$ is given by

$$G_0(t, s, \lambda) = \begin{cases} (u_1(t, \lambda) + M(\lambda) u_2(t, \lambda)) u_2(s, \lambda), & a < s < t, \\ u_2(t, \lambda)(u_1(s, \lambda) + M(\lambda) u_2(s, \lambda)), & t < s < b. \end{cases}$$

In particular, if $g \in L^2_r(a,b)$ has compact support, then

$$\begin{aligned}\big((A_0-\lambda)^{-1}g\big)(t) = u_2(t,\lambda)M(\lambda)\int_a^b u_2(s,\lambda)g(s)r(s)\,ds \\ + u_1(t,\lambda)\int_a^t u_2(s,\lambda)g(s)r(s)\,ds \\ + u_2(t,\lambda)\int_t^b u_1(s,\lambda)g(s)r(s)\,ds.\end{aligned} \tag{6.4.5}$$

Proof. As in the proof of Proposition 6.3.4, consider the function $f(\cdot,\lambda)$ given by the right-hand side in (6.4.4), which has the form

$$\begin{aligned}f(t,\lambda) = \big(u_1(t,\lambda)+M(\lambda)u_2(t,\lambda)\big)\int_a^t u_2(s,\lambda)g(s)r(s)\,ds \\ + u_2(t,\lambda)\int_t^b \big(u_1(s,\lambda)+M(\lambda)u_2(s,\lambda)\big)g(s)r(s)\,ds\end{aligned} \tag{6.4.6}$$

for $g \in L^2_r(a,b)$. Note that $f(\cdot,\lambda)$ is well defined, since $u_1(\cdot,\lambda)+M(\lambda)u_2(\cdot,\lambda)$ belongs to $L^2_r(a,b)$ by (6.4.3). A straightforward computation shows that $f(\cdot,\lambda)$ is a solution of the inhomogeneous differential equation $(L-\lambda)f = g$ satisfying the initial conditions

$$\begin{aligned}f(a,\lambda) &= 0,\\ (pf')(a,\lambda) &= (pu_2')(a,\lambda)\int_a^b (u_1(s,\lambda)+M(\lambda)u_2(s,\lambda))g(s)r(s)\,ds\\ &= \int_a^b g(s)\,\overline{(u_1(s,\bar\lambda)+M(\bar\lambda)u_2(s,\bar\lambda))}\,r(s)\,ds\\ &= (g,\gamma(\bar\lambda))_{L^2_r(a,b)}.\end{aligned}$$

On the other hand, since $A_0 \subset T_{\max}$, it is clear that the function $h = (A_0-\lambda)^{-1}g$ also satisfies the inhomogeneous differential equation $(L-\lambda)h = g$ and, moreover,

$$\begin{aligned}h(a) &= \Gamma_0 h = \Gamma_0(A_0-\lambda)^{-1}g = \big((A_0-\lambda)^{-1}g\big)(a) = 0,\\ (ph')(a) &= \Gamma_1 h = \Gamma_1(A_0-\lambda)^{-1}g = \gamma(\bar\lambda)^*g = (g,\gamma(\bar\lambda))_{L^2_r(a,b)},\end{aligned}$$

where Proposition 2.3.2 (iv) was used. Hence, $f = h$ by the uniqueness property for the initial value problem. This proves (6.4.4). The formula (6.4.5) follows for $g \in L^2_r(a,b)$ with compact support from (6.4.6). □

Proposition 6.4.4. *The minimal operator $T_{\min}$ is simple.*

Proof. It suffices to show that the defect spaces $\mathfrak{N}_\lambda(T_{\max})$ span the space $L^2_r(a,b)$; cf. Corollary 3.4.5. To see this, let $g \in L^2_r(a,b)$ and assume that

$$\int_a^b \big(u_1(t,\lambda)+M(\lambda)u_2(t,\lambda)\big)g(t)r(t)\,dt = 0$$

for all $\lambda \in \mathbb{C} \setminus \mathbb{R}$. Then it follows from (6.4.4) that

$$\begin{aligned}\big((A_0-\lambda)^{-1}g\big)(t) &= \big(u_1(t,\lambda)+M(\lambda)u_2(t,\lambda)\big)\int_a^t u_2(s,\lambda)g(s)r(s)\,ds \\ &\qquad + u_2(t,\lambda)\int_t^b \big(u_1(s,\lambda)+M(\lambda)u_2(s,\lambda)\big)g(s)r(s)\,ds \\ &= \big(u_1(t,\lambda)+M(\lambda)u_2(t,\lambda)\big)\int_a^t u_2(s,\lambda)g(s)r(s)\,ds \\ &\qquad - u_2(t,\lambda)\int_a^t \big(u_1(s,\lambda)+M(\lambda)u_2(s,\lambda)\big)g(s)r(s)\,ds \\ &= \int_a^t \big(u_1(t,\lambda)u_2(s,\lambda)-u_2(t,\lambda)u_1(s,\lambda)\big)g(s)r(s)\,ds\end{aligned}$$

and the right-hand side is entire in λ. Now consider a bounded interval $\delta \subset \mathbb{R}$ such that the endpoints of δ are not eigenvalues of A_0 and let $h \in L^2_r(a,b)$ be a function with compact support. It follows from Stone's formula (1.5.7) (see also Example A.1.4), Fubini's theorem, and dominated convergence that, for any $h \in L^2_r(a,b)$ with compact support,

$$\begin{aligned}&(E(\delta)g,h)_{L^2_r(a,b)} \\ &\quad = \lim_{\varepsilon\downarrow 0}\frac{1}{2\pi i}\int_\delta \big(\big((A_0-(\mu+i\varepsilon))^{-1}-(A_0-(\mu-i\varepsilon))^{-1}\big)g,h\big)_{L^2_r(a,b)}\,d\mu \\ &\quad = 0.\end{aligned}$$

This implies that $E(\delta)g=0$ for all bounded intervals $\delta\subset\mathbb{R}$ as above and letting δ expand to $\mathbb{R}$ one concludes that $g=E(\mathbb{R})g=0$. □

Let $\{\mathbb{C},\Gamma_0,\Gamma_1\}$ be the boundary triplet in Proposition 6.4.1 with γ-field and Weyl function given by

$$\gamma(\cdot,\lambda)=u_1(\cdot,\lambda)+M(\lambda)u_2(\cdot,\lambda) \quad\text{and}\quad M(\lambda)=\frac{(p\chi')(a,\lambda)}{\chi(a,\lambda)},$$

where $\chi(\cdot,\lambda)$ is a nontrivial element in $\mathfrak{N}_\lambda(T_{\max})$, $\lambda\in\rho(A_0)$. Since the operator $T_{\min}$ is simple by Proposition 6.4.4, Theorem 3.6.1 shows that the Weyl function M is analytic at λ if and only if $\lambda\in\rho(A_0)$, that $\lambda\in\sigma_{\rm p}(A_0)$ if and only if $\lim_{\varepsilon\downarrow 0} i\varepsilon M(\lambda+i\varepsilon)\neq 0$, the poles of M coincide with the isolated eigenvalues of A_0, and $\lambda\in\sigma_{\rm c}(A_0)$ if and only if $\lim_{\varepsilon\downarrow 0} i\varepsilon M(\lambda+i\varepsilon)=0$ and M does not admit an analytic continuation to λ. Furthermore, if Δ is an open interval in $\mathbb{R}$, then

$$\overline{\sigma_{\rm ac}(A_0)\cap\Delta}=\operatorname{clos}_{\rm ac}\big(\{\lambda\in\Delta : 0<\operatorname{Im} M(\lambda+i0)<+\infty\}\big).$$

In the special case $\Delta=\mathbb{R}$ one has

$$\sigma_{\rm ac}(A_0)=\operatorname{clos}_{\rm ac}\big(\{\lambda\in\mathbb{R} : 0<\operatorname{Im} M(\lambda+i0)<+\infty\}\big).$$

Furthermore, with the help of Corollary 3.6.9 one can exclude singular continuous spectrum as follows. If Δ is an open interval in $\mathbb{R}$ and there exist at most countably many $\lambda \in \Delta$ such that

$$\operatorname{Im} M(\lambda + i\varepsilon) \to +\infty, \quad \varepsilon M(\lambda + i\varepsilon) \to 0 \quad \text{as} \quad \varepsilon \downarrow 0,$$

then $\sigma_{\mathrm{sc}}(A_0) \cap \Delta = \emptyset$. For more results on the description of singular and singular continuous spectra of A_0 in this context see Section 3.6.

Now consider the self-adjoint (maximal dissipative, maximal accumulative) extensions of $T_{\min}$. According to Corollary 2.1.4 for $\tau \in \mathbb{R}$ ($\tau \in \mathbb{C}^+$, $\tau \in \mathbb{C}^-$), the realization A_τ of L with domain

$$\operatorname{dom} A_\tau = \bigl\{ f \in \operatorname{dom} T_{\max} : (pf')(a) = \tau f(a) \bigr\}$$

is self-adjoint (maximal dissipative, maximal accumulative), and the boundary condition $\tau = \infty$ is understood as $f(a) = 0$, which corresponds to $\ker \Gamma_0$. For $\tau \in \mathbb{R} \cup \{\infty\}$ introduce the following transformation of the boundary triplet in (6.4.2):

$$\begin{pmatrix} \Gamma_0^\tau \\ \Gamma_1^\tau \end{pmatrix} = \frac{1}{\sqrt{\tau^2+1}} \begin{pmatrix} \tau & -1 \\ 1 & \tau \end{pmatrix} \begin{pmatrix} \Gamma_0 \\ \Gamma_1 \end{pmatrix}. \tag{6.4.7}$$

Then $\{\mathbb{C}, \Gamma_0^\tau, \Gamma_1^\tau\}$ is a boundary triplet defined on $\operatorname{dom} T_{\max}$ with corresponding γ-field and Weyl function given by

$$\gamma_\tau(\lambda) = \frac{\gamma(\lambda)}{\tau - M(\lambda)} \sqrt{\tau^2+1} \quad \text{and} \quad M_\tau(\lambda) = \frac{1 + \tau M(\lambda)}{\tau - M(\lambda)}, \qquad \lambda \in \mathbb{C} \setminus \mathbb{R}; \tag{6.4.8}$$

cf. (2.5.19) and (2.5.20). Moreover, it is clear that

$$\ker \Gamma_0^\tau = \ker (\Gamma_1 - \tau \Gamma_0) = \operatorname{dom} A_\tau,$$

and again $\tau = \infty$ corresponds to the extension with boundary condition $f(a) = 0$. The spectrum of A_τ can now be characterized with the help of the Weyl function M_τ in the same way as the spectrum of the extension defined on $\ker \Gamma_0$ (that is, $\tau = \infty$) was characterized with the function M. E.g., λ is an eigenvalue of A_τ if and only if $\lim_{\varepsilon \downarrow 0} i\varepsilon M_\tau(\lambda + i\varepsilon) \neq 0$, and the absolutely continuous spectrum of A_τ is given by

$$\sigma_{\mathrm{ac}}(A_\tau) = \operatorname{clos}_{\mathrm{ac}}\bigl(\bigl\{\lambda \in \mathbb{R} : 0 < \operatorname{Im} M_\tau(\lambda + i0) < +\infty\bigr\}\bigr).$$

The transformation for the γ-field in (6.4.7) and (6.4.8) also shows up in a transformation of the fundamental solutions.

Lemma 6.4.5. *Let $\tau \in \mathbb{R} \cup \{\infty\}$. For $\lambda \in \mathbb{C} \setminus \mathbb{R}$ the γ-field $\gamma_\tau(\lambda)$ is of the form*

$$\gamma_\tau(\cdot, \lambda) = v_1(\cdot, \lambda) + M_\tau(\lambda) v_2(\cdot, \lambda).$$

Here the fundamental system $(v_1(\cdot,\lambda); v_2(\cdot,\lambda))$ *is given by*

$$\begin{pmatrix} v_1(\cdot,\lambda) \\ v_2(\cdot,\lambda) \end{pmatrix} = \frac{1}{\sqrt{\tau^2+1}} \begin{pmatrix} \tau & -1 \\ 1 & \tau \end{pmatrix} \begin{pmatrix} u_1(\cdot,\lambda) \\ u_2(\cdot,\lambda) \end{pmatrix}$$

and one has $W(v_1(\cdot,\lambda), v_2(\cdot,\lambda)) = 1$.

Proof. Recall that $\gamma(\cdot,\lambda) = u_1(\cdot,\lambda) + M(\lambda)u_2(\cdot,\lambda)$. If $v_1(\cdot,\lambda)$ and $v_2(\cdot,\lambda)$ are as above, then it is clear that

$$\begin{pmatrix} u_1(\cdot,\lambda) \\ u_2(\cdot,\lambda) \end{pmatrix} = \frac{1}{\sqrt{\tau^2+1}} \begin{pmatrix} \tau & 1 \\ -1 & \tau \end{pmatrix} \begin{pmatrix} v_1(\cdot,\lambda) \\ v_2(\cdot,\lambda) \end{pmatrix}.$$

Hence,

$$\begin{aligned} \gamma(\lambda) &= \frac{1}{\sqrt{\tau^2+1}} \big(\tau v_1(\cdot,\lambda) + v_2(\cdot,\lambda) + M(\lambda)(-v_1(\cdot,\lambda) + \tau v_2(\cdot,\lambda))\big) \\ &= \frac{1}{\sqrt{\tau^2+1}} \big((\tau - M(\lambda))v_1(\cdot,\lambda) + (1+\tau M(\lambda))v_2(\cdot,\lambda)\big) \\ &= \frac{\tau - M(\lambda)}{\sqrt{\tau^2+1}} \left(v_1(\cdot,\lambda) + \frac{1+\tau M(\lambda)}{\tau - M(\lambda)} v_2(\cdot,\lambda) \right), \end{aligned}$$

where it was used that $M(\lambda) \neq \tau \in \mathbb{R}$ for $\lambda \in \mathbb{C} \setminus \mathbb{R}$. This leads to

$$\frac{\gamma(\lambda)}{\tau - M(\lambda)} \sqrt{\tau^2+1} = v_1(\cdot,\lambda) + \frac{1+\tau M(\lambda)}{\tau - M(\lambda)} v_2(\cdot,\lambda).$$

Comparing with (6.4.8) one obtains the claimed form of $\gamma_\tau(\lambda)$. □

Note that the formal solution

$$v_2(\cdot,\lambda) = \frac{1}{\sqrt{\tau^2+1}} \big(u_1(\cdot,\lambda) + \tau u_2(\cdot,\lambda)\big) \tag{6.4.9}$$

satisfies the boundary condition $(pv_2')(a,\lambda) = \tau v_2(a,\lambda)$ which is connected with the self-adjoint realization A_τ defined on $\ker \Gamma_0^\tau = \ker(\Gamma_1 - \tau\Gamma_0)$. Observe that for $g \in L_r^2(a,b)$ and $\lambda \in \rho(A_\tau)$ the resolvent of A_τ has the form

$$\big((A_\tau - \lambda)^{-1} g\big)(t) = \int_a^b G_\tau(t,s,\lambda) g(s) r(s)\, ds, \quad g \in L_r^2(a,b), \tag{6.4.10}$$

where the Green function is given by

$$G_\tau(t,s,\lambda) = \begin{cases} (v_1(t,\lambda) + M_\tau(\lambda)v_2(t,\lambda))v_2(s,\lambda), & a < s < t, \\ v_2(t,\lambda)(v_1(s,\lambda) + M_\tau(\lambda)v_2(s,\lambda)), & t < s < b, \end{cases} \tag{6.4.11}$$

that is,

$$\begin{aligned}\big((A_\tau - \lambda)^{-1}g\big)(t) &= \big(v_1(t,\lambda) + M_\tau(\lambda)v_2(t,\lambda)\big)\int_a^t v_2(s,\lambda)g(s)r(s)\,ds \\ &\quad + v_2(t,\lambda)\int_t^b \big(v_1(s,\lambda) + M_\tau(\lambda)v_2(s,\lambda)\big)g(s)r(s)\,ds.\end{aligned}$$

This follows in the same way as in the proof of Proposition 6.4.3. In fact, a straightforward computation shows that the right-hand side (denoted by $f(\cdot,\lambda)$) satisfies the differential equation $(L-\lambda)f = g$ and that

$$\begin{aligned} f(a,\lambda) &= \frac{1}{\sqrt{\tau^2+1}}\int_a^b \big(v_1(s,\lambda) + M_\tau(\lambda)v_2(s,\lambda)\big)g(s)r(s)\,ds, \\ (pf')(a,\lambda) &= \frac{\tau}{\sqrt{\tau^2+1}}\int_a^b \big(v_1(s,\lambda) + M_\tau(\lambda)v_2(s,\lambda)\big)g(s)r(s)\,ds. \end{aligned}$$

Hence,

$$\frac{1}{\sqrt{\tau^2+1}}\big(\tau f(a,\lambda) - (pf')(a,\lambda)\big) = 0$$

and

$$\begin{aligned}\frac{1}{\sqrt{\tau^2+1}}\big(f(a,\lambda) + \tau(pf')(a,\lambda)\big) &= \int_a^b \big(v_1(s,\lambda) + M_\tau(\lambda)v_2(s,\lambda)\big)g(s)r(s)\,ds \\ &= (g, \gamma_\tau(\bar\lambda))_{L^2_r(a,b)}.\end{aligned}$$

Since $h = (A_\tau - \lambda)^{-1}g$ also satisfies the equation $(L-\lambda)h = g$ and the same boundary condition $\Gamma_0^\tau h = 0$ and $\Gamma_1^\tau h = (g,\gamma_\tau(\bar\lambda))_{L^2_r(a,b)}$, it follows that $f = h$.

In Theorem 6.4.7 below a unitary Fourier transform for the self-adjoint realization A_τ, $\tau \in \mathbb{R}\cup\{\infty\}$, will be provided, which takes A_τ into multiplication by the independent variable in the space $L^2_{d\sigma_\tau}(\mathbb{R})$. Here σ_τ denotes the nondecreasing function in the integral representation

$$M_\tau(\lambda) = \alpha_\tau + \int_{\mathbb{R}} \left(\frac{1}{t-\lambda} - \frac{t}{1+t^2}\right) d\sigma_\tau(t), \quad \lambda \in \mathbb{C}\setminus\mathbb{R}, \tag{6.4.12}$$

of the Weyl function M_τ. Observe that no linear term is present in the integral representation since A_τ is not multivalued (this follows, e.g., from Lemma A.4.3 and Proposition 3.5.7). Recall that α_τ is a real constant and that $\int_{\mathbb{R}}(1+t^2)^{-1}\,d\sigma_\tau(t)$ is finite; cf. Theorem A.2.5

The following preparatory lemma shows that the condition (B.1.2) in Appendix B for the Fourier transform is satisfied.

Lemma 6.4.6. *Let $\tau \in \mathbb{R} \cup \{\infty\}$ and let $E_\tau(\cdot)$ be the spectral measure of the self-adjoint operator A_τ. For $f \in L^2_r(a,b)$ with compact support define the Fourier transform $\widehat{f}$ by*

$$\widehat{f}(\mu) = \int_a^b v_2(s,\mu) f(s) r(s)\, ds, \quad \mu \in \mathbb{R},$$

where $v_2(\cdot,\mu)$ is the formal solution in (6.4.9). Let σ_τ be the function in the integral representation (6.4.12) of the Weyl function M_τ. Then for every bounded open interval $\delta \subset \mathbb{R}$ whose endpoints are not eigenvalues of A_τ one has

$$(E_\tau(\delta) f, f)_{L^2_r(a,b)} = \int_\delta \widehat{f}(\mu)\, \overline{\widehat{f}(\mu)}\, d\sigma_\tau(\mu).$$

Proof. Observe that for $f \in L^2_r(a,b)$ with compact support and $\lambda \in \rho(A_\tau)$ the resolvent of A_τ can be written as

$$\begin{aligned} \big((A_\tau - \lambda)^{-1} f\big)(t) &= M_\tau(\lambda) v_2(t,\lambda) \int_a^b v_2(s,\lambda) f(s) r(s)\, ds \\ &\quad + v_1(t,\lambda) \int_a^t v_2(s,\lambda) f(s) r(s)\, ds + v_2(t,\lambda) \int_t^b v_1(s,\lambda) f(s) r(s)\, ds. \end{aligned} \tag{6.4.13}$$

Now let $\delta \subset \mathbb{R}$ be a bounded open interval whose endpoints are not eigenvalues of A_τ. Then the spectral projection of A_τ corresponding to δ is given by Stone's formula

$$\begin{aligned} &(E_\tau(\delta) f, f)_{L^2_r(a,b)} \\ &\quad = \lim_{\varepsilon \downarrow 0} \frac{1}{2\pi i} \int_\delta \big(\big((A_\tau - (\mu + i\varepsilon))^{-1} - (A_\tau - (\mu - i\varepsilon))^{-1}\big) f, f\big)_{L^2_r(a,b)}\, d\mu. \end{aligned}$$

If $f \in L^2_r(a,b)$ has compact support, say in $[a',b'] \subset (a,b)$, then (6.4.13) and the fact that the function

$$\lambda \mapsto v_1(t,\lambda) \int_a^t v_2(s,\lambda) f(s) r(s)\, ds + v_2(t,\lambda) \int_t^b v_1(s,\lambda) f(s) r(s)\, ds$$

in (6.4.13) is entire imply that $(E_\tau(\delta) f, f)_{L^2_r(a,b)}$ has the form

$$\begin{aligned} \lim_{\varepsilon \downarrow 0} \frac{1}{2\pi i} \int_\delta \bigg(\int_a^b \int_a^b & \big[g_{t,s}(\mu + i\varepsilon) M_\tau(\mu + i\varepsilon) \\ & - g_{t,s}(\mu - i\varepsilon) M_\tau(\mu - i\varepsilon) \big] f(s) r(s) ds\, \overline{f(t)} r(t) dt \bigg) d\mu, \end{aligned} \tag{6.4.14}$$

where $g_{t,s}$ is defined by $g_{t,s}(\eta) = v_2(t,\eta) v_2(s,\eta)$. Note that for $t,s \in [a',b']$ the function $g_{t,s}$ is entire in η. For $\varepsilon_0 > 0$ and $A < B$ such that $\bar{\delta} \subset (A,B)$ consider the rectangle $R = [A, B] \times [-i\varepsilon_0, i\varepsilon_0]$. The function $\{t, s, \eta\} \mapsto g_{t,s}(\eta)$ is bounded on $[a', b'] \times [a', b'] \times R$ and hence for each fixed ε such that $0 < \varepsilon \le \varepsilon_0$ it follows that

$$
\begin{aligned}
&\int_\delta \bigg(\int_a^b \int_a^b \big[g_{t,s}(\mu + i\varepsilon) M_\tau(\mu + i\varepsilon) \\
&\qquad\qquad - g_{t,s}(\mu - i\varepsilon) M_\tau(\mu - i\varepsilon) \big] f(s) r(s) ds\, \overline{f(t)} r(t) dt \bigg) d\mu \\
&= \int_a^b \int_a^b \bigg(\int_\delta \big[g_{t,s}(\mu + i\varepsilon) M_\tau(\mu + i\varepsilon) \\
&\qquad\qquad - g_{t,s}(\mu - i\varepsilon) M_\tau(\mu - i\varepsilon) \big]\, d\mu \bigg) f(s) r(s) ds\, \overline{f(t)} r(t) dt.
\end{aligned}
$$

The Stieltjes inversion formula in Lemma A.2.7 shows that

$$
\lim_{\varepsilon \downarrow 0} \frac{1}{2\pi i} \int_\delta \big[(g_{t,s} M_\tau)(\mu + i\varepsilon) - (g_{t,s} M_\tau)(\mu - i\varepsilon) \big]\, d\mu = \int_\delta g_{t,s}(\mu)\, d\sigma_\tau(\mu)
$$

for all $t, s \in [a', b']$. To justify taking the limit $\varepsilon \downarrow 0$ inside the integral (6.4.14) one needs dominated convergence. Recall from Lemma A.2.7 that there exists a constant $m \ge 0$ such that for $0 < \varepsilon \le \varepsilon_0$ one has

$$
\begin{aligned}
&\left| \int_\delta \big[(g_{t,s} M_\tau)(\mu + i\varepsilon) - (g_{t,s} M_\tau)(\mu - i\varepsilon) \big]\, d\mu \right| \\
&\qquad\qquad \le m \sup \big\{ |g_{t,s}(\eta)|, |g'_{t,s}(\eta)| : t, s \in [a', b'],\ \eta \in R \big\},
\end{aligned}
\tag{6.4.15}
$$

where $R = [A, B] \times [-i\varepsilon_0, i\varepsilon_0]$. Since $\{t, s, \eta\} \mapsto g_{t,s}(\eta)$ and $\{t, s, \eta\} \mapsto g'_{t,s}(\eta)$ are bounded functions on $[a', b'] \times [a', b'] \times R$, it follows that the integral in (6.4.15) regarded as a function in $\{t, s\}$ on $[a', b'] \times [a', b']$ is bounded by some constant for all $0 < \varepsilon \le \varepsilon_0$. As $f \in L^2_r(a, b)$ has compact support, there is an integrable majorant for the integrands in (6.4.14). Dominated convergence and Fubini's theorem yield

$$
\begin{aligned}
(E_\tau(\delta) f, f)_{L^2_r(a,b)} &= \int_a^b \int_a^b \bigg(\int_\delta g_{t,s}(\mu)\, d\sigma_\tau(\mu) \bigg) f(s) r(s) ds\, \overline{f(t)} r(t) dt \\
&= \int_\delta \bigg(\int_a^b v_2(s, \mu) f(s) r(s)\, ds \bigg) \bigg(\int_a^b v_2(t, \mu) \overline{f(t)} r(t)\, dt \bigg) d\sigma_\tau(\mu)
\end{aligned}
$$

for every bounded open interval δ whose endpoints are not eigenvalues of A_τ. Now the assertion follows from the definition of $\widehat{f}$. $\square$

The next theorem is a consequence of Lemma 6.4.6 and Theorem B.1.4 in Appendix B.

Theorem 6.4.7. *Let $\tau \in \mathbb{R} \cup \{\infty\}$, let $v_2(\cdot,\mu)$ be the formal solution in (6.4.9), and let σ_τ be the function in the integral representation of the Weyl function M_τ. Then the Fourier transform*

$$f \mapsto \widehat{f}, \qquad \widehat{f}(\mu) = \int_a^b v_2(s,\mu) f(s) r(s)\, ds, \quad \mu \in \mathbb{R},$$

extends by continuity from compactly supported functions $f \in L^2_r(a,b)$ to a unitary mapping $\mathcal{F} : L^2_r(a,b) \to L^2_{d\sigma_\tau}(\mathbb{R})$, such that the self-adjoint operator A_τ in $L^2_r(a,b)$ is unitarily equivalent to multiplication by the independent variable in $L^2_{d\sigma_\tau}(\mathbb{R})$.

Proof. It follows from Lemma 6.4.6 that the condition (B.1.2) is satisfied. It is also clear that for every $\mu \in \mathbb{R}$ there exists $s \in (a,b)$ such that $v_2(s,\mu) \neq 0$ and hence (B.1.13) holds. Now the result follows from Theorem B.1.4. □

In the next lemma the Fourier transform $\mathcal{F}\gamma_\tau$ of the γ-field in (6.4.8) corresponding to the boundary triplet $\{\mathbb{C}, \Gamma_0^\tau, \Gamma_1^\tau\}$ is computed; this will be useful in identifying the model in Theorem 6.4.7 with the model for scalar Nevanlinna functions discussed in Section 4.3.

Lemma 6.4.8. *Let $\tau \in \mathbb{R} \cup \{\infty\}$ and let γ_τ be the γ-field in Lemma 6.4.5. Then for all $\lambda \in \mathbb{C} \setminus \mathbb{R}$ one has almost everywhere in the sense of $d\sigma_\tau$:*

$$\big[\mathcal{F}\gamma_\tau(\lambda)\big](\mu) = \frac{1}{\mu - \lambda}, \qquad \mu \in \mathbb{R},$$

where $\mathcal{F}$ is the Fourier transform from $L^2_r(a,b)$ onto $L^2_{d\sigma_\tau}(\mathbb{R})$ in Theorem 6.4.7.

Proof. Let $f \in L^2_r(a,b)$. From (6.4.10) one obtains for all $t \in (a,b)$ that

$$\big((A_\tau - \lambda)^{-1} f\big)(t) = \big(G_\tau(t,\cdot,\lambda), \bar{f}\big)_{L^2_r(a,b)},$$

where both terms are absolutely continuous in $t \in (a,b)$. Differentiation yields

$$p(t)\frac{d}{dt}\big((A_\tau - \lambda)^{-1} f\big)(t) = \big(p(t)\partial_t G_\tau(t,\cdot,\lambda), \bar{f}\big)_{L^2_r(a,b)},$$

where again both terms are absolutely continuous in $t \in (a,b)$. Since the Fourier transform $\mathcal{F}$ is unitary, these two formulas lead to

$$\big((A_\tau - \lambda)^{-1} f\big)(t) = \big(\mathcal{F}G_\tau(t,\cdot,\lambda), \mathcal{F}\bar{f}\big)_{L^2_{d\sigma_\tau}(\mathbb{R})} \tag{6.4.16}$$

and

$$p(t)\frac{d}{dt}\big((A_\tau - \lambda)^{-1} f\big)(t) = \big(\mathcal{F}p(t)\partial_t G_\tau(t,\cdot,\lambda), \mathcal{F}\bar{f}\big)_{L^2_{d\sigma_\tau}(\mathbb{R})}, \tag{6.4.17}$$

where all terms are absolutely continuous in $t \in (a,b)$.

With $f \in L^2_r(a,b)$ it follows from (B.1.8) that

$$\big((A_\tau - \lambda)^{-1} f\big)(t) = \int_{\mathbb{R}} \frac{v_2(t,\mu)}{\mu - \lambda} (\mathcal{F} f)(\mu)\, d\sigma_\tau(\mu) \tag{6.4.18}$$

for almost all $t \in (a,b)$ and that, in particular, the integrand on the right-hand side is integrable. In fact, if $\mathcal{F} f$ has compact support, then the right-hand side of (6.4.18) is absolutely continuous in $t \in (a,b)$ and hence in this case (6.4.18) holds for all $t \in (a,b)$. Moreover, if $\mathcal{F} f$ has compact support, one also has for all $t \in (a,b)$

$$p(t) \frac{d}{dt} \big((A_\tau - \lambda)^{-1} f\big)(t) = \int_{\mathbb{R}} \frac{(pv_2')(t,\mu)}{\mu - \lambda} (\mathcal{F} f)(\mu)\, d\sigma_\tau(\mu), \tag{6.4.19}$$

where again all terms are absolutely continuous in $t \in (a,b)$. Differentiation under the integral sign is now allowed as the integrand has compact support and the function $\mu \mapsto (pv_2')(t,\mu)$ is bounded.

Comparison of the integrals on the right-hand sides of (6.4.16) and (6.4.18) under the assumption that $\mathcal{F} f$ is an arbitrary function in $L^2_{d\sigma_\tau}(\mathbb{R})$ with compact support leads for each $t \in (a,b)$ to

$$\frac{v_2(t,\mu)}{\mu - \lambda} = \big[\mathcal{F} G_\tau(t,\cdot,\lambda)\big](\mu), \qquad \mu \in \mathbb{R}, \tag{6.4.20}$$

almost everywhere. Similarly, comparison of the integrals on the right-hand sides of (6.4.17) and (6.4.19) under the assumption that $\mathcal{F} f$ is an arbitrary function in $L^2_{d\sigma_\tau}(\mathbb{R})$ with compact support leads for each $t \in (a,b)$ to

$$\frac{(pv_2')(t,\mu)}{\mu - \lambda} = \mathcal{F}\big[p(t)\partial_t G_\tau(t,\cdot,\lambda)\big](\mu), \qquad \mu \in \mathbb{R}, \tag{6.4.21}$$

almost everywhere. The union of the exceptional sets in (6.4.20) and (6.4.21) is denoted by $\Omega(t)$ and it has measure 0 in the sense of $d\sigma_\tau$.

Let $t \in (a,b)$; then for all $\mu \in \mathbb{R} \setminus \Omega(t)$ it follows from the identities (6.4.20) and (6.4.21) that

$$\begin{aligned} &\frac{1}{\mu - \lambda}\big(v_1(t,\lambda)(pv_2')(t,\mu) - (pv_1')(t,\lambda)v_2(t,\mu)\big) \\ &\quad = v_1(t,\lambda)\mathcal{F}\big[p(t)\partial_t G_\tau(t,\cdot,\lambda)\big](\mu) - (pv_1')(t,\lambda)\big[\mathcal{F} G_\tau(t,\cdot,\lambda)\big](\mu) \\ &\quad = \mathcal{F}\big[v_1(t,\lambda)p(t)\partial_t G_\tau(t,\cdot,\lambda) - (pv_1')(t,\lambda)G_\tau(t,\cdot,\lambda)\big](\mu) \\ &\quad = \mathcal{F}\big[w(t,\cdot,\lambda)\big](\mu), \end{aligned} \tag{6.4.22}$$

where $w(t,s,\lambda)$ is defined by

$$w(t,s,\lambda) = \begin{cases} M_\tau(\lambda) v_2(s,\lambda), & a < s < t, \\ \gamma_\tau(s,\lambda), & t < s < b. \end{cases}$$

The form of $w(t,s,\lambda)$ follows from (6.4.11), Lemma 6.4.5, and a straightforward computation. First observe that according to this definition

$$\|\gamma_\tau(\cdot,\lambda) - w(t,\cdot,\lambda)\|^2_{L^2_r(a,b)} = \int_a^t |v_1(s,\lambda)|^2 \, r(s)\, ds \to 0 \quad \text{as} \quad t \to a,$$

and the continuity of $\mathcal{F}$ implies that

$$\big\|\mathcal{F}\big[\gamma_\tau(\cdot,\lambda)\big] - \mathcal{F}\big[w(t,\cdot,\lambda)\big]\big\|_{L^2_{d\sigma_\tau}(\mathbb{R})} \to 0 \quad \text{as} \quad t \to a.$$

Now approximate the endpoint a by a sequence (t_n). Then there exist a subsequence, again denoted by (t_n), and a set Ω of measure 0 in the sense of $d\sigma_\tau$, such that pointwise

$$\mathcal{F}\big[\gamma_\tau(\cdot,\lambda)\big](\mu) = \lim_{n\to\infty} \mathcal{F}\big[w(t_n,\cdot,\lambda)\big](\mu), \qquad \mu \in \mathbb{R}\setminus\Omega.$$

Observe that $\bigcup_{n=1}^\infty \Omega(t_n)$ is a set of measure 0 in the sense of $d\sigma_\tau$ and that via (6.4.22)

$$\mathcal{F}\big[w(t_n,\cdot,\lambda)\big](\mu) = \frac{1}{\mu-\lambda}\big(v_1(t_n,\lambda)(pv_2')(t_n,\mu) - (pv_1')(t_n,\lambda)v_2(t_n,\mu)\big)$$

for all $\mu \in \mathbb{R}\setminus\bigcup_{n=1}^\infty \Omega(t_n)$. The limit on the right-hand side as $n\to\infty$ gives

$$\frac{1}{\mu-\lambda}\big(v_1(a,\lambda)(pv_2')(a,\mu) - (pv_1')(a,\lambda)v_2(a,\mu)\big) = \frac{1}{\mu-\lambda},$$

which follows from the special form of the fundamental system $(v_1(\cdot,\lambda); v_2(\cdot,\lambda))$ in Lemma 6.4.5 and (6.4.1). Hence,

$$\mathcal{F}\big[\gamma_\tau(\cdot,\lambda)\big](\mu) = \frac{1}{\mu-\lambda}, \qquad \mu \in \mathbb{R}\setminus\left(\Omega \cup \bigcup_{n=1}^\infty \Omega(t_n)\right),$$

which completes the proof. □

Lemma 6.4.8 will be used to identify the model in Theorem 6.4.7 with the model for scalar Nevanlinna functions discussed in Chapter 4.3. The Weyl function M_τ of the boundary triplet $\{\mathbb{C},\Gamma_0^\tau,\Gamma_1^\tau\}$ for $T_{\max}$ has the integral representation (6.4.12). By Theorem 4.3.1, there is a closed simple symmetric operator S in $L^2_{d\sigma_\tau}(\mathbb{R})$ such that the Nevanlinna function M_τ in (6.4.12) is the Weyl function corresponding to the boundary triplet $\{\mathbb{C},\Gamma_0',\Gamma_1'\}$ for S^* in Theorem 4.3.1. The γ-field corresponding to $\{\mathbb{C},\Gamma_0',\Gamma_1'\}$ is denoted by γ' and is given by (4.3.8). Furthermore, the self-adjoint restriction A_0' corresponding to the boundary mapping Γ_0' is the maximal multiplication operator by the independent variable in $L^2_{d\sigma_\tau}(\mathbb{R})$. By comparing with (4.3.8) one sees that, according to Lemma 6.4.8, the Fourier transform $\mathcal{F}$ from $L^2_r(a,b)$ onto $L^2_{d\sigma_\tau}(\mathbb{R})$, being a unitary mapping, satisfies

$$\mathcal{F}\gamma_\tau(\lambda) = \gamma'(\lambda).$$

Hence, by Theorem 4.2.6, the boundary triplet $\{\mathbb{C}, \Gamma_0^\tau, \Gamma_1^\tau\}$ for the simple operator $T_{\rm min}$ and the boundary triplet $\{\mathbb{C}, \Gamma_0', \Gamma_1'\}$ for the simple operator S are unitarily equivalent under the Fourier transform $\mathcal{F}$. Thus, not only are A_0' and A_τ unitarily equivalent under $\mathcal{F}$,

$$A_0' = \mathcal{F} A_\tau \mathcal{F}^{-1},$$

as stated in Theorem 6.4.7, but in fact the complete boundary triplet structure is preserved under the Fourier transform $\mathcal{F}$.

Now assume that the coefficient functions satisfy (6.1.2) and that the endpoint a is in the limit-circle case, while the endpoint b is in the limit-point case. Let u, v be real solutions of $(L - \lambda_0)y = 0$, $\lambda_0 \in \mathbb{R}$, with $W(u,v) = 1$ with $u, v \in L^2_r(a, a')$. Let a fundamental system $(u_1(\cdot,\lambda); u_2(\cdot,\lambda))$ for the equation $(L - \lambda)y = 0$ be fixed by the initial conditions (6.2.11). The following proposition is proved along the same lines as Proposition 6.4.1.

Proposition 6.4.9. *Assume that the endpoint a is in the limit-circle case and that the endpoint b is in the limit-point case. Then $\{\mathbb{C}, \Gamma_0, \Gamma_1\}$, where*

$$\Gamma_0 f = f^{[0]}(a) \quad \textit{and} \quad \Gamma_1 f = f^{[1]}(a), \quad f \in \operatorname{dom} T_{\rm max}, \tag{6.4.23}$$

is a boundary triplet for the operator $(T_{\rm min})^* = T_{\rm max}$. The self-adjoint extension A_0 corresponding to Γ_0 is the restriction of $T_{\rm max}$ defined on

$$\operatorname{dom} A_0 = \bigl\{ f \in \operatorname{dom} T_{\rm max} : f^{[0]}(a) = 0 \bigr\}$$

and the minimal operator $T_{\rm min}$ is the restriction of $T_{\rm max}$ defined on

$$\operatorname{dom} T_{\rm min} = \bigl\{ f \in \operatorname{dom} T_{\rm max} : f^{[0]}(a) = f^{[1]}(a) = 0 \bigr\}.$$

Moreover, if $\lambda \in \mathbb{C} \setminus \mathbb{R}$ and $\chi(\cdot,\lambda)$ is a nontrivial element in $\mathfrak{N}_\lambda(T_{\rm max})$, then one has $\chi^{[0]}(a,\lambda) \neq 0$. For all $\lambda \in \mathbb{C} \setminus \mathbb{R}$ the corresponding γ-field and Weyl function are given by

$$\gamma(\cdot,\lambda) = u_1(\cdot,\lambda) + M(\lambda) u_2(\cdot,\lambda) \quad \textit{and} \quad M(\lambda) = \frac{\chi^{[1]}(a,\lambda)}{\chi^{[0]}(a,\lambda)}.$$

Proof. First it will be verified that the mapping $(\Gamma_0, \Gamma_1)^\top : \operatorname{dom} T_{\rm max} \to \mathbb{C}^2$ is surjective. Let $\alpha \in \mathbb{C}^2$; then there exists $f \in \operatorname{dom} T_{\rm max}$ such that $f^{[0]}(a) = \alpha_1$, $f^{[1]}(a) = \alpha_2$, and f vanishes in a neighborhood of b. To see this, define the function h on (a,b) by

$$h(x) = \alpha_1 u(x) + \alpha_2 v(x).$$

Then $h, ph' \in AC(a,b)$ and h satisfies $(L - \lambda_0)y = 0$, while $h \in L^2_r(a,a')$ by assumption. Now by cutting off the function h near b, one obtains a function f which satisfies $(L - \lambda_0)f = g$ for some $g \in L^2_r(a,b)$ and which vanishes in a neighborhood of b, see Proposition 6.1.3. Hence, $f \in \operatorname{dom} T_{\rm max}$ and at a one has

$$\Gamma_0 f = f^{[0]}(a) = h^{[0]}(a) = W_a(h,v) = \alpha_1$$

and

$$\Gamma_1 f = f^{[1]}(a) = h^{[1]}(a) = -W_a(h, u) = \alpha_2.$$

This proves the claim.

Now the abstract Green identity will be proved. The argument is the same as in the proof of Proposition 6.4.1. It will be shown first that $\lim_{x\to b} W_x(f, \bar{g}) = 0$ for all $f, g \in \operatorname{dom} T_{\max}$. In fact, since b is in the limit-point case, the minimal operator $T_{\min}$ has defect numbers $(1, 1)$ by Corollary 6.2.2. Now choose $h_1, h_2 \in \operatorname{dom} T_{\max}$ such that

$$h_1^{[0]}(a) = 1, \quad h_1^{[1]}(a) = 0, \quad h_2^{[0]}(a) = 0, \quad h_2^{[1]}(a) = 1,$$

and such that h_1 and h_2 vanish in a neighborhood of b; cf. Proposition 6.1.3. Then $h_1, h_2 \notin \operatorname{dom} T_{\min}$ follows in the same way as in the proof of Proposition 6.4.1 from (6.2.7) and Lemma 6.2.5. Thus, every function $f \in \operatorname{dom} T_{\max}$ can be written in the form

$$f = f_0 + c_1 h_1 + c_2 h_2, \qquad f_0 \in \operatorname{dom} T_{\min},$$

for some $c_1, c_2 \in \mathbb{C}$. Therefore,

$$W_x(f, \bar{g}) = W_x(f_0, \bar{g}) + W_x(c_1 h_1 + c_2 h_2, \bar{g})$$

for all $g \in \operatorname{dom} T_{\max}$ and since the last term vanishes in a neighborhood of b, one obtains

$$\lim_{x\to b} W(f, \bar{g}) = \lim_{x\to b} W(f_0, \bar{g}) = 0$$

for all $g \in \operatorname{dom} T_{\max}$. Hence, it follows from (6.2.6) and Lemma 6.2.5 that for $f, g \in \operatorname{dom} T_{\max}$ one has

$$\begin{aligned}(T_{\max} f, g)_{L^2_r(a,b)} - (f, T_{\max} g)_{L^2_r(a,b)} &= -\lim_{x\to a} W(f, \bar{g}) \\ &= f^{[1]}(a)\overline{g^{[0]}(a)} - f^{[0]}(a)\overline{g^{[1]}(a)},\end{aligned}$$

which implies that the abstract Green identity is satisfied with the choice of Γ_0 and Γ_1 in (6.4.23). Thus, (6.4.23) defines a boundary triplet for $(T_{\min})^* = T_{\max}$.

The forms of $\operatorname{dom} A_0$, $\operatorname{dom} T_{\min}$, the γ-field, and the Weyl function are verified in the same way as in the proof of Proposition 6.4.1. □

6.5 The case of two limit-point endpoints and interface conditions

Assume that the endpoints a and b of the interval (a, b) are both singular and that the differential expression L is in the limit-point case at a and at b. In this section interface conditions at an interior point $c \in (a, b)$ are discussed and the maximal operator $T_{\max}$ associated with L in $L^2_r(a, b)$ is identified as a natural extension of the coupling of the minimal operators on the subintervals (a, c) and (c, b). It turns out, in particular, that $T_{\rm max}$ is self-adjoint in $L^2_r(a,b)$ and hence $T_{\rm min} = T_{\rm max}$, so that the defect numbers of $T_{\rm min}$ are $(0,0)$; cf. Corollary 6.2.2.

Let $c \in (a,b)$ and consider the intervals (a,c) and (c,b) separately. The differential expression L will be restricted to the open intervals (a,c) and (c,b), so that the endpoint c is regular for L and the endpoints a and b are in the limit-point case. At the point c fix a fundamental system $(u_1(\cdot,\lambda); u_2(\cdot,\lambda))$ for the equation $(L-\lambda)y = 0$ by the conditions

$$\begin{pmatrix} u_1(c,\lambda) & u_2(c,\lambda) \\ (pu_1')(c,\lambda) & (pu_2')(c,\lambda) \end{pmatrix} = \begin{pmatrix} 1 & 0 \\ 0 & 1 \end{pmatrix}. \tag{6.5.1}$$

In the following the functions f on (a,b) will often be written in the vector form

$$\begin{pmatrix} f_+ \\ f_- \end{pmatrix}, \quad \text{where } f_+ = f\restriction_{(c,b)} \quad \text{and} \quad f_- = f\restriction_{(a,c)};$$

here the indices $+$ and $-$ stand for the restriction of a function on (a,b) to the subintervals (c,b) and (a,c), respectively.

Let $T^+_{\rm max}$ be the maximal operator generated by L on (c,b) and define

$$\Gamma_0^+ f_+ = f_+(c) \quad \text{and} \quad \Gamma_1^+ f_+ = (p_+ f_+')(c), \quad f_+ \in \operatorname{dom} T^+_{\rm max}.$$

According to Proposition 6.4.1, $\{\mathbb{C}, \Gamma_0^+, \Gamma_1^+\}$ is a boundary triplet for $T^+_{\rm max}$ with Weyl function m_+, so that on (c,b)

$$\gamma_+(\cdot,\lambda) = u_1(\cdot,\lambda) + m_+(\lambda) u_2(\cdot,\lambda) \in L^2_r(c,b),$$

where $u_1(\cdot,\lambda)$ and $u_2(\cdot,\lambda)$ are as in (6.5.1). Note that the operator A_0^+ with domain

$$\operatorname{dom} A_0^+ = \ker \Gamma_0^+ = \bigl\{ f_+ \in \operatorname{dom} T^+_{\rm max} : f_+(c) = 0 \bigr\}$$

is a self-adjoint extension of the minimal operator $T^+_{\rm min}$ in $L^2_{r_+}(c,b)$ defined on

$$\operatorname{dom} T^+_{\rm min} = \bigl\{ f_+ \in \operatorname{dom} T^+_{\rm max} : f_+(c) = (pf_+')(c) = 0 \bigr\}.$$

Likewise, let $T^-_{\rm max}$ be the maximal operator generated by L on (a,c) and define

$$\Gamma_0^- f_- = f_-(c) \quad \text{and} \quad \Gamma_1^- f_- = -(p_- f_-')(c), \quad f_- \in \operatorname{dom} T^-_{\rm max}.$$

Then $\{\mathbb{C}, \Gamma_0^-, \Gamma_1^-\}$ is a boundary triplet for $T^-_{\rm max}$ with Weyl function m_-, so that on (a,c)

$$\gamma_-(\cdot,\lambda) = u_1(\cdot,\lambda) - m_-(\lambda) u_2(\cdot,\lambda) \in L^2_r(a,c),$$

where again $u_1(\cdot,\lambda)$ and $u_2(\cdot,\lambda)$ are as in (6.5.1). Note that the operator A_0^- with domain

$$\operatorname{dom} A_0^- = \ker \Gamma_0^- = \bigl\{ f_- \in \operatorname{dom} T^-_{\rm max} : f_-(c) = 0 \bigr\}$$

is a self-adjoint extension of the minimal operator $T_{\min}^-$ in $L_r^2(a,c)$ defined on

$$\operatorname{dom} T_{\min}^- = \{f_- \in \operatorname{dom} T_{\max}^- : f_-(c) = (pf_-')(c) = 0\}.$$

The two maximal operators together give rise to the orthogonal coupling

$$T_{\max}^- \,\widehat{\oplus}\, T_{\max}^+ \quad \text{in} \quad L_r^2(a,b) = L_r^2(a,c) \oplus L_r^2(c,b).$$

It is clear from Proposition 1.3.13 that $T_{\max}^- \,\widehat{\oplus}\, T_{\max}^+$ is the adjoint of the orthogonal coupling of the corresponding minimal operators $T_{\min}^+ \,\widehat{\oplus}\, T_{\min}^-$, and that $T_{\min}^+ \,\widehat{\oplus}\, T_{\min}^-$ is a restriction of $T_{\max}^+ \,\widehat{\oplus}\, T_{\max}^-$ defined by the conditions

$$f_+(c) = 0 = f_-(c) \quad \text{and} \quad (p_+f_+')(c) = 0 = -(p_-f_-')(c)$$

on the functions $f \in \operatorname{dom}(T_{\max}^+ \,\widehat{\oplus}\, T_{\max}^-)$. In particular, these conditions force a smooth connection at c. Note that $T_{\min}^+ \,\widehat{\oplus}\, T_{\min}^-$ is a densely defined closed symmetric operator with defect numbers $(2,2)$ in $L_r^2(a,b) = L_r^2(a,c) \oplus L_r^2(c,b)$, which is simple since both operators $T_{\min}^+$ and $T_{\min}^-$ are simple by Proposition 6.4.4. The orthogonal coupling $T_{\max}^+ \,\widehat{\oplus}\, T_{\max}^-$ can also be identified with an operator associated with L when it is restricted to the domain

$$\{f \in L_r^2(a,b) : f, pf' \in AC((a,b)\backslash\{c\}),\, Lf \in L_r^2(a,b)\};$$

in other words, for the elements in this domain both f and pf' are allowed to have one-sided limits at c which need not be equal. Similarly, the orthogonal coupling $T_{\min}^+ \,\widehat{\oplus}\, T_{\min}^-$ can be identified as a restriction of the self-adjoint operator $T_{\max}$ in $L_r^2(a,b)$ defined by the interface conditions

$$f(c) = (pf')(c) = 0.$$

The following result is a direct consequence of the orthogonal coupling of boundary triplets; see Section 4.6.

Proposition 6.5.1. *A boundary triplet $\{\mathbb{C}^2, \widetilde{\Gamma}_0, \widetilde{\Gamma}_1\}$ for $T_{\max}^+ \,\widehat{\oplus}\, T_{\max}^-$ is given by*

$$\widetilde{\Gamma}_0 f = \begin{pmatrix} f_+(c) \\ f_-(c) \end{pmatrix} \quad \text{and} \quad \widetilde{\Gamma}_1 f = \begin{pmatrix} (p_+f_+')(c) \\ -(p_-f_-')(c) \end{pmatrix}, \tag{6.5.2}$$

where $f = (f_+, f_-)^\top \in \operatorname{dom}(T_{\max}^+ \,\widehat{\oplus}\, T_{\max}^-)$. The self-adjoint extension $\widetilde{A}_0$ corresponding to $\widetilde{\Gamma}_0$ is the orthogonal coupling of the operators A_0^+ and A_0^- with domain

$$\operatorname{dom} \widetilde{A}_0 = \left\{ f = \begin{pmatrix} f_+ \\ f_- \end{pmatrix} \in \operatorname{dom}(T_{\max}^+ \,\widehat{\oplus}\, T_{\max}^-) : f_+(c) = 0 = f_-(c) \right\}$$

and the orthogonal coupling of the minimal operators $T_{\min}^+ \,\widehat{\oplus}\, T_{\min}^-$ is the restriction of $T_{\max}^+ \,\widehat{\oplus}\, T_{\max}^-$ defined on

$$\begin{aligned} &\operatorname{dom}(T_{\min}^+ \,\widehat{\oplus}\, T_{\min}^-) \\ &\quad = \left\{ f = \begin{pmatrix} f_+ \\ f_- \end{pmatrix} \in \operatorname{dom}(T_{\max}^+ \,\widehat{\oplus}\, T_{\max}^-) : \begin{matrix} f_+(c) = 0 = f_-(c), \\ (pf_+')(c) = 0 = (pf_-')(c) \end{matrix} \right\}. \end{aligned}$$

Moreover, for all $\lambda \in \rho(\widetilde{A}_0) = \rho(A_0^+) \cap \rho(A_0^-)$ *the corresponding* γ*-field* $\widetilde{\gamma}$ *and Weyl function* $\widetilde{M}$ *are given by*

$$\widetilde{\gamma}(\lambda) = \begin{pmatrix} \gamma_+(\lambda) & 0 \\ 0 & \gamma_-(\lambda) \end{pmatrix} \quad \textit{and} \quad \widetilde{M}(\lambda) = \begin{pmatrix} m_+(\lambda) & 0 \\ 0 & m_-(\lambda) \end{pmatrix}.$$

Note that the self-adjoint operator $\widetilde{A}_0 = A_0^+ \,\widehat{\oplus}\, A_0^-$ is the orthogonal coupling of the self-adjoint realizations of L on (a,c) and (c,b) corresponding to Dirichlet boundary conditions at c. The resolvents of A_0^+ and A_0^- admit an integral representation as in Proposition 6.4.3 which extends in a natural form to the orthogonal coupling $\widetilde{A}_0$. Since the orthogonal coupling $T_{\min}^+ \,\widehat{\oplus}\, T_{\min}^-$ of the minimal operators is a densely defined closed simple symmetric operator with defect numbers $(2,2)$, the spectrum of $\widetilde{A}_0$ can be described with the help of the 2×2-matrix function $\widetilde{M}$ in Proposition 6.5.1 and the general results in Section 3.5 and Section 3.6; cf. the considerations below Proposition 6.4.4.

Recall for completeness that all self-adjoint extensions $\widetilde{A}_\Theta$ of the orthogonal coupling $T_{\min}^+ \,\widehat{\oplus}\, T_{\min}^-$ in $L_r^2(a,b) = L_r^2(a,c) \oplus L_r^2(c,b)$ are in one-to-one correspondence to the self-adjoint relations Θ in $\mathbb{C}^2$ via

$$\begin{aligned} \operatorname{dom} \widetilde{A}_\Theta &= \bigl\{ f \in \operatorname{dom}\bigl(T_{\max}^+ \,\widehat{\oplus}\, T_{\max}^-\bigr) : \{\widetilde{\Gamma}_0 f, \widetilde{\Gamma}_1 f\} \in \Theta \bigr\} \\ &= \left\{ f \in \operatorname{dom}\bigl(T_{\max}^+ \,\widehat{\oplus}\, T_{\max}^-\bigr) : \left\{ \begin{pmatrix} f_+(c) \\ f_-(c) \end{pmatrix}, \begin{pmatrix} (p_+ f_+')(c) \\ -(p_- f_-')(c) \end{pmatrix} \right\} \in \Theta \right\}. \end{aligned}$$

For $\lambda \in \rho(\widetilde{A}_\Theta) \cap \rho(\widetilde{A}_0)$ Kreĭn's formula in the present setting has the form

$$(\widetilde{A}_\Theta - \lambda)^{-1} = (\widetilde{A}_0 - \lambda)^{-1} + \widetilde{\gamma}(\lambda)\bigl(\Theta - \widetilde{M}(\lambda)\bigr)^{-1} \widetilde{\gamma}(\bar{\lambda})^*. \tag{6.5.3}$$

In the same way as in Section 6.3 and Section 6.4, the spectral properties of the self-adjoint realizations $\widetilde{A}_\Theta$ can be described in a convenient way with the help of transforms of the Weyl function $\widetilde{M}$; cf. Section 3.8.

Among all self-adjoint extensions of $T_{\min}^+ \,\widehat{\oplus}\, T_{\min}^-$ there is one of particular importance, namely the extension corresponding to the self-adjoint relation

$$\widetilde{\Theta} = \operatorname{span} \begin{pmatrix} 1 \\ 1 \end{pmatrix} \times \operatorname{span} \begin{pmatrix} 1 \\ -1 \end{pmatrix} = \left\{ \left\{ \begin{pmatrix} \varphi \\ \varphi \end{pmatrix}, \begin{pmatrix} \psi \\ -\psi \end{pmatrix} \right\} : \varphi, \psi \in \mathbb{C} \right\}, \tag{6.5.4}$$

which was also considered in the abstract context in Section 4.6; cf. Proposition 4.6.1.

Corollary 6.5.2. *Let* $\{\mathbb{C}^2, \widetilde{\Gamma}_0, \widetilde{\Gamma}_1\}$ *be the boundary triplet for* $T_{\max}^+ \,\widehat{\oplus}\, T_{\max}^-$ *as defined in* (6.5.2) *and let the self-adjoint relation* $\widetilde{\Theta}$ *be as in* (6.5.4)*. Then the corresponding self-adjoint extension* $\widetilde{A}_{\widetilde{\Theta}}$ *of* $T_{\min}^+ \,\widehat{\oplus}\, T_{\min}^-$ *satisfies*

$$\widetilde{A}_{\widetilde{\Theta}} = T_{\max}, \tag{6.5.5}$$

and for $\lambda \in \mathbb{C}\setminus\mathbb{R}$ *Kreĭn's formula in* (6.5.3) *reads*

$$(T_{\max} - \lambda)^{-1} = (\widetilde{A}_0 - \lambda)^{-1} - \frac{1}{m_+(\lambda)+m_-(\lambda)}\,\widetilde{\gamma}(\lambda)\begin{pmatrix}1 & 1\\ 1 & 1\end{pmatrix}\widetilde{\gamma}(\bar{\lambda})^*. \tag{6.5.6}$$

Proof. By definition, the self-adjoint extension $\widetilde{A}_{\widetilde{\Theta}}$ of $T^+_{\min} \,\widehat{\oplus}\, T^-_{\min}$ is determined by the boundary condition

$$\{\widetilde{\Gamma}_0 f, \widetilde{\Gamma}_1 f\} \in \widetilde{\Theta}, \quad f = \begin{pmatrix} f_+ \\ f_-\end{pmatrix} \in \operatorname{dom}(T^+_{\max} \,\widehat{\oplus}\, T^-_{\max}),$$

which, by the definition of the boundary mappings in Proposition 6.5.1 and (6.5.4), leads to

$$\operatorname{dom}\widetilde{A}_{\widetilde{\Theta}} = \left\{ f = \begin{pmatrix} f_+ \\ f_-\end{pmatrix} \in \operatorname{dom}(T^+_{\max} \,\widehat{\oplus}\, T^-_{\max}) : \begin{matrix} f_+(c) = f_-(c) \\ (p_+f_+')(c) = (p_-f_-')(c)\end{matrix} \right\}.$$

Observe that this domain coincides with $\operatorname{dom} T_{\max}$ and (6.5.5) follows. Kreĭn's formula in (6.5.6) follows from (6.5.3) and

$$\bigl(\widetilde{\Theta} - \widetilde{M}(\lambda)\bigr)^{-1} = -\frac{1}{m_+(\lambda)+m_-(\lambda)}\begin{pmatrix}1 & 1\\ 1 & 1\end{pmatrix};$$

cf. Proposition 4.6.1. □

Since $T_{\max}$ is self-adjoint, it follows from Corollary 6.5.2 that the defect numbers of $T_{\min}$ are $(0,0)$. This fact can be seen as a completion of the statements in Corollary 6.2.2.

The boundary triplet in (6.5.2) will now be transformed in order to interpret the self-adjoint extension $T_{\max}$ in (6.5.5) in a convenient way. The following proposition is a variant of Proposition 4.6.4 in the present situation.

Proposition 6.5.3. *A boundary triplet* $\{\mathbb{C}^2, \widehat{\Gamma}_0, \widehat{\Gamma}_1\}$ *for* $T^+_{\max} \,\widehat{\oplus}\, T^-_{\max}$ *is given by*

$$\widehat{\Gamma}_0 f = \begin{pmatrix} -(p_+f_+')(c) + (p_-f_-')(c) \\ f_+(c) - f_-(c)\end{pmatrix} \quad \text{and} \quad \widehat{\Gamma}_1 f = \begin{pmatrix} f_+(c) \\ (p_-f_-')(c)\end{pmatrix},$$

where $f = (f_+, f_-)^\top \in \operatorname{dom}(T^+_{\max} \,\widehat{\oplus}\, T^-_{\max})$. *Here the self-adjoint operator defined on* $\ker \widehat{\Gamma}_0$ *coincides with the maximal operator* $T_{\max}$ *associated with* L *in* $L^2_r(a,b)$. *For all* $\lambda \in \mathbb{C}\setminus\mathbb{R}$ *the Weyl function corresponding to the boundary triplet* $\{\mathbb{C}^2, \widehat{\Gamma}_0, \widehat{\Gamma}_1\}$ *is given by*

$$\widehat{M}(\lambda) = \begin{pmatrix} -\dfrac{1}{m_+(\lambda)+m_-(\lambda)} & \dfrac{m_-(\lambda)}{m_+(\lambda)+m_-(\lambda)} \\[2ex] \dfrac{m_-(\lambda)}{m_+(\lambda)+m_-(\lambda)} & \dfrac{m_-(\lambda)m_+(\lambda)}{m_+(\lambda)+m_-(\lambda)}\end{pmatrix}. \tag{6.5.7}$$

Assume that $\lambda \in \rho(\widetilde{A}_0) = \rho(A_0^+) \cap \rho(A_0^-)$. Then the functions m_+ and m_- are defined and analytic at λ. It is not difficult to check that $\lambda \in \sigma_p(T_{\max})$ if and only if $m_+(\lambda) + m_-(\lambda) = 0$; this also follows from Theorem 2.6.2, the special form of $\widetilde{M}$ in Proposition 6.5.1, and the choice of $\widetilde{\Theta}$. Since the resolvents of $T_{\max}$ and the orthogonal coupling $\widetilde{A}_0$ differ by a rank-one operator, it is also clear that a point $\lambda \in \rho(A_0^+) \cap \rho(A_0^-)$ is either an isolated eigenvalue of $\widetilde{A}$, or belongs to $\rho(T_{\max})$. Hence, the expression for the Weyl function $\widehat{M}$ in (6.5.7) remains valid for all $\lambda \in \rho(A_0^+) \cap \rho(A_0^-) \cap \rho(T_{\max})$. Note also that for $\lambda \in \mathbb{C} \setminus \mathbb{R}$ the Weyl function $\widehat{M}$ has the simple representation

$$\widehat{M}(\lambda) = -\begin{pmatrix} m_+(\lambda) & -1 \\ -1 & -\frac{1}{m_-(\lambda)} \end{pmatrix}^{-1}.$$

This representation remains valid for all $\lambda \in \rho(A_0^+) \cap \rho(A_1^-) \cap \rho(\widetilde{A})$, where A_1^- denotes the self-adjoint operator in $L_r^2(a,c)$ defined on $\ker \Gamma_1^-$.

As the orthogonal coupling $T_{\min}^+ \,\widehat{\oplus}\, T_{\min}^-$ is a simple symmetric operator with defect numbers $(2,2)$, the spectral properties of the self-adjoint extension $T_{\max}$ can be described by means of the Weyl function $\widehat{M}$ in Proposition 6.5.3 and the general results in Section 3.5 and Section 3.6. First of all, it is clear that the poles of $\widehat{M}$ coincide with the isolated eigenvalues of $T_{\max}$ and hence it follows from the representation (6.5.7) that $\lambda \in \sigma_p(T_{\max})$ is an isolated eigenvalue if and only if m_+ and m_- are holomorphic at λ and $m_+(\lambda) + m_-(\lambda) = 0$, or both m_+ and m_- have a pole at λ. Note that $\widehat{M}$ is holomorphic at λ if $m_\mp$ has a pole and $m_\pm$ is holomorphic at λ. For the description of the eigenvalues of $T_{\max}$ embedded in the continuous spectrum recall from Corollary 3.5.6 (see also Theorem 3.6.1) that $\lambda \in \sigma_p(T_{\max})$ if and only if $\widehat{\mathcal{R}}_\lambda \varphi = \lim_{\varepsilon \downarrow 0} i\varepsilon \widehat{M}(\lambda + i\varepsilon)\varphi \neq 0$ for some $\varphi \in \mathbb{C}^2$ and that the linear map

$$\widehat{\tau} : \ker (T_{\max} - \lambda) \to \operatorname{ran} \widehat{\mathcal{R}}_\lambda, \quad f(\cdot,\lambda) \mapsto \widehat{\Gamma}_1 f(\cdot,\lambda) = \begin{pmatrix} f_+(c,\lambda) \\ (p_- f_-')(c,\lambda) \end{pmatrix},$$

is bijective. The continuous, absolutely continuous, and singular continuous spectra are described as in Theorem 3.6.5 and Theorem 3.6.8.

Proposition 6.5.4. *For $\lambda \in \mathbb{C} \setminus \mathbb{R}$ the resolvent of the self-adjoint extension $T_{\max}$ is an integral operator of the form*

$$\big((T_{\max} - \lambda)^{-1} g\big)(t) = \int_a^b G(t,s,\lambda) g(s) r(s)\, ds, \quad g \in L_r^2(a,b), \tag{6.5.8}$$

where the Green function $G(t,s,\lambda)$ is given by

$$\begin{cases} \frac{-1}{m_+(\lambda)+m_-(\lambda)} (u_1(t,\lambda) + m_+(\lambda) u_2(t,\lambda))(u_1(s,\lambda) - m_-(\lambda) u_2(s,\lambda)), & a < s < t, \\ \frac{-1}{m_+(\lambda)+m_-(\lambda)} (u_1(t,\lambda) - m_-(\lambda) u_2(t,\lambda))(u_1(s,\lambda) + m_+(\lambda) u_2(s,\lambda)), & t < s < b. \end{cases}$$

In particular, if $g \in L^2_r(a,b)$ has compact support, then

$$
\begin{aligned}
\big((T_{\max} - \lambda)^{-1} g\big)(t) &= \big(u_1(t,\lambda) \;\; u_2(t,\lambda)\big)\, \widehat{M}(\lambda) \begin{pmatrix} \int_a^b u_1(s,\lambda) g(s) r(s)\, ds \\ \int_a^b u_2(s,\lambda) g(s) r(s)\, ds \end{pmatrix} \\
&\quad - u_1(t,\lambda) \int_t^b u_2(s,\lambda) g(s) r(s)\, ds - u_2(t,\lambda) \int_a^t u_1(s,\lambda) g(s) r(s)\, ds,
\end{aligned}
$$

where the 2×2 matrix $\widehat{M}(\lambda)$ is given by (6.5.7).

Proof. Observe that for $g \in L^2_r(a,b)$ and $\lambda \in \mathbb{C} \setminus \mathbb{R}$ the function $f(\cdot,\lambda)$ given by

$$
\begin{aligned}
&- \big(m_+(\lambda) + m_-(\lambda)\big) f(t,\lambda) \\
&\quad = \big(u_1(t,\lambda) + m_+(\lambda) u_2(t,\lambda)\big) \int_a^t \big(u_1(s,\lambda) - m_-(\lambda) u_2(s,\lambda)\big) g(s) r(s)\, ds \\
&\quad + \big(u_1(t,\lambda) - m_-(\lambda) u_2(t,\lambda)\big) \int_t^b \big(u_1(s,\lambda) + m_+(\lambda) u_2(s,\lambda)\big) g(s) r(s)\, ds
\end{aligned}
$$

is well defined. Moreover, it satisfies $(L - \lambda) f = g$ and it has the following initial values at c:

$$
f(c,\lambda) = - \frac{(g_-, \gamma_-(\bar\lambda))_{L^2_r(a,c)} + (g_+, \gamma_+(\bar\lambda))_{L^2_r(c,b)}}{m_+(\lambda) + m_-(\lambda)}
$$

and

$$
(pf')(c,\lambda) = - \frac{m_+(\lambda)(g_-, \gamma_-(\bar\lambda))_{L^2_r(a,c)} - m_-(\lambda)(g_+, \gamma_+(\bar\lambda))_{L^2_r(c,b)}}{m_+(\lambda) + m_-(\lambda)}.
$$

Recall that the function $h = (T_{\max} - \lambda)^{-1} g$ is given by the Kreĭn formula in (6.5.6). This leads to the following expressions for its components

$$
\begin{aligned}
\begin{pmatrix} h_+(\cdot,\lambda) \\ h_-(\cdot,\lambda) \end{pmatrix} &= \begin{pmatrix} (A_0^+ - \lambda)^{-1} g_+ \\ (A_0^- - \lambda)^{-1} g_- \end{pmatrix} \\
&\quad - \frac{1}{m_+(\lambda) + m_-(\lambda)} \begin{pmatrix} \gamma_+(\lambda) & 0 \\ 0 & \gamma_-(\lambda) \end{pmatrix} \begin{pmatrix} 1 & 1 \\ 1 & 1 \end{pmatrix} \begin{pmatrix} (g_+, \gamma_+(\bar\lambda))_{L^2_r(c,b)} \\ (g_-, \gamma_-(\bar\lambda))_{L^2_r(a,c)} \end{pmatrix} \\
&= \begin{pmatrix} (A_0^+ - \lambda)^{-1} g_+ \\ (A_0^- - \lambda)^{-1} g_- \end{pmatrix} \\
&\quad - \frac{1}{m_+(\lambda) + m_-(\lambda)} \begin{pmatrix} \gamma_+(\lambda) \big[(g_+, \gamma_+(\bar\lambda))_{L^2_r(c,b)} + (g_-, \gamma_-(\bar\lambda))_{L^2_r(a,c)}\big] \\ \gamma_-(\lambda) \big[(g_+, \gamma_+(\bar\lambda))_{L^2_r(c,b)} + (g_-, \gamma_-(\bar\lambda))_{L^2_r(a,c)}\big] \end{pmatrix}.
\end{aligned}
$$

To compute $h(c,\lambda)$ and $(ph')(c,\lambda)$, it suffices to compute $h_+(c,\lambda)$ and $(p_+ h_+')(c,\lambda)$ since $h \in \operatorname{dom} T_{\max}$ is smooth at c. Let $k_+(\cdot,\lambda) = (A_0^+ - \lambda)^{-1} g_+$, then clearly

$$
k_+(c,\lambda) = 0 \quad \text{and} \quad (p_+ k_+')(c,\lambda) = (g_+, \gamma_+(\bar\lambda))_{L^2_r(c,b)},
$$

using Proposition 2.3.2 (iv). Furthermore, observe that

$$\gamma_+(c,\lambda) = 1 \quad \text{and} \quad (p_+\gamma_+')(c,\lambda) = m_+(\lambda).$$

Hence, one sees that

$$h(c,\lambda) = h_+(c,\lambda) = -\frac{(g_+,\gamma_+(\bar\lambda))_{L^2_r(c,b)} + (g_-,\gamma_-(\bar\lambda))_{L^2_r(a,c)}}{m_+(\lambda)+m_-(\lambda)}$$

and that

$$\begin{aligned}(ph')(c,\lambda) &= (ph_+')(c,\lambda)\\ &= (g_+,\gamma_+(\bar\lambda))_{L^2_r(c,b)} - \frac{m_+(\lambda)\big[(g_+,\gamma_+(\bar\lambda))_{L^2_r(c,b)} + (g_-,\gamma_-(\bar\lambda))_{L^2_r(a,c)}\big]}{m_+(\lambda)+m_-(\lambda)}\\ &= \frac{m_-(\lambda)(g_+,\gamma_+(\bar\lambda))_{L^2_r(c,b)} - m_+(\lambda)(g_-,\gamma_-(\bar\lambda))_{L^2_r(a,c)}}{m_+(\lambda)+m_-(\lambda)}.\end{aligned}$$

Therefore, $f(\cdot,\lambda)$ and $h(\cdot,\lambda)$ satisfy the same differential equation and the same initial conditions. It follows that $f(\cdot,\lambda) = (T_{\max}-\lambda)^{-1}g$ and this yields (6.5.8).

Now assume that $g \in L^2_r(a,b)$ has compact support. Then writing out all the products on the right-hand side of $(T_{\max}-\lambda)^{-1}g$ is allowed, since each individual integral is well defined. This rewriting of the terms of the function $f(\cdot,\lambda)$ gives eight terms, which after adding and substracting of the terms

$$m_-(\lambda)u_1(t,\lambda)\int_t^b u_2(s,\lambda)g(s)r(s)\,ds$$

and

$$m_-(\lambda)u_2(t,\lambda)\int_a^t u_1(s,\lambda)g(s)r(s)\,ds$$

and regrouping leads to the desired result. □

Let $f \in L^2_r(a,b)$ have compact support and define the two-dimensional Fourier transform

$$\widehat f(\mu) = \begin{pmatrix} \int_a^b u_1(t,\mu)f(t)\,dt\\ \int_a^b u_2(t,\mu)f(t)\,dt\end{pmatrix}. \tag{6.5.9}$$

Consider the maximal operator $T_{\max}$ as the smooth extension of $T^+_{\min} \,\widehat\oplus\, T^-_{\min}$ and let $E(\lambda)$ be the corresponding spectral family. Then it follows from Proposition 6.5.4 in the same way as in the proof of Lemma 6.4.6 that

$$(E(\Delta)f,f)_{L^2_r(a,b)} = \int_\Delta \widehat f(x)^* d\Sigma(x)\widehat f(x),$$

where Σ denotes the 2×2-matrix function in the integral representation of the Weyl function $\widehat M$ in Proposition 6.5.3; cf. Theorem A.4.2. In the present context there is an analog of Theorem 6.4.7.

Theorem 6.5.5. *Let Σ be the 2×2-matrix function in the integral representation of the Weyl function $\widehat{M}$ in (6.5.7). Then the map $f \mapsto \widehat{f}$ in (6.5.9) extends by continuity from compactly supported functions $f \in L^2_r(a,b)$ to a unitary mapping $\mathcal{F} : L^2_r(a,b) \to L^2_{d\Sigma}(\mathbb{R})$, such that the self-adjoint operator $T_{\max}$ in $L^2_r(a,b)$ is unitarily equivalent to multiplication by the independent variable in $L^2_{d\Sigma}(\mathbb{R})$.*

The Fourier transform in the theorem is vector-valued. The details of the proof follow the scalar case, but this line of thought will not be pursued in the text.

Remark 6.5.6. The coupling technique in this section can also be applied in situations when the endpoints a or b are regular or in the limit-circle case. For simplicity a possible choice of the boundary triplets and Weyl functions will be made explicit when L is regular at a and b, and Dirichlet boundary conditions are imposed there. Let $c \in (a,b)$, consider the intervals (a,c) and (c,b) separately, and use the fundamental system $(u_1(\cdot,\lambda); u_2(\cdot,\lambda))$ in (6.5.1). As in Corollary 6.3.2 choose the operator $(T^+_{\min})'$ in $L^2_{r_+}(c,b)$ defined on

$$\operatorname{dom}(T^+_{\min})' = \bigl\{f \in \operatorname{dom} T^+_{\max} : f_+(c) = (p_+f'_+)(c) = f_+(b) = 0\bigr\}$$

and the boundary triplet $\{\mathbb{C}, \Gamma^+_0, \Gamma^+_1\}$ for the adjoint given by

$$\Gamma^+_0 f_+ = f_+(c) \quad \text{and} \quad \Gamma^+_1 f_+ = (p_+f'_+)(c), \quad f_+ \in \operatorname{dom}((T^+_{\min})')^*,$$

with corresponding Weyl function

$$m_+(\lambda) = -\frac{u_1(b,\lambda)}{u_2(b,\lambda)}. \tag{6.5.10}$$

Likewise, choose the operator $(T^-_{\min})'$ in $L^2_{r_-}(a,c)$ defined on

$$\operatorname{dom}(T^-_{\min})' = \bigl\{f \in \operatorname{dom} T^-_{\max} : f_-(c) = (p_-f'_-)(c) = f_-(b) = 0\bigr\}$$

and the boundary triplet $\{\mathbb{C}, \Gamma^-_0, \Gamma^-_1\}$ for the adjoint given by

$$\Gamma^-_0 f_- = f_-(c) \quad \text{and} \quad \Gamma^-_1 f_- = -(p_-f'_-)(c), \quad f_- \in \operatorname{dom}((T^-_{\min})')^*,$$

with corresponding Weyl function

$$m_-(\lambda) = \frac{u_1(a,\lambda)}{u_2(a,\lambda)}. \tag{6.5.11}$$

The earlier considerations in this section remain valid with the Weyl functions m_+ and m_- in (6.5.10) and (6.5.11), respectively.

6.6 Exit space extensions

In order to study boundary value problems where the spectral parameter appears in the boundary conditions one has to deal with self-adjoint extensions of a closed symmetric Sturm–Liouville operator in an exit space extending the original Hilbert space. In this section the exit space extensions will be investigated for a Sturm–Liouville expression L, which is regular at one endpoint and in the limit-point case at the other endpoint. Other situations may be studied in an analogous fashion.

At this stage observe that the construction in Section 6.5 may also be interpreted in the following way. The operator $T_{\min}^{+}$ in $L_r^2(c,b)$ has a self-adjoint extension $T_{\max}$ in the Hilbert space $L_r^2(a,b)$ when $L_r^2(a,c)$ is considered as an exit space: $L_r^2(a,b) = L_r^2(a,c) \oplus L_r^2(c,b)$. It follows from (6.5.6) in Corollary 6.5.2 that the compression of the resolvent of $T_{\max}$ from $L_r^2(a,b)$ onto $L_r^2(c,b)$ is of the form

$$P^{+}(T_{\max} - \lambda)^{-1}\iota_{+} = (A_0^{+} - \lambda)^{-1} - \gamma_{+}(\lambda)\big(m_{+}(\lambda) + m_{-}(\lambda)\big)^{-1}\gamma_{+}(\bar{\lambda})^{*},$$

where P^{+} denotes the orthogonal projection from $L_r^2(a,b)$ onto $L_r^2(c,b)$ and ι_{+} is the corresponding canonical embedding. It follows from Theorem 2.7.3 and Theorem 2.7.4 (see also Corollary 4.6.2) that the compressed resolvent of the self-adjoint operator $T_{\max}$ in $L_r^2(a,b)$ gives rise to the Štraus extensions of $T_{\min}^{+}$ in $L_r^2(c,b)$ corresponding to the boundary conditions

$$\Gamma_1^{+} f_{+} = -m_{-}(\lambda)\Gamma_0^{+} f \quad \text{or} \quad (pf_{+}')(c) = -m_{-}(\lambda) f_{+}(c). \tag{6.6.1}$$

Note that the family of Štraus extensions is defined via the Weyl function $m_{-}(\lambda)$ of the Sturm–Liouville operator on the interval (a,c); in particular, this family is described by the boundary conditions (6.6.1) in which the eigenvalue parameter appears.

In the present treatment one stays close to the above context. More precisely, one assumes that the Sturm–Liouville operator is defined on an interval (c,b) where the endpoint c is regular and the limit-point condition prevails at b. The maximal operator in $\mathfrak{H} = L_r^2(c,b)$ is denoted by $T_{\max}^{+}$ and the boundary triplet for the minimal operator $T_{\min}^{+}$ is denoted by $\{\mathbb{C}, \Gamma_0^{+}, \Gamma_1^{+}\}$, where $\Gamma_0^{+} f_{+} = f_{+}(c)$ and $\Gamma_1^{+} f_{+} = (p_{+} f_{+}')(c)$; cf. Proposition 6.4.1 and Section 6.4. The corresponding γ-field and Weyl function are denoted by γ_{+} and m_{+}, respectively. Now let τ be a scalar Nevanlinna function (which is not equal to a real constant). The interest is in boundary value problems of the form

$$\Gamma_1^{+} f_{+} = -\tau(\lambda)\Gamma_0^{+} f_{+} \quad \text{or, equivalently,} \quad (p_{+} f_{+}')(c) = -\tau(\lambda) f_{+}(c); \tag{6.6.2}$$

cf. (6.6.1). According to Theorem 4.2.4 (or Theorem 4.3.1), there exist a reproducing kernel Hilbert space $\mathfrak{H}' = \mathfrak{H}(\mathsf{N}_{\tau})$ (or an L^2-space $\mathfrak{H}'$, respectively), a closed simple symmetric operator T' in $\mathfrak{H}'$, and a boundary triplet $\{\mathbb{C}, \Gamma_0', \Gamma_1'\}$ for the adjoint $(T')^*$ with γ-field γ' and Weyl function τ. Observe that the closed symmetric operator T' is not necessarily densely defined and $(T')^*$ may not be an operator. Hence, to denote boundary triplets for the orthogonal sum of the maximal operators $T^+_{\max} \widehat{\oplus} (T')^*$ the graph notation will be used. In particular, instead of $\Gamma_0^+ f_+ = f_+(c)$ and $\Gamma_1^+ f_+ = (pf_+')(c)$ the notation $\Gamma_0^+ \widehat{f}_+ = f_+(c)$ and $\Gamma_1^+ \widehat{f}_+ = (pf_+')(c)$ for $\widehat{f}_+ = \{f_+, f_+'\} \in T^+_{\max}$ is used. Thus,

$$\widetilde{\Gamma}_0 \begin{pmatrix} \widehat{f}_+ \\ \widehat{k} \end{pmatrix} = \begin{pmatrix} \Gamma_0^+ \widehat{f}_+ \\ \Gamma_0' \widehat{k} \end{pmatrix} \quad \text{and} \quad \widetilde{\Gamma}_1 \begin{pmatrix} \widehat{f}_+ \\ \widehat{k} \end{pmatrix} = \begin{pmatrix} \Gamma_1^+ \widehat{f}_+ \\ \Gamma_1' \widehat{k} \end{pmatrix}$$

where $\widehat{f}_+ = \{f_+, f_+'\} \in T^+_{\max}$ and $\widehat{k} = \{k, k'\} \in (T')^*$, defines a boundary triplet $\{\mathbb{C}^2, \widetilde{\Gamma}_0, \widetilde{\Gamma}_1\}$ for $(T^+_{\min} \widehat{\oplus} T')^* = T^+_{\max} \widehat{\oplus} (T')^*$ and

$$\widetilde{A}_0 := A_0^+ \widehat{\oplus} A_0' = \ker \widetilde{\Gamma}_0,$$

where $A_0^+ = \ker \Gamma_0^+$ and $A_0' = \ker \Gamma_0'$, is a self-adjoint extension of $T^+_{\min} \widehat{\oplus} T'$ in $\mathfrak{H} \oplus \mathfrak{H}'$. It is clear that for $\lambda \in \rho(\widetilde{A}_0) = \rho(A_0^+) \cap \rho(A_0')$ the γ-field $\widetilde{\gamma}$ and Weyl function $\widetilde{M}$ corresponding to the boundary triplet $\{\mathbb{C}^2, \widetilde{\Gamma}_0, \widetilde{\Gamma}_1\}$ have the form

$$\widetilde{\gamma}(\lambda) = \begin{pmatrix} \gamma_+(\lambda) & 0 \\ 0 & \gamma'(\lambda) \end{pmatrix} \quad \text{and} \quad \widetilde{M}(\lambda) = \begin{pmatrix} m_+(\lambda) & 0 \\ 0 & \tau(\lambda) \end{pmatrix};$$

cf. Proposition 6.5.1. The self-adjoint extension $\widetilde{A}$ of $T^+_{\min} \widehat{\oplus} T'$ in the next proposition is of special interest since its compressed resolvent corresponds to the Nevanlinna function τ via the Kreĭn–Naĭmark formula. Proposition 6.6.1 is a special case of Theorem 2.7.4 and Corollary 4.6.2.

Proposition 6.6.1. *Let $T^+_{\min}$ and T' be the closed simple symmetric operators with boundary triplets $\{\mathbb{C}, \Gamma_0^+, \Gamma_1^+\}$ and $\{\mathbb{C}, \Gamma_0', \Gamma_1'\}$ as above. Then*

$$\widetilde{A} = \left\{ \begin{pmatrix} \widehat{f}_+ \\ \widehat{k} \end{pmatrix} : \widehat{f}_+ \in T^+_{\max},\, \widehat{k} \in (T')^*,\, \Gamma_0^+ \widehat{f}_+ = \Gamma_0' \widehat{k},\, \Gamma_1^+ \widehat{f}_+ = -\Gamma_1' \widehat{k} \right\} \tag{6.6.3}$$

is a self-adjoint relation in $\mathfrak{H} \oplus \mathfrak{H}'$ and for all $\lambda \in \mathbb{C} \setminus \mathbb{R}$ the resolvent of $\widetilde{A}$ has the form

$$(\widetilde{A} - \lambda)^{-1} = (\widetilde{A}_0 - \lambda)^{-1} - \frac{1}{m_+(\lambda) + \tau(\lambda)}\, \widetilde{\gamma}(\lambda) \begin{pmatrix} 1 & 1 \\ 1 & 1 \end{pmatrix} \widetilde{\gamma}(\bar{\lambda})^*.$$

The self-adjoint relation $\widetilde{A}$ satisfies the minimality condition

$$\mathfrak{H} \oplus \mathfrak{H}' = \overline{\operatorname{span}} \left\{ \mathfrak{H}, \operatorname{ran} (\widetilde{A} - \lambda)^{-1} \iota_{\mathfrak{H}} : \lambda \in \mathbb{C} \setminus \mathbb{R} \right\},$$

and for $\lambda \in \mathbb{C} \setminus \mathbb{R}$ the compression of the resolvent $(\widetilde{A} - \lambda)^{-1}$ to $\mathfrak{H}$ is given by

$$P_{\mathfrak{H}} (\widetilde{A} - \lambda)^{-1} \iota_{\mathfrak{H}} = (A_0^+ - \lambda)^{-1} - \gamma_+(\lambda) \big(m_+(\lambda) + \tau(\lambda) \big)^{-1} \gamma_+(\bar{\lambda})^*, \tag{6.6.4}$$

where $P_{\mathfrak{H}} : \mathfrak{H} \oplus \mathfrak{H}' \to \mathfrak{H}$ is the orthogonal projection from $\mathfrak{H} \oplus \mathfrak{H}'$ onto $\mathfrak{H}$ and $\iota_{\mathfrak{H}} : \mathfrak{H} \to \mathfrak{H} \oplus \mathfrak{H}'$ is the canonical embedding of $\mathfrak{H}$ into $\mathfrak{H} \oplus \mathfrak{H}'$.

The next result describes a particular boundary triplet $\{\mathbb{C}^2, \widehat{\Gamma}_0, \widehat{\Gamma}_1\}$ for which the self-adjoint relation $\widetilde{A}$ in (6.6.3) coincides with the kernel of the boundary mapping $\widehat{\Gamma}_0$; cf. Proposition 4.6.4.

Proposition 6.6.2. *Let $T^+_{\min}$ and T' be closed symmetric operators in the Hilbert spaces $\mathfrak{H}$ and $\mathfrak{H}'$ with boundary triplets $\{\mathbb{C}, \Gamma^+_0, \Gamma^+_1\}$ and $\{\mathbb{C}, \Gamma'_0, \Gamma'_1\}$ and corresponding Weyl functions m_+ and τ, respectively, as in the beginning of this section. Then $\{\mathbb{C}^2, \widehat{\Gamma}_0, \widehat{\Gamma}_1\}$, where*

$$\widehat{\Gamma}_0 \begin{pmatrix} \widehat{f}_+ \\ \widehat{k} \end{pmatrix} = \begin{pmatrix} -\Gamma^+_1 \widehat{f}_+ - \Gamma'_1 \widehat{k} \\ \Gamma^+_0 \widehat{f}_+ - \Gamma'_0 \widehat{k} \end{pmatrix} \quad \text{and} \quad \widehat{\Gamma}_1 \begin{pmatrix} \widehat{f}_+ \\ \widehat{k} \end{pmatrix} = \begin{pmatrix} \Gamma^+_0 \widehat{f}_+ \\ -\Gamma'_1 \widehat{k} \end{pmatrix},$$

with $\widehat{f}_+ \in T^+_{\max}$, $\widehat{k} \in (T')^*$, is a boundary triplet for $T^+_{\max} \widehat{\oplus} (T')^*$ such that the self-adjoint relation $\widetilde{A}$ in (6.6.3) corresponds to the boundary mapping $\widehat{\Gamma}_0$, that is,

$$\widetilde{A} = \ker \widehat{\Gamma}_0.$$

The Weyl function corresponding to $\{\mathbb{C}^2, \widehat{\Gamma}_0, \widehat{\Gamma}_1\}$ is given by

$$\widehat{M}(\lambda) = \begin{pmatrix} -\dfrac{1}{m_+(\lambda) + \tau(\lambda)} & \dfrac{\tau(\lambda)}{m_+(\lambda) + \tau(\lambda)} \\ \dfrac{\tau(\lambda)}{m_+(\lambda) + \tau(\lambda)} & \dfrac{m(\lambda)\tau(\lambda)}{m_+(\lambda) + \tau(\lambda)} \end{pmatrix}, \qquad \lambda \in \mathbb{C} \setminus \mathbb{R}. \tag{6.6.5}$$

Next it is shown that the Weyl function $\widehat{M}$ shows up in the integral representation of the compressed resolvent $R(\lambda) = P_{\mathfrak{H}}(\widetilde{A} - \lambda)^{-1}\iota_{\mathfrak{H}}$ of $\widetilde{A}$. The next result and its proof are similar to Proposition 6.5.4 and its proof.

Proposition 6.6.3. *For $\lambda \in \mathbb{C} \setminus \mathbb{R}$ the compressed resolvent of the self-adjoint extension $\widetilde{A}$ in (6.6.3) is an integral operator of the form*

$$(R(\lambda)g_+)(t) = \int_c^b G(t, s, \lambda) g_+(s) r(s)\, ds, \quad g_+ \in L^2_r(c, b), \tag{6.6.6}$$

where the Green function $G(t, s, \lambda)$ is given by

$$\begin{cases} \frac{-1}{m_+(\lambda)+\tau(\lambda)} (u_1(t,\lambda) + m_+(\lambda) u_2(t,\lambda))(u_1(s,\lambda) - \tau(\lambda) u_2(s,\lambda)), & c < s < t, \\ \frac{-1}{m_+(\lambda)+\tau(\lambda)} (u_1(t,\lambda) - \tau(\lambda) u_2(t,\lambda))(u_1(s,\lambda) + m_+(\lambda) u_2(s,\lambda)), & t < s < b. \end{cases}$$

In particular, if $g_+ \in L^2_r(c, b)$ has compact support, then

$$\begin{aligned} (R(\lambda)g_+)(t) &= \begin{pmatrix} u_1(t,\lambda) & u_2(t,\lambda) \end{pmatrix} \widehat{M}(\lambda) \begin{pmatrix} \int_c^b u_1(s,\lambda) g_+(s) r(s)\, ds \\ \int_c^b u_2(s,\lambda) g_+(s) r(s)\, ds \end{pmatrix} \\ &\quad - u_1(t,\lambda) \int_t^b u_2(s,\lambda) g_+(s) r(s)\, ds - u_2(t,\lambda) \int_c^t u_1(s,\lambda) g_+(s) r(s)\, ds, \end{aligned}$$

where the 2×2 matrix $\widehat{M}(\lambda)$ is given by (6.6.5).

Proof. Observe that for $g_+ \in L^2_r(c,b)$ and $\lambda \in \mathbb{C}\setminus\mathbb{R}$ the function $f_+(\cdot,\lambda)$ given by

$$\begin{aligned}
&-\big(m_+(\lambda)+\tau(\lambda)\big)f_+(t,\lambda)\\
&\quad=\big(u_1(t,\lambda)+m_+(\lambda)u_2(t,\lambda)\big)\int_c^t \big(u_1(s,\lambda)-\tau(\lambda)u_2(s,\lambda)\big)g_+(s)r(s)\,ds\\
&\qquad+\big(u_1(t,\lambda)-\tau(\lambda)u_2(t,\lambda)\big)\int_t^b \big(u_1(s,\lambda)+m_+(\lambda)u_2(s,\lambda)\big)g_+(s)r(s)\,ds
\end{aligned}$$

is well defined. Moreover, it satisfies $(L-\lambda)f_+ = g_+$ and it has the following initial values at c:

$$f_+(c,\lambda) = -\frac{(g_+,\gamma_+(\bar\lambda))_{L^2_r(c,b)}}{m_+(\lambda)+\tau(\lambda)} \quad\text{and}\quad (pf_+')(c,\lambda)=\frac{\tau(\lambda)(g_+,\gamma_+(\bar\lambda))_{L^2_r(c,b)}}{m_+(\lambda)+\tau(\lambda)}.$$

Now consider the function $h_+ = R(\lambda)g_+ = P_{\mathfrak{H}}(\widetilde{A}-\lambda)^{-1}\iota_{\mathfrak{H}} g_+$, which is given by the Kreĭn formula in (6.6.4). This leads to the equality

$$h_+(\cdot,\lambda) = (A_0^+-\lambda)^{-1}g_+ - \frac{1}{m_+(\lambda)+\tau(\lambda)}\gamma_+(\lambda)(g_+,\gamma_+(\bar\lambda))_{L^2_r(c,b)}.$$

Since $(L-\lambda)(A_0^+-\lambda)^{-1}g_+ = g_+$ and $(L-\lambda)\gamma_+(\lambda)=0$, one has $(L-\lambda)h_+ = g_+$. From $\gamma_+(c,\lambda)=1$ and $((A_0^+-\lambda)^{-1}g_+)(c)=0$ one concludes that

$$h_+(c,\lambda) = -\frac{(g_+,\gamma_+(\bar\lambda))_{L^2_r(c,b)}}{m_+(\lambda)+\tau(\lambda)}.$$

Furthermore, since $(p\gamma_+')(c,\lambda)=m_+(\lambda)$ and

$$\big(p((A_0^+-\lambda)^{-1}g_+)'\big)(c) = \Gamma_1(A_0^+-\lambda)^{-1}g_+ = \gamma_+(\bar\lambda)^*g_+ = (g_+,\gamma_+(\bar\lambda))_{L^2_r(c,b)},$$

it also follows that

$$\begin{aligned}
(ph_+')(c,\lambda) &= (g_+,\gamma_+(\bar\lambda))_{L^2_r(c,b)} - \frac{m_+(\lambda)(g_+,\gamma_+(\bar\lambda))_{L^2_r(c,b)}}{m_+(\lambda)+\tau(\lambda)}\\
&= \frac{\tau(\lambda)(g_+,\gamma_+(\bar\lambda))_{L^2_r(c,b)}}{m_+(\lambda)+\tau(\lambda)}.
\end{aligned}$$

Therefore, $f_+(\cdot,\lambda)$ and $h_+(\cdot,\lambda) = (R(\lambda)g_+)(\cdot)$ satisfy the same differential equation and the same initial conditions. Consequently, $f_+ = R(\lambda)g_+$, which yields (6.6.6).

Now assume that $g \in L^2_r(c,b)$ has compact support. Then writing out all the products in the Green function is allowed, as each individual integral is well defined. This rewriting of the terms of the function $f(\cdot,\lambda)$ gives eight terms, which after adding and substracting of the terms

$$\tau(\lambda)u_1(t,\lambda)\int_t^b u_2(s,\lambda)g(s)r(s)\,ds$$

and

$$\tau(\lambda)u_2(t,\lambda)\int_c^t u_1(s,\lambda)g(s)r(s)\,ds$$

and regrouping leads to the desired result. □

Returning to the boundary value problem (6.6.2) and taking into account Theorem 2.7.3 it is clear that for $\lambda \in \mathbb{C}\setminus\mathbb{R}$ and $g_+ \in L^2_r(c,b)$ the unique solution of

$$(L-\lambda)f_+ = g_+, \qquad (p_+f_+')(c) = -\tau(\lambda)f_+(c), \tag{6.6.7}$$

is given by $R(\lambda)g_+$ in Proposition 6.6.3. The last condition in (6.6.7) is an example of *λ-dependent boundary conditions.* If τ is the Weyl function corresponding to a Sturm–Liouville operator on (a,c) such that the endpoint a is in the limit-point case and c is regular, then the exit space extension $\widetilde{A}$ coincides with the maximal operator Sturm–Liouville on (a,b); this is the situation discussed in Section 6.5. For the special case where τ is a linear or rational Nevanlinna function the model space and hence the corresponding exit space extension $\widetilde{A}$ can be constructed explicitly; cf. Example 4.3.3.

6.7 Weyl functions and subordinate solutions

Consider the Sturm–Liouville equation on the interval (a,b) and assume that the endpoint a is regular and that the endpoint b is in the limit-point case (when replacing derivatives by quasi-derivatives the following discussion extends in a natural fashion to the situation where a is in the limit-circle case). Let $\{\mathbb{C},\Gamma_0,\Gamma_1\}$ be the boundary triplet for $T_{\max}$ in Proposition 6.4.1. The spectrum of the self-adjoint extension A_0 with $\operatorname{dom} A_0 = \ker\Gamma_0$ will be studied by means of *subordinate solutions* of the equation $(L-\lambda)y=0$.

Note that for each $x>a$ one can define the Hilbert space $L^2_r(a,x)$ with the inner product

$$(f,g)_{L^2_r(a,x)} = \int_a^x f(t)\overline{g(t)}r(t)\,dt, \quad f,g\in L^2_r(a,x).$$

For fixed $f,g\in L^2_r(a,c)$, $a<c<b$, the function $x\mapsto (f,g)_x$ is absolutely continuous and

$$\frac{d}{dx}(f,g)_{L^2_r(a,x)} = f(x)\overline{g(x)}r(x)$$

almost everywhere on (a,c). The norm corresponding to $(\cdot,\cdot)_{L^2_r(a,x)}$ will be denoted by $\|\cdot\|_{L^2_r(a,x)}$; it will play an important role in the estimates in this section.

Definition 6.7.1. Let $\xi\in\mathbb{R}$. A solution $v(\cdot,\xi)$ of $(L-\xi)y=0$ is said to be *subordinate* at b if

$$\lim_{x\to b}\frac{\|v(\cdot,\xi)\|_{L^2_r(a,x)}}{\|u(\cdot,\xi)\|_{L^2_r(a,x)}} = 0$$

for every solution $u(\cdot,\xi)$ of $(L-\xi)y=0$ which is not a scalar multiple of $v(\cdot,\xi)$.

The spectrum of A_0 will be studied in terms of solutions of the differential equation $(L-\xi)y=0$ which do not necessarily belong to $L^2_r(a,b)$. In fact, the interest will be in subordinate solutions which satisfy or do not satisfy the boundary condition $f(a)=0$ which characterizes $\operatorname{dom} A_0$. Observe that if a solution $v(\cdot,\xi)$ of $(L-\xi)y=0$ belongs to $L^2_r(a,b)$, then it is subordinate at b since b is in the limit-point case and hence any other solution which is not a scalar multiple of $v(\cdot,\xi)$ does not belong to $L^2_r(a,b)$.

Before the main result can be stated some preliminary considerations are necessary. Recall first the transformation of the boundary triplet in (6.4.7), where $\tau\in\mathbb{R}\cup\{\infty\}$. This results in a boundary triplet $\{\mathbb{C},\Gamma_0^\tau,\Gamma_1^\tau\}$, where

$$\begin{aligned}\Gamma_0^\tau f &= \frac{\tau}{\sqrt{\tau^2+1}}f(a)-\frac{1}{\sqrt{\tau^2+1}}(pf')(a),\\ \Gamma_1^\tau f &= \frac{1}{\sqrt{\tau^2+1}}f(a)+\frac{\tau}{\sqrt{\tau^2+1}}(pf')(a),\end{aligned}$$

for $f\in\operatorname{dom}T_{\max}$, with corresponding γ-field and Weyl function given by

$$\gamma_\tau(\lambda)=\frac{\gamma(\lambda)}{\tau-M(\lambda)}\sqrt{\tau^2+1}\quad\text{and}\quad M_\tau(\lambda)=\frac{1+\tau M(\lambda)}{\tau-M(\lambda)},\qquad \lambda\in\mathbb{C}\setminus\mathbb{R}. \tag{6.7.1}$$

It follows from Proposition 2.3.6 (iii) that

$$\frac{M_\tau(\lambda)-M_\tau(\mu)^*}{\lambda-\bar\mu}=\gamma_\tau(\mu)^*\gamma_\tau(\lambda),\qquad \lambda,\mu\in\mathbb{C}\setminus\mathbb{R}. \tag{6.7.2}$$

The fundamental system $(v_1(\cdot,\lambda);v_2(\cdot,\lambda))$, $\lambda\in\mathbb{C}$, in Lemma 6.4.5 given by

$$\begin{pmatrix}v_1(\cdot,\lambda)\\ v_2(\cdot,\lambda)\end{pmatrix}=\frac{1}{\sqrt{\tau^2+1}}\begin{pmatrix}\tau & -1\\ 1 & \tau\end{pmatrix}\begin{pmatrix}u_1(\cdot,\lambda)\\ u_2(\cdot,\lambda)\end{pmatrix}$$

satisfies the initial conditions

$$\begin{pmatrix}v_1(a,\lambda) & v_2(a,\lambda)\\ (pv_1')(a,\lambda) & (pv_2')(a,\lambda)\end{pmatrix}=\frac{1}{\sqrt{\tau^2+1}}\begin{pmatrix}\tau & -1\\ 1 & \tau\end{pmatrix}.$$

In terms of this fundamental system the γ-field $\gamma_\tau(\lambda)$ can then be expressed as

$$\gamma_\tau(\cdot,\lambda)=v_1(\cdot,\lambda)+M_\tau(\lambda)v_2(\cdot,\lambda); \tag{6.7.3}$$

cf. Lemma 6.4.5. Note that the formal solution

$$v_2(\cdot,\lambda)=\frac{1}{\sqrt{\tau^2+1}}\big(u_1(\cdot,\lambda)+\tau u_2(\cdot,\lambda)\big) \tag{6.7.4}$$

satisfies the boundary condition $(pf')(a)=\tau f(a)$ for the functions in the domain $\ker(\Gamma_1-\tau\Gamma_0)$ of the self-adjoint realization A_τ, $\tau\in\mathbb{R}\cup\{\infty\}$.

Using the fundamental system $(v_1(\cdot,\lambda); v_2(\cdot,\lambda))$, define for any $\lambda \in \mathbb{C}$ and $h \in L^2_r(a,x)$, $a < x < b$,

$$\begin{aligned}(\mathcal{H}(\lambda)h)(t) = v_1(t,\lambda)\int_a^t v_2(s,\lambda)h(s)r(s)\,ds \\ - v_2(t,\lambda)\int_a^t v_1(s,\lambda)h(s)r(s)\,ds, \quad t \in (a,x), \quad \lambda \in \mathbb{C}.\end{aligned}$$

Then $\mathcal{H}(\lambda)$ is a well-defined integral operator and one sees that for fixed $\lambda \in \mathbb{C}$ the function $f(t,\lambda) = (\mathcal{H}(\lambda)h)(t)$ is absolutely continuous and satisfies

$$(L-\lambda)f = h, \quad f(a) = 0,\ (pf')(a) = 0. \tag{6.7.5}$$

In particular, $\mathcal{H}(\lambda)$ maps $L^2_r(a,x)$ into itself. It follows directly that

$$v_i(\cdot,\lambda) - v_i(\cdot,\mu) = (\lambda-\mu)\mathcal{H}(\lambda)v_i(\cdot,\mu), \quad i = 1,2, \tag{6.7.6}$$

since the left-hand side and the right-hand side satisfy the same differential equation and the same initial conditions at a; cf. (6.7.5).

Lemma 6.7.2. *Let $a < x < b$ and let $h \in L^2_r(a,x)$. Then*

$$\|\mathcal{H}(\lambda)h\|^2_{L^2_r(a,x)} \le 2\|v_1(\cdot,\lambda)\|^2_{L^2_r(a,x)}\|v_2(\cdot,\lambda)\|^2_{L^2_r(a,x)}\|h\|^2_{L^2_r(a,x)}.$$

Proof. For $h \in L^2_r(a,x)$ define the functions $g_i(\cdot,\lambda)$, $i = 1,2$, by

$$g_i(t,\lambda) = \int_a^t v_i(s,\lambda)h(s)r(s)\,ds.$$

The Cauchy–Schwarz inequality then gives

$$|g_i(t,\lambda)|^2 \le \|v_i(\cdot,\lambda)\|^2_{L^2_r(a,t)}\|h\|^2_{L^2_r(a,t)}, \quad i = 1,2.$$

Hence, with $f(t,\lambda) = (\mathcal{H}(\lambda)h)(t)$, it follows from the definition of $\mathcal{H}(\lambda)$ that

$$\begin{aligned}|f(t,\lambda)|^2 &\le 2\big(|v_1(t,\lambda)|^2|g_2(t,\lambda)|^2 + |v_2(t,\lambda)|^2|g_1(t,\lambda)|^2\big) \\ &\le 2\big(|v_1(t,\lambda)|^2\|v_2(\cdot,\lambda)\|^2_{L^2_r(a,t)}\|h\|^2_{L^2_r(a,t)} \\ &\qquad + |v_2(t,\lambda)|^2\|v_1(\cdot,\lambda)\|^2_{L^2_r(a,t)}\|h\|^2_{L^2_r(a,t)}\big).\end{aligned}$$

Integration of this inequality yields

$$\begin{aligned}&\|f(\cdot,\lambda)\|^2_{L^2_r(a,x)} \\ &\le 2\int_a^x \big(|v_1(t,\lambda)|^2\|v_2(\cdot,\lambda)\|^2_{L^2_r(a,t)}\|h\|^2_{L^2_r(a,t)} \\ &\qquad + |v_2(t,\lambda)|^2\|v_1(\cdot,\lambda)\|^2_{L^2_r(a,t)}\|h\|^2_{L^2_r(a,t)}\big)r(t)\,dt \\ &= 2\int_a^x \left(\frac{d}{dt}\Big(\|v_1(\cdot,\lambda)\|^2_{L^2_r(a,t)}\|v_2(\cdot,\lambda)\|^2_{L^2_r(a,t)}\Big)\right)\|h\|^2_{L^2_r(a,t)}\,dt,\end{aligned}$$

since for $i = 1, 2$,

$$\frac{d}{dt}\|v_i(\cdot,\lambda)\|^2_{L^2_r(a,t)} = \frac{d}{dt}\int_a^t |v_i(s,\lambda)|^2 r(s)\,ds = |v_i(t,\lambda)|^2 r(t).$$

Therefore,

$$\|f(\cdot,\lambda)\|^2_{L^2_r(a,x)} \le 2\|h\|^2_{L^2_r(a,x)} \int_a^x \frac{d}{dt}\Big(\|v_1(\cdot,\lambda)\|^2_{L^2_r(a,t)}\|v_2(\cdot,\lambda)\|^2_{L^2_r(a,t)}\Big)\,dt,$$

which implies the assertion. □

Lemma 6.7.3. *Let $\xi \in \mathbb{R}$ be a fixed number. The function $x \mapsto \varepsilon_\tau(x,\xi)$ given by*

$$\sqrt{2}\,\varepsilon_\tau(x,\xi)\|v_1(\cdot,\xi)\|_{L^2_r(a,x)}\|v_2(\cdot,\xi)\|_{L^2_r(a,x)} = 1, \quad a < x < b,$$

is well defined, continuous, nonincreasing, and satisfies

$$\lim_{x\to b} \varepsilon_\tau(x,\xi) = 0.$$

Proof. For any $a < x < b$ the two functions

$$x \mapsto \|v_1(\cdot,\xi)\|_{L^2_r(a,x)} \quad \text{and} \quad x \mapsto \|v_2(\cdot,\xi)\|_{L^2_r(a,x)}$$

have positive values, so clearly $\varepsilon_\tau(x,\xi) > 0$ is well defined. Note that the mapping $x \mapsto \|v_1(\cdot,\xi)\|_{L^2_r(a,x)}\|v_2(\cdot,\xi)\|_{L^2_r(a,x)}$ is continuous and nondecreasing. The assumption that b is in the limit-point case implies that not both $v_1(\cdot,\xi)$ and $v_2(\cdot,\xi)$ belong to $L^2_r(a,b)$. Thus, the limit result follows. □

The function $x \mapsto \varepsilon_\tau(x,\xi)$ appears in the estimate in the following theorem.

Theorem 6.7.4. *Let M_τ be the Weyl function in (6.7.1) corresponding to the boundary triplet $\{\mathbb{C}, \Gamma_0^\tau, \Gamma_1^\tau\}$. Assume that $\xi \in \mathbb{R}$ and let $\varepsilon_\tau(x,\xi)$ be as in Lemma 6.7.3. Then for $a < x < b$*

$$\frac{1}{d_0} \le \frac{\|v_2(\cdot,\xi)\|_{L^2_r(a,x)}}{\|v_1(\cdot,\xi)\|_{L^2_r(a,x)}}\,|M_\tau(\xi + i\varepsilon_\tau(x,\xi))| \le d_0,$$

where $d_0 = 1 + 2\big(\sqrt{2} + \sqrt{2+\sqrt{2}}\big)$.

Proof. Assume that $\xi \in \mathbb{R}$ and let $\varepsilon > 0$. Define the function $\psi(\cdot,\xi,\varepsilon)$ by

$$\psi(\cdot,\xi,\varepsilon) = v_1(\cdot,\xi) + M_\tau(\xi + i\varepsilon)v_2(\cdot,\xi). \tag{6.7.7}$$

Then for $a < x < b$,

$$\Big|\,\|v_2(\cdot,\xi)\|_{L^2_r(a,x)}\,|M_\tau(\xi+i\varepsilon)| - \|v_1(\cdot,\xi)\|_{L^2_r(a,x)}\,\Big| \le \|\psi(\cdot,\xi,\varepsilon)\|_{L^2_r(a,x)}$$

or, equivalently,

$$\left| \frac{\|v_2(\cdot,\xi)\|_{L^2_r(a,x)}}{\|v_1(\cdot,\xi)\|_{L^2_r(a,x)}} |M_\tau(\xi+i\varepsilon)| - 1 \right| \leq \frac{\|\psi(\cdot,\xi,\varepsilon)\|_{L^2_r(a,x)}}{\|v_1(\cdot,\xi)\|_{L^2_r(a,x)}}. \tag{6.7.8}$$

The term on the right-hand side of (6.7.8) will now be estimated in a suitable way. In the definition (6.7.7) rewrite the right-hand side using the identity in (6.7.6) with $\lambda = \xi$ and $\mu = \xi + i\varepsilon$. Together with (6.7.3) this shows the identity

$$\psi(\cdot,\xi,\varepsilon) = \gamma_\tau(\cdot,\xi+i\varepsilon) - i\varepsilon\mathcal{H}(\xi)\gamma_\tau(\cdot,\xi+i\varepsilon),$$

expressing the function $\psi(\cdot,\xi,\varepsilon)$ directly in terms of the γ-field. It follows from Lemma 6.7.2 that

$$\begin{aligned}&\|\psi(\cdot,\xi,\varepsilon)\|_{L^2_r(a,x)}\\ &\qquad \leq \big(1+\sqrt{2}\,\varepsilon\,\|v_1(\cdot,\xi)\|_{L^2_r(a,x)}\|v_2(\cdot,\xi)\|_{L^2_r(a,x)}\big)\|\gamma_\tau(\cdot,\xi+i\varepsilon)\|_{L^2_r(a,x)}.\end{aligned}$$

Therefore, the right-hand side of (6.7.8) is estimated by

$$\begin{aligned}&\frac{\big(1+\sqrt{2}\,\varepsilon\,\|v_1(\cdot,\xi)\|_{L^2_r(a,x)}\|v_2(\cdot,\xi)\|_{L^2_r(a,x)}\big)\,\|\gamma_\tau(\cdot,\xi+i\varepsilon)\|_{L^2_r(a,x)}}{\|v_1(\cdot,\xi)\|_{L^2_r(a,x)}}\\ &\quad = \frac{1+\sqrt{2}\,\varepsilon\,\|v_1(\cdot,\xi)\|_{L^2_r(a,x)}\|v_2(\cdot,\xi)\|_{L^2_r(a,x)}}{\big(\|v_1(\cdot,\xi)\|_{L^2_r(a,x)}\|v_2(\cdot,\xi)\|_{L^2_r(a,x)}\big)^{\frac{1}{2}}}\,\frac{\|v_2(\cdot,\xi)\|^{\frac{1}{2}}_{L^2_r(a,x)}}{\|v_1(\cdot,\xi)\|^{\frac{1}{2}}_{L^2_r(a,x)}}\,\|\gamma_\tau(\cdot,\xi+i\varepsilon)\|_{L^2_r(a,x)}.\end{aligned}$$

Now observe that $\|\gamma_\tau(\cdot,\xi+i\varepsilon)\|_{L^2_r(a,x)} \leq \|\gamma_\tau(\cdot,\xi+i\varepsilon)\|_{L^2_r(a,b)}$ and it follows from (6.7.2) that

$$\|\gamma_\tau(\cdot,\xi+i\varepsilon)\|_{L^2_r(a,b)} \leq \sqrt{\frac{\operatorname{Im} M_\tau(\xi+i\varepsilon)}{\varepsilon}} \leq \sqrt{\frac{|M_\tau(\xi+i\varepsilon)|}{\varepsilon}}.$$

Thus, for any $\varepsilon > 0$,

$$\begin{aligned}&\left| \frac{\|v_2(\cdot,\xi)\|_{L^2_r(a,x)}}{\|v_1(\cdot,\xi)\|_{L^2_r(a,x)}} |M_\tau(\xi+i\varepsilon)| - 1 \right|\\ &\quad \leq \frac{1+\sqrt{2}\,\varepsilon\,\|v_1(\cdot,\xi)\|_{L^2_r(a,x)}\|v_2(\cdot,\xi)\|_{L^2_r(a,x)}}{\big(\varepsilon\,\|v_1(\cdot,\xi)\|_{L^2_r(a,x)}\|v_2(\cdot,\xi)\|_{L^2_r(a,x)}\big)^{\frac{1}{2}}}\left(\frac{\|v_2(\cdot,\xi)\|_{L^2_r(a,x)}}{\|v_1(\cdot,\xi)\|_{L^2_r(a,x)}}|M_\tau(\xi+i\varepsilon)|\right)^{\frac{1}{2}}.\end{aligned}$$

Now for $\xi \in \mathbb{R}$ choose $\varepsilon = \varepsilon_\tau(x,\xi)$ in this estimate. This choice minimizes the first factor on the right-hand side to $2^{5/4}$. Hence, the nonnegative quantity

$$Q = \frac{\|v_2(\cdot,\xi)\|_{L^2_r(a,x)}}{\|v_1(\cdot,\xi)\|_{L^2_r(a,x)}} |M_\tau(\xi+i\varepsilon_\tau(x,\xi))|$$

satisfies the inequality

$$|Q-1| \leq 2^{5/4} Q^{\frac{1}{2}}$$

or, equivalently, $Q^2 - 2Q + 1 \leq 4\sqrt{2}Q$. Therefore, $1/d_0 \leq Q \leq d_0$, which completes the proof. □

The following result is now a direct consequence of Theorem 6.7.4.

Theorem 6.7.5. *Let M be the Weyl function corresponding to the boundary triplet $\{\mathbb{C},\Gamma_0,\Gamma_1\}$ and let $\xi\in\mathbb{R}$. Then the following statements hold:*

(i) *If $\tau\in\mathbb{R}$, then the solution $u_1(\cdot,\xi)+\tau u_2(\cdot,\xi)$ of $(L-\xi)y=0$, $(py')(a)=\tau y(a)$ (which is unique up to scalar multiples) is subordinate if and only if*

$$\lim_{\varepsilon\downarrow 0} M(\xi+i\varepsilon)=\tau.$$

(ii) *If $\tau=\infty$, then the solution $u_2(\cdot,\xi)$ of $(L-\xi)y=0$, $y(a)=0$ (which is unique up to scalar multiples) is subordinate if and only if*

$$\lim_{\varepsilon\downarrow 0} |M(\xi+i\varepsilon)|=\infty.$$

Proof. Since $x\mapsto\varepsilon_\tau(x,\xi)$ is continuous, nonincreasing, and with limit 0 as $x\to b$, one has the identity

$$\lim_{\varepsilon\downarrow 0} M_\tau(\xi+i\varepsilon)=\lim_{x\to b} M_\tau(\xi+i\varepsilon_\tau(x,\xi)).$$

(i) Assume that $\tau\in\mathbb{R}$. It suffices to show that $|M_\tau(\xi+i\varepsilon)|\to\infty$ for $\varepsilon\downarrow 0$ if and only if the solution

$$v_2(\cdot,\xi)=\frac{1}{\sqrt{\tau^2+1}}\big(u_1(\cdot,\xi)+\tau u_2(\cdot,\xi)\big)$$

in (6.7.4) is subordinate. For this, assume first that $|M_\tau(\xi+i\varepsilon)|\to\infty$. Then, by Theorem 6.7.4,

$$\lim_{x\to b}\frac{\|v_2(\cdot,\xi)\|_{L^2_r(a,x)}}{\|v_1(\cdot,\xi)\|_{L^2_r(a,x)}}=0. \tag{6.7.9}$$

Hence, for any $c_1,c_2\in\mathbb{R}$, $c_1\neq 0$, one obtains from (6.7.9) that

$$\lim_{x\to b}\frac{\|v_2(\cdot,\xi)\|_{L^2_r(a,x)}}{\|c_1v_1(\cdot,\xi)+c_2v_2(\cdot,\xi)\|_{L^2_r(a,x)}}=0, \tag{6.7.10}$$

and therefore the solution $v_2(\cdot,\xi)$ is subordinate. Conversely, assume that $v_2(\cdot,\xi)$ is subordinate, so that (6.7.10) holds for all $c_1,c_2\in\mathbb{R}$, $c_1\neq 0$. Then clearly (6.7.9) holds, and from Theorem 6.7.4 it follows that $|M_\tau(\xi+i\varepsilon)|\to\infty$.

It is a consequence of (6.7.1) that for $\varepsilon\downarrow 0$ one has

$$|M_\tau(\xi+i\varepsilon)|\to\infty \quad\Leftrightarrow\quad M(\xi+i\varepsilon)\to\tau.$$

This establishes the assertion for $\tau\in\mathbb{R}$.

(ii) The case $\tau=\infty$ can be treated in the same way as (i). ☐

Let $\{\mathbb{C}, \Gamma_0, \Gamma_1\}$ be the boundary triplet for $T_{\max}$ in Proposition 6.4.1 and recall that A_0 corresponds to the Dirichlet boundary condition

$$f(a) = 0. \tag{6.7.11}$$

Let M be the Weyl function of $\{\mathbb{C}, \Gamma_0, \Gamma_1\}$ with the integral representation

$$M(\lambda) = \alpha + \int_{\mathbb{R}} \left(\frac{1}{t-\lambda} - \frac{t}{t^2+1} \right) d\tau(t), \tag{6.7.12}$$

where $\alpha \in \mathbb{R}$ and the measure τ satisfies

$$\int_{\mathbb{R}} \frac{1}{t^2+1}\, d\tau(t) < \infty;$$

cf. Theorem A.2.5. On the basis of Theorem 6.7.5 the following sets will be introduced.

Definition 6.7.6. With the Sturm–Liouville equation $(L - \xi)y = 0$, $\xi \in \mathbb{R}$, the following subsets of $\mathbb{R}$ are associated:

(i) $\mathcal{M}$ is the complement of the set of all $\xi \in \mathbb{R}$ for which a subordinate solution exists that does not satisfy (6.7.11);

(ii) $\mathcal{M}_{\mathrm{ac}}$ is the set of all $\xi \in \mathbb{R}$ for which no subordinate solution exists;

(iii) $\mathcal{M}_{\mathrm{s}}$ is the set of all $\xi \in \mathbb{R}$ for which a subordinate solution exists that satisfies (6.7.11);

(iv) $\mathcal{M}_{\mathrm{sc}}$ is the set of all $\xi \in \mathbb{R}$ for which a subordinate solution exists that satisfies (6.7.11) and does not belong to $L^2_r(a,b)$;

(v) $\mathcal{M}_{\mathrm{p}}$ is the set of all $\xi \in \mathbb{R}$ for which a (subordinate) solution exists that satisfies (6.7.11) and belongs to $L^2_r(a,b)$.

It is a direct consequence of Definition 6.7.6 that

$$\mathbb{R} = \mathcal{M}^c \sqcup \mathcal{M}_{\mathrm{ac}} \sqcup \mathcal{M}_{\mathrm{s}}, \quad \mathcal{M} = \mathcal{M}_{\mathrm{ac}} \sqcup \mathcal{M}_{\mathrm{s}}, \quad \text{and} \quad \mathcal{M}_{\mathrm{s}} = \mathcal{M}_{\mathrm{sc}} \sqcup \mathcal{M}_{\mathrm{p}}$$

hold, where $\sqcup$ stands for the disjoint union.

The following proposition is based on Corollary 3.1.8, where minimal supports for the various parts of the measure τ in the integral representation (6.7.12) of M are described in terms of the boundary behavior of the Nevanlinna function M.

Proposition 6.7.7. *Let M be the Weyl function associated with the boundary triplet $\{\mathbb{C}, \Gamma_0, \Gamma_1\}$ and let τ be the corresponding measure in (6.7.12). Then the sets*

$$\mathcal{M}, \mathcal{M}_{\mathrm{ac}}, \mathcal{M}_{\mathrm{s}}, \mathcal{M}_{\mathrm{sc}}, \mathcal{M}_{\mathrm{p}},$$

are minimal supports for the measures

$$\tau, \tau_{\mathrm{ac}}, \tau_{\mathrm{s}}, \tau_{\mathrm{sc}}, \tau_{\mathrm{p}},$$

respectively.

Proof. Step 1. It will be shown that the set $\mathcal{M}_{\rm ac}$ is a minimal support for the measure $\tau_{\rm ac}$. According to Theorem 6.7.5, $\mathcal{M}_{\rm ac}$ coincides with the set of all $\xi \in \mathbb{R}$ for which both conditions

$$\lim_{\varepsilon \downarrow 0} M(\xi + i\varepsilon) \in \mathbb{R} \quad \text{and} \quad \lim_{\varepsilon \downarrow 0} |M(\xi + i\varepsilon)| = \infty$$

are not satisfied. Hence, $\xi \notin \mathcal{M}_{\rm ac}$ if and only if

$$\lim_{\varepsilon \downarrow 0} M(\xi + i\varepsilon) \in \mathbb{R} \quad \text{or} \quad \lim_{\varepsilon \downarrow 0} |M(\xi + i\varepsilon)| = \infty.$$

Recall that the set $\mathcal{M}'_{\rm ac}$, defined by

$$\mathcal{M}'_{\rm ac} = \bigl\{\xi \in \mathbb{R} : 0 < \lim_{\varepsilon \downarrow 0} \operatorname{Im} M(\xi + i\varepsilon) < \infty\bigr\}, \tag{6.7.13}$$

is a minimal support for $\tau_{\rm ac}$; see Corollary 3.1.8. The following identity and inclusion are straightforward consequences of the definitions

$$\mathcal{M}'_{\rm ac} \setminus \mathcal{M}_{\rm ac} = \bigl\{\xi \in \mathcal{M}'_{\rm ac} : \lim_{\varepsilon \downarrow 0} |M(\xi + i\varepsilon)| = \infty\bigr\},$$

$$\mathcal{M}_{\rm ac} \setminus \mathcal{M}'_{\rm ac} \subset \bigl\{\xi \in \mathbb{R} : \lim_{\varepsilon \downarrow 0} M(\xi + i\varepsilon) \text{ does not exist in } \mathbb{C} \cup \{\infty\}\bigr\}.$$

Hence, it follows from Corollary 3.1.7 that

$$m(\mathcal{M}'_{\rm ac} \setminus \mathcal{M}_{\rm ac}) = 0 \quad \text{and} \quad m(\mathcal{M}_{\rm ac} \setminus \mathcal{M}'_{\rm ac}) = 0, \tag{6.7.14}$$

where m denotes the Lebesgue measure. However, since $\tau_{\rm ac}$ is absolutely continuous with respect to m, it also follows that $\tau_{\rm ac}(\mathcal{M}'_{\rm ac} \setminus \mathcal{M}_{\rm ac}) = 0$. Therefore, $\mathcal{M}_{\rm ac}$ is a minimal support for $\tau_{\rm ac}$; cf. Lemma 3.1.1.

Step 2. It will be shown that the set $\mathcal{M}_{\rm s}$ is a minimal support for the measure $\tau_{\rm s}$. According to Theorem 6.7.5, $\mathcal{M}_{\rm s}$ admits the following description

$$\mathcal{M}_{\rm s} = \bigl\{\xi \in \mathbb{R} : \lim_{\varepsilon \downarrow 0} |M(\xi + i\varepsilon)| = \infty\bigr\}.$$

Observe that the set

$$\mathcal{M}'_{\rm s} = \bigl\{\xi \in \mathbb{R} : \lim_{\varepsilon \downarrow 0} \operatorname{Im} M(\xi + i\varepsilon) = \infty\bigr\}$$

is a minimal support for the measure $\tau_{\rm s}$ by Corollary 3.1.8. Note that $\mathcal{M}'_{\rm s} \subset \mathcal{M}_{\rm s}$ and that

$$m(\mathcal{M}_{\rm s} \setminus \mathcal{M}'_{\rm s}) \leq m(\mathcal{M}_{\rm s}) = 0,$$

where the last identity follows from Corollary 3.1.7 Therefore, $\mathcal{M}_{\rm s}$ is a minimal support for $\tau_{\rm s}$ by Lemma 3.1.1.

Step 3. Here the remaining assertions will be proved. Since M is a Nevanlinna function, the limit value $\lim_{\varepsilon \downarrow 0} \varepsilon \operatorname{Im} M(\xi + i\varepsilon)$ exists, is finite, and is nonnegative; cf. (3.1.27) and (3.1.12). This allows to divide the minimal support $\mathcal{M}_s$ of τ_s into two disjoint subsets,

$$\mathcal{M}_{sc} = \left\{\xi \in \mathcal{M}_s : \lim_{\varepsilon \downarrow 0} \varepsilon \operatorname{Im} M(\xi + i\varepsilon) = 0\right\}$$

and

$$\mathcal{M}_p = \left\{\xi \in \mathcal{M}_s : \lim_{\varepsilon \downarrow 0} \varepsilon \operatorname{Im} M(\xi + i\varepsilon) > 0\right\}.$$

By Corollary 3.1.8 the set

$$\mathcal{M}'_{sc} = \left\{\xi \in \mathcal{M}'_s : \lim_{\varepsilon \downarrow 0} \varepsilon \operatorname{Im} M(\xi + i\varepsilon) = 0\right\}$$

is a minimal support for τ_{sc}. Since $\mathcal{M}'_{sc} \subset \mathcal{M}_{sc}$ and

$$m(\mathcal{M}_{sc} \setminus \mathcal{M}'_{sc}) \leq m(\mathcal{M}_{sc}) \leq m(\mathcal{M}_s) = 0$$

by Corollary 3.1.7, also $\mathcal{M}_{sc}$ is a minimal support for τ_{sc}.

On the other hand, clearly $\mathcal{M}_p \subset \mathcal{M}'_s$ and hence

$$\mathcal{M}_p = \left\{\xi \in \mathcal{M}'_s : \lim_{\varepsilon \downarrow 0} \varepsilon \operatorname{Im} M(\xi + i\varepsilon) > 0\right\},$$

which is a minimal support of τ_p by Corollary 3.1.8.

Finally, the assertion concerning the set $\mathcal{M}$ is a consequence of the other proved statements. □

The minimal supports in Proposition 6.7.7 are intimately connected with the spectrum of A_0. For the absolutely continuous spectrum one obtains the following result, where the notion of the absolutely continuous closure of a Borel set from Definition 3.2.4 is used. Similar statements (with an inclusion) can be formulated for the singular parts of the spectrum; cf. Section 3.6.

Theorem 6.7.8. *Let A_0 be the self-adjoint realization of L corresponding to the Dirichlet boundary condition at the regular endpoint a and let $\mathcal{M}_{ac}$ be as in Definition 6.7.6. Then*

$$\sigma_{ac}(A_0) = \operatorname{clos}_{ac}(\mathcal{M}_{ac}).$$

Proof. Since the minimal operator T_{min} is simple by Proposition 6.4.4, one can apply Theorem 3.6.5 with $\Delta = \mathbb{R}$, which yields

$$\sigma_{ac}(A_0) = \operatorname{clos}_{ac}\left\{\xi \in \mathbb{R} : 0 < \lim_{\varepsilon \downarrow 0} \operatorname{Im} M(\xi + i\varepsilon) < \infty\right\} = \operatorname{clos}_{ac}(\mathcal{M}'_{ac});$$

cf. (6.7.13). Since $m(\mathcal{M}'_{ac} \setminus \mathcal{M}_{ac}) = 0$ and $m(\mathcal{M}_{ac} \setminus \mathcal{M}'_{ac}) = 0$ by (6.7.14), it follows from Lemma 3.2.5 that

$$\operatorname{clos}_{ac}(\mathcal{M}'_{ac}) = \operatorname{clos}_{ac}(\mathcal{M}'_{ac} \cap \mathcal{M}_{ac}) = \operatorname{clos}_{ac}(\mathcal{M}_{ac}).$$

This leads to the result. □

6.8 Semibounded Sturm–Liouville expressions in the regular case

Let L be the Sturm–Liouville differential expression in (6.1.1),

$$L = \frac{1}{r}\left[-DpD + q\right], \quad D = d/dx,$$

and assume that the endpoints a and b are regular, that is, $[a,b]$ is a compact interval and the coefficient functions are real and satisfy

$$\begin{cases} p(x) \neq 0, \ r(x) > 0, \ \text{for almost all } x \in (a,b), \\ 1/p, q, r \in L^1(a,b). \end{cases} \tag{6.8.1}$$

Recall from Proposition 6.3.1 that $\{\mathbb{C}^2, \Gamma_0, \Gamma_1\}$, where

$$\Gamma_0 f = \begin{pmatrix} f(a) \\ f(b) \end{pmatrix} \quad \text{and} \quad \Gamma_1 f = \begin{pmatrix} (pf')(a) \\ -(pf')(b) \end{pmatrix}, \quad f \in \operatorname{dom} T_{\max}, \tag{6.8.2}$$

is a boundary triplet for $T_{\max}$. In the present section it will be assumed, in addition to (6.8.1), that the sign condition

$$p(x) > 0 \quad \text{for almost all} \quad x \in (a,b) \tag{6.8.3}$$

holds; cf. (6.1.26). This assumption will imply that the minimal operator and all self-adjoint realizations of L in $L^2_r(a,b)$ are semibounded from below. The main objective of this section is to provide a characterization of the closed semibounded forms and the corresponding semibounded self-adjoint realizations of L by using the abstract techniques developed in Section 5.6.

In order to apply the results from Section 5.6 a boundary pair will be constructed which is compatible with the boundary triplet in (6.8.2). As a first step, associate with the coefficient functions which satisfy (6.8.1) and (6.8.3) the quadratic form defined by

$$\mathfrak{t}[f,g] = \int_a^b \left((\sqrt{p}f')(x)\overline{(\sqrt{p}g')(x)} + q(x)f(x)\overline{g(x)} \right) dx \tag{6.8.4}$$

for $f, g \in \operatorname{dom} \mathfrak{t} = \mathfrak{D}$, where

$$\mathfrak{D} = \left\{ f \in L^2_r(a,b) : f \in AC(a,b), \ \sqrt{p}f' \in L^2(a,b) \right\}. \tag{6.8.5}$$

It turns out that $\mathfrak{t}$ is densely defined, closed, and semibounded. Hence, there exists a semibounded self-adjoint operator S_1 which corresponds to $\mathfrak{t}$ and it will be shown that S_1 extends $T_{\min}$ and that S_1 and the Friedrichs extension S_{F} are transversal. The next step is to define the mapping

$$\Lambda f = \begin{pmatrix} f(a) \\ f(b) \end{pmatrix}, \quad f \in \mathfrak{D} = \operatorname{dom} \mathfrak{t}. \tag{6.8.6}$$

In Lemma 6.8.4 it will be proved that $\{\mathbb{C}^2, \Lambda\}$ is a well-defined boundary pair for $T_{\min}$ on $\mathfrak{D}$ corresponding to S_1 that is compatible with the boundary triplet $\{\mathbb{C}^2, \Gamma_0, \Gamma_1\}$ in (6.8.2). In other words, the mapping Λ is an extension of Γ_0 and the self-adjoint operator S_1 corresponding to the form $\mathfrak{t}$ on $\mathfrak{D}$ coincides with the operator A_1 corresponding to Γ_1. Therefore, Theorem 5.6.13 and Corollary 5.6.14 can be applied, which leads to Theorem 6.8.5, the main result of this section.

Observe that the definition of the linear space $\mathfrak{D}$ in (6.8.5) does not involve the potential q in the differential expression L. Some properties concerning $\mathfrak{D}$ are collected in the following lemma.

Lemma 6.8.1. *Let $[a,b]$ be a compact interval and let the conditions (6.8.1) and (6.8.3) be satisfied. Then*

$$\operatorname{dom} T_{\max} \subset \big\{f \in AC[a,b] : pf' \in AC[a,b]\big\} \subset \mathfrak{D} \subset AC[a,b]. \tag{6.8.7}$$

In particular, $\mathfrak{D}$ is dense in $L^2_r(a,b)$ and for $f \in \mathfrak{D}$ both limits

$$f(a) = \lim_{x\to a} f(x) \quad \text{and} \quad f(b) = \lim_{x \to b} f(x) \tag{6.8.8}$$

exist.

Proof. The first inclusion in (6.8.7) is clear; cf. (6.2.1) and the beginning of Section 6.3. To see the second inclusion in (6.8.7), let $f \in AC[a,b]$ and $pf' \in AC[a,b]$. In particular, f is bounded so that $f \in L^2_r(a,b)$; and pf' is bounded so that $|\sqrt{p}f'| \le C\,1/\sqrt{p}$ for some positive C. It follows that $\sqrt{p}f' \in L^2(a,b)$ and thus $f \in \mathfrak{D}$.

To see the third inclusion in (6.8.7), it suffices to show that $f \in \mathfrak{D}$ implies $f' \in L^1(a,b)$. In fact, for $f \in \mathfrak{D}$ one has

$$\begin{aligned}\int_a^b |f'(x)|\,dx &= \int_a^b \frac{1}{\sqrt{p(x)}} |\sqrt{p(x)}f'(x)|\,dx \\ &\le \sqrt{\int_a^b \frac{1}{p(x)}\,dx}\,\sqrt{\int_a^b p(x)|f'(x)|^2\,dx} < \infty\end{aligned}$$

by the Cauchy–Schwarz inequality.

It follows from (6.8.7) that $\mathfrak{D}$ is a dense subspace of $L^2_r(a,b)$, since $\operatorname{dom} T_{\max}$ is dense in $L^2_r(a,b)$; cf. Theorem 6.2.1. Furthermore, the limits in (6.8.8) exist for $f \in \mathfrak{D}$ since $f \in AC[a,b]$ by (6.8.7). □

To study the properties of the form $\mathfrak{t}$ in (6.8.4) one uses the decomposition $\mathfrak{t} = \mathfrak{r} + \mathfrak{q}$, where the forms $\mathfrak{r}$ and $\mathfrak{q}$ are defined by

$$\mathfrak{r}[f,g] = \int_a^b (\sqrt{p}f')(x)\overline{(\sqrt{p}g')(x)}\,dx, \quad f,g \in \mathfrak{D}, \tag{6.8.9}$$

and

$$\mathfrak{q}[f,g] = \int_a^b q(x)f(x)\overline{g(x)}\,dx, \quad f,g \in \mathfrak{D}, \tag{6.8.10}$$

respectively. It follows directly from (6.8.5) that the form $\mathfrak{r}$ is well defined. To see that the form $\mathfrak{q}$ is well defined, note that $qf\overline{g} \in L^1(a,b)$ since $f,g \in \mathfrak{D} \subset AC[a,b]$ are bounded. Hence, $\mathfrak{D}$ is a natural domain of definition for the form $\mathfrak{t}$ in (6.8.4) for any real potential $q \in L^1(a,b)$. It will be shown that the form $\mathfrak{q}$ is a small perturbation of the form $\mathfrak{r}$; cf. Theorem 5.1.16. The properties of the unperturbed form $\mathfrak{r}$ will be studied first.

Associated with the function p is the first-order differential expression N on the interval (a,b) by $Nf = \sqrt{p}f'$, which is meaningful for $f \in AC[a,b]$. The differential expression N generates a linear operator R from $L^2_r(a,b)$ to $L^2(a,b)$ by

$$Rf := Nf, \qquad \operatorname{dom} R = \mathfrak{D}. \tag{6.8.11}$$

It is clear that the form $\mathfrak{r}$ in (6.8.9) and the operator R in (6.8.11) are connected by

$$\mathfrak{r}[f,g] = (Rf, Rg)_{L^2(a,b)}, \quad \operatorname{dom}\mathfrak{r} = \mathfrak{D}, \tag{6.8.12}$$

so that the form $\mathfrak{r}$ is closed if and only if R is closed as an operator from $L^2_r(a,b)$ to $L^2(a,b)$; cf. Lemma 5.1.21.

Lemma 6.8.2. *Let $[a,b]$ be a compact interval and assume that the conditions (6.8.1) and (6.8.3) are satisfied. Then $\mathfrak{r}$ in (6.8.9) is a densely defined closed nonnegative form in $L^2_r(a,b)$. Moreover, for $\varepsilon > 0$ there exists $C_\varepsilon > 0$ such that*

$$|f(x)|^2 \le C_\varepsilon \|f\|^2_{L^2_r(a,b)} + \varepsilon\mathfrak{r}[f], \qquad x \in [a,b], \tag{6.8.13}$$

holds for all $f \in \mathfrak{D}$.

Proof. It is clear from (6.8.12) that the form $\mathfrak{r}$ is nonnegative and densely defined; cf. Lemma 6.8.1. To show (6.8.13), observe that for $f \in \mathfrak{D}$ and $x,y \in [a,b]$ one has

$$\begin{aligned} |f(x)|^2 &\le \big(|f(y)| + |f(x) - f(y)|\big)^2 \\ &\le 2(|f(y)|^2 + |f(y) - f(x)|^2) \\ &\le 2\left(|f(y)|^2 + \left|\int_x^y f'(t)\,dt\right|^2\right) \\ &\le 2\left(|f(y)|^2 + \int_x^y p(t)|f'(t)|^2\,dt \int_x^y \frac{1}{p(t)}\,dt\right), \end{aligned} \tag{6.8.14}$$

where the Cauchy–Schwarz inequality and integrability of $1/p$ were used. Let $\varepsilon > 0$. Due to the absolute continuity of $x \mapsto \int_a^x 1/(p(t))\,dt$ on $[a,b]$, there exist $\delta > 0$ and $c_\delta > 0$ such that for all $x \in [a,b]$ and $J(x,\delta) = (x-\delta, x+\delta) \cap [a,b]$ one has

$$\left(\int_{J(x,\delta)} \frac{1}{p(t)}\,dt\right) \le \frac{\varepsilon}{2} \quad \text{and} \quad \int_{J(x,\delta)} r(y)\,dy \ge c_\delta. \tag{6.8.15}$$

For $x \in [a,b]$ and the corresponding interval $J(x,\delta)$ one observes from (6.8.14) that

$$
\begin{aligned}
|f(x)|^2 \int_{J(x,\delta)} r(y)\,dy &= \int_{J(x,\delta)} |f(x)|^2 r(y)\,dy \\
&\leq 2 \int_{J(x,\delta)} |f(y)|^2 r(y)\,dy \\
&\quad + 2 \int_{J(x,\delta)} \left(\int_x^y p(t)|f'(t)|^2\,dt \int_x^y \frac{1}{p(t)}\,dt \right) r(y)\,dy.
\end{aligned}
\tag{6.8.16}
$$

The first term on the right-hand side can be estimated by $2\|f\|^2_{L^2_r(a,b)}$. For the second term on the right-hand side note that

$$
\begin{aligned}
&2 \int_{J(x,\delta)} \left(\int_x^y p(t)|f'(t)|^2\,dt \int_x^y \frac{1}{p(t)}\,dt \right) r(y)\,dy \\
&\quad \leq 2 \left(\int_{J(x,\delta)} \frac{1}{p(t)}\,dt \right) \int_{J(x,\delta)} \left(\int_x^y p(t)|f'(t)|^2\,dt \right) r(y)\,dy \\
&\quad \leq \varepsilon \left(\int_a^b p(t)|f'(t)|^2\,dt \right) \int_{J(x,\delta)} r(y)\,dy.
\end{aligned}
$$

Use this with the estimate in (6.8.16), divide by $\int_{J(x,\delta)} r(y)\,dy$, and use (6.8.15) to conclude (6.8.13) with $C_\varepsilon = 2c_\delta^{-1}$ for $x \in [a,b]$.

To verify that the form $\mathfrak{r}$ is closed, it suffices by (6.8.12) to prove that the operator R in (6.8.11) is closed; cf. Lemma 5.1.21. Let $f_n \in \operatorname{dom} R = \mathfrak{D}$ and assume that $f_n \to f$ in $L^2_r(a,b)$ for some in $f \in L^2_r(a,b)$ and that $Rf_n = \sqrt{p} f_n' \to g$ in $L^2(a,b)$ for some $g \in L^2(a,b)$. It will be shown that $f \in \mathfrak{D}$ and $g = Rf$. In fact, the sequence $f_n(a)$ converges by (6.8.13) to some $\alpha \in \mathbb{C}$. Next define the function h on $[a,b]$ by

$$
h(x) := \alpha + \int_a^t \frac{g(t)}{\sqrt{p(t)}}\,dt.
$$

Since $1/p$ is integrable, the Cauchy–Schwarz inequality shows that

$$
\|g/\sqrt{p}\|_{L^1(a,b)} \leq \|p^{-1}\|^{1/2}_{L^1(a,b)} \|g\|_{L^2(a,b)}.
$$

Thus, it follows that h is well defined and absolutely continuous on $[a,b]$ and that $\sqrt{p}h' = g$ almost everywhere on $[a,b]$. Furthermore, for $a \leq x \leq b$ one obtains

$$
\begin{aligned}
|f_n(x) - h(x)| &\leq |f_n(a) - \alpha| + \left| \int_a^x f_n'(t)\,dt - \int_a^x \frac{g(t)}{\sqrt{p(t)}}\,dt \right| \\
&\leq |f_n(a) - \alpha| + \int_a^x \frac{1}{\sqrt{p(t)}} \left| \sqrt{p(t)} f_n'(t) - g(t) \right| dt
\end{aligned}
$$

$$\leq |f_n(a) - \alpha| + \left(\int_a^x |\sqrt{p(t)} f_n'(t) - g(t)|^2 \, dt \right)^{1/2} \left(\int_a^x \frac{1}{p(t)} \, dt \right)^{1/2}$$
$$\leq |f_n(a) - \alpha| + \|Rf_n - g\|_{L^2(a,b)} \|p^{-1}\|_{L^1(a,b)}^{1/2} \to 0$$

as $n \to \infty$. Thus, $f_n \to h$ uniformly on $[a, b]$. On the other hand, $f_n \to f$ in $L^2_r(a, b)$. Therefore, $f = h$ on $[a, b]$ and

$$\sqrt{p(x)} f'(x) = \sqrt{p(x)} h'(x) = g(x), \quad x \in [a, b].$$

Hence, one concludes $f \in \mathfrak{D}$ and $Rf = g$. □

For another appearance of the above lemma in a slightly more general setting, see Lemma 6.9.1. Now the main result about the perturbed form $\mathfrak{t}$ will be given.

Lemma 6.8.3. *Let $[a, b]$ be a compact interval and assume that the conditions (6.8.1) and (6.8.3) are satisfied. Then $\mathfrak{t}$ in (6.8.4) is a densely defined closed semi-bounded form in $L^2_r(a, b)$ and the corresponding semibounded self-adjoint operator S_1 is an extension of $T_{\min}$. Moreover, for $\varepsilon > 0$ there exists $C_\varepsilon > 0$ such that*

$$|f(x)|^2 \leq C_\varepsilon \|f\|^2_{L^2_r(a,b)} + \varepsilon \mathfrak{t}[f], \qquad x \in [a, b], \tag{6.8.17}$$

holds for all $f \in \mathfrak{D}$.

Proof. Recall the decomposition $\mathfrak{t} = \mathfrak{r} + \mathfrak{q}$ where $\mathfrak{r}$ and $\mathfrak{q}$ are defined in (6.8.9) and (6.8.10). According to Lemma 6.8.2, the form $\mathfrak{r}$ is nonnegative and closed in $L^2_r(a, b)$. Moreover, the form $\mathfrak{q}$ is a small perturbation of $\mathfrak{r}$. Indeed, $f \in \mathfrak{D}$ implies that $f \in AC[a, b]$, and hence

$$|\mathfrak{q}[f]| = \left| \int_a^b q(x) |f(x)|^2 \, dx \right| \leq \sup_{x \in [a,b]} |f(x)|^2 \int_a^b |q(x)| \, dx.$$

Therefore, with $\varepsilon > 0$ and $C_\varepsilon > 0$ as in Lemma 6.8.2, one concludes that

$$|\mathfrak{q}[f]| \leq C_\varepsilon \|q\|_{L^1(a,b)} \|f\|^2_{L^2_r(a,b)} + \varepsilon \|q\|_{L^1(a,b)} \mathfrak{r}[f],$$

and it follows that $\mathfrak{q}$ is form-bounded with respect to $\mathfrak{r}$ and the form bound is arbitrarily small. Now Theorem 5.1.16 implies that $\mathfrak{t} = \mathfrak{r} + \mathfrak{q}$ is a semibounded closed form in $L^2_r(a, b)$. Since $T_{\min}$ is equal to $\ker \Gamma_0 \cap \ker \Gamma_1$, integration by parts shows that

$$(T_{\min} f, g)_{L^2_r(a,b)} = \mathfrak{t}[f, g], \qquad f \in \operatorname{dom} T_{\min}, \ g \in \mathfrak{D},$$

and now the first representation theorem implies that $T_{\min} \subset S_1$. In particular, $T_{\min}$ is semibounded from below.

In order to show (6.8.17), decompose q as $q = q_+ - q_-$ into its positive part $q_+ = \max\{q, 0\}$ and its negative part $q_- = \max\{-q, 0\}$. Observe that for $f \in \mathfrak{D}$

$$\mathfrak{q}[f] = \int_a^b q(x)|f(x)|^2\,dx \geq -\int_a^b q_-(x)|f(x)|^2\,dx \tag{6.8.18}$$

and also that

$$\int_a^b q_-(x)|f(x)|^2\,dx \leq \sup_{x\in[a,b]} |f(x)|^2 \int_a^b q_-(x)\,dx. \tag{6.8.19}$$

By Lemma 6.8.2 one sees that for every $\delta > 0$ there exists C_δ such that

$$\int_a^b q_-(x)|f(x)|^2\,dx \leq C_\delta \|q_-\|_{L^1(a,b)}\|f\|^2_{L^2_r(a,b)} + \delta\|q_-\|_{L^1(a,b)}\mathfrak{r}[f].$$

Hence, it follows from (6.8.18), (6.8.19), and $\mathfrak{t} = \mathfrak{r} + \mathfrak{q}$ that

$$\mathfrak{t}[f] \geq \bigl(1 - \delta\|q_-\|_{L^1(a,b)}\bigr)\,\mathfrak{r}[f] - C_\delta\|q_-\|_{L^1(a,b)}\|f\|^2_{L^2_r(a,b)}.$$

If $\delta > 0$ is sufficiently small this means that there exist constants $\alpha, \beta > 0$ such that

$$\mathfrak{r}[f] \leq \alpha\mathfrak{t}[f] + \beta\|f\|^2_{L^2_r(a,b)}.$$

A further application of Lemma 6.8.2 yields the desired result. □

Now the theory developed in Chapter 5 concerning boundary pairs will be applied to the semibounded minimal operator $T_{\min}$. Recall the definition of the mapping in (6.8.6). A preparation for the main result is Lemma 6.8.4 below, which is based on Lemma 5.6.5.

Lemma 6.8.4. *Let $[a, b]$ be a compact interval, assume that the conditions (6.8.1) and (6.8.3) are satisfied, and let $\{\mathbb{C}^2, \Gamma_0, \Gamma_1\}$ be the boundary triplet in (6.8.2). Then $\{\mathbb{C}^2, \Lambda\}$ in (6.8.6) is a boundary pair for $T_{\min}$ corresponding to S_1 which is compatible with the boundary triplet $\{\mathbb{C}^2, \Gamma_0, \Gamma_1\}$. Moreover, one has*

$$(T_{\max} f, g)_{L^2_r(a,b)} = (\Gamma_1 f, \Lambda g) + \mathfrak{t}[f, g], \quad f \in \operatorname{dom} T_{\max},\ g \in \mathfrak{D}. \tag{6.8.20}$$

Proof. Consider the form $\mathfrak{t}$ in (6.8.4) defined on $\operatorname{dom}\mathfrak{t} = \mathfrak{D}$ and let S_1 be the corresponding semibounded self-adjoint operator in $L^2_r(a, b)$. By Lemma 6.8.1, $\operatorname{dom} T_{\max} \subset \mathfrak{D}$ and hence Λ in (6.8.6) is an extension of the boundary mapping Γ_0 in (6.8.2). Integration by parts shows that

$$\begin{aligned}(T_{\max} f, g)_{L^2_r(a,b)} &= \int_a^b \bigl(-(pf')'(x) + q(x)f(x)\bigr)\overline{g(x)}\,dx \\ &= (pf')(a)\overline{g(a)} - (pf')(b)\overline{g(b)} + \mathfrak{t}[f, g]\end{aligned} \tag{6.8.21}$$

for $f \in \operatorname{dom} T_{\max} \subset \mathfrak{D}$ and $g \in \mathfrak{D}$. This yields (6.8.20). It also follows from (6.8.21) that

$$(A_1 f, g)_{L^2_r(a,b)} = \mathfrak{t}[f,g]$$

for $f \in \operatorname{dom} A_1 = \ker \Gamma_1$ and $g \in \mathfrak{D}$, and the first representation theorem implies that $A_1 = S_1$. Let $\varepsilon > 0$ and $C_\varepsilon > 0$ be as in Lemma 6.8.3. It follows from the estimate (6.8.17) that for $\rho < m(S_1)$ there exists $C_{\rho,\varepsilon} > 0$ such that

$$\|\Lambda f\|^2_{\mathbb{C}^2} = |f(a)|^2 + |f(b)|^2 \leq 2C_\varepsilon \|f\|^2_{L^2_r(a,b)} + 2\varepsilon \mathfrak{t}[f] \leq C_{\rho,\varepsilon} \|f\|^2_{\mathfrak{t}_{S_1}-\rho}$$

for all $f \in \mathfrak{D}$. Therefore, $\Lambda \in \mathbf{B}(\mathfrak{H}_{\mathfrak{t}_{S_1}-\rho}, \mathbb{C}^2)$; recall that the Hilbert space $\mathfrak{H}_{\mathfrak{t}_{S_1}-\rho}$ was defined above Lemma 5.1.3. Now Lemma 5.6.5 implies that $\{\mathbb{C}^2, \Lambda\}$ is a boundary pair for $T_{\min}$ and since $A_1 = S_1$ one also sees that $\{\mathbb{C}^2, \Lambda\}$ and $\{\mathbb{C}^2, \Gamma_0, \Gamma_1\}$ are compatible. □

Recall that by means of the boundary triplet in (6.8.2) all the self-adjoint extensions of $T_{\min}$ are in a one-to-one correspondence to the self-adjoint relations Θ in $\mathbb{C}^2$ via

$$\operatorname{dom} A_\Theta = \bigl\{ f \in \operatorname{dom} T_{\max} : \{\Gamma_0 f, \Gamma_1 f\} \in \Theta \bigr\}. \tag{6.8.22}$$

The next result, which is an immediate consequence of Theorem 5.6.13 and Corollary 5.6.14, makes use of the compatible boundary pair in Lemma 6.8.4 and provides a characterization of all closed semibounded forms associated with the semibounded self-adjoint extensions A_Θ.

Theorem 6.8.5. *Let $\{\mathbb{C}^2, \Gamma_0, \Gamma_1\}$ be the boundary triplet in (6.8.2), let Θ be a self-adjoint relation in $\mathbb{C}^2$, and let A_Θ be the corresponding self-adjoint restriction of $T_{\max}$ in (6.8.22). Then A_Θ is semibounded from below and the corresponding densely defined closed semibounded form $\mathfrak{t}_\Theta$ in $L^2_r(a,b)$ such that*

$$(A_\Theta f, g)_{L^2_r(a,b)} = \mathfrak{t}_\Theta[f,g], \quad f \in \operatorname{dom} A_\Theta, \ g \in \operatorname{dom} \mathfrak{t}_\Theta,$$

is given as follows:

(i) *If Θ is a symmetric 2×2-matrix, then*

$$\mathfrak{t}_\Theta[f,g] = \mathfrak{t}[f,g] + \left(\Theta \begin{pmatrix} f(a) \\ f(b) \end{pmatrix}, \begin{pmatrix} g(a) \\ g(b) \end{pmatrix} \right), \quad \operatorname{dom} \mathfrak{t}_\Theta = \mathfrak{D}.$$

(ii) *If $\Theta = \Theta_{\rm op} \,\widehat{\oplus}\, \Theta_{\rm mul}$ with respect to the decomposition $\mathbb{C}^2 = \operatorname{dom} \Theta_{\rm op} \oplus \operatorname{mul} \Theta$ and $\dim \operatorname{dom} \Theta_{\rm op} = 1$, then*

$$\mathfrak{t}_\Theta[f,g] = \mathfrak{t}[f,g] + \Theta_{\rm op}\bigl(f(a)\overline{g(a)} + f(b)\overline{g(b)} \bigr),$$
$$\operatorname{dom} \mathfrak{t}_\Theta = \left\{ h \in \mathfrak{D} : \begin{pmatrix} h(a) \\ h(b) \end{pmatrix} \in \operatorname{dom} \Theta_{\rm op} \right\}.$$

(iii) *If* $\Theta = \{0\} \times \mathbb{C}^2$, *then* $A_\Theta = A_0$ *coincides with the Friedrichs extension* S_F *and*

$$\mathfrak{t}_\Theta[f,g] = \mathfrak{t}[f,g], \quad \operatorname{dom} \mathfrak{t}_\Theta = \bigl\{h \in \mathfrak{D} : h(a) = h(b) = 0\bigr\}.$$

Theorem 6.8.5 has an immediate corollary for the form $\mathfrak{t}_\Theta$, which is the analog of Lemma 6.8.3 for the form $\mathfrak{t}$.

Corollary 6.8.6. *Let* $[a,b]$ *be a compact interval and let the conditions* (6.8.1) *and* (6.8.3) *be satisfied. Let* Θ *be a self-adjoint relation in* $\mathbb{C}^2$ *and let the form* $\mathfrak{t}_\Theta$ *be as in Theorem* 6.8.5. *Then for* $\varepsilon > 0$ *there exists* $C_\varepsilon > 0$ *such that*

$$|f(x)|^2 \leq C_\varepsilon \|f\|^2_{L^2_r(a,b)} + \varepsilon \mathfrak{t}_\Theta[f], \qquad x \in [a,b],$$

holds for all $f \in \operatorname{dom} \mathfrak{t}_\Theta$.

Proof. Let $\mathfrak{t}_\Theta$ be given as in Theorem 6.8.5 (i); the case in Theorem 6.8.5 (ii) is treated in the same way and for $\mathfrak{t}_\Theta$ in Theorem 6.8.5 (iii) the result is clear from Lemma 6.8.3. Let $\mu(\Theta)$ be the smallest eigenvalue of the symmetric 2×2 matrix Θ and define $\mu \in \mathbb{R}$ by

$$\mu = \begin{cases} 0, & \mu(\Theta) \geq 0, \\ |\mu(\Theta)|, & \mu(\Theta) < 0. \end{cases}$$

Then $(\Theta \Lambda f, \Lambda f)_{\mathbb{C}^2} \geq -\mu(|f(a)|^2 + |f(b)|^2)$ for all $f \in \mathfrak{D} = \operatorname{dom} \mathfrak{t}_\Theta$ and using Lemma 6.8.3 one concludes that for $0 < \varepsilon' < 1$ there exists $C_{\varepsilon'} > 0$ such that

$$(\Theta \Lambda f, \Lambda f)_{\mathbb{C}^2} \geq -C_{\varepsilon'} \|f\|^2_{L^2_r(a,b)} - \varepsilon' \mathfrak{t}[f], \qquad f \in \mathfrak{D}.$$

The inequality

$$\mathfrak{t}_\Theta[f] = \mathfrak{t}[f] + (\Theta \Lambda f, \Lambda f)_{\mathbb{C}^2} \geq (1-\varepsilon')\mathfrak{t}[f] - C_{\varepsilon'} \|f\|^2_{L^2_r(a,b)}$$

shows that

$$\mathfrak{t}[f] \leq \frac{C_{\varepsilon'}}{1-\varepsilon'} \|f\|^2_{L^2_r(a,b)} + \frac{1}{1-\varepsilon'} \mathfrak{t}_\Theta[f], \qquad f \in \mathfrak{D},$$

and another application of Lemma 6.8.3 completes the proof. □

Finally, the Kreĭn type extensions $S_{\mathrm{K},x}$ from Definition 5.4.2 are provided for $x < m(S_\mathrm{F})$. Recall from Proposition 6.3.1 that

$$M(x) = \frac{1}{u_2(b,x)} \begin{pmatrix} -u_1(b,x) & 1 \\ 1 & -(pu_2')(b,x) \end{pmatrix}.$$

Hence, it follows from Theorem 5.5.1 that

$$\operatorname{dom} S_{\mathrm{K},x} = \left\{ f \in \operatorname{dom} T_{\max} : M(x) \begin{pmatrix} f(a) \\ f(b) \end{pmatrix} = \begin{pmatrix} (pf')(a) \\ -(pf')(b) \end{pmatrix} \right\}.$$

Note that the semibounded self-adjoint extensions of $T_{\min}$ in $L^2_r(a,b)$ with lower bound $x < m(S_F)$ are precisely those self-adjoint extensions A_Θ that satisfy the inequalities $S_{K,x} \leq A_\Theta \leq S_F$. These extensions and the corresponding closed semibounded forms can now be described explicitly using the results in Section 5.6. In particular, if $m(S_F) > 0$, then the Kreĭn–von Neumann extension $S_{K,0}$ is defined on

$$\operatorname{dom} S_{K,0} = \left\{ f \in \operatorname{dom} T_{\max} : M(0) \begin{pmatrix} f(a) \\ f(b) \end{pmatrix} = \begin{pmatrix} (pf')(a) \\ -(pf')(b) \end{pmatrix} \right\}$$

and Corollary 5.6.18 provides a one-to-one correspondence between all closed nonnegative forms corresponding to nonnegative self-adjoint extension A_Θ of $T_{\min}$ and all closed nonnegative forms corresponding to nonnegative self-adjoint relations Θ in $\mathbb{C}^2$.

6.9 Closed semibounded forms for Sturm–Liouville equations

Let L be the Sturm–Liouville differential expression given by (6.1.1) on the interval (a,b). It will be assumed that the coefficient functions satisfy the conditions in (6.1.2) and, in addition, that

$$p(x) > 0 \quad \text{for almost all } x \in (a,b). \tag{6.9.1}$$

In Section 6.8 it was assumed that L is regular at the endpoints, in which case the minimal operator $T_{\min}$ is semibounded from below and the form in (6.8.4) and the mapping in (6.8.6) give a boundary pair compatible with the boundary triplet in (6.8.2). The interest is now in the construction of a corresponding form when L is not necessarily regular at the endpoints, which implies that (6.8.4) is not adequate anymore. The key to defining an appropriate form in the general case is the condition (6.9.1) together with the assumption that there are nonoscillatory solutions of the Sturm–Liouville equation $(L-\lambda_0)y = 0$ for some $\lambda_0 \in \mathbb{R}$. It will be shown that these assumptions imply that $T_{\min}$ is semibounded from below. The main result in this section is Theorem 6.9.6. In Section 6.10 the properties of the nonoscillatory solutions will be further investigated.

The differential equation $(L-\lambda_0)y = 0$ with $\lambda_0 \in \mathbb{R}$ is said to be *nonoscillatory* at an endpoint a or b, if it has a real solution u whose zeros do not accumulate at a or b, respectively. Otherwise, the equation $(L-\lambda_0)y = 0$ is called *oscillatory*. If this is the case, then the zeros of any nontrivial real solution do not accumulate at that endpoint; cf. Lemma 6.1.8. Furthermore, Lemma 6.1.8 also implies that in this case the equation $(L-\lambda_0')y = 0$ with $\lambda_0' < \lambda_0$ is nonoscillatory. If the equation $(L-\lambda_0)y = 0$ is nonoscillatory at both endpoints, then there exist real solutions of this equation which do not vanish in neighborhoods of a and b, respectively. As a preparation for the general case there is first a closer look at the situation where the equation $(L - \lambda_0)y = 0$ has a real solution which does not vanish on a subinterval.

Assume that ϕ is a real solution of $(L - \lambda_0)y = 0$ with $\lambda_0 \in \mathbb{R}$ which does not vanish on the subinterval $(\alpha, \beta) \subset (a, b)$. Use ϕ to introduce the first-order differential expression N_ϕ

$$N_\phi f = \sqrt{p}\phi \left(\frac{f}{\phi}\right)' \tag{6.9.2}$$

for all functions $f \in AC(\alpha, \beta)$. The differential expression N_ϕ in (6.9.2) generates an operator R_ϕ from $L^2_r(\alpha, \beta)$ to $L^2(\alpha, \beta)$ defined on the linear subspace

$$\mathfrak{D}_\phi = \left\{ f \in L^2_r(\alpha, \beta) : f \in AC(\alpha, \beta), \, N_\phi f \in L^2(\alpha, \beta) \right\}$$

by

$$R_\phi f = N_\phi f, \quad \operatorname{dom} R_\phi = \mathfrak{D}_\phi. \tag{6.9.3}$$

In the special case $q = 0$, $\lambda_0 = 0$, and $\phi = 1$, the corresponding differential expression in (6.9.2) reduces to $N_\phi f = \sqrt{p} f'$, which played an important role in the proof of Lemma 6.8.2. The next lemma is an analog of Lemma 6.8.2. The interval (α, β) below is a possibly unbounded subinterval of (a, b) for which $\alpha = a$ or $\beta = b$ is allowed.

Lemma 6.9.1. *Let ϕ be a real solution of $(L - \lambda_0)y = 0$ with $\lambda_0 \in \mathbb{R}$, which does not vanish on a subinterval $(\alpha, \beta) \subset (a, b)$. Then the operator R_ϕ from $L^2_r(\alpha, \beta)$ to $L^2(\alpha, \beta)$ defined in (6.9.3) is closed. The associated form $\mathfrak{r}_\phi$ in $L^2_r(\alpha, \beta)$ defined by*

$$\mathfrak{r}_\phi[f, g] = (R_\phi f, R_\phi g)_{L^2(\alpha, \beta)}, \quad f, g \in \operatorname{dom} \mathfrak{r}_\phi = \operatorname{dom} R_\phi = \mathfrak{D}_\phi,$$

is nonnegative and closed.

Proof. The proof will be given in three steps. Observe that the functions $r\phi^2$ and $p\phi^2$ satisfy the integrability conditions

$$r\phi^2 \in L^1_{\mathrm{loc}}(\alpha, \beta), \quad \frac{1}{p\phi^2} \in L^1_{\mathrm{loc}}(\alpha, \beta), \tag{6.9.4}$$

while $r\phi^2$ and $p\phi^2$ are positive almost everywhere on (α, β).

Step 1. First take $q = 0$, $\lambda_0 = 0$, and $\phi = 1$, in which case $N_\phi f = \sqrt{p} f'$. Then the associated operator R_ϕ from $L^2_r(\alpha, \beta)$ to $L^2(\alpha, \beta)$ is closed. To see this, let $f_n \in \mathfrak{D}_\phi$ be such that $f_n \to f$ in $L^2_r(\alpha, \beta)$ and $R_\phi f_n \to g$ in $L^2(\alpha, \beta)$. Then clearly for every compact subinterval $[\alpha', \beta'] \subset (\alpha, \beta)$ one has that $f_n \to f$ in $L^2_r(\alpha', \beta')$ and $R_\phi f_n \to g$ in $L^2(\alpha', \beta')$. By Lemma 6.8.2, this implies that on $[\alpha', \beta']$ one has $f \in AC[\alpha', \beta']$ and $g = \sqrt{p} f'$. Since $[\alpha', \beta']$ is arbitrary, one concludes that $f \in AC(\alpha, \beta)$ and $g = \sqrt{p} f' \in L^2(\alpha, \beta)$. In other words, $f \in \mathfrak{D}_\phi$ and $g = R_\phi f$.

Step 2. Introduce the Hilbert space $L^2_{r\phi^2}(\alpha, \beta)$ and the linear space $\mathfrak{D}'$ by

$$\mathfrak{D}' = \left\{ f \in L^2_{r\phi^2}(\alpha, \beta) : f \in AC(\alpha, \beta), \, \sqrt{p}\phi f' \in L^2(\alpha, \beta) \right\}.$$

Note that

$$f \in L^2_r(\alpha,\beta) \quad \Leftrightarrow \quad \frac{f}{\phi} \in L^2_{r\phi^2}(\alpha,\beta),$$

and

$$f \in \mathfrak{D}_\phi \quad \Leftrightarrow \quad \frac{f}{\phi} \in \mathfrak{D}'.$$

Step 3. The operator R_ϕ from $L^2_r(\alpha,\beta)$ to $L^2(\alpha,\beta)$ is closed. Indeed, let $f_n \in \mathfrak{D}_\phi$ be such that $f_n \to f$ in $L^2_r(\alpha,\beta)$ and $R_\phi f_n \to g$ in $L^2(\alpha,\beta)$. Then it is clear that

$$\frac{f_n}{\phi} \in \mathfrak{D}', \quad \frac{f_n}{\phi} \to \frac{f}{\phi} \quad \text{in} \quad L^2_{r\phi^2}(\alpha,\beta),$$

while also

$$\sqrt{p}\phi \left(\frac{f_n}{\phi}\right)' \to g \quad \text{in} \quad L^2(\alpha,\beta).$$

Then Step 1, applied with p replaced by $p\phi^2$ and r replaced by $r\phi^2$ (and taking into account the integrability conditions (6.9.4)) shows that

$$\frac{f}{\phi} \in \mathfrak{D}' \quad \text{and} \quad \sqrt{p}\phi\left(\frac{f}{\phi}\right)' = g.$$

Thus, by Step 2 one obtains $f \in \mathfrak{D}_\phi$ and $g = R_\phi f$. Therefore, the operator R_ϕ is closed, as claimed. As a consequence, the associated nonnegative form $\mathfrak{t}_\phi$ is closed; cf. Lemma 5.1.21. □

The differential expression N_ϕ appears naturally when one considers the following form of the *first Green identity* for the differential expression L.

Lemma 6.9.2. *Let ϕ be a real solution of $(L-\lambda_0)y = 0$ with $\lambda_0 \in \mathbb{R}$, which does not vanish on a subinterval $(\alpha,\beta) \subset (a,b)$. Assume that $f, pf', g \in AC(\alpha,\beta)$. Then for any compact subinterval $[\alpha',\beta'] \subset (\alpha,\beta)$ one has*

$$\begin{aligned}\int_{\alpha'}^{\beta'} (Lf)(x)\overline{g(x)}r(x)\,dx &= W_x(f,\phi)\left(\frac{\overline{g}}{\phi}\right)(x)\Big|_{\alpha'}^{\beta'} \\ &+ \int_{\alpha'}^{\beta'} (N_\phi f)(x)\overline{(N_\phi g)(x)}\,dx + \lambda_0 \int_{\alpha'}^{\beta'} f(x)\overline{g(x)}r(x)\,dx.\end{aligned} \tag{6.9.5}$$

Proof. Since $f, pf' \in AC(\alpha,\beta)$, it follows from the definition of the Wronskian that

$$\big((L-\lambda_0)f\big)(x)r(x)\phi(x) = \frac{d}{dx}W_x(f,\phi);$$

cf. (6.1.9). Multiply this identity by $\overline{g(x)/\phi(x)}$; then integration by parts gives for any compact subinterval $[\alpha', \beta'] \subset (\alpha, \beta)$ that

$$\begin{aligned}\int_{\alpha'}^{\beta'} &((L-\lambda_0)f)(x)\overline{g(x)}r(x)\,dx \\ &= \int_{\alpha'}^{\beta'} (W_x(f,\phi))'(x)\overline{\frac{g(x)}{\phi(x)}}\,dx \\ &= W_x(f,\phi)\left(\frac{\overline{g}}{\phi}\right)(x)\Big|_{\alpha'}^{\beta'} - \int_{\alpha'}^{\beta'} W_x(f,\phi)\frac{d}{dx}\left(\overline{\frac{g(x)}{\phi(x)}}\right)dx.\end{aligned}$$

Now observe that the Wronskian $W_x(f,\phi)$ can be written in terms of the differential expressions $N_\phi f$ in (6.9.2) as

$$W_x(f,\phi) = -p\phi^2\left(\frac{f}{\phi}\right)' = -\sqrt{p}\phi(N_\phi f),$$

and so

$$-\int_{\alpha'}^{\beta'} W_x(f,\phi)\frac{d}{dx}\left(\overline{\frac{g(x)}{\phi(x)}}\right)dx = \int_{\alpha'}^{\beta'} (N_\phi f)(x)\overline{(N_\phi g)(x)}\,dx.$$

Hence, the result in (6.9.5) follows. □

In the first Green formula (6.9.5) there is an interplay between the out-integrated parts and the integrals involving the differential expression N_ϕ. In fact, assume, in addition, that $f, Lf \in L^2_r(\alpha,\beta)$ and $g = f$ in (6.9.5), then the limits

$$\lim_{x\to\alpha} W_x(f,\phi)\left(\frac{\overline{f}}{\phi}\right)(x) \quad \text{or} \quad \lim_{x\to\beta} W_x(f,\phi)\left(\frac{\overline{f}}{\phi}\right)(x), \tag{6.9.6}$$

exist in $\mathbb{C}$ if and only if $N_\phi f \in L^2(\alpha, c)$ or $N_\phi f \in L^2(c,\beta)$, respectively, since the corresponding limit on the left-hand side and the limit of the third term on the right-hand side in (6.9.5) exist.

Note that the integral terms on the right-hand side of (6.9.5) make sense for $f, g \in AC(\alpha,\beta)$. In the next lemma these terms are rewritten using the form in (6.8.4).

Lemma 6.9.3. *Let ϕ be a real solution of $(L-\lambda_0)y = 0$ with $\lambda_0 \in \mathbb{R}$, which does not vanish on a subinterval $(\alpha,\beta) \subset (a,b)$. Assume that $f, g \in AC(\alpha,\beta)$. Then for any compact subinterval $[\alpha',\beta'] \subset (\alpha,\beta)$ one has*

$$\begin{aligned}\int_{\alpha'}^{\beta'} &(N_\phi f)(x)\overline{(N_\phi g)(x)}\,dx + \lambda_0\int_{\alpha'}^{\beta'} f(x)\overline{g(x)}r(x)\,dx \\ &= \int_{\alpha'}^{\beta'} \big((\sqrt{p}f')(x)\overline{(\sqrt{p}g')(x)} + q(x)f(x)\overline{g(x)}\big)\,dx \\ &\qquad - \frac{(p\phi')(\beta')}{\phi(\beta')}f(\beta')\overline{g(\beta')} + \frac{(p\phi')(\alpha')}{\phi(\alpha')}f(\alpha')\overline{g(\alpha')}.\end{aligned} \tag{6.9.7}$$

Proof. Let $f, g \in AC(\alpha, \beta)$. Then

$$N_\phi f \overline{N_\phi g} = pf'\overline{g}' - (f\overline{g})' \frac{p\phi'}{\phi} + f\overline{g}\, p \left(\frac{\phi'}{\phi}\right)^2. \tag{6.9.8}$$

Observe that

$$\begin{aligned} -\int_{\alpha'}^{\beta'} (f\overline{g})'(x) \frac{(p\phi')(x)}{\phi(x)}\, dx = \int_{\alpha'}^{\beta'} (f\overline{g})(x) \left(\frac{p\phi'}{\phi}\right)'(x)\, dx \\ - \frac{(p\phi')(\beta')}{\phi(\beta')} f(\beta')\overline{g(\beta')} + \frac{(p\phi')(\alpha')}{\phi(\alpha')} f(\alpha')\overline{g(\alpha')}, \end{aligned}$$

and that

$$\left(\frac{p\phi'}{\phi}\right)' + p\left(\frac{\phi'}{\phi}\right)^2 = \frac{(p\phi')'}{\phi} = q - \lambda_0 r.$$

Now integration of (6.9.8) leads to the desired result. □

After this excursion to the case of a nonvanishing solution of the equation $(L - \lambda_0)y = 0$ on an arbitrary open subinterval $(\alpha, \beta) \subset (a, b)$, one returns to the general situation. Let L be the Sturm–Liouville expression (6.1.1) on the interval (a, b) and let the coefficient functions satisfy the conditions (6.1.2) and (6.9.1). Assume that there exist $\lambda_0^a \in \mathbb{R}$ and $\lambda_0^b \in \mathbb{R}$ for which the equations $(L - \lambda_0^a)y = 0$ and $(L - \lambda_0^b)y = 0$ are nonoscillatory at a and b, respectively. Then Lemma 6.1.8 implies that for $\lambda_0 \leq \min\{\lambda_0^a, \lambda_0^b\}$ the equation $(L - \lambda_0)y = 0$ is nonoscillatory at a and b. Thus, this equation has real solutions ϕ_a and ϕ_b which do not vanish on (a, a_0) and on (b_0, b), respectively. Denote the corresponding first-order differential expressions by N_{ϕ_a} and N_{ϕ_b}; cf. (6.9.2). To define a form associated with the differential expression L by means of N_{ϕ_a} and N_{ϕ_b}, let $[c, d] \subset (a, b)$ be a compact interval such that

$$a < c < a_0 < b_0 < d < b. \tag{6.9.9}$$

Define the linear subspace $\mathfrak{D} \subset L^2_r(a, b)$ by

$$\begin{aligned} \mathfrak{D} = \bigl\{ f \in L^2_r(a, b) : f \in AC(a, b),\ \sqrt{p} f' \in L^2(c, d), \\ N_{\phi_a} f \in L^2(a, c),\ N_{\phi_b} f \in L^2(d, b) \bigr\}, \end{aligned} \tag{6.9.10}$$

and define the form $\mathfrak{t}$ by

$$\begin{aligned} \mathfrak{t}[f, g] = \int_a^c (N_{\phi_a} f)(x) \overline{(N_{\phi_a} g)(x)}\, dx + \int_d^b (N_{\phi_b} f)(x) \overline{(N_{\phi_b} g)(x)}\, dx \\ + \lambda_0 \int_a^c f(x)\overline{g(x)} r(x)\, dx + \lambda_0 \int_d^b f(x)\overline{g(x)} r(x)\, dx \\ + \int_c^d \bigl((\sqrt{p} f')(x) \overline{(\sqrt{p} g')(x)} + q(x) f(x) \overline{g(x)} \bigr) dx \\ + \frac{(p\phi_a')(c)}{\phi_a(c)} f(c)\overline{g(c)} - \frac{(p\phi_b')(d)}{\phi_b(d)} f(d)\overline{g(d)}, \end{aligned} \tag{6.9.11}$$

where $f, g \in \mathfrak{D}$. Observe that here

$$\frac{(p\phi_a')(c)}{\phi_a(c)} \in \mathbb{R} \quad \text{and} \quad \frac{(p\phi_b')(d)}{\phi_b(d)} \in \mathbb{R}. \tag{6.9.12}$$

The basic properties of $\mathfrak{t}$ and its domain $\mathfrak{D}$ in (6.9.10) and (6.9.11) will be shown in the following lemma. It is clear that $\mathfrak{t}$ and $\mathfrak{D}$ depend on the choice of the nonoscillatory solutions ϕ_a and ϕ_b. However, they do not depend on the particular choice of the points c and d in (6.9.9).

Lemma 6.9.4. *Assume that ϕ_a and ϕ_b are real nonoscillatory solutions of the equation $(L-\lambda_0)y = 0$ which do not vanish on (a, a_0) and on (b_0, b), respectively. Then the form $\mathfrak{t}$ and its domain $\mathfrak{D}$ in (6.9.10) and (6.9.11) do not depend on the particular choice of the points $c < d$ in (6.9.9). Moreover, the form $\mathfrak{t}$ is closed and bounded from below in $L^2_r(a, b)$.*

Proof. *Step* 1. First one shows that $\mathfrak{t}$ and $\mathfrak{D}$ do not depend on the particular choice of the points $c < d$ in (a, b). Here only the case where the point d is replaced by some point d' with $b_0 < d' < b$ is considered. For the sake of definiteness assume that $d < d' < b$.

To see that in the definition (6.9.10) of $\mathfrak{D}$ the point d may be replaced by the point d' observe that for $f \in AC(a, b)$ one has on (d, b):

$$N_{\phi_b} f = \sqrt{p}\phi_b \left(\frac{f}{\phi_b}\right)' = \sqrt{p}f' - \frac{1}{\sqrt{p}}\left(f\frac{p\phi_b'}{\phi_b}\right). \tag{6.9.13}$$

Consider the compact interval $K = [d, d']$ and recall that the nonoscillatory solution ϕ_b does not vanish on K. Then the last term on the right-hand side of (6.9.13) belongs to $L^2(K)$ because $1/\sqrt{p} \in L^2(K)$, while the remaining factor is bounded because $f \in AC(a, b)$, $p\phi_b' \in AC(a, b)$, and $\phi_b \in AC(a, b)$. Hence, $N_{\phi_b} f \in L^2(K)$ if and only if $\sqrt{p}f' \in L^2(K)$. From this observation it follows directly that $\mathfrak{D}$ does not depend on the particular choice of the point d.

Next consider the right-hand side of (6.9.11) with the point d, subtract from it the right-hand side of (6.9.11) with d replaced by the point d', and observe that

$$\begin{aligned} &\int_d^{d'} (N_{\phi_b} f)(x)\overline{(N_{\phi_b} g)(x)}\, dx + \lambda_0 \int_d^{d'} f(x)\overline{g(x)} r(x)\, dx \\ &\quad = \int_d^{d'} \big((\sqrt{p}f')(x)\overline{(\sqrt{p}g')(x)} + q(x) f(x)\overline{g(x)} \big) dx \\ &\qquad + \frac{(p\phi_b')(d)}{\phi_b(d)} f(d)\overline{g(d)} - \frac{(p\phi_b')(d')}{\phi_b(d')} f(d')\overline{g(d')}, \end{aligned}$$

which follows from (6.9.7) in Lemma 6.9.3 with the nonvanishing solution ϕ_b on the interval $[d, d']$. This shows that $\mathfrak{t}$ in (6.9.11) does not depend on the choice of the point $d \in (b_0, b)$.

Step 2. Next as a preparation for the rest of the proof one defines forms on each of the disjoint intervals (a,c), (c,d), and (d,b). The properties of these forms are given in Theorem 6.8.5 and Lemma 6.9.1. In the next steps it will be shown how the information about each of the separate intervals can be pieced together for the form $\mathfrak{t}$. For the interval (a,c) define the form $\mathfrak{t}_{(a,c)}$ by

$$\mathfrak{t}_{(a,c)}[f,g] = \int_a^c (N_{\phi_a} f)(x)\overline{(N_{\phi_a} g)(x)}\, dx + \lambda_0 \int_a^c f(x)\overline{g(x)} r(x)\, dx,$$

on the domain

$$\mathfrak{D}_{(a,c)} = \{ f \in L^2_r(a,c) : f \in AC(a,c),\ N_{\phi_a} f \in L^2(a,c) \},$$

and, likewise, for the interval (d,b) define the form $\mathfrak{t}_{(d,b)}$ by

$$\mathfrak{t}_{(d,b)}[f,g] = \int_d^b (N_{\phi_b} f)(x)\overline{(N_{\phi_b} g)(x)}\, dx + \lambda_0 \int_d^b f(x)\overline{g(x)} r(x)\, dx,$$

on the domain

$$\mathfrak{D}_{(d,b)} = \{ f \in L^2_r(d,b) : f \in AC(d,b),\ N_{\phi_b} f \in L^2(d,b) \}.$$

The forms $\mathfrak{t}_{(a,c)}$ and $\mathfrak{t}_{(c,d)}$ are clearly bounded from below:

$$\mathfrak{t}_{(a,c)}[f] \geq \lambda_0 \int_a^c |f(x)|^2\, r(x)\, dx, \quad f \in \mathfrak{D}_{(a,c)},$$

and

$$\mathfrak{t}_{(d,b)}[f] \geq \lambda_0 \int_d^b |f(x)|^2\, r(x)\, dx, \quad f \in \mathfrak{D}_{(d,b)}.$$

It is a consequence of Lemma 6.9.1 that the forms $\mathfrak{t}_{(a,c)}$ and $\mathfrak{t}_{(d,b)}$ are closed. For the interval (c,d) define the form $\mathfrak{t}_{(c,d)}$ by

$$\begin{aligned}\mathfrak{t}_{(c,d)}[f,g] = \int_c^d &\big((\sqrt{p} f')(x)\overline{(\sqrt{p} g')(x)} + q(x) f(x)\overline{g(x)} \big)\, dx \\ &+ \frac{(p\phi_a')(c)}{\phi_a(c)} f(c)\overline{g(c)} - \frac{(p\phi_b')(d)}{\phi_b(d)} f(d)\overline{g(d)},\end{aligned}$$

on the domain

$$\mathfrak{D}_{(c,d)} = \{ f \in L^2_r(c,d) : f \in AC(c,d),\ \sqrt{p} f' \in L^2(c,d) \}.$$

It is a consequence of (6.9.12) and Theorem 6.8.5 with

$$\Theta = \begin{pmatrix} \frac{(p\phi_a')(c)}{\phi_a(c)} & 0 \\ 0 & -\frac{(p\phi_b')(d)}{\phi_b(d)} \end{pmatrix},$$

that in $L^2_r(c,d)$ the form $\mathfrak{t}_{(c,d)}$ is closed and bounded from below:

$$\mathfrak{t}_{(c,d)}[f] \geq C_{(c,d)} \int_c^d |f(x)|^2 \, r(x) \, dx, \quad f \in \mathfrak{D}_{(c,d)}.$$

Step 3. It will be shown that the form $\mathfrak{t}$ is bounded from below in $L^2_r(a,b)$. Let $f \in \mathfrak{D}$. Then the restrictions of f to the intervals (a,c), (c,d), and (d,b) belong to $\mathfrak{D}_{(a,c)}$, $\mathfrak{D}_{(c,d)}$, and $\mathfrak{D}_{(d,b)}$, respectively, and

$$\mathfrak{t}[f] = \mathfrak{t}_{(a,c)}[f] + \mathfrak{t}_{(c,d)}[f] + \mathfrak{t}_{(d,b)}[f].$$

This decomposition shows that

$$\mathfrak{t}[f] \geq \lambda_0 \left(\int_a^c + \int_d^b \right) |f(x)|^2 \, r(x) \, dx + C_{(c,d)} \int_c^d |f(x)|^2 \, r(x) \, dx.$$

Hence, it follows that

$$\mathfrak{t}[f] \geq M \int_a^b |f(x)|^2 \, r(x) \, dx, \quad M = \min \{\lambda_0, C_{(c,d)}\},$$

for all $f \in \mathfrak{D}$. Thus, the form $\mathfrak{t}$ is bounded from below in $L^2_r(a,b)$.

Step 4. It will be shown that the form $\mathfrak{t}$ is closed in $L^2_r(a,b)$. For this, let $f_n \in \mathfrak{D}$ be a sequence such that $f_n \to f$ in $L^2_r(a,b)$ and $\mathfrak{t}[f_n - f_m] \to 0$. One needs to establish that $f \in \mathfrak{D} = \operatorname{dom} \mathfrak{t}$ and $\mathfrak{t}[f_n - f] \to 0$. It is clear from the assumption that the restrictions to the intervals (a,c), (c,d), and (d,b) satisfy

$$\begin{aligned}
f_n \to f \quad &\text{in} \quad L^2_r(a,c) \quad \text{and} \quad \mathfrak{t}_{(a,c)}[f_n - f_m] \to 0,\\
f_n \to f \quad &\text{in} \quad L^2_r(c,d) \quad \text{and} \quad \mathfrak{t}_{(c,d)}[f_n - f_m] \to 0,\\
f_n \to f \quad &\text{in} \quad L^2_r(d,b) \quad \text{and} \quad \mathfrak{t}_{(d,b)}[f_n - f_m] \to 0,
\end{aligned}$$

respectively. It follows from Lemma 6.9.1 for the intervals (a,c) and (d,b), and from Theorem 6.8.5 for the interval (c,d), that

$$\begin{aligned}
&f \in AC(a,c), \quad N_{\phi_a} f \in L^2(a,c), \quad N_{\phi_a} f_n \to N_{\phi_a} f,\\
&f \in AC(c,d), \quad \sqrt{p} f' \in L^2(c,d), \quad \mathfrak{t}_{(c,d)}[f_n - f] \to 0,\\
&f \in AC(d,b), \quad N_{\phi_b} f \in L^2(d,b), \quad N_{\phi_b} f_n \to N_{\phi_b} f,
\end{aligned}$$

respectively. It remains to show that $f \in AC(a,b)$.

Recall from Step 1 that the definition of $\mathfrak{t}$ is independent of the choice of interval (c,d). By enlarging (c,d) to (c',d') with $c' < c$ and $d < d'$ one concludes that also $f \in AC(c',d')$. Hence, there is absolute continuity across the points c and d. Thus, $f \in AC(a,b)$ and consequently $f \in \operatorname{dom} \mathfrak{t}$ while $\mathfrak{t}[f_n - f] \to 0$. One concludes that the form $\mathfrak{t}$ is closed in $L^2_r(a,b)$. □

The Green formula in the following lemma will give the connection between the differential expression L and the form $\mathfrak{t}$ defined in (6.9.10) and (6.9.11).

Lemma 6.9.5. *Assume that ϕ_a and ϕ_b are real nonoscillatory solutions of the equation $(L-\lambda_0)y=0$ which do not vanish on (a,a_0) and on (b_0,b), respectively. Let $[c,d]$ be as in (6.9.9) and assume that $f, pf', g \in AC(a,b)$. Then for any choice of a' and b' with $a<a'<c$ and $d<b'<b$ one has*

$$\begin{aligned}
&\int_{a'}^{b'} (Lf)(x)\overline{g(x)}r(x)\,dx \\
&\quad = W_{b'}(f,\phi_b)\left(\frac{\bar g}{\phi_b}\right)(b') - W_{a'}(f,\phi_a)\left(\frac{\bar g}{\phi_a}\right)(a') \\
&\qquad + \int_{a'}^{c} (N_{\phi_a}f)(x)\overline{(N_{\phi_a}g)(x)}\,dx + \int_{d}^{b'} (N_{\phi_b}f)(x)\overline{(N_{\phi_b}g)(x)}\,dx \\
&\qquad + \lambda_0 \int_{a'}^{c} f(x)\overline{g(x)}r(x)\,dx + \lambda_0 \int_{d}^{b'} f(x)\overline{g(x)}r(x)\,dx \\
&\qquad + \int_{c}^{d} \bigl((\sqrt{p}f')(x)\overline{(\sqrt{p}g')(x)} + q(x)f(x)\overline{g(x)}\bigr)dx \\
&\qquad + \frac{(p\phi_a')(c)}{\phi_a(c)} f(c)\overline{g(c)} - \frac{(p\phi_b')(d)}{\phi_b(d)} f(d)\overline{g(d)}.
\end{aligned} \tag{6.9.14}$$

Proof. Split the integral on the left-hand side of (6.9.14) into three integrals over the subintervals (a',c), $[c,d]$, and (d,b'), and evaluate each integral by partial integration. Recall that the integral over the compact interval $[c,d]$ gives the usual formula

$$\begin{aligned}
\int_{c}^{d} (Lf)(x)\overline{g(x)}r(x)\,dx &= -(pf')(x)\overline{g(x)}\Big|_{c}^{d} \\
&\quad + \int_{c}^{d} \bigl((\sqrt{p}f')(x)\overline{(\sqrt{p}g')(x)} + q(x)f(x)\overline{g(x)}\bigr)dx.
\end{aligned}$$

As suggested by Lemma 6.9.2, the integral over the interval (a',c) can be written as

$$\begin{aligned}
\int_{a'}^{c} (Lf)(x)\overline{g(x)}r(x)\,dx &= W_x(f,\phi_a)\left(\frac{\bar g}{\phi_a}\right)(x)\Big|_{a'}^{c} \\
&\quad + \int_{a'}^{c} (N_{\phi_a}f)(x)\overline{(N_{\phi_a}g)(x)}\,dx + \lambda_0 \int_{a'}^{c} f(x)\overline{g(x)}r(x)\,dx
\end{aligned}$$

and the integral over the interval (d,b') can be written as

$$\begin{aligned}
\int_{d}^{b'} (Lf)(x)\overline{g(x)}r(x)\,dx &= W_x(f,\phi_b)\left(\frac{\bar g}{\phi_b}\right)(x)\Big|_{d}^{b'} \\
&\quad + \int_{d}^{b'} (N_{\phi_b}f)(x)\overline{(N_{\phi_b}g)(x)} + \lambda_0 \int_{d}^{b'} f(x)\overline{g(x)}r(x)\,dx.
\end{aligned}$$

Combining the out-integrated parts one obtains

$$\begin{aligned} W_x(f,\phi_a)\left(\frac{\bar g}{\phi_a}\right)(x)\Big|_{a'}^{c} - (pf')(x)\overline{g(x)}\Big|_c^d + W_x(f,\phi_b)\left(\frac{\bar g}{\phi_b}\right)(x)\Big|_d^{b'} \\ = W_{b'}(f,\phi_b)\left(\frac{\bar g}{\phi_b}\right)(b') - W_{a'}(f,\phi_a)\left(\frac{\bar g}{\phi_a}\right)(a') \\ + \frac{(p\phi_a')(c)}{\phi_a(c)} f(c)\overline{g(c)} - \frac{(p\phi_b')(d)}{\phi_b(d)} f(d)\overline{g(d)}, \end{aligned}$$

which yields the identity (6.9.14). □

A combination of Lemma 6.9.4 and Lemma 6.9.5 leads to the following theorem, which describes the interplay between $T_{\max}$ and the form $\mathfrak{t}$.

Theorem 6.9.6. *Assume the conditions in (6.1.2) and (6.9.1). Let ϕ_a and ϕ_b be real nonoscillatory solutions of $(L-\lambda_0)y=0$ for some $\lambda_0\in\mathbb{R}$ which do not vanish on (a,a_0) and on (b_0,b), respectively, and let $[c,d]$ be as in (6.9.9). Let $\mathfrak{t}$ and $\mathfrak{D}=\operatorname{dom}\mathfrak{t}$ be defined as in (6.9.10) and (6.9.11), so that $\mathfrak{t}$ is a closed semibounded form. Assume that $f\in\operatorname{dom}T_{\max}\cap\mathfrak{D}$ and $g\in\mathfrak{D}$. Then*

$$\begin{aligned} (T_{\max}f,g)_{L^2_r(a,b)} = \mathfrak{t}[f,g] + \lim_{b'\to b} W_{b'}(f,\phi_b)\left(\frac{\bar g}{\phi_b}\right)(b') \\ - \lim_{a'\to a} W_{a'}(f,\phi_a)\left(\frac{\bar g}{\phi_a}\right)(a'), \end{aligned} \tag{6.9.15}$$

where each of the limits exists in $\mathbb{C}$. Furthermore, the form $\mathfrak{t}$ is densely defined and $T_{\min}\subset S_1$, where S_1 is the semibounded self-adjoint operator corresponding to $\mathfrak{t}$ and, in fact,

$$(T_{\min}f,g)_{L^2_r(a,b)} = \mathfrak{t}[f,g], \quad f\in\operatorname{dom}T_{\min}\subset\mathfrak{D},\ g\in\mathfrak{D}. \tag{6.9.16}$$

In particular, $T_{\min}$ is bounded from below and the form $\mathfrak{t}_{S_{\mathrm{F}}}$ corresponding to the Friedrichs extension S_{F} of $T_{\min}$ satisfies

$$\mathfrak{t}_{S_{\mathrm{F}}}\subset\mathfrak{t}. \tag{6.9.17}$$

Proof. The assumptions $f\in\operatorname{dom}T_{\max}\cap\mathfrak{D}$ and $g\in\mathfrak{D}$ show that in (6.9.14) the term on the left-hand side and the integrals on the right-hand side which involve a' and b' have limits when $a'\to a$ or $b'\to b$. Therefore, the limits of

$$W_{a'}(f,\phi_a)\left(\frac{\bar g}{\phi_a}\right)(a') \quad\text{and}\quad W_{b'}(f,\phi_b)\left(\frac{\bar g}{\phi_b}\right)(b')$$

exist when $a'\to a$ or $b'\to b$. The definitions in (6.9.10) and (6.9.11) then lead to the identity (6.9.15).

To show that $\operatorname{dom} T_{\min} \subset \mathfrak{D}$ and (6.9.16) holds, take first $f \in \operatorname{dom} T_{\min}$ with compact support. In other words, $f \in \operatorname{dom} T_0$, where T_0 denotes the preminimal operator. Then $f, pf' \in AC(a,b)$ and hence $\sqrt{p}f' \in L^2(c,d)$ is clear. It is claimed that $N_{\phi_a} f \in L^2(a,c)$ and $N_{\phi_b} f \in L^2(d,b)$. Only the last inclusion will be shown. Consider the identity

$$N_{\phi_b} f = \sqrt{p}\phi_b \left(\frac{f}{\phi_b}\right)' = \frac{1}{\sqrt{p}}\left[pf' - p\phi_b' \frac{f}{\phi_b}\right],$$

and note that the functions $f, pf' \in AC(a,b)$ have compact support, $\phi_b \in AC(a,b)$ does not vanish on $[d,b)$, and $p\phi_b' \in AC(a,b)$. Hence, $pf' - p\phi_b'(f/\phi_b)$ vanishes in a neighborhood of b and is bounded on $[d,b)$. Since $1/p \in L^1_{\mathrm{loc}}(a,b)$ by the assumption (6.1.2) it follows that $N_{\phi_b} f \in L^2(d,b)$. Therefore,

$$\operatorname{dom} T_0 \subset \mathfrak{D}$$

and, in particular, $\mathfrak{t}$ is densely defined; cf. Theorem 6.2.1. For $f \in \operatorname{dom} T_0$ and $g \in \mathfrak{D}$ the identity

$$(T_0 f, g)_{L^2_r(a,b)} = \mathfrak{t}[f,g] \tag{6.9.18}$$

follows immediately from (6.9.15). By Lemma 6.9.4, the form $\mathfrak{t}$ is closed and bounded from below. Hence, by the first representation theorem, there exists a self-adjoint operator S_1 in $L^2_r(a,b)$ which is bounded from below such that

$$(S_1 f, g)_{L^2_r(a,b)} = \mathfrak{t}[f,g]$$

holds for all $f \in \operatorname{dom} T \subset \mathfrak{D}$ and $g \in \mathfrak{D}$. It follows from (6.9.18) and Theorem 5.1.18 that $T_0 \subset S_1$, and hence also $\overline{T_0} = T_{\min} \subset S_1$. In particular, $T_{\min}$ is bounded from below, one has $\operatorname{dom} T_{\min} \subset \mathfrak{D}$, and (6.9.16) holds.

In order to verify (6.9.17) consider the form $\mathfrak{t}_0[f,g] = (T_0 f, g)_{L^2_r(a,b)}$ defined on $\operatorname{dom} T_0$. Then one has $\mathfrak{t}_0 \subset \mathfrak{t}$ by (6.9.16). Since $\mathfrak{t}$ is closed, the closure of $\mathfrak{t}_0$, which coincides with the form $\mathfrak{t}_{S_F}$ corresponding to the Friedrichs extension S_F in Definition 5.3.2, is contained in $\mathfrak{t}$. This leads to (6.9.17). □

Note that the existence of nonoscillatory solutions implies the semiboundedness of $T_{\min}$. The following result will lead to a converse statement.

Lemma 6.9.7. *Assume the conditions in (6.1.2) and (6.9.1) and assume that $T_{\min}$ is bounded from below. Let u be a real solution of the equation $(L - \lambda_0)y = 0$ with $\lambda_0 \in \mathbb{R}$ and assume that u has at least two zeros in (a,b). Then $\lambda_0 \geq m(T_{\min})$. Consequently, for $\lambda_0 < m(T_{\min})$ any real solution of $(L - \lambda_0)y = 0$ has at most one zero on (a,b).*

Proof. Let $\lambda_0 \in \mathbb{R}$ and assume that u is a real solution of the equation $(L - \lambda_0)y = 0$ which has two zeros $\alpha < \beta$. Denote the maximal and minimal Sturm–Liouville operator in $L^2_r(\alpha,\beta)$ by $T_{\max}(\alpha,\beta)$ and $T_{\min}(\alpha,\beta)$. Likewise, the preminimal operator in $L^2_r(\alpha,\beta)$ is denoted by $T_0(\alpha,\beta)$. Denote the Friedrichs extension of

$T_{\min}(\alpha, \beta)$ by $S_F(\alpha, \beta)$. By Theorem 6.8.5 (iii), it is clear that the restriction of u to $[\alpha, \beta]$ is an eigenelement of $S_F(\alpha, \beta)$ with eigenvalue λ_0, and therefore $\lambda_0 \geq m(S_F(\alpha, \beta))$. Now recall from Lemma 5.3.1 and Definition 5.3.2 that

$$m(S_F(\alpha, \beta)) = m(T_{\min}(\alpha, \beta)) = m(T_0(\alpha, \beta)),$$

and obviously

$$m(T_0(\alpha, \beta)) \geq m(T_0) = m(T_{\min}).$$

Therefore, $\lambda_0 \geq m(T_{\min})$. □

Theorem 6.9.8. *Assume the conditions in (6.1.2) and (6.9.1). Then the operator $T_{\min}$ is bounded from below if and only if there exist $\lambda_a \in \mathbb{R}$ and $\lambda_b \in \mathbb{R}$ such that $(L - \lambda_a)y = 0$ and $(L - \lambda_b)y = 0$ are nonoscillatory at a and b, respectively.*

Proof. By Theorem 6.9.6, the existence of nonoscillatory solutions of $(L-\lambda_0)y = 0$ for some $\lambda_0 \in \mathbb{R}$ implies that $T_{\min}$ is bounded from below. Conversely, if $T_{\min}$ is bounded from below, then according to Lemma 6.9.7 the equation $(L - \lambda_0)y = 0$ is nonoscillatory at both endpoints for any $\lambda_0 < m(T_{\min})$. □

Remark 6.9.9. If one of the endpoints is regular, then the appearance of the form $\mathfrak{t}$ in (6.9.10) and (6.9.11) becomes somewhat simpler. Assume for instance that the endpoint a is regular and that ϕ_b is a real nonoscillatory solution of $(L-\lambda_0)y = 0$ that does not vanish on an open interval (b_0, b). Let $b_0 < d < b$ and define the linear space $\mathfrak{D}$ by

$$\mathfrak{D} = \left\{ f \in L^2_r(a, b) : f \in AC(a, b),\ \sqrt{p}f' \in L^2(a, d),\ N_{\phi_b} f \in L^2(d, b) \right\}, \tag{6.9.19}$$

and the form $\mathfrak{t}$ by

$$\begin{aligned} \mathfrak{t}[f, g] = &\int_d^b (N_{\phi_b} f)(x)\overline{(N_{\phi_b} g)(x)}\, dx + \lambda_0 \int_d^b f(x)\overline{g(x)} r(x)\, dx \\ &+ \int_a^d \big((\sqrt{p}f')(x)\overline{(\sqrt{p}g')}(x) + q(x)f(x)\overline{g(x)}\,\big)dx \\ &- \frac{(p\phi_b')(d)}{\phi_b(d)} f(d)\overline{g(d)}, \end{aligned} \tag{6.9.20}$$

where $f, g \in \mathfrak{D}$. The form $\mathfrak{t}$ and its domain $\mathfrak{D}$ in (6.9.19) and (6.9.20) do not depend on the particular choice of the point d in (b_0, b). For the corresponding Green formula, assume that $f, pf', g \in AC(a, b)$ and that $f, Lf \in L^2_r(a, b)$. Then for any choice of b' with $a < d < b' < b$ one has

$$\begin{aligned}
&\int_a^{b'} (Lf)(x)\overline{g(x)}r(x)\,dx \\
&\quad = W_{b'}(f,\phi_b)\left(\frac{\bar g}{\phi_b}\right)(b') + (pf')(a)\overline{g(a)} \\
&\qquad + \int_d^{b'} (N_{\phi_b}f)(x)\overline{(N_{\phi_b}g)(x)}\,dx + \lambda_0\int_d^{b'} f(x)\overline{g(x)}r(x)\,dx \\
&\qquad + \int_a^d \Big((\sqrt{p}f')(x)\overline{(\sqrt{p}g')}(x) + q(x)f(x)\overline{g(x)}\Big)\,dx \\
&\qquad - \frac{(p\phi_b')(d)}{\phi_b(d)}f(d)\overline{g(d)}.
\end{aligned}$$

Let $\mathfrak{t}$ and $\mathfrak{D}$ be as in (6.9.19) and (6.9.20), and assume that $f \in \operatorname{dom} T_{\max} \cap \mathfrak{D}$ and $g \in \mathfrak{D}$. Then the formula (6.9.15) becomes

$$(T_{\max} f, g) = \mathfrak{t}[f,g] + (pf')(a)\overline{g(a)} + \lim_{b' \to b} W_{b'}(f,\phi_b)\left(\frac{\bar g}{\phi_b}\right)(b'), \tag{6.9.21}$$

where the limit exists in $\mathbb{C}$.

6.10 Principal and nonprincipal solutions of Sturm–Liouville equations

This section contains a further treatment of the nonoscillatory solutions of a Sturm–Liouville equation. The forms in Section 6.9 were defined by means of solutions of $(L - \lambda_0)y = 0$ which are nonoscillatory near an endpoint. The nonoscillatory solutions ϕ will now be further classified as *nonprincipal* and *principal*, depending on the square-integrability of $(p\phi^2)^{-1}$ near an endpoint. The main result in this section is Theorem 6.10.9 which concerns the square-integrability of $N_\phi f$ near an endpoint when f is absolutely continuous, and it will be obtained by means of Lemma 6.10.1 and Theorem 6.10.4 below. In the following Section 6.11 and Section 6.12 there is a detailed treatment of the cases where the endpoint a and the endpoint b are either in the limit-circle case or in the limit-point case, respectively, and where the corresponding form in (6.9.10) and (6.9.11) will be defined in terms of nonprincipal and principal solutions.

The following auxiliary results concern a measurable function $P : (a,b) \to \mathbb{R}$ which satisfies

$$1/P \in L^1_{\mathrm{loc}}(a,b) \quad \text{and} \quad P(x) > 0 \text{ for almost all } x \in (a,b). \tag{6.10.1}$$

They are stated with respect to the endpoint a of the interval (a,b); the corresponding results for the other endpoint are clear.

Lemma 6.10.1. *Let the function P satisfy (6.10.1) and let $a < \alpha < b$. Assume that*

$$\varphi \in AC(a,\alpha) \quad \text{and} \quad \sqrt{P}\varphi' \in L^2(a,\alpha).$$

Then the following statements hold:

(i) *One has*

$$\lim_{x\to a} \frac{\varphi(x)^2}{\int_x^\alpha \frac{1}{P(t)}\,dt} \in \mathbb{C} \quad \text{and} \quad \int_a^{\alpha'} \frac{|\varphi(t)|^2}{P(t)(\int_t^\alpha \frac{1}{P(s)}\,ds)^2}\,dt < \infty$$

for any $a < \alpha' < \alpha$. Moreover,

$$\int_a^\alpha \frac{1}{P(t)}\,dt = \infty \quad \Rightarrow \quad \lim_{x\to a} \frac{\varphi(x)^2}{\int_x^\alpha \frac{1}{P(t)}\,dt} = 0.$$

(ii) *Assume that $\int_a^\alpha \frac{1}{P(t)}\,dt < \infty$. Then $\lim_{x\to a}\varphi(x)$ exists in $\mathbb{C}$ and one has*

$$\lim_{x\to a}\varphi(x) = 0 \quad \Rightarrow \quad \int_a^\alpha \frac{|\varphi(t)|^2}{P(t)(\int_a^t \frac{1}{P(s)}\,ds)^2}\,dt < \infty.$$

Moreover,

$$\lim_{x\to a}\varphi(x) = 0 \quad \Leftrightarrow \quad \lim_{x\to a} \frac{\varphi(x)^2}{\int_a^x \frac{1}{P(t)}\,dt} = 0 \quad \Leftrightarrow \quad \lim_{x\to a} \frac{\varphi(x)^2}{\int_a^x \frac{1}{P(t)}\,dt} \in \mathbb{C}.$$

Proof. In the following proof it will be assumed without loss of generality that the function φ is real.

(i) As abbreviation use the notation $H(t) = \int_t^\alpha \frac{1}{P(s)}\,ds$, $a < t < \alpha$. Then the identities

$$PH\left(\left(\frac{\varphi}{\sqrt{H}}\right)'\right)^2 = P\left(\varphi' + \frac{1}{2}\frac{\varphi}{PH}\right)^2 = P(\varphi')^2 + \frac{\varphi\varphi'}{H} + \frac{1}{4}\frac{\varphi^2}{PH^2}$$

and

$$\frac{1}{2}\left(\frac{\varphi^2}{H}\right)' = \frac{\varphi\varphi'}{H} + \frac{1}{2}\frac{\varphi^2}{PH^2}$$

yield

$$PH\left(\left(\frac{\varphi}{\sqrt{H}}\right)'\right)^2 = P(\varphi')^2 - \frac{1}{4}\frac{\varphi^2}{PH^2} + \frac{1}{2}\left(\frac{\varphi^2}{H}\right)'. \tag{6.10.2}$$

In particular, one sees that

$$P(\varphi')^2 \geq \frac{1}{4}\frac{\varphi^2}{PH^2} - \frac{1}{2}\left(\frac{\varphi^2}{H}\right)'.$$

Integrate this last inequality over $[x, \alpha']$ with $a < x < \alpha' < \alpha$. Then

$$\begin{aligned}\int_x^{\alpha'} P(t)\varphi'(t)^2\,dt &\geq \frac{1}{4}\int_x^{\alpha'} \frac{\varphi(t)^2}{P(t)H(t)^2}\,dt + \frac{1}{2}\frac{\varphi(x)^2}{H(x)} - \frac{1}{2}\frac{\varphi(\alpha')^2}{H(\alpha')} \\ &\geq \frac{1}{4}\int_x^{\alpha'} \frac{\varphi(t)^2}{P(t)H(t)^2}\,dt - \frac{1}{2}\frac{\varphi(\alpha')^2}{H(\alpha')}.\end{aligned}$$

Since $\sqrt{P}\varphi'$ is square-integrable, taking the limit $x \to a$ shows the integrability result in (i). Hence, both terms on the right-hand side of the estimate

$$PH\left(\left(\frac{\varphi}{\sqrt{H}}\right)'\right)^2 = P\left(\varphi' + \frac{1}{2}\frac{\varphi}{PH}\right)^2 \leq 2P(\varphi')^2 + \frac{1}{2}\frac{\varphi^2}{PH^2}$$

are integrable on (a, α'). Therefore, in view of (6.10.2), one sees that the limit $\lim_{x\to a} \frac{\varphi(x)^2}{H(x)}$ exists in $\mathbb{C}$. Thus, the first two statements in (i) have been proved.

In order to prove the last statement in (i) assume that $\int_a^\alpha \frac{1}{P(t)}\,dt = \infty$. Then

$$\begin{aligned}\int_a^{\alpha'} \frac{1}{P(t)H(t)}\,dt &= \lim_{x\to a}\left(\int_x^{\alpha'} \frac{1}{P(t)H(t)}\,dt\right) \\ &= \lim_{x\to a}\left(\log H(x) - \log H(\alpha')\right) = \infty;\end{aligned} \tag{6.10.3}$$

note that $(\log H)' = -1/PH$. In view of

$$\int_a^{\alpha'} \frac{1}{P(t)H(t)}\frac{\varphi(t)^2}{H(t)}\,dt = \int_a^{\alpha'} \frac{\varphi(t)^2}{P(t)H(t)^2}\,dt < \infty,$$

it follows from (6.10.3) that actually $\lim_{x\to a} \frac{\varphi(x)^2}{H(x)} = 0$, which is the limit result in assertion (i).

(ii) Assume that $\int_a^\alpha \frac{1}{P(t)}\,dt < \infty$. First observe that for $a < y < x < \alpha$ the identity

$$\varphi(x) - \varphi(y) = \int_y^x \sqrt{P(t)}\varphi'(t)\,\frac{1}{\sqrt{P(t)}}\,dt$$

together with the Cauchy–Schwarz inequality gives

$$|\varphi(x) - \varphi(y)|^2 \leq \left|\int_y^x \frac{1}{P(t)}\,dt\right|\left|\int_y^x P(t)\varphi'(t)^2\,dt\right|.$$

Due to the assumptions it is now clear that $\lim_{y\to a}\varphi(y)$ exists in $\mathbb{C}$. Moreover, if, in particular, $\lim_{y\to a}\varphi(y) = 0$, then the above inequality shows that

$$|\varphi(x)|^2 \leq \left|\int_a^x \frac{1}{P(t)}\,dt\right|\left|\int_a^x P(t)\varphi'(t)^2\,dt\right|.$$

Hence, the implications

$$\lim_{x\to a}\varphi(x)=0 \quad\Rightarrow\quad \lim_{x\to a}\frac{\varphi(x)^2}{\int_a^x \frac{1}{P(t)}\,dt}=0 \quad\Rightarrow\quad \lim_{x\to a}\frac{\varphi(x)^2}{\int_a^x \frac{1}{P(t)}\,dt}\in\mathbb{C}$$

are clear. Furthermore, since $\lim_{y\to a}\varphi(y)$ exists in $\mathbb{C}$,

$$\lim_{x\to a}\frac{\varphi(x)^2}{\int_a^x \frac{1}{P(t)}\,dt}\in\mathbb{C} \quad\Rightarrow\quad \lim_{x\to a}\varphi(x)=0.$$

This proves the equivalences in the last statement of (ii).

For the rest of the proof of (ii) use as abbreviation the notation

$$G(t)=\int_a^t \frac{1}{P(s)}\,ds, \qquad a<t<\alpha.$$

Note that the integral is well defined due to the assumption $\int_a^\alpha \frac{1}{P(t)}\,dt<\infty$. Then the identities

$$PG\left(\left(\frac{\varphi}{\sqrt{G}}\right)'\right)^2=P\left(\varphi'-\frac{1}{2}\frac{\varphi}{PG}\right)^2=P(\varphi')^2-\frac{\varphi\varphi'}{G}+\frac{1}{4}\frac{\varphi^2}{PG^2}$$

and

$$\frac{1}{2}\left(\frac{\varphi^2}{G}\right)'=\frac{\varphi\varphi'}{G}-\frac{1}{2}\frac{\varphi^2}{PG^2}$$

lead to

$$PG\left(\left(\frac{\varphi}{\sqrt{G}}\right)'\right)^2=P(\varphi')^2-\frac{1}{4}\frac{\varphi^2}{PG^2}-\frac{1}{2}\left(\frac{\varphi^2}{G}\right)'.$$

In particular, one sees that

$$P(\varphi')^2\geq\frac{1}{4}\frac{\varphi^2}{PG^2}+\frac{1}{2}\left(\frac{\varphi^2}{G}\right)'.$$

Integrate this last inequality over $[x,\alpha]$ with $a<x<\alpha$. Then

$$\int_x^\alpha P(t)\varphi'(t)^2\,dt\geq\frac{1}{4}\int_x^\alpha\frac{\varphi(t)^2}{P(t)G(t)^2}\,dt+\frac{1}{2}\frac{\varphi(\alpha)^2}{G(\alpha)}-\frac{1}{2}\frac{\varphi(x)^2}{G(x)}.$$

Recall that $\lim_{x\to a}\varphi(x)=0$ is equivalent to $\lim_{x\to a}\varphi(x)^2/G(x)$ exists in $\mathbb{C}$. Since $\sqrt{P}\varphi'$ is square-integrable, the integrability result in (ii) follows. □

Corollary 6.10.2. *Let the function P satisfy (6.10.1) and let $a<\alpha<b$. Assume that*

$$\varphi,\psi\in AC(a,\alpha) \quad \textit{and} \quad \sqrt{P}\varphi',\sqrt{P}\psi'\in L^2(a,\alpha).$$

Then

$$\liminf_{x\to a} \big|P(x)\varphi'(x)\psi(x)\big| = 0, \tag{6.10.4}$$

when either

$$\int_a^\alpha \frac{1}{P(t)}\,dt = \infty$$

or

$$\int_a^\alpha \frac{1}{P(t)}\,dt < \infty \quad \textit{and} \quad \lim_{x\to a} \psi(x) = 0.$$

Proof. According to Lemma 6.10.1 (i) applied to ψ one has with $H(t) = \int_t^\alpha \frac{1}{P(s)}\,ds$, $a < t < \alpha$, that

$$\frac{\psi}{\sqrt{PH}} \in L^2(a, \alpha')$$

for any $a < \alpha' < \alpha$. The assumption $\int_a^\alpha \frac{1}{P(t)}\,dt = \infty$ implies that

$$\int_a^{\alpha'} \frac{1}{P(t)H(t)}\,dt = \infty; \tag{6.10.5}$$

cf. (6.10.3). However, with $\sqrt{P}\varphi' \in L^2(a,\alpha)$ one also sees via the Cauchy–Schwarz inequality that

$$\int_a^{\alpha'} \big|P(t)\varphi'(t)\psi(t)\big| \frac{1}{P(t)H(t)}\,dt = \int_a^{\alpha'} \left|\sqrt{P(t)}\varphi'(t)\frac{\psi(t)}{\sqrt{P(t)H(t)}}\,dt\right| < \infty.$$

Therefore, it follows from (6.10.5) that (6.10.4) holds.

Next assume that

$$\int_a^\alpha \frac{1}{P(t)}\,dt < \infty \quad \text{and} \quad \lim_{x\to a} \psi(x) = 0.$$

Then with $G(t) = \int_a^t \frac{1}{P(s)}\,ds$, $a < t < \alpha$, the assumption $\int_a^\alpha \frac{1}{P(t)}\,dt < \infty$ implies that

$$\begin{aligned}\int_a^\alpha \frac{1}{P(t)G(t)}\,dt &= \lim_{x\to a} \int_x^\alpha \frac{1}{P(t)G(t)}\,dt \\ &= \lim_{x\to a} \big(\log G(\alpha) - \log G(x)\big) = \infty;\end{aligned}$$

note that one has $(\log G)' = 1/PG$. Thanks to the assumption $\lim_{x\to a} \psi(x) = 0$, Lemma 6.10.1 (ii) applied to ψ, shows that

$$\frac{\psi}{\sqrt{PG}} \in L^2(a, \alpha).$$

Combining this with the fact that $\sqrt{P}\varphi' \in L^2(a,\alpha)$ one sees in a similar way as above that (6.10.4) holds. □

Now return to the Sturm–Liouville differential expression L given by (6.1.1) on the interval (a, b). In addition to the conditions in (6.1.2), it will be assumed that

$$p(x) > 0 \quad \text{for almost all } x \in (a, b).$$

If the equation $(L - \lambda_0)y = 0$, $\lambda_0 \in \mathbb{R}$, is nonoscillatory at the endpoint a, then its real solutions may be distinguished by the following properties.

Definition 6.10.3. Let $(L-\lambda_0)y = 0$ with $\lambda_0 \in \mathbb{R}$ be nonoscillatory at the endpoint a and let u and v be real solutions of $(L - \lambda_0)y = 0$. Then u is said to be *principal* at a if $1/(pu^2)$ is not integrable at a and v is said to be *nonprincipal* at a if $1/(pv^2)$ is integrable at a.

It is clear that a real solution of $(L-\lambda_0)y = 0$ with $\lambda_0 \in \mathbb{R}$ is either principal or nonprincipal at a. In Theorem 6.10.4 and Corollary 6.10.5 below it turns out that a principal solution exists and is uniquely determined up to real multiples. Hence, every linearly independent solution is nonprincipal.

Theorem 6.10.4. *Let $(L - \lambda_0)y = 0$ with $\lambda_0 \in \mathbb{R}$ be nonoscillatory at the endpoint a. Then the following statements hold:*

(i) *Let u be a solution of $(L - \lambda_0)y = 0$ which is principal at a. Assume that u does not vanish on (a, a_0) and let $\alpha \in (a, a_0)$. Then v is a real solution of $(L - \lambda_0)y = 0$ with $W(u, v) = 1$ if and only if there exists $\gamma \in \mathbb{R}$ such that*

$$v(x) = -u(x)\left(\gamma + \int_x^\alpha \frac{ds}{p(s)u(s)^2}\right), \quad a < x < \alpha. \tag{6.10.6}$$

For each $\gamma \in \mathbb{R}$ the solution v is nonprincipal at a and

$$\int_a^x \frac{dt}{p(t)v(t)^2} = \frac{1}{\gamma + \int_x^\alpha \frac{dt}{p(t)u(t)^2}}, \quad a < x < a_v, \tag{6.10.7}$$

when v does not vanish on (a, a_v).

(ii) *Let v be a solution of $(L - \lambda_0)y = 0$ which is nonprincipal at a. Assume that v does not vanish on (a, a_0) and let $\alpha \in (a, a_0)$. Then w is a real solution of $(L - \lambda_0)y = 0$ with $W(v, w) = 1$ if and only if there exists $\gamma \in \mathbb{R}$ such that*

$$w(x) = v(x)\left(\gamma + \int_a^x \frac{ds}{p(s)v(s)^2}\right), \quad a < x < \alpha. \tag{6.10.8}$$

The solution w is principal at a if $\gamma = 0$ and nonprincipal at a if $\gamma \neq 0$, in which case

$$\int_a^x \frac{dt}{p(t)w(t)^2} = \frac{1}{\gamma} - \frac{1}{\gamma + \int_a^x \frac{1}{p(s)v(s)^2}\, ds}, \quad a < x < a_w, \tag{6.10.9}$$

when w does not vanish on (a, a_w).

Proof. (i) If v is given by (6.10.6), then it follows from the definition that

$$v'(x) = -u'(x)\left(\gamma + \int_x^\alpha \frac{ds}{p(s)u(s)^2}\right) + \frac{1}{p(x)u(x)}$$

and

$$(pv')'(x) = -(pu')'(x)\left(\gamma + \int_x^\alpha \frac{ds}{p(s)u(s)^2}\right).$$

Hence, v is a solution of $(L - \lambda_0)y = 0$ and $W(u, v) = 1$. Conversely, if v is a solution with $W(u, v) = 1$, then it follows from (6.1.28) that

$$\int_x^\alpha \frac{ds}{p(s)u(s)^2} = \frac{v(\alpha)}{u(\alpha)} - \frac{v(x)}{u(x)}, \quad a < x < \alpha,$$

which leads to (6.10.6). Observe that

$$\frac{d}{dt}\left(\frac{1}{\gamma + \int_t^\alpha \frac{ds}{p(s)u(s)^2}}\right) = \frac{1}{p(t)u(t)^2}\frac{1}{\left(\gamma + \int_t^\alpha \frac{ds}{p(s)u(s)^2}\right)^2} = \frac{1}{p(t)v(t)^2},$$

and consequently, for $a < y < x < a_v$:

$$\int_y^x \frac{dt}{p(t)v(t)^2} = \frac{1}{\gamma + \int_x^\alpha \frac{1}{p(s)u(s)^2}\,ds} - \frac{1}{\gamma + \int_y^\alpha \frac{1}{p(s)u(s)^2}\,ds}.$$

Since u is principal at a, the last term on the right-hand side goes to 0 as $y \to a$, so that v is nonprincipal at a, and (6.10.7) follows.

(ii) One verifies in the same way as in the proof of (i) that w is a real solution of $(L-\lambda_0)y = 0$ with $W(v, w) = 1$ if and only if there exists $\gamma \in \mathbb{R}$ such that (6.10.8) holds. In a similar way as above observe that

$$\frac{d}{dt}\left(\frac{1}{\gamma + \int_a^t \frac{ds}{p(s)v(s)^2}}\right) = -\frac{1}{p(t)v(t)^2}\frac{1}{\left(\gamma + \int_a^t \frac{ds}{p(s)v(s)^2}\right)^2} = -\frac{1}{p(t)w(t)^2},$$

and one obtains for $a < y < x < a_w$:

$$\int_y^x \frac{dt}{p(t)w(t)^2} = \frac{1}{\gamma + \int_a^y \frac{1}{p(s)v(s)^2}\,ds} - \frac{1}{\gamma + \int_a^x \frac{1}{p(s)v(s)^2}\,ds}.$$

Since v is nonprincipal at a, one sees for $\gamma \neq 0$ that w is nonprincipal at a, and (6.10.9) follows. For $\gamma = 0$ it follows that w is principal at a. □

Theorem 6.10.4 allows some flexibility. It is clear from the proof of Theorem 6.10.4 that one can choose $\alpha = a_0$ in (6.10.6) if u does not vanish on $(a, a_0]$ and one can choose $\alpha = a_0$ in (6.10.8) if v does not vanish on $(a, a_0]$. As to the nonoscillatory behavior of the solutions, observe that in (6.10.6) the factor

$$H_\alpha(x) = \gamma + \int_x^\alpha \frac{ds}{p(s)u(s)^2}, \quad a < x < a_0,$$

is a decreasing function with $\lim_{x\to a} H_\alpha(x) = \infty$, while $\lim_{x\to a_0} H_\alpha(x)$ is finite or $-\infty$, depending on whether u is nonprincipal or principal at a_0. In any case, H_α has at most one zero on (a, a_0). Note that $H_\alpha(\alpha) = \gamma$, so that $H_\alpha(x) \geq \gamma$ when $a < x \leq \alpha$.

If u and v are solutions which are principal and nonprincipal at a, respectively, and which are nonvanishing on (a, a_0), then, without loss of generality, one may assume that $W(u, v) = 1$. By choosing $a < \alpha < a_0$ it follows from Theorem 6.10.4 that v is of the form (6.10.6) for a unique $\gamma \in \mathbb{R}$. Moreover, by construction one sees that $\gamma > 0$. A similar observation can be made about (6.10.8).

Corollary 6.10.5. *Let $(L - \lambda_0)y = 0$ with $\lambda_0 \in \mathbb{R}$ be nonoscillatory at the endpoint a. Then there exists a nontrivial solution of $(L - \lambda_0)y = 0$ which is principal at a. This solution is unique up to real nonzero multiples. In fact, a nontrivial solution u is principal at a if and only if*

$$\lim_{x\to a} \frac{u(x)}{v(x)} = 0 \tag{6.10.10}$$

for all solutions v of $(L - \lambda_0)y = 0$ which are linearly independent of u.

Proof. A solution of $(L - \lambda_0)y = 0$ is either principal or nonprincipal at a, and by Theorem 6.10.4 any solution of $(L - \lambda_0)y = 0$ which is nonprincipal at a generates a solution which is principal at a. Thus, there exists a nontrivial solution of $(L - \lambda_0)y = 0$ which is principal at a. It also follows from Theorem 6.10.4 that a principal solution is unique up to real nonzero multiples.

If u is a solution of $(L - \lambda_0)y = 0$ which is principal at a, then it follows from Theorem 6.10.4 (i) that for every solution v of $(L - \lambda_0)y = 0$ with $W(u, v) = 1$ one has $u(x)/v(x) \to 0$ as $x \to a$, that is, (6.10.10) holds.

If u is a solution of $(L - \lambda_0)y = 0$ which is nonprincipal at a, then (6.10.10) does not hold. Indeed, Theorem 6.10.4 (ii) (with v replaced by u and w replaced by v) shows that for every solution v of $(L - \lambda_0)y = 0$ with $W(u, v) = 1$ there exists $\gamma \in \mathbb{R}$ such that

$$\frac{u(x)}{v(x)} = \frac{1}{\gamma + \int_a^x \frac{ds}{p(s)u(s)^2}}$$

for $x \in (a, a_v) \subset (a, a_0)$, where (a, a_v) is an interval on which v does not vanish. Hence, $u(x)/v(x) \to 1/\gamma$ as $x \to a$ when $\gamma \neq 0$ and $u(x)/v(x) \to \infty$ as $x \to a$ when $\gamma = 0$. Therefore, (6.10.10) does not hold. □

Let a be regular, which means that $a \in \mathbb{R}$ and that $\int_a^{a_0} \frac{1}{p(t)}\,dt < \infty$. For any real solution v of $(L - \lambda_0)y = 0$ with $v(a) \neq 0$ it follows that

$$\int_a^{a_0} \frac{1}{p(t)v(t)^2}\,dt < \infty.$$

Hence, the principal solution u corresponds to $u(a) = 0$ (and $(pu')(a) \neq 0$ as otherwise u would be trivial).

There is a refinement of the defining property of a principal solution when the endpoint a is in the limit-circle case.

Corollary 6.10.6. *Let $(L-\lambda_0)y=0$ with $\lambda_0\in\mathbb{R}$ be nonoscillatory at a and assume that a is in the limit-circle case. Let u be a principal solution and let v be a nonprincipal solution which do not vanish on the interval (a,a_0), and let $\alpha\in(a,a_0)$. Then*

$$\frac{\int_a^x u(t)^2r(t)\,dt}{\int_a^x v(t)^2r(t)\,dt}\leq\frac{u(x)^2}{v(x)^2},\quad a<x<\alpha. \tag{6.10.11}$$

Proof. By assumption, u is principal and v is nonprincipal. Hence, according to Theorem 6.10.4 (i) v may be written as (6.10.6). Since a is in the limit-circle case, $u,v\in L^2_r(a,a_0)$. It follows from (6.10.6) that for $a<x<\alpha$

$$\left(\frac{v(x)}{u(x)}\right)^2\int_a^x u(t)^2r(t)\,dt=\left(\gamma+\int_x^\alpha\frac{ds}{p(s)u(s)^2}\right)^2\int_a^x u(t)^2r(t)\,dt. \tag{6.10.12}$$

Moreover, since u and v do not vanish (6.10.6) also implies that

$$0\leq\gamma+\int_x^\alpha\frac{ds}{p(s)u(s)^2}\leq\gamma+\int_t^\alpha\frac{ds}{p(s)u(s)^2}$$

for $a<t<x$. Therefore, the right-hand side of (6.10.12) can be estimated by

$$\int_a^x u(t)^2\left(\gamma+\int_t^\alpha\frac{ds}{p(s)u(s)^2}\right)^2r(t)\,dt=\int_a^x v(t)^2r(t)\,dt,$$

and the assertion follows. □

Returning to the general case of an endpoint, observe that the connections between the solutions in Theorem 6.10.4 give also the following connections between the associated Wronskians.

Corollary 6.10.7. *Let $(L-\lambda_0)y=0$ with $\lambda_0\in\mathbb{R}$ be nonoscillatory at the endpoint a and assume that $f\in AC(a,a_0)$. Then the following statements hold:*

(i) *Let u be a principal solution at a which does not vanish on (a,a_0), let v be given by (6.10.6), and let $\alpha\in(a,a_0)$. Then*

$$W_x(f,v)=W_x(f,u)\,\frac{v(x)}{u(x)}+\frac{f(x)}{u(x)},\quad a<x<\alpha. \tag{6.10.13}$$

(ii) *Let v be a nonprincipal solution at a which does not vanish on (a,a_0), let w be given by (6.10.8), and let $\alpha\in(a,a_0)$. Then*

$$W_x(f,w)=W_x(f,v)\,\frac{w(x)}{v(x)}+\frac{f(x)}{v(x)},\quad a<x<\alpha. \tag{6.10.14}$$

The results about principal and nonprincipal solutions will now be applied in the context of the first-order differential expression (6.9.2) related to the Sturm–Liouville expression. Let ϕ be a real solution of $(L-\lambda_0)y=0$ which is nonoscillatory at the endpoint a and assume that ϕ does not vanish on (a,a_0). Recall that the first-order differential expression N_ϕ on (a,a_0) is given by

$$N_\phi f = \sqrt{p}\phi\left(\frac{f}{\phi}\right)' = -\frac{W(f,\phi)}{\sqrt{p}\phi} \tag{6.10.15}$$

for all functions $f \in AC(a,a_0)$; cf. (6.9.2). Note that for $q=0$ one may take $\lambda_0=0$ and $\phi=1$, in which case $N_\phi f=\sqrt{p}f$; cf. (6.8.11). One basic observation is contained in the following lemma.

Lemma 6.10.8. *Let $(L-\lambda_0)y=0$ with $\lambda_0\in\mathbb{R}$ be nonoscillatory at the endpoint a. Then the following statements hold:*

(i) *Let u be a principal solution of $(L-\lambda_0)y=0$ (which is unique up to real multiples) and let v be a nonprincipal solution of $(L-\lambda_0)y=0$ such that $W(u,v)=1$. Assume that u and v do not vanish on (a,a_0) and let $\alpha\in(a,a_0)$. Then*

$$(N_u f)(x) - (N_v f)(x) = \left(\frac{f}{\sqrt{puv}}\right)(x), \quad a<x<\alpha, \tag{6.10.16}$$

for $f\in AC(a,a_0)$.

(ii) *Let v and w be nonprincipal solutions of $(L-\lambda_0)y=0$ such that $W(v,w)=1$, assume that v and w do not vanish on (a,a_0), and let $\alpha\in(a,a_0)$. Then*

$$(N_v f)(x) - (N_w f)(x) = \left(\frac{f}{\sqrt{pvw}}\right)(x), \quad a<x<\alpha, \tag{6.10.17}$$

for $f\in AC(a,a_0)$.

Proof. (i) As v is assumed to be a real solution of $(L-\lambda_0)y=0$ with $W(u,v)=1$, it is given by (6.10.6) for some $\gamma\in\mathbb{R}$. Hence, (6.10.13) holds and it follows that

$$\frac{1}{\sqrt{pv}}W(f,v) = \frac{1}{\sqrt{pu}}W(f,u) + \frac{f}{\sqrt{puv}}.$$

Using (6.10.15) this implies (6.10.16).

(ii) Since w is assumed to be a real solution of $(L-\lambda_0)y=0$ with $W(v,w)=1$, it is given by (6.10.8) for some real $\gamma\neq 0$. Hence, (6.10.14) holds and it follows that

$$\frac{1}{\sqrt{pw}}W(f,w) = \frac{1}{\sqrt{pv}}W(f,v) + \frac{f}{\sqrt{pvw}}.$$

Using (6.10.15) this implies (6.10.17). □

The following theorem is based on a direct application of Lemma 6.10.1. It shows the usefulness of the various types of solutions.

Theorem 6.10.9. *Let $(L-\lambda_0)y=0$ with $\lambda_0\in\mathbb{R}$ be nonoscillatory at the endpoint a. Then the following statements hold:*

(i) *Let v be a solution which is nonprincipal at a and assume that v does not vanish on the subinterval (a,a_0). Let $f\in AC(a,a_0)$ and $N_v f\in L^2(a,c)$ for $a<c<a_0$. Then*

$$\lim_{x\to a}\frac{f(x)}{v(x)}$$

exists and is finite.

(ii) *Let v and w be nonprincipal solutions with $W(v,w)=1$, assume that both v and w do not vanish on (a,a_0), and let $f\in AC(a,a_0)$. Then*

$$N_v f\in L^2(a,c) \quad\Leftrightarrow\quad N_w f\in L^2(a,c)$$

for $a<c<a_0$.

(iii) *Let u be a principal solution, let v be a nonprincipal solution such that $W(u,v)=1$, assume that both u and v do not vanish on (a,a_0), and let $f\in AC(a,a_0)$. Then*

$$N_u f\in L^2(a,a') \quad\Leftrightarrow\quad N_v f\in L^2(a,a'') \quad \text{and} \quad \lim_{x\to a}\frac{f(x)}{v(x)}=0$$

for some $a<a',a''<a_0$.

(iv) *Let u be a principal solution which does not vanish on (a,a_0). If $f\in AC(a,a_0)$ and $N_u f\in L^2(a,c)$ for $a<c<a_0$, then*

$$\lim_{x\to a}\frac{f(x)}{u(x)}\left(\int_a^x\frac{dt}{p(t)v(t)^2}\right)^{1/2}=0$$

for any nonprincipal solution v at a.

Proof. (i) Let v be a nonprincipal solution and assume that $f\in AC(a,a_0)$ and $N_v f\in L^2(a,c)$. Now apply Lemma 6.10.1 (ii) with $P=pv^2$ and $\varphi=f/v$. In fact, in the present situation one has $\varphi=f/v\in AC(a,c)$ and $\sqrt{P}\varphi'=N_v f\in L^2(a,c)$, and $\int_a^c\frac{1}{P(t)}dt<\infty$, since v is nonprincipal at a. Therefore, Lemma 6.10.1 (ii) with $\alpha=c$ yields that the limit

$$\lim_{x\to a}\frac{f(x)}{v(x)}=\lim_{x\to a}\varphi(x)$$

exists and is finite.

(ii) By symmetry, it suffices to show the implication ($\Rightarrow$). Hence, assume that $f \in AC(a, a_0)$ and $N_v f \in L^2(a, c)$. Since v is nonprincipal at a it follows from (i) that the limit

$$\lim_{x \to a} \frac{f(x)}{v(x)}$$

exists and is finite. As v does not vanish in (a, a_0) one has $f/v \in AC(a, a_0)$ and for $c \in (a, a_0)$ there exists $M > 0$ such that

$$\left| \frac{f(x)}{v(x)} \right| \leq M, \quad x \in (a, c).$$

By Lemma 6.10.8 (ii),

$$(N_v f)(x) - (N_w f)(x) = \frac{1}{\sqrt{p(x)} w(x)} \left(\frac{f}{v} \right)(x). \tag{6.10.18}$$

Now it follows from (6.10.18) with $1/(pw^2)$ integrable on (a, c) that

$$\int_a^c |N_w f(s)|^2 ds \leq 2 \int_a^c |N_v f(s)|^2 ds + 2M^2 \int_a^c \frac{1}{p(s) w(s)^2} \, ds < \infty,$$

and hence $N_w f \in L^2(a, c)$.

(iii) ($\Rightarrow$) Assume that $f \in AC(a, a_0)$ and $N_u f \in L^2(a, a')$ for $a < a' < a_0$. Since the principal solution u does not vanish on (a, a_0), the function

$$H_{a'}(x) = \int_x^{a'} \frac{ds}{p(s) u(s)^2}, \quad a < x < a',$$

is well defined and $H_{a'}(x) \to \infty$ as $x \to a$. Then, according to Lemma 6.10.1 (i) with $P = pu^2$, $\varphi = f/u$, $\alpha = a'$, and $\alpha' = a''$ one obtains

$$\frac{1}{\sqrt{p} u} \left(\frac{f}{u} \right) \frac{1}{H_{a'}} \in L^2(a, a'') \tag{6.10.19}$$

for any $a < a'' < a'$, and

$$\left(\left(\frac{f}{u} \right)(x) \right)^2 \frac{1}{H_{a'}(x)} \to 0 \quad \text{as} \quad x \to a. \tag{6.10.20}$$

Since v is a nonprincipal solution it can be expressed in terms of u by means of Theorem 6.10.4 (i) with some $\gamma > 0$ as

$$v(x) = -u(x) \left(\gamma + H_{a'}(x) \right), \quad a < x < a',$$

and consequently

$$\frac{u(x)}{v(x)} H_{a'}(x) = -\frac{H_{a'}(x)}{\gamma + H_{a'}(x)}, \quad a < x < a'.$$

Since $H_{a'}(x) \to \infty$ as $x \to a$, it is clear that

$$\left| \frac{u(x)}{v(x)} H_{a'}(x) \right| \leq 1, \quad a < x < a'. \tag{6.10.21}$$

Write (6.10.16) in Lemma 6.10.8 on the interval (a, a') as

$$N_u f - N_v f = \frac{1}{\sqrt{pu}} \left(\frac{f}{u} \right) \frac{1}{H_{a'}} \frac{u}{v} H_{a'}. \tag{6.10.22}$$

Obviously, (6.10.19) and (6.10.21) imply that the right-hand side of (6.10.22) belongs to $L^2(a, a'')$. As $N_u f \in L^2(a, a')$ this gives $N_v f \in L^2(a, a'')$. To calculate $\lim_{x \to a} f(x)/v(x)$, observe that on (a, a_0)

$$\left(\frac{f}{v} \right)^2 = \left(\frac{f}{u} \right)^2 \frac{1}{H_{a'}} \left(\frac{u}{v} H_{a'} \right)^2 \frac{1}{H_{a'}}, \tag{6.10.23}$$

which in view of (6.10.20) and (6.10.21) shows that

$$\lim_{x \to a} \frac{f(x)}{v(x)} = 0.$$

($\Leftarrow$) For the converse implication assume that $f \in AC(a, a_0)$, $N_v f \in L^2(a, a'')$, and $\lim_{x \to a} f(x)/v(x) = 0$. By means of Theorem 6.10.4 (ii) one can express u in terms of v as

$$u(x) = v(x) K_a(x), \quad K_a(x) = \int_a^x \frac{ds}{p(s)v(s)^2} \, ds, \quad a < x < a''.$$

Now by Lemma 6.10.1 (ii) with $P = pv^2$, $\varphi = f/v$, and $\alpha = a''$ one has

$$\frac{1}{\sqrt{pv}} \left(\frac{f}{v} \right) \frac{1}{K_a} \in L^2(a, a''). \tag{6.10.24}$$

In order to show that the functions $N_u f$ is square-integrable near a, write (6.10.16) in Lemma 6.10.8 as

$$N_u f - N_v f = \frac{1}{\sqrt{pv}} \left(\frac{f}{v} \right) \cdot \frac{v}{u} = \frac{1}{\sqrt{pv}} \left(\frac{f}{v} \right) \frac{1}{K_a}. \tag{6.10.25}$$

It follows from (6.10.24) and (6.10.25) that $N_u f \in L^2(a, a'')$ and hence, in particular, $N_u f \in L^2(a, a')$ for $a < a' \leq a''$.

(iv) Let u be a principal solution which does not vanish on (a, a_0). Assume that $f \in AC(a, a_0)$ and $N_u f \in L^2(a, a')$, where $a' = c \in (a, a_0)$. Then it follows that (6.10.20) holds; cf. (iii). Recall that every solution v of $(L - \lambda_0)y = 0$ which is nonprincipal at a has the form (6.10.6) with $\alpha = a'$. Then v does not vanish on

$(a, a_v) \subset (a, a')$. Now observe that for $f \in AC(a, a_0)$ it then follows from (6.10.7) that for $a < x < a_v$

$$\frac{f(x)^2}{u(x)^2}\left(\int_a^x \frac{dt}{p(t)v(t)^2}\right) = \frac{f(x)^2}{u(x)^2}\frac{1}{\int_x^{a'} \frac{dt}{p(t)u(t)^2}} \frac{\int_x^{a'} \frac{dt}{p(t)u(t)^2}}{\gamma + \int_x^{a'} \frac{dt}{p(t)u(t)^2}}.$$

Therefore, the right-hand side has the limit 0 as $x \to a$. □

Let u and v be solutions of $(L - \lambda_0)y = 0$, $\lambda_0 \in \mathbb{R}$, which are principal and nonprincipal at a, respectively. If the endpoint a is in the limit-circle case, then one can say more about the limits of $f(x)/v(x)$ and $f(x)/u(x)$ as $x \to a$; cf. Theorem 6.10.9 (i) and (iv).

Lemma 6.10.10. *Let $(L - \lambda_0)y = 0$ with $\lambda_0 \in \mathbb{R}$ be nonoscillatory at a and assume that the endpoint a is in the limit-circle case. Let u and v be real solutions of $(L - \lambda_0)y = 0$ for $\lambda_0 \in \mathbb{R}$ which do not vanish on (a, a_0) and which are principal and nonprincipal at a, respectively, with $W(u, v) = 1$. Let $f, pf' \in AC(a, a_0)$ and assume that $f, Lf \in L^2_r(a, a_0)$. Then*

$$\lim_{x\to a} \frac{f(x)}{v(x)} = -\lim_{x\to a} W_x(f, u). \tag{6.10.26}$$

If, in addition,

$$\lim_{x\to a} W_x(f, u)\,\frac{v(x)}{u(x)} = 0, \tag{6.10.27}$$

then

$$\lim_{x\to a} \frac{f(x)}{u(x)} = \lim_{x\to a} W_x(f, v). \tag{6.10.28}$$

Proof. Note that under the present conditions both limits

$$\lim_{x\to a} W_x(f, u) \quad \text{and} \quad \lim_{x\to a} W_x(f, v)$$

exist; cf. Lemma 6.2.5. Recall that v can be written in terms of u as in (6.10.6). Hence, for $f \in AC(a, a_0)$ Corollary 6.10.7 (i) shows that

$$W_x(f, v) = W_x(f, u)\,\frac{v(x)}{u(x)} + \frac{f(x)}{u(x)}, \quad a < x < \alpha. \tag{6.10.29}$$

Multiplying the identity (6.10.29) by $\frac{u(x)}{v(x)}$ one obtains

$$\frac{u(x)}{v(x)}\,W_x(f, v) = W_x(f, u) + \frac{f(x)}{v(x)}, \quad a < x < \alpha.$$

Since, by Corollary 6.10.5, $\lim_{x\to a} \frac{u(x)}{v(x)} = 0$, it follows that (6.10.26) holds. Furthermore, if (6.10.27) is satisfied, then (6.10.28) follows from (6.10.29). □

The following result is a slight variation of Theorem 6.10.9 (iii) in the context of the limit-circle case.

Proposition 6.10.11. *Let $(L-\lambda_0)y=0$ with $\lambda_0\in\mathbb{R}$ be nonoscillatory at a and assume that the endpoint a is in the limit-circle case. Let u and v be real solutions of $(L-\lambda_0)y=0$ for some $\lambda_0\in\mathbb{R}$ which are principal and nonprincipal at a, respectively, and do not vanish on (a,a_0). Let $f,pf'\in AC(a,a_0)$ and assume that $f,Lf\in L^2_r(a,a_0)$. Then the following statements are equivalent:*

(i) $N_u f\in L^2(a,c)$ *for* $a<c<a_0$;

(ii) $\lim_{x\to a}\frac{f(x)}{v(x)}=0$;

(iii) $\lim_{x\to a}W_x(f,u)=0$;

(iv) $\lim_{x\to a}f(x)/u(x)$ *exists in* $\mathbb{C}$;

and in this case

$$\lim_{x\to a}\frac{f(x)}{u(x)}=\lim_{x\to a}W_x(f,v).$$

Furthermore, if the endpoint a is regular, then (i)–(iv) *are equivalent to*

(v) $f(a)=0$.

Proof. Let u and v be principal and nonprincipal at a, respectively, and assume without loss of generality that $W(u,v)=1$.

(i) $\Rightarrow$ (ii) If $N_u f\in L^2(a,c)$, then Theorem 6.10.9 (iii) implies $f(x)/v(x)\to 0$ as $x\to a$.

(ii) $\Rightarrow$ (iii) This follows from (6.10.26).

(iii) $\Rightarrow$ (iv) Assume that $\lim_{x\to a}W_x(f,u)=0$. Then the identity (6.1.9) shows that

$$W_x(f,u)=\int_a^x((L-\lambda_0)f)(t)u(t)r(t)\,dt,$$

which gives the estimate

$$|W_x(f,u)|\le\left(\int_a^x u(t)^2r(t)\,dt\right)^{\frac12}\left(\int_a^x|((L-\lambda_0)f)(t)|^2\,r(t)\,dt\right)^{\frac12}.$$

Combining this with the estimate (6.10.11) leads to

$$|W_x(f,u)|\left|\frac{v(x)}{u(x)}\right|\le\left(\int_a^x v(t)^2r(t)\,dt\right)^{\frac12}\left(\int_a^x|((L-\lambda_0)f)(t)|^2\,r(t)\,dt\right)^{\frac12}$$

for x sufficiently close to a, so that

$$W_x(f,u)\frac{v(x)}{u(x)}\to 0\quad\text{as}\quad x\to a.$$

Therefore, $\lim_{x\to a}f(x)/u(x)$ exists by Lemma 6.10.10.

(iv) $\Rightarrow$ (i) Since $\lim_{x\to a} W_x(f,u)$ exists, the assumption that $\lim_{x\to a} f(x)/u(x)$ exists implies that $N_u f \in L^2(a,c)$ for $a < c < a_0$; cf. (6.9.6).

(iv) $\Leftrightarrow$ (v) Observe that in the case of a regular endpoint one has $v(a) \neq 0$ for any nonprincipal solution v. This shows the equivalence of (ii) and (v). □

The importance of Theorem 6.10.9 and Proposition 6.10.11 will become clear in the treatment of the various forms associated with Sturm–Liouville expressions in Section 6.11 and Section 6.12. In fact, the dependence of the forms on the choice of nonprincipal and principal solutions will be indicated in Proposition 6.11.7 and Proposition 6.12.7.

6.11 Semibounded Sturm–Liouville operators and the limit-circle case

Let L be the Sturm–Liouville differential expression in (6.1.1) on the open interval (a,b):

$$L = \frac{1}{r}\left[-DpD + q\right], \quad D = d/dx,$$

and let the coefficient functions satisfy the conditions

$$\begin{cases} p(x) > 0, \ r(x) > 0, & \text{for almost all } x \in (a,b), \\ 1/p, q, r \in L^1_{\mathrm{loc}}(a,b). \end{cases} \tag{6.11.1}$$

In addition, it will be assumed that the equation $(L-\lambda_0)y = 0$ is nonoscillatory at the endpoints a and b for some $\lambda_0 \in \mathbb{R}$, which implies that the minimal operator $T_{\min}$ is bounded from below; cf. Theorem 6.9.6. Recall that if $T_{\min}$ is semibounded from below, then in any case for every $\lambda_0 < m(T_{\min})$ the equation $(L-\lambda_0)y = 0$ is nonoscillatory; cf. Theorem 6.9.8. Furthermore, it will be assumed that the endpoints a and b are in the limit-circle case.

Let v_a and v_b be solutions of $(L-\lambda_0)y = 0$, $\lambda_0 \in \mathbb{R}$, which are nonprincipal at a and b, respectively. Recall that v_a and v_b are real by Definition 6.10.3. Since it is assumed that a and b are in the limit-circle case, one sees that v_a and v_b belong to $L^2_r(a,b)$. These nonprincipal solutions v_a and v_b will be used to define a convenient boundary triplet.

Proposition 6.11.1. *Let v_a and v_b be solutions of $(L-\lambda_0)y = 0$, $\lambda_0 \in \mathbb{R}$, which are nonprincipal at a and b, respectively. Assume that the endpoints a and b are in the limit-circle case. Then $\{\mathbb{C}^2, \Gamma_0, \Gamma_1\}$, where*

$$\Gamma_0 f = \begin{pmatrix} \lim_{x\to a} \frac{f(x)}{v_a(x)} \\ \lim_{x\to b} \frac{f(x)}{v_b(x)} \end{pmatrix} \quad \text{and} \quad \Gamma_1 f = \begin{pmatrix} -\lim_{x\to a} W_x(f,v_a) \\ \lim_{x\to b} W_x(f,v_b) \end{pmatrix}, \quad f \in \operatorname{dom} T_{\max},$$

is a boundary triplet for $(T_{\min})^* = T_{\max}$.

Proof. Let u_a and u_b be solutions of $(L - \lambda_0)y = 0$, $\lambda_0 \in \mathbb{R}$, which are principal at a and b such that $W_x(u_a, v_a) = 1$ and $W_x(u_b, v_b) = 1$, respectively; cf. Theorem 6.10.4. By Lemma 6.10.10,

$$\lim_{x\to a} \frac{f(x)}{v_a(x)} = -\lim_{x\to a} W_x(f, u_a)$$

and in a similar way one obtains

$$\lim_{x\to b} \frac{f(x)}{v_b(x)} = -\lim_{x\to a} W_x(f, u_b).$$

Hence, the claim is that for $f \in \operatorname{dom} T_{\max}$ the mappings

$$\Gamma_0 f = \begin{pmatrix} \lim_{x\to a} W_x(f, -u_a) \\ \lim_{x\to b} W_x(f, -u_b) \end{pmatrix} \quad \text{and} \quad \Gamma_1 f = \begin{pmatrix} -\lim_{x\to a} W_x(f, v_a) \\ \lim_{x\to b} W_x(f, v_b) \end{pmatrix}$$

define a boundary triplet for $T_{\max}$. To see this, observe that

$$W_x(v_a, -u_a) = 1 \quad \text{and} \quad W_x(v_b, -u_b) = 1,$$

and apply Proposition 6.3.8. □

Choose a_0 and b_0 such that v_a does not vanish on (a, a_0) and v_b does not vanish on (b_0, b), and let c, d be as in (6.9.9). By means of the solutions v_a and v_b of $(L - \lambda_0)y = 0$, which are nonprincipal at a and b, one introduces the form $\mathfrak{t}$ by

$$\begin{aligned} \mathfrak{t}[f,g] = &\int_a^c (N_{v_a} f)(x)\overline{(N_{v_a} g)(x)}\, dx + \int_d^b (N_{v_b} f)(x)\overline{(N_{v_b} g)(x)}\, dx \\ &+ \lambda_0 \int_a^c f(x)\overline{g(x)} r(x)\, dx + \lambda_0 \int_d^b f(x)\overline{g(x)} r(x)\, dx \\ &+ \int_c^d \big((\sqrt{p} f')(x)\overline{(\sqrt{p} g')(x)} + q(x) f(x)\overline{g(x)}\,\big) dx \\ &+ \frac{(p v_a')(c)}{v_a(c)} f(c)\overline{g(c)} - \frac{(p v_b')(d)}{v_b(d)} f(d)\overline{g(d)} \end{aligned} \tag{6.11.2}$$

for $f, g \in \mathfrak{D}$, where

$$\begin{aligned} \mathfrak{D} = \big\{ f \in L^2_r(a,b) :\ &f \in AC(a,b),\ \sqrt{p} f' \in L^2(c,d), \\ &N_{v_a} f \in L^2(a,c),\ N_{v_b} f \in L^2(d,b) \big\}; \end{aligned} \tag{6.11.3}$$

cf. (6.9.10)–(6.9.11). The next corollary follows from Lemma 6.9.4 and Theorem 6.9.6 with $\phi_a = v_a$ and $\phi_b = v_b$.

Corollary 6.11.2. *Let v_a and v_b be solutions of $(L - \lambda_0)y = 0$, $\lambda_0 \in \mathbb{R}$, which are nonprincipal at a and b, respectively. Assume that the endpoints a and b are in* *the limit-circle case. Then* $\mathfrak{t}$ *in* (6.11.2)–(6.11.3) *is a densely defined closed semi-bounded form in* $L^2_r(a,b)$. *Moreover, if* S_1 *is the semibounded self-adjoint operator corresponding to* $\mathfrak{t}$, *then* $T_{\min} \subset S_1$ *and, in fact,*

$$(T_{\min} f, g)_{L^2_r(a,b)} = \mathfrak{t}[f,g]$$

holds for all $f \in \operatorname{dom} T_{\min} \subset \mathfrak{D}$ *and* $g \in \mathfrak{D}$.

Now one shows that also $\operatorname{dom} T_{\max} \subset \mathfrak{D}$. This property is important for the construction of a compatible boundary pair in Lemma 6.11.5.

Lemma 6.11.3. *Let* v_a *and* v_b *be solutions of* $(L - \lambda_0)y = 0$, $\lambda_0 \in \mathbb{R}$, *which are nonprincipal at* a *and* b, *respectively. Assume that the endpoints* a *and* b *are in the limit-circle case. Then*

$$\operatorname{dom} T_{\max} \subset \mathfrak{D}.$$

Proof. Let $f \in \operatorname{dom} T_{\max}$. Then $f, pf' \in AC(a,b)$ and hence $\sqrt{p}f' \in L^2(c,d)$. It follows from (6.10.15) with $\phi = v_a$ that

$$N_{v_a} f = -\frac{W(f, v_a)}{\sqrt{p}v_a}.$$

Since $f \in \operatorname{dom} T_{\max}$ and $v_a \in L^2_r(a,b)$, Lemma 6.2.5 shows that $\lim_{x\to a} W_x(f, v_a)$ exists. Hence, $x \mapsto W_x(f, v_a)$ is bounded in $(a, c]$. Consequently, $N_{v_a} f \in L^2(a,c)$, as v_a is nonprincipal at a and docs not vanish in (a, a_0). Likewise, it is clear that $N_{v_b} f \in L^2(d,b)$. Hence, $f \in \mathfrak{D}$. □

Introduce the mapping $\Lambda : \mathfrak{D} \to \mathbb{C}^2$ by

$$\Lambda f = \begin{pmatrix} \lim_{x\to a} \frac{f(x)}{v_a(x)} \\ \lim_{x\to b} \frac{f(x)}{v_b(x)} \end{pmatrix}, \quad f \in \mathfrak{D}. \tag{6.11.4}$$

Note that, by Theorem 6.10.9 (i), Λ is well defined.

Lemma 6.11.4. *Let* v_a *and* v_b *be solutions of* $(L - \lambda_0)y = 0$, $\lambda_0 \in \mathbb{R}$, *which are nonprincipal at* a *and* b, *respectively. Assume that the endpoints* a *and* b *are in the limit-circle case, and let* $\mathfrak{t}$ *be the form in* (6.11.2)–(6.11.3). *Then for every* $\varepsilon > 0$ *there exists* $D_\varepsilon > 0$ *such that*

$$\|\Lambda f\|^2_{\mathbb{C}^2} \le \varepsilon\, \mathfrak{t}[f] + D_\varepsilon \|f\|^2_{L^2_r(a,b)}, \quad f \in \mathfrak{D}.$$

Proof. Choose $\varepsilon > 0$ and write the form $\mathfrak{t}$ as the sum of

$$\begin{aligned} \mathfrak{t}_{[c,d]}[f,g] = &\int_c^d \big((\sqrt{p}f')(x)\overline{(\sqrt{p}g')(x)} + q(x)f(x)\overline{g(x)}\,\big)\,dx \\ &+ \frac{(pv_a)'(c)}{v_a(c)} f(c)\overline{g(c)} - \frac{(pv_b)'(d)}{v_b(d)} f(d)\overline{g(d)}, \end{aligned}$$

$$\mathfrak{t}_{(a,c)}[f,g] = \int_a^c (N_{v_a}f)(x)\overline{(N_{v_a}g)(x)}\,dx + \lambda_0 \int_a^c f(x)\overline{g(x)}r(x)\,dx,$$
$$\mathfrak{t}_{(b,d)}[f,g] = \int_d^b (N_{v_b}f)(x)\overline{(N_{v_b}g)(x)}\,dx + \lambda_0 \int_d^b f(x)\overline{g(x)}r(x)\,dx.$$

It has been shown in Lemma 6.9.4 that $\mathfrak{t}$ is independent of the choice of c and d, and hence $c \in (a, a_0)$ and $d \in (b_0, b)$ can be chosen suitably close to a and b, respectively; see below. First observe that for $f \in \mathfrak{D}$ and for all $d \le x < b$ one has

$$\frac{f(x)}{v_b(x)} = \frac{f(d)}{v_b(d)} + \int_d^x \left(\frac{f(t)}{v_b(t)}\right)' dt = \frac{f(d)}{v_b(d)} + \int_d^x \frac{1}{\sqrt{p(t)}v_b(t)} N_{v_b}f(t)\,dt$$

and hence

$$\begin{aligned}\left|\frac{f(x)}{v_b(x)}\right|^2 &\le 2\left|\frac{f(d)}{v_b(d)}\right|^2 + 2\int_d^x \frac{1}{p(t)v_b(t)^2}\,dt \int_d^x |N_{v_b}f(t)|^2\,dt \\ &\le 2\left|\frac{f(d)}{v_b(d)}\right|^2 + 2\int_d^b \frac{1}{p(t)v_b(t)^2}\,dt \int_d^b |N_{v_b}f(t)|^2\,dt.\end{aligned}$$

Now choose d so close to b that

$$2\int_d^b \frac{1}{p(t)v_b(t)^2}\,dt \le \varepsilon,$$

so that now for all $d \le x < b$ one has

$$\left|\frac{f(x)}{v_b(x)}\right|^2 \le 2\left|\frac{f(d)}{v_b(d)}\right|^2 + \varepsilon \int_d^b |N_{v_b}f(t)|^2\,dt.$$

Therefore,

$$\left|\lim_{x\to b} \frac{f(x)}{v_b(x)}\right|^2 \le 2\left|\frac{f(d)}{v_b(d)}\right|^2 + \varepsilon \int_d^b |N_{v_b}f(t)|^2\,dt$$

for all $f \in \mathfrak{D}$. Similarly, one can choose c so close to a that

$$\left|\lim_{x\to a} \frac{f(x)}{v_a(x)}\right|^2 \le 2\left|\frac{f(c)}{v_a(c)}\right|^2 + \varepsilon \int_a^c |N_{v_a}f(t)|^2\,dt$$

for all $f \in \mathfrak{D}$. An application of Corollary 6.8.6 shows that there exists $C_\varepsilon > 0$ such that for all $f \in \mathfrak{D}$ one has

$$\left|\frac{f(c)}{v_a(c)}\right|^2 + \left|\frac{f(d)}{v_b(d)}\right|^2 \le C_\varepsilon \|f\|_{L^2_r(c,d)} + \varepsilon \mathfrak{t}_{[c,d]}[f].$$

The assertion follows by combining the above inequalities. □

In order to apply the theory developed in Chapter 5 it will be shown that the map Λ in (6.11.4) leads to a boundary pair which is compatible with the boundary triplet $\{\mathbb{C}^2, \Gamma_0, \Gamma_1\}$ in Proposition 6.11.1. As usual, the self-adjoint operator defined on $\ker \Gamma_1$ is denoted by A_1.

Lemma 6.11.5. *Let v_a and v_b be solutions of $(L - \lambda_0)y = 0$, $\lambda_0 \in \mathbb{R}$, which are nonprincipal at a and b, respectively. Assume that the endpoints a and b are in the limit-circle case and let $\{\mathbb{C}^2, \Gamma_0, \Gamma_1\}$ be the boundary triplet in Proposition 6.11.1. Then $\{\mathbb{C}^2, \Lambda\}$ is a boundary pair for $T_{\min}$ corresponding to S_1 which is compatible with the boundary triplet $\{\mathbb{C}^2, \Gamma_0, \Gamma_1\}$. Moreover, one has*

$$(T_{\max} f, g)_{L^2_r(a,b)} = (\Gamma_1 f, \Lambda g) + \mathfrak{t}[f, g], \quad f \in \operatorname{dom} T_{\max}, \ g \in \mathfrak{D}. \tag{6.11.5}$$

Proof. Consider the form $\mathfrak{t}$ on $\operatorname{dom} \mathfrak{t} = \mathfrak{D}$ as in (6.11.2)–(6.11.3) and denote the corresponding semibounded self-adjoint operator in $L^2_r(a,b)$ by S_1; cf. Corollary 6.11.2. Let $\varepsilon > 0$ and $D_\varepsilon > 0$ be as in Lemma 6.11.4. It follows from the estimate in Lemma 6.11.4 that for $\rho < m(S_1)$ there exists $C_{\rho,\varepsilon} > 0$ such that

$$\|\Lambda f\|^2_{\mathbb{C}^2} \leq D_\varepsilon \|f\|^2_{L^2_r(a,b)} + \varepsilon \mathfrak{t}[f, f] \leq C_{\rho,\varepsilon} \|f\|^2_{\mathfrak{t}_{S_1} - \rho}$$

for all $f \in \mathfrak{D}$. Therefore, $\Lambda \in \mathbf{B}(\mathfrak{H}_{\mathfrak{t}_{S_1} - \rho}, \mathbb{C}^2)$. Moreover, according to Lemma 6.11.3 one has $\operatorname{dom} T_{\max} \subset \mathfrak{D}$ and hence Λ is an extension of the boundary mapping Γ_0 in Proposition 6.11.1. Now Lemma 5.6.5 implies that $\{\mathbb{C}^2, \Lambda\}$ is a boundary pair for $T_{\min}$ corresponding to S_1.

In order to conclude that $\{\mathbb{C}^2, \Lambda\}$ and $\{\mathbb{C}^2, \Gamma_0, \Gamma_1\}$ are compatible, it remains to show that $A_1 = S_1$, where A_1 is the self-adjoint operator defined on $\ker \Gamma_1$. In fact, since $\operatorname{dom} T_{\max} \subset \mathfrak{D}$, the Green formula (6.9.15) is valid for $f \in \operatorname{dom} T_{\max}$ and $g \in \mathfrak{D}$,

$$\begin{aligned}(T_{\max} f, g)_{L^2_r(a,b)} = \mathfrak{t}[f, g] &+ \lim_{b' \to b} W_{b'}(f, v_b) \left(\frac{\bar{g}}{v_b}\right)(b') \\ &- \lim_{a' \to a} W_{a'}(f, v_a) \left(\frac{\bar{g}}{v_a}\right)(a'),\end{aligned}$$

which, in the present context, is equivalent to (6.11.5). Hence,

$$(A_1 f, g)_{L^2_r(a,b)} = \mathfrak{t}[f, g]$$

for all $f \in \operatorname{dom} A_1$ and $g \in \operatorname{dom} \mathfrak{t}$. As A_1 is self-adjoint, the first representation theorem implies $A_1 = S_1$. □

Recall that by means of the boundary triplet in Proposition 6.11.1, all self-adjoint extensions of $T_{\min}$ are in a one-to-one correspondence to the self-adjoint relations Θ in $\mathbb{C}^2$ via

$$\operatorname{dom} A_\Theta = \bigl\{ f \in \operatorname{dom} T_{\max} : \{\Gamma_0 f, \Gamma_1 f\} \in \Theta \bigr\}. \tag{6.11.6}$$

The next result, which is an immediate consequence of Theorem 5.6.13 and Corollary 5.6.14, makes use of the compatible boundary pair in Lemma 6.11.5 and provides a characterization of all closed semibounded forms associated with the semibounded self-adjoint extensions A_Θ.

Theorem 6.11.6. *Let $\{\mathbb{C}^2, \Gamma_0, \Gamma_1\}$ be the boundary triplet in Proposition 6.11.1, let Θ be a self-adjoint relation in $\mathbb{C}^2$, and let A_Θ be the corresponding self-adjoint restriction of $T_{\max}$ in (6.11.6). Then A_Θ is semibounded from below and the corresponding densely defined closed semibounded form $\mathfrak{t}_\Theta$ in $L^2_r(a,b)$ such that*

$$(A_\Theta f, g)_{L^2_r(a,b)} = \mathfrak{t}_\Theta[f,g], \quad f \in \operatorname{dom} A_\Theta, \ g \in \operatorname{dom} \mathfrak{t}_\Theta,$$

is given in terms of $\mathfrak{t}$ in (6.11.2)–(6.11.3), and Λ in (6.11.4) as follows:

(i) *If Θ is a symmetric 2×2-matrix, then*

$$\mathfrak{t}_\Theta[f,g] = \mathfrak{t}[f,g] + (\Theta\Lambda f, \Lambda g), \quad \operatorname{dom} \mathfrak{t}_\Theta = \mathfrak{D}.$$

(ii) *If $\Theta = \Theta_{\rm op} \,\widehat{\oplus}\, \Theta_{\rm mul}$ with respect to the decomposition $\mathbb{C}^2 = \operatorname{dom} \Theta_{\rm op} \oplus \operatorname{mul} \Theta$ and $\dim \operatorname{dom} \Theta_{\rm op} = 1$, then*

$$\mathfrak{t}_\Theta[f,g] = \mathfrak{t}[f,g] + (\Theta_{\rm op}\Lambda f, \Lambda g), \quad \operatorname{dom} \mathfrak{t}_\Theta = \bigl\{h \in \mathfrak{D} : \Lambda h \in \operatorname{dom} \Theta_{\rm op}\bigr\}.$$

(iii) *If $\Theta = \{0\} \times \mathbb{C}^2$, then $A_\Theta = A_0$ coincides with the Friedrichs extension $S_{\rm F}$ and*

$$\mathfrak{t}_\Theta[f,g] = \mathfrak{t}[f,g], \quad \operatorname{dom} \mathfrak{t}_\Theta = \bigl\{h \in \mathfrak{D} : \Lambda h = 0\bigr\}.$$

The above description in Theorem 6.11.6 of the (automatically) semibounded self-adjoint extensions of $T_{\min}$ is in terms of the corresponding closed semibounded forms via the first representation theorem in Section 5.1, see also Theorem 5.6.13. The boundary triplet and the compatible boundary pair are provided by the choice of the solutions v_a and v_b of $(L - \lambda_0)y = 0$, which are nonprincipal at a and b, respectively; cf. Proposition 6.11.1.

It will now be shown what the results look like when there is a different choice of nonoscillatory solutions. First let w_a and w_b be nonprincipal solutions of $(L - \lambda_0)y = 0$ at a and b, respectively. Assume that v_a and w_a do not vanish on (a, a_0) and that v_b and w_b do not vanish on (b_0, b). Denote the form generated by the solutions w_a and w_b by $\mathfrak{t}'$; cf. (6.11.2)–(6.11.3). Then according to (ii) in Theorem 6.10.9, $\operatorname{dom} \mathfrak{t}' = \operatorname{dom} \mathfrak{t} = \mathfrak{D}$. To describe $\mathfrak{t}'$, let u_a and u_b be solutions of $(L - \lambda_0)y = 0$ which are principal at a and b, respectively, and which satisfy $W(u_a, v_a) = 1$ and $W(u_b, v_b) = 1$; cf. Theorem 6.10.4. Then clearly

$$w_a = \alpha_a v_a + \beta_a u_a \quad \text{and} \quad w_b = \alpha_b v_b + \beta_b u_b$$

for some $\alpha_a, \beta_a, \alpha_b, \beta_b \in \mathbb{R}$, where α_a, α_b are different from zero. Denote the boundary triplet generated by w_a and w_b by $\{\mathbb{C}^2, \Gamma_0', \Gamma_1'\}$ and let $\{\mathbb{C}^2, \Lambda'\}$ be the corresponding boundary pair; cf. Proposition 6.11.1 and (6.11.4).

Proposition 6.11.7. *The boundary triplet $\{\mathbb{C}^2, \Gamma_0', \Gamma_1'\}$ and the boundary pair $\{\mathbb{C}^2, \Lambda'\}$ generated by the nonprincipal solutions w_a and w_b are given by*

$$\Lambda' f = \begin{pmatrix} \frac{1}{\alpha_a} & 0 \\ 0 & \frac{1}{\alpha_b} \end{pmatrix} \Lambda f, \quad f \in \mathfrak{D},$$

and

$$\Gamma_1' f = \begin{pmatrix} \alpha_a & 0 \\ 0 & \alpha_b \end{pmatrix} \Gamma_1 f + \begin{pmatrix} \beta_a & 0 \\ 0 & -\beta_b \end{pmatrix} \Lambda f, \quad f \in \operatorname{dom} T_{\max}.$$

Moreover, the form $\mathfrak{t}'$ coincides with $\mathfrak{t}_\Theta$ as in Theorem 6.11.6, where the self-adjoint matrix Θ is given by

$$\Theta = \begin{pmatrix} -\frac{\beta_a}{\alpha_a} & 0 \\ 0 & \frac{\beta_b}{\alpha_b} \end{pmatrix}.$$

Proof. The following observations are for the endpoint a; the results are similar for the endpoint b. Since by Corollary 6.10.5 $u_a(x)/v_a(x) \to 0$ as $x \to a$, one has

$$\lim_{x \to a} \frac{f(x)}{w_a(x)} = \lim_{x \to a} \frac{f(x)}{\alpha_a v_a(x)} \cdot \frac{1}{1 + \frac{\beta_a}{\alpha_a} \frac{u_a(x)}{v_a(x)}} = \frac{1}{\alpha_a} \lim_{x \to a} \frac{f(x)}{v_a(x)}.$$

Furthermore, it is clear that

$$W(f, w_a) = \alpha_a W(f, v_a) + \beta_a W(f, u_a)$$

and recall that

$$\lim_{x \to a} W_x(f, u_a) = -\lim_{x \to a} \frac{f(x)}{v_a(x)};$$

cf. Lemma 6.10.10. Hence, one sees that

$$\lim_{x \to a} W_x(f, w_a) = \alpha_a \lim_{x \to a} W_x(f, v_a) - \beta_a \lim_{x \to a} \frac{f(x)}{v_a(x)}.$$

Thus, the results for the boundary triplet and boundary pair follow directly from Proposition 6.11.1 and (6.11.4). Analogous to the Green formula (6.11.5) one has

$$\begin{aligned} (T_{\max} f, g)_{L^2_r(a,b)} &= (\Gamma_1' f, \Lambda' g) + \mathfrak{t}'[f,g] \\ &= (\Gamma_1 f, \Lambda g) + \left(\begin{pmatrix} \frac{\beta_a}{\alpha_a} & 0 \\ 0 & -\frac{\beta_b}{\alpha_b} \end{pmatrix} \Lambda f, \Lambda g \right) + \mathfrak{t}'[f,g] \end{aligned}$$

for $f \in \operatorname{dom} T_{\max}$ and $g \in \mathfrak{D}$. Comparison of the right-hand sides gives

$$\mathfrak{t}[f,g] = \left(\begin{pmatrix} \frac{\beta_a}{\alpha_a} & 0 \\ 0 & -\frac{\beta_b}{\alpha_b} \end{pmatrix} \Lambda f, \Lambda g \right) + \mathfrak{t}'[f,g]$$

for $f \in \operatorname{dom} T_{\max}$ and $g \in \mathfrak{D}$. It is easily seen that in fact the last identity holds for $f, g \in \mathfrak{D}$, which completes the proof. □

Next let u_a and u_b be nontrivial solutions of $(L-\lambda_0)y=0$ which are principal at a and b, respectively, and assume that u_a does not vanish on (a,a_0) and that u_b does not vanish on (b_0,b). Denote the form generated by the solutions u_a and u_b by $\widetilde{\mathfrak{t}}$; cf. Theorem 6.9.6.

Proposition 6.11.8. *Let the form $\widetilde{\mathfrak{t}}$ be generated by the solutions u_a and u_b of $(L-\lambda_0)y=0$ which are principal at a and b, respectively. Then $\widetilde{\mathfrak{t}}$ coincides with $\mathfrak{t}_\Theta$ as in Theorem 6.11.6, where $\Theta=\{0\}\times\mathbb{C}^2$ or, equivalently, $\widetilde{\mathfrak{t}}$ is the form generated by the Friedrichs extension*

$$\mathfrak{t}_{S_{\mathrm{F}}}=\widetilde{\mathfrak{t}}.$$

Proof. Recall from Theorem 6.9.6 that

$$\mathfrak{t}_{S_{\mathrm{F}}}\subset\widetilde{\mathfrak{t}}. \tag{6.11.7}$$

Furthermore, let $\mathfrak{t}$ be the form in (6.11.2) defined on $\mathfrak{D}$ in (6.11.3) generated by the solutions v_a and v_b of $(L-\lambda_0)y=0$ which are nonprincipal at a and b, respectively. Now consider $f\in\mathfrak{D}$ and observe that, by Theorem 6.10.9 (iii), $N_{u_a}f$ is square-integrable at a if and only if $N_{v_a}f$ is square-integrable at a and $\lim_{x\to a}f(x)/v_a(x)=0$; an analogous statement holds at the endpoint b. Therefore, $f\in\operatorname{dom}\widetilde{\mathfrak{t}}$ if and only if $f\in\operatorname{dom}\mathfrak{t}$ and

$$\lim_{x\to a}\frac{f(x)}{v_a(x)}=0 \quad\text{and}\quad \lim_{x\to b}\frac{f(x)}{v_b(x)}=0. \tag{6.11.8}$$

Consider the boundary pair $\{\mathbb{C}^2,\Lambda\}$ in Lemma 6.11.5 with the boundary map Λ in (6.11.4). Then (6.11.8) is equivalent to $f\in\ker\Lambda$ and it follows that

$$\ker\Lambda=\operatorname{dom}\mathfrak{t}_{S_{\mathrm{F}}}\subset\operatorname{dom}\widetilde{\mathfrak{t}}=\operatorname{dom}\mathfrak{t}\cap\ker\Lambda\subset\ker\Lambda.$$

Hence, $\operatorname{dom}\mathfrak{t}_{S_{\mathrm{F}}}=\operatorname{dom}\widetilde{\mathfrak{t}}$ and from (6.11.7) one concludes $\mathfrak{t}_{S_{\mathrm{F}}}=\widetilde{\mathfrak{t}}$. □

For the sake of completeness the following equivalent characterizations of the Friedrichs extension of $T_{\min}$ are mentioned explicitly; cf. Proposition 6.10.11.

Corollary 6.11.9. *Let v_a and v_b be solutions of $(L-\lambda_0)y=0$, $\lambda_0\in\mathbb{R}$, which are nonprincipal at a and b, and let u_a and u_b be solutions of $(L-\lambda_0)y=0$, $\lambda_0\in\mathbb{R}$, which are principal at a and b. Assume that the endpoints a and b are in the limit-circle case. Then $f\in\operatorname{dom}T_{\max}$ is in the domain of the Friedrichs extension S_{F} of $T_{\min}$ if and only if one of the following equivalent conditions holds:*

(i) $N_{u_a}f\in L^2(a,a')$ *and* $N_{u_b}f\in L^2(b',b)$;

(ii) $\lim_{x\to a}\frac{f(x)}{v_a(x)}=0$ *and* $\lim_{x\to b}\frac{f(x)}{v_b(x)}=0$;

(iii) $\lim_{x\to a}W_x(f,u_a)=0$ *and* $\lim_{x\to b}W_x(f,u_b)=0$;

(iv) $\lim_{x\to a}f(x)/u_a(x)$ *and* $\lim_{x\to b}f(x)/u_b(x)$ *exist in* $\mathbb{C}$.

In the next remark the case of a regular endpoint is briefly discussed.

Remark 6.11.10. The considerations in this section simplify if one endpoint, say a, is regular. In that case one can choose the boundary triplet $\{\mathbb{C}^2, \Gamma_0, \Gamma_1\}$ where

$$\Gamma_0 f = \begin{pmatrix} f(a) \\ \lim_{x\to b} \frac{f(x)}{v_b(x)} \end{pmatrix} \quad \text{and} \quad \Gamma_1 f = \begin{pmatrix} (pf')(a) \\ \lim_{x\to b} W_x(f, v_b) \end{pmatrix}, \quad f \in \operatorname{dom} T_{\max};$$

cf. (6.9.21). The form $\mathfrak{t}$ and $\mathfrak{D}$ in (6.11.2) and (6.11.3) reduce to (6.9.19) and (6.9.20) with $\phi_b = v_b$ in Remark 6.9.9, respectively. The corresponding boundary pair $\Lambda : \mathfrak{D} \to \mathbb{C}^2$ in (6.11.4) has the form

$$\Lambda f = \begin{pmatrix} f(a) \\ \lim_{x\to b} \frac{f(x)}{v_b(x)} \end{pmatrix}, \quad f \in \mathfrak{D}.$$

6.12 Semibounded Sturm–Liouville operators and the limit-point case

Let L be the Sturm–Liouville differential expression in (6.1.1) on the open interval (a, b):

$$L = \frac{1}{r}\left[-DpD + q\right], \quad D = d/dx,$$

and let the coefficient functions satisfy the conditions (6.11.1). In addition, it will be assumed that the equation $(L - \lambda_0)y = 0$ is nonoscillatory at the endpoints a and b for some $\lambda_0 \in \mathbb{R}$; cf. Theorem 6.9.6 and Theorem 6.9.8. Let v_a be a solution of $(L-\lambda_0)y = 0$ which is nonprincipal at a and let u_b be a solution of $(L-\lambda_0)y = 0$ which is principal at b. Furthermore, it will be assumed that the endpoint a is in the limit-circle case and that b is in the limit-point case. Hence, $v_a \in L^2_r(a, a')$ for $a < a' < b$.

Proposition 6.12.1. *Assume that the endpoint a is in the limit-circle case and that the endpoint b is in the limit-point case. Let v_a be a solution of the equation $(L - \lambda_0)y = 0$, $\lambda_0 \in \mathbb{R}$, which is nonprincipal at a. Then $\{\mathbb{C}, \Gamma_0, \Gamma_1\}$, where*

$$\Gamma_0 f = \lim_{x\to a} \frac{f(x)}{v_a(x)} \quad \text{and} \quad \Gamma_1 f = -\lim_{x\to a} W_x(f, v_a), \quad f \in \operatorname{dom} T_{\max}, \tag{6.12.1}$$

is a boundary triplet for $(T_{\min})^* = T_{\max}$.

Proof. Let u_a be a solution of $(L - \lambda_0)y = 0$, $\lambda_0 \in \mathbb{R}$, which is principal at a such that $W_x(u_a, v_a) = 1$; cf. Theorem 6.10.4. By Lemma 6.10.10, one has that

$$\lim_{x\to a} \frac{f(x)}{v_a(x)} = -\lim_{x\to a} W_x(f, u_a),$$

and hence the claim is that for $f \in \operatorname{dom} T_{\max}$ the mappings

$$\Gamma_0 f = \lim_{x \to a} W_x(f, -u_a) \quad \text{and} \quad \Gamma_1 f = -\lim_{x \to a} W_x(f, v_a)$$

define a boundary triplet for $T_{\max}$. To see this, note that $W_x(v_a, -u_a) = 1$ and apply Proposition 6.4.9. □

Choose a_0 and b_0 such that v_a does not vanish on (a, a_0) and u_b does not vanish on (b_0, b), and let c, d be as in (6.9.9). By means of the solutions v_a and u_b of $(L - \lambda_0)y = 0$ the following form $\mathfrak{t}$ will be considered:

$$\begin{aligned} \mathfrak{t}[f,g] = &\int_a^c (N_{v_a} f)(x)\overline{(N_{v_a} g)(x)}\,dx + \int_d^b (N_{u_b} f)(x)\overline{(N_{u_b} g)(x)}\,dx \\ &+ \lambda_0 \int_a^c f(x)\overline{g(x)} r(x)\,dx + \lambda_0 \int_d^b f(x)\overline{g(x)} r(x)\,dx \\ &+ \int_c^d \bigl((\sqrt{p} f')(x)\overline{(\sqrt{p} g')(x)} + q(x) f(x)\overline{g(x)}\bigr)\,dx \\ &+ \frac{(p v_a')(c)}{v_a(c)} f(c)\overline{g(c)} - \frac{(p u_b')(d)}{u_b(d)} f(d)\overline{g(d)} \end{aligned} \tag{6.12.2}$$

for $f, g \in \mathfrak{D}$, where the domain of definition $\mathfrak{D}$ is given by

$$\begin{aligned} \mathfrak{D} = \bigl\{ f \in L^2_r(a,b) : \; &f \in AC(a,b),\ \sqrt{p} f' \in L^2(c,d), \\ &N_{v_a} f \in L^2(a,c),\ N_{u_b} f \in L^2(d,b) \bigr\}. \end{aligned} \tag{6.12.3}$$

The next corollary is the counterpart of Corollary 6.11.2 in the present situation; it follows from Lemma 6.9.4 and Theorem 6.9.6 with $\phi_a = v_a$ and $\phi_b = u_b$.

Corollary 6.12.2. *Let v_a and u_b be solutions of $(L - \lambda_0)y = 0$, $\lambda_0 \in \mathbb{R}$, which are nonprincipal at a and principal at b, respectively. Assume that the endpoint a is in the limit-circle case and the endpoint b is in the limit-point case. Then $\mathfrak{t}$ in (6.12.2)–(6.12.3) is a densely defined closed semibounded form in $L^2_r(a,b)$. Moreover, if S_1 is the semibounded self-adjoint operator corresponding to $\mathfrak{t}$, then $T_{\min} \subset S_1$ and, in fact,*

$$(T_{\min} f, g)_{L^2_r(a,b)} = \mathfrak{t}[f,g]$$

holds for all $f \in \operatorname{dom} T_{\min} \subset \mathfrak{D}$ and $g \in \mathfrak{D}$.

As in Lemma 6.11.3 one has $\operatorname{dom} T_{\max} \subset \mathfrak{D}$ when the endpoint b is in the limit-point case.

Lemma 6.12.3. *Let v_a and u_b be solutions of $(L - \lambda_0)y = 0$, $\lambda_0 \in \mathbb{R}$, which are nonprincipal at a and principal at b, respectively. Assume that the endpoint a is in the limit-circle case and the endpoint b is in the limit-point case. Then*

$$\operatorname{dom} T_{\max} \subset \mathfrak{D}.$$

Proof. Let $f \in \operatorname{dom} T_{\max}$. Then $f, pf' \in AC(a,b)$, and hence $\sqrt{p}f' \in L^2(c,d)$. It follows as in the proof of Lemma 6.11.3 that $N_{v_a} f \in L^2(a,c)$, as v_a is nonprincipal at a. For the behavior at b of $N_{u_b} f$ decompose $f \in \operatorname{dom} T_{\max}$ as

$$f = f_0 + h,$$

where $f_0 \in \operatorname{dom} T_{\min}$ and $h \in \operatorname{dom} T_{\max}$ is a function which vanishes in a neighborhood of b, say (b', b); cf. the proof of Proposition 6.4.9. Since $\operatorname{dom} T_{\min} \subset \mathfrak{D}$ by Corollary 6.12.2 it is clear that $N_{u_b} f_0 \in L^2(d,b)$. It follows from (6.10.15) with $\phi = u_b$ that

$$N_{u_b} h = -\frac{W(h, u_b)}{\sqrt{p} u_b}.$$

Since h vanishes on (b', b), it follows that $N_{u_b} h$ vanishes on (b', b), while on $[d, b']$ the function $x \to W_x(h, u_b)$ is bounded and the function u_b does not vanish. Consequently, one sees that $N_{u_b} h \in L^2(d,b)$ and thus $N_{u_b} f \in L^2(d,b)$. Hence, it follows that $f \in \mathfrak{D}$. □

Let the mapping $\Lambda : \mathfrak{D} \to \mathbb{C}$ be defined by

$$\Lambda f = \lim_{x \to a} \frac{f(x)}{v_a(x)}, \quad f \in \mathfrak{D}. \tag{6.12.4}$$

Note that Λ is well defined by Theorem 6.10.9 (i).

Lemma 6.12.4. *Let v_a and u_b be real solutions of $(L - \lambda_0)y = 0$, $\lambda_0 \in \mathbb{R}$, which are nonprincipal at a and principal at b, respectively. Assume that the endpoint a is in the limit-circle case and the endpoint b is in the limit-point case. Let $\mathfrak{t}$ be the form in (6.12.2)–(6.12.3). Then for every $\varepsilon > 0$ there exists $D_\varepsilon > 0$ such that*

$$|\Lambda f|^2 \le \varepsilon\, \mathfrak{t}[f] + D_\varepsilon \|f\|^2_{L^2_r(a,b)}, \quad f \in \mathfrak{D}.$$

Proof. Choose $\varepsilon > 0$ and write the form $\mathfrak{t}$ as the sum of

$$\begin{aligned}
\mathfrak{t}_{[c,d]}[f,g] &= \int_c^d \big((\sqrt{p}f')(x)\overline{(\sqrt{p}g')(x)} + q(x) f(x) \overline{g(x)} \big) dx \\
&\qquad + \frac{(pv_a)'(c)}{v_a(c)} f(c)\overline{g(c)} - \frac{(pv_b)'(d)}{u_b(d)} f(d)\overline{g(d)}, \\
\mathfrak{t}_{(a,c)}[f,g] &= \int_a^c (N_{v_a} f)(x) \overline{(N_{v_a} g)(x)}\, dx + \lambda_0 \int_a^c f(x)\overline{g(x)} r(x)\, dx, \\
\mathfrak{t}_{(b,d)}[f,g] &= \int_d^b (N_{u_b} f)(x) \overline{(N_{u_b} g)(x)}\, dx + \lambda_0 \int_d^b f(x)\overline{g(x)} r(x)\, dx.
\end{aligned}$$

It has been shown in Lemma 6.9.4 that $\mathfrak{t}$ is independent of the choice of c and d, and hence $c \in (a, a_0)$ and $d \in (b_0, b)$ can be chosen suitably close to a and b, respectively. In fact, choose c so close to a that

$$2 \int_a^c \frac{1}{p(t) v_a(t)^2}\, dt \le \varepsilon.$$

Then, as in the proof of Lemma 6.11.4, it follows that for all $f \in \mathfrak{D}$

$$\left|\lim_{x\to a}\frac{f(x)}{v_a(x)}\right|^2 \leq 2\left|\frac{f(c)}{v_a(c)}\right|^2 + \varepsilon\int_a^c |N_{v_a}f(t)|^2\,dt.$$

Consequently, with the above choice of c and any choice of $b_0 < d < b$ one sees that for all $f \in \mathfrak{D}$

$$\begin{aligned}\left|\lim_{x\to a}\frac{f(x)}{v_a(x)}\right|^2 &\leq 2\left|\frac{f(c)}{v_a(c)}\right|^2 + 2\left|\frac{f(d)}{u_b(d)}\right|^2\\ &\quad + \varepsilon\left(\int_a^c |N_{v_a}f(t)|^2\,dt + \int_d^b |N_{u_b}f(t)|^2\,dt\right).\end{aligned}$$

As in the proof of Lemma 6.11.4, an application of Corollary 6.8.6 shows that there exists $C_\varepsilon > 0$ such that for all $f \in \mathfrak{D}$ one has

$$\left|\frac{f(c)}{v_a(c)}\right|^2 + \left|\frac{f(d)}{v_b(d)}\right|^2 \leq C_\varepsilon\|f\|_{L^2_r(c,d)} + \varepsilon\mathfrak{t}_{[c,d]}[f].$$

The assertion follows by combining the above inequalities. □

The following lemma is the counterpart of Lemma 6.8.4 and Lemma 6.11.5 in the present situation.

Lemma 6.12.5. *Let v_a and u_b be solutions of $(L - \lambda_0)y = 0$, $\lambda_0 \in \mathbb{R}$, which are nonprincipal at a and principal at b, respectively. Assume that the endpoint a is in the limit-circle case and the endpoint b is in the limit-point case, and let $\{\mathbb{C}, \Gamma_0, \Gamma_1\}$ be the boundary triplet in Proposition 6.12.1. Then $\{\mathbb{C}, \Lambda\}$ is a boundary pair for $T_{\min}$ corresponding to S_1 which is compatible with the boundary triplet $\{\mathbb{C}, \Gamma_0, \Gamma_1\}$. Moreover, one has*

$$(T_{\max} f, g)_{L^2_r(a,b)} = (\Gamma_1 f, \Lambda g) + \mathfrak{t}[f,g], \quad f \in \operatorname{dom} T_{\max}, \ g \in \mathfrak{D}. \tag{6.12.5}$$

Proof. Consider the form $\mathfrak{t}$ defined on $\operatorname{dom}\mathfrak{t} = \mathfrak{D}$ as in (6.12.2)–(6.12.3) and denote the corresponding semibounded self-adjoint operator in $L^2_r(a,b)$ by S_1; cf. Corollary 6.12.2. Let $\varepsilon > 0$ and $D_\varepsilon > 0$ be as in Lemma 6.12.4. It follows from the estimate in Lemma 6.12.4 that for $\rho < m(S_1)$ there exists $C_{\rho,\varepsilon} > 0$ such that

$$|\Lambda f|^2 \leq D_\varepsilon\|f\|^2_{L^2_r(a,b)} + \varepsilon\mathfrak{t}[f] \leq C_{\rho,\varepsilon}\|f\|^2_{\mathfrak{t}_{S_1}-\rho}$$

for all $f \in \mathfrak{D}$. Therefore, $\Lambda \in \mathbf{B}(\mathfrak{H}_{\mathfrak{t}_{S_1}-\rho}, \mathbb{C})$. Moreover, by Lemma 6.12.3, one has $\operatorname{dom} T_{\max} \subset \mathfrak{D}$ and hence Λ is an extension of the boundary mapping Γ_0 in (6.12.1). Now Lemma 5.6.5 implies that $\{\mathbb{C}, \Lambda\}$ is a boundary pair for $T_{\min}$ corresponding to S_1.

In order to conclude that $\{\mathbb{C}, \Lambda\}$ and $\{\mathbb{C}, \Gamma_0, \Gamma_1\}$ are compatible it remains to show that $A_1 = S_1$, where A_1 is the self-adjoint operator defined on $\ker\Gamma_1$. In fact, due to $\operatorname{dom} T_{\max} \subset \mathfrak{D}$ the Green formula (6.9.15) is valid for $f \in \operatorname{dom} T_{\max}$ and $g \in \mathfrak{D}$

$$(T_{\max} f, g)_{L^2_r(a,b)} = \mathfrak{t}[f,g] + \lim_{b' \to b} W_{b'}(f, u_b) \left(\frac{\bar{g}}{u_b}\right)(b') - \lim_{a' \to a} W_{a'}(f, v_a) \left(\frac{\bar{g}}{v_a}\right)(a'),$$

where each of the limits exist. Now observe that

$$W(f, u_b) \left(\frac{\bar{g}}{u_b}\right) = -p u_b^2 \left(\frac{f}{u_b}\right)' \left(\frac{\bar{g}}{u_b}\right);$$

cf. (6.10.15). Since u_b is principal at b and $N_{u_b} f, N_{u_b} g \in L^2(d,b)$ for $b_0 < d < b$, it follows from Corollary 6.10.2 (applied to the endpoint b) with $P = p u_b^2$, $\varphi = f/u_b$, and $\psi = \bar{g}/u_b$, that

$$\lim_{b' \to b} W_{b'}(f, u_b) \left(\frac{\bar{g}}{u_b}\right)(b') = \liminf_{b' \to b} W_{b'}(f, u_b) \left(\frac{\bar{g}}{u_b}\right)(b') = 0.$$

Thus, in the present context, it follows that (6.12.5) holds. Hence,

$$(A_1 f, g)_{L^2_r(a,b)} = \mathfrak{t}[f,g]$$

holds for all $f \in \operatorname{dom} A_1$ and $g \in \operatorname{dom} \mathfrak{t}$. As A_1 is self-adjoint the first representation theorem implies $A_1 = S_1$. □

Recall that by means of the boundary triplet in Proposition 6.12.1, all the self-adjoint extensions of $T_{\min}$ are in a one-to-one correspondence to $\tau \in \mathbb{R} \cup \{\infty\}$ via

$$\operatorname{dom} A_\tau = \{ f \in \operatorname{dom} T_{\max} : \Gamma_1 f = \tau \Gamma_0 f \}, \tag{6.12.6}$$

where in case $\tau = \infty$ one means $\Gamma_0 f = 0$. The next result, which is an immediate consequence of Theorem 5.6.13 and Corollary 5.6.14, makes use of the compatible boundary pair in Lemma 6.12.5 and provides a characterization of all closed semibounded forms associated with the semibounded self-adjoint extensions A_τ. The boundary triplet and the compatible boundary pair are provided by the choice of the solution v_a of $(L - \lambda_0)y = 0$, which is nonprincipal at a.

Theorem 6.12.6. *Let $\{\mathbb{C}, \Gamma_0, \Gamma_1\}$ be the boundary triplet in Proposition 6.12.1, let $\tau \in \mathbb{R} \cup \{\infty\}$, and let A_τ be the corresponding self-adjoint restriction of $T_{\max}$ in (6.12.6). Then A_τ is semibounded from below and the corresponding densely defined closed semibounded form $\mathfrak{t}_\tau$ in $L^2_r(a,b)$ such that*

$$(A_\tau f, g)_{L^2_r(a,b)} = \mathfrak{t}_\tau[f,g], \quad f \in \operatorname{dom} A_\tau, \quad g \in \operatorname{dom} \mathfrak{t}_\tau,$$

is given in terms of $\mathfrak{t}$ in (6.12.2)–(6.12.3), and Λ in (6.12.4) as follows:

(i) *If* $\tau \in \mathbb{R}$, *then*

$$\mathfrak{t}_\tau[f,g] = \mathfrak{t}[f,g] + \tau(\Lambda f, \Lambda g), \quad \operatorname{dom} \mathfrak{t}_\tau = \mathfrak{D}.$$

(ii) *If* $\tau = \infty$, *then* $A_\tau = A_0$ *coincides with the Friedrichs extension* S_{F} *and*

$$\mathfrak{t}_\tau[f,g] = \mathfrak{t}[f,g], \quad \operatorname{dom} \mathfrak{t}_\tau = \{h \in \mathfrak{D} : \Lambda h = 0\}.$$

In the same way as in the previous section it will be discussed briefly what the results look like when there is a different choice for the nonoscillatory solution. Let w_a be a solution of $(L - \lambda_0)y = 0$ which is nonprincipal at a and assume that v_a and w_a do not vanish on (a, a_0). Denote the form generated by the solutions w_a and u_b by $\mathfrak{t}'$ and let u_a be a solution of $(L - \lambda_0)y = 0$ which is principal at a and which satisfies $W(u_a, v_a) = 1$. Then one has $\operatorname{dom} \mathfrak{t}' = \operatorname{dom} \mathfrak{t} = \mathfrak{D}$ and

$$w_a = \alpha_a v_a + \beta_a u_a$$

for some $\alpha_a, \beta_a \in \mathbb{R}$, where $\alpha_a \neq 0$. Denote the boundary triplet generated by w_a by $\{\mathbb{C}, \Gamma_0', \Gamma_1'\}$ and let $\{\mathbb{C}, \Lambda'\}$ be the corresponding boundary pair; cf. Proposition 6.12.1 and (6.12.4). Then the following result is clear; cf. Proposition 6.11.7.

Proposition 6.12.7. *The boundary triplet* $\{\mathbb{C}, \Gamma_0', \Gamma_1'\}$ *and the boundary pair* $\{\mathbb{C}, \Lambda'\}$ *generated by the nonprincipal solution* w_a *are given by*

$$\Lambda' f = \frac{1}{\alpha_a} \Lambda f, \quad f \in \mathfrak{D},$$

and

$$\Gamma_1' f = \alpha_a \Gamma_1 f + \beta_a \Lambda f, \quad f \in \operatorname{dom} T_{\max}.$$

Moreover, the form $\mathfrak{t}'$ *coincides with* $\mathfrak{t}_\tau$ *as in Theorem* 6.12.6, *where* $\tau \in \mathbb{R}$ *is given by*

$$\tau = -\frac{\beta_a}{\alpha_a}.$$

Next let u_a and u_b be nontrivial solutions of $(L-\lambda_0)y = 0$ which are principal at a and b, respectively, and assume that u_a does not vanish on (a, a_0) and that u_b does not vanish on (b_0, b). Denote the form generated by the solutions u_a and u_b by $\widetilde{\mathfrak{t}}$; cf. Theorem 6.9.6. Then the following analog of Proposition 6.11.8 holds.

Proposition 6.12.8. *The form* $\widetilde{\mathfrak{t}}$ *coincides with* $\mathfrak{t}_\infty$ *in Theorem* 6.12.6 *or, equivalently,* $\widetilde{\mathfrak{t}}$ *is the form generated by the Friedrichs extension*

$$\mathfrak{t}_{S_{\mathrm{F}}} = \widetilde{\mathfrak{t}}.$$

The following equivalent characterizations of the Friedrichs extension of $T_{\min}$ are mentioned for completeness; cf. Proposition 6.10.11 and Corollary 6.11.9.

Corollary 6.12.9. *Let* v_a *and* u_a *be solutions of* $(L - \lambda_0)y = 0$, $\lambda_0 \in \mathbb{R}$, *which are nonprincipal and principal at* a, *respectively. Assume that the endpoint* a *is in the* *limit-circle case and the endpoint b is in the limit-point case. Then* $f \in \operatorname{dom} T_{\max}$ *is in the domain of the Friedrichs extension* S_{F} *of* $T_{\min}$ *if and only if one of the following equivalent conditions holds:*

(i) $N_{u_a} f \in L^2(a, a')$;

(ii) $\lim_{x \to a} \frac{f(x)}{v_a(x)} = 0$;

(iii) $\lim_{x \to a} W_x(f, u_a) = 0$;

(iv) $\lim_{x \to a} f(x)/u_a(x)$ *exists in* $\mathbb{C}$.

Finally, the special case that the endpoint a is regular is briefly discussed.

Remark 6.12.10. The considerations in this section simplify if the endpoint a is regular. In that case one can choose the boundary triplet $\{\mathbb{C}, \Gamma_0, \Gamma_1\}$ in Proposition 6.4.1. The form $\mathfrak{t}$ and $\mathfrak{D}$ in (6.12.2) and (6.12.3) reduce to (6.9.19) and (6.9.20) with $\phi_b = u_b$ in Remark 6.9.9, respectively. The corresponding boundary pair $\Lambda : \mathfrak{D} \to \mathbb{C}$ in (6.11.4) has the form

$$\Lambda f = f(a), \quad f \in \mathfrak{D}.$$

6.13 Integrable potentials

In this section the Sturm–Liouville differential expression

$$L = -D^2 + q, \quad D = d/dx, \quad \text{with } q \in L^1(0, \infty) \text{ real},$$

is studied on the interval $(0, \infty)$; note that in this special case $r = p = 1$. It is clear that the endpoint 0 is regular. In the following lemma it will be shown that $-D^2 + q$ can be seen as a perturbation of the expression $-D^2$, in the sense that the fundamental solutions have the same asymptotic behavior as $x \mapsto e^{i\sqrt{\lambda}x}$ and $x \mapsto e^{-i\sqrt{\lambda}x}$; cf. Example 6.4.2. In this context it also follows that the endpoint ∞ is in the limit-point case. Throughout this section it will be tacitly assumed that the square root $\sqrt{\cdot}$ is fixed by the requirement that $\operatorname{Im} \sqrt{\lambda} > 0$ for all $\lambda \in \mathbb{C} \setminus [0, \infty)$ and $\sqrt{\lambda} \geq 0$ for $\lambda \in [0, \infty)$.

Lemma 6.13.1. *Assume that* $q \in L^1(0, \infty)$ *and that* $\lambda \in \mathbb{C} \setminus [0, \infty)$. *Then there is a fundamental system* $(e_1(\cdot, \lambda); e_2(\cdot, \lambda))$ *of the equation* $(L - \lambda)y = 0$ *such that*

$$e_1(x, \lambda) = e^{i\sqrt{\lambda}x}(1 + o(1)), \quad x \to \infty,$$
$$e_1'(x, \lambda) = i\sqrt{\lambda}e^{i\sqrt{\lambda}x}(1 + o(1)), \quad x \to \infty,$$

and

$$e_2(x, \lambda) = e^{-i\sqrt{\lambda}x}(1 + o(1)), \quad x \to \infty,$$
$$e_2'(x, \lambda) = -i\sqrt{\lambda}e^{-i\sqrt{\lambda}x}(1 + o(1)), \quad x \to \infty.$$

In particular, $e_1(\cdot, \lambda) \in L^2(0, \infty)$ *and* $e_2(\cdot, \lambda) \notin L^2(0, \infty)$ *for all* $\lambda \in \mathbb{C} \setminus [0, \infty)$.

Proof. Due to the integrability of q the nonnegative function

$$\sigma(x) = \int_x^\infty |q(t)|\,dt, \qquad x > 0,$$

is well defined, nonincreasing, and $\lim_{x\to\infty} \sigma(x) = 0$. The proof of the lemma will be given in three steps. In the first two steps each of the above solutions $e_1(\cdot, \lambda)$ and $e_2(\cdot, \lambda)$ is constructed; in the third step the linear independence is shown.

Step 1. Let $\lambda \in \mathbb{C} \setminus \{0\}$. It will be shown that there is a bounded function $\alpha(\cdot, \lambda)$ such that the integral equation

$$\alpha(x, \lambda) = 1 + \int_x^\infty \frac{e^{2i\sqrt{\lambda}(t-x)} - 1}{2i\sqrt{\lambda}}\, q(t)\, \alpha(t, \lambda)\, dt, \quad x > 0, \tag{6.13.1}$$

is satisfied. Note that for $t \geq x$ one has

$$|e^{2i\sqrt{\lambda}(t-x)} - 1| \leq 2.$$

Define the sequence of functions α_n, $n \in \mathbb{N} \cup \{0\}$, inductively by

$$\alpha_0(x, \lambda) = 1 \quad \text{and} \quad \alpha_{n+1}(x, \lambda) = \int_x^\infty \frac{e^{2i\sqrt{\lambda}(t-x)} - 1}{2i\sqrt{\lambda}}\, q(t)\, \alpha_n(t, \lambda)\, dt$$

for $x > 0$. Since σ is nonincreasing it is easily seen that

$$|\alpha_n(x, \lambda)| \leq \left(\frac{\sigma(x)}{|\sqrt{\lambda}|} \right)^n.$$

Now choose $\delta > 0$ such that $|\sqrt{\lambda}| \geq \delta$. Then there exists an $x_\delta > 0$ such that

$$\frac{\sigma(x_\delta)}{\delta} < 1, \tag{6.13.2}$$

and note that for all $x \geq x_\delta$ it follows that

$$\frac{\sigma(x)}{|\sqrt{\lambda}|} \leq \frac{\sigma(x_\delta)}{\delta} < 1.$$

For the function $\alpha(x, \lambda) = \sum_{n=0}^\infty \alpha_n(x, \lambda)$ defined for $x \geq x_\delta$ one has

$$|\alpha(x, \lambda)| \leq \sum_{n=0}^\infty |\alpha_n(x, \lambda)| \leq \sum_{n=0}^\infty \left(\frac{\sigma(x)}{|\sqrt{\lambda}|} \right)^n \leq \sum_{n=0}^\infty \left(\frac{\sigma(x_\delta)}{\delta} \right)^n = \frac{1}{1 - \frac{\sigma(x_\delta)}{\delta}}$$

for $x \geq x_\delta$. Hence, the function $\alpha(\cdot, \lambda)$ is well defined and bounded for $x \geq x_\delta$. By dominated convergence, it follows for $x \geq x_\delta$ that

$$\begin{aligned}\alpha(x,\lambda) &= 1 + \sum_{n=0}^{\infty} \alpha_{n+1}(x,\lambda)\\ &= 1 + \sum_{n=0}^{\infty} \int_x^\infty \frac{e^{2i\sqrt{\lambda}(t-x)} - 1}{2i\sqrt{\lambda}} q(t)\,\alpha_n(t,\lambda)\,dt\\ &= 1 + \int_x^\infty \frac{e^{2i\sqrt{\lambda}(t-x)} - 1}{2i\sqrt{\lambda}} q(t) \sum_{n=0}^{\infty} \alpha_n(t,\lambda)\,dt\\ &= 1 + \int_x^\infty \frac{e^{2i\sqrt{\lambda}(t-x)} - 1}{2i\sqrt{\lambda}} q(t)\,\alpha(t,\lambda)\,dt.\end{aligned}$$

Hence, the integral equation (6.13.1) is satisfied for all $x \geq x_\delta$. Note also that for $x > x_\delta$

$$\alpha'(x,\lambda) = -\int_x^\infty e^{2i\sqrt{\lambda}(t-x)}\,q(t)\,\alpha(t,\lambda)\,dt \tag{6.13.3}$$

and

$$\alpha''(x,\lambda) = 2i\sqrt{\lambda}\int_x^\infty e^{2i\sqrt{\lambda}(t-x)}\,q(t)\,\alpha(t,\lambda)\,dt + q(x)\alpha(x,\lambda). \tag{6.13.4}$$

It is clear that $|\alpha'(x,\lambda)| \to 0$ as $x \to \infty$.

Now consider the function $e_1(x,\lambda) = e^{i\sqrt{\lambda}x}\alpha(x,\lambda)$, defined for $x \geq x_\delta$. It follows from a straightforward computation and (6.13.3)–(6.13.4) that $e_1(\cdot,\lambda)$ satisfies the differential equation $(L-\lambda)e_1 = 0$ and the asymptotic properties of e_1 and e_1' for $x \to \infty$ are a consequence of the asymptotic properties of α in (6.13.1) and α' in (6.13.3) for $x \to \infty$. It remains to note that the solution e_1 on the interval (x_δ, ∞) can be extended to a solution on $(0,\infty)$.

Step 2. In this step it is assumed that $\lambda \in \mathbb{C} \setminus [0,\infty)$. As in Step 1, choose $\delta > 0$ and $x_\delta > 0$ such that $|\sqrt{\lambda}| \geq \delta$ and (6.13.2) hold. It will be shown that there exists a bounded function $\beta(\cdot,\lambda)$ such that the integral equation

$$\beta(x,\lambda) = 1 + \frac{1}{2i\sqrt{\lambda}}\int_{x_\delta}^x e^{2i\sqrt{\lambda}(x-t)}\,q(t)\,\beta(t,\lambda)\,dt + \frac{1}{2i\sqrt{\lambda}}\int_x^\infty q(t)\,\beta(t,\lambda)\,dt$$

is satisfied; in particular, then also the second integral is well defined.

Note first that for $x_\delta \leq t \leq x$ one has

$$|e^{2i\sqrt{\lambda}(x-t)}| \leq 1. \tag{6.13.5}$$

Define the sequence of functions β_n, $n \in \mathbb{N} \cup \{0\}$, inductively by $\beta_0(x,\lambda) = 1$ and

$$\beta_{n+1}(x,\lambda) = \frac{1}{2i\sqrt{\lambda}}\int_{x_\delta}^x e^{2i\sqrt{\lambda}(x-t)}\,q(t)\,\beta_n(t,\lambda)\,dt + \frac{1}{2i\sqrt{\lambda}}\int_x^\infty q(t)\,\beta_n(t,\lambda)\,dt$$

for $x > x_\delta$. In the same way as in Step 1 it follows that

$$|\beta_n(x,\lambda)| \le \left(\frac{\sigma(x_\delta)}{|\sqrt{\lambda}|}\right)^n \le \left(\frac{\sigma(x_\delta)}{\delta}\right)^n, \quad x \ge x_\delta,$$

and the function $\beta(x,\lambda) = \sum_{n=0}^\infty \beta_n(x,\lambda)$ is well defined for $x \ge x_\delta$, bounded by some constant $M_\beta \ge 0$, and solves the integral equation. For $x \ge x_\delta$ define the function $e_2(x,\lambda) = e^{-i\sqrt{\lambda}x}\beta(x,\lambda)$. Since for $x > x_\delta$ one has

$$\beta'(x,\lambda) = \int_{x_\delta}^x e^{2i\sqrt{\lambda}(x-t)}\, q(t)\, \beta(t,\lambda)\, dt$$

and

$$\beta''(x,\lambda) = 2i\sqrt{\lambda}\int_{x_\delta}^x e^{2i\sqrt{\lambda}(x-t)}\, q(t)\, \beta(t,\lambda)\, dt + q(x)\beta(x,\lambda),$$

it follows that $e_2(\cdot,\lambda)$ satisfies the differential equation $(L-\lambda)e_2 = 0$. For the asymptotic properties of e_2 and e_2' observe first that

$$\begin{aligned}&\int_{x_\delta}^x e^{2i\sqrt{\lambda}(x-t)}\, q(t)\, \beta(t,\lambda)\, dt \\ &\qquad = e^{i\sqrt{\lambda}x}\int_{x_\delta}^{x/2} e^{i\sqrt{\lambda}(x-2t)}\, q(t)\, \beta(t,\lambda)\, dt + \int_{x/2}^x e^{2i\sqrt{\lambda}(x-t)}\, q(t)\, \beta(t,\lambda)\, dt\end{aligned} \tag{6.13.6}$$

tends to 0 for $x \to \infty$. In fact, since $|e^{i\sqrt{\lambda}(x-2t)}| \le 1$ for $x_\delta \le t \le x/2$ the first term on the right-hand side in (6.13.6) satisfies the estimate

$$\left|e^{i\sqrt{\lambda}x}\int_{x_\delta}^{x/2} e^{i\sqrt{\lambda}(x-2t)}\, q(t)\, \beta(t,\lambda)\, dt\right| \le \left|e^{-\operatorname{Im}\sqrt{\lambda}x}\right| M_\beta \int_{x_\delta}^{x/2} |q(t)|\, dt,$$

and hence tends to 0 for $x \to \infty$ as $\operatorname{Im}\sqrt{\lambda} > 0$ and $q \in L^1(0,\infty)$. Similarly, the second term on the right-hand side in (6.13.6) tends to 0 for $x \to \infty$ by (6.13.5), $|\beta(t,\lambda)| \le M_\beta$, and $q \in L^1(0,\infty)$. Now the asymptotic properties of e_2 and e_2' for $x \to \infty$ follow from the asymptotic properties of β and β' for $x \to \infty$. Finally, the solution e_2 can be extended to a solution on $(0,\infty)$.

Step 3. Since the Wronskian $W(e_1(\cdot,\lambda), e_2(\cdot,\lambda))$ is constant, it follows from the asymptotic behavior of $e_1(\cdot,\lambda)$ and $e_2(\cdot,\lambda)$ that

$$W(e_1(\cdot,\lambda), e_2(\cdot,\lambda)) = -2i\sqrt{\lambda}.$$

Hence, $e_1(\cdot,\lambda)$ and $e_2(\cdot,\lambda)$ form a fundamental system. □

It is a direct consequence of Lemma 6.13.1 that the defect numbers of $T_{\min}$ are $(1,1)$. Hence, the endpoint ∞ is in the limit-point case and $\{\mathbb{C}, \Gamma_0, \Gamma_1\}$, where Γ_0 and Γ_1 are defined by

$$\Gamma_0 f = f(0) \quad \text{and} \quad \Gamma_1 f = f'(0), \quad f \in \operatorname{dom} T_{\max}, \tag{6.13.7}$$

is a boundary triplet for $T_{\max}$; cf. Proposition 6.4.1. The self-adjoint restriction A_0 of $T_{\max}$ is given by

$$A_0 = -D^2 + q, \quad \operatorname{dom} A_0 = \{f \in \operatorname{dom} T_{\max} : f(0) = 0\}.$$

Let $u_1(\cdot, \lambda)$ and $u_2(\cdot, \lambda)$ be the fundamental system corresponding to the usual initial conditions, that is, $u_1(\cdot, \lambda)$ and $u_2(\cdot, \lambda)$ satisfy

$$\begin{pmatrix} u_1(0,\lambda) & u_2(0,\lambda) \\ u_1'(0,\lambda) & u_2'(0,\lambda) \end{pmatrix} = \begin{pmatrix} 1 & 0 \\ 0 & 1 \end{pmatrix}. \tag{6.13.8}$$

Then the Weyl function belonging to the boundary triplet in (6.13.7) is uniquely defined by the property

$$u_1(\cdot, \lambda) + M(\lambda) u_2(\cdot, \lambda) \in L^2(0, \infty), \quad \lambda \in \mathbb{C} \setminus \mathbb{R}; \tag{6.13.9}$$

cf. Proposition 6.4.1. In order to determine the corresponding Weyl function one has to compare the fundamental system $(u_1(\cdot, \lambda); u_2(\cdot, \lambda))$ with the fundamental system $(e_1(\cdot, \lambda); e_2(\cdot, \lambda))$ from Lemma 6.13.1.

An important ingredient for the following considerations is *Gronwall's lemma.*

Lemma 6.13.2. *Let f be a continuous complex function on $[c, b)$.*

(i) *Assume that $\alpha, \beta \in L^1_{\mathrm{loc}}[c, b)$ are nonnegative functions and that α is nondecreasing. If*

$$|f(t)| \leq \alpha(t) + \int_c^t \beta(s)|f(s)|\, ds, \quad t \in (c, b), \tag{6.13.10}$$

then f satisfies the inequality

$$|f(t)| \leq \alpha(t)\, e^{\int_c^t \beta(u)\, du}, \quad t \in (c, b). \tag{6.13.11}$$

(ii) *Assume that $\beta \in L^1(c, b)$ is a nonnegative function and that $\alpha \geq 0$ is a constant. If $\beta f \in L^1(c, b)$ and*

$$|f(t)| \leq \alpha + \int_t^b \beta(s)|f(s)|\, ds, \quad t \in [c, b), \tag{6.13.12}$$

then f satisfies the inequality

$$|f(t)| \leq \alpha\, e^{\int_t^b \beta(u)\, du}, \quad t \in [c, b). \tag{6.13.13}$$

Proof. (i) It follows from the inequality in (6.13.10) that

$$\begin{aligned}\frac{d}{ds}\left(e^{-\int_c^s \beta(u)\,du}\int_c^s \beta(u)|f(u)|\,du\right) &= \left[|f(s)| - \int_c^s \beta(u)|f(u)|\,du\right]\beta(s)e^{-\int_c^s \beta(u)\,du} \\ &\leq \alpha(s)\beta(s)e^{-\int_c^s \beta(u)\,du}\end{aligned}$$

almost everywhere. Integration of this inequality over the interval $[c,t]$ leads to

$$e^{-\int_c^t \beta(u)\,du}\int_c^t \beta(u)|f(u)|\,du \leq \int_c^t \alpha(s)\beta(s)e^{-\int_c^s \beta(u)\,du}\,ds$$

or, equivalently,

$$\int_c^t \beta(u)|f(u)|\,du \leq \int_c^t \alpha(s)\beta(s)e^{\int_s^t \beta(u)\,du}\,ds.$$

Due to (6.13.10) and the assumption that α is nondecreasing one obtains

$$\begin{aligned}|f(t)| - \alpha(t) &\leq \int_c^t \beta(u)|f(u)|\,du \\ &\leq \alpha(t)\int_c^t \beta(s)e^{\int_s^t \beta(u)\,du}\,ds \\ &= -\alpha(t)\int_c^t \frac{d}{ds}e^{\int_s^t \beta(u)\,du}\,ds \\ &= -\alpha(t)\left(1 - e^{\int_c^t \beta(u)\,du}\right),\end{aligned}$$

which gives (6.13.11).

(ii) It follows from the inequality in (6.13.12) that

$$\begin{aligned}\frac{d}{ds}\left(e^{-\int_s^b \beta(u)\,du}\int_s^b \beta(u)|f(u)|\,du\right) &= \left[\int_s^b \beta(u)|f(u)|\,du - |f(s)|\right]\beta(s)e^{-\int_s^b \beta(u)\,du} \\ &\geq -\alpha\beta(s)e^{-\int_s^b \beta(u)\,du}\end{aligned}$$

almost everywhere. Integration of this inequality over the interval $[t,b]$ leads to

$$-e^{-\int_t^b \beta(u)\,du}\int_t^b \beta(u)|f(u)|\,du \geq -\alpha\int_t^b \beta(s)e^{-\int_s^b \beta(u)\,du}\,ds$$

or, equivalently,

$$\begin{aligned}\int_t^b \beta(u)|f(u)|\,du &\leq \alpha \int_t^b \beta(s)e^{\int_t^s \beta(u)\,du}\,ds \\ &= \alpha \int_t^b \frac{d}{ds}e^{\int_t^s \beta(u)\,du}\,ds \\ &= \alpha \left(e^{\int_t^b \beta(u)\,du} - 1\right).\end{aligned}$$

Due to (6.13.12) one obtains (6.13.13). □

The next lemma on the asymptotic properties of solutions of $(L-\lambda)u = 0$ is the first step to determine the Weyl function corresponding to the boundary triplet in (6.13.7).

Lemma 6.13.3. *Assume that $q \in L^1(0,\infty)$, let $\lambda \in \mathbb{C} \setminus [0,\infty)$ and $c_1, c_2 \in \mathbb{C}$. Let $u(\cdot,\lambda)$ be a solution of $(L-\lambda)u = 0$ satisfying*

$$(-D^2 + q)u = \lambda u, \quad u(0,\lambda) = c_1, \quad u'(0,\lambda) = c_2. \tag{6.13.14}$$

Then

$$|e^{i\sqrt{\lambda}x}u(x,\lambda)| \leq \left(|c_1| + \frac{|c_2|}{|\sqrt{\lambda}|}\right)\exp\left(\frac{1}{|\sqrt{\lambda}|}\int_0^x |q(t)|\,dt\right) \tag{6.13.15}$$

and

$$u(x,\lambda) = e^{-i\sqrt{\lambda}x}\left(\frac{c_1}{2} - \frac{c_2}{2i\sqrt{\lambda}} - \frac{1}{2i\sqrt{\lambda}}\int_0^\infty e^{i\sqrt{\lambda}t}\,q(t)u(t,\lambda)\,dt + o(1)\right)$$

as $x \to \infty$.

Proof. A simple computation shows that for $\lambda \in \mathbb{C} \setminus \{0\}$ the unique solution of (6.13.14) is given by

$$u(x,\lambda) = c_1 \cos\sqrt{\lambda}x + c_2\frac{\sin\sqrt{\lambda}x}{\sqrt{\lambda}} + \int_0^x \frac{\sin\sqrt{\lambda}(x-t)}{\sqrt{\lambda}}q(t)u(t,\lambda)\,dt. \tag{6.13.16}$$

It follows from (6.13.16) that $\varphi(x,\lambda) = e^{-(\operatorname{Im}\sqrt{\lambda})x}u(x,\lambda)$ satisfies

$$|\varphi(x,\lambda)| \leq |c_1| + \frac{|c_2|}{|\sqrt{\lambda}|} + \frac{1}{|\sqrt{\lambda}|}\int_0^x |q(t)|\,|\varphi(t,\lambda)|\,dt,$$

and hence Lemma 6.13.2 (i) leads to

$$|\varphi(x,\lambda)| \leq \left(|c_1| + \frac{|c_2|}{|\sqrt{\lambda}|}\right)\exp\left(\frac{1}{|\sqrt{\lambda}|}\int_0^x |q(t)|\,dt\right).$$

Since $|e^{i\sqrt{\lambda}x}u(x,\lambda)| = |e^{-(\operatorname{Im}\sqrt{\lambda})x}u(x,\lambda)| = |\varphi(x,\lambda)|$, the estimate (6.13.15) follows.

Furthermore, (6.13.16) yields that

$$\begin{aligned} u(x,\lambda) = e^{-i\sqrt{\lambda}\,x}\left(\frac{c_1}{2} - \frac{c_2}{2i\sqrt{\lambda}}\right) + e^{i\sqrt{\lambda}\,x}\left(\frac{c_1}{2} + \frac{c_2}{2i\sqrt{\lambda}}\right) \\ - \frac{1}{2i\sqrt{\lambda}}\int_0^x e^{-i\sqrt{\lambda}(x-t)}q(t)u(t,\lambda)\,dt \\ + \frac{1}{2i\sqrt{\lambda}}\int_0^x e^{i\sqrt{\lambda}(x-t)}q(t)u(t,\lambda)\,dt. \end{aligned} \tag{6.13.17}$$

For the second term on the right-hand side of (6.13.17) with $\lambda \in \mathbb{C} \setminus [0,\infty)$ one has

$$e^{i\sqrt{\lambda}\,x}\left(\frac{c_1}{2} + \frac{c_2}{2i\sqrt{\lambda}}\right) = e^{-i\sqrt{\lambda}x}o(1), \quad x \to \infty.$$

To estimate the third term on the right-hand side of (6.13.17) note first that this term is equal to

$$e^{-i\sqrt{\lambda}x}\left[-\frac{1}{2i\sqrt{\lambda}}\int_0^x e^{i\sqrt{\lambda}t}q(t)u(t,\lambda)\,dt\right].$$

Next one has

$$-\frac{1}{2i\sqrt{\lambda}}\int_x^\infty e^{i\sqrt{\lambda}t}q(t)u(t,\lambda)\,dt = o(1), \quad x \to \infty.$$

In fact, this holds since $t \mapsto e^{i\sqrt{\lambda}t}u(t,\lambda)$ is bounded by (6.13.15) and

$$\left|\int_x^\infty e^{i\sqrt{\lambda}t}q(t)u(t,\lambda)\,dt\right| \le C\int_x^\infty |q(t)|\,dt.$$

Therefore, the third term on the right-hand side of (6.13.17) has the form

$$e^{-i\sqrt{\lambda}x}\left(-\frac{1}{2i\sqrt{\lambda}}\int_0^\infty e^{i\sqrt{\lambda}t}\,q(t)u(t,\lambda)\,dt + o(1)\right).$$

For the fourth term on the right-hand side of (6.13.17) one uses again that $t \mapsto e^{i\sqrt{\lambda}t}u(t,\lambda)$ is bounded. Then

$$\left|\frac{1}{2i\sqrt{\lambda}}\int_0^x e^{i\sqrt{\lambda}(x-t)}q(t)u(t,\lambda)\,dt\right| \le \frac{C}{|\sqrt{\lambda}|}\left(\int_0^x e^{(\operatorname{Im}\sqrt{\lambda})(2t-x)}|q(t)|\,dt\right),$$

and splitting the interval of integration leads to

$$\begin{aligned}
&\int_0^x e^{(\mathrm{Im}\,\sqrt{\lambda})(2t-x)}|q(t)|\,dt\\
&\quad = \int_0^{x/2} e^{(\mathrm{Im}\,\sqrt{\lambda})(2t-x)}|q(t)|\,dt + \int_{x/2}^x e^{(\mathrm{Im}\,\sqrt{\lambda})(2t-x)}|q(t)|\,dt\\
&\quad \leq \int_0^{x/2} |q(t)|\,dt + e^{(\mathrm{Im}\,\sqrt{\lambda})x}\int_{x/2}^x |q(t)|\,dt\\
&\quad = e^{-i\sqrt{\lambda}x}o(1), \quad x\to\infty.
\end{aligned}$$

This completes the proof, as the last assertion in the lemma now follows from (6.13.17). □

The next proposition is a consequence of Lemma 6.13.1 and Lemma 6.13.3. Here the Weyl function M in (6.13.9) is specified.

Proposition 6.13.4. *Let M be the Weyl function corresponding to the boundary triplet $\{\mathbb{C},\Gamma_0,\Gamma_1\}$ in (6.13.7). Then there exists a countable (possibly empty) set $\mathcal{D}\subset(-\infty,0)$ which is bounded from below and may only accumulate at 0, such that M is holomorphic on $\mathbb{C}\setminus([0,\infty)\cup\mathcal{D})$ and*

$$M(\lambda)=\frac{i\sqrt{\lambda}-\int_0^\infty e^{i\sqrt{\lambda}t}q(t)u_1(t,\lambda)\,dt}{1+\int_0^\infty e^{i\sqrt{\lambda}t}q(t)u_2(t,\lambda)\,dt},\qquad \lambda\in\mathbb{C}\setminus([0,\infty)\cup\mathcal{D}). \tag{6.13.18}$$

Proof. In order to determine the Weyl function M corresponding to the boundary triplet $\{\mathbb{C},\Gamma_0,\Gamma_1\}$ observe first that for $\lambda\in\mathbb{C}\setminus[0,\infty)$ the fundamental systems $(u_1(\cdot,\lambda);u_2(\cdot,\lambda))$ and $(e_1(\cdot,\lambda);e_2(\cdot,\lambda))$ in Lemma 6.13.1 are connected by

$$\begin{aligned}
u_1(\cdot,\lambda)&=A_{11}(\lambda)e_1(\cdot,\lambda)+A_{12}(\lambda)e_2(\cdot,\lambda),\\
u_2(\cdot,\lambda)&=A_{21}(\lambda)e_1(\cdot,\lambda)+A_{22}(\lambda)e_2(\cdot,\lambda),
\end{aligned} \tag{6.13.19}$$

where $A_{ij}(\lambda)$, $i,j=1,2$, are connection coefficients. Since

$$\begin{aligned}
u_1(\cdot,\lambda)+M(\lambda)u_2(\cdot,\lambda)&=e_1(\cdot,\lambda)\big(A_{11}(\lambda)+M(\lambda)A_{21}(\lambda)\big)\\
&\quad+e_2(\cdot,\lambda)\big(A_{12}(\lambda)+M(\lambda)A_{22}(\lambda)\big)
\end{aligned}$$

and $e_1(\cdot,\lambda)\in L^2(0,\infty)$, $e_2(\cdot,\lambda)\notin L^2(0,\infty)$ for $\lambda\in\mathbb{C}\setminus[0,\infty)$ by Lemma 6.13.1, it follows that for $\lambda\in\mathbb{C}\setminus[0,\infty)$ the function $u_1(\cdot,\lambda)+M(\lambda)u_2(\cdot,\lambda)$ belongs to $L^2(0,\infty)$ if and only if $M(\lambda)$ satisfies the equation

$$A_{12}(\lambda)+M(\lambda)A_{22}(\lambda)=0. \tag{6.13.20}$$

Hence, it remains to compute the connection coefficients $A_{12}(\lambda)$ and $A_{22}(\lambda)$. It follows from Lemma 6.13.3 with $c_1=1$ and $c_2=0$ that

$$u_1(x,\lambda)=e^{-i\sqrt{\lambda}x}\left(\frac{1}{2}-\frac{1}{2i\sqrt{\lambda}}\int_0^\infty e^{i\sqrt{\lambda}t}q(t)u_1(t,\lambda)\,dt+o(1)\right),$$

and with $c_1 = 0$ and $c_2 = 1$ that

$$u_2(x,\lambda) = e^{-i\sqrt{\lambda}x}\left(-\frac{1}{2i\sqrt{\lambda}} - \frac{1}{2i\sqrt{\lambda}}\int_0^\infty e^{i\sqrt{\lambda}t}\,q(t)u_2(t,\lambda)\,dt + o(1)\right).$$

Comparing with (6.13.19) and Lemma 6.13.1, and taking care of the terms involving $o(1)$, this gives

$$A_{12}(\lambda) = \frac{1}{2} - \frac{1}{2i\sqrt{\lambda}}\int_0^\infty e^{i\sqrt{\lambda}t}\,q(t)u_1(t,\lambda)\,dt$$

and, likewise,

$$A_{22}(\lambda) = -\frac{1}{2i\sqrt{\lambda}} - \frac{1}{2i\sqrt{\lambda}}\int_0^\infty e^{i\sqrt{\lambda}t}\,q(t)u_2(t,\lambda)\,dt.$$

Hence, for $\lambda \in \mathbb{C}\setminus[0,\infty)$ such that $A_{22}(\lambda) \neq 0$ it follows from (6.13.20) that

$$M(\lambda) = -\frac{A_{12}(\lambda)}{A_{22}(\lambda)} = \frac{\frac{1}{2} - \frac{1}{2i\sqrt{\lambda}}\int_0^\infty e^{i\sqrt{\lambda}t}\,q(t)u_1(t,\lambda)\,dt}{\frac{1}{2i\sqrt{\lambda}} + \frac{1}{2i\sqrt{\lambda}}\int_0^\infty e^{i\sqrt{\lambda}t}\,q(t)u_2(t,\lambda)\,dt},$$

which leads to the expression for M in (6.13.18). Since the Weyl function is holomorphic on $\rho(A_0)$, it is clear that $A_{22}(\lambda) \neq 0$ for all $\lambda \in \mathbb{C}\setminus\mathbb{R}$. Note also that the functions A_{12} and A_{22} are both holomorphic in $\mathbb{C}\setminus[0,\infty)$, and that the zeros of A_{22} in $(-\infty,0)$ may only accumulate at 0 and $-\infty$. However, (6.13.15) shows that

$$|e^{i\sqrt{\lambda}x}u_2(x,\lambda)| \leq \frac{1}{|\sqrt{\lambda}|}\exp\left(\frac{1}{|\sqrt{\lambda}|}\int_0^x |q(t)|\,dt\right),$$

and hence there exists $C_- \in (-\infty,0)$ such that for all $\lambda \in (-\infty, C_-)$

$$\left|\int_0^\infty e^{i\sqrt{\lambda}t}q(t)u_2(t,\lambda)\,dt\right| < 1.$$

Therefore, $A_{22}(\lambda) \neq 0$ for all $\lambda \in (-\infty, C_-)$ and 0 is the only possible accumulation point of the zeros of A_{22} in the interval $(-\infty,0)$. This completes the proof of the proposition. □

It will turn out in Corollary 6.13.6 that the set $\mathbb{C}\setminus([0,\infty)\cup\mathcal{D})$ coincides with the resolvent set of the self-adjoint operator A_0, so that the form of the Weyl function M in Proposition 6.13.4 is valid for all $\lambda \in \rho(A_0)$.

In the following lemma, which complements Proposition 6.13.4, it will be shown that the Weyl function admits a continuation onto $(0,\infty)$ from $\mathbb{C}^+$ with a positive imaginary part.

Lemma 6.13.5. *Let M be the Weyl function corresponding to the boundary triplet $\{\mathbb{C},\Gamma_0,\Gamma_1\}$ in (6.13.7). Then the limits* $\lim_{\varepsilon\downarrow 0} M(\lambda+i\varepsilon)$ *and* $\lim_{\varepsilon\downarrow 0}\operatorname{Im} M(\lambda+i\varepsilon)$

exist for all $\lambda > 0$ *and are given by*

$$M(\lambda + i0) = -\frac{a_{11}(\lambda) + ia_{12}(\lambda)}{a_{21}(\lambda) + ia_{22}(\lambda)}, \qquad \lambda > 0, \tag{6.13.21}$$

and

$$\operatorname{Im} M(\lambda + i0) = \frac{1}{\sqrt{\lambda}} \frac{1}{a_{21}(\lambda)^2 + a_{22}(\lambda)^2} > 0, \qquad \lambda > 0, \tag{6.13.22}$$

respectively, where $a_{ij} : (0, \infty) \to \mathbb{R}$, $i, j = 1, 2$, *are the coefficients in the asymptotic formulas*

$$\begin{aligned} u_1(x, \lambda) &= a_{11}(\lambda) \cos \sqrt{\lambda} x + a_{12}(\lambda) \sin \sqrt{\lambda} x + o(1), \quad x \to \infty, \\ u_2(x, \lambda) &= a_{21}(\lambda) \cos \sqrt{\lambda} x + a_{22}(\lambda) \sin \sqrt{\lambda} x + o(1), \quad x \to \infty, \end{aligned} \tag{6.13.23}$$

for $\lambda > 0$. *The functions* a_{ij} *have the form*

$$\begin{aligned} a_{11}(\lambda) &= 1 - \frac{1}{\sqrt{\lambda}} \int_0^\infty \sin \sqrt{\lambda} t \, q(t) u_1(t, \lambda) \, dt, \\ a_{12}(\lambda) &= \frac{1}{\sqrt{\lambda}} \int_0^\infty \cos \sqrt{\lambda} t \, q(t) u_1(t, \lambda) \, dt, \\ a_{21}(\lambda) &= -\frac{1}{\sqrt{\lambda}} \int_0^\infty \sin \sqrt{\lambda} t \, q(t) u_2(t, \lambda) \, dt, \\ a_{22}(\lambda) &= \frac{1}{\sqrt{\lambda}} \left(1 + \int_0^\infty \cos \sqrt{\lambda} t \, q(t) u_2(t, \lambda) \, dt \right), \end{aligned} \tag{6.13.24}$$

and satisfy

$$a_{11}(\lambda) a_{22}(\lambda) - a_{12}(\lambda) a_{21}(\lambda) = \frac{1}{\sqrt{\lambda}}, \quad \lambda > 0. \tag{6.13.25}$$

Proof. It follows from Lemma 6.13.2 (i) that the solution

$$u(x, \lambda) = c_1 \cos \sqrt{\lambda} x + c_2 \frac{\sin \sqrt{\lambda} x}{\sqrt{\lambda}} + \int_0^x \frac{\sin \sqrt{\lambda}(x - t)}{\sqrt{\lambda}} q(t) u(t, \lambda) \, dt$$

of (6.13.14) is bounded for all $\lambda > 0$; cf. the proof of Lemma 6.13.3. Therefore,

$$\begin{aligned} u(x, \lambda) = c_1 \cos \sqrt{\lambda} x &+ c_2 \frac{\sin \sqrt{\lambda} x}{\sqrt{\lambda}} \\ &+ \int_0^\infty \frac{\sin \sqrt{\lambda}(x - t)}{\sqrt{\lambda}} q(t) u(t, \lambda) \, dt + o(1) \end{aligned} \tag{6.13.26}$$

and in the same way one obtains

$$\begin{aligned} u'(x, \lambda) = -c_1 \sqrt{\lambda} \sin \sqrt{\lambda} x &+ c_2 \cos \sqrt{\lambda} x \\ &+ \int_0^\infty \cos \sqrt{\lambda}(x - t) \, q(t) u(t, \lambda) \, dt + o(1). \end{aligned} \tag{6.13.27}$$

From (6.13.26) and

$$\int_0^\infty \frac{\sin\sqrt{\lambda}(x-t)}{\sqrt{\lambda}} q(t)u(t,\lambda)\,dt = \frac{\sin\sqrt{\lambda}x}{\sqrt{\lambda}} \int_0^\infty \cos\sqrt{\lambda}t\, q(t)u(t,\lambda)\,dt - \frac{\cos\sqrt{\lambda}x}{\sqrt{\lambda}} \int_0^\infty \sin\sqrt{\lambda}t\, q(t)u(t,\lambda)\,dt$$

one then derives for $\lambda > 0$ the asymptotic formulas (6.13.23), where the coefficient functions a_{ij} are as in (6.13.24). Similarly, from (6.13.27) and

$$\int_0^\infty \cos\sqrt{\lambda}(x-t)\, q(t)u(t,\lambda)\,dt = \sin\sqrt{\lambda}x \int_0^\infty \sin\sqrt{\lambda}t\, q(t)u(t,\lambda)\,dt + \cos\sqrt{\lambda}x \int_0^\infty \cos\sqrt{\lambda}t\, q(t)u(t,\lambda)\,dt$$

one obtains for the derivatives

$$\begin{aligned} u_1'(x,\lambda) &= -a_{11}(\lambda)\sqrt{\lambda}\sin\sqrt{\lambda}x + a_{12}(\lambda)\sqrt{\lambda}\cos\sqrt{\lambda}x + o(1), \quad x\to\infty, \\ u_2'(x,\lambda) &= -a_{21}(\lambda)\sqrt{\lambda}\sin\sqrt{\lambda}x + a_{22}(\lambda)\sqrt{\lambda}\cos\sqrt{\lambda}x + o(1), \quad x\to\infty. \end{aligned}$$

In view of the initial values of $u_1(\cdot,\lambda)$ and $u_2(\cdot,\lambda)$, their Wronskian satisfies

$$1 = W(u_1(\cdot,\lambda), u_2(\cdot,\lambda)) = \sqrt{\lambda}\big(a_{11}(\lambda)a_{22}(\lambda) - a_{12}(\lambda)a_{21}(\lambda)\big) + o(1)$$

as $x \to \infty$, and hence (6.13.25) follows.

To complete the proof of (6.13.21) and (6.13.22), it remains to note that for $\lambda > 0$ the limits

$$\lim_{\varepsilon\downarrow 0}\left(i\sqrt{\lambda+i\varepsilon} - \int_0^\infty e^{i\sqrt{\lambda+i\varepsilon}t} q(t)u_1(t,\lambda+i\varepsilon)\,dt\right)$$

and

$$\lim_{\varepsilon\downarrow 0}\left(1 + \int_0^\infty e^{i\sqrt{\lambda+i\varepsilon}t} q(t)u_2(t,\lambda+i\varepsilon)\,dt\right)$$

exist and are given by

$$i\sqrt{\lambda} - \int_0^\infty e^{i\sqrt{\lambda}t} q(t)u_1(t,\lambda)\,dt \quad \text{and} \quad 1 + \int_0^\infty e^{i\sqrt{\lambda}t} q(t)u_2(t,\lambda)\,dt,$$

respectively, so that the statements follow from the representation of M in Proposition 6.13.4 and the form of the coefficient functions a_{ij}. Note that $a_{21}(\lambda)$ and $a_{22}(\lambda)$ do not vanish simultaneously for any $\lambda > 0$ by (6.13.25). □

In the next corollary the spectral properties of the self-adjoint operator A_0 with Dirichlet boundary condition at 0 are discussed.

Corollary 6.13.6. *Let $\{\mathbb{C}, \Gamma_0, \Gamma_1\}$ be the boundary triplet in (6.13.7) and consider the self-adjoint operator*

$$A_0 = -D^2 + q, \quad \operatorname{dom} A_0 = \{f \in \operatorname{dom} T_{\max} : f(0) = 0\}.$$

Then the following holds for the spectrum of A_0:

(i) *$\sigma_{\mathrm{ac}}(A_0) = [0, \infty)$;*

(ii) *$\sigma_{\mathrm{p}}(A_0) \cap (0, \infty) = \emptyset$ and $\sigma_{\mathrm{sc}}(A_0) \cap (0, \infty) = \emptyset$;*

(iii) *$(\sigma(A_0) \cap (-\infty, 0)) \subset \sigma_{\mathrm{p}}(A_0)$ is at most countable, bounded from below, and may only accumulate at 0; each eigenvalue has multiplicity one.*

Proof. In order to apply the results from Chapter 3, recall first that, by Proposition 6.4.4, the minimal operator $T_{\min}$ is simple. It follows from Proposition 6.13.4 that the spectrum of A_0 in $(-\infty, 0)$ consists of at most countably many eigenvalues which are bounded from below and may only accumulate at 0. Since the singular endpoint ∞ is in the limit-point case, each eigenvalue has multiplicity one. This shows (iii). From Theorem 3.6.5 and Lemma 6.13.5 one then concludes that

$$\sigma_{\mathrm{ac}}(A_0) = \operatorname{clos}_{\mathrm{ac}}\big(\{\lambda \in \mathbb{R} : 0 < \operatorname{Im} M(\lambda + i0) < +\infty\}\big) = [0, \infty),$$

i.e., (i) holds. According to Lemma 6.13.5, $M(\lambda + i0)$ exists for all $\lambda > 0$ and hence $\mathcal{R}_\lambda = \lim_{\varepsilon \downarrow 0} i\varepsilon M(\lambda + i\varepsilon) = 0$ for all $\lambda > 0$. That $\sigma_{\mathrm{p}}(A_0) \cap (0, \infty) = \emptyset$ follows from Theorem 3.5.5 and Corollary 3.5.6 (see also Theorem 3.6.1). Finally, that $\sigma_{\mathrm{sc}}(A_0) \cap (0, \infty) = \emptyset$ follows from Theorem 3.6.8 (see also Corollary 3.6.9) and $0 < \operatorname{Im} M(\lambda + i0) < +\infty$ in Lemma 6.13.5. □

Recall from (6.4.8) that for $\tau \in \mathbb{R}$ the Weyl function corresponding to the boundary triplet $\{\mathbb{C}, \Gamma_0^\tau, \Gamma_1^\tau\}$ in (6.4.7) is given by

$$M_\tau(\lambda) = \frac{1 + \tau M(\lambda)}{\tau - M(\lambda)} \tag{6.13.28}$$

and that the self-adjoint restriction of $T_{\max}$ corresponding to $\ker \Gamma_0^\tau$ is

$$A_\tau = -D^2 + q, \qquad \operatorname{dom} A_\tau = \{f \in \operatorname{dom} T_{\max} : f'(0) = \tau f(0)\}.$$

For A_τ one obtains a statement similar to Corollary 6.13.6.

Proposition 6.13.7. *Let $\tau \in \mathbb{R}$ and let $\{\mathbb{C}, \Gamma_0^\tau, \Gamma_1^\tau\}$ be the boundary triplet in (6.4.7) with A_τ as above. Then the following holds for the spectrum of A_τ:*

(i) *$\sigma_{\mathrm{ac}}(A_\tau) = [0, \infty)$;*

(ii) *$\sigma_{\mathrm{p}}(A_\tau) \cap (0, \infty) = \emptyset$ and $\sigma_{\mathrm{sc}}(A_\tau) \cap (0, \infty) = \emptyset$;*

(iii) *$(\sigma(A_\tau) \cap (-\infty, 0)) \subset \sigma_{\mathrm{p}}(A_\tau)$ is at most countable, bounded from below, and may only accumulate at 0; each eigenvalue has multiplicity one.*

Proof. It follows from Proposition 6.13.4 that $1+\tau M$ and $\tau - M$ are holomorphic on $\mathbb{C} \setminus [0,\infty)$ with the possible exception of a countable set of poles in $(-\infty, 0)$ which may only accumulate at 0. Furthermore, since M is a Nevanlinna function, it is nondecreasing in between two consecutive poles in $(-\infty,0)$ and hence $\tau - M$ has at most countable many zeros in $(-\infty,0)$ which may only accumulate at 0. Therefore, the function M_τ in (6.13.28) has at most countably many poles in $(-\infty,0)$ which may only accumulate to 0.

By Lemma 6.13.5, the limit $M_\tau(\lambda+i0)$ exists for all $\lambda>0$. In fact, it is clear that the limits $1+\tau M(\lambda+i0)$ and $\tau - M(\lambda+i0)$ exist for all $\lambda>0$. Now assume that for some $\lambda>0$ one has $\tau = M(\lambda)$. Then it follows with the functions a_{ij} in Lemma 6.13.5 that

$$\tau\big(a_{21}(\lambda)+ia_{22}(\lambda)\big) = -a_{11}(\lambda) - ia_{12}(\lambda)$$

or, equivalently,

$$a_{11}(\lambda)+\tau a_{21}(\lambda) + i\big(a_{12}(\lambda)+\tau a_{22}(\lambda)\big) = 0.$$

Hence, $a_{11}(\lambda) = -\tau a_{21}(\lambda)$ and $a_{12}(\lambda) = -\tau a_{22}(\lambda)$ and (6.13.25) yields

$$\frac{1}{\sqrt{\lambda}} = -\tau a_{21}(\lambda)a_{22}(\lambda) + \tau a_{22}(\lambda)a_{21}(\lambda) = 0;$$

a contradiction. It follows that the limit $M_\tau(\lambda+i0)$ exists for all $\lambda>0$. A simple computation using $\operatorname{Im} M(\lambda+i0)>0$ for $\lambda>0$ shows that

$$\operatorname{Im} M_\tau(\lambda+i0) = \frac{(1+\tau^2)\operatorname{Im} M(\lambda+i0)}{|\tau - M(\lambda+i0)|^2} > 0, \qquad \lambda>0.$$

From these properties of M_τ the assertions (i)–(iii) follow in the same way as in the proof of Corollary 6.13.6. □

Example 6.13.8. Consider the integrable potential

$$q(x) = -\frac{2A^2}{\cosh^2(Ax+B)},$$

where $A \geq 0$ and $B \in \mathbb{R}$ is arbitrary. Define the smooth bounded function

$$T(x) = -A\tanh(Ax+B),$$

so that $q(x) = 2T'(x)$ and

$$T(x)^2 - T'(x) = A^2,$$

which gives $T''(x) = q(x)T(x)$. It is therefore clear that the functions

$$\begin{aligned}\varphi(x,\lambda) &= \cos\sqrt{\lambda}x + T(x)\,\frac{\sin\sqrt{\lambda}x}{\sqrt{\lambda}},\\ \psi(x,\lambda) &= \sqrt{\lambda}\sin\sqrt{\lambda}x - T(x)\,\cos\sqrt{\lambda}x,\end{aligned}$$

satisfy the equation $-y'' + qy = \lambda y$ with the initial values

$$\begin{aligned} \varphi(0,\lambda) &= 1, & \varphi'(0,\lambda) &= T(0), \\ \psi(0,\lambda) &= -T(0), & \psi'(0,\lambda) &= \lambda - T'(0). \end{aligned}$$

Hence, the following combinations of $\varphi(\cdot,\lambda)$ and $\psi(\cdot,\lambda)$, given by

$$u_1(x,\lambda) = \frac{1}{\lambda + A^2}\big((\lambda - T'(0))\varphi(x,\lambda) - T(0)\psi(x,\lambda)\big)$$

and

$$u_2(x,\lambda) = \frac{1}{\lambda + A^2}\big(T(0)\varphi(x,\lambda) + \psi(x,\lambda)\big),$$

form a fundamental system with the initial conditions in (6.13.8). It is clear that $\lim_{x\to\infty} T(x) = -A$ and hence the coefficients in Lemma 6.13.5 are given by the functions

$$\begin{pmatrix} a_{11}(\lambda) & a_{12}(\lambda) \\ a_{21}(\lambda) & a_{22}(\lambda) \end{pmatrix} = \frac{1}{\lambda + A^2} \begin{pmatrix} \lambda - T'(0) - AT(0) & \frac{-\lambda(A+T(0))+AT'(0)}{\sqrt{\lambda}} \\ T(0) + A & \frac{\lambda - AT(0)}{\sqrt{\lambda}} \end{pmatrix}.$$

Note that $a_{11}(\lambda)a_{22}(\lambda) - a_{12}(\lambda)a_{21}(\lambda) = 1/\sqrt{\lambda}$ by Lemma 6.13.5 and this also leads to

$$M(\lambda) = -\frac{\lambda^{3/2} - \sqrt{\lambda}(T'(0) + AT(0)) + i\big(AT'(0) - \lambda(A + T(0))\big)}{\sqrt{\lambda}(T(0) + A) + i(\lambda - AT(0))}, \qquad \lambda > 0, \tag{6.13.29}$$

and

$$\operatorname{Im} M(\lambda) = \frac{\sqrt{\lambda}\,(\lambda + A^2)}{\lambda + T(0)^2}, \qquad \lambda > 0.$$

Upon writing the solution $x \mapsto u_1(x,\lambda) + M(\lambda)u_2(x,\lambda)$ for $\lambda \in \mathbb{C}\setminus[0,\infty)$ in the form

$$d_1(x,\lambda)e^{i\sqrt{\lambda}x} + d_2(x,\lambda)e^{-i\sqrt{\lambda}x} \tag{6.13.30}$$

one observes that $x \mapsto d_1(x,\lambda)$ and $x \mapsto d_2(x,\lambda)$ are bounded with limits as $x \to b$. One concludes that $\lim_{x\to b} d_2(x,\lambda) = 0$ since the solution (6.13.30) belongs to $L^2(0,\infty)$. A computation of the coefficient $d_2(x,\lambda)$ shows that the expression for M in (6.13.29) remains valid also for $\lambda \in \mathbb{C}\setminus[0,\infty)$. Now one verifies that M has a pole at $\lambda < 0$ if and only if $B < 0$ and $\lambda = -T(0)^2$. Hence, the operator

$$A_0 = -D^2 + q, \quad \operatorname{dom} A_0 = \big\{f \in \operatorname{dom} T_{\max} : f(0) = 0\big\},$$

has one negative eigenvalue $-T(0)^2$ if $B < 0$ and

$$f(x) = e^{-T(0)x}\big(T(x) - T(0)\big)$$

is a corresponding eigenfunction; in the case $B \geq 0$ the operator A_0 has no negative eigenvalues.

Chapter 7

Canonical Systems of Differential Equations

Boundary value problems for regular and singular canonical systems of differential equations are investigated. After a brief introduction to Hilbert spaces of $\mathbb{C}^2$-valued vector functions which are square-integrable with respect to some 2×2 matrix-valued function in Section 7.1, the class of canonical systems to be studied here is introduced in Section 7.2. In Section 7.3 the notions of regular, quasiregular, and singular endpoints for canonical systems are explained. The number of square-integrable solutions at an endpoint of the interval is studied in Section 7.4. Together with a monotonicity principle from Chapter 5, this leads to a limit-circle/limit-point classification of singular endpoints in the same way as in Weyl's alternative in Chapter 6. The important concept of definiteness of canonical systems is defined and studied in Section 7.5, and a cut-off technique for solutions is provided. Afterwards, in Section 7.6, a symmetric minimal relation in the appropriate L^2-Hilbert space and its adjoint, the maximal relation, are associated with real definite canonical systems. The defect numbers of the minimal relation are specified for regular endpoints and for endpoints in the limit-circle or limit-point case. Boundary triplets and Weyl functions for canonical systems in the limit-circle case are constructed in Section 7.7, while the limit-point case is treated in Section 7.8. The connection between subordinate solutions and properties of the Weyl function, as well as the description of absolutely continuous and singular spectrum are studied in Section 7.9. Finally, in Section 7.10 some special classes of canonical systems of differential equations are discussed, among them weighted Sturm–Liouville equations.

J. Behrndt et al., *Boundary Value Problems, Weyl Functions, and Differential Operators*, Monographs in Mathematics 108,
https://doi.org/10.1007/978-3-030-36714-5_8

7.1 Classes of integrable functions

The purpose of this section is to introduce classes of vector functions which are locally square-integrable with respect to a measurable nonnegative matrix function and to collect some useful properties of such functions.

Let $\imath \subset \mathbb{R}$ be an interval, not necessarily bounded, with endpoints $a < b$, not necessarily belonging to $\imath$. In the following an integral of a vector function or a matrix function over $\imath$ or over a subinterval is always understood in the componentwise sense. The linear space $\mathcal{L}^1_{\mathrm{loc}}(\imath)$ of locally integrable $\mathbb{C}^2$-valued vector functions consists of all measurable $\mathbb{C}^2$-valued vector functions f defined almost everywhere on $\imath$ such that for each compact subinterval $K \subset \imath$

$$\int_K |f(s)|\, ds < \infty.$$

Here $|x|$ denotes the Euclidean norm of x in $\mathbb{C}^2$. Note that for $f \in \mathcal{L}^1_{\mathrm{loc}}(\imath)$ and each compact subinterval $K \subset \imath$ the norm inequality

$$\left|\int_K f(s)\, ds\right| \leq \int_K |f(s)|\, ds \tag{7.1.1}$$

holds. A $\mathbb{C}^2$-valued vector function $f \in \mathcal{L}^1_{\mathrm{loc}}(\imath)$ is said to be integrable at the left endpoint a of the interval $\imath$ or integrable at the right endpoint b of the interval $\imath$ if for some, and hence for all $c \in \mathbb{R}$ with $a < c < b$

$$\int_a^c |f(s)|\, ds < \infty \quad \text{or} \quad \int_c^b |f(s)|\, ds < \infty, \tag{7.1.2}$$

respectively. Similarly, a measurable 2×2 matrix function Φ is locally integrable on $\imath$ if for each compact subinterval $K \subset \imath$

$$\int_K |\Phi(s)|\, ds < \infty;$$

here and in the following $|A|$ stands for the operator norm of a 2×2 matrix A. In particular,

$$\left|\int_K \Phi(s)\, ds\right| \leq \int_K |\Phi(s)|\, ds. \tag{7.1.3}$$

The linear space consisting of all locally integrable 2×2 matrix functions on $\imath$ will also be denoted by $\mathcal{L}^1_{\mathrm{loc}}(\imath)$; it will be clear from the context if the values of the functions in $\mathcal{L}^1_{\mathrm{loc}}(\imath)$ are vectors in $\mathbb{C}^2$ or 2×2 matrices. A 2×2 matrix function $\Phi \in \mathcal{L}^1_{\mathrm{loc}}(\imath)$ is said to be integrable at the left endpoint a of the interval $\imath$ or integrable at the right endpoint b of the interval $\imath$ if (7.1.2) holds for some, and hence for all $c \in \mathbb{R}$ with $a < c < b$ and f replaced by Φ.

Note that all norms on the linear space of 2×2 matrices are equivalent and that the operator norm $|A|$ of a 2×2 matrix A can be estimated by the Hilbert–Schmidt matrix norm as follows:

$$|A| \leq \|A\|_2 \leq \sqrt{2}|A|, \quad \text{where } \|A\|_2 := \sqrt{\sum_{i,j=1}^{2} |a_{ij}|^2}. \tag{7.1.4}$$

For the product of 2×2 matrices A and B one has

$$|AB| \leq |A|\,|B| \quad \text{and} \quad \|AB\|_2 \leq \|A\|_2\,\|B\|_2. \tag{7.1.5}$$

In the case where the 2×2 matrix A is nonnegative the trace norm of A can be estimated by the Hilbert–Schmidt matrix norm:

$$\|A\|_2 \leq \operatorname{tr} A \leq \sqrt{2}\|A\|_2, \quad \text{where } \operatorname{tr} A = a_{11} + a_{22},$$

which also gives

$$|A| \leq \operatorname{tr} A \leq 2|A|. \tag{7.1.6}$$

The following definition introduces the semi-inner product space $\mathcal{L}^2_\Delta(\imath)$ of $\mathbb{C}^2$-valued functions which are square-integrable with respect to Δ; it is assumed that Δ is a 2×2 matrix function on $\imath$ and that $\Delta(s) \geq 0$ for almost every $s \in \imath$. In order to express the seminorm on $\mathcal{L}^2_\Delta(\imath)$ the notation

$$f(s)^*\Delta(s)f(s) = \big(\Delta(s)f(s), f(s)\big)$$

will be useful. Here $(\cdot,\cdot)$ denotes the standard scalar product in $\mathbb{C}^2$ and f is any $\mathbb{C}^2$-function defined on the interval $\imath$.

Definition 7.1.1. Let $\imath \subset \mathbb{R}$ be an interval and let Δ be a measurable 2×2 matrix function such that $\Delta(s) \geq 0$ for almost every $s \in \imath$. Then $\mathcal{L}^2_\Delta(\imath)$ denotes the linear space of all measurable functions f on $\imath$ with values in $\mathbb{C}^2$ which are *square-integrable with respect to* Δ, that is,

$$\|f\|^2_\Delta = \int_\imath f(s)^*\Delta(s)f(s)\,ds = \int_\imath |\Delta(s)^{\frac{1}{2}} f(s)|^2\,ds < \infty. \tag{7.1.7}$$

The semidefinite inner product $(\cdot,\cdot)_\Delta$ on $\mathcal{L}^2_\Delta(\imath)$ corresponding to the seminorm $\|\cdot\|_\Delta$ in (7.1.7) is given by

$$(f,g)_\Delta = \int_\imath g(s)^*\Delta(s)f(s)\,ds, \qquad f,g \in \mathcal{L}^2_\Delta(\imath). \tag{7.1.8}$$

Theorem 7.1.2. *Let $\imath \subset \mathbb{R}$ be an interval and let Δ be a measurable 2×2 matrix function such that $\Delta(s) \geq 0$ for almost every $s \in \imath$. Then the linear space $\mathcal{L}^2_\Delta(\imath)$ equipped with the seminorm* (7.1.8) *is complete.*

Proof. Since the 2×2 matrix function Δ is measurable and nonnegative almost everywhere, there are measurable nonnegative functions e_1 and e_2, and a measurable 2×2 matrix function U with unitary values, such that

$$\Delta(s) = U(s)^* \Xi(s) U(s), \quad \text{where } \Xi(s) = \begin{pmatrix} e_1(s) & 0 \\ 0 & e_2(s) \end{pmatrix},$$

for almost all $s \in \imath$. Hence, one has for all measurable functions f with values in $\mathbb{C}^2$ on $\imath$ that

$$\int_\imath f(s)^* \Delta(s) f(s)\, ds = \int_\imath (Uf)(s)^* \Xi(s) (Uf)(s)\, ds.$$

Written out in components this gives

$$\begin{aligned} \int_\imath f(s)^* \Delta(s) f(s)\, ds &= \int_\imath |(Uf)_1(s)|^2 e_1(s)\, ds + \int_\imath |(Uf)_2(s)|^2 e_2(s)\, ds \\ &= \int_\imath |(Uf)_1(s)|^2\, d\mu_1(s) + \int_\imath |(Uf)_2(s)|^2\, d\mu_2(s), \end{aligned} \tag{7.1.9}$$

where the measures μ_1 and μ_2 are absolutely continuous with respect to the Lebesgue measure m and their Radon–Nikodým derivatives are given by e_1 and e_2, respectively. Therefore, it is now clear that

$$f \in \mathcal{L}^2_\Delta(\imath) \quad \Leftrightarrow \quad (Uf)_1 \in \mathcal{L}^2_{d\mu_1}(\imath) \quad \text{and} \quad (Uf)_2 \in \mathcal{L}^2_{d\mu_2}(\imath).$$

This shows that the transformation U maps the space $\mathcal{L}^2_\Delta(\imath)$ bijectively onto $\mathcal{L}^2_{d\mu_1}(\imath) \times \mathcal{L}^2_{d\mu_2}(\imath)$ and from (7.1.9) one sees that the seminorms in $\mathcal{L}^2_\Delta(\imath)$ and $\mathcal{L}^2_{d\mu_1}(\imath) \times \mathcal{L}^2_{d\mu_2}(\imath)$ are preserved. Therefore, the completeness of $\mathcal{L}^2_\Delta(\imath)$ is a consequence of the completeness of $\mathcal{L}^2_{d\mu_1}(\imath)$ and $\mathcal{L}^2_{d\mu_2}(\imath)$. □

The space $\mathcal{L}^2_\Delta(\imath)$ has the following approximation property.

Lemma 7.1.3. *Each element of the seminormed space $\mathcal{L}^2_\Delta(\imath)$ can be approximated by functions in $\mathcal{L}^2_\Delta(\imath)$ which have compact support.*

Proof. Let $(K_n)_{n\in\mathbb{N}}$ be a sequence of nondecreasing compact intervals such that $\imath = \bigcup_{n=1}^\infty K_n$. For $f \in \mathcal{L}^2_\Delta(\imath)$ put $f_n(s) = f(s)$ for $s \in K_n$ and $f_n(s) = 0$ elsewhere. Then $f_n \in \mathcal{L}^2_\Delta(\imath)$, f_n has support in K_n, and

$$\|f - f_n\|^2_\Delta = \int_\imath (f(s) - f_n(s))^* \Delta(s) (f(s) - f_n(s))\, ds \to 0$$

as $n \to \infty$, by dominated convergence. □

The space $\mathcal{L}^2_{\Delta,\mathrm{loc}}(\imath)$ consists of all $\mathbb{C}^2$-valued functions which are square-integrable with respect to Δ for each compact subinterval $K \subset \imath$, i.e.,

$$\int_K f(s)^* \Delta(s) f(s)\, ds < \infty.$$

A function $f \in \mathcal{L}^2_{\Delta,\mathrm{loc}}(\imath)$ is said to be *square-integrable with respect to* Δ at the left endpoint a of the interval $\imath$ or *square-integrable with respect to* Δ at the right endpoint b of the interval $\imath$ if for some, and hence for all $c \in \mathbb{R}$ with $a < c < b$,

$$\int_a^c f(s)^* \Delta(s) f(s)\, ds < \infty \quad \text{or} \quad \int_c^b f(s)^* \Delta(s) f(s)\, ds < \infty,$$

respectively. A function $f \in \mathcal{L}^2_{\Delta,\mathrm{loc}}(\imath)$ belongs to $\mathcal{L}^2_\Delta(\imath)$ if and only if f is square-integrable with respect to Δ at both endpoints a and b of $\imath$.

Clearly, if Δ is a nonnegative matrix function and f is a vector function, then

$$|\Delta(s)f(s)| = |\Delta(s)^{\frac{1}{2}}\Delta(s)^{\frac{1}{2}}f(s)| \leq |\Delta(s)^{\frac{1}{2}}|\,|\Delta(s)^{\frac{1}{2}}f(s)|.$$

Now the statement in the next lemma is a consequence of the Cauchy–Schwarz inequality and the fact that

$$|\Delta(s)^{\frac{1}{2}}|^2 = |\Delta(s)^{\frac{1}{2}}\Delta(s)^{\frac{1}{2}}| = |\Delta(s)|.$$

Lemma 7.1.4. *Let Δ be a locally integrable nonnegative 2×2 matrix function on $\imath$ and let $K \subset \imath$ be compact. If $f \in \mathcal{L}^2_\Delta(K)$, then $\Delta f \in \mathcal{L}^1(K)$ and*

$$\int_K |\Delta(s)f(s)|\, ds \leq \left(\int_K |\Delta(s)|\, ds\right)^{\frac{1}{2}} \left(\int_K f(s)^*\Delta(s)f(s)\, ds\right)^{\frac{1}{2}}.$$

In particular, if $f \in \mathcal{L}^2_\Delta(\imath)$, then $\Delta f \in \mathcal{L}^1_{\mathrm{loc}}(\imath)$ and for all compact $K \subset \imath$

$$\int_K |\Delta(s)f(s)|\, ds \leq \left(\int_K |\Delta(s)|\, ds\right)^{\frac{1}{2}} \|f\|_\Delta.$$

Let $\mathfrak{N} = \{f \in \mathcal{L}^2_\Delta(\imath) : \|f\|_\Delta = 0\}$, so that $\mathfrak{N}$ is a linear space, and consider the quotient space

$$L^2_\Delta(\imath) := \mathcal{L}^2_\Delta(\imath)/\mathfrak{N}$$

equipped with the scalar product induced by (7.1.8), that is, $(f,g)_\Delta = (\widetilde{f},\widetilde{g})_\Delta$, where $\widetilde{f},\widetilde{g} \in \mathcal{L}^2_\Delta(\imath)$ are representatives in the equivalence classes $f, g \in L^2_\Delta(\imath)$. From Theorem 7.1.2 it is clear that $L^2_\Delta(\imath)$ is a Hilbert space. When no confusion can arise, the equivalence classes in $L^2_\Delta(\imath)$ will also be referred to as functions that are square-integrable with respect to Δ. Note that the compactly supported functions in $L^2_\Delta(\imath)$ are dense in $L^2_\Delta(\imath)$ by Lemma 7.1.3.

Recall that a $\mathbb{C}^2$-valued vector function f on an open interval $\imath$ is *absolutely continuous* if there exists a $\mathbb{C}^2$-valued vector function $h \in \mathcal{L}^1_{\mathrm{loc}}(\imath)$ such that

$$f(t) - f(s) = \int_s^t h(u)\, du \tag{7.1.10}$$

for all $s, t \in \imath$. In this case, f is differentiable and $f' = h$ almost everywhere. The space of absolutely continuous $\mathbb{C}^2$-valued vector functions is denoted by $AC(\imath)$. When $a \in \mathbb{R}$, then $AC[a,b)$ stands for the subclass of $f \in AC(a,b)$ for which $h \in \mathcal{L}^1_{\rm loc}(a,b)$ in (7.1.10) additionally belongs to $\mathcal{L}^1(a,a')$ for some, and hence for all $a < a' < b$, in which case

$$f(t) - f(a) = \int_a^t h(u)\, du$$

holds for all $t \in (a,b)$ and thus $f(a) = \lim_{t \to a} f(t)$. When $b \in \mathbb{R}$ there is a similar notation $AC(a,b]$ and for $f \in AC(a,b]$ one has $f(b) = \lim_{t \to b} f(t)$. The notation $AC[a,b]$ is analogous.

7.2 Canonical systems of differential equations

This section offers a brief review of so-called 2×2 canonical systems of differential equations. The existence and uniqueness result for linear systems of differential equations will be discussed and properties of the corresponding fundamental matrices will be derived.

Let $\imath = (a,b) \subset \mathbb{R}$ be an open, not necessarily bounded, interval and let H and Δ be 2×2 matrix functions defined almost everywhere on $\imath$ such that

$$H, \Delta \in \mathcal{L}^1_{\rm loc}(\imath), \quad H(t) = H(t)^*, \quad \text{and} \quad \Delta(t) \geq 0 \tag{7.2.1}$$

for almost every $t \in \imath$. Furthermore, let

$$J = \begin{pmatrix} 0 & -1 \\ 1 & 0 \end{pmatrix} \tag{7.2.2}$$

and note that $J^* = -J = J^{-1}$. A *canonical system* is a system of differential equations of the form

$$Jf'(t) - H(t)f(t) = \lambda \Delta(t) f(t) + \Delta(t) g(t), \quad t \in \imath, \quad \lambda \in \mathbb{C}, \tag{7.2.3}$$

where $g \in \mathcal{L}^2_{\Delta,{\rm loc}}(\imath)$ is a function that is locally square-integrable with respect to Δ with values in $\mathbb{C}^2$. The condition $g \in \mathcal{L}^2_{\Delta,{\rm loc}}(\imath)$ implies that Δg is locally integrable; cf. Lemma 7.1.4. In the general case of (7.2.3) one speaks of an *inhomogeneous system*, while if the term involving Δg is absent, that is,

$$Jf'(t) - H(t)f(t) = \lambda \Delta(t) f(t), \quad t \in \imath, \quad \lambda \in \mathbb{C}, \tag{7.2.4}$$

one speaks of the corresponding *homogeneous system*.

A function f on $\imath$ with values in $\mathbb{C}^2$ is said to be a *solution* of the canonical system (7.2.3) if f belongs to $AC(\imath)$ and the equation (7.2.3) holds for almost every

$t \in \imath$. Observe that if f is a solution of (7.2.3), then f is also a solution of (7.2.3) when $g \in \mathcal{L}^2_{\Delta,\mathrm{loc}}(\imath)$ is replaced by $\widetilde{g} \in \mathcal{L}^2_{\Delta,\mathrm{loc}}(\imath)$ with $\Delta(g - \widetilde{g}) = 0$. Furthermore, if f is a solution of (7.2.3) and h is a solution of (7.2.4), then $f + h$ is a solution of (7.2.3). In fact, the collection of all solutions of the homogeneous system (7.2.4) forms a linear space. The following result on the existence and uniqueness of solutions of initial value problems for inhomogeneous canonical systems will be useful.

Theorem 7.2.1. *Let $g \in \mathcal{L}^2_{\Delta,\mathrm{loc}}(\imath)$ and $\lambda \in \mathbb{C}$. Fix some $c_0 \in \imath = (a,b)$ and $\gamma \in \mathbb{C}^2$. Then the initial value problem*

$$Jf'(t) - H(t)f(t) = \lambda\Delta(t)f(t) + \Delta(t)g(t), \qquad f(c_0) = \gamma, \tag{7.2.5}$$

admits a unique solution $f \in AC(\imath)$. Moreover, the mapping $\lambda \mapsto f(t,\lambda)$ is entire for every fixed $t \in \imath$.

In order to prove this theorem one replaces the initial value problem (7.2.5) by an equivalent integral equation; recall that the functions H and Δ are locally integrable. The integral equation can be solved, for instance, by successive iterations, which also leads to the statement concerning the mapping $\lambda \mapsto f(t,\lambda)$ being entire, see, e.g., [754, Theorem 2.1].

In the next lemma a *Lagrange identity* for solutions of the inhomogeneous canonical system is obtained.

Lemma 7.2.2. *Assume that $\lambda, \mu \in \mathbb{C}$ and that $g, k \in \mathcal{L}^2_{\Delta,\mathrm{loc}}(\imath)$. Let f, h be solutions of the inhomogeneous equations*

$$\begin{aligned} Jf'(t) - H(t)f(t) &= \lambda\Delta(t)f(t) + \Delta(t)g(t), \\ Jh'(t) - H(t)h(t) &= \mu\Delta(t)h(t) + \Delta(t)k(t), \end{aligned}$$

respectively. Then for every compact interval $[\alpha,\beta] \subset \imath$,

$$\begin{aligned} h(\beta)^*Jf(\beta) - h(\alpha)^*Jf(\alpha) = \int_\alpha^\beta \big(h(s)^*\Delta(s)g(s) - k(s)^*\Delta(s)f(s)\big)\,ds \\ + (\lambda - \bar{\mu}) \int_\alpha^\beta h(s)^*\Delta(s)f(s)ds. \end{aligned}$$

Proof. The assumptions that J is skew-adjoint and that $H(t)$ and $\Delta(t)$ are self-adjoint almost everywhere on $\imath$ lead to the identities

$$\begin{aligned} (h^*Jf)' &= h^*(Jf') - (Jh')^*f \\ &= h^*(\lambda\Delta f + \Delta g + Hf) - (\mu\Delta h + \Delta k + Hh)^*f \\ &= h^*\Delta g - k^*\Delta f + (\lambda - \bar{\mu})h^*\Delta f, \end{aligned}$$

which are valid almost everywhere on $\imath$. Integration over the interval $[\alpha,\beta]$ completes the argument. □

Taking $\lambda = \mu = 0$ in Lemma 7.2.2, one obtains the following corollary. It provides the form of the Lagrange identity that will be studied in detail later in this chapter.

Corollary 7.2.3. *Assume that $g, k \in \mathcal{L}^2_{\Delta,\mathrm{loc}}(\imath)$. Let f, h be solutions of the inhomogeneous equations*

$$\begin{aligned} Jf'(t) - H(t)f(t) &= \Delta(t)g(t), \\ Jh'(t) - H(t)h(t) &= \Delta(t)k(t), \end{aligned} \tag{7.2.6}$$

respectively. Then for every compact interval $[\alpha, \beta] \subset \imath$,

$$h(\beta)^* Jf(\beta) - h(\alpha)^* Jf(\alpha) = \int_\alpha^\beta \big(h(s)^*\Delta(s)g(s) - k(s)^*\Delta(s)f(s)\big)\, ds.$$

There is also a corollary of Lemma 7.2.2 involving solutions of the corresponding homogeneous system. Let $Y_1(\cdot, \lambda)$ and $Y_2(\cdot, \lambda)$ be solutions of (7.2.4) and define the *solution matrix*

$$Y(\cdot, \lambda) = \begin{pmatrix} Y_1(\cdot, \lambda) & Y_2(\cdot, \lambda) \end{pmatrix}, \qquad \lambda \in \mathbb{C}, \tag{7.2.7}$$

which is a 2×2 matrix function for each $\lambda \in \mathbb{C}$. Then the matrix function $Y(\cdot, \lambda)$ solves the equation (7.2.4) in the sense that it actually solves the matrix version of (7.2.4),

$$JY'(t, \lambda) - H(t)Y(t, \lambda) = \lambda \Delta(t) Y(t, \lambda), \quad t \in \imath.$$

Corollary 7.2.4. *Let $Y(\cdot, \lambda)$ be a solution matrix of the homogeneous canonical system (7.2.4). Then for every compact interval $[\alpha, \beta] \subset \imath$ and all $\lambda, \mu \in \mathbb{C}$,*

$$Y(\beta, \mu)^* JY(\beta, \lambda) - Y(\alpha, \mu)^* JY(\alpha, \lambda) = (\lambda - \bar{\mu}) \int_\alpha^\beta Y(s, \mu)^* \Delta(s) Y(s, \lambda)\, ds.$$

In particular, for all $[\alpha, \beta] \subset \imath$ and all $\lambda \in \mathbb{C}$,

$$Y(\beta, \bar{\lambda})^* JY(\beta, \lambda) = Y(\alpha, \bar{\lambda})^* JY(\alpha, \lambda).$$

It is a consequence of Corollary 7.2.4 that for every solution matrix $Y(\cdot, \lambda)$ the function

$$t \mapsto Y(t, \bar{\lambda})^* JY(t, \lambda)$$

is constant on $\imath$. Hence, if for some $c_0 \in \imath$

$$Y(c_0, \bar{\lambda})^* JY(c_0, \lambda) = J, \tag{7.2.8}$$

then $Y(t, \bar{\lambda})^* JY(t, \lambda) = J$ for all $t \in \imath$. This shows that

$$Y(t, \lambda)^{-1} = -JY(t, \bar{\lambda})^* J \quad \text{and} \quad Y(t, \bar{\lambda})^{-*} = -JY(t, \lambda)J, \quad t \in \imath, \tag{7.2.9}$$

and thus it also follows that

$$Y(t,\lambda)JY(t,\bar{\lambda})^* = J, \quad t \in \imath. \tag{7.2.10}$$

Let X be an invertible matrix which does not depend on λ and assume that $X^*JX = J$. Let $Y(\cdot,\lambda)$ be the solution matrix which is fixed by the initial condition

$$Y(c_0,\lambda) = X \tag{7.2.11}$$

for some $c_0 \in \imath$ and all $\lambda \in \mathbb{C}$. Then (7.2.8) is valid and hence (7.2.9) and (7.2.10) are satisfied; the matrix $Y(\cdot,\lambda)$ is a *fundamental matrix*, that is, its columns are linearly independent solutions of the homogeneous canonical system (7.2.4) on $\imath$. Frequently the fundamental matrix $Y(\cdot,\lambda)$ will be fixed by the initial condition

$$Y(c_0,\lambda) = I \tag{7.2.12}$$

for some $c_0 \in \imath$.

According to Theorem 7.2.1, there is a unique solution of the initial value problem (7.2.5). It is possible to express this unique solution in terms of the fundamental matrix $Y(\cdot,\lambda)$ determined by the initial condition (7.2.12) (and in a similar way with the initial condition (7.2.11)). In fact, for any $\lambda \in \mathbb{C}$, any $\gamma \in \mathbb{C}^2$, and any $g \in \mathcal{L}^2_{\Delta,\mathrm{loc}}(\imath)$, the unique solution of the inhomogeneous initial value problem

$$Jf' - Hf = \lambda\Delta f + \Delta g, \quad f(c_0) = \gamma, \tag{7.2.13}$$

is provided by the *variation of constant formula*:

$$f(t) = Y(t,\lambda)\gamma + Y(t,\lambda)\int_{c_0}^{t} Y(s,\lambda)^{-1}J^{-1}\Delta(s)g(s)\,ds. \tag{7.2.14}$$

This can be seen by verifying that the second term on the right-hand side is a solution of the inhomogeneous equation that vanishes at c_0. Making use of (7.2.9), one recasts (7.2.14) as

$$f(t) = Y(t,\lambda)\gamma - Y(t,\lambda)\int_{c_0}^{t} JY(s,\bar{\lambda})^*\Delta(s)g(s)\,ds. \tag{7.2.15}$$

In terms of the notation (7.2.7) for the fundamental matrix $Y(\cdot,\lambda)$ fixed by (7.2.12) the unique solution (7.2.15) of (7.2.13) can be written as

$$\begin{aligned} f(t) = {} & Y_1(t,\lambda)\gamma_1 + Y_2(t,\lambda)\gamma_2 \\ & + Y_1(t,\lambda)\int_{c_0}^{t} Y_2(s,\bar{\lambda})^*\Delta(s)g(s)\,ds \\ & - Y_2(t,\lambda)\int_{c_0}^{t} Y_1(s,\bar{\lambda})^*\Delta(s)g(s)\,ds, \end{aligned} \tag{7.2.16}$$

where $\gamma = (\gamma_1,\gamma_2)^\top$. This form of the solution will be used later.

The general form of the inhomogeneous equation (7.2.3) can be simplified by a transformation of the system. This transformation will be employed in Corollary 7.4.8.

Lemma 7.2.5. *Let $\lambda_0 \in \mathbb{R}$, $c_0 \in \imath$, and let $U(\cdot,\lambda_0)$ be a solution matrix which satisfies*

$$JU'(\cdot,\lambda_0) - HU(\cdot,\lambda_0) = \lambda_0 \Delta U(\cdot,\lambda_0), \quad U(c_0,\lambda_0)^* JU(c_0,\lambda_0) = J. \tag{7.2.17}$$

Assume that $g \in \mathcal{L}^2_{\Delta,\mathrm{loc}}(\imath)$ and let f be a solution of the inhomogeneous equation (7.2.3). Define the functions $\widetilde{f}$, $\widetilde{g}$, and $\widetilde{\Delta}$ by

$$\widetilde{f} = U(\cdot,\lambda_0)^{-1} f, \quad \widetilde{g} = U(\cdot,\lambda_0)^{-1} g, \quad \widetilde{\Delta}(\cdot) = U(\cdot,\lambda_0)^* \Delta(\cdot) U(\cdot,\lambda_0). \tag{7.2.18}$$

Then $\widetilde{\Delta}$ is a locally integrable nonnegative measurable matrix function,

$$\widetilde{f}^* \widetilde{\Delta} \widetilde{f} = f^* \Delta f \quad \text{and} \quad \widetilde{g}^* \widetilde{\Delta} \widetilde{g} = g^* \Delta g, \tag{7.2.19}$$

and, in particular, $\widetilde{g} \in \mathcal{L}^2_{\widetilde{\Delta},\mathrm{loc}}(\imath)$. Moreover, the function $\widetilde{f}$ is a solution of the system of differential equations

$$J\widetilde{f}' = (\lambda - \lambda_0)\widetilde{\Delta}\widetilde{f} + \widetilde{\Delta}\widetilde{g}. \tag{7.2.20}$$

Conversely, if

$$\widetilde{f}, \widetilde{g} \in \mathcal{L}^2_{\widetilde{\Delta},\mathrm{loc}}(\imath) \quad \text{and} \quad \widetilde{\Delta}(\cdot) = U(\cdot,\lambda_0)^* \Delta(\cdot) U(\cdot,\lambda_0)$$

satisfy the equation (7.2.20), then $f = U(\cdot,\lambda_0)\widetilde{f}$ and $g = U(\cdot,\lambda_0)\widetilde{g}$ satisfy the inhomogeneous equation (7.2.3).

Proof. First observe that it is a direct consequence of (7.2.17) that the function $U(\cdot,\lambda_0)$ satisfies

$$U(\cdot,\lambda_0)^* JU(\cdot,\lambda_0) = J;$$

cf. (7.2.8). In particular, this shows that $U(t,\lambda_0)$ is invertible for each $t \in \imath$.

Let $\widetilde{f}$, $\widetilde{g}$, and $\widetilde{\Delta}$ be defined by (7.2.18). Then it is clear that (7.2.19) holds. Since $g \in \mathcal{L}^2_{\Delta,\mathrm{loc}}(\imath)$ it also follows that $\widetilde{g} \in \mathcal{L}^2_{\widetilde{\Delta},\mathrm{loc}}(\imath)$. Moreover,

$$Jf' - Hf = \lambda \Delta f + \Delta g \tag{7.2.21}$$

holds by assumption. Substituting $f = U(\cdot,\lambda_0)\widetilde{f}$ and $g = U(\cdot,\lambda_0)\widetilde{g}$ in (7.2.21), multiplying by $U(\cdot,\lambda_0)^*$ from the left, and using (7.2.17) a straightforward calculation leads to (7.2.20). Similarly, one verifies by a direct calculation that the converse statement holds. □

It follows from (7.2.19) that the functions $\widetilde{f}$ or $\widetilde{g}$ in (7.2.18) are square-integrable with respect to $\widetilde{\Delta}$ if and only if f or g are square-integrable with respect to Δ, respectively. The transformation in Lemma 7.2.5 implies that the boundary terms $h(x)^*Jf(x)$ in the Lagrange formula of the original equation in Lemma 7.2.2 can be written in terms of the boundary terms of the corresponding solutions of the transformed equation (7.2.20).

Corollary 7.2.6. *Assume that $\lambda, \mu \in \mathbb{C}$, $g, k \in \mathcal{L}^2_{\Delta,\mathrm{loc}}(\imath)$ and that f, h are solutions of the inhomogeneous equations*

$$\begin{aligned} Jf'(t) - H(t)f(t) &= \lambda\Delta(t)f(t) + \Delta(t)g(t), \\ Jh'(t) - H(t)h(t) &= \mu\Delta(t)h(t) + \Delta(t)k(t). \end{aligned}$$

Assume that $U(\cdot, \lambda_0)$ with $\lambda_0 \in \mathbb{R}$ is a solution matrix which satisfies (7.2.17) and define the functions $\widetilde{f} = U(\cdot, \lambda_0)^{-1}f$ and $\widetilde{h} = U(\cdot, \lambda_0)^{-1}h$ as in (7.2.18). Then for each $t \in \imath$

$$h(t)^*Jf(t) = \widetilde{h}(t)^*J\widetilde{f}(t).$$

Recall that the functions H and Δ were assumed to be 2×2 matrix functions with complex entries. When these functions are real the solutions enjoy a certain symmetry property.

Definition 7.2.7. The canonical system $Jf' - Hf = \lambda\Delta f$ is said to be *real* if the entries of the 2×2 matrix functions H and Δ in (7.2.1) are real functions.

To deal with real canonical systems the notion of conjugate matrices is useful. For a matrix T the *conjugate matrix* $\overline{T}$ is the matrix whose entries are the complex conjugates of the entries of T. Let T and S be matrices, not necessarily of the same size, for which the matrix product TS is defined. Then clearly

$$\overline{TS} = \overline{T}\,\overline{S}. \tag{7.2.22}$$

Lemma 7.2.8. *Assume that the canonical system (7.2.4) is real. Let $Y(\cdot, \lambda)$ be a solution matrix of (7.2.4) such that for all $\lambda \in \mathbb{C}$*

$$\overline{Y}(c_0, \overline{\lambda}) = Y(c_0, \lambda) \tag{7.2.23}$$

for some point $c_0 \in \imath$. Then

$$\overline{Y}(\cdot, \overline{\lambda}) = Y(\cdot, \lambda) \tag{7.2.24}$$

for all $\lambda \in \mathbb{C}$. In particular, (7.2.24) holds when $Y(\cdot, \lambda)$ is a fundamental matrix fixed by (7.2.11) or (7.2.12).

Proof. By definition, the solution matrix $Y(\cdot, \lambda)$ satisfies

$$JY'(\cdot, \lambda) - HY(\cdot, \lambda) = \lambda\Delta Y(\cdot, \lambda). \tag{7.2.25}$$

By assumption, the entries of J, H, and Δ are real; hence taking complex conjugates and using (7.2.22) one sees that

$$J\overline{Y}'(\cdot,\lambda) - H\overline{Y}(\cdot,\lambda) = \bar{\lambda}\Delta\overline{Y}(\cdot,\lambda).$$

Therefore, the matrix function $\overline{Y}(\cdot,\bar{\lambda})$ satisfies the same equation (7.2.25) as $Y(\cdot,\lambda)$ and by (7.2.23) these matrix functions satisfy the same initial condition at c_0. Now the uniqueness in Theorem 7.2.1 leads to (7.2.24). □

The following observation is an easy consequence of Lemma 7.2.8.

Corollary 7.2.9. *Let the system $Jf' - Hf = \lambda\Delta f$ be real and let a fundamental matrix $Y(\cdot,\lambda_0)$ be fixed by the initial condition $Y(c,\lambda_0) = I$ for some $a < c < b$. Then for every $u \in \mathbb{C}^2$*

$$\int_\imath u^*Y(s,\lambda)^*\Delta(s)Y(s,\lambda)u\,ds = \int_\imath \bar{u}^*Y(s,\bar{\lambda})^*\Delta(s)Y(s,\bar{\lambda})\bar{u}\,ds.$$

In particular,

$$Y(\cdot,\lambda)u \in \mathcal{L}^2_\Delta(\imath) \quad\Leftrightarrow\quad Y(\cdot,\bar{\lambda})\bar{u} \in \mathcal{L}^2_\Delta(\imath).$$

Proof. Clearly, for any $u \in \mathbb{C}^2$ and all $s \in \imath$ one has

$$u^*Y(s,\lambda)^*\Delta(s)Y(s,\lambda)u \geq 0.$$

Therefore,

$$\begin{aligned} u^*Y(s,\lambda)^*\Delta(s)Y(s,\lambda)u &= \overline{u^*Y(s,\lambda)^*\Delta(s)Y(s,\lambda)u} \\ &= \bar{u}^*Y(s,\bar{\lambda})^*\Delta(s)Y(s,\bar{\lambda})\bar{u}, \end{aligned}$$

which gives the assertion. □

7.3 Regular and quasiregular endpoints

In this section the notions of regular and quasiregular for an endpoint of the interval $\imath$ are introduced; this makes it possible to extend Theorem 7.2.1, so that one may solve an initial value problem in an endpoint.

The following definition gives a classification for the endpoints of the canonical system (7.2.3).

Definition 7.3.1. An endpoint of the interval $\imath$ is said to be a *quasiregular endpoint* of the canonical system (7.2.3) if the locally integrable functions H and Δ in (7.2.1) are integrable up to that endpoint. A finite quasiregular endpoint is called *regular*. An endpoint is said to be *singular* when it is not regular. The canonical system (7.2.3) is called *regular* if both endpoints are regular; otherwise it is called *singular*.

The main result in this section implies that if the term $g \in \mathcal{L}^2_{\Delta,\mathrm{loc}}(\imath)$ in (7.2.3) is square-integrable with respect to Δ at an endpoint which is regular or quasiregular, then every solution of the inhomogeneous equation has a continuous extension to that endpoint, so that it is square-integrable with respect to Δ there.

Proposition 7.3.2. *Assume that the endpoint a or b of $\imath = (a,b)$ is regular or quasiregular and that $g \in \mathcal{L}^2_{\Delta,\mathrm{loc}}(\imath)$ is square-integrable with respect to Δ at a or b, respectively. Then each solution f of (7.2.3) is square-integrable with respect to Δ at a or at b and the limits*

$$f(a) := \lim_{t\to a} f(t) \quad \text{or} \quad f(b) := \lim_{t\to b} f(t) \tag{7.3.1}$$

exist, respectively. Moreover, for each $\gamma \in \mathbb{C}^2$ there exists a unique solution f of (7.2.3) such that $f(a) = \gamma$ or $f(b) = \gamma$, respectively, and the corresponding function $\lambda \mapsto f(t,\lambda)$ is entire for every $t \in \imath$ and $t = a$ or $t = b$, respectively.

Proof. It suffices to consider the case of the endpoint b. So let b be a regular or quasiregular endpoint, let $\lambda \in \mathbb{C}$, and fix $c \in (a,b)$. The proof is split in three separate steps.

Step 1. Any solution f of (7.2.3) with $f(c) = \eta$ satisfies

$$f(t) = \eta + \int_c^t J^{-1}\left(\lambda\Delta(s) + H(s)\right) f(s)\, ds + \int_c^t J^{-1}\Delta(s)g(s)\, ds \tag{7.3.2}$$

with $t \in \imath$. Recall that, since g is square-integrable with respect to Δ at b, it follows that Δg is integrable on $[c,b)$; cf. Lemma 7.1.4. By definition also $\lambda\Delta + H$ is integrable on $[c,b)$. Hence, Gronwall's lemma in Section 6.13 (see Lemma 6.13.2) shows that

$$|f(t)| \leq \left(|\eta| + \int_c^t |\Delta(s)g(s)|\, ds\right) e^{\int_c^t |\lambda\Delta(s)+H(s)|\, ds}, \quad c \leq t < b. \tag{7.3.3}$$

Thus, the solution f is bounded on $[c,b)$: $|f(s)| \leq M$, $c < s < b$. In particular, this shows that

$$\int_c^b f(s)^*\Delta(s)f(s)\, ds \leq M^2 \int_c^b |\Delta(s)|\, ds < \infty,$$

and hence f is square-integrable with respect to Δ at b. Moreover, it is clear from (7.3.2) that the limit $f(b) = \lim_{t\to b} f(t)$ in (7.3.1) exists.

Step 2. In the special case where h is a solution of the homogeneous system (7.2.4) with $h(c) = \eta$ it follows from (7.3.2) that

$$h(b) = \eta + \int_c^b J^{-1}\left(\lambda\Delta(s) + H(s)\right) h(s)\, ds. \tag{7.3.4}$$

The solution $h(\cdot,\lambda)$ actually depends on λ and according to Theorem 7.2.1 for each $c \le t < b$ the function $\lambda \mapsto h(t,\lambda)$ is entire. It will be shown that also $\lambda \mapsto h(b,\lambda)$ is entire. In fact, from (7.3.4) it is clear that it suffices to prove that the mapping

$$\lambda \mapsto \int_c^b J^{-1}\left(\lambda\Delta(s) + H(s)\right) h(s,\lambda)\, ds \tag{7.3.5}$$

is entire. To see this note that (7.3.3) and the equality $h(b) = \lim_{t\to b} h(t)$ imply that, for each compact set $K \subset \mathbb{C}$,

$$|h(t,\lambda)| \le C_K\, e^{\int_c^b (|\Delta(s)|+|H(s)|)\, ds}$$

for all $c \le t \le b$ and for all $\lambda \in K$. Hence, by dominated convergence, the mapping in (7.3.5) is continuous, and an application of Morera's theorem implies that this mapping is holomorphic. Therefore, $\lambda \mapsto h(b,\lambda)$ is entire.

Step 3. Let $Z(\cdot,\lambda)$ be a fundamental matrix of the homogeneous equation (7.2.4) fixed by $Z(c,\lambda) = I$. Then, according to Step 1 and Step 2, one has

$$Z(t) = I + \int_c^t J^{-1}\left(\lambda\Delta(s) + H(s)\right) Z(s)\, ds$$

and Gronwall's lemma yields the estimate

$$|Z(t,\lambda)| \le e^{\int_c^t |\lambda\Delta(s)+H(s)|\, ds}, \quad c \le t < b. \tag{7.3.6}$$

Thus, $Z(b,\lambda) = \lim_{t\to b} Z(t,\lambda)$ exists and it follows from Step 2 that the mapping $\lambda \mapsto Z(b,\lambda)$ is entire. Moreover, from $Z(t,\bar\lambda)^* J Z(t,\lambda) = J$ for $c \le t < b$ one concludes by taking the limit $t \to b$ that the matrix $Z(b,\lambda)$ is invertible for all $\lambda \in \mathbb{C}$. It is also clear that $Z(b,\bar\lambda)^* J Z(b,\lambda) = J$. Thus, the function $U(\cdot,\lambda)$ defined by

$$U(t,\lambda) = Z(t,\lambda) Z(b,\lambda)^{-1}$$

is a fundamental matrix of the homogeneous equation which satisfies $U(b,\lambda) = I$ and $\lambda \mapsto U(t,\lambda)$ is entire for $c \le t \le b$. Therefore, if $\gamma \in \mathbb{C}^2$ is fixed one sees that

$$f(t) = U(t,\lambda)\gamma + U(t,\lambda)\int_t^b J U(s,\bar\lambda)^* \Delta(s) g(s)\, ds$$

is the unique solution of the inhomogeneous equation with $f(b) = \gamma$; cf. (7.2.15). It remains to verify that $\lambda \mapsto f(t,\lambda)$ is entire for $c \le t \le b$. For this it suffices to check that

$$\lambda \mapsto U(t,\lambda)\int_t^b J U(s,\bar\lambda)^* \Delta(s) g(s)\, ds = Z(t,\lambda)\int_t^b J Z(s,\bar\lambda)^* \Delta(s) g(s)\, ds$$

is entire for $c \le t \le b$, which can be seen with the help of (7.3.6) in the same way as in Step 2. □

Corollary 7.3.3. *Assume that the endpoints a and b of the canonical system (7.2.3) are regular or quasiregular and that $g \in \mathcal{L}^2_\Delta(\imath)$. Then each solution f of (7.2.3) belongs to $\mathcal{L}^2_\Delta(\imath)$ and both limits in (7.3.1) exist.*

The next statement follows from Corollary 7.2.3 and Corollary 7.3.3.

Corollary 7.3.4. *Assume that the endpoints a and b of the canonical system (7.2.3) are regular or quasiregular and that $g, k \in \mathcal{L}^2_\Delta(\imath)$. Let f, h be solutions of the inhomogeneous equations (7.2.6). Then*

$$h(b)^* J f(b) - h(a)^* J f(a) = \int_a^b \big(h(s)^* \Delta(s) g(s) - k(s)^* \Delta(s) f(s)\big)\, ds.$$

Finally, the next statement is a consequence of Proposition 7.3.2 and identity (7.2.10).

Corollary 7.3.5. *Assume that the endpoint a or b of the canonical system (7.2.3) is regular or quasiregular and let $Y(\cdot, \lambda)$ be a fundamental matrix of the canonical system (7.2.3). Then $Y(\cdot, \lambda)\phi$ is square-integrable with respect to Δ at a or b for every $\phi \in \mathbb{C}^2$ and $Y(\cdot, \lambda)$ admits a unique continuous extension to a or b such that $Y(a, \lambda)$ or $Y(b, \lambda)$ is invertible, respectively. In particular, the point c_0 in (7.2.12) can be chosen to be a or b, respectively.*

7.4 Square-integrability of solutions of real canonical systems

Let $\imath = (a, b)$ be an open interval and consider on this interval the homogeneous system $Jf' - Hf = \lambda \Delta f$. Recall that a solution f, depending on $\lambda \in \mathbb{C}$, is called square-integrable with respect to Δ at a or b if for some $c \in \imath$

$$\int_a^c f(s)^* \Delta(s) f(s)\, ds < \infty \quad \text{or} \quad \int_c^b f(s)^* \Delta(s) f(s)\, ds < \infty,$$

respectively. In this section the existence of such solutions is studied for *real* canonical systems; cf. Definition 7.2.7. The first main result asserts that if there are two linearly independent solutions which are square-integrable with respect to Δ at an endpoint for some $\lambda \in \mathbb{C}$, then for any $\lambda \in \mathbb{C}$ all solutions are square-integrable with respect to Δ at that endpoint. The second main result states that for any $\lambda \in \mathbb{C} \setminus \mathbb{R}$ there is at least one solution that is square-integrable with respect to Δ at an endpoint. A combination of these two results gives a general description of the existence of the solutions that are square-integrable with respect to Δ at an endpoint and leads to the limit-point and limit-circle classification.

In the rest of this section it will be assumed that the system (7.2.3) is real and the symmetry result in Corollary 7.2.9 will be used throughout.

Theorem 7.4.1. *Assume that for $\lambda_0 \in \mathbb{C}$ the equation $Jf' - Hf = \lambda_0 \Delta f$ has two linearly independent solutions which are square-integrable with respect to Δ at a or b. Then for any $\lambda \in \mathbb{C}$ each solution of $Jf' - Hf = \lambda \Delta f$ is square-integrable with respect to Δ at a or b, respectively.*

Proof. It is sufficient to show the result for one endpoint, say b. Assume without loss of generality that the endpoint a of the canonical system (7.2.3) is regular. Fix a fundamental solution $Y(\cdot, \lambda_0)$ by the initial condition $Y(a, \lambda_0) = I$. The columns $Y_1(\cdot, \lambda_0)$ and $Y_2(\cdot, \lambda_0)$ of $Y(\cdot, \lambda_0)$ belong to $\mathcal{L}^2_\Delta(\imath)$ by assumption. As the system is assumed to be real, one has

$$\int_a^b |\Delta(s)^{\frac{1}{2}} Y_i(s, \overline{\lambda}_0)|^2 \, ds = \int_a^b |\Delta(s)^{\frac{1}{2}} Y_i(s, \lambda_0)|^2 \, ds, \quad i = 1, 2; \tag{7.4.1}$$

cf. Corollary 7.2.9.

Let $\lambda \in \mathbb{C}$ and let $f(\cdot, \lambda)$ be any solution of $Jf' - Hf = \lambda \Delta f$. It will be shown that $f(\cdot, \lambda)$ is square-integrable with respect to Δ at b. Since the function $f(\cdot, \lambda)$ satisfies

$$Jf'(\cdot, \lambda) - Hf(\cdot, \lambda) = \lambda_0 \Delta f(\cdot, \lambda) + (\lambda - \lambda_0) \Delta f(\cdot, \lambda),$$

it follows from (7.2.16) (with $g = (\lambda - \lambda_0) f(\cdot, \lambda)$) that $f(\cdot, \lambda)$ can be written as

$$\begin{aligned} f(t, \lambda) &= Y_1(t, \lambda_0)\alpha_1 + Y_2(t, \lambda_0)\alpha_2 \\ &\qquad + (\lambda - \lambda_0)\big[Y_1(t, \lambda_0) y_2(t, \lambda) - Y_2(t, \lambda_0) y_1(t, \lambda)\big], \end{aligned} \tag{7.4.2}$$

where $f(a, \lambda) = (\alpha_1, \alpha_2)^\top$ and $y_i(\cdot, \lambda)$ is defined by

$$y_i(t, \lambda) = \int_a^t Y_i(s, \overline{\lambda}_0)^* \Delta(s) f(s, \lambda) \, ds, \quad i = 1, 2,$$

respectively. By applying the Cauchy–Schwarz inequality in the definition of $y_i(t, \lambda)$ and using (7.4.1) one obtains for $i = 1, 2$,

$$\begin{aligned} |y_i(t, \lambda)| &\leq \sqrt{\int_a^t |\Delta(s)^{\frac{1}{2}} Y_i(s, \overline{\lambda}_0)|^2 \, ds} \sqrt{\int_a^t |\Delta(s)^{\frac{1}{2}} f(s, \lambda)|^2 \, ds} \\ &\leq \sqrt{\int_a^b |\Delta(s)^{\frac{1}{2}} Y_i(s, \overline{\lambda}_0)|^2 \, ds} \sqrt{\int_a^t |\Delta(s)^{\frac{1}{2}} f(s, \lambda)|^2 \, ds} \\ &= \sqrt{\int_a^b |\Delta(s)^{\frac{1}{2}} Y_i(s, \lambda_0)|^2 \, ds} \sqrt{\int_a^t |\Delta(s)^{\frac{1}{2}} f(s, \lambda)|^2 \, ds}. \end{aligned}$$

Introduce the number $\alpha \geq 0$ and the nonnegative function φ by

$$\alpha = \max\{|\alpha_1|, |\alpha_2|\}, \quad \varphi(t) = \max\{|\Delta(t)^{\frac{1}{2}} Y_1(t, \lambda_0)|, |\Delta(t)^{\frac{1}{2}} Y_2(t, \lambda_0)|\},$$

so that $\varphi \in L^2(a,b)$. Multiply both sides in the identity in (7.4.2) from the left by $\Delta(t)^{\frac{1}{2}}$, then

$$
\begin{aligned}
&|\Delta(t)^{\frac{1}{2}} f(t,\lambda)| \\
&\qquad \le 2\alpha\varphi(t) + 2|\lambda-\lambda_0|\varphi(t)\sqrt{\int_a^b \varphi(s)^2\,ds}\,\sqrt{\int_a^t |\Delta(s)^{\frac{1}{2}} f(s,\lambda)|^2\,ds}.
\end{aligned}
$$

Therefore, one obtains that

$$
|\Delta(t)^{\frac{1}{2}} f(t,\lambda)|^2 \le \varphi(t)^2\left(A + B\int_a^t |\Delta(s)^{\frac{1}{2}} f(s,\lambda)|^2\,ds\right), \tag{7.4.3}
$$

where

$$
A = 8\alpha^2, \quad B = 8|\lambda-\lambda_0|^2\int_a^b \varphi(s)^2\,ds.
$$

It follows from (7.4.3) by means of Lemma 6.1.4 with $u(t) = |\Delta(t)^{\frac{1}{2}} f(t,\lambda)|$, φ as above, and $r=1$, that the function $f(\cdot,\lambda)$ is square-integrable with respect to Δ at b. □

Next it will be shown that for each endpoint and any $\lambda \in \mathbb{C}\setminus\mathbb{R}$ there is at least one solution of the homogeneous canonical system (7.2.4) which is square-integrable with respect to Δ at that endpoint. The proof of this fact is based on the monotonicity principle in Section 5.2; cf. Corollary 5.2.14. To apply this result, let $Y(\cdot,\lambda)$ be a fundamental matrix of the canonical system (7.2.3) fixed as in (7.2.12) and consider the 2×2 matrix function $D(\cdot,\lambda)$ on $\imath$ defined by

$$
D(t,\lambda) = Y(t,\lambda)^*(-iJ)Y(t,\lambda), \quad t \in \imath, \quad \lambda \in \mathbb{C}. \tag{7.4.4}
$$

Observe that the function $t \mapsto D(t,\lambda)$, $t \in \imath$, is absolutely continuous for every $\lambda \in \mathbb{C}$ and that the matrices $D(t,\lambda)$ are self-adjoint and invertible for all $t \in \imath$ and $\lambda \in \mathbb{C}$.

According to the following theorem, the matrix function in (7.4.4) admits self-adjoint limits at a and b, which may be either self-adjoint matrices or self-adjoint relations with a one-dimensional domain and a one-dimensional multivalued part. Furthermore, the dimensions of the domains of the limit relations are directly connected with the number of linearly independent solutions of the homogeneous canonical system (7.2.4) that are square-integrable with respect to Δ.

Theorem 7.4.2. *For $\lambda \in \mathbb{C}^+$ or $\lambda \in \mathbb{C}^-$ the 2×2 matrix function $t \mapsto D(t,\lambda)$ is nondecreasing or nonincreasing on $\imath$, respectively. There exist self-adjoint relations $D(b,\lambda)$ and $D(a,\lambda)$ in $\mathbb{C}^2$ such that*

$$
D(t,\lambda) \to D(a,\lambda) \quad \text{and} \quad D(t,\lambda) \to D(b,\lambda)
$$

in the (strong) resolvent sense when $t \to a$ and $t \to b$, respectively, and

$$
1 \le \dim\big(\operatorname{dom} D(a,\lambda)\big) \le 2 \quad \text{and} \quad 1 \le \dim\big(\operatorname{dom} D(b,\lambda)\big) \le 2.
$$

Furthermore, $\phi \in \operatorname{dom} D(a,\lambda)$ *or* $\phi \in \operatorname{dom} D(b,\lambda)$ *if and only if* $Y(\cdot,\lambda)\phi$ *is a solution of* (7.2.4) *that is square-integrable with respect to* Δ *at* a *or* b, *respectively.*

Proof. It follows from Corollary 7.2.4 that

$$D(\beta,\lambda) - D(\alpha,\lambda) = 2\operatorname{Im}\lambda \int_\alpha^\beta Y(s,\lambda)^*\Delta(s)Y(s,\lambda)\,ds, \quad \lambda \in \mathbb{C}, \tag{7.4.5}$$

holds for any compact interval $[\alpha,\beta] \subset \imath$. Hence, the matrix function $D(\cdot,\lambda)$ is nondecreasing for $\lambda \in \mathbb{C}^+$ and nonincreasing for $\lambda \in \mathbb{C}^-$. It follows from Corollary 5.2.14 that there exist self-adjoint relations $D(a,\lambda)$ and $D(b,\lambda)$ such that

$$\lim_{t\to a} \big(D(t,\lambda) - \mu\big)^{-1} = \big(D(a,\lambda) - \mu\big)^{-1}, \qquad \mu \in \mathbb{C}\setminus\mathbb{R},$$

and

$$\lim_{t\to b} \big(D(t,\lambda) - \mu\big)^{-1} = \big(D(b,\lambda) - \mu\big)^{-1}, \qquad \mu \in \mathbb{C}\setminus\mathbb{R}.$$

Next it will be shown that the dimension of the domains of the self-adjoint relations $D(a,\lambda)$ and $D(b,\lambda)$ is at least one. For this it is sufficient to prove that there exists at least one (finite) eigenvalue.

Note first that, by (7.4.4) and (7.2.12),

$$D(c_0,\lambda) = Y(c_0,\lambda)^*(-iJ)Y(c_0,\lambda) = -iJ$$

and hence the eigenvalues of $D(c_0,\lambda)$ are $\nu_-(c_0) = -1$ and $\nu_+(c_0) = 1$. As the function $D(\cdot,\lambda)$ is continuous on $\imath$, the same holds true for its eigenvalues $\nu_-(\cdot)$ and $\nu_+(\cdot)$. Since the matrices $D(t,\lambda)$ are self-adjoint and invertible for all $t \in \imath$ it follows that $\nu_-(t) < 0$ and $\nu_+(t) > 0$ for all $t \in \imath$. Recall that

$$\nu_-(t) = \inf_{|x|=1} (D(t,\lambda)x,x) \quad \text{and} \quad \nu_+(t) = \sup_{|x|=1} (D(t,\lambda)x,x),$$

and since $D(t_1,\lambda) \le D(t_2,\lambda)$, $t_1 \le t_2$, it follows that

$$\nu_-(t_1) \le \nu_-(t_2) \quad \text{and} \quad \nu_+(t_1) \le \nu_+(t_2), \quad t_1 \le t_2.$$

Therefore, it is clear that the limits of $\nu_-(t)$ and $\nu_+(t)$ exist and that

$$\nu_-(b) = \lim_{t\to b} \nu_-(t) \le 0 \quad \text{and} \quad 0 < \nu_+(b) = \lim_{t\to b} \nu_+(t) \le \infty.$$

In order to see the connection of these limits with the self-adjoint relation $D(b,\lambda)$ observe that for $\mu \in \mathbb{C}\setminus\mathbb{R}$

$$\frac{1}{\nu_-(t) - \mu} \quad \text{and} \quad \frac{1}{\nu_+(t) - \mu}$$

are the eigenvalues of the matrix $(D(t,\lambda) - \mu)^{-1}$. Therefore, again by continuity, one sees that

$$\frac{1}{\nu_-(b) - \mu} \quad \text{and} \quad \frac{1}{\nu_+(b) - \mu}$$

are the eigenvalues of the matrix $(D(b,\lambda)-\mu)^{-1}$. Hence, $\nu_-(b)$ is a nonpositive eigenvalue of the self-adjoint relation $D(b,\lambda)$, which implies $\dim(\operatorname{dom} D(b,\lambda)) \geq 1$. More precisely, if $\nu_+(b) < \infty$, then $\nu_+(b)$ is a positive eigenvalue of $D(b,\lambda)$, in which case $\dim(\operatorname{dom} D(b,\lambda)) = 2$, while if $\nu_+(b) = \infty$, then $D(b,\lambda)$ has a one-dimensional multivalued part and $\dim(\operatorname{dom} D(b,\lambda)) = 1$. Similar observations may be made for the self-adjoint relation $D(a,\lambda)$. In particular, it follows that $\dim(\operatorname{dom} D(a,\lambda)) \geq 1$.

Finally, it will be shown that $\phi \in \operatorname{dom} D(b,\lambda)$ if and only if the solution $Y(\cdot,\lambda)\phi$ of (7.2.4) is square-integrable with respect to Δ at b; the argument for the left endpoint a is the same. Suppose that $\lambda \in \mathbb{C}^+$, so that $D(\cdot,\lambda)$ is nondecreasing on $\imath$. In this case it follows from Corollary 5.2.13 and Corollary 5.2.14 that

$$\operatorname{dom} D(b,\lambda) = \left\{\phi \in \mathbb{C}^2 : \lim_{t\to b} \phi^* D(t,\lambda)\phi < \infty\right\}$$

and hence (7.4.5) implies that $\phi \in \operatorname{dom} D(b,\lambda)$ if and only if

$$\int_\alpha^b \phi^* Y(s,\lambda)^* \Delta(s) Y(s,\lambda)\phi \, ds < \infty,$$

that is, the solution $Y(\cdot,\lambda)\phi$ is square-integrable with respect to Δ at b. The case where $\lambda \in \mathbb{C}^-$ is dealt with in a similar way. □

A combination of Theorems 7.4.1 and 7.4.2 leads to the following observation.

Corollary 7.4.3. *If for some $\lambda_0 \in \mathbb{C}\setminus\mathbb{R}$ the equation $Jf' - Hf = \lambda_0\Delta f$ has, up to scalar multiples, only one nontrivial solution which is square-integrable with respect to Δ at a or b, then for any $\lambda \in \mathbb{C}\setminus\mathbb{R}$ the equation $Jf' - Hf = \lambda\Delta f$ has, up to scalar multiples, precisely one nontrivial solution which is square-integrable with respect to Δ at a or b, respectively.*

Proof. It is sufficient to consider the endpoint b. Assume that for some $\lambda_0 \in \mathbb{C}\setminus\mathbb{R}$ the equation $Jf' - Hf = \lambda_0\Delta f$ has, up to scalar multiples, only one nontrivial solution that is square-integrable with respect to Δ at b, and suppose that for some $\lambda \in \mathbb{C}\setminus\mathbb{R}$ with $\lambda \neq \lambda_0$ the equation $Jf' - Hf = \lambda\Delta f$ does not have, up to scalar multiples, only one nontrivial solution which is square-integrable with respect to Δ at b. Since

$$1 \leq \dim\big(\operatorname{dom} D(b,\lambda)\big) \leq 2$$

by Theorem 7.4.2 there exist two linearly independent solutions of $Jf' - Hf = \lambda\Delta f$ which are square-integrable with respect to Δ at b. But then Theorem 7.4.1 implies that there also exist two linearly independent solutions of $Jf' - Hf = \lambda_0\Delta f$ that are square-integrable with respect to Δ at b; a contradiction. □

Theorem 7.4.1 and Corollary 7.4.3 yield the limit-point and limit-circle classification for real canonical systems in the next definition and corollary. The terminology is inspired by the terminology for Sturm–Liouville equations in Section 6.1.

Definition 7.4.4. For a real canonical system the endpoint a or b of the interval $\imath$ is said to be in the *limit-circle case* if for some, and hence for all $\lambda \in \mathbb{C}$ there exist two linearly independent solutions of $Jf' - Hf = \lambda \Delta f$ that are square-integrable with respect to Δ at a or b, respectively. The endpoint a or b of the interval $\imath$ is said to be in the *limit-point case* if for some, and hence for all $\lambda \in \mathbb{C} \setminus \mathbb{R}$ there exists, up to scalar multiples, only one nontrivial solution of $Jf' - Hf = \lambda \Delta f$ that is square-integrable with respect to Δ at a or b, respectively.

Note that, by Theorem 7.4.2, at any endpoint of the interval $\imath$ there is at least one nontrivial solution that is square-integrable with respect to Δ and there are at most two linearly independent solutions that are square-integrable with respect to Δ. This leads to *Weyl's alternative* for canonical systems.

Corollary 7.4.5. *For a real canonical system each of the endpoints of the interval is either in the limit-circle case or in the limit-point case.*

For completeness also the special case of regular and quasiregular endpoints is briefly discussed. The next corollary is an immediate consequence of Corollary 7.3.5.

Corollary 7.4.6. *A regular or quasiregular endpoint of a real canonical system is in the limit-circle case.*

A simple but useful characterization of the limit-point case is given in the following corollary. It is stated for the endpoint b, but clearly there is a similar statement for the endpoint a.

Corollary 7.4.7. *Let the canonical system be real and assume that the endpoint a is regular or quasiregular. Then the following statements hold:*

(i) *If the endpoint b is in the limit-point case, then for all $\lambda \in \mathbb{R}$ the equation $Jf' - Hf = \lambda \Delta f$ has, up to scalar multiples, at most one nontrivial solution that is square-integrable with respect to Δ at b.*

(ii) *If there exists $\lambda_0 \in \mathbb{R}$ such that the equation $Jf' - Hf = \lambda_0 \Delta f$ has, up to scalar multiples, at most one nontrivial solution that is square-integrable with respect to Δ at b, then the endpoint b is in the limit-point case.*

Proof. (i) If there exists $\lambda \in \mathbb{R}$ for which the homogeneous equation has two linearly independent solutions that are square-integrable with respect to Δ at b, then by Theorem 7.4.1, for each $\lambda \in \mathbb{C}$ all nontrivial solutions are square-integrable with respect to Δ at b. Hence, b is in the limit-circle case; a contradiction.

(ii) If b is in the limit-circle case, then for all $\lambda \in \mathbb{C}$, and hence for $\lambda \in \mathbb{R}$, the homogeneous equation has two linearly independent solutions which are square-integrable with respect to Δ at b. This implies (ii). □

If, for instance, the endpoint b is regular or quasiregular, then any solution of (7.2.3) with g square-integrable with respect to Δ at b has a limit at b by

Proposition 7.3.2 and b is in the limit-circle case by Corollary 7.4.6. However, if b is in the limit-circle case, then the solutions of (7.2.3) are square-integrable with respect to Δ at b, but they do not necessarily have a limit at b. It will be shown in this case that there exists a natural transformation which turns the system into one where b is quasiregular; cf. Lemma 7.2.5.

Corollary 7.4.8. *Assume that a is regular and that b is in the limit-circle case. Let $g \in \mathcal{L}^2_\Delta(a,b)$ and let $f(\cdot,\lambda)$ be a solution of*

$$Jf' - Hf = \lambda \Delta f + \Delta g.$$

Let $U(\cdot,\lambda_0)$, $\lambda_0 \in \mathbb{R}$, be a matrix function as in (7.2.17). Then the limit

$$\widetilde{f}(b) = \lim_{t\to b} U(t,\lambda_0)^{-1} f(t) \tag{7.4.6}$$

exists in $\mathbb{C}^2$. Moreover, for each $\gamma \in \mathbb{C}^2$ there exists a unique solution $f(\cdot,\lambda)$ of (7.2.3) such that $\widetilde{f}(b) = \gamma$ and the corresponding function

$$\lambda \mapsto \lim_{t\to b} U(t,\lambda_0)^{-1} f(t,\lambda)$$

is entire.

Proof. Let $g \in \mathcal{L}^2_\Delta(a,b)$ and let f be a solution of (7.2.3). Since b is in the limit-circle case, there exists for $\lambda_0 \in \mathbb{R}$ and $c_0 \in [a,b)$ a matrix function $U(\cdot,\lambda_0)$ satisfying (7.2.17) that is square-integrable with respect to Δ at b. Thus, the function $\widetilde{\Delta}$ defined in (7.2.18) is integrable at b, which means that the endpoint b for the system in (7.2.20) is quasiregular. Since g is square-integrable with respect to Δ at b, the function $\widetilde{g}$ in Lemma 7.2.5 is square-integrable with respect to $\widetilde{\Delta}$ at b. Therefore, the assertion is clear from Proposition 7.3.2 as $\widetilde{f}$ is a solution of (7.2.20). □

Let the endpoint a be regular or quasiregular. Let $g, k \in \mathcal{L}^2_\Delta(\imath)$ and let f, h be solutions of the inhomogeneous equations (7.2.6) such that $f, h \in \mathcal{L}^2_\Delta(\imath)$. Then for $a \le t < b$ one has

$$h(t)^* Jf(t) - h(a)^* Jf(a) = \int_a^t \big(h(s)^*\Delta(s)g(s) - k(s)^*\Delta(s)f(s)\big)\, ds \tag{7.4.7}$$

by the Lagrange identity in Corollary 7.2.3. It follows from (7.4.7) that the limit

$$\lim_{t\to b} h(t)^* Jf(t)$$

exists. Of course, when b is regular or quasiregular, then the individual limits $\lim_{t\to b} f(t)$ and $\lim_{t\to b} h(t)$ exist by Proposition 7.3.2, see also Corollary 7.3.4. In general the existence of the individual limits $\lim_{t\to b} f(t)$ and $\lim_{t\to b} h(t)$ is not guaranteed. However, in the case where b is in the limit-circle case but not quasiregular the next corollary suggests to employ the limits in (7.4.6).

Corollary 7.4.9. *Assume that the endpoint a is regular and that b is in the limit-circle case. Let $g, k \in \mathcal{L}^2_\Delta(\imath)$ and let f, h be solutions of the inhomogeneous equations (7.2.6) such that $f, h \in \mathcal{L}^2_\Delta(\imath)$. Then*

$$\lim_{t \to b} h(t)^* J f(t) = \widetilde{h}(b)^* J \widetilde{f}(b), \tag{7.4.8}$$

where $\widetilde{f}(b)$ and $\widetilde{h}(b)$ are as in (7.4.6). Moreover,

$$\widetilde{h}(b)^* J \widetilde{f}(b) - h(a)^* J f(a) = \int_a^b \big(h(s)^* \Delta(s) g(s) - k(s)^* \Delta(s) f(s)\big)\, ds. \tag{7.4.9}$$

Proof. It follows by taking limits in (7.4.7) that

$$\lim_{t \to b} h(t)^* J f(t) - h(a)^* J f(a) = \int_a^b \big(h(s)^* \Delta(s) g(s) - k(s)^* \Delta(s) f(s)\big)\, ds.$$

Now apply Corollary 7.2.6 and Corollary 7.4.8. Take the limit $t \to b$ and (7.4.8) and (7.4.9) follow. □

7.5 Definite canonical systems

The general class of canonical differential equations as in (7.2.3) will now be narrowed down by imposing a definiteness condition; see Definition 7.5.5. This condition will be assumed in the rest of this chapter. In this section various equivalent formulations of the definiteness condition will be presented. Moreover, it will be shown that the solution of a definite canonical system (7.2.3) can be cut off near an endpoint of the interval $\imath$, in the sense that the solution is modified in such a way that it becomes trivial in a neighborhood of that endpoint.

It will be convenient to begin the discussion of definiteness of the canonical system (7.2.3) with the notion of definiteness when the system is restricted to an arbitrary subinterval $\jmath \subset \imath$.

Definition 7.5.1. Let $\jmath \subset \imath$ be a nonempty interval. The canonical system (7.2.3) is said to be *definite* on $\jmath$ if for each solution f of $Jf' - Hf = 0$ on $\jmath$ one has

$$\Delta(t) f(t) = 0, \ \ t \in \jmath \quad \Rightarrow \quad f(t) = 0, \ \ t \in \jmath.$$

Observe that if a solution f of the canonical system (7.2.3) vanishes on a nonempty subinterval $\jmath \subset \imath$, then $f(t) = 0$ for $t \in \imath$; cf. Theorem 7.2.1. Hence, it is clear that if the canonical system (7.2.3) is definite on $\jmath$, then it is also definite on every interval $\widetilde{\jmath}$ with the property that $\jmath \subset \widetilde{\jmath} \subset \imath$. Also observe that with the subinterval $\jmath \subset \imath$ and a continuous function f one has

$$\Delta(t) f(t) = 0, \ \ t \in \jmath \quad \Leftrightarrow \quad \int_\jmath f(s)^* \Delta(s) f(s)\, ds = 0. \tag{7.5.1}$$

Clearly, if $\Delta(t)$ has full rank for almost all $t \in \jmath$, then the canonical system is automatically definite on $\jmath$.

Lemma 7.5.2. *Let $\jmath \subset \imath$ be a nonempty interval. The canonical system (7.2.3) is definite on the interval $\jmath \subset \imath$ if and only if for all $\lambda \in \mathbb{C}$ and for each solution f of $Jf' - Hf = \lambda\Delta f$ on $\jmath$ one has*

$$\Delta(t)f(t) = 0, \ \ t \in \jmath \quad \Rightarrow \quad f(t) = 0, \ \ t \in \jmath.$$

Proof. Assume that the canonical system is definite on $\jmath$. Choose $\lambda \in \mathbb{C}$ and let f be a solution of $Jf' - Hf = \lambda\Delta f$ on $\jmath$ with $\Delta(t)f(t) = 0$ for almost all $t \in \jmath$. Thus, f is a solution of $Jf' - Hf = 0$ with $\Delta(t)f(t) = 0$ for almost all $t \in \jmath$. By assumption this implies that $f(t) = 0$ for $t \in \jmath$. The converse statement is trivial. □

The following result is an alternative useful version of Lemma 7.5.2 in terms of a fundamental matrix $Y(\cdot, \lambda)$.

Corollary 7.5.3. *Let $Y(\cdot, \lambda)$, $\lambda \in \mathbb{C}$, be a fundamental matrix for (7.2.3) and let $I \subset \imath$ be a compact interval. Then the system (7.2.3) is definite on I if and only if the 2×2 matrix*

$$\int_I Y(s,\lambda)^* \Delta(s) Y(s,\lambda)\, ds \tag{7.5.2}$$

is invertible for some, and hence for all $\lambda \in \mathbb{C}$.

Proof. Assume that (7.2.3) is definite on I. If the (nonnegative) matrix in (7.5.2) is not invertible, then there exists a nontrivial $\gamma \in \mathbb{C}^2$ for which

$$\gamma^* \left(\int_I Y(s,\lambda)^* \Delta(s) Y(s,\lambda)\, ds \right) \gamma = 0, \tag{7.5.3}$$

or alternatively $\Delta(t)Y(t,\lambda)\gamma = 0$ for $t \in I$; cf. (7.5.1). Since $Y(\cdot,\lambda)\gamma$ is a solution of $Jf' - Hf = \lambda\Delta f$, it follows from the definiteness that $Y(t,\lambda)\gamma = 0$ for $t \in I$, which implies $\gamma = 0$. This contradiction shows that the matrix in (7.5.2) is invertible.

Conversely, assume that the (nonnegative) matrix in (7.5.2) is invertible. In order to show that (7.2.3) is definite, let

$$Jf'(t) - H(t)f(t) = \lambda\Delta(t)f(t), \quad \Delta(t)f(t) = 0, \quad t \in I.$$

Since $Y(\cdot, \lambda)$ is a fundamental matrix of $Jf' - Hf = \lambda\Delta f$, every solution of this equation can be written in the form $f = Y(\cdot,\lambda)\gamma$ with a unique $\gamma \in \mathbb{C}^2$. The condition $\Delta(t)f(t) = 0$, $t \in I$, implies that (7.5.3) holds. Therefore, $\gamma = 0$ and thus the system (7.2.3) is definite. □

The next proposition shows that there is no difference between global definiteness and local definiteness.

Proposition 7.5.4. *The canonical system (7.2.3) is definite on $\imath$ if and only if there exists a compact interval $I \subset \imath$ such that the canonical system (7.2.3) is definite on the interval I.*

Proof. If the canonical system (7.2.3) is definite on the interval I, then it is clearly definite on the larger interval $\imath$.

To see the converse statement, let the canonical system (7.2.3) be definite on the interval $\imath$; in other words, assume that for each solution f of $Jf' - Hf = 0$ on $\imath$ one has

$$\Delta(t)f(t) = 0, \quad t \in \imath \quad \Rightarrow \quad f(t) = 0, \quad t \in \imath.$$

Introduce for each compact subinterval K of $\imath$ the subset $d(K)$ of $\mathbb{C}^2$ by

$$d(K) = \left\{ \phi \in \mathbb{C}^2 : |\phi| = 1, \int_K \phi^* Y(s,0)^* \Delta(s) Y(s,0)\phi \, ds = 0 \right\}.$$

Clearly, $d(K)$ is compact and $K \subset \widetilde{K}$ implies $d(\widetilde{K}) \subset d(K)$. Now choose an increasing sequence of compact intervals (K_n) such that their union equals the interval $\imath$. Then

$$\bigcap_{n \in \mathbb{N}} d(K_n) = \emptyset. \tag{7.5.4}$$

Indeed, assume that there exists an element $\phi \in \mathbb{C}^2$ with $|\phi| = 1$, such that

$$\int_{K_n} \phi^* Y(s,0)^* \Delta(s) Y(s,0)\phi \, ds = 0$$

for every $n \in \mathbb{N}$. Then, by monotone convergence,

$$\int_\imath \phi^* Y(s,0)^* \Delta(s) Y(s,0)\phi \, ds = 0.$$

As the canonical system (7.2.3) is definite, this implies by (7.5.1) that $Y(\cdot,0)\phi = 0$, which leads to $\phi = 0$; a contradiction. Therefore, the identity (7.5.4) is valid. Since each of the sets $d(K_n)$ in (7.5.4) is compact, it follows that there exists a compact interval K_m such that $d(K_m) = \emptyset$. Hence, $I = K_m$ satisfies the requirements. To see this, let $Jf' - Hf = 0$ on K_m and assume that $\Delta(t)f(t) = 0$, $t \in K_m$, or, equivalently, $\int_{K_m} f(s)^* \Delta(s) f(s) = 0$; cf. (7.5.1). Since $d(K_m) = \emptyset$ one concludes that $f = 0$. □

In the rest of the text one often speaks of definite systems in the following sense.

Definition 7.5.5. The canonical system (7.2.3) is said to be *definite* if it is *definite on $\imath$*.

The next result is about smoothly cutting off the solution of a definite canonical system (7.2.3) near an endpoint of the interval $\imath$, i.e., modifying the solution so that it becomes trivial in a neighborhood of that endpoint. The following proposition and corollary will be used in Section 7.6.

Proposition 7.5.6. *Let the canonical system (7.2.3) be definite and choose a compact interval $[\alpha, \beta] \subset \imath$ such that the system is definite on $[\alpha, \beta]$. Let $g \in \mathcal{L}^2_{\Delta,\mathrm{loc}}(\imath)$ and let $f \in AC(\imath)$ be a solution of the inhomogeneous equation (7.2.3) for some $\lambda \in \mathbb{C}$. Then there exist functions $f_a \in AC(\imath)$ and $g_a \in \mathcal{L}^2_{\Delta,\mathrm{loc}}(\imath)$ satisfying*

$$Jf_a'(t) - H(t)f_a(t) = \lambda\Delta(t)f_a(t) + \Delta(t)g_a(t)$$

such that

$$f_a(t) = \begin{cases} f(t), & t \in (a, \alpha], \\ 0, & t \in [\beta, b), \end{cases} \quad \textit{and} \quad g_a(t) = \begin{cases} g(t), & t \in (a, \alpha], \\ 0, & t \in [\beta, b). \end{cases}$$

Similarly, there exist functions $f_b \in AC(\imath)$ and $g_b \in \mathcal{L}^2_{\Delta,\mathrm{loc}}(\imath)$ satisfying

$$Jf_b'(t) - H(t)f_b(t) = \lambda\Delta(t)f_b(t) + \Delta(t)g_b(t)$$

such that

$$f_b(t) = \begin{cases} 0, & t \in (a, \alpha], \\ f(t), & t \in [\beta, b), \end{cases} \quad \textit{and} \quad g_b(t) = \begin{cases} 0, & t \in (a, \alpha], \\ g(t), & t \in [\beta, b). \end{cases}$$

Proof. Let the functions f and g be as indicated. The result will be proved for the functions f_b and g_b; the proof for the functions f_a and g_a is similar.

Let $[\alpha, \beta] \subseteq \imath$ be a compact interval on which the canonical system (7.2.3) is definite; cf. Proposition 7.5.4. Let $k \in \mathcal{L}^2_\Delta(\alpha, \beta)$ and fix a fundamental system $Y(\cdot, \lambda)$ by the initial condition $Y(\alpha, \lambda) = I$. According to (7.2.15), the function defined by

$$h(t) = -Y(t, \lambda) \int_\alpha^t JY(s, \bar{\lambda})^* \Delta(s) k(s)\, ds \tag{7.5.5}$$

satisfies the inhomogeneous equation

$$Jh'(t) - H(t)h(t) = \lambda\Delta(t)h(t) + \Delta(t)k(t), \quad \alpha < t < \beta,$$

and in the endpoints it has the values

$$h(\alpha) = 0 \quad \text{and} \quad h(\beta) = -Y(\beta, \lambda) \int_\alpha^\beta JY(s, \bar{\lambda})^* \Delta(s) k(s)\, ds.$$

It will be shown that there exists a function $k \in \mathcal{L}^2_\Delta(\alpha, \beta)$ such that $h(\beta) = f(\beta)$.

In order to verify this, observe that $Y(\beta, \lambda)$ is invertible and that the integral operator

$$\ell \mapsto \int_\alpha^\beta JY(s, \bar{\lambda})^* \Delta(s) \ell(s)\, ds$$

taking $\mathcal{L}^2_\Delta(\alpha,\beta)$ into $\mathbb{C}^2$ is surjective. To see this, assume that $\gamma \in \mathbb{C}^2$ is orthogonal to the range of this integral operator, that is,

$$0 = \gamma^* \int_\alpha^\beta JY(s,\bar\lambda)^*\Delta(s)\ell(s)\,ds = \int_\alpha^\beta (Y(s,\bar\lambda)J^*\gamma)^*\Delta(s)\ell(s)\,ds$$

for all $\ell \in \mathcal{L}^2_\Delta(\alpha,\beta)$. With $\ell(s) = Y(s,\bar\lambda)J^*\gamma$ it then follows from (7.5.1) and Lemma 7.5.2 that $\ell(s) = 0$ for $s \in (\alpha,\beta)$, which implies that $\gamma = 0$. Thus, the integral operator is surjective.

Now choose $k \in \mathcal{L}^2_\Delta(\alpha,\beta)$ as above, so that h defined by (7.5.5) satisfies $h(\alpha) = 0$ and $h(\beta) = f(\beta)$. Hence, the functions f_b and g_b defined by

$$f_b(t) = \begin{cases} 0, & t \in (a,\alpha], \\ h(t), & t \in (\alpha,\beta), \\ f(t), & t \in [\beta,b), \end{cases} \quad \text{and} \quad g_b(t) = \begin{cases} 0, & t \in (a,\alpha], \\ k(t), & t \in (\alpha,\beta), \\ g(t), & t \in [\beta,b), \end{cases}$$

satisfy the appropriate inhomogeneous canonical equations on (a,α), (α,β), and (β,b). Since $f_b(\alpha) = h(\alpha)$ and $f_b(\beta) = h(\beta)$ it follows that $f_b \in AC(\imath)$. □

In particular, if f is a solution of the homogeneous system (7.2.4), then f can be localized as indicated above. The following restatement of this fact in terms of matrix functions (groupings of column vector functions) is useful. Note that the modification of the solutions of the homogeneous equation involves a solution of the inhomogeneous equation.

Corollary 7.5.7. *Let the canonical system (7.2.3) be definite and choose a compact interval $[\alpha,\beta] \subset \imath$ such that the system is definite on $[\alpha,\beta]$. Let $Y(\cdot,\lambda)$ be a fundamental matrix of (7.2.4). Then there exist a 2×2 matrix function $Y_a(\cdot,\lambda) \in AC(\imath)$ and a 2×2 matrix function $Z_a(\cdot,\lambda)$ whose columns belong to $\mathcal{L}^2_\Delta(\imath)$, satisfying*

$$JY_a'(t,\lambda) - H(t)Y_a(t,\lambda) = \lambda\Delta(t)Y_a(t,\lambda) + \Delta(t)Z_a(t,\lambda)$$

such that

$$Y_a(t,\lambda) = \begin{cases} Y(t,\lambda), & t \in (a,\alpha], \\ 0, & t \in [\beta,b), \end{cases} \quad \text{and} \quad Z_a(t,\lambda) = \begin{cases} 0, & t \in (a,\alpha], \\ 0, & t \in [\beta,b). \end{cases}$$

Similarly, there exist a 2×2 matrix function $Y_b(\cdot,\lambda) \in AC(\imath)$ and a 2×2 matrix function $Z_b(\cdot,\lambda)$ whose columns belong to $\mathcal{L}^2_\Delta(\imath)$, satisfying

$$JY_b'(t,\lambda) - H(t)Y_b(t,\lambda) = \lambda\Delta(t)Y_b(t,\lambda) + \Delta(t)Z_b(t,\lambda)$$

such that

$$Y_b(t,\lambda) = \begin{cases} 0, & t \in (a,\alpha], \\ Y(t,\lambda), & t \in [\beta,b), \end{cases} \quad \text{and} \quad Z_b(t,\lambda) = \begin{cases} 0, & t \in (a,\alpha], \\ 0, & t \in [\beta,b). \end{cases}$$

With $\phi \in \mathbb{C}^2$ observe that the function $Y_a(\cdot,\lambda)\phi$ belongs to $\mathcal{L}^2_\Delta(\imath)$ if and only if $Y(\cdot,\lambda)\phi$ is square-integrable with respect to Δ at a, and, likewise, that the function $Y_b(\cdot,\lambda)\phi$ belongs to $\mathcal{L}^2_\Delta(\imath)$ if and only if $Y(\cdot,\lambda)\phi$ is square-integrable with respect to Δ at b.

It is useful to have a special notation for the elements that modify the pairs $\{Y(\cdot,\lambda), \lambda Y(\cdot,\lambda)\}$ in Corollary 7.5.7. Define the matrix functions $\mathcal{Y}_a(\cdot,\lambda)$ and $\mathcal{Y}_b(\cdot,\lambda)$ by

$$\begin{aligned} \mathcal{Y}_a(\cdot,\lambda) &:= \{Y_a(\cdot,\lambda), \lambda Y_a(\cdot,\lambda) + Z_a(\cdot,\lambda)\}, \\ \mathcal{Y}_b(\cdot,\lambda) &:= \{Y_b(\cdot,\lambda), \lambda Y_b(\cdot,\lambda) + Z_b(\cdot,\lambda)\}, \end{aligned} \tag{7.5.6}$$

that is, for $\phi \in \mathbb{C}^2$ one has

$$\begin{aligned} \mathcal{Y}_a(\cdot,\lambda)\phi &= \{Y_a(\cdot,\lambda)\phi, \lambda Y_a(\cdot,\lambda)\phi + Z_a(\cdot,\lambda)\phi\}, \\ \mathcal{Y}_b(\cdot,\lambda)\phi &= \{Y_b(\cdot,\lambda)\phi, \lambda Y_b(\cdot,\lambda)\phi + Z_b(\cdot,\lambda)\phi\}. \end{aligned}$$

Note that $\mathcal{Y}_a(\cdot,\lambda)$ and $\mathcal{Y}_b(\cdot,\lambda)$ satisfy

$$\begin{aligned} \mathcal{Y}_a(t,\lambda) &= \begin{cases} \{Y(t,\lambda), \lambda Y(t,\lambda)\}, & a < t \le \alpha, \\ \{0,0\}, & \beta \le t < b, \end{cases} \\ \mathcal{Y}_b(t,\lambda) &= \begin{cases} \{0,0\}, & a < t \le \alpha, \\ \{Y(t,\lambda), \lambda Y(t,\lambda)\}, & \beta \le t < b. \end{cases} \end{aligned} \tag{7.5.7}$$

It is clear from the construction that the columns of $\mathcal{Y}_a(\cdot,\lambda)$ or $\mathcal{Y}_b(\cdot,\lambda)$ are square-integrable on (a,b) with respect to Δ if and only if the corresponding columns of $Y(\cdot,\lambda)$ have this property at a or b, respectively.

7.6 Maximal and minimal relations for canonical systems

In this and later sections it will be assumed that the canonical system (7.2.3) is real as in Definition 7.2.7 and definite as in Definition 7.5.5: such systems will be called *real definite canonical systems*. In this context the central Hilbert space will be $L^2_\Delta(\imath)$, in which the maximal and minimal relations associated with the real definite canonical system (7.2.3) will be defined. In principle, both these relations may be multivalued. The results from Section 7.4 and Section 7.5 make it possible to consider the limit-circle case and the limit-point case from the point of view of the maximal and minimal relations.

The real definite canonical system (7.2.3) induces the *maximal relation* $T_{\max}$ in $L^2_\Delta(\imath)$ defined by

$$T_{\max} = \{\{f,g\} \in L^2_\Delta(\imath) \times L^2_\Delta(\imath) : Jf' - Hf = \Delta g\}.$$

Since the elements of $L^2_\Delta(\imath)$ are equivalence classes, the definition of $T_{\max}$ needs the following explanation: an element $\{f,g\} \in L^2_\Delta(\imath) \times L^2_\Delta(\imath)$ belongs to $T_{\max}$ if and only if the equivalence class f contains an absolutely continuous representative $\widetilde{f}$ such that the inhomogeneous equation $J\widetilde{f}'(t) - H(t)\widetilde{f}(t) = \Delta(t)\widetilde{g}(t)$ is satisfied for almost every $t \in \imath$. Here $\widetilde{g}$ is any representative of $g \in L^2_\Delta(\imath)$; observe that the function $\Delta(t)\widetilde{g}(t)$ is independent of the representative. The above argument also shows that the relation $T_{\max}$ is linear.

Since the canonical system (7.2.3) is assumed to be definite, the absolutely continuous representative is unique.

Lemma 7.6.1. *If $\{f,g\} \in T_{\max}$, then the equivalence class f has a unique absolutely continuous representative.*

Proof. Let $\{f,g\} \in T_{\max}$ and let $\widetilde{f}_1$ and $\widetilde{f}_2$ be absolutely continuous representatives of f. Then $J(\widetilde{f}_1 - \widetilde{f}_2)' - H(\widetilde{f}_1 - \widetilde{f}_2) = 0$ holds and

$$\Delta(t)(\widetilde{f}_1 - \widetilde{f}_2)(t) = 0, \quad t \in \imath.$$

Therefore, by Definition 7.5.5, it follows that $\widetilde{f}_1(t) = \widetilde{f}_2(t)$ for all $t \in \imath$. □

It will be shown that $T_{\max}$ is the adjoint of a symmetric relation whose defect numbers are equal and at most $(2,2)$. Let T_0 be the *preminimal relation*, i.e., the restriction of the maximal relation $T_{\max}$ to the elements where the first component has compact support in $\imath$:

$$T_0 = \big\{\{f,g\} \in T_{\max} : f \text{ has compact support}\big\}.$$

More precisely, an element $\{f,g\} \in L^2_\Delta(\imath) \times L^2_\Delta(\imath)$ belongs to T_0 if and only if the equivalence class f contains an absolutely continuous representative $\widetilde{f}$ with compact support such that the inhomogeneous equation $J\widetilde{f}'(t) - H(t)\widetilde{f}(t) = \Delta(t)\widetilde{g}(t)$ is satisfied for almost every $t \in \imath$. Here $\widetilde{g}$ is any representative of $g \in L^2_\Delta(\imath)$. The *minimal relation* $T_{\min}$ is defined as $T_{\min} = \overline{T}_0$.

Theorem 7.6.2. *The closure $T_{\min} = \overline{T}_0$ of T_0 is a closed symmetric relation in $L^2_\Delta(\imath)$ and it satisfies*

$$T_{\min} \subset (T_{\min})^* = T_{\max},$$

and, consequently, $T_{\min} = (T_{\max})^*$.

Proof. Step 1. It will be shown that

$$T_{\max} \subset (T_0)^*. \tag{7.6.1}$$

For this purpose, let $\{f,g\} \in T_{\max}$, $\{h,k\} \in T_0$, and choose an interval $[\alpha,\beta] \subset \imath$ containing the support h (and hence the support of Δk). Then

$$\begin{aligned}(g,h)_\Delta - (f,k)_\Delta &= \int_\alpha^\beta h(s)^*\Delta(s)g(s)\,ds - \int_\alpha^\beta k(s)^*\Delta(s)f(s)\,ds \\ &= h(\beta)^*Jf(\beta) - h(\alpha)^*Jf(\alpha) \\ &= 0\end{aligned}$$

by Corollary 7.2.3. Here $(\cdot,\cdot)_\Delta$ denotes the scalar product in $L^2_\Delta(\imath)$, f and h are the uniquely defined absolutely continuous representatives, while g and k are arbitrary representatives. Observe that the integral does not depend on the particular choice of g and k. This shows $\{f,g\} \in (T_0)^*$ and hence (7.6.1) follows.

Step 2. It will be shown that

$$(T_0)^* \subset T_{\max}. \tag{7.6.2}$$

For this, let $\{f,g\} \in (T_0)^*$. By Theorem 7.2.1, there exists a nontrivial absolutely continuous function u on $\imath$ such that $Ju' - Hu = \Delta g$. The aim is to show that for any representative f the difference $f - u$ is absolutely continuous modulo an element whose $\mathfrak{L}^2_\Delta(a,b)$-norm is zero. Recall that the system is assumed to be definite, and hence there exists a compact interval $[\alpha_0,\beta_0]$ on which it is definite; cf. Proposition 7.5.4. Choose an interval $[\alpha_1,\beta_1] \subset \imath$ which contains $[\alpha_0,\beta_0]$; then the system is also definite on $[\alpha_1,\beta_1]$.

It is convenient to introduce the subspace

$$\mathfrak{M}_1 := \left\{ k \in \mathcal{L}^2_\Delta(\alpha_1,\beta_1) : \begin{matrix} Jh' - Hh = \Delta k \text{ for some } h \in AC[\alpha_1,\beta_1] \\ \text{such that } h(\alpha_1) = h(\beta_1) = 0 \end{matrix} \right\}.$$

Let $k \in \mathfrak{M}_1$ and let $h \in AC[\alpha_1,\beta_1]$ be a solution of $Jh' - Hh = \Delta k$ for which $h(\alpha_1) = h(\beta_1) = 0$. It follows from (7.2.15) that

$$h(t) = Y(t,0)J^{-1} \int_{\alpha_1}^t Y(s,0)^*\Delta(s)k(s)\,ds, \quad t \in [\alpha_1,\beta_1], \tag{7.6.3}$$

where the fundamental matrix $Y(\cdot,\lambda)$ is fixed by $Y(\alpha_1,\lambda) = I$. Note that the condition $h(\beta_1) = 0$ implies

$$\int_{\alpha_1}^{\beta_1} Y(s,0)^*\Delta(s)k(s)\,ds = 0. \tag{7.6.4}$$

Conversely, if $k \in \mathcal{L}^2_\Delta(\alpha_1,\beta_1)$ satisfies (7.6.4), then $k \in \mathfrak{M}_1$ since $h \in AC[\alpha_1,\beta_1]$ in (7.6.3) satisfies $Jh' - Hh = \Delta k$ and $h(\alpha_1) = h(\beta_1) = 0$. In other words, one has

$$\mathfrak{M}_1 = \left\{ k \in \mathcal{L}^2_\Delta(\alpha_1,\beta_1) : \int_{\alpha_1}^{\beta_1} Y(s,0)^*\Delta(s)k(s)\,ds = 0 \right\}.$$

Now let $k \in \mathfrak{M}_1$ and let h be defined by (7.6.3). Then the pair of functions $\{h,k\}$ can be trivially extended to all of $\imath$ and the extended pair, which will also be denoted by $\{h,k\}$, belongs to T_0. As $\{f,g\} \in (T_0)^*$, one has $(h,g)_\Delta = (k,f)_\Delta$, and since the supports of h and k are inside $[\alpha_1,\beta_1]$ it follows that

$$\int_{\alpha_1}^{\beta_1} g(s)^*\Delta(s)h(s)\,ds = \int_{\alpha_1}^{\beta_1} f(s)^*\Delta(s)k(s)\,ds. \tag{7.6.5}$$

Consider the pair $\{u, g\}$ on $[\alpha_1, \beta_1]$. Note that on this interval u is absolutely continuous and g is square-integrable with respect to Δ. It follows from the Lagrange identity in Corollary 7.2.3 applied to the pairs $\{h, k\}$ and $\{u, g\}$, and $h(\alpha_1) = h(\beta_1) = 0$ that

$$\int_{\alpha_1}^{\beta_1} g(s)^* \Delta(s) h(s)\, ds = \int_{\alpha_1}^{\beta_1} u(s)^* \Delta(s) k(s)\, ds. \tag{7.6.6}$$

Combining (7.6.5) and (7.6.6), one obtains that

$$\int_{\alpha_1}^{\beta_1} (f(s) - u(s))^* \Delta(s) k(s)\, ds = 0$$

for all $k \in \mathfrak{M}_1$. In other words, the restriction of $f - u$ to $[\alpha_1, \beta_1]$ is orthogonal to $\mathfrak{M}_1$ in the semidefinite Hilbert space $\mathcal{L}^2_\Delta(\alpha_1, \beta_1)$. Furthermore, by (7.6.4) one has that $Y(\cdot, 0)\gamma$ is orthogonal to $\mathfrak{M}_1$ in $\mathcal{L}^2_\Delta(\alpha_1, \beta_1)$ for all $\gamma \in \mathbb{C}^2$, and since the same is true for $f - u$ it follows that $f - u - Y(\cdot, 0)\gamma$ is orthogonal to $\mathfrak{M}_1$ in $\mathcal{L}^2_\Delta(\alpha_1, \beta_1)$ for all $\gamma \in \mathbb{C}^2$.

Next it will be shown that for some $\gamma_1 \in \mathbb{C}^2$ the function $f - u - Y(\cdot, 0)\gamma_1$ belongs to $\mathfrak{M}_1$. In fact, first of all it is clear from (7.2.15) that for any $\gamma \in \mathbb{C}^2$

$$h(t) = Y(t, 0) J^{-1} \int_{\alpha_1}^{t} Y(s, 0)^* \Delta(s) \big(f(s) - u(s) - Y(s, 0)\gamma\big)\, ds$$

satisfies $Jh' - Hh = \Delta(f - u - Y(\cdot, 0)\gamma)$ and $h(\alpha_1) = 0$. To satisfy the boundary condition $h(\beta_1) = 0$, choose $\gamma = \gamma_1 \in \mathbb{C}^2$ such that

$$\int_{\alpha_1}^{\beta_1} Y(s, 0)^* \Delta(s)(f(s) - u(s))\, ds = \int_{\alpha_1}^{\beta_1} Y(s, 0)^* \Delta(s) Y(s, 0)\gamma_1\, ds;$$

this is possible since the system is definite on $[\alpha_1, \beta_1]$ and hence the matrix

$$\int_{\alpha_1}^{\beta_1} Y(s, 0)^* \Delta(s) Y(s, 0)\, ds$$

is invertible; cf. Corollary 7.5.3. Therefore, $f - u - Y(\cdot, 0)\gamma_1 \in \mathfrak{M}_1$. Since the element $f - u - Y(\cdot, 0)\gamma_1$ is orthogonal to $\mathfrak{M}_1$ in $\mathcal{L}^2_\Delta(\alpha_1, \beta_1)$, this yields

$$\int_{\alpha_1}^{\beta_1} \big(f(s) - u(s) - Y(s, 0)\gamma_1\big)^* \Delta(s) \big(f(s) - u(s) - Y(s, 0)\gamma_1\big)\, ds = 0,$$

and hence there exists a function ω_1 on $[\alpha_1, \beta_1]$ such that

$$f(s) = u(s) + Y(s, 0)\gamma_1 + \omega_1(s), \quad \Delta(s)\omega_1(s) = 0, \quad s \in [\alpha_1, \beta_1].$$

Likewise, on any interval $[\alpha_2, \beta_2]$ extending $[\alpha_1, \beta_1]$ the same argument shows that there exist $\gamma_2 \in \mathbb{C}^2$ and a function ω_2 such that

$$f(s) = u(s) + \widetilde{Y}(s,0)\gamma_2 + \omega_2(s), \quad \Delta(s)\omega_2(s) = 0, \quad s \in [\alpha_2, \beta_2];$$

here the fundamental matrix $\widetilde{Y}(\cdot, \lambda)$ is fixed by $\widetilde{Y}(\alpha_2, \lambda) = I$. Hence, on the smaller interval one obtains for $s \in [\alpha_1, \beta_1]$

$$\omega_1(s) - \omega_2(s) = \widetilde{Y}(s,0)\gamma_2 - Y(s,0)\gamma_1, \quad \Delta(s)(\omega_1(s) - \omega_2(s)) = 0.$$

Since the system is definite on the interval $[\alpha_1, \beta_1]$, this shows that $\omega_1(s) = \omega_2(s)$ and $Y(s,0)\gamma_1 = \widetilde{Y}(s,0)\gamma_2$ for $s \in [\alpha_1, \beta_1]$, and hence $Y(s,0)\gamma_1 = \widetilde{Y}(s,0)\gamma_2$ for $s \in \imath$. One concludes that there exists a function ω such that

$$f(s) = u(s) + Y(s,0)\gamma_1 + \omega(s), \quad \Delta(s)\omega(s) = 0, \quad s \in \imath.$$

Thus, the functions f and $u + Y(\cdot,0)\gamma_1$ belong to the same equivalence class in $L^2_\Delta(\imath)$. Since $J(u+Y(\cdot,0)\gamma_1)' - H(u+Y(\cdot,0)\gamma_1) = \Delta g$ it follows that $\{f, g\} \in T_{\max}$ and $u + Y(\cdot,0)\gamma_1$ is the unique absolutely continuous representative of f. This implies (7.6.2).

Step 3. It follows from (7.6.1) and (7.6.2) that $T_{\max} = (T_0)^*$ and, in particular, this implies that $T_{\max}$ is closed. Hence, the fact that $T_0 \subset T_{\max}$ and the definition $T_{\min} = \overline{T_0}$ imply that

$$T_{\min} = \overline{T_0} \subset T_{\max} = (T_0)^* = (T_{\min})^*.$$

Thus, $T_{\min}$ is a (closed) symmetric relation and $T_{\min} = (T_{\max})^*$. □

At this stage note that $T_{\min}$ is a closed symmetric relation which need not be densely defined in $L^2_\Delta(\imath)$. Consider the orthogonal decomposition

$$L^2_\Delta(\imath) = (\operatorname{mul} T_{\min})^\perp \oplus \operatorname{mul} T_{\min} = \overline{\operatorname{dom} T_{\max}} \oplus \operatorname{mul} T_{\min} \tag{7.6.7}$$

and recall from Theorem 1.4.11 that $T_{\min}$ admits the corresponding orthogonal sum decomposition

$$T_{\min} = (T_{\min})_{\rm op} \,\widehat{\oplus}\, (\{0\} \times \operatorname{mul} T_{\min}). \tag{7.6.8}$$

The operator part $(T_{\min})_{\rm op}$ is not necessarily densely defined in $\overline{\operatorname{dom} T_{\max}}$ and $\{0\} \times \operatorname{mul} T_{\min}$ is the purely multivalued self-adjoint relation in $\operatorname{mul} T_{\min}$.

Since by Theorem 7.6.2 the relation $T_{\min}$ is closed and symmetric, while $(T_{\min})^* = T_{\max}$, it follows from the von Neumann decomposition, as given in Theorem 1.7.11, that the relation $T_{\max}$ has the componentwise sum decomposition

$$T_{\max} = T_{\min} \,\widehat{+}\, \widehat{\mathfrak{N}}_\lambda(T_{\max}) \,\widehat{+}\, \widehat{\mathfrak{N}}_\mu(T_{\max}), \quad \lambda \in \mathbb{C}^+, \mu \in \mathbb{C}^-,$$

where the sums are direct. Now assume that $f \in \mathfrak{N}_\zeta(T_{\max})$ with $\zeta \in \mathbb{C}\setminus\mathbb{R}$. Then $\{f,\zeta f\} \in T_{\max}$, which is equivalent to $f \in L^2_\Delta(\imath)$ having an absolutely continuous representative $\widetilde{f}$ such that $J\widetilde{f}' - H\widetilde{f} = \zeta\Delta\widetilde{f}$. Since the system consists of 2×2 matrix functions, there are precisely two linearly independent solutions of this homogeneous equation and at most two linearly independent solutions that are square-integrable with respect to Δ. Furthermore, since the canonical system is assumed to be real, the number of solutions at $\zeta \in \mathbb{C}^+$ that are square-integrable with respect to Δ coincides with the number of solutions at $\bar{\zeta} \in \mathbb{C}^-$ that are square-integrable with respect to Δ by Corollary 7.2.9. Taking into account Corollary 7.4.5 and Corollary 7.4.6 one then obtains the following statement. The case where both endpoints of the interval $\imath$ are in the limit-point case will be dealt with in Corollary 7.6.9.

Corollary 7.6.3. *Let $T_{\min}$ be the minimal symmetric relation associated with the real definite canonical system* (7.2.3) *in $L^2_\Delta(\imath)$. Then the following statements hold:*

(i) *If both endpoints of $\imath$ are regular or quasiregular, then the defect numbers of $T_{\min}$ are $(2,2)$.*

(ii) *If one endpoint of $\imath$ is regular or quasiregular and one endpoint is in the limit-point case, then the defect numbers of $T_{\min}$ are $(1,1)$.*

Recall that elements $\{f,g\} \in T_{\max}$ satisfy the equation $Jf' - Hf = \Delta g$ and that the entries also satisfy the integrability condition $f,g \in L^2_\Delta(\imath)$. These two ingredients make it possible to extend the usual Lagrange identity in Corollary 7.2.3 on a compact subinterval to all of $\imath$. This new Lagrange identity for the elements in $T_{\max}$ will play an important role in the rest of this chapter.

Lemma 7.6.4. *Let $T_{\max}$ be the maximal relation associated with the real definite canonical system* (7.2.3) *in $L^2_\Delta(\imath)$. Then for all $\{f,g\},\{h,k\} \in T_{\max}$ one has*

$$(g,h)_\Delta - (f,k)_\Delta = \lim_{t\to b} h(t)^*Jf(t) - \lim_{t\to a} h(t)^*Jf(t), \tag{7.6.9}$$

where $f(t)$ and $h(t)$ denote the values of the unique absolutely continuous representatives of f and h, respectively.

Proof. First observe that for all elements $\{f,g\},\{h,k\} \in T_{\max}$ and every compact subinterval $[\alpha,\beta] \subset \imath$ one has

$$\int_\alpha^\beta \bigl(h(s)^*\Delta(s)g(s) - k(s)^*\Delta(s)f(s)\bigr)\,ds = h(\beta)^*Jf(\beta) - h(\alpha)^*Jf(\alpha)$$

by the Lagrange identity in Corollary 7.2.3; here $f(t)$ and $h(t)$ denote the values of the unique absolutely continuous representatives of f and h, and $g(t)$ and $k(t)$ are the values of some representatives of g and k. Observe that the integral on the left-hand side does not depend on the choice of the representatives of g and

k. The limit of the left-hand side exists as $\beta \to b$ and $\alpha \to a$, respectively, since $f, g, h, k \in L^2_\Delta(\imath)$. As a consequence, one sees that each of the limits

$$\lim_{t \to a} h(t)^* J f(t) \quad \text{and} \quad \lim_{t \to b} h(t)^* J f(t)$$

exists and hence the Lagrange identity takes the limit form (7.6.9). □

Remark 7.6.5. Observe that in (7.6.9) one uses the values $f(t)$ and $h(t)$ of the unique absolutely continuous representatives of f and h in $\operatorname{dom} T_{\max}$, respectively. For instance, if $\{0, g\} \in T_{\max}$ and $\{h, k\} \in T_{\max}$, then there exists an absolutely continuous function $\widetilde{f}$ such that $\Delta(t)\widetilde{f}(t) = 0$ and $J\widetilde{f}'(t) - H(t)\widetilde{f}(t) = \Delta(t)\widetilde{g}(t)$ is satisfied for almost every $t \in \imath$, where $\widetilde{g}$ is any representative of $g \in L^2_\Delta(\imath)$. In this situation the identity (7.6.9) has the form

$$(g, h)_\Delta - (0, k)_\Delta = \lim_{t \to b} h(t)^* J \widetilde{f}(t) - \lim_{t \to a} h(t)^* J \widetilde{f}(t).$$

The elements in the minimal symmetric relation $T_{\min} = \overline{T}_0$ can be easily characterized in terms of these limits.

Corollary 7.6.6. *Let $T_{\min}$ and $T_{\max}$ be the minimal and maximal relations associated with the real definite canonical system (7.2.3) in $L^2_\Delta(\imath)$ and let $\{f, g\} \in T_{\max}$. Then $\{f, g\} \in T_{\min}$ if and only if*

$$\lim_{t \to a} h(t)^* J f(t) = 0 \quad \textit{and} \quad \lim_{t \to b} h(t)^* J f(t) = 0 \tag{7.6.10}$$

for all $\{h, k\} \in T_{\max}$, where $f(t)$ and $h(t)$ denote the values of the unique absolutely continuous representatives of f and h, respectively.

Proof. Observe first that since $T_{\min} = (T_{\max})^*$ one has $\{f, g\} \in T_{\min}$ if and only if $(g, h)_\Delta = (f, k)_\Delta$ for all $\{h, k\} \in T_{\max}$. Hence, it follows from the Lagrange identity (7.6.9) that $\{f, g\} \in T_{\min}$ if and only if

$$\lim_{t \to b} h(t)^* J f(t) = \lim_{t \to a} h(t)^* J f(t) \tag{7.6.11}$$

for all $\{h, k\} \in T_{\max}$. To see that for $\{f, g\} \in T_{\min}$ each of the limits in (7.6.11) is zero, consider $\{h, k\} \in T_{\max}$ and use Proposition 7.5.6 (with $\lambda = 0$) to obtain an element $\{h_a, k_a\} \in T_{\max}$ that coincides with $\{h, k\}$ in a neighborhood of a and with $\{0, 0\}$ in a neighborhood of b. Then (7.6.11) implies

$$\lim_{t \to a} h(t)^* J f(t) = \lim_{t \to a} h_a(t)^* J f(t) = \lim_{t \to b} h_a(t)^* J f(t) = 0,$$

and hence (7.6.10) follows together with (7.6.11). Conversely, if (7.6.10) holds for some $\{f, g\} \in T_{\max}$ and all $\{h, k\} \in T_{\max}$, then the identity (7.6.11) holds for all $\{h, k\} \in T_{\max}$ and hence $\{f, g\} \in T_{\min}$. □

The main difficulty when dealing with the boundary value problems associated with the system (7.2.3) is to break the limits in (7.6.9) and in (7.6.10) into limits of the separate factors. The case where the endpoints are regular, quasiregular, or in the limit-circle case will be pursued in Section 7.7. If one of the endpoints is in the limit-point case, the situation is somewhat simpler, since one of the limits in (7.6.9) automatically vanishes, as will be shown now. A further discussion of the remaining limit will be pursued in Section 7.8.

Lemma 7.6.7. *Let $T_{\max}$ be the maximal relation associated with the real definite canonical system* (7.2.3) *in $L^2_\Delta(\imath)$, let $\lambda \in \mathbb{C}$, and let $\mathcal{Y}_a(\cdot,\lambda)$ and $\mathcal{Y}_b(\cdot,\lambda)$ be as in* (7.5.6)*. Then the following statements hold:*

(i) *Let a be a regular or quasiregular endpoint and let b be in the limit-point case. Then for $\lambda \in \mathbb{C}$*

$$T_{\max} = T_{\min} \,\widehat{+}\, \{\mathcal{Y}_a(\cdot,\lambda)\phi : \phi \in \mathbb{C}^2\}, \tag{7.6.12}$$

where the sum is direct.

(ii) *Let b be a regular or quasiregular endpoint and let a be in the limit-point case. Then for $\lambda \in \mathbb{C}$*

$$T_{\max} = T_{\min} \,\widehat{+}\, \{\mathcal{Y}_b(\cdot,\lambda)\phi : \phi \in \mathbb{C}^2\},$$

where the sum is direct.

Proof. It suffices to consider the case (i) since the case (ii) can be proved in a similar way. Let $Y(\cdot,\lambda)$ be a fundamental matrix of (7.2.4). Note that if a is regular or quasiregular, then it follows from Corollary 7.3.3 and (7.5.7) that $\mathcal{Y}_a(\cdot,\lambda)\phi \in L^2_\Delta(\imath) \times L^2_\Delta(\imath)$, and Corollary 7.5.7 implies that $\mathcal{Y}_a(\cdot,\lambda)\phi \in T_{\max}$ for all $\phi \in \mathbb{C}^2$.

As $T_{\min} \subset T_{\max}$, it is clear that the right-hand side of (7.6.12) is contained in $T_{\max}$. By assumption and Corollary 7.6.3 (ii) $T_{\max}$ is a two-dimensional extension of $T_{\min}$ and hence it suffices to show that the elements $\mathcal{Y}_a(\cdot,\lambda)\phi$, $\phi \in \mathbb{C}^2$, span a two-dimensional subspace of $T_{\max}$ which has a trivial intersection with $T_{\min}$. In other words, it remains to check that $\mathcal{Y}_a(\cdot,\lambda)\phi \in T_{\min}$ if and only if $\phi = 0$. Suppose that $\mathcal{Y}_a(\cdot,\lambda)\phi \in T_{\min}$ for some $\phi \in \mathbb{C}^2$. For all $\psi \in \mathbb{C}^2$ one has $\mathcal{Y}_a(\cdot,\bar\lambda)\psi \in T_{\max}$ and therefore, by Corollary 7.6.6,

$$0 = \lim_{t\to a} \psi^* Y_a(t,\bar\lambda)^* J Y_a(t,\lambda)\phi.$$

Since $Y_a(\cdot,\lambda) = Y(\cdot,\lambda)$ and $Y_a(\cdot,\bar\lambda) = Y(\cdot,\bar\lambda)$ in a neighborhood of a, it follows that

$$0 = \lim_{t\to a} \psi^* Y(t,\bar\lambda)^* J Y(t,\lambda)\phi$$

for all $\psi \in \mathbb{C}^2$. Fix the fundamental matrix $Y(\cdot,\lambda)$ by $Y(a,\lambda) = I$. This leads to $\psi^* J\phi = 0$ for all $\psi \in \mathbb{C}^2$, which implies $\phi = 0$. Hence, the right-hand side of (7.6.12) is a two-dimensional extension of $T_{\min}$ which is contained in $T_{\max}$, and therefore coincides with $T_{\max}$. □

In the next lemma the case of a singular endpoint in the limit-point case is discussed.

Lemma 7.6.8. *Let $T_{\max}$ be the maximal relation associated with the real definite canonical system (7.2.3) in $L^2_\Delta(\imath)$. Then the endpoint a or b of the interval $\imath$ is in the limit-point case if and only if for all $\{f,g\},\{h,k\} \in T_{\max}$ one has*

$$\lim_{t\to a} h(t)^* J f(t) = 0 \quad \text{or} \quad \lim_{t\to b} h(t)^* J f(t) = 0,$$

respectively. Here $f(t)$ and $h(t)$ denote the values of the unique absolutely continuous representatives of f and h, respectively.

Proof. It suffices to consider the case that the endpoint a is regular. The proof of the case where b is regular is similar. As usual the fundamental matrix is fixed by $Y(a,\lambda) = I$.

Assume that b is in the limit-point case. In this case one has the decomposition (7.6.12) in Lemma 7.6.7. Let $\{f,g\},\{h,k\} \in T_{\max}$ be decomposed in the form

$$\{f,g\} = \{f_0,g_0\} + \mathcal{Y}_a(\cdot,\lambda)\phi \quad \text{and} \quad \{h,k\} = \{h_0,k_0\} + \mathcal{Y}_a(\cdot,\lambda)\psi,$$

where $\{f_0,g_0\},\{h_0,k_0\} \in T_{\min}$ and $\phi,\psi \in \mathbb{C}^2$. Then it follows from (7.5.7) that

$$\lim_{t\to b} h(t)^* J f(t) = \lim_{t\to b} h_0(t)^* J f_0(t) = 0,$$

where Corollary 7.6.6 was used in the last step.

Conversely, assume that for all $\{f,g\},\{h,k\} \in T_{\max}$

$$\lim_{t\to b} h^*(t) J f(t) = 0. \tag{7.6.13}$$

Then b is in the limit-point case. To see this, assume that b is not in the limit-point case, so that b is in the limit-circle case by Corollary 7.4.5. It then follows that for $\lambda_0 \in \mathbb{R}$ the columns of the matrix function $\mathcal{Y}_b(\cdot,\lambda_0)$ are square-integrable with respect to Δ at b. Consider $\{f,g\} = \{h,k\} = \mathcal{Y}_b(\cdot,\lambda_0)\phi \in T_{\max}$ for some $\phi \in \mathbb{C}^2$ such that $\phi^* J\phi \neq 0$. Using (7.5.7) and $Y(t,\lambda_0)^* JY(t,\lambda_0)$ (see (7.2.8)), one computes

$$\lim_{t\to b} h^*(t) J f(t) = \lim_{t\to b} \phi^* Y(t,\lambda_0)^* J Y(t,\lambda_0)\phi = \phi^* J \phi \neq 0,$$

which contradicts (7.6.13). □

Corollary 7.6.9. *Let $T_{\min}$ be the minimal symmetric relation associated with the real definite canonical system (7.2.3) in $L^2_\Delta(\imath)$ and assume that both endpoints of $\imath$ are in the limit-point case. Then the defect numbers of $T_{\min}$ are $(0,0)$ and $T_{\min} = T_{\max}$ is self-adjoint in $L^2_\Delta(\imath)$.*

Proof. Assume that the system is definite on the interval $[\alpha, \beta] \subset \imath$. Then the system is also definite on the intervals (a, β') and (α', b), with $\beta' \in (\beta, b)$ and $\alpha' \in (a, \alpha)$, respectively. Denote the maximal relation in $L^2_\Delta(\alpha', b)$ associated with the canonical system by $T_{\max}(\alpha', b)$. It follows that the endpoint α' is regular and that the endpoint b is in the limit-point case for the canonical system on (α', b). To see the last assertion, assume that there are two linearly independent solutions on (α', b) which are square-integrable with respect to Δ at b. Since these solutions admit unique extensions to solutions on (a, b), one obtains a contradiction. In particular, for all $\{f_b, g_b\}, \{h_b, k_b\} \in T_{\max}(\alpha', b)$ one concludes from Lemma 7.6.8 that

$$\lim_{t \to b} h_b(t)^* J f_b(t) = 0.$$

Now consider $\{f, g\}, \{h, k\} \in T_{\max}$ and let f_b, g_b, h_b, k_b be the restrictions of f, g, h, k to the interval (α', b). Then one has $\{f_b, g_b\}, \{h_b, k_b\} \in T_{\max}(\alpha', b)$, and hence

$$\lim_{t \to b} h(t)^* J f(t) = \lim_{t \to b} h_b(t)^* J f_b(t) = 0.$$

A similar argument applies to the canonical system on (a, β') and shows that

$$\lim_{t \to a} h(t)^* J f(t) = 0$$

for all $\{f, g\}, \{h, k\} \in T_{\max}$. Therefore, Lemma 7.6.4 implies

$$(g, h)_\Delta - (f, k)_\Delta = \lim_{t \to b} h(t)^* J f(t) - \lim_{t \to a} h(t)^* J f(t) = 0$$

for all $\{f, g\}, \{h, k\} \in T_{\max}$, and hence $T_{\max} \subset T_{\max}^*$. From Theorem 7.6.2 one now concludes $T_{\min} = T_{\max}$ and thus it follows that the defect numbers of $T_{\min}$ are $(0, 0)$. □

7.7 Boundary triplets for the limit-circle case

Assume that the system (7.2.3) is real and definite, and assume that the endpoints of the system are both in the limit-circle case. A boundary triplet will be presented for $T_{\max} = (T_{\min})^*$ and the self-adjoint extensions of $T_{\min}$ will be described in terms of the boundary triplet. For a straightforward presentation the case where the endpoints are regular or quasiregular is discussed first. At the end of the section it will be explained what modifications are necessary for endpoints which are in the limit-circle case and which are not regular or quasiregular.

The symmetric relation $T_{\min} = \overline{T_0}$ will now be described when a and b are regular or quasiregular.

Lemma 7.7.1. *Assume that a and b are regular or quasiregular endpoints for the canonical system (7.2.3). Then the minimal relation $T_{\min}$ is given by*

$$T_{\min} = \big\{\{f, g\} \in T_{\max} : f(a) = f(b) = 0\big\},$$

where $f(a)$ and $f(b)$ denote the boundary values of the unique absolutely continuous representatives of f.

Proof. According to Corollary 7.6.6, the element $\{f,g\} \in T_{\max}$ belongs to $T_{\min}$ if and only if

$$\lim_{t\to a} h(t)^* Jf(t) = 0 \quad \text{and} \quad \lim_{t\to b} h(t)^* Jf(t) = 0$$

for all $\{h,k\} \in T_{\max}$. Since the endpoints are regular or quasiregular, these conditions are the same as

$$h(a)^* Jf(a) = 0 \quad \text{and} \quad h(b)^* Jf(b) = 0$$

for all $\{h,k\} \in T_{\max}$. Now observe that for any $\gamma \in \mathbb{C}^2$ and $k \in L^2_\Delta(\imath)$ there exists an element $h \in L^2_\Delta(\imath)$ such that $\{h,k\} \in T_{\max}$ and $h(a) = \gamma$ or $h(b) = \gamma$. Hence, it follows that $f(a) = 0$ and $f(b) = 0$. □

When the endpoints of the interval $\imath = (a,b)$ are regular or quasiregular for the canonical system, then the solutions of $Jf' - Hf = \lambda \Delta f$, $\lambda \in \mathbb{C}$, automatically belong to $L^2_\Delta(\imath)$ and thus $\dim\ker(T_{\max} - \lambda) = 2$, so that the defect numbers of $T_{\min}$ are $(2,2)$; cf. Corollary 7.6.3. In the next theorem a boundary triplet for $(T_{\min})^* = T_{\max}$ is provided and the corresponding γ-field and Weyl function are obtained in terms of an arbitrary fundamental matrix $Y(\cdot,\lambda)$ fixed by $Y(c,\lambda) = I$ for some $c \in [a,b]$.

Theorem 7.7.2. *Assume that a and b are regular or quasiregular endpoints for the canonical system (7.2.3) and let the fundamental matrix $Y(\cdot,\lambda)$ of (7.2.4) be fixed by $Y(c,\lambda) = I$ for some $c \in [a,b]$. Then $\{\mathbb{C}^2, \Gamma_0, \Gamma_1\}$, with*

$$\Gamma_0\{f,g\} = \frac{1}{\sqrt{2}}(f(a) + f(b)) \quad \textit{and} \quad \Gamma_1\{f,g\} = -\frac{J}{\sqrt{2}}(f(a) - f(b)),$$

where $\{f,g\} \in T_{\max}$, is a boundary triplet for $(T_{\min})^* = T_{\max}$; here $f(a)$ and $f(b)$ denote the boundary values of the unique absolutely continuous representative of f. The corresponding γ-field and Weyl function are given by

$$\gamma(\lambda) = \sqrt{2}Y(\cdot,\lambda)\big(Y(a,\lambda) + Y(b,\lambda)\big)^{-1}, \quad \lambda \in \rho(A_0),$$

and

$$M(\lambda) = -J\big(Y(a,\lambda) - Y(b,\lambda)\big)\big(Y(a,\lambda) + Y(b,\lambda)\big)^{-1}, \quad \lambda \in \rho(A_0).$$

Proof. Let $\{f,g\},\{h,k\} \in T_{\max}$. Since the endpoints a and b are regular or quasiregular, one has the Lagrange identity

$$\begin{aligned}(g,h)_\Delta - (f,k)_\Delta &= \int_a^b \big(h(s)^*\Delta(s)g(s) - k(s)^*\Delta(s)f(s)\big)\,ds \\ &= h(b)^*Jf(b) - h(a)^*Jf(a);\end{aligned}$$

cf. Corollary 7.3.4. On the other hand, a straightforward calculation shows that

$$\begin{aligned}&(\Gamma_1\{f,g\},\Gamma_0\{h,k\}) - (\Gamma_0\{f,g\},\Gamma_1\{h,k\})\\&\quad= -\frac{1}{2}\big(h(a)+h(b)\big)^*J\big(f(a)-f(b)\big) + \frac{1}{2}\big(h(a)-h(b)\big)^*J^*\big(f(a)+f(b)\big)\\&\quad= h(b)^*Jf(b) - h(a)^*Jf(a),\end{aligned}$$

and hence the boundary mappings Γ_0 and Γ_1 satisfy the abstract Green identity (2.1.1). Furthermore, the mapping $(\Gamma_0,\Gamma_1)^\top : T_{\max} \to \mathbb{C}^4$ is surjective. To see this, observe first that

$$\begin{pmatrix}\Gamma_0\{f,g\}\\ \Gamma_1\{f,g\}\end{pmatrix} = \frac{1}{\sqrt{2}}\begin{pmatrix} I & I\\ -J & J\end{pmatrix}\begin{pmatrix} f(a)\\ f(b)\end{pmatrix}, \quad \{f,g\}\in T_{\max},$$

and that the 4×4-matrix on the right-hand side is invertible. Hence, it suffices to check that for any $\gamma_a,\gamma_b\in\mathbb{C}^2$ there exists $\{f,g\}\in T_{\max}$ such that

$$\begin{pmatrix} f(a)\\ f(b)\end{pmatrix} = \begin{pmatrix}\gamma_a\\ \gamma_b\end{pmatrix}. \tag{7.7.1}$$

Choose a solution of the equation $Jh' - Hh = 0$ such that $h(a)=\gamma_a$ and modify h as in Proposition 7.5.6, so that it becomes a solution h_a of an inhomogeneous equation $Jh_a' - Hh_a = \Delta k_a$ which coincides with h in a neighborhood of a and vanishes in a neighborhood of the endpoint b. Then one has $\{h_a,k_a\}\in T_{\max}$ and $h_a(a)=\gamma_a$ and $h_a(b)=0$. The same argument shows that there exists an element $\{h_b,k_b\}\in T_{\max}$ such that $h_b(b)=\gamma_b$ and $h_b(a)=0$. Thus, for $f=h_a+h_b$ and $g=k_a+k_b$ one has $\{f,g\}\in T_{\max}$ and (7.7.1) holds. It follows that the mapping $(\Gamma_0,\Gamma_1)^\top : T_{\max}\to\mathbb{C}^4$ is surjective, as claimed. Therefore, $\{\mathbb{C}^2,\Gamma_0,\Gamma_1\}$ is a boundary triplet for $(T_{\min})^* = T_{\max}$.

To obtain the expressions for the associated γ-field and Weyl function, let $\lambda\in\rho(A_0)$, where $A_0=\ker\Gamma_0$, and note that

$$\mathfrak{N}_\lambda(T_{\max}) = \{Y(\cdot,\lambda)\phi : \phi\in\mathbb{C}^2\}, \qquad \lambda\in\mathbb{C}.$$

Hence, for $\widehat{f}_\lambda = \{Y(\cdot,\lambda)\phi, \lambda Y(\cdot,\lambda)\phi\}$, $\phi\in\mathbb{C}^2$, and $\lambda\in\rho(A_0)$ one has

$$\Gamma_0\widehat{f}_\lambda = \frac{1}{\sqrt{2}}\big(Y(a,\lambda)+Y(b,\lambda)\big)\phi \quad\text{and}\quad \Gamma_1\widehat{f}_\lambda = -\frac{J}{\sqrt{2}}\big(Y(a,\lambda)-Y(b,\lambda)\big)\phi,$$

which leads to

$$\gamma(\lambda) = \left\{\left\{\frac{1}{\sqrt{2}}\big(Y(a,\lambda)+Y(b,\lambda)\big)\phi, Y(\cdot,\lambda)\phi\right\} : \phi\in\mathbb{C}^2\right\}$$

and

$$M(\lambda) = \left\{\left\{\frac{1}{\sqrt{2}}\big(Y(a,\lambda)+Y(b,\lambda)\big)\phi, -\frac{J}{\sqrt{2}}\big(Y(a,\lambda)-Y(b,\lambda)\big)\phi\right\} : \phi\in\mathbb{C}^2\right\};$$

cf. Definition 2.3.1 and Definition 2.3.4. Now observe that for $\lambda \in \rho(A_0)$ the matrix $Y(a,\lambda)+Y(b,\lambda)$ is invertible, as otherwise $(Y(a,\lambda)+Y(b,\lambda))\psi = 0$ for some nontrivial $\psi \in \mathbb{C}^2$ would imply that λ is an eigenvalue of the self-adjoint relation $A_0 = \ker \Gamma_0$ with corresponding eigenfunction $Y(\cdot,\lambda)\psi$; a contradiction. Therefore, the formulas for the γ-field and the Weyl function follow from the above identities for $\gamma(\lambda)$ and $M(\lambda)$. □

Before formulating the next proposition some terminology is recalled. Let T be an integral operator of the form

$$Tf(t) = \int_a^b K(t,s)f(s)\,ds,$$

where f is a $\mathbb{C}^2$-valued function and K is a $\mathbb{C}^{2\times 2}$-valued measurable matrix kernel. If K is square-integrable with respect to the Lebesgue measure on $\imath \times \imath$, that is,

$$\int_a^b \int_a^b \|K(t,s)\|_2^2\,ds\,dt < \infty,$$

where $\|\cdot\|_2$ is the Hilbert–Schmidt matrix norm in (7.1.4), then T is a bounded linear operator from $L^2(\imath)$ into itself, which belongs to the Hilbert–Schmidt class. Recall that a bounded linear operator from $L^2(\imath)$ into itself belongs to the *Hilbert–Schmidt class* if for some, and hence for all orthonormal bases (φ_i) in $L^2(\imath)$ one has

$$\sum_{i,j} |(T\varphi_i,\varphi_j)|^2 < \infty.$$

Proposition 7.7.3. *Assume that a and b are regular or quasiregular endpoints for the canonical system (7.2.3) and let $\{\mathbb{C}^2,\Gamma_0,\Gamma_1\}$ be the boundary triplet for $T_{\max}$ in Theorem 7.7.2 with corresponding Weyl function M. Let the fundamental matrix $Y(\cdot,\lambda)$ be fixed by $Y(a,\lambda) = I$. Then the self-adjoint relation $A_0 = \ker\Gamma_0$ is given by*

$$A_0 = \ker \Gamma_0 = \bigl\{\{f,g\} \in T_{\max} : f(a)+f(b) = 0\bigr\},$$

where $f(a)$ and $f(b)$ denote the boundary values of the unique absolutely continuous representative of f. The resolvent of A_0 is an integral operator

$$((A_0-\lambda)^{-1}g)(t) = \int_a^b G_0(t,s,\lambda)\Delta(s)g(s)\,ds, \quad \lambda \in \rho(A_0), \tag{7.7.2}$$

which belongs to the Hilbert–Schmidt class. The Green function $G_0(t,s,\lambda)$ is given by

$$G_0(t,s,\lambda) = G_{0,\mathrm{e}}(t,s,\lambda) + G_{0,\mathrm{i}}(t,s,\lambda), \tag{7.7.3}$$

where the entire part $G_{0,\mathrm{e}}$ is given by

$$\begin{aligned} G_{0,\mathrm{e}}(t,s,\lambda) &= Y(t,\lambda)\left[\frac{1}{2}J\,\mathrm{sgn}\,(s-t)\right]Y(s,\bar\lambda)^* \\ &= \frac{1}{2}\begin{cases} -Y(t,\lambda)JY(s,\bar\lambda)^*, & s<t, \\ Y(t,\lambda)JY(s,\bar\lambda)^*, & s>t, \end{cases} \end{aligned} \tag{7.7.4}$$

and

$$G_{0,\mathrm{i}}(t,s,\lambda) = Y(t,\lambda)\left[-\frac{1}{2}JM(\lambda)J\right]Y(s,\bar{\lambda})^*. \tag{7.7.5}$$

Proof. Step 1. The resolvent of A_0 has the form (7.7.2) with G_0 as in (7.7.3). To see this, let $\lambda \in \rho(A_0)$ and $g \in L^2_\Delta(\imath)$, and define the function

$$f(t) = \int_a^b G_0(t,s,\lambda)\Delta(s)g(s)\,ds.$$

From the structure of the Green function in (7.7.3), (7.7.4), and (7.7.5), it follows that

$$\begin{aligned} f(t) &= \frac{1}{2}Y(t,\lambda)J\left(-\int_a^t Y(s,\bar{\lambda})^*\Delta(s)g(s)\,ds + \int_t^b Y(s,\bar{\lambda})^*\Delta(s)g(s)\,ds\right) \\ &\quad + Y(t,\lambda)E_0(\lambda)\int_a^b Y(s,\bar{\lambda})^*\Delta(s)g(s)\,ds, \end{aligned}$$

with $E_0(\lambda) = -\frac{1}{2}JM(\lambda)J$. Hence, f is well defined and absolutely continuous. A straightforward computation together with (7.2.10) shows that

$$\begin{aligned} Jf'(t) &= \frac{1}{2}JY'(t,\lambda)J\left(-\int_a^t Y(s,\bar{\lambda})^*\Delta(s)g(s)\,ds + \int_t^b Y(s,\bar{\lambda})^*\Delta(s)g(s)\,ds\right) \\ &\quad + \Delta(t)g(t) + JY'(t,\lambda)E_0(\lambda)\int_a^b Y(s,\bar{\lambda})^*\Delta(s)g(s)\,ds. \end{aligned}$$

This implies that

$$Jf' - Hf = \lambda\Delta f + \Delta g = \Delta(g+\lambda f),$$

and thus one has $\{f, g+\lambda f\} \in T_{\max}$. Furthermore, it is clear from the definition of f and $Y(a,\lambda) = I$ that

$$f(a) = \left[\frac{1}{2}J + E_0(\lambda)\right]\int_a^b Y(s,\bar{\lambda})^*\Delta(s)g(s)\,ds$$

and

$$f(b) = Y(b,\lambda)\left[-\frac{1}{2}J + E_0(\lambda)\right]\int_a^b Y(s,\bar{\lambda})^*\Delta(s)g(s)\,ds.$$

Since $E_0(\lambda) = -\frac{1}{2}JM(\lambda)J = -\frac{1}{2}(I - Y(b,\lambda))(I + Y(b,\lambda))^{-1}J$, observe that

$$\begin{aligned} \frac{1}{2}J + E_0(\lambda) &= Y(b,\lambda)(I+Y(b,\lambda))^{-1}J, \\ -\frac{1}{2}J + E_0(\lambda) &= -(I+Y(b,\lambda))^{-1}J. \end{aligned}$$

Thus,

$$\left[\frac{1}{2}J + E_0(\lambda)\right] + Y(b,\lambda)\left[-\frac{1}{2}J + E_0(\lambda)\right] = 0,$$

and hence $\Gamma_0\{f, g + \lambda f\} = \frac{1}{\sqrt{2}}(f(a) + f(b)) = 0$, which implies $\{f, g + \lambda f\} \in A_0$. Therefore, $f = (A_0 - \lambda)^{-1} g$ and the resolvent of A_0 is given by (7.7.2).

Step 2. The kernel $G_\Delta(\cdot, \cdot, \lambda)$ defined by

$$G_\Delta(t, s, \lambda) = \Delta(t)^{\frac{1}{2}} G_0(t, s, \lambda) \Delta(s)^{\frac{1}{2}}, \tag{7.7.6}$$

where the kernel $G_0(\cdot, \cdot, \lambda)$ is given in (7.7.3), satisfies

$$\int_a^b \int_a^b \left\| G_\Delta(t, s, \lambda) \right\|_2^2 ds\, dt < \infty; \tag{7.7.7}$$

here $\|\cdot\|_2$ is the Hilbert–Schmidt matrix norm. In fact, from (7.7.3), (7.7.4), (7.7.5), and (7.1.5) it follows that

$$\begin{aligned} \int_a^b \int_a^b \left\| G_\Delta(t, s, \lambda) \right\|_2^2 ds\, dt &= \int_a^b \int_a^b \left\| \Delta(t)^{\frac{1}{2}} G_0(t, s, \lambda) \Delta(s)^{\frac{1}{2}} \right\|_2^2 ds\, dt \\ &\leq C \int_a^b \int_a^b \left\| \Delta(t)^{\frac{1}{2}} Y(t, \lambda) \right\|_2^2 \left\| Y(s, \bar{\lambda})^* \Delta(s)^{\frac{1}{2}} \right\|_2^2 ds\, dt. \end{aligned}$$

To show that the right-hand side is finite, note that with $Y(\cdot,\lambda) = (Y_1(\cdot,\lambda)\, Y_2(\cdot,\lambda))$ one has

$$\int_a^b \left\| \Delta(t)^{\frac{1}{2}} Y(t, \lambda) \right\|_2^2 dt = \int_a^b \left| \Delta(t)^{\frac{1}{2}} Y_1(t, \lambda) \right|^2 dt + \int_a^b \left| \Delta(t)^{\frac{1}{2}} Y_2(t, \lambda) \right|^2 dt < \infty,$$

as the columns $Y_1(\cdot, \lambda)$ and $Y_2(\cdot, \lambda)$ of $Y(\cdot, \lambda)$ are square-integrable with respect to Δ. Due to the identity $\|A\|_2 = \|A^*\|_2$, it follows that

$$\int_a^b \left\| Y(s, \bar{\lambda})^* \Delta(s)^{\frac{1}{2}} \right\|_2^2 ds = \int_a^b \left\| \Delta(s)^{\frac{1}{2}} Y(s, \lambda) \right\|_2^2 dt < \infty.$$

Therefore, the kernel G_Δ is square-integrable with respect to the Lebesgue measure on $[a, b] \times [a, b]$ and hence (7.7.7) holds. Consequently, the integral operator T_Δ, defined by

$$T_\Delta f(t) = \int_a^b G_\Delta(t, s, \lambda) f(s)\, ds, \quad f \in L^2(\imath), \tag{7.7.8}$$

belongs to the Hilbert–Schmidt class in $L^2(\imath)$.

Step 3. The operator $(A_0 - \lambda)^{-1}$ belongs to the Hilbert–Schmidt class or, equivalently,

$$\sum_{i,j} \left| \left((A_0 - \lambda)^{-1} u_i, u_j \right) \right|^2 < \infty, \quad \lambda \in \rho(A_0), \tag{7.7.9}$$

for some, and hence for any orthonormal basis (u_i) in $L^2_\Delta(\imath)$. To see (7.7.9), observe that $(\Delta^{\frac{1}{2}} u_i)$ is an orthonormal system in $L^2(\imath)$ and that (7.7.2) and (7.7.6) give

$$\begin{aligned} \big((A_0-\lambda)^{-1}u_i, u_j\big)_\Delta &= \int_a^b u_j(t)^*\Delta(t)\left(\int_a^b G_0(t,s,\lambda)\Delta(s)u_i(s)\,ds\right)dt \\ &= \int_a^b\int_a^b \big(\Delta(t)^{\frac{1}{2}}u_j(t)\big)^* G_\Delta(t,s,\lambda)\big(\Delta(s)^{\frac{1}{2}}u_i(s)\big)\,ds\,dt \\ &= (T_\Delta\,\Delta^{\frac{1}{2}}u_i, \Delta^{\frac{1}{2}}u_j)_{L^2(\imath)}, \end{aligned}$$

where T_Δ is the Hilbert–Schmidt operator (7.7.8) in $L^2(\imath)$ whose kernel is given by (7.7.6). Hence, by Step 2 it follows that (7.7.9) holds, which implies that $(A_0-\lambda)^{-1}$ is a Hilbert–Schmidt operator. □

Since the resolvent of the self-adjoint relation A_0 is a Hilbert–Schmidt operator, the spectrum of A_0 is discrete. As the minimal relation $T_{\min}$ has no eigenvalues, the next statement follows immediately from Proposition 3.4.8.

Theorem 7.7.4. *Let $\{\mathbb{C}^2, \Gamma_0, \Gamma_1\}$ be the boundary triplet for $T_{\max}$ as in Theorem 7.7.2. Then the operator part $(T_{\min})_{\mathrm{op}}$ is simple in $L^2_\Delta(\imath) \ominus \operatorname{mul} T_{\min}$.*

Theorem 7.7.4 together with the considerations in Section 3.5 and Section 3.6 ensure that the Weyl function M in Theorem 7.7.2 contains the complete spectral data of A_0. In the present situation the eigenvalues of A_0 coincide with the poles of the Weyl function and the multiplicities of the eigenvalues of A_0 coincide with the multiplicities of the poles of M.

Let $\{\mathbb{C}^2, \Gamma_0, \Gamma_1\}$ be the boundary triplet in Theorem 7.7.2 with corresponding γ-field γ and Weyl function M. The self-adjoint (maximal dissipative, maximal accumulative) extensions $A_\Theta \subset T_{\max}$ of $T_{\min}$ are in a one-to-one correspondence to the self-adjoint (maximal dissipative, maximal accumulative) relations Θ in $\mathbb{C}^2$ via

$$\begin{aligned} A_\Theta &= \big\{\{f,g\} \in T_{\max} : \{\Gamma_0\{f,g\}, \Gamma_1\{f,g\}\} \in \Theta\big\} \\ &= \big\{\{f,g\} \in T_{\max} : \{f(a)+f(b), -Jf(a)+Jf(b)\} \in \Theta\big\}, \end{aligned} \tag{7.7.10}$$

where $f(a)$ and $f(b)$ denote the boundary values of the unique absolutely continuous representative of f. Recall from Theorem 2.6.1 that for $\lambda \in \rho(A_\Theta) \cap \rho(A_0)$ the Kreĭn formula for the corresponding resolvents reads

$$(A_\Theta - \lambda)^{-1} = (A_0-\lambda)^{-1} + \gamma(\lambda)\big(\Theta - M(\lambda)\big)^{-1}\gamma(\bar\lambda)^*. \tag{7.7.11}$$

Assume in the following that Θ is a self-adjoint relation in $\mathbb{C}^2$. Since the spectrum of A_0 is discrete and the difference of the resolvents of A_0 and A_Θ is an operator of rank ≤ 2, it is clear that the spectrum of the self-adjoint relation

A_Θ is also discrete. Note that $\lambda \in \rho(A_0)$ is an eigenvalue of A_Θ if and only if $\ker(\Theta - M(\lambda))$ is nontrivial, and that

$$\ker(A_\Theta - \lambda) = \gamma(\lambda)\ker(\Theta - M(\lambda)).$$

For the self-adjoint relation Θ one may use a parametric representation with the help of 2×2 matrices $\mathcal{A}$ and $\mathcal{B}$ as in Section 1.10 and give a complete description of the (discrete) spectrum of A_Θ via poles of a transform of the Weyl function M; cf. Section 3.8 and Section 6.3.

In the following paragraph and corollary it is assumed for simplicity that the relation Θ in (7.7.10) is a self-adjoint 2×2 matrix. In this case the self-adjoint relation A_Θ in (7.7.10) is given by

$$A_\Theta = \big\{\{f,g\} \in T_{\max} : \Theta(f(a) + f(b)) = -Jf(a) + Jf(b)\big\} \tag{7.7.12}$$

and according to Section 3.8 the spectral properties of A_Θ can also be described with the help of the function

$$\lambda \mapsto \big(\Theta - M(\lambda)\big)^{-1}; \tag{7.7.13}$$

that is, the poles of the matrix function (7.7.13) coincide with the (discrete) spectrum of A_Θ and the dimension of the eigenspace $\ker(A_\Theta - \lambda)$ coincides with the dimension of the range of the residue of the function in (7.7.13) at λ. Now fix a fundamental matrix $Y(\cdot,\lambda)$ by $Y(a,\lambda) = I$ as in Proposition 7.7.3. By Proposition 7.7.3, the resolvent $(A_0 - \lambda)^{-1}$ in the Kreĭn formula (7.7.11) is an integral operator. Since $I + JM(\lambda) = 2(I + Y(b,\lambda))^{-1}$, the γ-field and Weyl function in Theorem 7.7.2 are connected in the present situation via

$$\gamma(\cdot,\lambda) = \frac{1}{\sqrt{2}} Y(\cdot,\lambda)(I + JM(\lambda)), \quad \lambda \in \mathbb{C} \setminus \mathbb{R}.$$

One verifies that

$$\gamma(\bar\lambda)^* g = \frac{1}{\sqrt{2}}(I - M(\lambda)J) \int_a^b Y(s,\bar\lambda)^* \Delta(s) g(s)\, ds, \quad g \in L^2_\Delta(\imath),$$

and this implies that the second term on the right-hand side of (7.7.11) applied to $g \in L^2_\Delta(\imath)$ can be written as

$$\frac{1}{2} Y(\cdot,\lambda)(I + JM(\lambda))\big(\Theta - M(\lambda)\big)^{-1}(I - M(\lambda)J) \int_a^b Y(s,\bar\lambda)^* \Delta(s) g(s)\, ds.$$

Combining this expression with the entire part $G_{0,\mathrm{e}}$ in (7.7.4) and the part $G_{0,\mathrm{i}}$ in (7.7.5) for $(A_0 - \lambda)^{-1}$, one sees that the resolvent of the self-adjoint extension A_Θ is an integral operator in $L^2_\Delta(\imath)$ of the form

$$\big((A_\Theta - \lambda)^{-1} g\big)(t) = \int_a^b G_\Theta(t,s,\lambda)\Delta(s) g(s)\, ds, \quad \lambda \in \rho(A_\Theta) \cap \rho(A_0), \tag{7.7.14}$$

where $g \in L^2_\Delta(\imath)$. The Green function $G_\Theta(t,s,\lambda)$ in (7.7.14) is given by

$$G_\Theta(t,s,\lambda) = Y(t,\lambda)\left[\frac{1}{2}J\,\mathrm{sgn}\,(s-t) + E_\Theta(\lambda)\right]Y(s,\bar{\lambda})^*, \tag{7.7.15}$$

where

$$E_\Theta(\lambda) = -\frac{1}{2}J\left[M(\lambda) + (M(\lambda)-J)\big(\Theta - M(\lambda)\big)^{-1}(M(\lambda)+J)\right]J. \tag{7.7.16}$$

In the next corollary the Green function in (7.7.14) is further decomposed in the case that the self-adjoint relation Θ in $\mathbb{C}^2$ is a self-adjoint matrix.

Corollary 7.7.5. *Let a and b be regular or quasiregular endpoints for the canonical system (7.2.3) and let $\{\mathbb{C}^2, \Gamma_0, \Gamma_1\}$ be the boundary triplet for $T_{\max}$ in Theorem 7.7.2 with corresponding Weyl function M. Assume that the fundamental matrix $Y(\cdot,\lambda)$ is fixed by $Y(a,\lambda) = I$, let Θ be a self-adjoint matrix in $\mathbb{C}^2$, and let A_Θ be the self-adjoint extension in (7.7.12). Then the Green function $G_\Theta(t,s,\lambda)$ in (7.7.14) has the decomposition*

$$G_\Theta(t,s,\lambda) = G_{\Theta,\mathrm{e}}(t,s,\lambda) + G_{\Theta,\mathrm{i}}(t,s,\lambda),$$

where the entire part $G_{\Theta,\mathrm{e}}$ is given by

$$G_{\Theta,\mathrm{e}}(t,s,\lambda) = Y(t,\lambda)\left[\frac{1}{2}J\,\mathrm{sgn}\,(s-t) + \frac{1}{2}J\Theta J\right]Y(s,\bar{\lambda})^*,$$

and

$$G_{\Theta,\mathrm{i}}(t,s,\lambda) = Y_\Theta(t,\lambda)\left[\frac{1}{2}(\Theta - M(\lambda))^{-1}\right]Y_\Theta(s,\bar{\lambda})^*,$$

where $Y_\Theta(t,\lambda) = Y(t,\lambda)(I + J\Theta)$.

Proof. Since Θ is a self-adjoint 2×2-matrix, one sees that

$$\begin{aligned}&(M(\lambda)-J)\big(\Theta - M(\lambda)\big)^{-1}(M(\lambda)+J)\\ &\qquad = -M(\lambda) - \Theta + (\Theta - J)\big(\Theta - M(\lambda)\big)^{-1}(\Theta + J).\end{aligned}$$

Therefore, $E_\Theta(\lambda)$ in (7.7.16) has the form

$$\begin{aligned}E_\Theta(\lambda) &= -\frac{1}{2}J\left[-\Theta + (\Theta - J)\big(\Theta - M(\lambda)\big)^{-1}(\Theta + J)\right]J\\ &= \frac{1}{2}J\Theta J + (I + J\Theta)\left[\frac{1}{2}\big(\Theta - M(\lambda)\big)^{-1}\right](I - \Theta J).\end{aligned}$$

The assertion now follows from this identity combined with (7.7.15). □

At the end of this section the assumption is that the endpoints a and b are in the general limit-circle case, so that the assumption that a and b are regular or quasiregular is abandoned. The transformation in Lemma 7.2.5 will be useful as for any $\lambda_0 \in \mathbb{R}$ the solution matrix $U(\cdot,\lambda_0)$ is now square-integrable with respect to Δ. This implies that the transformed equation (7.2.20) is in the quasiregular case at a and b. Then most of the above results remain true once the (limit) values $f(a)$ and $f(b)$ are replaced by the limits in (7.4.6). The next proposition is the counterpart of Theorem 7.7.2.

Proposition 7.7.6. *Assume that a and b are in the limit-circle case and let $Y(\cdot,\lambda)$ be a fundamental matrix. Let λ_0 in $\mathbb{R}$, let $U(\cdot,\lambda_0)$ be a solution matrix as in Lemma 7.2.5, and consider the limits*

$$\widetilde{f}(a)=\lim_{t\to a}U(t,\lambda_0)^{-1}f(t) \quad \text{and} \quad \widetilde{f}(b)=\lim_{t\to b}U(t,\lambda_0)^{-1}f(t)$$

for $\{f,g\}\in T_{\max}$; cf. Corollary 7.4.8. Then $\{\mathbb{C}^2,\Gamma_0,\Gamma_1\}$, with

$$\Gamma_0\{f,g\}=\frac{1}{\sqrt{2}}(\widetilde{f}(a)+\widetilde{f}(b)) \quad \text{and} \quad \Gamma_1\{f,g\}=-\frac{J}{\sqrt{2}}(\widetilde{f}(a)-\widetilde{f}(b)),$$

where $\{f,g\}\in T_{\max}$, is a boundary triplet for $(T_{\min})^*=T_{\max}$. The corresponding γ-field and Weyl function are given by

$$\gamma(\lambda)=\sqrt{2}Y(\cdot,\lambda)\big(\widetilde{Y}(a,\lambda)+\widetilde{Y}(b,\lambda)\big)^{-1}, \quad \lambda\in\rho(A_0),$$

and

$$M(\lambda)=-J\big(\widetilde{Y}(a,\lambda)-\widetilde{Y}(b,\lambda)\big)\big(\widetilde{Y}(a,\lambda)+\widetilde{Y}(b,\lambda)\big)^{-1}, \quad \lambda\in\rho(A_0),$$

where $\widetilde{Y}(\cdot,\lambda)\phi=U(\cdot,\lambda_0)^{-1}Y(\cdot,\lambda)\phi$ for $\phi\in\mathbb{C}^2$ and $\lambda\in\rho(A_0)$.

Proof. Recall that due to Corollary 7.4.9 the Lagrange formula takes the form

$$\widetilde{h}(b)^*J\widetilde{f}(b)-\widetilde{h}(a)^*J\widetilde{f}(a)=\int_a^b\big(h(s)^*\Delta(s)g(s)-k(s)^*\Delta(s)f(s)\big)\,ds$$

for $\{f,g\},\{h,k\}\in T_{\max}$. Now the same computation as in the proof of Theorem 7.7.2 shows that the abstract Green identity (2.1.1) is satisfied. The surjectivity of the map $(\Gamma_0,\Gamma_1)^\top : T_{\max}\to\mathbb{C}^4$, and the form of the γ-field and Weyl function follow in the same way as in the proof of Theorem 7.7.2. □

7.8 Boundary triplets for the limit-point case

Assume that the system (7.2.3) is real and definite, and assume that the endpoint a is in the limit-circle case and the endpoint b is in the limit-point case. A boundary triplet will be presented for $T_{\max}=(T_{\min})^*$ and will be used to describe the self-adjoint extensions of $T_{\min}$. To make the presentation straightforward, the case where the endpoint a is regular or quasiregular is dealt with first. At the end of the section it will be explained what modifications are necessary if the endpoint a is in the limit-circle case.

The symmetric relation $T_{\min}=\overline{T_0}$ will now be described when a is a regular or quasiregular endpoint and b is in the limit-point case.

Lemma 7.8.1. *Assume that the endpoint a is regular or quasiregular and the endpoint b is in the limit-point case. Then the minimal relation $T_{\min}$ is given by*

$$T_{\min} = \{\{f,g\} \in T_{\max} : f(a) = 0\},$$

where $f(a)$ denotes the boundary value of the unique absolutely continuous representative of f.

Proof. According to Corollary 7.6.6 and Lemma 7.6.8, an element $\{f,g\} \in T_{\max}$ belongs to $T_{\min}$ if and only if

$$\lim_{t\to a} h(t)^* J f(t) = 0$$

for all $\{h,k\} \in T_{\max}$. Since the endpoint a is regular or quasiregular this condition is the same as

$$h(a)^* J f(a) = 0$$

for all $\{h,k\} \in T_{\max}$. Now observe that for any $\gamma \in \mathbb{C}^2$ there exists $\{h,k\} \in T_{\max}$ such that $h(a) = \gamma$. In fact, choose a solution $Ju' - Hu = 0$ such that $u(a) = \gamma$ and use Proposition 7.5.6 to modify u to a function $h \in L^2_\Delta(\imath)$ which coincides with u in a neighborhood of a, vanishes in a neighborhood of b, and satisfies $Jh' - Hh = \Delta k$ with some $k \in L^2_\Delta(\imath)$, that is, $\{h,k\} \in T_{\max}$. Since $\gamma^* J f(a) = 0$ for all $\gamma \in \mathbb{C}^2$, it follows that $f(a) = 0$. □

Let the endpoint a be regular or quasiregular and let b be in the limit-point case. Then there exists for some, and hence for all $\lambda \in \mathbb{C} \setminus \mathbb{R}$, up to scalar multiples, one nontrivial solution of $Jf' - Hf = \lambda \Delta f$, which is square-integrable with respect to Δ at b and thus $\dim \ker (T_{\max} - \lambda) = 1$ for $\lambda \in \mathbb{C} \setminus \mathbb{R}$. This implies that the defect numbers are $(1,1)$; cf. Corollary 7.6.3. In the next theorem a boundary triplet is provided in this case. To avoid confusion, recall that $Y_1(\cdot,\lambda)$ and $Y_2(\cdot,\lambda)$ are the columns of a fundamental matrix $Y(\cdot,\lambda)$, whereas f_1 and f_2 stand for the components of the 2×1 vector function f.

Theorem 7.8.2. *Assume that the endpoint a is regular or quasiregular and that the endpoint b is in the limit-point case. Let $Y(\cdot,\lambda)$ be a fundamental matrix fixed by $Y(a,\lambda) = I$. Then $\{\mathbb{C}, \Gamma_0, \Gamma_1\}$, where*

$$\Gamma_0\{f,g\} = f_1(a) \quad \text{and} \quad \Gamma_1\{f,g\} = f_2(a), \quad \{f,g\} \in T_{\max},$$

is a boundary triplet for $(T_{\min})^* = T_{\max}$; here $f_1(a)$ and $f_2(a)$ denote the boundary values of the components of the unique absolutely continuous representative of f. Moreover, if $\lambda \in \mathbb{C} \setminus \mathbb{R}$ and $\chi(\cdot,\lambda)$ is a nontrivial element in $\mathfrak{N}_\lambda(T_{\max})$, then one has $\chi_1(a,\lambda) \neq 0$. For all $\lambda \in \mathbb{C} \setminus \mathbb{R}$ the corresponding γ-field and Weyl function are given by

$$\gamma(\cdot,\lambda) = Y_1(\cdot,\lambda) + M(\lambda) Y_2(\cdot,\lambda) \quad \text{and} \quad M(\lambda) = \frac{\chi_2(a,\lambda)}{\chi_1(a,\lambda)}.$$

Proof. Since the endpoint a is assumed to be regular or quasiregular, the elements $\{f,g\},\{h,k\} \in T_{\max}$ have boundary values $f(a),h(a) \in \mathbb{C}^2$. Due to Lemma 7.6.8, the Lagrange identity in Corollary 7.3.4 takes the form

$$\begin{aligned}(g,h)_\Delta - (f,k)_\Delta &= \int_a^b \bigl(h(s)^*\Delta(s)g(s) - k(s)^*\Delta(s)f(s)\bigr)\,ds\\ &= -h(a)^* Jf(a)\\ &= f_2(a)\overline{h_1(a)} - f_1(a)\overline{h_2(a)}\\ &= \bigl(\Gamma_1\{f,g\},\Gamma_0\{h,k\}\bigr) - \bigl(\Gamma_0\{f,g\},\Gamma_1\{h,k\}\bigr).\end{aligned}$$

Hence, the boundary mappings Γ_0 and Γ_1 satisfy the abstract Green identity (2.1.1). In the proof of Lemma 7.8.1 it was shown that for $\gamma \in \mathbb{C}^2$ there exists $\{h,k\} \in T_{\max}$ such that $h(a) = \gamma$, and so the mapping $(\Gamma_0,\Gamma_1)^\top : T_{\max} \to \mathbb{C}^2$ is surjective. It follows that $\{\mathbb{C},\Gamma_0,\Gamma_1\}$ is a boundary triplet for $(T_{\min})^* = T_{\max}$.

Due to the assumption that the endpoint b is in the limit-point case, each eigenspace $\mathfrak{N}_\lambda(T_{\max})$, $\lambda \in \mathbb{C}\setminus\mathbb{R}$, has dimension 1. Hence, $\widehat{f}_\lambda \in \widehat{\mathfrak{N}}_\lambda(T_{\max})$ has the form $\widehat{f}_\lambda = \{\chi(\cdot,\lambda)c, \lambda\chi(\cdot,\lambda)c\}$ for some $c \in \mathbb{C}$, where $\chi(\cdot,\lambda)$ is a nontrivial element in $\mathfrak{N}_\lambda(T_{\max})$, $\lambda \in \mathbb{C}\setminus\mathbb{R}$. It follows from Definition 2.3.1 and Definition 2.3.4 that

$$\gamma(\lambda) = \bigl\{\{\chi_1(a,\lambda)c, \chi(\cdot,\lambda)c\} : c \in \mathbb{C}\bigr\}, \quad \lambda \in \mathbb{C}\setminus\mathbb{R},$$

and

$$M(\lambda) = \bigl\{\{\chi_1(a,\lambda)c, \chi_2(a,\lambda)c\} : c \in \mathbb{C}\bigr\}, \quad \lambda \in \mathbb{C}\setminus\mathbb{R}.$$

Observe that $\chi_1(a,\lambda) \neq 0$ for $\lambda \in \mathbb{C}\setminus\mathbb{R}$, as otherwise $\lambda \in \mathbb{C}\setminus\mathbb{R}$ would be an eigenvalue of the self-adjoint relation $A_0 = \ker\Gamma_0$ and $\chi(\cdot,\lambda)$ would be a corresponding eigenfunction. Thus, one concludes that

$$\gamma(\lambda) = \frac{\chi(\cdot,\lambda)}{\chi_1(a,\lambda)} \quad \text{and} \quad M(\lambda) = \frac{\chi_2(a,\lambda)}{\chi_1(a,\lambda)}, \quad \lambda \in \mathbb{C}\setminus\mathbb{R}.$$

Note that $\chi(\cdot,\lambda) = \alpha_1 Y_1(\cdot,\lambda) + \alpha_2 Y_2(\cdot,\lambda)$ for some $\alpha_1,\alpha_2 \in \mathbb{C}$ and that the assumption $Y(a,\lambda) = I$ yields

$$\begin{pmatrix}\chi_1(a,\lambda)\\ \chi_2(a,\lambda)\end{pmatrix} = \chi(a,\lambda) = \begin{pmatrix}\alpha_1\\ \alpha_2\end{pmatrix}.$$

This implies

$$\gamma(\lambda) = \frac{\alpha_1 Y_1(\cdot,\lambda) + \alpha_2 Y_2(\cdot,\lambda)}{\chi_1(a,\lambda)} = Y_1(\cdot,\lambda) + M(\lambda)Y_2(\cdot,\lambda), \quad \lambda \in \mathbb{C}\setminus\mathbb{R},$$

establishing the formulas for the γ-field and Weyl function. □

Note that the γ-field and Weyl function corresponding to the boundary triplet $\{\mathbb{C},\Gamma_0,\Gamma_1\}$ in Theorem 7.8.2 are defined and analytic on the resolvent set of the self-adjoint relation $A_0 = \ker \Gamma_0$. It follows in the same way as in Section 6.4 (see the discussion after the proof of Proposition 6.4.1) that the expressions for γ and M in Theorem 7.8.2 extend to points in $\rho(A_0) \cap \mathbb{R}$.

In the next proposition the resolvent of the self-adjoint relation A_0 is expressed as an integral operator.

Proposition 7.8.3. *Assume that the endpoint a is regular or quasiregular and that the endpoint b is in the limit-point case. Let $Y(\cdot, \lambda)$ be a fundamental matrix fixed by $Y(a, \lambda) = I$. Let $\{\mathbb{C}, \Gamma_0, \Gamma_1\}$ be the boundary triplet for $T_{\max}$ in Theorem 7.8.2 with corresponding Weyl function M. Then the self-adjoint relation $A_0 = \ker \Gamma_0$ is given by*

$$A_0 = \ker \Gamma_0 = \big\{\{f, g\} \in T_{\max} : f_1(a) = 0\big\},$$

where $f_1(a)$ denotes the boundary value of the first component of the unique absolutely continuous representative of f. The resolvent of A_0 is an integral operator

$$\big((A_0 - \lambda)^{-1} g\big)(t) = \int_a^b G_0(t, s, \lambda) \Delta(s) g(s)\, ds, \quad \lambda \in \mathbb{C} \setminus \mathbb{R}, \tag{7.8.1}$$

where $g \in L^2_\Delta(\imath)$. The Green function $G_0(t, s, \lambda)$ is given by

$$G_0(t, s, \lambda) = G_{0,\mathrm{e}}(t, s, \lambda) + G_{0,\mathrm{i}}(t, s, \lambda), \tag{7.8.2}$$

where the entire part $G_{0,\mathrm{e}}$ is given by

$$G_{0,\mathrm{e}}(t, s, \lambda) = \begin{cases} Y_1(t, \lambda) Y_2(s, \bar{\lambda})^*, & s < t, \\ Y_2(t, \lambda) Y_1(s, \bar{\lambda})^*, & s > t, \end{cases} \tag{7.8.3}$$

and

$$G_{0,\mathrm{i}}(t, s, \lambda) = Y_2(t, \lambda) M(\lambda) Y_2(s, \bar{\lambda})^*. \tag{7.8.4}$$

Proof. To prove the identity (7.8.1), consider $g \in L^2_\Delta(\imath)$ and define the function f by the right-hand side of (7.8.1) with G_0 as in (7.8.2). In view of (7.8.3) and (7.8.4), this means that

$$\begin{aligned} f(t) = \big(Y_1(t, \lambda) + Y_2(t, \lambda) M(\lambda)\big) \int_a^t Y_2(s, \bar{\lambda})^* \Delta(s) g(s)\, ds \\ + Y_2(t, \lambda) \int_t^b \big(Y_1(s, \bar{\lambda})^* + M(\bar{\lambda})^* Y_2(s, \bar{\lambda})^*\big) \Delta(s) g(s)\, ds. \end{aligned} \tag{7.8.5}$$

Observe that, indeed, the integral near b exists, since one has

$$\gamma(\cdot, \bar{\lambda}) = Y_1(\cdot, \bar{\lambda}) + M(\bar{\lambda}) Y_2(\cdot, \bar{\lambda}) \in L^2_\Delta(\imath)$$

and $g \in L^2_\Delta(\imath)$. It follows that the function f in (7.8.5) is well defined and absolutely continuous. Rewrite (7.8.5) in the form

$$\begin{aligned} f(t) = Y(t,\lambda)\begin{pmatrix} 1 \\ M(\lambda)\end{pmatrix}\int_a^t \begin{pmatrix} 0 & 1\end{pmatrix} Y(s,\bar\lambda)^*\Delta(s)g(s)\,ds \\ + Y(t,\lambda)\begin{pmatrix} 0 \\ 1\end{pmatrix}\int_t^b \begin{pmatrix} 1 & M(\bar\lambda)^*\end{pmatrix} Y(s,\bar\lambda)^*\Delta(s)g(s)\,ds. \end{aligned} \tag{7.8.6}$$

Then a straightforward calculation using the identity

$$\begin{pmatrix} 1 \\ M(\lambda)\end{pmatrix}\begin{pmatrix} 0 & 1\end{pmatrix} - \begin{pmatrix} 0 \\ 1\end{pmatrix}\begin{pmatrix} 1 & M(\bar\lambda)^*\end{pmatrix} = \begin{pmatrix} 0 & 1 \\ -1 & 0\end{pmatrix} = -J \tag{7.8.7}$$

and (7.2.10) shows that f satisfies the inhomogenenous equation

$$Jf' - Hf = \lambda\Delta f + \Delta g. \tag{7.8.8}$$

Moreover, one sees from (7.8.5) that f satisfies

$$f(a) = \begin{pmatrix} 0 \\ 1\end{pmatrix}\int_a^b \big(Y_1(s,\bar\lambda) + Y_2(s,\bar\lambda)M(\bar\lambda)\big)^*\Delta(s)g(s)\,ds = \begin{pmatrix} 0 \\ (g,\gamma(\bar\lambda))_\Delta\end{pmatrix}.$$

Now denote the function on the left-hand side of (7.8.1) by $h = (A_0 - \lambda)^{-1}g$. Then it is clear that

$$\{h, \lambda h + g\} = \{(A_0 - \lambda)^{-1}g, g + \lambda(A_0 - \lambda)^{-1}g\} \in A_0 \subset T_{\max}, \tag{7.8.9}$$

so that h also satisfies (7.8.8). Moreover, by (7.8.9) and Proposition 2.3.2 one obtains

$$\begin{aligned} h_1(a) &= \Gamma_0\{h, \lambda h + g\} = 0, \\ h_2(a) &= \Gamma_1\{h, \lambda h + g\} = \gamma(\bar\lambda)^* g = (g, \gamma(\bar\lambda))_\Delta. \end{aligned}$$

Thus, for a fixed $\lambda \in \mathbb{C}\setminus\mathbb{R}$ the functions h and f satisfy the same inhomogeneous equation (7.8.8) and they have the same initial value $f(a) = h(a)$. Since the solution is unique, $h = f$. One concludes that $(A_0 - \lambda)^{-1}g$ is given by the right-hand side of (7.8.1). □

Note that there exist canonical systems whose Weyl functions are of the form $M(\lambda) = \alpha + \beta\lambda$ where $\alpha \in \mathbb{R}$ and $\beta \geq 0$; cf. Example 7.10.4 for a special case. Hence, the functions M and $G_{0,\mathrm{i}}$ in (7.8.4) may be entire.

Theorem 7.8.4. *Assume that the endpoint a is regular or quasiregular and that the endpoint b is in the limit-point case. Then the operator part $(T_{\min})_{\mathrm{op}}$ is simple in the Hilbert space $L^2_\Delta(\imath) \ominus \mathrm{mul}\, T_{\min}$.*

Proof. Step 1. Let $g \in L^2_\Delta(\imath)$ and define $f = (A_0 - \lambda)^{-1} g$. Then it is clear that $\{f, g + \lambda f\} \in A_0$ and one has

$$f(t) = -Y(t,\lambda)J \int_a^t Y(s,\bar{\lambda})^* \Delta(s) g(s)\, ds + Y(t,\lambda) \begin{pmatrix} 0 \\ (g, \gamma(\bar{\lambda}))_\Delta \end{pmatrix}, \tag{7.8.10}$$

where $Y(\cdot,\lambda)$ is the fundamental matrix fixed by $Y(a,\lambda) = I$. In fact, it follows from Proposition 7.8.3 and its proof that f is given by (7.8.1) or, equivalently, by (7.8.6). Now on the right-hand side of (7.8.6) substract and add the term

$$Y(t,\lambda) \begin{pmatrix} 0 \\ 1 \end{pmatrix} \int_a^t \begin{pmatrix} 1 & M(\bar{\lambda})^* \end{pmatrix} Y(s,\bar{\lambda})^* \Delta(s) g(s)\, ds,$$

and use (7.8.7) and $\gamma(\cdot,\bar{\lambda}) = Y_1(\cdot,\bar{\lambda}) + M(\bar{\lambda}) Y_2(\cdot,\bar{\lambda})$. This yields (7.8.10).

Step 2. The multivalued part $\operatorname{mul} T_{\min}$ is given by

$$\big(\operatorname{span}\{\gamma(\lambda) : \lambda \in \mathbb{C} \setminus \mathbb{R}\}\big)^\perp = \operatorname{mul} T_{\min}, \tag{7.8.11}$$

which, in view of Corollary 3.4.6, is equivalent to $(T_{\min})_{\rm op}$ being simple in the Hilbert space $L^2_\Delta(\imath) \ominus \operatorname{mul} T_{\min}$.

The identity (7.8.11) will be verified by exhibiting the corresponding inclusions. For the inclusion ($\supset$) in (7.8.11), let $g \in \operatorname{mul} T_{\min}$. Since $\{0, g\} \in T_{\min}$ and $\{\gamma(\lambda), \lambda\gamma(\lambda)\} \in T_{\max}$ for all $\lambda \in \mathbb{C} \setminus \mathbb{R}$ one sees that

$$(g, \gamma(\lambda))_\Delta = (g, \gamma(\lambda))_\Delta - (0, \lambda\gamma(\lambda))_\Delta = 0.$$

Hence, $g \in \operatorname{mul} T_{\min}$ is orthogonal to all $\gamma(\lambda)$, $\lambda \in \mathbb{C} \setminus \mathbb{R}$.

For the inclusion ($\subset$) in (7.8.11), assume that $g \in L^2_\Delta(\imath)$ is orthogonal to all $\gamma(\lambda)$, $\lambda \in \mathbb{C} \setminus \mathbb{R}$. Then it follows from (7.8.10) that

$$f(t) = \big((A_0 - \lambda)^{-1} g\big)(t) = -Y(t,\lambda)J \int_a^t Y(s,\bar{\lambda})^* \Delta(s) g(s)\, ds. \tag{7.8.12}$$

Clearly, $f(a) = 0$, so that, in fact,

$$\{f, g + \lambda f\} \in T_{\min}. \tag{7.8.13}$$

Let $h \in L^2_\Delta(\imath)$ have compact support, say in $[a', b'] \subset \imath$. Then it follows from (7.8.12) that

$$\big((A_0 - \lambda)^{-1} g, h\big)_\Delta = -\int_a^b \left(\int_a^t h(t)^* \Delta(t) Y(t,\lambda) J Y(s,\bar{\lambda})^* \Delta(s) g(s)\, ds \right) dt$$

and due to the structure of the double integral the integration takes place only on the square $[a', b'] \times [a', b']$.

Now consider a bounded interval $\delta \subset \mathbb{R}$ such that the endpoints of δ are not eigenvalues of A_0. Then the spectral projection of A_0 corresponding to the interval δ is given by Stone's formula (1.5.7) (see also Example A.1.4),

$$(E(\delta)g,h)_\Delta = \lim_{\varepsilon\downarrow 0} \frac{1}{2\pi i} \int_\delta \big(((A_0-(\mu+i\varepsilon))^{-1} - (A_0-(\mu-i\varepsilon))^{-1})g,h\big)_\Delta \, d\mu.$$

Making use of the above integral for $((A_0-\lambda)^{-1}g,h)_\Delta$ one has that

$$\begin{aligned}(E(\delta)g,h)_\Delta &= \lim_{\varepsilon\downarrow 0} \frac{1}{2\pi i} \int_\delta \int_a^b \int_a^t h(t)^*\Delta(t)F_\varepsilon(t,s,\mu)\Delta(s)g(s)\,ds\,dt\,d\mu \\ &= \lim_{\varepsilon\downarrow 0} \frac{1}{2\pi i} \int_a^b \int_a^t h(t)^*\Delta(t)\left(\int_\delta F_\varepsilon(t,s,\mu)\,d\mu\right)\Delta(s)g(s)\,ds\,dt,\end{aligned}$$

where

$$F_\varepsilon(t,s,\mu) = Y(t,\mu-i\varepsilon)JY(s,\overline{\mu-i\varepsilon})^* - Y(t,\mu+i\varepsilon)JY(s,\overline{\mu+i\varepsilon})^*.$$

To justify the application of Fubini's theorem above note that each of the functions

$$s \mapsto \Delta(s)g(s) \quad \text{and} \quad t \mapsto \Delta(t)h(t)$$

is integrable on $[a',b']$, due to $g,h \in L^2_\Delta(a,b)$ and Lemma 7.1.4, and that the function

$$(s,t,\lambda) \mapsto Y(t,\lambda)JY(s,\bar\lambda)^* - Y(t,\bar\lambda)JY(s,\lambda)^*, \quad s,t\in[a',b'],\ \lambda\in K,$$

where $K\subset\mathbb{C}$ is some compact set, is continuous and hence bounded on the set $[a',b']\times[a',b']\times K$. Since the mapping $\lambda\mapsto Y(t,\lambda)$ is entire, it follows that

$$\lim_{\varepsilon\downarrow 0}\int_\delta F_\varepsilon(t,s,\mu)\,d\mu = 0$$

and dominated convergence implies that $(E(\delta)g,h)_\Delta = 0$ for any $h\in L^2_\Delta(\imath)$ with compact support. Therefore, $E(\delta)g=0$ for any bounded interval δ with endpoints not in $\sigma_p(A_0)$. With $\delta\to\mathbb{R}$ one concludes $E(\mathbb{R})g=0$ and this implies $g\in \operatorname{mul} A_0$. Since $\{f,g+\lambda f\}\in A_0$ and $\{0,g\}\in A_0$, it follows that $\{f,\lambda f\}\in A_0$ and hence $f=0$, as $\lambda\in\mathbb{C}\setminus\mathbb{R}$ is not an eigenvalue of A_0. Since $\{f,g+\lambda f\}\in T_{\min}$ by (7.8.13), one concludes $\{0,g\}\in T_{\min}$, that is, $g\in\operatorname{mul} T_{\min}$. This shows the inclusion ($\subset$) in (7.8.11). □

Let $\{\mathbb{C},\Gamma_0,\Gamma_1\}$ be the boundary triplet in Theorem 7.8.2. Then the self-adjoint extensions of $T_{\min}$ are in a one-to-one correspondence to the numbers $\tau\in\mathbb{R}\cup\{\infty\}$ via

$$A_\tau = \big\{\{f,g\}\in T_{\max} : \Gamma_1\{f,g\} = \tau\Gamma_0\{f,g\}\big\}. \tag{7.8.14}$$

Note that A_0 corresponds to $\tau = \infty$. For a given $\tau \in \mathbb{R} \cup \{\infty\}$ one can transform the boundary triplet $\{\mathbb{C}, \Gamma_0, \Gamma_1\}$ as follows:

$$\begin{pmatrix} \Gamma_0^\tau \\ \Gamma_1^\tau \end{pmatrix} = \frac{1}{\sqrt{\tau^2+1}} \begin{pmatrix} \tau & -1 \\ 1 & \tau \end{pmatrix} \begin{pmatrix} \Gamma_0 \\ \Gamma_1 \end{pmatrix} \tag{7.8.15}$$

(see (2.5.19)), so that $A_\tau = \ker \Gamma_0^\tau$. Then, by (2.5.20), the γ-field and Weyl function corresponding to the new boundary triplet are given by

$$M_\tau(\lambda) = \frac{1+\tau M(\lambda)}{\tau - M(\lambda)} \quad \text{and} \quad \gamma_\tau(\lambda) = \frac{\sqrt{\tau^2+1}}{\tau - M(\lambda)}\gamma(\lambda), \quad \lambda \in \mathbb{C} \setminus \mathbb{R}. \tag{7.8.16}$$

The Weyl function M_τ and the γ-field γ_τ are connected by

$$\frac{M_\tau(\lambda) - M_\tau(\mu)^*}{\lambda - \bar{\mu}} = \gamma_\tau(\mu)^*\gamma_\tau(\lambda), \quad \lambda, \mu \in \mathbb{C} \setminus \mathbb{R}.$$

Let $Y(\cdot, \lambda)$ be a fundamental matrix fixed by $Y(a, \lambda) = I$. In a similar fashion one can transform $Y(\cdot, \lambda)$ to a fundamental matrix $V(\cdot, \lambda)$ given by

$$\begin{pmatrix} V_1(\cdot,\lambda) & V_2(\cdot,\lambda) \end{pmatrix} = \begin{pmatrix} Y_1(\cdot,\lambda) & Y_2(\cdot,\lambda) \end{pmatrix} \frac{1}{\sqrt{\tau^2+1}} \begin{pmatrix} \tau & 1 \\ -1 & \tau \end{pmatrix}. \tag{7.8.17}$$

Note that $V(a, \bar{\lambda})^* J V(a, \lambda) = J$ holds; cf. (7.2.8) and (7.2.11). Due to this transformation the γ-field γ_τ can be written in terms of the new fundamental system as

$$\gamma_\tau(\lambda) = V_1(\cdot, \lambda) + M_\tau(\lambda) V_2(\cdot, \lambda),$$

which belongs to $L^2_\Delta(\imath)$, while the second column of $V(\cdot, \lambda)$ satisfies the (formal) boundary condition which determines $A_\tau = \ker \Gamma_0^\tau$:

$$V_2(a, \lambda) = \frac{1}{\sqrt{\tau^2+1}} \begin{pmatrix} 1 \\ \tau \end{pmatrix}. \tag{7.8.18}$$

The next proposition is the counterpart of Proposition 7.8.3 for the self-adjoint extensions A_τ.

Proposition 7.8.5. *Assume that the endpoint a is regular or quasiregular and that the endpoint b is in the limit-point case. Let $Y(\cdot, \lambda)$ be a fundamental matrix fixed by $Y(a, \lambda) = I$. Let $\{\mathbb{C}, \Gamma_0, \Gamma_1\}$ be the boundary triplet for $T_{\max}$ in Theorem 7.8.2. For $\tau \in \mathbb{R}$ let A_τ be a self-adjoint extension of $T_{\min}$ given by (7.8.14) and let M_τ be as in (7.8.16). Then the resolvent of A_τ is an integral operator*

$$\big((A_\tau - \lambda)^{-1} g\big)(t) = \int_a^b G_\tau(t, s, \lambda)\Delta(s) g(s)\, ds, \quad \lambda \in \mathbb{C} \setminus \mathbb{R}, \tag{7.8.19}$$

where $g \in L^2_\Delta(\imath)$. The Green's function $G_\tau(t, s, \lambda)$ is given by

$$G_\tau(t, s, \lambda) = G_{\tau,\mathrm{e}}(t, s, \lambda) + G_{\tau,\mathrm{i}}(t, s, \lambda), \tag{7.8.20}$$

where the entire part $G_{\tau,\mathrm{e}}$ is given by

$$G_{\tau,\mathrm{e}}(t,s,\lambda) = \begin{cases} V_1(t,\lambda)V_2(s,\bar{\lambda})^*, & s < t, \\ V_2(t,\lambda)V_1(s,\bar{\lambda})^*, & s > t, \end{cases} \tag{7.8.21}$$

and

$$G_{\tau,\mathrm{i}}(t,s,\lambda) = V_2(t,\lambda)M_\tau(\lambda)V_2(s,\bar{\lambda})^*. \tag{7.8.22}$$

Proof. The proof of Proposition 7.8.5 is similar to the proof of Proposition 7.8.3. In order to show the identity (7.8.19), consider $g \in L^2_\Delta(\imath)$ and define the function f by the right-hand side of (7.8.19). Then one has

$$\begin{aligned} f(t) &= \big(V_1(t,\lambda) + V_2(t,\lambda)M_\tau(\lambda)\big)\int_a^t V_2(s,\bar{\lambda})^*\Delta(s)g(s)\,ds \\ &\quad + V_2(t,\lambda)\int_t^b \big(V_1(s,\bar{\lambda})^* + M_\tau(\bar{\lambda})^*V_2(s,\bar{\lambda})^*\big)\Delta(s)g(s)\,ds \end{aligned} \tag{7.8.23}$$

and the same arguments as in the proof of Proposition 7.8.3 show that f is well defined, absolutely continuous, and satisfies the inhomogenenous equation

$$Jf' - Hf = \lambda\Delta f + \Delta g. \tag{7.8.24}$$

Moreover, one sees from (7.8.23) that f satisfies

$$\begin{aligned} f(a) &= \frac{1}{\sqrt{\tau^2+1}}\begin{pmatrix}1\\ \tau\end{pmatrix}\int_a^b \big(V_1(s,\bar{\lambda}) + V_2(s,\bar{\lambda})M_\tau(\bar{\lambda})\big)^*\Delta(s)g(s)\,ds \\ &= \frac{1}{\sqrt{\tau^2+1}}\begin{pmatrix}1\\ \tau\end{pmatrix}(g,\gamma_\tau(\bar{\lambda}))_\Delta, \end{aligned}$$

and hence

$$\frac{\tau f_1(a) - f_2(a)}{\sqrt{\tau^2+1}} = 0 \quad\text{and}\quad \frac{f_1(a) + \tau f_2(a)}{\sqrt{\tau^2+1}} = (g,\gamma_\tau(\bar{\lambda}))_\Delta.$$

Now denote the function on the left-hand side of (7.8.19) by $h = (A_\tau - \lambda)^{-1}g$. Then h also satisfies (7.8.24) and from (7.8.15) and Proposition 2.3.2 one obtains

$$\begin{aligned} \frac{\tau h_1(a) - h_2(a)}{\sqrt{\tau^2+1}} &= \Gamma_0^\tau\{h, \lambda h + g\} = 0, \\ \frac{h_1(a) + \tau h_2(a)}{\sqrt{\tau^2+1}} &= \Gamma_1^\tau\{h, \lambda h + g\} = \gamma_\tau(\bar{\lambda})^*g = (g,\gamma_\tau(\bar{\lambda}))_\Delta. \end{aligned}$$

Thus, for a fixed $\lambda \in \mathbb{C}\setminus\mathbb{R}$ the functions h and f satisfy the same initial value problem and hence it follows that $h = f$. Therefore, $(A_\tau - \lambda)^{-1}g$ is given by the right-hand side of (7.8.19). □

Let $\{\mathbb{C}, \Gamma_0, \Gamma_1\}$ be the boundary triplet for $T_{\max}$ in Theorem 7.8.2 and consider the self-adjoint extension $A_\tau = \ker \Gamma_\tau$; cf. (7.8.14). Assume that the Weyl function M_τ corresponding to $\{\mathbb{C}, \Gamma_0^\tau, \Gamma_1^\tau\}$ has the integral representation

$$M_\tau(\lambda) = \alpha_\tau + \beta_\tau \lambda + \int_{\mathbb{R}} \left(\frac{1}{t-\lambda} - \frac{t}{t^2+1} \right) d\sigma_\tau(t), \tag{7.8.25}$$

with $\alpha_\tau \in \mathbb{R}$, $\beta_\tau \geq 0$, and σ_τ a nondecreasing function with

$$\int_{\mathbb{R}} \frac{1}{t^2+1}\, d\sigma_\tau(t) < \infty.$$

Recall from Theorem 3.5.10 and Lemma A.2.6 that $\operatorname{mul} A_\tau \ominus \operatorname{mul} T_{\min}$ is nontrivial if and only if $\beta_\tau > 0$.

Lemma 7.8.6. *Let $\tau \in \mathbb{R} \cup \{\infty\}$ and let $E_\tau(\cdot)$ be the spectral measure of the self-adjoint relation A_τ. For $h \in L^2_\Delta(\imath)$ with compact support define the Fourier transform $\widehat{f}$ by*

$$\widehat{f}(\mu) = \int_a^b V_2(s,\mu)^* \Delta(s) f(s)\, ds, \quad \mu \in \mathbb{R},$$

where $V_2(\cdot, \mu)$ is the formal solution in (7.8.17). Let σ_τ be the function in the integral representation (7.8.25) of the Weyl function M_τ. Then for every bounded open interval $\delta \subset \mathbb{R}$ such that its endpoints are not eigenvalues of A_τ one has

$$(E_\tau(\delta) f, f)_\Delta = \int_\delta \widehat{f}(\mu)\, \overline{\widehat{f}(\mu)}\, d\sigma_\tau(\mu). \tag{7.8.26}$$

Proof. Recall that $(A_\tau - \lambda)^{-1}$ is given by (7.8.19), where the Green function $G_\tau(t,s,\lambda)$ in (7.8.20) is given by (7.8.21) and (7.8.22). Assume that the function $h \in L^2_\Delta(\imath)$ has compact support in $[a',b'] \subset \imath$. Then

$$\big((A_\tau - \lambda)^{-1} f, f\big)_\Delta = \int_a^b f(t)^* \Delta(t) \left(\int_a^b G_\tau(t,s,\lambda) \Delta(s) f(s)\, ds \right) dt$$

for each $\lambda \in \mathbb{C} \setminus \mathbb{R}$, where, in fact, the integration takes place only on the square $[a',b'] \times [a',b']$.

Let $\delta \subset \mathbb{R}$ be a bounded interval such that the endpoints of δ are not eigenvalues of A_τ. Then the spectral projection of A_τ corresponding to the interval δ is given by Stone's formula (1.5.7) (see also Example A.1.4)

$$\begin{aligned}
(E(\delta) f, f)_\Delta &= \lim_{\varepsilon \downarrow 0} \frac{1}{2\pi i} \int_\delta \big(((A_\tau - (\mu + i\varepsilon))^{-1} - (A_\tau - (\mu - i\varepsilon))^{-1}) f, f\big)_\Delta d\mu \\
&= \lim_{\varepsilon \downarrow 0} \frac{1}{2\pi i} \int_\delta \bigg(\int_a^b \int_a^b f(t)^* \Delta(t) \big(G_\tau(t,s,\mu + i\varepsilon) \\
&\qquad - G_\tau(t,s,\mu - i\varepsilon)\big) \Delta(s) f(s)\, ds\, dt \bigg) d\mu.
\end{aligned}$$

Decompose the Green function in (7.8.20) as in (7.8.21) and (7.8.22). Since the function $\lambda \mapsto V(t,\lambda)$ is entire, one verifies in the same way as in the proof of Theorem 7.8.4 that

$$\lim_{\varepsilon\downarrow 0}\frac{1}{2\pi i}\int_\delta\biggl(\int_a^b\int_a^b f(t)^*\Delta(t)\bigl(G_{\tau,\mathrm{e}}(t,s,\mu+i\varepsilon) \\ - G_{\tau,\mathrm{e}}(t,s,\mu-i\varepsilon)\bigr)\Delta(s)f(s)\,ds\,dt\biggr)d\mu = 0.$$

Therefore, it remains to consider the corresponding integral with $G_{\tau,\mathrm{i}}$, which takes the form

$$\begin{aligned}
&\lim_{\varepsilon\downarrow 0}\frac{1}{2\pi i}\int_\delta\biggl(\int_a^b\int_a^b f(t)^*\Delta(t)\bigl(V_2(t,\mu+i\varepsilon)M_\tau(\mu+i\varepsilon)V_2(s,\mu-i\varepsilon)^* \\
&\qquad - V_2(t,\mu-i\varepsilon)M_\tau(\mu-i\varepsilon)V_2(s,\mu+i\varepsilon)^*\bigr)\Delta(s)f(s)\,ds\,dt\biggr)d\mu \\
&= \lim_{\varepsilon\downarrow 0}\frac{1}{2\pi i}\int_\delta\biggl(\int_a^b\int_a^b f(t)^*\Delta(t)\bigl[(g_{t,s}M_\tau)(\mu+i\varepsilon) \\
&\qquad - (g_{t,s}M_\tau)(\mu-i\varepsilon)\bigr]\Delta(s)f(s)\,ds\,dt\biggr)d\mu,
\end{aligned}\tag{7.8.27}$$

where $g_{t,s}$ stands for the 2×2 matrix function

$$g_{t,s}(\eta) = V_2(t,\eta)\,V_2(s,\bar\eta)^*.$$

For $t,s\in[a',b']$ this function is entire in η. For $\varepsilon_0>0$ and $A<B$ such that $\bar\delta\subset(A,B)$ consider the rectangle $R=[A,B]\times[-i\varepsilon_0,i\varepsilon_0]$. Then the function $\{t,s,\eta\}\mapsto g_{t,s}(\eta)$ is bounded on $[a',b']\times[a',b']\times R$, and since $\Delta h\in L^1(a',b')$, it follows that for each fixed ε such that $0<\varepsilon\le\varepsilon_0$

$$\frac{1}{2\pi i}\int_\delta\biggl(\int_a^b\int_a^b f(t)^*\Delta(t)\bigl[(g_{t,s}M_\tau)(\mu+i\varepsilon)-(g_{t,s}M_\tau)(\mu-i\varepsilon)\bigr]\Delta(s)f(s)\,ds\,dt\biggr)d\mu$$

$$= \frac{1}{2\pi i}\int_a^b\int_a^b f(t)^*\Delta(t)\biggl(\int_\delta\bigl[(g_{t,s}M_\tau)(\mu+i\varepsilon)-(g_{t,s}M_\tau)(\mu-i\varepsilon)\bigr]\,d\mu\biggr)\Delta(s)f(s)\,ds\,dt.$$

By the Stieltjes inversion formula in Lemma A.2.7 and Remark A.2.10, one sees

$$\lim_{\varepsilon\downarrow 0}\frac{1}{2\pi i}\int_\delta\bigl[(g_{t,s}M_\tau)(\mu+i\varepsilon)-(g_{t,s}M_\tau)(\mu-i\varepsilon)\bigr]\,d\mu = \int_\delta g_{t,s,}(\mu)\,d\sigma_\tau(\mu)$$

for all $t,s\in[a',b']$. To justify taking the limit $\varepsilon\downarrow 0$ inside the integral (7.8.27) one needs dominated convergence. For this purpose recall from Lemma A.2.7 and

Remark A.2.10 that there exists $m \geq 0$ such that for $0 < \varepsilon \leq \varepsilon_0$ one has

$$\begin{aligned}\left| \int_\delta \left[(g_{t,s} M_\tau)(\mu + i\varepsilon) - (g_{t,s} M_\tau)(\mu - i\varepsilon) \right] d\mu \right| \\ \leq m \sup\{ |g_{t,s}(\eta)|, |g'_{t,s}(\eta)| : t, s \in [a', b'], \lambda \in R \},\end{aligned} \tag{7.8.28}$$

where $R = [A, B] \times [-i\varepsilon_0, i\varepsilon_0]$. Since the functions

$$\{t, s, \eta\} \mapsto g_{t,s}(\eta) \quad \text{and} \quad \{t, s, \eta\} \mapsto g'_{t,s}(\eta)$$

are bounded on $[a', b'] \times [a', b'] \times R$, it follows that the integral in (7.8.28), regarded as a function in $\{t, s\}$ on $[a', b'] \times [a', b']$, is bounded by some constant for all $0 < \varepsilon \leq \varepsilon_0$. Furthermore, $\Delta f \in L^1(a', b')$ implies that there is an integrable majorant for (7.8.27), and so dominated convergence and Fubini's theorem show that

$$\begin{aligned}(E(\delta)f, f)_\Delta &= \int_a^b \int_a^b f(t)^* \Delta(t) \int_\delta g_{t,s}(\mu)\, d\sigma_\tau(\mu)\, \Delta(s) f(s)\, ds\, dt \\ &= \int_\delta \left(\int_a^b \left(V_2(t, \mu)^* \Delta(t) f(t) \right)^* dt \right) \left(\int_a^b V_2(s, \mu)^* \Delta(s) f(s)\, ds \right) d\sigma_\tau(\mu)\end{aligned}$$

for every open interval δ such that the endpoints are not eigenvalues of A_τ. This gives the formula in (7.8.26). □

The next theorem is a consequence of Lemma 7.8.6 and Theorem B.2.3.

Theorem 7.8.7. *Let $\tau \in \mathbb{R} \cup \{\infty\}$, let $V_2(\cdot, \mu)$ be the formal solution in (7.8.17), and let σ_τ be the function in the integral representation of the Weyl function M_τ. Then the Fourier transform*

$$f \mapsto \widehat{f}, \qquad \widehat{f}(\mu) = \int_a^b V_2(s, \mu)^* \Delta(s) f(s)\, ds, \quad \mu \in \mathbb{R},$$

extends by continuity from compactly supported functions $f \in L^2_\Delta(\imath)$ to a surjective partial isometry $\mathcal{F}$ from $L^2_\Delta(\imath)$ to $L^2_{d\sigma_\tau}(\mathbb{R})$ with $\ker \mathcal{F} = \operatorname{mul} A_\tau$. The restriction $\mathcal{F}_{\rm op} : L^2_\Delta(\imath) \ominus \operatorname{mul} A_\tau \to L^2_{d\sigma_\tau}(\mathbb{R})$ is a unitary mapping, such that the self-adjoint operator $(A_\tau)_{\rm op}$ in $L^2_\Delta(\imath) \ominus \operatorname{mul} A_\tau$ is unitarily equivalent to multiplication by the independent variable in $L^2_{d\sigma_\tau}(\mathbb{R})$.

Proof. It follows from Lemma 7.8.6 that the condition (B.2.2) is satisfied. Furthermore, for every $\mu \in \mathbb{R}$ there exists a compactly supported function $f \in L^2_\Delta(\imath)$ such that

$$\widehat{f}(\mu) = \int_a^b V_2(s, \mu)^* \Delta(s) f(s)\, ds \neq 0.$$

To see this, assume that for some $\mu \in \mathbb{R}$

$$\int_a^b V_2(s, \mu)^* \Delta(s) f(s)\, ds = 0$$

for all compactly supported $f \in L^2_\Delta(\imath)$. This implies that $V_2(s,\mu)^*\Delta(s) = 0$ for a.e. $s \in (a,b)$. By definiteness one has $V_2(s,\mu) = 0$ for a.e. $s \in (a,b)$, which is a contradiction. Therefore, condition (B.2.9) is satisfied and the result follows from Theorem B.2.3. □

In the next lemma the Fourier transform $\mathcal{F}\gamma_\tau$ of the γ-field in (7.8.16) corresponding to the boundary triplet $\{\mathbb{C}, \Gamma_0^\tau, \Gamma_1^\tau\}$ is computed; this allows to identify the model in Theorem 7.8.7 with the model for scalar Nevanlinna functions in Section 4.3.

Lemma 7.8.8. *Let $\tau \in \mathbb{R} \cup \{\infty\}$ and let γ_τ be the γ-field in (7.8.16). Then for all $\lambda \in \mathbb{C} \setminus \mathbb{R}$ one has almost everywhere in the sense of $d\sigma_\tau$:*

$$\big[\mathcal{F}\gamma_\tau(\lambda)\big](\mu) = \frac{1}{\mu - \lambda}, \qquad \mu \in \mathbb{R},$$

where $\mathcal{F}$ is the Fourier transform from $L^2_\Delta(\imath)$ onto $L^2_{d\sigma_\tau}(\mathbb{R})$ in Theorem 7.8.7.

Proof. Recall first that for $g \in L^2_\Delta(\imath)$ and $\lambda \in \mathbb{C} \setminus \mathbb{R}$ the function $(A_\tau - \lambda)^{-1} g$ is given by the identity (7.8.19), which holds for all $t \in \imath$. In fact, the Green function G_τ in (7.8.19) is a 2×2 matrix function and the following notation will be useful:

$$G_\tau(t,s,\lambda) = \begin{pmatrix} G_{\tau,1}(t,s,\lambda) \\ G_{\tau,2}(t,s,\lambda) \end{pmatrix},$$

where each of these components is a 1×2 matrix function. The fundamental matrix $V(\cdot,\lambda)$ in (7.8.17) is written in the form

$$\begin{pmatrix} V_1(\cdot,\lambda) & V_2(\cdot,\lambda) \end{pmatrix} = \begin{pmatrix} V_{11}(\cdot,\lambda) & V_{12}(\cdot,\lambda) \\ V_{21}(\cdot,\lambda) & V_{22}(\cdot,\lambda) \end{pmatrix}.$$

Now observe that

$$G_{\tau,1}(t,s,\lambda) = \begin{cases} \big(V_{11}(t,\lambda) + M_\tau(\lambda) V_{12}(t,\lambda)\big) V_2(s,\bar\lambda)^*, & s < t, \\ V_{12}(t,\lambda)\gamma_\tau(s,\bar\lambda)^*, & s > t, \end{cases} \tag{7.8.29}$$

and

$$G_{\tau,2}(t,s,\lambda) = \begin{cases} \big(V_{21}(t,\lambda) + M_\tau(\lambda) V_{22}(t,\lambda)\big) V_2(s,\bar\lambda)^*, & s < t, \\ V_{22}(t,\lambda)\gamma_\tau(s,\bar\lambda)^*, & s > t, \end{cases} \tag{7.8.30}$$

which follows easily from (7.8.21) and (7.8.22); cf. (7.8.23). Note also that

$$V_{ij}(\cdot,\bar\lambda) = \overline{V_{ij}(\cdot,\lambda)}, \qquad i,j = 1,2,$$

since the system is real (see Lemma 7.2.8), and that (7.2.10) implies the useful identity

$$V_{11}(t,\lambda)V_{22}(t,\lambda) - V_{21}(t,\lambda)V_{12}(t,\lambda) = 1, \quad t \in \imath. \tag{7.8.31}$$

For $\lambda \in \mathbb{C} \setminus \mathbb{R}$ one has

$$((A_\tau - \lambda)^{-1} g)(t) = \begin{pmatrix} \int_a^b G_{\tau,1}(t,s,\lambda)\Delta(s)g(s)\,ds \\ \int_a^b G_{\tau,2}(t,s,\lambda)\Delta(s)g(s)\,ds \end{pmatrix} = \begin{pmatrix} (g, G_{\tau,1}(t,\cdot,\lambda)^*)_\Delta \\ (g, G_{\tau,2}(t,\cdot,\lambda)^*)_\Delta \end{pmatrix}.$$

Since the Fourier transform $\mathcal{F} : L^2_\Delta(\imath) \to L^2_{d\sigma_\tau}(\mathbb{R})$ in Theorem 7.8.7 is a partial isometry with $\ker \mathcal{F} = \operatorname{mul} A_\tau$, it follows from (1.1.10) that

$$\begin{aligned} ((A_\tau - \lambda)^{-1} g)(t) &= \begin{pmatrix} (\mathcal{F}g, \mathcal{F}G_{\tau,1}(t,\cdot,\lambda)^*)_{L^2_{d\sigma_\tau}(\mathbb{R})} \\ (\mathcal{F}g, \mathcal{F}G_{\tau,2}(t,\cdot,\lambda)^*)_{L^2_{d\sigma_\tau}(\mathbb{R})} \end{pmatrix} \\ &= \begin{pmatrix} \int_\mathbb{R} \mathcal{F}g(\mu)\, \overline{[\mathcal{F}G_{\tau,1}(t,\cdot,\lambda)^*](\mu)}\, d\sigma_\tau(\mu) \\ \int_\mathbb{R} \mathcal{F}g(\mu)\, \overline{[\mathcal{F}G_{\tau,2}(t,\cdot,\lambda)^*](\mu)}\, d\sigma_\tau(\mu) \end{pmatrix} \end{aligned} \tag{7.8.32}$$

is valid for all $g \in (\operatorname{mul} A_\tau)^\perp$. It is clear that (7.8.32) is also true for all $g \in \operatorname{mul} A_\tau$, since in this case $(A_\tau - \lambda)^{-1} g = 0$ and $\mathcal{F}g = 0$. Therefore, (7.8.32) is valid for all $g \in L^2_\Delta(\imath)$. Moreover, if $g \in L^2_\Delta(\imath)$ one has for almost all $t \in \imath$

$$\begin{aligned} ((A_\tau - \lambda)^{-1} g)(t) &= \int_\mathbb{R} \frac{V_2(t,\mu)}{\mu - \lambda} \mathcal{F}g(\mu)\, d\sigma_\tau(\mu) \\ &= \begin{pmatrix} \int_\mathbb{R} \frac{V_{12}(t,\mu)}{\mu-\lambda} \mathcal{F}g(\mu)\, d\sigma_\tau(\mu) \\ \int_\mathbb{R} \frac{V_{22}(t,\mu)}{\mu-\lambda} \mathcal{F}g(\mu)\, d\sigma_\tau(\mu) \end{pmatrix}; \end{aligned} \tag{7.8.33}$$

see (B.2.5). Furthermore, if $\mathcal{F}g$ has compact support, then the right-hand side of (7.8.33) is absolutely continuous, and hence in this case the equality holds for all $t \in \imath$. Therefore, when $\mathcal{F}g$ has compact support the right-hand sides of (7.8.32) and (7.8.33) are equal for all $t \in \imath$, and hence one has

$$\frac{V_{12}(t,\mu)}{\mu - \lambda} = \overline{[\mathcal{F}G_{\tau,1}(t,\cdot,\lambda)^*](\mu)} \quad \text{and} \quad \frac{V_{22}(t,\mu)}{\mu - \lambda} = \overline{[\mathcal{F}G_{\tau,2}(t,\cdot,\lambda)^*](\mu)}$$

for all $t \in \imath$. These identities hold for all $\mu \in \mathbb{R} \setminus \Omega(t)$, where the set $\Omega(t) \subset \mathbb{R}$ has $d\sigma_\tau$-measure 0. Hence, (by replacing λ with $\overline{\lambda}$ and taking conjugates) one has for all $t \in \imath$ and all $\mu \in \mathbb{R} \setminus \Omega(t)$

$$\frac{V_{12}(t,\mu)}{\mu - \lambda} = [\mathcal{F}G_{\tau,1}(t,\cdot,\overline{\lambda})^*](\mu) \quad \text{and} \quad \frac{V_{22}(t,\mu)}{\mu - \lambda} = [\mathcal{F}G_{\tau,2}(t,\cdot,\overline{\lambda})^*](\mu). \tag{7.8.34}$$

By means of (7.8.29), (7.8.30), (7.8.31), and (7.8.34) it is straightforward to verify that for all $t \in \imath$ and all $\mu \in \mathbb{R} \setminus \Omega(t)$,

$$\begin{aligned} &\frac{1}{\mu - \lambda}\big(V_{11}(t,\lambda)V_{22}(t,\mu) - V_{21}(t,\lambda)V_{12}(t,\mu)\big) \\ &\quad = V_{11}(t,\lambda)\big[\mathcal{F}G_{\tau,2}(t,\cdot,\overline{\lambda})^*\big](\mu) - V_{21}(t,\lambda)\big[\mathcal{F}G_{\tau,1}(t,\cdot,\overline{\lambda})^*\big](\mu) \\ &\quad = \mathcal{F}\big[V_{11}(t,\lambda)\, G_{\tau,2}(t,\cdot,\overline{\lambda})^* - V_{21}(t,\lambda)\, G_{\tau,1}(t,\cdot,\overline{\lambda})^*\big](\mu) \\ &\quad = \mathcal{F}\big[W(t,\cdot,\lambda)\big](\mu), \end{aligned} \tag{7.8.35}$$

where the 2×1 matrix function $W(\cdot, \cdot, \lambda)$ is given by

$$W(t, s, \lambda) = \begin{cases} M_\tau(\lambda)V_2(s, \lambda), & s < t, \\ \gamma_\tau(s, \lambda), & s > t. \end{cases} \tag{7.8.36}$$

The above identity and a limit process will give the desired result. In fact, first observe that according to the definition of $W(\cdot, \cdot, \lambda)$ in (7.8.36) one has

$$\|\gamma_\tau(\cdot, \lambda) - W(t, \cdot, \lambda)\|_\Delta^2 = \int_a^t |\Delta(s)^{\frac{1}{2}} V_1(s, \lambda)|^2 \, ds \to 0 \quad \text{as} \quad t \to a,$$

and hence the continuity of $\mathcal{F} : L^2_\Delta(\imath) \to L^2_{d\sigma_\tau}(\mathbb{R})$ implies that

$$\big\| [\mathcal{F}\gamma_\tau(\cdot, \lambda)] - [\mathcal{F}W(t, \cdot, \lambda)] \big\|_{L^2_{d\sigma_\tau}(\mathbb{R})} \to 0 \quad \text{as} \quad t \to a.$$

Now approximate a by a sequence $t_n \in \imath$. Then there exists a subsequence, again denoted by t_n, such that pointwise

$$\big[\mathcal{F}\gamma_\tau(\cdot, \lambda)\big](\mu) = \lim_{n\to\infty} \big[\mathcal{F}W(t_n, \cdot, \lambda)\big](\mu), \quad \mu \in \mathbb{R} \setminus \Omega,$$

where Ω is a set of measure 0 in the sense of $d\sigma_\tau$. Observe that (7.8.35) gives

$$\big[\mathcal{F}W(t_n, \cdot, \lambda)\big](\mu) = \frac{1}{\mu - \lambda} \big(V_{11}(t_n, \lambda)V_{22}(t_n, \mu) - V_{21}(t_n, \lambda)V_{12}(t_n, \mu)\big)$$

for all $\mu \in \mathbb{R} \setminus \Omega(t_n)$. The limit on the right-hand side as $n \to \infty$ gives

$$\frac{1}{\mu - \lambda} \big(V_{11}(a, \lambda)V_{22}(a, \mu) - V_{21}(a, \lambda)V_{12}(a, \mu)\big) = \frac{1}{\mu - \lambda},$$

which follows from the form of the fundamental matrix $(V_1(\cdot,\lambda)\, V_2(\cdot,\lambda))$ in (7.8.17). Hence,

$$\big[\mathcal{F}\gamma_\tau(\cdot, \lambda)\big](\mu) = \frac{1}{\mu - \lambda}, \quad \mu \in \mathbb{R} \setminus \left(\Omega \cup \bigcup_{n=1}^{\infty} \Omega(t_n) \right),$$

which completes the proof. □

Lemma 7.8.8 will be used to explain the model in Theorem 7.8.7 with the model for scalar Nevanlinna functions discussed in Section 4.3. Without loss of generality it is assumed that $T_{\min}$ is simple; cf. Section 3.4. The Weyl function M_τ of the boundary triplet $\{\mathbb{C}, \Gamma_0^\tau, \Gamma_1^\tau\}$ for $T_{\max}$ has the integral representation (7.8.25). If $\beta = 0$, then the discussion in Chapter 6 following Lemma 6.4.8 applies in this case as well. Hence, assume $\beta > 0$ in (7.8.25). Then by Theorem 4.3.4 there is a closed simple symmetric operator S in $L^2_{d\sigma_\tau}(\mathbb{R}) \oplus \mathbb{C}$ such that the Nevanlinna function M_τ in (7.8.25) is the Weyl function corresponding to the boundary triplet $\{\mathbb{C}, \Gamma_0', \Gamma_1'\}$ for S^* in Theorem 4.3.4. The γ-field corresponding to $\{\mathbb{C}, \Gamma_0', \Gamma_1'\}$ is denoted by γ' and it is given by (4.3.16). Furthermore, the restriction A_0' corresponding to the boundary mapping Γ_0' is a self-adjoint relation in $L^2_{d\sigma_\tau}(\mathbb{R}) \oplus \mathbb{C}$ whose operator part $(A_0')_{\rm op}$ is the maximal multiplication operator by the independent variable in $L^2_{d\sigma_\tau}(\mathbb{R})$. By comparing with (4.3.16) one sees that, according to Lemma 7.8.8, the Fourier transform $\mathcal{F}_{\rm op}$ from $L^2_\Delta(\imath) \ominus \operatorname{mul} A_\tau$ onto $L^2_{d\sigma_\tau}(\mathbb{R})$ as a unitary mapping satisfies

$$\mathcal{F}_{\rm op} P\gamma_\tau(\lambda) = P'\gamma'(\lambda),$$

where P and P' stand for the orthogonal projections from $L^2_\Delta(\imath)$ onto $(\operatorname{mul} A_\tau)^\perp$ and from $L^2_{d\sigma_\tau}(\mathbb{R}) \oplus \mathbb{C}$ onto $L^2_{d\sigma_\tau}(\mathbb{R}) = (\operatorname{mul} A_0')^\perp$. Recall that

$$(I-P)\gamma_\tau(\lambda) \quad \text{and} \quad (I-P')\gamma'(\lambda)$$

are independent of λ and belong to $\operatorname{mul} A_\tau$ and $\operatorname{mul} A_0'$, respectively; cf. Corollary 2.5.16. Hence, the mapping $\mathcal{F}_{\rm m}$ from $\operatorname{mul} A_\tau$ to $\operatorname{mul} A_0'$ defined by

$$\mathcal{F}_{\rm m}(I-P)\gamma_\tau(\lambda) = \beta^{\frac{1}{2}} = (I-P')\gamma'(\lambda)$$

is a one-to-one correspondence. In fact, $\mathcal{F}_{\rm m}$ is an isometry due to Proposition 3.5.7. Define the mapping U from the space $L^2_\Delta(\imath)$ to the model space $L^2_{d\sigma_\tau}(\mathbb{R}) \oplus \mathbb{C}$ by

$$U = \begin{pmatrix} \mathcal{F}_{\rm op} & 0 \\ 0 & \mathcal{F}_{\rm m} \end{pmatrix} : \begin{pmatrix} (\operatorname{mul} A_\tau)^\perp \\ \operatorname{mul} A_\tau \end{pmatrix} \to \begin{pmatrix} (\operatorname{mul} A_0')^\perp \\ \operatorname{mul} A_0' \end{pmatrix}.$$

Then it is clear that U is unitary and that

$$U\gamma_\tau(\lambda) = \gamma'(\lambda).$$

Hence, by Theorem 4.2.6, it follows that the boundary triplet $\{\mathbb{C}, \Gamma_0^\tau, \Gamma_1^\tau\}$ for $T_{\max}$ and the boundary triplet $\{\mathbb{C}, \Gamma_0', \Gamma_1'\}$ for S^* are unitarily equivalent under the mapping U, in particular, one has

$$(A_0')_{\rm op} = \mathcal{F}_{\rm op}(A_\tau)_{\rm op}\mathcal{F}_{\rm op}^{-1} \quad \text{and} \quad A_0' = UA_\tau U^{-1}.$$

At the end of this section the case where the endpoint a is in the limit-circle case and the endpoint b is in the limit-point case is briefly discussed. In a similar way as in the end of Section 7.7 one makes use of the transformation in Lemma 7.2.5. The next proposition is the counterpart of Theorem 7.8.2; it is proved in the same way.

Proposition 7.8.9. *Assume that a is in the limit-circle case and that b is in the limit-point case. Let λ_0 in $\mathbb{R}$, let $U(\cdot, \lambda_0)$ be a solution matrix as in Lemma 7.2.5, and consider the limit*

$$\widetilde{f}(a) = \lim_{t \to a} U(t, \lambda_0)^{-1} f(t)$$

for $\{f,g\} \in T_{\max}$*; cf. Corollary 7.4.8. Then* $\{\mathbb{C}, \Gamma_0, \Gamma_1\}$*, where*

$$\Gamma_0\{f,g\} = \widetilde{f}_1(a) \quad \text{and} \quad \Gamma_1\{f,g\} = \widetilde{f}_2(a), \quad \{f,g\} \in T_{\max},$$

is a boundary triplet for $(T_{\min})^* = T_{\max}$*. Let* $Y(\cdot,\lambda)$ *be a fundamental matrix fixed in such a way that* $\widetilde{Y}(\cdot,\lambda) = U(\cdot,\lambda_0)^{-1}Y(\cdot,\lambda)$ *satisfies* $\widetilde{Y}(a,\lambda) = I$*. Then for all* $\lambda \in \mathbb{C} \setminus \mathbb{R}$ *the* γ*-field* γ *and Weyl function* M *corresponding to* $\{\mathbb{C}, \Gamma_0, \Gamma_1\}$ *are given by*

$$\gamma(\lambda) = Y_1(\cdot,\lambda) + M(\lambda)Y_2(\cdot,\lambda) \quad \text{and} \quad M(\lambda) = \frac{\widetilde{\chi}_2(a,\lambda)}{\widetilde{\chi}_1(a,\lambda)},$$

where $\widetilde{\chi}(\cdot,\lambda) = U(\cdot,\lambda_0)^{-1}\chi(\cdot,\lambda)$ *and* $\chi(\cdot,\lambda)$ *is a nontrivial element in* $\mathfrak{N}_\lambda(T_{\max})$*.*

7.9 Weyl functions and subordinate solutions

Consider the real definite canonical system (7.2.3) on the interval $\imath = (a,b)$ and assume that the endpoint a is regular and that the endpoint b is in the limit-point case. Let $\{\mathbb{C}, \Gamma_0, \Gamma_1\}$ be the boundary triplet for $T_{\max}$ in Theorem 7.7.2 with γ-field γ and Weyl function M. The spectrum of the self-adjoint extension

$$A_0 = \ker \Gamma_0 = \big\{\{f,g\} \in T_{\max} : f_1(a) = 0\big\}$$

will be studied by means of *subordinate solutions* of the equation $Jy' - Hy = \lambda\Delta y$. The discussion in this section is parallel to the discussion in Section 6.7.

It is useful to take into account all self-adjoint extensions of $T_{\min}$. As in Section 7.8, there is a one-to-one correspondence to the numbers $\tau \in \mathbb{R} \cup \{\infty\}$ as restrictions of $T_{\max}$ via

$$A_\tau = \big\{\{f,g\} \in T_{\max} : \Gamma_1\{f,g\} = \tau\Gamma_0\{f,g\}\big\}, \tag{7.9.1}$$

with the understanding that A_0 corresponds to $\tau = \infty$. As before, let the boundary triplet $\{\mathbb{C}, \Gamma_0^\tau, \Gamma_1^\tau\}$ be defined by the transformation (7.8.15) with the Weyl function and γ-field given by

$$M_\tau(\lambda) = \frac{1 + \tau M(\lambda)}{\tau - M(\lambda)} \quad \text{and} \quad \gamma_\tau(\lambda) = \frac{\sqrt{\tau^2+1}}{\tau - M(\lambda)}\gamma(\lambda), \quad \lambda \in \mathbb{C} \setminus \mathbb{R}. \tag{7.9.2}$$

Recall that the Weyl function and the γ-field are connected via

$$\frac{M_\tau(\lambda) - M_\tau(\mu)^*}{\lambda - \bar{\mu}} = \gamma_\tau(\mu)^*\gamma_\tau(\lambda), \quad \lambda, \mu \in \mathbb{C} \setminus \mathbb{R}. \tag{7.9.3}$$

The transformation (7.8.15) also induces a transformation of the fundamental matrix $Y(\cdot,\lambda)$ with $Y(a,\lambda) = I$ as in (7.8.17):

$$\begin{pmatrix} V_1(\cdot,\lambda) & V_2(\cdot,\lambda) \end{pmatrix} = \begin{pmatrix} Y_1(\cdot,\lambda) & Y_2(\cdot,\lambda) \end{pmatrix} \frac{1}{\sqrt{\tau^2+1}} \begin{pmatrix} \tau & 1 \\ -1 & \tau \end{pmatrix}. \tag{7.9.4}$$

Recall that

$$\gamma_\tau(\lambda) = V_1(\cdot,\lambda) + M_\tau(\lambda)V_2(\cdot,\lambda)$$

is square-integrable with respect to Δ, while the second column of $V(\cdot,\lambda)$ satisfies the (formal) boundary condition which determines A_τ; cf. (7.8.18).

In the next estimates it is more convenient to work in the semi-Hilbert space $\mathcal{L}^2_\Delta(\imath)$ rather than in the Hilbert space $L^2_\Delta(\imath)$. In fact, for each $x > a$ the notation $\mathcal{L}^2_\Delta(a,x)$ stands for the semi-Hilbert space with the semi-inner product

$$(f,g)_x = \int_a^x g(t)^*\Delta(t)f(t)\,dt, \quad f,g \in \mathcal{L}^2_\Delta(a,x),$$

and the seminorm corresponding to $(\cdot,\cdot)_x$ will be denoted by $\|\cdot\|_x$; here the index Δ is omitted. Hence, for fixed $f,g \in \mathcal{L}^2_\Delta(a,c)$, $a < c < b$, the function $x \mapsto (f,g)_x$ is absolutely continuous and

$$\frac{d}{dx}(f,g)_x = g(x)^*\Delta(x)f(x) \tag{7.9.5}$$

holds almost everywhere on (a,c).

Definition 7.9.1. Let $\lambda \in \mathbb{C}$. Then a solution $h(\cdot,\lambda)$ of $Jh' - Hh = \lambda\Delta h$ is said to be *subordinate* at b if

$$\lim_{x\to b} \frac{\|h(\cdot,\lambda)\|_x}{\|k(\cdot,\lambda)\|_x} = 0$$

for every nontrivial solution $k(\cdot,\lambda)$ of $Jk' - Hk = \lambda\Delta k$ which is not a scalar multiple of $h(\cdot,\lambda)$.

The spectrum of the self-adjoint extension A_0 will be studied in terms of solutions of the canonical system $Jy' - Hy = \xi\Delta y$, $\xi \in \mathbb{R}$, which do not necessarily belong to $\mathcal{L}^2_\Delta(a,b)$. Observe that if a solution $h(\cdot,\xi)$ of $Jy' - Hy = \xi\Delta y$ belongs to $\mathcal{L}^2_\Delta(a,b)$, then it is subordinate at b since b is in the limit-point case, and hence any other nontrivial solution which is not a multiple does not belong to $\mathcal{L}^2_\Delta(a,b)$.

By means of the fundamental system $(V_1(\cdot,\lambda)\ V_2(\cdot,\lambda))$ in (7.8.17) define for any $\lambda \in \mathbb{C}$ and $h \in \mathcal{L}^2_\Delta(a,x)$, $a < x < b$,

$$\begin{aligned}(\mathcal{H}(\lambda)h)(t) &= V_1(t,\lambda)\int_a^t V_2(s,\bar\lambda)^*\Delta(s)h(s)\,ds \\ &\quad - V_2(t,\lambda)\int_a^t V_1(s,\bar\lambda)^*\Delta(s)h(s)\,ds, \quad t \in (a,x).\end{aligned}$$

Thus, $\mathcal{H}(\lambda)$ is a well-defined integral operator and it is clear that the function $\mathcal{H}(\lambda)h$ is absolutely continuous. Using the identity (7.2.10) for $V(\cdot,\lambda)$ (which holds because $V(a,\bar\lambda)^*JV(a,\lambda) = J$) one sees in the same way as in (7.2.14)–(7.2.16) that $f = \mathcal{H}(\lambda)h$ satisfies

$$Jf' - Hf = \lambda\Delta f + \Delta h, \quad f(a) = 0.$$

In particular, $\mathcal{H}(\lambda)$ maps $\mathcal{L}^2_\Delta(a,x)$ into itself. It follows directly that for $\lambda, \mu \in \mathbb{C}$ one has

$$V_i(\cdot,\lambda) - V_i(\cdot,\mu) = (\lambda - \mu)\mathcal{H}(\lambda)V_i(\cdot,\mu), \quad i = 1,2, \tag{7.9.6}$$

since the functions on the left-hand side and the right-hand side both satisfy the same equation $Jy' - Hy = \lambda\Delta y + (\lambda - \mu)\Delta V_i(\cdot,\mu)$ and both functions vanish at the endpoint a.

Lemma 7.9.2. *Let $a < x < b$ and let $h \in \mathcal{L}^2_\Delta(a,x)$. Then the operator $\mathcal{H}(\lambda)$ satisfies*

$$\|\mathcal{H}(\lambda)h\|_x^2 \le 2\|V_1(\cdot,\lambda)\|_x^2\,\|V_2(\cdot,\lambda)\|_x^2\,\|h\|_x^2$$

for each $x > a$.

Proof. The definition of $\mathcal{H}(\lambda)$ may be written as

$$(\mathcal{H}(\lambda)h)(t) = V_1(t,\lambda)g_2(t,\lambda) - V_2(t,\lambda)g_1(t,\lambda), \tag{7.9.7}$$

with the functions $g_i(\cdot,\lambda)$, $i = 1,2$, defined by

$$g_i(t,\lambda) = \int_a^t V_i(s,\bar\lambda)^*\Delta(s)h(s)\,ds.$$

The Cauchy–Schwarz inequality and Corollary 7.2.9 show that

$$|g_i(t,\lambda)|^2 \le \|V_i(\cdot,\bar\lambda)\|_t^2\|h\|_t^2 = \|V_i(\cdot,\lambda)\|_t^2\|h\|_t^2, \quad i = 1,2. \tag{7.9.8}$$

Multiplying $(\mathcal{H}(\lambda)h)(t)$ in (7.9.7) on the left by the matrix $\Delta(t)^{\frac12}$, using the inequality $|a+b|^2 \le 2(|a|^2 + |b|^2)$, and using (7.9.8) one obtains

$$\begin{aligned}
&|\Delta(t)^{\frac12}(\mathcal{H}(\lambda)h)(t)|^2\\
&\quad\le 2\big(|\Delta(t)^{\frac12}V_1(t,\lambda)|^2\,|g_2(t,\lambda)|^2 + |\Delta(t)^{\frac12}V_2(t,\lambda)|^2\,|g_1(t,\lambda)|^2\big)\\
&\quad\le 2\big(|\Delta(t)^{\frac12}V_1(t,\lambda)|^2\,\|V_2(\cdot,\lambda)\|_t^2\|h\|_t^2 + |\Delta(t)^{\frac12}V_2(t,\lambda)|^2\,\|V_1(\cdot,\lambda)\|_t^2\|h\|_t^2\big).
\end{aligned}$$

Integration of this inequality over (a,x) and (7.9.5) lead to

$$\begin{aligned}
\|\mathcal{H}(\lambda)h\|_x^2 &\le 2\int_a^x \Big(|\Delta(t)^{\frac12}V_1(t,\lambda)|^2\|V_2(\cdot,\lambda)\|_t^2\|h\|_t^2\\
&\qquad\qquad +|\Delta(t)^{\frac12}V_2(t,\lambda)|^2\|V_1(\cdot,\lambda)\|_t^2\|h\|_t^2\Big)\,dt\\
&= 2\int_a^x \left(\frac{d}{dt}\,\|V_1(\cdot,\lambda)\|_t^2\|V_2(\cdot,\lambda)\|_t^2\right)\|h\|_t^2\,dt\\
&\le 2\|h\|_x^2\int_a^x \left(\frac{d}{dt}\,\|V_1(\cdot,\lambda)\|_t^2\|V_2(\cdot,\lambda)\|_t^2\right)dt,
\end{aligned}$$

which implies the desired result. □

Since the system is assumed to be definite on $\imath = (a, b)$, there is a compact subinterval $[\alpha, \beta]$ such that the system is definite on $[\alpha, \beta]$ and hence on any interval (a, x) with $x > \beta$. This implies that for $x > \beta$ both functions

$$x \mapsto \|V_1(\cdot, \lambda)\|_x \quad \text{and} \quad x \mapsto \|V_2(\cdot, \lambda)\|_x$$

have positive values; cf. (7.5.1) and Lemma 7.5.2.

Lemma 7.9.3. *Let $\xi \in \mathbb{R}$ be a fixed number. The function $x \mapsto \varepsilon_\tau(x, \xi)$ given by*

$$\sqrt{2}\, \varepsilon_\tau(x, \xi) \|V_1(\cdot, \xi)\|_x \|V_2(\cdot, \xi)\|_x = 1, \quad x > \beta,$$

is well defined, continuous, nonincreasing, and satisfies

$$\lim_{x \to b} \varepsilon_\tau(x, \xi) = 0.$$

Proof. It is clear that $\varepsilon_\tau(x, \xi) > 0$ is well defined due to the assumption that $x > \beta$. Note that $x \mapsto \|V_1(\cdot, \xi)\|_x \|V_2(\cdot, \xi)\|_x$ is continuous and nondecreasing. The assumption that b is in the limit-point case implies that not both $V_1(\cdot, \xi)$ and $V_2(\cdot, \xi)$ belong to $L^2_\Delta(\imath)$. Thus, the limit result follows. □

The function $x \mapsto \varepsilon_\tau(x, \xi)$ appears in the estimate in the following theorem.

Theorem 7.9.4. *Let M_τ be the Weyl function in (7.9.2) corresponding to the boundary triplet $\{\mathbb{C}, \Gamma_0^\tau, \Gamma_1^\tau\}$. Assume that $\xi \in \mathbb{R}$ and let $\varepsilon_\tau(x, \xi)$ be as in Lemma 7.9.3. Then for $a < \beta < x < b$*

$$\frac{1}{d_0} \leq \frac{\|V_2(\cdot, \xi)\|_x}{\|V_1(\cdot, \xi)\|_x} |M_\tau(\xi + i\varepsilon_\tau(x, \xi))| \leq d_0,$$

where $d_0 = 1 + 2\big(\sqrt{2} + \sqrt{2 + \sqrt{2}}\big)$.

Proof. Assume that $\xi \in \mathbb{R}$ and let $\varepsilon > 0$. Define the function $\psi(\cdot, \xi, \varepsilon)$ by

$$\psi(\cdot, \xi, \varepsilon) = V_1(\cdot, \xi) + M_\tau(\xi + i\varepsilon) V_2(\cdot, \xi). \tag{7.9.9}$$

For any $a < x < b$ this leads to

$$\big|\|V_2(\cdot, \xi)\|_x |M_\tau(\xi + i\varepsilon)| - \|V_1(\cdot, \xi)\|_x\big| \leq \|\psi(\cdot, \xi, \varepsilon)\|_x$$

or, equivalently, when $\beta < x < a$

$$\left| \frac{\|V_2(\cdot, \xi)\|_x}{\|V_1(\cdot, \xi)\|_x} |M_\tau(\xi + i\varepsilon)| - 1 \right| \leq \frac{\|\psi(\cdot, \xi, \varepsilon)\|_x}{\|V_1(\cdot, \xi)\|_x}. \tag{7.9.10}$$

The term on the right-hand side of (7.9.10) will now be estimated in a suitable way. First note that it follows from (7.9.6) that for $\lambda \in \mathbb{C}$ and $\mu \in \mathbb{C} \setminus \mathbb{R}$ one obtains

$$V_1(\cdot, \lambda) + M_\tau(\mu) V_2(\cdot, \lambda) - \gamma_\tau(\mu) = (\lambda - \mu) \mathcal{H}(\lambda) \gamma_\tau(\cdot, \mu). \tag{7.9.11}$$

Applying the identity in (7.9.11) with $\lambda = \xi$ and $\mu = \xi + i\varepsilon$ one sees that

$$\psi(\cdot,\xi,\varepsilon) = \gamma_\tau(\cdot,\xi+i\varepsilon) - i\varepsilon\mathcal{H}(\xi)\gamma_\tau(\cdot,\xi+i\varepsilon),$$

which expresses the function $\psi(\cdot,\xi,\varepsilon)$ in (7.9.9) in terms of the γ-field γ_τ. Hence, it follows from Lemma 7.9.2 that

$$\|\psi(\cdot,\xi,\varepsilon)\|_x \leq \big(1+\sqrt{2}\,\varepsilon\,\|V_1(\cdot,\xi)\|_x\|V_2(\cdot,\xi)\|_x\big)\|\gamma_\tau(\cdot,\xi+i\varepsilon)\|_x.$$

Therefore, the right-hand side of (7.9.10) is estimated by

$$\begin{aligned}&\frac{\big(1+\sqrt{2}\,\varepsilon\,\|V_1(\cdot,\xi)\|_x\|V_2(\cdot,\xi)\|_x\big)\,\|\gamma_\tau(\cdot,\xi+i\varepsilon)\|_x}{\|V_1(\cdot,\xi)\|_x}\\ &\quad=\frac{1+\sqrt{2}\,\varepsilon\,\|V_1(\cdot,\xi)\|_x\|V_2(\cdot,\xi)\|_x}{(\|V_1(\cdot,\xi)\|_x\|V_2(\cdot,\xi)\|_x)^{\frac{1}{2}}}\,\frac{\|V_2(\cdot,\xi)\|_x^{\frac{1}{2}}}{\|V_1(\cdot,\xi)\|_x^{\frac{1}{2}}}\,\|\gamma_\tau(\cdot,\xi+i\varepsilon)\|_x.\end{aligned}$$

Now observe that $\|\gamma_\tau(\cdot,\xi+i\varepsilon)\|_x \leq \|\gamma_\tau(\cdot,\xi+i\varepsilon)\|_b$ and it follows from (7.9.3) that

$$\|\gamma_\tau(\cdot,\xi+i\varepsilon)\|_b \leq \sqrt{\frac{\operatorname{Im} M_\tau(\xi+i\varepsilon)}{\varepsilon}} \leq \sqrt{\frac{|M_\tau(\xi+i\varepsilon)|}{\varepsilon}}.$$

Thus, for any $\varepsilon > 0$ and $\beta < x < b$ one obtains the inequality

$$\begin{aligned}&\left|\frac{\|V_2(\cdot,\xi)\|_x}{\|V_1(\cdot,\xi)\|_x}\,|M_\tau(\xi+i\varepsilon)| - 1\right|\\ &\quad\leq \frac{1+\sqrt{2}\,\varepsilon\,\|V_1(\cdot,\xi)\|_x\|V_2(\cdot,\xi)\|_x}{(\varepsilon\,\|V_1(\cdot,\xi)\|_x\|V_2(\cdot,\xi)\|_x)^{\frac{1}{2}}}\left(\frac{\|V_2(\cdot,\xi)\|_x}{\|V_1(\cdot,\xi)\|_x}\,|M_\tau(\xi+i\varepsilon)|\right)^{\frac{1}{2}}.\end{aligned}$$

Now for $\xi \in \mathbb{R}$ and $\beta < x < b$ choose $\varepsilon = \varepsilon_\tau(x,\xi)$ in this estimate. This choice minimizes the first factor on the right-hand side to $2^{5/4}$. Hence, the nonnegative quantity

$$Q = \frac{\|V_2(\cdot,\xi)\|_x}{\|V_1(\cdot,\xi)\|_x}|M_\tau(\xi+i\varepsilon_\tau(x,\xi))|$$

satisfies the inequality

$$|Q-1| \leq 2^{5/4}Q^{\frac{1}{2}}$$

or, equivalently, $Q^2 - 2Q + 1 \leq 4\sqrt{2}Q$. Therefore, $1/d_0 \leq Q \leq d_0$, which completes the proof. □

The following result is now a direct consequence of Theorem 7.9.4.

Theorem 7.9.5. *Let M be the Weyl function corresponding to the boundary triplet $\{\mathbb{C},\Gamma_0,\Gamma_1\}$ and let $\xi \in \mathbb{R}$. Then the following statements hold:*

(i) *If* $\tau \in \mathbb{R}$, *then the solution* $Y_1(\cdot,\xi) + \tau Y_2(\cdot,\xi)$ *of the boundary value problem*

$$Jf' - Hf = \xi \Delta f, \quad f_2(a) = \tau f_1(a),$$

which is unique up to scalar multiples, is subordinate if and only if

$$\lim_{\varepsilon \downarrow 0} M(\xi + i\varepsilon) = \tau.$$

(ii) *If* $\tau = \infty$, *then the solution* $Y_2(\cdot,\xi)$ *of the boundary value problem*

$$Jf' - Hf = \xi \Delta f, \quad f_1(a) = 0,$$

which is unique up to scalar multiples, is subordinate if and only if

$$\lim_{\varepsilon \downarrow 0} M(\xi + i\varepsilon) = \infty.$$

Proof. Since $x \mapsto \varepsilon_\tau(x,\xi)$ is continuous, nonincreasing, and has limit 0 as $x \to b$, one obtains the identity

$$\lim_{\varepsilon \downarrow 0} M_\tau(\xi + i\varepsilon) = \lim_{x \to b} M_\tau(\xi + i\varepsilon_\tau(x,\xi)).$$

(i) Assume that $\tau \in \mathbb{R}$ and note that

$$V_2(\cdot,\xi) = \frac{1}{\sqrt{\tau^2+1}}\big(Y_1(\cdot,\xi) + \tau Y_2(\cdot,\xi)\big)$$

by (7.9.4). It will be shown that $|M_\tau(\xi + i\varepsilon)| \to \infty$ for $\varepsilon \downarrow 0$ if and only if the solution $V_2(\cdot,\xi)$ is subordinate. To see this, assume first that $|M_\tau(\xi + i\varepsilon)| \to \infty$. Then, by Theorem 7.9.4, it follows that

$$\lim_{x \to b} \frac{\|V_2(\cdot,\xi)\|_x}{\|V_1(\cdot,\xi)\|_x} = 0. \tag{7.9.12}$$

Hence, for any $c_1, c_2 \in \mathbb{R}$, $c_1 \neq 0$, one obtains from (7.9.12) that

$$\lim_{x \to b} \frac{\|V_2(\cdot,\xi)\|_x}{\|c_1 V_1(\cdot,\xi) + c_2 V_2(\cdot,\xi)\|_x} = 0, \tag{7.9.13}$$

and therefore the solution $V_2(\cdot,\xi)$ is subordinate. Conversely, assume that $V_2(\cdot,\xi)$ is subordinate, so that (7.9.13) holds for all $c_1, c_2 \in \mathbb{R}$, $c_1 \neq 0$. Then clearly (7.9.12) holds, and therefore it follows from Theorem 7.9.4 that $|M_\tau(\xi + i\varepsilon)| \to \infty$.

It is a consequence of (7.9.2) that for $\varepsilon \downarrow 0$ one has

$$|M_\tau(\xi + i\varepsilon)| \to \infty \quad \Leftrightarrow \quad M(\xi + i\varepsilon) \to \tau.$$

This equivalence leads to the assertion for $\tau \in \mathbb{R}$.

(ii) The case $\tau = \infty$ can be treated in the same way as (i). □

The self-adjoint extension $A_0 = \ker \Gamma_0$ of $T_{\min}$ is given by

$$A_0 = \{\{f, g\} \in T_{\max} : f_1(a) = 0\}; \tag{7.9.14}$$

cf. (7.9.1). The boundary condition

$$f_1(a) = 0 \tag{7.9.15}$$

plays a central role in the following definition, which is based on Theorem 7.9.5; cf. Definition 6.7.6.

Definition 7.9.6. With the canonical system $Jf' - Hf = \xi \Delta f$, $\xi \in \mathbb{R}$, the following subsets of $\mathbb{R}$ are associated:

(i) $\mathcal{M}$ is the complement of the set of all $\xi \in \mathbb{R}$ for which a subordinate solution exists that does not satisfy (7.9.15);

(ii) $\mathcal{M}_{\rm ac}$ is the set of all $\xi \in \mathbb{R}$ for which no subordinate solution exists;

(iii) $\mathcal{M}_{\rm s}$ is the set of all $\xi \in \mathbb{R}$ for which a subordinate solution exists that satisfies (7.9.15);

(iv) $\mathcal{M}_{\rm sc}$ is the set of all $\xi \in \mathbb{R}$ for which a subordinate solution exists that satisfies (7.9.15) and does not belong to $L^2_\Delta(\imath)$;

(v) $\mathcal{M}_{\rm p}$ is the set of all $\xi \in \mathbb{R}$ for which a subordinate solution exists that satisfies (7.9.15) and belongs to $L^2_\Delta(\imath)$.

It is a direct consequence of Definition 7.9.6 that

$$\mathbb{R} = \mathcal{M}^c \sqcup \mathcal{M}_{\rm ac} \sqcup \mathcal{M}_{\rm s}, \quad \mathcal{M} = \mathcal{M}_{\rm ac} \sqcup \mathcal{M}_{\rm s}, \quad \text{and} \quad \mathcal{M}_{\rm s} = \mathcal{M}_{\rm sc} \sqcup \mathcal{M}_{\rm p},$$

where $\sqcup$ stands for disjoint union.

Let the Weyl function M of the boundary triplet $\{\mathbb{C}, \Gamma_0, \Gamma_1\}$ have the integral representation

$$M(\lambda) = \alpha + \beta\lambda + \int_{\mathbb{R}} \left(\frac{1}{t-\lambda} - \frac{t}{t^2+1} \right) d\sigma(t), \tag{7.9.16}$$

where $\alpha \in \mathbb{R}$, $\beta \geq 0$, and the measure σ satisfies

$$\int_{\mathbb{R}} \frac{1}{t^2+1}\, d\sigma(t) < \infty;$$

cf. Theorem A.2.5. The following proposition is based on Corollary 3.1.8, where minimal supports for the various parts of the measure σ in the integral representation of M are described in terms of the boundary behavior of the Nevanlinna function M. The proof of Proposition 7.9.7 is the same as the proof of Proposition 6.7.7 and will not be repeated.

Proposition 7.9.7. *Let M be the Weyl function associated with the boundary triplet $\{\mathbb{C}, \Gamma_0, \Gamma_1\}$ and let σ be the corresponding measure in (7.9.16). Then the sets*

$$\mathcal{M}, \mathcal{M}_{\rm ac}, \mathcal{M}_{\rm s}, \mathcal{M}_{\rm sc}, \mathcal{M}_{\rm p},$$

are minimal supports for the measures

$$\sigma, \sigma_{\rm ac}, \sigma_{\rm s}, \sigma_{\rm sc}, \sigma_{\rm p},$$

respectively.

The minimal supports in Proposition 7.9.7 are intimately connected with the spectrum of A_0. For the absolutely continuous spectrum one obtains in the same way as in Theorem 6.7.8 the following result, where the notion of the absolutely continuous closure of a Borel set from Definition 3.2.4 is used. Similar statements (with an inclusion) can be formulated for the singular parts of the spectrum; cf. Section 3.6.

Theorem 7.9.8. *Let A_0 be the self-adjoint relation in (7.9.14) and let $\mathcal{M}_{\rm ac}$ be as in Definition 7.9.6. Then*

$$\sigma_{\rm ac}(A_0) = {\rm clos}_{\rm ac}(\mathcal{M}_{\rm ac}).$$

7.10 Special classes of canonical systems

In this section two particular types of canonical systems are studied. First it is shown how a class of Sturm–Liouville problems, which are slightly more general than the equations treated in Chapter 6, fit in the framework of canonical systems. In this context the results from the previous sections can be carried over to Sturm–Liouville equations. The second class of canonical systems which is discussed here consists of systems of the form (7.2.3) with $H = 0$. In this situation a simple limit-point/limit-circle criterion is provided.

Weighted Sturm–Liouville equations

Let $\imath \subset \mathbb{R}$ be an open interval. Let $1/p, q, r, s \in \mathcal{L}^1_{\rm loc}(\imath)$ be real functions, assume $r(t) \geq 0$ for almost all $t \in \imath$, and define the 2×2 matrix functions H and Δ by

$$H(t) = \begin{pmatrix} -q(t) & -s(t) \\ -s(t) & 1/p(t) \end{pmatrix} \quad \text{and} \quad \Delta(t) = \begin{pmatrix} r(t) & 0 \\ 0 & 0 \end{pmatrix}, \tag{7.10.1}$$

respectively. Let the 2×2 matrix J be as in (7.2.2). If the vector functions f and g satisfy the canonical system $Jf' - Hf = \Delta g$, then their first components $\mathfrak{f} = f_1$ and $\mathfrak{g} = g_1$ satisfy the weighted Sturm–Liouville equation

$$-(\mathfrak{f}^{[1]})' + s\mathfrak{f}^{[1]} + q\mathfrak{f} = r\mathfrak{g}, \quad \text{where} \quad \mathfrak{f}^{[1]} = p(\mathfrak{f}' + s\mathfrak{f}), \tag{7.10.2}$$

and, since $f \in AC(\imath)$, it follows that $\mathfrak{f}, \mathfrak{f}^{[1]} \in AC(\imath)$. Conversely, if $\mathfrak{f}, \mathfrak{f}^{[1]} \in AC(\imath)$ and $\mathfrak{f}, \mathfrak{g}$ satisfy (7.10.2), then the vector functions

$$f = \begin{pmatrix} \mathfrak{f} \\ \mathfrak{f}^{[1]} \end{pmatrix} \quad \text{and} \quad g = \begin{pmatrix} \mathfrak{g} \\ 0 \end{pmatrix},$$

satisfy the canonical system (7.2.3) with the coefficients given by (7.10.1). Since the functions $1/p, q, r, s$ are assumed to be real, the system $Jf' - Hf = \Delta g$ is real. Moreover, the canonical system corresponding to (7.10.1) is definite, when any solution f of $Jf' - Hf = 0$ which satisfies $\Delta f = 0$ vanishes or, equivalently,

$$-f_2' + qf_1 + sf_2 = 0, \quad f_1' + sf_1 - (1/p)f_2 = 0, \quad rf_1 = 0 \quad \Rightarrow \quad f = 0.$$

In accordance with the definition of definiteness for canonical equations, the Sturm–Liouville equation (7.10.2) is said to be *definite* if

$$-(\mathfrak{f}^{[1]})' + s\mathfrak{f}^{[1]} + q\mathfrak{f} = 0, \quad r\mathfrak{f} = 0 \quad \Rightarrow \quad \mathfrak{f} = 0.$$

In particular, the Sturm–Liouville equation (7.10.2) is definite if the weight function r is positive on an open interval.

The matrix function Δ in (7.10.1) induces the spaces $\mathcal{L}^2_\Delta(\imath)$ and $L^2_\Delta(\imath)$. With the weight r it is natural to introduce the space $\mathcal{L}^2_r(\imath)$ of all complex measurable functions φ for which

$$\int_\imath |\varphi(s)|^2 r(s)\, ds = \int_\imath \varphi(s)^* r(s) \varphi(s)\, ds < \infty.$$

The corresponding semi-inner product is denoted by $(\cdot,\cdot)_r$ and the corresponding Hilbert space of equivalence classes of elements from $\mathcal{L}^2_r(\imath)$ is denoted by $L^2_r(\imath)$. It is clear that the mapping R defined by

$$f = \begin{pmatrix} f_1 \\ f_2 \end{pmatrix} \in \mathcal{L}^2_\Delta(\imath) \mapsto f_1 \in \mathcal{L}^2_r(\imath)$$

is an isometry with respect to the semi-inner products, thanks to the identity

$$(f, f)_\Delta = \int_\imath \begin{pmatrix} f_1(s)^* & f_2(s)^* \end{pmatrix} \begin{pmatrix} r(s) & 0 \\ 0 & 0 \end{pmatrix} \begin{pmatrix} f_1(s) \\ f_2(s) \end{pmatrix} ds = (f_1, f_1)_r.$$

Furthermore, this mapping is onto, since each function in $\mathcal{L}^2_r(\imath)$ can be regarded as the first component of an element in $\mathcal{L}^2_\Delta(\imath)$ with the understanding that the second component can be any measurable function. Therefore, the mapping R induces a unitary operator, again denoted by R, from $L^2_\Delta(\imath)$ onto $L^2_r(\imath)$.

Assume now that the system or, equivalently, the Sturm–Liouville equation is definite. In the Hilbert space $L^2_\Delta(\imath)$ there are the preminimal relation T_0, minimal relation $T_{\min}$, and maximal relation $T_{\max}$ associated with the canonical system $Jf' - Hf = \Delta g$:

$$T_0 \subset \overline{T}_0 = T_{\min} \subset T_{\max} = (T_{\min})^*.$$

Likewise, one can define corresponding relations in the Hilbert space $L^2_r(\imath)$. The maximal relation $\mathfrak{T}_{\max}$ is defined as follows:

$$\mathfrak{T}_{\max} = \big\{\{\mathfrak{f},\mathfrak{g}\} \in L^2_r(\imath) \times L^2_r(\imath) : -(p\mathfrak{f}^{[1]})' + s\mathfrak{f}^{[1]} + q\mathfrak{f} = r\mathfrak{g}\big\},$$

in the sense that there exist representatives $\mathfrak{f}$ and $\mathfrak{g} \in \mathcal{L}^2_r(\imath)$ of $\mathfrak{f}$ and $\mathfrak{g}$, respectively, such that $\mathfrak{f} \in AC(\imath)$, $p\mathfrak{f}^{[1]} \in AC(\imath)$, and (7.10.2) holds. It is clear that the definiteness of the canonical system or, equivalently, of the equation (7.10.2) implies that each element $\mathfrak{f} \in \operatorname{dom}\mathfrak{T}_{\max}$ has a unique representative $\mathfrak{f}$ such that $\mathfrak{f} \in AC(\imath)$, $p\mathfrak{f}^{[1]} \in AC(\imath)$; cf. Lemma 7.6.1. The preminimal relation $\mathfrak{T}_0$ and the minimal relation $\mathfrak{T}_{\min}$ are defined by

$$\mathfrak{T}_0 = \big\{\{\mathfrak{f},\mathfrak{g}\} \in \mathfrak{T}_{\max} : \mathfrak{f} \text{ has compact support}\big\} \quad \text{and} \quad \mathfrak{T}_{\min} = \overline{\mathfrak{T}}_0.$$

It is not difficult to see that the mapping $\widehat{R}$ defined by

$$\widehat{R}\{f,g\} = \{Rf, Rg\}, \quad \{f,g\} \in L^2_\Delta(\imath) \times L^2_\Delta(\imath),$$

takes $T_{\max}$ one-to-one onto $\mathfrak{T}_{\max}$, including absolutely continuous representatives, and that with $\{\mathfrak{f},\mathfrak{g}\} = \widehat{R}\{f,g\}$ and $\{\mathfrak{h},\mathfrak{k}\} = \widehat{R}\{h,k\}$:

$$(g,h)_\Delta - (f,k)_\Delta = (\mathfrak{g},\mathfrak{h})_r - (\mathfrak{f},\mathfrak{k})_r, \quad \{f,g\},\{h,k\} \in T_{\max}. \tag{7.10.3}$$

Similarly, $\widehat{R}$ takes T_0 one-to-one onto $\mathfrak{T}_0$ and hence $\widehat{R}$ takes $T_{\min}$ one-to-one onto $\mathfrak{T}_{\min}$.

In the Hilbert space $L^2_r(\imath)$ the relations $\mathfrak{T}_0$, $\mathfrak{T}_{\min}$, and $\mathfrak{T}_{\max}$ associated with the Sturm–Liouville equation (7.10.2) satisfy

$$\mathfrak{T}_0 \subset \overline{\mathfrak{T}}_0 = \mathfrak{T}_{\min} \subset \mathfrak{T}_{\max} = (\mathfrak{T}_{\min})^*.$$

Furthermore, R maps $\ker(T_{\max} - \lambda)$ one-to-one onto $\ker(\mathfrak{T}_{\max} - \lambda)$. Since the functions p, q, s, and r are real, it follows that the defect numbers of $T_{\min}$ and $\mathfrak{T}_{\min}$ are equal. Let $\{\mathcal{G},\Gamma_0,\Gamma_1\}$ be a boundary triplet for $T_{\max}$ and let $\mathfrak{T}_{\max}$ be the corresponding maximal relation for the Sturm–Liouville operator. Then the mappings Γ_0' and Γ_1' from $\mathfrak{T}_{\max}$ to $\mathcal{G}$ given by

$$\Gamma_0'\{\mathfrak{f},\mathfrak{g}\} = \Gamma_0\{f,g\} \quad \text{and} \quad \Gamma_1'\{\mathfrak{f},\mathfrak{g}\} = \Gamma_1\{f,g\}, \quad \{\mathfrak{f},\mathfrak{g}\} = \widehat{R}\{f,g\}, \tag{7.10.4}$$

form a boundary triplet for $\mathfrak{T}_{\max}$; cf. (7.10.3). The boundary triplet $\{\mathcal{G},\Gamma_0,\Gamma_1\}$ and the one in (7.10.4) have the same Weyl function.

Via the above identification, the discussion and the results for canonical systems with regular, quasiregular, and singular endpoints in Section 7.7 and Section 7.8 remain valid for weighted Sturm–Liouville equations of the form (7.10.2). Note that in the special case where $s(t) = 0$ and $r(t) > 0$ for almost all $t \in \imath$ the Sturm–Liouville expression (7.10.2) coincides with the Sturm–Liouville expression studied in Chapter 6.

Special canonical systems

This subsection is devoted to the special class of canonical differential equations which have the form

$$Jf' = \lambda \Delta f + \Delta g \tag{7.10.5}$$

on an open interval $\imath = (a, b)$, i.e., the class of canonical systems of the form (7.2.3) with $H = 0$. It will be assumed that the system is real and definite on $\imath$. Here definiteness means that the identity $\Delta(t)e = 0$ for some $e \in \mathbb{C}^2$ and all $t \in \imath$ implies $e = 0$. Note that Lemma 7.2.5 shows that any real definite canonical system of the form (7.2.3) can be transformed into the form (7.10.5) with a possible real shift of the eigenvalue parameter. For this class of equations the limit-point and limit-circle classification at an endpoint can be characterized in terms of the integrability of the function Δ.

Theorem 7.10.1. *Let the canonical system (7.10.5) be real and definite, and let the endpoint a be regular. Then there is the alternative:*

(i) *Δ is integrable and the endpoint b is in the limit-circle case;*

(ii) *Δ is not integrable and the endpoint b is in the limit-point case.*

Proof. By assumption, the canonical system (7.10.5) is real and definite, and the endpoint a is regular. If Δ is integrable on $\imath$, then b is quasiregular, which implies that b is in the limit-circle case; see Corollary 7.4.6. Therefore, it suffices to show that if Δ not integrable, then b is in the limit-point case. Hence, assume that Δ is not integrable at b, so that

$$\infty = \int_a^b |\Delta(s)|\, ds \leq \int_a^b \operatorname{tr} \Delta(s)\, ds, \tag{7.10.6}$$

where the estimate $|\Delta(s)| \leq \operatorname{tr} \Delta(s)$ follows from (7.1.6). In order to show that the endpoint b is in the limit-point case one must verify that

$$\lim_{t \to b} h(t)^* J f(t) = 0$$

for all $\{f, g\}, \{h, k\} \in T_{\max}$; cf. Lemma 7.6.8. Since the limit on the left-hand side exists due to Lemma 7.6.4, it suffices to verify the weaker statement

$$\liminf_{t \to b} h(t)^* J f(t) = 0 \tag{7.10.7}$$

for all $\{f, g\}, \{h, k\} \in T_{\max}$. According to Corollary 7.6.6 and the von Neumann formula in Theorem 1.7.11, it then suffices to prove (7.10.7) for elements of the form $f = u_f + v_f$ and $h = u_h + v_h$, where

$$\{u_f, \lambda u_f\}, \{u_h, \lambda u_h\} \in T_{\max} \quad \text{and} \quad \{v_f, \mu v_f\}, \{v_h, \mu v_h\} \in T_{\max}$$

with λ and μ in different half-planes. Finally, by polarization, it is clearly sufficient to show that

$$\liminf_{t\to b} f(t)^* J f(t) = 0 \tag{7.10.8}$$

for $f = u + v$, where $\{u, \lambda u\}, \{v, \mu v\} \in T_{\max}$. The proof of (7.10.8) is carried out in five steps.

Step 1. Let u be a solution of the homogeneous equation $Jy' = \lambda\Delta y$ which satisfies $u \in \mathcal{L}^2_\Delta(\imath)$. Then

$$u(t) = u(a) + \int_a^t J^{-1}\lambda\Delta(s)u(s)\,ds. \tag{7.10.9}$$

Hence, it follows from (7.10.9) as in Lemma 7.1.4 that

$$\begin{aligned}
|u(t)| &\le |u(a)| + |\lambda| \int_a^t |\Delta(s)u(s)|\,ds \\
&\le |u(a)| + |\lambda| \left(\int_a^t |\Delta(s)|\,ds\right)^{\frac{1}{2}} \left(\int_a^t |\Delta(s)^{\frac{1}{2}}u(s)|^2\,ds\right)^{\frac{1}{2}} \\
&\le |u(a)| + |\lambda| \left(\int_a^t |\Delta(s)|\,ds\right)^{\frac{1}{2}} \|u\|_\Delta.
\end{aligned}$$

Due to the estimate $|\Delta(s)| \le \operatorname{tr}\Delta(s)$ one obtains

$$|u(t)| \le |u(a)| + |\lambda| \left(\int_a^t \operatorname{tr}\Delta(s)\,ds\right)^{\frac{1}{2}} \|u\|_\Delta. \tag{7.10.10}$$

It is clear that for a solution v of the homogeneous equation $Jy' = \mu\Delta y$ which satisfies $v \in \mathcal{L}^2_\Delta(\imath)$ one obtains the similar inequality

$$|v(t)| \le |v(a)| + |\mu| \left(\int_a^t \operatorname{tr}\Delta(s)\,ds\right)^{\frac{1}{2}} \|v\|_\Delta. \tag{7.10.11}$$

Step 2. Let u and v be as in Step 1 with λ and μ in different half-planes. Due to the assumption $\int_a^b \operatorname{tr}\Delta(s)\,ds = \infty$ in (7.10.6), one can choose $t_0 > a$ so large that $\int_a^t \operatorname{tr}\Delta(s)\,ds \ge 1$ for $t \ge t_0$. Then it follows from the estimates (7.10.10) and (7.10.11) that

$$|u(t)| \le \left(\int_a^t \operatorname{tr}\Delta(s)\,ds\right)^{\frac{1}{2}} \big(|u(a)| + |\lambda|\,\|u\|_\Delta\big), \quad t \ge t_0,$$

$$|v(t)| \le \left(\int_a^t \operatorname{tr}\Delta(s)\,ds\right)^{\frac{1}{2}} \big(|v(a)| + |\mu|\,\|v\|_\Delta\big), \quad t \ge t_0.$$

Consequently, for $f = u + v$,

$$|f(t)| \le C_f \left(\int_a^t \operatorname{tr} \Delta(s)\, ds \right)^{\frac{1}{2}}, \quad t \ge t_0, \tag{7.10.12}$$

where $C_f = |u(a)| + |\lambda|\, \|u\|_\Delta + |v(a)| + |\mu|\, \|v\|_\Delta$.

Step 3. Define the 2×2 matrix function Δ_0 by

$$\Delta_0(t) = \begin{cases} (\operatorname{tr} \Delta(t))^{-1} \Delta(t), & \operatorname{tr} \Delta(t) \ne 0, \\ \frac{1}{2} I, & \operatorname{tr} \Delta(t) = 0, \end{cases}$$

for almost every $t \in \imath$. Since $\operatorname{tr} \Delta(t) = 0$ implies $\Delta(t) = 0$ (see (7.1.6)), one has $\Delta(t) = (\operatorname{tr} \Delta(t)) \Delta_0(t)$. Then Δ_0 is nonnegative and

$$\Delta_0 = \begin{pmatrix} \alpha & \beta \\ \beta & \delta \end{pmatrix},$$

where the functions α and δ are nonnegative with $\alpha + \delta = 1$, and the function β is real. Define the matrix function Δ_1 by

$$\begin{aligned} \Delta_1 = &\left((\operatorname{sgn} \beta) \alpha^{\frac{1}{2}} \quad \delta^{\frac{1}{2}} \right)^* \left((\operatorname{sgn} \beta) \alpha^{\frac{1}{2}} \quad \delta^{\frac{1}{2}} \right) \\ &- \begin{pmatrix} \alpha & (\operatorname{sgn} \beta) \alpha^{\frac{1}{2}} \delta^{\frac{1}{2}} \\ (\operatorname{sgn} \beta) \alpha^{\frac{1}{2}} \delta^{\frac{1}{2}} & \delta \end{pmatrix}; \end{aligned} \tag{7.10.13}$$

then Δ_1 is nonnegative. Moreover, since $\alpha^{\frac{1}{2}} \delta^{\frac{1}{2}} \ge (\operatorname{sgn} \beta) \beta$ it follows that the matrix

$$2\Delta_0 - \Delta_1 = \begin{pmatrix} \alpha & 2\beta - (\operatorname{sgn} \beta) \alpha^{\frac{1}{2}} \delta^{\frac{1}{2}} \\ 2\beta - (\operatorname{sgn} \beta) \alpha^{\frac{1}{2}} \delta^{\frac{1}{2}} & \delta \end{pmatrix}$$

is nonnegative. Therefore, $\Delta_1(t) \le 2\Delta_0(t)$ and one has the estimate

$$(\operatorname{tr} \Delta(t)) \Delta_1(t) \le 2 (\operatorname{tr} \Delta(t)) \Delta_0(t) = 2\Delta(t) \tag{7.10.14}$$

for almost every $t \in \imath$.

Step 4. It will be shown that

$$\liminf_{t \to b} \left[f(t)^* \Delta_1(t) f(t) \int_a^t \operatorname{tr} \Delta(s)\, ds \right] = 0 \tag{7.10.15}$$

for $f \in \mathcal{L}^2_\Delta(\imath)$ and, in particular, for $f = u + v$ as in Step 2. In fact, assume that (7.10.15) does not hold. Then there exist $a < a' < b$ and $\varepsilon > 0$ such that for all $t \ge a'$

$$\varepsilon \le f(t)^* \Delta_1(t) f(t) \int_a^t \operatorname{tr} \Delta(s)\, ds$$

or, equivalently,

$$\varepsilon \frac{\operatorname{tr}\Delta(t)}{\int_a^t \operatorname{tr}\Delta(s)\,ds} \le f(t)^*\Delta_1(t)f(t)\operatorname{tr}\Delta(t). \tag{7.10.16}$$

Integration of the right-hand side of (7.10.16) together with (7.10.14) lead to

$$\int_{a'}^b f(t)^*\Delta_1(t)f(t)\operatorname{tr}\Delta(t)\,dt \le 2\int_{a'}^b f(t)^*\Delta(t)f(t)\,dt < \infty,$$

while integration of the left-hand side of (7.10.16) gives

$$\varepsilon \int_{a'}^b \frac{\operatorname{tr}\Delta(t)}{\int_a^t \operatorname{tr}\Delta(s)\,ds}\,dt = \varepsilon \int_{a'}^b \frac{d}{dt}\left(\log \int_a^t \operatorname{tr}\Delta(s)\,ds\right)dt = \infty,$$

due to (7.10.6). This contradiction shows that (7.10.15) is valid.

Step 5. It will be shown that for $f = u + v$ as in Step 2 the limit in (7.10.15) implies the limit in (7.10.8). It is helpful to introduce the notation

$$\varphi = \begin{pmatrix}(\operatorname{sgn}\beta)\alpha^{\frac12} & \delta^{\frac12}\end{pmatrix}\begin{pmatrix}f_1\\ f_2\end{pmatrix},$$

so that $|\varphi|^2 = f^*\Delta_1 f$; cf. (7.10.13). Then the limit result (7.10.15) can be written as

$$\liminf_{t\to b}\left[\,|\varphi(t)|^2 \int_a^t \operatorname{tr}\Delta(s)\,ds\right] = 0. \tag{7.10.17}$$

Observe that the term f^*Jf in (7.10.8) is given by

$$f^*Jf = 2i\operatorname{Im}(\overline{f}_2 f_1). \tag{7.10.18}$$

To estimate the term $|\operatorname{Im}(\overline{f}_2 f_1)|$ note that, by the definition of the function φ,

$$\overline{f}_1\varphi = (\operatorname{sgn}\beta)\alpha^{\frac12}\overline{f}_1 f_1 + \delta^{\frac12}\overline{f}_1 f_2 \quad\text{and}\quad \overline{f}_2\varphi = (\operatorname{sgn}\beta)\alpha^{\frac12}\overline{f}_2 f_1 + \delta^{\frac12}\overline{f}_2 f_2.$$

This yields the identities

$$\operatorname{Im}(\overline{f}_1\varphi) = \delta^{\frac12}\operatorname{Im}(\overline{f}_1 f_2) \quad\text{and}\quad \operatorname{Im}(\overline{f}_2\varphi) = (\operatorname{sgn}\beta)\alpha^{\frac12}\operatorname{Im}(\overline{f}_2 f_1).$$

Therefore, it is clear that

$$\delta^{\frac12}|\operatorname{Im}(\overline{f}_1 f_2)| = |\operatorname{Im}(\overline{f}_1\varphi)| \le |f_1|\,|\varphi| \tag{7.10.19}$$

and

$$\alpha^{\frac12}|\operatorname{Im}(\overline{f}_2 f_1)| = |\operatorname{Im}(\overline{f}_2\varphi)| \le |f_2|\,|\varphi|. \tag{7.10.20}$$

Since $\alpha + \delta = 1$, one has $\alpha < \frac12$ if and only if $\delta \ge \frac12$. Note that $x \ge \frac12$ if and only if $1/\sqrt{x} \le \sqrt{2}$, so it follows from (7.10.18) and (7.10.19)–(7.10.20) that

$$|f^*Jf| \le \begin{cases} 2\sqrt{2}\,|\varphi||f_2|, & \alpha \ge \frac12,\\ 2\sqrt{2}\,|\varphi||f_1|, & \delta \ge \frac12.\end{cases}$$

Therefore, if $t \geq t_0$, then (7.10.12) implies that

$$|f(t)^* J f(t)| \leq 2\sqrt{2}\, C_f |\varphi(t)| \left(\int_a^t \operatorname{tr} \Delta(s)\, ds \right)^{\frac{1}{2}}, \quad t \geq t_0.$$

Combined with (7.10.17) this shows that (7.10.8) is satisfied. □

The next corollary follows from Theorem 7.10.1 and (7.1.6).

Corollary 7.10.2. *Let the canonical system (7.10.5) be real and definite, and let the endpoint a be regular. Assume that Δ is trace-normed in the sense that $\operatorname{tr} \Delta = 1$. Then the following alternative holds:*

(i) *if $b \in \mathbb{R}$, then the limit-circle case prevails;*

(ii) *if $b = \infty$, then the limit-point case prevails.*

Next two simple examples for trace-normed canonical systems are discussed.

Example 7.10.3. Let $\imath = (-1, 1)$ and define the matrix function Δ by

$$\Delta(t) = \begin{pmatrix} 1 & 0 \\ 0 & 0 \end{pmatrix}, \quad t \in (-1, 0), \qquad \Delta(t) = \begin{pmatrix} 0 & 0 \\ 0 & 1 \end{pmatrix}, \quad t \in (0, 1).$$

A measurable function $f = (f_1, f_2)^\top$ belongs to $\mathcal{L}^2_\Delta(\imath)$ if and only if

$$\int_{-1}^{0} |f_1(t)|^2\, dt < \infty \quad \text{and} \quad \int_0^1 |f_2(t)|^2\, dt < \infty,$$

and for $f, g \in \mathcal{L}^2_\Delta(\imath)$ the semi-inner product is given by

$$(f, g)_\Delta = \int_{-1}^{0} \bar{g}_1(t) f_1(t)\, dt + \int_0^1 \bar{g}_2(t) f_2(t)\, dt.$$

Hence, an element $f \in \mathcal{L}^2_\Delta(\imath)$ has Δ-norm 0 if and only if

$$\begin{pmatrix} f_1(t) \\ f_2(t) \end{pmatrix} = \begin{pmatrix} 0 \\ f_2(t) \end{pmatrix} \qquad \text{for a.e. } t \in (-1, 0)$$

and

$$\begin{pmatrix} f_1(t) \\ f_2(t) \end{pmatrix} = \begin{pmatrix} f_1(t) \\ 0 \end{pmatrix} \qquad \text{for a.e. } t \in (0, 1),$$

where f_2 on $(-1, 0)$ and f_1 on $(0, 1)$ are completely arbitrary complex measurable functions.

It is straightforward to see that the regular canonical system $Jf' = \Delta g$ is definite. Hence, in the Hilbert space $L^2_\Delta(\imath)$ the maximal relation

$$T_{\max} = \{\{f, g\} \in L^2_\Delta(\imath) \times L^2_\Delta(\imath) : Jf' = \Delta g\}$$

is well defined and for each $\{f,g\} \in T_{\max}$ the equivalence class f contains a unique absolutely continuous representative such that $Jf' = \Delta g$; cf. Lemma 7.6.1. In fact, an absolutely continuous function $f = (f_1, f_2)^\top$ satisfies $Jf' = \Delta g$ with $g \in L^2_\Delta(\imath)$ if and only if

$$\begin{pmatrix} f_1(t) \\ f_2(t) \end{pmatrix} = \begin{pmatrix} \gamma_1 \\ \gamma_2 + \int_t^0 g_1(s)\,ds \end{pmatrix} \qquad \text{for a.e. } t \in (-1,0)$$

and

$$\begin{pmatrix} f_1(t) \\ f_2(t) \end{pmatrix} = \begin{pmatrix} \gamma_1 + \int_0^t g_2(s)\,ds \\ \gamma_2 \end{pmatrix} \qquad \text{for a.e. } t \in (0,1)$$

for some constants $\gamma_1, \gamma_2 \in \mathbb{C}$. From the equality

$$\int_{-1}^1 \left(f(t) - \begin{pmatrix} \gamma_1 \\ \gamma_2 \end{pmatrix} \right)^* \Delta(t) \left(f(t) - \begin{pmatrix} \gamma_1 \\ \gamma_2 \end{pmatrix} \right) dt = 0$$

it follows that $f = (f_1, f_2)^\top$ and $(\gamma_1, \gamma_2)^\top$ are in the same equivalence class in $L^2_\Delta(\imath)$. Therefore,

$$\dim \big(\operatorname{dom} T_{\max}\big) = 2,$$

and the functions $\varphi = (1,0)^\top$ and $\psi = (0,1)^\top$ form an orthonormal system in $\operatorname{dom} T_{\max}$. Furthermore, it follows from the representation of $T_{\min}$ in Lemma 7.7.1 that $\{f,g\} \in T_{\min}$ if and only if

$$\gamma_1 = 0, \quad \gamma_2 = 0, \quad \int_{-1}^0 g_1(t)\,dt = 0, \quad \int_0^1 g_2(t)\,dt = 0.$$

Hence,

$$\operatorname{dom} T_{\min} = \{0\} \quad \text{and} \quad \operatorname{mul} T_{\min} = (\operatorname{dom} T_{\max})^\perp,$$

and

$$\operatorname{mul} T_{\max} = (\operatorname{dom} T_{\min})^\perp = L^2_\Delta(\imath).$$

The boundary mappings in Theorem 7.7.2 are given by

$$\Gamma_0\{f,g\} = \frac{1}{\sqrt{2}} \begin{pmatrix} 2\gamma_1 + \int_0^1 g_2(t)\,dt \\ 2\gamma_2 + \int_{-1}^0 g_1(t)\,dt \end{pmatrix} \quad \text{and} \quad \Gamma_1\{f,g\} = \frac{1}{\sqrt{2}} \begin{pmatrix} \int_{-1}^0 g_1(t)\,dt \\ \int_0^1 g_2(t)\,dt \end{pmatrix}.$$

In order to compute the γ-field and Weyl function corresponding to the boundary triplet $\{\mathbb{C}^2, \Gamma_0, \Gamma_1\}$ fix a fundamental system by $Y(-1,\lambda) = I$. Then

$$Y(t,\lambda) = \begin{pmatrix} 1 & 0 \\ -\lambda t - \lambda & 1 \end{pmatrix} \qquad \text{for a.e. } t \in (-1,0),$$

$$Y(t,\lambda) = \begin{pmatrix} -\lambda^2 t + 1 & \lambda t \\ -\lambda & 1 \end{pmatrix} \qquad \text{for a.e. } t \in (0,1).$$

Hence, it follows from Theorem 7.7.2 that the γ-field γ and the Weyl function M are given by

$$\gamma(\cdot,\lambda) = Y(\cdot,\lambda)\frac{\sqrt{2}}{4-\lambda^2}\begin{pmatrix} 2 & -\lambda \\ \lambda & 2-\lambda^2 \end{pmatrix} \quad \text{and} \quad M(\lambda) = \frac{1}{4-\lambda^2}\begin{pmatrix} 2\lambda & -\lambda^2 \\ -\lambda^2 & 2\lambda \end{pmatrix}.$$

In particular, the poles of M are $\{-2, 2\}$ and hence the spectrum of A_0 consists of the eigenvalues 2 and -2 which both have multiplicity 1.

The next example is a variant of Example 7.10.3 in the limit-point case.

Example 7.10.4. Let $\imath = (-1, \infty)$ and define the matrix function Δ by

$$\Delta(t) = \begin{pmatrix} 1 & 0 \\ 0 & 0 \end{pmatrix}, \quad t \in (-1, 0), \qquad \Delta(t) = \begin{pmatrix} 0 & 0 \\ 0 & 1 \end{pmatrix}, \quad t \in (0, \infty).$$

As in Example 7.10.3, a measurable function $f = (f_1, f_2)^\top$ belongs to $\mathcal{L}^2_\Delta(\imath)$ if and only if

$$\int_{-1}^{0} |f_1(t)|^2\,dt < \infty \quad \text{and} \quad \int_{0}^{\infty} |f_2(t)|^2\,dt < \infty.$$

The semi-inner product and the elements with Δ-norm 0 are as in Example 7.10.3, except that the interval $(0, 1)$ has to be replaced by $(0, \infty)$. Furthermore, the canonical system $Jf' = \Delta g$ is definite and in the limit-point case; cf. Corollary 7.10.2. Hence, the maximal relation

$$T_{\max} = \big\{\{f, g\} \in L^2_\Delta(\imath) \times L^2_\Delta(\imath) : Jf' = \Delta g\big\}$$

is well defined in $L^2_\Delta(\imath)$. In a similar way as in Example 7.10.3 it follows that an absolutely continuous function $f = (f_1, f_2)^\top \in \mathcal{L}^2_\Delta(\imath)$ satisfies $Jf' = \Delta g$ with $g \in \mathcal{L}^2_\Delta(\imath)$ if and only if

$$\begin{pmatrix} f_1(t) \\ f_2(t) \end{pmatrix} = \begin{pmatrix} \gamma_1 \\ \int_t^0 g_1(s)\,ds \end{pmatrix} \qquad \text{for a.e. } t \in (-1, 0)$$

and

$$\begin{pmatrix} f_1(t) \\ f_2(t) \end{pmatrix} = \begin{pmatrix} \gamma_1 + \int_0^t g_2(s)\,ds \\ 0 \end{pmatrix} \qquad \text{for a.e. } t \in (0, \infty)$$

hold for some constant $\gamma_1 \in \mathbb{C}$. The functions $f = (f_1, f_2)^\top$ and $(\gamma_1, 0)^\top$ are in the same equivalence class in $L^2_\Delta(\imath)$ and therefore

$$\dim\big(\operatorname{dom} T_{\max}\big) = 1,$$

and $\operatorname{dom} T_{\max}$ is spanned by the function $\varphi = (1, 0)^\top$. It follows from Lemma 7.8.1 that $\{f, g\} \in T_{\min}$ if and only if

$$\gamma_1 = 0 \quad \text{and} \quad \int_{-1}^{0} g_1(t)\,dt = 0.$$

Hence, $\operatorname{dom} T_{\min} = \{0\}$, and $\operatorname{mul} T_{\min}$ and $\operatorname{mul} T_{\max}$ are related as in Example 7.10.3.

The boundary mappings in Theorem 7.8.2 are given by

$$\Gamma_0\{f,g\} = \gamma_1 \quad \text{and} \quad \Gamma_1\{f,g\} = \int_{-1}^{0} g_1(t)\,dt.$$

To compute the γ-field and Weyl function corresponding to the boundary triplet $\{\mathbb{C}, \Gamma_0, \Gamma_1\}$ use the fundamental system

$$Y(t,\lambda) = \begin{pmatrix} 1 & 0 \\ -\lambda t - \lambda & 1 \end{pmatrix} \qquad \text{for a.e. } t \in (-1,0),$$

$$Y(t,\lambda) = \begin{pmatrix} -\lambda^2 t + 1 & \lambda t \\ -\lambda & 1 \end{pmatrix} \qquad \text{for a.e. } t \in (0,\infty).$$

Clearly, not both colums of $Y(\cdot,\lambda)$ belong to $L^2_\Delta(\imath)$, but the function $\chi(\cdot,\lambda)$ given by

$$\begin{pmatrix} 1 \\ -\lambda t \end{pmatrix} \quad \text{for a.e. } t \in (-1,0), \qquad \begin{pmatrix} 1 \\ 0 \end{pmatrix} \quad \text{for a.e. } t \in (0,\infty),$$

belongs to $L^2_\Delta(\imath)$ and satisfies $J\chi'(\cdot,\lambda) = \lambda\Delta\chi(\cdot,\lambda)$. Hence, by Theorem 7.8.2, the γ-field γ and the Weyl function M are given by

$$\gamma(\cdot,\lambda) = Y(\cdot,\lambda)\begin{pmatrix} 1 \\ \lambda \end{pmatrix} \quad \text{and} \quad M(\lambda) = \lambda.$$

Chapter 8

Schrödinger Operators on Bounded Domains

For the multi-dimensional Schrödinger operator $-\Delta + V$ with a bounded real potential V on a bounded domain $\Omega \subset \mathbb{R}^n$ with a C^2-smooth boundary a boundary triplet and a Weyl function will be constructed. The self-adjoint realizations of $-\Delta + V$ in $L^2(\Omega)$ and their spectral properties will be investigated. One of the main difficulties here is to provide trace mappings on the domain of the maximal realization, in such a way that the second Green identity remains valid in an appropriate form. It is necessary to introduce and study Sobolev spaces on the domain Ω and its boundary $\partial\Omega$, which will be done in Section 8.2; in this context also the rigged Hilbert spaces from Section 8.1 arise as Sobolev spaces and their duals. The minimal and maximal operators, and the Dirichlet and Neumann trace maps on the maximal domain will be discussed in Section 8.3, and in Section 8.4 a boundary triplet and Weyl function for the maximal operator associated with $-\Delta + V$ is provided. The self-adjoint realizations, their spectral properties, and some natural boundary conditions are also discussed in Section 8.4. The class of semibounded self-adjoint realizations of $-\Delta + V$ in $L^2(\Omega)$ and the corresponding semibounded forms are studied in Section 8.5. For this purpose a boundary pair which is compatible with the boundary triplet in Section 8.4 is provided. Orthogonal couplings of Schrödinger operators are treated in Section 8.6 for the model problem in which $\mathbb{R}^n$ decomposes into a bounded C^2-domain Ω_+ and an unbounded component $\Omega_- = \mathbb{R}^n \setminus \overline{\Omega}_+$. Finally, in Section 8.7 the more general setting of Schrödinger operators on bounded Lipschitz domains is briefly discussed.

8.1 Rigged Hilbert spaces

In this preparatory section the notion of rigged Hilbert spaces or Gelfand triples is briefly recalled. For this, let $\mathfrak{G}$ and $\mathfrak{H}$ be Hilbert spaces and assume that $\mathfrak{G}$ is densely and continuously embedded in $\mathfrak{H}$, that is, one has $\mathfrak{G} \subset \mathfrak{H}$ and the

J. Behrndt et al., *Boundary Value Problems, Weyl Functions, and Differential Operators*, Monographs in Mathematics 108,
https://doi.org/10.1007/978-3-030-36714-5_9

embedding operator $\iota : \mathfrak{G} \hookrightarrow \mathfrak{H}$ is continuous with dense range and $\ker \iota = \{0\}$. In the following the dual space $\mathfrak{H}'$ is identified with $\mathfrak{H}$, but the dual space $\mathfrak{G}'$ of antilinear continuous functionals is not identified with $\mathfrak{G}$. Instead, the isometric isomorphism

$$\mathfrak{I} : \mathfrak{G}' \to \mathfrak{G}, \quad g' \mapsto \mathfrak{I}g', \quad \text{where} \quad (\mathfrak{I}g', g)_{\mathfrak{G}} = g'(g), \quad g \in \mathfrak{G}, \tag{8.1.1}$$

is written explicitly whenever used. In fact, in this section the continuous (antilinear) functionals on $\mathfrak{G}$ will be identified via the scalar product in $\mathfrak{H}$. In the following usually the notation

$$\langle g', g \rangle_{\mathfrak{G}' \times \mathfrak{G}} := g'(g) \tag{8.1.2}$$

is employed for the (antilinear) dual pairing in (8.1.1), and when no confusion can arise the index is suppressed, that is, one writes $\langle g', g \rangle = g'(g)$ for (8.1.2).

In the above setting the dual operator of the embedding operator $\iota : \mathfrak{G} \hookrightarrow \mathfrak{H}$ is given by

$$\iota' : \mathfrak{H} \hookrightarrow \mathfrak{G}', \quad (\iota' h)(g) = (h, \iota g)_{\mathfrak{H}}, \quad g \in \mathfrak{G}, \tag{8.1.3}$$

and in terms of the pairing $\langle \cdot, \cdot \rangle$ this means

$$\langle \iota' h, g \rangle = (h, \iota g)_{\mathfrak{H}}, \qquad h \in \mathfrak{H}, \ g \in \mathfrak{G}. \tag{8.1.4}$$

Since the scalar product $(\cdot, \cdot)_{\mathfrak{H}}$ is antilinear in the second argument one has $\langle \iota' h, \lambda g \rangle = \bar{\lambda} \langle \iota' h, g \rangle$ for $\lambda \in \mathbb{C}$, and hence $\iota' h$ is indeed antilinear. Observe that the dual operator ι' in (8.1.3) is continuous since ι is continuous. Moreover, from the identity $\ker \iota' = (\operatorname{ran} \iota)^{\perp_{\mathfrak{H}}}$ it follows that ι' is injective, and the range of ι' is dense in $\mathfrak{G}'$ since $\ker \iota'' = (\operatorname{ran} \iota')^{\perp_{\mathfrak{G}'}}$ and $\iota = \iota''$ as $\mathfrak{G}$ is reflexive. Thus,

$$\mathfrak{G} \overset{\iota}{\hookrightarrow} \mathfrak{H} \overset{\iota'}{\hookrightarrow} \mathfrak{G}' \quad \text{with } \operatorname{ran} \iota \subset \mathfrak{H} \text{ dense and } \operatorname{ran} \iota' \subset \mathfrak{G}' \text{ dense},$$

and since $\mathfrak{G}$ can be viewed as a subspace of $\mathfrak{H}$, and $\mathfrak{H}$ can be viewed as a subspace of $\mathfrak{G}'$, instead of (8.1.4) also the notation

$$\langle h, g \rangle = (h, g)_{\mathfrak{H}}, \qquad h \in \mathfrak{H}, \ g \in \mathfrak{G}, \tag{8.1.5}$$

will be used. The present situation will appear naturally in the context of Sobolev spaces later in this chapter. First the terminology will be fixed in the next definition.

Definition 8.1.1. Let $\mathfrak{G}$ and $\mathfrak{H}$ be Hilbert spaces such that $\mathfrak{G}$ is densely and continuously embedded in $\mathfrak{H}$. Then the triple $\{\mathfrak{G}, \mathfrak{H}, \mathfrak{G}'\}$ is a called a *Gelfand triple* or a *rigged Hilbert space.*

Assume now that $\{\mathfrak{G}, \mathfrak{H}, \mathfrak{G}'\}$ is a Gelfand triple. Since the embedding operator $\iota : \mathfrak{G} \hookrightarrow \mathfrak{H}$ is continuous, one has $\|g\|_{\mathfrak{H}} \le C \|g\|_{\mathfrak{G}}$ for all $g \in \mathfrak{G}$ with the constant

$C = \|\iota\| > 0$. Moreover, as $\mathfrak{G}$ is a Hilbert space it follows from Lemma 5.1.9 that the symmetric form

$$\mathfrak{t}[g_1, g_2] := (g_1, g_2)_{\mathfrak{G}}, \quad \operatorname{dom} \mathfrak{t} = \mathfrak{G},$$

is densely defined and closed in $\mathfrak{H}$ with a positive lower bound. Hence, by the first representation theorem (Theorem 5.1.18) there exists a unique self-adjoint operator T with the same positive lower bound in $\mathfrak{H}$, such that $\operatorname{dom} T \subset \operatorname{dom} \mathfrak{t}$ and

$$(g_1, g_2)_{\mathfrak{G}} = \mathfrak{t}[g_1, g_2] = (Tg_1, g_2)_{\mathfrak{H}}, \qquad g_1 \in \operatorname{dom} T, \; g_2 \in \mathfrak{G}.$$

Moreover, if $R := T^{\frac{1}{2}}$, then the second representation theorem (Theorem 5.1.23) implies $\operatorname{dom} R = \operatorname{dom} \mathfrak{t}$ and

$$(g_1, g_2)_{\mathfrak{G}} = \mathfrak{t}[g_1, g_2] = (Rg_1, Rg_2)_{\mathfrak{H}}, \qquad g_1, g_2 \in \operatorname{dom} R = \mathfrak{G}. \tag{8.1.6}$$

Note that R is a uniformly positive self-adjoint operator in $\mathfrak{H}$.

In the next lemma some more properties of the Gelfand triple $\{\mathfrak{G}, \mathfrak{H}, \mathfrak{G}'\}$ and the operator R are collected.

Lemma 8.1.2. *Let $\{\mathfrak{G}, \mathfrak{H}, \mathfrak{G}'\}$ be a Gelfand triple, let $\mathfrak{I} : \mathfrak{G}' \to \mathfrak{G}$ be the isometric isomorphism in (8.1.1), and let R be the uniformly positive self-adjoint operator in $\mathfrak{H}$ such that (8.1.6) holds. Then the following statements hold:*

(i) *The Hilbert space $\mathfrak{G}'$ coincides with the completion of $\mathfrak{H}$ equipped with the inner product $(R^{-1}\cdot, R^{-1}\cdot)_{\mathfrak{H}}$.*

(ii) *The operators $\iota_+ = R : \mathfrak{G} \to \mathfrak{H}$ and $\iota_- = R\mathfrak{I} : \mathfrak{G}' \to \mathfrak{H}$ are isometric isomorphisms such that*

$$(\iota_- g', \iota_+ g)_{\mathfrak{H}} = \langle g', g \rangle, \qquad g \in \mathfrak{G}, \; g' \in \mathfrak{G}'. \tag{8.1.7}$$

(iii) *For all $h \in \mathfrak{H}$ one has $\iota_- h = R^{-1} h$.*

(iv) *For all $h \in \mathfrak{H}$ and $g \in \mathfrak{G}$ one has $\iota_+ \iota_- h = h$ and $\iota_- \iota_+ g = g$.*

(v) *The operator R^{-2} can be extended by continuity to an isometric operator $\widetilde{R}^{-2} : \mathfrak{G}' \to \mathfrak{G}$ which coincides with the isometric isomorphism $\mathfrak{I} : \mathfrak{G}' \to \mathfrak{G}$.*

Proof. (i) Consider an element $g' \in \mathfrak{G}'$ and assume, in addition, that $g' \in \mathfrak{H}$. Then one has

$$\|g'\|_{\mathfrak{G}'} = \sup_{g \in \mathfrak{G} \setminus \{0\}} \frac{|g'(g)|}{\|g\|_{\mathfrak{G}}} = \sup_{g \in \mathfrak{G} \setminus \{0\}} \frac{|\langle g', g \rangle|}{\|g\|_{\mathfrak{G}}} = \sup_{g \in \mathfrak{G} \setminus \{0\}} \frac{|(g', g)_{\mathfrak{H}}|}{\|g\|_{\mathfrak{G}}},$$

where (8.1.2) was used in the second equality, and $g' \in \mathfrak{H}$ and (8.1.5) were used in the last step. Since R is uniformly positive, one has $R^{-1} \in \mathbf{B}(\mathfrak{H})$, and using (8.1.6) one obtains

$$\|g'\|_{\mathfrak{G}'} = \sup_{g \in \mathfrak{G} \setminus \{0\}} \frac{|(R^{-1}g', Rg)_{\mathfrak{H}}|}{\|Rg\|_{\mathfrak{H}}} = \sup_{h \in \mathfrak{H} \setminus \{0\}} \frac{|(R^{-1}g', h)_{\mathfrak{H}}|}{\|h\|_{\mathfrak{H}}} = \|R^{-1}g'\|_{\mathfrak{H}}.$$

Therefore, $\|g'\|_{\mathfrak{G}'} = \|R^{-1}g'\|_{\mathfrak{H}}$ for all $g' \in \mathfrak{H} \subset \mathfrak{G}'$ and as $\mathfrak{H}$ is dense in $\mathfrak{G}'$ with respect to the norm $\|\cdot\|_{\mathfrak{G}'}$, one concludes that $\mathfrak{G}'$ coincides with the completion of $\mathfrak{H}$ with respect to the norm $\|R^{-1}\cdot\|_{\mathfrak{H}}$.

(ii) Observe that by the definition of ι_+ and (8.1.6) one has

$$\|\iota_+ g\|_{\mathfrak{H}} = \|Rg\|_{\mathfrak{H}} = \|g\|_{\mathfrak{G}}, \qquad g \in \mathfrak{G} = \operatorname{dom}\iota_+ = \operatorname{dom} R,$$

and hence $\iota_+ : \mathfrak{G} \to \mathfrak{H}$ is isometric. Moreover, since R is bijective, it follows that ι_+ is an isometric isomorphism. Similarly, for $g' \in \mathfrak{G}'$ one has

$$\|\iota_- g'\|_{\mathfrak{H}} = \|R\mathfrak{I}g'\|_{\mathfrak{H}} = \|\mathfrak{I}g'\|_{\mathfrak{G}} = \|g'\|_{\mathfrak{G}'},$$

where in the last step it was used that $\mathfrak{I} : \mathfrak{G}' \to \mathfrak{G}$ is an isometric isomorphism. In order to check the identity (8.1.7), let $g' \in \mathfrak{G}'$ and $g \in \mathfrak{G}$. Then (8.1.6) and (8.1.1) imply

$$(\iota_- g', \iota_+ g)_{\mathfrak{H}} = (R\mathfrak{I}g', Rg)_{\mathfrak{H}} = (\mathfrak{I}g', g)_{\mathfrak{G}} = \langle g', g\rangle. \tag{8.1.8}$$

(iii) Let $g' \in \mathfrak{H} \subset \mathfrak{G}'$ and $g \in \mathfrak{G}$. By (8.1.5), one has

$$\langle g', g\rangle = (g', g)_{\mathfrak{H}} = (R^{-1}g', Rg)_{\mathfrak{H}} = (R^{-1}g', \iota_+ g)_{\mathfrak{H}}$$

and comparing this with (8.1.8) it follows that $R^{-1}g' = \iota_- g'$ for all $g' \in \mathfrak{H}$.

(iv) By the definition of ι_+ and (iii) it is clear that $\iota_+\iota_- h = RR^{-1}h = h$. Similarly, $\iota_-\iota_+ g = \iota_- Rg = R^{-1}Rg = g$ for $g \in \mathfrak{G}$ by (iii).

(v) For $h \in \mathfrak{H}$ one has $\|R^{-2}h\|_{\mathfrak{G}} = \|R^{-1}h\|_{\mathfrak{H}} = \|h\|_{\mathfrak{G}'}$ by (8.1.6) and (i), and since $\mathfrak{H}$ is dense in $\mathfrak{G}'$, it follows that R^{-2} admits an extension to an isometric operator $\widetilde{R}^{-2} : \mathfrak{G}' \to \mathfrak{G}$. Moreover, for $h \in \mathfrak{H}$ it follows from the definition of ι_- in (ii) and (iii) that

$$R\mathfrak{I}h = \iota_- h = R^{-1}h, \quad \text{and hence} \quad \mathfrak{I}h = R^{-2}h.$$

Thus $\mathfrak{I}$ and the restriction R^{-2} of $\widetilde{R}^{-2}$ coincide on the dense subspace $\mathfrak{H} \subset \mathfrak{G}'$. This implies $\mathfrak{I} = \widetilde{R}^{-2}$. □

Now a different point of view is taken on Gelfand triples. In the next lemma it is shown that the powers R^s for $s \geq 0$ of a uniformly positive self-adjoint operator R in $\mathfrak{H}$ give rise to Gelfand triples with certain compatibility properties.

Lemma 8.1.3. *Let $\mathfrak{H}$ be a Hilbert space and let R be a uniformly positive self-adjoint operator R in $\mathfrak{H}$. Let $s \geq 0$ and equip $\mathfrak{G}_s := \operatorname{dom} R^s$ with the inner product*

$$(h, k)_{\mathfrak{G}_s} := (R^s h, R^s k)_{\mathfrak{H}}, \qquad h, k \in \operatorname{dom} R^s. \tag{8.1.9}$$

Then $\mathfrak{G}_t \subset \mathfrak{G}_s$ for all $t \geq s \geq 0$ and the following statements hold:

(i) *$\{\mathfrak{G}_s, \mathfrak{H}, \mathfrak{G}_s'\}$ is a Gelfand triple and the assertions in Lemma 8.1.2 hold with R, $\mathfrak{G}$, and $\mathfrak{G}'$ replaced by R^s, $\mathfrak{G}_s$, and $\mathfrak{G}_s'$, respectively.*

(ii) *If $\iota_+ : \mathfrak{G}_1 \to \mathfrak{H}$ and $\iota_- : \mathfrak{G}_1' \to \mathfrak{H}$ denote the isometric isomorphisms corresponding to the Gelfand triple $\{\mathfrak{G}_1, \mathfrak{H}, \mathfrak{G}_1'\}$ such that*

$$(\iota_- g', \iota_+ g)_{\mathfrak{H}} = \langle g', g\rangle_{\mathfrak{G}_1' \times \mathfrak{G}_1}, \quad g' \in \mathfrak{G}_1', \, g \in \mathfrak{G}_1,$$

then their restrictions

$$\iota_+ = R : \mathfrak{G}_{s+1} \to \mathfrak{G}_s \quad \text{and} \quad \iota_- = R^{-1} : \mathfrak{G}_s \to \mathfrak{G}_{s+1}, \quad s \geq 0, \tag{8.1.10}$$

are isometric isomorphisms such that $\iota_+\iota_- g = g$ for $g \in \mathfrak{G}_s$ and $\iota_-\iota_+ l = l$ for $l \in \mathfrak{G}_{s+1}$.

Proof. (i) For $s \geq 0$ the self-adjoint operator R^s is uniformly positive in $\mathfrak{H}$ and hence $\mathfrak{G}_s = \operatorname{dom} R^s$ equipped with the inner product (8.1.9) is a Hilbert space which is dense in $\mathfrak{H}$. Moreover, from $R^{-s} \in \mathbf{B}(\mathfrak{H})$ and (8.1.9) one obtains that

$$\|g\|_{\mathfrak{H}} = \|R^{-s}R^s g\|_{\mathfrak{H}} \leq \|R^{-s}\| \|R^s g\|_{\mathfrak{H}} = \|R^{-s}\| \|g\|_{\mathfrak{G}_s}, \quad g \in \mathfrak{G}_s,$$

which shows that the embedding $\mathfrak{G}_s \hookrightarrow \mathfrak{H}$ is continuous. Therefore, if $\mathfrak{G}_s'$ denotes the dual of $\mathfrak{G}_s$, then $\{\mathfrak{G}_s, \mathfrak{H}, \mathfrak{G}_s'\}$ is a Gelfand triple. Comparing (8.1.9) with (8.1.6) shows that the operator R^s plays the same role as the representing operator of the inner product in (8.1.6). Hence, the assertions of Lemma 8.1.2 are valid with R, $\mathfrak{G}$, and $\mathfrak{G}'$ replaced by R^s, $\mathfrak{G}_s$, and $\mathfrak{G}_s'$, respectively.

(ii) Let $s \geq 0$ and consider $l \in \mathfrak{G}_{s+1} = \operatorname{dom} R^{s+1}$. It follows from (8.1.9) that

$$\|Rl\|_{\mathfrak{G}_s} = \|R^s Rl\|_{\mathfrak{H}} = \|R^{s+1} l\|_{\mathfrak{H}} = \|l\|_{\mathfrak{G}_{s+1}}$$

and hence $\iota_+ = R : \mathfrak{G}_{s+1} \to \mathfrak{G}_s$ is isometric. In order to verify that this mapping is onto let $k \in \mathfrak{G}_s$. Then $k \in \mathfrak{H}$, and as R is bijective, there exists $l \in \operatorname{dom} R$ such that $Rl = k$. Therefore, $l = R^{-1}k$ and as $k \in \mathfrak{G}_s = \operatorname{dom} R^s$ one concludes $l \in \operatorname{dom} R^{s+1} = \mathfrak{G}_{s+1}$. This shows that $\iota_+ = R : \mathfrak{G}_{s+1} \to \mathfrak{G}_s$ is an isometric isomorphism for $s \geq 0$. A similar reasoning shows that $\iota_- = R^{-1} : \mathfrak{G}_s \to \mathfrak{G}_{s+1}$ is an isometric isomorphism for $s \geq 0$. The remaining assertions $\iota_+\iota_- g = g$ for $g \in \mathfrak{G}_s$ and $\iota_-\iota_+ l = l$ for $l \in \mathfrak{G}_{s+1}$ follow immediately from (8.1.10). □

8.2 Sobolev spaces, C^2-domains, and trace operators

In this section Sobolev spaces on $\mathbb{R}^n$, open subsets $\Omega \subset \mathbb{R}^n$, and on the boundaries $\partial\Omega$ of C^2-domains are defined and some of their features are briefly recalled. Furthermore, the mapping properties of the Dirichlet and Neumann trace map on a C^2-domain Ω are recalled and the first Green identity is established.

For $s \geq 0$ the scale of L^2-based Sobolev spaces $H^s(\mathbb{R}^n)$ is defined with the help of the (classical) Fourier transform $\mathcal{F} \in \mathbf{B}(L^2(\mathbb{R}^n))$ by

$$H^s(\mathbb{R}^n) := \bigl\{ f \in L^2(\mathbb{R}^n) : (1 + |\cdot|^2)^{s/2}\mathcal{F}f \in L^2(\mathbb{R}^n) \bigr\}$$

and $H^s(\mathbb{R}^n)$ is equipped with the natural norm

$$\|f\|_{H^s(\mathbb{R}^n)} := \big\|(1+|\cdot|^2)^{s/2}\mathcal{F}f\big\|_{L^2(\mathbb{R}^n)}, \qquad f \in H^s(\mathbb{R}^n),$$

and corresponding scalar product

$$(f,g)_{H^s(\mathbb{R}^n)} := \big((1+|\cdot|^2)^{s/2}\mathcal{F}f, (1+|\cdot|^2)^{s/2}\mathcal{F}g\big)_{L^2(\mathbb{R}^n)}, \quad f,g \in H^s(\mathbb{R}^n).$$

Then the space $H^s(\mathbb{R}^n)$ is a separable Hilbert space for every $s \geq 0$ and one has $H^0(\mathbb{R}^n) = L^2(\mathbb{R}^n)$. It is also useful to note that the space $C_0^\infty(\mathbb{R}^n)$ is dense in $H^s(\mathbb{R}^n)$ for all $s \geq 0$. Since the Fourier transform is a unitary operator in $L^2(\mathbb{R}^n)$, it is clear that

$$\mathcal{R} = \mathcal{F}^{-1}(1+|\cdot|^2)^{1/2}\mathcal{F}$$

is a uniformly positive self-adjoint operator in $L^2(\mathbb{R}^n)$ such that $\operatorname{dom}\mathcal{R} = H^1(\mathbb{R}^n)$. Furthermore, for each $s \geq 0$ one has

$$\mathcal{R}^s = \mathcal{F}^{-1}(1+|\cdot|^2)^{s/2}\mathcal{F}$$

and hence $\mathcal{R}^s$ for $s \geq 0$ is also a uniformly positive self-adjoint operator in $L^2(\mathbb{R}^n)$ such that $\operatorname{dom}\mathcal{R}^s = H^s(\mathbb{R}^n)$. Note that the scalar product in $H^s(\mathbb{R}^n)$ satisfies

$$(f,g)_{H^s(\mathbb{R}^n)} = (\mathcal{R}^s f, \mathcal{R}^s g)_{L^2(\mathbb{R}^n)}, \qquad f,g \in H^s(\mathbb{R}^n),$$

for all $s \geq 0$. In particular, $\mathcal{R}$ plays the same role as the operator R in (8.1.6) and $\mathcal{R}^s$ plays the same role as the operator R^s in (8.1.9). Hence, $\mathcal{R}^s$, $s \geq 0$, gives rise to a Gelfand triple $\{H^s(\mathbb{R}^n), L^2(\mathbb{R}^n), H^{-s}(\mathbb{R}^n)\}$, where $H^{-s}(\mathbb{R}^n)$ denotes the dual space consisting of continuous antilinear functionals on $H^s(\mathbb{R}^n)$. From Lemma 8.1.3 it is now clear that the restrictions $\mathcal{R} : H^{s+1}(\mathbb{R}^n) \to H^s(\mathbb{R}^n)$ and $\mathcal{R}^{-1} : H^s(\mathbb{R}^n) \to H^{s+1}(\mathbb{R}^n)$ are isometric isomorphisms for $s \geq 0$.

For a nonempty open subset $\Omega \subset \mathbb{R}^n$ and $s \geq 0$ define

$$H^s(\Omega) := \big\{f \in L^2(\Omega) : \text{there exists } g \in H^s(\mathbb{R}^n) \text{ such that } f = g|_\Omega\big\}$$

and endow this space with the norm

$$\|f\|_{H^s(\Omega)} := \inf_{\substack{g \in H^s(\mathbb{R}^n)\\ f = g|_\Omega}} \|g\|_{H^s(\mathbb{R}^n)}, \qquad f \in H^s(\Omega). \tag{8.2.1}$$

The space $H^s(\Omega)$ is a separable Hilbert space; the corresponding scalar product will be denoted by $(\cdot,\cdot)_{H^s(\Omega)}$. For $s \geq 0$ the space $C^\infty(\overline{\Omega}) := \{\varphi|_\Omega : \varphi \in C_0^\infty(\mathbb{R}^n)\}$ is dense in $H^s(\Omega)$. The closure of $C_0^\infty(\Omega)$ in $H^s(\Omega)$ is a closed subspace of $H^s(\Omega)$; it is denoted by

$$H_0^s(\Omega) := \overline{C_0^\infty(\Omega)}^{\|\cdot\|_{H^s(\Omega)}}. \tag{8.2.2}$$

In order to define Sobolev spaces on the boundary $\partial\Omega$ of some domain $\Omega \subset \mathbb{R}^n$ assume first that $\phi : \mathbb{R}^{n-1} \to \mathbb{R}$ is a C^2-function. The vectors in $\mathbb{R}^{n-1}$ will be denoted by $x' = (x_1, \dots, x_{n-1})^\top \in \mathbb{R}^{n-1}$ and the notation $(x', x_n)^\top$ is used for $(x_1, \dots, x_n)^\top \in \mathbb{R}^n$. Then the domain

$$\Omega_\phi := \{(x', x_n)^\top \in \mathbb{R}^n : x_n < \phi(x')\} \tag{8.2.3}$$

is called a C^2-*hypograph* and its boundary is given by

$$\partial\Omega_\phi = \{(x', \phi(x'))^\top \in \mathbb{R}^n : x' \in \mathbb{R}^{n-1}\}.$$

For a measurable function $h : \partial\Omega_\phi \to \mathbb{C}$ the surface integral on $\partial\Omega_\phi$ is defined as

$$\int_{\partial\Omega_\phi} h\, d\sigma := \int_{\mathbb{R}^{n-1}} h(x', \phi(x'))\sqrt{1 + |\nabla\phi(x')|^2}\, dx'. \tag{8.2.4}$$

If $\mathbf{1}_B$ denotes the characteristic function of a Borel set $B \subset \partial\Omega_\phi$, then the surface integral in (8.2.4) induces a surface measure

$$\sigma(B) = \int_{\partial\Omega_\phi} \mathbf{1}_B\, d\sigma. \tag{8.2.5}$$

This surface measure also gives rise to the usual L^2-space on $\partial\Omega_\phi$, which will be denoted by $L^2(\partial\Omega_\phi)$. Furthermore, for $s \in [0, 2]$ define the Sobolev space of order s on $\partial\Omega_\phi$ by

$$H^s(\partial\Omega_\phi) := \{h \in L^2(\partial\Omega_\phi) : x' \mapsto h(x', \phi(x')) \in H^s(\mathbb{R}^{n-1})\}$$

and equip $H^s(\partial\Omega_\phi)$ with the corresponding Hilbert space scalar product

$$(h, k)_{H^s(\partial\Omega_\phi)} := \big(h(\cdot, \phi(\cdot)), k(\cdot, \phi(\cdot))\big)_{H^s(\mathbb{R}^{n-1})}, \quad h, k \in H^s(\partial\Omega_\phi). \tag{8.2.6}$$

Note that the operator $V_\phi : H^s(\partial\Omega_\phi) \to H^s(\mathbb{R}^{n-1})$ that maps $h \in H^s(\partial\Omega_\phi)$ to the function $x' \mapsto h(x', \phi(x')) \in H^s(\mathbb{R}^{n-1})$ is an isometric isomorphism.

In the next step the notion of C^2-hypograph is replaced by a bounded domain with a C^2-smooth boundary, that is, the boundary is locally the boundary of a C^2-hypograph.

Definition 8.2.1. A bounded nonempty open subset $\Omega \subset \mathbb{R}^n$ is called a C^2-*domain* if there exist open sets $U_1, \dots, U_l \subset \mathbb{R}^n$ and (possibly up to rotations of coordinates) C^2-hypographs $\Omega_1, \dots, \Omega_l \subset \mathbb{R}^n$ such that

$$\partial\Omega \subset \bigcup_{j=1}^{l} U_j \quad \text{and} \quad \Omega \cap U_j = \Omega_j \cap U_j, \quad j = 1, \dots, l.$$

Let $\Omega \subset \mathbb{R}^n$ be a bounded C^2-domain as in Definition 8.2.1. Then the boundary $\partial\Omega \subset \mathbb{R}^n$ is compact and there exists a partition of unity subordinate to the open cover $\{U_j\}$ of $\partial\Omega$, that is, there exist functions $\eta_j \in C_0^\infty(\mathbb{R}^n)$, $j = 1, \dots, l$, with $\operatorname{supp} \eta_j \subset U_j$ such that $0 \leq \eta_j(x) \leq 1$ for all $x \in \mathbb{R}^n$ and $\sum_{j=1}^l \eta_j(x) = 1$ for all $x \in \partial\Omega$. For a measurable function $h : \partial\Omega \to \mathbb{C}$ the surface integral on $\partial\Omega$ is defined as

$$\int_{\partial\Omega} h \, d\sigma := \sum_{j=1}^{l} \int_{\mathbb{R}^{n-1}} \eta_j(x', \phi_j(x'))h(x', \phi_j(x'))\sqrt{1 + |\nabla\phi_j(x')|^2} \, dx',$$

where the C^2-functions $\phi_j : \mathbb{R}^{n-1} \to \mathbb{R}$ define the C^2-hypographs Ω_j as in (8.2.3) and the possible rotation of coordinates is suppressed. This surface integral induces a surface measure and the notion of an L^2-space $L^2(\partial\Omega)$ in the same way as in (8.2.4) and (8.2.5). In the present setting the Sobolev space $H^s(\partial\Omega)$ for $s \in [0, 2]$ is now defined by

$$H^s(\partial\Omega) := \{h \in L^2(\partial\Omega) : \eta_j h \in H^s(\partial\Omega_j), \, j = 1, \dots, l\}$$

and is equipped with the corresponding Hilbert space scalar product

$$(h, k)_{H^s(\partial\Omega)} = \sum_{j=1}^{l} (\eta_j h, \eta_j k)_{H^s(\partial\Omega_j)}, \quad h, k \in H^s(\partial\Omega). \tag{8.2.7}$$

It follows from the construction that $H^s(\partial\Omega)$ is densely and continuously embedded in $L^2(\partial\Omega)$ for $s \in [0, 2]$. Furthermore, since $\partial\Omega$ is a compact subset of $\mathbb{R}^n$, the embedding

$$H^t(\partial\Omega) \hookrightarrow H^s(\partial\Omega), \qquad 0 \leq s < t \leq 2, \tag{8.2.8}$$

is compact; see, e.g., [774, Theorem 7.10].

For later purposes it is convenient to use an equivalent characterization of the spaces $H^s(\partial\Omega)$ via interpolation; cf. [573, Theorem B.11]. More precisely, as in (8.1.6) it follows that there exists a unique uniformly positive self-adjoint operator Q in $L^2(\partial\Omega)$ such that

$$\operatorname{dom} Q = H^2(\partial\Omega) \quad \text{and} \quad (h, k)_{H^2(\partial\Omega)} = (Qh, Qk)_{L^2(\partial\Omega)} \tag{8.2.9}$$

for all $h, k \in H^2(\partial\Omega)$. It can be shown that the spaces $H^s(\partial\Omega)$ coincide with the domains $\operatorname{dom} Q^{s/2}$ for $s \in [0, 2]$ and that $(Q^{s/2}\cdot, Q^{s/2}\cdot)_{L^2(\partial\Omega)}$ defines a scalar product and equivalent norm in $H^s(\partial\Omega)$. The dual space of the antilinear continuous functionals on $H^s(\partial\Omega)$ is denoted by $H^{-s}(\partial\Omega)$, $s \in [0, 2]$. Then one obtains the following statement from Lemma 8.1.2 and Lemma 8.1.3.

Corollary 8.2.2. *Let $\Omega \subset \mathbb{R}^n$ be a bounded C^2-domain and let $s \in [0, 2]$. Then the following statements hold:*

(i) *$\{H^s(\partial\Omega), L^2(\partial\Omega), H^{-s}(\partial\Omega)\}$ is a Gelfand triple and the assertions in Lemma 8.1.2 hold with R, $\mathfrak{G}$, and $\mathfrak{G}'$ replaced by $Q^{s/2}$, $H^s(\partial\Omega)$, and $H^{-s}(\partial\Omega)$, respectively.*

(ii) *If* $\iota_+ : H^{1/2}(\partial\Omega) \to L^2(\partial\Omega)$ *and* $\iota_- : H^{-1/2}(\partial\Omega) \to L^2(\partial\Omega)$ *denote the isometric isomorphisms from Lemma* 8.1.2 (ii) *corresponding to the Gelfand triple* $\{H^{1/2}(\partial\Omega), L^2(\partial\Omega), H^{-1/2}(\partial\Omega)\}$ *such that*

$$(\iota_-\varphi, \iota_+\psi)_{L^2(\partial\Omega)} = \langle\varphi, \psi\rangle_{H^{-1/2}(\partial\Omega)\times H^{1/2}(\partial\Omega)}$$

holds for $\varphi \in H^{-1/2}(\partial\Omega)$ *and* $\psi \in H^{1/2}(\partial\Omega)$, *then for* $s \in [0, 3/2]$ *their restrictions*

$$\iota_+ = Q^{1/4} : H^{s+1/2}(\partial\Omega) \to H^s(\partial\Omega)$$

and

$$\iota_- = Q^{-1/4} : H^s(\partial\Omega) \to H^{s+1/2}(\partial\Omega)$$

are isometric isomorphisms such that $\iota_+\iota_-\phi = \phi$ *for* $\phi \in H^s(\partial\Omega)$ *and* $\iota_-\iota_+\chi = \chi$ *for* $\chi \in H^{s+1/2}(\partial\Omega)$; *here* Q *is the uniformly positive self-adjoint operator in* (8.2.9).

Assume now that $\Omega \subset \mathbb{R}^n$ is a bounded C^2-domain as in Definition 8.2.1. The weak derivative of order $|\alpha|$ of an L^2-function f is denoted by $D^\alpha f$ in the following; as usual, here $\alpha \in \mathbb{N}_0^n$ stands for a multiindex and $|\alpha| = \alpha_1 + \cdots + \alpha_n$. Then for $k \in \mathbb{N}_0$ one has

$$H^k(\Omega) = \{f \in L^2(\Omega) : D^\alpha f \in L^2(\Omega) \text{ for all } \alpha \in \mathbb{N}_0^n \text{ with } |\alpha| \leq k\}$$

and

$$\|f\|_k := \sum_{|\alpha|\leq k} \|D^\alpha f\|_{L^2(\Omega)}, \quad f \in H^k(\Omega), \tag{8.2.10}$$

is equivalent to the norm on $H^k(\Omega)$ in (8.2.1); cf. [573, Theorem 3.30]. Recall also that for $k \in \mathbb{N}$ there exists $C_k > 0$ such that the *Poincaré inequality*

$$\|f\|_k \leq C_k \sum_{|\alpha|=k} \|D^\alpha f\|_{L^2(\Omega)}, \qquad f \in H_0^k(\Omega), \tag{8.2.11}$$

is valid. In particular, for $f \in C_0^\infty(\Omega)$ and $k = 2$, integration by parts and the Schwarz theorem give

$$\begin{aligned}\sum_{|\alpha|=2} \|D^\alpha f\|^2_{L^2(\Omega)} &= \sum_{|\alpha|=2} (D^\alpha f, D^\alpha f)_{L^2(\Omega)} \\ &= \sum_{j,k=1}^n (\partial_j\partial_k f, \partial_j\partial_k f)_{L^2(\Omega)} \\ &= \sum_{j,k=1}^n (\partial_j^2 f, \partial_k^2 f)_{L^2(\Omega)} \\ &= \|\Delta f\|^2_{L^2(\Omega)},\end{aligned}$$

and this equality extends to all $f \in H_0^2(\Omega)$ by (8.2.2). As a consequence one obtains the following useful fact.

Lemma 8.2.3. *The mapping $f \mapsto \|\Delta f\|_{L^2(\Omega)}$ is a norm on $H_0^2(\Omega)$ which is equivalent to the norms $\|\cdot\|_2$ and $\|\cdot\|_{H^2(\Omega)}$ in (8.2.10) and (8.2.1), respectively.*

Let again $\Omega \subset \mathbb{R}^n$ be a bounded C^2-domain and denote the unit normal vector field pointing outwards of $\partial\Omega$ by ν. The notion of *trace operator* or *trace map* and some of their properties are discussed next. Recall first that the mapping

$$C^\infty(\overline{\Omega}) \ni f \mapsto \left\{ f|_{\partial\Omega}, \frac{\partial f}{\partial \nu}\Big|_{\partial\Omega} \right\} \in H^{3/2}(\partial\Omega) \times H^{1/2}(\partial\Omega)$$

extends by continuity to a continuous operator

$$H^2(\Omega) \ni f \mapsto \{\tau_{\mathrm{D}} f, \tau_{\mathrm{N}} f\} \in H^{3/2}(\partial\Omega) \times H^{1/2}(\partial\Omega), \tag{8.2.12}$$

which is surjective; here

$$\tau_{\mathrm{D}} : H^2(\Omega) \to H^{3/2}(\partial\Omega) \tag{8.2.13}$$

denotes the *Dirichlet trace operator* and

$$\tau_{\mathrm{N}} : H^2(\Omega) \to H^{1/2}(\partial\Omega) \tag{8.2.14}$$

denotes the *Neumann trace operator*. In particular, for all $f \in C^\infty(\overline{\Omega})$ one has

$$\tau_{\mathrm{D}} f = f|_{\Omega} \quad \text{and} \quad \tau_{\mathrm{N}} f = \frac{\partial f}{\partial \nu}|_{\partial\Omega},$$

respectively. With the help of the trace operators one has another useful characterization of the space $H_0^2(\Omega)$ in (8.2.2), namely,

$$H_0^2(\Omega) = \{ f \in H^2(\Omega) : \tau_{\mathrm{D}} f = \tau_{\mathrm{N}} f = 0 \}. \tag{8.2.15}$$

It will also be used that the Dirichlet trace operator $\tau_{\mathrm{D}} : H^2(\Omega) \to H^{3/2}(\partial\Omega)$ admits a continuous surjective extension

$$\tau_{\mathrm{D}}^{(1)} : H^1(\Omega) \to H^{1/2}(\partial\Omega), \tag{8.2.16}$$

which, in analogy to (8.2.15), leads to the characterization

$$H_0^1(\Omega) = \{ f \in H^1(\Omega) : \tau_{\mathrm{D}}^{(1)} f = 0 \}. \tag{8.2.17}$$

Recall next that for $f \in H^2(\Omega)$ and $g \in H^1(\Omega)$ the *first Green identity*

$$(-\Delta f, g)_{L^2(\Omega)} = (\nabla f, \nabla g)_{L^2(\Omega;\mathbb{C}^n)} - \big(\tau_{\mathrm{N}} f, \tau_{\mathrm{D}}^{(1)} g\big)_{L^2(\partial\Omega)} \tag{8.2.18}$$

holds. Note that $\tau_{\mathrm{N}} f, \tau_{\mathrm{D}} g \in H^{1/2}(\partial\Omega)$ by (8.2.14) and (8.2.16). If, in addition, also $g \in H^2(\Omega)$, then one concludes from (8.2.18) the second Green identity

$$(-\Delta f, g)_{L^2(\Omega)} - (f, -\Delta g)_{L^2(\Omega)} = (\tau_{\mathrm{D}} f, \tau_{\mathrm{N}} g)_{L^2(\partial\Omega)} - (\tau_{\mathrm{N}} f, \tau_{\mathrm{D}} g)_{L^2(\partial\Omega)}, \quad (8.2.19)$$

which is valid for all $f, g \in H^2(\Omega)$.

In the next lemma it will be shown that the Neumann trace operator τ_{N} in (8.2.14) admits an extension to the subspace of $H^1(\Omega)$ consisting of all those functions $f \in H^1(\Omega)$ such that $\Delta f \in L^2(\Omega)$, and it turns out that the first Green identity (8.2.18) remains valid in an extended form. Here, and in the following, the expression Δf is understood in the sense of distributions. If, in addition, one has that $\Delta f \in L^2(\Omega)$, then Δf is a regular distribution generated by the function $\Delta f \in L^2(\Omega)$ via

$$(\Delta f)(\varphi) = \int_\Omega (\Delta f)(x)\, \overline{\varphi(x)}\, dx, \qquad \varphi \in C_0^\infty(\Omega). \quad (8.2.20)$$

Lemma 8.2.4. *For $f \in H^1(\Omega)$ with $\Delta f \in L^2(\Omega)$ there exists a unique element $\varphi \in H^{-1/2}(\partial\Omega)$ such that*

$$(-\Delta f, g)_{L^2(\Omega)} = (\nabla f, \nabla g)_{L^2(\Omega;\mathbb{C}^n)} - \langle \varphi, \tau_{\mathrm{D}}^{(1)} g \rangle_{H^{-1/2}(\partial\Omega)\times H^{1/2}(\partial\Omega)} \quad (8.2.21)$$

holds for all $g \in H^1(\Omega)$. In the following the notation $\tau_{\mathrm{N}}^{(1)} f := \varphi$ will be used.

Proof. Notice that there exists a bounded right inverse of $\tau_{\mathrm{D}}^{(1)}$ in (8.2.16), that is, there is a bounded operator $\eta : H^{1/2}(\partial\Omega) \to H^1(\Omega)$ with the property

$$\tau_{\mathrm{D}}^{(1)} \eta\psi = \psi, \qquad \psi \in H^{1/2}(\partial\Omega).$$

For a fixed $f \in H^1(\Omega)$ such that $\Delta f \in L^2(\Omega)$ define the antilinear functional $\varphi : H^{1/2}(\partial\Omega) \to \mathbb{C}$ by

$$\varphi(\psi) := (\nabla f, \nabla \eta\psi)_{L^2(\Omega;\mathbb{C}^n)} + (\Delta f, \eta\psi)_{L^2(\Omega)}, \quad \psi \in H^{1/2}(\partial\Omega). \quad (8.2.22)$$

Then one has

$$\begin{aligned} |\varphi(\psi)| &\leq \|\nabla f\|_{L^2(\Omega;\mathbb{C}^n)} \|\nabla \eta\psi\|_{L^2(\Omega;\mathbb{C}^n)} + \|\Delta f\|_{L^2(\Omega)} \|\eta\psi\|_{L^2(\Omega)} \\ &\leq C \|\eta\psi\|_{H^1(\Omega)} \\ &\leq C' \|\psi\|_{H^{1/2}(\partial\Omega)} \end{aligned}$$

with some constants $C, C' > 0$, and hence $\varphi \in H^{-1/2}(\partial\Omega)$. Thus, (8.2.22) can also be written in the form

$$\langle \varphi, \psi \rangle_{H^{-1/2}(\partial\Omega)\times H^{1/2}(\partial\Omega)} = (\nabla f, \nabla \eta\psi)_{L^2(\Omega;\mathbb{C}^n)} + (\Delta f, \eta\psi)_{L^2(\Omega)}, \quad (8.2.23)$$

where $\psi \in H^{1/2}(\partial\Omega)$. Now let $g \in H^1(\Omega)$ and set $g_0 := g - \eta\tau_{\mathrm{D}}^{(1)} g$. Then it follows from the characterization of the space $H_0^1(\Omega)$ in (8.2.17) that $g_0 \in H_0^1(\Omega)$, and hence (8.2.2) shows that there is a sequence $(g_m) \subset C_0^\infty(\Omega)$ such that $g_m \to g_0$ in $H^1(\Omega)$. It follows that

$$\begin{aligned}(\nabla f, \nabla g_0)_{L^2(\Omega;\mathbb{C}^n)} &= \lim_{m\to\infty} (\nabla f, \nabla g_m)_{L^2(\Omega;\mathbb{C}^n)} \\ &= -\lim_{m\to\infty} (\Delta f, g_m)_{L^2(\Omega)} \\ &= -(\Delta f, g_0)_{L^2(\Omega)}\end{aligned}$$

and one obtains, together with (8.2.23) (and with $\psi = \tau_{\mathrm{D}}^{(1)} g$), that

$$\begin{aligned}(\nabla f, \nabla g)_{L^2(\Omega;\mathbb{C}^n)} &= \bigl(\nabla f, \nabla\bigl(g_0 + \eta\tau_{\mathrm{D}}^{(1)} g\bigr)\bigr)_{L^2(\Omega;\mathbb{C}^n)} \\ &= -(\Delta f, g_0)_{L^2(\Omega)} + \bigl(\nabla f, \nabla\bigl(\eta\tau_{\mathrm{D}}^{(1)} g\bigr)\bigr)_{L^2(\Omega;\mathbb{C}^n)} \\ &= -(\Delta f, g_0)_{L^2(\Omega)} + \bigl\langle\varphi, \tau_{\mathrm{D}}^{(1)} g\bigr\rangle_{H^{-1/2}(\partial\Omega)\times H^{1/2}(\partial\Omega)} - \bigl(\Delta f, \eta\tau_{\mathrm{D}}^{(1)} g\bigr)_{L^2(\Omega)} \\ &= (-\Delta f, g)_{L^2(\Omega)} + \bigl\langle\varphi, \tau_{\mathrm{D}}^{(1)} g\bigr\rangle_{H^{-1/2}(\partial\Omega)\times H^{1/2}(\partial\Omega)}.\end{aligned}$$

This shows that $\varphi \in H^{-1/2}(\partial\Omega)$ in (8.2.22)–(8.2.23) satisfies (8.2.21).

It remains to check that $\varphi \in H^{-1/2}(\partial\Omega)$ in (8.2.21) is unique. Suppose that $\varphi_1, \varphi_2 \in H^{-1/2}(\partial\Omega)$ satisfy (8.2.21). Then

$$\bigl\langle\varphi_1 - \varphi_2, \tau_{\mathrm{D}}^{(1)} g\bigr\rangle_{H^{-1/2}(\partial\Omega)\times H^{1/2}(\partial\Omega)} = 0$$

for all $g \in H^1(\Omega)$. As $\tau_{\mathrm{D}}^{(1)} : H^1(\Omega) \to H^{1/2}(\partial\Omega)$ is surjective it follows that $\varphi_1 - \varphi_2 = 0$ and hence φ in (8.2.21) is unique. □

Remark 8.2.5. The assertion in Lemma 8.2.4 and its proof extend in a natural manner to all $f \in H^1(\Omega)$ such that $-\Delta f \in H^1(\Omega)^*$. In this situation there still exists a unique element $\varphi \in H^{-1/2}(\partial\Omega)$ such that (instead of (8.2.21)) one has the slightly more general first Green identity

$$\langle -\Delta f, g\rangle_{H^1(\Omega)^*\times H^1(\Omega)} = (\nabla f, \nabla g)_{L^2(\Omega;\mathbb{C}^n)} - \bigl\langle\varphi, \tau_{\mathrm{D}}^{(1)} g\bigr\rangle_{H^{-1/2}(\partial\Omega)\times H^{1/2}(\partial\Omega)}$$

for all $g \in H^1(\Omega)$; cf. [573, Lemma 4.3].

8.3 Trace maps for the maximal Schrödinger operator

The differential expression $-\Delta + V$ is considered on a bounded domain Ω, where the function $V \in L^\infty(\Omega)$ is assumed to be real. One then associates with $-\Delta + V$ a preminimal, minimal and maximal operator in $L^2(\Omega)$, which are adjoints of each other. Furthermore, the Dirichlet and Neumann operators are defined via the corresponding sesquilinear form and the first representation theorem, and some of their properties are collected. In the case where Ω is a bounded C^2-domain it is shown in Theorem 8.3.9 and Theorem 8.3.10 that the Dirichlet and Neumann trace operators in the previous section admit continuous extensions to the maximal domain; this is a key ingredient in the construction of a boundary triplet in the next section.

Let $n \geq 2$, let $\Omega \subset \mathbb{R}^n$ be a bounded domain, and assume that the function $V \in L^\infty(\Omega)$ is real. The *preminimal operator* associated to the differential expression $-\Delta + V$ is defined as

$$T_0 = -\Delta + V, \qquad \operatorname{dom} T_0 = C_0^\infty(\Omega).$$

It follows immediately from

$$(T_0 f, f)_{L^2(\Omega)} = (\nabla f, \nabla f)_{L^2(\Omega;\mathbb{C}^n)} + (Vf, f)_{L^2(\Omega)}, \quad f \in \operatorname{dom} T_0,$$

that T_0 is a densely defined symmetric operator in $L^2(\Omega)$ which is bounded from below with $v_- := \operatorname{essinf} V$ as a lower bound, so that $T_0 - v_-$ is nonnegative. Actually, for $f \in \operatorname{dom} T_0$ one has

$$\big((T_0 - v_-)f, f\big)_{L^2(\Omega)} = (\nabla f, \nabla f)_{L^2(\Omega;\mathbb{C}^n)} + \big((V - v_-)f, f\big)_{L^2(\Omega)} \geq \|\nabla f\|^2_{L^2(\Omega;\mathbb{C}^n)}$$

and hence, by the Poincaré inequality (8.2.11),

$$\big((T_0 - v_-)f, f\big)_{L^2(\Omega)} \geq C\|f\|_1^2 \geq C\|f\|^2_{L^2(\Omega)} \tag{8.3.1}$$

with some constant $C > 0$. This shows that $T_0 - v_-$ is uniformly positive.

The closure of T_0 in $L^2(\Omega)$ is the *minimal operator*

$$T_{\min} = -\Delta + V, \qquad \operatorname{dom} T_{\min} = H_0^2(\Omega). \tag{8.3.2}$$

In fact, using Lemma 8.2.3 and the fact that $V \in L^\infty(\Omega)$ one obtains that the graph norm

$$\|\cdot\|_{L^2(\Omega)} + \|T_{\min}\cdot\|_{L^2(\Omega)}$$

is equivalent to the H^2-norm on the closed subspace $H_0^2(\Omega)$ of $H^2(\Omega)$. Hence, $T_{\min}$ is a closed operator in $L^2(\Omega)$ and it follows from (8.2.2) that $\overline{T_0} = T_{\min}$. Therefore, $T_{\min}$ is a densely defined closed symmetric operator in $L^2(\Omega)$ and $T_{\min} - v_-$ is uniformly positive.

Besides the preminimal and minimal operator, also the *maximal operator* $T_{\max}$ associated with $-\Delta + V$ in $L^2(\Omega)$ will be important in the sequel; it is defined by

$$\begin{aligned} T_{\max} &= -\Delta + V, \\ \operatorname{dom} T_{\max} &= \big\{ f \in L^2(\Omega) : -\Delta f + Vf \in L^2(\Omega) \big\}. \end{aligned} \tag{8.3.3}$$

Here the expression Δf for $f \in L^2(\Omega)$ is understood in the distributional sense. Since $V \in L^\infty(\Omega)$, it is clear that $f \in L^2(\Omega)$ belongs to $\operatorname{dom} T_{\max}$ if and only if $\Delta f \in L^2(\Omega)$, that is, the (regular) distribution Δf is generated by an L^2-function; cf. (8.2.20). Observe that $H^2(\Omega) \subset \operatorname{dom} T_{\max}$, and it will also turn out that $H^2(\Omega) \neq \operatorname{dom} T_{\max}$.

Proposition 8.3.1. *Let T_0, $T_{\min}$, and $T_{\max}$ be the preminimal, minimal, and maximal operator associated with $-\Delta + V$ in $L^2(\Omega)$, respectively. Then $\overline{T}_0 = T_{\min}$, and*

$$(T_{\min})^* = T_{\max} \quad \text{and} \quad T_{\min} = (T_{\max})^*. \tag{8.3.4}$$

Proof. It has already been shown above that $\overline{T}_0 = T_{\min}$ holds. In particular, this implies $T_0^* = (T_{\min})^*$ and thus for the first identity in (8.3.4) it suffices to show $T_0^* = T_{\max}$. Furthermore, since multiplication by $V \in L^\infty(\Omega)$ is a bounded operator in $L^2(\Omega)$, it is no restriction to assume $V = 0$ in the following. Let $f \in \operatorname{dom} T_0^*$ and consider $T_0^* f \in L^2(\Omega)$ as a distribution. Then one has for all $\varphi \in C_0^\infty(\Omega) = \operatorname{dom} T_0$

$$(T_0^* f)(\varphi) = (T_0^* f, \overline{\varphi})_{L^2(\Omega)} = (f, T_0\overline{\varphi})_{L^2(\Omega)} = (f, -\Delta\overline{\varphi})_{L^2(\Omega)} = (-\Delta f)(\varphi),$$

and hence $-\Delta f = T_0^* f \in L^2(\Omega)$. Thus, $f \in \operatorname{dom} T_{\max}$ and $T_{\max} f = T_0^* f$. Conversely, for $f \in \operatorname{dom} T_{\max}$ and all $\varphi \in C_0^\infty(\Omega) = \operatorname{dom} T_0$ one has

$$(T_0\varphi, f)_{L^2(\Omega)} = (-\Delta\varphi, f)_{L^2(\Omega)} = (\varphi, -\Delta f)_{L^2(\Omega)},$$

that is, $f \in \operatorname{dom} T_0^*$ and $T_0^* f = -\Delta f = T_{\max} f$. Thus, the first identity in (8.3.4) has been shown. The second identity in (8.3.4) follows by taking adjoints. □

In the following the self-adjoint *Dirichlet realization* A_D and the self-adjoint *Neumann realization* A_N of $-\Delta + V$ in $L^2(\Omega)$ will play an important role. The operators A_D and A_N will be introduced via the corresponding sesquilinear forms using the first representation theorem. More precisely, consider the densely defined forms

$$\mathfrak{t}_D[f,g] = (\nabla f, \nabla g)_{L^2(\Omega;\mathbb{C}^n)} + (Vf, g)_{L^2(\Omega)}, \quad \operatorname{dom} \mathfrak{t}_D = H_0^1(\Omega),$$

and

$$\mathfrak{t}_N[f,g] = (\nabla f, \nabla g)_{L^2(\Omega;\mathbb{C}^n)} + (Vf, g)_{L^2(\Omega)}, \quad \operatorname{dom} \mathfrak{t}_N = H^1(\Omega),$$

in $L^2(\Omega)$. It is easy to see that both forms are semibounded from below and that $v_- = \operatorname{essinf} V$ is a lower bound. The same argument as in (8.3.1) using the Poincaré inequality (8.2.11) on $\operatorname{dom} \mathfrak{t}_D = H_0^1(\Omega)$ implies the stronger statement that the form $\mathfrak{t}_D - v_-$ is uniformly positive. Furthermore, it follows from the definitions that the form $(\nabla\cdot, \nabla\cdot)_{L^2(\Omega;\mathbb{C}^n)}$ defined on $H_0^1(\Omega)$ or $H^1(\Omega)$ is closed in $L^2(\Omega)$; cf. Lemma 5.1.9. Since $V \in L^\infty(\Omega)$, it is clear that the form $(V\cdot,\cdot)_{L^2(\Omega)}$ is bounded on $L^2(\Omega)$ and hence it follows from Theorem 5.1.16 that also the forms $\mathfrak{t}_D$ and $\mathfrak{t}_N$ are closed in $L^2(\Omega)$. Therefore, by the first representation theorem (Theorem 5.1.18), there exist unique semibounded self-adjoint operators A_D and A_N in $L^2(\Omega)$ associated with $\mathfrak{t}_D$ and $\mathfrak{t}_N$, respectively, such that

$$(A_D f, g)_{L^2(\Omega)} = \mathfrak{t}_D[f,g] \quad \text{for } f \in \operatorname{dom} A_D,\ g \in \operatorname{dom} \mathfrak{t}_D,$$

and

$$(A_N f, g)_{L^2(\Omega)} = \mathfrak{t}_N[f,g] \quad \text{for } f \in \operatorname{dom} A_N,\ g \in \operatorname{dom} \mathfrak{t}_N.$$

The self-adjoint operators A_{D} and A_{N} are called the *Dirichlet operator* and *Neumann operator*, respectively. In the next propositions some properties of these operators are discussed.

Proposition 8.3.2. *Assume that $\Omega \subset \mathbb{R}^n$ is a bounded domain. Then the Dirichlet operator A_{D} is given by*

$$\begin{aligned} A_{\mathrm{D}} f &= -\Delta f + Vf, \\ \operatorname{dom} A_{\mathrm{D}} &= \bigl\{ f \in H_0^1(\Omega) : -\Delta f + Vf \in L^2(\Omega) \bigr\}, \end{aligned} \tag{8.3.5}$$

and for all $\lambda \in \rho(A_{\mathrm{D}})$ the resolvent $(A_{\mathrm{D}} - \lambda)^{-1}$ is a compact operator in $L^2(\Omega)$. The Dirichlet operator A_{D} coincides with the Friedrichs extension S_{F} of the minimal operator $T_{\min}$ in (8.3.2). In particular, $A_{\mathrm{D}} - v_-$ is uniformly positive. Furthermore, if $\Omega \subset \mathbb{R}^n$ is a bounded C^2-domain, then the Dirichlet operator A_{D} is given by

$$\begin{aligned} A_{\mathrm{D}} f &= -\Delta f + Vf, \\ \operatorname{dom} A_{\mathrm{D}} &= \bigl\{ f \in H^1(\Omega) : -\Delta f + Vf \in L^2(\Omega),\ \tau_{\mathrm{D}}^{(1)} f = 0 \bigr\}. \end{aligned} \tag{8.3.6}$$

Proof. Observe that for $f \in \operatorname{dom} A_{\mathrm{D}}$ and $g \in C_0^\infty(\Omega) \subset \operatorname{dom} \mathfrak{t}_{\mathrm{D}}$ one has

$$(A_{\mathrm{D}} f, g)_{L^2(\Omega)} = \mathfrak{t}_{\mathrm{D}}[f, g] = (\nabla f, \nabla g)_{L^2(\Omega;\mathbb{C}^n)} + (Vf, g)_{L^2(\Omega)} = \bigl(-\Delta f + Vf\bigr)(\bar{g}),$$

where $-\Delta f + Vf$ is viewed as a distribution. Since this identity holds for all $g \in C_0^\infty(\Omega)$ and A_{D} is an operator in $L^2(\Omega)$ it follows that

$$-\Delta f + Vf = A_{\mathrm{D}} f \in L^2(\Omega).$$

Therefore, A_{D} is given by (8.3.5). In the case that $\Omega \subset \mathbb{R}^n$ has a C^2-smooth boundary the form of the domain of A_{D} in (8.3.6) follows from (8.3.5) and (8.2.17).

Next it will be shown that for $\lambda \in \rho(A_{\mathrm{D}})$ the resolvent $(A_{\mathrm{D}} - \lambda)^{-1}$ is a compact operator in $L^2(\Omega)$. For this observe first that

$$(A_{\mathrm{D}} - \lambda)^{-1} : L^2(\Omega) \to H_0^1(\Omega), \quad \lambda \in \rho(A_{\mathrm{D}}), \tag{8.3.7}$$

is everywhere defined and closed as an operator from $L^2(\Omega)$ into $H_0^1(\Omega)$. In fact, if $f_n \to f$ in $L^2(\Omega)$ and $(A_{\mathrm{D}} - \lambda)^{-1} f_n \to h$ in $H_0^1(\Omega)$, then $(A_{\mathrm{D}} - \lambda)^{-1} f_n \to h$ in $L^2(\Omega)$, and since the operator $(A_{\mathrm{D}} - \lambda)^{-1}$ is everywhere defined and continuous in $L^2(\Omega)$, it is clear that $(A_{\mathrm{D}} - \lambda)^{-1} f = h$. Hence, the operator in (8.3.7) is bounded by the closed graph theorem. By Rellich's theorem the embedding $H_0^1(\Omega) \hookrightarrow L^2(\Omega)$ is compact, and it follows that $(A_{\mathrm{D}} - \lambda)^{-1}$, $\lambda \in \rho(A_{\mathrm{D}})$, is a compact operator in $L^2(\Omega)$.

It remains to verify that A_{D} is the Friedrichs extension S_{F} of $T_{\min} = \overline{T_0}$ or, equivalenty, the Friedrichs extension of T_0; cf. Lemma 5.3.1 and Definition 5.3.2. For this, consider the form $\mathfrak{t}_{T_0}[f, g] = (T_0 f, g)_{L^2(\Omega)}$, defined for $f, g \in \operatorname{dom} T_0$, and note that

$$\mathfrak{t}_{T_0}[f, g] = (\nabla f, \nabla g)_{L^2(\Omega;\mathbb{C}^n)} + (Vf, g)_{L^2(\Omega)}, \quad \operatorname{dom} \mathfrak{t}_{T_0} = C_0^\infty(\Omega).$$

Observe that $f_m \to_{\mathfrak{t}_{T_0}} f$ if and only if $f_m \to f$ in $L^2(\Omega)$ and (∇f_m) is a Cauchy sequence in $L^2(\Omega;\mathbb{C}^n)$. Hence, $f_m \to_{\mathfrak{t}_{T_0}} f$ implies $f_m \to f$ in the norm of $H^1(\Omega)$, and so $f \in H_0^1(\Omega)$ by (8.2.2). Therefore, by (5.1.16), the closure of the form $\mathfrak{t}_{T_0}$ is given by

$$\widetilde{\mathfrak{t}}_{T_0}[f,g] = \lim_{m\to\infty} \mathfrak{t}_{T_0}[f_m,g_m] = (\nabla f,\nabla g)_{L^2(\Omega;\mathbb{C}^n)} + (Vf,g)_{L^2(\Omega)},$$

where $f,g \in H_0^1(\Omega)$ and $f_m \to_{\mathfrak{t}_{T_0}} f$, $g_m \to_{\mathfrak{t}_{T_0}} g$. Hence, $\widetilde{\mathfrak{t}}_{T_0} = \mathfrak{t}_{\mathrm{D}}$, and since by Definition 5.3.2 the Friedrichs extension of T_0 is the unique self-adjoint operator corresponding to the closed form $\widetilde{\mathfrak{t}}_{T_0}$, the assertion follows. □

In order to specify the Neumann operator A_{N}, the first Green identity and the trace operators $\tau_{\mathrm{D}}^{(1)} : H^1(\Omega) \to H^{1/2}(\partial\Omega)$ and $\tau_{\mathrm{N}}^{(1)} : H^1(\Omega) \to H^{-1/2}(\partial\Omega)$ will be used; cf. Lemma 8.2.4. For this reason in the next proposition it is assumed that $\Omega \subset \mathbb{R}^n$ is a bounded C^2-domain. It is also important to note that the Neumann operator A_{N} below differs from the Kreĭn–von Neumann extension and the Kreĭn type extensions in Definition 5.4.2; cf. Section 8.5 for more details.

Proposition 8.3.3. *Assume that $\Omega \subset \mathbb{R}^n$ is a bounded C^2-domain. Then the Neumann operator A_{N} is given by*

$$\begin{aligned} A_{\mathrm{N}} f &= -\Delta f + Vf, \\ \operatorname{dom} A_{\mathrm{N}} &= \bigl\{ f \in H^1(\Omega) : -\Delta f + Vf \in L^2(\Omega),\ \tau_{\mathrm{N}}^{(1)} f = 0 \bigr\}, \end{aligned} \tag{8.3.8}$$

and for all $\lambda \in \rho(A_{\mathrm{N}})$ the resolvent $(A_{\mathrm{N}} - \lambda)^{-1}$ is a compact operator in $L^2(\Omega)$.

Proof. In a first step it follows for $f \in \operatorname{dom} A_{\mathrm{N}} \subset H^1(\Omega)$ and all $g \in C_0^\infty(\Omega)$ in the same way as in the proof of Proposition 8.3.2 that

$$(A_{\mathrm{N}} f,g)_{L^2(\Omega)} = \mathfrak{t}_{\mathrm{N}}[f,g] = (\nabla f,\nabla g)_{L^2(\Omega;\mathbb{C}^n)} + (Vf,g)_{L^2(\Omega)} = (-\Delta f + Vf)(\bar g),$$

and hence $A_{\mathrm{N}} f = (-\Delta + V)f \in L^2(\Omega)$. In particular, for $f \in \operatorname{dom} A_{\mathrm{N}}$ one has $f \in H^1(\Omega)$ and $-\Delta f \in L^2(\Omega)$, so that Lemma 8.2.4 applies and yields

$$\begin{aligned} (A_{\mathrm{N}} f,g)_{L^2(\Omega)} &= \mathfrak{t}_{\mathrm{N}}[f,g] \\ &= (\nabla f,\nabla g)_{L^2(\Omega;\mathbb{C}^n)} + (Vf,g)_{L^2(\Omega)} \\ &= \bigl((-\Delta + V)f,g\bigr)_{L^2(\Omega)} + \bigl\langle \tau_{\mathrm{N}}^{(1)} f, \tau_{\mathrm{D}}^{(1)} g \bigr\rangle_{H^{-1/2}(\partial\Omega)\times H^{1/2}(\partial\Omega)} \end{aligned}$$

for all $g \in \operatorname{dom} \mathfrak{t}_{\mathrm{N}} = H^1(\Omega)$. As $A_{\mathrm{N}} f = (-\Delta + V)f$, one concludes that

$$\bigl\langle \tau_{\mathrm{N}}^{(1)} f, \tau_{\mathrm{D}}^{(1)} g \bigr\rangle_{H^{-1/2}(\partial\Omega)\times H^{1/2}(\partial\Omega)} = 0 \quad \text{for all } g \in H^1(\Omega).$$

Since $\tau_{\mathrm{D}}^{(1)} : H^1(\Omega) \to H^{1/2}(\partial\Omega)$ is surjective, it follows that $\tau_{\mathrm{N}}^{(1)} f = 0$. This implies the representation (8.3.8).

To show that the resolvent $(A_N - \lambda)^{-1}$ is a compact operator in $L^2(\Omega)$ one argues in the same way as in the proof of Proposition 8.3.2. In fact, the operator

$$(A_N - \lambda)^{-1} : L^2(\Omega) \to H^1(\Omega), \quad \lambda \in \rho(A_N),$$

is everywhere defined and closed, and hence bounded by the closed graph theorem. Since $\Omega \subset \mathbb{R}^n$ is a bounded C^2-domain, the embedding $H^1(\Omega) \hookrightarrow L^2(\Omega)$ is compact and this implies that $(A_N - \lambda)^{-1}$, $\lambda \in \rho(A_N)$, is a compact operator in $L^2(\Omega)$. □

It is known that functions f in $\operatorname{dom} A_D$ or $\operatorname{dom} A_N$ are locally H^2-regular, that is, for every compact subset $K \subset \Omega$ the restriction of f to K is in $H^2(K)$. The next theorem is an important elliptic regularity result which ensures H^2-regularity of the functions in $\operatorname{dom} A_D$ or $\operatorname{dom} A_N$ in (8.3.6) and (8.3.8), respectively, up to the boundary if the bounded domain Ω is C^2-smooth in the sense of Definition 8.2.1.

Theorem 8.3.4. *Assume that $\Omega \subset \mathbb{R}^n$ is a bounded C^2-domain. Then one has*

$$A_D f = -\Delta f + Vf, \quad \operatorname{dom} A_D = \{f \in H^2(\Omega) : \tau_D f = 0\},$$

and

$$A_N f = -\Delta f + Vf, \quad \operatorname{dom} A_N = \{f \in H^2(\Omega) : \tau_N f = 0\}.$$

Note that under the assumptions in Theorem 8.3.4 the domain of the Dirichlet operator A_D is $H^2(\Omega) \cap H^1_0(\Omega)$; cf. (8.2.17). The direct sum decompositions in the next corollary follow immediately from Theorem 1.7.1 when considering the operator $T = -\Delta + V$, $\operatorname{dom} T = H^2(\Omega)$, and taking into account that $A_D \subset T$ and $A_N \subset T$.

Corollary 8.3.5. *Assume that $\Omega \subset \mathbb{R}^n$ is a bounded C^2-domain and denote by $\tau_D : H^2(\Omega) \to H^{3/2}(\partial\Omega)$ and $\tau_N : H^2(\Omega) \to H^{1/2}(\partial\Omega)$ the Dirichlet and Neumann trace operator in (8.2.13) and (8.2.14), respectively. Then for $\lambda \in \rho(A_D)$ one has the direct sum decomposition*

$$\begin{aligned} H^2(\Omega) &= \operatorname{dom} A_D \dotplus \{f_\lambda \in H^2(\Omega) : (-\Delta + V)f_\lambda = \lambda f_\lambda\} \\ &= \ker \tau_D \dotplus \{f_\lambda \in H^2(\Omega) : (-\Delta + V)f_\lambda = \lambda f_\lambda\}, \end{aligned} \tag{8.3.9}$$

and for $\lambda \in \rho(A_N)$ one has the direct sum decomposition

$$\begin{aligned} H^2(\Omega) &= \operatorname{dom} A_N \dotplus \{f_\lambda \in H^2(\Omega) : (-\Delta + V)f_\lambda = \lambda f_\lambda\} \\ &= \ker \tau_N \dotplus \{f_\lambda \in H^2(\Omega) : (-\Delta + V)f_\lambda = \lambda f_\lambda\}. \end{aligned} \tag{8.3.10}$$

As a consequence of the decomposition (8.3.9) in Corollary 8.3.5 and (8.2.12) one concludes that the so-called Dirichlet-to-Neumann map in the next definition is a well-defined operator from $H^{3/2}(\partial\Omega)$ into $H^{1/2}(\partial\Omega)$.

Definition 8.3.6. Let $\Omega \subset \mathbb{R}^n$ be a bounded C^2-domain, let A_D be the self-adjoint Dirichlet operator, and let $\tau_D : H^2(\Omega) \to H^{3/2}(\partial\Omega)$ and $\tau_N : H^2(\Omega) \to H^{1/2}(\partial\Omega)$

be the Dirichlet and Neumann trace operator in (8.2.13) and (8.2.14), respectively. For $\lambda \in \rho(A_{\mathrm{D}})$ the *Dirichlet-to-Neumann map* is defined as

$$D(\lambda) : H^{3/2}(\partial\Omega) \to H^{1/2}(\partial\Omega), \qquad \tau_{\mathrm{D}} f_\lambda \mapsto \tau_{\mathrm{N}} f_\lambda,$$

where $f_\lambda \in H^2(\Omega)$ is such that $(-\Delta + V) f_\lambda = \lambda f_\lambda$.

Note that for $\lambda \in \rho(A_{\mathrm{D}}) \cap \rho(A_{\mathrm{N}})$ both decompositions (8.3.9) and (8.3.10) in Corollary 8.3.5 hold and together with (8.2.12) this implies that the Dirichlet-to-Neumann map $D(\lambda)$ is a bijective operator from $H^{3/2}(\partial\Omega)$ onto $H^{1/2}(\partial\Omega)$.

A further useful consequence of Theorem 8.3.4 is given by the following a priori estimates.

Corollary 8.3.7. *Assume that $\Omega \subset \mathbb{R}^n$ is a bounded C^2-domain and let A_{D} and A_{N} be the Dirichlet and Neumann operator, respectively. Then there exist constants $C_{\mathrm{D}} > 0$ and $C_{\mathrm{N}} > 0$ such that*

$$\|f\|_{H^2(\Omega)} \leq C_{\mathrm{D}} \big(\|f\|_{L^2(\Omega)} + \|A_{\mathrm{D}} f\|_{L^2(\Omega)} \big), \quad f \in \operatorname{dom} A_{\mathrm{D}},$$

and

$$\|g\|_{H^2(\Omega)} \leq C_{\mathrm{N}} \big(\|g\|_{L^2(\Omega)} + \|A_{\mathrm{N}} g\|_{L^2(\Omega)} \big), \quad g \in \operatorname{dom} A_{\mathrm{N}}.$$

Proof. One verifies in the same way as in the proof of Proposition 8.3.2 that for $\lambda \in \rho(A_{\mathrm{D}})$ the operator $(A_{\mathrm{D}} - \lambda)^{-1} : L^2(\Omega) \to H^2(\Omega)$ is everywhere defined and closed, and hence bounded. For $f \in \operatorname{dom} A_{\mathrm{D}}$ choose $h \in L^2(\Omega)$ such that $f = (A_{\mathrm{D}} - \lambda)^{-1} h$. Then

$$\|f\|_{H^2(\Omega)} = \|(A_{\mathrm{D}} - \lambda)^{-1} h\|_{H^2(\Omega)} \leq C \|h\|_{L^2(\Omega)} = C_{\mathrm{D}} \|(A_{\mathrm{D}} - \lambda) f\|_{L^2(\Omega)}$$

for some $C > 0$ and $C_{\mathrm{D}} > 0$, and the first estimate follows. The second estimate is proved in the same way. □

The next lemma is an important ingredient in the following.

Lemma 8.3.8. *Let $T_{\max}$ be the maximal operator associated to $-\Delta + V$ in (8.3.3). Then the space $C^\infty(\overline{\Omega})$ is dense in $\operatorname{dom} T_{\max}$ with respect to the graph norm.*

Proof. Since $V \in L^\infty(\Omega)$ is bounded, the graph norms

$$\big(\|\cdot\|^2_{L^2(\Omega)} + \|T_{\max} \cdot\|^2_{L^2(\Omega)} \big)^{1/2} \quad \text{and} \quad \big(\|\cdot\|^2_{L^2(\Omega)} + \|\Delta \cdot\|^2_{L^2(\Omega)} \big)^{1/2}$$

are equivalent on $\operatorname{dom} T_{\max}$, and hence it is no restriction to assume that $V = 0$. Now suppose that $f \in \operatorname{dom} T_{\max}$ is such that for all $g \in C^\infty(\overline{\Omega})$

$$0 = (f, g)_{L^2(\Omega)} + (\Delta f, \Delta g)_{L^2(\Omega)}. \tag{8.3.11}$$

Then (8.3.11) holds for all $g \in C_0^\infty(\Omega)$, so that $0 = (f + \Delta^2 f)(\bar{g})$, where $f + \Delta^2 f$ is viewed as a distribution. As $f \in L^2(\Omega)$, one concludes that

$$\Delta^2 f = -f \in L^2(\Omega). \tag{8.3.12}$$

Next it will be shown that

$$\Delta f \in H_0^2(\Omega). \tag{8.3.13}$$

In fact, choose an open ball B such that $\overline{\Omega} \subset B$ and let $h \in C_0^\infty(B)$. Let

$$\widetilde{A}_D = -\Delta, \qquad \operatorname{dom} \widetilde{A}_D = H^2(B) \cap H_0^1(B),$$

be the self-adjoint Dirichlet Laplacian in $L^2(B)$; cf. Theorem 8.3.4. Since B is bounded, one has $0 \in \rho(\widetilde{A}_D)$ by Proposition 8.3.2. As $h \in C_0^\infty(B)$, elliptic regularity yields $\widetilde{A}_D^{-1} h \in C^\infty(B)$ and hence $(\widetilde{A}_D^{-1} h)|_\Omega \in C^\infty(\overline{\Omega})$ for the restriction onto Ω. Denote by $\widetilde{f}$ and $\widetilde{\Delta f}$ the extension of f and Δf by zero to B. Then it follows with the help of (8.3.11) that

$$\begin{aligned}
(\widetilde{A}_D^{-1} \widetilde{f}, h)_{L^2(B)} &= (\widetilde{f}, \widetilde{A}_D^{-1} h)_{L^2(B)} \\
&= \bigl(f, (\widetilde{A}_D^{-1} h)|_\Omega\bigr)_{L^2(\Omega)} \\
&= -\bigl(\Delta f, \Delta(\widetilde{A}_D^{-1} h)|_\Omega\bigr)_{L^2(\Omega)} \\
&= (-\widetilde{\Delta f}, h)_{L^2(B)}
\end{aligned}$$

holds for $h \in C_0^\infty(B)$. This yields $-\widetilde{\Delta f} = \widetilde{A}_D^{-1} \widetilde{f} \in H^2(B)$. Moreover, as $\widetilde{\Delta f}$ vanishes outside of Ω it follows that $\Delta f \in H_0^2(\Omega)$, that is, (8.3.13) holds.

Now choose a sequence $(\psi_k) \subset C_0^\infty(\Omega)$ such that $\psi_k \to \Delta f$ in $H^2(\Omega)$. Then, by (8.3.12),

$$\begin{aligned}
0 \le (\Delta f, \Delta f)_{L^2(\Omega)} &= \lim_{k\to\infty} (\psi_k, \Delta f)_{L^2(\Omega)} = \lim_{k\to\infty} (\Delta \psi_k, f)_{L^2(\Omega)} \\
&= (\Delta^2 f, f)_{L^2(\Omega)} = -(f, f)_{L^2(\Omega)} \le 0,
\end{aligned}$$

that is, $f = 0$ in (8.3.11). Hence, $C^\infty(\overline{\Omega})$ is dense in $\operatorname{dom} T_{\max}$ with respect to the graph norm. □

The following result on the extension of the Dirichlet trace operator onto $\operatorname{dom} T_{\max}$ is essential for the construction of a boundary triplet for $T_{\max}$.

Theorem 8.3.9. *Assume that $\Omega \subset \mathbb{R}^n$ is a bounded C^2-domain. Then the Dirichlet trace operator $\tau_D : H^2(\Omega) \to H^{3/2}(\partial\Omega)$ in (8.2.13) admits a unique extension to a continuous surjective operator*

$$\widetilde{\tau}_D : \operatorname{dom} T_{\max} \to H^{-1/2}(\partial\Omega),$$

where $\operatorname{dom} T_{\max}$ is equipped with the graph norm. Furthermore,

$$\ker \widetilde{\tau}_D = \ker \tau_D = \operatorname{dom} A_D.$$

Proof. In the following fix $\lambda \in \rho(A_D)$ and consider the operator

$$\Upsilon := -\tau_N (A_D - \overline{\lambda})^{-1} : L^2(\Omega) \to H^{1/2}(\partial\Omega). \tag{8.3.14}$$

Since $(A_\mathrm{D} - \bar{\lambda})^{-1} : L^2(\Omega) \to H^2(\Omega)$ is everywhere defined and closed, it is clear that $(A_\mathrm{D} - \bar{\lambda})^{-1} : L^2(\Omega) \to H^2(\Omega)$ is continuous and maps onto $\operatorname{dom} A_\mathrm{D}$. Hence, it follows from Theorem 8.3.4 and (8.2.12) that $\Upsilon \in \mathbf{B}(L^2(\Omega), H^{1/2}(\partial\Omega))$ in (8.3.14) is a surjective operator.

Next it will be shown that

$$\ker \Upsilon = \mathfrak{N}_\lambda(T_{\max})^\perp, \tag{8.3.15}$$

where $\mathfrak{N}_\lambda(T_{\max}) = \ker (T_{\max} - \lambda)$. In fact, for the inclusion $(\subset)$ in (8.3.15), assume that

$$\Upsilon h = -\tau_\mathrm{N}(A_\mathrm{D} - \bar{\lambda})^{-1} h = 0$$

for some $h \in L^2(\Omega)$. Then it follows from Theorem 8.3.4 that

$$(A_\mathrm{D} - \bar{\lambda})^{-1} h \in \operatorname{dom} A_\mathrm{D} \cap \operatorname{dom} A_\mathrm{N}$$

and hence $(A_\mathrm{D} - \bar{\lambda})^{-1} h \in \operatorname{dom} T_{\min}$ by (8.2.15) and (8.3.2). For $f_\lambda \in \mathfrak{N}_\lambda(T_{\max})$ one concludes, together with Proposition 8.3.1, that

$$\begin{aligned}
(f_\lambda, h)_{L^2(\Omega)} &= \bigl(f_\lambda, (T_{\min} - \bar{\lambda})(A_\mathrm{D} - \bar{\lambda})^{-1} h\bigr)_{L^2(\Omega)} \\
&= \bigl((T_{\max} - \lambda) f_\lambda, (A_\mathrm{D} - \bar{\lambda})^{-1} h\bigr)_{L^2(\Omega)} \\
&= 0,
\end{aligned}$$

which shows $h \in \mathfrak{N}_\lambda(T_{\max})^\perp$. For the inclusion $(\supset)$ in (8.3.15), let $h \in \mathfrak{N}_\lambda(T_{\max})^\perp$. Then $h \in \operatorname{ran}(T_{\min} - \bar{\lambda})$, and hence there exists $k \in \operatorname{dom} T_{\min} = H^2_0(\Omega)$ such that $h = (T_{\min} - \bar{\lambda})k$. It follows that

$$\Upsilon h = -\tau_\mathrm{N}(A_\mathrm{D} - \bar{\lambda})^{-1} h = -\tau_\mathrm{N}(A_\mathrm{D} - \bar{\lambda})^{-1}(T_{\min} - \bar{\lambda})k = -\tau_\mathrm{N} k = 0,$$

which shows that $h \in \ker \Upsilon$. This completes the proof of (8.3.15).

From (8.3.14) and (8.3.15) it follows that the restriction of Υ to $\mathfrak{N}_\lambda(T_{\max})$ is an isomorphism from $\mathfrak{N}_\lambda(T_{\max})$ onto $H^{1/2}(\partial\Omega)$. This implies that the dual operator

$$\Upsilon' : H^{-1/2}(\partial\Omega) \to L^2(\Omega) \tag{8.3.16}$$

is bounded and invertible, and by the closed range theorem (see Theorem 1.3.5 for the Hilbert space adjoint) one has

$$\operatorname{ran} \Upsilon' = (\ker \Upsilon)^\perp = \mathfrak{N}_\lambda(T_{\max}).$$

The inverse $(\Upsilon')^{-1}$ is regarded as an isomorphism from $\mathfrak{N}_\lambda(T_{\max})$ onto $H^{-1/2}(\partial\Omega)$. Now recall the direct sum decomposition

$$\operatorname{dom} T_{\max} = \operatorname{dom} A_\mathrm{D} \dotplus \mathfrak{N}_\lambda(T_{\max})$$

from Theorem 1.7.1 or Corollary 1.7.5, and write the elements $f \in \operatorname{dom} T_{\max}$ accordingly,

$$f = f_\mathrm{D} + f_\lambda, \quad f_\mathrm{D} \in \operatorname{dom} A_\mathrm{D}, \quad f_\lambda \in \mathfrak{N}_\lambda(T_{\max}).$$

Define the mapping

$$\widetilde{\tau}_{\mathrm{D}} : \operatorname{dom} T_{\max} \to H^{-1/2}(\partial\Omega), \qquad f \mapsto \widetilde{\tau}_{\mathrm{D}} f = (\Upsilon')^{-1} f_\lambda. \tag{8.3.17}$$

Next it will be shown that $\widetilde{\tau}_{\mathrm{D}}$ is an extension of the Dirichlet trace operator $\tau_{\mathrm{D}} : H^2(\Omega) \to H^{3/2}(\partial\Omega)$. For this, consider $\varphi \in \operatorname{ran} \tau_{\mathrm{D}} = H^{3/2}(\partial\Omega) \subset H^{-1/2}(\partial\Omega)$ and note that by (8.3.9) and (8.2.12) there exists a unique $f_\lambda \in H^2(\Omega)$ such that

$$(-\Delta + V) f_\lambda = \lambda f_\lambda \quad \text{and} \quad \tau_{\mathrm{D}} f_\lambda = \varphi. \tag{8.3.18}$$

Let $h \in L^2(\Omega)$ and set $k := (A_{\mathrm{D}} - \overline{\lambda})^{-1} h$. Then, by (8.3.14), the fact that $\tau_{\mathrm{D}} k = 0$, and the second Green identity (8.2.19),

$$\begin{aligned}
(\Upsilon'\varphi, h)_{L^2(\Omega)} &= \langle \varphi, \Upsilon h \rangle_{H^{-1/2}(\partial\Omega)\times H^{1/2}(\partial\Omega)} \\
&= (\varphi, \Upsilon h)_{L^2(\partial\Omega)} \\
&= -\bigl(\varphi, \tau_{\mathrm{N}} (A_{\mathrm{D}} - \overline{\lambda})^{-1} h\bigr)_{L^2(\partial\Omega)} \\
&= -(\tau_{\mathrm{D}} f_\lambda, \tau_{\mathrm{N}} k)_{L^2(\partial\Omega)} + (\tau_{\mathrm{N}} f_\lambda, \tau_{\mathrm{D}} k)_{L^2(\partial\Omega)} \\
&= -\bigl((-\Delta + V) f_\lambda, k\bigr)_{L^2(\Omega)} + \bigl(f_\lambda, (-\Delta + V) k\bigr)_{L^2(\Omega)} \\
&= -(\lambda f_\lambda, k)_{L^2(\Omega)} + (f_\lambda, A_{\mathrm{D}} k)_{L^2(\Omega)} \\
&= \bigl(f_\lambda, (A_{\mathrm{D}} - \overline{\lambda}) k\bigr)_{L^2(\Omega)} \\
&= (f_\lambda, h)_{L^2(\Omega)},
\end{aligned}$$

and thus $\Upsilon'\varphi = f_\lambda$. Hence, the restriction of Υ' to $H^{3/2}(\partial\Omega)$ maps $\varphi \in H^{3/2}(\partial\Omega)$ to the unique $H^2(\Omega)$-solution f_λ of the boundary value problem (8.3.18), that is, to the unique element $f_\lambda \in \mathfrak{N}_\lambda(T_{\max}) \cap H^2(\Omega)$ such that $\tau_{\mathrm{D}} f_\lambda = \varphi$. Therefore, $(\Upsilon')^{-1}$ maps the elements in $\mathfrak{N}_\lambda(T_{\max}) \cap H^2(\Omega)$ onto their Dirichlet boundary values, that is,

$$(\Upsilon')^{-1} f_\lambda = \tau_{\mathrm{D}} f_\lambda \quad \text{for} \quad f_\lambda \in \mathfrak{N}_\lambda(T_{\max}) \cap H^2(\Omega).$$

By definition $\widetilde{\tau}_{\mathrm{D}} f_{\mathrm{D}} = 0 = \tau_{\mathrm{D}} f_{\mathrm{D}}$ for $f_{\mathrm{D}} \in \operatorname{dom} A_{\mathrm{D}}$. Therefore, if $f \in H^2(\Omega)$ is decomposed according to (8.3.9) as

$$f = f_{\mathrm{D}} + f_\lambda, \quad f_{\mathrm{D}} \in \operatorname{dom} A_{\mathrm{D}}, \quad f_\lambda \in \mathfrak{N}_\lambda(T_{\max}) \cap H^2(\Omega),$$

then

$$\widetilde{\tau}_{\mathrm{D}} f = \widetilde{\tau}_{\mathrm{D}} (f_{\mathrm{D}} + f_\lambda) = (\Upsilon')^{-1} f_\lambda = \tau_{\mathrm{D}} f_\lambda = \tau_{\mathrm{D}} f,$$

so that $\widetilde{\tau}_{\mathrm{D}}$ in (8.3.17) is an extension of τ_{D}. Note that by construction $\widetilde{\tau}_{\mathrm{D}}$ is surjective. Furthermore, the property $\ker \widetilde{\tau}_{\mathrm{D}} = \ker \tau_{\mathrm{D}}$ is clear from the definition.

It remains to show that $\widetilde{\tau}_{\mathrm{D}}$ in (8.3.17) is continuous with respect to the graph norm on $\operatorname{dom} T_{\max}$. For this, consider $f = f_{\mathrm{D}} + f_\lambda \in \operatorname{dom} T_{\max}$ with $f_{\mathrm{D}} \in \operatorname{dom} A_{\mathrm{D}}$ and $f_\lambda \in \mathfrak{N}_\lambda(T_{\max})$, and note that

$$\begin{aligned}
f_\lambda = f - f_{\mathrm{D}} &= f - (A_{\mathrm{D}} - \lambda)^{-1} (T_{\max} - \lambda) f_{\mathrm{D}} \\
&= f - (A_{\mathrm{D}} - \lambda)^{-1} (T_{\max} - \lambda) f.
\end{aligned}$$

Since $(\Upsilon')^{-1} : \mathfrak{N}_\lambda(T_{\max}) \to H^{-1/2}(\partial\Omega)$ is an isomorphism and hence, in particular, bounded, one has

$$\begin{aligned}\|\widetilde{\tau}_D f\|_{H^{-1/2}(\partial\Omega)} &= \|(\Upsilon')^{-1} f_\lambda\|_{H^{-1/2}(\partial\Omega)}\\ &\leq C\|f_\lambda\|_{L^2(\Omega)}\\ &\leq C\big(\|f\|_{L^2(\Omega)} + \|(A_D-\lambda)^{-1}(T_{\max}-\lambda)f\|_{L^2(\Omega)}\big)\\ &\leq C'\big(\|f\|_{L^2(\Omega)} + \|(T_{\max}-\lambda)f\|_{L^2(\Omega)}\big)\\ &\leq C''\big(\|f\|_{L^2(\Omega)} + \|T_{\max}f\|_{L^2(\Omega)}\big)\end{aligned}$$

with some constants $C, C', C'' > 0$. Thus, $\widetilde{\tau}_D$ is continuous. The proof of Theorem 8.3.9 is complete. □

The following result is parallel to Theorem 8.3.9 and can be proved in a similar way.

Theorem 8.3.10. *Assume that $\Omega \subset \mathbb{R}^n$ is a bounded C^2-domain. Then the Neumann trace operator $\tau_N : H^2(\Omega) \to H^{1/2}(\partial\Omega)$ in (8.2.14) admits a unique extension to a continuous surjective operator*

$$\widetilde{\tau}_N : \operatorname{dom} T_{\max} \to H^{-3/2}(\partial\Omega),$$

where $\operatorname{dom} T_{\max}$ *is equipped with the graph norm. Furthermore,*

$$\ker \widetilde{\tau}_N = \ker \tau_N = \operatorname{dom} A_N.$$

As a consequence of Theorem 8.3.9 and Theorem 8.3.10 one can also extend the second Green identity in (8.2.19) to elements $f \in \operatorname{dom} T_{\max}$ and $g \in H^2(\Omega)$.

Corollary 8.3.11. *Assume that $\Omega \subset \mathbb{R}^n$ is a bounded C^2-domain, and let*

$$\widetilde{\tau}_D : \operatorname{dom} T_{\max} \to H^{-1/2}(\partial\Omega) \quad \text{and} \quad \widetilde{\tau}_N : \operatorname{dom} T_{\max} \to H^{-3/2}(\partial\Omega)$$

be the unique continuous extensions of the Dirichlet and Neumann trace operators

$$\tau_D : H^2(\Omega) \to H^{3/2}(\partial\Omega) \quad \text{and} \quad \tau_N : H^2(\Omega) \to H^{1/2}(\partial\Omega)$$

from Theorem 8.3.9 and Theorem 8.3.10, respectively. Then the second Green identity in (8.2.19) extends to

$$\begin{aligned}&(T_{\max}f, g)_{L^2(\Omega)} - (f, T_{\max}g)_{L^2(\Omega)}\\ &\qquad = \langle \widetilde{\tau}_D f, \tau_N g\rangle_{H^{-1/2}(\partial\Omega)\times H^{1/2}(\partial\Omega)} - \langle \widetilde{\tau}_N f, \tau_D g\rangle_{H^{-3/2}(\partial\Omega)\times H^{3/2}(\partial\Omega)}\end{aligned}$$

for $f \in \operatorname{dom} T_{\max}$ and $g \in H^2(\Omega)$.

Proof. Let $f \in \operatorname{dom} T_{\max}$ and $g \in H^2(\Omega)$. Since $C^\infty(\overline{\Omega})$ is dense in $\operatorname{dom} T_{\max}$ with respect to the graph norm by Lemma 8.3.8 and $C^\infty(\overline{\Omega}) \subset H^2(\Omega) \subset \operatorname{dom} T_{\max}$, there exists a sequence $(f_n) \subset H^2(\Omega)$ such that $f_n \to f$ and $T_{\max} f_n \to T_{\max} f$

in $L^2(\Omega)$. Moreover, $\tau_D f_n \to \widetilde{\tau}_D f$ in $H^{-1/2}(\partial\Omega)$ and $\tau_N f_n \to \widetilde{\tau}_N f$ in $H^{-3/2}(\partial\Omega)$, because $\widetilde{\tau}_D$ and $\widetilde{\tau}_N$ are continuous with respect to the graph norm. Therefore, with the help of the second Green identity (8.2.19), one concludes that

$$\begin{aligned}
&(T_{\max} f, g)_{L^2(\Omega)} - (f, T_{\max} g)_{L^2(\Omega)} \\
&\quad = \lim_{n\to\infty} (T_{\max} f_n, g)_{L^2(\Omega)} - \lim_{n\to\infty} (f_n, T_{\max} g)_{L^2(\Omega)} \\
&\quad = \lim_{n\to\infty} \big[(\tau_D f_n, \tau_N g)_{L^2(\partial\Omega)} - (\tau_N f_n, \tau_D g)_{L^2(\partial\Omega)}\big] \\
&\quad = \lim_{n\to\infty} \big[\langle \tau_D f_n, \tau_N g\rangle_{H^{-1/2}(\partial\Omega)\times H^{1/2}(\partial\Omega)} - \langle \tau_N f_n, \tau_D g\rangle_{H^{-3/2}(\partial\Omega)\times H^{3/2}(\partial\Omega)}\big] \\
&\quad = \langle \widetilde{\tau}_D f, \tau_N g\rangle_{H^{-1/2}(\partial\Omega)\times H^{1/2}(\partial\Omega)} - \langle \widetilde{\tau}_N f, \tau_D g\rangle_{H^{-3/2}(\partial\Omega)\times H^{3/2}(\partial\Omega)},
\end{aligned}$$

which completes the proof. □

Note that, by construction, there exists a bounded right inverse for the extended Dirichlet trace operator $\widetilde{\tau}_D$ (see (8.3.16)–(8.3.17)) and similarly there exists a bounded right inverse for the extended Neumann trace operator $\widetilde{\tau}_N$. This also implies that the Dirichlet-to-Neumann map in Definition 8.3.6 admits a natural extension to a bounded mapping from from $H^{-1/2}(\partial\Omega)$ into $H^{-3/2}(\partial\Omega)$.

Corollary 8.3.12. *Assume that $\Omega \subset \mathbb{R}^n$ is a bounded C^2-domain and let $\widetilde{\tau}_D$ and $\widetilde{\tau}_N$ be the unique continuous extensions of the Dirichlet and Neumann trace operators from Theorem 8.3.9 and Theorem 8.3.10, respectively. Then for $\lambda \in \rho(A_D)$ the Dirichlet-to-Neumann map in Definition 8.3.6 admits an extension to a bounded operator*

$$\widetilde{D}(\lambda) : H^{-1/2}(\partial\Omega) \to H^{-3/2}(\partial\Omega), \qquad \widetilde{\tau}_D f_\lambda \mapsto \widetilde{\tau}_N f_\lambda,$$

where $f_\lambda \in \mathfrak{N}_\lambda(T_{\max})$.

For later purposes the following fact is provided.

Proposition 8.3.13. *The minimal operator $T_{\min}$ in (8.3.2) is simple.*

Proof. Since A_D is a self-adjoint extension of $T_{\min}$ with discrete spectrum, it suffices to check that $T_{\min}$ has no eigenvalues; cf. Proposition 3.4.8. For this, assume that $T_{\min} f = \lambda f$ for some $\lambda \in \mathbb{R}$ and some $f \in \operatorname{dom} T_{\min}$. Since $\operatorname{dom} T_{\min} = H^2_0(\Omega)$, there exist $(f_k) \in C_0^\infty(\Omega)$ such that $f_k \to f$ in $H^2(\Omega)$. Denote the zero extensions of f and f_k to all of $\mathbb{R}^n$ by $\widetilde{f}$ and $\widetilde{f}_k$, respectively. Then $\widetilde{f}_k \to \widetilde{f}$ in $L^2(\mathbb{R}^n)$ and for all $h \in C_0^\infty(\mathbb{R}^n)$ and $\alpha \in \mathbb{N}_0^n$ such that $|\alpha| \le 2$ one computes

$$\begin{aligned}
\int_{\mathbb{R}^n} \widetilde{f}(x) D^\alpha h(x)\,dx &= \lim_{k\to\infty} \int_{\mathbb{R}^n} \widetilde{f}_k(x) D^\alpha h(x)\,dx \\
&= (-1)^{|\alpha|} \lim_{k\to\infty} \int_{\mathbb{R}^n} (D^\alpha \widetilde{f}_k)(x) h(x)\,dx \\
&= (-1)^{|\alpha|} \lim_{k\to\infty} \int_{\Omega} (D^\alpha f_k)(x) h(x)\,dx
\end{aligned}$$

$$
\begin{aligned}
&= (-1)^{|\alpha|} \int_\Omega (D^\alpha f)(x) h(x)\,dx \\
&= (-1)^{|\alpha|} \int_{\mathbb{R}^n} \widetilde{(D^\alpha f)}(x) h(x)\,dx,
\end{aligned}
$$

where $\widetilde{(D^\alpha f)}$ denotes the zero extension of $D^\alpha f$ to all of $\mathbb{R}^n$. It follows from this computation that

$$
D^\alpha \widetilde{f} = \widetilde{(D^\alpha f)} \in L^2(\mathbb{R}^n), \qquad |\alpha| \le 2,
$$

and hence $\widetilde{f} \in H^2(\mathbb{R}^n)$. Furthermore, if $\widetilde{V} \in L^\infty(\mathbb{R}^n)$ denotes some real extension of V, then $(-\Delta + \widetilde{V})\widetilde{f} = \lambda \widetilde{f}$ and since $\widetilde{f}$ vanishes on an open subset of $\mathbb{R}^n$, the unique continuation principle (see, e.g., [652, Theorem XIII.63]) implies $\widetilde{f} = 0$, so that $f = 0$. Therefore, $T_{\min}$ has no eigenvalues and now Proposition 3.4.8 shows that $T_{\min}$ is simple. □

8.4 A boundary triplet for the maximal Schrödinger operator

In this section a boundary triplet $\{L^2(\partial\Omega), \Gamma_0, \Gamma_1\}$ for the maximal operator $T_{\max}$ in (8.3.3) is provided under the assumption that $\Omega \subset \mathbb{R}^n$ is a bounded C^2-domain. The corresponding Weyl function is closely connected to the extended Dirichlet-to-Neumann map in Corollary 8.3.12. As examples, Neumann and Robin type boundary conditions are discussed, and it is also explained that there exist self-adjoint realizations of $-\Delta + V$ in $L^2(\Omega)$ which are not semibounded and which may have essential spectrum of rather arbitrary form.

Recall from Corollary 8.2.2 that

$$
\{H^{1/2}(\partial\Omega),\, L^2(\partial\Omega), H^{-1/2}(\partial\Omega)\}
$$

is a Gelfand triple and there exist isometric isomorphisms $\iota_\pm : H^{\pm\frac{1}{2}}(\partial\Omega) \to L^2(\partial\Omega)$ such that

$$
\langle \varphi, \psi \rangle_{H^{-1/2}(\partial\Omega)\times H^{1/2}(\partial\Omega)} = (\iota_- \varphi, \iota_+ \psi)_{L^2(\partial\Omega)}
$$

holds for all $\varphi \in H^{-1/2}(\partial\Omega)$ and $\psi \in H^{1/2}(\partial\Omega)$. For the definition of the boundary mappings in the next proposition recall also the definition and the properties of the Dirichlet operator A_{D} (see Theorem 8.3.4), as well as the direct sum decomposition

$$
\operatorname{dom} T_{\max} = \operatorname{dom} A_{\mathrm{D}} \dotplus \mathfrak{N}_\eta(T_{\max}), \tag{8.4.1}
$$

which holds for all $\eta \in \rho(A_{\mathrm{D}})$. In particular, since A_{D} is semibounded from below, one may choose $\eta \in \rho(A_{\mathrm{D}}) \cap \mathbb{R}$ in (8.4.1). Further, let $\tau_{\mathrm{N}} : H^2(\Omega) \to H^{1/2}(\partial\Omega)$ be the Neumann trace operator in (8.2.12) and let $\widetilde{\tau}_{\mathrm{D}} : \operatorname{dom} T_{\max} \to H^{-1/2}(\partial\Omega)$ be the extension of the Dirichlet trace operator in Theorem 8.3.9.

Theorem 8.4.1. *Let $\Omega \subset \mathbb{R}^n$ be a bounded C^2-domain, let $A_{\rm D}$ be the self-adjoint Dirichlet realization of $-\Delta + V$ in $L^2(\Omega)$ in Theorem 8.3.4, fix $\eta \in \rho(A_{\rm D}) \cap \mathbb{R}$, and decompose $f \in \operatorname{dom} T_{\max}$ according to (8.4.1) in the form $f = f_{\rm D} + f_\eta$, where $f_{\rm D} \in \operatorname{dom} A_{\rm D}$ and $f_\eta \in \mathfrak{N}_\eta(T_{\max})$. Then $\{L^2(\partial\Omega), \Gamma_0, \Gamma_1\}$, where*

$$\Gamma_0 f = \iota_- \widetilde{\tau}_{\rm D} f \quad \textit{and} \quad \Gamma_1 f = -\iota_+ \tau_{\rm N} f_{\rm D}, \qquad f = f_{\rm D} + f_\eta \in \operatorname{dom} T_{\max},$$

is a boundary triplet for $(T_{\min})^* = T_{\max}$ such that

$$A_0 = A_{\rm D} \qquad \textit{and} \qquad A_1 = T_{\min} \,\widehat{+}\, \widehat{\mathfrak{N}}_\eta(T_{\max}). \tag{8.4.2}$$

Proof. Let $f, g \in \operatorname{dom} T_{\max}$ and decompose f and g in the form $f = f_{\rm D} + f_\eta$ and $g = g_{\rm D} + g_\eta$ with $f_{\rm D}, g_{\rm D} \in \operatorname{dom} A_{\rm D} \subset H^2(\Omega)$ and $f_\eta, g_\eta \in \mathfrak{N}_\eta(T_{\max})$. Since $A_{\rm D}$ is self-adjoint,

$$(T_{\max} f_{\rm D}, g_{\rm D})_{L^2(\Omega)} = (A_{\rm D} f_{\rm D}, g_{\rm D})_{L^2(\Omega)} = (f_{\rm D}, A_{\rm D}\, g_{\rm D})_{L^2(\Omega)} = (f_{\rm D}, T_{\max}\, g_{\rm D})_{L^2(\Omega)}$$

and since η is real, one also has

$$(T_{\max} f_\eta, g_\eta)_{L^2(\Omega)} = (\eta f_\eta, g_\eta)_{L^2(\Omega)} = (f_\eta, \eta g_\eta)_{L^2(\Omega)} = (f_\eta, T_{\max}\, g_\eta)_{L^2(\Omega)}.$$

Therefore, one obtains

$$\begin{aligned}
&(T_{\max} f, g)_{L^2(\Omega)} - (f, T_{\max} g)_{L^2(\Omega)} \\
&\quad = \big(T_{\max}(f_{\rm D} + f_\eta), g_{\rm D} + g_\eta\big)_{L^2(\Omega)} - \big(f_{\rm D} + f_\eta, T_{\max}(g_{\rm D} + g_\eta)\big)_{L^2(\Omega)} \\
&\quad = (T_{\max} f_\eta, g_{\rm D})_{L^2(\Omega)} + (T_{\max} f_{\rm D}, g_\eta)_{L^2(\Omega)} \\
&\qquad\quad - (f_\eta, T_{\max}\, g_{\rm D})_{L^2(\Omega)} - (f_{\rm D}, T_{\max}\, g_\eta)_{L^2(\Omega)}.
\end{aligned}$$

Let $\widetilde{\tau}_{\rm N}$ be the extension of the Neumann trace to $\operatorname{dom} T_{\max}$ from Theorem 8.3.10. Then it follows together with Corollary 8.3.11 and $\tau_{\rm D} f_{\rm D} = \tau_{\rm D} g_{\rm D} = 0$ that

$$\begin{aligned}
&(T_{\max} f_\eta, g_{\rm D})_{L^2(\Omega)} - (f_\eta, T_{\max}\, g_{\rm D})_{L^2(\Omega)} \\
&\quad = \langle \widetilde{\tau}_{\rm D} f_\eta, \tau_{\rm N} g_{\rm D}\rangle_{H^{-1/2}(\partial\Omega)\times H^{1/2}(\partial\Omega)} - \langle \widetilde{\tau}_{\rm N} f_\eta, \tau_{\rm D} g_{\rm D}\rangle_{H^{-3/2}(\partial\Omega)\times H^{3/2}(\partial\Omega)} \\
&\quad = \langle \widetilde{\tau}_{\rm D} f_\eta, \tau_{\rm N} g_{\rm D}\rangle_{H^{-1/2}(\partial\Omega)\times H^{1/2}(\partial\Omega)}
\end{aligned}$$

and

$$\begin{aligned}
&(T_{\max} f_{\rm D}, g_\eta)_{L^2(\Omega)} - (f_{\rm D}, T_{\max} g_\eta)_{L^2(\Omega)} \\
&\quad = \langle \tau_{\rm D} f_{\rm D}, \widetilde{\tau}_{\rm N} g_\eta\rangle_{H^{3/2}(\partial\Omega)\times H^{-3/2}(\partial\Omega)} - \langle \tau_{\rm N} f_{\rm D}, \widetilde{\tau}_{\rm D} g_\eta\rangle_{H^{1/2}(\partial\Omega)\times H^{-1/2}(\partial\Omega)} \\
&\quad = -\langle \tau_{\rm N} f_{\rm D}, \widetilde{\tau}_{\rm D} g_\eta\rangle_{H^{1/2}(\partial\Omega)\times H^{-1/2}(\partial\Omega)}.
\end{aligned}$$

Hence,

$$\begin{aligned}
&(T_{\max} f, g)_{L^2(\Omega)} - (f, T_{\max} g)_{L^2(\Omega)} \\
&\quad = \langle \widetilde{\tau}_{\rm D} f_\eta, \tau_{\rm N} g_{\rm D}\rangle_{H^{-1/2}(\partial\Omega)\times H^{1/2}(\partial\Omega)} - \langle \tau_{\rm N} f_{\rm D}, \widetilde{\tau}_{\rm D} g_\eta\rangle_{H^{1/2}(\partial\Omega)\times H^{-1/2}(\partial\Omega)} \\
&\quad = \big(\iota_- \widetilde{\tau}_{\rm D} f_\eta, \iota_+ \tau_{\rm N} g_{\rm D}\big)_{L^2(\partial\Omega)} - \big(\iota_+ \tau_{\rm N} f_{\rm D}, \iota_- \widetilde{\tau}_{\rm D} g_\eta\big)_{L^2(\partial\Omega)}
\end{aligned}$$

and, since $f_{\rm D}, g_{\rm D} \in \ker \tau_{\rm D} = \ker \widetilde{\tau}_{\rm D}$ according to Theorem 8.3.9, one sees that

$$\begin{aligned}(T_{\max} f, g)_{L^2(\Omega)} - (f, T_{\max} g)_{L^2(\Omega)} \\ &= \bigl(\iota_- \widetilde{\tau}_{\rm D} f, \iota_+ \tau_{\rm N} g_{\rm D}\bigr)_{L^2(\partial\Omega)} - \bigl(\iota_+ \tau_{\rm N} f_{\rm D}, \iota_- \widetilde{\tau}_{\rm D} g\bigr)_{L^2(\partial\Omega)} \\ &= \bigl(-\iota_+ \tau_{\rm N} f_{\rm D}, \iota_- \widetilde{\tau}_{\rm D} g\bigr)_{L^2(\partial\Omega)} - \bigl(\iota_- \widetilde{\tau}_{\rm D} f, -\iota_+ \tau_{\rm N} g_{\rm D}\bigr)_{L^2(\partial\Omega)} \\ &= (\Gamma_1 f, \Gamma_0 g)_{L^2(\partial\Omega)} - (\Gamma_0 f, \Gamma_1 g)_{L^2(\partial\Omega)}\end{aligned}$$

for all $f, g \in \operatorname{dom} T_{\max}$, that is, the abstract Green identity is satisfied. To verify the surjectivity of the mapping

$$\begin{pmatrix}\Gamma_0 \\ \Gamma_1\end{pmatrix} : \operatorname{dom} T_{\max} \to L^2(\partial\Omega) \times L^2(\partial\Omega), \tag{8.4.3}$$

let $\varphi, \psi \in L^2(\partial\Omega)$ and consider $\iota_-^{-1}\varphi \in H^{-1/2}(\partial\Omega)$ and $-\iota_+^{-1}\psi \in H^{1/2}(\partial\Omega)$. Observe that by (8.2.12) the Neumann trace operator $\tau_{\rm N}$ is a surjective mapping from $\{h \in H^2(\Omega) : \tau_{\rm D} h = 0\}$ onto $H^{1/2}(\partial\Omega)$, that is, $\tau_{\rm N} : \operatorname{dom} A_{\rm D} \to H^{1/2}(\partial\Omega)$ is onto, and hence there exists $f_{\rm D} \in \operatorname{dom} A_{\rm D}$ such that $\tau_{\rm N} f_{\rm D} = -\iota_+^{-1}\psi$. Next recall from Theorem 8.3.9 that the extended Dirichlet trace operator $\widetilde{\tau}_{\rm D}$ maps $\operatorname{dom} T_{\max}$ onto $H^{-1/2}(\partial\Omega)$ and that $\ker \widetilde{\tau}_{\rm D} = \ker \tau_{\rm D} = \operatorname{dom} A_{\rm D}$. Hence, it follows from the direct sum decomposition $\operatorname{dom} T_{\max} = \operatorname{dom} A_{\rm D} + \mathfrak{N}_\eta(T_{\max})$ that the restriction $\widetilde{\tau}_{\rm D} : \mathfrak{N}_\eta(T_{\max}) \to H^{-1/2}(\partial\Omega)$ is bijective, in particular, there exists $f_\eta \in \mathfrak{N}_\eta(T_{\max})$ such that $\widetilde{\tau}_{\rm D} f_\eta = \iota_-^{-1}\varphi$. Now it follows that $f := f_{\rm D} + f_\eta \in \operatorname{dom} T_{\max}$ satisfies

$$\Gamma_0 f = \iota_- \widetilde{\tau}_{\rm D} f = \iota_- \widetilde{\tau}_{\rm D} f_\eta = \iota_- \iota_-^{-1}\varphi = \varphi$$

and

$$\Gamma_1 f = -\iota_+ \tau_{\rm N} f_{\rm D} = \iota_+ \iota_+^{-1}\psi = \psi,$$

and hence the mapping in (8.4.3) is onto. Thus, $\{L^2(\partial\Omega), \Gamma_0, \Gamma_1\}$ is a boundary triplet for $(T_{\min})^* = T_{\max}$, as claimed.

From the definition of Γ_0 and $\ker \widetilde{\tau}_{\rm D} = \ker \tau_{\rm D} = \operatorname{dom} A_{\rm D}$ it is clear that $\operatorname{dom} A_{\rm D} = \ker \Gamma_0$, and hence the self-adjoint extension corresponding to Γ_0 coincides with the Dirichlet operator $A_{\rm D}$, that is, the first identity in (8.4.2) holds. It remains to check the second identity in (8.4.2). For this let $f = f_{\rm D} + f_\eta \in \ker \Gamma_1$, which means $\tau_{\rm N} f_{\rm D} = 0$. Thus, $f_{\rm D} \in \operatorname{dom} T_{\min}$ by (8.2.15) and it follows that $A_1 \subset T_{\min} \,\widehat{+}\, \widehat{\mathfrak{N}}_\eta(T_{\max})$. The inclusion $T_{\min} \,\widehat{+}\, \widehat{\mathfrak{N}}_\eta(T_{\max}) \subset A_1$ is clear from the definition of Γ_1. This leads to the second identity in (8.4.2). □

Remark 8.4.2. The boundary triplet $\{L^2(\partial\Omega), \Gamma_0, \Gamma_1\}$ in Theorem 8.4.1 is closely related to the boundary triplet $\{\mathfrak{N}_\eta(T_{\max}), \Gamma_0', \Gamma_1'\}$ in Corollary 5.5.12, where

$$\Gamma_0' f = f_\eta \quad \text{and} \quad \Gamma_1' f = P_{\mathfrak{N}_\eta(T_{\max})}(A_{\rm D} - \eta) f_{\rm D}, \quad f = f_{\rm D} + f_\eta \in \operatorname{dom} T_{\max}.$$

In fact, one has $\ker \Gamma_0 = \ker \Gamma_0'$ and $\ker \Gamma_1 = \ker \Gamma_1'$, and hence

$$\begin{pmatrix}\Gamma_0' \\ \Gamma_1'\end{pmatrix} = \begin{pmatrix} W_{11} & 0 \\ 0 & W_{22}\end{pmatrix} \begin{pmatrix}\Gamma_0 \\ \Gamma_1\end{pmatrix}$$

with some 2×2 operator matrix $\mathcal{W} = (W_{ij})_{i,j=1}^2$ as in Theorem 2.5.1, see also Corollary 2.5.5. In the present situation it follows from Theorem 8.3.9 and (8.4.1) that the restriction $\iota_- \widetilde{\tau}_{\mathrm{D}} : \mathfrak{N}_\eta(T_{\max}) \to L^2(\partial\Omega)$ is bijective and one concludes $W_{11} = (\iota_- \widetilde{\tau}_{\mathrm{D}})^{-1}$. Now the properties of $\mathcal{W}$ imply that $W_{22} = (\iota_- \widetilde{\tau}_{\mathrm{D}})^*$.

With the help of the extended Dirichlet-to-Neumann map in Corollary 8.3.12 one obtains a more explicit description of the domain of the self-adjoint operator A_1 in (8.4.2).

Proposition 8.4.3. *Let $\Omega \subset \mathbb{R}^n$ be a bounded C^2-domain, let A_{D} be the self-adjoint Dirichlet realization of $-\Delta + V$ in $L^2(\Omega)$, and fix $\eta \in \rho(A_{\mathrm{D}}) \cap \mathbb{R}$. Moreover, let $\widetilde{D}(\eta)$ be the extended Dirichlet-to-Neumann map in Corollary 8.3.12. Then the self-adjoint extension A_1 of $T_{\min}$ in (8.4.2) is defined on*

$$\operatorname{dom} A_1 = \bigl\{ f \in \operatorname{dom} T_{\max} : \widetilde{\tau}_{\mathrm{N}} f = \widetilde{D}(\eta) \widetilde{\tau}_{\mathrm{D}} f \bigr\}. \tag{8.4.4}$$

In the case that $\eta < m(A_{\mathrm{D}})$, where $m(A_{\mathrm{D}})$ denotes the lower bound of A_{D}, the operator A_1 coincides with the Kreĭn type extension $S_{\mathrm{K},\eta}$ of $T_{\min}$ in Definition 5.4.2. In particular, if $m(A_{\mathrm{D}}) > 0$ and $\eta = 0$, then $A_1 = S_{\mathrm{K},0}$ is the Kreĭn–von Neumann extension of $T_{\min}$.

Proof. It is clear from Theorem 8.4.1 that

$$\operatorname{dom} A_1 = \ker \Gamma_1 = \bigl\{ f = f_{\mathrm{D}} + f_\eta \in \operatorname{dom} T_{\max} : \tau_{\mathrm{N}} f_{\mathrm{D}} = 0 \bigr\}.$$

Let $\widetilde{\tau}_{\mathrm{N}}$ be the extension of the Neumann trace τ_{N} to the maximal domain in Theorem 8.3.10. Then the boundary condition $\tau_{\mathrm{N}} f_{\mathrm{D}} = 0$ can be rewritten as $\widetilde{\tau}_{\mathrm{N}} f = \widetilde{\tau}_{\mathrm{N}} f_\eta$, where $f = f_{\mathrm{D}} + f_\eta \in \operatorname{dom} T_{\max}$. With the help of the extended Dirichlet-to-Neumann map

$$\widetilde{D}(\eta) : H^{-1/2}(\partial\Omega) \to H^{-3/2}(\partial\Omega), \quad \widetilde{\tau}_{\mathrm{D}} f_\eta \mapsto \widetilde{\tau}_{\mathrm{N}} f_\eta, \qquad f_\eta \in \mathfrak{N}_\eta(T_{\max}),$$

one obtains $\widetilde{\tau}_{\mathrm{N}} f_\eta = \widetilde{D}(\eta) \widetilde{\tau}_{\mathrm{D}} f_\eta = \widetilde{D}(\eta) \widetilde{\tau}_{\mathrm{D}} f$, which implies (8.4.4).

If $\eta \in \mathbb{R}$ is chosen smaller than the lower bound $m(A_{\mathrm{D}})$ of A_{D}, then it follows from the second identity in (8.4.2), Lemma 5.4.1, and Definition 5.4.2 that the Kreĭn type extension $S_{\mathrm{K},\eta} = T_{\min} \widehat{+} \widehat{\mathfrak{N}}_\eta(T_{\max})$ of $T_{\min}$ and A_1 coincide. In the special case $m(A_{\mathrm{D}}) > 0$ and $\eta = 0$ one has $A_1 = S_{\mathrm{K},0}$, which is the Kreĭn–von Neumann extension of $T_{\min}$; cf. Definition 5.4.2. □

In the next proposition the γ-field and the Weyl function corresponding to the boundary triplet $\{L^2(\partial\Omega), \Gamma_0, \Gamma_1\}$ in Theorem 8.4.1 are provided. Note that for $f = f_{\mathrm{D}} + f_\eta$ decomposed as in (8.4.1) one has

$$\Gamma_0 f = \iota_- \widetilde{\tau}_{\mathrm{D}} f = \iota_- \widetilde{\tau}_{\mathrm{D}} f_\eta,$$

as $\ker \widetilde{\tau}_{\mathrm{D}} = \ker \tau_{\mathrm{D}} = \operatorname{dom} A_{\mathrm{D}}$ by Theorem 8.3.9. It is also clear from (8.4.1) that Γ_0 is a bijective mapping from $\mathfrak{N}_\eta(T_{\max})$ onto $L^2(\partial\Omega)$.

Proposition 8.4.4. *Let $\{L^2(\partial\Omega), \Gamma_0, \Gamma_1\}$ be the boundary triplet for $(T_{\min})^* = T_{\max}$ in Theorem 8.4.1 and let $f_\eta(\varphi)$ be the unique element in $\mathfrak{N}_\eta(T_{\max})$ such that $\Gamma_0 f_\eta(\varphi) = \varphi$. Then for all $\lambda \in \rho(A_D)$ the γ-field corresponding to the boundary triplet $\{L^2(\partial\Omega), \Gamma_0, \Gamma_1\}$ is given by*

$$\gamma(\lambda)\varphi = \bigl(I + (\lambda - \eta)(A_D - \lambda)^{-1}\bigr) f_\eta(\varphi), \quad \varphi \in L^2(\partial\Omega), \tag{8.4.5}$$

and $f_\lambda(\varphi) := \gamma(\lambda)\varphi$ is the unique element in $\mathfrak{N}_\lambda(T_{\max})$ such that $\Gamma_0 f_\lambda(\varphi) = \varphi$. Furthermore, one has

$$\gamma(\lambda)^* = -\iota_+ \tau_N (A_D - \bar{\lambda})^{-1}, \quad \lambda \in \rho(A_D). \tag{8.4.6}$$

The Weyl function M corresponding to the boundary triplet $\{L^2(\partial\Omega), \Gamma_0, \Gamma_1\}$ is given by

$$M(\lambda)\varphi = (\eta - \lambda)\iota_+ \tau_N (A_D - \lambda)^{-1} f_\eta(\varphi), \quad \varphi \in L^2(\partial\Omega).$$

In particular, $\gamma(\eta)\varphi = f_\eta(\varphi)$ and $M(\eta)\varphi = 0$ for all $\varphi \in L^2(\partial\Omega)$.

Proof. Since by definition $\gamma(\eta)$ is the inverse of the restriction of Γ_0 to $\mathfrak{N}_\eta(T_{\max})$, it is clear that $\gamma(\eta)\varphi = f_\eta(\varphi)$, where $f_\eta(\varphi)$ is the unique element in $\mathfrak{N}_\eta(T_{\max})$ such that $\Gamma_0 f_\eta(\varphi) = \varphi$. Both (8.4.5) and (8.4.6) are consequences of Proposition 2.3.2. In order to compute the Weyl function note that

$$\gamma(\lambda)\varphi = f_\eta(\varphi) + (\lambda - \eta)(A_D - \lambda)^{-1} f_\eta(\varphi)$$

is decomposed in $(\lambda - \eta)(A_D - \lambda)^{-1} f_\eta(\varphi) \in \operatorname{dom} A_D$ and $f_\eta(\varphi) \in \mathfrak{N}_\eta(T_{\max})$, and hence by the definition of Γ_1 it follows that

$$\begin{aligned} M(\lambda)\varphi = \Gamma_1 \gamma(\lambda)\varphi &= -\iota_+ \tau_N \bigl[(\lambda - \eta)(A_D - \lambda)^{-1} f_\eta(\varphi)\bigr] \\ &= (\eta - \lambda)\iota_+ \tau_N (A_D - \lambda)^{-1} f_\eta(\varphi). \end{aligned}$$

The assertion $M(\eta)\varphi = 0$ for all $\varphi \in L^2(\partial\Omega)$ is clear from the above. □

The Weyl function M in Proposition 8.4.4 is closely connected with the Dirichlet-to-Neumann map $D(\lambda)$ and its extension $\widetilde{D}(\lambda)$, $\lambda \in \rho(A_D)$, in Definition 8.3.6 and Corollary 8.3.12. This connection will be made explicit in the next lemma. First, consider $f = f_D + f_\eta \in \operatorname{dom} T_{\max}$ as in (8.4.1). In the present situation one has

$$\tau_N f_D = \widetilde{\tau}_N f_D = \widetilde{\tau}_N f - \widetilde{\tau}_N f_\eta.$$

Hence, making use of $\widetilde{D}(\eta)\widetilde{\tau}_D f_\eta = \widetilde{\tau}_N f_\eta$ (see Corollary 8.3.12) and the identity $\ker \widetilde{\tau}_D = \operatorname{dom} A_D$, it follows that

$$\tau_N f_D = \widetilde{\tau}_N f - \widetilde{D}(\eta)\widetilde{\tau}_D f_\eta = \widetilde{\tau}_N f - \widetilde{D}(\eta)\widetilde{\tau}_D f. \tag{8.4.7}$$

Lemma 8.4.5. *Let M be the Weyl function corresponding to the boundary triplet in Theorem 8.4.1 and let $\widetilde{D}(\lambda)$, $\lambda \in \rho(A_\mathrm{D})$, be the extended Dirichlet-to-Neumann map in Corollary 8.3.12. Then the regularization property*

$$\operatorname{ran}\big(\widetilde{D}(\eta) - \widetilde{D}(\lambda)\big) \subset H^{1/2}(\partial\Omega) \tag{8.4.8}$$

holds and one has

$$M(\lambda)\varphi = \iota_+\big(\widetilde{D}(\eta) - \widetilde{D}(\lambda)\big)\iota_-^{-1}\varphi, \qquad \varphi \in L^2(\partial\Omega), \tag{8.4.9}$$

and

$$M(\lambda)\varphi = \iota_+\big(D(\eta) - D(\lambda)\big)\iota_-^{-1}\varphi, \qquad \varphi \in H^2(\partial\Omega), \tag{8.4.10}$$

Proof. For $\psi \in H^{-1/2}(\partial\Omega)$ choose $f_\lambda \in \mathfrak{N}_\lambda(T_{\max})$ such that $\widetilde{\tau}_\mathrm{D} f_\lambda = \psi$ or, equivalently, $\Gamma_0 f_\lambda = \iota_-\psi$. Decompose f_λ in the form $f_\lambda = f_\mathrm{D}^\lambda + f_{\lambda,\eta}$ with $f_\mathrm{D}^\lambda \in \operatorname{dom} A_\mathrm{D}$ and $f_{\lambda,\eta} \in \mathfrak{N}_\eta(T_{\max})$. Then one computes

$$\big(\widetilde{D}(\eta) - \widetilde{D}(\lambda)\big)\psi = \widetilde{D}(\eta)\widetilde{\tau}_\mathrm{D} f_\lambda - \widetilde{\tau}_\mathrm{N} f_\lambda = -\tau_\mathrm{N} f_\mathrm{D}^\lambda, \tag{8.4.11}$$

where (8.4.7) was used in the last step for $f = f_\lambda$. Since $f_\mathrm{D}^\lambda \in \operatorname{dom} A_\mathrm{D} \subset H^2(\Omega)$, the regularization property (8.4.8) follows from (8.2.12). From (8.4.11) one also concludes that

$$\iota_+\big(\widetilde{D}(\eta) - \widetilde{D}(\lambda)\big)\iota_-^{-1}\Gamma_0 f_\lambda = -\iota_+\tau_\mathrm{N} f_\mathrm{D}^\lambda = \Gamma_1 f_\lambda,$$

and since $M(\lambda)\Gamma_0 f_\lambda = \Gamma_1 f_\lambda$ by the definition of the Weyl function, this shows (8.4.9).

It remains to prove the second assertion (8.4.10). For this note that the restriction of $\iota_-^{-1} : L^2(\partial\Omega) \to H^{-1/2}(\partial\Omega)$ to $H^2(\partial\Omega)$ is an isometric isomorphism from $H^2(\partial\Omega)$ onto $H^{3/2}(\partial\Omega)$ by Corollary 8.2.2. Furthermore, it follows from the definition that the extended Dirichlet-to-Neumann map $\widetilde{D}(\lambda)$ coincides with the Dirichlet-to-Neumann map $D(\lambda)$ on $H^{3/2}(\partial\Omega)$. With these observations it is clear that (8.4.10) follows when restricting (8.4.9) to $H^2(\partial\Omega)$. □

Remark 8.4.6. The boundary mappings in Theorem 8.4.1 and the corresponding γ-field and Weyl function depend on the choice of $\eta \in \rho(A_\mathrm{D}) \cap \mathbb{R}$ and the decomposition of $f \in \operatorname{dom} T_{\max}$ as $f = f_\mathrm{D}^\eta + f_\eta$; observe that also $f_\mathrm{D} = f_\mathrm{D}^\eta \in \operatorname{dom} A_\mathrm{D}$ depends on η. Suppose now that the boundary mappings are defined with respect to some other $\eta' \in \rho(A_\mathrm{D}) \cap \mathbb{R}$ and decompose f accordingly as $f = f_\mathrm{D}^{\eta'} + f_{\eta'}$. If $\Gamma_0^\eta, \Gamma_1^\eta$ denote the boundary mappings in Theorem 8.4.1 with respect to η, and $\Gamma_0^{\eta'}, \Gamma_1^{\eta'}$ denote the boundary mappings in Theorem 8.4.1 with respect to η', then one has

$$\begin{pmatrix} \Gamma_0^{\eta'} \\ \Gamma_1^{\eta'} \end{pmatrix} = \begin{pmatrix} I & 0 \\ -M(\eta') & I \end{pmatrix} \begin{pmatrix} \Gamma_0^{\eta} \\ \Gamma_1^{\eta} \end{pmatrix}. \tag{8.4.12}$$

In fact, that $\Gamma_0^{\eta'} f = \Gamma_0^{\eta} f$ for $f \in \operatorname{dom} T_{\max}$ is clear from Theorem 8.4.1, and for the remaining identity in (8.4.12) it follows from Lemma 8.4.5 that

$$\begin{aligned}
-M(\eta')\Gamma_0^{\eta} f + \Gamma_1^{\eta} f &= \iota_+\big(\widetilde{D}(\eta') - \widetilde{D}(\eta)\big)\iota_-^{-1}\Gamma_0^{\eta} f + \Gamma_1^{\eta} f \\
&= \iota_+\big(\widetilde{D}(\eta')\widetilde{\tau}_{\mathrm{D}} f - \widetilde{D}(\eta)\widetilde{\tau}_{\mathrm{D}} f\big) - \iota_+\tau_{\mathrm{N}} f_{\mathrm{D}}^{\eta} \\
&= \iota_+\big(\widetilde{D}(\eta')\widetilde{\tau}_{\mathrm{D}} f_{\eta'} - \widetilde{D}(\eta)\widetilde{\tau}_{\mathrm{D}} f_{\eta}\big) - \iota_+\tau_{\mathrm{N}} f_{\mathrm{D}}^{\eta} \\
&= \iota_+\big(\widetilde{\tau}_{\mathrm{N}} f_{\eta'} - \widetilde{\tau}_{\mathrm{N}} f_{\eta}\big) - \iota_+\tau_{\mathrm{N}} f_{\mathrm{D}}^{\eta} \\
&= \iota_+\tau_{\mathrm{N}}\big(f_{\mathrm{D}}^{\eta} - f_{\mathrm{D}}^{\eta'}\big) - \iota_+\tau_{\mathrm{N}} f_{\mathrm{D}}^{\eta} \\
&= \Gamma_1^{\eta'} f.
\end{aligned}$$

Finally, note that the γ-fields and Weyl functions of the boundary triplets in Theorem 8.4.1 for different η and η' transform accordingly; cf. Proposition 2.5.3.

Next some classes of extensions of $T_{\min}$ and their spectral properties are briefly discussed. Let $\{L^2(\partial\Omega), \Gamma_0, \Gamma_1\}$ be the boundary triplet in Theorem 8.4.1 with corresponding γ-field γ and Weyl function M in Proposition 8.4.4. According to Corollary 2.1.4, the self-adjoint (maximal dissipative, maximal accumulative) extensions $A_\Theta \subset T_{\max}$ of $T_{\min}$ are in a one-to-one correspondence to the self-adjoint (maximal dissipative, maximal accumulative) relations Θ in $L^2(\partial\Omega)$ via

$$\begin{aligned}
\operatorname{dom} A_\Theta &= \big\{ f \in \operatorname{dom} T_{\max} : \{\Gamma_0 f, \Gamma_1 f\} \in \Theta \big\} \\
&= \big\{ f \in \operatorname{dom} T_{\max} : \{\iota_-\widetilde{\tau}_{\mathrm{D}} f, -\iota_+\tau_{\mathrm{N}} f_{\mathrm{D}}\} \in \Theta \big\}.
\end{aligned} \tag{8.4.13}$$

If Θ is an operator in $L^2(\partial\Omega)$, then the domain of A_Θ is given by

$$\operatorname{dom} A_\Theta = \big\{ f \in \operatorname{dom} T_{\max} : \Theta\iota_-\widetilde{\tau}_{\mathrm{D}} f = -\iota_+\tau_{\mathrm{N}} f_{\mathrm{D}} \big\}. \tag{8.4.14}$$

Let Θ be a self-adjoint relation in $L^2(\partial\Omega)$ and let A_Θ be the corresponding self-adjoint realization of $-\Delta + V$ in $L^2(\Omega)$. By Corollary 1.10.9, Θ can be represented in terms of bounded operators $\mathcal{A}, \mathcal{B} \in \mathbf{B}(L^2(\partial\Omega))$ satisfying the conditions $\mathcal{A}^*\mathcal{B} = \mathcal{B}^*\mathcal{A}$, $\mathcal{A}\mathcal{B}^* = \mathcal{B}\mathcal{A}^*$, and $\mathcal{A}^*\mathcal{A} + \mathcal{B}^*\mathcal{B} = I = \mathcal{A}\mathcal{A}^* + \mathcal{B}\mathcal{B}^*$ such that

$$\Theta = \big\{ \{\mathcal{A}\varphi, \mathcal{B}\varphi\} : \varphi \in L^2(\partial\Omega) \big\} = \big\{ \{\psi, \psi'\} : \mathcal{A}^*\psi' = \mathcal{B}^*\psi \big\}.$$

In this case one has

$$\operatorname{dom} A_\Theta = \big\{ f \in \operatorname{dom} T_{\max} : -\mathcal{A}^*\iota_+\tau_{\mathrm{N}} f_{\mathrm{D}} = \mathcal{B}^*\iota_-\widetilde{\tau}_{\mathrm{D}} f \big\},$$

and for $\lambda \in \rho(A_\Theta) \cap \rho(A_{\mathrm{D}})$ the Kreĭn formula for the corresponding resolvents

$$\begin{aligned}
(A_\Theta - \lambda)^{-1} &= (A_{\mathrm{D}} - \lambda)^{-1} + \gamma(\lambda)\big(\Theta - M(\lambda)\big)^{-1}\gamma(\bar{\lambda})^* \\
&= (A_{\mathrm{D}} - \lambda)^{-1} + \gamma(\lambda)\mathcal{A}\big(\mathcal{B} - M(\lambda)\mathcal{A}\big)^{-1}\gamma(\bar{\lambda})^*
\end{aligned} \tag{8.4.15}$$

holds by Theorem 2.6.1 and Corollary 2.6.3. Recall that in the present situation the spectrum of $A_D = A_0$ is discrete by Proposition 8.3.2. According to Theorem 2.6.2, $\lambda \in \rho(A_D)$ is an eigenvalue of A_Θ if and only if $\ker(\Theta - M(\lambda))$ or, equivalently, $\ker(\mathcal{B} - M(\lambda)\mathcal{A})$ is nontrivial, and that

$$\ker(A_\Theta - \lambda) = \gamma(\lambda)\ker\bigl(\Theta - M(\lambda)\bigr) = \gamma(\lambda)\mathcal{A}\ker\bigl(\mathcal{B} - M(\lambda)\mathcal{A}\bigr).$$

Although Ω is a bounded C^2-domain, it will turn out in Example 8.4.9 that the spectrum of A_Θ is in general not discrete, and thus continuous spectrum may be present. It then follows from Theorem 2.6.2 and Theorem 2.6.5 that $\lambda \in \rho(A_D)$ belongs to the continuous spectrum $\sigma_c(A_\Theta)$ (essential spectrum $\sigma_{ess}(A_\Theta)$ or discrete spectrum $\sigma_d(A_\Theta)$) of A_Θ if and only if 0 belongs to $\sigma_c(\Theta - M(\lambda))$ ($\sigma_{ess}(\Theta - M(\lambda))$ or $\sigma_d(\Theta - M(\lambda))$).

For a complete description of the spectrum of A_Θ recall that the symmetric operator T_{min} is simple according to Proposition 8.3.13 and make use of a transform of the boundary triplet $\{L^2(\partial\Omega), \Gamma_0, \Gamma_1\}$ as in Chapter 3.8. This reasoning implies that λ is an eigenvalue of A_Θ if and only if λ is a pole of the function

$$\lambda \mapsto M_\Theta(\lambda) = \bigl(\mathcal{A}^* + \mathcal{B}^* M(\lambda)\bigr)\bigl(\mathcal{B}^* - \mathcal{A}^* M(\lambda)\bigr)^{-1}.$$

It is important to note in this context that the multiplicity of the eigenvalues of A_Θ is not necessarily finite and that the dimension of the eigenspace $\ker(A_\Theta - \lambda)$ of an isolated eigenvalue λ of A_Θ coincides with the dimension of the range of the residue of M_Θ at λ. Furthermore, the continuous and absolutely continuous spectrum of A_Θ can be characterized as in Section 3.8, e.g., one has

$$\sigma_{ac}(A_\Theta) = \overline{\bigcup_{\varphi \in L^2(\partial\Omega)} \operatorname{clos}_{ac}\bigl(\bigl\{x \in \mathbb{R} : 0 < \operatorname{Im}\bigl(M_\Theta(x + i0)\varphi, \varphi\bigr)_{L^2(\partial\Omega)} < \infty\bigr\}\bigr)}.$$

In the special case that the self-adjoint relation Θ in $L^2(\partial\Omega)$ is a bounded operator the boundary condition reads as in (8.4.14) and according to Section 3.8 the spectral properties of the self-adjoint operator A_Θ can also be described with the help of the function

$$\lambda \mapsto \bigl(\Theta - M(\lambda)\bigr)^{-1}.$$

The general boundary conditions in (8.4.13) and (8.4.14) contain also typical classes of boundary conditions that are treated in spectral problems for partial differential operators, as, e.g., Neumann or Robin type boundary conditions. In the following example the standard Neumann boundary conditions are discussed. Note that the Neumann operator does not coincide with the Kreĭn type extension $S_{K,\eta}$ or the Kreĭn–von Neumann extension $S_{K,0}$ of T_{min} in Proposition 8.4.3.

Example 8.4.7. Let $\{L^2(\partial\Omega), \Gamma_0, \Gamma_1\}$ be the boundary triplet in Theorem 8.4.1 and choose $\eta \in \rho(A_D) \cap \mathbb{R}$ in (8.4.1) in such a way that also $\eta \in \rho(A_N)$, where

$A_{\rm N}$ denotes the Neumann realization of $-\Delta + V$ in Proposition 8.3.3 and Theorem 8.3.4. Since both self-adjoint operators $A_{\rm D}$ and $A_{\rm N}$ are semibounded from below (or both have discrete spectrum), such an η exists. In this situation it follows that the Dirichlet-to-Neumann map

$$D(\eta) : H^{3/2}(\partial\Omega) \to H^{1/2}(\partial\Omega)$$

in Definition 8.3.6 is a bijective mapping. Furthermore, $\iota_+ : H^{1/2}(\partial\Omega) \to L^2(\partial\Omega)$ is bijective and the restriction of $\iota_-^{-1} : L^2(\partial\Omega) \to H^{-1/2}(\partial\Omega)$ to $H^2(\partial\Omega)$ is an isometric isomorphism from $H^2(\partial\Omega)$ onto $H^{3/2}(\partial\Omega)$ according to Corollary 8.2.2. Hence, it is clear that

$$\Theta_{\rm N} := \iota_+ D(\eta)\iota_-^{-1}, \qquad \operatorname{dom}\Theta_{\rm N} := H^2(\partial\Omega), \tag{8.4.16}$$

is a densely defined bijective operator in $L^2(\partial\Omega)$. Furthermore, for $\varphi \in H^2(\partial\Omega)$ and $\psi = \iota_-^{-1}\varphi \in H^{3/2}(\partial\Omega)$ it follows from Corollary 8.2.2 that

$$\bigl(\iota_+ D(\eta)\iota_-^{-1}\varphi, \varphi\bigr)_{L^2(\partial\Omega)} = \bigl(\iota_+ D(\eta)\psi, \iota_-\psi\bigr)_{L^2(\partial\Omega)} = (D(\eta)\psi, \psi)_{L^2(\partial\Omega)}. \tag{8.4.17}$$

Now choose $f_\eta \in H^2(\Omega)$ such that $(-\Delta + V)f_\eta = \eta f_\eta$ and $\tau_{\rm D} f_\eta = \psi$, which is possible by (8.3.9) and (8.2.12). Then it follows from Definition 8.3.6 and the first Green identity in (8.2.18) that

$$\begin{aligned}
(D(\eta)\psi, \psi)_{L^2(\partial\Omega)} &= \bigl(D(\eta)\tau_{\rm D} f_\eta, \tau_{\rm D} f_\eta\bigr)_{L^2(\partial\Omega)} \\
&= (\tau_{\rm N} f_\eta, \tau_{\rm D} f_\eta)_{L^2(\partial\Omega)} \\
&= \|\nabla f_\eta\|^2_{L^2(\Omega;\mathbb{C}^n)} + (\Delta f_\eta, f_\eta)_{L^2(\Omega)} \\
&= \|\nabla f_\eta\|^2_{L^2(\Omega;\mathbb{C}^n)} + ((V-\eta) f_\eta, f_\eta)_{L^2(\Omega)},
\end{aligned} \tag{8.4.18}$$

so that $(D(\eta)\psi, \psi)_{L^2(\partial\Omega)} \in \mathbb{R}$ and hence $(\Theta_{\rm N}\varphi, \varphi)_{L^2(\partial\Omega)} \in \mathbb{R}$ by (8.4.16)–(8.4.17) for all $\varphi \in H^2(\partial\Omega)$. It follows that the bijective operator $\Theta_{\rm N}$ is symmetric in $L^2(\partial\Omega)$, and hence $\Theta_{\rm N}$ is an unbounded self-adjoint operator in $L^2(\partial\Omega)$ such that $0 \in \rho(\Theta_{\rm N})$.

The self-adjoint realization of $-\Delta + V$ in $L^2(\Omega)$ corresponding to the self-adjoint operator $\Theta_{\rm N}$ in (8.4.16) is denoted by $A_{\Theta_{\rm N}}$. A function $f \in \operatorname{dom} T_{\max}$ belongs to $\operatorname{dom} A_{\Theta_{\rm N}}$ if and only if

$$\Gamma_0 f = \iota_- \widetilde{\tau}_{\rm D} f \in \operatorname{dom}\Theta_{\rm N} \quad \text{and} \quad \Gamma_1 f = \Theta_{\rm N}\Gamma_0 f.$$

Note that $\iota_-\widetilde{\tau}_{\rm D} f \in \operatorname{dom}\Theta_{\rm N}$ forces $\widetilde{\tau}_{\rm D} f \in H^{3/2}(\partial\Omega)$ and hence $f \in H^2(\Omega)$ and $\widetilde{\tau}_{\rm D} f = \tau_{\rm D} f$ by (8.2.12) and Theorem 8.3.9. It then follows from (8.4.7) and (8.4.16) that the boundary condition $\Gamma_1 f = \Theta_{\rm N}\Gamma_0 f$ takes on the form

$$\iota_+ D(\eta)\tau_{\rm D} f - \iota_+\tau_{\rm N} f = -\iota_+\tau_{\rm N} f_{\rm D} = \Gamma_1 f = \Theta_{\rm N}\Gamma_0 f = \iota_+ D(\eta)\tau_{\rm D} f,$$

that is, $\tau_{\mathrm{N}} f = 0$. Hence, it has been shown that $\operatorname{dom} A_{\Theta_{\mathrm{N}}} \subset H^2(\Omega)$ and that $\tau_{\mathrm{N}} f = 0$ for all $f \in \operatorname{dom} A_{\Theta_{\mathrm{N}}}$. Therefore, $A_{\Theta_{\mathrm{N}}} \subset A_{\mathrm{N}}$ and since both operators are self-adjoint one concludes that $A_{\Theta_{\mathrm{N}}} = A_{\mathrm{N}}$.

Note also that by (8.4.16) and Lemma 8.4.5 one has

$$\big(\Theta_{\mathrm{N}} - M(\lambda)\big)\varphi = \iota_+ D(\eta)\iota_-^{-1}\varphi - \iota_+\big(D(\eta) - D(\lambda)\big)\iota_-^{-1}\varphi = \iota_+ D(\lambda)\iota_-^{-1}\varphi$$

for $\varphi \in H^2(\partial\Omega)$ and $\lambda \in \rho(A_{\mathrm{D}})$. Hence, it follows that $\Theta_{\mathrm{N}} - M(\lambda)$ is a bijective operator in $L^2(\partial\Omega)$ for all $\lambda \in \rho(A_{\mathrm{D}}) \cap \rho(A_{\mathrm{N}})$ which is defined on $H^2(\partial\Omega)$. Therefore, (8.4.15) implies that the resolvents of A_{D} and A_{N} are related via

$$(A_{\mathrm{N}} - \lambda)^{-1} = (A_{\mathrm{D}} - \lambda)^{-1} + \gamma(\lambda)\iota_- D(\lambda)^{-1}\iota_+^{-1}\gamma(\bar{\lambda})^*,$$

where γ is the γ-field corresponding to the boundary triplet $\{L^2(\partial\Omega), \Gamma_0, \Gamma_1\}$ in Proposition 8.4.4.

The next example is a generalization of the previous example from Neumann to local and nonlocal Robin boundary conditions.

Example 8.4.8. Let $\{L^2(\partial\Omega), \Gamma_0, \Gamma_1\}$ be as in the previous example and fix some $\eta \in \rho(A_{\mathrm{D}}) \cap \rho(A_{\mathrm{N}}) \cap \mathbb{R}$. Then the operator $\Theta_{\mathrm{N}} = \iota_+ D(\eta)\iota_-^{-1}$ in (8.4.16) is an unbounded self-adjoint operator in $L^2(\partial\Omega)$ with domain $H^2(\partial\Omega)$, and $0 \in \rho(\Theta_{\mathrm{N}})$. Assume that

$$B : H^{3/2}(\partial\Omega) \to H^{1/2}(\partial\Omega) \tag{8.4.19}$$

is compact as an operator from $H^{3/2}(\partial\Omega)$ into $H^{1/2}(\partial\Omega)$ and that B is symmetric in $L^2(\partial\Omega)$, that is, $(B\psi, \psi)_{L^2(\partial\Omega)} \in \mathbb{R}$ for all $\psi \in \operatorname{dom} B = H^{3/2}(\partial\Omega)$. Then it follows that

$$\iota_+ B\iota_-^{-1} : H^2(\partial\Omega) \to L^2(\partial\Omega)$$

is compact as an operator from $H^2(\partial\Omega)$ into $L^2(\partial\Omega)$ and as in (8.4.17) one sees that $\iota_+ B\iota_-^{-1}$ is symmetric in $L^2(\partial\Omega)$. Consider the operator

$$\Theta_B := \iota_+\big(D(\eta) - B\big)\iota_-^{-1} = \Theta_{\mathrm{N}} - \iota_+ B\iota_-^{-1}, \quad \operatorname{dom}\Theta_B = H^2(\partial\Omega), \tag{8.4.20}$$

and observe that the symmetric operator $\iota_+ B\iota_-^{-1}$ is a relative compact perturbation of the self-adjoint operator Θ_{N} in $L^2(\partial\Omega)$, that is, the operator

$$\iota_+ B\iota_-^{-1}\Theta_{\mathrm{N}}^{-1}$$

is compact in $L^2(\partial\Omega)$. It is well known from standard perturbation results (see, e.g., [652, Corollary 2 of Theorem XIII.14]) that in this case the perturbed operator Θ_B is self-adjoint in $L^2(\partial\Omega)$.

The self-adjoint realization of $-\Delta + V$ in $L^2(\Omega)$ corresponding to the self-adjoint operator Θ_B in (8.4.20) is denoted by A_{Θ_B}. It is clear that a function $f \in \operatorname{dom} T_{\max}$ belongs to $\operatorname{dom} A_{\Theta_B}$ if and only if

$$\Gamma_0 f = \iota_-\widetilde{\tau}_{\mathrm{D}} f \in \operatorname{dom}\Theta_B \quad \text{and} \quad \Gamma_1 f = \Theta_B \Gamma_0 f. \tag{8.4.21}$$

In the same way as in Example 8.4.7 the fact that $\iota_-\widetilde{\tau}_{\mathrm{D}}f \in \operatorname{dom}\Theta_B$ implies that $f \in H^2(\Omega)$ and $\widetilde{\tau}_{\mathrm{D}}f = \tau_{\mathrm{D}}f$, and the boundary condition $\Gamma_1 f = \Theta_B\Gamma_0 f$ takes the explicit form

$$\iota_+ D(\eta)\tau_{\mathrm{D}}f - \iota_+\tau_{\mathrm{N}}f = \Gamma_1 f = \Theta_B\Gamma_0 f = \iota_+\bigl(D(\eta) - B\bigr)\tau_{\mathrm{D}}f,$$

that is, $\tau_{\mathrm{N}}f = B\tau_{\mathrm{D}}f$. Conversely, if $f \in H^2(\Omega)$ is such that $\tau_{\mathrm{N}}f = B\tau_{\mathrm{D}}f$, then f satisfies (8.4.21) and hence $f \in \operatorname{dom} A_{\Theta_B}$. Thus, it has been shown that the self-adjoint operator A_{Θ_B} is defined on

$$\operatorname{dom} A_{\Theta_B} = \bigl\{f \in H^2(\Omega) : \tau_{\mathrm{N}}f = B\tau_{\mathrm{D}}f\bigr\}.$$

In the same way as in the previous example one obtains

$$\Theta_B - M(\lambda) = \iota_+\bigl(D(\lambda) - B\bigr)\iota_-^{-1}$$

and hence for all $\lambda \in \rho(A_{\mathrm{D}}) \cap \rho(A_{\Theta_B})$ one has

$$(A_{\Theta_B} - \lambda)^{-1} = (A_{\mathrm{D}} - \lambda)^{-1} + \gamma(\lambda)\iota_-\bigl(D(\lambda) - B\bigr)^{-1}\iota_+^{-1}\gamma(\bar{\lambda})^*.$$

Finally, note that a sufficient condition for the operator B in (8.4.19) to be compact is that $B : H^{3/2}(\partial\Omega) \to H^{1/2+\varepsilon}(\partial\Omega)$ is bounded for some $\varepsilon > 0$, or that $B : H^{3/2-\varepsilon'}(\partial\Omega) \to H^{1/2}(\partial\Omega)$ is bounded for some $\varepsilon' > 0$, since the embeddings $H^{1/2+\varepsilon}(\partial\Omega) \hookrightarrow H^{1/2}(\partial\Omega)$ and $H^{3/2}(\partial\Omega) \hookrightarrow H^{3/2-\varepsilon'}(\partial\Omega)$ are compact by (8.2.8).

In the next example it is shown that the (essential) spectrum of a self-adjoint realization A_Θ of $-\Delta + V$ can be very general, depending on the properties of the parameter Θ. In particular, the self-adjoint realization A_Θ may not be semi-bounded.

Example 8.4.9. Let $\eta \in \rho(A_{\mathrm{D}}) \cap \mathbb{R}$, consider an arbitrary self-adjoint operator Ξ in the Hilbert space $\mathfrak{N}_\eta(T_{\max}) = \ker(T_{\max} - \eta)$, and assume that $\eta \in \rho(\Xi)$. Denote by $P_{\mathfrak{N}_\eta}$ the orthogonal projection in $L^2(\Omega)$ onto $\mathfrak{N}_\eta(T_{\max})$ and let $\iota_{\mathfrak{N}_\eta}$ be the natural embedding of $\mathfrak{N}_\eta(T_{\max})$ into $L^2(\Omega)$.

Let $\{L^2(\partial\Omega), \Gamma_0, \Gamma_1\}$ be the boundary triplet in Theorem 8.4.1 with corresponding γ-field and Weyl function in Proposition 8.4.4. Note that $M(\eta) = 0$ and that both

$$P_{\mathfrak{N}_\eta}\gamma(\eta) : L^2(\partial\Omega) \to \mathfrak{N}_\eta(T_{\max}) \quad \text{and} \quad \gamma(\eta)^*\iota_{\mathfrak{N}_\eta} : \mathfrak{N}_\eta(T_{\max}) \to L^2(\partial\Omega)$$

are isomorphisms. It follows that

$$\Theta := \bigl(\gamma(\eta)^*\iota_{\mathfrak{N}_\eta}\bigr)(\Xi - \eta)\bigl(P_{\mathfrak{N}_\eta}\gamma(\eta)\bigr)$$

is a self-adjoint operator in $L^2(\partial\Omega)$ with $0 \in \rho(\Theta)$ and

$$\Theta^{-1} = \bigl(P_{\mathfrak{N}_\eta}\gamma(\eta)\bigr)^{-1}(\Xi - \eta)^{-1}\bigl(\gamma(\eta)^*\iota_{\mathfrak{N}_\eta}\bigr)^{-1}. \tag{8.4.22}$$

Let A_Θ be the corresponding self-adjoint realization of $-\Delta+V$ in (8.4.13)–(8.4.14) defined on

$$\operatorname{dom} A_\Theta = \bigl\{ f \in \operatorname{dom} T_{\max} : \Theta\iota_- \widetilde{\tau}_{\mathrm{D}} f = -\iota_+ \tau_{\mathrm{N}} f_{\mathrm{D}} \bigr\}.$$

Since $M(\eta) = 0$ and $\eta \in \mathbb{R}$, Kreĭn's formula in (8.4.15) takes the form

$$\begin{aligned}(A_\Theta - \eta)^{-1} &= (A_{\mathrm{D}} - \eta)^{-1} + \gamma(\eta)\Theta^{-1}\gamma(\eta)^* \\ &= (A_{\mathrm{D}} - \lambda)^{-1} + \begin{pmatrix} P_{\mathfrak{N}_\eta}\gamma(\eta)\Theta^{-1}\gamma(\eta)^*\iota_{\mathfrak{N}_\eta} & 0 \\ 0 & 0 \end{pmatrix},\end{aligned}$$

where the block operator matrix is acting with respect to the space decomposition $L^2(\Omega) = \mathfrak{N}_\eta(T_{\max}) \oplus (\mathfrak{N}_\eta(T_{\max}))^\perp$. Using (8.4.22) one then concludes that

$$(A_\Theta - \eta)^{-1} = (A_{\mathrm{D}} - \eta)^{-1} + \begin{pmatrix} (\Xi - \eta)^{-1} & 0 \\ 0 & 0 \end{pmatrix}.$$

In particular, since $(A_{\mathrm{D}} - \eta)^{-1}$ is compact, well-known perturbation results show that

$$\sigma_{\mathrm{ess}}\bigl((A_\Theta - \eta)^{-1}\bigr) = \sigma_{\mathrm{ess}}\bigl((\Xi - \eta)^{-1}\bigr) \cup \{0\},$$

and hence $\sigma_{\mathrm{ess}}(A_\Theta) = \sigma_{\mathrm{ess}}(\Xi)$.

8.5 Semibounded Schrödinger operators

The semibounded self-adjoint realizations of $-\Delta + V$, where $V \in L^\infty(\Omega)$ is real, and the corresponding densely defined closed semibounded forms in $L^2(\Omega)$ are described in this section. For this purpose it is convenient to construct a boundary pair which is compatible with the boundary triplet in Theorem 8.4.1 and to apply the general results from Section 5.6. Under the additional assumption that $V \geq 0$, the nonnegative realizations of $-\Delta + V$ and the corresponding nonnegative forms in $L^2(\Omega)$ are discussed as a special case. In this situation the Kreĭn–von Neumann extension appears as the smallest nonnegative extension.

Let $\Omega \subset \mathbb{R}^n$ be a bounded C^2-domain and let A_{D} be the self-adjoint Dirichlet realization of $-\Delta + V$. It is clear from Proposition 8.3.2 that A_{D} coincides with the Friedrichs extension of the minimal operator $T_{\min}$ in (8.3.2) and that A_{D} is bounded from below with lower bound $m(A_{\mathrm{D}}) > v_-$, where $v_- = \operatorname{essinf} V$. Furthermore, the resolvent of A_{D} is compact since the domain Ω is bounded. Therefore, the following description of the semibounded self-adjoint extensions of $T_{\min}$ is an immediate consequence of Proposition 5.5.6 and Proposition 5.5.8.

Proposition 8.5.1. *Let $\Omega \subset \mathbb{R}^n$ be a bounded C^2-domain, let $\{L^2(\partial\Omega), \Gamma_0, \Gamma_1\}$ be the boundary triplet for $(T_{\min})^* = T_{\max}$ from Theorem 8.4.1, and let*

$$\begin{aligned} A_\Theta &= -\Delta + V, \\ \operatorname{dom} A_\Theta &= \bigl\{ f \in \operatorname{dom} T_{\max} : \{\Gamma_0 f, \Gamma_1 f\} \in \Theta \bigr\}, \end{aligned}$$

be a self-adjoint extension of $T_{\min}$ in $L^2(\Omega)$ corresponding to a self-adjoint relation Θ in $L^2(\partial\Omega)$ as in (8.4.13). Then

$$A_\Theta \text{ is semibounded} \quad \Leftrightarrow \quad \Theta \text{ is semibounded.}$$

Recall also from Section 8.3 that the densely defined closed semibounded form $\mathfrak{t}_{A_\mathrm{D}}$ corresponding to A_D is defined on $H_0^1(\Omega)$. Now fix some $\eta < m(A_\mathrm{D})$, use the direct sum decomposition

$$\operatorname{dom} T_{\max} = \mathfrak{N}_\eta(T_{\max}) \dotplus \operatorname{dom} A_\mathrm{D} = \mathfrak{N}_\eta(T_{\max}) \dotplus \big(H^2(\Omega) \cap H_0^1(\Omega)\big) \tag{8.5.1}$$

from (8.4.1) and Proposition 8.3.2, and consider the corresponding boundary triplet $\{L^2(\partial\Omega), \Gamma_0, \Gamma_1\}$ for $(T_{\min})^* = T_{\max}$ in Theorem 8.4.1 given by

$$\Gamma_0 f = \iota_- \widetilde{\tau}_\mathrm{D} f \quad \text{and} \quad \Gamma_1 f = -\iota_+ \tau_\mathrm{N} f_\mathrm{D}, \tag{8.5.2}$$

where $f = f_\eta + f_\mathrm{D} \in \operatorname{dom} T_{\max}$ with $f_\eta \in \mathfrak{N}_\eta(T_{\max})$ and $f_\mathrm{D} \in \operatorname{dom} A_\mathrm{D}$; cf. (8.5.1). It is clear that $A_0 = A_\mathrm{D}$ coincides with the Friedrichs extension of $T_{\min}$ and $A_1 = T_{\min} \,\widehat{+}\, \widehat{\mathfrak{N}}_\eta(T_{\max})$ coincides with the Kreĭn type extension $S_{\mathrm{K},\eta}$ of $T_{\min}$; cf. Definition 5.4.2. In order to define a boundary pair for $T_{\min}$ corresponding to $A_1 = S_{\mathrm{K},\eta}$, consider the densely defined closed semibounded form $\mathfrak{t}_{S_{\mathrm{K},\eta}}$ associated with $S_{\mathrm{K},\eta}$ and recall from Corollary 5.4.16 the direct sum decomposition

$$\operatorname{dom} \mathfrak{t}_{S_{\mathrm{K},\eta}} = \mathfrak{N}_\eta(T_{\max}) \dotplus \operatorname{dom} \mathfrak{t}_{A_\mathrm{D}} = \mathfrak{N}_\eta(T_{\max}) \dotplus H_0^1(\Omega) \tag{8.5.3}$$

of $\operatorname{dom} \mathfrak{t}_{S_{\mathrm{K},\eta}}$. Comparing (8.5.1) and (8.5.3) one sees that $\operatorname{dom} T_{\max} \subset \operatorname{dom} \mathfrak{t}_{S_{\mathrm{K},\eta}}$ and that the domain of the Dirichlet operator A_D in (8.5.1) is replaced by the corresponding form domain in (8.5.3). The functions $f \in \operatorname{dom} \mathfrak{t}_{S_{\mathrm{K},\eta}}$ will be written in the form $f = f_\eta + f_\mathrm{F}$, where $f_\eta \in \mathfrak{N}_\eta(T_{\max})$ and $f_\mathrm{F} \in \operatorname{dom} \mathfrak{t}_{A_\mathrm{D}} = H_0^1(\Omega)$. Now define the mapping

$$\Lambda : \operatorname{dom} \mathfrak{t}_{S_{\mathrm{K},\eta}} \to L^2(\partial\Omega), \quad f \mapsto \Lambda f = \iota_- \widetilde{\tau}_\mathrm{D} f_\eta. \tag{8.5.4}$$

It will be shown next that $\{L^2(\partial\Omega), \Lambda\}$ is a boundary pair that is compatible with the boundary triplet $\{L^2(\partial\Omega), \Gamma_0, \Gamma_1\}$ in the sense of Definition 5.6.4; although the main part of the proof of Lemma 8.5.2 is similar to Example 5.6.9, the details are provided.

Lemma 8.5.2. *Let $\Omega \subset \mathbb{R}^n$ be a bounded C^2-domain and let A_D be the self-adjoint Dirichlet realization of $-\Delta + V$ with lower bound $m(A_\mathrm{D})$. Fix $\eta < m(A_\mathrm{D})$, let $\{L^2(\partial\Omega), \Gamma_0, \Gamma_1\}$ be the corresponding boundary triplet for $(T_{\min})^* = T_{\max}$ from Theorem 8.4.1, and let Λ be the mapping in (8.5.4). Then $\{L^2(\partial\Omega), \Lambda\}$ is a boundary pair for $T_{\min}$ corresponding to the Kreĭn type extension $S_{\mathrm{K},\eta}$ which is compatible with the boundary triplet $\{L^2(\partial\Omega), \Gamma_0, \Gamma_1\}$. Moreover, one has*

$$(T_{\max} f, g)_{L^2(\Omega)} = \mathfrak{t}_{S_{\mathrm{K},\eta}}[f, g] + (\Gamma_1 f, \Lambda g)_{L^2(\partial\Omega)} \tag{8.5.5}$$

for all $f \in \operatorname{dom} T_{\max}$ and $g \in \operatorname{dom} \mathfrak{t}_{S_{\mathrm{K},\eta}}$.

Proof. According to Lemma 5.6.5 (ii), it suffices to show that for some $a < \eta$ the mapping Λ in (8.5.4) is bounded from the Hilbert space

$$\mathfrak{H}_{\mathfrak{t}_{S_{K,\eta}}-a} = \big(\operatorname{dom}\mathfrak{t}_{S_{K,\eta}}, (\cdot,\cdot)_{\mathfrak{t}_{S_{K,\eta}}-a}\big)$$

to $L^2(\partial\Omega)$ and that Λ extends the mapping Γ_0 in (8.5.2). In the present situation it is clear that the compatibility condition $A_1 = S_{K,\eta}$ is satisfied.

In order to show that Λ is bounded fix some $a < \eta$, recall first from (5.1.7) that the Hilbert space norm on $\mathfrak{H}_{\mathfrak{t}_{S_{K,\eta}}-a}$ is given by

$$\|f\|^2_{\mathfrak{t}_{S_{K,\eta}}-a} = \mathfrak{t}_{S_{K,\eta}}[f] - a\|f\|^2_{L^2(\Omega)}, \qquad f \in \operatorname{dom}\mathfrak{t}_{S_{K,\eta}} = \mathfrak{H}_{\mathfrak{t}_{S_{K,\eta}}-a}.$$

It follows from Theorem 8.3.9 that the restriction $\iota_-\widetilde{\tau}_D : \mathfrak{N}_\eta(T_{\max}) \to L^2(\partial\Omega)$ is bounded and hence for $f = f_\eta + f_F \in \operatorname{dom}\mathfrak{t}_{S_{K,\eta}}$, decomposed according to (8.5.3) in $f_\eta \in \mathfrak{N}_\eta(T_{\max})$ and $f_F \in \operatorname{dom}\mathfrak{t}_{A_D}$, one has the estimate

$$\|\Lambda f\|^2_{L^2(\partial\Omega)} = \|\iota_-\widetilde{\tau}_D f_\eta\|^2_{L^2(\partial\Omega)} \leq C\|f_\eta\|^2_{L^2(\Omega)}. \tag{8.5.6}$$

Now the orthogonal sum decomposition

$$\operatorname{dom}\mathfrak{t}_{S_{K,\eta}} = \mathfrak{N}_a(T_{\max}) \oplus_{\mathfrak{t}_{S_{K,\eta}}-a} \operatorname{dom}\mathfrak{t}_{A_D} \tag{8.5.7}$$

from Corollary 5.4.15 will be used. To this end, define

$$f_a := \big(I + (a-\eta)(A_D - a)^{-1}\big)f_\eta$$

and note that $f = f_a + h_F$ with $f_a \in \mathfrak{N}_a(T_{\max})$ and $h_F = f_\eta - f_a + f_F \in \operatorname{dom}\mathfrak{t}_{A_D}$. Then one has

$$f_\eta = \big(I + (\eta - a)(A_D - \eta)^{-1}\big)f_a$$

and Proposition 1.4.6 leads to the estimate

$$\|f_\eta\|_{L^2(\Omega)} \leq \frac{m(A_D) - a}{m(A_D) - \eta}\|f_a\|_{L^2(\Omega)}. \tag{8.5.8}$$

Furthermore, it follows from (5.1.9) and the orthogonal sum decomposition (8.5.7) that

$$(\eta - a)\|f_a\|^2_{L^2(\Omega)} \leq \|f_a\|^2_{\mathfrak{t}_{S_{K,\eta}}-a} \leq \|f_a\|^2_{\mathfrak{t}_{S_{K,\eta}}-a} + \|h_F\|^2_{\mathfrak{t}_{S_{K,\eta}}-a} = \|f\|^2_{\mathfrak{t}_{S_{K,\eta}}-a}.$$

From this estimate, (8.5.6), and (8.5.8) one concludes that $\Lambda : \mathfrak{H}_{\mathfrak{t}_{S_{K,\eta}}-a} \to L^2(\partial\Omega)$ is bounded.

From the definition of Λ in (8.5.4) and the decompositions (8.5.1) and (8.5.3) it is clear that Λ is an extension of the mapping Γ_0 in (8.5.2). Moreover, by construction, the condition $A_1 = S_{K,\eta}$ is satisfied. Therefore, Lemma 5.6.5 (ii) shows that $\{L^2(\partial\Omega), \Lambda\}$ is a boundary pair for $T_{\min}$ corresponding to $S_{K,\eta}$ which is compatible with the boundary triplet $\{L^2(\partial\Omega), \Gamma_0, \Gamma_1\}$. The identity (8.5.5) follows from Corollary 5.6.7. □

The next theorem is a variant of Theorem 5.6.13 in the present situation.

Theorem 8.5.3. *Let $\Omega \subset \mathbb{R}^n$ be a bounded C^2-domain, let A_{D} be the self-adjoint Dirichlet realization of $-\Delta + V$ with lower bound $m(A_{\mathrm{D}})$, and fix $\eta < m(A_{\mathrm{D}})$. Let $\{L^2(\partial\Omega), \Gamma_0, \Gamma_1\}$ be the boundary triplet for $(T_{\min})^* = T_{\max}$ from Theorem 8.4.1 and let $\{L^2(\partial\Omega), \Lambda\}$ be the compatible boundary pair in Lemma 8.5.2. Furthermore, let Θ be a semibounded self-adjoint relation in $L^2(\partial\Omega)$ and let A_Θ be the corresponding semibounded self-adjoint extension of $T_{\min}$ in Proposition 8.5.1. Then the closed semibounded form ω_Θ in $L^2(\partial\Omega)$ corresponding to Θ and the densely defined closed semibounded form $\mathfrak{t}_{A_\Theta}$ corresponding to A_Θ are related by*

$$\begin{aligned} \mathfrak{t}_{A_\Theta}[f,g] &= \mathfrak{t}_{S_{\mathrm{K},\eta}}[f,g] + \omega_\Theta[\Lambda f, \Lambda g], \\ \operatorname{dom} \mathfrak{t}_{A_\Theta} &= \bigl\{ f \in \operatorname{dom} \mathfrak{t}_{S_{\mathrm{K},\eta}} : \Lambda f \in \operatorname{dom} \omega_\Theta \bigr\}. \end{aligned} \tag{8.5.9}$$

For completeness, the form $\mathfrak{t}_{A_\Theta}$ in Theorem 8.5.3 will be made more explicit using Corollary 5.6.14. First note that, by the definition of the boundary map Λ in (8.5.4) and the decomposition (8.5.3), one can rewrite (8.5.9) as

$$\begin{aligned} \mathfrak{t}_{A_\Theta}[f,g] &= \mathfrak{t}_{S_{\mathrm{K},\eta}}[f,g] + \omega_\Theta[\iota_-\widetilde{\tau}_{\mathrm{D}} f_\eta, \iota_-\widetilde{\tau}_{\mathrm{D}} g_\eta], \\ \operatorname{dom} \mathfrak{t}_{A_\Theta} &= \bigl\{ f = f_\eta + f_{\mathrm{F}} \in \operatorname{dom} \mathfrak{t}_{S_{\mathrm{K},\eta}} : \iota_-\widetilde{\tau}_{\mathrm{D}} f_\eta \in \operatorname{dom} \omega_\Theta \bigr\}. \end{aligned} \tag{8.5.10}$$

If $m(\Theta)$ denotes the lower bound of the semibounded self-adjoint relation Θ and $\mu \leq m(\Theta)$ is fixed, then the closed semibounded form $\mathfrak{t}_{A_\Theta}$ in (8.5.9)–(8.5.10) corresponding to A_Θ is given by

$$\begin{aligned} \mathfrak{t}_{A_\Theta}[f,g] &= \mathfrak{t}_{S_{\mathrm{K},\eta}}[f,g] + \bigl((\Theta_{\mathrm{op}} - \mu)^{\frac{1}{2}} \iota_-\widetilde{\tau}_{\mathrm{D}} f_\eta, (\Theta_{\mathrm{op}} - \mu)^{\frac{1}{2}} \iota_-\widetilde{\tau}_{\mathrm{D}} g_\eta \bigr)_{L^2(\partial\Omega)} \\ &\qquad + \mu \bigl(\iota_-\widetilde{\tau}_{\mathrm{D}} f_\eta, \iota_-\widetilde{\tau}_{\mathrm{D}} g_\eta \bigr)_{L^2(\partial\Omega)}, \\ \operatorname{dom} \mathfrak{t}_{A_\Theta} &= \bigl\{ f = f_\eta + f_{\mathrm{F}} \in \operatorname{dom} \mathfrak{t}_{S_{\mathrm{K},\eta}} : \iota_-\widetilde{\tau}_{\mathrm{D}} f_\eta \in \operatorname{dom} (\Theta_{\mathrm{op}} - \mu)^{\frac{1}{2}} \bigr\}; \end{aligned}$$

as usual, here Θ_{op} denotes the semibounded self-adjoint operator part of Θ acting in $L^2(\partial\Omega)_{\mathrm{op}} = \overline{\operatorname{dom} \Theta}$. In the special case where $\Theta_{\mathrm{op}} \in \mathbf{B}(L^2(\partial\Omega)_{\mathrm{op}})$ one has

$$\begin{aligned} \mathfrak{t}_{A_\Theta}[f,g] &= \mathfrak{t}_{S_{\mathrm{K},\eta}}[f,g] + \bigl(\Theta_{\mathrm{op}} \iota_-\widetilde{\tau}_{\mathrm{D}} f_\eta, \iota_-\widetilde{\tau}_{\mathrm{D}} g_\eta \bigr)_{L^2(\partial\Omega)}, \\ \operatorname{dom} \mathfrak{t}_{A_\Theta} &= \bigl\{ f = f_\eta + f_{\mathrm{F}} \in \operatorname{dom} \mathfrak{t}_{S_{\mathrm{K},\eta}} : \iota_-\widetilde{\tau}_{\mathrm{D}} f_\eta \in \operatorname{dom} \Theta_{\mathrm{op}} \bigr\}, \end{aligned}$$

and if $\Theta \in \mathbf{B}(L^2(\partial\Omega))$, then

$$\mathfrak{t}_{A_\Theta}[f,g] = \mathfrak{t}_{S_{\mathrm{K},\eta}}[f,g] + \bigl(\Theta \iota_-\widetilde{\tau}_{\mathrm{D}} f_\eta, \iota_-\widetilde{\tau}_{\mathrm{D}} g_\eta \bigr)_{L^2(\partial\Omega)}, \qquad \operatorname{dom} \mathfrak{t}_{A_\Theta} = \operatorname{dom} \mathfrak{t}_{S_{\mathrm{K},\eta}}.$$

Recall also from Corollary 5.4.15 that the form $\mathfrak{t}_{S_{\mathrm{K},\eta}}$ can be expressed in terms of the form $\mathfrak{t}_{A_{\mathrm{D}}}$ and the resolvent of A_{D}.

Finally, the special case $V \geq 0$ will be briefly considered. In this situation the minimal operator $T_{\min}$ and the Dirichlet operator A_{D} are both uniformly positive and hence in the above construction of a boundary triplet and corresponding boundary pair one may choose $\eta = 0$. More precisely, Theorem 8.4.1 has the following form.

Corollary 8.5.4. *Let $\Omega \subset \mathbb{R}^n$ be a bounded C^2-domain, let A_{D} be the self-adjoint Dirichlet realization of $-\Delta+V$ in $L^2(\Omega)$ with $V \geq 0$, and decompose $f \in \operatorname{dom} T_{\max}$ according to (8.4.1) with $\eta = 0$ in the form $f = f_{\mathrm{D}} + f_0$, where $f_{\mathrm{D}} \in \operatorname{dom} A_{\mathrm{D}}$ and $f_0 \in \ker T_{\max}$. Then $\{L^2(\partial\Omega), \Gamma_0, \Gamma_1\}$, where*

$$\Gamma_0 f = \iota_- \widetilde{\tau}_{\mathrm{D}} f \quad \text{and} \quad \Gamma_1 f = -\iota_+ \tau_{\mathrm{N}} f_{\mathrm{D}}, \qquad f = f_{\mathrm{D}} + f_0 \in \operatorname{dom} T_{\max},$$

is a boundary triplet for $(T_{\min})^* = T_{\max}$ such that

$$A_0 = A_{\mathrm{D}} \qquad \text{and} \qquad A_1 = T_{\min} \,\widehat{+}\, \widehat{\mathfrak{N}}_0(T_{\max})$$

coincide with the Friedrichs extension and the Kreĭn–von Neumann extension of $T_{\min}$, respectively.

It is clear from Proposition 8.4.4 that for all $\lambda \in \rho(A_{\mathrm{D}})$ the γ-field and Weyl function corresponding to the boundary triplet $\{L^2(\partial\Omega), \Gamma_0, \Gamma_1\}$ in Corollary 8.5.4 have the form

$$\gamma(\lambda)\varphi = \bigl(I + \lambda(A_{\mathrm{D}} - \lambda)^{-1}\bigr) f_0(\varphi), \quad \varphi \in L^2(\partial\Omega),$$

and

$$M(\lambda)\varphi = -\iota_+ \tau_{\mathrm{N}} \lambda (A_{\mathrm{D}} - \lambda)^{-1} f_0(\varphi), \quad \varphi \in L^2(\partial\Omega), \tag{8.5.11}$$

respectively, where $f_0(\varphi)$ is the unique element in $\mathfrak{N}_0(T_{\max})$ with the property that $\Gamma_0 f_0(\varphi) = \iota_- \widetilde{\tau}_{\mathrm{D}} f_0(\varphi) = \varphi$.

The next proposition is a variant of Proposition 8.5.1 for nonnegative extensions.

Proposition 8.5.5. *Let $\Omega \subset \mathbb{R}^n$ be a bounded C^2-domain, assume that $V \geq 0$, and let $\{L^2(\partial\Omega), \Gamma_0, \Gamma_1\}$ be the boundary triplet for $(T_{\min})^* = T_{\max}$ from Corollary 8.5.4. Let*

$$\begin{aligned} A_\Theta &= -\Delta + V, \\ \operatorname{dom} A_\Theta &= \bigl\{ f \in \operatorname{dom} T_{\max} : \{\Gamma_0 f, \Gamma_1 f\} \in \Theta \bigr\}, \end{aligned}$$

be a self-adjoint extension of $T_{\min}$ in $L^2(\Omega)$ corresponding to a self-adjoint relation Θ in $L^2(\partial\Omega)$ as in (8.4.13). Then

$$A_\Theta \text{ is nonnegative} \quad \Leftrightarrow \quad \Theta \text{ is nonnegative.}$$

Proof. Note that the Weyl function M in (8.5.11) satisfies $M(0) = 0$ and that $T_{\min}$ is uniformly positive. Therefore, if A_Θ is a nonnegative self-adjoint extension of $T_{\min}$, then Proposition 5.5.6 with $x = 0$ shows that the self-adjoint relation Θ in $L^2(\partial\Omega)$ is nonnegative. Conversely, if Θ is a nonnegative self-adjoint relation in $L^2(\partial\Omega)$, then it follows from Corollary 5.5.15 and $A_1 = S_{\mathrm{K},0} \geq 0$ that A_Θ is a nonnegative self-adjoint extension of $T_{\min}$. □

In the nonnegative case the boundary mapping Λ in (8.5.4) is given by

$$\Lambda : \operatorname{dom} \mathfrak{t}_{S_{\mathrm{K},0}} \to L^2(\partial\Omega), \quad f \mapsto \Lambda f = \iota_- \widetilde{\tau}_{\mathrm{D}} f_0, \tag{8.5.12}$$

where one has the direct sum decomposition

$$\operatorname{dom} \mathfrak{t}_{S_{K,0}} = \mathfrak{N}_0(T_{\max}) + \operatorname{dom} \mathfrak{t}_{A_{\mathrm{D}}} = \mathfrak{N}_0(T_{\max}) + H_0^1(\Omega),$$

and according to Lemma 8.5.2 $\{L^2(\partial\Omega), \Lambda\}$ is a boundary pair that is compatible with the boundary triplet $\{L^2(\partial\Omega), \Gamma_0, \Gamma_1\}$ in Corollary 8.5.4.

In the nonnegative case a description of the nonnegative extensions and their form domains is of special interest. In the present situation Corollary 5.6.18 reads as follows.

Corollary 8.5.6. *Let $\Omega \subset \mathbb{R}^n$ be a bounded C^2-domain, let A_{D} be the self-adjoint Dirichlet realization of $-\Delta + V$ with $V \geq 0$, let $\{L^2(\partial\Omega), \Gamma_0, \Gamma_1\}$ be the boundary triplet for $(T_{\min})^* = T_{\max}$ from Corollary 8.5.4, and let $\{L^2(\partial\Omega), \Lambda\}$ be the compatible boundary pair in (8.5.12). Then the formula*

$$\begin{aligned}
\mathfrak{t}_{A_\Theta}[f,g] &= \mathfrak{t}_{S_{\mathrm{K},0}}[f,g] + \bigl(\Theta_{\mathrm{op}}^{\frac{1}{2}} \iota_- \widetilde{\tau}_{\mathrm{D}} f_0, \Theta_{\mathrm{op}}^{\frac{1}{2}} \iota_- \widetilde{\tau}_{\mathrm{D}} g_0\bigr)_{L^2(\partial\Omega)}, \\
\operatorname{dom} \mathfrak{t}_{A_\Theta} &= \bigl\{ f = f_0 + f_{\mathrm{F}} \in \operatorname{dom} \mathfrak{t}_{S_{\mathrm{K},0}} : \iota_- \widetilde{\tau}_{\mathrm{D}} f_0 \in \operatorname{dom} \Theta_{\mathrm{op}}^{\frac{1}{2}} \bigr\},
\end{aligned}$$

establishes a one-to-one correspondence between all closed nonnegative forms $\mathfrak{t}_{A_\Theta}$ corresponding to nonnegative self-adjoint extension A_Θ of $T_{\min}$ in $L^2(\Omega)$ and all closed nonnegative forms ω_Θ corresponding to nonnegative self-adjoint relations Θ in $L^2(\partial\Omega)$.

8.6 Coupling of Schrödinger operators

The aim of this section is to interpret the *natural* self-adjoint Schrödinger operator

$$A = -\Delta + V, \qquad \operatorname{dom} A = H^2(\mathbb{R}^n), \tag{8.6.1}$$

in $L^2(\mathbb{R}^n)$ with a real potential $V \in L^\infty(\mathbb{R}^n)$ as a coupling of Schrödinger operators on a bounded C^2-domain and its complement, that is, A is identified as a self-adjoint extension of the orthogonal sum of the minimal Schrödinger operators on the subdomains and its resolvent is expressed in a Kreĭn type resolvent formula. The present treatment is a multidimensional variant of the discussion in Section 6.5 and is based on the abstract coupling construction in Section 4.6.

Let $\Omega_+ \subset \mathbb{R}^n$ be a bounded C^2-domain and let $\Omega_- := \mathbb{R}^n \setminus \overline{\Omega}_+$ be the corresponding exterior domain. Since $\mathcal{C} := \partial\Omega_- = \partial\Omega_+$ is C^2-smooth in the sense of Definition 8.2.1, the term C^2-domain will be used here for Ω_-, although Ω_- is unbounded. In the following the common boundary $\mathcal{C}$ is sometimes referred to as an interface, linking the two domains Ω_+ and Ω_-. Note that one has the identification

$$L^2(\mathbb{R}^n) = L^2(\Omega_+) \oplus L^2(\Omega_-). \tag{8.6.2}$$

Consider the Schrödinger operator $A = -\Delta + V$, $\operatorname{dom} A = H^2(\mathbb{R}^n)$, in (8.6.1) with $V \in L^\infty(\mathbb{R}^n)$ real. Since the Laplacian $-\Delta$ defined on $H^2(\mathbb{R}^n)$ is unitarily equivalent in $L^2(\mathbb{R}^n)$ via the Fourier transform to the maximal multiplication operator with the function $x \mapsto |x|^2$, it is clear that $-\Delta$, and hence A in (8.6.1), is self-adjoint in $L^2(\mathbb{R}^n)$. Moreover, for $f \in C_0^\infty(\mathbb{R}^n)$ integration by parts shows that

$$(Af, f)_{L^2(\mathbb{R}^n)} = (\nabla f, \nabla f)_{L^2(\mathbb{R}^n;\mathbb{C}^n)} + (Vf, f)_{L^2(\mathbb{R}^n)} \geq v_- \|f\|^2_{L^2(\mathbb{R}^n)},$$

where $v_- = \operatorname{essinf} V$. As $C_0^\infty(\mathbb{R}^n)$ is dense in $H^2(\mathbb{R}^n)$, this estimate extends to $H^2(\mathbb{R}^n)$. Therefore, A is semibounded from below and v_- is a lower bound.

The restriction of the real function $V \in L^\infty(\Omega)$ to $\Omega_\pm$ is denoted by $V_\pm$ and the same $\pm$-index notation will be used for the restriction $f_\pm \in L^2(\Omega_\pm)$ of an element $f \in L^2(\mathbb{R}^n)$. The minimal and maximal operator associated with $-\Delta + V_+$ in $L^2(\Omega_+)$ will be denoted by $T^+_{\min}$ and $T^+_{\max}$, respectively, and the self-adjoint Dirichlet realization in $L^2(\Omega_+)$ will be denoted by A^+_{D}; cf. Proposition 8.3.1, Proposition 8.3.2, and Theorem 8.3.4. For the minimal operator

$$T^-_{\min} = -\Delta + V_-, \qquad \operatorname{dom} T^-_{\min} = H^2_0(\Omega_-),$$

and the maximal operator

$$\begin{aligned} T^-_{\max} &= -\Delta + V_-, \\ \operatorname{dom} T^-_{\max} &= \bigl\{ f_- \in L^2(\Omega_-) : -\Delta f_- + V_- f_- \in L^2(\Omega_-) \bigr\}, \end{aligned}$$

on the unbounded C^2-domain one can show in the same way as in the proof of Proposition 8.3.1 that $(T^-_{\min})^* = T^-_{\max}$ and $T^-_{\min} = (T^-_{\max})^*$. Furthermore, since Ω_- has a compact C^2-smooth boundary, it follows by analogy to Theorem 8.3.4 that the self-adjoint Dirichlet realization A^-_{D} corresponding to the densely defined closed semibounded form

$$\mathfrak{t}^-_{\mathrm{D}}[f_-, g_-] = (\nabla f_-, \nabla g_-)_{L^2(\Omega_-;\mathbb{C}^n)} + (V_- f_-, g_-)_{L^2(\Omega_-)}, \quad \operatorname{dom} \mathfrak{t}^-_{\mathrm{D}} = H^1_0(\Omega_-),$$

via the first representation theorem (Theorem 5.1.18) is given by

$$A^-_{\mathrm{D}} = -\Delta + V_-, \quad \operatorname{dom} A^-_{\mathrm{D}} = \bigl\{ f_- \in H^2(\Omega_-) : \tau^-_{\mathrm{D}} f_- = 0 \bigr\},$$

where τ^-_{D} denotes the Dirichlet trace operator on Ω_-; cf. (8.2.13). The operator A^-_{D} is semibounded from below and $v_- = \operatorname{essinf} V$ is a lower bound. In contrast to the Dirichlet operator A^+_{D}, the resolvent of A^-_{D} is not compact since Rellich's theorem is not valid on the unbounded domain Ω_-; cf. the proof of Proposition 8.3.2. Note also that the Dirichlet trace operator $\tau^-_{\mathrm{D}} : H^2(\Omega_-) \to H^{3/2}(\mathcal{C})$ and Neumann trace operator $\tau_N^- : H^2(\Omega_-) \to H^{1/2}(\mathcal{C})$ have the same mapping properties as on a bounded domain. Moreover, both trace operators admit continuous extensions to $\operatorname{dom} T_{\max}^-$ as in Theorem 8.3.9 and Theorem 8.3.10. With the identification (8.6.2) it is clear that the orthogonal sum

$$\widetilde{A}_D = \begin{pmatrix} A_D^+ & 0 \\ 0 & A_D^- \end{pmatrix} \tag{8.6.3}$$

is a self-adjoint operator in $L^2(\mathbb{R}^n)$ with Dirichlet boundary conditions on $\mathcal{C}$. The goal of the following considerations is to identify the self-adjoint Schrödinger operator A in (8.6.1) as a self-adjoint extension of the orthogonal sum of the minimal operators $T_{\min}^\pm$ and to compare A with the orthogonal sum $\widetilde{A}_D$ in (8.6.3) using a Kreĭn type resolvent formula.

From now on it is assumed that $\eta < \operatorname{essinf} V$ is fixed, so that, in particular, $\eta \in \rho(A_D^+) \cap \rho(A_D^-) \cap \mathbb{R}$. Consider the boundary triplet $\{L^2(\mathcal{C}), \Gamma_0^+, \Gamma_1^+\}$ for $T_{\max}^+$ in Theorem 8.4.1, that is,

$$\Gamma_0^+ f_+ = \iota_- \widetilde{\tau}_D^+ f_+ \quad \text{and} \quad \Gamma_1^+ f_+ = -\iota_+ \tau_N^+ f_{D,+}, \quad f_+ \in \operatorname{dom} T_{\max}^+,$$

where $f_+ = f_{D,+} + f_{\eta,+}$ with $f_{D,+} \in \operatorname{dom} A_D^+$ and $f_{\eta,+} \in \mathfrak{N}_\eta(T_{\max}^+)$. In the same way as in the proof of Theorem 8.4.1 one verifies that $\{L^2(\mathcal{C}), \Gamma_0^-, \Gamma_1^-\}$, where

$$\Gamma_0^- f_- = \iota_- \widetilde{\tau}_D^- f_- \quad \text{and} \quad \Gamma_1^- f_- = -\iota_+ \tau_N^- f_{D,-}, \quad f_- \in \operatorname{dom} T_{\max}^-,$$

where $f_- = f_{D,-} + f_{\eta,-}$ with $f_{D,-} \in \operatorname{dom} A_D^-$ and $f_{\eta,-} \in \mathfrak{N}_\eta(T_{\max}^-)$, is a boundary triplet for $T_{\max}^-$ such that $\operatorname{dom} A_D^- = \ker \Gamma_0^-$. The γ-fields and Weyl functions in Proposition 8.4.4 corresponding to the boundary triplets $\{L^2(\mathcal{C}), \Gamma_0^\pm, \Gamma_1^\pm\}$ are denoted by $\gamma_\pm$ and $M_\pm$, respectively.

In analogy to Section 4.6, the orthogonal coupling of the boundary triplets $\{L^2(\mathcal{C}), \Gamma_0^+, \Gamma_1^+\}$ and $\{L^2(\mathcal{C}), \Gamma_0^-, \Gamma_1^-\}$ leads to the boundary triplet

$$\{L^2(\mathcal{C}) \oplus L^2(\mathcal{C}), \widetilde{\Gamma}_0, \widetilde{\Gamma}_1\} \tag{8.6.4}$$

for the orthogonal sum $T_{\max} := T_{\max}^+ \mathbin{\widehat{\oplus}} T_{\max}^-$ of the maximal operator $T_{\max}^\pm$, where

$$\widetilde{\Gamma}_0 f = \begin{pmatrix} \Gamma_0^+ f_+ \\ \Gamma_0^- f_- \end{pmatrix} = \begin{pmatrix} \iota_- \widetilde{\tau}_D^+ f_+ \\ \iota_- \widetilde{\tau}_D^- f_- \end{pmatrix}, \quad f = \begin{pmatrix} f_+ \\ f_- \end{pmatrix}, \quad f_\pm \in \operatorname{dom} T_{\max}^\pm, \tag{8.6.5}$$

and

$$\widetilde{\Gamma}_1 f = \begin{pmatrix} \Gamma_1^+ f_+ \\ \Gamma_1^- f_- \end{pmatrix} = \begin{pmatrix} -\iota_+ \tau_N^+ f_{D,+} \\ -\iota_+ \tau_N^- f_{D,-} \end{pmatrix}, \quad f = \begin{pmatrix} f_+ \\ f_- \end{pmatrix}, \quad f_\pm \in \operatorname{dom} T_{\max}^\pm. \tag{8.6.6}$$

It is clear that

$$\operatorname{dom} A_D^+ \times \operatorname{dom} A_D^- = \ker \widetilde{\Gamma}_0,$$

and hence the self-adjoint operator in (8.6.3) coincides with the self-adjoint extension of $T_{\min} := T^+_{\min} \,\widehat{\oplus}\, T^-_{\min}$ corresponding to the boundary condition $\ker \widetilde{\Gamma}_0$. Note also that the corresponding γ-field $\widetilde{\gamma}$ and Weyl function $\widetilde{M}$ have the form

$$\widetilde{\gamma}(\lambda) = \begin{pmatrix} \gamma_+(\lambda) & 0 \\ 0 & \gamma_-(\lambda) \end{pmatrix} \quad \text{and} \quad \widetilde{M}(\lambda) = \begin{pmatrix} M_+(\lambda) & 0 \\ 0 & M_-(\lambda) \end{pmatrix} \tag{8.6.7}$$

for $\lambda \in \rho(A^+_{\mathrm{D}}) \cap \rho(A^-_{\mathrm{D}})$.

In Lemma 8.6.2 it will be shown that a certain relation $\widetilde{\Theta}$ is self-adjoint in $L^2(\mathcal{C}) \oplus L^2(\mathcal{C})$. This relation will turn out to be the boundary parameter that corresponds to the Schrödinger operator A in (8.6.1) via the boundary triplet (8.6.4). The following lemma on the sum of the Dirichlet-to-Neumann maps is preparatory.

Lemma 8.6.1. *Let $\eta < \operatorname{essinf} V$ and let $D_\pm(\lambda) : H^{3/2}(\mathcal{C}) \to H^{1/2}(\mathcal{C})$ be the Dirichlet-to-Neumann maps as in Definition 8.3.6 corresponding to $-\Delta + V_\pm$. Then for all $\lambda \in \mathbb{C} \setminus [\eta, \infty)$ the operator*

$$D_+(\lambda) + D_-(\lambda) : H^{3/2}(\mathcal{C}) \to H^{1/2}(\mathcal{C}) \tag{8.6.8}$$

is bijective.

Proof. First it will be shown that the operator in (8.6.8) is injective. Assume that $(D_+(\lambda) + D_-(\lambda))\varphi = 0$ for some $\varphi \in H^{3/2}(\mathcal{C})$ and some $\lambda \in \mathbb{C} \setminus [\eta, \infty)$. Then there exist $f_{\lambda,\pm} \in H^2(\Omega_\pm)$ such that

$$(-\Delta + V_\pm) f_{\lambda,\pm} = \lambda f_{\lambda,\pm}, \qquad \tau^+_{\mathrm{D}} f_{\lambda,+} = \tau^-_{\mathrm{D}} f_{\lambda,-} = \varphi, \tag{8.6.9}$$

and

$$0 = \big(D_+(\lambda) + D_-(\lambda)\big)\varphi = \big(D_+(\lambda) + D_-(\lambda)\big)\tau^\pm_{\mathrm{D}} f_{\lambda,\pm} = \tau^+_{\mathrm{N}} f_{\lambda,+} + \tau^-_{\mathrm{N}} f_{\lambda,-}.$$

As $\tau^+_{\mathrm{D}} f_{\lambda,+} = \tau^-_{\mathrm{D}} f_{\lambda,-}$ and $\tau^+_{\mathrm{N}} f_{\lambda,+} = -\tau^-_{\mathrm{N}} f_{\lambda,-}$ this implies that

$$f_\lambda = \begin{pmatrix} f_{\lambda,+} \\ f_{\lambda,-} \end{pmatrix} \in H^2(\mathbb{R}^n). \tag{8.6.10}$$

In fact, for each $h = (h_+, h_-)^\top \in \operatorname{dom} A = H^2(\mathbb{R}^n)$ one also has $\tau^+_{\mathrm{D}} h_+ = \tau^-_{\mathrm{D}} h_-$ and $\tau^+_{\mathrm{N}} h_+ = -\tau^-_{\mathrm{N}} h_-$ (note that the different signs are due to the fact that the Neumann trace on each domain is taken with respect to the outward normal vector) and hence

$$\begin{aligned}
(Ah, f_\lambda)_{L^2(\mathbb{R}^n)} &- (h, T_{\max} f_\lambda)_{L^2(\mathbb{R}^n)} \\
&= (T^+_{\max} h_+, f_{\lambda,+})_{L^2(\Omega_+)} - (h_+, T^+_{\max} f_{\lambda,+})_{L^2(\Omega_+)} \\
&\qquad + (T^-_{\max} h_-, f_{\lambda,-})_{L^2(\Omega_-)} - (h_-, T^-_{\max} f_{\lambda,-})_{L^2(\Omega_-)} \\
&= (\tau^+_{\mathrm{D}} h_+, \tau^+_{\mathrm{N}} f_{\lambda,+})_{L^2(\mathcal{C})} - (\tau^+_{\mathrm{N}} h_+, \tau^+_{\mathrm{D}} f_{\lambda,+})_{L^2(\mathcal{C})} \\
&\qquad + (\tau^-_{\mathrm{D}} h_-, \tau^-_{\mathrm{N}} f_{\lambda,-})_{L^2(\mathcal{C})} - (\tau^-_{\mathrm{N}} h_-, \tau^-_{\mathrm{D}} f_{\lambda,-})_{L^2(\mathcal{C})} \\
&= 0.
\end{aligned}$$

As the operator A is self-adjoint this shows, in particular, $f_\lambda \in \operatorname{dom} A$ and hence (8.6.10) holds. Furthermore, from (8.6.9) it follows that

$$Af_\lambda = (-\Delta + V)f_\lambda = \lambda f_\lambda.$$

Since $\sigma(A) \subset [v_-, \infty) \subset [\eta, \infty)$ and $\lambda \in \mathbb{C} \setminus [\eta, \infty)$, this implies that $f_\lambda = 0$ and hence $\varphi = \tau_{\mathrm{D}}^{\pm} f_{\lambda,\pm} = 0$. Thus, it has been shown that the operator in (8.6.8) is injective for all $\lambda \in \mathbb{C} \setminus [\eta, \infty)$.

Next it will be shown that the operator in (8.6.8) is surjective. For this consider the space

$$H_{\mathcal{C}}(\mathbb{R}^n) := \left\{ f = \begin{pmatrix} f_+ \\ f_- \end{pmatrix} : f_\pm \in H^2(\Omega_\pm),\ \tau_{\mathrm{D}}^+ f_+ = \tau_{\mathrm{D}}^- f_- \right\}$$

and observe that as a consequence of (8.2.12) the mapping

$$\tau_{\mathrm{N}}^{\mathcal{C}} : H_{\mathcal{C}}(\mathbb{R}^n) \to H^{1/2}(\mathcal{C}), \qquad f \mapsto \tau_{\mathrm{N}}^{\mathcal{C}} f := \tau_{\mathrm{N}}^+ f_+ + \tau_{\mathrm{N}}^- f_-,$$

is surjective. For $\lambda \in \mathbb{C} \setminus [\eta, \infty)$ it will be shown now that the direct sum decomposition

$$H_{\mathcal{C}}(\mathbb{R}^n) = \operatorname{dom} A \dotplus \left\{ f_\lambda = \begin{pmatrix} f_{\lambda,+} \\ f_{\lambda,-} \end{pmatrix} : \begin{matrix} f_{\lambda,\pm} \in H^2(\Omega_\pm),\ \tau_{\mathrm{D}}^+ f_{\lambda,+} = \tau_{\mathrm{D}}^- f_{\lambda,-}, \\ (-\Delta + V_\pm) f_{\lambda,\pm} = \lambda f_{\lambda,\pm} \end{matrix} \right\}$$

holds. In fact, the inclusion ($\supset$) is clear since $\operatorname{dom} A = H^2(\mathbb{R}^n)$ and the second summand on the right-hand side is obviously contained in $H_{\mathcal{C}}(\mathbb{R}^n)$. The inclusion ($\subset$) follows from Theorem 1.7.1 applied to $T = -\Delta + V$, $\operatorname{dom} T = H_{\mathcal{C}}(\mathbb{R}^n)$, after observing that the space

$$\left\{ f_\lambda = \begin{pmatrix} f_{\lambda,+} \\ f_{\lambda,-} \end{pmatrix} : \begin{matrix} f_{\lambda,\pm} \in H^2(\Omega_\pm),\ \tau_{\mathrm{D}}^+ f_{\lambda,+} = \tau_{\mathrm{D}}^- f_{\lambda,-}, \\ (-\Delta + V_\pm) f_{\lambda,\pm} = \lambda f_{\lambda,\pm} \end{matrix} \right\} \tag{8.6.11}$$

coincides with $\mathfrak{N}_\lambda(T) = \ker(T - \lambda)$ and $\lambda \in \rho(A)$. Note also that $\lambda \in \rho(A)$ implies that the sum is direct.

Next observe that for $f \in \operatorname{dom} A$ one has $\tau_{\mathrm{N}}^{\mathcal{C}} f = 0$ and hence also the restriction of $\tau_{\mathrm{N}}^{\mathcal{C}}$ to the space (8.6.11) maps onto $H^{1/2}(\mathcal{C})$. Therefore, for $\psi \in H^{1/2}(\mathcal{C})$ there exists $f_\lambda = (f_{\lambda,+}, f_{\lambda,-})^\top$ such that $f_{\lambda,\pm} \in H^2(\Omega_\pm)$, $(-\Delta + V_\pm) f_{\lambda,\pm} = \lambda f_{\lambda,\pm}$,

$$\tau_{\mathrm{D}}^+ f_{\lambda,+} = \tau_{\mathrm{D}}^- f_{\lambda,-} =: \varphi \in H^{3/2}(\mathcal{C}) \quad \text{and} \quad \tau_{\mathrm{N}}^{\mathcal{C}} f_\lambda = \tau_{\mathrm{N}}^+ f_{\lambda,+} + \tau_{\mathrm{N}}^- f_{\lambda,-} = \psi.$$

It follows that

$$\big(D_+(\lambda) + D_-(\lambda)\big)\varphi = D_+(\lambda)\tau_{\mathrm{D}}^+ f_{\lambda,+} + D_-(\lambda)\tau_{\mathrm{D}}^- f_{\lambda,-} = \tau_{\mathrm{N}}^+ f_{\lambda,+} + \tau_{\mathrm{N}}^- f_{\lambda,-} = \psi,$$

and hence the operator in (8.6.8) is surjective.

Consequently, it has been shown that (8.6.8) is a bijective operator for all $\lambda \in \mathbb{C} \setminus [\eta, \infty)$. □

Lemma 8.6.1 will be used to prove the following lemma on the self-adjointness of a particular relation $\widetilde{\Theta}$ in $L^2(\mathcal{C}) \oplus L^2(\mathcal{C})$.

Lemma 8.6.2. *Let $\eta < \operatorname{essinf} V$ and let $D_\pm(\eta) : H^{3/2}(\mathcal{C}) \to H^{1/2}(\mathcal{C})$ be the Dirichlet-to-Neumann maps as in Definition 8.3.6 corresponding to $-\Delta + V_\pm$. Then the relation*

$$\widetilde{\Theta} = \left\{ \left\{ \begin{pmatrix} \xi \\ \xi \end{pmatrix}, \begin{pmatrix} \varphi \\ \psi \end{pmatrix} \right\} : \xi \in H^2(\mathcal{C}),\ \varphi + \psi = \iota_+ \big(D_+(\eta) + D_-(\eta)\big) \iota_-^{-1} \xi \right\}$$

is self-adjoint in $L^2(\mathcal{C}) \oplus L^2(\mathcal{C})$.

Proof. Recall first from Example 8.4.7 that $\iota_+ D_+(\eta) \iota_-^{-1}$ and $\iota_+ D_-(\eta) \iota_-^{-1}$ are both unbounded bijective self-adjoint operators in $L^2(\mathcal{C})$ with domain $H^2(\mathcal{C})$. Since $\eta < \operatorname{essinf} V$, one also sees from (8.4.18) that these operators are nonnegative. It follows, in particular, that $\iota_+(D_+(\eta) + D_-(\eta))\iota_-^{-1}$ is a symmetric operator in $L^2(\mathcal{C})$. Since

$$D_+(\eta) + D_-(\eta) : H^{3/2}(\mathcal{C}) \to H^{1/2}(\mathcal{C})$$

is bijective by Lemma 8.6.1 and the restricted operators $\iota_-^{-1} : H^2(\mathcal{C}) \to H^{3/2}(\mathcal{C})$ and $\iota_+ : H^{1/2}(\mathcal{C}) \to L^2(\mathcal{C})$ are also bijective, one concludes that

$$\iota_+(D_+(\eta) + D_-(\eta))\iota_-^{-1} \tag{8.6.12}$$

is a uniformly positive self-adjoint operator in $L^2(\mathcal{C})$ defined on $H^2(\mathcal{C})$.

To show that $\widetilde{\Theta} \subset \widetilde{\Theta}^*$, consider two arbitrary elements

$$\left\{ \begin{pmatrix} \xi \\ \xi \end{pmatrix}, \begin{pmatrix} \varphi \\ \psi \end{pmatrix} \right\}, \left\{ \begin{pmatrix} \xi' \\ \xi' \end{pmatrix}, \begin{pmatrix} \varphi' \\ \psi' \end{pmatrix} \right\} \in \widetilde{\Theta},$$

that is, $\xi, \xi' \in H^2(\mathcal{C})$,

$$\varphi + \psi = \iota_+ \big(D_+(\eta) + D_-(\eta)\big) \iota_-^{-1} \xi \quad \text{and} \quad \varphi' + \psi' = \iota_+ \big(D_+(\eta) + D_-(\eta)\big) \iota_-^{-1} \xi'.$$

Then one computes

$$\begin{aligned} &\left(\begin{pmatrix} \xi \\ \xi \end{pmatrix}, \begin{pmatrix} \varphi' \\ \psi' \end{pmatrix} \right)_{(L^2(\mathcal{C}))^2} - \left(\begin{pmatrix} \varphi \\ \psi \end{pmatrix}, \begin{pmatrix} \xi' \\ \xi' \end{pmatrix} \right)_{(L^2(\mathcal{C}))^2} \\ &\quad = (\xi, \varphi' + \psi')_{L^2(\mathcal{C})} - (\varphi + \psi, \xi')_{L^2(\mathcal{C})} \\ &\quad = \big(\xi, \iota_+ \big(D_+(\eta) + D_-(\eta)\big) \iota_-^{-1} \xi'\big)_{L^2(\mathcal{C})} - \big(\iota_+ \big(D_+(\eta) + D_-(\eta)\big) \iota_-^{-1} \xi, \xi'\big)_{L^2(\mathcal{C})} \\ &\quad = 0, \end{aligned}$$

where in the last step it was used that (8.6.12) is a symmetric operator in $L^2(\mathcal{C})$. Hence, the relation $\widetilde{\Theta}$ is symmetric in $L^2(\mathcal{C})$. For the opposite inclusion $\widetilde{\Theta}^* \subset \widetilde{\Theta}$ consider an element

$$\left\{ \begin{pmatrix} \alpha \\ \beta \end{pmatrix}, \begin{pmatrix} \gamma \\ \delta \end{pmatrix} \right\} \in \widetilde{\Theta}^*, \tag{8.6.13}$$

As the operator A is self-adjoint this shows, in particular, $f_\lambda \in \operatorname{dom} A$ and hence (8.6.10) holds. Furthermore, from (8.6.9) it follows that

$$Af_\lambda = (-\Delta + V)f_\lambda = \lambda f_\lambda.$$

Since $\sigma(A) \subset [v_-, \infty) \subset [\eta, \infty)$ and $\lambda \in \mathbb{C} \setminus [\eta, \infty)$, this implies that $f_\lambda = 0$ and hence $\varphi = \tau_D^\pm f_{\lambda,\pm} = 0$. Thus, it has been shown that the operator in (8.6.8) is injective for all $\lambda \in \mathbb{C} \setminus [\eta, \infty)$.

Next it will be shown that the operator in (8.6.8) is surjective. For this consider the space

$$H_{\mathcal{C}}(\mathbb{R}^n) := \left\{ f = \begin{pmatrix} f_+ \\ f_- \end{pmatrix} : f_\pm \in H^2(\Omega_\pm),\ \tau_D^+ f_+ = \tau_D^- f_- \right\}$$

and observe that as a consequence of (8.2.12) the mapping

$$\tau_N^{\mathcal{C}} : H_{\mathcal{C}}(\mathbb{R}^n) \to H^{1/2}(\mathcal{C}), \qquad f \mapsto \tau_N^{\mathcal{C}} f := \tau_N^+ f_+ + \tau_N^- f_-,$$

is surjective. For $\lambda \in \mathbb{C} \setminus [\eta, \infty)$ it will be shown now that the direct sum decomposition

$$H_{\mathcal{C}}(\mathbb{R}^n) = \operatorname{dom} A \dotplus \left\{ f_\lambda = \begin{pmatrix} f_{\lambda,+} \\ f_{\lambda,-} \end{pmatrix} : \begin{matrix} f_{\lambda,\pm} \in H^2(\Omega_\pm),\ \tau_D^+ f_{\lambda,+} = \tau_D^- f_{\lambda,-}, \\ (-\Delta + V_\pm) f_{\lambda,\pm} = \lambda f_{\lambda,\pm} \end{matrix} \right\}$$

holds. In fact, the inclusion ($\supset$) is clear since $\operatorname{dom} A = H^2(\mathbb{R}^n)$ and the second summand on the right-hand side is obviously contained in $H_{\mathcal{C}}(\mathbb{R}^n)$. The inclusion ($\subset$) follows from Theorem 1.7.1 applied to $T = -\Delta + V$, $\operatorname{dom} T = H_{\mathcal{C}}(\mathbb{R}^n)$, after observing that the space

$$\left\{ f_\lambda = \begin{pmatrix} f_{\lambda,+} \\ f_{\lambda,-} \end{pmatrix} : \begin{matrix} f_{\lambda,\pm} \in H^2(\Omega_\pm),\ \tau_D^+ f_{\lambda,+} = \tau_D^- f_{\lambda,-}, \\ (-\Delta + V_\pm) f_{\lambda,\pm} = \lambda f_{\lambda,\pm} \end{matrix} \right\} \tag{8.6.11}$$

coincides with $\mathfrak{N}_\lambda(T) = \ker(T - \lambda)$ and $\lambda \in \rho(A)$. Note also that $\lambda \in \rho(A)$ implies that the sum is direct.

Next observe that for $f \in \operatorname{dom} A$ one has $\tau_N^{\mathcal{C}} f = 0$ and hence also the restriction of $\tau_N^{\mathcal{C}}$ to the space (8.6.11) maps onto $H^{1/2}(\mathcal{C})$. Therefore, for $\psi \in H^{1/2}(\mathcal{C})$ there exists $f_\lambda = (f_{\lambda,+}, f_{\lambda,-})^\top$ such that $f_{\lambda,\pm} \in H^2(\Omega_\pm)$, $(-\Delta + V_\pm) f_{\lambda,\pm} = \lambda f_{\lambda,\pm}$,

$$\tau_D^+ f_{\lambda,+} = \tau_D^- f_{\lambda,-} =: \varphi \in H^{3/2}(\mathcal{C}) \quad \text{and} \quad \tau_N^{\mathcal{C}} f_\lambda = \tau_N^+ f_{\lambda,+} + \tau_N^- f_{\lambda,-} = \psi.$$

It follows that

$$\big(D_+(\lambda) + D_-(\lambda)\big)\varphi = D_+(\lambda)\tau_D^+ f_{\lambda,+} + D_-(\lambda)\tau_D^- f_{\lambda,-} = \tau_N^+ f_{\lambda,+} + \tau_N^- f_{\lambda,-} = \psi,$$

and hence the operator in (8.6.8) is surjective.

Consequently, it has been shown that (8.6.8) is a bijective operator for all $\lambda \in \mathbb{C} \setminus [\eta, \infty)$. □

Lemma 8.6.1 will be used to prove the following lemma on the self-adjointness of a particular relation $\widetilde{\Theta}$ in $L^2(\mathcal{C}) \oplus L^2(\mathcal{C})$.

Lemma 8.6.2. *Let $\eta < \operatorname{essinf} V$ and let $D_\pm(\eta) : H^{3/2}(\mathcal{C}) \to H^{1/2}(\mathcal{C})$ be the Dirichlet-to-Neumann maps as in Definition 8.3.6 corresponding to $-\Delta + V_\pm$. Then the relation*

$$\widetilde{\Theta} = \left\{ \left\{ \begin{pmatrix} \xi \\ \xi \end{pmatrix}, \begin{pmatrix} \varphi \\ \psi \end{pmatrix} \right\} : \xi \in H^2(\mathcal{C}),\ \varphi + \psi = \iota_+ \big(D_+(\eta) + D_-(\eta) \big) \iota_-^{-1} \xi \right\}$$

is self-adjoint in $L^2(\mathcal{C}) \oplus L^2(\mathcal{C})$.

Proof. Recall first from Example 8.4.7 that $\iota_+ D_+(\eta) \iota_-^{-1}$ and $\iota_+ D_-(\eta) \iota_-^{-1}$ are both unbounded bijective self-adjoint operators in $L^2(\mathcal{C})$ with domain $H^2(\mathcal{C})$. Since $\eta < \operatorname{essinf} V$, one also sees from (8.4.18) that these operators are nonnegative. It follows, in particular, that $\iota_+(D_+(\eta) + D_-(\eta))\iota_-^{-1}$ is a symmetric operator in $L^2(\mathcal{C})$. Since

$$D_+(\eta) + D_-(\eta) : H^{3/2}(\mathcal{C}) \to H^{1/2}(\mathcal{C})$$

is bijective by Lemma 8.6.1 and the restricted operators $\iota_-^{-1} : H^2(\mathcal{C}) \to H^{3/2}(\mathcal{C})$ and $\iota_+ : H^{1/2}(\mathcal{C}) \to L^2(\mathcal{C})$ are also bijective, one concludes that

$$\iota_+(D_+(\eta) + D_-(\eta))\iota_-^{-1} \tag{8.6.12}$$

is a uniformly positive self-adjoint operator in $L^2(\mathcal{C})$ defined on $H^2(\mathcal{C})$.

To show that $\widetilde{\Theta} \subset \widetilde{\Theta}^*$, consider two arbitrary elements

$$\left\{ \begin{pmatrix} \xi \\ \xi \end{pmatrix}, \begin{pmatrix} \varphi \\ \psi \end{pmatrix} \right\}, \left\{ \begin{pmatrix} \xi' \\ \xi' \end{pmatrix}, \begin{pmatrix} \varphi' \\ \psi' \end{pmatrix} \right\} \in \widetilde{\Theta},$$

that is, $\xi, \xi' \in H^2(\mathcal{C})$,

$$\varphi + \psi = \iota_+ \big(D_+(\eta) + D_-(\eta) \big) \iota_-^{-1} \xi \quad \text{and} \quad \varphi' + \psi' = \iota_+ \big(D_+(\eta) + D_-(\eta) \big) \iota_-^{-1} \xi'.$$

Then one computes

$$\begin{aligned}
&\left(\begin{pmatrix} \xi \\ \xi \end{pmatrix}, \begin{pmatrix} \varphi' \\ \psi' \end{pmatrix} \right)_{(L^2(\mathcal{C}))^2} - \left(\begin{pmatrix} \varphi \\ \psi \end{pmatrix}, \begin{pmatrix} \xi' \\ \xi' \end{pmatrix} \right)_{(L^2(\mathcal{C}))^2} \\
&\quad = (\xi, \varphi' + \psi')_{L^2(\mathcal{C})} - (\varphi + \psi, \xi')_{L^2(\mathcal{C})} \\
&\quad = \big(\xi, \iota_+ \big(D_+(\eta) + D_-(\eta) \big) \iota_-^{-1} \xi' \big)_{L^2(\mathcal{C})} - \big(\iota_+ \big(D_+(\eta) + D_-(\eta) \big) \iota_-^{-1} \xi, \xi' \big)_{L^2(\mathcal{C})} \\
&\quad = 0,
\end{aligned}$$

where in the last step it was used that (8.6.12) is a symmetric operator in $L^2(\mathcal{C})$. Hence, the relation $\widetilde{\Theta}$ is symmetric in $L^2(\mathcal{C})$. For the opposite inclusion $\widetilde{\Theta}^* \subset \widetilde{\Theta}$ consider an element

$$\left\{ \begin{pmatrix} \alpha \\ \beta \end{pmatrix}, \begin{pmatrix} \gamma \\ \delta \end{pmatrix} \right\} \in \widetilde{\Theta}^*, \tag{8.6.13}$$

that is,

$$\left(\begin{pmatrix}\alpha\\ \beta\end{pmatrix},\begin{pmatrix}\varphi\\ \psi\end{pmatrix}\right)_{(L^2(\mathcal{C}))^2}=\left(\begin{pmatrix}\gamma\\ \delta\end{pmatrix},\begin{pmatrix}\xi\\ \xi\end{pmatrix}\right)_{(L^2(\mathcal{C}))^2} \tag{8.6.14}$$

holds for all

$$\left\{\begin{pmatrix}\xi\\ \xi\end{pmatrix},\begin{pmatrix}\varphi\\ \psi\end{pmatrix}\right\}\in\widetilde{\Theta}.$$

The special choice $\xi = 0$ yields $\varphi + \psi = 0$ by the definition of $\widetilde{\Theta}$ and hence $(\alpha - \beta, \varphi)_{L^2(\mathcal{C})} = 0$ for all $\varphi \in L^2(\mathcal{C})$. This shows $\alpha = \beta$ and therefore (8.6.14) becomes

$$\big(\alpha, \iota_+\big(D_+(\eta) + D_-(\eta)\big)\iota_-^{-1}\xi\big)_{L^2(\mathcal{C})} = (\alpha, \varphi + \psi)_{L^2(\mathcal{C})} = (\gamma + \delta, \xi)_{L^2(\mathcal{C})}$$

for all $\xi \in H^2(\mathcal{C})$. Since $\iota_+(D_+(\eta) + D_-(\eta))\iota_-^{-1}$ is a self-adjoint operator in $L^2(\mathcal{C})$ defined on $H^2(\mathcal{C})$, it follows that $\alpha \in H^2(\mathcal{C})$ and

$$\iota_+\big(D_+(\eta) + D_-(\eta)\big)\iota_-^{-1}\alpha = \gamma + \delta.$$

This implies that the element in (8.6.13) belongs to $\widetilde{\Theta}$. Thus, $\widetilde{\Theta}$ is a self-adjoint relation in $L^2(\mathcal{C}) \oplus L^2(\mathcal{C})$. □

The following theorem is the main result in this section. It turns out that the self-adjoint operator corresponding to $\widetilde{\Theta}$ in Lemma 8.6.2 coincides with the Schrödinger operator A.

Theorem 8.6.3. *Let $\{L^2(\mathcal{C})\oplus L^2(\mathcal{C}),\widetilde{\Gamma}_0,\widetilde{\Gamma}_1\}$ be the boundary triplet for $T_{\max}^+ \,\widehat{\oplus}\, T_{\max}^-$ from (8.6.4) with γ-field $\widetilde{\gamma}$, let $\widetilde{\Theta}$ be the self-adjoint relation in Lemma 8.6.2, and let $D_\pm(\lambda)$ be the Dirichlet-to-Neumann maps corresponding to $-\Delta + V_\pm$. Then the self-adjoint operator $\widetilde{A}_{\widetilde{\Theta}}$ corresponding to the parameter $\widetilde{\Theta}$ coincides with the Schrödinger operator A in (8.6.1) and for all $\lambda \in \mathbb{C}\setminus[\eta,\infty)$ one has the resolvent formula*

$$(A-\lambda)^{-1} = (\widetilde{A}_{\rm D} - \lambda)^{-1} + \widetilde{\gamma}(\lambda)\widetilde{\Lambda}(\lambda)\widetilde{\gamma}(\bar{\lambda})^*,$$

where $\widetilde{\Lambda}(\lambda) \in \mathbf{B}(L^2(\mathcal{C}) \oplus L^2(\mathcal{C}))$ has the form

$$\widetilde{\Lambda}(\lambda) = \begin{pmatrix}\iota_-(D_+(\lambda)+D_-(\lambda))^{-1}\iota_+^{-1} & \iota_-(D_+(\lambda)+D_-(\lambda))^{-1}\iota_+^{-1}\\ \iota_-(D_+(\lambda)+D_-(\lambda))^{-1}\iota_+^{-1} & \iota_-(D_+(\lambda)+D_-(\lambda))^{-1}\iota_+^{-1}\end{pmatrix}.$$

Proof. First it will be shown that the self-adjoint extension $\widetilde{A}_{\widetilde{\Theta}}$ and the self-adjoint Schrödinger operator A in (8.6.1) coincide. Since both operators are self-adjoint, it suffices to verify the inclusion $A \subset \widetilde{A}_{\widetilde{\Theta}}$. For this, consider $f \in \operatorname{dom} A = H^2(\mathbb{R}^n)$ and note that $f = (f_+, f_-)^\top$ satisfies $\tau_{\rm D}^+ f_+ = \tau_{\rm D}^- f_-$ and $\tau_{\rm N}^+ f_+ = -\tau_{\rm N}^- f_-$. It will be shown that $\{\widetilde{\Gamma}_0 f, \widetilde{\Gamma}_1 f\} \in \widetilde{\Theta}$. By the definition of the boundary mappings $\widetilde{\Gamma}_0$ and $\widetilde{\Gamma}_1$ in (8.6.5)–(8.6.6), one has

$$\widetilde{\Gamma}_0 f = \begin{pmatrix}\iota_-\widetilde{\tau}_{\rm D}^+ f_+\\ \iota_-\widetilde{\tau}_{\rm D}^- f_-\end{pmatrix} \quad\text{and}\quad \widetilde{\Gamma}_1 f = \begin{pmatrix}-\iota_+\tau_{\rm N}^+ f_{{\rm D},+}\\ -\iota_+\tau_{\rm N}^- f_{{\rm D},-}\end{pmatrix} =: \begin{pmatrix}\varphi\\ \psi\end{pmatrix}$$

and, as $f \in H^2(\mathbb{R}^n)$, it follows that

$$\xi := \iota_- \tau_{\rm D}^+ f_+ = \iota_- \widetilde{\tau}_{\rm D}^+ f_+ = \iota_- \widetilde{\tau}_{\rm D}^- f_- = \iota_- \tau_{\rm D}^- f_- \in H^2(\mathcal{C}).$$

Since $f_\pm = f_{{\rm D},\pm} + f_{\eta,\pm}$ with $f_{{\rm D},\pm} \in \operatorname{dom} A_{\rm D}^\pm$ and $f_{\eta,\pm} \in \mathfrak{N}_\eta(T_{\max}^\pm)$, one has $\tau_{\rm D}^\pm f_\pm = \tau_{\rm D}^\pm f_{\eta,\pm}$ and one concludes that

$$\begin{aligned}
\iota_+\big(D_+(\eta) + D_-(\eta)\big)\iota_-^{-1}\xi &= \iota_+\big(D_+(\eta)\tau_{\rm D}^+ f_{\eta,+} + D_-(\eta)\tau_{\rm D}^- f_{\eta,-}\big)\\
&= \iota_+\big(\tau_{\rm N}^+ f_{\eta,+} + \tau_{\rm N}^- f_{\eta,-}\big)\\
&= \iota_+\big(\tau_{\rm N}^+ f_+ + \tau_{\rm N}^- f_- - \tau_{\rm N}^+ f_{{\rm D},+} - \tau_{\rm N}^- f_{{\rm D},-}\big)\\
&= -\iota_+\tau_{\rm N}^+ f_{{\rm D},+} - \iota_+\tau_{\rm N}^- f_{{\rm D},-}\\
&= \varphi + \psi,
\end{aligned}$$

where the property $\tau_{\rm N}^+ f_+ = -\tau_{\rm N}^- f_-$ for $f \in \operatorname{dom} A$ was used. These considerations imply $\{\widetilde{\Gamma}_0 f, \widetilde{\Gamma}_1 f\} \in \widetilde{\Theta}$ and thus $f \in \operatorname{dom} \widetilde{A}_{\widetilde{\Theta}}$. Therefore, $\operatorname{dom} A = H^2(\mathbb{R}^n)$ is contained in $\operatorname{dom} \widetilde{A}_{\widetilde{\Theta}}$, and since both operators are self-adjoint, it follows that they coincide, that is, $A = \widetilde{A}_{\widetilde{\Theta}}$.

As a consequence of Theorem 2.6.1 one has for $\lambda \in \rho(A) \cap \rho(\widetilde{A}_{\rm D})$ that

$$(A - \lambda)^{-1} = (\widetilde{A}_{\rm D} - \lambda)^{-1} + \widetilde{\gamma}(\lambda)\big(\widetilde{\Theta} - \widetilde{M}(\lambda)\big)^{-1}\widetilde{\gamma}(\bar{\lambda})^*,$$

where $\widetilde{\gamma}$ and $\widetilde{M}$ are the γ-field and Weyl function, respectively, of the boundary triplet $\{L^2(\mathcal{C}) \oplus L^2(\mathcal{C}), \widetilde{\Gamma}_0, \widetilde{\Gamma}_1\}$ in (8.6.7). Here it is also clear from Theorem 2.6.1 that

$$\big(\widetilde{\Theta} - \widetilde{M}(\lambda)\big)^{-1} \in \mathbf{B}(L^2(\mathcal{C}) \oplus L^2(\mathcal{C})), \qquad \lambda \in \rho(A) \cap \rho(\widetilde{A}_{\rm D}).$$

From now consider only $\lambda \in \mathbb{C} \setminus [\eta, \infty)$. It follows from Lemma 8.6.2 and (8.6.7) that

$$\begin{aligned}
&\big(\widetilde{\Theta} - \widetilde{M}(\lambda)\big)^{-1}\\
&= \left\{\left\{\begin{pmatrix} \varphi - M_+(\lambda)\xi \\ \psi - M_-(\lambda)\xi \end{pmatrix}, \begin{pmatrix} \xi \\ \xi \end{pmatrix}\right\} : \begin{array}{c} \xi \in H^2(\mathcal{C}), \\ \varphi + \psi = \iota_+\big(D_+(\eta) + D_-(\eta)\big)\iota_-^{-1}\xi \end{array}\right\},
\end{aligned}$$

and setting $\vartheta_1 = \varphi - M_+(\lambda)\xi$ and $\vartheta_2 = \psi - M_-(\lambda)\xi$ one obtains

$$\begin{aligned}
\vartheta_1 + \vartheta_2 &= \varphi + \psi - M_+(\lambda)\xi - M_-(\lambda)\xi\\
&= \iota_+\big(D_+(\eta) + D_-(\eta)\big)\iota_-^{-1}\xi - M_+(\lambda)\xi - M_-(\lambda)\xi.
\end{aligned}$$

Since $M_\pm(\lambda)\xi = \iota_+(D_\pm(\eta) - D_\pm(\lambda))\iota_-^{-1}\xi$ for $\xi \in H^2(\mathcal{C})$ by Lemma 8.4.5, it follows that

$$\vartheta_1 + \vartheta_2 = \iota_+\big(D_+(\lambda) + D_-(\lambda)\big)\iota_-^{-1}\xi.$$

Lemma 8.6.1 implies that $\iota_+(D_+(\lambda) + D_-(\lambda))\iota_-^{-1}$ is a bijective operator in $L^2(\mathcal{C})$ for $\lambda \in \mathbb{C} \setminus [\eta, \infty)$ and hence

$$\iota_-\big(D_+(\lambda) + D_-(\lambda)\big)^{-1}\iota_+^{-1}\vartheta_1 + \iota_-\big(D_+(\lambda) + D_-(\lambda)\big)^{-1}\iota_+^{-1}\vartheta_2 = \xi.$$

Therefore, one has

$$(\widetilde{\Theta} - \widetilde{M}(\lambda))^{-1} = \begin{pmatrix} \iota_-(D_+(\lambda) + D_-(\lambda))^{-1}\iota_+^{-1} & \iota_-(D_+(\lambda) + D_-(\lambda))^{-1}\iota_+^{-1} \\ \iota_-(D_+(\lambda) + D_-(\lambda))^{-1}\iota_+^{-1} & \iota_-(D_+(\lambda) + D_-(\lambda))^{-1}\iota_+^{-1} \end{pmatrix}.$$

This completes the proof of Theorem 8.6.3. □

Finally, the boundary triplet in (8.6.4) is modified in the same way as in Proposition 4.6.4 to interpret the Schrödinger operator A as the self-adjoint extension corresponding to the boundary mapping $\widehat{\Gamma}_0$. More precisely, the boundary triplets $\{L^2(\mathcal{C}), \Gamma_0^+, \Gamma_1^+\}$ and $\{L^2(\mathcal{C}), \Gamma_0^-, \Gamma_1^-\}$ lead to the boundary triplet

$$\{L^2(\mathcal{C}) \oplus L^2(\mathcal{C}), \widehat{\Gamma}_0, \widehat{\Gamma}_1\} \tag{8.6.15}$$

for $T_{\max} = T_{\max}^+ \widehat{\oplus}\, T_{\max}^-$, where

$$\widehat{\Gamma}_0 f = \begin{pmatrix} -\Gamma_1^+ f_+ - \Gamma_1^- f_- \\ \Gamma_0^+ f_+ - \Gamma_0^- f_- \end{pmatrix} = \begin{pmatrix} \iota_+(\tau_{\mathrm{N}}^+ f_{\mathrm{D},+} + \tau_{\mathrm{N}}^- f_{\mathrm{D},-}) \\ \iota_-(\widetilde{\tau}_{\mathrm{D}}^+ f_+ - \widetilde{\tau}_{\mathrm{D}}^- f_-) \end{pmatrix}$$

and

$$\widehat{\Gamma}_1 f = \begin{pmatrix} \Gamma_0^+ f_+ \\ -\Gamma_1^- f_- \end{pmatrix} = \begin{pmatrix} \iota_- \widetilde{\tau}_{\mathrm{D}}^+ f_+ \\ \iota_+ \tau_{\mathrm{N}}^- f_{\mathrm{D},-} \end{pmatrix}$$

for $f = (f_+, f_-)^\top$ with $f_\pm \in \operatorname{dom} T_{\max}^\pm$. It follows from Proposition 4.6.4 that the Schrödinger operator $A = -\Delta + V$ in (8.6.1) coincides with the self-adjoint extension defined on $\ker \widehat{\Gamma}_0$ and that the Weyl function corresponding to the boundary triplet in (8.6.15) is given by

$$\widehat{M}(\lambda) = -\begin{pmatrix} M_+(\lambda) & -I \\ -I & -M_-(\lambda)^{-1} \end{pmatrix}^{-1}, \qquad \lambda \in \mathbb{C} \setminus \mathbb{R},$$

where $M_\pm(\lambda) = \iota_+(\widetilde{D}_\pm(\eta) - \widetilde{D}_\pm(\lambda))\iota_-^{-1}$ is the Weyl function corresponding to the boundary triplet $\{L^2(\mathcal{C}), \Gamma_0^\pm, \Gamma_1^\pm\}$; cf. Proposition 8.4.4 and Lemma 8.4.5. In particular, the results in Section 3.5 and Section 3.6 can be used to describe the isolated and embedded eigenvalues, continuous, and absolutely continuous spectrum of A with the help of the limit properties of the Dirichlet-to-Neumann maps $\widetilde{D}_\pm$. For this, however, one has to ensure that the underlying minimal operator $T_{\min} = T_{\min}^+ \widehat{\oplus}\, T_{\min}^-$ is simple, which follows from Proposition 8.3.13 and [120, Proposition 2.2].

8.7 Bounded Lipschitz domains

In this last section Schrödinger operators $-\Delta + V$ with a real function $V \in L^\infty(\Omega)$ on bounded Lipschitz domains are briefly discussed. This situation is more general than the setting of bounded C^2-domains treated in the previous sections. The main objective here is to highlight the differences to the C^2-case and to indicate which methods have to be adapted in order to obtain results of similar nature as above.

The notions of a Lipschitz hypograph and a bounded Lipschitz domain are defined in the same way as C^2-hypographs and bounded C^2-domains in Section 8.2. More precisely, for a Lipschitz continuous function $\phi : \mathbb{R}^{n-1} \to \mathbb{R}$ the domain

$$\Omega_\phi := \{(x', x_n)^\top \in \mathbb{R}^n : x_n < \phi(x')\}$$

is called a *Lipschitz hypograph* with boundary $\partial\Omega$. The surface integral and surface measure on $\partial\Omega_\phi$ are defined in the same way as in (8.2.4), and this leads to the L^2-space $L^2(\partial\Omega_\phi)$ on $\partial\Omega_\phi$. For $s \in [0,1]$ define the Sobolev space of order s on $\partial\Omega_\phi$ by

$$H^s(\partial\Omega_\phi) := \{h \in L^2(\partial\Omega_\phi) : x' \mapsto h(x', \phi(x')) \in H^s(\mathbb{R}^{n-1})\}$$

and equip $H^s(\partial\Omega_\phi)$ with the corresponding scalar product (8.2.6).

Definition 8.7.1. A bounded nonempty open subset $\Omega \subset \mathbb{R}^n$ is called a *Lipschitz domain* if there exist open sets $U_1, \dots, U_l \subset \mathbb{R}^n$ and (possibly up to rotations of coordinates) Lipschitz hypographs $\Omega_1, \dots, \Omega_l \subset \mathbb{R}^n$, such that

$$\partial\Omega \subset \bigcup_{j=1}^{l} U_j \quad \text{and} \quad \Omega \cap U_j = \Omega_j \cap U_j, \quad j = 1, \dots, l.$$

For a bounded Lipschitz domain $\Omega \subset \mathbb{R}^n$ the boundary $\partial\Omega \subset \mathbb{R}^n$ is compact. Using a partition of unity subordinate to the open cover $\{U_j\}$ of $\partial\Omega$ one defines the surface integral, surface measure, and the L^2-space $L^2(\partial\Omega)$ in the same way as in Section 8.2. The Sobolev space $H^s(\partial\Omega)$ for $s \in [0,1]$ is then defined by

$$H^s(\partial\Omega) := \{h \in L^2(\partial\Omega) : \eta_j h \in H^s(\partial\Omega_j),\ j = 1, \dots, l\}$$

and equipped with the corresponding Hilbert space scalar product (8.2.7). It follows that $H^s(\partial\Omega)$, $s \in [0,1]$, is densely and continuously embedded in $L^2(\partial\Omega)$, and the embedding $H^t(\partial\Omega) \hookrightarrow H^s(\partial\Omega)$ is compact for $s < t \leq 1$. As in Section 8.2, the spaces $H^s(\partial\Omega)$, $s \in [0,1]$, can be defined in an equivalent way via interpolation. The dual space of the antilinear continuous functionals on $H^s(\partial\Omega)$ is denoted by $H^{-s}(\partial\Omega)$, $s \in [0,1]$.

For a bounded Lipschitz domain Ω define the spaces

$$H^s_\Delta(\Omega) := \{f \in H^s(\Omega) : \Delta f \in L^2(\Omega)\}, \qquad s \geq 0,$$

and equip them with the Hilbert space scalar product

$$(f,g)_{H^s_\Delta(\Omega)} := (f,g)_{H^s(\Omega)} + (\Delta f, \Delta g)_{L^2(\Omega)}, \qquad f, g \in H^s_\Delta(\Omega). \tag{8.7.1}$$

It is clear that $H^s(\Omega) = H^s_\Delta(\Omega)$ for $s \geq 2$ and that $H^0_\Delta(\Omega) = \operatorname{dom} T_{\max}$ for $s = 0$, with (8.7.1) as the graph norm; cf. (8.3.3). The unit normal vector field pointing outwards on $\partial\Omega$ will again be denoted by ν. It is known that the Dirichlet trace mapping $C^\infty(\overline{\Omega}) \ni f \mapsto f|_{\partial\Omega}$ extends by continuity to a continuous surjective mapping

$$\tau_{\mathrm{D}} : H^s_\Delta(\Omega) \to H^{s-1/2}(\partial\Omega), \qquad \frac{1}{2} \leq s \leq \frac{3}{2},$$

and that the Neumann trace mapping $C^\infty(\overline{\Omega}) \ni f \mapsto \nu \cdot \nabla f|_{\partial\Omega}$ extends by continuity to a continuous surjective mapping

$$\tau_{\mathrm{N}} : H^s_\Delta(\Omega) \to H^{s-3/2}(\partial\Omega), \qquad \frac{1}{2} \leq s \leq \frac{3}{2};$$

cf. [92, 326]. For the present purposes it is particularly useful to note that the mappings

$$\tau_{\mathrm{D}} : H^{3/2}_\Delta(\Omega) \to H^1(\partial\Omega) \quad \text{and} \quad \tau_{\mathrm{N}} : H^{3/2}_\Delta(\Omega) \to L^2(\partial\Omega) \tag{8.7.2}$$

are both continuous and surjective. Furthermore, the first and second Green identities remain true in the natural form, that is,

$$(-\Delta f, g)_{L^2(\Omega)} = (\nabla f, \nabla g)_{L^2(\Omega;\mathbb{C}^n)} - \big\langle \tau_{\mathrm{N}} f, \tau_{\mathrm{D}} g \big\rangle_{H^{-1/2}(\partial\Omega)\times H^{1/2}(\partial\Omega)}$$

and

$$\begin{aligned}(-\Delta f, g)_{L^2(\Omega)} &- (f, -\Delta g)_{L^2(\Omega)} \\ &= \big\langle \tau_{\mathrm{D}} f, \tau_{\mathrm{N}} g \big\rangle_{H^{1/2}(\partial\Omega)\times H^{-1/2}(\partial\Omega)} - \big\langle \tau_{\mathrm{N}} f, \tau_{\mathrm{D}} g \big\rangle_{H^{-1/2}(\partial\Omega)\times H^{1/2}(\partial\Omega)}\end{aligned}$$

hold for all $f, g \in H^1_\Delta(\Omega)$.

The minimal operator $T_{\min}$ and maximal operator $T_{\max}$ associated with $-\Delta + V$ on a bounded Lipschitz domain are defined in exactly the same way as in the beginning of Section 8.3. The assertions $T^*_{\min} = T_{\max}$ and $T_{\min} = T^*_{\max}$ in Proposition 8.3.1 remain valid in the present situation. Furthermore, the Dirichlet realization A_{D} and Neumann realization A_{N} of $-\Delta + V$ are defined as in Section 8.3, and their properties are the same as in Proposition 8.3.2 and Proposition 8.3.3. The first remarkable and substantial difference for Schrödinger operators on a bounded Lipschitz domain appears in connection with the regularity of the domains of A_{D} and A_{N} when comparing with Theorem 8.3.4. In the present case one has the following regularity result from [431, 432], see also [92, 323].

Theorem 8.7.2. *Assume that $\Omega \subset \mathbb{R}^n$ is a bounded Lipschitz domain. Then one has*

$$A_{\mathrm{D}} f = -\Delta f + V f, \quad \operatorname{dom} A_{\mathrm{D}} = \big\{ f \in H^{3/2}_\Delta(\Omega) : \tau_{\mathrm{D}} f = 0 \big\},$$

and

$$A_{\mathrm{N}} f = -\Delta f + V f, \quad \operatorname{dom} A_{\mathrm{N}} = \big\{ f \in H^{3/2}_\Delta(\Omega) : \tau_{\mathrm{N}} f = 0 \big\}.$$

The same reasoning as in Section 8.3 one obtains the following useful decomposition of the space $H^{3/2}_\Delta(\Omega)$.

Corollary 8.7.3. *Assume that $\Omega \subset \mathbb{R}^n$ is a bounded Lipschitz domain. Then for $\lambda \in \rho(A_{\rm D})$ one has the direct sum decomposition*

$$\begin{aligned} H^{3/2}_{\Delta}(\Omega) &= \operatorname{dom} A_{\rm D} \dotplus \bigl\{f_\lambda \in H^{3/2}_{\Delta}(\Omega) : (-\Delta + V)f_\lambda = \lambda f_\lambda\bigr\} \\ &= \ker \tau_{\rm D} \dotplus \bigl\{f_\lambda \in H^{3/2}_{\Delta}(\Omega) : (-\Delta + V)f_\lambda = \lambda f_\lambda\bigr\}, \end{aligned}$$

and for $\lambda \in \rho(A_{\rm N})$ one has the direct sum decomposition

$$\begin{aligned} H^{3/2}_{\Delta}(\Omega) &= \operatorname{dom} A_{\rm N} \dotplus \bigl\{f_\lambda \in H^{3/2}_{\Delta}(\Omega) : (-\Delta + V)f_\lambda = \lambda f_\lambda\bigr\} \\ &= \ker \tau_{\rm N} \dotplus \bigl\{f_\lambda \in H^{3/2}_{\Delta}(\Omega) : (-\Delta + V)f_\lambda = \lambda f_\lambda\bigr\}. \end{aligned}$$

For a bounded Lipschitz domain and $\lambda \in \rho(A_{\rm D})$ the Dirichlet-to-Neumann map is defined as

$$D(\lambda) : H^1(\partial\Omega) \to L^2(\partial\Omega), \qquad \tau_{\rm D} f_\lambda \mapsto \tau_{\rm N} f_\lambda, \tag{8.7.3}$$

where $f_\lambda \in H^{3/2}_{\Delta}(\Omega)$ is such that $(-\Delta+V)f_\lambda = \lambda f_\lambda$. This definition is the natural analog of Definition 8.3.6, taking into account the decomposiion in (8.7.3). As before, it follows that for $\lambda \in \rho(A_{\rm D})\cap\rho(A_{\rm N})$ the Dirichlet-to-Neumann map (8.7.3) is a bijective operator.

For completeness the following a priori estimates are stated.

Corollary 8.7.4. *Assume that $\Omega \subset \mathbb{R}^n$ is a bounded Lipschitz domain. Then there exist constants $C_{\rm D} > 0$ and $C_{\rm N} > 0$ such that*

$$\|f\|_{H^{3/2}_{\Delta}(\Omega)} \le C_{\rm D}\bigl(\|f\|_{L^2(\Omega)} + \|A_{\rm D} f\|_{L^2(\Omega)}\bigr), \quad f \in \operatorname{dom} A_{\rm D},$$

and

$$\|g\|_{H^{3/2}_{\Delta}(\Omega)} \le C_{\rm N}\bigl(\|g\|_{L^2(\Omega)} + \|A_{\rm N} g\|_{L^2(\Omega)}\bigr) \quad g \in \operatorname{dom} A_{\rm N}.$$

Next a variant of Theorem 8.3.9 and Theorem 8.3.10 on the extensions of the Dirichlet and Neumann trace operators to $\operatorname{dom} T_{\max} = H^0_{\Delta}(\Omega)$ for bounded Lipschitz domains is formulated. For this consider the spaces

$$\mathscr{G}_0 := \bigl\{\tau_{\rm D} f : f \in \operatorname{dom} A_{\rm N}\bigr\} \quad \text{and} \quad \mathscr{G}_1 := \bigl\{\tau_{\rm N} g : g \in \operatorname{dom} A_{\rm D}\bigr\}, \tag{8.7.4}$$

and note that for the special case of a bounded C^2-domain the spaces $\mathscr{G}_0$ and $\mathscr{G}_1$ coincide with the spaces $H^{3/2}(\partial\Omega)$ and $H^{1/2}(\partial\Omega)$, respectively. The spaces $\mathscr{G}_0$ and $\mathscr{G}_1$ are dense in $L^2(\partial\Omega)$ and, equipped with the scalar products

$$\begin{aligned} (\varphi,\psi)_{\mathscr{G}_0} &:= (\Sigma^{-1/2}\varphi, \Sigma^{-1/2}\psi)_{L^2(\partial\Omega)}, \qquad &\Sigma = \operatorname{Im}\bigl(D(i)^{-1}\bigr), \\ (\varphi,\psi)_{\mathscr{G}_1} &:= (\Lambda^{-1/2}\varphi, \Lambda^{-1/2}\psi)_{L^2(\partial\Omega)}, \qquad &\Lambda = -\overline{\operatorname{Im} D(i)}, \end{aligned} \tag{8.7.5}$$

they are Hilbert spaces, as was shown in [92, 115]; here both $\Sigma^{-1/2}$ and $\Lambda^{-1/2}$ are unbounded nonnegative self-adjoint operators in $L^2(\partial\Omega)$. The corresponding dual spaces of antilinear continuous functionals are denoted by $\mathscr{G}_0'$ and $\mathscr{G}_1'$, respectively, and one obtains Gelfand triples $\{\mathscr{G}_i, L^2(\partial\Omega), \mathscr{G}_i'\}$, $i = 0, 1$, which serve as the counterparts of $\{H^s(\partial\Omega), L^2(\partial\Omega), H^s(\partial\Omega)\}$, $s = 1/2, 3/2$. Now one can prove the variant of Theorem 8.3.9 and Theorem 8.3.10 alluded to above.

Theorem 8.7.5. *Assume that $\Omega \subset \mathbb{R}^n$ is a bounded Lipschitz domain. Then the Dirichlet and Neumann trace operators in (8.7.2) admit unique extensions to continuous surjective operators*

$$\widetilde{\tau}_{\mathrm{D}} : \operatorname{dom} T_{\max} \to \mathscr{G}_1' \quad \text{and} \quad \widetilde{\tau}_{\mathrm{N}} : \operatorname{dom} T_{\max} \to \mathscr{G}_0',$$

where $\operatorname{dom} T_{\max}$ *is equipped with the graph norm. Furthermore,*

$$\ker \widetilde{\tau}_{\mathrm{D}} = \ker \tau_{\mathrm{D}} = \operatorname{dom} A_{\mathrm{D}} \quad \text{and} \quad \ker \widetilde{\tau}_{\mathrm{N}} = \ker \tau_{\mathrm{N}} = \operatorname{dom} A_{\mathrm{N}}.$$

By analogy to Corollary 8.3.11, the second Green identity extends to elements $f \in \operatorname{dom} T_{\max}$ and $g \in \operatorname{dom} A_{\mathrm{D}}$ in the form

$$(T_{\max} f, g)_{L^2(\Omega)} - (f, T_{\max} g)_{L^2(\Omega)} = \langle \widetilde{\tau}_{\mathrm{D}} f, \tau_{\mathrm{N}} g \rangle_{\mathscr{G}_1' \times \mathscr{G}_1},$$

and for $f \in \operatorname{dom} T_{\max}$ and $g \in \operatorname{dom} A_{\mathrm{N}}$ the second Green identity reads

$$(T_{\max} f, g)_{L^2(\Omega)} - (f, T_{\max} g)_{L^2(\Omega)} = -\langle \widetilde{\tau}_{\mathrm{N}} f, \tau_{\mathrm{D}} g \rangle_{\mathscr{G}_0' \times \mathscr{G}_0}.$$

It will also be used that for $\lambda \in \rho(A_{\mathrm{D}})$ the Dirichlet-to-Neumann map in (8.7.3) admits an extension to a bounded operator

$$\widetilde{D}(\lambda) : \mathscr{G}_1' \to \mathscr{G}_0', \qquad \widetilde{\tau}_{\mathrm{D}} f_\lambda \mapsto \widetilde{\tau}_{\mathrm{N}} f_\lambda, \tag{8.7.6}$$

where $f_\lambda \in \mathfrak{N}_\lambda(T_{\max})$.

With the preparations above one can now follow the strategy in Section 8.4 and construct a boundary triplet for the maximal operator $T_{\max}$ under the assumption that $\Omega \subset \mathbb{R}^n$ is a bounded Lipschitz domain. Consider the Gelfand triple $\{\mathscr{G}_1, L^2(\partial\Omega), \mathscr{G}_1'\}$ and the corresponding isometric isomorphisms $\iota_+ : \mathscr{G}_1 \to L^2(\partial\Omega)$ and $\iota_- : \mathscr{G}_1' \to L^2(\partial\Omega)$ such that

$$\langle \varphi, \psi \rangle_{\mathscr{G}_1' \times \mathscr{G}_1} = (\iota_- \varphi, \iota_+ \psi)_{L^2(\partial\Omega)}, \qquad \varphi \in \mathscr{G}_1', \ \psi \in \mathscr{G}_1;$$

cf. Lemma 8.1.2. When comparing (8.1.6) and (8.7.5) it is clear that $\iota_+ = \Lambda^{-1/2}$ and ι_- is the extension of $\Lambda^{1/2}$ onto $\mathscr{G}_1'$. Recall also the definition and the properties of the Dirichlet operator A_{D} in Theorem 8.7.2 and the direct sum decomposition (8.4.1).

Theorem 8.7.6. *Let $\Omega \subset \mathbb{R}^n$ be a bounded Lipschitz domain and let A_{D} be the self-adjoint Dirichlet realization of $-\Delta + V$ in $L^2(\Omega)$ in Theorem 8.7.2. Fix a number $\eta \in \rho(A_{\mathrm{D}}) \cap \mathbb{R}$ and decompose $f \in \operatorname{dom} T_{\max}$ according to (8.4.1) in the form*

$f = f_{\rm D} + f_\eta$, where $f_{\rm D} \in \operatorname{dom} A_{\rm D}$ and $f_\eta \in \mathfrak{N}_\eta(T_{\max})$. Then $\{L^2(\partial\Omega), \Gamma_0, \Gamma_1\}$, where

$$\Gamma_0 f = \iota_- \widetilde{\tau}_{\rm D} f \quad \textit{and} \quad \Gamma_1 f = -\iota_+ \tau_{\rm N} f_{\rm D}, \qquad f = f_{\rm D} + f_\eta \in \operatorname{dom} T_{\max},$$

is a boundary triplet for $(T_{\min})^* = T_{\max}$ such that

$$A_0 = A_{\rm D} \qquad \textit{and} \qquad A_1 = T_{\min} \,\widehat{+}\, \widehat{\mathfrak{N}}_\eta(T_{\max}).$$

The γ-field and Weyl function corresponding to the boundary triplet in Theorem 8.7.6 are formally the same as in Proposition 8.4.4. In fact, if $f_\eta(\varphi)$ denotes the unique element in $\mathfrak{N}_\eta(T_{\max})$ such that $\Gamma_0 f_\eta(\varphi) = \varphi$, then for all $\lambda \in \rho(A_{\rm D})$ the γ-field is given by

$$\gamma(\lambda)\varphi = \bigl(I + (\lambda - \eta)(A_{\rm D} - \lambda)^{-1}\bigr) f_\eta(\varphi), \quad \varphi \in L^2(\partial\Omega),$$

where $f_\lambda(\varphi) := \gamma(\lambda)\varphi$ is the unique element in $\mathfrak{N}_\lambda(T_{\max})$ such that $\Gamma_0 f_\lambda(\varphi) = \varphi$. As in Proposition 8.4.4 one also has

$$\gamma(\lambda)^* = -\iota_+ \tau_{\rm N} (A_{\rm D} - \bar{\lambda})^{-1}, \quad \lambda \in \rho(A_{\rm D}).$$

Moreover, the Weyl function M is given by

$$M(\lambda)\varphi = (\eta - \lambda)\iota_+ \tau_{\rm N} (A_{\rm D} - \lambda)^{-1} f_\eta(\varphi), \quad \varphi \in L^2(\partial\Omega).$$

As in the case of bounded C^2-domains, the Weyl function can be expressed via the Dirichlet-to-Neumann map; here the extended mapping $\widetilde{D}(\lambda)$ in (8.7.6) is used. In the same way as in Lemma 8.4.5 one verifies the relation

$$M(\lambda) = \iota_+ \bigl(\widetilde{D}(\eta) - \widetilde{D}(\lambda)\bigr)\iota_-^{-1}.$$

With the boundary triplet $\{L^2(\partial\Omega), \Gamma_0, \Gamma_1\}$ in Theorem 8.7.6 and the corresponding γ-field and Weyl function the self-adjoint realizations of $-\Delta + V$ on a bounded Lipschitz domain $\Omega \subset \mathbb{R}^n$ can be parametrized and the spectral properties can be described in a similar form as in Section 8.4. The discussion of the semibounded extensions and of the corresponding sesquilinear forms with the help of a compatible boundary pair is parallel to the considerations in Section 8.5 and is not provided here. Finally, the coupling technique of Schrödinger operators from Section 8.6 also extends under appropriate modifications to the general situation of Lipschitz domains.

Appendix A

Integral Representations of Nevanlinna Functions

Operator-valued Nevanlinna functions and their integral representations are presented in this appendix. First the case of scalar Nevanlinna functions is considered. Then follows a short introduction to operator-valued integrals; by interpreting these integrals as improper integrals the methods are kept as simple as possible. The general operator-valued Nevanlinna functions are treated based on the previous notions. Special operator-valued Nevanlinna functions such as Kac functions, Stieltjes functions, and inverse Stieltjes functions are discussed in detail.

A.1 Borel transforms and their Stieltjes inversion

This preparatory section contains a brief discussion of the Stieltjes inversion formula for the Borel transform. The form of the transform and the conditions have been chosen so that the results are easy to apply. In particular, with the inversion formula one can prove a weak form of the Stone inversion formula, a useful denseness property, and a general form of the Stieltjes inversion formula for Nevanlinna functions.

Let $\tau : \mathbb{R} \to \mathbb{R}$ be a nondecreasing function and let $g : \mathbb{R} \to \mathbb{C}$ be a measurable function such that

$$\int_{\mathbb{R}} \frac{|g(t)|}{|t|+1} \, d\tau(t) < \infty. \tag{A.1.1}$$

The *Borel transform* $G : \mathbb{C} \setminus \mathbb{R} \to \mathbb{C}$ of the combination $g\, d\tau$ is defined by

$$G(\lambda) = \int_{\mathbb{R}} \frac{1}{t-\lambda} \, g(t) \, d\tau(t), \quad \lambda \in \mathbb{C} \setminus \mathbb{R}. \tag{A.1.2}$$

Observe that the Borel transform G of $g\, d\tau$ in (A.1.2) is well defined and holomorphic on $\mathbb{C} \setminus \mathbb{R}$. The following result is the *Stieltjes inversion formula* for the Borel transform in (A.1.1).

J. Behrndt et al., *Boundary Value Problems, Weyl Functions, and Differential Operators*, Monographs in Mathematics 108,
https://doi.org/10.1007/978-3-030-36714-5

Proposition A.1.1. *Let $\tau : \mathbb{R} \to \mathbb{R}$ be a nondecreasing function, let $g : \mathbb{R} \to \mathbb{C}$ be a measurable function which satisfies (A.1.1), and let G be the Borel transform of $g\, d\tau$. Then*

$$\lim_{\varepsilon \downarrow 0} \frac{1}{2\pi i} \int_a^b \big(G(s+i\varepsilon) - G(s-i\varepsilon)\big)\, ds = \frac{1}{2}\int_{\{a\}} g(t)\, d\tau(t) + \int_{a+}^{b-} g(t)\, d\tau(t) + \frac{1}{2}\int_{\{b\}} g(t)\, d\tau(t)$$

holds for each compact interval $[a,b] \subset \mathbb{R}$. Furthermore, if $g \in L^1_{d\tau}(\mathbb{R})$, then for $0 < \varepsilon < 1$:

$$\frac{1}{2\pi}\left| \int_a^b \big(G(s+i\varepsilon) - G(s-i\varepsilon)\big)\, ds \right| \leq \int_{\mathbb{R}} |g(t)|\, d\tau(t). \tag{A.1.3}$$

Proof. For $\varepsilon > 0$ and $s \in \mathbb{R}$ one has

$$\frac{G(s+i\varepsilon) - G(s-i\varepsilon)}{2\pi i} = \frac{1}{\pi}\int_{\mathbb{R}} \frac{\varepsilon}{(s-t)^2+\varepsilon^2}\, g(t)\, d\tau(t).$$

Integration of the left-hand side over the interval $[a,b]$ and Fubini's theorem lead to

$$\begin{aligned} &\frac{1}{2\pi i}\int_a^b \big(G(s+i\varepsilon) - G(s-i\varepsilon)\big)\, ds \\ &\quad = \frac{1}{\pi}\int_{\mathbb{R}} \left(\int_a^b \frac{\varepsilon}{(s-t)^2+\varepsilon^2}\, ds \right) g(t)\, d\tau(t) \\ &\quad = \frac{1}{\pi}\int_{\mathbb{R}} \left(\arctan\left(\frac{b-t}{\varepsilon}\right) - \arctan\left(\frac{a-t}{\varepsilon}\right)\right) g(t)\, d\tau(t). \end{aligned} \tag{A.1.4}$$

In order to justify the use of Fubini's theorem in (A.1.4) note first that the functions

$$\arctan\left(\frac{b-t}{\varepsilon}\right) - \arctan\left(\frac{a-t}{\varepsilon}\right), \qquad 0 < \varepsilon < 1, \tag{A.1.5}$$

are nonnegative and bounded on $\mathbb{R}$. Furthermore, observe that for $0 < \varepsilon < 1$ and $x > 0$ one has $\arctan x < \arctan x/\varepsilon < \pi/2$ and hence the functions in (A.1.5) have the upper bound

$$k(t) = \begin{cases} \frac{\pi}{2} - \arctan(a-t), & t \leq a, \\ \pi, & a < t < b, \\ \arctan(b-t) + \frac{\pi}{2}, & t \geq b. \end{cases}$$

Since

$$t\left(\frac{\pi}{2} - \arctan(a-t)\right) \to -1, \quad t \to -\infty,$$
$$t\left(\arctan(b-t) + \frac{\pi}{2}\right) \to 1, \quad t \to \infty,$$

one has $kg \in L^1_{d\tau}(\mathbb{R})$ by (A.1.1), and it follows that the integral on the right-hand side in (A.1.4) is finite. Thus, the interchange of integration in (A.1.4) is justified.

Now the dominated convergence theorem will be applied to (A.1.4). For $\varepsilon \downarrow 0$ one has

$$\arctan\left(\frac{b-t}{\varepsilon}\right) - \arctan\left(\frac{a-t}{\varepsilon}\right) \to \begin{cases} 0, & t < a, \\ \pi/2, & t = a, \\ \pi, & a < t < b, \\ \pi/2, & t = b, \\ 0, & t > b, \end{cases}$$

and as an integrable majorant one can use $k|g|$. The right-hand side of (A.1.4) then shows that

$$\begin{aligned} &\lim_{\varepsilon \downarrow 0} \frac{1}{\pi} \int_{\mathbb{R}} \left(\arctan\left(\frac{b-t}{\varepsilon}\right) - \arctan\left(\frac{a-t}{\varepsilon}\right)\right) g(t)\, d\tau(t) \\ &\quad = \frac{1}{2} \int_{\{a\}} g(t)\, d\tau(t) + \int_{a+}^{b-} g(t)\, d\tau(t) + \frac{1}{2} \int_{\{b\}} g(t)\, d\tau(t), \end{aligned}$$

which leads to the assertion.

Finally, assume that $g \in L^1_{d\tau}(\mathbb{R})$ and $0 < \varepsilon < 1$. Then the estimate (A.1.3) follows due to (A.1.4) and (A.1.5). □

Observe that when the function g in (A.1.1) is real, then the function G in (A.1.2) satisfies the symmetry property $\overline{G(\lambda)} = G(\bar{\lambda})$, in which case one has that

$$\frac{1}{2\pi i} \int_a^b \big(G(s+i\varepsilon) - G(s-i\varepsilon)\big)\, ds = \frac{1}{\pi} \int_a^b \operatorname{Im} G(s+i\varepsilon)\, ds.$$

The special case $g(t) = 1$ in Proposition A.1.1 is of particular interest; see for instance Chapter 3.

Corollary A.1.2. *Let $\tau : \mathbb{R} \to \mathbb{R}$ be a nondecreasing function which satisfies the integrability condition*

$$\int_{\mathbb{R}} \frac{1}{|t|+1}\, d\tau(t) < \infty \tag{A.1.6}$$

and let G be given by

$$G(\lambda) = \int_{\mathbb{R}} \frac{1}{t-\lambda}\, d\tau(t), \quad \lambda \in \mathbb{C} \setminus \mathbb{R}.$$

Then the inversion formula

$$\lim_{\varepsilon\downarrow 0}\frac{1}{2\pi i}\int_a^b \big(G(s+i\varepsilon)-G(s-i\varepsilon)\big)\,ds=\frac{\tau(b+)+\tau(b-)}{2}-\frac{\tau(a+)+\tau(a-)}{2} \tag{A.1.7}$$

holds for every compact interval $[a,b]\subset\mathbb{R}$. In particular, if $\tau:\mathbb{R}\to\mathbb{R}$ is a nondecreasing function which is bounded, then (A.1.6) is satisfied and (A.1.7) holds.

The Stieltjes inversion result in Proposition A.1.1 and Corollary A.1.2 has a number of interesting consequences. A first observation concerns functions of bounded variation. Recall that any function of bounded variation on $\mathbb{R}$ is a linear combination of four bounded nondecreasing functions. Hence, the following corollary is straightforward.

Corollary A.1.3. *Let $\tau:\mathbb{R}\to\mathbb{C}$ be a function of bounded variation. Then the function*

$$H(\lambda)=\int_{\mathbb{R}}\frac{1}{t-\lambda}\,d\tau(t),\quad \lambda\in\mathbb{C}\setminus\mathbb{R},$$

is well defined and holomorphic, and for each compact interval $[a,b]\subset\mathbb{R}$ one has

$$\lim_{\varepsilon\downarrow 0}\frac{1}{2\pi i}\int_a^b \big(H(s+i\varepsilon)-H(s-i\varepsilon)\big)\,ds=\frac{\tau(b+)+\tau(b-)}{2}-\frac{\tau(a+)+\tau(a-)}{2}.$$

Proposition A.1.1 and Corollary A.1.2 can also be used to compute the spectral projection of a self-adjoint relation via Stone's formula; see Chapter 1.

Example A.1.4. Let H be a self-adjoint relation in a Hilbert space $\mathfrak{H}$ and let $E(\cdot)$ be the corresponding spectral measure. For $f\in\mathfrak{H}$ consider the function

$$G(\lambda)=\big((H-\lambda)^{-1}f,f\big),\quad \lambda\in\mathbb{C}\setminus\mathbb{R}.$$

It is clear that $\tau(t)=(E(-\infty,t)f,f)$, $t\in\mathbb{R}$, is a bounded nondecreasing function and that

$$G(\lambda)=\int_{\mathbb{R}}\frac{1}{t-\lambda}\,d\tau(t).$$

By Proposition A.1.1 with $g(t)=1$, $t\in\mathbb{R}$, one has for $[a,b]\subset\mathbb{R}$

$$\begin{aligned}\lim_{\varepsilon\downarrow 0}\frac{1}{2\pi i}\int_a^b &\big(\big((H-(s+i\varepsilon))^{-1}-(H-(s-i\varepsilon))^{-1}\big)f,f\big)\,ds\\ &=\frac{1}{2}(E(\{a\})f,f)+(E((a,b))f,f)+\frac{1}{2}(E(\{b\})f,f),\end{aligned}$$

which is Stone's formula in the weak sense; cf. (1.5.4) and (1.5.7) in Section 1.5.

Another consequence is the Stieltjes inversion formula for Nevanlinna functions; see Lemma A.2.7 and Corollary A.2.8. Moreover, there is the following denseness statement for a space of the form $L^2_{d\sigma}(\mathbb{R})$ which is used in Section 4.3.

Corollary A.1.5. *Assume that the function $\sigma : \mathbb{R} \to \mathbb{R}$ is nondecreasing and that it satisfies*

$$\int_{\mathbb{R}} \frac{1}{t^2+1}\, d\sigma(t) < \infty. \tag{A.1.8}$$

Let g be an element in $L^2_{d\sigma}(\mathbb{R})$ such that

$$\int_{\mathbb{R}} \frac{1}{t-\lambda}\, g(t)\, d\sigma(t) = 0, \qquad \lambda \in \mathbb{C} \setminus \mathbb{R}. \tag{A.1.9}$$

Then $g = 0$ in $L^2_{d\sigma}(\mathbb{R})$.

Proof. The conditions (A.1.8) and $g \in L^2_{d\sigma}(\mathbb{R})$ show, using the Cauchy–Schwarz inequality, that

$$\int_{\mathbb{R}} \frac{|g(t)|}{\sqrt{t^2+1}}\, d\sigma(t) \leq \left(\int_{\mathbb{R}} |g(t)|^2\, d\sigma(t)\right)\left(\int_{\mathbb{R}} \frac{1}{t^2+1}\, d\sigma(t)\right) < \infty,$$

which implies that the condition (A.1.1) is satisfied. It follows from (A.1.9) and Proposition A.1.1 that for each compact interval $[a,b] \subset \mathbb{R}$ one has

$$\frac{1}{2}\int_{\{a\}} g(t)\, d\sigma(t) + \int_{a+}^{b-} g(t)\, d\sigma(t) + \frac{1}{2}\int_{\{b\}} g(t)\, d\sigma(t) = 0. \tag{A.1.10}$$

Next it will be shown that the contribution of the endpoints a and b in (A.1.10) is trivial. To see this, suppose that a is a point mass of $d\sigma$ and choose $\eta_k > 0$, $k = 1, 2, \dots$, such that $\eta_k \to 0$, $k \to \infty$, and $a \pm \eta_k$ are not point masses of $d\sigma$, which is possible since the point masses of $d\sigma$ form a countable subset of $\mathbb{R}$. Now Proposition A.1.1 for the compact interval $[a-\eta_k, a+\eta_k] \subset \mathbb{R}$, (A.1.9), and dominated convergence lead to

$$0 = \int_{a-\eta_k}^{a+\eta_k} g(t)\, d\sigma(t) = \lim_{k\to\infty} \int_{a-\eta_k}^{a+\eta_k} g(t)\, d\sigma(t) = \int_{\{a\}} g(t)\, d\sigma(t).$$

The same argument shows that the contribution of the endpoint b in (A.1.10) is trivial, and hence (A.1.10) reduces to

$$\int_a^b g(t)\, d\sigma(t) = 0 \quad \text{for all } \ a < b.$$

In other words, g is orthogonal to all characteristic functions in $L^2_{d\sigma}(\mathbb{R})$ and hence $g = 0$ in $L^2_{d\sigma}(\mathbb{R})$. □

Remark A.1.6. Sometimes it is useful to have a matrix-valued version of the Borel transform in (A.1.2) and of the Stieltjes inversion formula in Proposition A.1.1; see also Remark A.2.10. More precisely, if g is a measurable $n \times n$ matrix function on $\mathbb{R}$ such that the entries of g satisfy the integrability condition (A.1.1), then Proposition A.1.1 remains valid with the integrals interpreted in the matrix sense.

A.2 Scalar Nevanlinna functions

This section contains a brief treatment of the integral representation of scalar Nevanlinna functions and its consequences.

Lemma A.2.1. *Let $f : \mathbb{D} \to \mathbb{C}$ be a holomorphic function and let $r \in (0,1)$. Then the representation*

$$f(z) = i\mathrm{Im}\, f(0) + \frac{1}{2\pi}\int_0^{2\pi} \frac{re^{it}+z}{re^{it}-z}\,\mathrm{Re}\, f(re^{it})\,dt$$

holds for all $z \in \mathbb{D}$ with $|z| < r$.

Proof. Let $r \in (0,1)$ and $\mathbb{T}_r = \{z \in \mathbb{C} : |z| = r\}$. Then for $z \in \mathbb{D}$, $|z| < r$, one obtains by Cauchy's integral formula

$$f(z) = \frac{1}{2\pi i}\int_{\mathbb{T}_r} \frac{f(w)}{w-z}\,dw = \frac{1}{2\pi}\int_0^{2\pi} \frac{re^{it}}{re^{it}-z} f(re^{it})\,dt, \tag{A.2.1}$$

and, since $|r^2/\bar{z}| > r$, one obtains in a similar way

$$f(0) = \frac{1}{2\pi i}\int_{\mathbb{T}_r} \frac{f(w)}{w(1-w(\bar{z}/r^2))}\,dw = \frac{1}{2\pi}\int_0^{2\pi} \frac{re^{-it}}{re^{-it}-\bar{z}}\, f(re^{it})\,dt,$$

and, by taking complex conjugates,

$$\overline{f(0)} = \frac{1}{2\pi}\int_0^{2\pi} \frac{re^{it}}{re^{it}-z}\,\overline{f(re^{it})}\,dt.$$

Furthermore, it is clear from (A.2.1) that

$$\mathrm{Re}\, f(0) = \frac{1}{2\pi}\int_0^{2\pi} \mathrm{Re}\, f(re^{it})\,dt.$$

Therefore,

$$\begin{aligned}
&\frac{1}{2\pi}\int_0^{2\pi} \frac{re^{it}+z}{re^{it}-z}\,\mathrm{Re}\, f(re^{it})\,dt \\
&\quad= \frac{1}{2\pi}\int_0^{2\pi} \left(\frac{2re^{it}}{re^{it}-z} - 1\right) \frac{f(re^{it}) + \overline{f(re^{it})}}{2}\,dt \\
&\quad= \frac{1}{2\pi}\int_0^{2\pi} \frac{re^{it}}{re^{it}-z} f(re^{it})\,dt + \frac{1}{2\pi}\int_0^{2\pi} \frac{re^{it}}{re^{it}-z}\,\overline{f(re^{it})}\,dt - \mathrm{Re}\, f(0) \\
&\quad= f(z) + \overline{f(0)} - \mathrm{Re}\, f(0) = f(z) - i\mathrm{Im}\, f(0),
\end{aligned}$$

which implies the assertion of the lemma. □

If $f : \mathbb{D} \to \mathbb{C}$ is a holomorphic function such that $\operatorname{Re} f(re^{it}) \geq 0$, then the expression

$$\operatorname{Re} f(re^{it})\, dt/2\pi$$

in Lemma A.2.1 leads to a measure. This observation is important in the proof of the next lemma.

Lemma A.2.2. *Let $f : \mathbb{D} \to \mathbb{C}$ be a function. Then the following statements are equivalent:*

(i) *f has an integral representation of the form*

$$f(z) = ic + \int_0^{2\pi} \frac{e^{it} + z}{e^{it} - z}\, d\tau(t), \qquad z \in \mathbb{D},$$

with $c \in \mathbb{R}$ and a bounded nondecreasing function $\tau : [0, 2\pi] \to \mathbb{R}$.

(ii) *f is holomorphic on $\mathbb{D}$ and $\operatorname{Re} f(z) \geq 0$ for all $z \in \mathbb{D}$.*

Proof. (i) $\Rightarrow$ (ii) It is clear that f is holomorphic on $\mathbb{D}$. For $z \in \mathbb{D}$ a straightforward calculation shows that

$$\operatorname{Re} f(z) = \int_0^{2\pi} \frac{1 - |z|^2}{|e^{it} - z|^2}\, d\tau(t) \geq 0.$$

(ii) $\Rightarrow$ (i) Since $\operatorname{Re} f(z) \geq 0$, $z \in \mathbb{D}$, it is obvious that for any $r \in (0, 1)$ the function

$$\tau_r : [0, 2\pi] \to \mathbb{R}, \qquad t \mapsto \frac{1}{2\pi} \int_0^t \operatorname{Re} f(re^{is})\, ds,$$

is nondecreasing. Furthermore, Cauchy's integral formula shows that

$$\tau_r(2\pi) = \frac{1}{2\pi} \int_0^{2\pi} \operatorname{Re} f(re^{is})\, ds = \operatorname{Re} \left(\frac{1}{2\pi i} \int_{\mathbb{T}_r} \frac{f(w)}{w}\, dw \right) = \operatorname{Re} f(0),$$

and so

$$0 = \tau_r(0) \leq \tau_r(t) \leq \tau_r(2\pi) = \operatorname{Re} f(0) < \infty \tag{A.2.2}$$

for $t \in (0, 2\pi)$ and $r \in (0, 1)$. Therefore, the Borel measure induced by τ_r on $[0, 2\pi]$ is finite and since $\operatorname{Re} f$ is continuous, it follows that τ_r, $r \in (0, 1)$, is a regular Borel measure on $[0, 2\pi]$. Observe that, by Lemma A.2.1,

$$\begin{aligned} f(z) &= i\operatorname{Im} f(0) + \frac{1}{2\pi} \int_0^{2\pi} \frac{re^{it} + z}{re^{it} - z} \operatorname{Re} f(re^{it})\, dt \\ &= i\operatorname{Im} f(0) + \int_0^{2\pi} \frac{re^{it} + z}{re^{it} - z}\, d\tau_r(t) \end{aligned} \tag{A.2.3}$$

for all $z \in \mathbb{D}$, $|z| < r$. Next it will be verified that the above formula remains valid when r tends to 1.

By the Helly selection principle (cf. [763, Theorem 16.2]) and (A.2.2), there exists a nonnegative nondecreasing sequence (r_k), $k = 1, 2, \dots$, tending to 1 and a function τ such that $\tau_{r_k}(t) \to \tau(t)$, $0 \leq t \leq 2\pi$. Moreover, according to the Helly–Bray theorem (cf. [763, Theorem 16.4]), one has

$$\lim_{k\to\infty} \int_0^{2\pi} h(t)\, d\tau_{r_k}(t) = \int_0^{2\pi} h(t)\, d\tau(t)$$

for all continuous functions $h : [0, 2\pi] \to \mathbb{C}$. Observe that, by (A.2.2), in particular

$$\tau(2\pi) - \tau(0) = \operatorname{Re} f(0).$$

Therefore, using the Helly–Bray theorem and the fact that $t \mapsto \frac{r_k e^{it}+z}{r_k e^{it}-z}$ converges uniformly to $t \mapsto \frac{e^{it}+z}{e^{it}-z}$, one finds that

$$\begin{aligned} &\lim_{k\to\infty} \int_0^{2\pi} \frac{r_k e^{it} + z}{r_k e^{it} - z}\, d\tau_{r_k}(t) \\ &\quad = \int_0^{2\pi} \frac{e^{it}+z}{e^{it}-z}\, d\tau(t) + \lim_{k\to\infty} \int_0^{2\pi} \left(\frac{r_k e^{it} + z}{r_k e^{it} - z} - \frac{e^{it}+z}{e^{it}-z} \right) d\tau_{r_k}(t) \\ &\quad = \int_0^{2\pi} \frac{e^{it}+z}{e^{it}-z}\, d\tau(t). \end{aligned}$$

Hence, (A.2.3) yields

$$f(z) = ic + \int_0^{2\pi} \frac{e^{it}+z}{e^{it}-z}\, d\tau(t) \quad \text{and} \quad c = \operatorname{Im} f(0),$$

as needed. □

Here is the definition of a scalar Nevanlinna function. The operator-valued version will be considered in Definition A.4.1.

Definition A.2.3. A function $F : \mathbb{C} \setminus \mathbb{R} \to \mathbb{C}$ is called a *Nevanlinna function* if

(i) F is holomorphic on $\mathbb{C} \setminus \mathbb{R}$;
(ii) $\overline{F(\lambda)} = F(\bar{\lambda})$, $\lambda \in \mathbb{C} \setminus \mathbb{R}$;
(iii) $\operatorname{Im} F(\lambda) / \operatorname{Im} \lambda \geq 0$, $\lambda \in \mathbb{C} \setminus \mathbb{R}$.

The next result provides an integral representation for Nevanlinna functions.

Theorem A.2.4. *Let $F : \mathbb{C} \setminus \mathbb{R} \to \mathbb{C}$. Then the following statements are equivalent:*

(i) *F has an integral representation of the form*

$$F(\lambda) = \alpha + \beta\lambda + \int_{\mathbb{R}} \frac{1 + t\lambda}{t - \lambda}\, d\theta(t), \quad \lambda \in \mathbb{C} \setminus \mathbb{R}, \tag{A.2.4}$$

with $\alpha \in \mathbb{R}$, $\beta \geq 0$, and a bounded nondecreasing function $\theta : \mathbb{R} \to \mathbb{R}$.

(ii) *F is a Nevanlinna function.*

Proof. (i) ⇒ (ii) It is clear from the representation (A.2.4) that F is holomorphic on $\mathbb{C}\setminus\mathbb{R}$ and that $\overline{F(\lambda)} = F(\bar{\lambda})$. Moreover it follows that

$$\frac{\operatorname{Im} F(\lambda)}{\operatorname{Im}\lambda} = \beta + \int_{\mathbb{R}} \frac{t^2+1}{|t-\lambda|^2}\, d\theta(t), \quad \lambda\in\mathbb{C}\setminus\mathbb{R}.$$

Hence, F is a Nevanlinna function.

(ii) ⇒ (i) Assume that F is a Nevanlinna function and consider the following transformations

$$\lambda = i\,\frac{1+z}{1-z}, \quad z\in\mathbb{D}, \qquad F(\lambda) = if(z).$$

Note that $z\mapsto\lambda$ is a bijective mapping from $\mathbb{D}$ onto $\mathbb{C}^+$ and that the function f is holomorphic on $\mathbb{D}$. Furthermore, observe that

$$\operatorname{Im} F(\lambda)\geq 0 \quad\Rightarrow\quad \operatorname{Re} f(z)\geq 0.$$

Hence, according to Lemma A.2.2, there exist $c\in\mathbb{R}$ and a bounded nondecreasing function $\tau:[0,2\pi]\to\mathbb{R}$, such that the function f has the representation

$$\begin{aligned} f(z) &= ic + \int_0^{2\pi} \frac{e^{is}+z}{e^{is}-z}\, d\tau(s) \\ &= ic + \int_{0+}^{2\pi-} \frac{e^{is}+z}{e^{is}-z}\, d\tau(s) \\ &\quad + \frac{1+z}{1-z}\big(\tau(2\pi)-\tau(2\pi-)+\tau(0+)-\tau(0)\big) \\ &= ic + \beta\,\frac{1+z}{1-z} + \int_{0+}^{2\pi-} \frac{e^{is}+z}{e^{is}-z}\, d\tau(s), \end{aligned}$$

where $\beta = \tau(2\pi)-\tau(2\pi-)+\tau(0+)-\tau(0)\geq 0$. Thus, the function F has the integral representation

$$F(\lambda) = -c + \beta\lambda + i\int_{0+}^{2\pi-} \frac{e^{is}+z}{e^{is}-z}\, d\tau(s).$$

Since $z = (\lambda-i)/(\lambda+i)$, one sees that

$$F(\lambda) = -c + \beta\lambda + \int_{0+}^{2\pi-} \frac{\lambda\cot s/2 - 1}{\cot s/2 + \lambda}\, d\tau(s).$$

With the substitutions $\alpha = -c$, $-\cot s/2 = t$, and the function θ defined by $\tau(s) = \theta(t)$, one finds that

$$F(\lambda) = \alpha + \beta\lambda + \int_{\mathbb{R}} \frac{1+t\lambda}{t-\lambda}\, d\theta(t).$$

Note that the function $\theta:\mathbb{R}\to\mathbb{R}$ is bounded since the function $\tau:[0,2\pi]\to\mathbb{R}$ is bounded. □

There is an equivalent formulation of this theorem involving the possibly unbounded measure $d\sigma(t) = (t^2+1)d\theta(t)$ defined by

$$\sigma(t) = \int_0^t (s^2+1)d\theta(s),$$

which is equivalent to

$$\theta(b) - \theta(a) = \int_a^b \frac{d\sigma(t)}{t^2+1}$$

for every compact interval $[a,b] \subset \mathbb{R}$. Hence, one obtains the following variant of Theorem A.2.4.

Theorem A.2.5. *Let $F : \mathbb{C} \setminus \mathbb{R} \to \mathbb{C}$. Then the following statements are equivalent:*

(i) *F has an integral representation of the form*

$$F(\lambda) = \alpha + \beta\lambda + \int_{\mathbb{R}} \left(\frac{1}{t-\lambda} - \frac{t}{t^2+1} \right) d\sigma(t), \quad \lambda \in \mathbb{C} \setminus \mathbb{R}, \tag{A.2.5}$$

with $\alpha \in \mathbb{R}$, $\beta \geq 0$, and a nondecreasing function $\sigma : \mathbb{R} \to \mathbb{R}$ such that

$$\int_{\mathbb{R}} \frac{d\sigma(t)}{t^2+1} < \infty.$$

(ii) *F is a Nevanlinna function.*

It follows from the integral representation (A.2.5) that the imaginary part of the function F satisfies

$$\frac{\operatorname{Im} F(\lambda)}{\operatorname{Im} \lambda} = \beta + \int_{\mathbb{R}} \frac{1}{|t-\lambda|^2} d\sigma(t), \quad \lambda \in \mathbb{C} \setminus \mathbb{R}. \tag{A.2.6}$$

With (A.2.5) and (A.2.6) it is possible to recover the ingredients in the integral formula (A.2.5) directly in terms of the function F. These results are used in Chapter 3.

Lemma A.2.6. *Let F be a Nevanlinna function as in Theorem A.2.5. Then*

$$\alpha = \operatorname{Re} F(i) \quad \textit{and} \quad \beta = \lim_{y \to \infty} \frac{F(iy)}{iy} = \lim_{y \to \infty} \frac{\operatorname{Im} F(iy)}{y}. \tag{A.2.7}$$

Moreover, for all $x \in \mathbb{R}$

$$\lim_{y \downarrow 0} y \operatorname{Im} F(x+iy) = \sigma(x+) - \sigma(x-) \tag{A.2.8}$$

and

$$\lim_{y \downarrow 0} y \operatorname{Re} F(x+iy) = 0. \tag{A.2.9}$$

Proof. The statement concerning α in (A.2.7) is clear. It follows from (A.2.5) that

$$\frac{1}{iy}F(iy) = \frac{1}{iy}\alpha + \beta + \int_{\mathbb{R}} \frac{1}{iy}\frac{1+iyt}{(t-iy)}\frac{d\sigma(t)}{t^2+1}. \tag{A.2.10}$$

An application of the dominated convergence theorem shows the first identity for β in (A.2.7). The second identity in (A.2.7) follows from (A.2.6). The integral representation (A.2.6) also shows that

$$y\mathrm{Im}\, F(x+iy) = \beta y^2 + \int_{\mathbb{R}} \frac{y^2}{(t-x)^2+y^2}\, d\sigma(t),$$

and the identity (A.2.8) follows from the dominated convergence theorem. One also sees from (A.2.5) that

$$y\,\mathrm{Re}\, F(x+iy) = y\,(\alpha+\beta x) + \int_{\mathbb{R}} \frac{y[(t-x)(1+xt)-y^2t]}{((t-x)^2+y^2)(t^2+1)}\, d\sigma(t). \tag{A.2.11}$$

By writing $xt = (t-x)x + x^2$ it follows that the numerator of the integrand is equal to

$$\begin{aligned} &y\big[(t-x)\big(1+(t-x)x+x^2\big) - y^2t\big] \\ &\qquad = y\big[(t-x)\big(1+(t-x)x+x^2-y^2\big) - y^2x\big] \\ &\qquad = y(t-x)[1+x^2-y^2] + (t-x)^2xy - y^3x. \end{aligned}$$

Now assume that $|y| \leq 1$. Then one sees that for a fixed $x \in \mathbb{R}$ the integrand in (A.2.11) is dominated by

$$\frac{x^2+4|x|+2}{2(t^2+1)}.$$

Thus, the identity in (A.2.9) follows from the dominated convergence theorem. □

In addition to Lemma A.2.6, the following *Stieltjes inversion formula* helps to recover the essential parts of the function σ.

Lemma A.2.7. *Let $F : \mathbb{C}\setminus\mathbb{R} \to \mathbb{C}$ be a Nevanlinna function with the integral representation (A.2.5). Let $\mathfrak{U}$ be an open neighborhood in $\mathbb{C}$ of $[a,b] \subset \mathbb{R}$ and let $g : \mathfrak{U} \to \mathbb{C}$ be holomorphic. Then*

$$\begin{aligned} &\lim_{\varepsilon\downarrow 0} \frac{1}{2\pi i}\int_a^b \big[(gF)(s+i\varepsilon) - (gF)(s-i\varepsilon)\big]\, ds \\ &\qquad = \frac{1}{2}\int_{\{a\}} g(t)\, d\sigma(t) + \int_{a+}^{b-} g(t)\, d\sigma(t) + \frac{1}{2}\int_{\{b\}} g(t)\, d\sigma(t). \end{aligned} \tag{A.2.12}$$

For any rectangle $R = [A,B]\times[-i\varepsilon_0, i\varepsilon_0] \subset \mathfrak{U}$ with $A < a < b < B$ there exists $M \geq 0$ such that, for $0 < \varepsilon \leq \varepsilon_0$,

$$\left|\int_a^b \big[(gF)(s+i\varepsilon) - (gF)(s-i\varepsilon)\big]\, ds\right| \leq M \sup\big\{|g(\lambda)|, |g'(\lambda)| : \lambda \in R\big\}, \tag{A.2.13}$$

where g' stands for the derivative of g.

Proof. Consider the Nevanlinna function F given by

$$F(\lambda) = \alpha + \beta\lambda + \int_{\mathbb{R}} \left(\frac{1}{t-\lambda} - \frac{t}{t^2+1} \right) d\sigma(t), \quad \lambda \in \mathbb{C} \setminus \mathbb{R},$$

where $\alpha \in \mathbb{R}$, $\beta \geq 0$, and σ is a nondecreasing function satisfying the integrability condition

$$\int_{\mathbb{R}} \frac{1}{t^2+1} \, d\sigma(t);$$

cf. Theorem A.2.5. Choose an interval $(A, B) \subset \mathbb{R}$ such that

$$[a, b] \subset (A, B) \subset [A, B] \subset \mathfrak{U}$$

and choose $\varepsilon_0 > 0$ such that $R = [A, B] \times [-i\varepsilon_0, i\varepsilon_0] \subset \mathfrak{U}$. Observe that the choice of A and B leads to the decomposition

$$g(\lambda)F(\lambda) = G(\lambda) + H(\lambda) + g(\lambda)K(\lambda), \quad \lambda \in (\mathbb{C} \setminus \mathbb{R}) \cap \mathfrak{U}, \tag{A.2.14}$$

where the functions G and H are given by

$$G(\lambda) = \int_A^B \frac{1}{t-\lambda} g(t) \, d\sigma(t), \quad H(\lambda) = \int_A^B \frac{g(\lambda) - g(t)}{t-\lambda} \, d\sigma(t),$$

while the factor K is given by

$$\begin{aligned} K(\lambda) = \Bigg[\alpha + \beta\lambda - \int_A^B \frac{t}{t^2+1} \, d\sigma(t) \\ + \left(\int_{-\infty}^A + \int_B^\infty \right) \left(\frac{1}{t-\lambda} - \frac{t}{t^2+1} \right) d\sigma(t) \Bigg]. \end{aligned}$$

The contributions of the functions G, H, and K will be considered separately.

Denote by $\widetilde{g}$ the extension of g on $[A, B]$ by zero to all of $\mathbb{R}$. Then G can be written as

$$G(\lambda) = \int_{\mathbb{R}} \frac{1}{t-\lambda} \widetilde{g}(t) \, d\sigma(t), \quad \text{where} \quad \int_{\mathbb{R}} \frac{|\widetilde{g}(t)|}{|t|+1} \, d\sigma(t) < \infty.$$

Hence, one concludes from Proposition A.1.1 that

$$\begin{aligned} \lim_{\varepsilon \downarrow 0} \frac{1}{2\pi i} \int_a^b \big(G(s+i\varepsilon) - G(s-i\varepsilon) \big) \, ds \\ = \frac{1}{2} \int_{\{a\}} \widetilde{g}(t) \, d\sigma(t) + \int_{a+}^{b-} \widetilde{g}(t) \, d\sigma(t) + \frac{1}{2} \int_{\{b\}} \widetilde{g}(t) \, d\sigma(t) \\ = \frac{1}{2} \int_{\{a\}} g(t) \, d\sigma(t) + \int_{a+}^{b-} g(t) \, d\sigma(t) + \frac{1}{2} \int_{\{b\}} g(t) \, d\sigma(t). \end{aligned} \tag{A.2.15}$$

Moreover, since $\widetilde{g} \in L^1_{d\sigma}(\mathbb{R})$, it follows from the estimate (A.1.3) in Proposition A.1.1 that

$$\begin{aligned}\left|\int_a^b \big(G(s+i\varepsilon) - G(s-i\varepsilon)\big)\,ds\right| &\leq M' \sup\big\{|g(\lambda)| : \lambda \in [A,B]\big\}\\ &\leq M' \sup\big\{|g(\lambda)| : \lambda \in R\big\}.\end{aligned} \tag{A.2.16}$$

The function H is defined for $\lambda \in (\mathbb{C} \setminus \mathbb{R}) \cap \mathfrak{U}$ and can be extended to $\mathfrak{U}$ by setting

$$H(\lambda) = \int_A^B h(t,\lambda)\,d\sigma(t), \quad \text{where} \quad h(t,\lambda) = \begin{cases} \frac{g(\lambda)-g(t)}{t-\lambda}, & t \neq \lambda,\\ -g'(t), & t = \lambda.\end{cases} \tag{A.2.17}$$

Clearly, the function H in (A.2.17) is bounded on the rectangle R by

$$L \sup\big\{|g'(\lambda)| : \lambda \in R\big\},$$

where L is a constant. Note that for all $s \in (a,b)$

$$H(s+i\varepsilon) - H(s-i\varepsilon) \to 0, \quad \varepsilon \to 0,$$

and hence dominated convergence yields

$$\lim_{\varepsilon \downarrow 0} \frac{1}{2\pi i}\int_a^b \big(H(s+i\varepsilon) - H(s-i\varepsilon)\big)\,ds = 0. \tag{A.2.18}$$

Note that it also follows from the above that there exists $M'' \geq 0$ such that for all $0 < \varepsilon \leq \varepsilon_0$

$$\left|\int_a^b \big(H(s+i\varepsilon) - H(s-i\varepsilon)\big)\,ds\right| \leq M'' \sup\big\{|g'(\lambda)| : \lambda \in R\big\}. \tag{A.2.19}$$

The function K has a holomorphic extension to the set $(\mathbb{C} \setminus \mathbb{R}) \cup (A,B)$ and this extension is uniformly continuous on the rectangle $[a,b] \times [-i\varepsilon, i\varepsilon_0]$. It is clear that

$$(gK)(s+i\varepsilon) - (gK)(s-i\varepsilon) \to 0, \quad \varepsilon \to 0,$$

holds for all $s \in (a,b)$. Since $|g|$ is also bounded on $[a,b] \times [-i\varepsilon, i\varepsilon_0]$, dominated convergence shows that

$$\lim_{\varepsilon \downarrow 0} \frac{1}{2\pi i}\int_a^b \big((gK)(s+i\varepsilon) - (gK)(s-i\varepsilon)\big)\,ds = 0. \tag{A.2.20}$$

Furthermore, one sees that there exists $M''' \geq 0$ such that, for all $0 < \varepsilon \leq \varepsilon_0$,

$$\begin{aligned}&\left|\int_a^b \big((gK)(s+i\varepsilon) - (gK)(s-i\varepsilon)\big)\,ds\right|\\ &\qquad \leq M''' \sup\big\{|g(\lambda)| : \lambda \in [a,b] \times [-i\varepsilon_0, i\varepsilon_0]\big\}\\ &\qquad \leq M''' \sup\big\{|g(\lambda)| : \lambda \in R\big\}.\end{aligned} \tag{A.2.21}$$

Now the assertion (A.2.12) follows from (A.2.14), (A.2.15), (A.2.18), and (A.2.20); in a similar way the assertion (A.2.13) follows from (A.2.14), (A.2.16), (A.2.19), and (A.2.21). □

For the special case $g(t) = 1$ Lemma A.2.7 has the following form.

Corollary A.2.8. *Let $F : \mathbb{C} \setminus \mathbb{R} \to \mathbb{C}$ be a Nevanlinna function with the integral representation (A.2.5). Then*

$$\lim_{\varepsilon \downarrow 0} \frac{1}{\pi} \int_a^b \operatorname{Im} F(s + i\varepsilon)\, ds = \frac{\sigma(b+) + \sigma(b-)}{2} - \frac{\sigma(a+) + \sigma(a-)}{2}.$$

It may happen that a Nevanlinna function has an analytic continuation to a subinterval of $\mathbb{R}$.

Proposition A.2.9. *Let F be a Nevanlinna function as in Theorem A.2.5 and let $(c, d) \subset \mathbb{R}$ be an open interval. Then the following statements are equivalent:*

(i) *F is holomorphic on $(\mathbb{C} \setminus \mathbb{R}) \cup (c, d)$;*

(ii) *σ is constant on (c, d).*

In this case

$$F(x) = \alpha + x\beta + \int_{\mathbb{R}\setminus(c,d)} \left(\frac{1}{t - x} - \frac{t}{t^2 + 1} \right) d\sigma(t), \quad x \in (c, d), \tag{A.2.22}$$

and F is a real nondecreasing function on (c, d).

Proof. (i) ⇒ (ii) By assumption, $F(x)$ is real for every $x \in (c, d)$. Now apply Corollary A.2.8 to any compact subinterval of (c, d). This implies that σ is constant on every compact subinterval of (c, d).

(ii) ⇒ (i) This is a direct consequence of Theorem A.2.5, since

$$F(\lambda) = \alpha + \lambda\beta + \int_{\mathbb{R}\setminus(c,d)} \left(\frac{1}{t - \lambda} - \frac{t}{t^2 + 1} \right) d\sigma(t), \quad \lambda \in \mathbb{C} \setminus \mathbb{R}.$$

If either (i) or (ii) holds, then (A.2.22) follows by a limit process. In particular, F is real on (c, d). From (A.2.22) one also concludes that for all $x \in (c, d)$ the function F is differentiable and

$$F'(x) = \beta + \int_{\mathbb{R}\setminus(c,d)} \frac{1}{(t - x)^2}\, d\sigma(t), \quad x \in (c, d),$$

is nonnegative. Hence, F is nondecreasing on (c, d). This completes the proof. □

Assume that $\operatorname{Im} F(\mu) = 0$ for some $\mu \in \mathbb{C} \setminus \mathbb{R}$. Then it follows from (A.2.6) that in the integral representation (A.2.5) $\beta = 0$ and $\sigma(t) = 0$, $t \in \mathbb{R}$. In other words, $\operatorname{Im} F(\lambda) = 0$ for all $\lambda \in \mathbb{C} \setminus \mathbb{R}$ and $F(\lambda) = \alpha$ for all $\lambda \in \mathbb{C} \setminus \mathbb{R}$. If F

admits an analytic continuation to $(c,d) \subset \mathbb{R}$ one concludes in the same way that $F(x) = \alpha$ for $x \in (c,d)$. If $F'(x_0) = 0$ for some $x_0 \in (c,d)$, then $F'(x) = 0$ for all $x \in (c,d)$ and $F(\lambda) = \alpha$ for all $\lambda \in \mathbb{C} \setminus \mathbb{R} \cup (c,d)$. Finally, observe that if $\operatorname{Im} F(\mu) \neq 0$ for some $\mu \in \mathbb{C} \setminus \mathbb{R}$, then

$$\operatorname{Im}\left(-\frac{1}{F(\lambda)}\right) = \frac{\operatorname{Im} F(\lambda)}{|F(\lambda)|^2}, \quad \lambda \in \mathbb{C} \setminus \mathbb{R},$$

and hence $-1/F$ is also a Nevanlinna function.

Remark A.2.10. This remark is a continuation of Remark A.1.6. If the function g in Lemma A.2.7 is a holomorphic $n \times n$ matrix function on $\mathcal{U}$, then the results (A.2.12) and (A.2.13) remain valid with the integrals interpreted in the matrix sense; cf. Remark A.1.6. Furthermore, the function g may be defined on $K \times \mathcal{U}$ with some compact space K such that $x \mapsto g_x(\lambda)$ is continuous for all $\lambda \in \mathcal{U}$ and $\lambda \mapsto g_x(\lambda)$ is holomorphic for all $x \in K$. In this case (A.2.12) remains valid for all $x \in K$, while the upper bound in (A.2.13) must be replaced by

$$M \sup\big\{|g_x(\lambda)|, |g_x'(\lambda)| : x \in K, \lambda \in R\big\}.$$

A.3 Operator-valued integrals

This section is concerned with operator-valued integrals which will be used in the integral representation of operator-valued Nevanlinna functions. For this purpose the notion of an improper Riemann–Stieltjes integral of bounded continuous functions is carried over to the case of operator-valued distribution functions.

In order to treat the Riemann–Stieltjes integral in the operator-valued case one needs the following preparatory observations. A function $\Theta : \mathbb{R} \to \mathbf{B}(\mathcal{G})$ whose values are self-adjoint operators is said to be *nondecreasing* if $t_1 \leq t_2$ implies

$$(\Theta(t_1)\varphi, \varphi) \leq (\Theta(t_2)\varphi, \varphi), \quad \varphi \in \mathcal{G}.$$

In general, such a function has limits as $t \to \pm\infty$, that are self-adjoint relations; cf. Chapter 5. In the following Θ will be called a *self-adjoint nondecreasing operator function*. Furthermore, Θ will be called *uniformly bounded* if there exists M such that for all $t \in \mathbb{R}$

$$|(\Theta(t)\varphi, \varphi)| \leq M\|\varphi\|^2, \quad \varphi \in \mathcal{G}.$$

Lemma A.3.1. *Let $\Theta : \mathbb{R} \to \mathbf{B}(\mathcal{G})$ be a self-adjoint nondecreasing operator function. Then the one-sided limits*

$$\Theta(t\pm), \quad t \in \mathbb{R},$$

exist in the strong sense and are bounded self-adjoint operators. If, in addition, $\Theta : \mathbb{R} \to \mathbf{B}(\mathcal{G})$ is uniformly bounded, then $\Theta(\pm\infty)$ exist in the strong sense and are bounded self-adjoint operators.

Proof. Let $t \in \mathbb{R}$ and choose an increasing sequence $t_n \to t-$. It is no restriction to assume that $\Theta(t_n) \geq 0$ for all $n \in \mathbb{N}$. Then for every $\varphi \in \mathcal{G}$ the sequence $(\Theta(t_n)\varphi, \varphi)$ is nondecreasing and bounded by $(\Theta(t)\varphi, \varphi) \leq \|\Theta(t)\|\|\varphi\|^2$. Hence, $(\Theta(t_n)\varphi, \varphi)$ converges to some $\nu_{\varphi,\varphi} \in \mathbb{R}$. Via polarization one finds that $(\Theta(t_n)\varphi, \psi)$ converges for all $\varphi, \psi \in \mathcal{G}$ and therefore

$$\mathcal{G} \times \mathcal{G} \ni \varphi \times \psi \mapsto \lim_{n\to\infty} (\Theta(t_n)\varphi, \psi)$$

is a symmetric sesquilinear form which is continuous, because

$$\begin{aligned}\Big|\lim_{n\to\infty} (\Theta(t_n)\varphi, \psi)\Big| &\leq \lim_{n\to\infty} (\Theta(t_n)\varphi, \varphi)^{1/2}(\Theta(t_n)\psi, \psi)^{1/2} \\ &\leq \|\Theta(t)\|\|\varphi\|\|\psi\|,\end{aligned}$$

where the Cauchy–Schwarz inequality was used for the nonnegative sesquilinear form $(\Theta(t_n)\cdot, \cdot)$. Thus, there exists a self-adjoint operator $\Omega \in \mathbf{B}(\mathcal{G})$ such that

$$\lim_{n\to\infty} (\Theta(t_n)\varphi, \psi) = (\Omega\varphi, \psi)$$

for all $\varphi, \psi \in \mathcal{G}$. Then $\|\Omega - \Theta(t_n)\| \leq \|\Omega\| + \|\Theta(t)\|$, while $\Omega - \Theta(t_n) \geq 0$. Recall the Cauchy–Schwarz inequality $\|Af\|^2 \leq \|A\|(Af, f)$, $f \in \mathcal{G}$, for nonnegative operators $A \in \mathbf{B}(\mathcal{G})$. Thus, one obtains

$$\|(\Omega - \Theta(t_n))\varphi\|^2 \leq (\|\Omega\| + \|\Theta(t)\|)\big((\Omega - \Theta(t_n))\varphi, \varphi\big) \to 0$$

for $n \to \infty$ and $\varphi \in \mathcal{G}$, i.e., $\Theta(t-)$ exists in the strong sense. Similar arguments show that $\Theta(t+)$, $t \in \mathbb{R}$, exists in the strong sense. If, in addition, Θ is uniformly bounded one verifies in the same way that the limits $\Theta(\pm\infty)$ exist in the strong sense and are bounded self-adjoint operators. □

Corollary A.3.2. *Let $\Theta : \mathbb{R} \to \mathbf{B}(\mathcal{G})$ be a self-adjoint nondecreasing operator function which is uniformly bounded. Then for every compact interval $[a, b]$ one has*

$$0 \leq \big((\Theta(b) - \Theta(a))\varphi, \varphi\big) \leq \big((\Theta(+\infty) - \Theta(-\infty))\varphi, \varphi\big), \quad \varphi \in \mathcal{G},$$

and consequently

$$\|(\Theta(b) - \Theta(a)\| \leq \|\Theta(+\infty) - \Theta(-\infty)\|.$$

After these preliminaries the operator-valued integrals will be introduced. Let $[a, b]$ be a compact interval and let $\Theta : [a, b] \to \mathbf{B}(\mathcal{G})$ be a self-adjoint nondecreasing operator function. Let $f : [a, b] \to \mathbb{C}$ be a continuous function. For a finite partition $a = t_0 < t_1 < \cdots < t_n = b$ of the interval $[a, b]$, define the Riemann–Stieltjes sum

$$S_n := \sum_{i=1}^{n} f(t_i)\big(\Theta(t_i) - \Theta(t_{i-1})\big). \tag{A.3.1}$$

The bounded operators S_n converge in $\mathbf{B}(\mathcal{G})$ if $\max |t_i - t_{i-1}|$ tends to zero. The limit will be called the *operator Riemann–Stieltjes integral* of f with respect to Θ and will be denoted by

$$\int_a^b f(t)\, d\Theta(t) \in \mathbf{B}(\mathcal{G}). \tag{A.3.2}$$

Lemma A.3.3. *Let $[a,b]$ be a compact interval and let $\Theta : [a,b] \to \mathbf{B}(\mathcal{G})$ be a self-adjoint nondecreasing operator function. Let $f : [a,b] \to \mathbb{C}$ be a continuous function. Then for all $\varphi, \psi \in \mathcal{G}$*

$$\left(\left(\int_a^b f(t)\, d\Theta(t)\right)\varphi, \psi\right) = \int_a^b f(t)\, d(\Theta(t)\varphi, \psi). \tag{A.3.3}$$

Moreover, for all $\varphi \in \mathcal{G}$,

$$\left\|\left(\int_a^b f(t)\, d\Theta(t)\right)\varphi\right\|^2 \leq \|\Theta(b) - \Theta(a)\| \int_a^b |f(t)|^2\, d(\Theta(t)\varphi, \varphi). \tag{A.3.4}$$

In particular, for all $\varphi \in \mathcal{G}$,

$$\left\|\left(\int_a^b f(t)\, d\Theta(t)\right)\varphi\right\| \leq \left(\sup_{t\in[a,b]} |f(t)|\right) \|\Theta(b) - \Theta(a)\| \|\varphi\|. \tag{A.3.5}$$

Proof. It is clear from the definition involving the Riemann–Stieltjes sums in (A.3.1) that the identity (A.3.3) holds.

To see that (A.3.4) holds, observe first that

$$\|T_1\varphi_1 + \cdots + T_n\varphi_n\|^2 \leq \|T_1T_1^* + \cdots + T_nT_n^*\| \left(\|\varphi_1\|^2 + \cdots + \|\varphi_n\|^2\right), \tag{A.3.6}$$

where $T_1, \dots, T_n \in \mathbf{B}(\mathcal{G})$ and $\varphi_1, \dots, \varphi_n \in \mathcal{G}$. One verifies (A.3.6) by interpreting the row $(T_1 \dots T_n)$ as a bounded operator A from $\mathcal{G} \times \cdots \times \mathcal{G}$ to $\mathcal{G}$ and recalling that $\|A\|^2 = \|AA^*\|$. Now rewrite the following Riemann–Stieltjes sum as indicated:

$$\begin{aligned}&\sum_{i=1}^n f(t_i)\left(\Theta(t_i) - \Theta(t_{i-1})\right)\varphi \\ &\qquad = \sum_{i=1}^n \left(\Theta(t_i) - \Theta(t_{i-1})\right)^{\frac{1}{2}} f(t_i)\left(\Theta(t_i) - \Theta(t_{i-1})\right)^{\frac{1}{2}}\varphi.\end{aligned}$$

The right-hand side may be written as $T_1\varphi_1 + \cdots + T_n\varphi_n$, where

$$T_i = \left(\Theta(t_i) - \Theta(t_{i-1})\right)^{\frac{1}{2}} \in \mathbf{B}(\mathcal{G}), \ \ \varphi_i = f(t_i)\left(\Theta(t_i) - \Theta(t_{i-1})\right)^{\frac{1}{2}}\varphi \in \mathcal{G}$$

for $i = 1, \dots, n$. Furthermore, one has

$$T_1T_1^* + \cdots + T_nT_n^* = \sum_{i=1}^n \left(\Theta(t_i) - \Theta(t_{i-1})\right) = \Theta(b) - \Theta(a).$$

Hence, the general estimate (A.3.6) above gives

$$\begin{aligned}\left\|\sum_{i=1}^{n} f(t_i)\left(\Theta(t_i)-\Theta(t_{i-1})\right)\varphi\right\|^2 &= \|T_1\varphi_1+\cdots+T_n\varphi_n\|^2\\ &\leq \|\Theta(b)-\Theta(a)\|\left(\|\varphi_1\|^2+\cdots+\|\varphi_n\|^2\right).\end{aligned}$$

Since

$$\begin{aligned}\|\varphi_1\|^2+\cdots+\|\varphi_n\|^2 &= \sum_{i=1}^{n}|f(t_i)|^2\left\|(\Theta(t_i)-\Theta(t_{i-1}))^{\frac{1}{2}}\varphi\right\|^2\\ &= \sum_{i=1}^{n}|f(t_i)|^2\big((\Theta(t_i)-\Theta(t_{i-1}))\varphi,\varphi\big)\end{aligned}$$

one concludes (A.3.4) with a limit argument. Finally, (A.3.5) is an immediate consequence of (A.3.4) and

$$\int_a^b d(\Theta(t)\varphi,\varphi) = \big((\Theta(b)-\Theta(a))\varphi,\varphi\big) \leq \|\Theta(b)-\Theta(a)\|\|\varphi\|^2.$$

This completes the proof. □

The integral in (A.3.2) enjoys the usual linearity properties. The nonnegativity property

$$f(t)\geq 0,\ t\in[a,b] \quad\Rightarrow\quad \int_a^b f(t)\,d\Theta(t)\geq 0$$

is a direct consequence of (A.3.3) of the previous lemma. Moreover, if c is a point of the open interval (a,b), then

$$\int_a^b f(t)\,d\Theta(t) = \int_a^c f(t)\,d\Theta(t) + \int_c^b f(t)\,d\Theta(t). \tag{A.3.7}$$

It follows from (A.3.7) that the integral defined in (A.3.2) and the properties in Lemma A.3.3 remain valid for functions $f:[a,b]\to\mathbb{C}$ that are piecewise continuous.

Now the integral in (A.3.2) will be extended to an improper Riemann–Stieltjes integral on $\mathbb{R}$ under the assumption that the self-adjoint nondecreasing operator function $\Theta:\mathbb{R}\to\mathbf{B}(\mathcal{G})$ is uniformly bounded; cf. Lemma A.3.1. Note that the restriction to bounded continuous functions guarantees the existence of the improper integral. However, it is clear that the results remain valid for bounded functions f that are continuous up to finitely many points.

Proposition A.3.4. *Let $\Theta : \mathbb{R} \to \mathbf{B}(\mathcal{G})$ be a self-adjoint nondecreasing operator function which is uniformly bounded and let $f : \mathbb{R} \to \mathbb{C}$ be a bounded continuous function. Then there exists a unique linear operator*

$$\int_{\mathbb{R}} f(t)\, d\Theta(t) \in \mathbf{B}(\mathcal{G}) \tag{A.3.8}$$

such that

$$\left(\int_{\mathbb{R}} f(t)\, d\Theta(t)\right)\varphi = \lim_{a\to-\infty}\lim_{b\to\infty}\left(\int_a^b f(t)\, d\Theta(t)\right)\varphi \tag{A.3.9}$$

for all $\varphi \in \mathcal{G}$, and

$$\left(\left(\int_{\mathbb{R}} f(t)\, d\Theta(t)\right)\varphi, \psi\right) = \int_{\mathbb{R}} f(t)\, d(\Theta(t)\varphi, \psi) \tag{A.3.10}$$

for all $\varphi, \psi \in \mathcal{G}$. Moreover,

$$\left\|\left(\int_{\mathbb{R}} f(t)\, d\Theta(t)\right)\varphi\right\|^2 \leq \|\Theta(+\infty) - \Theta(-\infty)\| \int_{\mathbb{R}} |f(t)|^2\, d(\Theta(t)\varphi, \varphi) \tag{A.3.11}$$

for all $\varphi \in \mathcal{G}$. In particular,

$$\left\|\left(\int_{\mathbb{R}} f(t)\, d\Theta(t)\right)\varphi\right\| \leq \sup_{t\in\mathbb{R}} |f(t)| \|\Theta(+\infty) - \Theta(-\infty)\| \|\varphi\| \tag{A.3.12}$$

for all $\varphi \in \mathcal{G}$.

Proof. By assumption, there exists some $M > 0$ such that for all $\varphi \in \mathcal{G}$

$$|(\Theta(t)\varphi, \varphi)| \leq M\|\varphi\|^2, \qquad t \in \mathbb{R}; \tag{A.3.13}$$

cf. Lemma A.3.1. First the existence of the limit in (A.3.9) will be verified. With the estimate (A.3.13) the inequality (A.3.4) may be written as

$$\left\|\left(\int_a^b f(t)\, d\Theta(t)\right)\varphi\right\|^2 \leq 2M \int_a^b |f(t)|^2\, d(\Theta(t)\varphi, \varphi). \tag{A.3.14}$$

Now consider two compact intervals $[a, b] \subset [a', b']$ and observe from (A.3.7) that

$$\int_{a'}^{b'} f(t)\, d\Theta(t) - \int_a^b f(t)\, d\Theta(t) = \int_{a'}^{a} f(t)\, d\Theta(t) + \int_b^{b'} f(t)\, d\Theta(t).$$

Hence, for every $\varphi \in \mathcal{G}$ one has

$$\begin{aligned}
&\left\|\int_{a'}^{b'} f(t)\, d\Theta(t)\varphi - \int_a^b f(t)\, d\Theta(t)\varphi\right\|^2 \\
&\quad\leq 2\left\|\int_{a'}^{a} f(t)\, d\Theta(t)\varphi\right\|^2 + 2\left\|\int_b^{b'} f(t)\, d\Theta(t)\varphi\right\|^2 \qquad\qquad (A.3.15)\\
&\quad\leq 4M \int_{a'}^{a} |f(t)|^2\, d(\Theta(t)\varphi, \varphi) + 4M \int_b^{b'} |f(t)|^2\, d(\Theta(t)\varphi, \varphi),
\end{aligned}$$

where the estimate (A.3.14) has been used. The right-hand side of (A.3.15) gives a Cauchy sequence for $a, a' \to -\infty$ and $b, b' \to +\infty$, since for all $\varphi \in \mathcal{G}$

$$\begin{aligned}\int_{\mathbb{R}} |f(t)|^2 \, d(\Theta(t)\varphi, \varphi) &\leq \|f\|_\infty^2 \int_{\mathbb{R}} d(\Theta(t)\varphi, \varphi) \\ &= \|f\|_\infty^2 \big((\Theta(+\infty)\varphi, \varphi) - (\Theta(-\infty)\varphi, \varphi)\big) \\ &\leq 2M\|f\|_\infty^2 \|\varphi\|^2 < \infty.\end{aligned}$$

Therefore, the strong limit on the right-hand side of (A.3.9) exists. This limit is denoted by the left-hand side of (A.3.9).

To verify (A.3.10), observe that the left-hand side of (A.3.10) is given by

$$\lim_{a\to-\infty} \lim_{b\to\infty} \left(\left(\int_a^b f(t) \, d\Theta(t) \right) \varphi, \psi \right) = \lim_{a\to-\infty} \lim_{b\to\infty} \int_a^b f(t) \, d(\Theta(t)\varphi, \psi),$$

where (A.3.3) was used. The statement now follows from the dominated convergence theorem.

Finally, as to (A.3.11) and (A.3.8), recall from (A.3.4) that for every $\varphi \in \mathcal{G}$ and for every compact interval $[a, b]$ one has the estimate

$$\begin{aligned}\left\| \left(\int_a^b f(t) \, d\Theta(t) \right) \varphi \right\|^2 &\leq \|\Theta(b) - \Theta(a)\| \int_a^b |f(t)|^2 \, d(\Theta(t)\varphi, \varphi) \\ &\leq \|\Theta(+\infty) - \Theta(-\infty)\| \int_{\mathbb{R}} |f(t)|^2 \, d(\Theta(t)\varphi, \varphi),\end{aligned} \tag{A.3.16}$$

where in the last inequality Corollary A.3.2 and the dominated convergence theorem have been used. Clearly, (A.3.11) follows from (A.3.16). This also leads to (A.3.12), which implies (A.3.8). □

The linearity and nonnegativity properties are preserved for the improper Riemann–Stieltjes integral. The adjoint of $\int_{\mathbb{R}} f(t) \, d\Theta(t)$ is given by

$$\left(\int_{\mathbb{R}} f(t) \, d\Theta(t) \right)^* = \int_{\mathbb{R}} \overline{f(t)} \, d\Theta(t). \tag{A.3.17}$$

Another immediate consequence is the following limit result.

Corollary A.3.5. *Let $\Theta : \mathbb{R} \to \mathbf{B}(\mathcal{G})$ be a self-adjoint nondecreasing operator function which is uniformly bounded. Let $f_n : \mathbb{R} \to \mathbb{C}$ be a sequence of continuous functions which is uniformly bounded. Let $f : \mathbb{R} \to \mathbb{C}$ be a bounded continuous function such that* $\lim_{n\to\infty} f_n(t) = f(t)$ *for all $t \in \mathbb{R}$. Then for all $\varphi \in \mathcal{G}$*

$$\left(\int_{\mathbb{R}} f_n(t) \, d\Theta(t) \right) \varphi \to \left(\int_{\mathbb{R}} f(t) \, d\Theta(t) \right) \varphi.$$

Proof. Use the linearity property and Proposition A.3.4 to conclude that for all $\varphi \in \mathcal{G}$

$$\left\| \left(\int_{\mathbb{R}} f(t)\, d\Theta(t) \right) \varphi - \left(\int_{\mathbb{R}} f_n(t)\, d\Theta(t) \right) \varphi \right\|^2 \leq \|\Theta(+\infty) - \Theta(-\infty)\| \int_{\mathbb{R}} |f(t) - f_n(t)|^2 \, d(\Theta(t)\varphi, \varphi).$$

Now apply the dominated convergence theorem. □

The next goal is to extend the operator-valued Riemann–Stieltjes integral to the case where the self-adjoint nondecreasing $\mathbf{B}(\mathcal{G})$-valued function appearing in the integral is not uniformly bounded. In principle this more general situation will be reduced to the case discussed above. In the following consider a (not necessarily uniformly bounded) self-adjoint nondecreasing operator function $\Sigma : \mathbb{R} \to \mathbf{B}(\mathcal{G})$. Let $\omega : \mathbb{R} \to \mathbb{R}$ be a continuous positive function with a positive lower bound, and define the operator function $\Theta : \mathbb{R} \to \mathbf{B}(\mathcal{G})$ by

$$\Theta(t) = \int_0^t \frac{d\Sigma(s)}{\omega(s)}, \quad t \in \mathbb{R}. \tag{A.3.18}$$

The function Θ can be used to extend the definition of the integral in Proposition A.3.4. First a preliminary lemma is needed.

Lemma A.3.6. *Let $\Sigma : \mathbb{R} \to \mathbf{B}(\mathcal{G})$ be a self-adjoint nondecreasing operator function and let $\omega : \mathbb{R} \to \mathbb{R}$ be a continuous positive function with a positive lower bound. Then Θ in (A.3.18) defines a self-adjoint nondecreasing operator function from $\mathbb{R}$ to $\mathbf{B}(\mathcal{G})$. Moreover, for every bounded continuous function $f : \mathbb{R} \to \mathbb{C}$ and for every compact interval $[a, b]$ one has*

$$\int_a^b f(t) \frac{d\Sigma(t)}{\omega(t)} = \int_a^b f(t)\, d\Theta(t) \in \mathbf{B}(\mathcal{G}). \tag{A.3.19}$$

Proof. It follows from Lemma A.3.3 that $\Theta(t)$, $t \in \mathbb{R}$, in (A.3.18) is well defined and that $\Theta(t) \in \mathbf{B}(\mathcal{G})$. Moreover,

$$(\Theta(t)\varphi, \varphi) = \int_0^t \frac{d(\Sigma(s)\varphi, \varphi)}{\omega(s)}, \quad t \in \mathbb{R}, \tag{A.3.20}$$

for all $\varphi \in \mathcal{G}$; cf. Lemma A.3.3. Thus, $\Theta(t)$ is self-adjoint and it follows that

$$(\Theta(t_2)\varphi, \varphi) - (\Theta(t_1)\varphi, \varphi) = \int_{t_1}^{t_2} \frac{d(\Sigma(s)\varphi, \varphi)}{\omega(s)}, \quad t_1 \leq t_2,$$

so that Θ is a nondecreasing operator function. Since the functions f/ω and f are continuous on $[a, b]$, Lemma A.3.3 shows that both integrals

$$\int_a^b f(t) \frac{d\Sigma(t)}{\omega(t)} \quad \text{and} \quad \int_a^b f(t)\, d\Theta(t)$$

belong to $\mathbf{B}(\mathcal{G})$. Observe that, for all $\varphi \in \mathcal{G}$,

$$\begin{aligned}\left(\left(\int_a^b f(t)\frac{d\Sigma(t)}{\omega(t)}\right)\varphi,\varphi\right) &= \int_a^b f(t)\frac{d(\Sigma(t)\varphi,\varphi)}{\omega(t)}\\ &= \int_a^b f(t)\,d(\Theta(t)\varphi,\varphi)\\ &= \left(\left(\int_a^b f(t)d\Theta(t)\right)\varphi,\varphi\right).\end{aligned}$$

Here the first and the third equality follow from Lemma A.3.3, while the second equality uses the Radon–Nikodým derivative in (A.3.20). By polarization,

$$\left(\left(\int_a^b f(t)\frac{d\Sigma(t)}{\omega(t)}\right)\varphi,\psi\right) = \left(\left(\int_a^b f(t)d\Theta(t)\right)\varphi,\psi\right)$$

for all $\varphi,\psi \in \mathcal{G}$, and hence

$$\left(\int_a^b f(t)\frac{d\Sigma(t)}{\omega(t)}\right)\varphi = \left(\int_a^b f(t)d\Theta(t)\right)\varphi \tag{A.3.21}$$

holds for all $\varphi \in \mathcal{G}$. This implies (A.3.19). □

As a consequence of Lemma A.3.6 one may recover the function Σ from the function Θ, since for every compact interval $[a,b] \subset \mathbb{R}$ one has

$$\int_a^b d\Sigma(t) = \int_a^b \omega(t)\,d\Theta(t).$$

Now the definition of the integral in Proposition A.3.4 is extended to "unbounded" measures by means of a "Radon–Nikodým" derivative.

Proposition A.3.7. *Let $\Sigma : \mathbb{R} \to \mathbf{B}(\mathcal{G})$ be a self-adjoint nondecreasing operator function, let $\omega : \mathbb{R} \to \mathbb{R}$ be a continuous positive function with a positive lower bound, and let $\Theta : \mathbb{R} \to \mathbf{B}(\mathcal{G})$ be defined by (A.3.18). Then the following statements are equivalent:*

(i) *For all $\varphi \in \mathcal{G}$ one has*

$$\int_{\mathbb{R}} \frac{d(\Sigma(s)\varphi,\varphi)}{\omega(s)} < \infty. \tag{A.3.22}$$

(ii) *In the sense of strong limits one has*

$$\int_{\mathbb{R}} \frac{d\Sigma(t)}{\omega(t)} \in \mathbf{B}(\mathcal{G}).$$

(iii) *The nondecreasing operator function $\Theta : \mathbb{R} \to \mathbf{B}(\mathcal{G})$ in (A.3.18) is uniformly bounded.*

Assume that either of the above conditions is satisfied. Let $f : \mathbb{R} \to \mathbb{C}$ be a bounded continuous function. Then

$$\int_{\mathbb{R}} f(t)\, \frac{d\Sigma(t)}{\omega(t)} = \int_{\mathbb{R}} f(t)\, d\Theta(t), \tag{A.3.23}$$

where each side belongs to $\mathbf{B}(\mathcal{G})$ and is meant as the strong limit of the correponding operators in the identity (A.3.19), respectively.

Proof. (i) $\Rightarrow$ (iii) Assume the condition (A.3.22). Then it follows from (A.3.18) and the monotone convergence theorem that

$$\|((\Theta(b) - \Theta(a))^{\frac{1}{2}}\varphi\|^2 = \int_a^b \frac{d(\Sigma(s)\varphi, \varphi)}{\omega(s)} \leq \int_{\mathbb{R}} \frac{d(\Sigma(s)\varphi, \varphi)}{\omega(s)}$$

for every compact interval $[a, b]$. The assumption (A.3.22) and the uniform boundedness principle show the existence of a constant M such that

$$\|(\Theta(b) - \Theta(a))^{\frac{1}{2}}\| \leq M$$

for every compact interval $[a, b]$. In particular, this leads to the inequality

$$|(\Theta(b) - \Theta(a))\varphi, \varphi)| \leq M^2 \|\varphi\|^2,$$

which implies that the nondecreasing operator function Θ is uniformly bounded. This gives (iii).

(iii) $\Rightarrow$ (ii) Assume that $\Theta : \mathbb{R} \to \mathbf{B}(\mathcal{G})$ in (A.3.18) is uniformly bounded. It follows from (A.3.19) that for every compact interval $[a, b] \subset \mathbb{R}$

$$\int_a^b \frac{d\Sigma(t)}{\omega(t)} = \int_a^b d\Theta(t) = \Theta(b) - \Theta(a),$$

where each side belongs to $\mathbf{B}(\mathcal{G})$. The result now follows by taking strong limits for $[a, b] \to \mathbb{R}$.

(ii) $\Rightarrow$ (i) The assumption means that for all $\varphi \in \mathcal{G}$

$$\left(\int_{\mathbb{R}} \frac{d\Sigma(t)}{\omega(t)}\right)\varphi = \lim_{a\to-\infty} \lim_{b\to\infty} \left(\int_a^b \frac{d\Sigma(t)}{\omega(t)}\right)\varphi,$$

with convergence in $\mathcal{G}$. In particular, this gives that

$$\begin{aligned} \left(\left(\int_{\mathbb{R}} \frac{d\Sigma(t)}{\omega(t)}\right)\varphi, \varphi\right) &= \lim_{a\to-\infty} \lim_{b\to\infty} \left(\left(\int_a^b \frac{d\Sigma(t)}{\omega(t)}\right)\varphi, \varphi\right) \\ &= \lim_{a\to-\infty} \lim_{b\to\infty} \int_a^b \frac{d(\Sigma(t)\varphi, \varphi)}{\omega(t)} \\ &= \int_{\mathbb{R}} \frac{d(\Sigma(t)\varphi, \varphi)}{\omega(t)}, \end{aligned}$$

where the second step is justified by Lemma A.3.3 and the last step by the monotone convergence theorem. This leads to (A.3.22).

By Proposition A.3.4, the integral on the right-hand side of (A.3.21) converges strongly to

$$\int_{\mathbb{R}} f(t) d\Theta(t) \in \mathbf{B}(\mathcal{G}).$$

Together with the identity (A.3.19) this leads to the assertion in (ii). □

For the reader's convenience the following facts are mentioned in terms of $\Sigma : \mathbb{R} \to \mathbf{B}(\mathcal{G})$ and $\omega : \mathbb{R} \to \mathbb{R}$ from Proposition A.3.7. They can be easily verified via the identity (A.3.23). Let $f : \mathbb{R} \to \mathbb{C}$ be a bounded continuous function. Then

$$\left(\int_{\mathbb{R}} f(t) \frac{d\Sigma(t)}{\omega(t)} \right)^* = \int_{\mathbb{R}} \overline{f(t)}\, \frac{d\Sigma(t)}{\omega(t)} \in \mathbf{B}(\mathcal{G}); \tag{A.3.24}$$

cf. (A.3.17), and for all $\varphi \in \mathcal{G}$

$$\left(\int_{\mathbb{R}} f(t) \frac{d\Sigma(t)}{\omega(t)} \varphi, \varphi \right) = \int_{\mathbb{R}} f(t) \frac{d(\Sigma(t)\varphi, \varphi)}{\omega(t)}; \tag{A.3.25}$$

cf. (A.3.10). Furthermore, if $f_n : \mathbb{R} \to \mathbb{C}$ is a sequence of continuous functions which is uniformly bounded and $f : \mathbb{R} \to \mathbb{C}$ is a bounded continuous function such that $\lim_{n\to\infty} f_n(t) = f(t)$ for all $t \in \mathbb{R}$, then for all $\varphi \in \mathcal{G}$

$$\left(\int_{\mathbb{R}} f_n(t) \frac{d\Sigma(t)}{\omega(t)} \right) \varphi \to \left(\int_{\mathbb{R}} f(t) \frac{d\Sigma(t)}{\omega(t)} \right) \varphi; \tag{A.3.26}$$

cf. Corollary A.3.5. It is also clear that Proposition A.3.7 and the above properties of the integral remain true for bounded functions $f : \mathbb{R} \to \mathbb{C}$ with finitely many discontinuities.

The following notation will be used later: Consider an interval $I \subset \mathbb{R}$, let χ_I be the corresponding characteristic function, and let $f : \mathbb{R} \to \mathbb{C}$ be a bounded continuous function. Then $\chi_I f$ is a piecewise continuous function and one defines

$$\int_I f(t) \frac{d\Sigma(t)}{\omega(t)} = \int_{\mathbb{R}} \chi_I(t) f(t) \frac{d\Sigma(t)}{\omega(t)}. \tag{A.3.27}$$

In particular, for $c \in \mathbb{R}$ integrals of the form

$$\int_{[c,\infty)} f(t) \frac{d\Sigma(t)}{\omega(t)} = \int_{\mathbb{R}} \chi_{[c,\infty)}(t) f(t) \frac{d\Sigma(t)}{\omega(t)} \tag{A.3.28}$$

will appear in the context of Nevanlinna functions that admit an analytic continuation to $(-\infty, c)$; see Section A.6.

A.4 Operator-valued Nevanlinna functions

The notion of scalar Nevanlinna function in Definition A.2.3 is easily carried over to the operator-valued case. The present section is concerned with developing the corresponding operator-valued integral representations. The main tool is provided by Proposition A.3.7.

Definition A.4.1. Let $\mathcal{G}$ be a Hilbert space and let $F : \mathbb{C} \setminus \mathbb{R} \to \mathbf{B}(\mathcal{G})$ be an operator function. Then F is called a $\mathbf{B}(\mathcal{G})$-valued *Nevanlinna function* if

(i) F is holomorphic on $\mathbb{C} \setminus \mathbb{R}$;

(ii) $F(\lambda) = F(\bar{\lambda})^*$, $\lambda \in \mathbb{C} \setminus \mathbb{R}$;

(iii) $\operatorname{Im} F(\lambda)/\operatorname{Im} \lambda \geq 0$, $\lambda \in \mathbb{C} \setminus \mathbb{R}$.

Operator-valued Nevanlinna functions admit integral representations as in the scalar case; cf. Theorem A.2.5.

Theorem A.4.2. *Let $\mathcal{G}$ be a Hilbert space and let $F : \mathbb{C} \setminus \mathbb{R} \to \mathbf{B}(\mathcal{G})$ be an operator function. Then the following statements are equivalent:*

(i) *F has an integral representation of the form*

$$F(\lambda) = \alpha + \lambda\beta + \int_{\mathbb{R}} \left(\frac{1}{t-\lambda} - \frac{t}{t^2+1} \right) d\Sigma(t), \quad \lambda \in \mathbb{C} \setminus \mathbb{R}, \tag{A.4.1}$$

with self-adjoint operators $\alpha, \beta \in \mathbf{B}(\mathcal{G})$, $\beta \geq 0$, and a nondecreasing self-adjoint operator function $\Sigma : \mathbb{R} \to \mathbf{B}(\mathcal{G})$ such that

$$\int_{\mathbb{R}} \frac{d\Sigma(t)}{t^2+1} \in \mathbf{B}(\mathcal{G}), \tag{A.4.2}$$

where the integrals in (A.4.1) and (A.4.2) converge in the strong topology.

(ii) *F is a Nevanlinna function.*

Note that for $\lambda \in \mathbb{C} \setminus \mathbb{R}$ the identity (A.4.1) can be rewritten as

$$F(\lambda) = \alpha + \lambda\beta + \int_{\mathbb{R}} f_\lambda(t)\, d\Theta(t), \quad \lambda \in \mathbb{C} \setminus \mathbb{R}, \tag{A.4.3}$$

where the bounded continuous function $f_\lambda : \mathbb{R} \to \mathbb{C}$ and the "bounded measure" Θ are given by

$$f_\lambda(t) = \frac{1+\lambda t}{t-\lambda} \quad \text{and} \quad \Theta(t) = \int_0^t \frac{d\Sigma(s)}{s^2+1}, \quad t \in \mathbb{R}, \tag{A.4.4}$$

respectively; cf. (A.3.18) and (A.3.23) with $\omega(t) = t^2 + 1$.

Proof. (i) $\Rightarrow$ (ii) Due to the condition (A.4.2) it follows that $F(\lambda)$ in (A.4.1) is well defined and represents an element in $\mathbf{B}(\mathcal{G})$; cf. (A.4.3), (A.4.4), and Proposition A.3.7. Moreover, it follows from (A.4.3) and (A.3.25) that for $\varphi \in \mathcal{G}$

$$(F(\lambda)\varphi,\varphi) = (\alpha\varphi,\varphi) + \lambda(\beta\varphi,\varphi) + \int_{\mathbb{R}} f_\lambda(t)\, d(\Theta(t)\varphi,\varphi), \quad \lambda \in \mathbb{C}\setminus\mathbb{R}.$$

Therefore, one sees that the function F is holomorphic on $\mathbb{C}\setminus\mathbb{R}$. It is clear that $F(\lambda)^* = F(\bar{\lambda})$; cf. (A.3.24). Since $(\alpha\varphi,\varphi) \in \mathbb{R}$ and $(\beta\varphi,\varphi) \in \mathbb{R}$, one also sees that

$$\operatorname{Im}(F(\lambda)\varphi,\varphi) = (\operatorname{Im}\lambda)\left[(\beta\varphi,\varphi) + \int_{\mathbb{R}} \frac{t^2+1}{|t-\lambda|^2} d(\Theta(t)\varphi,\varphi)\right], \quad \lambda \in \mathbb{C}\setminus\mathbb{R}.$$

Therefore, $\operatorname{Im}(F(\lambda)\varphi,\varphi)/\operatorname{Im}\lambda \geq 0$ for all $\lambda \in \mathbb{C}\setminus\mathbb{R}$. This implies that F is a Nevanlinna function.

(ii) $\Rightarrow$ (i) Let $F : \mathbb{C}\setminus\mathbb{R} \to \mathbf{B}(\mathcal{G})$ be an operator-valued Nevanlinna function. Then $M(\lambda) = F(\lambda) - \operatorname{Re} F(i)$ is also a Nevanlinna function and $\operatorname{Re} M(i) = 0$. For $\varphi \in \mathcal{G}$ the function $(M(\lambda)\varphi,\varphi)$ is a scalar Nevanlinna function with the integral representation

$$(M(\lambda)\varphi,\varphi) = \lambda\beta_{\varphi,\varphi} + \int_{\mathbb{R}} \frac{1+\lambda t}{t-\lambda}\, d\theta_{\varphi,\varphi}(t), \quad \lambda \in \mathbb{C}\setminus\mathbb{R},$$

where $\beta_{\varphi,\varphi} \geq 0$ and $\theta_{\varphi,\varphi} : \mathbb{R} \to \mathbb{R}$ is a nondecreasing function; cf. Theorem A.2.4 and Lemma A.2.6. Then it follows that

$$\operatorname{Im}(M(i)\varphi,\varphi) = \beta_{\varphi,\varphi} + \int_{\mathbb{R}} d\theta_{\varphi,\varphi}(t),$$

where each term on the right-hand side is nonnegative, so that

$$0 \leq \beta_{\varphi,\varphi} + \int_{\mathbb{R}} d\theta_{\varphi,\varphi}(t) = \operatorname{Im}(M(i)\varphi,\varphi) \leq \|M(i)\|\|\varphi\|^2, \quad \varphi \in \mathcal{G}.$$

Without loss of generality it will be assumed that

$$\theta_{\varphi,\varphi}(-\infty) = 0 \quad \text{and} \quad \theta_{\varphi,\varphi}(t) = \frac{\theta_{\varphi,\varphi}(t+) - \theta_{\varphi,\varphi}(t-)}{2}, \ t \in \mathbb{R}, \tag{A.4.5}$$

(see also the proof of Theorem A.2.4) and thus

$$0 \leq \beta_{\varphi,\varphi} \leq \|M(i)\|\|\varphi\|^2 \quad \text{and} \quad 0 \leq \theta_{\varphi,\varphi}(t) \leq \|M(i)\|\|\varphi\|^2 \tag{A.4.6}$$

for all $t \in \mathbb{R}$. For $\varphi, \psi \in \mathcal{G}$ one defines by polarization

$$\beta_{\varphi,\psi} = \frac{1}{4}\big(\beta_{\varphi+\psi,\varphi+\psi} - \beta_{\varphi-\psi,\varphi-\psi} + i\beta_{\varphi+i\psi,\varphi+i\psi} - i\beta_{\varphi-i\psi,\varphi-i\psi}\big),$$

and similarly

$$\theta_{\varphi,\psi} = \frac{1}{4}\big(\theta_{\varphi+\psi,\varphi+\psi} - \theta_{\varphi-\psi,\varphi-\psi} + i\theta_{\varphi+i\psi,\varphi+i\psi} - i\theta_{\varphi-i\psi,\varphi-i\psi}\big). \tag{A.4.7}$$

Then it follows that

$$(M(\lambda)\varphi,\psi) = \lambda\beta_{\varphi,\psi} + \int_{\mathbb{R}} \frac{1+\lambda t}{t-\lambda}\, d\theta_{\varphi,\psi}(t), \quad \lambda \in \mathbb{C}\setminus\mathbb{R}, \tag{A.4.8}$$

where $\beta_{\varphi,\psi}$ is determined by

$$\beta_{\varphi,\psi} = \lim_{y\to\infty} \frac{(M(iy)\varphi,\psi)}{iy}$$

(see (A.2.10) in the proof of Lemma A.2.6) and $\theta_{\varphi,\psi} : \mathbb{R} \to \mathbb{C}$ is a function of bounded variation.

Now the representation (A.4.8) will be used to verify that the bounded forms

$$\{\varphi,\psi\} \mapsto \beta_{\varphi,\psi}, \quad \{\varphi,\psi\} \mapsto \theta_{\varphi,\psi}(t), \tag{A.4.9}$$

are sesquilinear. For instance, with $\varphi,\varphi',\psi \in \mathcal{G}$ it follows that

$$\begin{aligned}
\lambda\beta_{\varphi+\varphi',\psi} &+ \int_{\mathbb{R}} \frac{1+\lambda t}{t-\lambda}\, d\theta_{\varphi+\varphi',\psi}(t) \\
&= \big(M(\lambda)(\varphi+\varphi'),\psi\big) \\
&= (M(\lambda)\varphi,\psi) + (M(\lambda)\varphi',\psi) \\
&= \lambda\beta_{\varphi,\psi} + \int_{\mathbb{R}} \frac{1+\lambda t}{t-\lambda}\, d\theta_{\varphi,\psi}(t) + \lambda\beta_{\varphi',\psi} + \int_{\mathbb{R}} \frac{1+\lambda t}{t-\lambda}\, d\theta_{\varphi',\psi}(t) \\
&= \lambda(\beta_{\varphi,\psi} + \beta_{\varphi',\psi}) + \int_{\mathbb{R}} \frac{1+\lambda t}{t-\lambda}\, d(\theta_{\varphi,\psi}(t) + \theta_{\varphi',\psi}(t)).
\end{aligned}$$

After dividing this equality by $\lambda = iy$ and letting $y \to \infty$, one concludes that

$$\beta_{\varphi+\varphi',\psi} = \beta_{\varphi,\psi} + \beta_{\varphi',\psi}. \tag{A.4.10}$$

Setting $\widetilde{\theta} = \theta_{\varphi+\varphi',\psi} - \theta_{\varphi,\psi} - \theta_{\varphi',\psi}$, one obtains

$$0 = \int_{\mathbb{R}} \frac{1+\lambda t}{t-\lambda}\, d\widetilde{\theta}(t) = \int_{\mathbb{R}} \left(\lambda + \frac{1+\lambda^2}{t-\lambda}\right) d\widetilde{\theta}(t), \qquad \lambda \in \mathbb{C}\setminus\mathbb{R},$$

and hence

$$\int_{\mathbb{R}} \frac{1}{t-\lambda}\, d\widetilde{\theta}(t) = -\int_{\mathbb{R}} \frac{\lambda}{1+\lambda^2}\, d\widetilde{\theta}(t), \qquad \lambda \in \mathbb{C}\setminus\mathbb{R}.$$

Now it follows from Corollary A.1.3 that for every compact subinterval $[a,b] \subset \mathbb{R}$ one has

$$0 = \frac{\widetilde{\theta}(b+) + \widetilde{\theta}(b-)}{2} - \frac{\widetilde{\theta}(a+) + \widetilde{\theta}(a-)}{2} = \widetilde{\theta}(b) - \widetilde{\theta}(a),$$

where (A.4.5) and (A.4.7) were used in the second equality. With the help of (A.4.5) and (A.4.7) via polarization one also concludes that $\widetilde{\theta}(a) \to 0$ for $a \to -\infty$ and therefore $\widetilde{\theta}(b) = 0$ for all $b \in \mathbb{R}$. This implies

$$\theta_{\varphi+\varphi',\psi}(t) = \theta_{\varphi,\psi}(t) + \theta_{\varphi',\psi}(t), \qquad t \in \mathbb{R}. \tag{A.4.11}$$

It follows from (A.4.10), (A.4.11), and similar considerations that the forms in (A.4.9) are both sesquilinear. Furthermore, these forms are nonnegative and from the Cauchy–Schwarz inequality and (A.4.6) it follows that they are bounded. Hence, there exists a uniquely determined nonnegative operator $\beta \in \mathbf{B}(\mathcal{G})$ such that

$$\beta_{\varphi,\psi} = (\beta\varphi, \psi), \qquad \varphi, \psi \in \mathcal{G},$$

and for each fixed $t \in \mathbb{R}$ there exists a uniquely determined bounded operator $\Theta(t) \in \mathbf{B}(\mathcal{G})$ such that

$$\theta_{\varphi,\psi}(t) = (\Theta(t)\varphi, \psi) \qquad \varphi, \psi \in \mathcal{G}.$$

From

$$(\Theta(t)\varphi, \varphi) = \theta_{\varphi,\varphi}(t) = \overline{\theta_{\varphi,\varphi}(t)} = (\varphi, \Theta(t)\varphi), \quad \varphi \in \mathcal{G},$$

one concludes that $(\Theta(t)\varphi, \psi) = (\varphi, \Theta(t)\psi)$, so that $\Theta(t)$ is a self-adjoint operator in $\mathbf{B}(\mathcal{G})$. Furthermore, for $t \leq t'$ one sees that

$$\theta_\varphi(t) \leq \theta_\varphi(t') \quad \Rightarrow \quad \Theta(t) \leq \Theta(t')$$

and therefore $\Theta : \mathbb{R} \to \mathbf{B}(\mathcal{G})$, $t \mapsto \Theta(t)$, is a nondecreasing self-adjoint operator function. Thus, one obtains for all $\varphi, \psi \in \mathcal{G}$

$$\begin{aligned}\left(\left(\lambda\beta + \int_{\mathbb{R}} \frac{1+\lambda t}{t-\lambda}\, d\Theta(t)\right)\varphi, \psi\right) &= \lambda\beta_{\varphi,\psi} + \int_{\mathbb{R}} \frac{1+\lambda t}{t-\lambda}\, d\theta_{\varphi,\psi}(t) \\ &= \big((F(\lambda) - \operatorname{Re} F(i))\varphi, \psi\big),\end{aligned}$$

which gives (A.4.3) with $\alpha = \operatorname{Re} F(i)$. □

The imaginary part $\operatorname{Im} F(\lambda) \in \mathbf{B}(\mathcal{G})$ of the Nevanlinna function N in Theorem A.4.2 admits the representation

$$\frac{\operatorname{Im} F(\lambda)}{\operatorname{Im} \lambda} = \beta + \int_{\mathbb{R}} \frac{1}{|t-\lambda|^2}\, d\Sigma(t), \quad \lambda \in \mathbb{C} \setminus \mathbb{R}, \tag{A.4.12}$$

where the integral exists in the strong sense. In particular, one has for all $\varphi \in \mathcal{G}$:

$$\frac{(\operatorname{Im} F(\lambda)\varphi, \varphi)}{\operatorname{Im} \lambda} = (\beta\varphi, \varphi) + \int_{\mathbb{R}} \frac{1}{|t-\lambda|^2}\, d(\Sigma(t)\varphi, \varphi), \quad \lambda \in \mathbb{C} \setminus \mathbb{R};$$

cf. (A.3.25). Hence, $\operatorname{Im} F(\lambda)/\operatorname{Im} \lambda$ is a nonnegative operator in $\mathbf{B}(\mathcal{G})$.

The next lemma is the counterpart of Lemma A.2.6 for operator-valued Nevanlinna functions.

Lemma A.4.3. *Let F be a $\mathbf{B}(\mathcal{G})$-valued Nevanlinna function with integral representation (A.4.1). The operators α, β, and the nondecreasing self-adjoint operator function $\Sigma : \mathbb{R} \to \mathbf{B}(\mathcal{G})$ are related to the function F by the following identities:*

$$\alpha = \operatorname{Re} F(i), \tag{A.4.13}$$

$$\beta\varphi = \lim_{y\to\infty} \frac{F(iy)}{iy}\varphi = \lim_{y\to\infty} \frac{\operatorname{Im} F(iy)}{y}\varphi, \quad \varphi \in \mathcal{G}. \tag{A.4.14}$$

Moreover, for all $x \in \mathbb{R}$

$$\lim_{y\downarrow 0} y\big(\operatorname{Im} F(x+iy)\varphi, \varphi\big) = (\Sigma(x+)\varphi, \varphi) - (\Sigma(x-)\varphi, \varphi), \quad \varphi \in \mathcal{G}. \tag{A.4.15}$$

Proof. It is clear from (A.4.1) that (A.4.13) holds. The identities for β in (A.4.14) follow with the help of (A.3.26) in the same was as in the proof of Lemma A.2.6. Finally, note that the statement (A.4.15) is a direct consequence of Lemma A.2.6 and Lemma A.3.1. □

In the present context the Stieltjes inversion formula has the following form; cf. Lemma A.2.7.

Lemma A.4.4. *Let $F : \mathbb{C}\setminus\mathbb{R} \to \mathbb{C}$ be a $\mathbf{B}(\mathcal{G})$-valued Nevanlinna function with the integral representation (A.4.1) and let $[a,b] \subset \mathbb{R}$. Assume that $\mathcal{U}$ is an open neighborhood of $[a,b]$ in $\mathbb{C}$ and that $g : \mathcal{U} \to \mathbb{C}$ is holomorphic. Then*

$$\begin{aligned}
&\lim_{\varepsilon\downarrow 0} \frac{1}{2\pi i} \int_a^b \big(((gF)(s+i\varepsilon) - (gF)(s-i\varepsilon))\varphi, \varphi\big)\, ds \\
&= \frac{1}{2}\int_{\{a\}} g(t)\, d(\Sigma(t)\varphi,\varphi) + \int_{a+}^{b-} g(t)\, d(\Sigma(t)\varphi,\varphi) + \frac{1}{2}\int_{\{b\}} g(t)\, d(\Sigma(t)\varphi,\varphi)
\end{aligned} \tag{A.4.16}$$

holds for all $\varphi \in \mathcal{G}$. If the function g is entire, then (A.4.16) is valid for any compact interval $[a,b]$. In particular,

$$\begin{aligned}
&\lim_{\varepsilon\downarrow 0} \frac{1}{\pi} \int_a^b (\operatorname{Im} F(s+i\varepsilon)\varphi, \varphi)\, ds \\
&= \frac{(\Sigma(b+)\varphi,\varphi) + (\Sigma(b-)\varphi,\varphi)}{2} - \frac{(\Sigma(a+)\varphi,\varphi) + (\Sigma(a-)\varphi,\varphi)}{2}
\end{aligned}$$

for all $\varphi \in \mathcal{G}$.

For operator-valued Nevanlinna functions Proposition A.2.9 has the following form.

Proposition A.4.5. *Let F be a Nevanlinna function as in Theorem A.4.2 and let $(c,d) \subset \mathbb{R}$ be an open interval. Then the following statements are equivalent:*

(i) *F is holomorphic on $(\mathbb{C}\setminus\mathbb{R}) \cup (c,d)$.*

(ii) *For every $\varphi \in \mathcal{G}$ the function $t \mapsto \Sigma(t)\varphi$ is constant on (c,d).*

In this case

$$F(x) = \alpha + x\beta + \int_{\mathbb{R}\setminus(c,d)} \left(\frac{1}{t-x} - \frac{t}{t^2+1} \right) d\Sigma(t), \quad x \in (c,d), \tag{A.4.17}$$

is self-adjoint; the integral in (A.4.17) converges in the strong topology. Moreover, F is nondecreasing on (c,d):

$$(F(x_1)\varphi,\varphi) \le (F(x_2)\varphi,\varphi), \quad c < x_1 < x_2 < d.$$

Proof. The implication (i) $\Rightarrow$ (ii) follows from the Stieltjes inversion formula in Lemma A.4.4, which implies that $t \mapsto (\Sigma(t)\varphi,\varphi)$ is constant for $t \in (c,d)$ and $\varphi \in \mathcal{G}$. For $c < t_1 < t_2 < d$ one concludes from the inequality

$$\big|((\Sigma(t_2)-\Sigma(t_1))\varphi,\psi)\big| \le ((\Sigma(t_2)-\Sigma(t_1))\varphi,\varphi)((\Sigma(t_2)-\Sigma(t_1))\psi,\psi) = 0$$

that the function $t \mapsto \Sigma(t)\varphi$ is constant on (c,d) for all $\varphi \in \mathcal{G}$. For the implication (ii) $\Rightarrow$ (i) consider the integral representation (A.4.1) in Theorem A.4.2, where the integral converges in the strong sense. Since $t \mapsto \Sigma(t)\varphi$ is constant on (c,d), this representation takes the form

$$\begin{aligned} F(\lambda) &= \alpha + \lambda\beta + \int_{\mathbb{R}} \chi_{\mathbb{R}\setminus(c,d)}(t) \left(\frac{1}{t-\lambda} - \frac{t}{t^2+1} \right) d\Sigma(t) \\ &= \alpha + \lambda\beta + \int_{\mathbb{R}\setminus(c,d)} \left(\frac{1}{t-\lambda} - \frac{t}{t^2+1} \right) d\Sigma(t) \end{aligned} \tag{A.4.18}$$

for all $\lambda \in \mathbb{C}\setminus\mathbb{R}$; cf. (A.3.27). It follows that $\lambda \mapsto (F(\lambda)\varphi,\varphi)$ is holomorphic on $(\mathbb{C}\setminus\mathbb{R}) \cup (c,d)$ for all $\varphi \in \mathcal{G}$ and hence F is holomorphic on $(\mathbb{C}\setminus\mathbb{R}) \cup (c,d)$.

To obtain (A.4.17), observe that for $x \in (c,d)$ one has

$$\chi_{\mathbb{R}\setminus(c,d)}(t) \left(\frac{1}{t-\lambda} - \frac{t}{t^2+1} \right) \to \chi_{\mathbb{R}\setminus(c,d)}(t) \left(\frac{1}{t-x} - \frac{t}{t^2+1} \right)$$

when $\lambda \to x$, $\lambda \in \mathbb{C}\setminus\mathbb{R}$, and the functions are uniformly bounded in t. Hence, (A.3.26) and (A.4.18) imply (A.4.17), where the integral in (A.4.17) converges in the strong topology by Proposition A.3.7. Furthermore, the bounded operators $F(x)$, $x \in (c,d)$, are self-adjoint by (A.3.24) and Proposition A.2.9 shows that $x \mapsto (F(x)\varphi,\varphi)$ is nondecreasing on (c,d). □

If F has an analytic continuation to $(c,d) \subset \mathbb{R}$, then it follows from (A.4.17) that F is differentiable on (c,d) and that

$$F'(x) = \beta + \int_{\mathbb{R}} \frac{1}{|t-x|^2} d\Sigma(t), \quad x \in (c,d), \tag{A.4.19}$$

where the integral exists in the strong sense. In particular, one has for all $\varphi \in \mathcal{G}$:

$$(F'(x)\varphi, \varphi) = (\beta\varphi, \varphi) + \int_{\mathbb{R}} \frac{1}{|t-x|^2}\, d(\Sigma(t)\varphi, \varphi), \quad x \in (c,d);$$

cf. (A.3.25). It is clear that $F'(x)$ is a nonnegative operator in $\mathbf{B}(\mathcal{G})$.

The next observation on isolated singularities of F is useful. If $(c,d) \subset \mathbb{R}$ and F admits an analytic continuation to $(c,d) \setminus \{y\}$ for some $y \in (c,d)$, then Proposition A.4.5 implies

$$F(\lambda) = \alpha + \lambda\beta + \int_{\mathbb{R}\setminus((c,y)\cup(y,d))} \left(\frac{1}{t-\lambda} - \frac{t}{t^2+1}\right) d\Sigma(t)$$

for all $\lambda \in (\mathbb{C}\setminus\mathbb{R}) \cup (c,y) \cup (y,d)$. Dominated convergence shows that

$$-\lim_{\lambda\to y} (\lambda - y)(F(\lambda)\varphi, \psi) = \big(\Sigma(y+)\varphi, \psi\big) - \big(\Sigma(y-)\varphi, \psi\big), \quad \varphi, \psi \in \mathcal{G},$$

whence

$$\lim_{\lambda\to y} \big\|(\lambda - y)F(\lambda)\big\| = \big\|\Sigma(y+) - \Sigma(y-)\big\|.$$

The following result establishes an important property of operator-valued Nevanlinna functions which will be characteristic for the Weyl functions in this text; see Chapter 4. The proof given here depends on the integral representation of a Nevanlinna function.

Proposition A.4.6. *Let F be a $\mathbf{B}(\mathcal{G})$-valued Nevanlinna function. Then the following statements hold:*

(i) *If $\operatorname{Im} F(\mu)$ is boundedly invertible for some $\mu \in \mathbb{C}\setminus\mathbb{R}$, then $\operatorname{Im} F(\lambda)$ is boundedly invertible for all $\lambda \in \mathbb{C}\setminus\mathbb{R}$.*

(ii) *Assume that F has an analytic continuation to the interval $(c,d) \subset \mathbb{R}$. If either $\operatorname{Im} F(\mu)$ is boundedly invertible for some $\mu \in \mathbb{C}\setminus\mathbb{R}$ or $F'(x_0)$ is boundedly invertible for some $x_0 \in (c,d)$, then $\operatorname{Im} F(\lambda)$ is boundedly invertible for all $\lambda \in \mathbb{C}\setminus\mathbb{R}$ and $F'(x)$ is boundedly invertible for all $x \in (c,d)$.*

Proof. The proof will be given in three steps and involves the nonnegativity of the operators in (A.4.12) and (A.4.19).

Step 1. Assume for some $\mu \in \mathbb{C}\setminus\mathbb{R}$ that $0 \in \sigma_{\mathrm{p}}(\operatorname{Im} F(\mu))$. Then there exists a nontrivial element $\varphi \in \ker(\operatorname{Im} F(\mu))$. Since $\beta \geq 0$, it follows from (A.4.12) that

$$(\beta\varphi, \varphi) = 0 \quad \text{and} \quad \int_{\mathbb{R}} \frac{1}{|t-\mu|^2}\, d(\Sigma(t)\varphi, \varphi) = 0,$$

and therefore $\varphi \in \ker\beta$ and $(\Sigma(t)\varphi, \varphi) = 0$ for all $t \in \mathbb{R}$. Hence, due to (A.4.12) one sees that $(\operatorname{Im} F(\lambda)\varphi, \varphi) = 0$ for all $\lambda \in \mathbb{C}\setminus\mathbb{R}$. For $\lambda \in \mathbb{C}^+$ one concludes by the nonnegativity of $\operatorname{Im} F(\lambda)$ that $\operatorname{Im} F(\lambda)\varphi = 0$. For $\lambda \in \mathbb{C}^-$ one has that $\operatorname{Im} F(\lambda)\varphi = \operatorname{Im} F(\bar{\lambda})\varphi = 0$. Therefore, it follows that $0 \in \sigma_{\mathrm{p}}(\operatorname{Im} F(\lambda))$ for all $\lambda \in \mathbb{C} \setminus \mathbb{R}$. Moreover, in case there is an analytic continuation to (c,d), (A.4.19) implies that $(F'(x)\varphi, \varphi) = 0$ for all $x \in (c,d)$. For $x \in (c,d)$ one now concludes by the nonnegativity of $F'(x)$ that $F'(x)\varphi = 0$. Hence, it follows that $0 \in \sigma_{\mathrm{p}}(F'(x))$ for all $x \in (c,d)$.

If F admits an analytic continuation to (c,d), then the above arguments show that the assumption $0 \in \sigma_{\mathrm{p}}(F'(x_0))$ for some $x_0 \in (c,d)$ leads to $0 \in \sigma_{\mathrm{p}}(\operatorname{Im} F(\lambda))$ for all $\lambda \in \mathbb{C} \setminus \mathbb{R}$ and $0 \in \sigma_{\mathrm{p}}(F'(x))$ for all $x \in (c,d)$.

Step 2. Assume for some $\mu \in \mathbb{C} \setminus \mathbb{R}$ that $0 \in \sigma_{\mathrm{c}}(\operatorname{Im} F(\mu))$. Then there exists a sequence $\varphi_n \in \mathcal{G}$, $\|\varphi_n\| = 1$, such that $\operatorname{Im} F(\mu)\varphi_n \to 0$ for $n \to \infty$. It follows from (A.4.12) that

$$(\beta\varphi_n, \varphi_n) \to 0 \quad \text{and} \quad \int_{\mathbb{R}} \frac{1}{|t-\mu|^2}\, d(\Sigma(t)\varphi_n, \varphi_n) \to 0, \qquad n \to \infty,$$

and hence also $(\operatorname{Im} F(\lambda)\varphi_n, \varphi_n) \to 0$ for any $\lambda \in \mathbb{C} \setminus \mathbb{R}$. Thus, for $\lambda \in \mathbb{C}^+$ one concludes from the nonnegativity of $\operatorname{Im} F(\lambda)$ that $\operatorname{Im} F(\lambda)\varphi_n \to 0$. Hence, one concludes that $0 \in \sigma_{\mathrm{c}}(\operatorname{Im} F(\lambda))$ for all $\lambda \in \mathbb{C} \setminus \mathbb{R}$. Moreover, if there is an analytic continuation to (c,d), then it follows from (A.4.19) that

$$(\beta\varphi_n, \varphi_n) \to 0 \quad \text{and} \quad \int_{\mathbb{R}} \frac{1}{|t-x|^2}\, d(\Sigma(t)\varphi_n, \varphi_n) \to 0, \qquad n \to \infty,$$

and hence also $(F'(x)\varphi_n, \varphi_n) \to 0$ for any $x \in (c,d)$. Thus, for $x \in (c,d)$ one concludes by the nonnegativity of $F'(x)$ that $F'(x)\varphi_n \to 0$. Hence, $x \in \sigma_{\mathrm{c}}(F'(x))$ for all $x \in (c,d)$.

Therefore, it has been shown that the assumption $0 \in \sigma_{\mathrm{c}}(\operatorname{Im} F(\mu))$ for some $\mu \in \mathbb{C} \setminus \mathbb{R}$ leads to

$$0 \in \sigma_{\mathrm{c}}(\operatorname{Im} F(\lambda)), \quad \lambda \in \mathbb{C} \setminus \mathbb{R}, \quad \text{and} \quad 0 \in \sigma_{\mathrm{c}}(F'(x)), \quad x \in (c,d).$$

It is clear that if F admits an analytic continuation to (c,d), the above arguments show that the assumption $0 \in \sigma_{\mathrm{c}}(F'(x_0))$ for some $x_0 \in (c,d)$ leads to $0 \in \sigma_{\mathrm{c}}(\operatorname{Im} F(\lambda))$ for all $\lambda \in \mathbb{C} \setminus \mathbb{R}$ and $0 \in \sigma_{\mathrm{c}}(F'(x))$ for all $x \in (c,d)$.

Step 3. Now assume that $\operatorname{Im} F(\mu)$ is boundedly invertible for some $\mu \in \mathbb{C} \setminus \mathbb{R}$ or, if there is an analytic continuation to (c,d), that $F'(x_0)$ is boundedly invertible for some $x_0 \in (c,d)$, in other words, $0 \in \rho(\operatorname{Im} F(\mu))$ or $0 \in \rho(F'(x_0))$. Since $\operatorname{Im} F(\lambda)$, $\lambda \in \mathbb{C} \setminus \mathbb{R}$, and $F'(x)$, $x \in (c,d)$, are self-adjoint operators in $\mathbf{B}(\mathcal{G})$, it follows that $\sigma_{\mathrm{r}}(\operatorname{Im} F(\lambda)) = \emptyset$ and $\sigma_{\mathrm{r}}(F'(x)) = \emptyset$. The assertions (i) and (ii) of the proposition now follow from Step 1 and Step 2. □

Next the notion of a uniformly strict Nevanlinna function will be defined. It can be used in conjunction with Proposition A.4.6.

Definition A.4.7. Let $\mathcal{G}$ be a Hilbert space and let $F : \mathbb{C} \setminus \mathbb{R} \to \mathbf{B}(\mathcal{G})$ be a Nevanlinna function. The function F is said to be *uniformly strict* if its imaginary part $\operatorname{Im} F(\lambda)$ is boundedly invertible for some, and hence for all $\lambda \in \mathbb{C} \setminus \mathbb{R}$.

In this context it is helpful to recall the following facts. Let $A \in \mathbf{B}(\mathcal{G})$ with $\operatorname{Im} A \geq 0$. Then clearly $\operatorname{Im} A$ is boundedly invertible if and only if $\operatorname{Im} A \geq \varepsilon$ for some $\varepsilon > 0$. Note that if $\operatorname{Im} A \geq \varepsilon$ for some $\varepsilon > 0$, then

$$\varepsilon \|\varphi\|^2 \leq (\operatorname{Im} A \, \varphi, \varphi) \leq |(A\varphi, \varphi)| \leq \|A\varphi\| \|\varphi\|, \quad \varphi \in \mathcal{G},$$

yields that $\varepsilon\|\varphi\| \leq \|A\varphi\|$, $\varphi \in \mathcal{G}$. Therefore, if $\operatorname{Im} A$ is boundedly invertible, then so is A itself, i.e., $A^{-1} \in \mathbf{B}(\mathcal{G})$, and furthermore

$$\operatorname{Im} (-A^{-1}) = A^{-1} (\operatorname{Im} A) A^{-*},$$

so that $\operatorname{Im} (-A^{-1}) \geq 0$, and in fact $\operatorname{Im} (-A^{-1}) \geq \varepsilon'$ for some $\varepsilon' > 0$. The next lemma is now clear.

Lemma A.4.8. *Let $\mathcal{G}$ be a Hilbert space and let $F : \mathbb{C} \setminus \mathbb{R} \to \mathbf{B}(\mathcal{G})$ be a Nevanlinna function. If the function F is uniformly strict, then its inverse $-F^{-1}$ is a uniformly strict $\mathbf{B}(\mathcal{G})$-valued Nevanlinna function and*

$$\operatorname{Im} (-F(\lambda)^{-1}) = F(\lambda)^{-1} (\operatorname{Im} F(\lambda)) F(\lambda)^{-*}, \quad \lambda \in \mathbb{C} \setminus \mathbb{R}.$$

In general, for an operator-valued Nevanlinna function F with values in $\mathbf{B}(\mathcal{G})$ the values of the inverse $-F^{-1}$ need not be bounded operators.

A.5 Kac functions

The notion of operator-valued Nevanlinna function in Definition A.4.1 gave rise to the integral representation of Nevanlinna functions in Theorem A.4.2. Next special subclasses of Nevanlinna functions with corresponding integral representations will be considered. Whenever one deals with "unbounded measures" the interpretation of the integrals is again via Proposition A.3.7.

Definition A.5.1. Let $\mathcal{G}$ be a Hilbert space and let $F : \mathbb{C} \setminus \mathbb{R} \to \mathbf{B}(\mathcal{G})$ be a Nevanlinna function. Then F is said to belong to the class of *Kac functions* if for all $\varphi \in \mathcal{G}$

$$\int_1^\infty \frac{\operatorname{Im} (F(iy)\varphi, \varphi)}{y} \, dy < \infty. \tag{A.5.1}$$

Note that every $\mathbf{B}(\mathcal{G})$-valued Nevanlinna function F that satisfies

$$\sup_{y>0} y \big(\operatorname{Im} (F(iy)\varphi, \varphi) \big) < \infty, \quad \varphi \in \mathcal{G}, \tag{A.5.2}$$

is a Kac function.

Theorem A.5.2. *Let $\mathcal{G}$ be a Hilbert space and let $F : \mathbb{C} \setminus \mathbb{R} \to \mathbf{B}(\mathcal{G})$ be an operator function. Then the following statements are equivalent:*

(i) *F has an integral representation of the form*

$$F(\lambda) = \gamma + \int_{\mathbb{R}} \frac{1}{t-\lambda}\, d\Sigma(t), \quad \lambda \in \mathbb{C} \setminus \mathbb{R}, \tag{A.5.3}$$

with a self-adjoint operator $\gamma \in \mathbf{B}(\mathcal{G})$ and a nondecreasing self-adjoint operator function $\Sigma : \mathbb{R} \to \mathbf{B}(\mathcal{G})$ such that

$$\int_{\mathbb{R}} \frac{d\Sigma(t)}{|t|+1} \in \mathbf{B}(\mathcal{G}), \tag{A.5.4}$$

where the integrals in (A.5.3) and (A.5.4) converge in the strong topology.

(ii) *F is a Kac function.*

If either of these equivalent conditions is satisfied, then $\gamma = \lim_{y\to\infty} F(iy)$ in the strong sense.

Proof. It is helpful to recall the identity

$$\int_1^\infty \frac{1}{t^2+y^2}\, dy = \frac{1}{|t|}\left(\frac{\pi}{2} - \arctan\frac{1}{|t|}\right), \quad t \neq 0, \tag{A.5.5}$$

and the fact that

$$\lim_{|t|\to 0} \frac{1}{|t|}\left(\frac{\pi}{2} - \arctan\frac{1}{|t|}\right) = 1. \tag{A.5.6}$$

(i) $\Rightarrow$ (ii) Write the representation (A.5.3) as

$$F(\lambda) = \gamma + \int_{\mathbb{R}} g_\lambda(t)\, \frac{d\Sigma(t)}{|t|+1}, \quad \lambda \in \mathbb{C} \setminus \mathbb{R},$$

where the bounded continuous function $g_\lambda : \mathbb{R} \to \mathbb{C}$ is given by

$$g_\lambda(t) = \frac{|t|+1}{t-\lambda}.$$

Due to the condition (A.5.4) it follows that $F(\lambda)$ in (A.5.3) is well defined and represents an element in $\mathbf{B}(\mathcal{G})$; cf. Proposition A.3.7. Moreover, (A.5.3) and (A.3.25) imply that for $\varphi \in \mathcal{G}$

$$(F(\lambda)\varphi, \varphi) = (\gamma\varphi, \varphi) + \int_{\mathbb{R}} g_\lambda(t)\, \frac{d(\Sigma(t)\varphi, \varphi)}{|t|+1}, \quad \lambda \in \mathbb{C} \setminus \mathbb{R}.$$

Therefore, one sees that the function F is holomorphic on $\mathbb{C} \setminus \mathbb{R}$. Furthermore, it is clear that $F(\lambda)^* = F(\bar{\lambda})$; cf. (A.3.24). Since $(\gamma\varphi, \varphi) \in \mathbb{R}$, one also sees that

$$\frac{\operatorname{Im}(F(\lambda)\varphi, \varphi)}{\operatorname{Im}\lambda} = \int_{\mathbb{R}} \frac{1}{|t-\lambda|^2}\, d(\Sigma(t)\varphi, \varphi).$$

Thus, F is a Nevanlinna function. Furthermore, integration of this identity shows, after a change of the order of integration, that

$$\int_1^\infty \frac{\operatorname{Im}(F(iy)\varphi,\varphi)}{y}\,dy = \int_{\mathbb{R}} \frac{1}{|t|}\left(\frac{\pi}{2} - \arctan\frac{1}{|t|}\right) d(\Sigma(t)\varphi,\varphi) < \infty.$$

Here the identity (A.5.5) was used in the first equality and (A.5.4), (A.5.6), and $\arctan 1/|t| \to 0$ for $|t| \to \infty$ were used to conclude that the last integral is finite; cf. Proposition A.3.7. This shows that F is a Kac function.

(ii) $\Rightarrow$ (i) Since F is a Nevanlinna function, it follows from the integral representation (A.4.1) and (A.3.25) that for all $\varphi \in \mathcal{G}$

$$\frac{\operatorname{Im}(F(iy)\varphi,\varphi)}{y} = (\beta\varphi,\varphi) + \int_{\mathbb{R}} \frac{1}{t^2+y^2}\,d(\Sigma(t)\varphi,\varphi).$$

Each of the terms on the right-hand side is nonnegative. Hence, the integrability condition (A.5.1) implies that $\beta = 0$ and furthermore

$$\int_1^\infty \left(\int_{\mathbb{R}} \frac{1}{t^2+y^2}\,d(\Sigma(t)\varphi,\varphi)\right) dy < \infty.$$

Changing the order of integration and using (A.5.5) gives

$$\int_{\mathbb{R}} \frac{1}{|t|}\left(\frac{\pi}{2} - \arctan\frac{1}{|t|}\right) d(\Sigma(t)\varphi,\varphi) < \infty,$$

and hence $\int_{\mathbb{R}}(1+|t|)^{-1}\,d(\Sigma(t)\varphi,\varphi) < \infty$. By Proposition A.3.7, this implies that (A.5.4) is satisfied. Observe that for each compact interval $[a,b] \subset \mathbb{R}$ one has the identity

$$\int_a^b \frac{1+\lambda t}{t-\lambda}\frac{d\Sigma(t)}{t^2+1} = \int_a^b \frac{|t|+1}{t-\lambda}\frac{d\Sigma(t)}{|t|+1} - \int_a^b \frac{|t|+1}{t^2+1}\frac{d\Sigma(t)}{|t|+1},$$

and now all integrals have strong limits as $[a,b] \to \mathbb{R}$ by Proposition A.3.7. Hence, one obtains from (A.4.1) with $\beta = 0$ that

$$F(\lambda) = \alpha + \int_{\mathbb{R}} \frac{1+\lambda t}{t-\lambda}\frac{d\Sigma(t)}{t^2+1} = \alpha + \int_{\mathbb{R}} \frac{d\Sigma(t)}{t-\lambda} - \int_{\mathbb{R}} \frac{t}{t^2+1}\,d\Sigma(t).$$

Observe that

$$\int_{\mathbb{R}} \frac{t}{t^2+1}\,d\Sigma(t) = \int_{\mathbb{R}} \frac{t(|t|+1)}{t^2+1}\frac{d\Sigma(t)}{|t|+1}$$

is a self-adjoint operator in $\mathbf{B}(\mathcal{G})$. Hence, with γ defined by

$$\gamma = \alpha - \int_{\mathbb{R}} \frac{t}{t^2+1}\,d\Sigma(t)$$

one sees that $\gamma \in \mathbf{B}(\mathcal{G})$ is self-adjoint and the assertion (i) follows.

Finally, observe that by means of (A.3.26) one has

$$\lim_{y\to\infty}(F(iy)-\gamma)\varphi = \lim_{y\to\infty}\left(\int_{\mathbb{R}}\frac{1}{t-iy}\,d\Sigma(t)\right)\varphi = 0$$

for all $\varphi \in \mathcal{G}$. This gives the last assertion. □

A subclass of the Kac functions in Theorem A.5.2 concerns the case with bounded "measures".

Proposition A.5.3. *Let $\mathcal{G}$ be a Hilbert space and let $F : \mathbb{C}\setminus\mathbb{R} \to \mathbf{B}(\mathcal{G})$ be an operator function. Then the following statements are equivalent:*

(i) *F has an integral representation of the form*

$$F(\lambda) = \gamma + \int_{\mathbb{R}}\frac{1}{t-\lambda}\,d\Sigma(t), \quad \lambda \in \mathbb{C}\setminus\mathbb{R},$$

with a self-adjoint operator $\gamma \in \mathbf{B}(\mathcal{G})$ and a nondecreasing self-adjoint operator function $\Sigma : \mathbb{R} \to \mathbf{B}(\mathcal{G})$ such that $\int_{\mathbb{R}} d\Sigma(t) \in \mathbf{B}(\mathcal{G})$.

(ii) *F is a Kac function satisfying (A.5.2).*

If either of these equivalent conditions is satisfied, then $\gamma = \lim_{y\to\infty} F(iy)$ in the strong sense, and furthermore, for all $\varphi \in \mathcal{G}$

$$\int_{\mathbb{R}} d(\Sigma(t)\varphi,\varphi) = \sup_{y>0} y(\operatorname{Im} F(iy)\varphi,\varphi) < \infty. \tag{A.5.7}$$

Proof. (i) ⇒ (ii) It follows from Theorem A.5.2 that F is a Kac function. Moreover,

$$\begin{aligned}\sup_{y>0} y\big(\operatorname{Im} F(iy)\varphi,\varphi\big) &= \sup_{y>0}\int_{\mathbb{R}}\frac{y^2}{t^2+y^2}\,d(\Sigma(t)\varphi,\varphi)\\ &= \int_{\mathbb{R}} d(\Sigma(t)\varphi,\varphi) - \inf_{y>0}\int_{\mathbb{R}}\frac{t^2}{t^2+y^2}\,d(\Sigma(t)\varphi,\varphi)\\ &= \int_{\mathbb{R}} d(\Sigma(t)\varphi,\varphi),\end{aligned} \tag{A.5.8}$$

which shows that the condition (A.5.2) is satisfied.

(ii) ⇒ (i) Since F is a Kac function, the integral representation follows from Theorem A.5.2 with a nondecreasing self-adjoint operator function $\Sigma : \mathbb{R} \to \mathbf{B}(\mathcal{G})$ which satisfies the integrability condition (A.5.4). Now it follows from the assumption (A.5.2) and (A.5.8) that $\int_{\mathbb{R}} d\Sigma(t) \in \mathbf{B}(\mathcal{G})$.

Moreover, the identity $\gamma = \lim_{y\to\infty} F(iy)$ in the strong sense is clear from Theorem A.5.2 and (A.5.7) was shown in (A.5.8). □

The previous proposition has an interesting consequence for a further subclass of the Kac functions which satisfy (A.5.2). The following result is connected with the characterization of generalized resolvents; see Chapter 4, where also the Sz.-Nagy dilation theorem is treated. Note that the conditions $\gamma = \gamma^*$ and $\int_{\mathbb{R}} d\Sigma(t) \in \mathbf{B}(\mathcal{G})$ in Proposition A.5.3 are now specialized to the conditions $\gamma = 0$ and $\| \int_{\mathbb{R}} d\Sigma(t)\| \leq 1$.

Proposition A.5.4. *Let $\mathcal{G}$ be a Hilbert space and let $F : \mathbb{C} \setminus \mathbb{R} \to \mathbf{B}(\mathcal{G})$ be an operator function. Then the following statements are equivalent:*

(i) *F has an integral representation of the form*

$$F(\lambda) = \int_{\mathbb{R}} \frac{1}{t-\lambda}\, d\Sigma(t), \quad \lambda \in \mathbb{C} \setminus \mathbb{R},$$

with a nondecreasing self-adjoint operator function $\Sigma : \mathbb{R} \to \mathbf{B}(\mathcal{G})$ such that $\int_{\mathbb{R}} d\Sigma(t) \in \mathbf{B}(\mathcal{G})$ and $\| \int_{\mathbb{R}} d\Sigma(t)\| \leq 1$.

(ii) *F is a Nevanlinna function which satisfies*

$$\frac{\operatorname{Im} F(\lambda)}{\operatorname{Im} \lambda} - F(\lambda)^* F(\lambda) \geq 0, \quad \lambda \in \mathbb{C} \setminus \mathbb{R}.$$

Proof. (i) $\Rightarrow$ (ii) It is clear from Proposition A.5.3 that F is a Kac function and hence a Nevanlinna function. Moreover, for all $\lambda \in \mathbb{C} \setminus \mathbb{R}$ one has

$$\frac{(\operatorname{Im} F(\lambda)\varphi, \varphi)}{\operatorname{Im} \lambda} = \int_{\mathbb{R}} \frac{1}{|t-\lambda|^2}\, d(\Sigma(t)\varphi, \varphi), \quad \varphi \in \mathcal{G}.$$

Since $\|\Theta(+\infty) - \Theta(-\infty)\| \leq 1$, it follows from Proposition A.3.4 that

$$\begin{aligned}(F(\lambda)^* F(\lambda)\varphi, \varphi) = \|F(\lambda)\varphi\|^2 &= \left\| \left(\int_{\mathbb{R}} \frac{1}{t-\lambda}\, d\Sigma(t) \right) \varphi \right\|^2 \\ &\leq \int_{\mathbb{R}} \frac{1}{|t-\lambda|^2}\, d(\Sigma(t)\varphi, \varphi), \quad \varphi \in \mathcal{G},\end{aligned}$$

which gives the desired result.

(ii) $\Rightarrow$ (i) Let F be a Nevanlinna function which satisfies

$$|\operatorname{Im} \lambda|\, \|F(\lambda)\varphi\|^2 \leq |\operatorname{Im} (F(\lambda)\varphi, \varphi)|, \quad \varphi \in \mathcal{G}, \quad \lambda \in \mathbb{C} \setminus \mathbb{R}.$$

Then

$$|\operatorname{Im} \lambda|\, \|F(\lambda)\varphi\| \leq \|\varphi\|, \quad \varphi \in \mathcal{G}, \quad \lambda \in \mathbb{C} \setminus \mathbb{R},$$

which leads to

$$|\operatorname{Im} \lambda|\, |(F(\lambda)\varphi, \varphi)| \leq \|\varphi\|^2, \quad \varphi \in \mathcal{G}, \quad \lambda \in \mathbb{C} \setminus \mathbb{R}.$$

In particular, one has

$$|\mathrm{Im}\,\lambda|\,|\mathrm{Im}\,(F(\lambda)\varphi,\varphi)| \leq \|\varphi\|^2, \quad \varphi \in \mathcal{G}, \quad \lambda \in \mathbb{C}\setminus\mathbb{R}, \tag{A.5.9}$$

and

$$|\mathrm{Im}\,\lambda|\,|\mathrm{Re}\,(F(\lambda)\varphi,\varphi)| \leq \|\varphi\|^2, \quad \varphi \in \mathcal{G}, \quad \lambda \in \mathbb{C}\setminus\mathbb{R}. \tag{A.5.10}$$

The inequality (A.5.9) and Proposition A.5.3 imply that F admits the integral representation

$$F(\lambda) = \gamma + \int_{\mathbb{R}} \frac{1}{t-\lambda}\, d\Sigma(t), \quad \lambda \in \mathbb{C}\setminus\mathbb{R},$$

with $\int_{\mathbb{R}} d\Sigma(t) \in \mathbf{B}(\mathcal{G})$, and hence $\int_{\mathbb{R}} d(\Sigma(t)\varphi,\varphi) < \infty$. Moreover, it follows from (A.5.9) and (A.5.7) that

$$\int_{\mathbb{R}} d(\Sigma(t)\varphi,\varphi) = \sup_{y>0} y\big(\mathrm{Im}\,F(iy)\varphi,\varphi\big) \leq \|\varphi\|^2.$$

Hence, one sees that $\| \int_{\mathbb{R}} d\Sigma(t)\| \leq 1$. Finally, note that

$$\mathrm{Re}\,(F(iy)\varphi,\varphi) = (\gamma\varphi,\varphi) + \int_{\mathbb{R}} \frac{t}{t^2+y^2}\, d(\Sigma(t)\varphi,\varphi), \quad y>0.$$

Now (A.5.10) and the dominated convergence theorem show that $\gamma = 0$. □

A.6 Stieltjes and inverse Stieltjes functions

Let F be a $\mathbf{B}(\mathcal{G})$-valued Nevanlinna function F with the integral representation (A.4.1) as in Theorem A.4.2. Recall from Proposition A.4.5 that F is holomorphic on $\mathbb{C}\setminus[c,\infty)$ if and only if the function $t \mapsto \Sigma(t)\varphi$ is constant on $(-\infty,c)$ for all $\varphi \in \mathcal{G}$, and in this case one has the integral representation

$$F(\lambda) = \alpha + \lambda\beta + \int_{[c,\infty)} \left(\frac{1}{t-\lambda} - \frac{t}{t^2+1}\right) d\Sigma(t), \tag{A.6.1}$$

where the integral converges in the strong topology for all $\lambda \in \mathbb{C}\setminus[c,\infty)$; cf. (A.3.28). Note that (A.6.1) implies

$$\beta\varphi = \lim_{x\downarrow -\infty} \frac{F(x)}{x}\varphi, \qquad \varphi \in \mathcal{G}. \tag{A.6.2}$$

This identity can be verified in the same way as in the proof of Lemma A.2.6 using (A.3.26); cf. Lemma A.4.3. Moreover, $x \mapsto F(x)$ is a $\mathbf{B}(\mathcal{G})$-valued operator function on $(-\infty,c)$ with self-adjoint operators as values and one has

$$F(x_1) \leq F(x_2), \quad x_1 < x_2 < c, \tag{A.6.3}$$

by Proposition A.4.5. In the present section this class of Nevanlinna functions F will now be further specified by requiring, in addition to holomorphy of F on $\mathbb{C}\setminus[c,\infty)$, a sign condition for the values $F(x)$ for $x \in (-\infty,c)$.

Definition A.6.1. Let $\mathcal{G}$ be a Hilbert space and let $c \in \mathbb{R}$ be fixed. A Nevanlinna function $F : \mathbb{C} \setminus [c, \infty) \to \mathbf{B}(\mathcal{G})$ is said to belong to the class $\mathbf{S}_{\mathcal{G}}(-\infty, c)$ of *Stieltjes functions* if

(i) F is holomorphic on $\mathbb{C} \setminus [c, \infty)$;
(ii) $F(x) \geq 0$ for all $x < c$.

Similarly, a Nevanlinna function $F : \mathbb{C} \setminus [c, \infty) \to \mathbf{B}(\mathcal{G})$ is said to belong to the class $\mathbf{S}^{-1}_{\mathcal{G}}(-\infty, c)$ of *inverse Stieltjes functions* if

(i) F is holomorphic on $\mathbb{C} \setminus [c, \infty)$;
(ii) $F(x) \leq 0$ for all $x < c$.

Thus, if $F \in \mathbf{S}_{\mathcal{G}}(-\infty, c)$, then $x \mapsto F(x)$ is an operator function on $(-\infty, c)$ and

$$0 \leq F(x_1) \leq F(x_2), \quad x_1 < x_2 < c.$$

In particular, $\lim_{x \downarrow -\infty} F(x)$ exists in the strong sense and defines a nonnegative self-adjoint operator in $\mathbf{B}(\mathcal{G})$; cf. Lemma A.3.1. There is also a limit as $x \uparrow c$ in the sense of relations, see for details Chapter 5. Similarly, one sees that if $F \in \mathbf{S}^{-1}_{\mathcal{G}}(-\infty, c)$, then $x \mapsto F(x)$ is an operator function on $(-\infty, c)$ and

$$F(x_1) \leq F(x_2) \leq 0, \quad x_1 < x_2 < c.$$

In particular, $\lim_{x \uparrow c} F(x)$ exists in the strong sense and defines a nonnegative self-adjoint operator in $\mathbf{B}(\mathcal{G})$; cf. Lemma A.3.1. There is also a limit as $x \downarrow -\infty$ in the sense of relations, see for details Chapter 5.

The characterization of Nevanlinna functions in Theorem A.4.2 can be specialized for Stieltjes functions and, in fact, the following result also shows that the Stieltjes functions form a subclass of the Kac functions; cf. Theorem A.5.2.

Theorem A.6.2. *Let $\mathcal{G}$ be a Hilbert space, let $F : \mathbb{C} \setminus \mathbb{R} \to \mathbf{B}(\mathcal{G})$ be an operator function, and let $c \in \mathbb{R}$. Then the following statements are equivalent:*

(i) *F has an integral representation of the form*

$$F(\lambda) = \gamma + \int_{[c,\infty)} \frac{1}{t - \lambda}\, d\Sigma(t), \quad \lambda \in \mathbb{C} \setminus [c, \infty), \tag{A.6.4}$$

with a nonnegative self-adjoint operator $\gamma \in \mathbf{B}(\mathcal{G})$, a nondecreasing self-adjoint operator function $\Sigma : \mathbb{R} \to \mathbf{B}(\mathcal{G})$ such that $t \mapsto \Sigma(t)\varphi$, $\varphi \in \mathcal{G}$, is constant on $(-\infty, c)$, and

$$\int_{[c,\infty)} \frac{d(\Sigma(t)\varphi, \varphi)}{|t| + 1} < \infty, \quad \varphi \in \mathcal{G}, \tag{A.6.5}$$

where the integral in the representation (A.6.4) is interpreted in the weak topology for all $\lambda \in \mathbb{C} \setminus [c, \infty)$.

(ii) *F is a Stieltjes function in $\mathbf{S}_{\mathcal{G}}(-\infty, c)$.*

Proof. (i) $\Rightarrow$ (ii) Assume that the function F has the integral representation (A.6.4) with $\gamma \in \mathbf{B}(\mathcal{G})$ and $\Sigma : \mathbb{R} \to \mathbf{B}(\mathcal{G})$ as above, such that (A.6.5) holds. The condition (A.6.5) ensures that the integral in (A.6.4) is well defined in the weak sense. From the integral representation (A.6.4) one sees that F is holomorphic on $\mathbb{C} \setminus [c, \infty)$,

$$\frac{(\operatorname{Im} F(\lambda)\varphi, \varphi)}{\operatorname{Im} \lambda} = \int_{[c,\infty)} \frac{1}{|t-\lambda|^2}\, d(\Sigma(t)\varphi, \varphi) \geq 0, \quad \lambda \in \mathbb{C} \setminus \mathbb{R}, \quad \varphi \in \mathcal{G},$$

and $F(\lambda)^* = F(\overline{\lambda})$ for $\lambda \in \mathbb{C} \setminus \mathbb{R}$. Hence, F is a $\mathbf{B}(\mathcal{G})$-valued Nevanlinna function. Moreover, $\gamma \geq 0$ and

$$(F(x)\varphi, \varphi) = (\gamma\varphi, \varphi) + \int_{[c,\infty)} \frac{1}{t-x}\, d(\Sigma(t)\varphi, \varphi), \quad x \in (-\infty, c), \quad \varphi \in \mathcal{G},$$

imply $F(x) \geq 0$ for all $x < c$. Thus, $F \in \mathbf{S}_{\mathcal{G}}(-\infty, c)$.

(ii) $\Rightarrow$ (i) Assume that F is a Nevanlinna function in $\mathbf{S}_{\mathcal{G}}(-\infty, c)$. Then F has the integral representation (A.6.1), the operators $F(x)$ defined for $-\infty < x < c$ satisfy (A.6.3), and they are all nonnegative. Hence, the limit

$$\gamma = \lim_{x \downarrow -\infty} F(x) \in \mathbf{B}(\mathcal{G}) \tag{A.6.6}$$

exists in the strong sense and one has $\gamma \geq 0$ by Lemma A.3.1. Note that (A.6.2) implies $\beta = 0$ in (A.6.1). Therefore, (A.6.1) implies that for all $x < c$:

$$-\int_{[c,\infty)} \frac{1+tx}{t-x} \frac{d(\Sigma(t)\varphi, \varphi)}{t^2+1} \leq (\alpha\varphi, \varphi), \quad \varphi \in \mathcal{G}.$$

Letting $x \downarrow -\infty$ and using the monotone convergence theorem one obtains

$$\int_{[c,\infty)} \frac{t}{t^2+1}\, d(\Sigma(t)\varphi, \varphi) \leq (\alpha\varphi, \varphi), \quad \varphi \in \mathcal{G}. \tag{A.6.7}$$

Since

$$\frac{d(\Sigma(t)\varphi, \varphi)}{|t|+1} = \frac{t^2+1}{t(|t|+1)} \frac{t}{t^2+1}\, d(\Sigma(t)\varphi, \varphi)$$

for large t, one concludes from (A.6.7) that (A.6.5) holds. Moreover, one obtains from (A.6.6) and (A.6.1) that

$$\gamma = \alpha - \int_{[c,\infty)} \frac{t}{t^2+1}\, d\Sigma(t).$$

Thus, one may rewrite the integral representation (A.6.1) in the form (A.6.4). $\square$

The inverse Stieltjes functions can also be characterized by integral representations. Here is the analog of Theorem A.6.2.

Theorem A.6.3. *Let $\mathcal{G}$ be a Hilbert space, let $F : \mathbb{C} \setminus \mathbb{R} \to \mathbf{B}(\mathcal{G})$ be an operator function, and let $c \in \mathbb{R}$. Then the following statements are equivalent:*

(i) *F has an integral representation of the form*

$$F(\lambda) = L+\beta(\lambda-c)+\int_{[c,\infty)} \left(\frac{1}{t-\lambda} - \frac{1}{t-c}\right) d\Sigma(t), \quad \lambda \in \mathbb{C}\setminus[c,\infty), \tag{A.6.8}$$

with $L, \beta \in \mathbf{B}(\mathcal{G})$, $L \leq 0$, $\beta \geq 0$, a nondecreasing self-adjoint operator function $\Sigma : \mathbb{R} \to \mathbf{B}(\mathcal{G})$ such that $t \mapsto \Sigma(t)\varphi$, $\varphi \in \mathcal{G}$, is constant on $(-\infty, c)$, and

$$\int_{[c,\infty)} \frac{d(\Sigma(t)\varphi,\varphi)}{(t-c)(|t|+1)} < \infty, \qquad \varphi \in \mathcal{G}, \tag{A.6.9}$$

where the integral in the representation (A.6.8) is interpreted in the weak topology for all $\lambda \in \mathbb{C} \setminus [c,\infty)$.

(ii) *F is an inverse Stieltjes function in $\mathbf{S}_{\mathcal{G}}^{-1}(-\infty, c)$.*

Proof. (i) $\Rightarrow$ (ii) Assume that the function F has the integral representation (A.6.8) with $L, \beta \in \mathbf{B}(\mathcal{G})$ and $\Sigma : \mathbb{R} \to \mathbf{B}(\mathcal{G})$ as above such that the condition (A.6.9) holds. First observe that

$$\int_{[c,\infty)} \left(\frac{1}{t-\lambda} - \frac{1}{t-c}\right) d(\Sigma(t)\varphi,\varphi) = \int_{[c,\infty)} \frac{(\lambda-c)(|t|+1)}{t-\lambda} \frac{d(\Sigma(t)\varphi,\varphi)}{(t-c)(|t|+1)}$$

for $\varphi \in \mathcal{G}$, and hence condition (A.6.9) ensures that the integral in (A.6.8) converges in the weak topology for all $\lambda \in \mathbb{C} \setminus [c,\infty)$. It also follows from (A.6.8) that F is holomorphic on $\mathbb{C} \setminus [c,\infty)$,

$$\frac{(\operatorname{Im} F(\lambda)\varphi,\varphi)}{\operatorname{Im}\lambda} = (\beta\varphi,\varphi) + \int_{[c,\infty)} \frac{1}{|t-\lambda|^2} d(\Sigma(t)\varphi,\varphi) \geq 0$$

for $\lambda \in \mathbb{C} \setminus \mathbb{R}$ and $\varphi \in \mathcal{G}$, and $F(\lambda)^* = F(\overline{\lambda})$ for $\operatorname{Im}\lambda \neq 0$. This shows that F is a $\mathbf{B}(\mathcal{G})$-valued Nevanlinna function. Now if $x < c$, then $\beta(x-c) \leq 0$ and the integrand in (A.6.8) is nonpositive. Since also $L \leq 0$, one concludes that $F(x) \leq 0$ for all $x < c$. Thus, $F \in \mathbf{S}_{\mathcal{G}}^{-1}(-\infty, c)$.

(ii) $\Rightarrow$ (i) Assume that the function F is a Nevanlinna function in $\mathbf{S}_{\mathcal{G}}^{-1}(-\infty, c)$. Then F has the integral representation (A.6.1) and the operators $F(x)$ defined for $-\infty < x < c$ satisfy (A.6.3) and they are all nonpositive. Hence, the limit

$$L = \lim_{x \uparrow c} F(x) \in \mathbf{B}(\mathcal{G})$$

exists in the strong sense and one has $L \leq 0$ by Lemma A.3.1. Furthermore, for $\varphi \in \mathcal{G}$ and $-\infty < x < c$ one has

$$(F(x)\varphi, \varphi) = (\alpha\varphi, \varphi) + x(\beta\varphi, \varphi) + \int_{[c,\infty)} \left(\frac{1}{t-x} - \frac{t}{t^2+1} \right) d(\Sigma(t)\varphi, \varphi),$$

so that for $x \uparrow c$ the monotone convergence theorem gives

$$(L\varphi, \varphi) = (\alpha\varphi, \varphi) + c(\beta\varphi, \varphi) + \int_{[c,\infty)} \left(\frac{1}{t-c} - \frac{t}{t^2+1} \right) d(\Sigma(t)\varphi, \varphi). \quad \text{(A.6.10)}$$

In particular, the integral on the right-hand side of (A.6.10) exists for all $\varphi \in \mathcal{G}$. From

$$\frac{d(\Sigma(t)\varphi, \varphi)}{(t-c)(|t|+1)} = \frac{t^2+1}{(1+tc)(|t|+1)} \left(\frac{1}{t-c} - \frac{t}{t^2+1} \right) d(\Sigma(t)\varphi, \varphi)$$

one then concludes that (A.6.9) holds. Using (A.6.10) the integral representation (A.6.1) can be rewritten as

$$(F(\lambda)\varphi, \varphi) = (L\varphi, \varphi) + (\lambda - c)(\beta\varphi, \varphi) + \int_{[c,\infty)} \left(\frac{1}{t-\lambda} - \frac{1}{t-c} \right) d(\Sigma(t)\varphi, \varphi)$$

for $\lambda \in \mathbb{C} \setminus [c, \infty)$ and $\varphi \in \mathcal{G}$. This proves (A.6.8). □

Remark A.6.4. The integral representations in (A.6.4) and (A.6.8) are understood in the weak sense. Using Proposition A.3.7 one can verify that the integral representation for Stieltjes functions in (A.6.4) remains valid in the strong sense and that the integrability condition (A.6.5) can be replaced by the condition

$$\int_{[c,\infty)} \frac{d\Sigma(t)}{|t|+1} \in \mathbf{B}(\mathcal{G}),$$

where the integral exists in the strong sense. Within the theory of operator-valued integrals developed in Section A.3, the integral representation (A.6.8) for inverse Stieltjes functions and the integrability condition (A.6.9) cannot be interpreted directly in the strong sense.

Let F be a Nevanlinna function and let $c \in \mathbb{R}$. Define the functions F_c and F_c^- by

$$F_c(\lambda) = (\lambda - c)F(\lambda), \quad \lambda \in \mathbb{C} \setminus \mathbb{R},$$

and

$$F_c^-(\lambda) = (\lambda - c)^{-1}F(\lambda), \quad \lambda \in \mathbb{C} \setminus \mathbb{R}.$$

The class of Nevanlinna functions is not stable under either of the mappings

$$F \mapsto F_c \quad \text{or} \quad F \mapsto F_c^-.$$

In fact, the next results show that the set of Nevanlinna functions that is stable under these mappings coincides with $\mathbf{S}_{\mathcal{G}}(-\infty, c)$ or $\mathbf{S}_{\mathcal{G}}^{-1}(-\infty, c)$, respectively.

Proposition A.6.5. *Let $\mathcal{G}$ be a Hilbert space, let F be a $\mathbf{B}(\mathcal{G})$-valued Nevanlinna function F, and let $c \in \mathbb{R}$. Then the following statements are equivalent:*

(i) *F belongs to the class $\mathbf{S}_{\mathcal{G}}(-\infty, c)$;*

(ii) *F_c is a Nevanlinna function.*

In fact, if $F \in \mathbf{S}_{\mathcal{G}}(-\infty, c)$, then $F_c \in \mathbf{S}_{\mathcal{G}}^{-1}(-\infty, c)$.

Proof. (i) $\Rightarrow$ (ii) Assume that F belongs to the class $\mathbf{S}_{\mathcal{G}}(-\infty, c)$. Then, by Theorem A.6.2, the function F has the integral representation (A.6.4) interpreted in the weak sense. This implies that F_c has the representation

$$F_c(\lambda) = \gamma(\lambda - c) + \int_{[c,\infty)} \frac{\lambda - c}{t - \lambda}\, d\Sigma(t), \quad \lambda \in \mathbb{C} \setminus [c, \infty), \tag{A.6.11}$$

in the weak sense. It follows that F_c is holomorphic on $\mathbb{C} \setminus [c, \infty)$,

$$\frac{(\operatorname{Im} F_c(\lambda)\varphi, \varphi)}{\operatorname{Im} \lambda} = (\gamma\varphi, \varphi) + \int_{[c,\infty)} \frac{t - c}{|t - \lambda|^2}\, d(\Sigma(t)\varphi, \varphi) \geq 0$$

for $\lambda \in \mathbb{C} \setminus \mathbb{R}$ and $\varphi \in \mathcal{G}$, and $F_c(\lambda)^* = F_c(\overline{\lambda})$ for $\lambda \in \mathbb{C} \setminus \mathbb{R}$. Therefore, F_c is a $\mathbf{B}(\mathcal{G})$-valued Nevanlinna function. It is also clear from (A.6.11) that $F_c(x) \leq 0$ for $x < c$, and hence $F_c \in \mathbf{S}_{\mathcal{G}}^{-1}(-\infty, c)$.

(ii) $\Rightarrow$ (i) Assume that with F also F_c is a Nevanlinna function. Let $\varphi \in \mathcal{G}$ and express the scalar Nevanlinna function $f_\varphi(\lambda) = (F(\lambda)\varphi, \varphi)$ with its integral representation

$$f_\varphi(\lambda) = \alpha_\varphi + \beta_\varphi \lambda + \int_{\mathbb{R}} \left(\frac{1}{t - \lambda} - \frac{t}{t^2 + 1} \right) d\sigma_\varphi(t), \quad \lambda \in \mathbb{C} \setminus \mathbb{R},$$

in Theorem A.2.5. The function $g(\lambda) = \lambda - c$, $c \in \mathbb{R}$, is entire and real on $\mathbb{R}$. According to the Stieltjes inversion formula in Lemma A.2.7, for any compact subinterval $[a, b] \subset (-\infty, c)$ one obtains

$$\begin{aligned} &\lim_{\varepsilon \downarrow 0} \frac{1}{2\pi i} \int_a^b \big[(gf_\varphi)(s + i\varepsilon) - (gf_\varphi)(s - i\varepsilon) \big]\, ds \\ &\quad = \frac{1}{2} \int_{\{a\}} (t - c)\, d\sigma_\varphi(t) + \int_{a+}^{b-} (t - c)\, d\sigma_\varphi(t) + \frac{1}{2} \int_{\{b\}} (t - c)\, d\sigma_\varphi(t). \end{aligned} \tag{A.6.12}$$

Since the function $g(\lambda) f_\varphi(\lambda) = (F_c(\lambda)\varphi, \varphi)$ is also a Nevanlinna function, the limit in (A.6.12) is nonnegative. However, since $t - c < 0$ for all $t \in [a, b]$ one concludes that the right-hand side in (A.6.12) is nonpositive and consequently $\sigma(t)$ must take a constant value on the whole interval $[a, b]$. Proposition A.2.9 implies that f_φ is holomorphic on $\mathbb{C} \setminus [c, \infty)$ for all $\varphi \in \mathcal{G}$. This shows that F is holomorphic on $\mathbb{C} \setminus [c, \infty)$.

Finally, to see that F takes nonnegative values on the interval $(-\infty, c)$ Proposition A.2.9 will be applied. Again consider the two functions $f_\varphi(\lambda)$ and $(\lambda - c)f_\varphi(\lambda)$. Both of them are differentiable and nondecreasing on the interval $(-\infty, c)$. In particular, for all $x < c$ one has $f'_\varphi(x) \geq 0$ and

$$f_\varphi(x) + (x - c)f'_\varphi(x) = \frac{d}{dx}\left((x - c)f_\varphi(x)\right) \geq 0.$$

Since $x < c$, this implies that

$$f_\varphi(x) \geq (c - x)f'_\varphi(x) \geq 0,$$

and hence $(F(x)\varphi, \varphi) \geq 0$ for all $\varphi \in \mathcal{G}$. Therefore, $F(x) \geq 0$ for all $x < c$ and the claim $F \in \mathbf{S}_\mathcal{G}(-\infty, c)$ is proved. □

The next proposition is a variant of Proposition A.6.5 for inverse Stieltjes functions.

Proposition A.6.6. *Let $\mathcal{G}$ be a Hilbert space, let F be a $\mathbf{B}(\mathcal{G})$-valued Nevanlinna function F, and let $c \in \mathbb{R}$. Then the following statements are equivalent:*

- (i) *F belongs to the class $\mathbf{S}_\mathcal{G}^{-1}(-\infty, c)$;*
- (ii) *F_c^- is a Nevanlinna function.*

In fact, if $F \in \mathbf{S}_\mathcal{G}^{-1}(-\infty, c)$, then $F_c^- \in \mathbf{S}_\mathcal{G}(-\infty, c)$.

Proof. (i) $\Rightarrow$ (ii) Assume that F belongs to the class $\mathbf{S}_\mathcal{G}^{-1}(-\infty, c)$. Then, by Theorem A.6.3, the function F has the integral representation (A.6.8) interpreted in the weak sense. This implies that F_c^- has the representation

$$F_c^-(\lambda) = \frac{-L}{c - \lambda} + \beta + \int_{(c,\infty)} \frac{d\Sigma(t)}{(t - c)(t - \lambda)}, \quad \lambda \in \mathbb{C} \setminus [c, \infty), \tag{A.6.13}$$

in the weak sense. It follows that F_c^- is holomorphic on $\mathbb{C} \setminus [c, \infty)$,

$$\frac{(\operatorname{Im} F_c^-(\lambda)\varphi, \varphi)}{\operatorname{Im} \lambda} = \frac{(-L\varphi, \varphi)}{|c - \lambda|^2} + \int_{[c,\infty)} \frac{1}{(t - c)|t - \lambda|^2}\, d(\Sigma(t)\varphi, \varphi) \geq 0$$

for $\lambda \in \mathbb{C} \setminus \mathbb{R}$ and $\varphi \in \mathcal{G}$, and $F_c^-(\lambda)^* = F_c^-(\overline{\lambda})$ for $\lambda \in \mathbb{C} \setminus \mathbb{R}$. Therefore, F_c^- is a $\mathbf{B}(\mathcal{G})$-valued Nevanlinna function. It is also clear from (A.6.13) that $F_c^-(x) \geq 0$ for $x < c$, and hence $F_c^- \in \mathbf{S}_\mathcal{G}(-\infty, c)$.

(ii) $\Rightarrow$ (i) Assume that with F also F_c^- is a Nevanlinna function. Let $\varphi \in \mathcal{G}$ and express the scalar Nevanlinna function $f_\varphi(\lambda) = (F(\lambda)\varphi, \varphi)$ with its integral representation in Theorem A.2.5. Since the function $g(\lambda) = (\lambda - c)^{-1}$ is holomorphic when $\lambda \neq c$ and real on $\mathbb{R} \setminus \{c\}$, the same argument as in the proof of Theorem A.6.5 shows that f_φ is holomorphic on $\mathbb{C} \setminus [c, \infty)$ for all $\varphi \in \mathcal{G}$, and hence F is holomorphic on $\mathbb{C} \setminus [c, \infty)$.

To see that F takes nonnegative values on the interval $(-\infty, c)$ one applies Proposition A.2.9. Consider the functions $f_\varphi(\lambda)$ and $(\lambda - c)^{-1} f_\varphi(\lambda)$. Both of them are differentiable and nondecreasing on the interval $(-\infty, c)$. In particular, for all $x < c$ one has $f'_\varphi(x) \geq 0$ and

$$\frac{f'_\varphi(x)}{x - c} - \frac{f_\varphi(x)}{(x - c)^2} = \frac{d}{dx}\left((x - c)^{-1} f_\varphi(x)\right) \geq 0.$$

Since $x < c$, this implies that

$$f_\varphi(x) \leq (x - c) f'_\varphi(x) \leq 0$$

and hence $(F(x)\varphi, \varphi) \leq 0$ for all $\varphi \in \mathcal{G}$. Therefore, $F(x) \leq 0$ for all $x < c$ and the claim $F \in \mathbf{S}^{-1}_{\mathcal{G}}(-\infty, c)$ is proved. □

The classes of Stieltjes functions and inverse Stieltjes functions are also connected to each other by inversion. For an operator-valued Nevanlinna function F with values in $\mathbf{B}(\mathcal{G})$ the values of the inverse $-F^{-1}$ need not be bounded operators. For simplicity, it is assumed here that the relevant functions are uniformly strict, see Definition A.4.7. Recall that if F is a uniformly strict $\mathbf{B}(\mathcal{G})$-valued Nevanlinna function, then so is the function $-F^{-1}$; cf. Lemma A.4.8.

Proposition A.6.7. *Let $\mathcal{G}$ be a Hilbert space, let F be a $\mathbf{B}(\mathcal{G})$-valued Nevanlinna function, and assume that F is uniformly strict. Then the following statements hold:*

(i) *if $F \in \mathbf{S}_{\mathcal{G}}(-\infty, c)$, then $-F^{-1} \in \mathbf{S}^{-1}_{\mathcal{G}}(-\infty, c)$;*

(ii) *if $F \in \mathbf{S}^{-1}_{\mathcal{G}}(-\infty, c)$, then $-F^{-1} \in \mathbf{S}_{\mathcal{G}}(-\infty, c)$.*

Proof. (i) Let $F \in \mathbf{S}_{\mathcal{G}}(-\infty, c)$ and note first that in the integral representation (A.6.1) one has $\beta = 0$, as follows from (A.6.2); cf. the proof of Theorem A.6.2. Hence, one concludes from (A.6.1) that

$$\operatorname{Im} F(i) = \int_{[c,\infty)} \frac{d\Sigma(t)}{t^2 + 1},$$

which by assumption is a nonnegative boundedly invertible operator. Recall that F has the integral representation (A.6.4) with the same Σ as in (A.6.1). It is straightforward to see that for every $x < c$ there exists a constant $C_x > 0$ such that

$$\frac{t - x}{t^2 + 1} \leq C_x \quad \text{for all} \quad t \geq c.$$

Thus, one concludes that

$$\begin{aligned}(F(x)\varphi, \varphi) &= (\gamma\varphi, \varphi) + \int_{[c,\infty)} \frac{1}{t - x}\, d(\Sigma(t)\varphi, \varphi) \\ &\geq (\gamma\varphi, \varphi) + \frac{1}{C_x} \int_{[c,\infty)} \frac{d(\Sigma(t)\varphi, \varphi)}{t^2 + 1}\end{aligned}$$

for all $\varphi \in \mathcal{G}$. It follows that $(F(x)\varphi, \varphi) \geq \delta_x \|\varphi\|^2$, $\varphi \in \mathcal{G}$, for some $\delta_x > 0$. Therefore, $F(x)^{-1} \in \mathbf{B}(\mathcal{G})$ for all $x < c$. Thus, the function $-F^{-1}$ is holomorphic in the region $\mathbb{C} \setminus [c, \infty)$. Since $F(x) \geq 0$, it is clear that $-F^{-1}(x) \leq 0$ for all $x < c$. Therefore, $-F^{-1} \in \mathbf{S}_{\mathcal{G}}^{-1}(-\infty, c)$.

(ii) Let $F \in \mathbf{S}_{\mathcal{G}}^{-1}(-\infty, c)$. Then it follows from Theorem A.6.3 that for all $x < c$

$$F(x) = L + (x - c)\left(\beta + \int_{[c,\infty)} \frac{d\Sigma(t)}{(t-x)(t-c)}\right) \leq 0, \tag{A.6.14}$$

where the integral is interpreted in the weak sense. Since F is uniformly strict, Proposition A.4.6 asserts that for all $x < c$

$$F'(x) = \beta + \int_{[c,\infty)} \frac{d\Sigma(t)}{(t-x)^2}$$

is a nonnegative boundedly invertible operator. It follows that

$$(\beta\varphi, \varphi) + \int_{[c,\infty)} \frac{d(\Sigma(t)\varphi, \varphi)}{(t-x)(t-c)} \geq (\beta\varphi, \varphi) + \int_{[c,\infty)} \frac{d(\Sigma(t)\varphi, \varphi)}{(t-x)^2} \geq \delta_x \|\varphi\|^2$$

for all $\varphi \in \mathcal{G}$ and for some $\delta_x > 0$, $x < c$. This implies that $F(x)$ in (A.6.14) has a bounded inverse for every $x < c$. Thus, $-F^{-1}$ is holomorphic in the region $\mathbb{C} \setminus [c, \infty)$. Since $F(x) \leq 0$, it is clear that $0 \leq -F(x)^{-1}$ for all $x < c$. Therefore, $-F^{-1} \in \mathbf{S}_{\mathcal{G}}(-\infty, c)$. □

Appendix B

Self-adjoint Operators and Fourier Transforms

Let A be a self-adjoint operator in the Hilbert space $\mathfrak{H}$ and let $E(\cdot)$ be the corresponding spectral measure. The operator A will be diagonalized by means of a self-adjoint operator Q in a Hilbert space $L^2_{d\rho}(\mathbb{R})$, where ρ is a nondecreasing function on $\mathbb{R}$. Here Q stands for multiplication by the independent variable in $L^2_{d\rho}(\mathbb{R})$. By means of the spectral measure one constructs an integral transform $\mathcal{F}$ which maps $\mathfrak{H}$ unitarily onto $L^2_{d\rho}(\mathbb{R})$ such that $A = \mathcal{F}^* Q \mathcal{F}$. This transform shares the diagonalization property with the classical Fourier transform and hence, for convenience, it will be referred to as Fourier transform in the following. There are two cases of interest in the present text.

The first (scalar) case is where $\mathfrak{H} = L^2_r(a,b)$ and r is a locally integrable function that is positive almost everywhere, and where A is a self-adjoint operator in $\mathfrak{H}$ with spectral measure $E(\cdot)$. The Fourier transform is initially defined on the compactly supported scalar functions in $L^2_r(a,b)$ by

$$\widehat{f}(x) = \int_a^b \omega(t,x) f(t)\, r(t)\, dt,$$

where $(t,x) \mapsto \omega(t,x)$ is a continuous real function defined on the square $(a,b) \times \mathbb{R}$. The spectral measure of A and the Fourier transform are assumed to be connected by

$$(E(\delta)f,f) = \int_\delta \widehat{f}(x)\overline{\widehat{f}(x)}\, d\rho(x), \quad \delta \subset \mathbb{R}, \tag{B.0.1}$$

for all $f \in L^2_r(a,b)$ with compact support. Due to this condition the Fourier transform can be extended to an isometry on all of $L^2_r(a,b)$. Under an additional assumption on ω it is shown that the Fourier transform is a unitary map.

The second (vector) case is where $\mathfrak{H} = L^2_\Delta(a,b)$ and Δ is a nonnegative 2×2 matrix function with locally integrable coefficients and where A is a self-adjoint

J. Behrndt et al., *Boundary Value Problems, Weyl Functions, and Differential Operators*, Monographs in Mathematics 108,
https://doi.org/10.1007/978-3-030-36714-5

relation. Then A_{op} is a self-adjoint operator in $\mathfrak{H} \ominus \operatorname{mul} A$ with the spectral measure $E(\cdot)$. The Fourier transform is initially defined on the compactly supported 2×1 vector functions in $L^2_\Delta(a,b)$ by

$$\widehat{f}(x) = \int_a^b \omega(t,x)\Delta(t)f(t)\,dt,$$

where $(t,x) \mapsto \omega(t,x)$ is a continuous 1×2 matrix function defined on the square $(a,b) \times \mathbb{R}$. The spectral measure and the Fourier transform are assumed to be connected by (B.0.1) for all $f \in L^2_\Delta(a,b)$ with compact support. Due to this condition, the Fourier transform can be extended to a partial isometry on all of $L^2_\Delta(a,b)$. Under an additional assumption on ω it is shown that the Fourier transform is onto.

For the scalar case the theory will be developed in detail; the results in the vector case will be only briefly explained, as the details are very much the same.

B.1 The scalar case

Let (a,b) be an open interval and let r be a locally integrable function that is positive almost everywhere. Let A be a self-adjoint operator in the Hilbert space $L^2_r(a,b)$ and let $E(\cdot)$ be the spectral measure of A. The basic ingredient is a continuous real function $(t,x) \mapsto \omega(t,x)$ on $(a,b)\times\mathbb{R}$. Hence, the *Fourier transform* $\widehat{f}$ of a compactly supported function $f \in L^2_r(a,b)$ defined by

$$\widehat{f}(x) = \int_a^b \omega(t,x)f(t)r(t)\,dt, \quad x \in \mathbb{R}, \tag{B.1.1}$$

is a well-defined complex function which is continuous on $\mathbb{R}$. The main assumption is that there exists a nondecreasing function ρ on $\mathbb{R}$ such the identity

$$(E(\delta)f,g) = \int_\delta \widehat{f}(x)\overline{\widehat{g}(x)}\,d\rho(x), \quad \delta \subset \mathbb{R}, \tag{B.1.2}$$

holds for all $f,g \in L^2_r(a,b)$ with compact support and all bounded open intervals $\delta \subset \mathbb{R}$, whose endpoints are not eigenvalues of A. Since the eigenvalues of A are discrete, an approximation argument and the dominated convergence theorem show that the identity (B.1.2) is in fact true for all $f,g \in L^2_r(a,b)$ with compact support and all bounded open intervals $\delta \subset \mathbb{R}$, regardless of whether their endpoints being eigenvalues or not.

It follows from the assumption (B.1.2) that for every function $f \in L^2_r(a,b)$ with compact support the continuous function $\widehat{f}$ belongs to $L^2_{d\rho}(\mathbb{R})$. In fact, the following result is valid.

Lemma B.1.1. *The Fourier transform $f \to \widehat{f}$ in (B.1.1) extends by continuity from the compactly supported functions in $L^2_r(a,b)$ to an isometric mapping*

$$\mathcal{F} : L^2_r(a,b) \to L^2_{d\rho}(\mathbb{R})$$

such that for all $f \in L^2_r(a,b)$

$$\lim_{\alpha \to a, \beta \to b} \int_{\mathbb{R}} \left| (\mathcal{F}f)(x) - \int_{\alpha}^{\beta} \omega(t,x) f(t) r(t)\, dt \right|^2 d\rho(x) = 0. \tag{B.1.3}$$

The Fourier transform and the spectral measure $E(\cdot)$ are related via

$$(E(\delta)f, g) = \int_{\delta} (\mathcal{F}f)(x)\overline{(\mathcal{F}g)(x)}\, d\rho(x) \tag{B.1.4}$$

for all $f, g \in L^2_r(a,b)$, where $\delta \subset \mathbb{R}$ is any bounded open interval.

Proof. Step 1. The mapping $f \mapsto \widehat{f}$ is a contraction on the functions in $L^2_r(a,b)$ that have compact support. To see this, let $f \in L^2_r(a,b)$ have compact support. Then it follows from the assumption (B.1.2) that

$$\int_{\delta} \widehat{f}(x)\overline{\widehat{f}(x)}\, d\rho(x) = (E(\delta)f, f) \leq (f, f),$$

where δ is an arbitrary bounded open interval. The monotone convergence theorem shows that $\widehat{f}$ belongs to $L^2_{d\rho}(\mathbb{R})$ and

$$(\widehat{f}, \widehat{f})_{\rho} \leq (f, f),$$

when $f \in L^2_r(a,b)$ has compact support.

Step 2. The mapping $f \mapsto \widehat{f}$, defined on the functions in $L^2_r(a,b)$ that have compact support, can be extended as a contraction to all of $L^2_r(a,b)$. For this, let $f \in L^2_r(a,b)$ and approximate f in $L^2_r(a,b)$ by functions $f_n \in L^2_r(a,b)$ with compact support. Then $f_n - f_m$ is a Cauchy sequence in $L^2_r(a,b)$ and since

$$\|\widehat{f_n} - \widehat{f_m}\|_{\rho} \leq \|f_n - f_m\|,$$

there is an element $\varphi \in L^2_{d\rho}(\mathbb{R})$ such that $\widehat{f_n} \to \varphi$ in $L^2_{d\rho}(\mathbb{R})$. It follows from $\|\widehat{f_n}\|_{\rho} \leq \|f_n\|$ that, in fact,

$$\|\varphi\|_{\rho} \leq \|f\|.$$

In particular, the mapping $f \mapsto \varphi$ from $L^2_r(a,b)$ to $L^2_{d\rho}(\mathbb{R})$ is a contraction. Hence, the operator $\mathcal{F}$ given by $\mathcal{F}f = \varphi$ is well defined and takes $L^2_r(a,b)$ contractively to $L^2_{d\rho}(\mathbb{R})$. The assertion in (B.1.3) is clear by multiplying $f \in L^2_r(a,b)$ with appropriately chosen characteristic functions.

Step 3. The operator $\mathfrak{F}$ is an isometry. To see this, let $f, g \in L^2_r(a,b)$ and approximate them by compactly supported functions $f_n, g_n \in L^2_r(a,b)$. Then by the assumption (B.1.2) one obtains

$$(E(\delta)f_n, g_n) = \int_\delta (\mathfrak{F}f_n)(x)\overline{(\mathfrak{F}g_n)(x)}\, d\rho(x). \tag{B.1.5}$$

Taking limits in (B.1.5) yields (B.1.4). It follows from (B.1.4) and the dominated convergence theorem that

$$(f,g) = \int_\delta (\mathfrak{F}f)(x)\overline{(\mathfrak{F}g)(x)}\, d\rho(x)$$

holds for all functions $f, g \in L^2_r(a,b)$. Therefore, $\mathfrak{F}$ is an isometry. □

The following simple observation is a direct consequence of Lemma B.1.1.

Corollary B.1.2. *For any bounded open interval $\delta \subset \mathbb{R}$ one has*

$$\mathfrak{F}(E(\delta)f) = \chi_\delta\, \mathfrak{F}f, \quad f \in L^2_r(a,b).$$

Proof. It follows from (B.1.4) that

$$(E(\delta)f, g) = \int_\delta (\mathfrak{F}f)(x)\overline{(\mathfrak{F}g)(x)} d\rho(x) = (\chi_\delta \mathfrak{F}f, \mathfrak{F}g)_\rho$$

for all $f, g \in L^2_r(a,b)$. Consequently,

$$\begin{aligned} \|E(\delta)f - g\|^2 &= (E(\delta)f - g, E(\delta)f - g) \\ &= (E(\delta)f, f) - (E(\delta)f, g) - (g, E(\delta)f) + (g,g) \\ &= (\chi_\delta\mathfrak{F}f, \mathfrak{F}f)_\rho - (\chi_\delta\mathfrak{F}f, \mathfrak{F}g)_\rho \\ &\qquad - (\mathfrak{F}g, \chi_\delta\mathfrak{F}f)_\rho + (\mathfrak{F}g, \mathfrak{F}g)_\rho \\ &= \|\chi_\delta\mathfrak{F}f - \mathfrak{F}g\|^2_\rho. \end{aligned}$$

Setting $g = E(\delta)f$, the desired result follows. □

Let $\delta \subset \mathbb{R}$ be a bounded open interval. The identity (B.1.4) is now written in the equivalent form

$$\int_\mathbb{R} \chi_\delta(t) d(E(t)f, g) = \int_\mathbb{R} \chi_\delta(x)(\mathfrak{F}f)(x)\overline{(\mathfrak{F}g)(x)}\, d\rho(x)$$

for all $f, g \in L^2_r(a,b)$. Let Ξ be a bounded Borel measurable function on $\mathbb{R}$. An approximation argument involving characteristic functions shows that then also

$$\int_\mathbb{R} \Xi(t) d(E(t)f, g) = \int_\mathbb{R} \Xi(x)(\mathfrak{F}f)(x)\overline{(\mathfrak{F}g)(x)}\, d\rho(x) \tag{B.1.6}$$

for all $f, g \in L^2_r(a,b)$. As a particular case of (B.1.6) one has

$$((A-\lambda)^{-1}f, g) = \int_{\mathbb{R}} \frac{\mathcal{F}f(x)\overline{\mathcal{F}g(x)}}{x-\lambda}\, d\rho(x) \tag{B.1.7}$$

for $f, g \in L^2_r(a,b)$ and $\lambda \in \rho(A)$. Moreover, if $g \in L^2_r(a,b)$ has compact support one obtains

$$\begin{aligned}\int_a^b ((A-\lambda)^{-1}f)(t)\overline{g(t)}\, r(t)\, dt &= \int_{\mathbb{R}} \frac{\mathcal{F}f(x)}{x-\lambda}\left(\int_a^b \omega(t,x)\overline{g(t)}\, r(t)\, dt\right) d\rho(x)\\ &= \int_a^b \left(\int_{\mathbb{R}} \frac{\mathcal{F}f(x)}{x-\lambda}\,\omega(t,x)\, d\rho(x)\right)\overline{g(t)}\, r(t)\, dt.\end{aligned}$$

The Fubini theorem now implies that

$$((A-\lambda)^{-1}f)(t) = \int_{\mathbb{R}} \frac{\omega(t,x)}{x-\lambda}\,\mathcal{F}f(x)\, d\rho(x) \tag{B.1.8}$$

for almost all $t \in (a,b)$ and, in particular, the integrand on the right-hand side is integrable for almost all $t \in (a,b)$.

Parallel to the Fourier transform one may also introduce a reverse Fourier transform acting on the space $L^2_{d\rho}(\mathbb{R})$. Let $\varphi \in L^2_{d\rho}(\mathbb{R})$ have compact support and define the *reverse Fourier transform* $\breve{\varphi}$ by

$$\breve{\varphi}(t) = \int_{\mathbb{R}} \omega(t,x)\varphi(x)\, d\rho(x), \quad t \in (a,b). \tag{B.1.9}$$

Then $\breve{\varphi}$ is a well-defined function which is continuous on (a,b). By means of Lemma B.1.1 one can now prove the following result.

Lemma B.1.3. *The reverse Fourier transform $\varphi \to \breve{\varphi}$ in (B.1.9) extends by continuity from the compactly supported functions in $L^2_{d\rho}(\mathbb{R})$ to a contractive mapping $\mathcal{G} : L^2_{d\rho}(\mathbb{R}) \to L^2_r(a,b)$ such that for all $\varphi \in L^2_{d\rho}(\mathbb{R})$*

$$\lim_{\eta \to \mathbb{R}} \int_a^b \left| \mathcal{G}\varphi(t) - \int_\eta \omega(t,x)\varphi(x)\, d\rho(x)\right|^2 r(t) dt = 0. \tag{B.1.10}$$

In fact, the extension $\mathcal{G}$ satisfies $\mathcal{G}\mathcal{F}f = f$ for all $f \in L^2_r(a,b)$.

Proof. Step 1. The mapping $\varphi \mapsto \breve{\varphi}$ takes the compactly supported functions in $L^2_{d\rho}(\mathbb{R})$ contractively into $L^2_r(a,b)$. To see this, let $\varphi \in L^2_{d\rho}(\mathbb{R})$ have compact support and let $[\alpha, \beta] \subset (a, b)$ be a compact interval. Then

$$\begin{aligned}\int_\alpha^\beta |\breve{\varphi}(t)|^2 r(t)\, dt &= \int_\alpha^\beta \breve{\varphi}(t)\Big(\int_{\mathbb{R}} \omega(t,x)\overline{\varphi(x)}\, d\rho(x)\Big) r(t)\, dt \\ &= \int_{\mathbb{R}}\Big(\int_\alpha^\beta \omega(t,x)\breve{\varphi}(t) r(t)\, dt\Big)\overline{\varphi(x)}\, d\rho(x) \\ &= \int_{\mathbb{R}} (\mathcal{F}(\chi_{[\alpha,\beta]}\breve{\varphi}))(x)\overline{\varphi(x)}\, d\rho(x) \\ &\le \|\mathcal{F}(\chi_{[\alpha,\beta]}\breve{\varphi})\|_\rho \|\varphi\|_\rho = \|\chi_{[\alpha,\beta]}\breve{\varphi}\| \|\varphi\|_\rho,\end{aligned}$$

where Lemma B.1.1 was used in the last step. This estimate gives that for any compact interval $[\alpha, \beta] \subset (a, b)$,

$$\sqrt{\int_\alpha^\beta |\breve{\varphi}(t)|^2 r(t)\, dt} \le \|\varphi\|_\rho.$$

By the monotone convergence theorem this leads to the inequality

$$\|\breve{\varphi}\| \le \|\varphi\|_\rho.$$

Hence, the mapping $\varphi \mapsto \breve{\varphi}$ takes the compactly supported functions in $L^2_{d\rho}(\mathbb{R})$ contractively into $L^2_r(a, b)$.

Step 2. The mapping $\varphi \mapsto \breve{\varphi}$ defined on the functions in $L^2_{d\rho}(\mathbb{R})$ that have compact support can be contractively extended to all of $L^2_{d\rho}(\mathbb{R})$. For this, let $\varphi \in L^2_{d\rho}(\mathbb{R})$ and approximate φ in $L^2_{d\rho}(\mathbb{R})$ by functions $\varphi_n \in L^2_{d\rho}(\mathbb{R})$ with compact support. Then $\varphi_n - \varphi_m$ is a Cauchy sequence in $L^2_{d\rho}(\mathbb{R})$ and since

$$\|\breve{\varphi}_n - \breve{\varphi}_m\| \le \|\varphi_n - \varphi_m\|_\rho,$$

there is an element $f \in L^2_r(a, b)$ such that $\breve{\varphi}_n \to f$ in $L^2_r(a, b)$. It follows from $\|\breve{\varphi}_n\| \le \|\varphi_n\|_\rho$ that, in fact,

$$\|f\| \le \|\varphi\|_\rho.$$

In particular, the mapping $\varphi \mapsto f$ from $L^2_{d\rho}(\mathbb{R})$ to $L^2_r(a, b)$ is a contraction. Hence, the operator $\mathcal{G}$ given by $\mathcal{G}\varphi = f$ is well defined and takes $L^2_{d\rho}(\mathbb{R})$ contractively to $L^2_r(a, b)$. The assertion in (B.1.10) is clear by multiplying $\varphi \in L^2_{d\rho}(\mathbb{R})$ by appropriately chosen characteristic functions.

Step 3. The extended mapping $\mathcal{G}$ is a left inverse of the Fourier transform $\mathcal{F}$. For this, observe that

$$(f, g) = \int_{\mathbb{R}} (\mathcal{F}f)(x)\overline{(\mathcal{F}g)(x)}\, d\rho(x)$$

for all $f, g \in L^2_r(a,b)$, since $\mathcal{F}$ is isometric by Lemma B.1.1. Thus, by means of the Fubini theorem one concludes that for $f, g \in L^2_r(a,b)$ and g having compact support,

$$\begin{aligned}(f,g) &= \lim_{\eta \to \mathbb{R}} \int_\eta (\mathcal{F}f)(x) \left(\int_a^b \omega(t,x) \overline{g(t)} r(t) dt \right) d\rho(x) \\ &= \lim_{\eta \to \mathbb{R}} \int_a^b \left(\int_\eta \omega(t,x) (\mathcal{F}f)(x)\, d\rho(x) \right) \overline{g(t)} r(t) dt \\ &= (\mathcal{G}\mathcal{F}f, g),\end{aligned}$$

where, in the last step, (B.1.10) and the continuity of the inner product have been used. Since the functions with compact support are dense in $L^2_r(a,b)$, one obtains $\mathcal{G}\mathcal{F}f = f$ for all $f \in L^2_r(a,b)$. □

Recall that multiplication by the independent variable in the Hilbert space $L^2_{d\rho}(\mathbb{R})$ generates a self-adjoint operator Q whose resolvent is given by

$$(Q-\lambda)^{-1} = \frac{1}{x-\lambda}, \qquad \lambda \in \mathbb{C} \setminus \mathbb{R}.$$

Hence, by (B.1.7), one sees that

$$\big((A-\lambda)^{-1} f, g\big) = \big((Q-\lambda)^{-1} \mathcal{F}f, \mathcal{F}g\big)_\rho = \big(\mathcal{F}^*(Q-\lambda)^{-1}\mathcal{F}f, g\big)$$

for all $f, g \in L^2_r(a,b)$, which leads to

$$(A-\lambda)^{-1} = \mathcal{F}^*(Q-\lambda)^{-1}\mathcal{F}, \qquad \lambda \in \mathbb{C} \setminus \mathbb{R}. \tag{B.1.11}$$

In general, here the Fourier transform $\mathcal{F}$ is an isometry, which is not necessarily onto.

The above results have been proved under the assumption that ω is a continuous function on $(a,b) \times \mathbb{R}$ which satisfies (B.1.2). For the following theorem one needs the additional condition:

$$\begin{array}{c}\text{for each } x_0 \in \mathbb{R} \text{ there exists a compactly} \\ \text{supported function } f \in L^2_r(a,b) \text{ such that } (\mathcal{F}f)(x_0) \neq 0.\end{array} \tag{B.1.12}$$

In the present situation this condition is equivalent to the following:

$$\text{for each } x_0 \in (a,b) \text{ there exists } t_0 \in \mathbb{R} \text{ such that } \omega(t_0, x_0) \neq 0. \tag{B.1.13}$$

Theorem B.1.4. *Let ω be continuous on $(a,b)\times\mathbb{R}$ and assume that (B.1.2) and one of the equivalent conditions (B.1.12) or (B.1.13) are satisfied. Then the Fourier transform*

$$f \mapsto \widehat{f}, \quad \widehat{f}(x) = \int_a^b \omega(t,x) f(t)\, r(t)\, dt, \quad x \in \mathbb{R},$$

extends by continuity from the compactly supported functions $f \in L^2_r(a,b)$ *to a unitary mapping* $\mathcal{F} : L^2_r(a,b) \to L^2_{d\rho}(\mathbb{R})$. *Moreover, the self-adjoint operator* A *in* $L^2_r(a,b)$ *is unitarily equivalent to the multiplication operator* Q *by the independent variable in* $L^2_{d\rho}(\mathbb{R})$ *via the Fourier transform* $\mathcal{F}$:

$$A = \mathcal{F}^* Q \mathcal{F}. \tag{B.1.14}$$

Proof. It is clear from Lemma B.1.1 that $\mathcal{F} : L^2_r(a,b) \to L^2_{d\rho}(\mathbb{R})$ is isometric and hence it remains to show for the first part of the theorem that $\mathcal{F}$ is surjective.

Assume that $\psi \in L^2_{d\rho}(\mathbb{R})$, $\psi \neq 0$, is orthogonal to $\operatorname{ran} \mathcal{F}$. Note that

$$(\psi, \mathcal{F} f)_\rho = 0 \quad \Rightarrow \quad (\overline{\psi}, \mathcal{F}\overline{f})_\rho = 0$$

and thus it follows that $\operatorname{Re}\psi$ and $\operatorname{Im}\psi$ are also orthogonal to $\operatorname{ran}\mathcal{F}$; hence it is no restriction to assume that $\psi \in L^2_{d\rho}(\mathbb{R})$ is real. Since $\psi \neq 0$, there exist an interval $I = [\alpha, \beta]$ and a Borel set $B \subset I$ with positive ρ-measure, such that $\psi(x) > 0$ (or $\psi(x) < 0$) for all $x \in B$. By the condition (B.1.12), there exists for each $x_0 \in I$ a compactly supported function $f_{x_0} \in L^2_r(a,b)$ such that $(\mathcal{F} f_{x_0})(x_0) > 0$. Since $\mathcal{F} f_{x_0}$ is continuous, there exists an open interval I_{x_0} containing x_0 such that $(\mathcal{F} f_{x_0})(x) > 0$ for all $x \in I_{x_0}$. As I is compact there exist finitely many points $x_1, \dots, x_n \in I$ such that

$$I \subset \bigcup_{i=1}^{n} I_{x_i}.$$

Hence, $B \cap I_{x_j}$ has positive ρ-measure for some $j \in \{1, \dots, n\}$ and therefore

$$(\chi_{B\cap I_{x_j}} \psi, \mathcal{F} f_{x_j})_\rho = \int_{\mathbb{R}} \chi_{B\cap I_{x_j}}(x)\psi(x)(\mathcal{F} f_{x_j})(x)\, d\rho(x) \neq 0. \tag{B.1.15}$$

On the other hand, since $\psi \perp \operatorname{ran}\mathcal{F}$, Corollary B.1.2 implies that

$$(\chi_\delta \psi, \mathcal{F} f)_\rho = (\psi, \chi_\delta \mathcal{F} f)_\rho = (\psi, \mathcal{F} E(\delta) f)_\rho = 0$$

for all $f \in L^2_r(a,b)$ and all bounded open intervals $\delta \subset \mathbb{R}$. Hence, by the regularity of the Borel measure ρ, also $(\chi_{B\cap I_{x_j}} \psi, \mathcal{F} f)_\rho = 0$ for all $f \in L^2_r(a,b)$; this contradicts (B.1.15). Therefore, $\operatorname{ran}\mathcal{F}$ is dense in $L^2_{d\rho}(\mathbb{R})$, and since $\mathcal{F} : L^2_r(a,b) \to L^2_{d\rho}(\mathbb{R})$ is isometric, one obtains that $\mathcal{F}$ is surjective.

The identity (B.1.14) follows from (B.1.11), the fact that $\mathcal{F}$ is unitary, and Lemma 1.3.8. □

In the situation of Theorem B.1.4 the inverse of the Fourier transform $\mathcal{F}$ is actually given by the reverse Fourier transform $\mathcal{G}$ in Lemma B.1.3, which is now a unitary mapping from $L^2_{d\rho}(\mathbb{R})$ to $L^2_r(a,b)$. Thus, for all $\varphi \in L^2_{d\rho}(\mathbb{R})$ one has

$$\lim_{\eta \to \mathbb{R}} \int_a^b \left| \mathcal{F}^{-1}\varphi(t) - \int_\eta \omega(t,x)\varphi(x)\, d\rho(x) \right|^2 r(t)dt = 0.$$

B.2 The vector case

Let (a,b) be an open interval and let Δ be a measurable nonnegative 2×2 matrix function on (a,b). Let A be a self-adjoint relation in the corresponding Hilbert space $L^2_\Delta(a,b)$ and assume that its multivalued part is at most finite-dimensional. Let $E(\cdot)$ be the spectral measure of $A_{\rm op}$, where $A_{\rm op}$ is the orthogonal operator part of A: it is a self-adjoint operator in $L^2_\Delta(a,b)\ominus \operatorname{mul} A$. In this case the basic ingredient is a continuous 1×2 matrix function $(t,x)\mapsto \omega(t,x)$ on $(a,b)\times\mathbb{R}$ whose entries are real. Hence, the *Fourier transform* $\widehat{f}$ of a compactly supported function $f\in L^2_\Delta(a,b)$ given by

$$\widehat{f}(x)=\int_a^b \omega(t,x)\Delta(t)f(t)\,dt,\quad x\in\mathbb{R}, \tag{B.2.1}$$

is a well-defined complex function that is continuous on $\mathbb{R}$. The main assumption is that there exists a nondecreasing function ρ on $\mathbb{R}$ such the identity

$$(E(\delta)f,g)=\int_\delta \widehat{f}(x)\overline{\widehat{g}(x)}\,d\rho(x),\quad \delta\subset\mathbb{R}, \tag{B.2.2}$$

holds for all $f,g\in L^2_\Delta(a,b)$ with compact support and all bounded open intervals $\delta\subset\mathbb{R}$, whose endpoints are not eigenvalues of $A_{\rm op}$. An approximation argument shows that (B.2.2) remains valid for all $f,g\in L^2_\Delta(a,b)$ with compact support and all bounded open intervals $\delta\subset\mathbb{R}$.

It follows from the assumption (B.2.2) that for every function $f\in L^2_\Delta(a,b)$ with compact support the continuous function $\widehat{f}$ belongs to $L^2_{d\rho}(\mathbb{R})$. In the present situation Lemma B.1.1 remains valid in a slightly modified form; here $\mathcal{F}$ is a partial isometry with $\ker\mathcal{F}=\operatorname{mul} A$.

Lemma B.2.1. *The Fourier transform $f\mapsto\widehat{f}$ in (B.2.1) extends by continuity from the compactly supported functions in $L^2_\Delta(a,b)$ to a partial isometry*

$$\mathcal{F}: L^2_\Delta(a,b)\to L^2_{d\rho}(\mathbb{R})$$

with $\ker\mathcal{F}=\operatorname{mul} A$ such that for all $f\in L^2_\Delta(a,b)$

$$\lim_{\alpha\to a,\beta\to b}\int_{\mathbb{R}}\left|(\mathcal{F}f)(x)-\int_\alpha^\beta \omega(t,x)\Delta(t)f(t)\,dt\right|^2 d\rho(x)=0.$$

The Fourier transform $\mathcal{F}$ and the spectral measure $E(\cdot)$ are related via

$$(E(\delta)f,g)=\int_\delta(\mathcal{F}f)(x)\overline{(\mathcal{F}g)(x)}\,d\rho(x) \tag{B.2.3}$$

for all $f,g\in L^2_\Delta(a,b)$, where $\delta\subset\mathbb{R}$ is any bounded open interval.

Next one verifies as in the proof of Corollary B.1.2 that the identity

$$\mathcal{F}(E(\delta)f) = \chi_\delta \mathcal{F}f, \quad f \in L^2_\Delta(a,b),$$

holds for bounded open intervals $\delta \subset \mathbb{R}$. Further, (B.2.3) and an approximation argument using characteristic functions show that

$$\big((A-\lambda)^{-1}f, g\big) = \int_{\mathbb{R}} \frac{\mathcal{F}f(x)\overline{\mathcal{F}g(x)}}{x-\lambda}\, d\rho(x) \tag{B.2.4}$$

for all $f, g \in L^2_\Delta(a,b)$ and $\lambda \in \rho(A)$. Moreover,

$$\big((A-\lambda)^{-1}f\big)(t) = \int_{\mathbb{R}} \frac{\omega(t,x)^*}{x-\lambda} \mathcal{F}f(x)\, d\rho(x), \quad \lambda \in \mathbb{C}\setminus\mathbb{R}, \tag{B.2.5}$$

for almost all $t \in (a,b)$ and, in particular, the integrand on the right-hand side is integrable for almost all $t \in (a,b)$.

The *reverse Fourier transform* of a function $\varphi \in L^2_{d\rho}(\mathbb{R})$ with compact support is defined by

$$\breve{\varphi}(t) = \int_{\mathbb{R}} \omega(t,x)^* \varphi(x)\, d\rho(x), \quad t \in (a,b). \tag{B.2.6}$$

Then $\breve{\varphi}$ is a well-defined 2×1 matrix function which is continuous on (a,b). By means of Lemma B.2.1 one now obtains the following result, which is similar to Lemma B.1.3. There are some slight differences in the proof for the vector case, which is provided here for completeness.

Lemma B.2.2. *The reverse Fourier transform $\varphi \mapsto \breve{\varphi}$ in (B.2.6) extends by continuity from the compactly supported functions in $L^2_{d\rho}(\mathbb{R})$ to a contractive mapping $\mathcal{G} : L^2_{d\rho}(\mathbb{R}) \to L^2_\Delta(a,b)$ such that for all $\varphi \in L^2_{d\rho}(\mathbb{R})$*

$$\lim_{\eta \to \mathbb{R}} \int_a^b \left| \mathcal{G}\varphi(t) - \int_\eta \omega(t,x)^*\varphi(x)\, d\rho(x) \right|^2 dt = 0. \tag{B.2.7}$$

In fact, the extension $\mathcal{G}$ satisfies $\mathcal{G}\mathcal{F}f = f$ for all $f \in L^2_\Delta(a,b) \ominus \operatorname{mul} A$.

Proof. Step 1. The mapping $\varphi \mapsto \breve{\varphi}$ takes the compactly supported functions in $L^2_{d\rho}(\mathbb{R})$ contractively into $L^2_\Delta(a,b)$. For this, let $\varphi \in L^2_{d\rho}(\mathbb{R})$ have compact support and let $[\alpha,\beta] \subset (a,b)$ be a compact interval. First observe that $\chi_{[\alpha,\beta]}\breve{\varphi} \in L^2_\Delta(a,b)$, so that indeed

$$\int_\alpha^\beta \omega(t,x)\Delta(t)\breve{\varphi}(t)\, dt = (\mathcal{F}(\chi_{[\alpha,\beta]}\breve{\varphi}))(x).$$

Therefore, one obtains

$$\begin{aligned}\int_\alpha^\beta \breve{\varphi}(t)^*\Delta(t)\breve{\varphi}(t)\,dt &= \int_\alpha^\beta \left(\int_{\mathbb{R}} \omega(t,x)^*\varphi(x)\,d\rho(x)\right)^* \Delta(t)\breve{\varphi}(t)\,dt \\ &= \int_{\mathbb{R}} \left(\int_\alpha^\beta \omega(t,x)\Delta(t)\breve{\varphi}(t)\,dt\right) \overline{\varphi(x)}\,d\rho(x) \\ &= \int_{\mathbb{R}} (\mathcal{F}(\chi_{[\alpha,\beta]}\breve{\varphi}))(x)\overline{\varphi(x)}\,d\rho(x) \\ &\leq \|\mathcal{F}(\chi_{[\alpha,\beta]}\breve{\varphi})\|_\rho \|\varphi\|_\rho \leq \|\chi_{[\alpha,\beta]}\breve{\varphi}\|\|\varphi\|_\rho,\end{aligned}$$

where Lemma B.2.1 was used in the last step. The above estimate gives for any compact interval $[\alpha, \beta] \subset (a, b)$

$$\sqrt{\int_\alpha^\beta \breve{\varphi}(t)^*\Delta(t)\breve{\varphi}(t)\,dt} \leq \|\varphi\|_\rho.$$

By the monotone convergence theorem this leads to the inequality

$$\|\breve{\varphi}\| \leq \|\varphi\|_\rho.$$

Hence, the mapping $\varphi \mapsto \breve{\varphi}$ takes the compactly supported functions in $L^2_{d\rho}(\mathbb{R})$ contractively into $L^2_\Delta(a, b)$.

Step 2. The mapping $\varphi \mapsto \breve{\varphi}$ defined on the functions in $L^2_{d\rho}(\mathbb{R})$ that have compact support can be contractively extended to all of $L^2_{d\rho}(\mathbb{R})$. To see this, let $\varphi \in L^2_{d\rho}(\mathbb{R})$ and approximate φ in $L^2_{d\rho}(\mathbb{R})$ by functions $\varphi_n \in L^2_{d\rho}(\mathbb{R})$ with compact support. Then $\varphi_n - \varphi_m$ is a Cauchy sequence in $L^2_{d\rho}(\mathbb{R})$ and since

$$\|\breve{\varphi}_n - \breve{\varphi}_m\| \leq \|\varphi_n - \varphi_m\|_\rho,$$

there is an element $f \in L^2_\Delta(a, b)$ such that $\breve{\varphi}_n \to f$ in $L^2_\Delta(a, b)$. It follows from $\|\breve{\varphi}_n\| \leq \|\varphi_n\|_\rho$ that in fact

$$\|f\| \leq \|\varphi\|_\rho.$$

In particular, the mapping $\varphi \mapsto f$ from $L^2_{d\rho}(\mathbb{R})$ to $L^2_\Delta(a, b)$ is a contraction. Hence, the operator $\mathcal{G}$ given by $\mathcal{G}\varphi = f$ is well defined and takes $L^2_{d\rho}(\mathbb{R})$ contractively to $L^2_\Delta(a, b)$. The assertion in (B.2.7) is clear by multiplying $\varphi \in L^2_{d\rho}(\mathbb{R})$ by appropriately chosen characteristic functions.

Step 3. The extended mapping $\mathcal{G}$ is a left inverse of the Fourier transform $\mathcal{F}$. For this, first note that it follows from (B.2.3) and dominated convergence that

$$(f, g) = \int_{\mathbb{R}} (\mathcal{F}f)(x)\overline{(\mathcal{F}g)(x)}\,d\rho(x)$$

for all $f \in L^2_\Delta(a,b) \ominus \operatorname{mul} A$ and $g \in L^2_\Delta(a,b)$. Thus, by means of the Fubini theorem, one concludes for $f \in L^2_\Delta(a,b) \ominus \operatorname{mul} A$ and $g \in L^2_\Delta(a,b)$, each having compact support, that

$$\begin{aligned}(f,g) &= \lim_{\eta\to\mathbb{R}} \int_\eta (\mathcal{F}f)(x) \left(\int_a^b g(t)^* \Delta(t) \omega(t,x)^* \, dt \right) d\rho(x) \\ &= \lim_{\eta\to\mathbb{R}} \int_a^b g(t)^* \Delta(t) \left(\int_\eta \omega(t,x)^* (\mathcal{F}f)(x) \, d\rho(x) \right) dt \\ &= (\mathcal{G}\mathcal{F}f, g),\end{aligned}$$

where, in the last step, (B.2.7) and the continuity of the inner product have been used. This implies

$$\mathcal{G}\mathcal{F}f = f$$

for all $f \in L^2_\Delta(a,b) \ominus \operatorname{mul} A$ with compact support. Since the functions in $L^2_\Delta(a,b)$ with compact support are dense in $L^2_\Delta(a,b)$ and $\operatorname{mul} A$ is finite-dimensional by assumption, it follows that the functions with compact support are also dense in $L^2_\Delta(a,b) \ominus \operatorname{mul} A$. Therefore, one obtains from the contractivity of $\mathcal{F}$ and $\mathcal{G}$ that $\mathcal{G}\mathcal{F}f = f$ for all $f \in L^2_\Delta(a,b) \ominus \operatorname{mul} A$. □

Recall that multiplication by the independent variable in the Hilbert space $L^2_{d\rho}(\mathbb{R})$ generates a self-adjoint operator Q whose resolvent is given by

$$(Q-\lambda)^{-1} = \frac{1}{x-\lambda}, \qquad \lambda \in \mathbb{C}\setminus\mathbb{R}.$$

Hence, by (B.2.4), one sees that

$$\big((A-\lambda)^{-1}f, g\big) = \big((Q-\lambda)^{-1}\mathcal{F}f, \mathcal{F}g\big)_\rho = \big(\mathcal{F}^*(Q-\lambda)^{-1}\mathcal{F}f, g\big)$$

for all $f \in L^2_\Delta(a,b)$ and $g \in L^2_\Delta(a,b)$, which leads to

$$(A-\lambda)^{-1} = \mathcal{F}^*(Q-\lambda)^{-1}\mathcal{F}, \qquad \lambda \in \mathbb{C}\setminus\mathbb{R}. \tag{B.2.8}$$

Consider the restriction $\mathcal{F}_{\rm op} : L^2_\Delta(a,b) \ominus \operatorname{mul} A \to L^2_{d\rho}(\mathbb{R})$ of the partial isometry $\mathcal{F}$ onto $L^2_\Delta(a,b) \ominus \operatorname{mul} A$ and recall that

$$\ker \mathcal{F} = \operatorname{mul} A = \ker\,(A-\lambda)^{-1}, \qquad \lambda \in \mathbb{C}\setminus\mathbb{R}.$$

Then $\mathcal{F}_{\rm op}$ is an isometry and (B.2.8) leads to

$$(A_{\rm op}-\lambda)^{-1} = \mathcal{F}_{\rm op}^*(Q-\lambda)^{-1}\mathcal{F}_{\rm op}, \qquad \lambda \in \mathbb{C}\setminus\mathbb{R}.$$

In general, here the restricted Fourier transform $\mathcal{F}_{\rm op}$ is not necessarily onto.

In order to prove that $\mathcal{F}_{\rm op}$ is surjective one needs an additional condition:

$$\begin{gathered}\text{for each } x_0 \in \mathbb{R} \text{ there exists a compactly} \\ \text{supported function } f \in L^2_\Delta(a,b) \text{ such that } (\mathcal{F}_{\rm op}f)(x_0) \neq 0.\end{gathered} \tag{B.2.9}$$

For the following result, recall that it is assumed that mul A is finite-dimensional.

Theorem B.2.3. *Let ω be a continuous 1×2 matrix function $(t,x) \mapsto \omega(t,x)$ on $(a,b) \times \mathbb{R}$ whose entries are real, and assume that the conditions (B.2.2) and (B.2.9) are satisfied. Then the Fourier transform*

$$f \mapsto \widehat{f}, \quad \widehat{f}(x) = \int_a^b \omega(t,x)\Delta(t)f(t)\,dt, \quad x \in \mathbb{R},$$

extends by continuity from the compactly supported functions $f \in L^2_\Delta(a,b)$ to a surjective partial isometry $\mathcal{F}$ from $L^2_\Delta(a,b)$ to $L^2_{d\rho}(\mathbb{R})$ with $\ker \mathcal{F} = \operatorname{mul} A$, that is, the restriction

$$\mathcal{F}_{\rm op} : L^2_\Delta(a,b) \ominus \operatorname{mul} A \to L^2_{d\rho}(\mathbb{R})$$

is a unitary mapping. Moreover, the self-adjoint operator $A_{\rm op}$ in $L^2_\Delta(a,b) \ominus \operatorname{mul} A$ is unitarily equivalent to the multiplication operator Q by the independent variable in $L^2_{d\rho}(\mathbb{R})$ via the restricted Fourier transform $\mathcal{F}_{\rm op}$:

$$A_{\rm op} = \mathcal{F}_{\rm op}^* Q \mathcal{F}_{\rm op}. \tag{B.2.10}$$

Proof. Is is clear from Lemma B.2.1 that $\mathcal{F} : L^2_\Delta(a,b) \to L^2_{d\rho}(\mathbb{R})$ is a partial isometry with $\ker \mathcal{F} = \operatorname{mul} A$. One verifies in the same way as in the proof of Theorem B.1.4 that $\mathcal{F}$ is surjective, and hence $\mathcal{F}_{\rm op}$ is unitary. The identity (B.2.10) follows from (B.2.8). □

In the situation of Theorem B.2.3 the inverse of the Fourier transform $\mathcal{F}_{\rm op}$ is actually given by the reverse Fourier transform $\mathcal{G}$ in Lemma B.2.2, which is now a unitary mapping from $L^2_{d\rho}(\mathbb{R})$ to $L^2_\Delta(a,b) \ominus \operatorname{mul} A$. Thus, for all $\varphi \in L^2_{d\rho}(\mathbb{R})$ one has

$$\lim_{\eta \to \mathbb{R}} \int_a^b \left| \Delta(t)^{\frac{1}{2}} \left(\mathcal{F}_{\rm op}^{-1}\varphi(t) - \int_\eta \omega(t,x)^* \varphi(x)\,d\rho(x) \right) \right|^2 dt = 0.$$

Appendix C

Sums of Closed Subspaces in Hilbert Spaces

In this appendix the sum of closed (linear) subspaces in a Hilbert space is discussed and, in particular, conditions are given so that sums of closed subspaces are closed. There is also a brief review of the opening and gap of closed subspaces.

In the following $\mathfrak{M}$ and $\mathfrak{N}$ will be closed subspaces of a Hilbert space $\mathfrak{H}$. The first lemma on the sum of closed subspaces is preliminary.

Lemma C.1. *Let $\mathfrak{M}$ and $\mathfrak{N}$ be closed subspaces of $\mathfrak{H}$. Then the following statements are equivalent:*

(i) $\mathfrak{M} + \mathfrak{N}$ *is closed and* $\mathfrak{M} \cap \mathfrak{N} = \{0\}$*;*

(ii) *there exists* $0 < \rho < 1$ *such that*

$$\rho\sqrt{\|f\|^2 + \|g\|^2} \leq \|f + g\|, \quad f \in \mathfrak{M}, \; g \in \mathfrak{N}. \tag{C.1}$$

Proof. (i) $\Rightarrow$ (ii) Assume that $\mathfrak{M} + \mathfrak{N}$ is closed and $\mathfrak{M} \cap \mathfrak{N} = \{0\}$. The projection from the Hilbert space $\mathfrak{M} + \mathfrak{N}$ onto $\mathfrak{M}$ parallel to $\mathfrak{N}$ is a closed, everywhere defined operator. Hence, by the closed graph theorem, the projection is bounded and there exists $C > 0$ such that

$$\|f\| \leq C\|f + g\|, \quad f \in \mathfrak{M}, \; g \in \mathfrak{N}.$$

Likewise, there exists $D > 0$ such that

$$\|g\| \leq D\|f + g\|, \quad f \in \mathfrak{M}, \; g \in \mathfrak{N}.$$

A combination of these inequalities leads to

$$\|f\|^2 + \|g\|^2 \leq (C^2 + D^2)\|f + g\|^2, \quad f \in \mathfrak{M}, \; g \in \mathfrak{N}.$$

By eventually enlarging $C^2 + D^2$ the inequality (C.1) follows with $0 < \rho < 1$.

J. Behrndt et al., *Boundary Value Problems, Weyl Functions, and Differential Operators*, Monographs in Mathematics 108,
https://doi.org/10.1007/978-3-030-36714-5

(ii) $\Rightarrow$ (i) Let (h_n) be a sequence in $\mathfrak{M}+\mathfrak{N}$ converging to $h \in \mathfrak{H}$. Decompose each element h_n as

$$h_n = f_n + g_n, \quad f_n \in \mathfrak{M}, \quad g_n \in \mathfrak{N}.$$

Since (h_n) is a Cauchy sequence in $\mathfrak{H}$ it follows from (C.1) that (f_n) and (g_n) are Cauchy sequences in $\mathfrak{M}$ and $\mathfrak{N}$, respectively. Therefore, there exist elements $f \in \mathfrak{M}$ and $g \in \mathfrak{N}$, so that $f_n \to f$ in $\mathfrak{M}$ and $g_n \to g$ in $\mathfrak{N}$. Hence, $h = f + g \in \mathfrak{M}+\mathfrak{N}$. Thus, $\mathfrak{M}+\mathfrak{N}$ is closed. To see that the sum is direct, assume that $h \in \mathfrak{M}\cap\mathfrak{N}$. Then, in particular, $h \in \mathfrak{M}$ and $-h \in \mathfrak{N}$ and the inequality (C.1) with $f = h$ and $g = -h$ implies $h = 0$. □

Let $\mathfrak{M}$ and $\mathfrak{N}$ be closed subspaces of $\mathfrak{H}$. Then the intersection $\mathfrak{M}\cap\mathfrak{N}$ is a closed subspace which generates an orthogonal decomposition of the Hilbert space:

$$\mathfrak{H} = (\mathfrak{M}\cap\mathfrak{N})^\perp \oplus (\mathfrak{M}\cap\mathfrak{N}).$$

In order to study properties of the sums $\mathfrak{M}+\mathfrak{N}$ and $\mathfrak{M}^\perp+\mathfrak{N}^\perp$ it is sometimes useful to reduce to a direct sum.

Lemma C.2. *Let $\mathfrak{M}$ and $\mathfrak{N}$ be closed subspaces of $\mathfrak{H}$. Then the subspaces*

$$\mathfrak{M}\cap(\mathfrak{M}\cap\mathfrak{N})^\perp \quad \text{and} \quad \mathfrak{N}\cap(\mathfrak{M}\cap\mathfrak{N})^\perp$$

have a trivial intersection and due to

$$(\mathfrak{M}\cap\mathfrak{N})^\perp = [\mathfrak{M}\cap(\mathfrak{M}\cap\mathfrak{N})^\perp]\oplus\mathfrak{M}^\perp = [\mathfrak{N}\cap(\mathfrak{M}\cap\mathfrak{N})^\perp]\oplus\mathfrak{N}^\perp \tag{C.2}$$

their orthogonal complements in the subspace $(\mathfrak{M}\cap\mathfrak{N})^\perp$ coincide with $\mathfrak{M}^\perp$ and $\mathfrak{N}^\perp$, respectively. Moreover,

$$\begin{aligned}\mathfrak{M} &= [\mathfrak{M}\cap(\mathfrak{M}\cap\mathfrak{N})^\perp]\oplus(\mathfrak{M}\cap\mathfrak{N}),\\ \mathfrak{N} &= [\mathfrak{N}\cap(\mathfrak{M}\cap\mathfrak{N})^\perp]\oplus(\mathfrak{M}\cap\mathfrak{N}),\end{aligned} \tag{C.3}$$

and, consequently, $\mathfrak{M}+\mathfrak{N}$ has the decomposition

$$\mathfrak{M}+\mathfrak{N} = \big[\mathfrak{M}\cap(\mathfrak{M}\cap\mathfrak{N})^\perp + \mathfrak{N}\cap(\mathfrak{M}\cap\mathfrak{N})^\perp\big]\oplus(\mathfrak{M}\cap\mathfrak{N}). \tag{C.4}$$

Proof. It is clear that $\mathfrak{M}\cap(\mathfrak{M}\cap\mathfrak{N})^\perp$ and $\mathfrak{N}\cap(\mathfrak{M}\cap\mathfrak{N})^\perp$ have a trivial intersection. To see the first identity in (C.2) note that

$$\mathfrak{M}\cap(\mathfrak{M}\cap\mathfrak{N})^\perp \subset (\mathfrak{M}\cap\mathfrak{N})^\perp \quad \text{and} \quad \mathfrak{M}^\perp \subset (\mathfrak{M}\cap\mathfrak{N})^\perp,$$

so that the right-hand side is contained in the left-hand side. To see the other inclusion, decompose $\mathfrak{H}$ as

$$\mathfrak{H} = \mathfrak{M}\oplus\mathfrak{M}^\perp.$$

Let $f \in (\mathfrak{M}\cap\mathfrak{N})^\perp$. Then according to this decomposition one has

$$f = g + h, \quad g \in \mathfrak{M}, \quad h \in \mathfrak{M}^\perp.$$

Since $\mathfrak{M}^\perp \subset (\mathfrak{M} \cap \mathfrak{N})^\perp$ one sees that $g \in \mathfrak{M} \cap (\mathfrak{M} \cap \mathfrak{N})^\perp$. Hence, the left-hand side is contained in the right-hand side. The second identity in (C.2) follows by interchanging $\mathfrak{M}$ and $\mathfrak{N}$.

Since $\mathfrak{M} \cap \mathfrak{N}$ is a closed subspace, it is clear that $\mathfrak{H}$ has the following orthogonal decomposition

$$\mathfrak{H} = (\mathfrak{M} \cap \mathfrak{N})^\perp \oplus (\mathfrak{M} \cap \mathfrak{N}).$$

Hence, in an analogous way as above, the decompositions (C.3) are clear. Thus, (C.4) follows. □

The above reduction process will play a role in the following theorem.

Theorem C.3. *Let $\mathfrak{M}$ and $\mathfrak{N}$ be closed subspaces of $\mathfrak{H}$. Then the following statements are equivalent:*

(i) $\mathfrak{M} + \mathfrak{N}$ *is closed;*

(ii) $\mathfrak{M}^\perp + \mathfrak{N}^\perp$ *is closed.*

Furthermore, the following statements are equivalent:

(iii) $\mathfrak{M} + \mathfrak{N}$ *is closed and* $\mathfrak{M} \cap \mathfrak{N} = \{0\}$*;*

(iv) $\mathfrak{M}^\perp + \mathfrak{N}^\perp = \mathfrak{H}$*,*

and the following statements are equivalent:

(v) $\mathfrak{M} + \mathfrak{N} = \mathfrak{H}$ *and* $\mathfrak{M} \cap \mathfrak{N} = \{0\}$*;*

(vi) $\mathfrak{M}^\perp + \mathfrak{N}^\perp = \mathfrak{H}$ *and* $\mathfrak{M}^\perp \cap \mathfrak{N}^\perp = \{0\}$.

Proof. (iii) ⇒ (iv) Assume that $\mathfrak{M} \cap \mathfrak{N} = \{0\}$. It will be shown that $\mathfrak{H} = \mathfrak{M}^\perp + \mathfrak{N}^\perp$ or, equivalently, that $\mathfrak{H} \subset \mathfrak{M}^\perp + \mathfrak{N}^\perp$. To see this, choose $h \in \mathfrak{H}$. The element $h \in \mathfrak{H}$ induces two linear functionals F and G on $\mathfrak{M} + \mathfrak{N}$ by

$$F(\gamma) = (\alpha, h), \quad G(\gamma) = (\beta, h), \quad \gamma = \alpha + \beta, \quad \alpha \in \mathfrak{M}, \ \beta \in \mathfrak{N}. \tag{C.5}$$

Since $\mathfrak{M} + \mathfrak{N}$ is closed and $\mathfrak{M} \cap \mathfrak{N} = \{0\}$, it follows from the inequality (C.1) that the functionals F and G are bounded on $\mathfrak{M} + \mathfrak{N}$. Extend the functionals F and G trivially to bounded linear functionals on all of $\mathfrak{H}$, and denote the extensions by F and G, respectively. By the Riesz representation theorem there exist unique elements $f, g \in \mathfrak{H}$ such that

$$F(\gamma) = (\gamma, f), \quad G(\gamma) = (\gamma, g), \quad \gamma \in \mathfrak{H}. \tag{C.6}$$

The definition in (C.5) implies that $F(\gamma) = 0$, $\gamma \in \mathfrak{N}$, and $G(\gamma) = 0$, $\gamma \in \mathfrak{M}$. For the corresponding elements f and g in (C.6) this means that $f \in \mathfrak{N}^\perp$ and $g \in \mathfrak{M}^\perp$. Now for $\gamma = \alpha + \beta$, $\alpha \in \mathfrak{M}$, $\beta \in \mathfrak{N}$, it follows from (C.5) and (C.6) that

$$(\gamma, h) = (\alpha, h) + (\beta, h) = F(\gamma) + G(\gamma) = (\gamma, f) + (\gamma, g).$$

This implies that $k = h - f - g \in (\mathfrak{M} + \mathfrak{N})^\perp \subset \mathfrak{M}^\perp$ and hence $h = f + g + k$ with $f \in \mathfrak{N}^\perp$ and $g + k \in \mathfrak{M}^\perp$. Therefore, $\mathfrak{H} \subset \mathfrak{M}^\perp + \mathfrak{N}^\perp$, which completes the proof.

(i) $\Rightarrow$ (ii) Use the reduction process from Lemma C.2. Since $\mathfrak{M} + \mathfrak{N}$ is assumed to be closed it follows from (C.4) that

$$\mathfrak{M} \cap (\mathfrak{M} \cap \mathfrak{N})^\perp + \mathfrak{N} \cap (\mathfrak{M} \cap \mathfrak{N})^\perp$$

is closed in $(\mathfrak{M} \cap \mathfrak{N})^\perp$. Since this is a direct sum, the sum of their orthogonal complements is closed in $(\mathfrak{M}\cap\mathfrak{N})^\perp$ by the implication (iii) $\Rightarrow$ (iv). Recall that the sum of their orthogonal complements coincides with $\mathfrak{M}^\perp + \mathfrak{N}^\perp$.

(ii) $\Rightarrow$ (i) This follows from the previous implication by symmetry.

(iv) $\Rightarrow$ (iii) Apply (ii) $\Rightarrow$ (i) to conclude that $\mathfrak{M} + \mathfrak{N}$ is closed. This sum is direct since $\{0\} = (\mathfrak{M}^\perp + \mathfrak{N}^\perp)^\perp$ by assumption.

(v) $\Leftrightarrow$ (vi) It suffices to show (v) $\Rightarrow$ (vi). Note that it follows from (iii) $\Rightarrow$ (iv) that $\mathfrak{M}^\perp + \mathfrak{N}^\perp = \mathfrak{H}$. Moreover, $\mathfrak{M}^\perp \cap \mathfrak{N}^\perp = (\mathfrak{M} + \mathfrak{N})^\perp = \{0\}$ by assumption, which completes the argument. □

Let again $\mathfrak{M}$ and $\mathfrak{N}$ be closed subspaces of $\mathfrak{H}$ and assume that $\mathfrak{M} + \mathfrak{N}$ is closed. It follows from (C.3) that $\mathfrak{M} + \mathfrak{N}$ can be written as

$$\begin{aligned} \mathfrak{M} + \mathfrak{N} &= \big([\mathfrak{M} \cap (\mathfrak{M} \cap \mathfrak{N})^\perp] \oplus (\mathfrak{M} \cap \mathfrak{N})\big) + \mathfrak{N} \\ &= [\mathfrak{M} \cap (\mathfrak{M} \cap \mathfrak{N})^\perp] + \mathfrak{N}. \end{aligned} \tag{C.7}$$

Since $\mathfrak{M} + \mathfrak{N}$ is closed, the orthogonal decomposition $\mathfrak{H} = (\mathfrak{M} + \mathfrak{N})^\perp \oplus (\mathfrak{M} + \mathfrak{N})$ and (C.7) show that

$$\begin{aligned} \mathfrak{H} &= (\mathfrak{M} + \mathfrak{N})^\perp \oplus \big([\mathfrak{M} \cap (\mathfrak{M} \cap \mathfrak{N})^\perp] + \mathfrak{N}\big) \\ &= \big((\mathfrak{M} + \mathfrak{N})^\perp \oplus [\mathfrak{M} \cap (\mathfrak{M} \cap \mathfrak{N})^\perp]\big) + \mathfrak{N}. \end{aligned} \tag{C.8}$$

The sum in the identity (C.8) is direct. To see this, assume that

$$f + g + \varphi = 0, \quad f \in (\mathfrak{M} + \mathfrak{N})^\perp, \quad g \in \mathfrak{M} \cap (\mathfrak{M} \cap \mathfrak{N})^\perp, \quad \varphi \in \mathfrak{N}.$$

Then $f = -g - \varphi$, where $-g - \varphi \in \mathfrak{M} + \mathfrak{N}$. Hence, $f = 0$ and $g = -\varphi$ implies that $g = 0$ and $\varphi = 0$.

The decompositions (C.7) and (C.8) will be used in the proof of the next lemma.

Lemma C.4. *Let $B \in \mathbf{B}(\mathfrak{H}, \mathfrak{K})$ have a closed range and let $\mathfrak{M}$ be a closed subspace of $\mathfrak{H}$. Then $\mathfrak{M} + \ker B$ is closed if and only if $B(\mathfrak{M})$ is closed.*

Proof. Assume that $\mathfrak{M} + \ker B$ is closed and set $\mathfrak{N} = \ker B$. It follows from the direct sum decomposition (C.8) that B maps

$$(\mathfrak{M} + \mathfrak{N})^\perp \oplus [\mathfrak{M} \cap (\mathfrak{M} \cap \mathfrak{N})^\perp]$$

bijectively onto the closed subspace $\operatorname{ran} B$ and hence B also provides a bijection between the closed subspaces of $(\mathfrak{M}+\mathfrak{N})^\perp \oplus [\mathfrak{M} \cap (\mathfrak{M}\cap\mathfrak{N})^\perp]$ and those of $\operatorname{ran} B$. In particular, it follows from this observation and (C.7) that

$$B(\mathfrak{M}) = B\big(\mathfrak{M} \cap (\mathfrak{M}\cap\mathfrak{N})^\perp\big)$$

is closed in $\mathfrak{K}$.

For the converse statement assume that $B(\mathfrak{M})$ is closed. In order to show that $\mathfrak{M} + \ker B$ is closed, let $f_n \in \mathfrak{M}$ and $\varphi_n \in \ker B$ have the property that

$$f_n + \varphi_n \to \chi$$

for some $\chi \in \mathfrak{H}$. In particular, this shows that $Bf_n \to B\chi$. The assumption that $B(\mathfrak{M})$ is closed implies that $Bf_n \to Bf$ for some $f \in \mathfrak{M}$. Hence,

$$\chi - f = \varphi \in \ker B \quad \text{or} \quad \chi = f + \varphi \in \mathfrak{M} + \ker B.$$

It follows that $\mathfrak{M} + \ker B$ is closed. □

There is another way to approach the topic of sums of closed subspaces, namely via various notions of opening or gap between closed subspaces.

Definition C.5. Let $\mathfrak{M}$ and $\mathfrak{N}$ be closed subspaces of $\mathfrak{H}$ with corresponding orthogonal projections $P_\mathfrak{M}$ and $P_\mathfrak{N}$, respectively. The *opening* $\omega(\mathfrak{M},\mathfrak{N})$ between $\mathfrak{M}$ and $\mathfrak{N}$ is defined as

$$\omega(\mathfrak{M},\mathfrak{N}) = \|P_\mathfrak{M} P_\mathfrak{N}\|.$$

It is clear that $0 \le \omega(\mathfrak{M},\mathfrak{N}) \le 1$ and that $\omega(\mathfrak{M},\mathfrak{N}) = \omega(\mathfrak{N},\mathfrak{M})$, since one has $\|A\| = \|A^*\|$ for any $A \in \mathbf{B}(\mathfrak{H})$.

Proposition C.6. *Let $\mathfrak{M}$ and $\mathfrak{N}$ be closed subspaces of $\mathfrak{H}$. Then the following statements are equivalent:*

(i) $\omega(\mathfrak{M},\mathfrak{N}) < 1$;

(ii) $\mathfrak{M}+\mathfrak{N}$ *is closed and* $\mathfrak{M}\cap\mathfrak{N} = \{0\}$.

Proof. (i) ⇒ (ii) Assume that $\omega(\mathfrak{M},\mathfrak{N}) < 1$. Observe that for $f \in \mathfrak{M}$ and $g \in \mathfrak{N}$ one has

$$|(f,g)| = |(P_\mathfrak{M} f, P_\mathfrak{N} g)| \le \omega(\mathfrak{M},\mathfrak{N})\|f\|\|g\|.$$

It follows that for all $f \in \mathfrak{M}$ and $g \in \mathfrak{N}$

$$\begin{aligned} \|f\|^2 + \|g\|^2 &= \|f+g\|^2 - 2\mathrm{Re}\,(f,g) \\ &\le \|f+g\|^2 + 2|(f,g)| \\ &\le \|f+g\|^2 + 2\omega(\mathfrak{M},\mathfrak{N})\|f\|\|g\| \\ &\le \|f+g\|^2 + \omega(\mathfrak{M},\mathfrak{N})\big(\|f\|^2 + \|g\|^2\big). \end{aligned}$$

In particular, this shows that

$$(1 - \omega(\mathfrak{M}, \mathfrak{N}))\big(\|f\|^2 + \|g\|^2\big) \le \|f + g\|^2, \quad f \in \mathfrak{M}, \quad g \in \mathfrak{N}.$$

Hence, Lemma C.1 implies that $\mathfrak{M} \cap \mathfrak{N} = \{0\}$ and that $\mathfrak{M} + \mathfrak{N}$ is closed, which gives (ii).

(ii) $\Rightarrow$ (i) Assume that $\mathfrak{M} + \mathfrak{N}$ is closed and $\mathfrak{M} \cap \mathfrak{N} = \{0\}$. By Lemma C.1, the inequality (C.1) holds for some $0 < \rho < 1$, hence for all $f \in \mathfrak{M}$ and $g \in \mathfrak{N}$

$$\rho^2 \left(\|f\|^2 + \|g\|^2\right) \le \|f\|^2 + 2\mathrm{Re}\,(f, g) + \|g\|^2$$

or, equivalently,

$$-\,2\mathrm{Re}\,(f, g) \le (1 - \rho^2)\left(\|f\|^2 + \|g\|^2\right). \tag{C.9}$$

Now let $h, k \in \mathfrak{H}$ with $\|h\| \le 1$ and $\|k\| \le 1$ and choose $\theta \in \mathbb{R}$ such that

$$e^{i\theta}(P_{\mathfrak{M}}h, P_{\mathfrak{N}}k) = -|(P_{\mathfrak{M}}h, P_{\mathfrak{N}}k)|.$$

Then $(e^{i\theta}P_{\mathfrak{M}}h, P_{\mathfrak{N}}k) \in \mathbb{R}$ and (C.9) with $f = e^{i\theta}P_{\mathfrak{M}}h$ and $g = P_{\mathfrak{N}}k$ yields

$$\begin{aligned} |(h, P_{\mathfrak{M}}P_{\mathfrak{N}}k)| &= |(P_{\mathfrak{M}}h, P_{\mathfrak{N}}k)| \\ &= -\mathrm{Re}\,(e^{i\theta}P_{\mathfrak{M}}h, P_{\mathfrak{N}}k) \\ &\le \frac{1-\rho^2}{2}\big(\|e^{i\theta}P_{\mathfrak{M}}h\|^2 + \|P_{\mathfrak{N}}k\|^2\big) \\ &\le 1 - \rho^2. \end{aligned}$$

This implies $\|P_{\mathfrak{M}}P_{\mathfrak{N}}\| \le 1 - \rho^2 < 1$ and hence (i) holds. □

Corollary C.7. *Let $\mathfrak{M}$ and $\mathfrak{N}$ be closed subspaces of $\mathfrak{H}$ such that $\mathfrak{M} + \mathfrak{N}$ is closed and $\mathfrak{M} \cap \mathfrak{N} = \{0\}$. Then for closed subspaces $\mathfrak{M}_1 \subset \mathfrak{M}$ and $\mathfrak{N}_1 \subset \mathfrak{N}$ also the subspace $\mathfrak{M}_1 + \mathfrak{N}_1$ is closed.*

Proof. This statement follows immediately from Proposition C.6 by noting that $\omega(\mathfrak{M}_1, \mathfrak{N}_1) \le \omega(\mathfrak{M}, \mathfrak{N}) < 1$. □

Proposition C.6 and the reduction of closed subspaces in Lemma C.2 leads to the following corollary.

Corollary C.8. *Let $\mathfrak{M}$ and $\mathfrak{N}$ be closed subspaces of $\mathfrak{H}$. Then the following statements are equivalent:*

(i) $\omega\big(\mathfrak{M} \cap (\mathfrak{M} \cap \mathfrak{N})^\perp, \mathfrak{N} \cap (\mathfrak{M} \cap \mathfrak{N})^\perp\big) < 1$;

(ii) $\mathfrak{M} + \mathfrak{N}$ *is closed.*

Proof. Since the closed subspaces $\mathfrak{M} \cap (\mathfrak{M} \cap \mathfrak{N})^\perp$ and $\mathfrak{N} \cap (\mathfrak{M} \cap \mathfrak{N})^\perp$ have trivial intersection, it follows from Proposition C.6 that (i) is equivalent to

$$\mathfrak{M} \cap (\mathfrak{M} \cap \mathfrak{N})^\perp + \mathfrak{N} \cap (\mathfrak{M} \cap \mathfrak{N})^\perp \tag{C.10}$$

is closed. Furthermore, the decomposition (C.4) shows that the space in (C.10) is closed if and only if $\mathfrak{M} + \mathfrak{N}$ is closed. □

The notion of opening is now supplemented with the notion of gap between closed subspaces.

Definition C.9. Let $\mathfrak{M}$ and $\mathfrak{N}$ be closed subspaces of $\mathfrak{H}$ with corresponding orthogonal projections $P_\mathfrak{M}$ and $P_\mathfrak{N}$, respectively. The *gap* $g(\mathfrak{M}, \mathfrak{N})$ between $\mathfrak{M}$ and $\mathfrak{N}$ is defined as

$$g(\mathfrak{M}, \mathfrak{N}) = \|P_\mathfrak{M} - P_\mathfrak{N}\|.$$

Note that it follows directly from the definition that

$$g(\mathfrak{M}, \mathfrak{N}) = g(\mathfrak{N}, \mathfrak{M}) \quad \text{and} \quad g(\mathfrak{M}^\perp, \mathfrak{N}^\perp) = g(\mathfrak{M}, \mathfrak{N}). \tag{C.11}$$

The connection between the gap and the opening between closed subspaces is contained in the following proposition.

Proposition C.10. *Let $\mathfrak{M}$ and $\mathfrak{N}$ be closed subspaces in $\mathfrak{H}$. Then*

$$g(\mathfrak{M}, \mathfrak{N}) = \max\left\{\omega(\mathfrak{M}, \mathfrak{N}^\perp), \omega(\mathfrak{M}^\perp, \mathfrak{N})\right\} \tag{C.12}$$

and, in particular, $g(\mathfrak{M}, \mathfrak{N}) \leq 1$.

Proof. First observe that

$$P_{\mathfrak{N}^\perp} P_\mathfrak{M} = (I - P_\mathfrak{N}) P_\mathfrak{M} = (P_\mathfrak{M} - P_\mathfrak{N}) P_\mathfrak{M},$$

so that $\omega(\mathfrak{M}, \mathfrak{N}^\perp) = \omega(\mathfrak{N}^\perp, \mathfrak{M}) \leq g(\mathfrak{M}, \mathfrak{N}) \|P_\mathfrak{M}\| \leq g(\mathfrak{M}, \mathfrak{N})$. Therefore, also

$$\omega(\mathfrak{M}^\perp, \mathfrak{N}) \leq g(\mathfrak{M}^\perp, \mathfrak{N}^\perp) = g(\mathfrak{M}, \mathfrak{N})$$

and hence

$$\max\left\{\omega(\mathfrak{M}, \mathfrak{N}^\perp), \omega(\mathfrak{M}^\perp, \mathfrak{N})\right\} \leq g(\mathfrak{M}, \mathfrak{N}).$$

Furthermore, observe that for all $h \in \mathfrak{H}$

$$\begin{aligned}
\|(P_\mathfrak{M} - P_\mathfrak{N})h\|^2 &= \|(I - P_\mathfrak{N}) P_\mathfrak{M} h - P_\mathfrak{N}(I - P_\mathfrak{M})h\|^2 \\
&= \|P_{\mathfrak{N}^\perp} P_\mathfrak{M} h\|^2 + \|P_\mathfrak{N} P_{\mathfrak{M}^\perp} h\|^2 \\
&= \|P_{\mathfrak{N}^\perp} P_\mathfrak{M} P_\mathfrak{M} h\|^2 + \|P_\mathfrak{N} P_{\mathfrak{M}^\perp} P_{\mathfrak{M}^\perp} h\|^2 \\
&\leq \omega(\mathfrak{M}, \mathfrak{N}^\perp)^2 \|P_\mathfrak{M} h\|^2 + \omega(\mathfrak{M}^\perp, \mathfrak{N})^2 \|P_{\mathfrak{M}^\perp} h\|^2,
\end{aligned}$$

and the last term is majorized by

$$\max\left\{\omega(\mathfrak{M},\mathfrak{N}^\perp)^2,\omega(\mathfrak{M}^\perp,\mathfrak{N})^2\right\}\left(\|P_\mathfrak{M}h\|^2+\|P_{\mathfrak{M}^\perp}h\|^2\right).$$

It follows that

$$g(\mathfrak{M},\mathfrak{N})\le\max\left\{\omega(\mathfrak{M},\mathfrak{N}^\perp),\omega(\mathfrak{M}^\perp,\mathfrak{N})\right\}.$$

Therefore, (C.12) has been established. □

Proposition C.11. *Let* $\mathfrak{M}$ *and* $\mathfrak{N}$ *be closed subspaces of* $\mathfrak{H}$. *Then*

$$g(\mathfrak{M},\mathfrak{N})<1 \quad\Rightarrow\quad \dim\mathfrak{M}=\dim\mathfrak{N}.$$

Proof. If $g(\mathfrak{M},\mathfrak{N})=\|P_\mathfrak{M}-P_\mathfrak{N}\|<1$, then the bounded operator

$$I-(P_\mathfrak{M}-P_\mathfrak{N})$$

maps $\mathfrak{H}$ onto itself with bounded inverse. Hence,

$$P_\mathfrak{M}P_\mathfrak{N}\,\mathfrak{H}=P_\mathfrak{M}\left[I-(P_\mathfrak{M}-P_\mathfrak{N})\right]\mathfrak{H}=P_\mathfrak{M}\,\mathfrak{H},$$

so that $P_\mathfrak{M}$ maps $\operatorname{ran}P_\mathfrak{N}$ onto $\operatorname{ran}P_\mathfrak{M}$. Observe that for $h\in\operatorname{ran}P_\mathfrak{N}$

$$\|P_\mathfrak{M}h\|=\|h+(P_\mathfrak{M}-P_\mathfrak{N})h\|\ge\left(1-\|P_\mathfrak{M}-P_\mathfrak{N}\|\right)\|h\|$$

and hence $P_\mathfrak{M}$ maps $\operatorname{ran}P_\mathfrak{N}$ boundedly and boundedly invertible onto $\operatorname{ran}P_\mathfrak{M}$. This implies that the dimensions of $\mathfrak{M}=\operatorname{ran}P_\mathfrak{M}$ and $\mathfrak{N}=\operatorname{ran}P_\mathfrak{N}$ coincide. □

Theorem C.12. *Let* $\mathfrak{M}$ *and* $\mathfrak{N}$ *be closed subspaces of* $\mathfrak{H}$. *Then the following statements are equivalent:*

(i) $g(\mathfrak{M},\mathfrak{N}^\perp)<1$;

(ii) $\mathfrak{M}+\mathfrak{N}=\mathfrak{H}$ *and* $\mathfrak{M}\cap\mathfrak{N}=\{0\}$,

and in this case $\dim\mathfrak{M}=\dim\mathfrak{N}^\perp$ *and* $\dim\mathfrak{M}^\perp=\dim\mathfrak{N}$.

Proof. (i) ⇒ (ii) Proposition C.10 implies $\omega(\mathfrak{M},\mathfrak{N})<1$ and $\omega(\mathfrak{M}^\perp,\mathfrak{N}^\perp)<1$. By Proposition C.6, the condition $\omega(\mathfrak{M},\mathfrak{N})<1$ implies that $\mathfrak{M}+\mathfrak{N}$ is closed and that $\mathfrak{M}\cap\mathfrak{N}=\{0\}$, and in the same way $\omega(\mathfrak{M}^\perp,\mathfrak{N}^\perp)<1$ yields that $\mathfrak{M}^\perp+\mathfrak{N}^\perp$ is closed and that $\mathfrak{M}^\perp\cap\mathfrak{N}^\perp=\{0\}$. The identity $\mathfrak{M}^\perp\cap\mathfrak{N}^\perp=\{0\}$ implies that $\mathfrak{M}+\mathfrak{N}$ is dense in $\mathfrak{H}$. Therefore, $\mathfrak{M}+\mathfrak{N}=\mathfrak{H}$.

(ii) ⇒ (i) It follows from Proposition C.6 that $\omega(\mathfrak{M},\mathfrak{N})<1$. Moreover, the equivalence of (v) and (vi) in Theorem C.3 shows $\mathfrak{M}^\perp+\mathfrak{N}^\perp=\mathfrak{H}$ and $\mathfrak{M}^\perp\cap\mathfrak{N}^\perp=\{0\}$, and therefore $\omega(\mathfrak{M}^\perp,\mathfrak{N}^\perp)<1$ by Proposition C.6. Now Proposition C.10 implies that $g(\mathfrak{M},\mathfrak{N}^\perp)<1$.

It is clear from (i) and Proposition C.11 that $\dim\mathfrak{M}=\dim\mathfrak{N}^\perp$. Furthermore, as $g(\mathfrak{M}^\perp,\mathfrak{N})=g(\mathfrak{M},\mathfrak{N}^\perp)$ by (C.11), the same argument gives $\dim\mathfrak{M}^\perp=\dim\mathfrak{N}$. □

Appendix D

Factorization of Bounded Linear Operators

This appendix contains a number of results pertaining to the factorization of bounded linear operators based on range inclusions or norm inequalities. These results will be useful in conjunction with range inclusions or norm inequalities for relations.

Let $\mathfrak{H}$ and $\mathfrak{K}$ be Hilbert spaces and let $H \in \mathbf{B}(\mathfrak{H}, \mathfrak{K})$. The restriction of H to $(\ker H)^\perp \subset \mathfrak{H}$,

$$H : (\ker H)^\perp \to \operatorname{ran} H,$$

is a bijective mapping between $(\ker H)^\perp$ and $\operatorname{ran} H$. The inverse of this restriction is a (linear) operator from $\operatorname{ran} H$ onto $(\ker H)^\perp$ and is called the *Moore–Penrose inverse* of the operator H, which in general is an unbounded operator. Since $\mathfrak{H} = \ker H \oplus (\ker H)^\perp$, one sees immediately that

$$H^{(-1)} H = P_{(\ker H)^\perp}. \tag{D.1}$$

Moreover, the closed graph theorem shows that $\operatorname{ran} H$ is closed if and only if $H^{(-1)}$ takes $\operatorname{ran} H$ boundedly into $(\ker H)^\perp$. Hence, if $\operatorname{ran} H$ is closed, then

$$H^{(-1)} \in \mathbf{B}\big(\operatorname{ran} H, (\ker H)^\perp\big) \quad \text{and} \quad (H^{(-1)})^\times \in \mathbf{B}\big((\ker H)^\perp, \operatorname{ran} H\big), \tag{D.2}$$

where $^\times$ denotes the adjoint with respect to the scalar products in $\operatorname{ran} H$ and $(\ker H)^\perp$.

Lemma D.1. *Let $\mathfrak{H}$ and $\mathfrak{K}$ be Hilbert spaces and let $H \in \mathbf{B}(\mathfrak{H}, \mathfrak{K})$. Then* $\operatorname{ran} H$ *is closed if and only if* $\operatorname{ran} H^*$ *is closed.*

Proof. Since $H \in \mathbf{B}(\mathfrak{H}, \mathfrak{K})$, it suffices to show that if $\operatorname{ran} H$ is closed in $\mathfrak{K}$, then $\operatorname{ran} H^*$ is closed in $\mathfrak{H}$. Since $\overline{\operatorname{ran}}\, H^* = (\ker H)^\perp$, one only needs to show

$$(\ker H)^\perp \subset \operatorname{ran} H^*.$$

J. Behrndt et al., *Boundary Value Problems, Weyl Functions, and Differential Operators*, Monographs in Mathematics 108,
https://doi.org/10.1007/978-3-030-36714-5

For this let $h \in (\ker H)^\perp$ and $f \in \mathfrak{H}$. Then it follows from (D.1) and (D.2) that with $(H^{(-1)})^\times \in \mathbf{B}((\ker H)^\perp, \operatorname{ran} H)$ in (D.2) one has

$$\begin{aligned}
(h,f)_{\mathfrak{H}} &= (h, P_{(\ker H)^\perp} f)_{(\ker H)^\perp} \\
&= \bigl(h, H^{(-1)} H f\bigr)_{(\ker H)^\perp} \\
&= \bigl((H^{(-1)})^\times h, Hf\bigr)_{\operatorname{ran} H} \\
&= \bigl((H^{(-1)})^\times h, Hf\bigr)_{\mathfrak{K}} \\
&= \bigl(H^*(H^{(-1)})^\times h, f\bigr)_{\mathfrak{H}}
\end{aligned}$$

and it follows that $h = H^*(H^{(-1)})^\times h$. Hence, $h \in \operatorname{ran} H^*$ and therefore $\operatorname{ran} H^*$ is closed. □

Lemma D.1 has a direct consequence, which is useful.

Lemma D.2. *For $H \in \mathbf{B}(\mathfrak{H}, \mathfrak{K})$ there are the inclusions*

$$\operatorname{ran} HH^* \subset \operatorname{ran} H \subset \overline{\operatorname{ran}}\, H = \overline{\operatorname{ran}}\, HH^*. \tag{D.3}$$

Moreover, the following statements are equivalent:

(i) $\operatorname{ran} HH^*$ *is closed;*

(ii) $\operatorname{ran} H$ *is closed;*

(iii) $\operatorname{ran} HH^* = \operatorname{ran} H$,

in which case all inclusions in (D.3) are identities.

Proof. The chain of inclusions in (D.3) is clear; the last equality is a consequence of the identity $\ker H^* = \ker HH^*$.

(i) ⇒ (iii) Assume that $\operatorname{ran} HH^*$ is closed. Then all inclusions in (D.3) are equalities and, in particular, (iii) follows.

(iii) ⇒ (ii) Assume that $\operatorname{ran} HH^* = \operatorname{ran} H$. It will be sufficient to show that $\overline{\operatorname{ran}}\, H^* \subset \operatorname{ran} H^*$, which implies that $\operatorname{ran} H^*$ is closed and hence also $\operatorname{ran} H$ is closed by Lemma D.1. Assume that $h \in \overline{\operatorname{ran}}\, H^* = (\ker H)^\perp$. By the assumption, it follows that $Hh = HH^*k$ for some k, which gives $H(h - H^*k) = 0$. Note that $h \in (\ker H)^\perp$ and $H^*k \in \operatorname{ran} H^* \subset \overline{\operatorname{ran}}\, H^* = (\ker H)^\perp$, and hence $h = H^*k$. Thus, $\overline{\operatorname{ran}}\, H^* \subset \operatorname{ran} H^*$ which implies that $\operatorname{ran} H^*$ is closed, so that also $\operatorname{ran} H$ is closed by Lemma D.1.

(ii) ⇒ (i) Assume that $\operatorname{ran} H$ is closed. Then it follows from (D.3) that

$$\overline{\operatorname{ran}}\, HH^* = \operatorname{ran} H. \tag{D.4}$$

It will suffice to show that $\overline{\operatorname{ran}}\, HH^* \subset \operatorname{ran} HH^*$. Assume that $k \in \overline{\operatorname{ran}}\, HH^*$. Then $k = Hh$ for some $h \in (\ker H)^\perp$ by (D.4). As $(\ker H)^\perp = \overline{\operatorname{ran}}\, H^* = \operatorname{ran} H^*$ by the assumption (ii) and Lemma D.1, one has $h \in \operatorname{ran} H^*$ and therefore $k \in \operatorname{ran} HH^*$. This shows that $\overline{\operatorname{ran}}\, HH^* \subset \operatorname{ran} HH^*$ which implies that $\operatorname{ran} HH^*$ is closed. □

Let $\mathfrak{F}$, $\mathfrak{G}$, and $\mathfrak{H}$ be Hilbert spaces, let $A \in \mathbf{B}(\mathfrak{F},\mathfrak{H})$, $B \in \mathbf{B}(\mathfrak{G},\mathfrak{H})$, and $C \in \mathbf{B}(\mathfrak{F},\mathfrak{G})$, and assume that the following factorization holds:

$$A = BC. \tag{D.5}$$

Then it follows from (D.5) that $\ker C \subset \ker A$. Note that the orthogonal decomposition $\mathfrak{G} = \ker B \oplus \overline{\operatorname{ran}}\, B^*$ allows one to write $A = BPC$, where P is the orthogonal projection in $\mathfrak{G}$ onto $\overline{\operatorname{ran}}\, B^*$. Hence, one may always assume that $\operatorname{ran} C \subset \overline{\operatorname{ran}}\, B^*$. In this case $\ker C = \ker A$ and hence C maps $(\ker A)^\perp = \overline{\operatorname{ran}}\, A^*$ into $\overline{\operatorname{ran}}\, B^*$ injectively. Moreover, in this case the operator C in (D.5) is uniquely determined. The following proposition is a version of the well-known *Douglas lemma.*

Proposition D.3. *Assume that $A \in \mathbf{B}(\mathfrak{F},\mathfrak{H})$ and $B \in \mathbf{B}(\mathfrak{G},\mathfrak{H})$. Let $\rho > 0$, then the following statements are equivalent:*

(i) *$A = BC$ for some $C \in \mathbf{B}(\mathfrak{F},\mathfrak{G})$ with $\operatorname{ran} C \subset \overline{\operatorname{ran}}\, B^*$ and $\|C\| \leq \rho$;*

(ii) *$\operatorname{ran} A \subset \operatorname{ran} B$ and $\|B^{(-1)}\varphi\| \leq \rho\, \|A^{(-1)}\varphi\|$, $\varphi \in \operatorname{ran} A$;*

(iii) *$AA^* \leq \rho^2\, BB^*$.*

Moreover, if $\operatorname{ran} A \subset \operatorname{ran} B$, then $\|B^{(-1)}\varphi\| \leq \rho\, \|A^{(-1)}\varphi\|$, $\varphi \in \operatorname{ran} A$, for some $\rho > 0$.

Proof. (i) $\Rightarrow$ (ii) It is clear that $\operatorname{ran} A \subset \operatorname{ran} B$ and therefore also

$$B^{(-1)}A = B^{(-1)}BC = P_{(\ker B)^\perp} C = P_{\overline{\operatorname{ran}}\, B^*} C = C.$$

From the last identity one sees that

$$\|B^{(-1)}A\psi\| \leq \|C\|\, \|\psi\|, \quad \psi \in \mathfrak{F}.$$

Now let $\varphi \in \operatorname{ran} A$. Then there exists $\psi \perp \ker A$ such that $\varphi = A\psi$. Therefore, $\psi = A^{(-1)}\varphi$ and it follows immediately that

$$\|B^{(-1)}\varphi\| \leq \|C\|\, \|A^{(-1)}\varphi\|, \quad \varphi \in \operatorname{ran} A.$$

Hence, (ii) has been shown.

(ii) $\Rightarrow$ (i) It follows from (ii) that for each $h \in \mathfrak{F}$ there exists a uniquely determined element $k \in (\ker B)^\perp = \overline{\operatorname{ran}}\, B^*$ such that $Ah = Bk$. Hence, the mapping $h \mapsto k$ from $\mathfrak{F}$ to $\overline{\operatorname{ran}}\, B^* \subset \mathfrak{G}$ defines a linear operator C with $\operatorname{dom} C = \mathfrak{F}$ and one has $A = BC$. To show that the operator C is closed, assume that

$$h_n \to h \quad \text{and} \quad k_n = Ch_n \to k,$$

where $k_n \in \overline{\operatorname{ran}}\, B^*$. Since A and B are bounded linear operators, it follows from from $Ah_n = Bk_n$ that $Ah = Bk$. Furthermore, $\overline{\operatorname{ran}}\, B^*$ is closed, which implies that $k \in \overline{\operatorname{ran}}\, B^*$. Hence, $k = Ch$ and thus C is closed. By the closed graph theorem, it follows that $C \in \mathbf{B}(\mathfrak{F}, \mathfrak{G})$. The property $\operatorname{ran} C \subset \overline{\operatorname{ran}}\, B^*$ holds by construction. Furthermore, one sees that for $\psi \in \mathfrak{F}$

$$\begin{aligned}\|C\psi\| &= \|P_{\overline{\operatorname{ran}}\, B^*} C\psi\| \\ &= \|P_{(\ker B)^\perp} C\psi\| \\ &= \|B^{(-1)} BC\psi\| \\ &\leq \rho\, \|A^{(-1)} A\psi\| \\ &= \rho\, \|P_{(\ker A)^\perp}\psi\| \\ &\leq \rho\, \|\psi\|,\end{aligned}$$

and so $\|C\| \leq \rho$.

(i) $\Rightarrow$ (iii) It is clear that $AA^* = BCC^*B^*$. For $\psi \in \mathfrak{F}$ one has

$$\big(AA^*\psi, \psi\big) = \|C^*B^*\psi\|^2 \leq \|C^*\|^2\|B^*\psi\|^2 = \|C^*\|^2\big(BB^*\psi, \psi\big)$$

and hence $AA^* \leq \|C^*\|^2 BB^*$. Since $\|C^*\| = \|C\| \leq \rho$ this leads to (iii).

(iii) $\Rightarrow$ (i) The mapping D from $\operatorname{ran} B^*$ to $\operatorname{ran} A^*$ given by

$$B^*h \mapsto A^*h, \quad h \in \mathfrak{H},$$

is well defined and linear. To see that it is well defined observe that $B^*h = 0$ implies $AA^*h = 0$ and hence $A^*h = 0$. Then $DB^*h = A^*h$ for $h \in \mathfrak{H}$ and one has $D \in \mathbf{B}(\operatorname{ran} B^*, \operatorname{ran} A^*)$ with $\|DB^*h\| \leq \rho^2\|B^*h\|$. Due to the boundedness, this operator has a unique extension in $\mathbf{B}(\overline{\operatorname{ran}}\, B^*, \overline{\operatorname{ran}}\, A^*)$, denoted again by D, and $\|D\| \leq \rho^2$. Finally, this mapping has a trivial extension $D \in \mathbf{B}(\mathfrak{G}, \mathfrak{F})$, satisfying $DB^*h = A^*h$ for $h \in \mathfrak{H}$ and again $\|D\| \leq \rho^2$. With $C = D^*$ one obtains that $C \in \mathbf{B}(\mathfrak{F}, \mathfrak{G})$, $A = BC$, and $\|C\| \leq \rho^2$. In view of the construction of D one has $\ker B = (\overline{\operatorname{ran}}\, B^*)^\perp \subset \ker D$, so that $\operatorname{ran} C = \operatorname{ran} D^* \subset \overline{\operatorname{ran}}\, B^*$.

For the last statement it suffices to observe that the inclusion $\operatorname{ran} A \subset \operatorname{ran} B$ implies $A = BC$ for some $C \in \mathbf{B}(\mathfrak{F}, \mathfrak{G})$ (see (ii) $\Rightarrow$ (i)). However, this implies $AA^* \leq \rho^2\, BB^*$ for some $\rho > 0$ (see (i) $\Rightarrow$ (iii)). □

The next result contains a strengthening of Proposition D.3. Recall that $\operatorname{ran} C \subset \overline{\operatorname{ran}}\, B^*$ implies that $\ker C = \ker A$. Hence, if C maps $\overline{\operatorname{ran}}\, A^*$ onto $\overline{\operatorname{ran}}\, B^*$, then C maps $\overline{\operatorname{ran}}\, A^*$ onto $\operatorname{ran} C = \overline{\operatorname{ran}}\, B^*$ bijectively.

Corollary D.4. *Assume that $A \in \mathbf{B}(\mathfrak{F}, \mathfrak{H})$ and $B \in \mathbf{B}(\mathfrak{G}, \mathfrak{H})$. Let $\rho > 0$. Then the following statements are equivalent:*

(a) *$A = BC$ for some $C \in \mathbf{B}(\mathfrak{F}, \mathfrak{G})$ with $\operatorname{ran} C = \overline{\operatorname{ran}}\, B^*$ and $\|C\| \leq \rho$;*
(b) *$\operatorname{ran} A = \operatorname{ran} B$ and $\|B^{(-1)}\varphi\| \leq \rho\, \|A^{(-1)}\varphi\|$, $\varphi \in \operatorname{ran} A$;*
(c) *$\varepsilon^2 BB^* \leq AA^* \leq \rho^2\, BB^*$ for some $0 < \varepsilon < \rho$.*

Moreover, if $\operatorname{ran} A = \operatorname{ran} B$, then $\|B^{(-1)}\varphi\| \leq \rho\, \|A^{(-1)}\varphi\|$, $\varphi \in \operatorname{ran} A$, for some $\rho > 0$.

Proof. The assertions in Proposition D.3 will be freely used in the arguments below.

(a) $\Rightarrow$ (c) Recall that C maps $\overline{\operatorname{ran}}\, A^*$ bijectively onto $\overline{\operatorname{ran}}\, B^*$. Since

$$\overline{\operatorname{ran}}\, C^* = \overline{\operatorname{ran}}\, A^* \quad \text{and} \quad \ker C^* = \ker B,$$

one sees that C^* maps $\overline{\operatorname{ran}}\, B^*$ bijectively onto $\overline{\operatorname{ran}}\, A^*$, as $\operatorname{ran} C^*$ is closed by Lemma D.1. Thus, CC^* is a bijective mapping from $\overline{\operatorname{ran}}\, B^*$ onto itself and hence

$$(AA^*h, h) = (CC^*B^*h, B^*h) \geq \varepsilon^2(B^*h, B^*h) = \varepsilon^2(BB^*h, h), \quad h \in \mathfrak{H}.$$

Therefore, (c) follows.

(c) $\Rightarrow$ (b) The inequality $BB^* \leq \varepsilon^{-2}AA^*$ implies $\operatorname{ran} B \subset \operatorname{ran} A$.

(b) $\Rightarrow$ (a) It suffices to show that $\overline{\operatorname{ran}}\, B^* \subset \operatorname{ran} C$. Let $k \in \overline{\operatorname{ran}}\, B^*$. The assumption $\operatorname{ran} B \subset \operatorname{ran} A$ implies that $Bk = Ah$ for some $h \in \mathfrak{F}$. Since $A = BC$, one sees that $Bk = BCh$, and hence $k - Ch \in \ker B$. On the other hand, $k \in \overline{\operatorname{ran}}\, B^*$ and $Ch \in \operatorname{ran} C \subset \overline{\operatorname{ran}}\, B^*$ yield $k - Ch \in \overline{\operatorname{ran}}\, B^* = (\ker B)^\perp$. Therefore, $k = Ch$ and it follows that $\overline{\operatorname{ran}}\, B^* \subset \operatorname{ran} C$ holds. $\square$

An operator $T \in \mathbf{B}(\mathfrak{F}, \mathfrak{G})$ is said to be a *partial isometry* if the restriction of T to $(\ker T)^\perp$ is an isometry. The *initial space* is $(\ker T)^\perp$ and the *final space* is $\operatorname{ran} T$, which is automatically closed since $(\ker T)^\perp$ is closed and the restriction of T is isometric. Let $T \in \mathbf{B}(\mathfrak{F}, \mathfrak{G})$, then the following statements are equivalent:

(i) T is a partial isometry with initial space $\mathfrak{M}$ and final space $\mathfrak{N}$;
(ii) T^*T is an orthogonal projection in $\mathfrak{F}$ onto $\mathfrak{M}$;
(iii) T^* is a partial isometry with initial space $\mathfrak{N}$ and final space $\mathfrak{M}$;
(vi) TT^* is an orthogonal projection in $\mathfrak{G}$ onto $\mathfrak{N}$.

The next result can be considered to be a special case of Proposition D.3 and Corollary D.4.

Corollary D.5. *Assume that $A \in \mathbf{B}(\mathfrak{F}, \mathfrak{H})$ and $B \in \mathbf{B}(\mathfrak{G}, \mathfrak{H})$. Then the following statements are equivalent:*

(i) *$A = BC$ for some $C \in \mathbf{B}(\mathfrak{F}, \mathfrak{G})$, where C is a partial isometry with initial space $\overline{\operatorname{ran}}\, A^*$ and final space $\overline{\operatorname{ran}}\, B^*$;*
(ii) *$AA^* = BB^*$.*

Proof. (i) $\Rightarrow$ (ii) Observe that $AA^* = BCC^*B^*$. By assumption, C is a partial isometry with initial space $\overline{\operatorname{ran}}\, A^*$ and final space $\overline{\operatorname{ran}}\, B^*$. Therefore, C^* is a partial isometry with initial space $\overline{\operatorname{ran}}\, B^*$ and final space $\overline{\operatorname{ran}}\, A^*$, and it follows that CC^* is the orthogonal projection onto the final space $\overline{\operatorname{ran}}\, B^*$ of C. Hence, $AA^* = BB^*$.

(ii) $\Rightarrow$ (i) It follows from Corollary D.4 that $A = BC$ for some $C \in \mathbf{B}(\mathfrak{F}, \mathfrak{G})$ with $\operatorname{ran} C = \overline{\operatorname{ran}}\, B^*$ and $\|C\| \leq 1$. Moreover

$$\|C^*B^*h\| = \|A^*h\| = \|B^*h\|, \qquad h \in \mathfrak{H},$$

shows that C^* is a partial isometry with initial space $\overline{\operatorname{ran}}\, B^*$ and final space $\overline{\operatorname{ran}}\, A^*$. Therefore, C is a partial isometry with initial space $\overline{\operatorname{ran}}\, A^*$ and final space $\overline{\operatorname{ran}}\, B^*$. □

The following *polar decomposition* of $H \in \mathbf{B}(\mathfrak{F}, \mathfrak{H})$ can be seen as a consequence of the previous corollaries. Recall that $|H^*| \in \mathbf{B}(\mathfrak{H})$ is a nonnegative operator in $\mathfrak{H}$ defined by $|H^*| = (HH^*)^{\frac{1}{2}}$.

Corollary D.6. *Let $H \in \mathbf{B}(\mathfrak{F}, \mathfrak{H})$. Then $H = |H^*|C$ and $|H^*| = HC^*$ for some partial isometry $C \in \mathbf{B}(\mathfrak{F}, \mathfrak{H})$ with initial space $\overline{\operatorname{ran}}\, H^*$ and final space $\overline{\operatorname{ran}}\, |H^*|$. In addition, one has $\operatorname{ran} H = \operatorname{ran} |H^*|$.*

The following corollary is a direct consequence of Lemma D.2.

Corollary D.7. *Let $H \in \mathbf{B}(\mathfrak{H})$ be self-adjoint and nonnegative and let $H^{\frac{1}{2}}$ be its square root. Then there are the inclusions*

$$\operatorname{ran} H \subset \operatorname{ran} H^{\frac{1}{2}} \subset \overline{\operatorname{ran}}\, H^{\frac{1}{2}} = \overline{\operatorname{ran}}\, H. \tag{D.6}$$

Moreover, the following statements are equivalent:

(i) $\operatorname{ran} H$ *is closed;*

(ii) $\operatorname{ran} H^{\frac{1}{2}}$ *is closed;*

(iii) $\operatorname{ran} H = \operatorname{ran} H^{\frac{1}{2}}$,

in which case all the inclusions in (D.6) are identities.

Recall that the Moore–Penrose inverse of $H^{\frac{1}{2}}$ is the uniquely defined inverse of the restriction

$$H^{\frac{1}{2}} : (\ker H^{\frac{1}{2}})^{\perp} = (\ker H)^{\perp} \to \operatorname{ran} H^{\frac{1}{2}}.$$

In the sequel the Moore–Penrose inverse of $H^{\frac{1}{2}}$ will be denoted by $H^{(-\frac{1}{2})}$, i.e.,

$$H^{(-\frac{1}{2})} := (H^{\frac{1}{2}})^{(-1)}.$$

It is clear that

$$H^{(-\frac{1}{2})} \in \mathbf{B}(\operatorname{ran} H^{\frac{1}{2}}, (\ker H)^{\perp}) \quad \Leftrightarrow \quad \operatorname{ran} H^{\frac{1}{2}} \text{ is closed.}$$

Here is the version of Proposition D.3 for nonnegative operators.

Proposition D.8. *Let $A, B \in \mathbf{B}(\mathfrak{H})$ be self-adjoint and nonnegative, and let $\rho > 0$. Then the following statements are equivalent:*

(i) $A^{\frac{1}{2}} = B^{\frac{1}{2}}C$ *for some* $C \in \mathbf{B}(\mathfrak{H})$ *with* $\operatorname{ran} C \subset \overline{\operatorname{ran}}\, B^{\frac{1}{2}}$ *and* $\|C\| \le \rho$;

(ii) $\operatorname{ran} A^{\frac{1}{2}} \subset \operatorname{ran} B^{\frac{1}{2}}$ *and* $\|B^{(-\frac{1}{2})}\varphi\| \le \rho\, \|A^{(-\frac{1}{2})}\varphi\|$, $\varphi \in \operatorname{ran} A^{\frac{1}{2}}$;

(iii) $A \le \rho^2 B$.

Morover, if $\operatorname{ran} A^{\frac{1}{2}} \subset \operatorname{ran} B^{\frac{1}{2}}$, then $\|B^{(-\frac{1}{2})}\varphi\| \le \rho\, \|A^{(-\frac{1}{2})}\varphi\|$, $\varphi \in \operatorname{ran} A^{\frac{1}{2}}$, for some $\rho > 0$.

Notes

These pages contain a number of comments about the various results in this monograph and some further developments. We also give some historical remarks without any claim of completeness; the history of this area is complicated, also because of the limited exchange of ideas that has existed between East and West. The main setting of the book is that of operators and relations in Hilbert spaces. For an introduction and comprehensive treatment of linear operators in Hilbert spaces and related topics we refer the reader to the standard textbooks [2, 133, 141, 272, 276, 341, 348, 598, 649, 650, 651, 652, 673, 691, 723, 738, 743, 752, 756, 757]. We shall occasionally address topics that fall outside the scope of the monograph itself, e.g., the role of operators and relations in spaces with an indefinite inner product is indicated with proper references.

Notes on Chapter 1

Linear relations in Hilbert spaces go back at least to Arens [42]; they attracted attention because of their usefulness in the extension theory of not necessarily densely defined operators; see for instance [82, 128, 202, 218, 266, 297, 523, 619, 638]. Linear relations also appear in a natural way in the description of boundary conditions; cf. [2, 660]. Related material can be found in [74, 298, 406, 512, 513, 514, 515, 516, 576, 683, 685, 686]. The description of self-adjoint, maximal dissipative, or accumulative extensions of a symmetric operator (see Sections 1.4, 1.5, and 1.6) in terms of operators between the defect spaces in Theorem 1.7.12 and Theorem 1.7.14 goes back to von Neumann and Štraus [610, 731] in the densely defined case and was worked out in the general case by Coddington [202, 203]; see also [266, 490, 733]. These descriptions may be seen as building stones for the extension theory via boundary triplets in Chapter 2.

Sections 1.1, 1.2, and 1.3 present elementary facts. The identity (1.1.10) goes back to Štraus [726]. The notions of spectrum and points of regular type for a linear relation in Definition 1.2.1 and Definition 1.2.3 are introduced as in the operator case [268, 269, 404], while the adjoint of a linear relation in Definition 1.3.1 is introduced as in [611], see also [42, 659] and [396, 599]. Note that in Proposition 1.2.9 one sees a "pseudo-resolvent" which already shows that there is a relation in the background; cf. [421]. In Theorem 1.3.14 we present the notion of

J. Behrndt et al., *Boundary Value Problems, Weyl Functions, and Differential Operators*, Monographs in Mathematics 108,
https://doi.org/10.1007/978-3-030-36714-5

the operator part of a relation and the corresponding Hilbert space decomposition in the spirit of the Lebesgue decomposition of a relation [397, 398]; see also [439, 621, 622, 623] for the operator case. Operator parts have a connection with generalized inverses (see for instance [127, 609]), such as the Moore–Penrose inverse in Definition 1.3.17. Note that the more general definition $P_{\ker \overline{H}} H^{-1}$ gives a Moore–Penrose inverse that is a closable operator. The special classes of relations and their transforms are developed in the usual way; see [660, 705, 706]. The presentation is influenced by personal communication with McKelvey (see also [572]) and lecture notes by Kaltenbäck [451] and Woracek [775, 776]; see also [365, 556, 662] for more references. Although we assume a working knowledge of the spectral theory of self-adjoint operators in Hilbert spaces, we highlight in Section 1.5 a couple of useful facts, with special attention paid to the semibounded case. Lemma 1.5.7 goes back to [210], Proposition 1.5.11 is a consequence of the Douglas lemma in Appendix D, and Lemma 1.5.12 is taken from [95]. The extension theory in the sense of von Neumann can be found in Section 1.7. Note that extensions of a symmetric relation may be disjoint or transversal. Our present characterization of these notions in Theorem 1.7.3 seems to be new; see also [391].

For the development of boundary triplets we only need a few basic facts concerning spaces with an indefinite inner product, see Section 1.8. A typical result in this respect is the automatic boundedness property in Lemma 1.8.1; see [73, 77, 235, 236]. The transform in Definition 1.8.4 goes back to Shmuljan [506, 705, 706]. The notions of strong graph convergence and strong resolvent convergence of relations are treated in Section 1.9 (see [650] for the case of self-adjoint operators). Here we prove the equivalence of the two notions when there is a uniform bound. For Corollary 1.9.5 see [755]. The parametric representation of linear relations in Section 1.10 is closely connected with the theory of operator ranges. Here we only consider such representations from the point of view of boundary value problems; see also [2, 660] and Chapter 2. However, these parametric representations also play an important role in the above mentioned Lebesgue decompositions of relations and, more generally, in the Lebesgue type decompositions of relations; cf. [397]. Most of the material in the beginning of Section 1.11 consists of straightforward generalizations of properties of the resolvent. Lemma 1.11.4 is well known for the usual resolvent and seems to be new in the generalized context; it is used in Section 2.8. Lemma 1.11.5 was inspired by [523]. Nevanlinna families and pairs are generalizations of operator-valued Nevanlinna functions; cf. Appendix A. The notion of Nevanlinna pair goes back to [617]. The notion of Nevanlinna family appears already in [523] (and in a different guise in [266]) and was later more systematically studied in [20, 21, 22, 89, 94, 95, 234, 246].

Although in this monograph we will restrict ourselves mainly to Hilbert spaces, we will make some comments in the notes concerning the setting of indefinite inner product spaces; see for instance [31, 77, 79, 142, 430, 452] and also [775, 776] for Pontryagin spaces, almost Pontryagin spaces, Kreĭn spaces, and almost Kreĭn spaces. The spectral theory of operators and relations in such spaces has been studied extensively; in Kreĭn spaces the operators are often required to have (locally) finitely many negative squares or to be (locally) definitizable, that is, roughly speaking some polynomial (or rational function) of the operator or relation under consideration is (locally) nonnegative in the Kreĭn space sense. Here we only mention work related to the present topics [75, 78, 79, 99, 220, 222, 253, 260, 264, 268, 269, 274, 275, 434, 435, 436, 437, 497, 498, 499, 500, 501, 502, 503, 519, 522, 631, 632, 633, 634, 674, 717, 718, 719, 720, 721, 722, 764, 765].

Notes on Chapter 2

Boundary triplets were originally introduced in the works of Bruk [176, 177, 178] and Kochubeĭ [466] and were used in the study of operator differential equations in the monograph [346]; see also [509]. The notion of boundary triplet appeared implicitly in a different form already much earlier in the work of Calkin [186, 187] (see also [276, Chapter XII] and [410]). The Weyl function was introduced by Derkach and Malamud in [243, 244]; it can also be interpreted as the Q-function from the works of Kreĭn [491, 492] (see also [679]) and, e.g., the papers [499, 500, 501, 502, 503, 504] of Kreĭn and Langer, and [523] of Langer and Textorius. We also mention the more recent monographs [359, 691], where boundary triplets and Weyl functions are briefly discussed, as well as the very recent monograph by Derkach and Malamud [247] (in Russian), in which a more detailed exposition and more than 400 references (also many older papers published in Russian) can be found. From our point of view the most influential general papers on boundary triplets and Weyl functions area are [245, 246] and [184], the latter also has an introduction to the basic notions.

While a large part of the material in this chapter can be found in the literature, the presentation here is given in a unified form for general symmetric relations. The connection between the abstract Green identity and the appropriate indefinite inner products (see Section 1.8) is used in Section 2.1 when deriving some of the basic properties of boundary triplets and their transforms. The inverse result for boundary triplets in Theorem 2.1.9 is taken from [103, 104]. The discussion in Section 2.2 on parametric representations of boundary conditions is given to establish a connection between abstract and concrete boundary value problems. The definitions and the central properties of γ-fields and Weyl functions are presented in Section 2.3 and these results can be found in [245, 246]. In particular, Proposition 2.3.2 (part (ii)) and Proposition 2.3.6 (parts (iii), (v)) show the connection with the notion of Q-function (see [184, 245, 246]), while the formula in part (vi) of Proposition 2.3.6 is taken from [265]. Without specification of the γ-field and Weyl function, constructions of boundary triplets appear already in the early literature; see [176, 466, 577] and the monographs [346, 509]. Here the constructions are based on decompositions that are closely related to the von Neumann formulas in Section 1.7. The Weyl function in Theorem 2.4.1 coincides with the abstract Donoghue type M-function that was studied, for instance, in [317, 320, 327]. Transformations of boundary triplets and the corresponding

γ-fields and Weyl functions have been treated in [234, 237, 245, 246]. In particular, the description of boundary triplets in Theorem 2.5.1 is taken from [246]. We make the precise connection with Q-functions in Corollary 2.5.8. The present notion of unitary equivalence of boundary triplets in Definition 2.5.14 is slightly more general than the definition used in [382].

Section 2.6 is devoted to Kreĭn's formula for intermediate extensions. The Kreĭn type formula (for intermediate extensions) phrased in terms of boundary triplets in Theorem 2.6.1 is adapted to the present setting from [233]. This yields new simple proofs for the description of various parts of the spectra of intermediate extensions in Theorem 2.6.2. This description appears in a similar form in [243, 244, 245, 246] and is treated for dissipative extensions in [346, 468] by means of characteristic functions. The same remark applies to the proof of Theorem 2.6.5, which can be found in a different form in [184]. These results indicate the importance of Kreĭn's formula for the spectral analysis of self-adjoint extensions of symmetric operators. Closely related is the completion problem, briefly touched upon in Remark 2.4.4, which is investigated in [383], where also an analogous boundary triplet appears for the nonnegative case. This case of extensions of a bounded symmetric operator was studied earlier in, for instance, [735, 736].

Section 2.7 is devoted to the Kreĭn–Naĭmark formula, which is our terminology for the Kreĭn formula for compressed resolvents of self-adjoint exit space extensions, in short, Kreĭn's formula for exit space extensions. Actually, it was Naĭmark [605, 606, 607] who investigated different types of extensions (with exit) and a proper interpretation then yields the corresponding resolvent formula. For the notion of Štraus family and for Kreĭn's formula for exit space extensions we refer to [491, 492, 679] and to [726, 727, 728, 729, 730, 731, 732, 733]; some other related papers are [51, 121, 126, 219, 234, 237, 254, 266, 320, 497, 523, 554, 595, 596, 624]. The Štraus families are restrictions of the adjoint in terms of "λ-dependent boundary conditions" given by a Nevanlinna family or corresponding Nevanlinna pair. When the exit space is a Pontryagin space, the same mechanism is in force and the boundary conditions are now given by a generalized Nevanlinna family or corresponding Nevanlinna pair (generalized in the sense that the family or pair has negative squares); we mention [76, 100, 111, 253, 257, 258, 259, 260, 261, 263, 290, 521, 523] and [664] for the special case of a finite-dimensional exit space. Boundary triplets for symmetric relations in Pontryagin and even in Kreĭn spaces were introduced by Derkach (see [101, 227, 228, 229, 230]) and for isometric operators in Pontryagin spaces see [80]; cf. Chapter 6 for concrete λ-dependent boundary conditions.

The perturbation problems in Section 2.8 and Kreĭn's formula are closely related. The $\mathfrak{S}_p$-perturbation result in Theorem 2.8.3 appears in a similar form in [245]. Standard (additive) perturbations of an unbounded self-adjoint operator yield an analogous situation, where the symmetric operator is maximally nondensely defined [405].

Rigged Hilbert spaces offer a framework where extension theory of unbounded symmetric operators can be developed in a somewhat analogous manner as in the bounded case; see [371, 400, 401, 746]. A closely related approach to extensions of symmetric relations relies on the concept of graph perturbations studied in [203, 204, 205, 206, 238, 264, 267, 270, 399, 403]. There has been a considerable interest in the related concept of singular perturbations, where perturbation elements belong to rigged Hilbert spaces with negative indices; see [10], which contains an extensive list of earlier literature, and see also the notes to Chapter 6 and Chapter 8 in the context of differential operators with singular potentials. General operator models for highly singular perturbations involve lifting of operators in Pontryagin spaces [117, 239, 240, 241, 252, 256, 640, 641, 707].

The interest in boundary triplets and Weyl functions has substantially grown in the last decade, so a complete list of references is beyond the scope of these notes. However, for a selection of papers in which boundary triplet techniques were applied to differential operators and other related problems we refer to [28, 50, 86, 87, 97, 112, 113, 123, 143, 144, 155, 169, 170, 173, 174, 184, 185, 241, 288, 307, 342, 343, 344, 377, 461, 475, 477, 478, 483, 511, 529, 551, 562, 563, 564, 565, 589, 590, 591, 626, 627, 642, 644, 677, 750, 751]. For boundary triplets and similar techniques in the analysis of quantum graphs see [110, 134, 172, 183, 196, 287, 289, 294, 536, 625, 637, 645, 646].

Boundary triplets and their Weyl functions for symmetric operators have been further generalized in [103, 105, 109, 110, 115, 233, 236, 237, 246] by relaxing some of the conditions in the definition of a boundary triplet. Moreover, in the setting of dual pairs of operators (see [525]) boundary triplets have been introduced in [559, 560] and applied, e.g., in [170, 173, 300, 302, 381]; they were specialized to the case of isometric operators in [561]. Boundary triplets and their extensions also occur naturally in a system-theoretic environment, where the underlying operator is often isometric, contractive, or skew-symmetric; cf. [65, 66, 67, 102, 394, 750, 751].

By applying a transform of the boundary triplet, resulting in a Cayley transform of the Weyl function, one gets a connection with the notion of characteristic function, which was used to elaborate an alternative analytic tool for studying extension theory for symmetric operators, for some literature in this direction, see, e.g., [165, 166, 246, 346, 468, 509, 603, 728].

Notes on Chapter 3

Most of the material in Section 3.1 and Section 3.2 is working knowledge from measure theory. Typical classical textbooks providing a more detailed exposition on Borel measures are [272, 335, 416, 676, 682] and the papers [61, 60, 271] are listed here for symmetric derivatives and the limit behavior of the Borel transform (see Appendix A). We also recommend the more recent monograph [738]. Section 3.3 is a brief exposition of the notions of absolutely continuous and singular spectrum of self-adjoint operators (see [649, 691, 738, 757]) in the context of self-adjoint relations; cf. Section 1.5 and Theorem 1.5.1.

Simplicity of symmetric operators and the decomposition of a symmetric operator into a self-adjoint and a simple symmetric part in Theorem 3.4.4 go back to [496]; cf. [523] for a version of this result for symmetric relations. The local version of simplicity in Definition 3.4.9 appears in papers of Jonas (see, e.g., [437]) and was more recently considered in [119].

The characterization of the spectrum via the limit properties of the Weyl function in Section 3.5 and Section 3.6 is well known for the special case of singular Sturm–Liouville differential operators and the Titchmarsh–Weyl m-function; cf. Chapter 6 and the classics [209, 740, 741] and, e.g., [60, 97, 175, 224, 321, 328, 335, 333, 334, 426, 479, 711, 712] and also [195]. This description was extended in [155] to the abstract setting of boundary triplets and Weyl functions in the case where the underlying symmetric operator is densely defined and simple (on $\mathbb{R}$). These ideas where further generalized in [119, 120] and applied to elliptic partial differential operators. The present treatment in the context of linear relations seems to be new; cf. [265] for Theorem 3.6.1 (i).

Our presentation of the material in Section 3.7 is strongly inspired by the contribution [523] of Langer and Textorius. Due to the limit properties in this section there are important connections between analytic properties of the Weyl functions and the geometric properties of the associated self-adjoint extensions. The characterizations stated in Proposition 3.7.1 and Proposition 3.7.4 have been established initially in [371, 374, 379], where they were used to introduce the notions of generalized Friedrichs and Kreĭn–von Neumann extensions for non-semibounded symmetric operators or relations; see the notes on Chapter 5.

Finally, the translation of the results in Section 3.5 and Section 3.6 to arbitrary self-adjoint extensions in Section 3.8 based on the corresponding transformation of the Weyl function (see Section 2.5) is immediate.

Notes on Chapter 4

One of the central themes in this chapter is the construction of a Hilbert space via a nonnegative reproducing kernel; such a construction has been studied intensively after Aronszjan [59]. For some introductory texts on this topic we refer the reader to [32, 272, 277, 278, 680, 737]. It should be mentioned that there are different approaches avoiding the mechanism of reproducing kernel Hilbert spaces; see for instance Brodskiĭ [165] and for another method Sz.-Nagy and Korányi [604] (sometimes called the ε-method).

Section 4.1 gives a short review of all the facts about reproducing kernel Hilbert spaces that are needed in the monograph. We remind the reader that special reproducing kernels were considered by de Branges and Rovnyak for the setting of Nevanlinna functions or Schur functions [146, 147, 148, 149, 150, 151, 152, 153]; see also [24, 25, 33].

In Section 4.2 the case of operator-valued Nevanlinna functions is taken up first. Theorem 4.2.1 states that the Nevanlinna kernel of a Nevanlinna function M is nonnegative; the proof can be found in [450]. In this case the reproducing kernel approach leads to the realization of the function $-(M(\lambda)+\lambda)^{-1}$ in terms of compressed resolvents in Theorem 4.2.2; see [231, Theorem 2.5 and Remark 2.6] and also [409]. The unitary equivalence of different models in Theorem 4.2.3 is based on a standard argument. If the Nevanlinna function is uniformly strict, then it can be regarded as a Weyl function; see Theorem 4.2.4 as follows by an inspection of the proof of Theorem 4.2.2. The uniqueness of the model in Theorem 4.2.6 is obtained by a further specialization of the proof of Theorem 4.2.3.

Section 4.3 presents a reproducing kernel approach for scalar Nevanlinna functions via their integral representation (see Appendix A) as can be found in [246]; cf. [19, 379, 530]. In the present monograph only the case of scalar Nevanlinna functions is treated in this way; the matrix-valued case is a straightforward generalization. For a rigorous treatment involving operator-valued Nevanlinna functions one should apply [558].

Parallel to the above treatment of operator-valued Nevanlinna functions the case of Nevanlinna families or Nevanlinna pairs is taken up in Section 4.4. The nonnegativity of the Nevanlinna kernel for a Nevanlinna family or Nevanlinna pair in Theorem 4.4.1 is proved via a simple reduction argument. The representation model for Nevanlinna pairs in Theorem 4.4.2 is again given along the lines of Derkach's work [231]; see also [265]. In fact, the proof can be carried forward to show that any Nevanlinna family is the Weyl family of a boundary relation (see the notes for Chapter 2). Another reproducing kernel approach was followed in [95, Theorem 6.1] by a reduction to the corresponding case of Schur functions; cf. [152, 153]. Closely connected is the notion of generalized resolvents given in Definition 4.4.5. It was formalized by McKelvey [572], see also [207]. Its representation in Corollary 4.4.7 is equivalent to the representation of Nevanlinna pairs in Theorem 4.4.2. In Corollary 4.4.9 we present Naĭmark's dilation result [606, 607] as a straightforward consequence of the characterization of generalized resolvents in Theorem 4.4.8; cf. [659].

Our construction of the exit space extension in Theorem 4.5.2 uses the above representation of generalized resolvents; cf. [497]. The identity in (4.5.3) gives the precise connection between the exit space and the Nevanlinna pair leading to the exit space. Closely related is an interpretation via the coupling method in Section 4.6. By taking an "orthogonal sum of two boundary triplets" as in Proposition 4.6.1 one obtains an exit space extension for one of the original symmetric relations whose compressed resolvent is described in Corollary 4.6.3. This leads to an interpretation of the Kreĭn formula when the parameter family coincides with a uniformly strict Nevanlinna function; see [373, 375, 376, 630]. However, a similar interpretation exists when the parameter family is a general Nevanlinna family, once it is interpreted as the Weyl family of a boundary relation; see [236, 237, 238]. It has already been explained in the notes on Chapter 2 that it is quite natural to have exit spaces which may be indefinite, in which case the Nevanlinna family must be replaced by a more general object.

And, indeed, more general reproducing kernel spaces can be constructed. For the Pontryagin space situation and reproducing kernels with finitely many negative squares one may consult the monograph [23] and the papers [89, 232, 409, 497, 499, 500]; see also [454, 455, 456, 457, 458, 459]. For the setting of almost Pontryagin spaces, see [452, 775, 777]. Finally, for Kreĭn spaces we refer to [15, 16, 17, 262, 434, 435, 437, 778].

Notes on Chapter 5

The approach to semibounded forms in Section 5.1 follows the lines of Kato's work [462], where densely defined semibounded and sectorial forms are treated; see also [2, 49, 246, 346, 359, 552, 649, 650, 690, 691]. An arbitrary form is not necessarily closable, but there is a Lebesgue decomposition into the sum of a closable form and a singular form; see [395, 397, 404, 476, 710]. Versions of the representation theorems for nondensely defined closed forms appear in [14, 44, 393, 661, 710]; see Theorem 5.1.18 and Theorem 5.1.23. Nondensely defined forms appear, for instance, when treating sums of densely defined forms, leading to a representation problem for the form sum; see [295, 296, 389, 390, 392]. The ordering of semibounded forms and semibounded self-adjoint relations in Section 5.2 results in Proposition 5.2.7; the present simple treatment via the Douglas lemma extends [390]; see Proposition 1.5.11. The characterization in terms of the corresponding resolvent operators goes back to Rellich [654]; see also [210, 390, 412, 462]. For the monotonicity principle (Theorem 5.2.11) see [96], and for similar statements [83, 245, 246, 618, 661, 709, 710, 753].

The main properties of the Friedrichs extension, namely Theorem 5.3.3 and Proposition 5.3.6, follow from the present approach via forms; cf. [202, 370, 734]. The square root decomposition approach involving resolvents to describe those semibounded self-adjoint extensions which are transversal with the Friedrichs extension extends [391] (for the nonnegative case). The transversality criterion in Theorem 5.3.8 is an extension of Malamud's earlier result [553]. The introduction of Kreĭn type extensions in Definition 5.4.2 is inspired by [210]. The minimality of the Kreĭn type extension is inherited from the maximality of the Friedrichs extension and this yields a characterization of all semibounded extensions whose lower bounds belong to a certain prescribed interval in Theorem 5.4.6. This is the semibounded analog of a characterization of nonnegative self-adjoint extensions in the nonnegative case [210]. The interpretation of the Friedrichs and Kreĭn type extensions as strong resolvent limits in Theorem 5.4.10 goes back in the operator case to [34] and in the general case to [383]. The present treatment is based on the general monotonicity principle in Theorem 5.2.11. The concept of a positively closable operator (see (iii) of Corollary 5.4.8) was introduced in [34] to detect the nonnegative self-adjoint operator extensions. This notion is closely connected with [490], where self-adjoint operator extensions were determined. Closely related to the Friedrichs and the Kreĭn–von Neumann extensions is the study of all extremal extensions of a nonnegative or a sectorial operator in [43, 44, 45, 46, 47, 53, 55, 383, 389, 391, 392, 393, 510, 648, 702, 703].

If a symmetric relation is semibounded, then a boundary triplet can be chosen so that A_0 is the Friedrichs extension and A_1 is a semibounded self-adjoint extension (for instance, a Kreĭn type extension, when transversality is satisfied). This type of boundary triplet serves as an extension of the notions of positive boundary triplets introduced in [43, 467]. The main part of the results in Section 5.5 when specialized to the nonnegative case, reduce to the results in [245, 246]. A result similar to Proposition 5.5.8 (ii) can be found in [349, 355], see also the more recent contribution [362] that deals with elliptic partial differential operators on exterior domains. We remark that the sufficient condition in Lemma 5.5.7 is also necessary for the extension A_Θ to be semibounded for every semibounded self-adjoint relation Θ in $\mathcal{G}$; see [245, 246]. For an example where Θ is a nonzero bounded operator with arbitrary small operator norm $\|\Theta\|$, while A_Θ is not semibounded from below, see [378]. In the context of elliptic partial differential operators on bounded domains the boundary triplet in Example 5.5.13 appears implicitly already in Grubb [352]; see also [169, 359, 557] and the notes to Chapter 8 for more details.

The first Green identity appearing in Theorem 5.5.14 establishes a link with the notion of boundary pairs for semibounded operators and the corresponding semibounded forms in Section 5.6. Definition 5.6.1 of a boundary pair for a semibounded relation S is adapted from Arlinskiĭ [44], see also [50] for the context of generalized boundary triplets. This notion makes it possible to describe the closed forms associated with all semibounded self-adjoint extensions of a given semibounded relation S by means of closed semibounded forms in the parameter space. In particular, Theorem 5.6.11 contains the description of all nonnegative closed forms generated by the nonnegative self-adjoint extensions in [53]; for a similar result in the case of maximal sectorial extensions see [44]. By connecting boundary pairs with compatible boundary triplets (see Theorem 5.6.6 and Theorem 5.6.10) one obtains an explicit description of the forms associated with all semibounded self-adjoint extensions by means of the boundary conditions, see Theorem 5.6.13.

We remind the reader of the study of nonnegative self-adjoint extensions of a densely defined nonnegative operator as initiated by von Neumann [610]; see [280, 309, 310, 724, 773] and the papers by Kreĭn [491, 492, 493, 494]. Kreĭn's analysis was complemented by Vishik [747] and Birman [139]; see also [2, 14, 58, 157, 217, 295, 346, 347, 352, 354, 372, 464, 714]. For an operator matrix completion approach see [383, 472], see also [81] for an extension in a Pontryagin space setting. This approach has its origin in the famous paper of Shmuljan [704]. The notion of boundary pairs for forms can also be traced back to Kreĭn [493, 494], Vishik [747], and Birman [139]. Some further developments can be found, e.g., in [36, 43, 44, 47, 48, 54, 56, 57, 248, 301, 357, 359, 363, 553, 555, 556], where also accretive and sectorial extensions, and associated closed forms are treated, and see also [552, 580, 581, 582, 636, 725]. A related concept of boundary pairs designed for elliptic partial differential operators and corresponding quadratic forms was recently proposed by Post [647].

A study of symmetric forms which are not semibounded and concepts of generalized Friedrichs and Kreĭn–von Neumann extensions goes beyond the scope of the present text. However, an interested reader may consult in this direction, e.g., [125, 215, 216, 303, 306, 364, 371, 374, 379, 570, 571].

Notes on Chapter 6

The main textbooks in this area are [2, 26, 72, 208, 281, 284, 414, 420, 438, 489, 542, 574, 608, 663, 724, 740, 741, 754, 757, 782]. The Sturm–Liouville differential expressions that we consider here have coefficients which are assumed to satisfy some weak integrability conditions giving rise to quasiderivatives (where under stronger smoothness conditions ordinary derivatives would suffice); see for instance [724] where already such quasiderivatives were used. Most of the material concerning singular Sturm–Liouville equations has been stimulated by the work of Weyl [758, 759, 760]; see also [761]. For later work on operators of higher order see, for instance, [200, 201, 465, 469, 470].

Section 6.1 and Section 6.2 contain standard material, where we follow parts of [414, 757]; for the quasiderivatives in the limit-circle case we refer to [312]. The treatment of the regular and limit-circle cases in Section 6.3 is straightforward; see [312] and also [12] for a boundary triplet in the limit-circle case. For an alternative description of boundary conditions and Weyl functions in the regular case we mention the recent papers [197, 199, 332] using the concept of boundary data maps. The limit-point case is treated in Section 6.4. In the proof of Proposition 6.4.4 concerning the simplicity of the minimal operator we follow the arguments of R.C. Gilbert [336]; cf. [263]. The treatment of the Fourier transform in Lemma 6.4.6 and Theorem 6.4.7 uses the theory in Appendix B. The surjectivity is direct in this case. The statement in Lemma 6.4.8 that the Fourier transform provides a model for the Weyl function uses an argument provided in [281].

In general, interface conditions are a tool to paste together several boundary value problems. The interface conditions which we consider in Section 6.5 make it possible in the case of two endpoints in the limit-point case to consider minimal operators [339, 740] and to apply the coupling of boundary triplets as explained in Section 4.6. This automatically leads to some kind of exit space extensions. As this goes beyond the present text, we only refer to [336, 338, 441, 713].

The present theory of subordinate solutions in Section 6.7 goes back to D.J. Gilbert and D.B. Pearson [335]; see also [333, 334, 463]. Further contributions can be found in [433, 533, 655, 656, 657]; see also [738]. Our presentation is modelled on [388].

If the minimal operator is semibounded one can apply the form methods from Chapter 5. In the regular case the treatment of the forms associated with the Sturm–Liouville expression in Section 6.8 is based on the inequality in Lemma 6.8.2 and Theorem 5.1.16; see [212, 493, 494, 757]. This makes it possible to introduce boundary pairs which are compatible with the given boundary triplet. Section 6.9 and Section 6.10 contain preparatory material so that boundary pairs can be introduced also when the endpoints are singular. Section 6.9 on Dirichlet forms is inspired by Rellich [654] and Kalf [448]. The proof of Lemma 6.9.7 seems to be new. Section 6.10 is concerned with principal and nonprincipal solutions; cf. [198, 368, 369, 535] and, in particular, [616]. The Hardy type inequalities in Lemma 6.10.1 go back to [449]. The treatment is in some sense folklore, but Theorem 6.10.9 seems to be new. Our handling of the regular case in Section 6.8 serves as model for the semibounded singular cases in Section 6.11 and Section 6.12. These two sections are influenced by Kalf [448]; see also [671, 672] and a recent contribution [168]. The determination of the Friedrichs extension in our treatment goes hand in hand with the boundary pair; see also [310, 311, 508, 597, 600, 601, 615, 616, 654, 661, 671, 672, 779, 782].

The case of an integrable potential on a half-line is treated in Section 6.13. It appears already in [740], where the corresponding spectral measure is determined. The construction of solutions with a given asymptotic behavior can be found for instance in [542, 740]. The example with the Pöschl–Teller potential at the very end of Section 6.13 can be found in, e.g., [2].

There is a large amount of literature devoted to special topics. For λ-dependent boundary conditions we just refer to [117, 313, 314, 675, 687, 688, 689, 749] and [261, 263] for Pontryagin exit spaces. The determination of the Kreĭn–von Neumann extension and other nonnegative extensions can be found in [212, 350]. For singular perturbations associated with Sturm–Liouville operators, see for instance [331] and the later papers [9, 28, 29, 124, 171, 182, 282, 315, 316, 479, 480, 481, 482, 507, 531, 532, 549, 550]; for δ-point interactions we refer to [8, 293]. Special properties of the Titchmarsh–Weyl coefficient have been studied in many papers; we just mention [130, 292, 366, 367]. Already early in the 20th century there was an interest in boundary value problems with conditions at interior points and, more general, integral boundary conditions; for instance, see [135, 583] and for a later review [762]. It was pointed out by Coddington [203, 204] that such conditions can be described in tems of relation extensions of a restriction of the usual minimal operator; see for a brief selection also [188, 189, 190, 203, 204, 205, 206, 238, 264, 267, 270, 399, 403, 484, 485, 486, 697, 784].

The topic of semibounded self-adjoint extensions in Section 5.5 and Section 5.6 is closely related to problems that are called left-definite in the literature; we only mention [13, 131, 473, 474, 488, 545, 698, 699, 708, 748] and the references therein. Another case of interest is when the weight function in the Sturm–Liouville equation changes sign, in which case Kreĭn spaces come up naturally: the following is just a limited selection [116, 122, 123, 138, 223, 225, 304, 305]. Under certain circumstances the Weyl function in the limit-point case (of a usual Sturm–Liouville operator) belongs to the Kac class; see for instance [420]. For further results in this direction, see [384, 385, 386, 387]; in these cases there is a distinguished self-adjoint extension, namely the generalized Friedrichs extension; see the notes on Chapter 5. Sturm–Liouville equations with vector-valued coefficients are beyond the scope of this text; see [345, 346] and for a recent contribution [330].

Notes on Chapter 7

Canonical systems of differential equations are discussed in the monographs [72, 340, 653, 681]. The spectral theory for such systems has been developed in varying degrees of generality in an abundance of papers [62, 63, 64, 85, 161, 162, 163,

213, 263, 265, 279, 422, 423, 424, 425, 427, 428, 487, 575, 592, 612, 613, 614, 620, 667, 668, 669, 687, 688, 689, 692, 693, 694, 695, 696]. In [520] there is a general procedure to reduce systems to a canonical form. Many investigations concerning boundary problems can be written in terms of canonical systems; see for a general procedure [619, 693], so that for instance [129, 211] can be subsumed. As a particular example we mention the system of second-order differential equations studied in the dissertation of a student of Titchmarsh [191, 192, 193, 194]. Frequently conditions are imposed on the canonical system so that there is a full analogy with the usual operator treatment. However, in general one is confronted with relations.

The first treatment of canonical systems in terms of relations is due to Orcutt [619]; see also [97, 408, 471, 524, 538] and [11, 593, 594]. In our opinion the present case of 2×2 systems already serves as a good illustration of the various phenomena that may occur. The interest in 2×2 systems is justified by the work of de Branges [146, 147, 148, 149]; see also [278, 504, 528, 715, 716, 766, 767, 768, 769, 770, 771, 772]. Via these systems one may also approach Sturm–Liouville equations with distributional coefficients as, for instance, in [281, 283]. In Remling [658] and the forthcoming book [666] by Romanov our systems are treated with the de Branges results in mind. For a number of topics we rely on the lecture notes by Kaltenbäck and Woracek [460].

Sections 7.1, 7.2, and 7.3 contain preparatory material. The inequalites (7.1.1) and (7.1.3) are standard for Bochner integrals. The construction of the Hilbert space $\mathcal{L}^2_\Delta(\imath)$ follows the treatment in [460]; see also [276, 670]. For further information concerning these and more general spaces, see [440] and the expositions in [2, 276]. The treatment of the square-integrable solutions in Section 7.4 is based on the monotonicity principle in Theorem 5.2.11; cf. [97]. For a different treatment, see for instance [612]. Section 7.5 is devoted to definite systems. The present notion of definiteness can be found in [340, 619]; in the literature sometimes a more restrictive form of definiteness is used. Proposition 7.5.4 can be found in [614, Hilfsatz (3.1)] and [471]; for a more abstract treatment, see [98]. The modification of solutions is avoided in [619]; the present argument in Proposition 7.5.6 seems to be new. The minimal and maximal (multivalued) relations associated with canonical systems can be found in Section 7.6. They were originally introduced by Orcutt [619]; see also Kac [442, 443, 444, 445] and [408]. The extension theory for them naturally involves (multivalued) relations. In [97] it is indicated how all such systems fit in the boundary triplet scheme; see also [471, 538].

In Section 7.7 it is assumed that the endpoints are (quasi)regular, in which case a boundary triplet is constructed in Theorem 7.7.2. The resolvent of the self-adjoint extension $\ker \Gamma_0$ is an integral operator whose kernel belongs to the Hilbert–Schmidt class. In this way we can show that the operator part of the minimal relation is simple. In Section 7.8 it is assumed that one of the endpoints is in the limit-point case. We prove the simplicity of the operator part of the minimal relation along the lines of [337], cf. [263] (see also Chapter 6 for the Sturm–Liouville case). The treatment of the Fourier transform in the limit-point case uses

Appendix B and parallels the treatment in Chapter 6. Subordinate solutions for canonical systems are introduced in Section 7.9; cf. [388]. Here again we follow the results for the Sturm–Liouville case in Section 6.7; see also the corresponding notes in Chapter 5, where the appropriate references can be found.

The discussion of the special cases in Section 7.10 is just an indication of the possibilities; we could also pay attention to, for instance, the so-called Dirac systems. The connection with the Sturm–Liouville case and its generalization in [281] is only briefly indicated. The special case in Theorem 7.10.1 is modelled on [460]. For the connection of canonical systems with strings see, for instance, [453]. Finally, we would like to mention that the approach via boundary triplets allows also λ-dependent boundary conditions; for some related papers we refer to [221, 263, 265, 675, 687, 688, 689, 742].

Notes on Chapter 8

The notion of Gelfand triples or riggings of Hilbert spaces in Section 8.1 is often used in the treatment of partial differential equations and Sobolev spaces, and can be found in a similar form in Berezanskiĭ's monograph [133] (see also the textbook [774]); cf. [132] and the contributions [534, 537] by Lax and Leray. There are many well-known textbooks on Sobolev spaces, among which we mention here only the monographs [3, 5, 164, 136, 284, 291, 351, 569, 584, 744, 783]. In our opinion a very useful source for trace maps, the Green identities, and similar related results (also for nonsmooth domains) is the monograph [573] by McLean, see also the list of references therein. For the description of the spaces $H^s(\partial\Omega)$ in Corollary 8.2.2 as domains of powers of the Laplace–Beltrami operator on $\partial\Omega$ see [322, 544, 567]. In some cases it is also convenient to use powers of a Dirichlet-to-Neumann map, as in [114].

The discussion of the minimal and maximal operators in Section 8.3 is standard, e.g., Proposition 8.3.1 can be found in Triebel's textbook [745]. The H^2-regularity in Theorem 8.3.4 for smooth domains can be found in [167] and [84], see also [4, 308, 544, 517] or [1, Theorem 7.2] for a recent very general result on the H^2-regularity of the Neumann operator (and more general) on certain classes of nonsmooth bounded and unbounded domains. As explained in Section 8.7, for the class of Lipschitz domains the H^2-regularity of the Dirichlet and Neumann Laplacian up to the boundary fails and has to be replaced by the weaker $H^{3/2}$-regularity; cf. [431, 432] and [92, 115, 323, 324, 325, 326] for some recent closely related works dealing with Schrödinger operators on Lipschitz domains. In this context we also mention the papers [585, 586, 587, 588] by Mitrea and Taylor for general layer potential methods in Lipschitz domains on Riemannian manifolds. It is worth mentioning that the resolvent of the Neumann Laplacian in Proposition 8.3.3 is not necessarily compact if the boundary of the bounded domain Ω is not of class C^2 (or not Lipschitz), see [415] for a well-known counterexample using a rooms-and-passages domain. For similar *unusual* spectral properties we also mention Example 8.4.9 on the essential spectrum of self-adjoint realizations of the

Laplacian on a bounded domain, which however is very different from a technical point of view. In a related context the existence of self-adjoint extensions with prescribed point spectrum, absolutely continuous, and singular continuous spectrum in spectral gaps of a fixed underlying symmetric operator was also discussed in [6, 7, 154, 156, 158, 159, 160]. For our purposes Theorem 8.3.9 and Theorem 8.3.10 play an important role; for the case of C^∞-smooth domains such results can be found in [544], see also [352, 354]. The present versions of the extension theorems are inspired by slightly different considerations in [115]; the variant for Lipschitz domains in Theorem 8.7.5 can be proved by means of a similar technique (see also [92] for a comprehensive discussion). We remark that the definition and topologies on the spaces $\mathscr{G}_0$ and $\mathscr{G}_1$ in (8.7.4)–(8.7.5) from [92, 115] are partly inspired by abstract considerations in [246] for generalized boundary triplets. Concerning Section 8.3, as a final comment we mention that for more general second-order elliptic operators in bounded and unbounded domains Proposition 8.3.13 can be found in [118, 119, 120].

The boundary triplet in Theorem 8.4.1 can also be found in, e.g., [105, 169, 173, 359, 557, 643] and extends with some simple modifications to second-order and $2m$-th order elliptic operators with variable coefficients. In a different form this boundary triplet is already essentially contained in the well-known work of Grubb [352], where all closed extensions of a minimal operator elliptic partial differential operator were characterised by nonlocal boundary conditions; see also the early contribution [747] by Vishik and the fundamental paper [140] by Birman. In fact, it seems that the more recent paper [27] by Amrein and Pearson on a generalisation of Weyl–Titchmarsh theory for Schrödinger operators inspired many operator theorists to investigate partial differential operators from an extension theory point of view. Various papers based on boundary triplet techniques and related methods were published in the last decade; besides those papers listed before we mention as a selection here [18, 90, 91, 103, 107, 108, 251, 323, 324, 360, 429, 566, 568, 635, 647, 678] in which, e.g., Dirichlet-to-Neumann maps and Kreĭn type resolvent formulas are treated. For self-adjoint realizations of the Laplacians, Schrödinger operators, and more general second-order elliptic differential operators in nonsmooth domains we refer to, e.g., [1, 92, 115, 326, 358]. Particular attention has been paid to the spectral properties of realizations with local and nonlocal Robin boundary conditions in [41, 106, 179, 180, 226, 325, 361, 413, 543, 628, 629, 665]. Furthermore, the recent contributions [35, 36, 37, 38, 39, 40] by Arendt, ter Elst, and coauthors form an interesting series of papers on elliptic differential operators and Dirichlet-to-Neumann operators based mainly on form methods.

The present treatment of semibounded Schrödinger operator in Section 8.5 with the help of boundary pairs and boundary triplets seems to be new. However, the problem of lower boundedness was also discussed with different methods in [349, 355, 362]. The Kreĭn–von Neumann extension which is of special interest in this context was investigated in, e.g., [14, 68, 69, 70, 71, 93, 181, 356, 362, 578, 579]. For coupling methods for elliptic differential operators based on boundary triplet techniques in the spirit of Section 8.6 we refer to the recent paper [88], where also an abstract version of the third Green identity was proved. Finally, we mention that various classes of λ-dependent boundary value problems for elliptic operators can be treated in such a context, see, e.g., [87, 103, 137, 285, 286].

Notes on Appendices A–D

The appendices were included for the convenience of the reader. Here we collect some notes for each of the appendices A–D.

In Appendix A we present the basic properties of Nevanlinna functions that are needed in the text. The Stieltjes inversion formula for the Borel transform is folklore. The integral representation of scalar Nevanlinna functions is derived in a classical way involving the Helly and Helly–Bray theorems; see [72, 763]. The inversion formula in Lemma A.2.7 is standard; see [272, 446]. For the present purposes it suffices to approach the integration with respect to operator-valued measures in terms of improper Riemann integrals. For the operator measure version see for instance [691]. For Kac functions, Stieltjes functions, and inverse Stieltjes functions see [49, 52, 329, 407, 446, 447, 505]. As far as we know, the proof of Proposition A.5.4 is new; cf. [572]. For related work see [318, 319, 329, 602]. The case of generalized Nevanlinna functions was initiated in the works of Kreĭn and Langer [499, 500]; see [145, 225, 232, 239, 242, 255, 380, 411, 434, 435, 437, 546, 547, 548] for further developments.

For the general notion of Fourier transforms in Appendix B associated with Sturm–Liouville equations, see [539, 540, 542, 700, 701, 740, 780, 781]. Our basic idea here is inspired by the treatment in [208]. The arguments establishing the surjectivity of the Fourier transform are also inspired by [281, 715, 716]. Fourier transforms that are partially isometric go back at least to [204] in a treatment connected with multivalued operators. The discussion in this section has connections with the treatment of Kreĭn's directing functionals in [495, 518, 526, 527].

Necessary and sufficient conditions (as in Appendix C) for the sum of closed subspaces have a long history. The concept of opening of a pair of subspaces goes back to Friedrichs and Dixmier; see for instance [214, 249, 250]. Lemma C.4 goes back to [246].

The main reference for Appendix D is the paper by Douglas [273]; see also [30, 299]. There have been many generalizations of the Douglas paper. We only mention a particular direction of extension, namely [639, 684]; see also [402]. The application in Proposition 1.5.11 in Chapter 1 has the same flavor, but is obtained by reduction to the case in this appendix; see [390].

Bibliography

[1] H. Abels, G. Grubb, and I. Wood. Extension theory and Krein-type resolvent formulas for nonsmooth boundary value problems. J. Functional Analysis 266 (2014) 4037–4100.

[2] N.I. Achieser and I.M. Glasman. *Theorie der linearen Operatoren im Hilbertraum.* 8th edition. Akademie Verlag, 1981.

[3] R.A. Adams and J.J.F. Fournier. *Sobolev spaces.* 2nd edition. Academic Press, 2003.

[4] S. Agmon, A. Douglis, and L. Nirenberg. Estimates near the boundary for solutions of elliptic partial differential equations satisfying general boundary conditions I. Comm. Pure Appl. Math. 12 (1959) 623–727.

[5] M.S. Agranovich. *Sobolev spaces, their generalizations and elliptic problems in smooth and Lipschitz domains.* Monographs in Mathematics, Springer, 2015.

[6] S. Albeverio, J.F. Brasche, M.M. Malamud, and H. Neidhardt. Inverse spectral theory for symmetric operators with several gaps: scalar-type Weyl functions. J. Functional Analysis 228 (2005) 144–188.

[7] S. Albeverio, J.F. Brasche, and H. Neidhardt. On inverse spectral theory for self-adjoint extensions: mixed types of spectra. J. Functional Analysis 154 (1998) 130–173.

[8] S. Albeverio, F. Gesztesy, R. Hoegh-Krohn, and H. Holden. *Solvable models in quantum mechanics.* 2nd edition. AMS Chelsea Publishing, Amer. Math. Soc., 2005.

[9] S. Albeverio, A. Kostenko, and M.M. Malamud. Spectral theory of semibounded Sturm–Liouville operators with local interactions on a discrete set. J. Math. Physics 51 (2010) 1–24.

[10] S. Albeverio and P. Kurasov. *Singular perturbations of differential operators and Schrödinger type operators.* Cambridge University Press, 2000.

[11] S. Albeverio, M.M. Malamud, and V.I. Mogilevskiĭ. On Titchmarsh–Weyl functions and eigenfunction expansions of first-order symmetric systems. Integral Equations Operator Theory 77 (2013) 303–354.

[12] B.P. Allakhverdiev. On the theory of dilatation and on the spectral analysis of dissipative Schrödinger operators in the case of the Weyl limit circle (Russian). Izv. Akad. Nauk SSSR Ser. Mat. 54 (1990) 242–257 [Math. USSR-Izv. 36 (1991) 247–262].

J. Behrndt et al., *Boundary Value Problems, Weyl Functions, and Differential Operators*, Monographs in Mathematics 108,
https://doi.org/10.1007/978-3-030-36714-5

[13] W. Allegretto and A.B. Mingarelli. Boundary problems of the second order with an indefinite weight function. J. Reine Angew. Math. 398 (1989) 1–24.

[14] A. Alonso and B. Simon. The Birman–Kreĭn–Vishik theory of self-adjoint extensions of semibounded operators. J. Operator Theory 4 (1980) 251–270; 6 (1981) 407.

[15] D. Alpay. *Reproducing kemel Krein spaces of analytic functions and inverse scattering.* Doctoral dissertation, The Weizmann Institute of Science, Rehovot, 1985.

[16] D. Alpay. Some remarks on reproducing kernel Krein spaces. Rocky Mountain J. Math. 21 (1991) 1189–1205.

[17] D. Alpay. A theorem on reproducing kernel Hilbert spaces of pairs. Rocky Mountain J. Math. 22 (1992) 1243–1258.

[18] D. Alpay and J. Behrndt. Generalized Q-functions and Dirichlet-to-Neumann maps for elliptic differential operators. J. Functional Analysis 257 (2009) 1666–1694.

[19] D. Alpay, P. Bruinsma, A. Dijksma, and H.S.V. de Snoo. A Hilbert space associated with a Nevanlinna function. In: *Signal processing, scattering and operator theory, and numerical methods.* Birkhäuser, 1990, pp. 115–122.

[20] D. Alpay, P. Bruinsma, A. Dijksma, and H.S.V. de Snoo. Interpolation problems, extensions of operators and reproducing kernel spaces I. Oper. Theory Adv. Appl. 50 (1990) 35–82.

[21] D. Alpay, P. Bruinsma, A. Dijksma, and H.S.V. de Snoo. Interpolation problems, extensions of symmetric operators and reproducing kernel spaces II. Integral Equations Operator Theory 14 (1991) 465–500.

[22] D. Alpay, P. Bruinsma, A. Dijksma, and H.S.V. de Snoo. Interpolation problems, extensions of symmetric operators and reproducing kernel spaces II (missing Section 3). Integral Equations Operator Theory 15 (1992) 378–388.

[23] D. Alpay, A. Dijksma, J. Rovnyak, and H.S.V. de Snoo. *Schur functions, operator colligations, and Pontryagin spaces.* Oper. Theory Adv. Appl., Vol. 96, Birkhäuser, 1997.

[24] D. Alpay and H. Dym. Hilbert spaces of analytic functions, inverse scattering and operator models I. Integral Equations Operator Theory 7 (1984) 589–641.

[25] D. Alpay and H. Dym. Hilbert spaces of analytic functions, inverse scattering and operator models II. Integral Equations Operator Theory 8 (1985) 145–180.

[26] W.O. Amrein, A.M. Hinz, and D.B. Pearson (editors). *Sturm–Liouville theory: past and present.* Birkhäuser, 2005.

[27] W.O. Amrein and D.B. Pearson. M-operators: a generalisation of Weyl–Titchmarsh theory. J. Comp. Appl. Math. 171 (2004) 1–26.

[28] A.Yu. Ananieva and V. Budyika. To the spectral theory of the Bessel operator on finite interval and half-line. J. Math. Sciences 211 (2015) 624–645.

[29] A.Yu. Ananieva and V.S. Budyika. On the spectral theory of the Bessel operator on a finite interval and the half-line (Russian). Differ. Uravn. 52 (2016) 1568–1572 [Differential Equations 52 (2016) 1517–1522].

[30] W.N. Anderson Jr. and G.E. Trapp. Shorted operators. II. SIAM J. Appl. Math. 28 (1975) 60–71.

[31] T. Ando. *Linear operators on Krein spaces.* Division of Applied Mathematics, Research Institute of Applied Electricity, Hokkaido University, Sapporo, 1979.

[32] T. Ando. *Reproducing kernel spaces and quadratic inequalities.* Division of Applied Mathematics, Research Institute of Applied Electricity, Hokkaido University, Sapporo, 1987.

[33] T. Ando. *De Branges spaces and analytic operator functions.* Division of Applied Mathematics, Research Institute of Applied Electricity, Hokkaido University, Sapporo, 1990.

[34] T. Ando and K. Nishio. Positive selfadjoint extensions of positive symmetric operators. Tohoku Math. J. 22 (1970) 65–75.

[35] W. Arendt and A.F.M. ter Elst. The Dirichlet-to-Neumann operator on rough domains. J. Differential Equations 251 (2011) 2100–2124.

[36] W. Arendt and A.F.M. ter Elst. Sectorial forms and degenerate differential operators. J. Operator Theory 67 (2012) 33–72.

[37] W. Arendt and A.F.M. ter Elst. From forms to semigroups. Oper. Theory Adv. Appl. 221 (2012) 47–69.

[38] W. Arendt and A.F.M. ter Elst. The Dirichlet-to-Neumann operator on exterior domains. Potential Analysis 43 (2015) 313–340.

[39] W. Arendt, A.F.M. ter Elst, J.B. Kennedy, and M. Sauter. The Dirichlet-to-Neumann operator via hidden compactness. J. Functional Analysis 266 (2014) 1757–1786.

[40] W. Arendt, A.F.M. ter Elst, and M. Warma. Fractional powers of sectorial operators via the Dirichlet-to-Neumann operator. Comm. Partial Differential Equations 43 (2018) 1–24.

[41] W. Arendt and M. Warma. The Laplacian with Robin boundary conditions on arbitrary domains. Potential Analysis 19 (2003) 341–363.

[42] R. Arens. Operational calculus of linear relations. Pacific J. Math. 11 (1961) 9–23.

[43] Yu.M. Arlinskiĭ. Positive spaces of boundary values and sectorial extensions of nonnegative symmetric operators. Ukrainian Math. J. 40 (1988) 8–15.

[44] Yu.M. Arlinskiĭ. Maximal sectorial extensions and closed forms associated with them. Ukrainian Math. J. 48 (1996) 723–739.

[45] Yu.M. Arlinskiĭ. Extremal extensions of sectorial linear relations. Mat. Studii 7 (1997) 81–96.

[46] Yu.M. Arlinskiĭ. On a class of nondensely defined contractions and their extensions. J. Math. Sciences 79 (1999) 4391–4419.

[47] Yu.M. Arlinskiĭ. Abstract boundary conditions for maximal sectorial extensions of sectorial operators. Math. Nachr. 209 (2000) 5–35.

[48] Yu.M. Arlinskiĭ. Boundary triplets and maximal accretive extensions of sectorial operators. In: *Operator methods for boundary value problems.* London Math. Soc. Lecture Note Series, Vol. 404, 2012, pp. 35–72.

[49] Yu.M. Arlinskiĭ, S. Belyi, and E.R. Tsekanovskiĭ. *Conservative realizations of Herglotz-Nevanlinna functions*. Oper. Theory Adv. Appl., Vol. 217, Birkhäuser, 2011.

[50] Yu.M. Arlinskiĭ and S. Hassi. Q-functions and boundary triplets of nonnegative operators. Oper. Theory Adv. Appl. 244 (2015) 89–130.

[51] Yu.M. Arlinskiĭ and S. Hassi. Compressed resolvents of selfadjoint contractive exit space extensions and holomorphic operator-valued functions associated with them. Methods Funct. Anal. Topology 21 (2015) 199–224.

[52] Yu.M. Arlinskiĭ and S. Hassi. Stieltjes and inverse Stieltjes holomorphic families of linear relations and their representations. Studia Math. To appear.

[53] Yu.M. Arlinskiĭ, S. Hassi, Z. Sebestyén, and H.S.V. de Snoo. On the class of extremal extensions of a nonnegative operator. Oper. Theory Adv. Appl. 127 (2001) 41–81.

[54] Yu.M. Arlinskiĭ, Yu. Kovalev, and E.R. Tsekanovskiĭ. Accretive and sectorial extensions of nonnegative symmetric operators. Complex Anal. Oper. Theory 6 (2012) 677–718.

[55] Yu. M. Arlinskiĭ and E.R. Tsekanovskiĭ. Quasiselfadjoint contractive extensions of a Hermitian contraction (Russian). Teor. Funktsii Funktsional. Anal. i Prilozhen (1988) 9–16 [J. Soviet Math. 49 (1990) 1241–1247].

[56] Yu.M. Arlinskiĭ and E.R. Tsekanovskiĭ. On von Neumann's problem in extension theory of nonnegative operators. Proc. Amer. Math. Soc. 131 (2003) 3143–3154.

[57] Yu.M. Arlinskiĭ and E.R. Tsekanovskiĭ. The von Neumann problem for nonnegative symmetric operators. Integral Equations Operator Theory 51 (2005) 319–356.

[58] Yu.M. Arlinskiĭ and E.R. Tsekanovskiĭ. M. Kreĭn's research on semi-bounded operators, its contemporary developments, and applications. Oper. Theory Adv. Appl. 190 (2009) 65–112.

[59] N. Aronszajn. Theory of reproducing kernels. Trans. Amer. Math. Soc. 68 (1950) 337–404.

[60] N. Aronszajn. On a problem of Weyl in the theory of singular Sturm–Liouville equations. Amer. J. Math. 79 (1957) 597–610.

[61] N. Aronszajn and W.F. Donoghue. On exponential representations of analytic functions. J. Anal. Math. 5 (1956/1957) 321–388.

[62] D.Z. Arov and H. Dym. J-inner matrix functions, interpolation and inverse problems for canonical systems, I. Foundations. Integral Equations Operator Theory 29 (1997) 373–454.

[63] D.Z. Arov and H. Dym. J-inner matrix functions, interpolation and inverse problems for canonical systems, II. The inverse monodromy problem. Integral Equations Operator Theory 36 (2000) 11–70.

[64] D.Z. Arov and H. Dym. J-inner matrix functions, interpolation and inverse problems for canonical systems, III. More on the inverse monodromy problem. Integral Equations Operator Theory 36 (2000) 127–181.

[65] D.Z. Arov, M. Kurula, and O.J. Staffans. Boundary control state/signal systems and boundary triplets. In: *Operator methods for boundary value problems.* London Math. Soc. Lecture Note Series, Vol. 404, 2012, pp. 73–85.

[66] D.Z. Arov, M. Kurula, and O.J. Staffans. Passive state/signal systems and conservative boundary relations. In: *Operator methods for boundary value problems.* London Math. Soc. Lecture Note Series, Vol. 404, 2012, pp. 87–119.

[67] D.Z. Arov and O.J. Staffans. State/signal linear time-invariant systems theory. Part I: discrete time systems. Oper. Theory Adv. Appl. 161 (2006) 115–177.

[68] M.S. Ashbaugh, F. Gesztesy, A. Laptev, M. Mitrea, and S. Sukhtaiev. A bound for the eigenvalue counting function for Krein–von Neumann and Friedrichs extensions. Adv. Math. 304 (2017) 1108–1155.

[69] M.S. Ashbaugh, F. Gesztesy, M. Mitrea, R. Shterenberg, and G. Teschl. The Krein–von Neumann extension and its connection to an abstract buckling problem. Math. Nachr. 210 (2010) 165–179.

[70] M.S. Ashbaugh, F. Gesztesy, M. Mitrea, R. Shterenberg, and G. Teschl. A survey on the Krein-von Neumann extension, the corresponding abstract buckling problem, and Weyl-type spectral asymptotics for perturbed Krein Laplacians in nonsmooth domains. Oper. Theory Adv. Appl. 232 (2013) 1–106.

[71] M.S. Ashbaugh, F. Gesztesy, M. Mitrea, and G. Teschl. Spectral theory for perturbed Krein Laplacians in nonsmooth domains. Adv. Math. 223 (2010) 1372–1467.

[72] F.V. Atkinson. *Discrete and continuous boundary problems.* Academic Press, 1964.

[73] T.Ya. Azizov. Extensions of J-isometric and J-symmetric operators (Russian). Funktsional. Anal. i Prilozhen 18 (1984) 57–58 [Funct. Anal. Appl. 18 (1984) 46–48].

[74] T.Ya. Azizov, J. Behrndt, P. Jonas, and C. Trunk. Compact and finite rank perturbations of linear relations in Hilbert spaces. Integral Equations Operator Theory 63 (2009) 151–163.

[75] T.Ya. Azizov, J. Behrndt, P. Jonas, and C. Trunk. Spectral points of definite type and type π for linear operators and relations in Krein spaces. J. London Math. Soc. 83 (2011) 768–788.

[76] T.Ya. Azizov, B. Ćurgus, and A. Dijksma. Standard symmetric operators in Pontryagin spaces: a generalized von Neumann formula and minimality of boundary coefficients. J. Functional Analysis 198 (2003) 361–412.

[77] T.Ya. Azizov and I.S. Iokhvidov. *Linear operators in spaces with indefinite metric.* John Wiley and Sons, 1989.

[78] T.Ya. Azizov, P. Jonas, and C. Trunk. Spectral points of type π_+ and π_- of self-adjoint operators in Krein spaces. J. Functional Analysis 226 (2005) 114–137.

[79] T.Ya. Azizov and L.I. Soukhotcheva. Linear operators in almost Krein spaces. Oper. Theory Adv. Appl. 175 (2007) 1–11.

[80] D. Baidiuk. Boundary triplets and generalized resolvents of isometric operators in a Pontryagin space (Russian). Ukr. Mat. Visn. 10 (2013), no. 2, 176–200, 293 [J. Math. Sciences 194 (2013) 513–531].

[81] D. Baidiuk and S. Hassi. Completion, extension, factorization, and lifting of operators. Math. Ann. 364 (2016) 1415–1450.

[82] A.G. Baskakov and K.I. Chernyshov. Spectral analysis of linear relations and degenerate operator semigroups. Mat. Sb. 193 (2002) 3–42 [Sb. Math. 193 (2002) 1573–1610].

[83] C.J.K. Batty and A.F.M. ter Elst. On series of sectorial forms. J. Evolution Equations 14 (2014) 29–47.

[84] R. Beals. Non-local boundary value problems for elliptic operators. Amer. J. Math. 87 (1965) 315–362.

[85] H. Behncke and D. Hinton. Two singular point linear Hamiltonian systems with an interface condition. Math. Nachr. 283 (2010) 365–378.

[86] J. Behrndt. Boundary value problems with eigenvalue depending boundary conditions. Math. Nachr. 282 (2009) 659–689.

[87] J. Behrndt. Elliptic boundary value problems with λ-dependent boundary conditions. J. Differential Equations 249 (2010) 2663–2687.

[88] J. Behrndt, V.A. Derkach, F. Gesztesy, and M. Mitrea. Coupling of symmetric operators and the third Green identity. Bull. Math. Sciences 8 (2018) 49–80.

[89] J. Behrndt, V.A. Derkach, S. Hassi, and H.S.V. de Snoo. A realization theorem for generalized Nevanlinna pairs. Operators Matrices 5 (2011) 679–706.

[90] J. Behrndt and A.F.M. ter Elst. Dirichlet-to-Neumann maps on bounded Lipschitz domains. J. Differential Equations 259 (2015) 5903–5926.

[91] J. Behrndt, F. Gesztesy, H. Holden, and R. Nichols. Dirichlet-to-Neumann maps, abstract Weyl–Titchmarsh M-functions, and a generalized index of unbounded meromorphic operator-valued functions. J. Differential Equations 261 (2016) 3551–3587.

[92] J. Behrndt, F. Gesztesy, and M. Mitrea. Sharp boundary trace theory and Schrödinger operators on bounded Lipschitz domains. In preparation

[93] J. Behrndt, F. Gesztesy, M. Mitrea, and T. Micheler. The Krein-von Neumann realization of perturbed Laplacians on bounded Lipschitz domains. Oper. Theory Adv. Appl. 255 (2014) 49–66.

[94] J. Behrndt, S. Hassi, and H.S.V. de Snoo. Functional models for Nevanlinna families. Opuscula Mathematica 28 (2008) 233–245.

[95] J. Behrndt, S. Hassi, and H.S.V. de Snoo. Boundary relations, unitary colligations, and functional models. Complex Anal. Oper. Theory 3 (2009) 57–98.

[96] J. Behrndt, S. Hassi, H.S.V. de Snoo, and H.L. Wietsma. Monotone convergence theorems for semibounded operators and forms with applications. Proc. Royal Soc. Edinburgh 140A (2010) 927–951.

[97] J. Behrndt, S. Hassi, H.S.V. de Snoo, and H.L. Wietsma. Square-integrable solutions and Weyl functions for singular canonical systems. Math. Nachr. 284 (2011) 1334–1384.

[98] J. Behrndt, S. Hassi, H.S.V. de Snoo, and H.L. Wietsma. Limit properties of monotone matrix functions. Lin. Alg. Appl. 436 (2012) 935–953.

[99] J. Behrndt and P. Jonas. On compact perturbations of locally definitizable self-adjoint relations in Krein spaces. Integral Equations Operator Theory 52 (2005) 17–44.

[100] J. Behrndt and P. Jonas. Boundary value problems with local generalized Nevanlinna functions in the boundary condition. Integral Equations Operator Theory 55 (2006) 453–475.

[101] J. Behrndt and H.-C. Kreusler. Boundary relations and generalized resolvents of symmetric relations in Krein spaces. Integral Equations Operator Theory 59 (2007) 309–327.

[102] J. Behrndt, M. Kurula, A.J. van der Schaft, and H. Zwart. Dirac structures and their composition on Hilbert spaces. J. Math. Anal. Appl. 372 (2010) 402–422.

[103] J. Behrndt and M. Langer. Boundary value problems for elliptic partial differential operators on bounded domains. J. Functional Analysis 243 (2007) 536–565.

[104] J. Behrndt and M. Langer. On the adjoint of a symmetric operator. J. London Math. Soc. 82 (2010) 563–580.

[105] J. Behrndt and M. Langer. Elliptic operators, Dirichlet-to-Neumann maps and quasi boundary triples. In: *Operator methods for boundary value problems.* London Math. Soc. Lecture Note Series, Vol. 404, 2012, pp. 121–160.

[106] J. Behrndt, M. Langer, I. Lobanov, V. Lotoreichik, and I.Yu. Popov. A remark on Schatten–von Neumann properties of resolvent differences of generalized Robin Laplacians on bounded domains. J. Math. Anal. Appl. 371 (2010) 750–758.

[107] J. Behrndt, M. Langer, and V. Lotoreichik. Spectral estimates for resolvent differences of self-adjoint elliptic operators. Integral Equations Operator Theory 77 (2013) 1–37.

[108] J. Behrndt, M. Langer, and V. Lotoreichik. Schrödinger operators with δ and δ'-potentials supported on hypersurfaces. Ann. Henri Poincaré 14 (2013) 385–423.

[109] J. Behrndt, M. Langer, V. Lotoreichik, and J. Rohleder. Quasi boundary triples and semibounded self-adjoint extensions. Proc. Roy. Soc. Edinburgh 147A (2017) 895–916.

[110] J. Behrndt, M. Langer, V. Lotoreichik, and J. Rohleder. Spectral enclosures for non-self-adjoint extensions of symmetric operators. J. Functional Analysis 275 (2018) 1808–1888.

[111] J. Behrndt and A. Luger. An analytic characterization of the eigenvalues of self-adjoint extensions. J. Functional Analysis 242 (2007) 607–640.

[112] J. Behrndt, M.M. Malamud, and H. Neidhardt. Scattering theory for open quantum systems with finite rank coupling. Math. Phys. Anal. Geom. 10 (2007) 313–358.

[113] J. Behrndt, M.M. Malamud, and H. Neidhardt. Scattering matrices and Weyl functions. Proc. London Math. Soc. 97 (2008) 568–598.

[114] J. Behrndt, M.M. Malamud, and H. Neidhardt. Scattering matrices and Dirichlet-to-Neumann maps. J. Functional Analysis 273 (2017) 1970–2025.

[115] J. Behrndt and T. Micheler. Elliptic differential operators on Lipschitz domains and abstract boundary value problems. J. Functional Analysis 267 (2014) 3657–3709.

[116] J. Behrndt and F. Philipp. Spectral analysis of singular ordinary differential operators with indefinite weights. J. Differential Equations 248 (2010) 2015–2037.

[117] J. Behrndt and F. Philipp. Finite rank perturbations in Pontryagin spaces and a Sturm–Liouville problem with λ-rational boundary conditions. Oper. Theory Adv. Appl. 263 (2018) 163–189.

[118] J. Behrndt and J. Rohleder. An inverse problem of Calderón type with partial data. Comm. Partial Differential Equations 37 (2012) 1141–1159.

[119] J. Behrndt and J. Rohleder. Spectral analysis of self-adjoint elliptic differential operators, Dirichlet-to-Neumann maps, and abstract Weyl functions. Adv. Math. 285 (2015) 1301–1338.

[120] J. Behrndt and J. Rohleder. Titchmarsh–Weyl theory for Schrödinger operators on unbounded domains. J. Spectral Theory 6 (2016) 67–87.

[121] J. Behrndt and H.S.V. de Snoo. On Kreĭn's formula. J. Math. Anal. Appl. 351 (2009) 567–578.

[122] J. Behrndt and C. Trunk. Sturm–Liouville operators with indefinite weight functions and eigenvalue depending boundary conditions. J. Differential Equations 222 (2006) 297–324.

[123] J. Behrndt and C. Trunk. On the negative squares of indefinite Sturm–Liouville operators. J. Differential Equations 238 (2007) 491–519.

[124] A. Beigl, J. Eckhardt, A. Kostenko, and G. Teschl. On spectral deformations and singular Weyl functions for one-dimensional Dirac operators. J. Math. Phys. 56 (2015), no. 1, 012102, 11 pp.

[125] S. Di Bella and C. Trapani. Some representation theorems for sesquilinear forms. J. Math. Anal. Appl. 451 (2017) 64–83.

[126] S. Belyi, G. Menon, and E.R. Tsekanovskiĭ. On Krein's formula in the case of non-densely defined symmetric operators. J. Math. Anal. Appl. 264 (2001) 598–616.

[127] A. Ben-Israel and T.N.E. Greville. *Generalized inverses: theory and applications.* CMS Books in Mathematics, Springer, 2003.

[128] C. Bennewitz. Symmetric relations on a Hilbert space. Springer Lecture Notes Math. 280 (1972) 212–218.

[129] C. Bennewitz. Spectral theory for pairs of differential operators. Ark. Mat. 15 (1977) 33–61.

[130] C. Bennewitz. Spectral asymptotics for Sturm–Liouville equations. Proc. London Math. Soc. (3) 59 (1989) 294–338.

[131] C. Bennewitz and W.N. Everitt. On second order left definite boundary value problems. Springer Lecture Notes Math. 1032 (1983) 31–67.

[132] Yu.M. Berezanskiĭ. Spaces with negative norm (Russian). Uspehi Mat. Nauk 18 no. 1 (1963) 63–96.

[133] Yu.M. Berezanskiĭ. *Expansions in eigenfunctions of selfadjoint operators* (Russian). Naukova Dumka, Kiev, 1965 [Transl. Math. Monographs, Vol. 17, Amer. Math. Soc., 1968].

[134] G. Berkolaiko and P. Kuchment. *Introduction to quantum graphs.* Mathematical Surveys and Monographs, Vol. 186, Amer. Math. Soc., 2013.

[135] F. Betschler. *Über Integraldarstellungen welche aus speziellen Randwertproblemen bei gewöhnlichen linearen inhomogenen Differentialgleichungen entspringen.* Doctoral dissertation, Julius-Maximilians-Universität, Würzburg, 1912.

[136] P.K. Bhattacharyya. *Distributions. Generalized functions with applications in Sobolev spaces.* de Gruyter Textbook, 2012.

[137] P. Binding, R. Hryniv, H. Langer, and B. Najman. Elliptic eigenvalue problems with eigenparameter dependent boundary conditions. J. Differential Equations 174 (2001) 30–54.

[138] P. Binding, H. Langer, and M. Möller. Oscillation results for Sturm–Liouville problems with an indefinite weight function. J. Comput. Appl. Math. 171 (2004) 93–101.

[139] M.S. Birman. On the self-adjoint extensions of positive definite operators (Russian). Mat. Sb. 38 (1956) 431–450.

[140] M.S. Birman. Perturbations of the continuous spectrum of a singular elliptic operator by varying the boundary and the boundary conditions (Russian). Vestnik Leningrad Univ. 17 (1962) 22–55 [Amer. Math. Soc. Transl. 225 (2008), 19–53].

[141] M.S. Birman and M.Z. Solomjak. *Spectral theory of self-adjoint operators in Hilbert space.* Kluwer, 1987.

[142] J. Bognar. *Indefinite inner product spaces.* Ergebnisse der Mathematik und ihrer Grenzgebiete, Vol. 78, Springer, 1974.

[143] A.A. Boitsev. Boundary triplets approach for Dirac operator. In: *Mathematical results in quantum mechanics.* World Scientific, 2015, pp. 213–219.

[144] A.A. Boitsev, J.F. Brasche, M.M. Malamud, H. Neidhardt, and I.Yu. Popov. Boundary triplets, tensor products and point contacts to reservoirs. Ann. Henri Poincaré 19 (2018) 2783–2837.

[145] M. Borogovac and A. Luger. Analytic characterizations of Jordan chains by pole cancellation functions of higher order. J. Functional Analysis 267 (2014) 4499–4518.

[146] L. de Branges. Some Hilbert spaces of entire functions. Trans. Amer. Math. Soc. 96 (1960) 259–295.

[147] L. de Branges. Some Hilbert spaces of entire functions II. Trans. Amer. Math. Soc. 99 (1961) 118–152.

[148] L. de Branges. Some Hilbert spaces of entire functions III. Trans. Amer. Math. Soc. 100 (1961) 73–115.

[149] L. de Branges. Some Hilbert spaces of entire functions IV. Trans. Amer. Math. Soc. 105 (1962) 43–83.

[150] L. de Branges. The expansion theorem for Hilbert spaces of entire functions. In: *Entire functions and related parts of analysis.* Proc. Symp. Pure Math., Vol. 11, Amer. Math. Soc., 1968, pp. 79-148.

[151] L. de Branges, *Hilbert spaces of entire functions.* Prentice-Hall, 1968.

[152] L. de Branges and J. Rovnyak. *Square summable power series.* Holt, Rinehart and Winston, 1966.

[153] L. de Branges and J. Rovnyak. Appendix on square summable power series, Canonical models in quantum scattering theory. In: *Perturbation theory and its applications in quantum mechanics.* John Wiley and Sons, 1966, pp. 295–392.

[154] J.F. Brasche. Spectral theory for self-adjoint extensions. In: *Spectral theory of Schrödinger operators.* Contemp. Math., Vol. 340, Amer. Math. Soc., 2004, pp. 51–96.

[155] J.F. Brasche, M.M. Malamud, and H. Neidhardt. Weyl function and spectral properties of selfadjoint extensions. Integral Equations Operator Theory 43 (2002) 264–289.

[156] J.F. Brasche, M.M. Malamud, and H. Neidhardt. Selfadjoint extensions with several gaps: finite deficiency indices. Oper. Theory Adv. Appl. 162 (2006) 85–101.

[157] J.F. Brasche and H. Neidhardt. Some remarks on Krein's extension theory. Math. Nachr. 165 (1994) 159–181.

[158] J.F. Brasche and H. Neidhardt. On the absolutely continuous spectrum of self-adjoint extensions. J. Functional Analysis 131 (1995) 364–385.

[159] J.F. Brasche and H. Neidhardt. On the singular continuous spectrum of self-adjoint extensions. Math. Z. 222 (1996) 533–542.

[160] J.F. Brasche, H. Neidhardt, and J. Weidmann. On the point spectrum of self-adjoint extensions. Math. Z. 214 (1993) 343–355.

[161] F. Brauer. Singular self-adjoint boundary value problems for the differential equation $Lx = \lambda Mx$. Trans. Amer. Math. Soc. 88 (1958) 331–345.

[162] F. Brauer. Spectral theory for the differential equation $Lu = \lambda Mu$. Can. J. Math. 10 (1958) 431–446.

[163] F. Brauer. Spectral theory for linear systems of differential equations. Pacific J. Math. 10 (1960) 17–34.

[164] H. Brezis. *Functional analysis, Sobolev spaces and partial differential equations.* Universitext, Springer, 2011.

[165] M.S. Brodskiĭ. *Triangular and Jordan representations of linear operators.* Transl. Math. Monographs, Vol. 32, Amer. Math. Soc., 2006.

[166] M.S. Brodskiĭ. Unitary operator colligations and their characteristic functions (Russian). Uspekhi Mat. Nauk 33 (1978) 141–168 [Russian Math. Surveys 33 (1978) 159–191].

[167] F.E. Browder, On the spectral theory of elliptic differential operators. I. Math. Ann. 142 (1961) 22–130.

[168] B.M. Brown and W.D. Evans. Selfadjoint and m-sectorial extensions of Sturm–Liouville operators. Integral Equations Operator Theory 85 (2016) 151–166.

[169] B.M. Brown, G. Grubb, and I. Wood. M-functions for closed extensions of adjoint pairs of operators with applications to elliptic boundary problems. Math. Nachr. 282 (2009) 314–347.

[170] B.M. Brown, J. Hinchcliffe, M. Marletta, S. Naboko, and I. Wood. The abstract Titchmarsh–Weyl M-function for adjoint operator pairs and its relation to the spectrum. Integral Equations Operator Theory 63 (2009) 297–320.

[171] B.M. Brown, H. Langer, and M. Langer. Bessel-type operators with an inner singularity. Integral Equations Operator Theory 75 (2013) 257–300.

[172] B.M. Brown, H. Langer, and C. Tretter. Compressed resolvents and reduction of spectral problems on star graphs. Complex Anal. Oper. Theory 13 (2019) 291–320.

[173] B.M. Brown, M. Marletta, S. Naboko, and I. Wood. Boundary triplets and M-functions for non-selfadjoint operators, with applications to elliptic PDEs and block operator matrices. J. London Math. Soc. (2) 77 (2008) 700–718.

[174] B.M. Brown, M. Marletta, S. Naboko, and I. Wood. Inverse problems for boundary triples with applications. Studia Mathematica 237 (2017) 241–275.

[175] B.M. Brown, D.K.R. McCormack, W.D. Evans, and M. Plum. On the spectrum of second-order differential operators with complex coefficients. Proc. Roy. Soc. London A 455 (1999) 1235–1257.

[176] V.M. Bruk. A certain class of boundary value problems with a spectral parameter in the boundary condition (Russian). Mat. Sb. 100 (1976) 210–216 [Sb. Math. 29 (1976) 186–192].

[177] V.M. Bruk. Extensions of symmetric relations (Russian). Mat. Zametki 22 (1977) 825–834 [Math. Notes 22 (1977) 953–958].

[178] V.M. Bruk. Linear relations in a space of vector functions (Russian). Mat. Zametki 24 (1978) 499–511 [Math. Notes 24 (1978) 767–773].

[179] V. Bruneau, K. Pankrashkin, and N. Popoff. Eigenvalue counting function for Robin Laplacians on conical domains. J. Geom. Anal. 28 (2018) 123–151.

[180] V. Bruneau and N. Popoff. On the negative spectrum of the Robin Laplacian in corner domains. Anal. PDE. 9 (2016) 1259–1283.

[181] V. Bruneau and G. Raikov. Spectral properties of harmonic Toeplitz operators and applications to the perturbed Krein Laplacian. Asymptotic Analysis 109 (2018) 53–74.

[182] R. Brunnhuber, J. Eckhardt, A. Kostenko, and G. Teschl. Singular Weyl–Titchmarsh–Kodaira theory for one-dimensional Dirac operators. Monatsh. Math. 174 (2014) 515–547.

[183] J. Brüning, V. Geyler, and K. Pankrashkin. Cantor and band spectra for periodic quantum graphs with magnetic fields. Commun. Math. Phys. 269 (2007) 87–105.

[184] J. Brüning, V. Geyler, and K. Pankrashkin. Spectra of self-adjoint extensions and applications to solvable Schrödinger operators. Rev. Math. Phys. 20 (2008) 1–70.

[185] C. Cacciapuoti, D. Fermi, and A. Posilicano. On inverses of Kreĭn's Q-functions. Rend. Mat. Appl. (7) 39 (2018) 229–240.

[186] J.W. Calkin. Abstract symmetric boundary conditions. Trans. Amer. Math. Soc. 45 (1939) 369–442.

[187] J.W. Calkin. General self-adjoint boundary conditions for certain partial differential operators. Proc. Nat. Acad. Sci. U.S.A. 25 (1939) 201–206.

[188] R. Carlson. Eigenfunction expansions for self-adjoint integro-differential operators. Pacific J. Math. 81 (1979) 327–347.

[189] R. Carlson. Expansions associated with non-self-adjoint boundary-value problems. Proc. Amer. Math. Soc. 73 (1979) 173–179.

[190] E.A. Catchpole. An integro-differential operator. J. London Math. Soc. (2) 6 (1973) 513–523.

[191] N.K. Chakravarty. Some problems in eigenfunction expansions I. Quart. J. Math. 16 (1965) 135–150.

[192] N.K. Chakravarty. Some problems in eigenfunction expansions II. Quart. J. Math. 19 (1965) 213–224.

[193] N.K. Chakravarty. Some problems in eigenfunction expansions III. Quart. J. Math. 19 (1968) 397–415.

[194] N.K. Chakravarty. Some problems in eigenfunction expansions IV. Indian J. Pure Appl. Math. 1 (1970) 347–353.

[195] J. Chaudhuri and W.N. Everitt. On the spectrum of ordinary second-order differential operators. Proc. Royal Soc. Edinburgh 68A (1969) 95–119.

[196] K.D. Cherednichenko, A.V. Kiselev, and L.O. Silva. Functional model for extensions of symmetric operators and applications to scattering theory. Netw. Heterog. Media 13 (2018) 191–215.

[197] S. Clark, F. Gesztesy, and M. Mitrea. Boundary data maps for Schrödinger operators on a compact interval. Math. Model. Nat. Phenom. 5 (2010) 73–121.

[198] S. Clark, F. Gesztesy, and R. Nichols. Principal solutions revisited. In: *Stochastic and infinite dimensional analysis.* Trends in Mathematics, Birkhäuser, 2016, pp. 85–117.

[199] S. Clark, F. Gesztesy, R. Nichols, and M. Zinchenko. Boundary data maps and Krein's resolvent formula for Sturm–Liouville operators on a finite interval. Operators Matrices 8 (2014) 1–71.

[200] E.A. Coddington. The spectral representation of ordinary self-adjoint differential operators. Ann. Math. (2) 60 (1954) 192–211.

[201] E.A. Coddington. Generalized resolutions of the identity for symmetric ordinary differential operators. Ann. Math. (2) 68 (1958) 378–392.

[202] E.A. Coddington. Extension theory of formally normal and symmetric subspaces. Mem. Amer. Math. Soc. 134 (1973).

[203] E.A Coddington. Self-adjoint subspace extensions of non-densely defined symmetric operators. Adv. Math. 14 (1974) 309–332.

[204] E.A. Coddington. Self-adjoint problems for nondensely defined ordinary differential operators and their eigenfunction expansions. Adv. Math. 15 (1975) 1–40.

[205] E.A. Coddington and A. Dijksma. Self-adjoint subspaces and eigenfunction expansions for ordinary differential subspaces. J. Differential Equations 20 (1976) 473–526.

[206] E.A. Coddington and A. Dijksma. Adjoint subspaces in Banach spaces, with applications to ordinary differential subspaces. Annali di Matematica Pura ed Applicata 118 (1978) 1–118.

[207] E.A. Coddington and R.C. Gilbert. Generalized resolvents of ordinary differential operators. Trans. Amer. Math. Soc. 93 (1959) 216–241.

[208] E.A. Coddington and N. Levinson. On the nature of the spectrum of singular second order linear differential equations. Canadian J. Math. 3 (1951) 335–338.

[209] E.A. Coddington and N. Levinson. *Theory of ordinary differential equations.* McGraw-Hill, 1955.

[210] E.A. Coddington and H.S.V. de Snoo. Positive selfadjoint extensions of positive subspaces. Math. Z. 159 (1978) 203–214.

[211] E.A. Coddington and H.S.V. de Snoo. Differential subspaces associated with pairs of ordinary differential expressions. J. Differential Equations 35 (1980) 129–182.

[212] E.A. Coddington and H.S.V. de Snoo. *Regular boundary value problems associated with pairs of ordinary differential expressions.* Lecture Notes in Mathematics, Vol. 858, Springer, 1981.

[213] S.D. Conte and W.C. Sangren. An expansion theorem for a pair of first order equations. Canadian J. Math. 6 (1954) 554–560.

[214] G. Corach and A. Maestripieri. Redundant decompositions, angles between subspaces and oblique projections. Publicacions Matemàtiques 54 (2010) 461–484.

[215] R. Corso. A Kato's second type representation theorem for solvable sesquilinear forms. J. Math. Anal. Appl. 462 (2018) 982–998.

[216] R. Corso and C. Trapani. Representation theorems for solvable sesquilinear forms. Integral Equations Operator Theory 89 (2017) 43–68.

[217] M.G. Crandall. Norm preserving extensions of linear transformations on Hilbert spaces. Proc. Amer. Math. Soc. 21 (1969) 335–340.

[218] R. Cross. *Multi-valued linear operators.* Marcel Dekker, 1998.

[219] M.E. Cumakin. Generalized resolvents of isometric operators (Russian). Sibirsk. Mat. Z. 8 (1967) 876–892.

[220] B. Ćurgus. Definitizable extensions of positive symmetric operators in a Kreĭn space. Integral Equations Operator Theory 12 (1989) 615–631.

[221] B. Ćurgus, A. Dijksma, and T. Read. The linearization of boundary eigenvalue problems and reproducing kernel Hilbert spaces. Lin. Alg. Appl. 329 (2001) 97–136.

[222] B. Ćurgus, A. Dijksma, H. Langer, and H.S.V. de Snoo. Characteristic functions of unitary colligations and of bounded operators in Krein spaces. Oper. Theory Adv. Appl. 41 (1989) 125–152.

[223] B. Ćurgus and H. Langer. A Krein-space approach to symmetric ordinary differential operators with an indefinite weight function. J. Differential equations 79 (1989) 31–61.

[224] P. Deift and R. Killip. On the absolutely continuous spectrum of one-dimensional Schrödinger operators with square summable potentials. Comm. Math. Phys. 203 (1999) 341–347.

[225] K. Daho and H. Langer. Sturm–Liouville operators with indefinite weight function. Proc. Royal Soc. Edinburgh 78A (1977) 161–191.

[226] D. Daners. Robin boundary value problems on arbitrary domains. Trans. Amer. Math. Soc. 352 (2000) 4207–4236.

[227] V.A. Derkach. Extensions of a Hermitian operator in a Kreĭn space (Russian). Dokl. Akad. Nauk Ukrain. SSR Ser. A 1988, no. 5, 5–9.

[228] V.A. Derkach. Extensions of a Hermitian operator that is not densely defined in a Kreĭn space (Russian). Dokl. Akad. Nauk Ukrain. SSR Ser. A 1990, no. 10, 15–19.

[229] V.A. Derkach. On Weyl function and generalized resolvents of a Hermitian operator in a Krein space. Integral Equations Operator Theory 23 (1995) 387–415.

[230] V.A. Derkach. On generalized resolvents of Hermitian relations in Krein spaces. J. Math. Sciences 97 (1999) 4420–4460.

[231] V.A. Derkach. Abstract interpolation problem in Nevanlinna classes. Oper. Theory Adv. Appl. 190 (2009) 197–236.

[232] V.A. Derkach and S. Hassi. A reproducing kernel space model for N_κ-functions. Proc. Amer. Math. Soc. 131 (2003) 3795–3806.

[233] V.A. Derkach, S. Hassi, and M.M. Malamud. Generalized boundary triples, Weyl functions, and inverse problems. arXiv:1706.0794824 (2017), 104 pp.

[234] V.A. Derkach, S. Hassi, M.M. Malamud, and H.S.V. de Snoo. Generalized resolvents of symmetric operators and admissibility. Methods Funct. Anal. Topology 6 (2000) 24–55.

[235] V.A. Derkach, S. Hassi, M.M. Malamud, and H.S.V. de Snoo. Boundary relations and their Weyl families. Dokl. Russian Akad. Nauk 39 (2004) 151–156.

[236] V.A. Derkach, S. Hassi, M.M. Malamud, and H.S.V. de Snoo. Boundary relations and their Weyl families. Trans. Amer. Math. Soc. 358 (2006) 5351–5400.

[237] V.A. Derkach, S. Hassi, M.M. Malamud, and H.S.V. de Snoo. Boundary relations and generalized resolvents of symmetric operators. Russian J. Math. Phys. 16 (2009) 17–60.

[238] V.A. Derkach, S. Hassi, M.M. Malamud, and H.S.V. de Snoo. Boundary triplets and Weyl functions. Recent developments. In: *Operator methods for boundary value problems*. London Math. Soc. Lecture Note Series, Vol. 404, 2012, pp. 161–220.

[239] V.A. Derkach, S. Hassi, and H.S.V. de Snoo. Operator models associated with singular perturbations. Methods Funct. Anal. Topology 7 (2001) 1–21.

[240] V.A. Derkach, S. Hassi, and H.S.V. de Snoo. Rank one perturbations in a Pontryagin space with one negative square. J. Functional Analysis 188 (2002) 317–349.

[241] V.A. Derkach, S. Hassi, and H.S.V. de Snoo. Singular perturbations of self-adjoint operators. Math. Phys. Anal. Geom. 6 (2003) 349–384.

[242] V.A. Derkach, S. Hassi, and H.S.V. de Snoo. Asymptotic expansions of generalized Nevanlinna functions and their spectral properties. Oper. Theory Adv. Appl. 175 (2007) 51–88.

[243] V.A. Derkach and M.M. Malamud. Weyl function of Hermitian operator and its connection with characteristic function (Russian). Preprint 85-9 (104) Donetsk Fiz-Techn. Institute AN Ukrain. SSR, 1985.

[244] V.A. Derkach and M.M Malamud. On the Weyl function and Hermitian operators with gaps (Russian). Soviet Math. Dokl. 35 (1987) 393–398 [Dokl. Akad. Nauk SSSR 293 (1987) 1041–1046].

[245] V.A. Derkach and M.M. Malamud, Generalized resolvents and the boundary value problems for Hermitian operators with gaps. J. Functional Analysis 95 (1991) 1–95.

[246] V.A. Derkach and M.M. Malamud. The extension theory of Hermitian operators and the moment problem. J. Math. Sciences 73 (1995) 141–242.

[247] V.A. Derkach and M.M. Malamud. *Extension theory of symmetric operators and boundary value problems.* Proceedings of Institute of Mathematics of NAS of Ukraine, Vol. 104, 2017.

[248] V.A. Derkach, M.M. Malamud, and E.R. Tsekanovskiĭ. Sectorial extensions of positive operator (Russian). Ukr. Mat. Zh. 41 (1989) 151–158 [Ukrainian Math. J. 41 (1989) 136–142].

[249] F. Deutsch. The angle between subspaces of a Hilbert space. In: *Approximation theory, wavelets and applications.* Kluwer, 1995, pp. 107–130.

[250] F. Deutsch. *Best approximation in inner product spaces.* CMS Books in Mathematics, Springer, 2001.

[251] N.C. Dias, A. Posilicano, and J.N. Prata. Self-adjoint, globally defined Hamiltonian operators for systems with boundaries. Commun. Pure Appl. Anal. 10 (2011), no. 6, 1687–1.

[252] J.F. van Diejen and A. Tip. Scattering from generalized point interaction using self-adjoint extensions in Pontryagin spaces. J. Math. Phys. 32 (1991) 631–641.

[253] A. Dijksma and H. Langer. Operator theory and ordinary differential operators. In: *Lectures on operator theory and its applications.* Fields Institute Monographs, Vol. 3, Amer. Math. Soc., 1996, pp. 73–139.

[254] A. Dijksma and H. Langer. Compressions of self-adjoint extensions of a symmetric operator and M.G. Krein's resolvent formula. Integral Equations Operator Theory 90 (2018) 41 30 pp.

[255] A. Dijksma, H. Langer, A. Luger, and Yu. Shondin. A factorization result for generalized Nevanlinna functions of the class N_κ. Integral Equations Operator Theory 36 (2000) 121–125.

[256] A. Dijksma, H. Langer, Yu. Shondin, and C. Zeinstra. Self-adjoint operators with inner singularities and Pontryagin spaces. Oper. Theory Adv. Appl. 118 (2000) 105–175.

[257] A. Dijksma, H. Langer, and H.S.V. de Snoo. Selfadjoint Π_κ-extensions of symmetric subspaces: an abstract approach to boundary problems with spectral parameter in the boundary condition. Integral Equations Operator Theory 90 (1984) 459–515.

[258] A. Dijksma, H. Langer, and H.S.V. de Snoo. Unitary colligations in Π_κ-spaces, characteristic functions and Štraus extensions. Pacific J. Math. 125 (1986) 347–362.

[259] A. Dijksma, H. Langer, and H.S.V. de Snoo. Characteristic functions of unitary operator colligations in Π_κ-spaces. Oper. Theory Adv. Appl. 19 (1986) 125–194.

[260] A. Dijksma, H. Langer, and H.S.V. de Snoo. Unitary colligations in Krein spaces and their role in the extension theory of isometries and symmetric linear relations in Hilbert spaces. Springer Lecture Notes Math. 1242 (1987) 1–42.

[261] A. Dijksma, H. Langer, and H.S.V. de Snoo. Symmetric Sturm–Liouville operator with eigenvalue depending boundary conditions. Canadian Math. Soc. Conf. Proc. 8 (1987) 87–116.

[262] A. Dijksma, H. Langer, and H.S.V. de Snoo. Representations of holomorphic operator functions by means of resolvents of unitary or selfadjoint operators in Krein spaces. Oper. Theory Adv. Appl. 24 (1987) 123–143.

[263] A. Dijksma, H. Langer, and H.S.V. de Snoo. Hamiltonian systems with eigenvalue depending boundary conditions. Oper. Theory Adv. Appl. 35 (1988) 37–83.

[264] A. Dijksma, H. Langer, and H.S.V. de Snoo. Generalized coresolvents of standard isometric operators and generalized resolvents of standard symmetric relations in Kreĭn spaces. Oper. Theory Adv. Appl. 48 (1990) 261–274.

[265] A. Dijksma, H. Langer, and H.S.V. de Snoo. Eigenvalues and pole functions of Hamiltonian systems with eigenvalue depending boundary conditions. Math. Nachr. 161 (1993) 107–154.

[266] A. Dijksma and H.S.V. de Snoo. Self-adjoint extensions of symmetric subspaces. Pacific J. Math. 54 (1974) 71–100.

[267] A. Dijksma and H.S.V. de Snoo. Eigenfunction expansions for non-densely defined differential operators. J. Differential Equations 60 (1975) 21–56.

[268] A. Dijksma and H.S.V. de Snoo. Symmetric and selfadjoint relations in Kreĭn spaces I. Oper. Theory Adv. Appl. 24 (1987) 145–166.

[269] A. Dijksma and H.S.V. de Snoo. Symmetric and selfadjoint relations in Kreĭn spaces II. Anal. Acad. Sci. Fenn. Ser. A I Math. 12 (1987) 199–216.

[270] A. Dijksma, H.S.V. de Snoo, and A.A. El Sabbagh. Selfadjoint extensions of regular canonical systems with Stieltjes boundary conditions. J. Math. Anal. Appl. 152 (1990) 546–583.

[271] W.F. Donoghue. On the perturbation of spectra. Comm. Pure Appl. Math. 18 (1965) 559–579.

[272] W.F. Donoghue. *Monotone matrix functions and analytic continuation.* Springer, 1974.

[273] R.G. Douglas. On majorization, factorization, and range inclusion of operators on Hilbert space. Proc. Amer. Math. Soc. 17 (1966) 413–416.

[274] M.A. Dritschel and J. Rovnyak. Extension theorems for contraction operators on Krein spaces. Oper. Theory Adv. Appl. 47 (1990) 221–305.

[275] M.A. Dritschel and J. Rovnyak. Operators on indefinite inner product spaces. In: *Lectures on operator theory and its applications.* Fields Institute Monographs, Vol. 3, Amer. Math. Soc., 1996, pp. 141–232.

[276] N. Dunford and J.T. Schwarz. *Linear operators, Part II: Spectral theory.* John Wiley and Sons, 1963.

[277] H. Dym. *J-contractive matrix functions, reproducing kernel Hilbert spaces and interpolation.* Regional conference series in mathematics, Vol. 71, Amer. Math. Soc., 1989.

[278] H. Dym and McKean. *Gaussian processes, function theory, and the inverse spectral problem.* Academic Press, 1976.

[279] H. Dym and A. Iacob. Positive definite extensions, canonical equations and inverse problems. Oper. Theory Adv. Appl. 12 (1984) 141–240.

[280] W.F. Eberlein. Closure, convexity and linearity in Banach spaces. Ann. Math. 47 (1946) 688–703.

[281] J. Eckhardt, F. Gesztesy, R. Nichols, and G. Teschl. Weyl–Titchmarsh theory for Sturm–Liouville operators with distributional potentials. Opuscula Math. 33 (2013) 467–563.

[282] J. Eckhardt, F. Gesztesy, R. Nichols, and G. Teschl. Supersymmetry and Schrödinger-type operators with distributional matrix-valued potentials. J. Spectral Theory 4 (2014) 715–768.

[283] J. Eckhardt, A. Kostenko, and G. Teschl. Spectral asymptotics for canonical systems. J. Reine Angew. Math. 736 (2018) 285–315.

[284] D.E. Edmunds and W.D. Evans. *Spectral theory and differential operators.* Oxford University Press, 1987.

[285] J. Ercolano and M. Schechter. Spectral theory for operators generated by elliptic boundary problems with eigenvalue parameter in boundary conditions. I. Comm. Pure Appl. Math. 18 (1965) 83–105.

[286] J. Ercolano and M. Schechter. Spectral theory for operators generated by elliptic boundary problems with eigenvalue parameter in boundary conditions. II. Comm. Pure Appl. Math. 18 (1965) 397–414.

[287] Y. Ershova, I. Karpenko, and A. Kiselev. Isospectrality for graph Laplacians under the change of coupling at graph vertices. J. Spectral Theory 6 (2016) 43–66.

[288] Y. Ershova and A. Kiselev. Trace formulae for graph Laplacians with applications to recovering matching conditions. Methods Funct. Anal. Topology 18 (2012) 343–359.

[289] Y. Ershova and A. Kiselev. Trace formulae for Schrödinger operators on metric graphs with applications to recovering matching conditions. Methods Funct. Anal. Topology 20 (2014) 134–148.

[290] A. Etkin. On an abstract boundary value problem with the eigenvalue parameter in the boundary condition. Fields Inst. Commun. 25 (2000) 257–266.

[291] L.C. Evans. *Partial differential equations.* Graduate Series in Mathematics, Vol. 19, Amer. Math. Soc., 2010.

[292] W.N. Everitt. On a property of the m-coefficient of a second-order linear differential equation. J. London Math. Soc. (2) 4 (1972) 443–457.

[293] P. Exner. Seize ans après. In: Appendix K to *Solvable models in quantum mechanics.* 2nd edition. AMS Chelsea Publishing, Amer. Math. Soc., 2005.

[294] P. Exner, A. Kostenko, M.M. Malamud, and H. Neidhardt. Spectral theory of infinite quantum graphs. Ann. Henri Poincaré 19 (2018) 3457–3510.

[295] W. Faris. *Self-adjoint operators.* Lecture Notes in Mathematics, Vol. 437, Springer, 1975.

[296] B. Farkas and M. Matolcsi. Commutation properties of the form sum of positive, symmetric operators. Acta Sci. Math. (Szeged) 67 (2001) 777–790.

[297] A. Favini and A. Yagi. *Degenerate differential equations in Banach spaces.* CRC Press, 1998.

[298] M. Fernandez-Miranda and J.Ph. Labrousse. The Cayley transform of linear relations. Proc. Amer. Math. Soc. 133 (2005) 493–498.

[299] P. Fillmore and J. Williams. On operator ranges. Adv. Math. 7 (1971) 254–281.

[300] C. Fischbacher. The nonproper dissipative extensions of a dual pair. Trans. Amer. Math. Soc. 370 (2018) 8895–8920.

[301] C. Fischbacher. A Birman–Krein–Vishik–Grubb theory for sectorial operators. Complex Anal. Oper. Theory. To appear.

[302] C. Fischbacher, S. Naboko, and I. Wood. The proper dissipative extensions of a dual pair. Integral Equations Operator Theory 85 (2016) 573–599.

[303] A. Fleige, S. Hassi, and H.S.V. de Snoo. A Kreĭn space approach to representation theorems and generalized Friedrichs extensions. Acta Sci. Math. 66 (2000) 633–650.

[304] A. Fleige, S. Hassi, H.S.V. de Snoo, and H. Winkler. Generalized Friedrichs extensions associated with interface conditions for Sturm–Liouville operators. Oper. Theory Adv. Appl. 163 (2005) 135–145.

[305] A. Fleige, S. Hassi, H.S.V. de Snoo, and H. Winkler. Sesquilinear forms corresponding to a non-semibounded Sturm–Liouville operator. Proc. Royal Soc. Edinburgh 140A (2010) 291–318.

[306] A. Fleige, S. Hassi, H.S.V. de Snoo, and H. Winkler. Non-semibounded closed symmetric forms associated with a generalized Friedrichs extension. Proc. Roy. Soc. Edinburgh 144A (2014) 731–745.

[307] K.-H. Förster and M.M. Nafalska. A factorization of extremal extensions with applications to block operator matrices. Acta Math. Hungar. 129 (2010) 112–141.

[308] R. Freeman. Closed operators and their adjoints associated with elliptic differential operators. Pacific J. Math. 22 (1967) 71–97.

[309] H. Freudenthal. Über die Friedrichssche Fortsetzung halb-beschränkter hermitescher Operatoren. Proc. Akad. Amsterdam 39 (1936) 832–833.

[310] K.O. Friedrichs. Spektraltheorie halbbeschränkter Operatoren und Anwendung auf die Spektralzerlegung von Differentialoperatoren. Math. Ann. 109 (1933/1934) 465–487, 685–713; 110 (1935) 777–779.

[311] K.O. Friedrichs. Über die ausgezeichnete Randbedingung in der Spektraltheorie der halbbeschränkten gewöhnlichen Differentialoperatoren zweiter Ordnung. Math. Ann. 112 (1935/6) 1–23.

[312] C.T. Fulton. Parametrizations of Titchmarsh's $m(\lambda)$-functions in the limit circle case. Trans. Amer. Math. Soc. 229 (1977) 51–63.

[313] C.T. Fulton. Two-point boundary value problems with eigenvalue parameter contained in the boundary condition. Proc. Roy. Soc. Edinburgh 77A (1977) 293–308.

[314] C.T. Fulton. Singular eigenvalue problems with eigenvalue parameter contained in the boundary conditions. Proc. Roy. Soc. Edinburgh 87A (1980/81) 1–34.

[315] C.T. Fulton and H. Langer. Sturm–Liouville operators with singularities and generalized Nevanlinna functions. Complex Anal. Oper. Theory 4 (2010) 179–243.

[316] C.T. Fulton, H. Langer, and A. Luger. Mark Krein's method of directing functionals and singular potentials. Math. Nachr. 285 (2012) 1791–1798.

[317] F. Gesztesy, N.J. Kalton, K.A. Makarov, and E.R. Tsekanovskiĭ. Some applications of operator-valued Herglotz functions. Oper. Theory Adv. Appl. 123 (2001) 271–321.

[318] F. Gesztesy and K.A. Makarov. $SL_2(\mathbb{R})$, exponential Herglotz representations, and spectral averaging. St. Petersburg Math. J. 15 (2004) 393–418.

[319] F. Gesztesy, K.A. Makarov, and S.N. Naboko. The spectral shift operator. Oper. Theory Adv. Appl. 108 (1999) 59–90.

[320] F. Gesztesy, K.A. Makarov, and E.R. Tsekanovskiĭ. An addendum to Krein's formula. J. Math. Anal. Appl. 222 (1998) 594–606.

[321] F. Gesztesy, K.A. Makarov, and M. Zinchenko. Essential closures and ac spectra for reflectionless CMV, Jacobi, and Schrödinger operators revisited. Acta Applicandae Math. 103 (2008) 315–339.

[322] F. Gesztesy, D. Mitrea, I. Mitrea, and M. Mitrea. On the nature of the Laplace–Beltrami operator on Lipschitz manifolds. J. Math. Sci. 172 (2011) 279–346.

[323] F. Gesztesy and M. Mitrea. Generalized Robin boundary conditions, Robin-to-Dirichlet maps, and Krein-type resolvent formulas for Schrödinger operators on bounded Lipschitz domains. In: *Perspectives in partial differential equations, harmonic analysis and applications*. Proc. Sympos. Pure Math., Vol. 79, Amer. Math. Soc., 2008, pp. 105–173.

[324] F. Gesztesy and M. Mitrea, Robin-to-Robin maps and Krein-type resolvent formulas for Schrödinger operators on bounded Lipschitz domains. Oper. Theory Adv. Appl. 191 (2009) 81–113.

[325] F. Gesztesy and M. Mitrea. Nonlocal Robin Laplacians and some remarks on a paper by Filonov on eigenvalue inequalities. J. Differential Equations 247 (2009) 2871–2896.

[326] F. Gesztesy and M. Mitrea. A description of all self-adjoint extensions of the Laplacian and Krein-type resolvent formulas on non-smooth domains. J. Math. Anal. Appl. 113 (2011) 53–172.

[327] F. Gesztesy, S.N. Naboko, R. Weikard, and M. Zinchenko. Donoghue-type m-functions for Schrödinger operators with operator-valued potentials. J. Anal. Math. 137 (2019) 373–427.

[328] F. Gesztesy, R. Nowell, and W. Pötz. One-dimensional scattering theory for quantum systems with nontrivial spatial asymptotics. Differential Integral Equations 10 (1997) 521–546.

[329] F. Gesztesy and E.R. Tsekanovskiĭ. On matrix-valued Herglotz functions. Math. Nachr. 218 (2000) 61–138.

[330] F. Gesztesy, R. Weikard, and M. Zinchenko. On spectral theory for Schrödinger operators with operator-valued potentials. J. Differential Equations 255 (2013) 1784–1827.

[331] F. Gesztesy and M. Zinchenko. On spectral theory for Schrödinger operators with strongly singular potentials. Math. Nachr. 279 (2006) 1041–1082.

[332] F. Gesztesy and M. Zinchenko. Symmetrized perturbation determinants and applications to boundary data maps and Krein-type resolvent formulas. Proc. Lond. Math. Soc. (3) 104 (2012) 577–612.

[333] D.J. Gilbert. On subordinacy and analysis of the spectrum of Schrödinger operators with two singular endpoints. Proc. Roy. Soc. Edinburgh 112A (1989) 213–229.

[334] D.J. Gilbert. On subordinacy and spectral multiplicity for a class singular differential operators. Proc. Roy. Soc. Edinburgh 128A (1998) 549–584.

[335] D.J. Gilbert and D.B. Pearson. On subordinacy and analysis of the spectrum of one-dimensional Schrödinger operators. J. Math. Anal. Appl. 128 (1987) 30–56.

[336] R.C. Gilbert. Spectral multiplicity of selfadjoint dilations. Proc. Amer. Math. Soc. 19 (1968) 477–482.

[337] R.C. Gilbert. Simplicity of linear ordinary differential operators. J. Differential Equations 11 (1972) 672–681.

[338] R.C. Gilbert. Spectral representation of selfadjoint extensions of a symmetric operator. Rocky Mountain J. Math. 2 (1972) 75–96.

[339] I.M. Glazman. On the theory of singular differential operators. Uspekhi Mat. Nauk 5 (1950) 102–135.

[340] I. Gohberg and M.G. Kreĭn. The basic propositions on defect numbers, root numbers and indices of linear operators (Russian). Uspekhi Mat. Nauk. 12 (1957) 43–118 [Transl. Amer. Math. Soc. (2) 13 (1960) 185–264].

[341] I. Gohberg and M.G. Kreĭn. *Theory and applications of Volterra operators in Hilbert space.* Transl. Math. Monographs, Vol. 24, Amer. Math. Soc., 1970.

[342] N. Goloshchapova. On the multi-dimensional Schrödinger operators with point interactions. Methods Funct. Anal. Topology 17 (2011) 126–143.

[343] N. Goloshchapova, M.M. Malamud, and V. Zastavnyi. Radial positive definite functions and spectral theory of the Schrödinger operators with point interactions. Math. Nachr. 285 (2012) 1839–1859.

[344] N. Goloshchapova and L. Oridoroga. On the negative spectrum of one-dimensional Schrödinger operators with point interactions. Integral Equations Operator Theory 67 (2010) 1–14.

[345] M.L. Gorbachuk. Self-adjoint boundary value problems for differential equation of the second order with unbounded operator coefficient (Russian). Funktsional. Anal. i Prilozhen 5 (1971) 10–21 [Funct. Anal. Appl. 5 (1971) 9–18].

[346] V.I. Gorbachuk and M.L. Gorbachuk. *Boundary value problems for operator differential equations* (Russian). Naukova Dumka, Kiev, 1984 [Kluwer, 1991].

[347] V.I. Gorbachuk, M.L. Gorbachuk, and A.N. Kochubei. Extension theory for symmetric operators and boundary value problems for differential equations (Russian). Ukr. Mat. Zh. 41 (1989) 1299–1313 [Ukrainian Math. J. 41 (1989) 1117–1129].

[348] V.I. Gorbachuk and M.L. Gorbachuk. *M.G. Krein's lectures on entire operators.* Oper. Theory Adv. Appl., Vol. 97, Birkhäuser, 1991.

[349] M.L. Gorbachuk and V.A. Mikhailets. Semibounded selfadjoint extensions of symmetric operators (Russian). Dokl. Akad. Nauk SSSR 226 (1976) 765–768 [Sov. Math. Dokl. 17 (1976) 185–186].

[350] Y.I. Granovskyi and L.L. Oridoroga. Krein–von Neumann extension of an even order differential operator on a finite interval. Opuscula Math. 38 (2018) 681–698.

[351] P. Grisvard. *Elliptic problems in nonsmooth domains.* Pitman, 1985.

[352] G. Grubb. A characterization of the non-local boundary value problems associated with an elliptic operator. Ann. Scuola Norm. Sup. Pisa (3) 22 (1968) 425–513.

[353] G. Grubb. Les problems aux limites généraux d'un opérateur elliptique, provenant de la théorie variationnelle. Bull. Sci. Math. 94 (1970) 113–157.

[354] G. Grubb. On coerciveness and semiboundedness of general boundary problems. Israel J. Math. 10 (1971) 32–95.

[355] G. Grubb. Properties of normal boundary problems for elliptic even-order systems. Ann. Scuola Norm. Sup. Pisa (4) 1 (1974) 1–61.

[356] G. Grubb. Spectral asymptotics for the "soft" selfadjoint extension of a symmetric elliptic differential operator. J. Operator Theory 10 (1983) 9–20.

[357] G. Grubb. Known and unknown results on elliptic boundary problems. Bull. Amer. Math. Soc. 43 (2006) 227–230.

[358] G. Grubb. Krein resolvent formulas for elliptic boundary problems in nonsmooth domains. Rend. Semin. Mat. Univ. Politec. Torino 66 (2008) 13–39.

[359] G. Grubb. *Distributions and operators.* Graduate Texts in Mathematics, Vol. 252, Springer, 2009.

[360] G. Grubb. The mixed boundary value problem, Krein resolvent formulas and spectral asymptotic estimates. J. Math. Anal. Appl. 382 (2011) 339–363.

[361] G. Grubb. Spectral asymptotics for Robin problems with a discontinuous coefficient. J. Spectral Theory 1 (2011) 155–177.

[362] G. Grubb. Krein-like extensions and the lower boundedness problem for elliptic operators. J. Differential Equations 252 (2012) 852–885.

[363] G. Grubb. Extension theory for elliptic partial differential operators with pseudodifferential methods. In: *Operator methods for boundary value problems.* London Math. Soc. Lecture Note Series, Vol. 404, 2012, pp. 221–258.

[364] L. Grubišić, V. Kostrykin, K.A. Makarov, and K. Veselić. Representation theorems for indefinite quadratic forms revisited. Mathematika 59 (2013) 169–189.

[365] M. Haase. *The functional calculus for sectorial operators.* Oper. Theory Adv. Appl., Vol. 169, Birkhäuser, 2006.

[366] B.J. Harris. The asymptotic form of the Titchmarsh–Weyl m-function. J. London Math. Soc. (2) 30 (1984) 110–118.

[367] B.J. Harris. The asymptotic form of the Titchmarsh–Weyl m-function associated with a second order differential equation with locally integrable coefficient. Proc. Roy. Soc. Edinburgh 102A (1986) 243–251.

[368] P. Hartman. Self-adjoint, non-oscillatory systems of ordinary, second order, linear differential equations. Duke Math. J. 24 (1957) 25–35.

[369] P. Hartman. *Ordinary differential equations.* 2nd edition. SIAM Classics in Mathematics, 1982.

[370] S. Hassi. On the Friedrichs and the Kreĭn–von Neumann extension of nonnegative relations. Acta Wasaensia 122 (2004) 37–54.

[371] S. Hassi, M. Kaltenbäck, and H.S.V. de Snoo. Triplets of Hilbert spaces and Friedrichs extensions associated with the subclass N_1 of Nevanlinna functions. J. Operator Theory 37 (1997) 155–181.

[372] S. Hassi, M. Kaltenbäck, and H.S.V. de Snoo. A characterization of semibounded selfadjoint operators. Proc. Amer. Math. Soc. 124 (1997) 2681–2692.

[373] S. Hassi, M. Kaltenbäck, and H.S.V. de Snoo. Selfadjoint extensions of the orthogonal sum of symmetric relations. I. 16th Oper. Theory Conf. Proc. (1997) 163–178.

[374] S. Hassi, M. Kaltenbäck, and H.S.V. de Snoo. Generalized Kreĭn–von Neumann extensions and associated operator models. Acta Sci. Math. (Szeged) 64 (1998) 627–655.

[375] S. Hassi, M. Kaltenbäck, and H.S.V. de Snoo. Selfadjoint extensions of the orthogonal sum of symmetric relations. II. Oper. Theory Adv. Appl. 106 (1998) 187–200.

[376] S. Hassi, M. Kaltenbäck, and H.S.V. de Snoo. The sum of matrix Nevanlinna functions and selfadjoint exit space extensions. Oper. Theory Adv. Appl. 103 (1998) 137–154.

[377] S. Hassi and S. Kuzhel. On J-self-adjoint operators with stable C-symmetries. Proc. Roy. Soc. Edinburgh 143A (2013) 141–167.

[378] S. Hassi, J.Ph. Labrousse, and H.S.V. de Snoo. Selfadjoint extensions of relations whose domain and range are orthogonal. Methods Funct. Anal. Topology. To appear.

[379] S. Hassi, H. Langer, and H.S.V. de Snoo. Selfadjoint extensions for a class of symmetric operators with defect numbers (1,1). 15th Oper. Theory Conf. Proc. (1995) 115–145.

[380] S. Hassi and A. Luger. Generalized zeros and poles of N_κ-functions: on the underlying spectral structure. Methods Funct. Anal. Topology 12 (2006) 131–150.

[381] S. Hassi, M.M. Malamud, and V.I. Mogilevskiĭ. Generalized resolvents and boundary triplets for dual pairs of linear relations. Methods Funct. Anal. Topology 11 (2005) 170–187.

[382] S. Hassi, M.M. Malamud, and V.I. Mogilevskiĭ. Unitary equivalence of proper extensions of a symmetric operator and the Weyl function. Integral Equations Operator Theory 77 (2013) 449–487.

[383] S. Hassi, M.M. Malamud, and H.S.V. de Snoo. On Kreĭn's extension theory of nonnegative operators. Math. Nachr. 274/275 (2004) 40–73.

[384] S. Hassi, M. Möller, and H.S.V. de Snoo. Sturm–Liouville operators and their spectral functions. J. Math. Anal. Appl. 282 (2003) 584–602.

[385] S. Hassi, M. Möller, and H.S.V. de Snoo. Singular Sturm–Liouville equations whose coefficients depend rationally on the eigenvalue parameter. J. Math. Anal. Appl. 295 (2004) 258–275.

[386] S. Hassi, M. Möller, and H.S.V. de Snoo. A class of Nevanlinna functions associated with Sturm–Liouville operators. Proc. Amer. Math. Soc. 134 (2006) 2885–2893.

[387] S. Hassi, M. Möller, and H.S.V. de Snoo. Limit-point/limit-circle classification for Hain-Lüst type equations. Math. Nachr. 291 (2018) 652–668.

[388] S. Hassi, C. Remling, and H.S.V. de Snoo. Subordinate solutions and spectral measures of canonical systems. Integral Equations Operator Theory 37 (2000) 48–63.

[389] S. Hassi, A. Sandovici, and H.S.V. de Snoo. Factorized sectorial relations, their maximal sectorial extensions, and form sums. Banach J. Math. Anal. 13 (2019) 538–564.

[390] S. Hassi, A. Sandovici, H.S.V. de Snoo, and H. Winkler. Form sums of nonnegative selfadjoint operators. Acta Math. Hungar. 111 (2006) 81–105.

[391] S. Hassi, A. Sandovici, H.S.V. de Snoo, and H. Winkler. A general factorization approach to the extension theory of nonnegative operators and relations. J. Operator Theory 58 (2007) 351–386.

[392] S. Hassi, A. Sandovici, H.S.V. de Snoo, and H. Winkler. Extremal extensions for the sum of nonnegative selfadjoint relations. Proc. Amer. Math. Soc. 135 (2007) 3193–3204.

[393] S. Hassi, A. Sandovici, H.S.V. de Snoo, and H. Winkler. Extremal maximal sectorial extensions of sectorial relations. Indag. Math. 28 (2017) 1019–1055.

[394] S. Hassi, A.J. van der Schaft, H.S.V. de Snoo, and H.J. Zwart. Dirac structures and boundary relations. In: *Operator methods for boundary value problems*. London Math. Soc. Lecture Note Series, Vol. 404, 2012, pp. 259–274.

[395] S. Hassi, Z. Sebestyén, and H.S.V. de Snoo. Lebesgue type decompositions for nonnegative forms. J. Functional Analysis 257 (2009) 3858–3894.

[396] S. Hassi, Z. Sebestyén, and H.S.V. de Snoo. Domain and range descriptions for adjoint relations, and parallel sums and differences of forms. Oper. Theory Adv. Appl. 198 (2009) 211–227.

[397] S. Hassi, Z. Sebestyén, and H.S.V. de Snoo. Lebesgue type decompositions for linear relations and Ando's uniqueness criterion. Acta Sci. Math. (Szeged) 84 (2018) 465–507.

[398] S. Hassi, Z. Sebestyén, H.S.V. de Snoo, and F.H. Szafraniec. A canonical decomposition for linear operators and linear relations. Acta Math. Hungar. 115 (2007) 281–307.

[399] S. Hassi and H.S.V. de Snoo. One-dimensional graph perturbations of selfadjoint relations. Ann. Acad. Sci. Fenn. Ser. A I Math. 22 (1997) 123–164.

[400] S. Hassi and H.S.V. de Snoo. On rank one perturbations of selfadjoint operators. Integral Equations Operator Theory 29 (1997) 288–300.

[401] S. Hassi and H.S.V. de Snoo. Nevanlinna functions, perturbation formulas and triplets of Hilbert spaces. Math. Nachr. 195 (1998) 115–138.

[402] S. Hassi and H.S.V. de Snoo. Factorization, majorization, and domination for linear relations. Annales Univ. Sci. Budapest 58 (2015) 53–70.

[403] S. Hassi, H.S.V. de Snoo, A.E. Sterk, and H.Winkler. Finite-dimensional graph perturbations of selfadjoint Sturm–Liouville operators. In: *Operator theory, structured matrices, and dilations*. Theta Foundation, 2007, pp. 205–226.

[404] S. Hassi, H.S.V. de Snoo, and F.H. Szafraniec. Componentwise and Cartesian decompositions of linear relations. Dissertationes Mathematicae 465 (2009) 59 pp.

[405] S. Hassi, H.S.V. de Snoo, and F.H. Szafraniec. Infinite-dimensional perturbations, maximally nondensely defined symmetric operators, and some matrix representations. Indag. Math. 23 (2012) 1087–1117.

[406] S. Hassi, H.S.V. de Snoo, and F.H. Szafraniec. Normal intermediate extensions of symmetric relations. Acta Math. Szeged 80 (2014) 195–232.

[407] S. Hassi, H.S.V. de Snoo, and A.D.I. Willemsma. Smooth rank one perturbations of selfadjoint operators. Proc. Amer. Math. Soc. 126 (1998) 2663–2675.

[408] S. Hassi, H.S.V. de Snoo, and H. Winkler. Boundary-value problems for two-dimensional canonical systems. Integral Equations Operator Theory 36 (2000) 445–479.

[409] S. Hassi, H.S.V. de Snoo, and H. Woracek. Some interpolation problems of Nevanlinna-Pick type. The Kreĭn–Langer method. Operator Theory Adv. Appl. 106 (1998) 201–216.

[410] S. Hassi and H.L. Wietsma, On Calkin's abstract symmetric boundary conditions. In: *Operator methods for boundary value problems*. London Math. Soc. Lecture Note Series, Vol. 404, 2012, pp. 3–34.

[411] S. Hassi and H.L. Wietsma. Products of generalized Nevanlinna functions with symmetric rational functions. J. Functional Analysis 266 (2014) 3321–3376.

[412] E. Heinz. Halbbeschränktheit gewöhnlicher Differentialoperatoren höherer Ordnung. Math. Ann. 135 (1958) 1–49.

[413] B. Helffer and A. Kachmar. Eigenvalues for the Robin Laplacian in domains with variable curvature. Trans. Amer. Math. Soc. 369 (2017) 3253–3287.

[414] G. Hellwig. *Differential operators of mathematical physics*. Addison-Wesley, 1967.

[415] R. Hempel, L. Seco, and B. Simon. The essential spectrum of Neumann Laplacians on some bounded, singular domains. J. Functional Analysis 102 (1991) 448–483.

[416] E. Hewitt and K. Stromberg. *Real and abstract analysis*. Springer, 1965.

[417] E. Hilb. Über Integraldarstellungen willkürlicher Funktionen. Math. Ann. 66 (1909) 1–66.

[418] E. Hilb. Über Reihentwicklungen nach den Eigenfunktionen linearer Differentialgleichungen 2ter Ordnung. Math. Ann. 71 (1911) 76–87.

[419] E. Hilb. Über Reihentwicklungen, welche aus speziellen Randwertproblemen bei gewöhnlichen linearen inhomogenen Differentialgleichungen entspringen. J. Reine Angew. Math. 140 (1911) 205–229.

[420] E. Hille. *Lectures on ordinary differential equations.* Addison-Wesley, 1969.

[421] E. Hille and R.S. Phillips. *Functional analysis and semi-groups.* Amer. Math. Soc. Colloquium Publications, Volume 31, 1996.

[422] D.B. Hinton and A. Schneider. On the Titchmarsh–Weyl coefficients for singular S-Hermitian systems I. Math. Nachr. 163 (1993) 323–342.

[423] D.B. Hinton and A. Schneider. On the Titchmarsh–Weyl coefficients for singular S-Hermitian systems II. Math. Nachr. 185 (1997) 67–84.

[424] D.B. Hinton and A. Schneider. Titchmarsh–Weyl coefficients for odd-order linear Hamiltonian systems. J. Spectral Math. Appl. 1 (2006) 1–36.

[425] D.B. Hinton and J.K. Shaw. On Titchmarsh–Weyl $m(\lambda)$-functions for linear Hamiltonian systems. J. Differential Equations 40 (1981) 316–342.

[426] D.B. Hinton and J.K. Shaw. Titchmarsh–Weyl theory for Hamiltonian systems. In: *Spectral theory and differential operators.* North-Holland Math. Studies, Vol. 55, 1981, pp. 219–231.

[427] D.B. Hinton and J.K. Shaw. Hamiltonian systems of limit point or limit circle type with both endpoints singular. J. Differential Equations 50 (1983) 444–464.

[428] D.B. Hinton and J.K. Shaw. On boundary value problems for Hamiltonian systems with two singular endpoints. SIAM J. Math. Anal. 15 (1984) 272–286.

[429] A. Ibort, F. Lledó, and J.M. Pérez-Pardo. Self-adjoint extensions of the Laplace–Beltrami operator and unitaries at the boundary. J. Functional Analysis 268 (2015) 634–670.

[430] I.S. Iohvidov, M.G. Krein, and H. Langer. *Introduction to the spectral theory of operators in spaces with an indefinite metric.* Mathematische Forschung, Band 9, Akademie Verlag, 1982.

[431] D. Jerison and C. Kenig. The Neumann problem in Lipschitz domains. Bull. Amer. Math. Soc. 4 (1981) 203–207.

[432] D. Jerison and C. Kenig. The inhomogeneous Dirichlet problem in Lipschitz domains. J. Functional Analysis 113 (1995) 161–219.

[433] S.Y. Jitomirskaya, Y. Last, R. del Rio, and B. Simon. Operators with singular continuous spectrum. IV. Hausdorff dimensions, rank one perturbations, and localization. J. Anal. Math. 69 (1996) 153–200.

[434] P. Jonas. A class of operator-valued meromorphic functions on the unit disc. Ann. Acad. Sci. Fenn. Math. 17 (1992) 257–284.

[435] P. Jonas. Operator representations of definitizable functions. Ann. Acad. Sci. Fenn. Math. 25 (2000) 41–72.

[436] P. Jonas. On locally definite operators in Krein spaces. In: *Spectral analysis and its applications.* Theta Foundation, 2003, pp. 95–127.

[437] P. Jonas. On operator representations of locally definitizable functions. Oper. Theory Adv. Appl. 162 (2005) 165–190.

[438] K. Jörgens and F. Rellich. *Eigenwerttheorie gewöhnlicher Differentialgleichungen.* Springer, 1976.

[439] P.E.T. Jorgensen. Unbounded operators; perturbations and commutativity problems. J. Functional Analysis 39 (1980) 281–307.

[440] I.S. Kac. On the Hilbert spaces generated by monotone Henrmitian matrix functions. Zap. Mat. Otd. Fiz.-Mat. Fak. i Har'kov. Mat. Obšč 22 (1950) 95–113 (1951).

[441] I.S. Kac. Spectral multiplicity of a second-order differential operator and expansion in eigenfunctions. Izv. Akad Nauk. SSSR Ser. Mat 27 (1963) 1081–1112.

[442] I.S. Kac. Linear relations generated by canonical differential equations (Russian). Funktsional. Anal. i Prilozhen 17 (1983) 86-87.

[443] I.S. Kac. Linear relations, generated by a canonical differential equation on an interval with regular endpoints, and the expansibility in eigenfunctions (Russian). Deposited Paper, Odessa, 1984.

[444] I.S. Kac. Expansibility in eigenfunctions of a canonical differential equation on an interval with singular endpoints and associated linear relations (Russian). Deposited in Ukr NIINTI, No. 2111, 1986. (VINITI Deponirovannye Nauchnye Raboty, No. 12 (282), b.o. 1536, 1986).

[445] I.S. Kac. Linear relations, generated by a canonical differential equation and expansibility in eigenfunctions (Russian). Algebra i Analiz 14 (2002) 86–120 [St. Petersburg Math. J. 14 (2003) 429–452].

[446] I.S. Kac and M.G. Kreĭn. R-functions – analytic functions mapping the upper halfplane into itself. Supplement I to the Russian edition of F.V. Atkinson, *Discrete and continuous boundary problems* (Russian). Mir, 1968 [Amer. Math. Soc. Transl. (2) 103 (1974) 1–18].

[447] I.S. Kac and M.G. Kreĭn. On the spectral functions of the string. Supplement II to the Russian edition of F.V. Atkinson, *Discrete and continuous boundary problems* (Russian). Mir, 1968 [Amer. Math. Soc. Transl. (2) 103 (1974), 19–102].

[448] H. Kalf. A characterization of the Friedrichs extension of Sturm–Liouville operators. J. London Math. Soc. (2) 17 (1978) 511–521.

[449] H. Kalf and J. Walter. Strongly singular potentials and essential self-adjointness of singular elliptic operators in $C_0^\infty(\mathbb{R}^n \setminus \{0\})$. J. Functional Analysis 10 (1972) 114–130.

[450] M. Kaltenbäck. *Ausgewählte Kapitel Operatortheorie*. Lecture Notes, TU Wien.

[451] M. Kaltenbäck. *Funktionalanalysis* 2. Lecture Notes, TU Wien.

[452] M. Kaltenbäck, H. Winkler, and H. Woracek. Almost Pontryagin spaces. Oper. Theory Adv. Appl. 160 (2005) 253–271.

[453] M. Kaltenbäck, H. Winkler, and H. Woracek. Strings, dual strings, and related canonical systems. Math. Nachr. 280 (2007) 1518–1536.

[454] M. Kaltenbäck and H. Woracek. Pontryagin spaces of entire functions I. Integral Equations Operator Theory 33 (1999) 34–97.

[455] M. Kaltenbäck and H. Woracek. Pontryagin spaces of entire functions II. Integral Equations Operator Theory 33 (1999) 305–380.

[456] M. Kaltenbäck and H. Woracek. Pontryagin spaces of entire functions III. Acta Sci. Math. (Szeged) 69 (2003) 241–310.

[457] M. Kaltenbäck and H. Woracek. Pontryagin spaces of entire functions IV. Acta Sci. Math. (Szeged) 72 (2006) 709–835.

[458] M. Kaltenbäck and H. Woracek. Pontryagin spaces of entire functions V. Acta Sci. Math. (Szeged) 77 (2011) 223–336.

[459] M. Kaltenbäck and H. Woracek. Pontryagin spaces of entire functions VI. Acta Sci. Math. (Szeged) 76 (2010) 511–560.

[460] M. Kaltenbäck and H. Woracek. *Kanonische Systeme: Direkte Spektraltheorie*. Winter School, Reichenau, 2015.

[461] I.M. Karabash, A. Kostenko, and M.M. Malamud. The similarity problem for J-nonnegative Sturm–Liouville operators. J. Differential Equations 246 (2009) 964–997.

[462] T. Kato. *Perturbation theory for linear operators*. 1980 edition. Classics in Mathematics, Springer, 1995.

[463] S. Khan and D.B. Pearson. Subordinacy and spectral theory for infinite matrices. Helv. Phys. Acta 65 (1992) 505–527.

[464] Y. Kilpi. Über selbstadjungierte Fortsetzungen symmetrischer Transformationen im Hilbert-Raum. Ann. Acad. Sci. Fenn. Ser. A I 264 (1959) 23 pp.

[465] T. Kimura and M. Takahashi. Sur les opérateurs différentiels ordinaires linéaires formellement autoadjoints. I. Funkcial. Ekvac. 7 (1965) 35–90.

[466] A.N. Kochubeĭ. On extensions of symmetric operators and symmetric binary relations (Russian). Mat. Zametki 17 (1975) 41–48 [Math. Notes 17 (1975), 25–28].

[467] A.N. Kochubeĭ. Extensions of a positive definite symmetric operator (Russian). Dokl. Akad. Nauk Ukrain. SSR Ser. A 3 (1979) 168–171, 237 [Amer. Math. Soc. Transl. (2) 124 (1984) 139–142].

[468] A.N. Kochubeĭ. Characteristic functions of symmetric operators and their extensions (Russian). Izv. Akad. Nauk Armyan. SSR Ser. Mat. 15 (1980) 219–232 [Soviet J. Contemporary Math. Anal. 15 (1980)].

[469] K. Kodaira. The eigenvalue problem for ordinary differential equations of the second order and Heisenberg's theory of S-matrices. Amer. J. Math. 71 (1949) 921–945.

[470] K. Kodaira. On ordinary differential equations of any even order and the corresponding eigenfunction expansions. Amer. J. Math. 72 (1950) 501–544.

[471] V.I. Kogan and F.S. Rofe-Beketov. On square-integrable solutions of symmetric systems of differential equations of arbitrary order. Proc. Roy. Soc. Edinburgh 74A (1976) 5–40.

[472] V.Yu. Kolmanovich and M.M. Malamud. Extensions of sectorial operators and dual pairs of contractions (Russian). Deposited at VINITI No. 4428–85, Donetsk, 1985.

[473] Q. Kong, M. Möller, H. Wu, and A. Zettl. Indefinite Sturm–Liouville problems. Proc. Roy. Soc. Edinburgh 133A (2003) 639–652.

[474] Q. Kong, H. Wu, and A. Zettl. Singular left-definite Sturm–Liouville problems. J. Differential Equations 206 (2004) 1–29.

[475] N.D. Kopachevskiĭ and S.G. Kreĭn. Abstract Green formula for a triple of Hilbert spaces, abstract boundary-value and spectral problems (Russian). Ukr. Mat. Visn. 1 (2004) 69–97 [Ukr. Math. Bull. 1 (2004), 77–105].

[476] K. Koshmanenko. *Singular quadratic forms in perturbation theory*. Kluwer, 1999.

[477] A. Kostenko and M.M. Malamud. 1-D Schrödinger operators with local point interactions on a discrete set. J. Differential Equations 249 (2010) 253–304.

[478] A. Kostenko, M.M. Malamud, and D.D. Natyagajlo. The matrix Schrödinger operator with δ-interactions (Russian). Mat. Zametki 100 (2016) 59–77 [Math. Notes 100 (2016) 49–65].

[479] A. Kostenko, A. Sakhnovich, and G. Teschl. Weyl–Titchmarsh theory for Schrödinger operators with strongly singular potentials. Int. Math. Res. Not. IMRN 2012 (8) 1699–1747.

[480] A. Kostenko, A. Sakhnovich, and G. Teschl. Commutation methods for Schrödinger operators with strongly singular potentials. Math. Nachr. 285 (2012) 392–410.

[481] A. Kostenko and G. Teschl. On the singular Weyl–Titchmarsh function of perturbed spherical Schrödinger operators. J. Differential Equations 250 (2011) 3701–3739.

[482] A. Kostenko and G. Teschl. Spectral asymptotics for perturbed spherical Schrödinger operators and applications to quantum scattering. Comm. Math. Phys. 322 (2013) 255–275.

[483] Yu.G. Kovalev. 1-D nonnegative Schrödinger operators with point interactions. Mat. Stud. 39 (2013) 150–163.

[484] A.M. Krall. Differential-boundary operators. Trans. Amer. Math. Soc. 154 (1971) 429–458.

[485] A.M. Krall. The development of general differential and general differential boundary systems. Rocky Mountain J. Math. 5 (1975) 493–542.

[486] A.M. Krall. nth Order Stieltjes differential boundary operators and Stieltjes differential boundary systems. J. Differential Equations 24 (1977) 253–267.

[487] A.M. Krall. $M(\lambda)$-theory for singular Hamiltonian systems with one singular endpoint. SIAM J. Math. Anal. 20 (1989) 664–700.

[488] A.M. Krall. Left-definite regular Hamiltonian systems. Math. Nachr. 174 (1995) 203–217.

[489] A.M. Krall. *Hilbert space, boundary value problems and orthogonal polynomials*. Oper. Theory Adv. Appl., Vol. 133, Birkhäuser, 2002.

[490] M.A. Krasnoselskiĭ. On selfadjoint extensions of Hermitian operators. Ukr. Mat. Zh. 1 no. 1 (1949) 21–38.

[491] M.G. Kreĭn. On Hermitian operators whose deficiency indices are 1. Dokl. Akad. Nauk URSS 43 (1944) 323–326.

[492] M.G. Kreĭn. Concerning the resolvents of an Hermitian operator with the deficiency-index (m, m). Dokl. Akad. Nauk URSS 52 (1946) 651–654.

[493] M.G. Kreĭn. The theory of selfadjoint extensions of semibounded Hermitian operators and its applications I. Mat. Sb. 20 (62) (1947) 431–495.

[494] M.G. Kreĭn. The theory of selfadjoint extensions of semibounded Hermitian operators and its applications II. Mat. Sb. 21 (63) (1947) 366–404.

[495] M.G. Kreĭn. On Hermitian operators with directing functionals. Sbirnik Praz Inst. Math. AN URSS 10 (1948) 83–106.

[496] M.G. Kreĭn. The fundamental propositions of the theory of representations of Hermitian operators with deficiency index (m, m) (Russian). Ukr. Mat. Zh. 1 no. 2 (1949) 3–66.

[497] M.G. Kreĭn and H. Langer. On defect subspaces and generalized resolvents of Hermitian operator in Pontryagin space. Funct. Anal. Appl. 5 (1971) 136–146; 217–228.

[498] M.G. Kreĭn and H. Langer. Über die verallgemeinerten Resolventen und die charakteristische Funktion eines isometrischen Operators in Raume Π_κ. In: *Colloquia Math. Soc. Janos Bolyai* 5. North Holland, 1972, pp. 353–399.

[499] M.G. Kreĭn and H. Langer. Über die Q-Funktion eines π-hermiteschen Operators im Raume Π_κ. Acta Sci. Math. (Szeged) 34 (1973) 191–230.

[500] M.G. Kreĭn and H. Langer. Über einige Fortsetzungsprobleme, die eng mit der Theorie hermitescher Operatoren im Raume Π_κ zusammenhängen. I. Einige Funktionenklassen und ihre Darstellungen. Math. Nachr. 77 (1977) 187–236.

[501] M.G. Kreĭn and H. Langer. Über einige Fortsetzungsprobleme, die eng mit der Theorie hermitescher Operatoren im Raume Π_κ zusammenhängen. II. Verallgemeinerte Resolventen, u-Resolventen und ganze Operatoren. J. Functional Analysis 30 (1978) 390–447.

[502] M.G. Kreĭn and H. Langer. On some extension problems which are closely connected with the theory of Hermitian operators in a space Π_κ. III. Indefinite analogues of the Hamburger and Stieltjes moment problems. Part (I). Beiträge Analysis 14 (1979) 25–40. Part (II). Beiträge Analysis 15 (1980) 27–45.

[503] M.G. Kreĭn and H. Langer. On some continuation problems which are closely related to the theory of operators in spaces Π_κ. IV. Continuous analogues of orthogonal polynomials on the unit circle with respect to an indefinite weight and related continuation problems for some classes of functions. J. Operator Theory 13 (1985) 299–417.

[504] M.G. Kreĭn and H. Langer. Continuation of Hermitian positive definite functions and related questions. Integral Equations Operator Theory 78 (2014) 1–69.

[505] M.G. Kreĭn and A.A. Nudelman. *Markov moment problem and extremal problems.* Transl. Mathematical Monographs, Vol. 50, Amer. Math. Soc., 1977.

[506] M.G. Kreĭn and Yu.L. Shmuljan. On linear fractional transformations with operator coefficients (Russian). Mat. Issled 2 (1967) 64–96 [Amer. Math. Soc. Transl. (2) 103 (1974) 125–152].

[507] P. Kurasov and A. Luger. An operator theoretic interpretation of the generalized Titchmarsh–Weyl coefficient for a singular Sturm–Liouville problem. Math. Phys. Anal. Geom. 14 (2011) 115–151.

[508] T. Kusche, R. Mennicken, and M. Möller. Friedrichs extension and essential spectrum of systems of differential operators of mixed order. Math. Nachr. 278 (2005) 1591–1606.

[509] A. Kuzhel and S. Kuzhel. *Regular extensions of Hermitian operators.* Kluwer, 1998.

[510] S. Kuzhel. On some properties of unperturbed operators. Methods Funct. Anal. Topology 2 (1996) 78–84.

[511] S. Kuzhel and M. Znojil. Non-self-adjoint Schrödinger operators with nonlocal one-point interactions. Banach J. Math. Anal. 11 (2017) 923–944.

[512] J.Ph. Labrousse. Geodesics in the space of linear relations in a Hilbert space. 18th Oper. Theory Conf. Proc. (2000) 213–234.

[513] J.Ph. Labrousse. Idempotent linear relations. In: *Spectral theory and its applications.* The Theta Foundation, 2003, pp. 121–141.

[514] J.Ph. Labrousse. Bisectors, isometries and connected components in Hilbert spaces. Oper. Theory Adv. Appl. 198 (2009) 239–258.

[515] J.Ph. Labrousse, A. Sandovici, H.S.V. de Snoo, and H. Winkler. The Kato decomposition of quasi-Fredholm relations. Operators Matrices 4 (2010) 1–51.

[516] J.Ph. Labrousse, A. Sandovici, H.S.V. de Snoo, and H. Winkler. Closed linear relations and their regular points. Operators Matrices 6 (2012) 681–714.

[517] O.A. Ladyzhenskaya and N.N. Uraltseva. *Linear and quasilinear elliptic equations.* Mathematics in Science and Engineering, Vol. 46, Academic Press, 1968.

[518] H. Langer. Über die Methode der richtenden Funktionale von M.G. Kreĭn. Acta Math. Acad. Sci. Hung. 21 (1970) 207–224.

[519] H. Langer. Spectral functions of definitizable operators in Krein spaces. Springer Lecture Notes Math. 948 (1982) 1–46.

[520] H. Langer and R. Mennicken. A transformation of right-definite S-hermitian systems to canonical systems. Differential Integral Equations 3 (1990) 901–908.

[521] H. Langer and M. Möller. Linearization of boundary eigenvalue problems. Integral Equations Operator Theory 14 (1991) 105–119.

[522] H. Langer and P. Sorjonen. Verallgemeinerte Resolventen hermitischer und isometrischer Operatoren im Pontryaginraum. Ann. Acad. Sci. Fenn. Ser. A I561 (1974) 45 pp.

[523] H. Langer and B. Textorius. On generalized resolvents and Q-functions of symmetric linear relations (subspaces) in Hilbert space. Pacific J. Math. 72 (1977) 135–165.

[524] H. Langer and B. Textorius. L-resolvent matrices of symmetric linear relations with equal defect numbers; applications to canonical differential relations. Integral Equations Operator Theory 5 (1982) 208–243.

[525] H. Langer and B. Textorius. Generalized resolvents of dual pairs of contractions. Oper. Theory Adv. Appl. 6 (1982) 103–118.

[526] H. Langer and B. Textorius. Spectral functions of a symmetric linear relation with a directing mapping, I. Proc. Roy. Soc. Edinburgh 97A (1984) 165–176.

[527] H. Langer and B. Textorius. Spectral functions of a symmetric linear relation with a directing mapping, II. Proc. Roy. Soc. Edinburgh 101A (1985) 111–124.

[528] H. Langer and H. Winkler. Direct and inverse spectral problems for generalized strings. Integral Equations Operator Theory 30 (1998) 409–431.

[529] M. Langer. A general HELP inequality connected with symmetric operators. Proc. Roy. Soc. London 462A (2006) 587–606.

[530] M. Langer and A. Luger. Scalar generalized Nevanlinna functions: realizations with block operator matrices. Oper. Theory Adv. Appl. 162 (2006) 253–276.

[531] M. Langer and H. Woracek. Dependence of the Weyl coefficient on singular interface conditions. Proc. Edinburgh Math. Soc. 52 (2009) 445–487.

[532] M. Langer and H. Woracek. A local inverse spectral theorem for Hamiltonian systems. Inverse Problems 27 (2011), no. 5, 055002, 17 pp.

[533] Y. Last. Quantum dynamics and decompositions of singular continuous spectra. J. Functional Analysis 142 (1996) 406–445.

[534] P.D. Lax. On Cauchy's problem for hyperbolic equations and the differentiability of solutions of elliptic equations. Comm. Pure Appl. Math. 8 (1955) 615–633.

[535] W. Leighton and M. Morse. Singular quadratic functions. Trans. Amer. Math. Soc. 40 (1936) 252–288.

[536] D. Lenz, C. Schubert, and I. Veselić. Unbounded quantum graphs with unbounded boundary conditions. Math. Nachr. 287 (2014) 962–979.

[537] J. Leray. *Lectures on hyperbolic equations with variable coefficients.* Institute for Advanced Studies, Princeton, 1952.

[538] M. Lesch and M.M. Malamud. On the deficiency indices and self-adjointness of symmetric Hamiltonian systems. J. Differential Equations 189 (2003) 556–615.

[539] N. Levinson. A simplified proof of the expansion theorem for singular second order linear differential equations. Duke Math. J. 18 (1951) 57–71.

[540] N. Levinson. The expansion theorem for singular self-adjoint differential operators. Ann. Math. (2) 59 (1954) 300–315.

[541] B.M. Levitan. Proof of the theorem on the expansion in the eigenfunctions of selfadjoint differential equations. Dokl. Akad. Nauk SSSR 73 (1950) 651–654.

[542] B.M. Levitan and I.S. Sargsjan. *Sturm–Liouville and Dirac operators.* Kluwer, 1991.

[543] M. Levitin and L. Parnovski. On the principal eigenvalue of a Robin problem with a large parameter. Math. Nachr. 281 (2008) 272–281.

[544] J.L. Lions and E. Magenes. *Non-homogeneous boundary value problems and applications, Vol.* 1. Springer, 1972.

[545] L. Littlejohn and R. Wellman. On the spectra of left-definite operators. Complex Anal. Oper. Theory 7 (2011) 437–455.

[546] A. Luger. A factorization of regular generalized Nevanlinna functions. Integral Equations Operator Theory 43 (2002) 326–345.

[547] A. Luger. Generalized poles and zeros of generalized Nevanlinna functions, a survey. In: *Proc. algorithmic information theory conference.* Vaasan Yliop. Julk. Selvityksiä Rap., Vol. 124, Vaasan Yliopisto, 2005, pp. 85–94.

[548] A. Luger. A characterization of generalized poles of generalized Nevanlinna functions. Math. Nachr. 279 (2006) 891–910.

[549] A. Luger and C. Neuner. An operator theoretic interpretation of the generalized Titchmarsh–Weyl function for perturbed spherical Schrödinger operators. Complex Anal. Oper. Theory 9 (2015) 1391–1410.

[550] A. Luger and C. Neuner. On the Weyl solution of the 1-dim Schrödinger operator with inverse fourth power potential. Monatsh. Math. 180 (2016) 295–303.

[551] A. Lunyov. Spectral functions of the simplest even order ordinary differential operator. Methods Funct. Anal. Topology 19 (2013) 319–326.

[552] V.E. Lyantse and O.G. Storozh. *Methods of the theory of unbounded operators.* Naukova Dumka, Kiev, 1983.

[553] M.M. Malamud. Some classes of extensions of a Hermitian operator with gaps (Russian). Ukr. Mat. Zh. 44 (1992) 215–233 [Ukrainian Math. J. 44 (1992) 190–204].

[554] M.M. Malamud. On a formula for the generalized resolvents of a non-densely defined Hermitian operator (Russian). Ukr. Mat. Zh. 44 (1992) 1658–1688 [Ukrainian Math. J. 44 (1992) 1522–1547].

[555] M.M. Malamud. On some classes of extensions of sectorial operators and dual pairs of contractions. Oper. Theory Adv. Appl. 124 (2001) 401–449.

[556] M.M. Malamud. Operator holes and extensions of sectorial operators and dual pairs of contractions. Math. Nachr. 279 (2006) 625–655.

[557] M.M. Malamud. Spectral theory of elliptic operators in exterior domains. Russ. J. Math. Phys. 17 (2010) 96–125.

[558] M.M. Malamud and S.M. Malamud. Spectral theory of operator measures in a Hilbert space (Russian). Algebra i Analiz 15 (2003) 1–77 [St. Petersburg Math. J. 15 (2004) 323–373].

[559] M.M. Malamud and V.I. Mogilevskiĭ. On Weyl functions and Q-functions of dual pairs of linear relations. Dopov. Nats. Akad. Nauk Ukr. Mat. Prirodozn. Tekh. Nauki 1999, no. 4, 32–37.

[560] M.M. Malamud and V.I. Mogilevskiĭ. Krein type formula for canonical resolvents of dual pairs of linear relations. Methods Funct. Anal. Topology 8 (2002) 72–100.

[561] M.M. Malamud and V.I. Mogilevskiĭ. Generalized resolvents of an isometric operator (Russian). Mat. Zametki 73 (2003) 460–465 [Math. Notes 73 (2003) 429–435].

[562] M.M. Malamud and H. Neidhardt. On the unitary equivalence of absolutely continuous parts of self-adjoint extensions. J. Functional Analysis 260 (2011) 613–638.

[563] M.M. Malamud and H. Neidhardt. Perturbation determinants for singular perturbations. Russ. J. Math. Phys. 21 (2014) 55–98.

[564] M.M. Malamud and H. Neidhardt. Trace formulas for additive and non-additive perturbations. Adv. Math. 274 (2015) 736–832.

[565] M.M. Malamud and K. Schmüdgen. Spectral theory of Schrödinger operators with infinitely many point interactions and radial positive definite functions. J. Functional Analysis 263 (2012) 3144–3194.

[566] J. Malinen and O. J. Staffans. Impedance passive and conservative boundary control systems. Complex Anal. Oper. Theory 1 (2007) 279–300.

[567] A. Mantile, A. Posilicano, and M. Sini. Self-adjoint elliptic operators with boundary conditions on not closed hypersurfaces. J. Differential Equations 261 (2016) 1–55.

[568] M. Marletta. Eigenvalue problems on exterior domains and Dirichlet to Neumann maps. J. Comput. Appl. Math. 171 (2004) 367–391.

[569] V.G. Mazya. *Sobolev spaces with applications to elliptic partial differential equations.* Second, revised and augmented edition. Grundlehren der Mathematischen Wissenschaften, Vol. 342, Springer, 2011.

[570] A.G.R. McIntosh. Hermitian bilinear forms which are not semibounded. Bull. Amer. Math. Soc. 76 (1970) 732–737.

[571] A.G.R. McIntosh. Bilinear forms in Hilbert space. J. Math. Mech. 19 (1970) 1027–1045.

[572] R. McKelvey. Spectral measures, generalized resolvents, and functions of positive type. J. Math. Anal. Appl. 11 (1965) 447–477.

[573] W. McLean. *Strongly elliptic systems and boundary integral equations.* Cambridge University Press, 2000.

[574] R. Mennicken and M. Möller. *Non-self-adjoint boundary eigenvalue problems.* North-Holland Mathematics Studies, Vol. 192, North-Holland Publishing Co., 2003.

[575] R. Mennicken and H.-D. Niessen. $(S, \tilde{S})$-adjungierte Randwertoperatoren. Math. Nachr. 44 (1970) 259–284.

[576] Y. Mezroui. Projection orthogonale sur le graphe d'une relation linéaire fermée. Trans. Amer. Math. Soc. 352 (2000) 2789–2800.

[577] V.A. Mikhailets. Spectra of operators and boundary value problems (Russian). In: *Spectral analysis of differential operators.* Akad. Nauk Ukrain. SSR, Inst. Mat., Kiev, 1980, pp. 106–131, 134.

[578] V.A. Mikhailets. Distribution of the eigenvalues of finite multiplicity of Neumann extensions of an elliptic operator (Russian). Differentsial'nye Uravneniya 30 (1994) 178–179 [Differential Equations 30 (1994) 167–168].

[579] V.A. Mikhailets. Discrete spectrum of the extreme nonnegative extension of the positive elliptic differential operator. In: *Proceedings of the Ukrainian Mathematical Congress* 2001*, Section* 7*, Nonlinear Analysis.* Kyiv, 2006, pp. 80–94.

[580] O.Ya. Milyo and O.G. Storozh. On general form of maximal accretive extension of positive definite operator (Russian). Dokl. Akad. Nauk Ukr. SSR 6 (1991) 19–22.

[581] O.Ya. Milyo and O.G. Storozh. Maximal accretive extensions of positive definite operator with finite defect number (Russian). Lviv University, 31 pages, Deposited in GNTB of Ukraine 28.10.93, no 2139 Uk93.

[582] O.Ya. Milyo and O.G. Storozh. Maximal accretiveness and maximal nonnegativeness conditions for a class of finite-dimensional perturbations of a positive definite operator. Mat. Studii 12 (1999) 90–100.

[583] R. von Mises. Beitrag zum Oszillationsproblem. *Festschrift Heinrich Weber zu seinem siebzigsten Geburtstag am* 5. *März* 1912 *gewidmet von Freunden und Schülern.* B.G. Teubner, 1912, pp. 252-282.

[584] D. Mitrea. *Distributions, partial differential equations, and harmonic analysis.* Universitext, Springer, 2013.

[585] M. Mitrea and M. Taylor. Boundary layer methods for Lipschitz domains in Riemannian manifolds. J. Functional Analysis 163 (1999) 181–251.

[586] M. Mitrea and M. Taylor. Potential theory on Lipschitz domains in Riemannian manifolds: Sobolev–Besov space results and the Poisson problem. J. Functional Analysis 176 (2000) 1–79.

[587] M. Mitrea and M. Taylor. Potential theory on Lipschitz domains in Riemannian manifolds: Hölder continuous metric tensors. Comm. Partial Differential Equations 25 (2000) 1487–1536.

[588] M. Mitrea and M. Taylor. Potential theory on Lipschitz domains in Riemannian manifolds: L^p, Hardy and Hölder type estimates. Commun. Anal. Geometry 9 (2001) 369–421.

[589] V.I. Mogilevskiĭ. Boundary triplets and Krein type resolvent formula for symmetric operators with unequal defect numbers. Methods Funct. Anal. Topology 12 (2006) 258–280.

[590] V.I. Mogilevskiĭ. Boundary triplets and Titchmarsh–Weyl functions of differential operators with arbitrary deficiency indices. Methods Funct. Anal. Topology 15 (2009) 280–300.

[591] V.I. Mogilevskiĭ. Minimal spectral functions of an ordinary differential operator. Proc. Edinburgh Math. Soc. 55 (2012) 731–769.

[592] V.I. Mogilevskiĭ. Boundary pairs and boundary conditions for general (not necessarily definite) first-order symmetric systems with arbitrary deficiency indices. Math. Nachr. 285 (2012) 1895–1931.

[593] V.I. Mogilevskiĭ. On eigenfunction expansions of first-order symmetric systems and ordinary differential operators of an odd order. Integral Equations Operator Theory 82 (2015) 301–337.

[594] V.I. Mogilevskiĭ. Spectral and pseudospectral functions of Hamiltonian systems: development of the results by Arov–Dym and Sakhnovich. Methods Funct. Anal. Topology 21 (2015) 370–402.

[595] V.I. Mogilevskiĭ. Symmetric extensions of symmetric linear relations (operators) preserving the multivalued part. Methods Funct. Anal. Topology 24 (2018) 152–177.

[596] V.I. Mogilevskiĭ. On compressions of self-adjoint extensions of a symmetric linear relation. Integral Equations Operator Theory 91 (2019) 9 30 pp.

[597] M. Möller. On the unboundedness below of the Sturm–Liouville operator. Proc. Roy. Soc. Edinburgh 129A (1999) 1011–1015.

[598] M. Möller and V. Pivovarchik. *Spectral theory of operator pencils, Hermite–Biehler functions, and their applications.* Oper. Theory Adv. Appl., Vol. 246, Birkhäuser, 2015.

[599] M. Möller and F.H. Szafraniec. Adjoints and formal adjoints of matrices of unbounded operators. Proc. Amer. Math. Soc. 136 (2008) 2165–2176.

[600] M. Möller and A. Zettl. Semi-boundedness of ordinary differential operators. J. Differential Equations 115 (1995) 24–49.

[601] M. Möller and A. Zettl. Symmetric differential operators and their Friedrichs extension. J. Differential Equations 115 (1995) 50–69.

[602] S.N. Naboko. Nontangential boundary values of operator R-functions in a half-plane (Russian). Algebra i Analiz 1 (1989) 197–222 [Leningrad Math. J. 1 (1990) 1255–1278].

[603] B. Sz.-Nagy, C. Foias, H. Bercovici, and L. Kérchy. *Harmonic analysis of operators on Hilbert space.* 2nd edition. Universitext, Springer, 2010 [enlarged and revised edition of 1st edition by B. Sz.-Nagy and C. Foias].

[604] B. Sz.-Nagy and A. Koranyi. Operatortheoretische Behandlung und Verallgemeinerung eines Problemkreises in der komplexen Funktionentheorie. Acta Math. 100 (1958) 171–202.

[605] M.A. Naĭmark. Self-adjoint extensions of the second kind of a symmetric operator. Bull. Acad. Sci. URSS Ser. Math. 4 (1940) 53–104.

[606] M.A. Naĭmark. Spectral functions of a symmetric operator. Izv. Akad. Nauk SSSR. Ser. Mat. 4 (1940) 277–318.

[607] M.A. Naĭmark. On spectral functions of a symmetric operator. Izv. Akad. Nauk SSSR. Ser. Mat. 7 (1943) 285–296.

[608] M.A. Naĭmark. *Linear differential operators II.* Ungar Pub. Co., 1967.

[609] M.Z. Nashed (editor). *General inverses and applications.* Academic Press, 1976.

[610] J. von Neumann. Allgemeine Eigenwerttheorie Hermitescher Funktionaloperatoren. Math. Ann. 102 (1930) 49–131.

[611] J. von Neumann. Über adjungierte Operatoren. Ann. Math. 33 (1932) 294–310.

[612] H.-D. Niessen. Singuläre S-hermitesche Rand-Eigenwertprobleme. Manuscripta Math. 3 (1970) 35–68.

[613] H.-D. Niessen. Zum verallgemeinerten zweiten Weylschen Satz. Arch. Math. 22 (1971) 648–656.

[614] H.-D. Niessen. Greensche Matrix und die Formel von Titchmarsh–Kodaira für singuläre S-hermitesche Eigenwertprobleme. J. Reine Angew. Math. 261 (1973) 164–193.

[615] H.-D. Niessen and A. Zettl. The Friedrichs extension of regular ordinary differential operators. Proc. Roy. Soc. Edinburgh 114A (1990) 229–236.

[616] H.-D. Niessen and A. Zettl. Singular Sturm–Liouville problems: The Friedrichs extension and comparison of eigenvalues. Proc. London Math. Soc. 64 (1992) 545–578.

[617] A.A. Nudelman. On a new problem of moment problem type (Russian) Dokl. Akad. Nauk SSSR 233 (1977) 792–795 [Soviet Math. Dokl. 18 (1977) 507–510].

[618] E.-M. Ouhabaz. Second order elliptic operators with essential spectrum $[0, \infty)$ on L^p. Comm. Partial Differential Equations 20 (1995) 763–773.

[619] B.C. Orcutt. *Canonical differential equations.* Dissertation, University of Virginia, 1969.

[620] S.A. Orlov. Nested matrix disks depending analytically on a parameter, and the theorem on invariance of the ranks of the radii of limiting matrix disks. Izv. Akad. Nauk SSSR Ser. Mat 40 (1967) 593–644.

[621] S. Ôta. Decomposition of unbounded derivations in operator algebras. Tôhoku Math. J. 33 (1981) 215–225.

[622] S. Ôta. Closed linear operators with domain containing their range. Proc. Edinburgh Math. Soc. 27 (1984) 229–233.

[623] S. Ôta. On a singular part of an unbounded operator. Zeitschrift für Analysis und ihre Anwendungen 7 (1987) 15–18.

[624] K. Pankrashkin. Resolvents of self-adjoint extensions with mixed boundary conditions. Rep. Math. Phys. 58 (2006) 207–221.

[625] K. Pankrashkin. Spectra of Schrödinger operators on equilateral quantum graphs. Lett. Math. Phys. 77 (2006) 139–154.

[626] K. Pankrashkin. Unitary dimension reduction for a class of self-adjoint extensions with applications to graph-like structures. J. Math. Anal. Appl. 396 (2012) 640–655.

[627] K. Pankrashkin. An example of unitary equivalence between self-adjoint extensions and their parameters. J. Functional Analysis 265 (2013) 2910–2936.

[628] K. Pankrashkin and N. Popoff. Mean curvature bounds and eigenvalues of Robin Laplacians. Calculus Variations Partial Differential Equations 54 (2015) 1947–1961.

[629] K. Pankrashkin and N. Popoff. An effective Hamiltonian for the eigenvalue asymptotics of the Robin Laplacian with a large parameter. J. Math. Pures Appl. 106 (2016) 615–650.

[630] B.S. Pavlov. The theory of extensions and explicitly-solvable models. Russian Math. Surveys 42 (1987) 127–168.

[631] R.S. Phillips. Dissipative operators and hyperbolic systems of partial differential equations. Trans. Amer. Math. Soc. (1959) 193–254.

[632] R.S. Phillips. Dissipative operators and parabolic partial diferential equations. Commun. Pure Appl. Math. 12 (1959) 249–276.

[633] R.S. Phillips. The extension of dual subspaces invariant under an algebra. In: *Proceedings international symposium linear algebra, Israel*, 1960. Academic Press, 1961, pp. 366–398.

[634] R.S. Phillips. On dissipative operators. In: *Lecture series in differential equations, Vol.* 2. Van Nostrand, 1969, pp. 65–113.

[635] R. Picard, S. Seidler, S. Trostorff, and M. Waurick. On abstract grad-div systems. J. Differential Equations 260 (2016) 4888–4917.

[636] H.M. Pipa and O.G. Storozh. Accretive perturbations of some proper extensions of a positive definite operator. Mat. Studii 25 (2006) 181–190.

[637] V. Pivovarchik and H. Woracek. Sums of Nevanlinna functions and differential equations on star-shaped graphs. Operators Matrices 3 (2009) 451–501.

[638] A. Pleijel. Spectral theory for pairs of formally self-adjoint ordinary differential operators. J. Indian Math. Soc. 34 (1970) 259–268.

[639] D. Popovici and Z. Sebestyén. Factorizations of linear relations. Adv. Math. 233 (2013) 40–55.

[640] A. Posilicano. A Krein-like formula for singular perturbations of selfadjoint operators and applications. J. Functional Analysis 183 (2001) 109–147.

[641] A. Posilicano. Boundary triples and Weyl functions for singular perturbations of self-adjoint operators. Methods Funct. Anal. Topology 10 (2004) 57–63.

[642] A. Posilicano. Self-adjoint extensions of restrictions. Operators Matrices 2 (2008) 483–506.

[643] A. Posilicano and L. Raimondi. Krein's resolvent formula for self-adjoint extensions of symmetric second-order elliptic differential operators. J. Phys. A 42 (2009) 015204, 11 pp.

[644] O. Post. First-order operators and boundary triples. Russ. J. Math. Phys. 14 (2007), 482–492.

[645] O. Post. Equilateral quantum graphs and boundary triples. In: *Analysis on graphs and its applications*. Proc. Sympos. Pure Math., Vol. 77, Amer. Math. Soc., 2008, pp. 469–490.

[646] O. Post. Spectral analysis on graph-like spaces. Lecture Notes in Mathematics, Vol. 2039, Springer, 2012.

[647] O. Post. Boundary pairs associated with quadratic forms. Math. Nachr. 289 (2016) 1052–1099.

[648] V. Prokaj and Z. Sebestyén. On Friedrichs extensions of operators. Acta Sci. Math. 62 (1996) 243–246.

[649] M.C. Reed and B. Simon. *Methods of modern mathematical physics I. Functional analysis*. Academic Press, 1972.

[650] M.C. Reed and B. Simon. *Methods of modern mathematical physics II. Fourier analysis, self-adjointness*. Academic Press, 1975.

[651] M.C. Reed and B. Simon. *Methods of modern mathematical physics III. Scattering theory*. Academic Press, 1978.

[652] M.C. Reed and B. Simon. *Methods of modern mathematical physics IV. Analysis of operators*. Academic Press, 1977.

[653] W.T. Reid. *Ordinary differential equations*. John Wiley and Sons, 1971.

[654] F. Rellich. Halbbeschränkte gewöhnliche Differentialoperatoren zweiter Ordnung. Math. Ann. 122 (1950) 343–368.

[655] C. Remling. Relationships between the m-function and subordinate solutions of second order differential operators. J. Math. Anal. Appl. 206 (1997) 352–363.

[656] C. Remling. Spectral analysis of higher order differential operators I. General properties of the m-function. J. London Math. Soc. 58 (1998) 367–380.

[657] C. Remling. Schrödinger operators and de Branges spaces. J. Functional Analysis 196 (2002) 323–394.

[658] C. Remling. *Spectral theory of canonical systems*. Studies in Mathematics, Vol. 70, de Gruyter, 2018.

[659] F. Riesz and B. Sz.-Nagy. *Leçons d'analyse fonctionelle.* 6th edition. Gauthier-Villars, 1972.

[660] F.S. Rofe-Beketov. On selfadjoint extensions of differential operators in a space of vector-functions (Russian). Teor. Funktsii Funktsional. Anal. i Prilozhen 8 (1969) 3–24.

[661] F.S. Rofe-Beketov. Perturbations and Friedrichs extensions of semibounded operators on variable domains. Soviet Math. Dokl. 22 (1980) 795–799.

[662] F.S. Rofe-Beketov. The numerical range of a linear relation and maximum relations (Russian). Teor. Funktsii Funktsional. Anal. i Prilozhen 44 (1985) 103–112 [J. Math. Sciences 48 (1990) 329–336].

[663] F.S. Rofe-Beketov and A.M. Kholkin. *Spectral analysis of differential operators.* World Scientific Monograph Series in Math., Vol. 7, World Scientific, 2005.

[664] H. Röh. Self-adjoint subspace extensions satisfying λ-linear boundary conditions. Proc. Roy. Soc. Edinburgh 90A (1981) 107–124.

[665] J. Rohleder. Strict inequality of Robin eigenvalues for elliptic differential operators on Lipschitz domains. J. Math. Anal. Appl. 418 (2014) 978–984.

[666] R. Romanov. *Canonical systems and de Branges spaces.* arXiv:1408.6022.

[667] B.W. Roos and W.C. Sangren. Spectra for a pair of singular first order differential equations. Proc. Amer. Math. Soc. 12 (1961) 468–476.

[668] B.W. Roos and W.C. Sangren. Three spectral theorems for a pair of singular first order differential equations. Pacific J. Math. 12 (1962) 1047–1055.

[669] B.W. Roos and W.C. Sangren. Spectral theory of Dirac's radial relativistic wave equation. J. Math. Phys. 3 (1962) 882–890.

[670] M. Rosenberg. The square-integrability of matrix-valued functions with respect to a non-negative Hermitian measure. Duke Math. J. 31 (1964) 291–298.

[671] R. Rosenberger. *Charakterisierungen der Friedrichsfortsetzung von halbbeschränkten Sturm–Liouville Operatoren.* Doctoral dissertation, Technische Hochschule Darmstadt, 1984.

[672] R. Rosenberger. A new characterization of the Friedrichs extension of semibounded Sturm–Liouville operators. J. London Math. Soc. 31 (1985) 501–510.

[673] M. Rosenblum and J. Rovnyak. *Hardy classes and operator theory.* Oxford University Press, 1985.

[674] J. Rovnyak. Methods of Kreĭn space operator theory. Oper. Theory Adv. Appl. 134 (2002) 31–66.

[675] E.M. Russakovskiĭ. The matrix Sturm–Liouville problem with spectral parameter in the boundary condition. Algebraic and operator aspects. Trans. Moscow Math. Soc. 57 (1996) 159–184.

[676] W. Rudin. *Real and complex analysis.* 3rd edition. McGraw-Hill, 1986.

[677] V. Ryzhov. A general boundary value problem and its Weyl function. Opuscula Math. 27 (2007) 305–331.

[678] V. Ryzhov. Weyl–Titchmarsh function of an abstract boundary value problem, operator colligations, and linear systems with boundary control. Complex Anal. Oper. Theory 3 (2009) 289–322.

[679] Sh.N. Saakyan. Theory of resolvents of a symmetric operator with infinite defect numbers (Russian). Akad. Nauk Armjan. SSR Dokl. 41 (1965) 193–198.

[680] S. Saitoh. *Theory of reproducing kernels and its applications.* Longman Scientific and Technical, 1988.

[681] L.A. Sakhnovich. *Spectral theory of canonical differential systems. Method of operator identities.* Oper. Theory Adv. Appl., Vol. 107, Birkhäuser, 1999.

[682] S. Saks. *Theory of the integral.* 2nd revised edition. Dover Publications, 1964.

[683] A. Sandovici. Von Neumann's theorem for linear relations. Linear Multilinear Algebra 66 (2018) 1750–1756.

[684] A. Sandovici and Z. Sebestyén. On operator factorization of linear relations. Positivity 17 (2013) 1115–1122.

[685] A. Sandovici, H.S.V. de Snoo, and H. Winkler. The structure of linear relations in Euclidean spaces. Lin. Alg. Appl. 397 (2005) 141–169.

[686] A. Sandovici, H.S.V. de Snoo, and H. Winkler. Ascent, descent, nullity, and defect for linear relations in linear spaces. Lin. Alg. Appl. 423 (2007) 456–497.

[687] F.W. Schäfke and A. Schneider. S-hermitesche Rand-Eigenwertprobleme I. Math. Ann. 162 (1965) 9–26.

[688] F.W. Schäfke and A. Schneider. S-hermitesche Rand-Eigenwertprobleme II. Math. Ann. 165 (1966) 236–260.

[689] F.W. Schäfke and A. Schneider. S-hermitesche Rand-Eigenwertprobleme III. Math. Ann. 177 (1968) 67–94.

[690] M. Schechter. *Principles of functional analysis.* 2nd edition. Graduate Studies in Mathematics, Vol. 36, Amer. Math. Soc., 2001.

[691] K. Schmüdgen. *Unbounded self-adjoint operators on Hilbert space.* Graduate Texts in Mathematics, Vol. 265, Springer, 2012.

[692] A. Schneider. Untersuchungen über singuläre reelle S-hermitesche Differentialgleichungssysteme im Normalfall. Math. Z. 107 (1968) 271–296.

[693] A. Schneider. Zur Einordnung selbstadjungierter Rand-Eigenwertprobleme bei gewöhnlichen Differentialgleichungen in die Theorie S-hermitescher Rand-Eigenwertprobleme. Math. Ann. 178 (1968) 277–294.

[694] A. Schneider. Weitere Untersuchungen über singuläre reelle S-hermitesche Differentialgleichungssysteme im Normalfall. Math. Z. 109 (1969) 153–168.

[695] A. Schneider. Geometrische Bedeutung eines Satzes vom Weylschen Typ für S-hermitesche Differentialgleichungssysteme im Normalfall. Arch. Math. 20 (1969) 147–154.

[696] A. Schneider. Die Greensche Matrix S-hermitescher Rand-Eigenwertprobleme im Normalfall. Math. Ann. 180 (1969) 307–312.

[697] A. Schneider. On spectral theory for the linear selfadjoint equation $Fy = \lambda Gy$. Springer Lecture Notes Math. 846 (1981) 306–322.

[698] A. Schneider and H.-D. Niessen. Linksdefinite singuläre kanonische Eigenwertprobleme I. J. Reine Angew. Math. 281 (1976) 13–52.

[699] A. Schneider and H.-D. Niessen. Linksdefinite singuläre kanonische Eigenwertprobleme II. J. Reine Angew. Math. 289 (1977) 62–84.

[700] D.B. Sears. Integral transforms and eigenfunction theory. Quart. J. Math. (Oxford) 5 (1954) 47–58.

[701] D.B. Sears. Integral transforms and eigenfunction theory II. Quart. J. Math. (Oxford) 6 (1955) 213–217.

[702] Z. Sebestyén and E. Sikolya. On Kreĭn–von Neumann and Friedrichs extensions. Acta Sci. Math. (Szeged) 69 (2003) 323–336.

[703] Z. Sebestyén and J. Stochel. Restrictions of positive selfadjoint operators. Acta Sci. Math. Szeged 55 (1991) 149–154.

[704] Yu.L. Shmuljan. A Hellinger operator integral (Russian). Mat. Sb. (N.S.) 49 (91) (1959) 381–430 [Amer. Math. Soc. Transl. (2) 22 (1962) 289–337].

[705] Yu.L. Shmuljan. Theory of linear relations, and spaces with indefinite metric (Russian). Funktsional. Anal. i Prilozhen 10 (1976) 67–72.

[706] Yu.L. Shmuljan. Transformers of linear relations in J-spaces (Russian). Funktsional. Anal. i Prilozhen 14 (1980) 39–44 [Functional Anal. Appl. 14 (1980) 110–113].

[707] Yu.G. Shondin. Quantum-mechanical models in $\mathbb{R}^n$ associated with extensions of the energy operator in Pontryagin space (Russian). Teor. Mat. Fiz. 74 (1988) 331–344 [Theor. Math. Phys. 74 (1988) 220–230].

[708] D.A. Shotwell. *Boundary problems for the differential equation $Lu = \lambda\sigma u$ and associated eigenfunction-expansions.* Doctoral dissertation, University of Colorado, 1965.

[709] B. Simon. Lower semicontinuity of positive quadratic forms. Proc. Roy. Soc. Edinburgh 79A (1977) 267–273.

[710] B. Simon. A canonical decomposition for quadratic forms with applications to monotone convergence theorems. J. Functional Analysis 28 (1978) 377–385.

[711] B. Simon. m-functions and the absolutely continuous spectrum of one-dimensional almost periodic Schrödinger operators. In: *Differential equations.* North-Holland Math. Studies, Vol. 92, North-Holland, 1984, pp. 519.

[712] B. Simon. Bounded eigenfunctions and absolutely continuous spectra for one-dimensional Schrödinger operators. Proc. Amer. Math. Soc. 124 (1996) 3361–3369.

[713] S. Simonov and H. Woracek. Spectral multiplicity of selfadjoint Schrödinger operators on star-graphs with standard interface conditions. Integral Equations Operator Theory 78 (2014) 523–575.

[714] C.F. Skau. Positive self-adjoint extensions of operators affiliated with a von Neumann algebra. Math. Scand. 44 (1979) 171–195.

[715] H.S.V. de Snoo and H. Winkler. Canonical systems of differential equations with selfadjoint interface conditions on graphs. Proc. Roy. Soc. Edinburgh 135A (2005) 297–315.

[716] H.S.V. de Snoo and H. Winkler. Two-dimensional trace-normed canonical systems of differential equations and selfadjoint interface conditions. Integral Equations Operator Theory 51 (2005) 73–108.

[717] H.S.V. de Snoo and H. Woracek. Sums, couplings, and completions of almost Pontryagin spaces. Lin. Alg. Appl. 437 (2012) 559–580.

[718] H.S.V. de Snoo and H. Woracek. Restriction and factorization for isometric and symmetric operators in almost Pontryagin spaces. Oper. Theory Adv. Appl. 252 (2016) 123–170.

[719] H.S.V. de Snoo and H. Woracek. Compressed resolvents, Q-functions and h_0-resolvents in almost Pontryagin spaces. Oper. Theory Adv. Appl. 263 (2018) 425–483.

[720] H.S.V. de Snoo and H. Woracek. The Krein formula in almost Pontryagin spaces. A proof via orthogonal coupling. Indag. Math. 29 (2018) 714–729.

[721] P. Sorjonen. On linear relations in an indefinite inner product space. Ann. Acad. Sci. Fenn. Ser. A I Math. 4 (1978/1979) 169–192.

[722] P. Sorjonen. Extensions of isometric and symmetric linear relations in a Krein space. Ann. Acad. Sci. Fenn. Ser. A I Math. 5 (1980) 355–375.

[723] J. Stochel and F.H. Szafraniec. *Unbounded operators and subnormality.* To appear.

[724] M.H. Stone. *Linear transformations in Hilbert space and their applications to analysis.* Amer. Math. Soc. Colloquium Publications, Vol. 15, 1932.

[725] O.G. Storozh. Extremal extensions of a nonnegative operator, and accretive boundary value problems (Russian). Ukr. Mat. Zh. 42 (1990) 858–860 [Ukrainian Math. J. 42 (1990) 758–760 (1991)].

[726] A.V. Štraus. Generalized resolvents of symmetric operators (Russian). Izv. Akad. Nauk. SSSR Ser. Mat. 18 (1954) 51–86 [Math. USSR-Izv. 4 (1970) 179–208].

[727] A.V. Štraus. On spectral functions of differential operators (Russian). Izv. Akad. Nauk SSSR Ser. Mat. 19 (1955) 201–220.

[728] A.V. Štraus. Characteristic functions of linear operators (Russian). Izv. Akad. Nauk SSSR Ser. Mat. 24 (1960) 43–74 [Amer. Math. Soc. Transl. (2) 40 (1964) 1–37].

[729] A.V. Štraus. On selfadjoint operators in the orthogonal sum of Hilbert spaces (Russian). Dokl. Akad. Nauk SSSR 144 (1962) 512–515.

[730] A.V. Štraus. On the extensions of symmetric operators depending on a parameter (Russian). Izv. Akad. Nauk SSSR Ser. Mat. 29 (1965) 1389–1416 [Amer. Math. Soc. Transl. (2) 61 (1967) 113–141].

[731] A.V. Štraus. On one-parameter families of extensions of a symmetric operator (Russian). Izv. Akad. Nauk SSSR Ser. Mat. 30 (1966) 1325–1352 [Amer. Math. Soc. Transl. (2) 90 (1970) 135–164].

[732] A.V. Štraus. Extensions and characteristic function of a symmetric operator (Russian). Izv. Akad. Nauk SSSR Ser. Mat. 32 (1968) 186–207 [Math. USSR-Izv. 2 (1968) 181–204].

[733] A.V. Štraus. Extensions and generalized resolvents of a symmetric operator which is not densely defined (Russian). Izv. Akad. Nauk. SSSR Ser. Mat. 34 (1970) 175–202 [Math. USSR-Izv. 4 (1970) 179–208].

[734] A.V. Štraus. On extensions of semibounded operators (Russian). Dokl. Akad. Nauk SSSR 211 (1973) 543–546 [Soviet Math. Dokl. 14 (1973) 1075–1079].

[735] A.V. Štraus. On the theory of extremal extensions of a bounded positive operator. Funkts. Anal. 18 (1982) 115–126.

[736] A.V. Štraus. Generalized resolvents of nondensely defined bounded symmetric operators. Amer. Math. Soc. Transl. (2) 154 (1992) 189–195.

[737] F.H. Szafraniec. *The reproducing property: spaces and operators.* To appear.

[738] G. Teschl. *Mathematical methods in quantum mechanics with applications to Schrödinger operators.* Graduate studies in mathematics, Vol. 99, Amer. Math. Soc., 2009.

[739] E.C. Titchmarsh. On expansions in eigenfunctions IV. Quart. J. Math. Oxford Ser. 12 (1941) 33–50.

[740] E.C. Titchmarsh. *Eigenfunction expansions associated with second order differential equations I.* 2nd ed. Oxford University Press, Oxford, 1962.

[741] E.C. Titchmarsh. *Eigenfunction expansions associated with second order differential equations II.* Oxford University Press, Oxford, 1958.

[742] C. Tretter. Boundary eigenvalue problems for differential equations $N\eta = \lambda P\eta$ and λ-polynomial boundary conditions. J. Differential Equations 170 (2001) 408–471.

[743] C. Tretter. *Spectral theory of block operator matrices and applications.* Imperial College Press, 2008.

[744] H. Triebel. *Interpolation theory, function spaces, differential operators.* 2nd revised and enlarged edition. Johann Ambrosius Barth Verlag, 1995.

[745] H. Triebel. *Höhere Analysis.* 2nd edition. Harri Deutsch, 1980.

[746] E.R. Tsekanovskiĭ and Yu.L. Shmuljan. The theory of biextensions of operators in rigged Hilbert spaces. Unbounded operator colligations and characteristic functions. Russian Math. Surveys 32 (1977) 73–131.

[747] M.I. Vishik. On general boundary conditions for elliptic differential equations. Trudy Moskov. Mat. Obsc. 1 (1952) 187–246 [Amer. Math. Soc. Transl. Ser. (2) 24 (1963) 107–172].

[748] R. Vonhoff. *Spektraltheoretische Untersuchung linksdefiniter Differentialgleichungen im singulären Fall.* Doctoral dissertation, Universität Dormund, 1995.

[749] J. Walter. Regular eigenvalue problems with eigenvalue parameter in the boundary condition. Math. Z. 133 (1973) 301–312.

[750] M. Waurick and S.A. Wegner. Some remarks on the notions of boundary systems and boundary triple(t)s. Math. Nachr. 291 (2018) 2489–2497.

[751] S.A. Wegner. Boundary triplets for skew-symmetric operators and the generation of strongly continuous semigroups. Anal. Math. 43 (2017) 657–686.

[752] J. Weidmann. *Linear operators in Hilbert spaces.* Graduate Texts in Mathematics, Vol. 68, Springer, 1980.

[753] J. Weidmann. Stetige Abhängigkeit der Eigenwerte und Eigenfunktionen elliptischer Differentialoperatoren vom Gebiet. Math. Scand. 54 (1984) 51–69.

[754] J. Weidmann. *Spectral theory of ordinary differential operators.* Lecture Notes in Mathematics, Vol. 1258, Springer, 1987.

[755] J. Weidmann. Strong operator convergence and spectral theory of ordinary differential operators. Universitatis Iagellonicae Acta Math. 34 (1997) 153–163.

[756] J. Weidmann. *Lineare Operatoren im Hilberträumen. Teil I: Grundlagen.* B.G. Teubner, 2000.

[757] J. Weidmann. *Lineare Operatoren im Hilberträumen. Teil II: Anwendungen.* B.G. Teubner, 2003.

[758] H. Weyl. Über gewöhnliche lineare Differentialgleichungen mit singulären Stellen und ihre Eigenfunktionen. Göttinger Nachr. (1909) 37–64.

[759] H. Weyl. Über gewöhnliche Differentialgleichungen mit Singularitäten und die zugehörigen Entwicklungen willkürlicher Funktionen. Math. Ann. 68 (1910) 220–269.

[760] H. Weyl. Über gewöhnliche lineare Differentialgleichungen mit singulären Stellen und ihre Eigenfunktionen. Göttinger Nachr. (1910) 442–467.

[761] H. Weyl. Ramifications, old and new, of the eigenvalue problem. Bull. Amer. Math. Soc. 56 (1950) 115–139.

[762] W.M. Whyburn. Differential equations with general boundary conditions. Bull. Amer. Math. Soc. 48 (1942) 692–704.

[763] D.V. Widder. *The Laplace transform.* Princeton Mathematical Series, Vol. 6, Princeton University Press, 1941.

[764] H.L. Wietsma. *On unitary relations between Kreĭn spaces.* Doctoral dissertation, University of Vaasa, 2012.

[765] H.L. Wietsma. Representations of unitary relations between Kreĭn spaces. Integral Equations Operator Theory 72 (2012) 309–344.

[766] H. Winkler. The inverse spectral problem for canonical systems. Integral Equations Operator Theory 22 (1995) 360–374.

[767] H. Winkler. On transformations of canonical systems. Oper. Theory Adv. Appl. 80 (1995) 276–288.

[768] H. Winkler. Canonical systems with a semibounded spectrum. Oper. Theory Adv. Appl. 106 (1998) 397–417.

[769] H. Winkler. Spectral estimations for canonical systems. Math. Nachr. 220 (2000) 115–141.

[770] H. Winkler. On generalized Friedrichs and Krein–von Neumann extensions and canonical systems. Math. Nachr. 236 (2002) 175–191.

[771] H. Winkler and H. Woracek. On semibounded canonical systems. Lin. Alg. Appl. 429 (2008) 1082–1092.

[772] H. Winkler and H. Woracek. Reparametrizations of non trace-normed Hamiltonians. Oper. Theory Adv. Appl. 221 (2012) 667–690.

[773] A. Wintner. *Spektraltheorie unendlicher Matrizen.* Hirzel, 1929.

[774] J. Wloka. *Partial differential equations.* Cambridge University Press, 1987.

[775] H. Woracek. *Spektraltheorie im Pontryaginraum.* Lecture Notes, TU Wien.

[776] H. Woracek. *Operatortheorie im Krein Raum 1+2.* Lecture Notes, TU Wien.

[777] H. Woracek. Reproducing kernel almost Pontryagin spaces. Lin. Alg. Appl. 461 (2014) 271–317.

[778] A. Yang. *A construction of Krein spaces of analytic functions.* Doctoral dissertation, Purdue University, 1990.

[779] S. Yao, J. Sun, and A. Zettl. The Sturm–Liouville Friedrichs extension. Appl. Math. 60 (2015) 299–320.

[780] K. Yosida. On Titchmarsh–Kodaira's formula concerning Weyl–Stone's eigenfunction expansion. Nagoya Math. J. 1 (1950) 49–58.

[781] K. Yosida. Correction. Nagoya Math. J. 6 (1953) 187–188.

[782] A. Zettl. *Sturm–Liouville theory.* Mathematical Surveys and Monographs, Vol. 121, Amer. Math. Soc., 2005.

[783] W.P. Ziemer. *Weakly differentiable functions.* Graduate Texts in Mathematics, Vol. 120, Springer, 1989.

[784] H.J. Zimmerberg. Linear integro-differential-boundary-parameter problems. Ann. Mat. Pura Appl. 105 (1975) 241–256.

List of Symbols

$\mathbb{N}$	natural numbers,
$\mathbb{R}$	real line,
$\mathbb{C}$	complex plane,
$\mathbb{C}^+$	upper complex half-plane,
$\mathbb{C}^-$	lower complex half-plane,
$\mathbb{D}$	open unit disk.

For a linear relation H the following notation is used:

$\operatorname{dom} H$	domain,
$\overline{\operatorname{dom}}\, H$	closure of the domain,
$\ker H$	kernel,
$\operatorname{ran} H$	range,
$\overline{\operatorname{ran}}\, H$	closure of the range,
$\operatorname{mul} H$	multivalued part.

J. Behrndt et al., *Boundary Value Problems, Weyl Functions, and Differential Operators*, Monographs in Mathematics 108,
https://doi.org/10.1007/978-3-030-36714-5

Index

J. Behrndt et al., *Boundary Value Problems, Weyl Functions, and Differential Operators*, Monographs in Mathematics 108,
https://doi.org/10.1007/978-3-030-36714-5

www.ingramcontent.com/pod-product-compliance
Lightning Source LLC
LaVergne TN
LVHW062151200926
848484LV00007B/1898
9781013273339